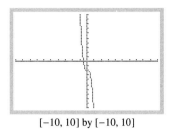

[−10, 10] by [−10, 10]

A graph of $y = x^4 − 12x^3 + x^2 − 2$ in the standard viewing window. How do we know this gives the complete picture?

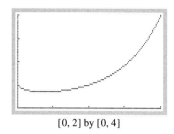

[0, 2] by [0, 4]

Determining $\lim_{x \to 0^+} x^x$ is difficult. From the graph on the graphing calculator, it appears this limit is 1.

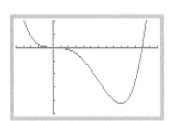

[−5, 14] by [−2500, 1000]

Another graph of $y = x^4 − 12x^3 + x^2 − 2$, showing behavior missed on the previous screen. Does this give the complete picture? Only by using calculus can we be sure that this is the complete picture.

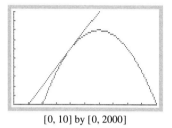

[0, 10] by [0, 2000]

It is easy to draw tangent lines using our graphing calculators.

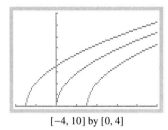

[−4, 10] by [0, 4]

The graph of $y = f(x) = \sqrt{x}$ is in the middle. The graph of $y = f(x + 3) = \sqrt{x + 3}$ is the graph of $y = f(x)$ shifted 3 units to the left. The graph of $y = f(x − 3) = \sqrt{x − 3}$ is the graph of $y = f(x)$ shifted 3 units to the right.

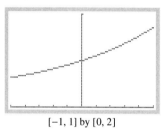

[−1, 1] by [0, 2]

A graph of $y = \dfrac{e^x − 1}{x}$ indicates that $\lim_{x \to 0} \dfrac{e^x − 1}{x} = 1$. This is an important limit.

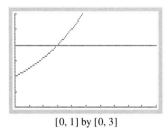

[0, 1] by [0, 3]

The solution of $10^x = 2$ can be found by finding the x-coordinate of the point of intersection of the graphs of the functions $y_1 = 10^x$ and $y_2 = 2$. The solution is called log 2.

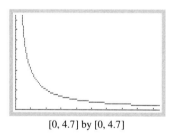

[0, 4.7] by [0, 4.7]

Let $f(x) = \ln x$. Then $f'(x) \approx \dfrac{\ln(x + 0.001) − \ln x}{0.001} = g(x)$. Shown is the graph of $y = g(x)$. We can readily see that $g(x) \approx 1/x$. Thus, we suspect that $f'(x) = 1/x$.

[0.75, 1.25] by [0.5, 1.5]

The graph of $y_1 = x^2$ is nearly the same as the graph of the line tangent to $y_1 = x^2$ at $x = 1$. Thus, we see that near a point where the function is differentiable, the graph of the function is approximately the same as the graph of the tangent line.

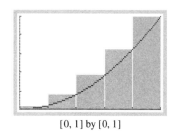

[0, 1] by [0, 1]

Using the program RECT we can easily draw a graph of $y = x^2$ and the rectangles associated with the right-hand sums.

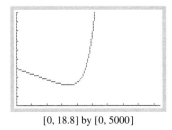

[0, 18.8] by [0, 5000]

A graph of marginal cost indicates marginal cost initially decreases, as the firm gains efficiency through increased production, but then marginal cost increases as too much production overwhelms the firm's capacity.

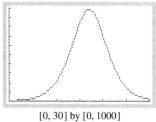

[0, 30] by [0, 1000]

The graph of $y = 3960(e^{0.2x-3.4} + e^{-0.2x+3.4})^{-2}$ models the spread of the Bombay plaque of 1905.

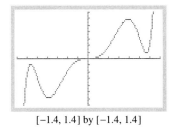

[−1.4, 1.4] by [−1.4, 1.4]

A graph of the complicated function $y = 1.6x^9 - 8x^5 + 7x^3$ is shown. Without using calculus, however, we will never know if this is the complete picture.

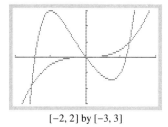

[−2, 2] by [−3, 3]

We want to find the area between the graphs of $y = x^5 + x^2 - 3x$ and $y = 0.5x^3$.

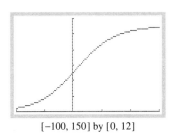

[−100, 150] by [0, 12]

A graph of the logistic curve $y = \dfrac{11.5}{1 + 1.2e^{-0.03x}}$ giving the population of the world over the last 200 years and projecting the population over the next 200 years. Logistic curves model population and sales growth and the spread of a technology. There is a rapid, exponential, increase at first, and then a leveling off later.

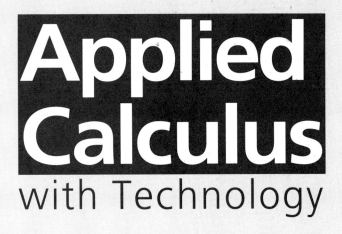

Applied Calculus
with Technology

by
Edmond C. Tomastik
University of Connecticut

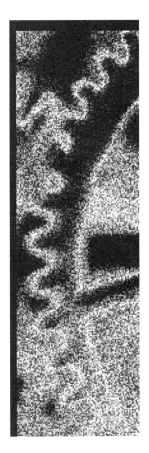

SAUNDERS COLLEGE PUBLISHING
Harcourt Brace College Publishers

Forth Worth   Philadelphia   San Diego   New York   Orlando   Austin
San Antonio   Toronto   Montreal   London   Sydney   Tokyo

Text Typeface: Times Roman
Compositor: University Graphics, Inc.
Executive Editor: Jay Ricci
Developmental Editor: Patrick Farace
Managing Editor: Carol Field
Project Editor: Laura Shur
Copy Editor: Linda Davoli
Manager of Art and Design: Carol Bleistine
Art Director: Joan S. Wendt
Cover Designer: Louis Fuiano
Text Artwork: Techsetters, Inc.
Director of EDP: Tim Frelick
Production Manager: Joanne Cassetti
Marketing Manager: Nick Agnew

Cover Credit: Chris Ferebee/Photonica

Printed in the United States of America

Applied Calculus with Technology

ISBN: 0-03-006863-0

Library of Congress Catalog Card Number: 95-071906

78   053   9876543

To Nancy

PREFACE

Applied Calculus With Technology is designed to be used in a two semester calculus course aimed at students majoring in business, management, economics, or life/social sciences. The text is written for students with two years of high school algebra. A wide range of topics is included, giving the instructor considerable flexibility in designing a course.

Since the text uses technology as a major tool, the reader is required to use a graphing calculator. The Technology Resource Manual, available with the text, gives all the details in user-friendly terms needed to use a graphing calculator in conjunction with the text. Instructors and students unfamiliar with modern graphing calculators will be surprised at how easy they are to use, and how powerful they have become. This text, together with the accompanying Technology Resource Manual, constitutes a completely organized, self-contained, and easy to use set of material, even for those without any knowledge of graphing calculators.

PHILOSOPHY

The writing of this text has been guided by four basic principles, all of which are consistent with the movement by national mathematics organizations for reform in calculus teaching and learning.

1. **The Rule of Three:** Every topic should be presented graphically, numerically, and algebraically.
2. **Technology:** Incorporate technology into the calculus instruction.
3. **The Way of Archimedes:** Formal definitions and procedures should evolve from the investigation of practical problems.
4. **Teaching Method:** Teach calculus using the investigative, exploratory approach.

The Rule of Three

By always bringing graphical and numerical, as well as algebraic, viewpoints to bear on each topic, the text presents a conceptual understanding of the calculus that is deep and useful in accommodating diverse applications. Sometimes a problem is done algebraically, then *supported* numerically and/or graphically (with a graphing calculator). Sometimes a problem is done numerically and/or graphically (with a graphing calculator), then *confirmed* algebraically. Other times a problem is done numerically or graphically because the algebra is too time consuming or impossible.

Technology

Technology permits more time to be spent on concepts, problem solving, and applications. The technology is used to assist the student to think about the geometric and numerical meaning of the calculus, without undermining the algebraic aspects. In this process, a balanced approach is presented. The text clearly points out that the graphing calculator may not give the whole story, motivating the need to learn the calculus. On the other hand, the text also stresses common situations where exact solutions are impossible, requiring an approximation technique using the graphing calculator. Thus, the graphing calculator is just another tool needed, along with the calculus, if we are to solve a variety of problems in the applications.

Applications and the Way of Archimedes

The text is written for *users* of mathematics. Applications play a central role and are woven into the development of the material. Practical problems are always investigated first, then used to motivate, to maintain interest, and to use as a basis for developing definitions and procedures. Here too, technology plays a natural role, allowing the forbidding and time-consuming difficulties associated with real applications to be overcome.

The Investigative, Exploratory Approach

The text also emphasizes an investigative and exploratory approach to teaching. Whenever practical, the text gives students the opportunity to explore and discover for themselves the basic calculus concepts. Again, technology plays an important role. For example, using their graphing calculators, students discover for themselves the derivatives of x^2, x^3, and x^4, and then generalize to x^n. They also discover the derivatives of $\ln x$ and e^x. None of this is realistically possible without technology.

Student response in the classroom has been exciting. My students enjoy using their graphing calculators in class and feel engaged and part of the learning process. I find students much more receptive to answering questions concerning their observations and more ready to ask questions.

A particularly effective technique is to take 15 or 20 minutes of class time and have students work in small groups to do an exploration or make a discovery. By walking around the classroom and talking with each group, lively discussions arise, even from students who do not normally participate. After such a minilab, the whole class is ready to discuss the insights gained.

Fully in sync with current goals in teaching and learning mathematics, every section in the text includes an Enrichment Exercise Set that encourages exploration, investigation, critical thinking, writing, and verbalization.

Which Graphing Calculator?

Any user of this text faces the immediate problem of what ''technology'' to use. The TI-82 graphing calculator made by Texas Instruments is highly recommended. The text does not require the additional functions found on the Texas Instruments TI-85, which is, in general, more difficult to use. The student should spend a minimum of time mastering the technology, leaving a maximum of time to learn calculus. In this regard the TI-82 strikes a good balance, powerful while still being user-friendly.

Other graphing calculators can be used. The Technology Resource Manual available with this text covers the TI-81, TI-82, TI-85, the Casio fx-7700G and fx-8700G, and the Sharp EL-9200C and EL-9300C. The TI-82, TI-85, and EL-9300C have a significant advantage in this group. Programs used in the text can be readily transferred electronically from calculator to calculator, eliminating time consuming and error prone programming by hand. About 10 or 15 seconds is required to link two of these calculators, and another second or two to transfer all the needed programs. The TI-82 and TI-85 can communicate with a personal computer and transfer programs. (A special cable and software are needed.) To make this process easier there are two disks available to adopters of the text, one for IBM compatible computers and one for Macintoshes, that contain the many programs used in this text. For more information see your Saunders' sales representative.

Why Graphing Calculators?

The modern graphing calculator is a more effective practical tool than computers in one-dimensional calculus. Computers, unfortunately, are stuck in a laboratory, graphing calculators are completely mobile. Every student can have one instantly ready for use at any time in the classroom. The graphing calculator can be used at the precise moment in the course when needed, with minilabs of 10 minutes being very practical.

Computers are expensive to purchase and maintain, and become obsolete all too soon. Computers also require rooms and monitors. On the other hand, the expense of graphing calculators can be shouldered by the student, not the institution.

IMPORTANT FEATURES

Cost. We understand the financial burden of buying a text and a graphing calculator. To help ease the student burden, the price of this text is less than half the usual price. Costs have been cut by printing with one color and using a soft cover. We believe that the creative layout of the text makes it visually appealing and user-friendly.

Style. The text is designed to implement the philosophy stated earlier. Every section opens by posing an interesting and relevant applied problem using familiar vocabulary, which is later solved in the section after the appropriate mathematics has been developed. Concepts are always introduced intuitively, evolve gradually from the investigation of practical problems or particular cases, and culminate in a definition or result. Students are given the opportunity to investigate and discover concepts for themselves, by using the graphing calculator to create the screens in the text or by doing the Explorations. Topics are presented graphically, numerically, and algebraically to give the reader a deep and conceptual understanding. Scattered throughout the text are historical and anecdotal comments. The historical comments are not only interesting in themselves, but also indicate that mathematics is a continually developing subject. The anecdotal comments relate the material to contemporary real life situations.

Applications. The text includes many meaningful applications drawn from a variety of fields, including numerous referenced examples extracted from current journals. Applications are given for all the mathematics that is presented and are used to motivate the student.

Worked Examples. Over 400 worked examples, including self-help examples mentioned below, have been carefully selected to take the reader progressively from the simplest idea to the most complex. All the steps needed for the complete solutions are included.

Screens. There are over 100 screens shown in the text. In almost all cases, they represent opportunities for the instructor to have the students reproduce these on their graphing calculators at the point in the lecture when they are needed. This allows the student to be an active partner in the learning process, emphasizes the point being made, and makes the classroom more exciting. A majority of the other graphs can also be done on the graphing calculator.

Explorations. These explorations are designed to further make the student an active partner in the learning process. Some of these explorations can be done in class, some can be done outside class, as group or individual projects. Not all of these explorations use the graphing calculator, some ask to solve a problem or make a discovery using pencil and paper.

Self-Help Exercises. Immediately preceding each exercise set is a set of Self-Help Exercises. These exercises have been very carefully selected to bridge the gap between the exposition in the chapter and the regular exercise set. By doing these exercises and checking the complete solutions provided, students will be able to test or check their comprehension of the material. This, in turn, will better prepare them to do the exercises in the regular exercise set.

Exercises. The book contains over 3300 exercises. The exercises in each set gradually increase in difficulty, concluding with the Enrichment Exercises mentioned below. The exercise sets also include an extensive array of realistic applications from diverse disciplines, including numerous referenced examples extracted from current journals.

Enrichment Exercises. Fully in line with current goals in teaching and learning mathematics, every section in the text includes an Enrichment Exercise Set that encourages exploration, investigation, critical thinking, writing, and verbalization.

End-of-Chapter Projects. These projects, found at the end of each chapter, are especially good for group assignments. These projects are interesting and will serve to motivate the mathematics student.

Comment on the Trigonometric Functions. Before going through a chapter by chapter content overview some comments are needed concerning the trigonometric functions. Many instructors will simply omit the trigonometric functions. For those who include the trigonometric functions, some will wait until the second semester to cover these functions as presented in Chapter 9. Others, however, will want to include the trigonometric functions beginning in Chapter 1, and use them throughout the first semester. The text is written with enough flexibility to accommodate these later instructors.

For example, Sections 9.1 and 9.2 can be presented after Section 1.6 or after Section 1.8. Exercises on derivatives and integrals of the trigonometric functions have been included in many of the Enrichment Exercise sets. The derivatives of $\sin x$ and $\cos x$ are considered in Exercises 55 through 60 in Section 2.4. Notice that the derivatives of these two functions are found graphically in Exercise 59 and 60. Exercise 33 and 34 in Section 2.5 involve the tangent line approximation of these two functions. In Section 3.2, Exercises 49 through 54 involve $\sin x$ and $\cos x$ with the product and quotient rule, while Exercises 55 and 56 ask for the derivation of the derivatives of $\tan x$ and $\cot x$. Further exercise on the derivatives of powers of $\sin x$ and $\cos x$ are given at the end of Section 3.3. Exercises using the chain rule are given at the end of Section 3.4. Exercises in Chapter 5 derive and use the integrals of $\sin \mu$ and $\cos \mu$. Naturally, where appropriate, exercises from Section 9.3, 9.4, and 9.5 can be added to the above list. Thus, the material in Chapter 9 can be easily be integrated into the first five chapters.

STUDENT AIDS

- **Boldface** is used when defining new terms.
- **Boxes** are used to highlight definitions, theorems, results, and procedures.
- **Remarks** are used to draw attention to important points that might otherwise be overlooked.
- **Warnings** alert students against making common mistakes.
- **Titles** for worked examples help to identify the subject.
- **Chapter summary outlines**, at the end of each chapter, conveniently summarize all the definitions, theorems, and procedures in one place.

- **Review exercises** are found at the end of each chapter.
- **Chapter projects** are found at the end of each chapter.
- **Answers** to odd-numbered exercises and to all the review exercises are provided in an appendix.
- The **Technology Resource Manual** available with this text has all the details, in user-friendly terms, on how to carry out any of the graphing calculator operations used in the text.
- A **Student's Solution Manual** that contains completely worked solutions to all odd-numbered exercises and to all chapter review exercises is available.

INSTRUCTOR AIDS

- An **Instructor's Solution Manual** with completely worked solutions to the even-numbered exercises and to all the Explorations is available free to adopters. The **Student's Solution Manual** is free to adopters and contains the completely worked solutions to all odd-numbered and to all chapter review exercises. Between the two manuals all exercises are covered.
- The **Technology Resource Manual**, available with this text has all the details, in user-friendly terms, on how to carry out any of the graphing calculator operations used in the text. The manual includes the Texas Instrument TI-81, TI-82, and TI-85, the Casio fx-7700G and fx-8700G, and the Sharp EL-9200C and EL-9300C.
- A **TI Graphing Calculator Program Disk** is available free to all adopters and contains all the programs found in the text. This disk allows you to download the programs from your PC or Macintosh to a Texas Instrument TI-82 or TI-85, with the proper hardware.
- A **Test Bank** written by Joan Van Glabek (Collier County College) contains 100 questions per chapter and is set up like the exercise sets.
- A **Computerized Test Bank** allows instructors to quickly create, edit and print tests or different versions of tests from the set of test questions accompanying the text. It is available free to adopters and is available in IBM or Mac versions.
- **Graph 2D/3D**, a software package by George Bergeman, Northern Virginia Community College, is available free to users. This software graphs functions in one variable and graphs surfaces of functions in two variables. It also provides computational support for solving calculus problems and investigating concepts. It is available for IBM (or IBM-compatible) computers.

Custom Publishing

Courses in business calculus are structured in various ways, differing in length, content, and organization. To cater to these differences, Saunders College Publishing is offering **Applied Calculus with Technology** in a custom-publishing format. Instructors can rearrange, add, or cut chapters to produce a text that best meets their needs.

CHAPTER DEPENDENCIES

The diagram below shows chapter dependencies in Applied Calculus which instructors should consider. Beyond these dependencies, instructors, with custom publishing, are free to choose the topics they want to cover in the order they want to cover them, thereby creating a text that follows their course syllabi.

x   Preface

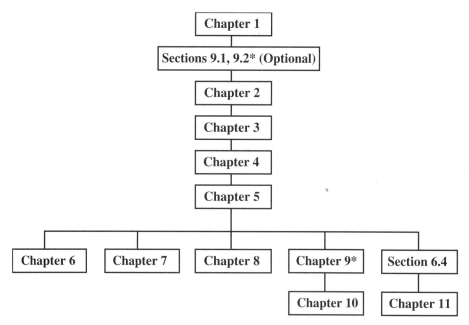

* See the two paragraphs following "Comments on the Trigonometric Functions" below.

Saunders College Publishing is working hard to provide the highest quality service and product for your courses. If you have any questions about custom publishing, please contact your local Saunders sales representative.

CONTENT OVERVIEW

Chapter 1. Section 1.0 contains some examples that clearly indicate instances when the graphing calculator fails to tell the whole story, and therefore motivates the need to learn the calculus. Chapter 4 features examples for which our current mathematical knowledge is inadequate to find the exact values of critical points, requiring us to use some approximation technique on our graphing calculators. This theme of needing both mathematical analysis and technology to solve important problems continues throughout the text. The rest of Chapter 1 presents a review of algebra topics. Depending on the preparation of the students, some of the material can be omitted. The first section presents standard coverage (with technology) of lines, the second presents linear models, including an introduction to the theory of the firm with some necessary economics background. The third section introduces functions, and the remainder of the chapter considers a variety of functions, together with graphing techniques and combinations of functions.

Chapter 2. Chapter 2 begins the study of calculus. The first section introduces limits intuitively, lending support with many geometric and numerical examples. The second section continues with continuity, limits at infinity, and a derivation of the natural exponential function from the idea of continuous compounding. The next two sections cover rates of change, slope of the tangent line to a curve, and the derivative. Using a program provided, students can use their graphing calculators to see secant lines converging to a tangent line to a curve. In the section on derivatives, graphing calculators are used to find the derivative of $f(x) = \ln x$. From the limit definition of derivative we know that for h small, $f'(x) \approx \dfrac{f(x + h) - f(x)}{h}$. We then take $h = 0.001$ and graph the function $g(x) = \dfrac{\ln(x + 0.001) - \ln x}{0.001}$. We see on our graphing calculator screens that $g(x) \approx 1/x$. Since $f'(x) \approx g(x)$, we then have strong evidence that $f'(x) = 1/x$. This is confirmed algebraically later in Chapter 3. The chapter ends with the tangent line approximation and local linearity near a point where the derivative exists.

Chapter 3. The first section of the chapter begins with some rules for derivatives. In this section we also discover the derivatives of a number of functions using graphing calculators. Just as we found the derivative of $\ln x$ in the preceding chapter, we graph $g(x) = \dfrac{f(x + 0.001) - f(x)}{0.001}$ for the functions $f(x) = x^2$, x^3, and x^4, and then discover from our graphing calculator screens what particular function $g(x)$ is in each case. Since $f'(x) \approx g(x)$, we then discover $f'(x)$. We then generalize to x^n. In the same way we find the derivative of $f(x) = e^x$. This is an exciting and innovative way for students to find these derivatives. Now that the derivatives of $\ln x$ and e^x are known, these functions can be used in conjunction with the product and quotient rules found in the second section, making this material more interesting and compelling. The third section covers the chain rule, and the fourth section derives the derivatives of the exponential and logarithmic functions in the standard fashion. The last section presents applications to business and economics.

Chapter 4. Graphing and curve sketching are introduced in this chapter. Section 1 describes the importance of the first derivative in graphing. We show clearly that our graphing calculator can fail to give a complete picture of the graph of a function, demonstrating the need for the calculus. We also consider examples for which the exact values of the critical points cannot be determined, and thus need to resort to using an approximation technique on our graphing calculators. Section 2 considers important and timely applications to Laffer curves in tax policy and to the harvesting of renewable natural resources. Section 3 presents the second derivative and its use in graphing, while Section 4 continues with more curve sketching. Extensive applications are given, including population growth, radioactive decay, and the logistic equation with derived estimates of the limiting human population of the earth. The chapter ends with optimization, implicit differentiation, related rates, and Newton's method.

Chapter 5. The first two sections of this chapter present antiderivatives and substitution. Section 3 lays the groundwork for the definite integral by considering left- and right-hand Riemann sums. Here again the graphing calculator plays a vital role. Using programs provided, students can easily graph the rectangles associated with these Riemann sums, and see graphically and numerically what happens as $n \to \infty$. The chapter continues with the definite integral, fundamental theorem of calculus, area between curves, and presents a number of additional applications of the integral, including average value, density, consumer's and producer's surplus, Lorentz's curves, and money flow.

Chapter 6. This chapter contains material on integration by parts, integration using tables, numerical integration, and improper integrals.

Chapter 7. The first sections presents an introduction to functions of several variables, including cost and revenue curves, Cobb-Douglas production functions, and level curves. The second section then introduces partial derivatives with applications that include competitive and complementary demand relations. The third section gives the second derivative test for functions of several variables and applied application on optimization. The fourth section covers Lagrange multipliers and carefully avoids algebraic complications. The method of least squares and correlation is presented in Section 5 and the tangent plane approximation is presented in the sixth section. The last section on double integrals covers double integrals over general domains, Riemann sums, and applications to average value and density. A program is given for the graphing calculator to compute Riemann sums over rectangular regions.

Chapter 8. This chapter is a brief introduction to differential equations and includes the technique of separation of variables, approximate solutions using Euler's method, some qualitative analysis, and mathematical problems involving the harvesting of a renewable natural resource. The graphing calculator is used to graph approximate solutions and to do some experimentation.

Chapter 9. This chapter covers an introduction to the trigonometric functions. The first section starts with angles and the next three section covers the sine and cosine functions, including differentiation and integration. The last section then covers the remaining trigonometric functions. Notice that these sections include extensive business applications, including models by Samuelson [137] and Phillips [138]. Notice in Section 9.3 that the derivatives of $\sin x$ and $\cos x$ are found using the graphing calculator and that the graphing calculator is used throughout this chapter.

Chapter 10. This chapter covers Taylor polynomials and infinite series. Sections 10.1, 10.2, and 10.7 constitute a subchapter on Taylor polynomials. Section 10.7 is written so that the reader can go from Section 10.2 directly to Section 10.7. The first section introduces Taylor polynomials and the second section considers the errors in Taylor polynomial approximation. The graphing calculator is used extensively to compare the Taylor polynomial with the approximated function. The last section looks at Taylor series, where the interval of convergence is found analytically in the simpler cases, while graphing calculator experiments cover the more difficult cases. Section 10.3 introduces infinite sequences and Sections 10.4 through 10.6 are on infinite series and includes a variety of test for convergence and divergence.

Chapter 11. This chapter is on probability. The first section is a brief review of discrete probability. The second section then considers continuous probability density functions and the third section presents the expected value and variance of these functions. The fourth section covers the normal distribution, arguably the most important probability density function. The graphing calculator is used to do many of the arithmetic calculations and to draw histograms.

ACKNOWLEDGMENTS

I appreciate Elizabeth Widdicombe, Publisher, and Jay Ricci, Executive Editor, for their generous support of this project. I greatly appreciate the very important help provided by my Developmental Editor, Anita Fallon.

My thanks to Laura Shur, Project Editor, Joan Wendt, Senior Art Director, Linda Davoli, Copy Editor, and Joanne Cassetti, Production Manager for a great job.

I wish to thank the Department of Mathematics here at the University of Connecticut for their collective support, and a particular thanks to Jeffrey Tollefson for his encouragement.

I wish to express my sincere appreciation to each of the reviewers for their many helpful suggestions. Rich Cambell, Butte College, Bob Denton, Orange Coast College, Gudryn Doherty, Community College of Denver, Michael Dutko, University of Scranton, Harvey Greenwald, California State Polytechnic University, Robert Goad, Sam Houston State University, Linda Halligan, Mohawk Valley Community College, Yvette Hester, Texas A & M University, Miles Hubbard, St. Cloud State University, John Lawlor, University of Vermont, Jaclyn LeFebvre, Illinois Central College, Joyce Longman, Villanova University, Mark Palko, University of Arkansas, Don Pierce, Western Oregon State College, Georgia Pyrros, University of Delaware, Geetha Ramanchandra, California State University, Sacramento, Deborah Ritchie, Moorpark College, Dale Rohm, University of Wisconsin—Stevens Point, Arlene Sherburn, Montgomery College, Steven Terry, Ricks College, Stuart Thomas, University of Oregon, Richard Witt, University of Wisconsin—Eau Claire, Judith Wolbert, Michigan State University, Cathleen Zucco, LeMoyne College.

Thanks also to the people listed below for taking the time to fill out a detailed questionnaire about their business calculus courses. Their responses were invaluable. Tom Adamson, Phoenix College, Keith Alford, Alcorn State University, Dan Anderson, Parkland College, Chris Barker, DeAnza College, Arlene Blasius, SUNY, College at Old Westbury, Bob Branch, Spokane Community College, Frank Caldwell, York Technical College, Connie Campbell, Millsaps College, Rich Campbell, Butte College, Roger Cooke, University of Vermont, Richard A. Didio, LaSalle University, Diane Doyle, Adirondack Community College, Margaret Ehringer, Indiana University, Southeast, Betty Fein, Oregon State University, James O. Friel, California State University, Fullerton, Deborah Garner, Umpqua Community College, Debbie Garrison, Valencia Community College, East, A. Karen Gragg, New Mexico State University, Carlsbad, Thomas Gruszka, Western New Mexico University, R. Guralnick, University of Southern California, Chris Haddock, Bentley College, Daniel L. Hansen, Northeastern Oklahoma State University, Lonnie Hass, North Dakota State University, Barbara A. Honhart, Baker College, Linda Jones, University of California, Davis, Gerald R. Krusinski, College of Du Page, Larry R. Lance, Columbus State Community College, C. Lando, University of Alaska, Fairbanks, Mary Ann Lee, Mankato State University, Jaclyn LeFebvre, Illinois Central College, M. Lehmann, University of San Francisco, Ricardo A. Martinez, Foothill College, Roger Maurer, Linn-Benton Community College, Ruth A. Meyer, Western Michigan University, David Meyers, Bakersfield College, Syed Moiz, Galveston College, Lorraine Edson-Perone, Cerritos College, Dennis Parker, University of the Pacific, Bob Pawlowski, Lansing Community College, William B. Peirce, Cape Cod Community College, Don Philley, Monterey Pennisula College, Wallace Pye, University of Southern Mississippi, Rosalind Reichard, Elon College, Kathy V. Rodgers, University of Southern Indiana, Howard Rolf, Baylor University, P. Rosnick, Greenfield Community College, Edward Rozema, University of Tennessee, Chattanooga, Frederick M. Russell, Charles County Community College, Tami Ryan, Valencia Community College, Helen E. Salzberg, Rhode Island College, Nancy Sattler, Terra Community College, Dan Schapiro, Yakima Valley Community College, M. Scott, Western Illinois University, Arlene Sherburne, Montgomery College, Toby Shook, Asheville-Buncombe Technical Community College, Minnie W. Shuler, Gulf Coast Community College, Brenda M. Shryock, Wake Technical Community College, Marlene Sims, Kennesaw State College, Gerald Skidmore, Alvin Community College, M. Smith, Georgia State University, Carol Soos, Beleville Area College, Richard Tebbs, Southern Utah University, Steven Terry, Ricks College, Gwen Terwilliger, University of Toledo, Anthony D. Thomas, University of Wisconsin, Platteville, Carolyn R. Thomas, San Diego City College, Stuart Thomas, University of Oregon, Joseph A. Tovissi, Cañada College, Paul J. Welsh, Pima Community College East, Randy Westhoff, Bemidji State University, June White, St. Petersburg Junior College, Jim Wooland, Florida State University, Ben F. Zirkle, Virginia Western Community College.

A particular thanks to Jack Porter and Joan Van Glabek (Collier County College) for checking the accuracy of the manuscript.

On a personal level, I am grateful to my wife Nancy for her love, patience and support.

Edmond Tomastik

October 1995

TABLE OF CONTENTS

APPLICATIONS INDEX

Life Sciences

Physical Sciences

Social Sciences

1

Functions

After an introduction to the graphing calculator, this chapter introduces lines and mathematical models. The remainder of the chapter covers functions, which form the basis of calculus. We introduce the notion of a function and then explore the properties and graphs of a variety of functions.

1.0 INTRODUCTION TO THE GRAPHING CALCULATOR

■ The Viewing Window   ■ Complete Graphs   ■ Applications

APPLICATION: BUDGET EQUATION

A person has $6 to purchase oranges and apples. Oranges cost $0.50 and apples $0.40. If x is the number of oranges and y the number of apples, write an equation that x and y must satisfy. How would you graph this equation on your graphing calculator? See Example 4 for the answer.

APPLICATION: PRICE OF OIL

Chapman [1] estimated that the relationship between the price x of a barrel of oil and the number y of billions of barrels of oil used in the world per year is $y = 30 - 0.4x$. Using a graphing calculator how can you determine the price of a barrel of oil if 22 billions of barrels of oil are used in the world per year? For the answer see Example 5.

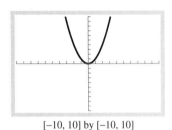

[−10, 10] by [−10, 10]

Screen 1.1
A graph of $y = x^2$ in the standard viewing window.

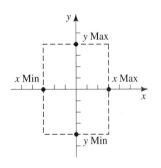

Figure 1.1
The dashed line outlines the viewing window. The dimensions of the window are [x_{min}, x_{max}] by [y_{min}, y_{max}]. The scale units are determined by x_{scl} and y_{scl} found among the range variables.

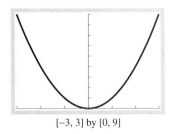

[−3, 3] by [0, 9]

Screen 1.2
Another graph of $y = x^2$.

The Viewing Window

This section is an introduction on how to use the graphing calculator. We begin by examining the viewing window and then, appropriately, by drawing a graph. We then see how to use many of the functions on the graphing calculator, use these functions to help us analyze and explore graphs, and use our grapher to solve problems. Using the grapher leads to a deeper understanding of the mathematics we are learning and lets us explore and ''do'' mathematics in exciting and creative ways.

For additional information, consult your owner's guide or the technology resource manual for your particular calculator.

EXAMPLE 1   The Viewing Window

Graph $y = x^2$ on your graphing calculator.

SOLUTION ■ First enter the equation $y = x^2$, and then graph. You probably see something like that shown in Screen 1.1. Your graph may not be connected, but in this text we will usually draw connected graphs. ■

The dashed rectangle in Figure 1.1 is the boundary of the viewing screen. The x-axis is marked from xMin to xMax and the y-axis from yMin to yMax. We shall say that this viewing window has dimensions [xMin, xMax] by [yMin, yMax]. Check the RANGE variables to see what they are. If your viewing window does not have the dimensions [−10, 10] by [−10, 10], change the RANGE variables so that it has these dimensions and graph. You then obtain Screen 1.1. The viewing window with dimensions [−10, 10] by [−10, 10] is referred to as the *standard viewing window*.

Notice in Screen 1.1 that the graph only takes up a small portion of the standard viewing window. Set the dimensions of your viewing window using the RANGE variables to [−3, 3] by [0, 9], graph again, and obtain Screen 1.2. Notice that we now are using all of the available screen.

EXPLORATION **1**

The Scale Marks

 (a) Graph $y = x^2$ using a viewing window with dimension $[-4, 4]$ by $[0, 10]$. Make sure the scale units are one.

(b) Now change the scale unit on the y-axis to 2 and graph. What is different?

(c) Now also change the scale unit on the x-axis to 2 and graph. What has changed?

Often we want to evaluate a function $y = f(x)$ at integer values or some other specific values. For example, we might wish to evaluate $y = f(x) = x^2$ in increments of tenths. Using TRACE on Screen 1.2 try to evaluate $y = x^2$ at $x = 1.5$. What happens? As we move the cursor on the TI-82, for example, using the last screen, the x-coordinate jumps from a value of about 1.468 to about 1.532. How can we get the cursor on exactly $x = 1.5$? Now we detail how to do this.

On your graphing calculator remove $y = x^2$ and set $x_{\min} = 0$, $x_{\max} = 10$, $y_{\min} = 0$, and $y_{\max} = 10$. Use the cursor controls to place the cursor at the point at the extreme lower left corner of the window. Notice that the point is given by $(0, 0) = (x_{\min}, y_{\min})$.

Now use the cursor control to move the cursor one unit to the right. Note the x coordinate and call it x_{step}. This is the size of each step. You can obtain the number N_x of these steps for your calculator by calculating (and rounding off to the nearest integer) the number $N_x = 10/x_{\text{step}}$, where 10 is the width of the current screen.

Now use the cursor control to move the cursor one unit up. Note the y-coordinate and call it y_{step}. This is the size of each step. You can obtain the number N_y of these steps for your calculator by calculating (and rounding off to the nearest integer) the number $N_y = 10/y_{\text{step}}$, where 10 is the height of the current screen.

The TI-82 graphing calculator has $N_x = 94$. Suppose we want the cursor to move in steps of size 0.1. Then $x_{\max} - x_{\min}$ must satisfy

$$\frac{x_{\max} - x_{\min}}{x_{\text{step}}} = N_x$$

On the TI-82 recall that $N_x = 94$. Thus

$$\frac{x_{\max} - x_{\min}}{x_{\text{step}}} = N_x$$

$$\frac{x_{\max} - x_{\min}}{0.1} = 94$$

$$x_{\max} - x_{\min} = 9.4$$

If $x_{\min} = 0$, then we must have $x_{\max} = 9.4$. Try it. Taking $x_{\max} = 4.7$ results in the cursor moving in steps of 0.05. Try it.

EXPLORATION **2**

The Step Size

 Find N_x on your calculator, and do something similar to what was done in the previous paragraph.

Complete Graphs

We (informally) call a graph *complete* if the portion of the graph we see in the viewing window suggests all the important features of the graph. For example if some interesting feature occurs beyond the viewing window, then the graph is not complete. If the graph has some important wiggle that does not show on the viewing window because the scale of the graph is too large, then again the graph is not complete. Unfortunately, no matter how large or small the scale of the graph, we can never be certain that some interesting behavior may be occurring outside the viewing window or some interesting wiggles hidden within the curve that we see. Thus, if we only use a graphing utility on a graphing calculator or computer, we may overlook important discoveries. This is one reason why we need to carefully do a *mathematical* analysis.

EXAMPLE 2   Complete Graphs

Graph $y = x^4 - 12x^3 + x^2 - 2$ using the standard viewing window on your graphing calculator.

SOLUTION ■ The graph is shown in Screen 1.3. Now set the dimensions of your viewing window to $[-5, 14]$ by $[-2500, 1000]$ and obtain Screen 1.4. Notice the missing behavior we have now discovered. ■

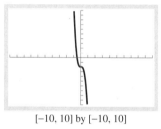

[−10, 10] by [−10, 10]

Screen 1.3
A graph of $y = x^4 - 12x^3 + x^2 - 2$ in the standard viewing window.

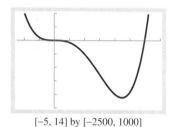

[−5, 14] by [−2500, 1000]

Screen 1.4
A graph of $y = x^4 - 12x^3 + x^2 - 2$ showing some hidden behavior.

EXAMPLE 3   Complete Graphs

Graph $y = x^4 - 2x^3 + x^2 - 2$ using the standard viewing window on your graphing calculator.

SOLUTION ■ The graph is shown in Screen 1.5. Now use the ZOOM feature of your graphing calculator and ZOOM about $(0, -2)$ and obtain Screen 1.6. Notice the missing behavior in the form of a wiggle that we have now discovered. ■

The last two examples indicate the shortfalls of using a graphing utility on a graphing calculator or computer. You can never know if some interesting behavior is taking place just outside the viewing screen, no matter how large the screen. Also, using

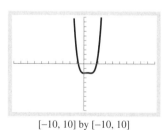

[−10, 10] by [−10, 10]

Screen 1.5
A graph of $y = x^4 - 2x^3 + x^2 - 2$ in the standard viewing window.

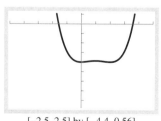

[−2.5, 2.5] by [−4.4, 0.56]

Screen 1.6
A graph of $y = x^4 - 2x^3 + x^2 - 2$ showing some hidden behavior.

only a graphing utility how can we ever know if there are some hidden wiggles somewhere in the graph? We cannot ZOOM everywhere and forever!

We will be able to determine complete graphs by using calculus. In Chapter 4 we will use calculus to find all the wiggles and hidden behavior of a graph.

Applications

EXAMPLE 4   Budget Equation

A person has $6 to purchase oranges and apples. Oranges cost $0.50 and apples $0.40. Let x be the number of oranges and y the number of apples.

(a) Write an equation that x and y must satisfy.

(b) Graph using a window with dimensions $[-1, 22.5]$ by $[-15, 15]$. (Change the dimensions of the viewing window so that on your graphing calculator the x-coordinate of the cursor can be placed at integer values. Consult the technology resource manual for your calculator.)

(c) Use the TRACE function to determine how much y changes when x changes by 1.

(d) Use TRACE to find the point on the graph with x-coordinate equal to 4. What is the y-coordinate and what does the point (x, y) mean in terms of oranges and apples?

(e) Find the x-coordinate of the point where the graph intersects the x-axis.

SOLUTION ■

(a) Because x is the number of oranges and each orange costs $0.50, $0.50x$ is the total number of dollars spent on oranges. In a similar fashion $0.40y$ is the total number of dollars spent on apples. The total spent on oranges and apples is $6, so we must have $0.50x + 0.40y = 6$.

(b) To graph on our graphing calculator we must solve the equation for y. Solving for y gives

$$0.50x + 0.40y = 6$$

$$0.40y = -0.50x + 6$$

$$y = -1.25x + 15$$

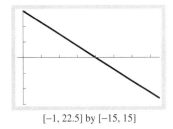

$[-1, 22.5]$ by $[-15, 15]$

Screen 1.7
A graph of $y = -1.25x + 15$.

Graphing $y = -1.25x + 15$ using a window with the requested dimensions gives Screen 1.7.

(c) Use the TRACE function to place the x-coordinate at 1. Assuming the x-coordinate is at 1, the y-coordinate is at 13.75. Moving the x-coordinate to 2 gives the y-coordinate as 12.5. Thus, y has changed by -1.25.

(d) Using the TRACE function, place the x-coordinate at 4. The y-coordinate reads 10. This means that we can purchase 4 oranges and 10 apples with $6.

(e) Move the cursor to a point as near as possible to the point where the line crosses the x-axis. It is possible that you are able to place the cursor so that the y-coordinate reads exactly 0. The corresponding x-coordinate is the number we are seeking and should be equal to 12. If as you move the cursor from left to right the y-coordinate jumps from positive to negative and you are unable to find a place where the y-coordinate is exactly 0, then you will need to use the ZOOM feature. By ZOOMing repeatedly you will eventually be able to place the y-coordinate at 0 or very close to 0, with the x-coordinate reading 12. Further ZOOMing will not change the x-coordinate of 12. ■

EXAMPLE 5   Intersection of Graphs

Chapman [1] estimated that the relationship between the price x of a barrel of oil and the number y of billions of barrels of oil used in the world per year is $y = 30 - 0.4x$. Graph using a graphing utility. By also graphing a certain horizontal line and using

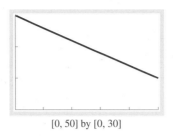

[0, 50] by [0, 30]

Screen 1.8
A graph of $y = 30 - 0.4x$.

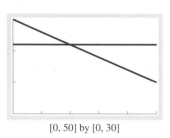

[0, 50] by [0, 30]

Screen 1.9
A graph of $y_1 = 30 - 0.4x$ and $y_2 = 22$.

the ZOOM feature, determine the price of a barrel of oil if 22 billions of barrels of oil are used in the world per year. Confirm algebraically.

S O L U T I O N ■ We graph $y_1 = 30 - 0.4x$ using a window with dimensions [0, 50] by [0, 30] and obtain Screen 1.8. Now in the same window also graph $y_2 = 22$ and obtain Screen 1.9. By ZOOMing a number of times we find that the x-coordinate of the point of intersection is 20.

To confirm algebraically, we need to solve for x in the equation $22 = 30 - 0.4x$. We have

$$22 = 30 - 0.4x$$

$$0.4x = 30 - 22 = 8$$

$$x = \frac{8}{0.4} = 20 \quad ■$$

SELF-HELP EXERCISE SET 1.0

1. If d is the twig diameter in inches of bitter bush, then it is estimated that the length l in inches of the twig is given by $l = 1.25 + 89.83d$. Use your grapher to find the twig diameter to four decimal places if the length is 10 inches.

EXERCISE SET 1.0

In Exercises 1 through 10 graph using your graphing calculator and using the standard viewing window.

1. $y = -2x + 1$ **2.** $y = -2x - 2$

3. $y = 3x - 2$ **4.** $y = 2x + 3$

5. $y = 3$ **6.** $x = 4$

7. $y = 3x^2$ **8.** $y = -x^2$

9. $y = x^3$ **10.** $y = \sqrt{x}$

In Exercises 11 through 14 determine which points lie in the given viewing window.

11. Viewing window: $[-10, 10]$ by $[-10, 10]$.
Points: $(8, 9)$, $(-7, -8)$, $(9, 11)$, $(-13, 2)$

12. Viewing window: $[-10, 10]$ by $[-10, 10]$
Points: $(7, -9)$, $(-7, 4)$, $(11, 11)$, $(3, -12)$

13. Viewing window: $[0, 10]$ by $[0, 20]$.
Points: $(7, -9)$, $(7, 14)$, $(11, 19)$, $(9, 22)$

14. Viewing window: $[0, 5]$ by $[0, 10]$.
Points: $(3, -1)$, $(7, 4)$, $(1, 9)$, $(4, 2)$

In Exercises 15 through 18 determine the smallest viewing window that contains the given points.

15. $(6, 1)$, $(5, -3)$, $(-1, 2)$ **16.** $(-4, -1)$, $(-5, 3)$, $(-1, 1)$

17. $(4, 5)$, $(6, -2)$, $(1, -1)$ **18.** $(1, 6)$, $(5, 3)$, $(-1, 5)$

In Exercises 19 through 26 determine which viewing window gives the better graph of the indicated equation.

19. $y = -x + 20$. (a) $[-10, 10]$ by $[-10, 10]$ (b) $[-10, 30]$ by $[-10, 30]$

20. $y = x + 20$. (a) $[-10, 10]$ by $[-10, 10]$ (b) $[-30, 10]$ by $[-10, 30]$

21. $y = x^2 - 50x + 100$. (a) $[-10, 10]$ by $[-10, 10]$ (b) $[-30, 60]$ by $[-600, 200]$

22. $y = -2x^2 + 30x + 20$. (a) $[-10, 10]$ by $[-10, 10]$ (b) $[-5, 20]$ by $[-100, 200]$

23. $y = x^3 - x + 10$. (a) $[-10, 10]$ by $[-10, 10]$ (b) $[-5, 5]$ by $[-10, 20]$

24. $y = -x^3 + 30x$. (a) $[-10, 10]$ by $[-10, 10]$ (b) $[-10, 10]$ by $[-100, 100]$

25. $y = x^4 - x^2$. (a) $[-10, 10]$ by $[-10, 10]$ (b) $[-3, 3]$ by $[-1, 3]$

26. $y = 0.001x^2 + x + 1$. (a) $[-10, 10]$ by $[-10, 10]$ (b) $[-1500, 500]$ by $[-300, 200]$

Applications

27. Profits. A furniture store sells chairs at a profit of $100 each and sofas at a profit of $150 each. Let x be the number of chairs sold each week and y the number of sofas sold each week.
  (a) If the profit in one week is $3000, write an equation that x and y must satisfy. Graph using your graphing calculator.
  (b) Use the TRACE function to determine how much y changes when x changes by 1.
  (c) Use TRACE to find the point on the graph with x-coordinate equal to 3. What is the y-coordinate, and what does the point (x, y) mean in terms of chairs and sofas?
  (d) Find the x-coordinate of the point where the graph intersects the x-axis.

28. Revenue. A restaurant has two specials: steak and chicken. The steak special is $13, and the chicken special is $10. Let x be the number of steak specials served and y the number of chicken specials.
  (a) If the restaurant took in $390 in total sales on these specials, find the equation that x and y must satisfy. Graph using your graphing calculator.
  (b) Use the TRACE function to determine how much y changes when x changes by 1.
  (c) Use TRACE to find the point on the graph with x-coordinate equal to 3. What is the y-coordinate, and what does the point (x, y) mean in terms of the steak and chicken specials?
  (d) Find the x-coordinate of the point where the graph intersects the x-axis.

29. Nutrition. An individual needs about 820 mg of calcium daily in his diet but is unable to eat any dairy products due to the large amounts of cholesterol in such products. This individual does however enjoy eating canned sardines and steamed broccoli. Let x be the number of ounces of canned sardines consumed each day and let y be the number of cups of steamed broccoli.
  (a) If there is 125 mg of calcium in each ounce of canned sardines and 190 mg in each cup of steamed broccoli, what equation must x and y satisfy if this person is to obtain his daily need of calcium from these two sources? Graph using your graphing calculator.
  (b) Use the TRACE function to determine how much y changes when x changes by 1.
  (c) Use TRACE to find the point on the graph with x-coordinate equal to 2. What is the y-coordinate, and what does the point (x, y) mean in terms of the amount of calcium and canned sardines?
  (d) Find the x-coordinate of the point where the graph intersects the x-axis.

30. Costs. A contractor builds ranch and split-level style homes. The ranch costs $130,000 to build and the split-level $150,000. Let x be the number of ranch-style homes built and y the number of split-level style homes.
  (a) If the contractor has $1,360,000 to build these homes, find the equation that x and y must satisfy. Graph using your graphing calculator.
  (b) Use the TRACE function to determine how much y changes when x changes by 1.
  (c) Use TRACE to find the point on the graph with x-coordinate equal to 2. What is the y-coordinate, and what does the point (x, y) mean in terms of the cost of each style home?
  (d) Find the x-coordinate of the point where the graph intersects the x-axis.

31. Sociology. Leigh [2] showed that the equation $y = -2011.44 + 1.06x$ was approximately true, where x is the calendar year (from 1958–1967) and y the percentage of whites who would not move if blacks came to live next door. Graph using your graphing calculator. Graphically find the year in which, according to the given equation, 69.34% of whites would not move if blacks came to live next door.

32. Forestry. Ewel [3] showed that the equation $T_s = 8.71 + 0.533T_a$ approximates the relationship between the soil and ground temperatures in a Florida pine plantation, where T_s is soil temperature and T_a is air temperature in degrees Celsius. Graph using your graphing calculator. Graphically find the air temperature if the soil temperature is 19.37°.

SOLUTIONS TO SELF-HELP EXERCISE SET 1.0

1. Taking $y_1 = 1.25 + 89.83x$ and $y_2 = 10$ and using a window with dimensions $[0, 0.5]$ by $[0, 50]$ we obtained the accompanying screen. By ZOOMing repeatedly we then obtained $x = 0.0974$ to four decimal places.

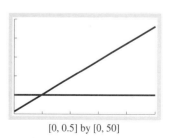

[0, 0.5] by [0, 50]

1.1 LINES

■ Slope ■ Equations of Lines ■ Applications

APPLICATION: PRODUCTION OF VACUUM CLEANERS

A firm produces two models of vacuum cleaners: regular and deluxe. The regular model requires 8 hours of labor to assemble, and the deluxe requires 10 hours of labor. Let x be the number of regular models produced and y the number of deluxe models. If the firm wishes to use 500 hours of labor to assemble these cleaners, find an equation that the two quantities x and y must satisfy. If the number of regular models to be assembled is increased by 5, how many fewer deluxe models can be assembled? See Example 8 for the answer.

APPLICATION: TROPICAL RAIN FOREST DESTRUCTION

According to data collected by Aiken [4] the equation $p = -1.10t + 2229$ approximates the percentage of rain forest cover in Peninsular Malaysia, where t is the calendar year. What is the significance of the constant -1.10? For the answer see Example 9.

Slope

To describe a straight line, one must first describe the "slant," or **slope**, of the line. See Figure 1.2.

Slope of a Line

Let (x_1, y_1) and (x_2, y_2) be two points on a line L. If the line is nonvertical ($x_1 \neq x_2$), the **slope** m of the line L is defined as

$$m = \frac{y_2 - y_1}{x_2 - x_1}$$

If L is vertical ($x_2 = x_1$), then the slope is said to be **undefined**.

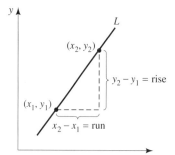

Figure 1.2
The slope of the line is $m = (y_2 - y_1)/(x_2 - x_1)$.

Remark. The term $x_2 - x_1$ is called the **run**, and $y_2 - y_1$ is called the **rise**. See Figure 1.2. Thus, the slope m can be thought of as

$$m = \frac{y_2 - y_1}{x_2 - x_1} = \frac{\text{rise}}{\text{run}}$$

If (x_1, y_1) and (x_2, y_2) are two points on a vertical line L, then we must have $x_1 = x_2$. (Why?) Using the formula for slope yields

$$m = \frac{y_2 - y_1}{x_2 - x_1} = \frac{y_2 - y_1}{0}$$

which is undefined. This is why a vertical line is said to have an undefined slope.

The slope of the line does not depend on which two points on the line are chosen to be used in the above-mentioned formula. To see this, notice in Figure 1.3 that the two right triangles are similar because the corresponding angles are equal. Thus, the ratio of the corresponding sides must be equal. That is,

$$\frac{b}{a} = \frac{d}{c}$$

where the ratio on the left is the slope using the points A and D, and the ratio on the right is the slope using the points B and C.

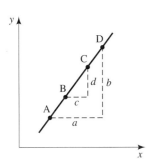

Figure 1.3
The same slope is obtained by using any two points on the line.

EXAMPLE 1 Finding the Slope of Lines

Find the slope (if it exists) of the line through each pair of points. Sketch the line.
(a) (1, 2), (3, 6) (b) (−3, 1), (3, −2) (c) (−1, 3), (2, 3) (d) (3, 4), (3, 1)

SOLUTIONS ■
(a) $(x_1, y_1) = (1, 2)$ and $(x_2, y_2) = (3, 6)$. So

$$m = \frac{y_2 - y_1}{x_2 - x_1} = \frac{(6) - (2)}{(3) - (1)} = \frac{4}{2} = 2$$

See Figure 1.4. Notice that for each unit we move to the right, the line moves up $m = 2$ units.
(b) $(x_1, y_1) = (−3, 1)$ and $(x_2, y_2) = (3, −2)$. So

$$m = \frac{y_2 - y_1}{x_2 - x_1} = \frac{(−2) - (1)}{(3) - (−3)} = \frac{−3}{6} = -\frac{1}{2}$$

See Figure 1.5. Notice that for each unit we move to the right, the line moves down $\frac{1}{2}$ unit.
(c) $(x_1, y_1) = (−1, 3)$ and $(x_2, y_2) = (2, 3)$. So

$$m = \frac{y_2 - y_1}{x_2 - x_1} = \frac{(3) - (3)}{(2) - (−1)} = \frac{0}{3} = 0$$

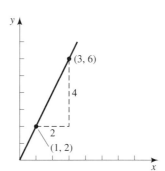

Figure 1.4
The slope is $m = 4/2 = 2$.

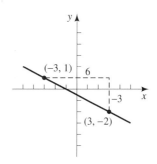

Figure 1.5
The slope is $m = -3/6 = -1/2$.

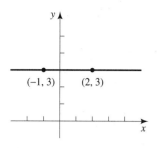

Figure 1.6
The slope is $m = 0/3 = 0$.

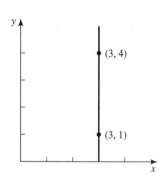

Figure 1.7
The slope is undefined.

See Figure 1.6. Notice that this line is horizontal.

(d) $(x_1, y_1) = (3, 4)$ and $(x_2, y_2) = (3, 1)$. So

$$m = \frac{y_2 - y_1}{x_2 - x_1} = \frac{(1) - (4)}{(3) - (3)} = \frac{-3}{0} = \text{undefined}$$

See Figure 1.7. Notice that this line is vertical. ∎

EXPLORATION *1*

Slope

  (a) (For calculators with Grid On Mode.) Set the MODE of your graphing calculator to Grid On. Use a window with dimensions [−4.7, 4.7] by [0, 5], and set $x_{sc\ell} = y_{sc\ell} = 1$. Use a square window, if possible. Graph the following four equations on the same standard viewing window:

$$y = x + 3 \qquad y = 2x + 3 \qquad y = -x + 3 \qquad y = -2x + 3$$

Using the grid, estimate the slope of each line.

(b) Using TRACE, move along the graph of each of the lines $y_1 = 2x + 3$ and $y_2 = -2x + 3$ in turn. Verify that as the cursor moves to the right, the y-coordinate on the first line changes twice as much as the x-coordinate, and the y-coordinate on the second line changes negatively twice as much as the x-coordinate.

EXPLORATION *2*

Equations of Lines

  (For calculators with Grid On Mode.) Set the MODE of your graphing calculator to Grid On. Use a square window, if possible. Graph $y_1 = 2x$ and $y_2 = -0.5x - 0.5$ on a screen with dimensions [−4.7, 4.7] by [−2, 6] and with $x_{sc\ell} = y_{sc\ell} = 0.5$. Verify that the graphs are the line shown in Figures 1.4 and 1.5, and go through the points shown in these figures. By using the grid, notice the change in the y-coordinate when the x-coordinate is increased by 1. What relationship does this change in y-coordinate have with the slope? Establish using algebra.

Consider any line L that is not vertical, and let $P_1(x_1, y_1)$ be one point on the line. We noted earlier that in determining the slope of a line, *any* two points on the line can be used. We now take a second point $P_2(x_2, y_2)$ on the line where P_2 is chosen with $x_2 = x_1 + 1$. Thus, we have run $= x_2 - x_1 = 1$, and therefore

$$m = \frac{y_2 - y_1}{x_2 - x_1} = \frac{y_2 - y_1}{1} = y_2 - y_1 = \text{rise} \qquad [1]$$

This says that if the run is taken as 1, then the slope equals the rise.

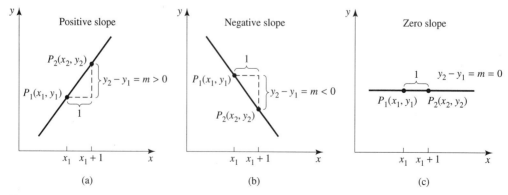

Figure 1.8
If the run is 1, the slope is $y_2 - y_1$. (a) Positive slope. (b) Negative slope. (c) Zero slope.

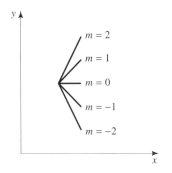

Figure 1.9
The slopes of different lines.

Now, if m is positive, as in Figure 1.8a, then moving 1 unit to the right results in moving *up* m units. Thus, the line *rises* (moving from left to right), and the larger the value of m the steeper the rise.

If m is negative, as in Figure 1.8b, then moving 1 unit to the right results in moving *down* $|m| = -m$ units. Thus, the line *falls*, and the more negative the value of m, the steeper the fall.

If m is zero, as in Figure 1.8c, then moving over 1 unit results in moving up *no* units. Thus, the line is *horizontal*.

The first three parts of Example 1 give specific examples of these three general cases.

Figure 1.9 shows several lines through the same point with different slopes.

EXAMPLE 2   Using the Slope of a Line

Find the y-coordinate of point P if P has x-coordinate 5 and P is on the line through $(2, 7)$ with slope -2.

SOLUTION ■ See Figure 1.10. Using the definition of slope, we have

$$-2 = \frac{y - 7}{5 - 2}$$

$$-2 = \frac{y - 7}{3}$$

$$-6 = y - 7$$

$$y = 1 \quad ■$$

Equations of Lines

Suppose we have a line through the point (x_1, y_1) with slope m as shown in Figure 1.11. If (x, y) is any other point on the line, then we must have

$$m = \frac{y - y_1}{x - x_1}$$

or

$$y - y_1 = m(x - x_1)$$

This equation is the **point-slope equation**.

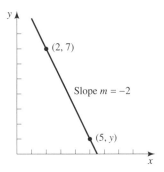

Figure 1.10
The line through $(2, 7)$ with slope $m = -2$.

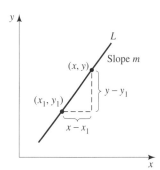

Figure 1.11
The line through the points (x_1, y_1) and (x, y).

The Point-Slope Equation of a Line

> The equation of the line through the point (x_1, y_1) with slope m is given by
>
> $$y - y_1 = m(x - x_1)$$

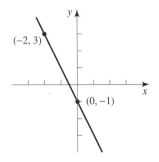

(−2, 3)

(0, −1)

Figure 1.12
The line through (−2, 3) with
slope $m = -2$.

EXAMPLE 3   Finding the Point-slope Equation of a Line

Find an equation of the line through $(-2, 3)$ with slope $m = -2$. Sketch the graph.

SOLUTION ■ Because $(x_1, y_1) = (-2, 3)$, the point-slope equation is

$$y - y_1 = m(x - x_1)$$
$$y - 3 = -2[x - (-2)]$$
$$y = -2x - 1$$

One quick way of finding the graph is to notice that when $x = 0$, $y = -2(0) - 1 = -1$. Thus $(0, -1)$ is a second point on the graph. See Figure 1.12.   ■

A point (x, y) is on the graph of a vertical line such as the one shown in Figure 1.13, if and only if $x = a$, where $(a, 0)$ is the point where the vertical line crosses the x-axis. Thus, we have the following.

Vertical Lines

> A vertical line has the equation
>
> $$x = a$$
>
> where $(a, 0)$ is the point at which the vertical line crosses the x-axis.

Given any line that is not vertical, then the line has a slope m and by Figure 1.14 must cross the y-axis at some point $(0, b)$. The number b is called the **y-intercept**. Thus, we can use the point-slope equation, where $(x_1, y_1) = (0, b)$ is a point on the line. This gives

$$y - b = m(x - 0)$$
$$y = mx + b$$

This is the **slope-intercept equation** of a nonvertcal line.

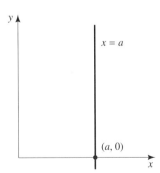

Figure 1.13
A vertical line through (a, 0) has
the equation $x = a$.

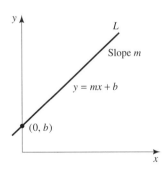

Figure 1.14
The line through the point (0, b)
with slope m has equation $y = mx + b$.

The Slope-Intercept Equation of a Line

The equation of the line with slope m and y-intercept b is given by

$$y = mx + b$$

Slope	Graph	Equation
$m > 0$	rises ($\nearrow$)	$y = mx + b$
$m < 0$	falls ($\searrow$)	$y = mx + b$
$m = 0$	horizontal ($\leftrightarrow$)	$y = $ constant
m undefined	vertical ($\updownarrow$)	$x = $ constant

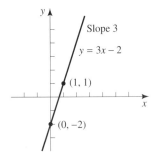

Figure 1.15
The graph of the equation with slope 3 and y-intercept -2.

EXAMPLE 4 Finding the Slope-Intercept Equation of a Line

Find the equation of a line with slope 3 and y-intercept -2. Draw a graph.

SOLUTION ■ We have $m = 3$ and $b = -2$. Thus,

$$y = mx + b = 3x - 2$$

To draw a graph, we first notice that since the y-intercept is -2, the point $(0, -2)$ is on the graph (see Figure 1.15). Since the slope is 3, if we move over 1 unit to the right, the line moves up 3 units so that the point $(1, 1)$ is also on the line. ■

If a and b are not both zero, the equation $ax + by = c$ is called a **linear equation**. The following can be proven. (See Exercise 59.)

The General Form of the Equation of a Line

If a and b are not both zero, then the graph of a linear equation $ax + by = c$ is a straight line, and, conversely, every line is the graph of a linear equation.

We already defined the y-intercept. We now need to define the x-intercept. Any line that is not horizontal must cross the x-axis at some point $(a, 0)$. The number a is called the **x-intercept**. To find the x-intercept of a line, simply set $y = 0$ in the equation of the line and solve for x. Similarly, to find the y-intercept of a line, set $x = 0$ in the equation of the line and solve for y. The intercepts are very useful in graphing, as the next example illustrates.

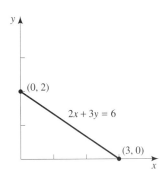

Figure 1.16
The graph of $2x + 3y = 6$ has x-intercept 3 and y-intercept 2.

EXAMPLE 5 Graphing an Equation of a Line

Sketch a graph of $2x + 3y = 6$ by finding the x- and y-intercepts. Support with a grapher.

SOLUTION ■ Since the equation is in the form $ax + by = c$, we know that the graph of this equation is a line. Setting $x = 0$, yields $3y = 6$, or $y = 2$, as the y-intercept. Setting $y = 0$, yields $2x = 6$, or $x = 3$, as the x-intercept. We then obtain Figure 1.16.

To support with a grapher, we must first solve for y and obtain $y = -\frac{2}{3}x + 2$. Using a window with dimension $[-1, 3.7]$ by $[-1, 4]$, we obtain the graph shown in Screen 1.10. This agrees with Figure 1.16 and we also verify the intercepts. ■

$[-1, 3.7]$ by $[-1, 4]$

Screen 1.10
A graph of $y = -\frac{2}{3}x + 2$.

In this text we use technology in two important ways. First, we solve a problem mathematically, and then support our answer graphically, as we did in Example 5.

Second, we solve a problem graphically and then confirm the solution mathematically, as we do in the next exploration.

EXPLORATION **3**

Finding an Intercept Graphically

 Find the *y*-intercept of $2y - 4x = -15$ graphically. Confirm algebraically.
(a) First find a complete graph on your graphing calculator.
(b) Now use the ZOOM feature to estimate the *y*-intercept.
(c) Find the *y*-intercept algebraically to confirm your answer.

EXPLORATION **4**

Parallel Lines

 Graph the following four lines on your graphing calculator. Change the grapher format to *simultaneous*, if you can.

$$y = -2x + 10 \qquad y = -2x + 8 \qquad y = -2x + 6 \qquad y = -2x + 4$$

Do these line appear parallel? Speculate on what is common about the equations that might indicate they are parallel.

Because slope is a number that indicates the direction or slant of a line, the following theorem can be proven.

Slope and Parallel Lines

Two lines are parallel if and only if they have the same slope.

EXAMPLE 6   Finding the Equation of a Line

Let L_1 be the line that goes through the two points $(-3, 1)$ and $(-1, 7)$. Find the equation of the line L through the point $(2, 1)$ and parallel to L_1. See Figure 1.17.

SOLUTION ■ We already have the point $(2, 1)$ on the line L. To use the point-slope equation, we now need to find the slope m of L. Because L is parallel to L_1, however, m equals the slope of L_1. The slope of L_1 is

$$m = \frac{y_2 - y_1}{x_2 - x_1} = \frac{(7) - (1)}{(-1) - (-3)} = \frac{6}{2} = 3$$

Now let us use the point-slope equation for L where $(2, 1)$ is a point on the line L with slope $m = 3$. Then we obtain

$$y - y_1 = m(x - x_1)$$
$$y - 1 = 3(x - 2)$$
$$y = 3x - 5 \quad ■$$

We also have the following theorem. (See Exercise 60.)

Slope and Perpendicular Lines

Two nonvertical lines with slope m_1 and m_2 are perpendicular if and only if

$$m_2 = -\frac{1}{m_1}$$

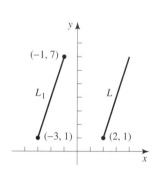

Figure 1.17
Since *L* is parallel to L_1, *L* has the same slope as L_1.

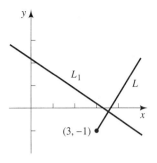

Figure 1.18
Since L is perpendicular to L_1, the slope of L is the negative reciprocal of the slope of L_1.

EXAMPLE 7   Finding the Equation of a Line

Find the equation of the line L through the point $(3, -1)$ and perpendicular to the line L_1 given by $2x + 3y = 6$.

SOLUTION ■ See Figure 1.18. We rewrite $2x + 3y = 6$ as

$$y = -\frac{2}{3}x + 2$$

This indicates that the slope of L_1 is $-\frac{2}{3}$. Thus, the slope m of L is the negative reciprocal or $m = \frac{3}{2}$. Now use the point-slope equation for L to give

$$y - y_1 = m(x - x_1)$$

$$y - (-1) = \frac{3}{2}(x - 3)$$

$$y = \frac{3}{2}x - \frac{11}{2}$$ ■

Applications

EXAMPLE 8   Production of Vacuum Cleaners

A firm produces two models of vacuum cleaners: regular and deluxe. The regular model requires 8 hours of labor to assemble, and the deluxe requires 10 hours of labor. Let x be the number of regular models produced and y the number of deluxe models. If the firm wishes to use 500 hours of labor to assemble these cleaners, find an equation that the two quantities x and y must satisfy. If the number of the regular models that are assembled is increased by 5, how many fewer deluxe models are assembled?

SOLUTION ■ To find the required equation, construct a table with the given information (see Table 1.1).

The number of work hours needed to assemble all of the regular models is the number of work hours required to produce one of these models times the number produced, or $8x$. In a similar fashion we see that the number of work hours needed to produce all of the deluxe models is $10y$. The number of hours allocated to assemble the regular model plus the number allocated to assemble the deluxe must equal 500. That is,

$$8x + 10y = 500$$

This is the required equation. Solving for y yields

$$y = -\frac{4}{5}x + 50$$

This is a line with slope $-\frac{4}{5}$. See Figure 1.19. Thus, increasing the number x of regular models produced by 5 decreases the number y of deluxe models produced by 4. ■

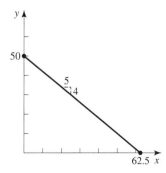

Figure 1.19
Since the slope is $-\frac{4}{5}$, moving 5 units to the right requires moving down 4 units to be on the line.

TABLE 1.1

Type	Regular	Deluxe	Total
Work hours for each cleaner	8	10	
Number of each model produced	x	y	
Total work hours for each model	$8x$	$10y$	500

EXPLORATION **5**

Slope

 Graph the line given in Example 8 on your grapher using a window with dimensions [0, 94] by [0, 50]. Use the TRACE feature to move the *x*-coordinate to the right by 5 units and note the change in the *y*-coordinate. Did you get 4?

■ **Rain Forests**

On a global scale, the rain forests are on the verge of extinction, a serious concern because of the amount of life tropical forests support. The rain forests account for only 7% of the earth's total land area, yet it is estimated that 5 to 10 million plant and animal species live within these regions. Today, one third to one fourth of the original tropical rain forests are gone. Deforestation for crop cultivation accounts for the loss of 20,000 to 30,000 square miles a year, an area the size of West Virginia. Experts believe that at this rate of destruction the tropical rain forests will be completely extinct in 200 years.

Source: Encyclopedia Americana 1994 ed.

EXAMPLE 9   Tropical Rain Forest Destruction

According to data collected by Aiken [4] the equation $p = -1.10t + 2229$ approximates the percentage of rain forest cover in Peninsular Malaysia, where t is the calendar year. What is the significance of the constant -1.10? According to this equation what was the percentage of rain forest cover in 1990 in Peninsular Malaysia?

SOLUTION ■ The line $p = -1.10t + 2229$ has slope $m = -1.10$. This implies that each year brings a 1.10% loss of rain forest cover in Peninsular Malaysia. Using this equation, we have $-1.10(1990) + 2229 = 40$. Thus, according to this equation, the percentage of rain forest cover was 40% in Peninsular Malaysia in 1990. Refer to Figure 1.20. ■

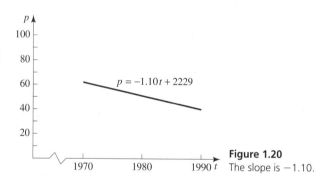

Figure 1.20
The slope is -1.10.

SELF-HELP EXERCISE SET 1.1

1. Find the equation of the line through the two points (3, 1) and (−2, −3) and then find the *y*-intercept. Sketch.

2. Find the equation of the line through the point (2, −1) and perpendicular to the line in the previous exercise. Sketch.

3. A household has $100 to purchase quarts of milk for $0.80 a quart or candy bars at $0.40 a bar. Let *x* be the number of quarts of milk and *y* the number of candy bars purchased. (**a**) Write a linear equation that *x* and *y* must satisfy. (This equation is called the **budget equation** in household economics.) (**b**) Plot this line. (**c**) What does the slope of this line mean?

EXERCISE SET 1.1

In Exercises 1 through 8 find the slope (if it exists) of the line through the given pair of points. Sketch the line.

1. (1, 3), (2, 7)   **2.** (3, −2), (2, 5)

3. (−2, −2), (−3, 1)   **4.** (−3, 2), (5, 4)

5. (−1, 3), (2, 3)   **6.** (−1, 2), (−3, −4)

7. (−3, −1), (−3, −7)   **8.** (−5, −2), (−3, −4)

In Exercises 9 through 20 find the slope (if it exists) of the line given by each of the equations. Also find the *x*- and *y*-intercepts if they exist. Sketch a graph.

9. $y = 2x - 1$ **10.** $y = -3x + 2$

11. $x + y = 1$ **12.** $x - 2y - 3 = 0$

13. $-x + 2y - 4 = 0$ **14.** $x + y = 0$

15. $x = 3$ **16.** $x = 0$

17. $y = 4$ **18.** $y = 0$

19. $x = 2y + 1$ **20.** $x = -3y + 2$

In Exercises 21 through 24 find the *x*- and *y*-intercepts and use these intercepts to sketch a graph of the line. Support your sketch with a graphing utility.

21. $3x + 5y = 15$ **22.** $5x + 3y = 15$

23. $4x - 5y = 20$ **24.** $-6x + 5y = 30$

In Exercises 25 through 46 find the equation of the line, given the indicated information.

25. Through (2, 1) with slope −3

26. Through (−3, 4) with slope 3

27. Through $(-1, -3)$ with slope 0

28. Through $(2, 3)$ with slope undefined

29. Through $(-2, 4)$ and $(-4, 6)$

30. Through $(-1, 3)$ and $(-1, 6)$

31. Through $(-2, 4)$ and $(-2, 6)$

32. Through $(-1, 4)$ and $(-3, 2)$

33. Through $(a, 0)$ and $(0, b)$, $ab \neq 0$

34. Through (a, b) and (b, a), $a \neq b$

35. Through $(-2, 3)$ and parallel to the line $2x + 5y = 3$

36. Through $(2, -3)$ and parallel to the line $2x - 3y = 4$

37. Through $(0, 0)$ and parallel to the line $y = 4$

38. Through $(0, 0)$ and parallel to the line $x = 4$

39. Through $(1, 2)$ and perpendicular to the line $y = 4$

40. Through $(1, 2)$ and perpendicular to the x-axis

41. Through $(-1, 3)$ and perpendicular to the line $x + 2y - 1 = 0$

42. Through $(2, -4)$ and perpendicular to the line $x - 3y = 4$

43. Through $(3, -1)$ and with a y-intercept of 4

44. Through $(-3, -1)$ and with a y-intercept of 0

45. A y-intercept of 4 and parallel to $2x + 5y = 6$

46. A y-intercept of -2 and perpendicular to $x - 3y = 4$

47. Are the two lines $2x + 3y = 6$ and $3x + 2y = 6$ perpendicular?

48. A person has $2.35 in change consisting entirely of dimes and quarters. If x is the number of dimes and y is the number of quarters, write an equation that x and y must satisfy.

Applications

49. **Investment.** A person invests some money in a bond that yields 8% a year and some money in a money market fund that yields 5% a year, and obtains $576 of interest in the first year. If x is the amount invested in the bond and y the amount invested in the money market fund, find an equation that x and y must satisfy. If the amount of money invested in bonds is increased by $100, by how much must the money invested in the money market decrease if the total interest is to stay the same?

50. **Transportation Costs.** New cars are transported from Detroit to Boston and also from Detroit to New York City. It costs $90 to transport each car to Boston and $80 to transport each car to New York City. Suppose $4000 is spent on transporting these cars. If x is the number transported to Boston and y is the number transported to New York City, find an equation that x and y must satisfy. If the number of cars transported to Boston is increased by 8, how many fewer cars must be transported to New York City if the total transportation costs remained the same?

51. **Size of Deer Antlers.** Clow [5] showed that the size of a deer's antlers depends primarily on the age of the deer. Mule deer in Northern Colorado have no antlers at age 10 months, but after that the weight of the antlers increases by 0.12 kilogram for every 10 months of additional age. Let x be the age of these mule deer in months and y the weight of their antlers in kilograms. Assuming a linear relationship, write an equation that x and y must satisfy. Graph using your graphing calculator. Using the TRACE function determine the weight in kilograms of antlers carried by a 40-month-old mule deer.

52. **Scheduling.** An attorney has two types of documents to process. The first document takes 3 hours and the second takes 4 hours to process. If x is the number of the first type that she processes and y is the number of the second type, find an equation that x and y must satisfy if she works 45 hours on these documents. If she processes 8 additional of the first type of document, how many fewer of the second type of document could she process?

53. **Milk Production.** The amount of milk produced by a cow depends on the quality and therefore the cost of feed. According to a study in Wisconsin the annual feed cost c in dollars per cow per year is approximately related to the annual pounds x of milk per cow per year by $c = 201.38 + 0.03x$. What is the significance of the number 0.03 in this equation, and what is the increase in cost for a 1000-pound increase in milk production per year?

54. **Biology.** In an extensive study of 82 species of desert lizards from three continents, Pianka [6] found that the body temperature T_b of the lizards in degrees Celsius was given approximately by $T_b = 38.8 + (1 - \beta)(T_a - 38.8)$, where T_a is the air temperature in degrees Celsius and β is a constant between 0 and 1 that depends on the specific species. Explain the significance of the number $1 - \beta$. What is the significance if $\beta = 1$?

55. **Biology.** Simpson [7] reports that in females of the snake *Lampropeltis polyzona* the total length L is linear in the tail length l. Find the linear equation relating L to l if for one of these snakes $L = 140$ mm (millimeters) when $l = 60$ mm and for another $L = 1050$ mm when $l = 455$ mm. Find the slope and its significance.

56. **Biology.** Clarke and McKenzie [8] report that weight w in milligrams (mg) of the sheep blowfly pupa is approximated by $w = 41.61 - 0.25T$, where T is the temperature in degrees Celsius. Explain the significance of the number -0.25. Graph.

57. **Veterinary Entomology.** Hribar and coworkers [9] found that the number y of flies on a bull was approximated by $y = 10.95 + 2.27x$, where x is the number of flies on the front legs of the bull. What is the significance of the number 2.27? 10.95? Graph.

58. **Environmental Quality.** Pierzynski and colleagues [10] found that the relative yield y of soybeans was approximated by $y = 1.54 - 0.00056x$, where x is the concentration in milligrams per kilogram (mg/kg) of zinc in the soybean tissue. What is the significance of the fact that the coefficient of x is *negative*? Graph for $0 \leq x \leq 1000$.

Enrichment Exercises

59. Sketch a proof of the following. If a and b are not both zero, then the graph of a linear equation $ax + by = c$ is a straight line, and, conversely, every line is the graph of a linear equation.

60. Establish the following theorem: Two lines with slope m_1 and m_2 are perpendicular if, and only if,

$$m_2 = -\frac{1}{m_1}$$

Give a proof in the special case shown in the figure in which the two lines intersect at the origin. (The general proof is very similar). *Hint:* The two lines L_1 and L_2 are perpendicular if and only if the triangle AOB is a right triangle. By the Pythagorean theorem this is the case if, and only if,

$$[d(A, B)]^2 = [d(A, O)]^2 + [d(B, O)]^2$$

Now use the distance formula for each of the three terms in the previous equation and simplify.

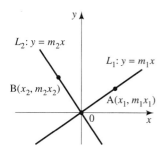

SOLUTIONS TO SELF-HELP EXERCISE SET 1.1

1. To find the equation of the line through the two points $(3, 1)$ and $(-2, -3)$, we first must find the slope of the line through the two points. This is

$$m = \frac{y_2 - y_1}{x_2 - x_1} = \frac{1 - (-3)}{3 - (-2)} = \frac{4}{5}$$

We now have the slope of the line and *two* points on the line. Pick one of the points, say $(3, 1)$, and use the point-slope form of the equation of the line and obtain

$$y - y_1 = m(x - x_1)$$

$$y - 1 = \frac{4}{5}(x - 3)$$

$$y = \frac{4}{5}x - \frac{7}{5}$$

This last equation is in the form of $y = mx + b$ with $b = -\frac{7}{5}$. Thus, the y-intercept is $-\frac{7}{5}$.

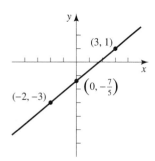

2. The equation through the point $(2, -1)$ and perpendicular to the line in the previous exercise must have a slope equal to the negative reciprocal of the slope of that line. Then the slope of the line whose equation we are seeking is $-\frac{5}{4}$. Using the point-slope form of the equation yields

$$y - y_1 = m(x - x_1)$$

$$y + 1 = -\frac{5}{4}(x - 2)$$

$$y = -\frac{5}{4}x + \frac{3}{2}$$

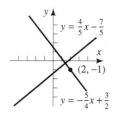

3. We organize the information as shown in the table.

Item	Quarts of Milk	Candy Bars	Total
Number of each item purchased	x	y	
Price of each item	0.80	0.40	
Total amount spent on each item	$0.80x$	$0.40y$	100

(a) Since $0.80x$ is spent on milk and $0.40y$ on candy, the budget equation is given by

$$0.80x + 0.40y = 100$$

(b) To find the y-intercept we set $x = 0$ and obtain $100 = 0.40y$ or $y = 250$. To find the x-intercept we set $y = 0$ and obtain $100 = 0.80x$ or $x = 125$. The graph is shown in the figure.

y

250
200
150
100
50

$0.80x + 0.40y = 100$

50 100 150 200 250 x

(c) To find the slope of the line we have

$$0.40y = -0.80x + 100$$

$$y = -2x + 250$$

We see that the slope is $m = -2$. This means that for each additional quart of milk purchased, 2 fewer candy bars can be purchased.

1.2 LINEAR MODELS

■ Mathematical Models ■ Cost, Revenue, and Profits ■ Supply and Demand
■ Straight-Line Depreciation

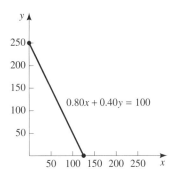

(Historical Pictures Collection/Stock Montage, Inc.)

Augustin Cournot 1801–1877

The first significant work dealing with the application of mathematics to economics was Cournot's *Researches into the Mathematical Principles of the Theory of Wealth*, published in 1836. It was Cournot who originated the supply and demand curves that are discussed in this section. Irving Fisher, a prominent economics professor at Yale and one of the first exponents of mathematical economics in the United States, wrote that Cournot's book "seemed a failure when first published. It was far in advance of the times. Its methods were too strange, its reasoning too intricate for the crude and confident notions of political economy then current."

APPLICATION: RAIL FREIGHT

In a report of the Federal Trade Commission (FTC) [11] an example is given in which the Portland Oregon mill price of 50,000 board square feet of plywood is $3525 and the rail freight is $0.3056 per mile.

(a) If a customer is located x rail miles from this mill, write an equation that gives the total freight f charged to this customer in terms of x for delivery of 50,000 board square feet of plywood.

(b) Write a (linear) equation that gives the total c charged to a customer x rail miles from the mill for delivery of 50,000 board square feet of plywood. Graph this equation.

(c) In the FTC report, a delivery of 50,000 board square feet of plywood from this mill is made to New Orleans, Louisiana, 2500 miles from the mill. What is the total charge? See Example 6 for the answers.

Mathematical Models

Mathematical modeling is an attempt to describe some part of the real world in mathematical terms. There are three steps in mathematical modeling: formulation, mathematical manipulation, and evaluation.

Formulation

First, based on observations, we must state a question or formulate a hypothesis. If the question or hypothesis is too vague, we need to make it precise. If it is too ambitious, we need to restrict the question or hypothesis or subdivide it into manageable parts.

Second, we need to identify important factors. We must decide which quantities and relationships are important to answer the question and which can be ignored. We then need to formulate a *mathematical* description. For example, each important quantity should be represented by a variable. Each relationship should be represented by an equation, inequality, or other mathematical construct.

Mathematical Manipulation

After the mathematical formulation we then need to do some mathematical manipulation to obtain the answer to our original question. We may need to do a calculation, solve an equation, or prove a theorem. Sometimes the mathematical formulation gives us a mathematical problem that is impossible to solve. In such a case, we need to reformulate the question in a less ambitious manner.

Evaluation

Naturally we need to check the answers given by the model with real data. We normally only expect the mathematical model to describe a very limited aspect of the world and to only give approximate answers. If the answers are wrong or not accurate enough for our purposes, we then need to identify the sources of the shortcomings of the model. Perhaps we need to change the model entirely or just make some refinements. In any case, this requires a new mathematical manipulation and evaluation. Thus, modeling often involves repeating the three steps of formulation, mathematical manipulation, and evaluation.

We now create mathematical equations that relate cost, revenue, and profits of a manufacturing firm to the number of items produced and sold. In this section we restrict ourselves to linear models, in the next section we consider quadratic models.

Cost, Revenue, and Profits

Any manufacturing firm has two types of costs: fixed costs and variable costs. **Fixed costs** are costs that do not depend on the amount of production. These costs include real estate taxes, interest on loans, some management salaries, certain minimal maintenance, and protection of plant and equipment. **Variable costs** depend on the amount of production. They include cost of material and labor. Total cost, or simply **cost**, is the sum of fixed and variable costs:

$$\text{Cost} = \text{variable cost} + \text{fixed cost}$$

Let x denote the number of units of a given product or commodity produced by a firm. In the **linear cost model** we assume that the cost m of manufacturing one item is the same no matter how many items are produced. Thus, the variable cost is the number of items produced times the cost of each item:

$$\text{Variable cost} = \text{cost per item} \times \text{number of items produced}$$
$$= mx$$

If b is the fixed cost and C is the cost, we then have the following:

$$C = \text{cost}$$
$$= \text{variable cost} + \text{fixed cost}$$
$$= mx + b$$

In the graph shown in Figure 1.21 note that the y-intercept is the fixed cost and the slope is the cost per item.

In the **linear revenue model** we assume that the price p of an item sold by a firm is the same no matter how many items are sold. (This is a reasonable assumption if the number sold by the firm is small compared with the total number sold by the entire industry.) Thus, the revenue is the price per item times the number of items sold. Let x be the number of items sold. (For convenience, we always assume that *the number*

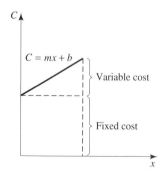

Figure 1.21
Cost is fixed cost plus variable cost.

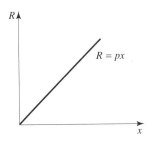

Figure 1.22
Revenue is the price of each item times the number of items sold.

of items sold equals the number of items produced.) Then, if we denote the revenue by R,

$$R = \text{revenue}$$
$$= \text{price per item} \times \text{number sold}$$
$$= px$$

Notice in Figure 1.22 that the straight line goes through $(0, 0)$ because nothing sold results in no revenue. The slope is the price per item.

No matter if our models of cost and revenue are linear or not, **profit** P is always revenue less cost. Thus,

$$P = \text{profit}$$
$$= \text{revenue} - \text{cost}$$
$$= R - C$$

We now determine the cost, revenue, and profit equations for a hypothetical firm.

E X A M P L E   1   Finding Cost, Revenue, and Profit Equations

---

A firm has weekly fixed costs of $80,000 associated with the manufacture of dresses that cost $25 per dress to produce. The firm sells all the dresses it produces at $75 per dress. Find the cost, revenue, and profit equations if x is the number of dresses produced per week.

S O L U T I O N   ■
(a) $C = \text{variable cost} + \text{fixed cost}$
$$= mx + b$$
$$= 25x + 80{,}000$$

(b) $R = \text{price per item} \times \text{number sold}$
$$= px$$
$$= 75x$$

(c) $P = \text{revenue} - \text{cost} = R - C$
$$= 75x - (25x + 80{,}000)$$
$$= 50x - 80{,}000 \quad ■$$

Given the number x of items produced and sold, we can use these equations to find the corresponding cost, revenue, and profit. For example, if $x = 1000$, then

$$C = 25(1000) + 80{,}000 = 105{,}000$$
$$R = 75(1000) = 75{,}000$$
$$P = 75{,}000 - 105{,}000 = -30{,}000$$

Thus, if 1000 items are produced and sold, the cost is $105,000, the revenue is $75,000, and there is a *loss* of $30,000.

Doing the same for some other values of x we have Table 1.2.

TABLE 1.2

Number Made and Sold	1000	1500	2000
Cost	105,000	117,500	130,000
Revenue	75,000	112,500	150,000
Profit (or loss)	−30,000	−5,000	20,000

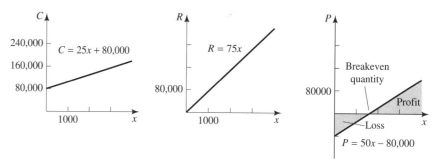

Figure 1.23
A linear profit function shows losses if only a small number of items are sold, but profits if a larger number are sold.

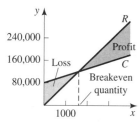

Figure 1.24
Profit is revenue less cost.

The cost, revenue, and profit equations are graphed in Figure 1.23.

One can see in Figure 1.23c or in Table 1.2, that for smaller values of x, P is *negative*, that is, the firm has losses. For larger values of x, P turns positive, and the firm has (positive) profits. The value of x at which the profit is zero is called the **break-even quantity**. Figure 1.24 shows that the break-even quantity is the value of x at which cost equals revenue. This is true since $P = R - C$, and $P = 0$ if, and only if, $R = C$.

EXAMPLE 2   Finding the Break-Even Quantity

Find the break-even quantity in the previous example.

SOLUTION ■ Set $P = 0$ and solve for x.

$$0 = P$$
$$= 50x - 80{,}000$$
$$50x = 80{,}000$$
$$x = 1600$$

Thus, the firm needs to manufacture and sell 1600 dresses to break even (i.e., for profits to be zero). ■

EXPLORATION 1

Break-Even Quantity

Lend graphical support to Example 2 by doing the following.

(a) On your graphing calculator graph $y_1 = 25x + 80{,}000$, $y_2 = 75x$, and $y_3 = y_2 - y_1$ using a window with dimensions [0, 2500] by [−100,000, 200,000].

(b) Now ZOOM in repeatedly on the point of intersection of the cost and revenue curves.

(c) Take note of the y-intercept found in (b). Confirm this number algebraically.

Supply and Demand

In the previous discussion we assumed that the number of items produced and sold by the given firm was small compared with the number sold by the industry. Under this assumption it was then reasonable to conclude that the price was constant and did not vary with the number x sold. But if the number of items sold by the firm represented a *large* percentage of the number sold by the entire industry, then trying to sell significantly more items could only be accomplished by *lowering* the price of each item. Thus,

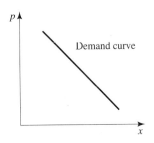

Figure 1.25
A typical demand curve slopes
downward.

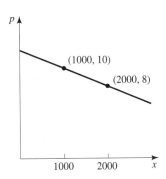

Figure 1.26
A linear demand curve through
the two points.

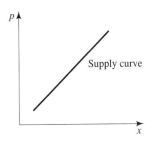

Figure 1.27
A typical supply curve slopes
upward.

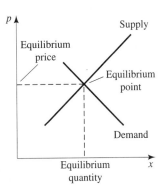

Figure 1.28
The equilibrium point is the point
of intersection of the supply and
demand curves.

under these assumptions, the price p of each item would depend on the number sold. If, in addition, the price is assumed to be linear, that is, $p = mx + b$, then the graph of this equation is a straight line that must slope down as shown in Figure 1.25.

We assume that x is the number of a certain item produced and sold by the entire industry during a given period and that $p = mx + b$ is the price of the item if x are sold, that is, $p = mx + b$ is the price of the xth item sold. We call $p = mx + b$ the **demand equation** and the graph the **demand curve**.

Estimating the demand equation is a fundamental problem for the management of any company or business. In the next example we consider the situation when just two data points are available and the demand equation is assumed to be linear. More complicated models are found later in the text.

EXAMPLE 3   Finding the Demand Equation

Suppose that 1000 units of a certain item are sold per day by the entire industry at a price of \$10 per item and that 2000 units can be sold per day by the same industry at a price of \$8 per item. (a) Find the demand equation $p = mx + b$, assuming the demand curve is a straight line. (b) If 1500 items are sold, what do you expect the price of each item to be?

SOLUTIONS ■

(a) Figure 1.26 shows the two points (1000, 10) and (2000, 8) that lie on the demand curve. Since we are assuming that the demand curve is a straight line, we need only find the straight line through these two points. The slope m is

$$m = \frac{8 - 10}{2000 - 1000} = -0.002$$

Notice that the slope is *negative*, indicating the line slants down moving from left to right. This is what we expect of a reasonable demand curve.

Now using the point-slope equation for a line with (1000, 10) as the point on the line, we have

$$p - 10 = m(x - 1000)$$
$$p - 10 = -0.002(x - 1000)$$
$$p = -0.002x + 12$$

(b) We can now use this equation to find the price of each item if 1500 items are made and sold. We set $x = 1500$ in this equation and obtain

$$p = -0.002(1500) + 12$$
$$= -3 + 12 = 9$$

Thus, according to this model, if 1500 items are made and sold, the price of each item must be \$9. ■

The **supply equation** $p = mx + b$ gives the price p necessary for suppliers to make available x items to the market. The graph of this equation is called the **supply curve**. Figure 1.27 shows a *linear* supply curve. A reasonable supply curve rises, moving from left to right, because the suppliers of an item naturally want to sell more items if the price per item is higher. (See Shea [12] who recently looked at a large number of industries and determined that the supply curve does indeed slope up.)

The best known law of economics is the law of supply and demand. Figure 1.28 shows a demand equation and a supply equation that intersect. The point of intersection, the point at which supply equals demand, is called the **equilibrium point**. The x-coordinate of the equilibrium point is called the **equilibrium quantity** and the p-coordinate is called the **equilibrium price**.

24 CHAPTER 1 Functions

EXAMPLE 4   Finding the Equilibrium Point

Suppose the supply curve is given by $p = 0.008x + 5$, and the demand curve is the same as in the previous example. Find the equilibrium point.

SOLUTION ■ The demand and supply equations are given by

Demand equation: $\quad p = -0.002x + 12$

Supply equation: $\quad p = 0.008x + 5$

These are shown in Figure 1.29. The equilibrium point must be on *both* curves, and thus the coordinates of this point must satisfy both equations. Thus

$$-0.002x + 12 = p = 0.008x + 5$$

$$0.01x = 7$$

$$x = 700$$

Then $x = 700$ is the equilibrium quantity, and when $x = 700$,

$$p = -0.002(700) + 12 = 10.6$$

is the equilibrium price in dollars. Thus, the equilibrium point is (700, 10.6). ■

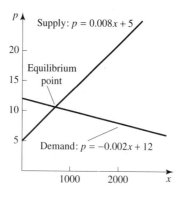

Figure 1.29
The equilibrium point is the point of intersection of the supply and demand curves.

EXPLORATION 2

Finding the Equilibrium Point

 Support Example 4 graphically by doing the following.
(a) Graph the supply and demand curves on your graphing calculator using a window of dimensions [0, 1000] by [0, 20].
(b) Now ZOOM repeatedly to find the equilibrium point.

Straight-Line Depreciation

Many assets, such as machines or buildings, have a *finite* useful life and furthermore *depreciate* in value from year to year. For purposes of determining profits and taxes, various methods of depreciation can be used. In **straight-line depreciation** we assume that the value V of the asset is given by a *linear* equation in time t, say, $V = mt + b$. The slope m must be *negative* since the value of the asset *decreases* over time.

EXAMPLE 5   Straight-Line Depreciation

A company has purchased a new grinding machine for $100,000 with a useful life of 10 years, after which it is assumed that the scrap value of the machine is $5000. If

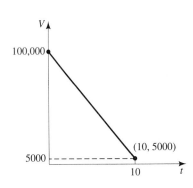

Figure 1.30
A graph of straight-line depreciation.

we use straight-line depreciation, write an equation for the value V of the machine where t is measured in years. What will be the value of the machine after the first year? After the second year? After the ninth year?

S O L U T I O N ■ We assume that $V = mt + b$, where m is the slope and b is the V-intercept. We then must find both m and b. We are told that the machine is initially worth \$100,000, that is, when $t = 0$, $V = 100,000$. Thus, the point $(0, 100,000)$ is on the line, and thus $100,000$ is the V-intercept b (see Figure 1.30).

Since the value of the machine in 10 years will be \$5000, this means that when $t = 10$, $V = 5000$. Thus, $(10, 5000)$ is also on the line. From Figure 1.30, the slope can then be calculated since we now know that the two points $(0, 100,000)$ and $(10, 5000)$ are on the line. Then

$$m = \frac{5000 - 100{,}000}{10 - 0} = -9500$$

Thus,

$$V = -9500t + 100{,}000$$

Then

$$\text{When } t = 1, \ V = -9500(1) + 100{,}000 = 90{,}500$$

$$\text{When } t = 2, \ V = -9500(2) + 100{,}000 = 81{,}000$$

$$\text{When } t = 9, \ V = -9500(9) + 100{,}000 = 14{,}500 \quad ■$$

E X A M P L E 6   Rail Freight
_____

In a report of the Federal Trade Commission [11] an example is given in which the Portland Oregon mill price of 50,000 board square feet of plywood is \$3525 and the rail freight is \$0.3056 per mile.

(a) If a customer is located x rail miles from this mill, write an equation that gives the total freight f charged to this customer in terms of x for delivery of 50,000 board square feet of plywood.

(b) Write a (linear) equation that gives the total c charged to a customer x rail miles from the mill for delivery of 50,000 board square feet of plywood. Graph this equation.

(c) In the FTC report, a delivery of 50,000 board square feet of plywood from this mill is made to New Orleans, Louisiana, 2500 miles from the mill. What is the total charge?

S O L U T I O N ■
(a) Since each mile costs \$0.3056, the freight charge for x miles is $0.3056x$ dollars.

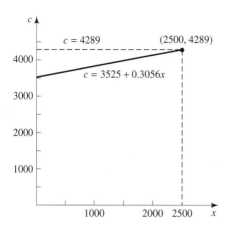

Figure 1.31
A graph of the cost of the delivery.

(b) Since the mill price is \$3525, the total charge is $c = 3525 + 0.3056x$. The graph of this equation is a line with c-intercept 3525 and slope 0.3056. The graph is shown in Figure 1.31.

(c) If $x = 2500$, then

$$c = 3525 + 0.3056(2500) = 4289$$

and the total charge is \$4289.   ■

SELF-HELP EXERCISE SET 1.2

1. A manufacturer has fixed costs of \$10,000 per week and variable costs of \$20 per item. The item is sold for \$30. Let x denote the number of items manufactured and sold.
   (a) Find the cost equation.
   (b) Find the revenue equation.
   (c) Find the profit equation.
   (d) Find the break-even quantity.

2. A manufacturer notes that it can sell 10,000 items at \$6 per item and 8000 items at \$8 per item. Find the demand curve, assuming it is a straight line.

3. Suppose the supply equation for the manufacturer in the previous exercise was $p = 0.002x + 4$. Find the equilibrium quantity.

EXERCISE SET 1.2

In Exercises 1 and 2 you are given the cost per item and the fixed costs over a certain period. Assuming a linear cost model, find the cost equation, where C is cost and x is the number produced.

1. Cost per item = \$3, fixed cost = \$10,000

2. Cost per item = \$6, fixed cost = \$14,000

In Exercises 3 and 4 you are given the price of each item assumed to be independent of the number of items sold. Find the revenue equation, where R is revenue and x is the number sold, assuming the revenue is linear in x.

3. Price per item = \$5   4. Price per item = \$0.1

5. Using the cost equation found in Exercise 1 and the revenue equation found in Exercise 3, find the profit equation for P, assuming the number produced equals the number sold.

6. Using the cost equation found in Exercise 2 and the revenue equation found in Exercise 4, find the profit equation for P, assuming the number produced equals the number sold.

In Exercises 7 through 10 linear cost and revenue equations are given. Find the break-even quantity.

7. $C = 2x + 4,$ $R = 4x$

8. $C = 3x + 10,$ $R = 6x$

9. $C = 0.1x + 2,$ $R = 0.2x$

10. $C = 0.03x + 1,$ $R = 0.04x$

In Exercises 11 through 14 you are given a demand equation and a supply equation. Sketch the demand and supply curves, and find the equilibrium point.

11. Demand: $p = -x + 6$;   supply: $p = x + 3$

12. Demand: $p = -3x + 12$;   supply: $p = 2x + 5$

13. Demand: $p = -10x + 25$;   supply: $p = 5x + 10$

14. Demand: $p = -0.1x + 2$:   supply: $p = 0.2x + 1$

Applications

15. Straight-Line Depreciation. A new machine that costs $50,000 has a useful life of nine years and a scrap value of $5000. Using a straight-line depreciation, find the equation for the value V in terms of t where t is in years. Find the value after one year. After five years.

16. Straight-Line Depreciation. A new building that costs $1,100,000 has a useful life of 50 years and a scrap value of $100,000. Using straight-line depreciation, find the equation for the value V in terms of t where t is in years. Find the value after one year. After two years. After 40 years.

17. Demand. Suppose that 500 units of a certain item are sold per day by the entire industry at a price of $20 per item and that 1500 units can be sold per day by the same industry at a price of $15 per item. Find the demand equation for p, assuming the demand curve to be a straight line.

18. Demand. Suppose that 10,000 units of a certain item are sold per day by the entire industry at a price of $150 per item and that 8000 units can be sold per day by the same industry at a price of $200 per item. Find the demand equation for p, assuming the demand curve to be a straight line.

19. Machine Allocation. A furniture manufacturer makes bookcases and small desks, each requiring use of a cutting machine. Each bookcase requires 3 hours on the machine, and each desk requires 4 hours on the machine. The machine is available 120 hours each week.
(a) If x is the number of bookcases and y the number of desks that are manufactured each week, find the equation that x and y must satisfy if all available time is used on the machine.
(b) The graph of the equation in the previous part is a straight line. Find the slope of this line and interpret its meaning.

20. Labor Allocation. A company manufactures 3-speed and 10-speed bikes. Each 3-speed bike requires 6 hours of labor, and each 10-speed bike requires 12 hours of labor. There are 240 hours of labor available each day.
(a) If x is the number of 3-speed bikes and y the number of 10-speed bikes that are manufactured each day, find the equation that x and y must satisfy if all available hours of labor are used.
(b) The graph of the equation in the previous part is a straight line. Find the slope of this line and interpret its meaning.

21. Sales Commissions. A salesperson is offered two salary plans. The first gives a base salary of $1000 a month and a 5% commission on all sales; the second plan gives a base salary of $1500 a month and a 4% commission.
(a) If x is the amount of sales, and y the monthly salary, find the relationship between x and y for each plan.
(b) How many sales are required so that each plan yields the same total salary?

22. Inventory. A store has 1000 units of a certain item in stock and sells 20 units a day.
(a) If y is the number of items in stock at the end of day t, find the relationship between y and t.
(b) How long until the inventory is exhausted?

23. Medicine. Two rules have been suggested by Cowling and by Friend to adjust adult drug dosage levels for young children. Let a denote the adult dosage and let t be the age (in years) of the child. The two rules are given respectively by

$$C = \frac{a}{24}(t+1) \quad \text{and} \quad F = \frac{2}{25}at$$

If, for a particular drug, $a = 100$ in the appropriate units, graph both functions. Find the age for which the two rules give the same dosage.

24. Temperature. Let C denote the temperature measured on the Celsius scale and F denote the temperature measured on the Fahrenheit scale. The freezing and boiling points of water are 0°C and 100°C, respectively, on the Celsius scale and 32°F and 212°F, respectively, on the Fahrenheit scale. If F is a linear equation in C, find the equation that relates F to C.

25. Elections. A candidate for the Senate of the United States obtained 41% of the vote in the previous election and spent $3 million in advertisements. The candidate's advisors estimate that each additional $1 million in advertisement expenditures will obtain an additional 6% of the vote for the next election. Write an equation that relates the advertisement expenditures in millions of dollars with the percentage of the vote obtained, assuming a linear relationship. How much must be spent on advertising, according to this analysis, to obtain 50% of the vote in the next election?

26. Weight of Whales. It is difficult to weigh large whales, but easy to measure their length. The International Whaling Commission adopted the formula $w = 3.51l - 192$ as a reliable way of relating the length l in feet of adult blue whales with their weight w in British tons. Graph this equation. According to this formula, what must be the increase in weight of these whales for every 5-foot increase in their length?

27. Make or Buy Decision. A company includes a manual with each piece of software it sells and is trying to decide whether to contract with an outside supplier or produce the manual in house. The lowest bid of any outside supplier is $0.75 per manual. The company estimates that producing the manuals in house will require fixed costs of $10,000 and variable costs of $0.50 per manual. Which alternative has the lower total cost if demand is 20,000 manuals?

28. Make or Buy Decision. Repeat the previous exercise for a demand of 50,000 manuals.

29. Rail Freight. In the FTC report mentioned in Example 6 suppose another mill exists in Holden, Louisiana. Suppose the Holden mill has a mill price of $4000 for 50,000 board square feet of plywood and can deliver by truck for $1.3333 per mile.
(a) If a customer is located x truck miles from this mill, write an equation that gives the total freight charged to this customer in terms of x.
(b) Write a (linear) equation that gives the total charged to a customer x truck miles from the mill for delivery of 50,000 board square feet of plywood.

30. Rail Freight. A customer is located on the straight line connecting the mill in Holden mentioned in the previous exercise and the mill in Portland mentioned in Example 6. Assuming rail distances and truck distances are the same, how far from Holden does a customer have to be so that the total charge (mill price plus freight charge) is the same (to the nearest mile) for both mills?

31. Supply and Demand. The excess supply and demand curves for wheat worldwide were recently estimated by Doering [13] to be

$$\text{Supply:} \quad x = 57.4 + 0.143p$$

$$\text{Demand:} \quad x = 142.8 - 0.284p$$

where p is price per metric ton and x is in millions of metric tons. Excess demand refers to the excess of wheat that producer countries have over their own consumption. Find the equilibrium point using your graphing calculator. Confirm your answer algebraically.

32. Supply and Demand. The demand and supply equations for potatoes has been given by Christ [14] as

$$\text{Demand:} \quad p = -1.425x + 2.425$$

$$\text{Supply:} \quad p = 1.222x - 0.222$$

Using the ZOOM feature of your graphing calculator find the equilibrium point to two decimal places.

Enrichment Exercises

33. Assume the linear cost model applies and fixed costs are $1000. If the *total* cost of producing 800 items is $5000, find the cost equation.

34. Assume the linear revenue model applies. If the *total* revenue from producing 1000 items is $8000, find the revenue equation.

35. Assume the linear cost model applies. If the *total* cost of producing 1000 items at $3 each is $5000, find the cost equation.

36. Assume the linear cost and revenue models applies. An item that costs $3 to make sells for $6. If profits of $5000 are made when 1000 items are made and sold, find the cost equation.

37. Assume the linear cost and revenue models applies. An item costs $3 to make. If fixed costs are $1000 and profits of $7000 are made when 1000 items are made and sold, find the revenue equation.

38. Assume the linear cost and revenue models applies. An item sells for $3. If fixed costs are $2000 and profits of $9000 are made when 1000 items are made and sold, find the cost equation.

39. Process Selection and Capacity. A machine shop needs to drill holes in a certain plate. An inexpensive manual drill press could be purchased that requires large labor costs to operate, or an expensive automatic press can be purchased that requires small labor costs to operate. The table summarizes the options. If these are the only fixed and variable costs, determine which machine should be purchased by filling in the second table if the number produced is (**a**) 1000 (**b**) 10,000.

| | Annual | Variable | Production |
Machine	Fixed Costs	Labor Costs	Rate
Manual	$1000	$16.00/hour	10 plates/hour
Automatic	$8000	$2.00/hour	100 plates/hour

| | Total Costs | |
Volume	Manual	Automatic
1,000		
10,000		

40. Process Selection and Capacity. If x is the number of plates in Exercise 39 produced and C is the total cost using the manual drill press, find an equation that x and C must satisfy. Repeat if C is the total cost using the automatic drill press. Graph both equations on your graphing calculator. Use the ZOOM feature to find the number of plates produced per hour for which both the manual and automatic drill presses cost the same.

41. Cost Per Unit. If C is the total cost to produce x items, then the cost per unit is $\overline{C} = C/x$. Fill in the table with the costs per unit for the indicated volumes for each of the machines in Exercise 39. Explain in words what happens to the costs per unit as the number of items made becomes larger. Find the equation that $\overline{C}$ and x must satisfy and graph this equation on your graphing calculator. What happens to the cost per unit as x becomes larger?

| | Costs Per Unit | |
Volume	Manual	Automatic
1,000		
10,000		
100,000		

42. Supply and Demand. (**a**) Explain what you think the consumers and suppliers will do if the price is at p_1 in the diagram. (See Taylor [15].) (**b**) Explain what you think the consumers and suppliers will do if the price is at p_2 in the diagram.

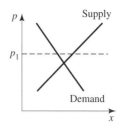

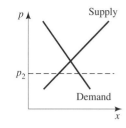

43. Cost, Revenue, and Profit Equations. Assuming a linear cost and revenue model, explain in complete sentences where you expect the *y*-intercepts to be for the cost, revenue, and profit equations. Give reasons for your answers.

44. In the linear cost model we assumed that the cost of manufacturing one item is the same no matter how many items are produced. Explain a situation that you know about when this is not the case. For example, there may be economies of scale. How might the graph of the cost equation look?

SOLUTIONS TO SELF-HELP EXERCISE SET 1.2

1. (a) A manufacturer has fixed costs of $10,000 per week and variable costs of $20 per item, and the number of items manufactured is x. The variable costs are the cost per item times the number of items, or $20x$. Thus, the cost in dollars is $C = 20x + 10,000$.

(b) The item is sold for $30. Thus, the revenue equation is $R = 30x$.

(c) Profit is always revenue less costs. Thus

$$P = R - C = 30x - (20x + 10,000)$$
$$= 10x - 10,000$$

(d) The break-even quantity is the quantity for which profits are zero. Setting $P = 0$ then gives

$$0 = 10x - 10,000$$
$$x = 1000$$

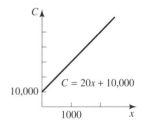

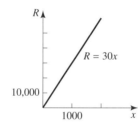

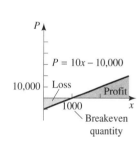

2. If a manufacturer sells 10,000 items at $6 per item and sells 8000 items at $8 per item and the demand curve is assumed to be a straight line, then the slope of this line is given by

$$m = \frac{6 - 8}{10,000 - 8000} = -0.001$$

(See the figure.) Then the demand equation is given by the point-slope equation of the line as

$$p - 8 = -0.001(x - 8000)$$
$$p = -0.001x + 16$$

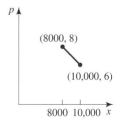

3. If the supply equation for the manufacturer in the previous exercise is $p = 0.002x + 4$, then to find the equilibrium quantity, we have

$$0.002x + 4 = -0.001x + 16$$
$$0.003x = 12$$
$$x = 4000$$

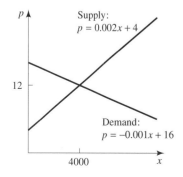

1.3 FUNCTIONS

■ Definition of Function   ■   Graphs of Functions   ■   Applications

(The Bettman Archive/ Hutton)

Lejeuné Dirichlet   1805–1859

Dirichlet was one of the mathematical giants of the 19th century. He formulated the notion of function that is currently used. He is also known for the Dirichlet series, the Dirichlet function, the Dirichlet principle, and the Dirichlet problem. The Dirichlet problem is fundamental to thermodynamics and electrodynamics. Although described as noble, sincere, humane, and possessing a modest disposition, he was known as a dreadful teacher. Dirichlet was also a failure as a family correspondent. When his first child was born, he neglected to inform his father-in-law. When the father-in-law found out about the event, he commented that Dirichlet might have at least written a note saying "2 + 1 = 3".

APPLICATION: ALFALFA YIELDS

Generally, alfalfa yields are highest in the second year of production and then decline as the stands thin out and grow less vigorously. This yield decline was estimated by Knapp [16] as

$$y = \begin{cases} 4.93, & \text{if } t = 1 \\ 7.47 - 0.584(t - 2), & \text{if } t \geq 2 \end{cases}$$

where y gives the yield in tons per acre per year and t denotes the age of the crop in years. Determine the yield in each of the first three years. See Example 3 for the answer.

APPLICATION: VOLUME OF AN OPEN BOX

From all four corners of a rectangular piece of cardboard of dimensions 10 inches by 20 inches squares are cut with dimensions x by x. The sides are turned up to form an open box. What is the volume V of the box as a function of x? See Example 9 for the answer.

TABLE 1.3

Company	1991 Sales (billions $)
GM	124
Exxon	103
Ford	89
IBM	65
GE	60
Mobile	57
Philip Morris	48
Du Pont	38
Texaco	38
Chevron	37

TABLE 1.4

Year	Debt (trillions $)
1982	1.1
1983	1.4
1984	1.6
1985	1.8
1986	2.1
1987	2.4
1988	2.6
1989	2.9
1990	3.2
1991	3.6

Definition of Function

We are all familiar with the correspondence between an element in one set and an element in another set. For example, to each house there corresponds a house number, to each automobile there corresponds a license number, to each individual there corresponds a name.

Table 1.3 lists the ten companies in the United States with the largest sales in 1991. The table indicates that to each company there corresponds a number that gives the sales for 1991. Notice that there is one and only one number associated with each company. Table 1.4 gives the federal debt in trillions of dollars for each of 10 recent years. Again, there is one and only one number (the debt) associated with each year.

We call any rule that assigns or corresponds to each element in one set precisely one element in another set a **function**. Thus, the correspondences indicated in Tables 1.3 and 1.4 are functions.

As we have seen, a table can represent a function. Functions can also be represented by formulas. For example, suppose you are going a steady 40 miles per hour in a car. Then in 1 hour you will travel 40 miles. In 2 hours you will travel 80 miles, and so on. The distance you travel depends on (corresponds to) the time. Indeed the equation relating distance (*d*), velocity (*v*), and time (*t*), is $d = v \cdot t$. In our example, we have $d = 40 \cdot t$. We can view this as a correspondence, or rule: given the time *t* in hours, the rule gives a distance *d* in miles according to $d = 40 \cdot t$. Thus, given $t = 3$, $d = 40 \cdot 3 = 120$. Notice carefully how this rule is *unambiguous*. That is, given any time *t*, the rule specifies one and only one distance *d*. This rule is therefore a function. The correspondence is between time and distance.

Often the letter *f* is used to denote a function. Thus, in the previous example, we can write $d = f(t) = 40 \cdot t$. The symbol $f(t)$ is read "*f* of *t*." One can think of *t* as the "input" and the value of $d = f(t)$ as the "output." For example an input of $t = 4$ results in an output of $d = f(4) = 40 \cdot 4 = 160$.

The following gives a general definition of function.

Definition of Function

Let *D* and *R* be two nonempty sets. A **function** *f* from *D* to *R* is a rule that assigns to each element *x* in *D* one and only one element $y = f(x)$ in *R* (see Figure 1.32).

Figure 1.32
The point *x* is sent to the unique point $y = f(x)$ for *f* to be a function.

The set *D* in the definition is called the **domain** of *f*. We might think of the domain as the set of inputs. We then can think of the values $f(x)$ as outputs.

Another helpful way to think of a function is shown in Figure 1.33. Here the function *f* accepts the input *x* from the conveyor belt, operates on *x*, and outputs (assigns) the new value $f(x)$.

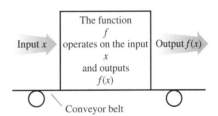

Figure 1.33

The set of all possible outputs is called the **range** of *f*. The letter representing elements in the domain is called the **independent variable**, and the letter representing the elements in the range is called the **dependent variable**. Thus, if $y = f(x)$, *x* is the independent variable, and *y* is the dependent variable since *y* *depends* on *x*. In the equation $d = 40t$, we can write $d = f(t) = 40t$ with *t* as the independent variable. We are free to set the independent variable *t* equal to any number of values. The dependent variable is *d*. Notice that *d* *depends* on the particular value of *t* that is used.

EXAMPLE 1   Determining When Rules are Functions

Which of the rules in Figure 1.34 are functions? Find the range of each function.

SOLUTION ■ The rule indicated in Figure 1.34a is a function from $A = \{a, b, c\}$ to $B = \{p, q, r, s\}$ since each element in *A* is assigned a unique element in *B*. The domain of this function is *A*, and the range is $\{p, r\}$. The rule indicated in Figure 1.34b is not a function since the element *a* has been assigned more than one element in *B*. The rule indicated in Figure 1.34c is a function with domain *A* and with range the single point $\{q\}$. ■

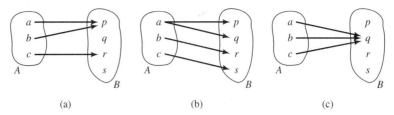

Figure 1.34
Which correspondences are functions?

Notice that it is possible for the same output to result from different inputs. For example, inputting a and b into the function in part (a) results in the same output p. In part (c), the output is always the same regardless of the input.

Functions are often given by equations, as we saw for the function $d = f(t) = 40 \cdot t$ at the beginning of this section. As another example, let both the sets D and R be the set of real numbers, and let f be the function from D to R defined by $f(x) = 3x^2 + 1$. In words, the output is obtained by taking three times the square of the input and adding one to the result. Thus, the formula

$$f(x) = 3x^2 + 1$$

can also be viewed as

$$f(\text{input}) = 3(\text{input})^2 + 1$$

If we want $f(2)$ then we replace x (or the input) with 2 and obtain

$$f(2) = 3(2)^2 + 1 = 13$$

If we want $f(-4)$ then we replace x (or the input) with -4 and obtain

$$f(-4) = 3(-4)^2 + 1 = 49$$

If we want $f(s + 1)$ then we replace x (or the input) with $s + 1$ and obtain

$$f(s + 1) = 3(s + 1)^2 + 1 = 3(s^2 + 2s + 1) + 1 = 3s^2 + 6s + 4$$

If we want $f(x + 2)$ then we replace x (or the input) with $x + 2$ and obtain

$$f(x + 2) = 3(x + 2)^2 + 1 = 3(x^2 + 4x + 4) + 1 = 3x^2 + 12x + 13$$

Convention

When the domain of a function f is not stated explicitly, we shall agree to let the domain be all values x for which $f(x)$ makes sense and is real.

For example, if $f(x) = 1/x$, the domain is assumed to be all real numbers other than zero.

EXAMPLE 2   Finding the Domain of a Function

Let $f(x) = \sqrt{2x - 4}$. Find the domain of f. Evaluate $f(2)$, $f(4)$, $f(2t + 2)$.

SOLUTIONS ■ Since $f(x)$ must be real, we must have $2x - 4 \geq 0$ or $x \geq 2$. Thus, the domain is $[2, \infty)$.

$$f(2) = \sqrt{2(2) - 4} = \sqrt{0} = 0$$

$$f(4) = \sqrt{2(4) - 4} = \sqrt{4} = 2$$

$$f(2t + 2) = \sqrt{2(2t + 2) - 4} = \sqrt{4t} = 2\sqrt{t}, \quad t \geq 0 \quad ■$$

In applications we often encounter functions with domains divided into two or more parts with a different rule applied to each part. We call such functions **piecewise-defined functions**. The following is such an example.

EXAMPLE 3   A Piecewise-Defined Function

Generally, alfalfa yields are highest in the second year of production and then decline as the stands thin out and grow less vigorously. This yield decline was estimated by Knapp [16] as

$$y = \begin{cases} 4.93, & \text{if } t = 1 \\ 7.47 - 0.584(t - 2), & \text{if } t \geq 2 \end{cases}$$

where y gives the yield in tons per acre per year and t denotes the age of the crop in years. Determine the yield in each of the first three years.

SOLUTION ■ For the first year yield we have $t = 1$ and must use the first line of the above formula. This gives $y = 4.93$, or 4.93 tons.

For the yield in the second year we have $t = 2$ and must use the second line of the above formula. This gives

$$y = 7.47 - 0.584(2 - 2) = 7.47$$

or 7.47 tons.

For the yield in the third year we have $t = 3$, and since $t = 3 \geq 2$, we must use the second line of the above formula. This gives

$$y = 7.47 - 0.584(3 - 2) = 6.886$$

or 6.886 tons. ■

Graphs of Functions

When the domain and range of a function are sets of real numbers, the function can be graphed.

Graph of a Function

The **graph** of a function f consists of all points (x, y) such that x is in the domain of f and $y = f(x)$.

EXAMPLE 4   Graphing a Function

Construct a graph of the function represented in Table 1.4.

SOLUTION ■ Let us label the x-axis in years from 1980 and the y-axis in trillions of dollars. Thus, for example, the year 1982 corresponds to 2 on the x-axis. Corresponding to the year 1982, we see a debt of $1.1 trillion. Thus, the point $(2, 1.1)$ is on the graph. In a similar fashion, the points $(3, 1.4)$ and $(4, 1.6)$ are on the graph. A complete graph is shown in Figure 1.35. ■

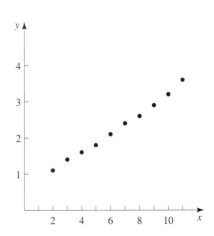

Figure 1.35
A graph of the function indicated in Table 1.4.

BABY BOOM AND BUST; 1940–90
(Births per 1,000 U.S. women aged 15–44)

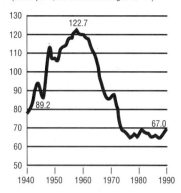

Figure 1.36

Notice the graph in Figure 1.35 gives a *picture* of the function that is represented in Table 1.4.

We have already seen that a function can be represented by a table or formula. A function can also be represented by a graph. Figure 1.36 represents the function that gives the births per 1000 U.S. women aged 15 through 44. We see that the birth rate peaked in 1960 at 122.7. If the name of the function represented by the graph in Figure 1.36 is f, then this means that $f(1960) = 122.7$.

EXPLORATION *1*

A Function Given by a Graph

 Your graphing calculator has a SIN key. Using this key graph $y_1 = \sin x$ using a window with dimensions $[-4.7, 4.7]$ by $[-1, 1]$. (Your graphing calculator must be in radian mode.) This graph determines a function. Let us call the function f, so $f(x) = \sin x$. Now use the TRACE function on your calculator to find a number of values of this function. For example, find $f(0)$, $f(1)$, $f(3.1)$, $f(-1)$, $f(-3.1)$.

We now determine the graphs of two important functions.

EXAMPLE 5   Graphing a Function

TABLE 1.5

x	−3	−2	−1	0	1	2	3
$f(x)$	9	4	1	0	1	4	9

Graph the functions f and g defined as follows. (a) $y = f(x) = x^2$ (b) $y = g(x) = |x|$.

SOLUTIONS ■ In both cases the domain is $(-\infty, \infty)$.

(a) To graph $y = f(x) = x^2$, we evaluate the function at a number of points as indicated by Table 1.5. The graph is then shown in Figure 1.37.

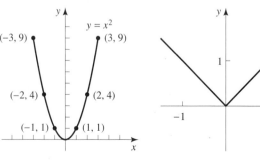

Figure 1.37
A graph of $y = x^2$.

Figure 1.38
A graph of $y = |x|$.

(b) If $x \geq 0$, then $g(x) = |x| = x$. This is a line through the origin with slope 1. This line is graphed in the first quadrant in Figure 1.38. Also if $x \leq 0$, then $g(x) = |x| = -x$, which is a line through the origin with slope -1. This is shown in Figure 1.38. ∎

Do not forget the graphs of these two functions, because we will have need to use them throughout this text.

EXPLORATION *2*

Finding a Domain Graphically

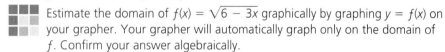 Estimate the domain of $f(x) = \sqrt{6 - 3x}$ graphically by graphing $y = f(x)$ on your grapher. Your grapher will automatically graph only on the domain of f. Confirm your answer algebraically.

We now see how to determine if a graph is the graph of a function.

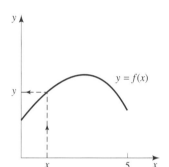

Figure 1.39
The function $y = f(x)$ takes x to y.

E X A M P L E 6   Determining If a Graph Is the Graph of a Function

A graph is shown in Figure 1.39. Does this represent the graph of a function of x?

S O L U T I O N ∎ Figure 1.39 indicates that given any number x there is one and only one value of y. You can see that starting at any value of x and moving vertically until you strike the graph, then moving horizontally until you strike the y-axis will never result in more than one value of y. Thus, this process assigns unambiguously one single value y to any value x on $[0, 5]$. Thus, this graph describes a function. ∎

E X A M P L E 7   Determining If a Graph Is the Graph of a Function

A graph is shown in Figure 1.40. Does this represent the graph of a function of x?

S O L U T I O N ∎ Notice that for the value a indicated in Figure 1.40, drawing a vertical line from a results in striking the graph in *two* places resulting in assigning two values to x when $x = a$. This cannot then be the graph of a function of x. ∎

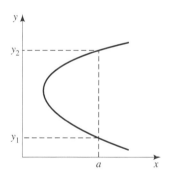

Figure 1.40
Not a function since a vertical line strikes the graph in two places.

This works in general.

Vertical Line Test

A graph in the xy-plane represents a function of x, if and only if, every vertical line intersects the graph in at most one point.

E X A M P L E 8   A Graph That Fails to Be the Graph of a Function

Since $\sqrt{x^2 + y^2}$ is the distance from the point (x, y) to the origin, the graph of the equation $x^2 + y^2 = 1$ is a circle centered at the origin with radius 1 (see Figure 1.41a). Does this represent a function of x?

S O L U T I O N ∎ As Figure 1.41b indicates, the vertical line test fails, so this cannot be the graph of a function of x. ∎

Remark. In the previous example, if we try to solve for y we obtain $y = \pm\sqrt{1 - x^2}$. Thus, for any value of x in the interval $(-1, 1)$, there are *two* corresponding values

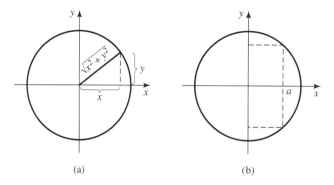

(a)   (b)

Figure 1.41
Not a function since a vertical line strikes the graph in two places.

of y. We are not told which of these two values to take. Thus, the rule is ambiguous and not a function.

Applications

To solve any applied problem, we must first take the problem and translate it into mathematics. This requires us to create equations and functions. We already did this at the beginning of the section when we related distance traveled with velocity and time. We also have created many other (linear) equations earlier. In the next example, we find a *nonlinear* function that gives the volume of a certain package. Later in the text we use this function to find important information such as the dimensions that yield the maximum volume. For now, we just find the function.

EXAMPLE 9   Finding a Function

From all four corners of a rectangular piece of cardboard of dimensions 10 inches by 20 inches squares are cut with dimensions x by x. The sides are turned up to form an open box (see Figure 1.42) that is used in packaging. Find the volume V of the box as a function of x.

SOLUTION ■ The volume of a box is given by the product of the width (w), the length (l), and the height (h), that is $V = w \cdot l \cdot h$. The width of the box is $w = 10 - 2x$, the length is $20 - 2x$, and the height is x. Thus, the volume is

$$V = V(x) = (10 - 2x)(20 - 2x)x \quad ■$$

Piecewise-defined functions occur naturally, as the following example illustrates.

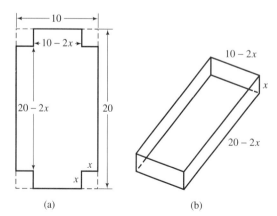

(a)   (b)

Figure 1.42
Fold the cut cardboard on the left to obtain the box without top on right.

EXAMPLE 10   A Piecewise-Defined Function

Taxable Income ($)	What You Pay
0.00–21,450	15% of sum over $0.00
21,450–51,900	$3217.50 + 28% of sum over $21,450
51,900 and over	$11,743.50 + 31% of sum over $51,900

The table gives the federal income tax table in 1994 for single individuals. Let $T(x)$ be the federal income tax you pay based on taxable income of x dollars if you are single. Write $T(x)$ as a piecewise-defined function. Draw a graph of $y = T(x)$. Find $T(20,000)$, $T(30,000)$, and $T(60,000)$.

SOLUTION ■ The first line of the table implies that for $x \le 21,450$, $T(x) = 0.15x$.

The second line of the table implies that for $21,450 \le x \le 51,900$, $T(x) = 3217.50 + 0.28(x - 21,450)$.

The third line of the table implies that for $x \ge 51,900$, $T(x) = 11,743.50 + 0.31(x - 51,900)$.

Putting this all together gives the piecewise-defined function

$$T(x) = \begin{cases} 0.15x, & \text{if } x \le 21,450 \\ 3217.50 + 0.28(x - 21,450), & \text{if } 21,450 \le x \le 51,900 \\ 11,743.50 + 0.31(x - 51,900), & \text{if } x \ge 51,900 \end{cases}$$

The first formula gives $y = 0.15x$, the graph of which is a line with slope 0.15 and y-intercept 0. You can easily check that the line also goes through the point $(21, 450, 3217.50)$, as shown in Figure 1.43a.

The second formula gives $y = 3217.50 + 0.28(x - 21,450)$, the graph of which is a line through the point $(21,450, 3217.50)$ with slope 0.28. You can easily check that the line also goes through the point $(51,900, 11,743.50)$. This is shown in Figure 1.43b.

The third formula gives $y = 11,743.50 + 0.31(x - 51,900)$, the graph of which is a line through the point $(51,900, 11,743.50)$ with slope 0.31. This is shown in Figure 1.43c.

Putting this all together gives the graph of $y = T(x)$ shown in Figure 1.43d.

Notice that the line segments become steeper. This shows graphically that the income tax rates, which are the slopes of the lines, are higher for larger incomes.

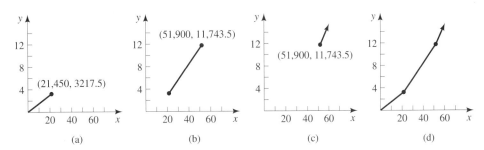

Figure 1.43
The graph of a piecewise-defined function.

The taxes on $20,000 are $T(20,000)$ dollars. Since $20,000 \le 21,450$, we use the formula in the first line and obtain

$$T(20,000) = 0.15(20,000) = 3000$$

that is, $3000.

The taxes on $30,000 are $T(30,000)$ dollars. Since $21,450 \le 30,000 \le 51,900$, we use the formula in the second line and obtain

$$T(30,000) = 3217.50 + 0.28(30,000 - 21,450) = 5611.50$$

that is, $5611.50.

The taxes on $60,000 are $T(60,000)$ dollars. Since $60,000 \ge 51,900$, we use the formula in the third line and obtain

$$T(60,000) = 11,743.50 + 0.31(60,000 - 51,900) = 14,254.50$$

that is, $14,254.50. ∎

EXPLORATION *3*

Graphing Piecewise-Defined Functions

 Graph $y = f(x)$ on your grapher where

$$f(x) = \begin{cases} x^2 + 1, & x \le 0 \\ 3, & 0 < x < 3 \\ x - 3 & x \ge 3 \end{cases}$$

Hint: (a) First graph $y = (x < 0)$ where "$<$" is a function (see the technology manual). The expression $(x < 0)$ is 1 if true and 0 otherwise. Thus, $y = (x < 0)$ is 1 if $x < 0$ and 0 if $x \ge 0$. Does your graph indicate this?
    (b) Now graph $y = (x^2 + 1)(x < 0)$.
    (c) Considering (a) what do you think the graph of $y = (x > 0)(x < 3)$ looks like? Confirm your conjecture by graphing on your grapher. Then graph $y = 3 (x > 0)(x < 3)$.
    (d) Do you now think the graph of

$$y = (x^2 + 1)(x < 0) + 3 (x > 0)(x < 3) + (x - 3)(x \ge 3)$$

is the same as the graph of $y = f(x)$? Graph and see.

SELF-HELP EXERCISE SET 1.3

1. Which of the graphs in the figures represents the graph of a function?

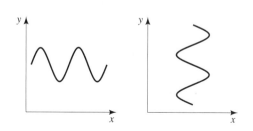

2. Find the domain of $f(x) = (\sqrt{x + 1})/x$.

3. For the function in the previous exercise evaluate, (**a**) $f(-1)$ (**b**) $f(3)$ (**c**) $f(a + 2)$ (**d**) $f(x + 2)$.

4. A telephone company charges $1.25 for the first minute for a certain long-distance call and $0.25 for each additional minute or fraction thereof. If C is the cost of a call and x is the length of the call in minutes, find the cost as a function of x, and sketch a graph. Assume $x \le 3$.

For Exercises 1 through 6, determine which of the rules in the figures represent functions.

1.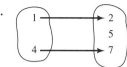

2.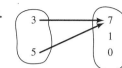

3.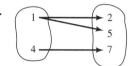

4.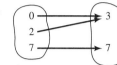

5.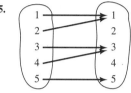

6.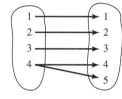

7. Specify the domain and range for each rule that represents a function in Exercises 1, 2, and 3.

8. Specify the domain and range for each rule that represents a function in Exercises 4, 5, and 6.

For Exercises 9 through 24, find the domain of each function.

9. $\dfrac{x}{x-1}$

10. $\dfrac{x-1}{x+2}$

11. $\sqrt{x}$

12. $\sqrt{x-1}$

13. $\left|\dfrac{1}{\sqrt{x-1}}\right|$

14. $\sqrt{2-x}$

15. $\dfrac{x-1}{x^2-5x+6}$

16. $\dfrac{x}{x^2-1}$

17. $\sqrt{|x-1|}$

18. $|\sqrt{x-1}|$

19. $\dfrac{\sqrt{x+2}}{x}$

20. $\dfrac{x}{\sqrt{x-1}}$

21. $\sqrt{x^2+1}$

22. $\dfrac{1}{x^2+1}$

23. $\dfrac{1}{\sqrt{x^2+1}}$

24. $\sqrt{x^2-1}$

For Exercises 25 through 30, find the domain and range of each function.

25.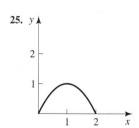

26.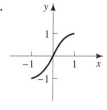

27.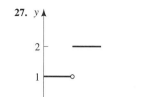

28.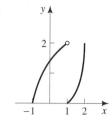

29.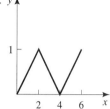

30.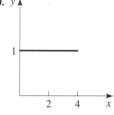

In Exercises 31 through 38 evaluate the given function at the given values.

31. $f(x) = 2x^3 + 3$, $f(1)$, $f(-1)$, $f(0)$, $f(\sqrt{3})$, $f(x^2)$, $f\left(\dfrac{1}{x}\right)$

32. $f(x) = 2x^4 - 1$, $f(0)$, $f(2)$, $f\left(-\dfrac{1}{2}\right)$, $f(-\sqrt{2})$, $f(x^2)$, $f(\sqrt{x})$

33. $f(x) = \dfrac{1}{x+1}$, $f(1)$, $f\left(\dfrac{1}{2}\right)$, $f(-2)$, $f(\sqrt{2}-1)$, $f(x+2)$, $f\left(\dfrac{1}{x+1}\right)$

34. $f(x) = \dfrac{1}{x^2+1}$, $f(2)$, $f\left(\dfrac{1}{2}\right)$, $f(-3)$, $f(2\sqrt{2})$, $f(\sqrt{x})$, $f(x-1)$

35. $f(x) = \sqrt{2x-1}$, $f(5)$, $f\left(\dfrac{1}{2}\right)$, $f\left(x^2-\dfrac{1}{2}\right)$, $f\left(\dfrac{1}{x}\right)$, $f(-x)$, $-f(x)$

36. $f(x) = \sqrt{2-3x}$, $f(-1)$, $f(0)$, $f(x+1)$, $f\left(\dfrac{1}{x}\right)$, $f(-x)$, $-f(x)$

37. $f(x) = |x+1|$, $f(-2)$, $-f(-2)$, $f(x-1)$, $f(x^2-1)$

38. $f(x) = \dfrac{x-1}{x+1}$, $f(1)$, $f(-1)$, $f(x+1)$, $f(x^2)$

39. Let $f(x)$ be the function given by the graph in Exercise 27. Find $f(0)$, $f(1)$, $f(1.5)$, $f(2)$.

40. Let $f(x)$ be the function given by the graph in Exercise 28. Find $f(-1)$, $f(1)$, $f(2)$.

41. Let $f(x)$ be the function given by the graph in Exercise 29. Find $f(0)$, $f(2)$, $f(4)$, $f(6)$.

42. Your graphing calculator has a COS key. Using this key graph $y_1 = \cos x$ using a window with dimensions $[-4.7, 4.7]$ by $[-1,1]$. (Your graphing calculator must be in radian mode.) This graph determines a function. Let us call the function f, so $f(x) = \cos x$. Now use the TRACE

function on your calculator to find a number of values of this function. For example, estimate $f(0)$, $f(1)$, $f(3.1)$, $f(-1)$, $f(-3.1)$.

For Exercises 43 through 50, determine whether each graph represents a function of x.

43.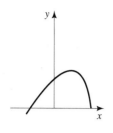

44.

45.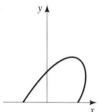

46.

47.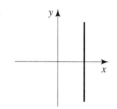

48.

49. **50.**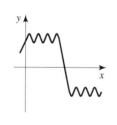

In Exercises 51 through 58 an equation involving x and y is given. Determine if y can be written as a function of x. If it can, find it.

51. $2x + 5y = 15$

52. $x - y = 2$

53. $x^2 + y^2 = 4$

54. $(x - 1)^2 + (y - 2)^2 = 1$

55. $y^3 = x$

56. $y^4 = x$

57. $x + y^2 - 1 = 0$

58. $x^2 + y - 1 = 0$

For Exercises 59 through 66 graph the indicated function.

59. $f(x) = \begin{cases} 3x - 1 & \text{if } x \le -1 \\ x & \text{if } x > -1 \end{cases}$

60. $f(x) = \begin{cases} x & \text{if } x \le 1 \\ -x & \text{if } x > 1 \end{cases}$

61. $f(x) = \begin{cases} x^2 & \text{if } x \le 0 \\ x & \text{if } x > 0 \end{cases}$

62. $f(x) = \begin{cases} -1 & \text{if } x < 0 \\ 1 & \text{if } x \ge 0 \end{cases}$

63. $f(x) = \begin{cases} x^2 & \text{if } x \le 0 \\ x & \text{if } 0 < x \le 1 \\ 2x & \text{if } x > 1 \end{cases}$

64. $f(x) = \begin{cases} 1 & \text{if } x < 0 \\ 0 & \text{if } 0 \le x \le 1 \\ -1 & \text{if } x > 1 \end{cases}$

65. $f(x) = \begin{cases} 1 & \text{if } 0 \le x < 1 \\ 2 & \text{if } 1 \le x < 2 \\ 3 & \text{if } 2 \le x \le 3 \end{cases}$

66. $f(x) = \begin{cases} -x & \text{if } x \le -1 \\ 1 & \text{if } -1 < x < 1 \\ x & \text{if } x \ge 1 \end{cases}$

In Exercises 67 through 72 graph using your graphing calculator and estimate the domain of each function. Confirm algebraically.

67. $f(x) = |-x^2 + 8|$

68. $f(x) = \sqrt{x^2 + 4}$

69. $f(x) = \sqrt{9 - x^2}$

70. $f(x) = \sqrt{x^2 - 9}$

71. $f(x) = \dfrac{\sqrt{x}}{x - 5}$

72. $f(x) = \begin{cases} \dfrac{\sqrt{1 - x}}{x + 4}, & x \le -2 \\ \sqrt{4 - x}, & x > 2 \end{cases}$

73. Find the area A of a circle as a function of the radius r.

74. Find the area A of a square as a function of the length s of a side.

Applications

75. Packaging. A box has a square base with each side of length x and height equal to $3x$. Find the volume V as a function of x.

76. Packaging. Find the surface area S of the box in the previous exercise as a function of x.

77. Velocity. A car travels at a steady 60 miles per hour. Write the distance d in miles that the car travels as a function of time t in hours.

78. Navigation. Two ships leave port at the same time. The first ship heads due north at 5 miles per hour, and the second heads due west at 3 miles per hour. Let d be the distance between the ships in miles and t the time in hours since they left port. Find d as a function of t.

79. Revenue. A company sells a certain style of shoe for $60. If x is the number of shoes sold, what is the revenue R as a function of x?

80. Delivery Fees. A store will deliver a sofa for free if you are within 10 miles of the store and charges $2 per mile (including fractions thereof) beyond 10 miles. Let x be the miles from the store and C the cost of delivery. Write C as a function of x and sketch a graph of this function.

81. Sales Commission. A salesman receives a commission of $1 per square yard for the first 500 yards of carpeting sold in a month and $2 per square yard for any additional carpet sold during the same month. If x is the number of yards of carpet sold and C is the commission, find C as a function of x, and sketch a graph of this function.

82. Taxes. A certain state has a tax on electricity of 1% of the monthly electricity bill with the first $50 of the bill exempt from tax. Let $P(x)$ be the percentage paid in taxes and x the amount of the bill in dollars. Sketch a graph of this function.

83. Taxes. In the previous problem, let $T(x)$ be the total amount of dollars paid on the tax, where x is the amount of the monthly bill. Sketch a graph of this function.

84. Production. A production function that often appears in the literature (see Kim and Mohtadi [17]) is

$$y = \begin{cases} 0 & \text{if } x < M \\ x - M & \text{if } x \geq M \end{cases}$$

where y is production (output) and x is total labor. Sketch a graph of this function.

85. Demand Equation. A demand equation that often appears in the literature (see Guesnerie [18]) is

$$D(p) = \begin{cases} A - Bp & \text{if } p \leq \dfrac{A}{B} \\ 0 & \text{if } p > \dfrac{A}{B} \end{cases}$$

Sketch a graph of this equation.

86. Interest Rates. The figure shows a graph of the prime rate (the rate that banks charge to their best customers) given in percent over a recent period of time [19]. Let $p(x)$ be this function, where x is the year. What is $p(1990.3)$? $P(1993.5)$?

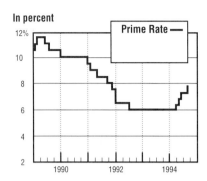

87. Biology. In an experiment Senanayake and coworkers [20] showed that the yield y of potato flea beetles treated with an insecticide given as a percentage of untreated beetles was approximated by the piecewise linear function $y = 100 + 0.24x$ when $x \leq 59$ and $y = 114.3 - 0.219x$ when $x > 59$, where x is the peak number of potato flea beetles per plant. What was the yield when $x = 40$? 200? Sketch a graph on $[0, 500]$.

88. Biology. Hardman [21] found that the survival rate S of the summer eggs of the European red mite was approximated by the piecewise linear function given by $S = 0.998 - (18 - T)/0.0019$ if $T < 18$ and $S = 0.998$ if $T \geq 18$, where T is temperature in degrees Celsius. Determine the survival rate if $T = 15$, $T = 20$. Sketch a graph.

Enrichment Exercises

In Exercises 89 through 94 find $[f(x + h) - f(x)]/h$ for the indicated functions.

89. $f(x) = 3x + 1$

90. $f(x) = x^2 - 1$

91. $f(x) = 3x^2 + 1$

92. $f(x) = |x|$

93. $\dfrac{1}{x}$

94. $\sqrt{x}$

95. Using your graphing calculator, graph on the same screen $y_1 = f(x) = |x|$ and $y_2 = g(x) = \sqrt{x^2}$. How many graphs do you see? What does this say about the two functions?

96. Using your graphing calculator, graph on the same screen $y_1 = f(x) = x - 3$ and $y_2 = g(x) = (2x^2 - 9x + 9)/(2x - 3)$. How many graphs do you see? What does this say about the two functions?

97. Let $f(x) = mx$. Show $f(a + b) = f(a) + f(b)$.

98. Let $f(x) = x^2$. Find x, such that $f(x + 1) = f(x + 2)$.

99. Let $f(x) = 3^x$. Show that $3f(x) = f(x + 1)$ and that $f(a + b) = f(a) \cdot f(b)$.

100. Let $f(x) = x^2 + 1$. Find $f[f(1)]$ and $f[f(x)]$.

SOLUTIONS TO SELF-HELP EXERCISE SET 1.3

1. The vertical line test indicates that the first figure represents a function but the second does not.

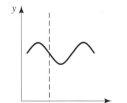
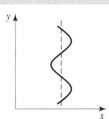

2. The function $f(x) = (\sqrt{x + 1})/x$ is defined if $x + 1 \geq 0$, or $x \geq -1$, and if $x \neq 0$. Thus the domain is all points on $[-1, 0)$ together with all points on $(0, \infty)$.

3. For the function in the previous exercise

(a) $f(-1) = \dfrac{\sqrt{-1 + 1}}{-1} = 0$

(b) $f(3) = \dfrac{\sqrt{3 + 1}}{3} = \dfrac{2}{3}$

(c) $f(a + 2) = \dfrac{\sqrt{a + 2 + 1}}{a + 2} = \dfrac{\sqrt{a + 3}}{a + 2}$

(d) $f(x + 2) = \dfrac{\sqrt{x + 2 + 1}}{x + 2} = \dfrac{\sqrt{x + 3}}{x + 2}$

4. For the first minute the cost is a constant \$1.25. During the second minute the cost is a constant \$1.25 + \$0.25 = \$1.50. During the third minute the cost is \$1.75. We can write the function as

$$C(x) = \begin{cases} 1.25 & \text{if } 0 < x \leq 1 \\ 1.50 & \text{if } 1 < x \leq 2 \\ 1.75 & \text{if } 2 < x \leq 3 \end{cases}$$

The graph is shown in the figure. You can graph this function on a graphing calculator that has a greatest integer function. Set the MODE of your calculator to Dot. If int(x) is the greatest integer function, let $y_1 = -0.25\text{int}(-x) + 1$ on a window with dimensions [0, 4.7] by [0, 1.9]. Use TRACE to check your answer. Notice what happens as x goes from 0.95 to 1.00 to 1.05.

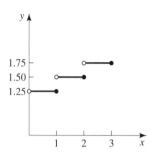

1.4 QUADRATIC FUNCTIONS

■ Quadratic Functions ■ Models Using Quadratics

APPLICATION: MAXIMIZING REVENUE

Suppose the demand equation for a certain commodity is given by $p = -2x + 12$, where x is given in thousands per week. For what value of x is the revenue at maximum? The answer can be found in Example 5.

APPLICATION: MAXIMIZING REVENUE

Assuming cotton prices of \$0.75 per pound, Chakravorty and Roumasset [22] showed that the revenue R in dollars for cotton in California is approximated by the function $R(w) = -0.2224 + 1.0944w - 0.5984w^2$, where w is the amount of irrigation water in appropriate units paid for and used. What happens to the revenue if only a small amount of water is paid for and used? A large amount? What is the optimal amount of water to use? For the answers to these questions see Example 7.

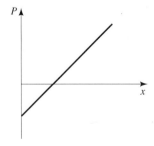

Figure 1.44
A linear profit functions implies large profits no matter how large the number of items sold. Is this realistic?

Quadratic Functions

In Section 1.2 we assumed a mathematical model that was *linear* for both the cost function and the revenue function. As a consequence the profit function was linear. This is adequate if the number sold is not too large. If the profit function is linear, as in Figure 1.44, the graph implies that profits become large without bound. This we know is not what actually happens. If, for example, we try to flood the market with pens, we most likely have to set the price per pen so low as to result in losses, not

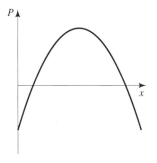

Figure 1.45
This profit function indicates that profits begin to fall if we try to flood the market with our items. Some quadratics behave like this.

profits. Thus, a more reasonable mathematical model should yield a profit curve that rises at first and then falls, such as shown in Figure 1.45. Some quadratic functions behave like this. Quadratics can also be used in other applications and are often more realistic than a linear model. Thus, we need to study quadratic functions.

The function

$$f(x) = ax^2 + bx + c, \quad a \neq 0$$

is a **quadratic function**. In this section we graph quadratic functions, find where each one takes either a maximum or minimum, and consider models based on quadratic functions.

EXPLORATION *1*

Shape of the Graph of a Quadratic

 Graph the following on the same screen. Note carefully if the graph opens upward or downward and if the coefficient of x^2 is positive or negative.

1. $y = x^2 - 3x + 1$ 2. $y = 2x^2 - 3x + 1$
3. $y = -x^2 - 3x + 1$ 4. $y = -2x^2 - 3x + 1$

What have you discovered about the relationship between the sign of the coefficient of x^2 and how the graph bends? Confirm your conjecture by graphing other quadratic functions.

We shall see that all quadratics have graphs that look basically like the graphs of ax^2. Thus, we begin with the following example.

EXAMPLE 1   Graphs of Some Simple Quadratics

Graph the quadratics

$$y = \frac{1}{2}x^2 \qquad y = x^2 \qquad y = 2x^2 \qquad y = -x^2$$

SOLUTION ■ Using the values found in Table 1.6, we obtain the graphs given in Figure 1.46.

As Figure 1.46 indicates, the graphs of $f(x) = ax^2$ when $a > 0$ all look similar; they open upward and are narrower when a is larger. The graphs of ax^2 when $a < 0$ are similar except that they open downward. ■

EXPLORATION *2*

More on the Shape of the Graph of a Quadratic

 Reproduce Figure 1.46 on your graphing calculator using a window with dimensions $[-4.7, 4.7]$ by $[-20, 20]$. Use TRACE and the cursor to hop up or down from one curve to the other. How does the y-coordinate change?

We now see how to effectively graph some quadratics by shifting the graph of $y = x^2$ up, down, left, or right.

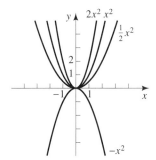

Figure 1.46
The graphs of some simple quadratics.

TABLE 1.6

x	-3	-2	-1	0	1	2	3
$0.5x^2$	4.5	2	0.5	0	0.5	2	4.5
x^2	9	4	1	0	1	4	9
$2x^2$	18	8	2	0	2	8	18
$-x^2$	-9	-4	-1	0	-1	-4	-9

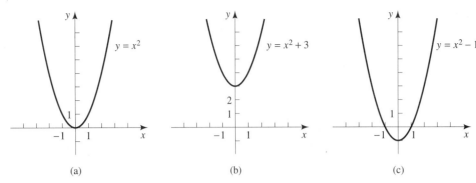

Figure 1.47
Adding 3 shifts the graph up 3 units. Subtracting 1 shifts the graph down 1 unit.

TABLE 1.7

x	x^2	$(x-2)^2$
-5	25	49
-4	16	36
-3	9	25
-2	4	16
-1	1	9
0	0	4
1	1	1
2	4	0
3	9	1
4	16	4
5	25	9

TABLE 1.8

x	x^2	$(x+1)^2$
-5	25	16
-4	16	9
-3	9	4
-2	4	1
-1	1	0
0	0	1
1	1	4
2	4	9
3	9	16
4	16	25
5	25	36

EXAMPLE 2   Shifting a Graph Up or Down

Graph $y = x^2 + 3$ and $y = x^2 - 1$.

SOLUTION ■ In Figure 1.47a we give the graph of $y = x^2$ found in Figure 1.46. To graph $y = x^2 + 3$ in Figure 1.47b, we simply add 3 to every y value obtained in Figure 1.47a. This results in the graph of $y = x^2$ shifted upward 3 units to obtain the graph of $y = x^2 + 3$. To obtain the graph of $y = x^2 - 1$ in Figure 1.47c, we shift the graph of $y = x^2$ down 1 unit. ■

EXPLORATION 3

Vertical Shift

On a screen with dimensions [−4.7, 4.7] by [−20, 20] graph

$$y = x^2 - 3 \qquad y = x^2 \qquad y = x^2 + 3 \qquad y = x^2 + 6$$

Use TRACE and the cursor to hop up or down from one curve to the other. How does the y-coordinate change?

EXAMPLE 3   Shifting a Graph Left or Right

Graph $y = (x - 2)^2$ and $y = (x + 1)^2$.

SOLUTION ■ First notice that these are indeed quadratics since

$$(x - 2)^2 = x^2 - 4x + 4 \qquad (x + 1)^2 = x^2 + 2x + 1$$

Table 1.7 compares the values of x^2 with $(x - 2)^2$. Notice that the y values of $y = (x - 2)^2$ are the same as the y values of $y = x^2$ except that they occur 2 units to the right on the x-axis. This is seen in Figure 1.48a.

Table 1.8 compares the values of x^2 with $(x + 1)^2$. Notice that the y values of $y = (x + 1)^2$ are the same as the y values of $y = x^2$ except that they occur 1 unit to the left on the x-axis. This is seen in Figure 1.48b. ■

EXPLORATION 4

Horizontal Shift

On the same screen graph

$$y = (x + 3)^2 \qquad y = x^2 \qquad y = (x - 3)^2 \qquad y = (x - 6)^2$$

How do these graphs differ?

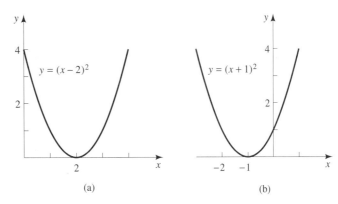

Figure 1.48
Replacing x with $x - 2$ shifts the graph 2 units to the right. Replacing x with $x + 1$ shifts the graph 1 unit to the left.

We now see how to graph a general quadratic by shifting the graph of a simple quadratic of the form $y = ax^2$ first to the left or right and then up or down. Before we can do this we must first do some algebraic manipulation to the quadratic (called completing the square) to put the quadratic expression into a convenient form.

EXAMPLE 4   Graphing a Quadratic

Sketch the graph of $y = 2x^2 - 16x + 14$.

S O L U T I O N ■ We complete the square.

STEP 1. Factor out the coefficient of x^2

$$2x^2 - 16x + 14 = 2(x^2 - 8x + 7)$$

STEP 2. Complete the square inside the bracket. Do this by adding and subtracting the square of one half of the coefficient of x.

$$\left(\frac{-8}{2}\right)^2 = (-4)^2 = 16$$

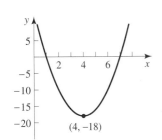

Figure 1.49
The vertex of this quadratic is at $(4, -18)$.

$$\begin{aligned}
2(x^2 - 8x + 7) &= 2(x^2 - 8x + 16 - 16 + 7) && \text{Add and subtract 16.} \\
&= 2([x^2 - 8x + 16] - 9) && \text{Collect terms.} \\
&= 2([x - 4]^2 - 9) && \text{Complete square.} \\
&= 2(x - 4)^2 - 18 && \text{Multiply by 2.}
\end{aligned}$$

In view of the previous examples, the graph of $2(x - 4)^2 - 18$ is obtained by shifting the graph of $2x^2$ four units to the right and 18 units down, as shown in Figure 1.49. ■

EXPLORATION 5

Do Graphically, Confirm Algebraically

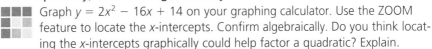 Graph $y = 2x^2 - 16x + 14$ on your graphing calculator. Use the ZOOM feature to locate the x-intercepts. Confirm algebraically. Do you think locating the x-intercepts graphically could help factor a quadratic? Explain.

The same procedure can be used for any quadratic $ax^2 + bx + c$, with $a \neq 0$.

$$\begin{aligned}
f(x) &= ax^2 + bx + c \\
&= a\left(x^2 + \frac{b}{a}x + \frac{c}{a}\right) && \text{Factor out } a.
\end{aligned}$$

$$= a\left(\left[x^2 + \frac{b}{a}x + \frac{b^2}{4a^2}\right] - \frac{b^2}{4a^2} + \frac{c}{a}\right) \qquad \text{Add and subtract } \left(\frac{b}{2a}\right)^2$$

$$= a\left(\left[x + \frac{b}{2a}\right]^2 + \frac{c}{a} - \frac{b^2}{4a^2}\right) \qquad \text{Complete the square.}$$

$$= a\left(x + \frac{b}{2a}\right)^2 + \left(c - \frac{b^2}{4a}\right) \qquad \text{Multiply by } a$$

If we set

$$h = -\frac{b}{2a} \qquad \text{and} \qquad k = c - \frac{b^2}{4a}$$

the quadratic can be written in **standard form**

$$f(x) = a(x - h)^2 + k$$

The point (h, k) is called the **vertex**. With the quadratic in the standard form $y = a(x - h)^2 + k$, and $a > 0$, notice that y is smallest when the term $a(x - h)^2 = 0$. This can only happen if $x = h$. Thus, when $a > 0$, $y = a(x - h)^2 + k$ takes a *minimum* when $x = h$. The minimum is $y = k$.

With the quadratic in the standard form $y = a(x - h)^2 + k$, and $a < 0$, notice that y is largest when the term $a(x - h)^2 = 0$. This can only happen if $x = h$. Thus, when $a < 0$, $y = a(x - h)^2 + k$ takes a *maximum* when $x = h$. The maximum is $y = k$.

The graphs, which are called **parabolas**, are given in Figure 1.50 in the two cases $a > 0$ and $a < 0$. Notice that when $a > 0$, the parabola opens upward and $y = k$ is the *minimum* value of y. When $a < 0$, the parabola opens downward and $y = k$ is the *maximum* value of y.

The Quadratic $ax^2 + bx + c$

Every quadratic with $a \neq 0$ can be put into the standard form

$$ax^2 + bx + c = a(x - h)^2 + k$$

where

$$h = -\frac{b}{2a} \qquad \text{and} \qquad k = c - \frac{b^2}{4a}$$

The graphs of the quadratics are shown in Figure 1.50.

When $a > 0$, the parabola opens upward and $y = k$ is the *minimum* value of y.
When $a < 0$, the parabola opens downward and $y = k$ is the *maximum* value of y.

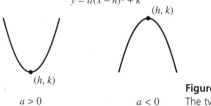

$y = a(x - h)^2 + k$

(h, k)

(h, k)

$a > 0$ $a < 0$

Figure 1.50
The two types of quadratics.

Models Using Quadratics

We now use quadratic functions to model the revenue equation. The exercise set includes quadratics that model cost and profit equations.

EXAMPLE 5   A Quadratic Revenue Function

Suppose the demand equation for a certain commodity is given by $p = -2x + 12$, where x is given in thousands per week and p is the price of one item. Find the value of x for which the revenue is at maximum. Find the value(s) of x for which the revenue is zero.

SOLUTION ■   If $R(x)$ is the revenue in thousands of dollars per week, then

$$R(x) = xp = x(-2x + 12) = -2x^2 + 12x$$

We can now proceed in either of two ways: complete the square or use the formulas listed earlier. We illustrate both methods.

First, we complete the square. We obtain

$$
\begin{aligned}
R(x) &= -2x^2 + 12x \\
&= -2(x^2 - 6x) & \text{Factor out } -2. \\
&= -2([x^2 - 6x + 9] - 9) & \text{Add and subtract } (-6/2)^2 = 9. \\
&= -2([x - 3]^2 - 9) & \text{Complete the square.} \\
&= -2(x - 3)^2 + 18 & \text{Multiply by } -2.
\end{aligned}
$$

The quadratic is now in standard form. Thus, we see that the maximum revenue occurs when $x = 3$ or when 3000 of the items are sold. See Figure 1.51.

With the second method, we can use the formula given earlier

$$h = -\frac{b}{2a} = -\frac{12}{2(-2)} = 3$$

To answer the second question we factor $R(x)$ to obtain

$$R(x) = 2x(6 - x)$$

This shows that the zeros of R occur at $x = 0$ and $x = 6$. Thus, revenue is zero when no items are sold and also when 6000 items are sold. ■

We see that there are two ways of graphing a quadratic. You can remember the process of completing the square or remember the formulas.

The graph of the revenue function shown in Figure 1.51 is typical. To sell an unusually large number of an item to consumers, the price needs to be lowered to such an extent as to begin lowering the (total) revenue.

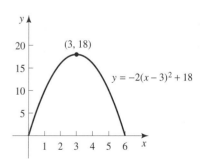

Figure 1.51
This quadratic has vertex at (3, 18) and opens downward since $a = -2 < 0$.

EXPLORATION 6

Revenue, Cost, Profit

 Suppose the revenue function is the same as found in Example 5 and the cost function is $C(x) = 2x + 5$. Using a window with dimensions $[0, 8]$ by $[-5, 20]$ graph $y_1 = R(x)$ and $y_2 = C(x)$.

    (a) Identify area(s) bounded by the revenue and cost graphs that correspond to profits. To losses.

(b) Now add the profit equation $y_3 = y_1 - y_2 = P(x)$. Note the region(s) where the profit curve is below the x-axis and make this correspond to the region(s) of profit found in (a). Note the region(s) where the profit curve is above the x-axis and make this correspond to the region(s) of loss found in (a).

(c) Does the profit curve peak at the same point as the revenue curve? What implication does this have for business?

We now work through an example that illustrates that finding the zeros of a quadratic is equivalent to factoring the quadratic.

EXAMPLE 6   Factoring a Quadratic Graphically

A profit function is given as $P(x) = -x^2 + 8.8x - 14.95$. Graph on your graphing calculator and use the ZOOM feature repeatedly to determine the break-even quantities. Now use this to factor the quadratic and confirm your answer algebraically.

SOLUTION ■   The graph using the standard viewing window is shown in Screen 1.11. By repeatedly using the ZOOM feature on each x-intercept, we see that the x-intercepts are 2.3 and 6.5. If this is correct, then we must have

$$-x^2 + 8.8x - 14.95 = -(x^2 - 8.8x + 14.95) = -(x - 2.3)(x - 6.5)$$

Because this is indeed correct, we have confirmed our answer algebraically. ■

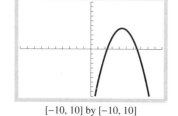

[−10, 10] by [−10, 10]

Screen 1.11
A graph of $y = -x^2 + 8.8x - 14.95$. Use the ZOOM feature to find the break-even quantities.

The graphical method we just used to find the zeros of a quadratic is an alternative to using the quadratic formula. As we shall see there is no corresponding formula for finding the zeros of many other types of important functions. In these instances the graphical method is the only way of finding the x-intercepts.

EXPLORATION 7

Finding x-intercepts Graphically

Find the x-intercepts of $y = 1.1x^2 - 9.68x + 12.045$ graphically. Confirm your answer using the quadratic formula.

EXAMPLE 7   Maximizing Revenue

Assuming cotton prices of $0.75 per pound, Chakravorty and Roumasset [22] showed that the revenue R in dollars for cotton in California is approximated by the function $R(w) = -0.2224 + 1.0944w - 0.5984w^2$, where w is the amount of irrigation water in appropriate units paid for and used. What happens to the revenue if only a small amount of water is paid for and used? A large amount? What is the optimal amount of water to use?

SOLUTION ■   Notice that $R(0) = -0.2224$. Thus, using little water results in little revenue. But $R(1) = 0.2736$, which represents positive revenue. However, $R(2) < 0$, so using (and paying for) too much water drives revenue down.

To see where revenue is maximum first recognize the expression $-0.2224 + 1.0944w - 0.5984w^2$ as a quadratic $aw^2 + bw + c$, where $a = -0.5984$, $b = 1.0944$, and $c = -0.2224$. Since $a < 0$, the graph opens downward. If the vertex is (h, k), then

$$h = -\frac{b}{2a} = -\frac{1.0944}{2(-0.5984)} = 0.914$$

$$k = c - \frac{b^2}{4a} = -0.2224 - \frac{(1.0944)^2}{4(-0.5984)} \approx 0.278$$

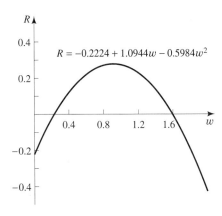

Figure 1.52
The revenue is negative for small w and larger w.

For this model 0.914 units of water yields a maximum revenue of about 0.278 (Fig. 1.52).

A graph using a graphing utility with window dimensions of $[0, 2]$ by $[-0.5, 0.5]$ is shown in Screen 1.12. ZOOMing several times, indicates that the peak is at $(0.914, 0.278)$.   ∎

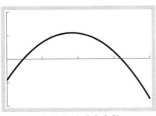

[0, 2] by [−0.5, 0.5]

Screen 1.12
A graph of $y = -0.2224 + 1.0944x - 0.5984x^2$. Use the ZOOM feature to find the vertex at about $(0.914, 0.278)$.

We close this section with a final observation about using a quadratic versus a linear model. Screen 1.13 shows a graph of the profit equation given by the quadratic $y = -x^2 + 22x - 40$ on a window with dimensions $[0, 20]$ by $[-50, 100]$. Notice that on the interval $[0, 5]$ the graph of the quadratic looks almost like the graph of a straight line. We can check this further by graphing using a window with dimensions $[0, 5]$ by $[-50, 100]$, as shown in Screen 1.14. Notice that the graph looks linear in this window! Thus, if we were only interested in the profits on the interval $[0, 5]$ we might replace the quadratic function with a linear function. This would make any mathematical analysis easier and would lead to only a small error.

EXPLORATION 8

Approximating a Quadratic Function with a Linear Function

 By experimentation find a linear equation whose graph approximates the graph of the quadratic $y = -x^2 + 22x - 40$ on $[0, 5]$. Graph this linear function together with the quadratic. Can you therefore say that a quadratic profit function is nearly linear over small intervals of x?

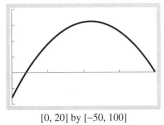

[0, 20] by [−50, 100]

Screen 1.13
A graph of $y = -x^2 + 22x - 40$. On the interval $[0, 5]$ the graph is approximately a straight line.

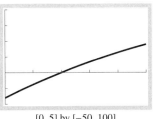

[0, 5] by [−50, 100]

Screen 1.14
A graph of $y = -x^2 + 22x - 40$ on the interval $[0, 5]$ is approximately a straight line.

SELF-HELP EXERCISE SET 1.4

1. Graph $y = (x + 3)^2 - 8$ and locate the vertex.

2. Suppose in Example 5 the cost function, in thousands of dollars per week, is given by $C(x) = 4x + 6$. Find the x for which the profits are maximized. Compare this value of x to the value that maximizes revenue. Find any break-even quantities.

3. A bus company charges $70 per person for a sightseeing trip if 40 people travel in a group. If for each person above 40 the company reduces the charge per person by $1, how many people will maximize the total revenue for the bus company? What is the maximum revenue?

EXERCISE SET 1.4

In Exercises 1 through 12 graph the given quadratic function and locate the vertex.

1. $y = -0.5x^2$
2. $y = -x^2$
3. $y = (x - 1)^2$
4. $y = (x - 3)^2$
5. $y = (x - 1)^2 + 2$
6. $y = 2(x + 2)^2 + 3$
7. $y = 2(x + 2)^2 - 3$
8. $y = (x + 1)^2 - 4$
9. $y = -(x - 1)^2 + 3$
10. $y = -2(x - 2)^2 + 1$
11. $y = -(x + 1)^2 - 2$
12. $y = -3(x + 2)^2 - 1$

In Exercises 13 through 16 complete the square and put in standard form. Locate the vertex and graph.

13. $y = x^2 - 4x + 4$
14. $y = x^2 - 8x + 16$
15. $y = -2x^2 + 12x - 20$
16. $y = -x^2 + 2x - 4$

In Exercises 17 through 20 locate the vertex and graph.

17. $y = x^2 + 2x + 1$
18. $y = x^2 + 6x + 9$
19. $y = -4x^2 - 8x - 3$
20. $y = -5x^2 - 10x - 4$

In Exercises 21 through 24 locate the vertex and graph. Support your answer using your graphing calculator.

21. $y = x^2 - 2x + 4$
22. $y = 2x^2 - 12x + 20$
23. $y = -x^2 - 2x - 2$
24. $y = -3x^2 - 6x - 4$

In Exercises 25 through 28 use your graphing calculator to locate the vertex. Confirm algebraically.

25. $y = x^2 - 2x$
26. $y = 2x^2 - 8x + 7$
27. $y = -x^2 + 4x - 2$
28. $y = -2x^2 + 4x + 1$

In Exercises 29 through 32 find where the revenue function is maximized using the given linear demand function.

29. $p = -x + 10$
30. $p = -2x + 3$
31. $p = -5x + 15$
32. $p = -0.5x + 20$

In Exercises 33 and 34 use your graphing calculator to find where the revenue function is maximized using the given linear demand function. Confirm your answer algebraically.

33. $p = -0.01x + 10$
34. $p = -0.2x + 12$

In Exercises 35 through 38 find the profit function using the given revenue and cost functions. Find the value of x that maximizes the profit. Find the break-even quantities (if they exist), that is, the value of x for which the profits are zero. Sketch a graph of the profit function.

35. $R(x) = -2x^2 + 30x$, $C(x) = 10x + 42$
36. $R(x) = -3x^2 + 20x$, $C(x) = 2x + 15$
37. $R(x) = -x^2 + 30x$, $C(x) = 15x + 36$
38. $R(x) = -2x^2 + 15x$, $C(x) = 5x + 8$

In Exercises 39 through 42 use your graphing calculator to find the break-even quantities for the given profit functions and the value of x that maximizes the profit.

39. $P(x) = -x^2 + 3.1x - 1.98$
40. $P(x) = -x^2 + 5.9x - 7.92$
41. $P(x) = -4x^2 + 38.4x - 59.36$
42. $P(x) = -2.4x^2 + 14.64x - 14.112$

43. The sum of two numbers is 9. What is the largest possible value for their product?

44. The difference between two numbers is 2. Find the smallest possible value for the sum of their squares.

45. If 200 feet of fence is used to fence in a rectangular field and x is the width of the field, the enclosed area is given by $A = x(100 - x)$. Determine the width and length that give the maximum area.

Applications

46. **Cost.** A farmer wishes to enclose a rectangular field of area 200 square feet using an existing wall as one of the sides. The cost of the fence to go on the other three sides is $1 per foot. Find the dimensions of the rectangular field that minimizes the cost of the fence.

47. **Revenue.** An apple orchard produces annual revenue of $60 per tree when planted with 100 trees. Due to overcrowding, the annual revenue per tree is reduced by $0.50 for each additional tree planted. How many trees should be planted to maximize the revenue from the orchard?

48. Profit. If in the previous problem the cost of maintaining each tree is $5 per year, how many trees should be planted to maximize the profit from the orchard?

49. Spread of Rumors. It has been estimated that a rumor spreads at a rate R that is proportional both to the ratio r of individuals who have heard the rumor and to the ratio $(1 - r)$ who have not. Thus, $R = kr(1 - r)$, where k is a positive constant. For what value of r does the rumor spread fastest?

50. Biology. When predators are not stimulated to search very hard for prey, the rate of change R of the number N of prey attacked per unit of area with respect to the number n of prey is given by $R(x) = kx(ax - N)$, where x is the number of predators, k is the per capita effectiveness of the predators, and a is the maximum per capita effectiveness of the predators over an extended period. Find the number of predators that minimizes R.

51. Advertising. Let x be a measure (in percent) of the degree of concentration in an industry. Sutton [23] noted that the advertising intensity y, defined as the advertising/sales ratio (in percent), rises to a peak at intermediate levels of concentration x and declines again for the most concentrated sectors. One economist noted from a sample of consumer industries that $y = -3.1545 + 0.1914x - 0.0015x^2$ approximately modeled this situation. Sketch a graph, find the value of the concentration ratio for which the advertising intensity is largest, and find the maximum value of this intensity. Confirm graphically.

52. Physics. If an object is initially at a height above the ground of s_0 feet and is thrown straight upward with an initial velocity of v_0 feet per second, then from physics it can be shown that the height in feet above the ground is given by $s(t) = -16t^2 + v_0 t + s_0$, where t is in seconds. Find how long it takes for the object to reach maximum height. Find when the object hits the ground.

53. Production Function. Spencer [24] found that the relative yield y of vegetables on farms in Alabama was related to the number x of hundreds of pounds of phosphate (fertilizer) per acre by $y = 0.15 + 0.63x - 0.11x^2$. How much

phosphate per acre, to the nearest pound, would you fertilize? Why?

54. Irrigation. Caswell and coauthors [25] indicated that the cotton yield y in pounds per acre in the San Joaquin Valley in California was given approximately by $y = -1589 + 3211x - 462x^2$, where x is the annual acre-feet per acre of water application. Determine the annual acre-feet per acre of water application that maximizes the yield and determine the maximum yield.

55. Political Action Committees (PAC). PACs are formed by corporations to funnel political contributions. Grier and collaborators [26] showed that the percentage P of firms within a manufacturing industry with PACs was given approximately by $P = -23.21 + 0.014x - 0.0000006x^2$, where x is the average industry sales in millions of dollars. Determine the sales that result in the maximum percentage of PACs and find this maximum.

56. Political Action Committees. In the same study mentioned in the previous exercise, P was also given approximately as $P = -23.21 + 0.659x - 0.008x^2$, where this time x represents the concentration ratio of the industry. (This ratio is defined as the proportion of total sales of the four largest firms to total industry sales.) Find the concentration ratio that maximizes the percentage of PACs and find this maximum.

57. Biology. Kaakeh and colleagues [27] showed that the amount of photosynthesis in apple leaves in appropriate units was approximated by $y = 19.8 + 0.28x - 0.004x^2$, where x is the number of days from an aphid infestation. Determine the number of days after the infestation until photosynthesis peaked.

58. Biology. Hardman [21] showed the survival rate S from insect predation of European red mite eggs in an apple orchard was approximated by

$$S = \begin{cases} 1, & t \le 0 \\ 1 - 0.01t - 0.01t^2, & t > 0 \end{cases}$$

where t is the number of days after June 1. Determine the predation rate on May 15. June 15.

Enrichment Exercises

59. Is the sum of two quadratic functions always a quadratic function? Explain carefully.

60. Suppose you have two quadratic functions, $y = f(x)$ and $y = g(x)$, both of which have graphs that open downward. Consider the function $y = f(x) + g(x)$. Show that this function is quadratic and its graph opens downward.

61. Give an intuitive geometric argument (not an algebraic one) that shows that the third quadratic in the previous exercise peaks at a value in between the values for which the first two quadratics peak.

62. Can the graph of a quadratic function cross the x-axis in three different places? Why or why not?

63. Profit Function. If you use the quadratic function $P(x) = ax^2 + bx + c$ to model profits on a very large interval, what sign should the coefficient a have? Explain carefully.

64. Cost Function. If you use the quadratic function $C(x) = ax^2 + bx + c$ to model costs on a very large interval, what sign should the coefficient a have? Explain carefully.

65. Revenue Function. Suppose we assume that the demand equation for a commodity is given by $p = mx + e$, where x is the number sold and p is the price. Explain carefully why the resulting revenue function is of the form $R(x) =$

$ax^2 + bx$, with the sign of a being negative and the sign of b positive.

66. Profit Function. Suppose the cost and revenue functions are quadratic functions. Must the profit function be a quadratic function? Explain carefully.

67. Peak Profit. Suppose the demand equation for a commodity is of the form $p = mx + b$, where $m < 0$ and $b > 0$. Suppose the cost function is of the form $C = dx + e$, $d > 0$ and $e < 0$. Show that profit peaks before revenue peaks.

SOLUTIONS TO SELF-HELP EXERCISE SET 1.4

1. The quadratic $y = (x + 3)^2 - 8$ is in standard form $y = a(x - h)^2 + k$ with $a = 1$, $h = -3$, and $k = -8$. Since $a > 0$, the parabola opens upward and the vertex $(h, k) = (-3, -8)$ is the point on the graph where the function attains its minimum.

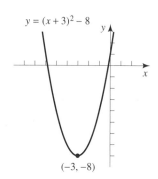

$y = (x+3)^2 - 8$

$(-3, -8)$

2. First complete the square.

$$P(x) = R(x) - C(x)$$
$$= [-2x^2 + 12x] - [4x + 6]$$
$$= -2x^2 + 8x - 6$$
$$= -2(x^2 - 4x + 3)$$
$$= -2([x^2 - 4x + 4] - 4 + 3) \qquad \left(\frac{-4}{2}\right)^2 = 4$$
$$= -2([x - 2]^2 - 1)$$
$$= -2(x - 2)^2 + 2$$

The quadratic is now in standard form. Thus, we see that profits are maximized at $x = 2$, or 2000 units.

We could have used the formula

$$h = -\frac{b}{2a} = -\frac{8}{2(-2)} = 2$$

To find the break-even quantities, we must set $P = 0$, which gives

$$0 = P(x) = -2x^2 + 8x - 6$$
$$= -2(x^2 - 4x + 3) = -2(x - 1)(x - 3)$$

From this factorization it is apparent that $x = 1$ and $x = 3$ are the break-even quantities (see the figure).

Notice the point where profit is maximized is not the same point for which revenue is maximized. This figure is typical of profit functions. As more items are produced and

sold, initial losses turn into profits. But if too many items are made and sold, this flooding of the market results in a downturn in profits.

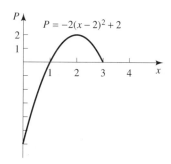

$P = -2(x - 2)^2 + 2$

3. The revenue R for the bus company is the number n of people who take the trip times the price p each person pays, that is, $R = n \cdot p$. Let x be the number of people in excess of 40 that take the trip. Then the total number that take the trip is $n = 40 + x$. For each person in excess of 40 who takes the trip, the company lowers the price by \$1. Thus, the price per person is $p = 70 - x$. Revenue is then

$$R = n \cdot p = (40 + x)(70 - x) = 2800 + 30x - x^2$$

This is a quadratic. Completing the square yields

$$2800 + 30x - x^2 = -(x^2 - 30x - 2800)$$
$$= -(x^2 - 30x + 225 - 225 - 2800)$$
$$\text{since } \left(\frac{-30}{2}\right)^2 = 225$$
$$= -(x^2 - 30x + 225 - 3025)$$
$$= -(x - 15)^2 + 3025$$

The quadratic is now in standard form. The maximum occurs at $x = 15$, and the maximum is 3025. Thus, the bus company has maximum revenue if the number that take the trip is 40 + x = 40 + 15 = 55. The maximum revenue is \$3025.

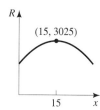

$(15, 3025)$

1.5 SOME SPECIAL FUNCTIONS AND GRAPHING TECHNIQUES

■ Some Special Functions   ■ Graphing Techniques

APPLICATION: LEARNING CURVE

Cost engineers have noted that the efficiency of workers in a plant or on a machine increases over time, as the work force gains efficiency and skill in performing the repeated manual tasks. Hirsch [28] reported that the relationship between the cumulative average amount of direct labor input y on a textile machine and the cumulative number of units produced x is approximated by $y = ax^{-0.175}$, where a is a constant. If production doubles, what happens to the average amount of needed labor? What does this mean in terms of learning? See Example 1 for the answer.

APPLICATION: COMPARING TWO MATHEMATICAL MODELS

Researchers have noted that the equation $y_1 = 0.0516L^{2.437}$ approximated the respiration rate of *Daphnia pulex* as a function of length. They have no scientific explanation of the meaning of the power 2.437. On the other hand, Richman [29] reasoned, based on certain biological facts, that the respiration rate should be a constant times the surface area plus another constant times the volume. Thus, he reasoned that the respiration rate should be given by a function of the form $aL^2 + bL^3$. He then discovered that the function $y_2 = 0.0336L^2 + 0.01845L^3$ was a good approximation. Graph these two functions on your grapher using a window of dimensions [0, 2] by [0, 0.25]. What can you say about the graphs? Which model do you prefer? See Example 2 for the answer.

TABLE 1.9

x	-3	-2	-1	0	1	2	3
x^3	-27	-8	-1	0	1	8	27

TABLE 1.10

x	0	1	4	9	16	25	36
$\sqrt{x}$	0	1	2	3	4	5	6

TABLE 1.11

x	-27	-8	-1	0	1	8	27
$\sqrt[3]{x}$	-3	-2	-1	0	1	2	3

Some Special Functions

The area of a square with sides of length s is $A = f(s) = s^2$. The volume of a cube with sides of length s is $V = g(s) = s^3$. The volume of a sphere of radius r is $V = h(r) = \frac{4}{3}\pi r^3$. All of these functions are *power functions*. Functions such as $f(x) = x^{1/2}$ and $f(x) = 1/x^2 = x^{-2}$ are also **power functions**. In general we have the following definition.

Power Functions

A **power function** is a function of the form

$$f(x) = kx^r$$

where r and k are any real numbers.

Let us consider the domains of some power functions. The domain of $y = x^3$ and $y = \sqrt[3]{x}$ is the entire real line, whereas the domain of $y = \sqrt{x}$ is [0, ∞). From Tables 1.9, 1.10, and 1.11, which show a number of values for these functions, the graphs are constructed in Figures 1.53, 1.54, and 1.55.

If $r > 0$, the domains of power functions are either all real numbers or else all nonnegative numbers, depending on r. For example, the two rational functions $f(x) =$

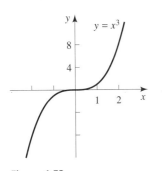

Figure 1.53
A graph of $y = x^3$.

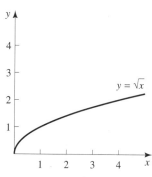

Figure 1.54
A graph of $y = \sqrt{x}$.

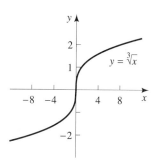

Figure 1.55
A graph of $y = \sqrt[3]{x}$.

$x^{1/2}$ and $g(x) = x^{5/4}$ have $[0, \infty)$ as their domains, whereas $f(x) = x^{1/3}$ and $g(x) = x^{4/5}$ have all real numbers as their domains. If $r < 0$, then $x = 0$ cannot be in the domain. For example, if $f(x) = x^{-1/3} = 1/x^{1/3}$, then the domain is the real line except for $x = 0$. If $f(x) = x^{-1/2} = 1/x^{1/2}$, then the domain is $(0, \infty)$.

Remark. Be careful when using your grapher to graph $y = x^r$ when r is a rational number $r = m/n$. For example, the domain of $f(x) = x^{2/3}$ is the set of all real numbers. Your grapher may graph this function incorrectly, however. If your grapher does give an incorrect graph, rewrite $f(x)$ as $f(x) = (x^{1/3})^2$ or $(x^2)^{1/3}$.

EXAMPLE 1   Learning Curve

Cost engineers have noted that the efficiency of workers in a plant or on a machine increases over time, as the work force gains efficiency and skill in performing the repeated manual tasks. Hirsch [28] reported that the relationship between the cumulative average amount of direct labor input y on a textile machine and the cumulative number of units produced x is approximated by $y = f(x) = ax^{-0.175}$, where a is a constant. Graph. If production doubles, what happens to the average amount of needed labor? What does this mean in terms of learning?

SOLUTION ■ A graph in the case that $a = 1$ is shown in Screen 1.15 using a window of dimensions $[0, 4.7]$ by $[0, 2]$. We see that the curve drops, indicating that an increase in the number of units produced results in less average labor input. In particular, suppose production is doubled, then

$$f(2x) = a(2x)^{-0.175}$$
$$= 2^{-0.175}ax^{-0.175}$$
$$= 2^{-0.175}f(x)$$
$$\approx 0.886f(x)$$

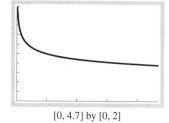

[0, 4.7] by [0, 2]

Screen 1.15
The graph of $y = x^{-0.175}$ drops.

Since $1.00 - 0.886 = 0.114$, average labor input is decreased by about 11.4%. ■

EXPLORATION 1

Support for Example 1

Use TRACE to evaluate $f(1)$ and $f(2)$ where $y = f(x) = x^{-0.175}$ is graphed in Screen 1.15. Does this support the equation $f(2x) = 0.886f(x)$ found above? Why?

We next consider polynomial functions.

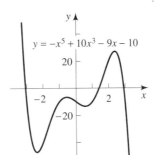

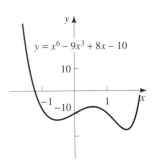

Figure 1.56
Graphs of typical polynomial functions.

Definition of Polynomial Function

A **polynomial function** of degree n is a function of the form

$$f(x) = a_n x^n + a_{n-1} x^{n-1} + \cdots + a_1 x + a_0 \quad (a_n \neq 0)$$

where $a_n, a_{n-1}, \ldots, a_1, a_0$, are constants and n is a nonnegative integer. The domain is all real numbers. The coefficient a_n is called the **leading coefficient**.

A polynomial function of degree 1 ($n = 1$), $f(x) = a_1 x + a_0$, ($a_1 \neq 0$) is a linear function and has already been encountered earlier. A polynomial function of degree 2 ($n = 2$), $f(x) = a_2 x^2 + a_1 x + a_0$ ($a_2 \neq 0$), is a quadratic function. The functions

$$f(x) = -x^5 + 10x^3 - 9x - 10 \qquad \text{and} \qquad g(x) = x^6 - 9x^3 + 8x - 10$$

are polynomials of degree 5 and 6, respectively, with leading coefficients of -1 and 1, respectively. The graphs are shown in Figure 1.56. In Chapter 4 we will see how to find the graphs of polynomial functions.

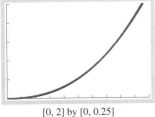

[0, 2] by [0, 0.25]

Screen 1.16
The graphs of $y_1 = 0.0516x^{2.437}$ and $y_2 = 0.0336x^2 + 0.01845x^3$ appear to be essentially the same.

EXAMPLE 2   Comparing Two Mathematical Models

Researchers have noted that the equation $y_1 = 0.0516L^{2.437}$ approximates the respiration rate of *Daphnia pulex* as a function of length. They have no scientific explanation of the meaning of the power 2.437. On the other hand, Richman [29] reasoned, based on certain biological facts, that the respiration rate should be a constant times the surface area plus another constant times the volume. Thus, he reasoned that the respiration rate should be given by a polynomial function of the form $aL^2 + bL^3$. He then discovered that the function $y_2 = 0.0336L^2 + 0.01845L^3$ was a good approximation. Graph these two functions on your grapher using a window of dimensions [0, 2] by [0, 0.25]. What can you say about the graphs? Which model do you prefer?

SOLUTION ■   The graphs of the two equations are shown in Screen 1.16. Notice that the graphs are indistinguishable! The equation $y_1 = 0.0516L^{2.437}$ was obtained using abstract curve-fitting techniques. That is, researchers simply looked for the curve that seemed to fit the data best. As a result we can refer to this curve as a *description*. On the other hand, the second equation was obtained as the consequence of biological reasoning. As such, this equation seems to offer an *explanation*, not just a description. Therefore, we prefer to use the second equation.   ■

We again look at abstract curve fitting versus a curve based on scientific reasoning in Project 2 at the end of Chapter 3.

We continue with the definition of *rational function*.

Rational Function

A **rational function** $R(x)$ is the quotient of two polynomial functions and thus is of the form

$$R(x) = \frac{f(x)}{g(x)}$$

where $f(x)$ and $g(x)$ are polynomial functions. The domain of $R(x)$ is all real numbers for which $g(x) \neq 0$.

Thus,

$$R(x) = \frac{x^5 + x - 2}{x - 3} \quad \text{and} \quad Q(x) = \frac{x + 1}{x^2 - 1}$$

are rational functions with the domain of $R(x)$ all real numbers other than $x = 3$ and the domain of $Q(x)$ all real numbers other than $x = \pm 1$.

Graphing Techniques

We now develop a number of techniques to aid in graphing.

In the previous section we noted that the graph of $y = x^2 + 3$ is the graph of $y = x^2$ shifted upward by 3 units, and the graph of $y = x^2 - 1$ is the graph of $y = x^2$ shifted downward by 1 unit. This is an example of the following general graphing principle.

Vertical Shift

For $k > 0$ the graph of $y = f(x) + k$ is the graph of $y = f(x)$ shifted upward by k units, and the graph of $y = f(x) - k$ is the graph of $y = f(x)$ shifted downward by k units.

EXAMPLE 3 Vertical Shift

Graph $y = \sqrt{x}$, $y = \sqrt{x} + 2$, and $y = \sqrt{x} - 1$ on the same graph.

SOLUTION ■ The graph of $y = \sqrt{x} + 2$ is the graph of $y = \sqrt{x}$ shifted upward by 2 units, and the graph of $y = \sqrt{x} - 1$ is the graph of $y = \sqrt{x}$ shifted downward by 1 unit (Fig. 1.57). ■

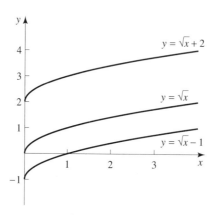

Figure 1.57
Adding 2 shifts the graph up 2 units, whereas subtracting 1 shifts the graph down 1 unit.

EXPLORATION *2*

Seeing Vertical Shift

 On a screen with dimensions [0, 9.4] by [−3, 10], graph

$$y = \sqrt{x} - 3 \qquad y = \sqrt{x} \qquad y = \sqrt{x} + 3 \qquad y = \sqrt{x} + 6$$

Use TRACE and the cursor to hop up or down from one curve to the other. Check $x = 1$, 4 and 9. How does the y-coordinate change?

We also noted in the previous section that the graph of $y = (x - 2)^2$ is the graph of $y = x^2$ shifted to the right by 2 units, and the graph of $y = (x + 1)^2$ is the graph of $y = x^2$ shifted to the left by 1 unit. This is an example of the following general graphing principle.

Horizontal Shift

> For $c > 0$ the graph of $y = f(x - c)$ is the graph of $y = f(x)$ shifted to the right by c units, whereas the graph of $y = f(x + c)$ is the graph of $y = f(x)$ shifted to the left by c units.

EXAMPLE 4   Horizontal Shift

Graph $y = x^3$, $y = (x - 2)^3$, and $y = (x + 1)^3$ on the same graph.

SOLUTION ■   The graph of $y = (x - 2)^3$ is the graph of $y = x^3$ shifted to the right by 2 units, whereas the graph of $y = (x + 1)^3$ is the graph of $y = x^3$ shifted to the left by 1 unit (Fig. 1.58). ■

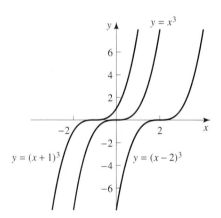

Figure 1.58
Replacing x with $x + 1$ shifts the graph 1 unit to the left, whereas replacing x with $x - 2$ shifts the graph 2 units to the right.

EXPLORATION *3*

Seeing Horizontal Shift

 On a screen with dimensions [−9.5, 37.5] by [0, 7] graph

$$y = \sqrt{x + 3} \qquad y = \sqrt{x} \qquad y = \sqrt{x - 3} \qquad y = \sqrt{x - 6}$$

How do these graphs differ?

In the previous section we compared the graphs of $y = f(x) = x^2$ with that of $y = -f(x) = -x^2$. The graphs are shown in Figure 1.59. As we see, the graph of $y = -f(x) = -x^2$ is the reflection of the graph of $y = f(x) = x^2$ across the x-axis. This is an example of the following general graphing principle.

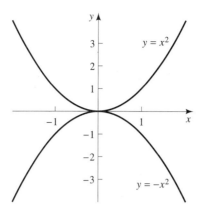

Figure 1.59
The graph of $y = -x^2$ is a reflection across the x-axis of the graph of $y = x^2$.

Reflection

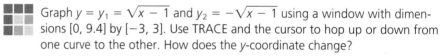

The graph of $y = -f(x)$ is the reflection of the graph of $y = f(x)$ across the x-axis.

EXAMPLE 5   Reflection

Graph $y = x^3$ and $y = -x^3$ on the same graph.

SOLUTION ■ The graphs are shown in Figure 1.60. ■

EXPLORATION 4

Seeing a Reflection

Graph $y = y_1 = \sqrt{x - 1}$ and $y_2 = -\sqrt{x - 1}$ using a window with dimensions [0, 9.4] by [−3, 3]. Use TRACE and the cursor to hop up or down from one curve to the other. How does the y-coordinate change?

In the previous section we compared the graphs of $y = ax^2 = af(x)$ for various values of a with that of $y = x^2 = f(x)$. Figure 1.61 shows some examples. We note that for $a = 2$, the graph of $y = 2x^2$ expands the graph of $y = x^2$ by a factor of 2, whereas the graph of $y = 0.5x^2$ contracts the graph of $y = x^2$ by a factor of 0.5. This is an example of the following general graphing principle.

Expansion and Contraction

For $a > 0$, the graph of $y = af(x)$ is an expansion of the graph of $y = f(x)$ if $a > 1$ and a contraction if $0 < a < 1$.

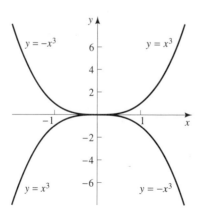

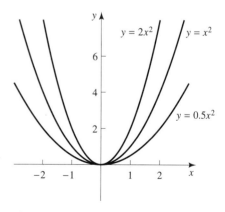

Figure 1.60
The graph of $y = -x^3$ is a reflection across the x-axis of the graph of $y = x^3$.

Figure 1.61
The graph of $y = ax^2$ for $a > 0$ expands or contracts the graph of $y = x^2$ depending on whether $a > 1$ or $a < 1$.

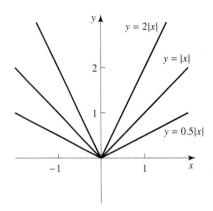

Figure 1.62
The graph of $y = a|x|$ for $a > 0$ expands or contracts the graph of $y = |x|$ depending on whether $a > 1$ or $a < 1$.

■ **E X A M P L E 6**   Expansion and Contraction

Draw the graphs of $y = |x|$, $y = 2|x|$, and $y = 0.5|x|$ on the same graph.

S O L U T I O N ■ The graphs are shown in Figure 1.62. ■

EXPLORATION *5*

Seeing Expansions and Contractions

Draw the graphs of $y = |x|$, $y = 0.25|x|$, and $y = 0.1|x|$ using a window with dimensions $[-4.7, 4.7]$ by $[0, 5]$. Use TRACE and the cursor to hop up or down from one curve to the other. How does the y-coordinate change?

We can combine several graphing principles to graph a function.

■ **E X A M P L E 7**   Combining Graphing Principles

Graph $y = g(x) = \sqrt{x - 1} + 2$.

S O L U T I O N ■ Let $y = f(x) = \sqrt{x}$. The graph is shown in Figure 1.63a. Then we know that the graph of $y = \sqrt{x - 1}$ is the graph of $y = f(x) = \sqrt{x}$ shifted to the right by 1 unit (see Fig. 1.63b). The graph of $y = g(x) = \sqrt{x - 1} + 2$ is then the graph of $y = \sqrt{x - 1}$ shifted upward by 2 units (see Fig. 1.63c). ■

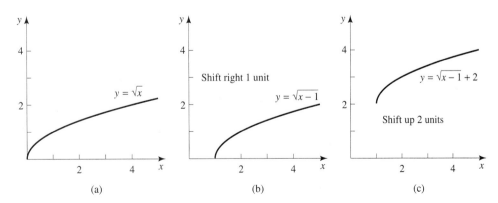

(a)   (b)   (c)

Figure 1.63
Replacing x with $x - 1$ shifts the graph 1 unit to the right. Adding 2 then shifts the graph up by 2 units.

SELF-HELP EXERCISE SET 1.5

1. Using the basic graphing principles sketch graphs of (**a**) $y = f(x) + 2$; (**b**) $y = f(x - 1)$; (**c**) $y = 2f(x)$, where $y = f(x)$ is given in the following figure.

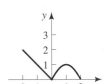

2. Graph $y = |x^3|$; $y = |(x - 1)^3|$; $y = 0.5|(x - 1)^3|$

EXERCISE SET 1.5

In Exercises 1 through 4 determine the degree of each of the given polynomial functions, their domains, and the leading coefficients.

1. $p(x) = x^8 - x + 2$

2. $p(x) = 0.001x^6 + x^5 + x - 3$

3. $p(x) = 1000x^9 - 10x^7 + x^2$

4. $p(x) = 100x^{101} + x^{50} - x^9 + 2$

In Exercises 5 through 10 determine the domains of each of the given functions.

5. $\dfrac{x + 1}{x - 2}$

6. $\dfrac{x^3 - 1}{x^2 - 4}$

7. $\dfrac{x^4 + 1}{x^2 + 1}$

8. $x^{1/4}$

9. $x^{-1/5}$

10. $x^{-3/4}$

In Exercises 11 through 24 refer to the two graphs.

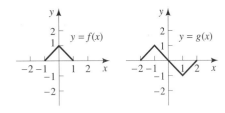

Use the graph of f or g and the basic graphing principles to graph each of the following:

11. $y = f(x) + 2$

12. $y = f(x) - 2$

13. $y = g(x) - 1$

14. $y = g(x) + 1$

15. $y = f(x + 2)$

16. $y = f(x - 2)$

17. $y = g(x - 1)$

18. $y = g(x + 1)$

19. $y = -f(x)$

20. $y = -g(x)$

21. $y = 2f(x)$

22. $y = 2g(x)$

23. $y = 0.5f(x)$

24. $y = 0.5g(x)$

In Exercises 25 through 32 graph using the graphing principles.

25. $y = x^3$ $y = x^3 - 1$ $y = x^3 - 2$

26. $y = |x|$ $y = |x| - 1$ $y = |x| - 2$

27. $y = \sqrt[3]{x}$ $y = \sqrt[3]{x - 1}$ $y = \sqrt[3]{x - 2}$

28. $y = |x|$ $y = |x - 1|$ $y = |x - 2|$

29. $y = |x|$ $y = -|x|$

30. $y = \sqrt[3]{x}$ $y = -\sqrt[3]{x}$

31. $y = \sqrt{x}$ $y = \sqrt{x - 1}$ $y = \sqrt{x - 2}$

32. $y = x^3$ $y = 2x^3$ $y = 0.5x^3$

In Exercises 33 through 40 graph using the graphing principles.

33. $y = |x - 1| + 2$

34. $y = (x - 1)^3 - 1$

35. $y = \sqrt{x + 1} - 2$

36. $y = \sqrt[3]{x + 1} + 2$

37. $y = (x + 2)^3 - 1$

38. $y = (x + 2)^3 + 1$

39. $y = -|x - 1| + 2$

40. $y = -(x - 1)^3 - 1$

In Exercises 41 through 46 transformations are given in a specified order to be applied to the graph of the given function. Give an equation for the function associated with the transformed graph. Support your answer with a grapher.

41. $y = f(x) = |x|$, vertical shrink by 0.5, shift up by 3

42. $y = f(x) = |x|$, vertical stretch by 2, shift down by 3

43. $y = f(x) = |x^3|$, vertical stretch by 2, shift down by 3

44. $y = f(x) = |x^3|$, vertical shrink by 0.5, shift up by 2

45. $y = f(x) = |x|$, reflect across x-axis, shift up by 3, shift right by 2

46. $y = f(x) = |x|$, reflect across x-axis, shift down by 3, shift left by 2

47. Reverse the order of the transformations in Exercise 45. Do you obtain the same function?

48. Reverse the order of the transformations in Exercise 46. Do you obtain the same function?

Applications

49. Forestry. McNeil and associates [30] showed that for small loblolly pine trees $V(H) = 0.0000837H^{3.191}$, where V is the volume in cubic meters and H is the tree height in meters. Using this formula find V when $H = 10, 20, 40$. What happens to V when H doubles? Graph $V = V(H)$.

50. Agriculture. The following equation, developed from data from sites in Minnesota [31] gives the soybean yield y in bushels per acre as $y(x) = 46.6x^{1.067}$, where $x = T/T_p$ with T as the amount of moisture released by the plant to the atmosphere and T_p the amount of moisture the plant can give off if the soil moisture is unlimited. Find y when $x = 0.1, 0.2, 0.4$. Graph $y = y(x)$.

51. Diversity of Species. Usher and coworkers [32] found that for certain farm woodlands the equation $S(A) = 1.81A^{0.284}$ describes the numbers of species of plants as a function of the area A of woodland in square meters. Graph. What happens to the number of species if the area is reduced by one-half?

Enrichment Exercises

52. On the same screen (with dimension [0, 1] by [0, 1]) graph $y = x^2$, $y = x^3$, $y = x^4$, $y = x^5$. What can you say about how these graphs compare with each other on the interval [0, 1]? What do you think is true about the graphs of $y = x^n$ on [0, 1] for n an integer?

53. Refer to Exercise 52. Change the dimension of the screen to [0, 4] by [0, 64]. What can you say about how these graphs compare with each other on the interval [1, 4]? What do you think is true about the graphs of $y = x^n$ on [1, ∞) for n an integer?

54. Speculate on how the graphs of $y = x^{1/n}$ for n an integer are related on [0, 1]. Confirm your suspicions by graphing $y = x^{1/2}$, $y = x^{1/3}$, $y = x^{1/4}$, $y = x^{1/5}$.

55. Speculate on how the graphs of $y = x^{1/n}$ for n an integer are related on [1, ∞). Confirm your suspicions by graphing $y = x^{1/2}$, $y = x^{1/3}$, $y = x^{1/4}$, $y = x^{1/5}$.

56. On your graphing calculator graph $y = x^n$ for some *even* powers n. Then graph $y = x^n$ for some *odd* powers n. What is a major distinguishing feature of the graphs when n is even and when n is odd? Explain why this must be so.

57. Two graphs are shown. One is the graph of $y = x^{11}$ and the other is the graph of $y = x^{12}$. Which is which? Explain why you think so.

58. How would you compare the graph of $y = f(cx)$ for $c > 1$ with the graph of $y = f(x)$? To answer, consider specific examples such as $f(x) = |x|$ and $f(x) = x^2$.

59. Repeat the previous exercise if $0 < c < 1$.

60. Cost Functions. The values of two functions that give variable costs are shown in the table. One function is of the form $y = ax^2$ and the other is of the form $y = bx^3$. Which function is which?

x	$f(x)$	$g(x)$
1	0.2	2
5	25.0	50
15	675.0	450

61. Cost Functions. The figure shows two graphs of variable cost functions. One is of the form $y_1 = ax^{0.75}$ and the other is of the form $y_2 = bx^{1.25}$. Which function is which? *Hint:* Determine what happens to y_1 and y_2 when x is doubled.

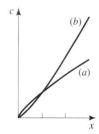

62. Cost Functions. Based on your answer to the previous exercise and to further experimentation on your graphing calculator, what is a distinguishing feature of the graphs of $y = ax^c$ with $0 < c < 1$ and $y = ax^c$ with $c > 1$?

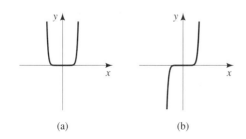

(a)   (b)

SOLUTIONS TO SELF-HELP EXERCISE SET 1.5

1.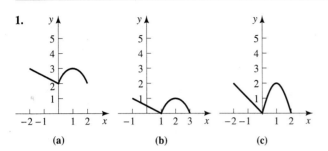

(a) (b) (c)

2.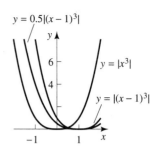

1.6 COMBINATIONS OF FUNCTIONS

■ The Algebra of Functions   ■   The Composition of Functions

---

APPLICATION: AREA OF FOREST FIRE

Suppose a forest fire is spreading in a circular manner with the radius of the burnt area increasing at a steady rate of 3 feet per minute. What is the area of the circular burnt region as a function of the time? See Example 7 for the answer.

The Algebra of Functions

A function such as $h(x) = x^4 + \sqrt{x + 1}$ can be viewed as the sum of two simpler functions: $f(x) = x^4$ and $g(x) = \sqrt{x + 1}$. We have always written profit as revenue less costs, that is, $P(x) = R(x) - C(x)$. A function such as

$$r(x) = \frac{1 + 10x^3}{1 + x^2}$$

can be viewed as the quotient of the two polynomials $p(x) = 1 + 10x^3$ and $q(x) = 1 + x^2$. The two polynomials are much simpler than the original rational function. Breaking a complicated function into less complicated parts can yield insight into the behavior of the original function.

As we noted above, functions can be added, subtracted, multiplied, and divided. We now state the formal definitions of these concepts.

The Algebra of Functions

Let f and g be two functions and define

$$D = \{x \mid x \in (\text{domain of } f), x \in (\text{domain of } g)\}$$

Then for all $x \in D$, we define

The sum $f + g$ by $(f + g)(x) = f(x) + g(x)$

The difference $f - g$ by $(f - g)(x) = f(x) - g(x)$

The product $f \cdot g$ by $(f \cdot g)(x) = f(x) \cdot g(x)$

The quotient f/g by $\left(\dfrac{f}{g}\right)(x) = \dfrac{f(x)}{g(x)}$ for x with $g(x) \neq 0$.

EXAMPLE 1 Examples of Combining Functions

Let $f(x) = \sqrt{x}$ and $g(x) = 1 - x$. Find $(f + g)(x)$; $(f - g)(x)$; $(f \cdot g)(x)$; $(f/g)(x)$; and find the domains of each of them.

SOLUTION ■ The domain of f is $[0, +\infty)$ and the domain of g is all real numbers. Thus, the common domain of the first three functions is $D = [0, +\infty)$.

$$(f + g)(x) = f(x) + g(x) = \sqrt{x} + (1 - x) = \sqrt{x} + 1 - x$$

$$(f - g)(x) = f(x) - g(x) = \sqrt{x} - (1 - x) = \sqrt{x} - 1 + x$$

$$(f \cdot g)(x) = f(x) \cdot g(x) = \sqrt{x}(1 - x)$$

$$\left(\frac{f}{g}\right)(x) = \frac{f(x)}{g(x)} = \frac{\sqrt{x}}{1 - x}$$

The domain of f/g is all nonnegative numbers other than 1. ■

EXAMPLE 2 Combining Functions on a Grapher

On the same screen graph $y_1 = f(x) = \sqrt{x + 5}$; $y_2 = g(x) = \sqrt{7 - x}$; $y_3 = y_1 + y_2$. From this obtain an indication of the domain of $f + g$. Verify analytically.

SOLUTION ■ The graphs are shown in Screen 1.17. Notice how the graph gives us an indication of the domain of $f + g$. The domain appears to be $[-5, 7]$. Using algebra we obtain the domain of f as $[-5, +\infty)$, and the domain of g as $(-\infty, 7]$. Thus the domain of $f + g$ is $[-5, 7]$. ■

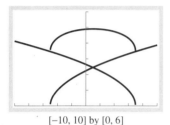

[−10, 10] by [0, 6]

Screen 1.17
Graphs of $y_1 = \sqrt{x + 5}$, $y_2 = \sqrt{7 - x}$, and $y_3 = \sqrt{x + 5} + \sqrt{7 - x}$. From these graphs we can estimate the domains.

EXPLORATION 1

Profit

 We know that $P(x) = R(x) - C(x)$, where $P(x)$ is profit, $R(x)$ is revenue, and $C(x)$ is cost. Graph on the same screen $y_1 = R(x) = 12x - x^2$; $y_2 = C(x) = 5x + 4$; $y_3 = y_1 - y_2$; and $y_4 = P(x) = -x^2 + 7x - 4$. What happened to the graph of the fourth function? Does this indicate that $y_3 = y_4$?

The Composition of Functions

Consider the function h given by the equation

$$h(x) = \sqrt{25 - x^2}$$

To evaluate $h(3)$, for example, we first take $25 - (3)^2 = 16$ and then take the square root of 16 to obtain $h(3) = 4$. Thus, we can view this function as the combination of two simpler functions as follows. The first is $g(x) = 25 - x^2$, and the second is the square root function. Let

$$u = g(x) = 25 - x^2$$

$$y = f(u) = \sqrt{u}$$

then

$$y = \sqrt{u} = \sqrt{25 - x^2}$$

In terms of the two functions f and g,

$$h(x) = \sqrt{25 - x^2} = \sqrt{g(x)} = f[g(x)]$$

The function $h(x)$ is called the **composition** of the two functions f and g. We use the notation $h(x) = (f \circ g)(x)$.

The domain of h consists of those values in the domain of g for which $g(x)$ is in the domain of f. In the above-mentioned case the domain of g is $(-\infty, \infty)$, and the domain of f is the set of nonnegative numbers. Thus, the domain of h consists of those

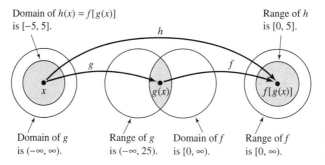

Domain of $h(x) = f[g(x)]$ is $[-5, 5]$.

Range of h is $[0, 5]$.

Domain of g is $(-\infty, \infty)$.

Range of g is $(-\infty, 25)$.

Domain of f is $[0, \infty)$.

Range of f is $[0, \infty)$.

Figure 1.64
A pictorial of the composite function where $f(x) = \sqrt{x}$ and $g(x) = 25 - x^2$.

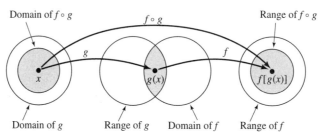

Domain of $f \circ g$

Range of $f \circ g$

Domain of g

Range of g

Domain of f

Range of f

Figure 1.65
Notice that $g(x)$ must be in the domain of f for $f[g(x)]$ to be defined.

number x for which $g(x) = 25 - x^2$ is nonnegative. Thus, the domain of h consists of all numbers for which $x^2 \le 25$ or the set $[-5, 5]$ (Fig. 1.64).

We now define in general terms the composition of two functions (Fig. 1.65).

Definition of Composite Function

Let f and g be two functions. The composite function $f \circ g$ is defined by

$$(f \circ g)(x) = f[g(x)].$$

The domain of $f \circ g$ is the set of all x in the domain of g for which $g(x)$ is in the domain of f.

Sometimes all of the range of g is in the domain of f, as shown in Figure 1.66. In this special case the domain of $f \circ g$ is just the domain of g.

Another way of looking at the composition is shown in Figure 1.67. Here the conveyor belt inputs x into the function g, which outputs $g(x)$. The conveyor belt then inputs the value $g(x)$ into the function f, which then outputs $f[g(x)]$.

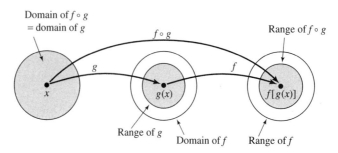

Domain of $f \circ g$ = domain of g

Range of $f \circ g$

Range of g

Domain of f

Range of f

Figure 1.66
This is the case when the range of g is in the domain of f.

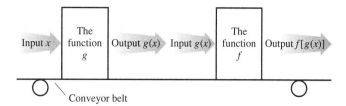

Figure 1.67

EXAMPLE 3   Finding a Composite Function

Let $f(x) = x + 1$ and $g(x) = x^3$. Find the domains of $f \circ g$ and $g \circ f$. Find $(f \circ g)(2)$ and $(g \circ f)(2)$. Find $(f \circ g)(x)$ and $(g \circ f)(x)$.

SOLUTIONS ■ Since the domains of f and g are all the real numbers, the domains of $f \circ g$ and $g \circ f$ are also all the real numbers.

$$(f \circ g)(2) = f[g(2)] = f[2^3] = f[8] = 8 + 1 = 9$$

$$(g \circ f)(2) = g[f(2)] = g[2 + 1] = g[3] = 3^3 = 27$$

$$(f \circ g)(x) = f[g(x)] = f[x^3] = x^3 + 1$$

$$(g \circ f)(x) = g[f(x)] = g[x + 1] = (x + 1)^3 \quad ■$$

Remark. Notice that $f \circ g$ and $g \circ f$ are two very different functions.

EXPLORATION 2

Compositions on Your Grapher

(a) Let $y_1 = f(x) = \sqrt{x}$, $y_2 = g(x) = 25 - x^2$, $y_3 = \sqrt{y_2}$, and $y_4 = 25 - y_1^2$. Does y_3 correspond to $f \circ g$ or $g \circ f$? What about y_4?

(b) Confirm your conjectures by graphing y_3 and either $y = (f \circ g)(x)$ or $y = (g \circ f)(x)$ together on the same screen. Do the same with y_4.

In the next example we need to be more careful about the domain of the composite function.

EXAMPLE 4   Finding a Composite Function and Its Domain

Let $g(x) = \sqrt{x}$ and let $f(x) = x^2 + 5$. Find $(f \circ g)(x)$ and its domain.

SOLUTION ■

$$(f \circ g)(x) = f[g(x)] = f[\sqrt{x}] = [\sqrt{x}]^2 + 5 = x + 5$$

One might naturally be inclined to think (incorrectly) that since one can take $x + 5$ for any x, that the domain of $f \circ g$ is all real numbers. But notice that the domain of $f \circ g$ is all x in the *domain of g* for which $g(x)$ is in the domain of f. But the domain of g is $[0, \infty)$, and the domain of f is all real numbers. Thus, the domain of $f \circ g$ is $[0, \infty)$. ■

Remark. To avoid possible mistakes keep in mind that the domain of $f \circ g$ is always a subset of the domain of g and that the range of $f \circ g$ is always a subset of the range of f.

EXPLORATION 3

The Domain of the Composite Function

Confirm Example 4 on your grapher by graphing $y = (\sqrt{x})^2 + 5$.

EXAMPLE 5   Writing a Function as the Composition of Simpler Functions

Let $h(x) = (3x + 1)^9$. Find two functions f and g such that $(f \circ g)(x) = h(x)$.

SOLUTION ■ There are many possibilities. But perhaps the simplest is to notice that we are taking the ninth power of something. The something is $3x + 1$. Thus, we might set $g(x) = 3x + 1$ and $f(x) = x^9$. Then

$$(f \circ g)(x) = f[g(x)] = f[3x + 1] = (3x + 1)^9 = h(x) \quad ■$$

Our last two examples are applications.

EXAMPLE 6   Measurement Conversions

Let x be the number of bushels, q the number of quarts, and c the number of cups. Given that there are 32 quarts to a bushel, find the function h such that $q = h(x)$. Given that there are 4 cups to a quart, find the function f such that $c = f(q)$. Now determine c as a function of x and relate this to the composition of two functions. Explain your formula in words.

SOLUTION ■ We have $q = 32x = h(x)$ and $c = 4q = f(q)$. Then

$$c = 4q = 4(32x) = 128x$$

In terms of the functions f and h this is the same as

$$c = (f \circ h)(x) = f[h(x)] = f[32x] = 4[32x] = 128x$$

This says that there are $128x$ cups in x bushels. ■

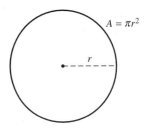

Figure 1.68
The area of a circle of radius r is πr^2.

EXAMPLE 7   Area of a Forest Fire

Suppose a forest fire is spreading in a circular manner with the radius of the burnt area increasing at a steady rate of 3 feet per minute. Write the area of the circular burnt region as a function of the time.

SOLUTION ■ The area of a circle of radius r is given by $A(r) = \pi r^2$ (Fig. 1.68). If r is the radius of the burnt region, then $r(t) = 3t$, where t is measured in minutes and r in feet. Thus

$$A = A[r(t)] = A[3t] = \pi(3t)^2 = 9\pi t^2 \quad ■$$

SELF-HELP EXERCISE SET 1.6

1. If $f(x) = x^2 + 1$ and $g(x) = x - 3$, find **(a)** $(f + g)(x)$; **(b)** $(f - g)(x)$; **(c)** $(f \cdot g)(x)$; and **(d)** $(f/g)(x)$, and the domains of each.

2. If $f(x) = x^2 + 1$ and $g(x) = \sqrt{x - 3}$, find $(f \circ g)(x)$ and the domain.

EXERCISE SET 1.6

In Exercises 1 through 2 let $f(x) = 3x + 1$ and $g(x) = x^2$ and find the indicated quantity.

1. (a) $(f + g)(2)$ **(b)** $(f - g)(2)$

   (c) $(f \cdot g)(2)$ **(d)** $\left(\dfrac{f}{g}\right)(2)$

2. (a) $(g - f)(-3)$ **(b)** $(g \cdot f)(-3)$

   (c) $\left(\dfrac{g}{f}\right)(-3)$ **(d)** $(g + f)(-3)$

In Exercises 3 through 10 you are given a pair of functions, f and g. In each case find $(f + g)(x)$; $(f - g)(x)$; $(f \cdot g)(x)$; and $(f/g)(x)$, and the domains of each.

3. $f(x) = 2x + 3$; $g(x) = x^2 + 1$

4. $f(x) = x^3$; $g(x) = 3$

5. $f(x) = 2x + 3$; $g(x) = x + 1$

6. $f(x) = x^3$; $g(x) = x - 1$

7. $f(x) = \sqrt{x + 1}$; $g(x) = x + 2$

8. $f(x) = \sqrt{x + 1}$; $g(x) = x - 3$

9. $f(x) = 2x + 1$; $g(x) = 1/x$

10. $f(x) = \sqrt{x + 2}$; $g(x) = 1/x$

In Exercises 11 through 14 you are given a pair of functions, f and g. In each case estimate the domain of $(f + g)(x)$ on your graphing calculator. Confirm analytically.

11. $f(x) = \sqrt{x + 3}$; $g(x) = \sqrt{x + 1}$

12. $f(x) = \sqrt{1 - x}$; $g(x) = \sqrt{3 - x}$

13. $f(x) = \sqrt{2x + 9}$; $g(x) = \sqrt{2x - 5}$

14. $f(x) = \sqrt{2x - 7}$; $g(x) = \sqrt{4x + 15}$

In Exercises 15 and 16 let $f(x) = 2x + 3$ and $g(x) = x^3$. Find the indicated quantity.

15. (a) $(g \circ f)(1)$ (b) $(g \circ f)(-2)$

16. (a) $(f \circ g)(1)$ (b) $(f \circ g)(-2)$

In Exercises 17 through 20 you are given a pair of functions, f and g. In each case find $(f \circ g)(x)$ and $(g \circ f)(x)$ and the domains of each.

17. $f(x) = 2x + 1$; $g(x) = 3x - 2$

18. $f(x) = 2x + 3$; $g(x) = x^3$

19. $f(x) = x^3$; $g(x) = \sqrt[3]{x}$

20. $f(x) = x^2 + x + 1$; $g(x) = x^2$

In Exercises 21 and 22 let $f(x) = 2x + 1$ and $g(x) = 2x^3$. Find the indicated quantity.

21. (a) $(f \circ g)(1)$ (b) $(g \circ f)(1)$
    (c) $(f \circ f)(1)$ (d) $(g \circ g)(1)$

22. (a) $(f \circ g)(x)$ (b) $(g \circ f)(x)$
    (c) $(f \circ f)(x)$ (d) $(g \circ g)(x)$

In Exercises 23 through 30 express each of the given functions as the composition of two functions. Find the two functions that seem the simplest.

23. $(x + 5)^5$ 24. $\sqrt{x^3 + 1}$

25. $\sqrt[3]{x + 1}$ 26. $|2x - 3|$

27. $|x^2 - 1|$ 28. $1/(3x + 2)$

29. $1/(x^2 + 1)$ 30. $\sqrt[3]{x^3 + 1}$

31. Let the two functions f and g be given by the graphs in the figure. Find
    (a) $(f \circ g)(0)$ (b) $(f \circ g)(1)$
    (c) $(g \circ f)(0)$ (d) $(g \circ f)(1)$

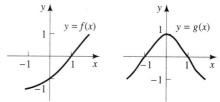

32. Let the two functions f and g be given by the graphs in the figure. Find the domains of $(f \circ g)(x)$ and $(g \circ f)(x)$.

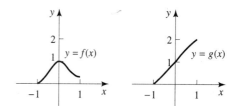

Applications

33. **Cost.** A manufacturing firm has a daily cost function of $C(x) = 3x + 10$, where x is the number of thousands of computer disks produced and C is in thousands of dollars. Suppose the number of disks that can be manufactured is given by $x = n(t) = 3t$, where t is measured in hours. Find $(C \circ n)(t)$, and state what this means.

34. **Revenue.** A firm has a revenue function given by $R(p) = 10p - p^2$, where p is the price of a chocolate bar sold in dollars and R is measured in thousands of dollars per day. Suppose the firm is able to increase the price of each bar by 5 cents each year (without affecting demand). If t is time measured in years, write an equation for the revenue as a function of t if the price of a candy bar starts out at 25 cents.

35. **Revenue and Cost.** If $R(x)$ is the revenue function and $C(x)$ is the cost function, what does the function $(R - C)(x)$ stand for?

36. **Distance.** Two ships leave the same port at the same time. The first travels due north at 4 miles per hour and the second travels due east at 5 miles per hour. Find the distance d

between the two ships as a function of time t measured in hours. *Hint:* $d = \sqrt{x^2 + y^2}$. Now find both x and y as functions of t.

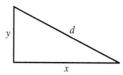

37. **Medicine.** A cancerous spherical tumor originally 30 millimeters in radius is decreasing at the rate of 2 millimeters per month after treatment. Write an equation for the volume of the tumor as a function of time t measured in months. Note that the volume V of a sphere of radius r is given by $V(r) = \frac{4}{3}\pi r^3$.

38. **Revenue.** Suppose a firm's annual revenue function is given by $R(x) = 20x + 0.01x^2$, where x is the number of items sold and R is in dollars. The firm sells 1000 items now and anticipates its sales to increase by 100 each of the

next number of years. If t is the number of years from now, write the number of sales as a function of t and also write the revenue as a function of t.

39. Measurement Conversion. Let x be the length of an object in furlongs, r the length in rods, and y the length in yards. Given that there are 40 rods to a furlong, find the function g such that $r = g(x)$. Given that there are 5.5 yards to a rod, find the function f such that $y = f(r)$. Now determine y as a function of x, and relate this to the

composition of two functions. Explain your formula in words.

40. Measurement Conversion. Let x be the size of a field in hectares, a the size in acres, and y the size in square yards. Given that there are 2.471 acres to a hectare, find the function g such that $a = g(x)$. Given that there are 4840 square yards to an acre, find the function f such that $y = f(a)$. Now determine y as a function of x, and relate this to the composition of two functions. Explain your formula in words.

Enrichment Exercises

In Exercises 41 through 46 you are given a pair of functions, f and g. In each case use your grapher to estimate the domain of $(g \circ f)(x)$. Confirm algebraically.

41. $f(x) = x^2$; $g(x) = \sqrt{x}$

42. $f(x) = |x|$; $g(x) = \sqrt{x}$

43. $f(x) = \sqrt{x + 5}$; $g(x) = x^2$

44. $f(x) = \sqrt{x - 5}$; $g(x) = x^2$

45. $f(x) = \sqrt{5 - x}$; $g(x) = x^4$

46. $f(x) = \sqrt{5 - x}$; $g(x) = x^3$

SOLUTIONS TO SELF-HELP EXERCISE SET 1.6

1. If $f(x) = x^2 + 1$ and $g(x) = x - 3$, then
  (a) $(f + g)(x) = f(x) + g(x) = (x^2 + 1) + (x - 3) = x^2 + x - 2$
  (b) $(f - g)(x) = f(x) - g(x) = (x^2 + 1) - (x - 3) = x^2 - x + 4$
  (c) $(f \cdot g)(x) = f(x) \cdot g(x) = (x^2 + 1)(x - 3) = x^3 - 3x^2 + x - 3$
  (d) $\left(\dfrac{f}{g}\right)(x) = \dfrac{f(x)}{g(x)} = \dfrac{x^2 + 1}{x - 3}$

The domains of both f and g are the set of all real numbers. The domains of the first three functions are also

the set of all real numbers. The domain of the quotient is the set of all real numbers except 3.

2. If $f(x) = x^2 + 1$ and $g(x) = \sqrt{x - 3}$, then
$$(f \circ g)(x) = f[g(x)] = (\sqrt{x - 3})^2 + 1$$
$$= (x - 3) + 1 = x - 2$$

The domain of f is $(-\infty, \infty)$ and the domain of g is $[3, \infty)$. The domain of $(f \circ g)$ is then $[3, \infty)$.

1.7 EXPONENTIAL FUNCTIONS

- Graphs of Exponential Functions
- Compound Interest
- Effective Yield
- Present Value

APPLICATION: A POPULATION MODEL

Table 1.12 gives the population in thousands of the United States for some selected years. Based on the data given for the years 1790 to 1794 find a mathematical model that approximately describes the annual population. What do you predict the population will be in 1804? An answer is given in Example 1.

TABLE 1.12

Year	1790	1791	1792	1793	1794	1804	1805
Population	3926	4056	4194	4332	4469	6065	6258

APPLICATION: COMPARING INVESTMENTS

> One bank advertises an annual rate of 9.1% compounded semiannually. A second bank advertises an annual rate of 9% compounded daily. In which bank will your money grow faster? The answer is given in Example 7.

APPLICATION: PRESENT VALUE OF A FUTURE BALANCE

> How much money must grandparents set aside at the birth of their grandchild if they wish to have $20,000 when the grandchild reaches its 18th birthday? They can earn 9% compounded quarterly. The answer is given in Example 8.

Graphs of Exponential Functions

We shall see that the graphs of functions of the form a^x for a a positive number fall into three categories: $a > 1$, $a = 1$, $0 < a < 1$. The case when $a = 1$ is not very interesting since $1^x = 1$ for all real numbers x. We now consider the other cases.

EXAMPLE 1   Exponential Functions

Table 1.12 at the beginning of this section gives the population in thousands of the United States for some selected years. Based on the data given for the years 1790 to 1794 find a mathematical model that approximately describes the annual population. Use the model to predict the population in 1804. Check the actual data to determine the accuracy of the prediction.

SOLUTION ■ We first check the annual increases in population. We obtain

$$\frac{\text{Population in 1791}}{\text{Population in 1790}} = \frac{4056}{3926} \approx 1.033$$

$$\frac{\text{Population in 1792}}{\text{Population in 1791}} = \frac{4194}{4056} \approx 1.034$$

$$\frac{\text{Population in 1793}}{\text{Population in 1792}} = \frac{4332}{4194} \approx 1.033$$

$$\frac{\text{Population in 1794}}{\text{Population in 1793}} = \frac{4469}{4332} \approx 1.032$$

Thus, for these early years, the population is increasing at about 3.3% per year. Taking 1794 to be our initial year, we thus expect

Population (1 year later) in 1795 = $4469(1.033)$

Population (2 years later) in 1796 = $4469(1.033)(1.033) = 4469(1.033)^2$

Population (3 years later) in 1797 = $4469(1.033)^2(1.033) = 4469(1.033)^3$

Thus, we expect the population n years after 1794 to be $P(n) = 4469(1.033)^n$. Based on this model, we predict the population (in thousands) in 1804 to be

$$P(10) = 4469(1.033)^{10} \approx 6183$$

As we can see from Table 1.12 this is close. Thus, we can say that during these early years the population of the United States was growing exponentially. Figure 1.69 is a graph of $y = P(t) = 4469(1.033)^t$. ■

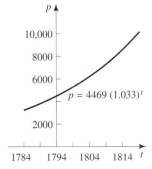

Figure 1.69
A graph of an exponential model of the population of the United States.

$p = 4469\,(1.033)^t$

Remark. Notice the graph in Figure 1.69 extends in the *negative* direction. Thus, for example, $t = -1$ corresponds to the year 1793.

The population of the United States has *not* been increasing exponentially in recent years. Thus, the population model given above is only adequate for the earlier years. Later in this text we use a more complex mathematical model that yields rather accurate estimates of the population over the entire last 200 years.

Biologists have observed from actual data on many species that populations tend to grow exponentially when the population has abundant resources available. Under these circumstances, there exists a positive constant k, called the **growth constant**, such that the population $P(t)$ at time t satisfies

$$P(t) = P_0(1 + k)^t$$

Since $P(0) = P_0(1 + k)^0 = P_0$, P_0 is the initial population, or amount at time $t = 0$. We see from Example 1 that the growth constant for the U.S. population in the early years was 0.033.

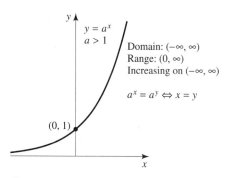

Figure 1.70

EXPLORATION *1*

Graph of Exponential Function

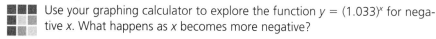

Use your graphing calculator to explore the function $y = (1.033)^x$ for negative x. What happens as x becomes more negative?

EXPLORATION *2*

Solving for Time

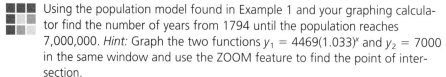

Using the population model found in Example 1 and your graphing calculator find the number of years from 1794 until the population reaches 7,000,000. *Hint:* Graph the two functions $y_1 = 4469(1.033)^x$ and $y_2 = 7000$ in the same window and use the ZOOM feature to find the point of intersection.

The graph of $y = (1.033)^t$ is similar to that shown in Figure 1.69 and is typical of the graphs of functions of the form $y = a^t$ for $a > 1$. Figure 1.70 shows the graph of $y = a^x$ for $a > 1$ and indicates some properties for this function.

We now turn to the case of $y = a^x$ when $0 < a < 1$. An important example is radioactive decay. Some elements exhibit radioactive decay. For these elements, on occasion, one nucleus spontaneously divides into two or more other nuclei. Thus, over time, the amount of the substance decreases (decays). The rate at which decay takes place depends on the substance. In the next example we see a way in which this property can be exploited.

TABLE 1.13

Day	0	1	2	3	4
Number of units	20.00	18.34	16.82	15.42	14.14

EXAMPLE 2　Radioactive Decay

Iodine-131 is a radioactive substance used as a tracer for medical diagnosis. Table 1.13 gives the amount of iodine-131 observed in a laboratory over a number of days. Based on this data, determine a mathematical model that gives the amount of iodine-131 at any time. Then use this model and your graphing calculator to find how long until this substance decays to half its original amount.

S O L U T I O N ■ Taking ratios of the amount of substance at the end of a day with the amount at the beginning we obtain

$$\frac{18.34}{20} \approx 0.917$$

$$\frac{16.82}{18.34} \approx 0.917$$

$$\frac{15.42}{16.82} \approx 0.917$$

$$\frac{14.14}{15.42} \approx 0.917$$

Thus we have

Initial amount $= 20$

Amount after the first day $= 20(0.917)$

Amount after the second day $= 20(0.917)(0.917) = 20(0.917)^2$

Amount after the third day $= 20(0.917)^2(0.917) = 20(0.917)^3$

Thus the amount after the nth day is $A(n) = 20(0.917)^n$ (Fig. 1.71).

To find the time until the substance decays to half its original amount, we graph the two functions $y_1 = 20(0.917)^x$ and $y_2 = 10$ using a window of dimensions [0, 10] by [0, 20]. We obtain Screen 1.18. Now using the ZOOM feature several

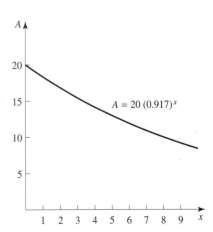

Figure 1.71
A graph of an exponential decay model.

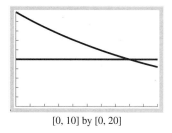

[0, 10] by [0, 20]

Screen 1.18
Graphs of $y_1 = 20(0.917)^x$ and $y_2 = 10$.

times, we find that the x-coordinate of the point of intersection is 8. Thus, it takes 8 days until the 20 units of iodine-131 decays to 10 units. ∎

For elements that exhibit radioactive decay, let $A(t)$ be the amount at time t. Then there is a positive constant k, called the **decay constant**, so that

$$A(t) = A_0(1 - k)^t$$

Since $A(0) = A_0(1 - k)^0 = A_0$, A_0 is the initial amount, or amount at time $t = 0$.

EXPLORATION 3

Graph of Exponential Function

 Use your graphing calculator to explore the function $y = (0.917)^x$ for negative x. What happens as x becomes more negative?

The graph of $y = (0.917)^x$ is similar to that shown in Screen 1.18 and is typical of the graphs of functions of the form $y = a^t$ for $0 < a < 1$. Figure 1.72 shows the graph of $y = a^x$ for $0 < a < 1$ and indicates some properties for this function.

Properties of a^x

Property 1: If $a > 1$, the function a^x is increasing.
Property 2: If $0 < a < 1$, the function a^x is decreasing.
Property 3: If $a \neq 1$, then $a^x = a^y$ if and only if $x = y$.
Property 4: If $0 < a < 1$, then the graph of $y = a^x$ approaches the x-axis as x becomes large. If $a > 1$, then the graph of $y = a^x$ approaches the x-axis as x becomes negatively large.

We now solve some special equations with the unknown as the exponent. We must wait until the next section to see how to solve for an unknown exponent in general.

EXAMPLE 3   Solving Equations with Exponents

Solve (a) $2^x = 8$ (b) $9^x = 27^{4x-10}$

SOLUTIONS ∎

(a) First rewrite $2^x = 8$ as $2^x = 2^3$. Then by property 3, $x = 3$.
(b) We first have $(3^2)^x = (3^3)^{4x-10}$. Then

$$3^{2x} = 3^{12x-30}$$

$$2x = 12x - 30 \quad \text{Property 3}$$

$$x = 3 \quad ∎$$

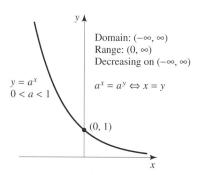

Domain: $(-\infty, \infty)$
Range: $(0, \infty)$
Decreasing on $(-\infty, \infty)$

$y = a^x$
$0 < a < 1$

$a^x = a^y \Leftrightarrow x = y$

$(0, 1)$

Figure 1.72

Compound Interest

The exponential function also arises in the most common type of interest, **compound interest**.

If the principal is invested for a period of time given by a fraction t of a year at an annual interest rate of r, where r is given as a decimal, then the amount at the end of the first period is

$$\text{Principal} + \text{interest} = P + Pi = P(1 + i)$$

with $i = rt$. The interest $i = rt$ is called the **interest per period**.

If, for example, the annual interest for a bank account is 6% and the time period is one month, then $i = 0.06\,(1/12) = 0.005$ is the interest per month.

If the interest and principal are left in the account for more than one period and interest is calculated not only on the principal but also on the previous interest earned, we say that the interest is being **compounded**.

If $1000 is deposited in a bank account earning interest at 6% a year and compounding monthly, then

Amount at end of month 1 is $1000(1 + 0.005) = \$1000(1.005)$

Amount at end of month 2 is $1000(1.005)(1.005) = \$1000(1.005)^2$

Amount at end of month 3 is $1000(1.005)^2(1.005) = \$1000(1.005)^3$

Continuing in this manner, the future amount of money F in the account at the end of n months is

$$F = \$1000(1.005)^n$$

In the same way if a principal P earns interest at the rate per period of i, expressed as a decimal, and interest is compounded, then the amount F after n periods is

$$F = P(1 + i)^n$$

EXAMPLE 4　　Finding Compound Interest

Suppose $1000 is deposited into an account with an annual interest rate of 8% compounded quarterly. Find the amount in the account at the end of 5 years, 10 years, 20 years, 30 years, and 40 years.

SOLUTIONS ■　We have $P = \$1000$, and the interest per quarter is $i = 0.08(1/4) = 0.02$. Thus, $F = \$1000(1 + 0.02)^n$. Now using the y^x key on a calculator, we can obtain

Years (t)	Periods ($n = 4t$)	Future Value
5	20	$1000(1 + 0.02)^{20} = \$1485.95$
10	40	$1000(1 + 0.02)^{40} = \$2208.04$
20	80	$1000(1 + 0.02)^{80} = \$4875.44$
30	120	$1000(1 + 0.02)^{120} = \$10,765.16$
40	160	$1000(1 + 0.02)^{160} = \$23,769.91$

Notice that during the first 10 years the account grows by about $1200, but during the last 10 years it grows by about $13,000. In fact, each year the account grows by more than in the previous year (Fig. 1.73).

Remark. We use the letter i to designate the interest rate for any period, whether annual or not. But we reserve the letter r to always designate an annual rate.

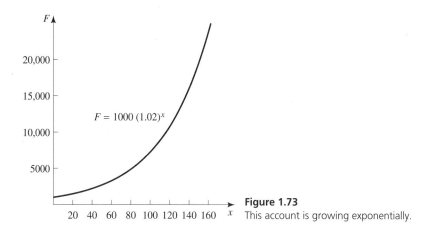

Figure 1.73
This account is growing exponentially.

Suppose r is the annual interest rate expressed as a decimal, and interest is compounded m times a year. Then the interest rate per period is $i = r/m$. If the compounding goes on for t years, then there are $n = mt$ periods, and the amount F after t years is

$$F = P(1 + i)^n = P\left(1 + \frac{r}{m}\right)^{mt}$$

Compound Interest

Suppose a principal P earns interest at the annual rate of r, expressed as a decimal, and interest is compounded m times a year. Then the amount F after t years is

$$F = P(1 + i)^n = P\left(1 + \frac{r}{m}\right)^{mt}$$

where $n = mt$ is the number of time periods, and $i = r/m$ is the interest per period.

EXAMPLE 5   Finding Compound Interest

Suppose $1000 is deposited into an account that yields 9% annually. Find the amount in the account at the end of the fifth year if the compounding is (a) quarterly (b) monthly.

SOLUTIONS ■

(a) Here $P = \$1000$, $r = 0.09$, $m = 4$, and $t = 5$. Using the y^x on your calculator gives

$$F = P\left(1 + \frac{r}{m}\right)^{mt} = \$1000\left(1 + \frac{0.09}{4}\right)^{4 \cdot 5} = \$1560.51$$

(b) Here $P = \$1000$, $r = 0.09$, $m = 12$, and $t = 5$. Using the y^x on your calculator gives

$$F = P\left(1 + \frac{r}{m}\right)^{mt} = \$1000\left(1 + \frac{0.09}{12}\right)^{12 \cdot 5} = \$1565.68 \quad ■$$

Remark. Notice that the amount at the end of the period is larger if the compounding is done more often.

EXPLORATION 4

Solving for Time

Using your graphing calculator, find the time for the amount in the account in Example 5b to reach $1400. Graph on the same screen $y_1 =$

1000(1.0075)^x and $y_2 = 1400$. Use screen dimensions of [0, 60] by [1000, 1500]. Then use the ZOOM feature to find the point of intersection.

EXAMPLE 6   Calculating Interest over a Long Period

Suppose after the Canarsie Indians sold Manhattan for $24 in 1626, the money was deposited into a Dutch Guilder account that yielded an annual rate of 6% compounded quarterly. How much would be in this account in 1994?

SOLUTION ■ We have

$$F = P\left(1 + \frac{r}{m}\right)^{mt}$$

$$= \$24\left(1 + \frac{0.06}{4}\right)^{368\cdot4}$$

$$\approx \$79 \text{ billion} \quad \blacksquare$$

The table indicates what the value of this account would have been at some intermediate times.

Year		Future Value ($)
1626		24
1650	$24\left(1 + \dfrac{0.06}{4}\right)^{4\cdot24} \approx$	100
1700	$24\left(1 + \dfrac{0.06}{4}\right)^{4\cdot74} \approx$	2,000
1750	$24\left(1 + \dfrac{0.06}{4}\right)^{4\cdot124} \approx$	39,000
1800	$24\left(1 + \dfrac{0.06}{4}\right)^{4\cdot174} \approx$	760,000
1850	$24\left(1 + \dfrac{0.06}{4}\right)^{4\cdot224} \approx$	15,000,000
1900	$24\left(1 + \dfrac{0.06}{4}\right)^{4\cdot274} \approx$	293,000,000
1950	$24\left(1 + \dfrac{0.06}{4}\right)^{4\cdot324} \approx$	6,000,000,000
1994	$24\left(1 + \dfrac{0.06}{4}\right)^{4\cdot368} \approx$	79,000,000,000

Remark. It is, of course, very unlikely that any investment could have survived through the upheavals of wars and financial crises that occurred during this 368-year span of time. Nonetheless, it is examples such as this one that inspire some to use the phrase "the wonders of compounding."

Effective Yield

If $1000 is invested at an annual rate of 9% compounded monthly, then at the end of a year there is

$$F = \$1000\left(1 + \frac{0.09}{12}\right)^{12} = \$1093.81$$

in the account. This is the same amount obtainable if the same principal of $1000 was invested for one year at an *annual* rate of 9.381% (or 0.09381 expressed as a decimal). We call the rate 9.381% the **effective annual yield**. The 9% annual rate is often referred to as the **nominal rate**.

Suppose r is the annual interest rate expressed as a decimal, and interest is compounded m times a year. If the compounding goes on for one year, then the amount F after one year is

$$F = P\left(1 + \frac{r}{m}\right)^m$$

If we let r_e be the effective annual yield then r_e must satisfy

$$P\left(1 + \frac{r}{m}\right)^m = P(1 + r_e)$$

Solving for r_e, we obtain

$$r_e = \left(1 + \frac{r}{m}\right)^m - 1$$

Effective Yield

Suppose a sum of money is invested at an annual rate of r expressed as a decimal and is compounded m times a year. The effective yield r_e is

$$r_e = \left(1 + \frac{r}{m}\right)^m - 1$$

EXAMPLE 7   Comparing Investments

One bank advertises a nominal rate of 9.1% compounded semiannually. A second bank advertises a nominal rate of 9% compounded daily. What are the effective yields? In which bank would you deposit your money?

SOLUTIONS ■ For the first bank $r = 0.091$ and $m = 2$. Then

$$r_e = \left(1 + \frac{0.091}{2}\right)^2 - 1 = 0.0931$$

or as a percent, 9.31%.

For the second bank $r = 0.09$ and $m = 365$. Then

$$r_e = \left(1 + \frac{0.09}{365}\right)^{365} - 1 = 0.0942$$

or as a percent, 9.42%.

Despite the higher nominal rate given by the first bank, the effective yield for the second bank is higher than the first. Thus, money deposited in the second bank grows faster than money deposited in the first bank.   ■

Present Value

If we have an account with an initial amount P and earning interest at an annual rate of r expressed as a decimal, and interest is compounded m times a year, then the amount F in the account after t years is

$$F = P\left(1 + \frac{r}{m}\right)^{mt}$$

If we wish to know how many dollars P to set aside now in this account so that we will have a future amount of F dollars after t years, we simply solve the preceding expression for P. Thus,

$$P = \frac{F}{(1 + [r/m])^{mt}}$$

This is called the **present value**.

Present Value

Suppose an account earns an annual rate of r expressed as a decimal and is compounded m times a year. Then the amount P, called the **present value**, needed presently in this account so that a future amount of F will be attained in t years is given by

$$P = \frac{F}{(1 + [r/m])^{mt}}$$

E X A M P L E 8   Finding the Present Value of a Future Balance

How much money must grandparents set aside at the birth of their grandchild if they wish to have \$20,000 when the grandchild reaches its 18th birthday? They can earn 9% compounded quarterly.

S O L U T I O N ■ Here $r = 0.09$, $m = 4$, $t = 18$, and $A = \$20,000$. Thus,

$$P = \frac{F}{(1 + [r/m])^{mt}}$$

$$= \frac{\$20,000}{(1 + [0.09/4])^{4(18)}} \approx \$4029.69 \quad ■$$

SELF-HELP EXERCISE SET 1.7

1. Solve $3^{2x+1} = 1/3^x$

2. An account with \$1000 earns interest at an annual rate of 8% compounded monthly. Find the amount in this account after 10 years.

3. Find the effective yield if the annual rate is 8% and the compounding is weekly.

4. How much money should be deposited in a bank account earning the annual interest rate of 8% compounded quarterly in order that there be \$10,000 in the account at the end of 10 years?

EXERCISE SET 1.7

In Exercises 1 through 8 sketch a graph of each of the functions without using your grapher. Then support your answer with your grapher.

1. $y = 5^x$

2. $y = 7^x$

3. $y = 3^{x^2}$

4. $y = 3^{-x^2}$

5. $y = \left(\frac{1}{2}\right)^x$

6. $y = (0.1)^x$

7. $y = 10^x$

8. $y = 5^{-x}$

In Exercises 9 through 16 solve for x.

9. $3^x = 9$

10. $4^x = 8$

11. $16^x = 8$

12. $\left(\frac{1}{2}\right)^x = \frac{1}{16}$

13. $\left(\frac{1}{16}\right)^x = \frac{1}{8}$

14. $\left(\frac{1}{4}\right)^x = 8$

15. $4^{2x} = 8^{9x+15}$

16. $5^{3x} = 125^{4x-4}$

In Exercises 17 through 24 graph each of the functions without using your graphing calculator. Then support your answer using your graphing calculator.

17. $y = 2^{3x}$

18. $y = 3^{-2x}$

19. $y = 3^{x^2}$

20. $y = 5^{-x^2}$

21. $y = 2 - 3^{-x}$

22. $y = 2 + 3^{-x}$

23. $y = \frac{1}{2}(3^x + 3^{-x})$

24. $y = \frac{1}{2}(2^x - 2^{-x})$

In Exercises 25 through 30 use the basic graphing principles to sketch the graphs. Then support your answers using your graphing calculator.

25. $y = 3^x$, $y = 3^{x+3}$, $y = 3^{x-3}$

26. $y = 2^x$, $y = 2 \cdot 2^x$, $y = 0.25 \cdot 2^x$

27. $y = 5^{-x}$, $y = 2 \cdot 5^{-x}$, $y = 0.25 \cdot 5^{-x}$

28. $y = 7^{-x}$, $y = 7^{-(x+3)}$, $y = 7^{-(x-3)}$

29. $y = 3^x$, $y = -3^x$

30. $y = 2^{-x}$, $y = -2^{-x}$

In Exercises 31 through 44 solve for x.

31. $3^x = \dfrac{1}{3}$ **32.** $5^x = \dfrac{1}{25}$

33. $2^x = \dfrac{1}{8}$ **34.** $2^{5x} = 2^{x+8}$

35. $(3^{x+1})^2 = 3$ **36.** $5^{2x-1}5^x = \dfrac{1}{5^x}$

37. $3^{x+1} = \dfrac{1}{3^x}$ **38.** $7^{x^2+2x} = 7^{-x}$

39. $3^{-x^2+2x} = 3$ **40.** $x2^x = 2^x$

41. $3^{-x}(x + 1) = 0$ **42.** $x^25^{2x} = 5^{2x}$

43. $x5^{-x} = x^25^{-x}$ **44.** $(x^2 + x - 2)2^x = 0$

Applications

In Exercises 45 through 48 find how much is in the accounts after the given years where P is the initial principal, r is the annual rate given as a percent with the indicated compounding.

45. After 1 year where $P = \$1000$, $r = 8\%$, compounded (**a**) annually (**b**) quarterly (**c**) monthly (**d**) weekly (**e**) daily

46. After 40 years where $P = \$1000$, $r = 8\%$, compounded (**a**) annually (**b**) quarterly (**c**) monthly (**d**) weekly (**e**) daily

47. After 40 years where $P = \$1000$, compounded annually, r equal to (**a**) 3% (**b**) 5% (**c**) 7% (**d**) 9% (**e**) 12% (**f**) 15%

48. $P = \$1000$, $r = 9\%$, compounded annually, after (**a**) 5 (**b**) 10 (**c**) 15 (**d**) 30 years.

In Exercises 49 and 50 find the effective yield given the annual rate r and the indicated compounding.

49. $r = 8\%$, compounded (**a**) semiannually (**b**) quarterly (**c**) monthly (**d**) weekly (**e**) daily

50. $r = 10\%$, compounded (**a**) semiannually (**b**) quarterly (**c**) monthly (**d**) weekly (**e**) daily

In Exercises 51 and 52 find the present value of the given amounts F with the indicated annual rate of return r, the number of years, t, and the indicated compounding.

51. $F = \$10,000$, $r = 9\%$, $t = 20$, compounded (**a**) annually (**b**) monthly (**c**) weekly

52. $F = \$10,000$, $r = 10\%$, $t = 20$, compounded (**a**) annually (**b**) quarterly (**c**) daily

53. Your rich uncle has just given you your high school graduation present of ''$\$1,000,000$.'' The present is in the form of a 40-year bond with an annual interest rate of 9% compounded annually. The bond says it will be worth $\$1,000,000$ in 40 years. What is this ''$\$1,000,000$'' gift worth at the present time?

54. Redo the previous exercise if the annual interest rate is 6%.

55. Your second rich uncle gives you a high school graduation present of ''$\$2,000,000$.'' The present is in the form of a 50-year bond with an annual interest rate of 9%. The bond says it will be worth $\$2,000,000$ in 50 years. What is this ''$\$2,000,000$'' gift worth at the present time? Compare your answer with that for Exercise 53.

56. Redo the previous exercise with an annual interest rate of 6%.

57. In Example 6 find the amount in 1994 if the annual interest was 7% compounded quarterly. Compare your answer with that for Example 6 in the text.

58. In Example 6 find the amount in 1994 if the annual interest was 5% compounded quarterly. Compare your answer with that for Example 6 in the text.

59. Real Estate Appreciation. The United States paid about 4 cents an acre for the Louisiana Purchase in 1803. Suppose the value of this property grew at an annual rate of 5.5% compounded annually. What would an acre be worth in 1994? Does this seem realistic?

60. Real Estate Appreciation. Redo the previous problem using a rate of 6% instead of 5.5%. Compare your answer with the answer to Exercise 59.

61. Comparing Rates at Banks. One bank advertises a nominal rate of 6.5% compounded quarterly. A second bank advertises a nominal rate of 6.6% compounded daily. What are the effective yields? In which bank would you deposit your money?

62. Comparing Rates at Banks. One bank advertises a nominal rate of 8.1% compounded semiannually. A second bank advertises a nominal rate of 8% compounded weekly. What are the effective yields? In which bank would you deposit your money?

63. Saving for Machinery. How much money should a company deposit in an account with a nominal rate of 8% compounded quarterly in order to have $\$100,000$ for a certain piece of machinery in 5 years?

64. Saving for Machinery. Repeat Exercise 63 with an annual rate of 7% and monthly compounding.

65. Suppose $\$1000$ is invested at an annual rate of 6% compounded monthly. Using your graphing calculator, determine the time (to two decimal places) it takes for this account to double.

66. One account with an initial amount of $\$1$ grows at 100% a year compounded quarterly and a second account with an initial amount of $\$2$ grows at 50% a year compounded quarterly. Graph two appropriate functions in the same viewing window and using the ZOOM feature find the time (to two decimal places) for the two accounts to have equal amounts.

67. An account earns an annual rate of r, expressed as a decimal, and is compounded quarterly. The account has $1000 initially and five years later has $1500. What is r?

68. Radioactive Decay. Plutonium-239 is a product of nuclear reactors with a half-life of about 24,000 years. What percentage of a given sample will remain after 10,000 years?

69. Medicine. As we noted in the text, radioactive tracers are used for medical diagnosis. Suppose 20 units of iodine-131 are shipped and take two days to arrive. Referring to Example 2 of the text determine how much of the substance arrives.

70. Medicine. Suppose 20 units of iodine-131 are needed in two days. Determine how much should be ordered.

Enrichment Exercises

71. Let $y_1 = 2^x$ and $y_2 = x^3$.
(a) Which is larger when $x = 2$?
(b) On a screen with dimensions [0, 12] by [0, 3000] graph and determine which is larger for large values of x.

72. Given any positive integer n, speculate on whether $y_1 = 2^x$ or $y_2 = x^n$ is larger for large x. Experiment on your graphing calculator to decide.

73. Let $y_1 = (1.1)^x$ and $y_2 = x^3$.
(a) Which is larger when $x = 2$?
(b) On a screen with dimensions [100, 180] by [0, 30,000,000] graph and determine which is larger for large values of x.

74. Given any positive integer n, speculate on whether $y_1 = (1.1)^x$ or $y_2 = x^n$ is larger for large x. Experiment on your graphing calculator to decide.

75. On a screen with dimensions $[-2, 2]$ by [0, 5] graph 2^x, 3^x, 5^x. Determine the interval for which $2^x < 3^x < 5^x$ and the interval for which $2^x > 3^x > 5^x$.

76. Decide whether $y = 1/x^2$ or 2^{-x} is smaller for large values of x.

SOLUTIONS TO SELF-HELP EXERCISE SET 1.7

1. Write the equation as $3^{2x+1} = 3^{-x}$. Then from property 3, we must have $2x + 1 = -x$ and $x = -1/3$.

2. Here $P = \$1000$, $r = 0.08$, and $n = 10$, $m = 12$. Thus,
$$F = P\left(1 + \frac{r}{m}\right)^{mt} = \$1000\left(1 + \frac{0.08}{12}\right)^{12 \cdot 10} = \$2219.64$$

3. If the annual rate is 8%, the effective yield is given by
$$r_e = \left(1 + \frac{r}{m}\right)^m - 1 = \left(1 + \frac{0.08}{52}\right)^{52} - 1 \approx 0.0832$$

4. The present value of $10,000 if the annual interest rate is 8% compounded quarterly for 10 years is
$$P = \frac{F}{(1 + [r/m])^{mt}} = \frac{10,000}{(1 + [0.08/4])^{4 \cdot 10}} = 4528.90$$

Thus, a person must deposit $4528.90 in an account earning 8% compounded quarterly so there will be $10,000 in the account after 10 years.

1.8 LOGARITHMS

■ Basic Properties of Logarithms ■ Solving Equations with Exponents and Logarithms

John Napier 1550–1617

John Napier, Laird of Merchiston, was a Scottish nobleman who invented logarithms around 1594 in order to ease the burden of numerical calculations needed in astronomy. Most of his time was spent engaged in the political and religious controversies of his day. He published an extremely widely read attack on the Roman Catholic Church, which he sincerely believed was his primary achievement. For relaxation he amused himself with the study of mathematics and science.

(The Granger Collection)

APPLICATION: TIME NEEDED TO DOUBLE AN ACCOUNT

A person has $1000 and can invest it in an account that earns an annual rate of 8% compounded quarterly. In five years this investor needs $1400. Will this account have $1400 within five years? See Example 4 for the answer.

In the previous section we considered the radioactive substance iodine-131, used in medical diagnosis. A hospital has ordered 20 grams of iodine-131 and will need 10 grams within a week of ordering. Did they order enough? See Example 7 for the answer.

Basic Properties of Logarithms

In the previous section we faced a number of problems in which the unknown quantity in an equation was an exponent. For example, when we wanted the time for $1000 in an account growing at an annual rate of 9% compounded monthly to reach $1400 we needed to solve for t in the equation

$$1000(1.0075)^t = 1400$$

Our only available method was to find an approximate solution using our grapher.

We now wish to develop an analytical method of solving such an equation. As a simplified example suppose we want to solve for x in the equation $2 = 10^x$. To do this analytically, we need to invent a function, that we temporarily call $l(x)$, that "undoes" the exponential. That is, we want $l(x)$ to have the property that for all x, $l(10^x) = x$. Thus, in Figure 1.74, the conveyor belt inputs x into the exponential function, which outputs 10^x. The term 10^x is then inputed into the l function, which outputs x. An equation such as $2 = 10^x$ can then be solved by applying l to each side, obtaining

$$l(2) = l(10^x) = x$$

The values of the function $l(x)$ can be tabulated in a table or obtained on a calculator. In this way we solve for x.

You are already familiar with a similar process. For example, to solve the equation $2 = x^3$, we apply the cube root function $g(x) = \sqrt[3]{x}$ to each side. The cube root function $g(x) = \sqrt[3]{x}$ "undoes" the cube function $f(x) = x^3$, that is, $g(x^3) = \sqrt[3]{x^3} = x$. Thus, we can solve the equation $2 = x^3$ for x by applying g to each side, obtaining

$$\sqrt[3]{2} = \sqrt[3]{x^3} = x$$

A table or calculator gives the value of $\sqrt[3]{2}$. In this way we solve for x.

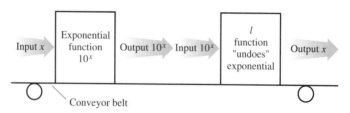

Figure 1.74

EXPLORATION 1

Undoing a Function

On a graphing calculator solve for x to three decimal places in the equation $2 = x^3$ by finding the point of intersection of the graphs of $y_1 = x^3$ and $y_2 = 2$ in a window of dimension [0, 2] by [0, 3]. Check your answer.

When given any x, Figure 1.75 indicates geometrically how we can find $y = 10^x$. Since the function l is to "undo" the exponential function $y = 10^x$, then, given any

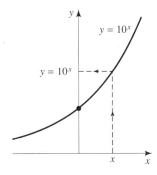

Figure 1.75
We see how to obtain 10^x graphically.

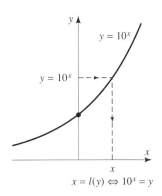

$x = l(y) \Leftrightarrow 10^x = y$

Figure 1.76
Given y we see graphically how to find the x for which $10^x = y$.

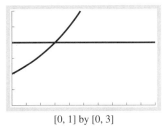

[0, 1] by [0, 3]

Screen 1.19
Graphs of $y_1 = 10^x$ and $y_2 = 2$.

$y > 0$, we see in Figure 1.76 how to obtain $l(y)$ by "reversing the arrows." That is, given any $y > 0$,

$$l(y) = x \qquad \text{if, and only if,} \qquad y = 10^x$$

Thus, $l(y)$ is that exponent to which 10 must be raised to obtain y. We normally write our functions with x as the independent variable. So interchanging x and y in the preceding equation gives for any $x > 0$

$$y = l(x) \qquad \text{if, and only if,} \qquad x = 10^y$$

We now see how to graphically evaluate $l(x)$.

EXAMPLE 1   Evaluating $l(x)$

Use a graphing calculator to solve for x to three decimal places in the equation $2 = 10^x$. Check your answer.

SOLUTION ■ We first graph $y_1 = 10^x$ and $y_2 = 2$ in a window of dimension [0, 1] by [0, 3] (Screen 1.19). This corresponds to Figure 1.76. Now repeatedly using the ZOOM feature we can obtain $x = 0.301$ to three decimal places. Thus, we have that $l(2) \approx 0.301$. In fact $10^{l(2)} = 10^{0.301} \approx 1.99986$. ■

The function $l(x)$ is usually written as $l(x) = \log_{10} x$ and is called the **logarithm to the base 10 of x**. In terms of this notation we have the following.

Logarithm to the Base 10

If $x > 0$, the **logarithm to the base 10 of x**, denoted by $\log_{10} x$, is defined as follows:

$$y = \log_{10} x \qquad \text{if, and only if,} \qquad x = 10^y$$

Instead of using the number 10, we can use any positive number $a \neq 1$.

Logarithm to the Base a

Let a be a positive number with $a \neq 1$. If $x > 0$, the **logarithm to the base a of x**, denoted by $\log_a x$, is defined as follows:

$$y = \log_a x \qquad \text{if, and only if} \qquad x = a^y$$

EXAMPLE 2   Using the Definition of Logarithm

Solve for x in (a) $\log_{10} x = -\frac{3}{2}$ (b) $\log_3 3^2 = x$.

SOLUTION ■
(a) $\log_{10} x = -\frac{3}{2}$ if, and only if, $x = 10^{-3/2}$.
(b) $\log_3 3^2 = x$ if, and only if, $3^x = 3^2$. This is true if and only if $x = 2$. ■

Remark. Notice that we can only have $x > 0$ in the definition of the logarithm since $x = a^y$ is always positive. Thus, the domain of the function $y = \log_a x$ is (0, ∞).

Remark. Notice that $y = \log_a x$ means that $x = a^y = a^{\log_a x}$. Thus, $\log_a x$ is an *exponent*. In fact $\log_a x$ is that exponent to which a must be raised to obtain x.

According to the definition of logarithm, for any $x > 0$, $y = \log_a a^x$ if, and only if, $a^x = a^y$. But this implies that $x = y = \log_a a^x$. This shows that the function $\log_a x$ "undoes" the exponential a^x.

Basic Properties of the Logarithmic Function

1. $a^{\log_a x} = x$ if $x > 0$.
2. $\log_a a^x = x$ for all x.

Remark. It has already been mentioned that Property 2 means that the logarithmic function $\log_a x$ "undoes" the exponential function a^x. Property 1 says that the exponential function a^x "undoes" the logarithmic function $\log_a x$.

Remark. To find any intercepts of the graph of $y = \log_a x$, set $0 = \log_a x$. Then we must have $x = a^0 = 1$. Thus, $\log_a 1 = 0$ and $(1, 0)$ is the only intercept of the graph of $y = \log_a x$.

EXPLORATION 2

An Overloaded Calculator

By Property 2 we know that $\log_{10} 10^{10000} = 10000$. Try to find $\log_{10} 10^{10000}$ on your calculator. What went wrong? Explain.

EXAMPLE 3   Using the Basic Properties of Logarithms

Find (a) $\log_3 3^{-2}$ (b) $10^{\log_{10}\pi}$

SOLUTIONS ■

(a) Use Property 2 to obtain $\log_3 3^{-2} = -2$.
(b) Use Property 1 to obtain $10^{\log_{10}\pi} = \pi$. ■

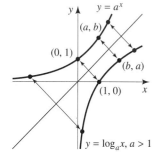

Figure 1.77
The graph of $y = \log_a x$ is the reflection about the line $y - x$ of the graph of $y = a^x$.

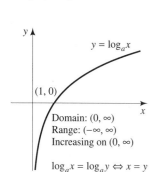

Domain: $(0, \infty)$
Range: $(-\infty, \infty)$
Increasing on $(0, \infty)$

$\log_a x = \log_a y \Leftrightarrow x = y$

Figure 1.78

We now consider the graphs of the functions $y = \log_a x$. The graphs are different depending on whether $a > 1$ or $0 < a < 1$. We are only interested in the case $a > 1$, since this covers the two important cases. The first of these is $a = 10$. This case is historically important. The logarithm $\log_{10} x$ is called the **common logarithm** and is also denoted by $\log x$. The second important case is $a = 2.71828$ (to five decimal places) and also denoted by the letter e. This case is actually critically important and will be encountered in the next chapter and throughout the remainder of the text. We call the logarithm $\log_e x$ the **natural logarithm** and also denote it by $\ln x$.

First notice that whether $a > 1$ or $0 < a < 1$, a point (u, v) is on the graph of $y = a^x$ if, and only if, $v = a^u$. But according to the definition of the natural logarithm, $v = a^u$ if, and only if, $u = \log_a v$. This says that (v, u) is on the graph of $y = \log_a x$. Thus, (u, v) lies on the graph of $y = a^x$ if, and only if, (v, u) lies on the graph of $y = \log_a x$. This means that the graphs of $y = a^x$ and $y = \log_a x$ are symmetrical about the line $y = x$. Figure 1.77 shows the case $a > 1$. In this way we are able to construct the graph of $y = \log_a x$ shown in Figures 1.77 and 1.78 where the properties of $y = \log_a x$ are also given in the case $a > 1$.

EXPLORATION 3

Symmetry

(a) On a screen of dimension $[-1, 3.7]$ by $[-2, 4]$ graph $y_1 = 10^x$ and $y_2 = \log_{10} x$.

(b) Now on the same screen graph $y = -x + 2$ or $x + y = 2$. Notice that (u, v) is on the graph of this line if and only if (v, u) is. Now use TRACE to place the blinking cursor on the graph of the line, and use the arrow keys to move the blinking cursor to approximately the point of intersection of this line with the graph of $y = 10^x$. Note carefully the x- and y-coordinate of this point. Now use the arrow keys to continue down the line until you find the point of intersection of this line with the graph of $y = \log_{10} x$. Note carefully the x- and y-coordinate. What have you found?

    (c) Repeat this procedure for the line $x + y = 1.5$.

For any positive $a \neq 1$ the logarithm $\log_a x$ obeys the following rules.

Rules for the Logarithm

1. $\log_a xy = \log_a x + \log_a y$
2. $\log_a \dfrac{x}{y} = \log_a x - \log_a y$
3. $\log_a x^c = c \log_a x$

To establish these rules, we need to use $x = a^{\log_a x}$, $y = a^{\log_a y}$, $xy = a^{\log_a xy}$, $x/y = a^{\log_a x/y}$.

To establish Rule 1, notice that

$$a^{\log_a xy} = xy = a^{\log_a x} a^{\log_a y} = a^{\log_a x + \log_a y}$$

Then equating exponents gives Rule 1.

We can establish Rule 2 in a similar manner. Notice that

$$a^{\log_a x/y} = \frac{x}{y} = \frac{a^{\log_a x}}{a^{\log_a y}} = a^{\log_a x - \log_a y}$$

Then equating exponents gives Rule 2.

To establish Rule 3, notice that

$$a^{\log_a x^c} = x^c = (a^{\log_a x})^c = a^{c \log_a x}$$

Equating exponents gives Rule 3.

EXPLORATION *4*

Determining Equalities by Graphing

Decide if any of the following equations is true by finding the graph of each side of each equation in a window of dimensions $[-4, 4]$ by $[-2, 2]$.

    (a) $\log_{10}(x + 3) = \log_{10} x + \log_{10} 3$

    (b) $\log_{10} x^2 = (\log_{10} x)^2$

    (c) $\log_{10} \dfrac{x}{2} = \dfrac{\log_{10} x}{\log_{10} 2}$

Solving Equations with Exponents and Logarithms

We now consider an important case in which we must solve for an unknown exponent.

EXAMPLE 4   Solving an Equation with an Unknown Exponent

A person has \$1000 and can invest it in an account that earns an annual rate of 8% compounded quarterly. In five years the investor needs \$1400. Will this account have \$1400 within five years?

S O L U T I O N ■ We need to find the time t for this account to grow from $1000 to $1400. This requires us to solve for t in the equation $1400 = 1000(1 + 0.02)^t$.

$$1400 = 1000(1.02)^t$$

$$1.4 = (1.02)^t$$

$$\log 1.4 = \log(1.02)^t = t \log 1.02 \qquad \text{Rule 3}$$

$$t = \frac{\log 1.4}{\log 1.02}$$

$$\approx \frac{0.1461}{0.0086} \approx 17$$

Thus, it takes about 17 quarters, or 4.25 years, for this account to reach $1400. This is less than the required five years (Fig. 1.79). ■

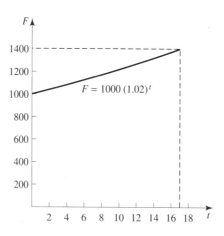

Figure 1.79
This account grows to $1400 in about 17 quarters.

If we have an equation with a logarithm, then at an appropriate point we can use the definition of logarithm as in the following example.

EXAMPLE 5 Solving an Equation with a Logarithm

Solve $2 \log(2x + 5) + 6 = 0$.

S O L U T I O N ■

$$2 \log(2x + 5) + 6 = 0$$

$$\log(2x + 5) = -3$$

$$2x + 5 = 10^{-3} = 0.001 \qquad \text{definition of logarithm}$$

$$x = \frac{1}{2}(-4.999) = -2.4995 \qquad ■$$

Remark. In Example 5, we had $\log(2x + 5) = -3$ at one point. We could have applied the exponential function to each side of this equation to "undo" the logarithm. Thus,

$$10^{-3} = 10^{\log(2x+5)} = 2x + 5$$

by Property 1.

No Analytical Solution

An equation with exponents cannot always be solved by taking logarithms. For example, the equation $1 + x = 5^x$ cannot be solved by taking logarithms. Try to! Use your graphing calculator to approximate all solutions.

E X A M P L E 6   Doubling Time
_____

Determine the time it takes for an account earning interest at an annual rate of 4% compounded annually to double.

S O L U T I O N ■ Let t be the time in years for the account to double and P the initial amount. Then we must solve for t in the equation $2P = P(1 + 0.04)^t$.

$$2P = P(1.04)^t$$

$$2 = (1.04)^t$$

$$\log 2 = \log(1.04)^t = t \log 1.04 \quad \text{Rule 3}$$

$$t = \frac{\log 2}{\log 1.04}$$

$$\approx \frac{0.30103}{0.017} \approx 17.67$$

Thus, it takes nearly 18 years for this account to double (Fig. 1.80). ■

The table gives the doubling times for various annual interest rates for accounts that are compounding annually.

Interest Rate	Doubling Times in Years
4	17.67
5	14.21
6	11.90
7	10.24
8	9.01
9	8.04
10	7.27
15	4.96
20	3.80

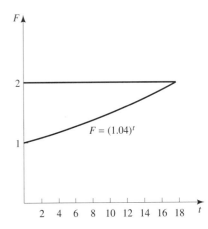

Figure 1.80
This account doubles in about 18 years.

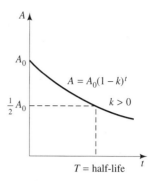

Figure 1.81
The half-life T is the time needed for the substance to decay to half its original size.

■ Half-Life

Half-life is the period of time for one-half of a radioactive or degradable substance to disappear. Half-lives of radioactive isotopes vary widely from seconds to billions of years. For example, the half-life of polonium-210 is 164 seconds, whereas thorium-232 has a half-life of 14,000,000,000 years. Dioxins, one of the most controversial chemicals discovered in the past 50 years, have a half-life of 10 years. Not much is known about the amount of dioxin in the environment because federal law doesn't require companies to report the chemical in annual toxic-release inventory reports. But the EPA estimates that about 30 pounds—20 pounds in the air and 10 pounds in water—are emitted nationwide each year. Dioxins are listed as a probable cause of human cancer and affect immune and reproductive systems.

Source: Melvin D. Joesten, David O. Johnston, John T. Netterville, et al., *World of Chemistry.* Saunders College Publishing, Philadelphia, PA, 1991; Linda Dono Reeves, "Dioxins Getting into Air, Water, Incinerators Emitting Possible Carcinogens," *Cincinnati Enquirer* September 17, 1994.

The interest rate i determines the rate at which the account is growing. The doubling times give an alternative way to assess the rate of growth.

In the previous section we considered the radioactive substance iodine-131 used in medical diagnosis. We found that if $A(t)$ is the amount after t days, then $A(t) = A_0(0.917)^t$, where A_0 is the initial amount. The decay constant is $k = 1 - 0.917 = 0.083$ and determines the rate at which the substance decays. An alternative way to assess the rate of decay is to determine the time until the substance decays to half of its original amount. This is called the **half-life** (Fig. 1.81).

EXAMPLE 7   Half-Life

A hospital orders 20 grams of iodine-131 and will need 10 grams within a week of ordering. Did they order enough? Use the half-life of iodine-131 to answer the question.

SOLUTION   ■   We can answer this question by finding the half-life of iodine-131. Let T be the half-life of iodine-131. Then $0.5A_0 = A(T) = A_0(0.917)^T$. We must solve for T in this equation. We do this by taking common logarithms at an appropriate point. We have

$$0.5A_0 = A_0(0.917)^T$$
$$0.5 = (0.917)^T$$
$$\log 0.5 = \log(0.917)^T$$
$$= T \log 0.917$$
$$T = \frac{\log 0.5}{\log 0.917}$$
$$\approx \frac{-0.30103}{-0.03763} \approx 8$$

Thus, the half-life of iodine-131 is about 8 days. As a consequence, 10 grams of the original 20 grams will be left in 8 days. Thus, more than 10 grams still remains in 7 days or less. So the hospital has ordered enough (Fig. 1.82).   ■

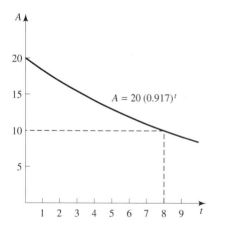

Figure 1.82
The half-life is about 8 days.

SELF-HELP EXERCISE SET 1.8

1. Solve for x in the equation $10^{x^3} = 5$.

2. Suppose inflation causes the value of a dollar to decrease by 3% a year. How long does it take for a dollar to be worth $0.75?

EXERCISE SET 1.8

In Exercises 1 through 6 solve for x.

1. $\log x = 2$

2. $\log x = -3$

3. $\log x = -\pi$

4. $\log x = -\dfrac{3}{4}$

5. $3 \log_3 x = 5$

6. $2 \log_2 x = 9$

In Exercises 7 through 16 simplify.

7. $\log 10^4$

8. $\log \sqrt{10}$

9. $\log \dfrac{1}{10}$

10. $\log \dfrac{1}{\sqrt{10}}$

11. $10^{\log 2\pi}$

12. $10^{\log \sqrt{2}}$

13. $5^{3 \log_5 2}$

14. $3^{0.5 \log_3 9}$

15. $(5^{\log_5 3})^4$

16. $\sqrt{3}^{\log_3 2}$

In Exercises 17 through 22 write the given quantity in terms of $\log x$, $\log y$, and $\log z$.

17. $\log x^2 \sqrt{yz}$

18. $\log \sqrt{xyz}$

19. $\log \dfrac{\sqrt{xy}}{z}$

20. $\log \dfrac{xy^2}{z^3}$

21. $\log \sqrt{\dfrac{xy}{z}}$

22. $\log \dfrac{x^2 y^3}{\sqrt{z}}$

In Exercises 23 through 28 write the given quantity as one logarithm.

23. $2 \log x + \log y$

24. $2 \log x - \log y$

25. $\dfrac{1}{2} \log x - \dfrac{1}{3} \log y$

26. $2 \log x - \dfrac{1}{2} \log y + \log z$

27. $3 \log x + \log y - \dfrac{1}{3} \log z$

28. $\sqrt{2} \log x - \log y$

In Exercises 29 through 40 solve the equation for x.

29. $5 \cdot 10^{5x} = 3$

30. $2 \cdot 10^{3x} = 5$

31. $6 = 2 \cdot 10^{-2x}$

32. $3 = 4 \cdot 10^{-0.5x}$

33. $2 \cdot 10^{3x-1} = 1$

34. $3 \cdot 10^{2-5x} = 4$

35. $10^{x^2} = 4$

36. $10^{\sqrt{x}} = 4$

37. $2 \log(x + 7) + 3 = 0$

38. $\log x^2 = 9$

39. $\log(\log 4x) = 0$

40. $\log 3x = \log 6$

In Exercises 41 through 43 use the basic graphing principles to sketch the graphs. Then support your answers using your graphing calculator.

41. $y = \log x$, $y = \log(x + 3)$, $y = \log(x - 3)$

42. $y = \log x$, $y = 2 \log x$, $y = 0.25 \log x$

43. $y = \log x$, $y = -\log x$

44. Solve the equation $10^x = -\log x$ to two decimal places by graphing $y = 10^x$ and $y = -\log x$ on the same screen and using the ZOOM feature.

45. Using your graphing calculator and the ZOOM feature find all solutions to $\log x = x^2 - 2$.

Applications

46. Doubling Time. If an account has an annual interest rate of 8% compounded monthly, how long is it before the account doubles?

47. Tripling Time. A population grows according to $P(t) = P_0(1.05)^t$. How long is it before the population triples?

48. Compounding. How long does it take for an account earning 7% per year and compounding annually to go from $1000 to $1800?

49. Compounding. An individual deposits $2000 into an account with an annual interest rate of 6% compounded annually and $1000 into an account with an annual interest rate of 9% compounded annually. How long will it take for the amount in the second account to equal the amount in the first account?

Enrichment Exercises

50. Establish the change of base theorem:

$$\log_a x = \frac{\log_b x}{\log_b a}$$

Hint: Start with $x = a^{\log_a x}$ and take a logarithm to the base b of each side.

51. Use the change of base theorem to find $\log_3 x$ in terms of $\log_{10} x$.

52. Determine the graph of $y = \log_a x$ in the case that $0 < a < 1$ in the same way as was done in the text for the case

$a > 1$. That is, use the fact that the graphs of $y = a^x$ and $y = \log_a x$ are symmetrical about the line $y = x$.

53. Can you define $\log_a x$ when $a = 1$? Explain why or why not.

54. Let $y_1 = \log x$ and $y_2 = x^{1/5}$. Determine which is larger for large values of x using your grapher.

55. Given any positive integer n, speculate on whether $y_1 = \log x$ or $y_2 = x^{1/n}$ is larger for large x. Experiment on your grapher to decide. What does this say about how slow $\log x$ is increasing?

SOLUTIONS TO SELF-HELP EXERCISE SET 1.8

1. Taking the logarithm of each side gives

$$\log 5 = \log 10^{x^3} = x^3$$

Thus, $x = \sqrt[3]{\log 5} \approx \sqrt[3]{0.69897} \approx 0.8875$.

2. The value $V(t)$ of the dollar is $V(t) = (0.97)^t$, where t is given in years. We must find t so that $0.75 = V(t) = (0.97)^t$. We have

$$0.75 = (0.97)^t$$
$$\log 0.75 = \log(0.97)^t$$
$$= t \log 0.97$$
$$t = \frac{\log 0.75}{\log 0.97} \approx 9.4$$

Thus, at this rate, a dollar will be reduced in value to $0.75 in about 9.4 years.

CHAPTER **1**

SUMMARY OUTLINE

- **Linear Cost, Revenue, and Profit Equations.** p. 20–21 Let x be the number of items made and sold.

   variable cost = cost per item × number of items produced = mx

   C = **cost** = variable cost + fixed cost = $mx + b$

   R = **revenue** = price per item × number sold = px

   P = **profit** = revenue − cost = $R - C$

- The quantity at which the profit is zero is called the **break-even quantity**. p. 22.

- Let x be the number of items made and sold and p the price of each item. A **linear demand equation**, which governs the behavior of the consumer, is of the form $p = mx + b$, where m must be negative. A **linear supply equation**, which governs the behavior of the producer, is of the form $p = mx + b$, where m must be positive. p. 23.

- The point at which supply equals demand is called the **equilibrium point**. The x-coordinate of the equilibrium point is called the **equilibrium quantity**, and the p-coordinate is called the **equilibrium price**. p. 23.

- In **straight-line depreciation** we assume that the value V of the asset is given by a *linear* equation in time t, say, $V = mt + b$, where the slope m must be *negative*. p. 24.

- A **function** f from the set D to the set R is a rule that assigns to each element x in D one and only one element $y = f(x)$ in R. p. 31.

- The set D above is called the **domain**.

- One thinks of the domain as the set of inputs and the values $y = f(x)$ as the outputs. p. 31.

- The set of all possible outputs is called the **range**. p. 31.

- The letter representing the elements in the domain is called the **independent variable**. p. 31.

- The letter representing the elements in the range is called the **dependent variable**. p. 31.

- The **graph** of the function f consists of all points (x, y) such that x is in the domain of f and $y = f(x)$. p. 33.

- **Vertical line test**: A graph in the xy-plane represents a function, if and only if every vertical line intersects the graph in at most one place. p. 35.

- The function $f(x) = ax^2 + bx + c$, $a \neq 0$, is called a **quadratic function**. p. 43.

- The graph of a quadratic function is called a **parabola**. p. 46.

- Every quadratic function $f(x) = ax^2 + bx + c$ with $a \neq 0$ can be put in **standard form** $f(x) = a(x - h)^2 + k$, where $h = -b/2a$ and $k = c - (b^2/4a)$ by **completing the square**. p. 46.

- The point (h, k) is called the **vertex**. p. 46.
- When $a < 0$, the graph of the quadratic $y = ax^2 + bx + c$ looks like ⌢ and assumes the maximum value k when $x = h$. p. 46.
- When $a > 0$, the graph of the quadratic $y = ax^2 + bx + c$ looks like ⌣ and assumes the minimum value k when $x = h$. p. 46.
- A **polynomial function** of degree n is a function of the form

$$f(x) = a_n x^n + a_{n-1} x^{n-1} + \cdots + a_1 x + a_0, \ a_n \neq 0.$$

  The domain is the set of all real numbers. The coefficient a_n is called the **leading coefficient**. p. 55.

- A **rational function** is the quotient of two polynomials. The domain is the set of all real numbers for which the denominator is not zero. p. 56.
- A **power function** is a function of the form $y = kx^r$, where r and k are any real numbers. p. 53.
- The **sum** $f + g$ is defined by $(f + g)(x) = f(x) + g(x)$. p. 62.
- The **difference** $f - g$ is defined by $(f - g)(x) = f(x) - g(x)$. p. 62.
- The **product** $f \cdot g$ is defined by $(f \cdot g)(x) = f(x) \cdot g(x)$. p. 62.
- The **quotient** f/g is defined by $(f/g)(x) = [f(x)/g(x)]$, if $g(x) \neq 0$. p. 62.
- The **composition** $f \circ g$ is defined by $(f \circ g)(x) = f[g(x)]$ for all x in the domain of g for which $g(x)$ is in the domain of f. p. 64.
- **Graphing Principles.** For $k > 0$ the graph of $y = f(x) + k$ is the graph of $y = f(x)$ shifted upward by k units, whereas the graph of $y = f(x) - k$ is the graph of $y = f(x)$ shifted downward by k units.

    For $c > 0$ the graph of $y = f(x - c)$ is the graph of $y = f(x)$ shifted to the right by c units, and the graph of $y = f(x + c)$ is the graph of $y = f(x)$ shifted to the left by c units.

    The graph of $y = -f(x)$ is the reflection of the graph of $y = f(x)$ across the x-axis. For $a > 0$, the graph of $y = af(x)$ is an expansion of the graph of $y = f(x)$ if $a > 1$ and a contraction if $0 < a < 1$.

- An **exponential** function is a function of the form $y = a^x$. p. 70.
- **Properties of a^x.** p. 70 and 72.

$a > 1$	$0 < a < 1$
Increasing on $(-\infty, \infty)$	Decreasing on $(-\infty, \infty)$
Domain $= (-\infty, \infty)$	
Range $= (0, \infty)$	
$a^x = a^y$ if and only if $x = y$	

- **Compound Interest.** Suppose a principal P earns interest at the annual rate of r, expressed as a decimal, and interest is compounded m times a year. Then the amount F after t years is $F = P(1 + i)^n = P(1 + [r/m])^{mt}$, where $n = mt$ is the number of time periods and $i = r/m$ is the interest per period. p. 74.
- **Effective Yield.** Suppose a sum of money is invested at an annual rate of r expressed as a decimal and is compounded m times a year. The effective yield r_e is $r_e = (1 + [r/m])^m - 1$. p. 76.
- **Present Value.** Suppose an account earns an annual rate of r expressed as a decimal and is compounded m times a year. Then the amount P, called the **present value**, needed presently in this account so that a future amount of F will be attained in t years is given by $P = \dfrac{F}{(1 + [r/m])^{mt}}$. p. 77.

- **Logarithm to Base a.** Let a be a positive number with $a \neq 1$. If $x > 0$, the logarithm to the base a of x, denoted by $\log_a x$, is defined as $y = \log_a x$ if, and only if, $x = a^y$. p. 81.

- **Properties of Logarithms** p. 82 and 83.

$$a^{\log_a x} = x \text{ if } x > 0 \qquad \log_a a^x = x \text{ for all } x$$

$$\log_a 1 = 0 \qquad \log_a xy = \log_a x + \log_a y$$

$$\log_a \frac{x}{y} = \log_a x - \log_a y \qquad \log_a x^c = c \log_a x$$

- If a quantity grows according to $P(t) = P_0(1 + k)^t$, where $k > 0$, k is called the **growth constant**. p. 70.

- If a quantity decays according to $A(t) = A_0(1 - k)^t$, where $k > 0$, k is called the **decay constant**. p. 72.

- The **half-life** of a radioactive substance is the time that must elapse for half of the material to decay. p. 86.

C H A P T E R **1**

REVIEW EXERCISES

In Exercises 1 through 4 plot the given point in the xy-plane.

1. $(2, 3)$ **2.** $(-2, -5)$

3. $(-3, 2)$ **4.** $(4, -5)$

5. Find the points (x, y) in the xy-plane that satisfy $xy \leq 0$.

6. Find the distance between the two points $(2, -3)$ and $(-3, -6)$.

7. Sketch the graph of $y^2 = 9x$.

8. Which of the points $(8, 9)$, $(-7, -6)$, $(9, 15)$, $(-11, 2)$ lie in the viewing window: $[-10, 10]$ by $[-5, 20]$.

9. Determine the smallest viewing window that contains the points $(8, 1)$, $(5, -5)$, $(-1, 3)$.

10. Which viewing window gives the better graph of $y = -x + 18$. (a) $[-10, 10]$ by $[-10, 10]$, (b) $[-10, 30]$ by $[-10, 30]$

11. Which viewing window gives the better graph of $y = e^{x^2 - x + 2}$. (a) $[-10, 10]$ by $[-10\ 10]$, (b) $[-2, 2]$ by $[0, 30]$

In Exercises 12 through 14 find the slope (if it exists) of the line through each pair of points. Sketch the line. Support your answer with a graphing utility in the case(s) when the slope exists.

12. $(-2, 4)$, $(1, -3)$ **13.** $(-2, -3)$, $(2, 3)$

14. $(-2, 5)$, $(-2, -3)$

In Exercises 15 through 17 find the slope (if it exists) of the lines. Find the x- and y-intercepts if they exist.

15. $y = 3x + 2$ **16.** $y = -2$

17. $x = 3$

18. Find the equation of the line through the point $(-2, 5)$ with slope -2.

19. Find the equation of the line through the two points $(-4, 2)$ and $(3, -5)$. Support your answer with a graphing utility.

20. Find the equation of the line through the point $(-3, -5)$ and parallel to the line $y = 3x - 2$.

21. Find the equation of the line through the point $(-3, 7)$ and perpendicular to the line $y = -4x + 5$.

22. Find the x- and y-intercepts of $(x/a) + (y/b) = 1$. If a and b are positive, sketch a graph.

In Exercises 23 through 28 determine which graphs are graphs of functions.

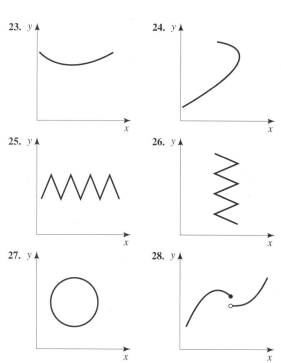

In Exercises 29 and 30 find the domain of each of the functions. Support your answer using a graphing calculator.

29. $\dfrac{x + 1}{x - 3}$

30. $\dfrac{x + 1}{\sqrt{x - 5}}$

31. Let $f(x) = \dfrac{1}{x^2 + 1}$. Evaluate $f(0)$, $f(\sqrt{2})$, $f(a - 1)$, $f(x + 1)$.

32. Graph

$$f(x) = \begin{cases} x^2 & \text{if } x \le 0 \\ x + 1 & \text{if } 0 < x \le 1 \\ -x + 2 & \text{if } x > 1 \end{cases}$$

In Exercises 33 through 39 sketch a graph and identify which graphing principles you are using. Support your sketch using a graphing calculator.

33. $y = -(x + 3)^2 + 5$

34. $y = |x - 2| - 1$

35. $y = 2(x - 1)^3 + 1$

36. $y = 0.5\sqrt[3]{x + 1} - 2$

37. $y = -(x - 1)^2$

38. $y = -\log(x - 4)$

39. $y = \log 10^x$

In Exercises 40 through 43 let $f(x) = x^3 + 4$ and $g(x) = 2x - 1$. Find the indicated quantity and the domains for each.

40. $(f + g)(x)$

41. $(f - g)(x)$

42. $(f \cdot g)(x)$

43. $\left(\dfrac{f}{g}\right)(x)$

In Exercises 44 and 45 find $(f \circ g)(x)$ and $(g \circ f)(x)$ for the given f and g, and find the domains for each.

44. $f(x) = x^2 + 2$; $g(x) = 3x + 1$

45. $f(x) = \sqrt{x - 1}$; $g(x) = x + 1$

Graph Exercises 46 through 51.

46. $y = 3^x$

47. $y = 3^{-x}$

48. $y = 10^{2x}$

49. $y = 10^{-2x}$

50. $y = 10^{|x|}$

51. $y = \log |x|$

In Exercises 52 through 57 solve for x.

52. $9^x = 27$

53. $\left(\dfrac{1}{3}\right)^x = \dfrac{1}{27}$

54. $\log x = 1$

55. $10^{4x} = 10^{8x-2}$

56. $\log 10^x = 4$

57. $10^{\log x} = 5$

58. Solve $10^{-x} = \log x$ to two decimal places using the ZOOM feature of your graphing calculator.

59. Cost. Assuming a linear cost model, find the equation for the cost C, where x is the number produced, the cost per item is $6, and the fixed costs are $2000.

60. Revenue. Assuming a linear revenue equation, find the revenue equation for R, where x is the number sold and the price per item is $10.

61. Profit. Assuming the cost and revenue equations in the previous two problems, find the profit equation. Also find the break-even quantity.

62. Profit. Given that the cost equation is $C = 5x + 3000$ and the revenue equation is $R = 25x$, find the break-even quantity.

63. Supply and Demand. Given the demand equation is $p = -2x + 4000$ and the supply equation is $p = x + 1000$, find the equilibrium point.

64. Transportation. A plane travels a straight path from O to A and then another straight path from A to B. How far does the plane travel? Refer to the figure.

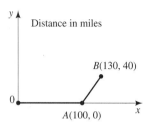

65. Transportation. If the plane in the previous exercise travels 200 miles per hour from O to A and 250 miles per hour from A to B how long does the trip take?

66. Demand. A company notes from experience that it can sell 100,000 pens at $1 each and 120,000 of the same pens at $0.90 each. Find the demand equation, assuming it is linear.

67. Profit. It costs a publisher $2 to produce each copy of a weekly magazine. The magazine sells for $2.50 a copy, and the publisher obtains advertising revenue equal to 30% of the revenue from sales. How many copies must be sold to obtain a profit of $15,000?

68. Nutrition. A certain woman needs 15 milligrams (mg) of iron in her diet each day but cannot eat fish or liver. She plans to obtain all of her iron from kidney beans (4.5 mg/cup) and soybeans (5.5 mg/cup). If x is the number of cups of kidney beans and y the number of cups of soybeans she consumes each day, what linear equation must x and y satisfy? Interpret the meaning of the slope of this equation.

69. Facility Location. A company is trying to decide whether to locate a new plant in Houston or Boston. Information on the two possible locations is given in the table. The initial investment is in land, buildings, and equipment. Suppose 10,000,000 items are produced each year.

	Houston	Boston
Variable cost ($)	0.25 per item	0.22 per item
Annual fixed costs ($)	4,000,000	4,200,000
Initial investment ($)	16,000,000	20,000,000

(a) Find which city has the lowest annual costs not counting the initial investment.

(b) Which city has the lowest total cost over five years counting the initial investment?

70. Quadruped Size. Let l denote the length of a quadruped measured from hip to shoulder and h denote the average height of the trunk of the body. From physics it is known that for a uniform bar with similar dimensions the ratio $l{:}h^{2/3}$ is limited by some value. If the ratio exceeds a certain value, the rod breaks. The table lists quadrupeds whose ratios are among the highest. Find the ratio and estimate the limiting value.

Quadruped	*l* (cm)	*h* (cm)	$l{:}h^{2/3}$
Ermine	12	4	
Dashshund	35	12	
Indian tiger	90	45	
Llama	122	73	
Indian elephant	153	135	

71. Loudness. Loudness is measured using the formula $L = 10 \log(I/I_0)$, where I_0 is the lowest intensity that can be heard. How much would L increase if we multiplied the intensity I by 10? By 100? By 1000?

72. Forestry. Klopzig [33] noted that a common symptom of red pine decline is a large circular area of dead trees (pocket), ringed by trees showing reduced diameter and height growth. The equation $y = 0.305 - 0.023x$, approximates this effect, where y is the proportion of dead roots and x is the distance from the pocket margin. Interpret what the slope of this line means.

73. Forestry. Harmer [34] indicated that the equation $y = 5.78 + 0.0275x$ approximated the relationship between shoot length x and bud number y for the plant *Quercus petraea*. Interpret what the slope of this line means.

74. Cost of Irrigation Water. Using an argument too complex to give here Tolley and Hastings [35] argued that if c is the cost in 1960 dollars per acre-foot of water in the area of Nebraska and x is the acre-feet of water available, then $c = 12$ when $x = 0$. They also noted that farms used about 2 acre-feet of water per acre in the Ainsworth area when this water was free. If we assume (as they did) that the relationship between c and x is linear, then find the equation that c and x must satisfy.

75. Agriculture. Tronstad and Gum [36] indicated that the weight w in pounds of a calf (at eight months) was related to the age t in years of its mother according to $w = 412.34 + 28.05t - 2.05t^2$. Determine the age of the mother that results in a maximum calf weight and determine this maximum. Support your answer with a graphing utility.

76. Agriculture. The yield response of Texas coastal bend grain sorghum to nitrogen fertilizer has been given by SriRamaratnam [37] as $y = 2102 + 42.4x - 0.205x^2$, where y is in pounds per acre and x is pounds of nitrogen per acre. Using a graphing utility determine the amount of nitrogen that results in a maximum yield, and find this maximum. Confirm your answer algebraically.

77. Forestry. To monitor the changes in forest growth, some estimate of leaf surface area y is needed. Blodwyn [38] gives the estimate $y = 0.6249x^{1.8280}$, where y is leaf surface area in square meters, and x is the (easy to measure) diameter at breast height in centimeters. Find y if $x = 20, 30, 45$. Graph.

78. Packaging. A box with top has length twice the width and height equal to the width. If x denotes the width and S the surface area, find S as a function of x.

79. Postage. The postage on a first class letter is \$0.32 for the first ounce, and \$0.23 for each additional ounce or any fraction thereof. If x denotes the number of ounces that a letter weighs and $C(x)$ the first class postage, graph the function $C(x)$ for $0 < x \le 3$. Plot $y = C(x)$ on your graphing calculator using the greatest integer function, if your calculator has this function. *Hint*: Set the MODE of your calculator to DOT. If int(x) is the greatest integer function, let $y_1 = -0.23\text{int}(-x) + 0.09$ on a window with dimensions [0, 4.7] by [0, 1]. Use TRACE to check your answer. Notice what happens as x goes from 0.95 to 1.00 to 1.05. Replace y_1 with $y_2 = 0.23\text{int}(x) + 0.32$. Graph. What is the difference between y_2 and y_1?

80. Spread of Infestation. A certain pine disease is spreading in a circular fashion through a pine forest with the radius of the infested region increasing at a rate of 2 miles per year. Write the area of the infested circular region as a function of time t measured in years.

81. Population. According to the U.S. Bureau of Statistics, the developing world is growing at a rate of 2% a year. If we assume exponential growth $P_0(1.02)^t$, how long until this population increases by 30%?

82. Population. According to the U.S. Bureau of Statistics, the population of India will double in 35 years. If we assume exponential growth $P_0(1 + k)^t$, what must be the growth constant k?

83. Population Growth. According to the U.S. Bureau of the Census, the population of Indonesia was 181 million in 1990 and growing at about 1.9% a year, whereas the population of the United States was 249 million in 1990 and growing at about 0.9% a year. Find when the populations of these two countries will be the same, assuming the two growth rates continue at the given values.

CHAPTER **1**

PROJECTS

Project 1
Price Discrimination

Economists have done extensive studies of price discrimination. Aside from the understanding of the workings of markets that these studies have advanced, additional knowledge is needed in order to enforce laws against illegal types of discrimination. In the type of price discrimination we consider here we assume that the seller can divide customers into two groups, each of which has its own demand function.

Suppose in Market A the demand equation is given by $p = 12 - 3x$ and in Market B by $p = 6 - x$, where x is a measure of the units of the commodity and p is the price in dollars of each unit (see the figure). The seller wants to ensure that the last unit of output sold in Market A adds the same amount to total revenue as the last unit sold in Market B. To accomplish this, we must sum the two demand curves horizontally to obtain the combined curve shown in the figure.

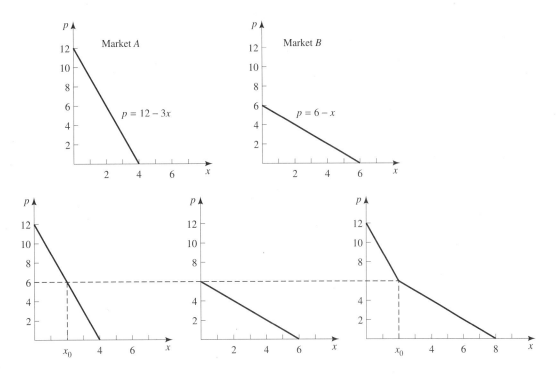

(a) If the seller wants to sell x_0 units indicated in the figure, then as the figure shows no units will be sold in Market B and all x_0 will be sold in Market A. Find x_0. Then for $0 \leq x \leq x_0$, the combined demand curve is $p = 12 - 3x$, the same as in Market A.

(b) Find the combined demand for $x \geq x_0$ by adding horizontally and then solving for p.

(c) Now suppose the seller wishes to sell three units in the combined market. What prices should be charged in each of the two markets?

Project 2
Portfolio Management

A fundamental tenet in the modern theory of finance is that an investor must assume greater risks to obtain larger returns. In general, the greater the risk, the larger the return. In the theory of finance the "risk" of an asset is defined to be a measure of the variability of the asset. The basis of this definition rests on the observation that the greater the risk of an asset the greater the daily fluctuations in the value of the asset.

If everyone has a great deal of confidence in the future value of an investment, then the value of the investment should vary little. For example, everyone has essentially complete confidence in the future performance of money in the bank. Thus, the value of this asset (the principal) does not vary in time. This reflects the fact that this investment is (essentially) risk-free. Everyone has very high confidence in the future performance of a high-quality bond. There is some variability in price, however, because investors change their assessment somewhat from week to week of the economy, inflation, and other factors that may affect the price of the bond. This reflects the fact that a bond does have some risk.

Moving on to stocks, the future earnings and dividend performance of a very large and regulated company such as AT&T is estimated by many analysts with reasonable confidence, and so the price of this security does not vary significantly. The future earnings of a small biotech firm is likely to be very unpredictable, however. Will it be able to come up with new products? Will it be able to fend off competitors? Will some key personnel leave? Thus, the weekly assessment of such a firm is likely to change substantially more than that of AT&T. As a consequence, the weekly price changes of this company will vary significantly more, reflecting the fact that an investment in the small company is riskier.

In the figure the letter M refers to the Market, defined as, say, all stocks listed on the New York Stock Exchange. The number r_M is the risk of the Market portfolio. The height R_T of the point T indicates the riskless rate of return of Treasury bills. Long-term studies indicate this rate is about 4% per year on average and the return R_M on the Market portfolio is about 9% per year on average. Thus, the point T is placed lower than the point M.

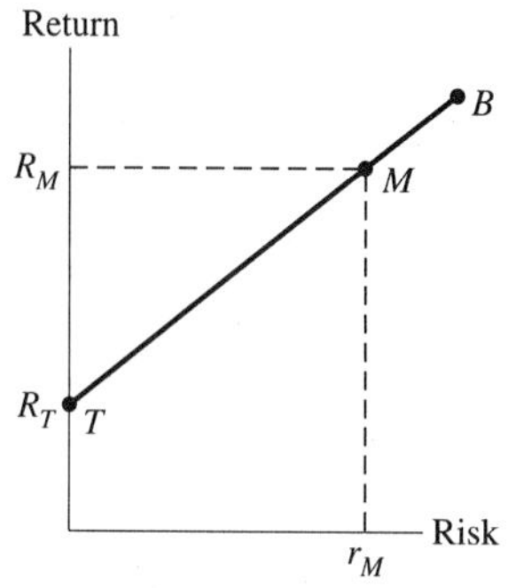

The line TMB is called the **capital market line** and plays a critical role in portfolio analysis. A person can construct a portfolio by investing a fraction of the funds in the Market portfolio and the remainder in Treasury bills. Such a portfolio is called a *lending portfolio.* Also a person can place all the funds in the Market portfolio, then using this as collateral, borrow money, and invest the borrowed funds in the Market portfolio. Such a portfolio is called a *leverage portfolio.* For convenience we assume that the rate charged on the borrowed money in the leveraged portfolio is R_T.

It turns out that the curve that relates the risk to the return of lending and leveraged portfolios is a straight line through the points T and M. To understand this fully requires a more careful definition of risk and additional mathematics beyond the scope of this text. Interested readers can refer to *The Stock Market: Theories and Evidence* by Lorie and Hamilton, an excellent text that presents the results of the academic research into the stock market.

(a) Label the risk axis as r and the return axis as R, and find the equation of the line TMB.
(b) Find R if $R_M = 0.09$, $R_T = 0.04$, and $r = 2r_M$.
(c) Find R if $R_M = 0.09$, $R_T = 0.04$, and $r = 0.5r_M$.
(d) Find R if $R_M = -0.10$, $R_T = 0.04$, and $r = 2r_M$.
(e) Find R if $R_M = -0.10$, $R_T = 0.04$, and $r = 0.5r_M$.

What conclusion can you make for lending and leveraged portfolios in rising markets? Falling markets?

Project 3
Diversity

Buoniorno and associates [39] gave the index of tree size diversity

$$H = -[p_1 \ln p_1 + p_2 \ln p_2 + \cdots + p_n \ln p_n]$$

where p_i is the proportion of trees in the ith size class.

(a) Is this index negative? Why or why not?
(b) Explain why this index does or does not equal

$$p_1 \ln \frac{1}{p_1} + p_2 \ln \frac{1}{p_2} + \cdots + p_n \ln \frac{1}{p_n}$$

(c) What is the value of this index if all the p_i's are equal?
(d) Suppose $n = 2$, $p_1 = p$, and $p_2 = 1 - p$. What happens to H if p is close to 0? To 1? For what value of p is the diversity a minimum? Use your graphing calculator to answer these questions. Explain why your answers make sense in view of the fact that H measures diversity.

Project 4
Island Biogeography

A fundamental result in island biogeography is that the number of species N in island faunas as a whole is given by the equation $N = aA^b$, where A is the area of the island. One of the first pieces of empirical evidence was given by Preston in 1960. The following figure gives his data in graphical form and was taken from MacArthur and Wilson [40]. The vertical axis gives the number of species of herpetofauna (amphibians plus reptiles). Notice that each axis is marked off in *powers* of 10. The indicated line does not go exactly through each data point, but is judged to be the line ''closest'' to the data points. Clearly, the dots very nearly fall on a line. (See Section 7.5 on least squares to see how this line is found.) From the graph, determine the equation of the indicated line. Use this to determine the parameters a and b in the equation $N = aA^b$. You will need to take logarithms of this last equation.

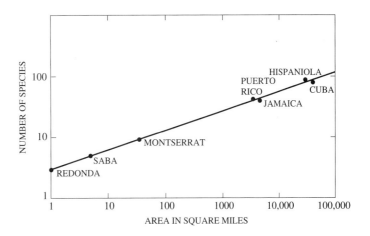

CHAPTER 2

Limits and the Derivative

This chapter begins the study of calculus. The notions of limits and continuity are discussed first. Then the derivative is introduced with applications.

2.0 INTRODUCTION TO CALCULUS

Isaac Newton   1642–1727

Newton is considered the inventor of the calculus. He also made very significant discoveries in the sciences and is often considered the most accomplished scientific thinker of recent centuries. He attended public schools in England, where he was undistinguished. Indeed he was reported to be idle and inattentive. He blossomed quickly, however, and, while confining himself to his room for two years from 1665–1667 in order to avoid the plague that was raging at that time, he laid the foundations for his future accomplishments. During these extremely productive years he set down the basis for calculus, discovered that white light can be split into different colors, and discovered the law of universal gravitation. Any one of these discoveries would have made him an outstanding figure in the history of science.

Gottfried Leibniz   1646–1716

Leibniz is considered the coinventor of the calculus. He was a precocious child and mentioned once that at age 14 he would go for walks in the woods, "comparing and contrasting the principles of Aristotle with those of Democritus." He invented a calculating machine that could multiply and divide. He commented on the machine that "It is unworthy of excellent men to lose hours like slaves in the labor of calculation." He received a degree in law and served in the diplomatic service, only later taking up mathematics. He developed much of the currently used notation in the calculus and a number of the elementary formulas.

Except possibly for the creation of Euclidean geometry, the calculus represents the greatest creation in all of mathematics. Armed with the calculus, major scientific problems of the 17th century were solved. Even today, the use of the calculus continues to spread into new areas, such as business, economics, and biology.

Four major 17th-century problems were solved by the calculus. The first was to find a means of calculating the (instantaneous) velocity of a moving body. To everyone who walks, runs, rides a horse, or flies in a plane (Fig. 2.1), it is physically apparent that a moving body has an instantaneous velocity. If the function $s = s(t)$ gives the distance a body has moved in the time t, how can one calculate the (instantaneous) velocity if the velocity is different at different times? One can calculate the average velocity as distance divided by time, but since no distance or time elapses in an instant, this formula results in the meaningless expression $\dfrac{0}{0}$ if used to calculate the instantaneous velocity.

The second problem was to find the slope of the line that is tangent to a curve at a given point (Fig. 2.2). This problem was not just a problem in pure geometry but was also of considerable scientific interest in the 17th century. At that time men such as Pierre de Fermat and Christiaan Huygens were interested in the construction of lenses, which required a knowledge of the angle between the ray of light and the normal line to the lens shown in Figure 2.3. Since the normal line is perpendicular to the tangent line, knowledge of the tangent line immediately gives the normal line.

Actually, a serious problem arose about exactly what tangent meant. For example, looking at Figure 2.4, it is easy to agree that a tangent to a figure, such as a circle, is that line that touches the figure at only one point and lies entirely on one side of the figure. This was known by the Greeks. But more complex curves such as that shown in Figure 2.5 were being considered in the 17th century. Is the line shown in this figure a tangent? (We will see later that it is.)

The third problem was to find the maximum or minimum value of a function on some given interval. A contemporary example of such a problem is finding the point at which maximum profits occur (Fig. 2.6).

The fourth type of problem was to find the area under a curve (Fig. 2.7). We will see in a later chapter why this is so important in the applications.

(North Wind Picture Archive)

Figure 2.1
The velocity of an airplane may change from instant to instant.

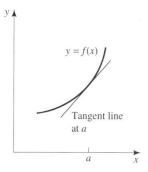

Figure 2.2
Finding the slope of the tangent line was a problem of major importance.

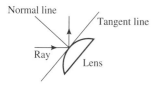

Figure 2.3
Making lenses requires knowledge of the tangent line.

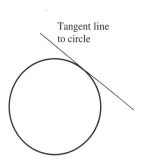

Figure 2.4
It is clear what a tangent line to a circle is.

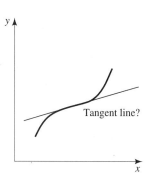

Figure 2.5
Do we want to call this line a tangent line? (We do.)

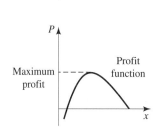

Figure 2.6
We want to find maxima and minima of functions.

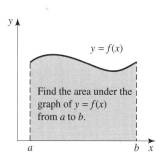

Figure 2.7
We want to find the area between the graph of $y = f(x)$ and the x-axis.

In the first problem considered here, velocity measures the rate at which distance is changing with respect to a change in time. In exact analogy, the calculus enables us to determine, among other things, the rate at which profits are changing with respect to the changes in the number of items produced and sold, the rate at which a population is growing with respect to changes in the time, and the rate at which demand for an item drops with respect to increases in its price. Thus, we will see that the calculus can be used not only to solve the great scientific problems of the past centuries but also to solve many important problems in modern business, economics, and biology.

2.1 LIMITS

■ Introduction to Limits ■ Some Limits That Do Not Exist ■ Rules for Limits

Augustin Louis Cauchy 1789–1857

It is interesting to note that Newton and Leibniz, who are given credit for the discovery of the calculus, did *not* develop the theory on the basis of a clear and logical foundation. Indeed they, and the many individuals that expanded the calculus in the subsequent 150 years, used arguments that were intuitively plausible rather than mathematically exact. From the beginning, the calculus, as presented, evoked considerable controversy, with many critics pointing out (correctly) that several fundamental concepts were vague. The success of the calculus in solving many physical problems is what gave individuals the confidence that the mathematics must somehow be correct. It was Cauchy who finally placed the calculus on a sound footing. He developed an acceptable theory of limits and used this theory as the logical underpinnings of the calculus. Thus, Cauchy has become known as the creator of the modern calculus.

(The Bettman Archive)

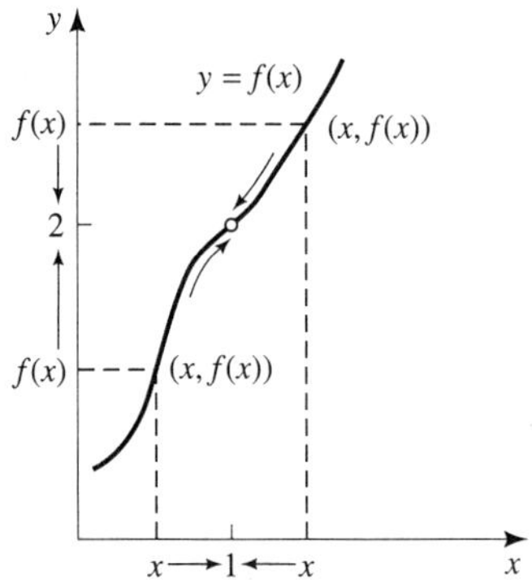

Figure 2.8
No matter how $x \to 1$, $f(x) \to 2$.
Thus, $\lim\limits_{x \to 1} f(x) = 2$.

APPLICATION: ADVERTISING COST AND MARKET SHARE

A large corporation already has cornered 30% of the toothpaste market and now wishes to undertake an extensive advertising campaign to increase its share of this market. It is estimated that the advertising cost $C(x)$, in billions of dollars, of attaining an $x\%$ share of this market is given by $C(x) = \dfrac{10}{100 - x}$. What does it cost to attain a 90% share? 99%? 99.9%? 99.99%? What happens to $C(x)$ as x approaches 100? Does this make sense? See Example 7 for the answer.

Introduction to Limits

The foundation of the calculus rests on the notion of limit. To understand the fundamental ideas in the calculus we need an intuitive understanding of limits.

Toward this end consider the graph shown in Figure 2.8 of some function $y = f(x)$. Notice that the function is not defined for the value $x = 1$. We seek to understand what happens to $f(x)$ as x approaches (gets closer and closer to) 1 without x actually attaining the value 1.

First notice that it is meaningless to have x actually attain the value 1 since $f(1)$ is not even defined. In Figure 2.8 we can see that as x approaches 1 from the right, the point $(x, f(x))$ slides along the graph and approaches the point $(1, 2)$ and $f(x)$ approaches the number 2. Also, we can see that as x approaches 1 from the left, the point $(x, f(x))$ slides along the graph and approaches the point $(1, 2)$, and $f(x)$ again approaches the number 2. We then see that no matter how x approaches 1, $f(x)$ approaches the same number $L = 2$.

In such a case we say that $L = 2$ is the **limit** of $f(x)$ as x approaches 1 and write

$$\lim_{x \to 1} f(x) = 2$$

We now give the following intuitive definition of limit of a function.

Definition of Limit of a Function

Suppose that the function $f(x)$ is defined for all values of x near a, but not necessarily at a. If as x approaches a (without actually attaining the value a), $f(x)$ approaches the number L, then we say that L is the limit of $f(x)$ as x approaches a, and write

$$\lim_{x \to a} f(x) = L$$

We say that this definition is intuitive because the definition does not define the term ''approaches'' in a strictly mathematical sense.

Remark. Notice that in this definition, $f(a)$ need not be defined. Furthermore, even if $f(a)$ is defined, the limit process never requires knowledge of $f(a)$, since this process only requires knowledge of $f(x)$ for x near a but not *equal* to a. Thus, even if $f(a)$ is defined, this value has nothing to do with the limit as x approaches a.

In determining the limit of $f(x)$ as x approaches 1, we find it convenient to consider the limit of $f(x)$ as x approaches 1 from the left and also the limit of $f(x)$ as x approaches 1 from the right. These two limits are written as

$$\lim_{x \to 1^-} f(x) \qquad \lim_{x \to 1^+} f(x)$$

respectively. In the preceding example, we discovered that both of these limits exist and equal the very same number, $L = 2$. We then conclude that the limit of $f(x)$ as x

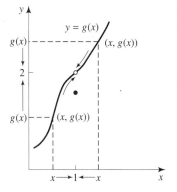

Figure 2.9
$\lim\limits_{x \to 1} g(x) = 2$ even though $g(1) =$ 1.5. The value of $g(x)$ at $x = 1$ is irrelevant to the limit as $x \to 1$.

approaches 1 exists and equals $L = 2$. This reasoning is true in general. We then have the following theorem:

Existence of a Limit

If the function $f(x)$ is defined near $x = a$ but not necessarily at $x = a$, then

$$\lim_{x \to a} f(x) = L$$

if and only if both the limits

$$\lim_{x \to a^-} f(x) \qquad \text{and} \qquad \lim_{x \to a^+} f(x)$$

exist and are equal to the same number L.

The following example should further clarify the idea of the limit:

EXAMPLE 1   Finding the Limit Graphically

Define the function $g(x)$ to be the very same function as $f(x)$, except that $g(1) = 1.5$ (Fig. 2.9). Find $\lim\limits_{x \to 1} g(x)$ if it exists.

SOLUTION ■ Notice from Figure 2.9 that we still have

$$\lim_{x \to 1^-} g(x) = 2 \qquad \text{and} \qquad \lim_{x \to 1^+} g(x) = 2$$

Thus, $\lim\limits_{x \to 1} g(x) = 2$. ■

Remark. Notice that this last statement is true despite the fact that $g(1) = 1.5$, since the limit process does not involve the value of the function at $x = 1$.

The two functions $f(x)$ and $g(x)$ given here are not actually the same function because they differ for the single value $x = 1$. They are equal for all other values, however, and therefore the limits as x approaches 1 of these two functions must be equal.

We now look at two limits that we can find using our intuitive knowledge of limits. In each case we support our answer both numerically and graphically.

EXAMPLE 2   Finding the Limit Intuitively, Supporting It Numerically and Graphically

Find $\lim\limits_{x \to 2} (3x + 1)$ if it exists.

SOLUTION ■ It seems intuitive that as x approaches 2, $3x$ approaches 6, and $3x + 1$ approaches $6 + 1 = 7$. This is supported by the calculations shown in Table 2.1.

Furthermore, the graph $y = 3x + 1$ is a straight line and is shown in Figure 2.10. We can see that as x approaches 2, y seems to approach 7. Thus $\lim\limits_{x \to 2} (3x + 1) = 7$. ■

TABLE 2.1

x	1.9	1.99	1.999	$\to$	2	$\leftarrow$	2.001	2.01	2.1
3x + 1	6.7	6.97	6.997	$\to$	7	$\leftarrow$	7.003	7.03	7.3

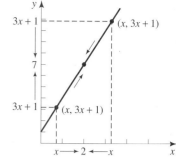

Figure 2.10
As $x \to 2$, $3x + 1 \to 3(2) + 1 = 7$. Thus, $\lim\limits_{x \to 2} (3x + 1) = 7$.

[−10, 10] by [−10, 10]

Screen 2.1
From the graph of $y = 3x^2 - 5x - 2$ we see that $y \to 0$ as $x \to 2$.

EXAMPLE 3 Finding the Limit Intuitively, Supporting It Graphically and Numerically

Find $\lim\limits_{x \to 2} (3x^2 - 5x - 2)$ based on your intuitive understanding of limits. Support your answer graphically and numerically.

SOLUTION ■ It seems intuitive that as x approaches 2, $x^2 = x \cdot x$ approaches $2 \cdot 2 = 4$, $3x^2$ approaches $3 \cdot 4 = 12$, and $3x^2 - 5x - 2$ approaches $12 - 5(2) - 2 = 0$.

The graph $y = 3x^2 - 5x - 2$ is shown in Screen 2.1, using the standard viewing window. Using the TRACE and the ZOOM features, if we wish, we can see that as x approaches 2, y seems to approach 0.

This is also supported by the numerical calculations shown in Table 2.2.
Thus, $\lim\limits_{x \to 2} (3x^2 - 5x - 2) = 0$. ■

The limits in Examples 3 and 4 were rather transparent. The following limit is not so obvious.

EXAMPLE 4 Finding the Limit Using Cancellation

Find $\lim\limits_{x \to 2} \dfrac{3x^2 - 5x - 2}{x - 2}$ if it exists.

SOLUTION ■ First notice that the function $f(x)$ is not defined at $x = 2$, because setting $x = 2$ in the fraction yields 0/0. We saw in Example 3 that the numerator approaches 0 as x approaches 2. On the other hand, the denominator, $x - 2$, also approaches 0 as x approaches 2. So the fraction approaches 0/0. At this point it is not clear if the limit exists.

To get some idea of what is happening as x approaches 2, we should evaluate the function for various values of x. The results are shown in Table 2.3.

TABLE 2.2

x	1.9	1.99	1.999	→	2	←	2.001	2.01	2.1
$3x^2 - 5x - 2$	−0.670	−0.070	−0.007	→	0	←	0.007	0.0703	0.730

TABLE 2.3

x	1.9	1.99	1.999	→	2	←	2.001	2.01	2.1
$f(x)$	6.7	6.97	6.997	→	7	←	7.003	7.03	7.3

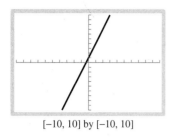

[−10, 10] by [−10, 10]

Screen 2.2
From the graph of $y = (3x^2 - 5x - 2)/(x - 2)$ we see that $y \to 7$ as $x \to 2$.

As Table 2.3 indicates, $f(x)$ seems to approach 7 as x approaches $x = 2$ from the right and also seems to approach 7 as x approaches $x = 2$ from the left.

Furthermore, Screen 2.2 shows a graph of $y = \dfrac{3x^2 - 5x - 2}{x - 2}$, using the standard window. From this graph it appears also that as x approaches 2, y approaches 7. Thus, we have every reason to suspect that the limit is equal to 7.

We can confirm this suspicion algebraically. First notice that the numerator can be factored as

$$3x^2 - 5x - 2 = (3x + 1)(x - 2)$$

Therefore,

$$f(x) = \frac{(3x + 1)(x - 2)}{x - 2} = 3x + 1 \quad \text{for } x \neq 2$$

Then

$$\lim_{x \to 2} \frac{3x^2 - 5x - 2}{x - 2} = \lim_{x \to 2} \frac{(3x + 1)(x - 2)}{x - 2} \quad \text{factor}$$

$$= \lim_{x \to 2} (3x + 1) \quad \text{cancel}$$

$$= 7 \quad \text{Example 2}$$

where we can cancel because the limit does not involve the value $x = 2$.   ∎

EXPLORATION 1

Find Graphically, Confirm Algebraically

Find $\lim\limits_{x \to 1} \left(\dfrac{2x^2 - x - 1}{x - 1} \right)$ graphically, if it exists. Confirm algebraically.

We present another limit that is not at all transparent. Although we will be able to give strong graphical and numerical support for the limit in the next example, we do not have the necessary mathematical tools at this time to confirm our answer algebraically.

EXAMPLE 5   Finding a Limit Graphically, Supporting It Numerically

Find $\lim\limits_{x \to 0^+} x^x$.

SOLUTION ∎ You probably do not have the vaguest idea of what the limit might be, or even if it exists. Using a screen with dimensions [0, 2] by [0, 4] we obtain Screen 2.3, which seems to indicate that 1 is the limit of $y = x^x$ as x goes to 0^+. You can use the ZOOM feature to give further support to this conclusion.

The table lends numerical support. Thus, we very strongly suspect, but are unable to prove, that $\lim\limits_{x \to 0^+} x^x = 1$.   ∎

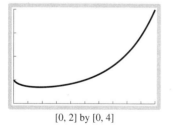

[0, 2] by [0, 4]

Screen 2.3
From the graph of $y = x^x$ we see that $y \to 1$ as $x \to 0^+$.

x	0	←	0.0001	0.001	0.01	0.1	1
y	1	←	0.99908	0.99312	0.95499	0.79433	1

Some Limits That Do Not Exist

We now look at two cases for which limits do not exist.

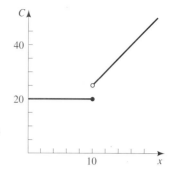

Figure 2.11

$\lim\limits_{x\to10^-} C(x) = 20$ while $\lim\limits_{x\to10^+} C(x) = 25$. Because these two limits are not equal, $\lim\limits_{x\to10} C(x)$ does not exist.

EXAMPLE 6   Different Limits from Each Side

A furniture store charges \$20 to deliver a couch to your house if you live 10 miles or less from the store and charges \$2.50 per mile from the store if you live more than 10 miles from the store. Find the function $C(x)$ that gives the delivery cost, where x is measured in miles. Find $\lim\limits_{x\to10} C(x)$, if the limit exists.

SOLUTION ■ The function $C(x)$ is given by

$$C(x) = \begin{cases} 20 & \text{if } x \le 10 \\ 2.5x & \text{if } x > 10 \end{cases}$$

This function is graphed in Figure 2.11. Notice that

$$\lim\limits_{x\to10^-} C(x) = \lim\limits_{x\to10^-} 20$$
$$= 20$$

$$\lim\limits_{x\to10^+} C(x) = \lim\limits_{x\to10^+} 2.5x$$
$$= 25$$

Because the limits from each side are not the same, $\lim\limits_{x\to10} C(x)$ does not exist. ■

EXAMPLE 7   Limit Does Not Exist if the Function Is Unbounded

A large corporation already has cornered 30% of the toothpaste market and now wishes to undertake an extensive advertising campaign to increase its share of this market. It is estimated that the advertising cost $C(x)$, in billions of dollars, of attaining an $x\%$ share of this market is given by $C(x) = \dfrac{10}{100 - x}$. What does it cost to attain a 90% share? 99%? 99.9%? 99.99%? What happens to $C(x)$ as $x \to 100^-$? Does this make sense?

SOLUTION ■ In Table 2.4, we see the result of evaluating $C(x)$ at values of x that approach 100 from the left. The values of the function $C(x)$ get large without bound and therefore do not approach any particular number L. From Table 2.4 we can construct Figure 2.12. Thus, $\lim\limits_{x\to100^-} C(x)$ does not exist.

TABLE 2.4

x	90	99	99.9	99.99	$\to$	100
$C(x)$	1	10	100	1000	$\to$	?

■ **Superbowl Advertising**

The 1996 Superbowl marks the 12th year of Superbowls as advertising events. No other advertising promotions cost as much as those done during the Superbowl. Production costs over a \$1 million and air time for the 1994 Superbowl was listed at \$900,000 for each of the 56 30-second spots. The most notorious campaign failure in marketing history was kicked off during the 1986 Superbowl. This 6-week promotional campaign featuring Burger King's "Herb the Nerd" cost \$40 million and produced no increase in sales.

Source: Stuart Elliott "Super Triumphs and Super Flops," *New York Times*, January 30, 1994.

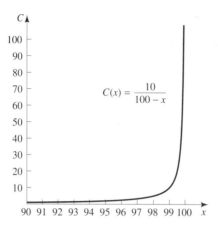

Figure 2.12
As $x \to 100^-$, $C(x)$ becomes large without bound. Thus, $\lim\limits_{x \to 100^-} C(x)$ does not exist.

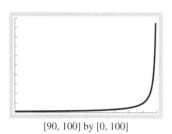

[90, 100] by [0, 100]

Screen 2.4
From the graph of $y = 10/(100 - x)$ we see that as $x \to 100^-$, y becomes large without bound.

Using a screen with dimensions [90, 100] by [0, 100] we obtain the graph shown in Screen 2.4, which supports the previous conclusion. The graph makes sense. Some people are fiercely loyal to other brands of toothpaste: for example, the employees of the competing manufacturers of toothpaste. The amount of advertising needed to win over these holdouts will be therefore astronomical. ∎

Whenever the graph of a curve, such as that in Figure 2.12, approaches a vertical line $x = x_0$, we say that $x = x_0$ is a **vertical asymptote**.

Remark. We have now seen two basic ways in which a limit at $x = a$ may not exist.

1. If a function "blows up" near $x = a$, such as in Figure 2.12 for $a = 100$.
2. If the limits from each side of $x = a$ exist but are unequal, as in Figure 2.11 at $a = 10$.

Rules for Limits

We have just seen that graphs and tables can be used to find limits or to determine that limits do not exist. Rules concerning limits can be used to evaluate limits efficiently, however. We now give a number of such rules. For convenience, let us use the notation $x \to a$ to represent the phrase "x approaches a" and $f(x) \to L$ to represent the phrase "$f(x)$ approaches L."

First consider $\lim\limits_{x \to a} p(x)$, where $p(x)$ is a polynomial. Take, for example, the polynomial in Example 3, $p(x) = 3x^2 - 5x - 2$, $a = 2$. It seems intuitive that as $x \to 2$, $p(x) \to 3(2)^2 - 5(2) - 2$, which is $p(2)$. Thus, $\lim\limits_{x \to 2} p(x) = p(2)$, which means we can evaluate $\lim\limits_{x \to 2} p(x)$ by *substituting* 2 for x in $p(x)$. This same reasoning applies to any polynomial function.

The Limit of a Polynomial Function

If $p(x)$ is any polynomial and a is any number, then $\lim\limits_{x \to a} p(x) = p(a)$.

Let us now consider some other rules. Let $f(x)$ and $g(x)$ be two functions, and suppose that as $x \to a$, $f(x) \to L$ and $g(x) \to M$. Then it is reasonable that $cf(x) \to cL$, $f(x) \pm g(x) \to L \pm M$, $f(x)g(x) \to LM$, $\dfrac{f(x)}{g(x)} \to L/M$ (if $M \neq 0$), and $[f(x)]^n \to L^n$ when L^n makes sense and $L \neq 0$. This then gives the following five rules:

Assume that $\lim_{x \to a} f(x) = L$ and $\lim_{x \to a} g(x) = M$. Then

Rule 1. $\lim_{x \to a} cf(x) = c \lim_{x \to a} f(x) = cL.$

Rule 2. $\lim_{x \to a} (f(x) \pm g(x)) = \lim_{x \to a} f(x) \pm \lim_{x \to a} g(x) = L \pm M.$

Rule 3. $\lim_{x \to a} (f(x) \cdot g(x)) = \left(\lim_{x \to a} f(x) \right) \cdot \left(\lim_{x \to a} g(x) \right) = L \cdot M.$

Rule 4. $\lim_{x \to a} \dfrac{f(x)}{g(x)} = \dfrac{\lim_{x \to a} f(x)}{\lim_{x \to a} g(x)} = \dfrac{L}{M}$ if $\lim_{x \to a} g(x) = M \neq 0.$

Rule 5. $\lim_{x \to a} (f(x))^n = L^n,$ n any real number, L^n defined, $L \neq 0.$

Remark. The following is a verbal interpretation of the above-mentioned rules:

1. The limit of a constant times a function is the constant times the limit of the function.
2. The limit of the sum or difference equals the sum or difference of the limits.
3. The limit of the product is the product of the limits.
4. The limit of a quotient is the quotient of the limits if the limit of the denominator is not zero.
5. The limit of a function raised to a power is the limit of the function raised to the power, provided the terms make sense.

Remark. In Rule 5, L^n is not defined if for example $L < 0$ and $n = \frac{1}{2}, \frac{1}{4},$ and so forth, or if $L = 0$ and $n \leq 0$. Refer to Exercise 52 for further comments on Rule 5.

Consider also the following, If $f(x) = c$, then since $f(x)$ is always the constant c, the limit as $x \to a$ must also be c. Thus, $\lim_{x \to a} c = c$. Furthermore, it is intuitively clear that $\lim_{x \to a} x = a$.

EXAMPLE 8   Using the Rules of Limits

Evaluate

(a) $\lim_{x \to 3} \dfrac{x^2 + 11}{x^2 - 4}$ (b) $\lim_{x \to 1} (x^2 - 9)^{1/3}$ (c) $\lim_{x \to 4} \dfrac{x^2 - 16}{x^2 - 2x - 8}$

SOLUTIONS ▪

(a) $\lim_{x \to 3} \dfrac{x^2 + 11}{x^2 - 4} = \dfrac{\lim_{x \to 3} (x^2 + 11)}{\lim_{x \to 3} (x^2 - 4)}$ Rule 4

$= \dfrac{(3)^2 + 11}{(3)^2 - 4} = \dfrac{20}{5} = 4$ limit of polynomial functions

(b) $\lim_{x \to 1} (x^2 - 9)^{1/3} = \left(\lim_{x \to 1} (x^2 - 9) \right)^{1/3}$ Rule 5

$= (-8)^{1/3} = -2$ limit of polynomial functions

(c) Using Rule 4 and substitution gives $\dfrac{0}{0}$. We therefore must try to factor the numerator and denominator to find a common factor.

$$\lim_{x \to 4} \dfrac{x^2 - 16}{x^2 - 2x - 8} = \lim_{x \to 4} \dfrac{(x + 4)(x - 4)}{(x + 2)(x - 4)}$$ factor

$$= \lim_{x \to 4} \frac{(x + 4)}{(x + 2)} \qquad \text{cancel}$$

$$= \frac{\lim_{x \to 4} (x + 4)}{\lim_{x \to 4} (x + 2)} \qquad \text{Rule 4}$$

$$= \frac{4 + 4}{4 + 2} = \frac{8}{6} = \frac{4}{3} \qquad \text{limit of polynomial functions} \qquad \blacksquare$$

EXAMPLE 9   Finding a Limit Graphically

Graphically find $\lim_{x \to 1} (\sqrt{x} - 1)/(x - 1)$.

SOLUTION ■ First notice that both the numerator and denominator are heading for 0 as $x \to 1$. Thus, we cannot use Rule 4. Using a window of dimension [0, 2] by [0, 1], we obtain Screen 2.5, from which it appears that the limit is 0.5. ■

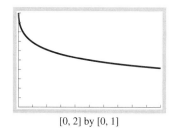

[0, 2] by [0, 1]

Screen 2.5
From the graph of $y = (\sqrt{x} - 1)/(x - 1)$ we see that $y \to 0.5$ as $x \to 1$.

EXPLORATION 2

Confirm Example 9 Algebraically

Confirm algebraically the limit found in Example 9. *Hint:* Write

$$\frac{\sqrt{x} - 1}{x - 1} = \frac{\sqrt{x} - 1}{x - 1} \cdot \frac{\sqrt{x} + 1}{\sqrt{x} + 1}$$

simplify the right-hand side, and use the rules of limits.

SELF-HELP EXERCISE SET 2.1

1. For the function in the figure, find $\lim_{x \to a^-} f(x)$, $\lim_{x \to a^+} f(x)$, $\lim_{x \to a} f(x)$ for $a = 1, 2, 3, 4$.

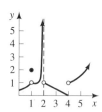

2. Find $\lim_{x \to 1^-} f(x)$, $\lim_{x \to 1^+} f(x)$, $\lim_{x \to 1} f(x)$ if

$$f(x) = \begin{cases} x^4 - x + 1 & \text{if } x < 1 \\ x^2 & \text{if } x > 1 \end{cases}$$

3. Find $\lim_{x \to 3} \dfrac{x^2 + x - 12}{x - 3}$.

4. Find $\lim_{x \to 2} \sqrt{\dfrac{x^2 - 8}{x^3 - 9}}$

EXERCISE SET 2.1

In Exercises 1 through 6 refer to the figure. Find $\lim_{x \to a^-} f(x)$, $\lim_{x \to a^+} f(x)$, $\lim_{x \to a} f(x)$ at the indicated value for the function given in the figure.

1. $a = -1$

2. $a = 0$

3. $a = 1$

4. $a = 2$

5. $a = 3$

6. $a = 4$

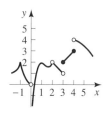

In Exercises 7 through 13, fill in the table, and then estimate the limit of the given function at the given value or decide the limit does not exist. Support your answer on a graphing calculator.

7. $a = 1$ and $f(x) = 3x + 2$.

x	0.9	0.99	0.999	→	1	←	1.001	1.01	1.1
f(x)				→	?	←			

8. $a = 2$ and $f(x) = 2x + 3$.

x	1.9	1.99	1.999	→	2	←	2.001	2.01	2.1
f(x)				→	?	←			

9. $a = 2$ and $f(x) = x^3 - 4$.

x	1.9	1.99	1.999	→	2	←	2.001	2.01	2.1
f(x)				→	?	←			

10. $a = -1$ and $f(x) = \dfrac{|x + 1|}{x + 1}$.

x	-1.1	-1.01	-1.001	→	-1	←	-0.999	-0.99	-0.9
f(x)				→	?	←			

11. $a = 2$ and $f(x) = \dfrac{1}{x - 2}$.

x	1.9	1.99	1.999	→	2	←	2.001	2.01	2.1
f(x)				→	?	←			

12. $a = 4$ and $f(x) = \dfrac{\sqrt{x} - 2}{x - 2}$.

x	3.9	3.99	3.999	→	4	←	4.001	4.01	4.1
f(x)				→	?	←			

13. $a = 4$ and $f(x) = \dfrac{x - 4}{\sqrt{x} - 2}$.

x	3.9	3.99	3.999	→	4	←	4.001	4.01	4.1
f(x)				→	?	←			

In Exercises 14 through 25 find $\lim_{x \to a^-} f(x)$, $\lim_{x \to a^+} f(x)$, and $\lim_{x \to a} f(x)$, at the indicated value for the indicated function. Do *not* use a graphing calculator.

14. $a = 1$, $f(x) = \begin{cases} x & \text{if } x < 1 \\ 2 & \text{if } x = 1 \\ x - 1 & \text{if } x > 1 \end{cases}$

15. $a = 0$, $f(x) = \begin{cases} x^4 - x + 1 & \text{if } x < 0 \\ x^2 - x & \text{if } x \geq 0 \end{cases}$

16. $a = 1$, $f(x) = \begin{cases} x^3 - x + 1 & \text{if } x < 1 \\ x^4 + x^2 - 1 & \text{if } x > 1 \end{cases}$

17. $a = 1$, $f(x) = \begin{cases} x^2 & \text{if } x < 1 \\ 2 & \text{if } x = 1 \\ x & \text{if } x > 1 \end{cases}$

18. $a = 1$, $f(x) = \begin{cases} -x + 2 & \text{if } x < 1 \\ 0 & \text{if } x = 1 \\ x^2 & \text{if } x > 1 \end{cases}$

19. $a = 0$, $f(x) = \begin{cases} x & \text{if } x < 0 \\ \dfrac{1}{x} & \text{if } x > 0 \end{cases}$

20. $a = 2$, $f(x) = \dfrac{x - 2}{|x - 2|}$ **21.** $a = -1$, $f(x) = \dfrac{x + 1}{|x + 1|}$

22. $a = 0$, $f(x) = \dfrac{1}{x^2}$ **23.** $a = 1$, $f(x) = \dfrac{1}{x - 1}$

24. $a = 1$, $f(x) = \begin{cases} -x + 1 & \text{if } x < 1 \\ \dfrac{1}{1 - x} & \text{if } x > 1 \end{cases}$

25. $a = 1$, $f(x) = \begin{cases} x^5 + x^4 + x^2 + 1 & \text{if } x < 1 \\ \dfrac{1}{x - 1} & \text{if } x > 1 \end{cases}$

In Exercises 26 through 36 use the rules of limits to find the indicated limits if they exist. Support your answer using a graphing calculator.

26. $\lim_{x \to 1} \pi$ **27.** $\lim_{x \to 0} 13$

28. $\lim_{x \to -1} (3x - 2)$ **29.** $\lim_{x \to 2} (3x^3 - 2x^2 - 2x + 1)$

30. $\lim_{x \to 0} \dfrac{x}{x^2 + x + 1}$ **31.** $\lim_{x \to 1} \dfrac{x^2 + 2}{x^2 - 2}$

32. $\lim_{x \to 2} \dfrac{x + 3}{x - 3}(2x^2 - 1)$ **33.** $\lim_{x \to 1} \dfrac{\sqrt{x}}{x - 3} \cdot \dfrac{x + 2}{x^2 + 1}$

34. $\lim_{x \to 2} \dfrac{x^2 + 3}{x - 1}$ **35.** $\lim_{x \to 2} \sqrt{x^2 - 3}$

36. $\lim_{x \to 0} \sqrt[3]{x - 8}$

In Exercises 37 through 46 find the limits graphically, then confirm algebraically.

37. $\lim_{x \to 2} \dfrac{x^2 - 4}{x - 2}$ **38.** $\lim_{x \to 2} \dfrac{x^2 - x - 2}{x - 2}$

39. $\lim_{x \to 1} \dfrac{x^2 - 1}{x - 1}$ **40.** $\lim_{x \to -1} \dfrac{x^2 + 3x + 2}{x + 1}$

41. $\lim_{x \to -1} \dfrac{x^2 - 1}{x + 1}$ **42.** $\lim_{x \to -2} \dfrac{x^2 + 4}{x + 2}$

43. $\lim_{x \to 0} \dfrac{x + 1}{x^2 + x}$ **44.** $\lim_{x \to 0} \dfrac{x^3 + x}{x}$

45. $\lim_{x \to 4} \dfrac{x - 4}{\sqrt{x} - 2}$ **46.** $\lim_{h \to 0} \dfrac{(2 + h)^2 - 4}{h}$

Applications

47. Postal Rates. First-class postage in 1995 was \$0.32 for the first ounce (or any fraction thereof) and \$0.23 for each additional ounce (or fraction thereof). If $C(x)$ is the cost of postage for a letter weighing x ounces, then for $x \le 3$,

$$C(x) = \begin{cases} 0.32 & \text{if } 0 < x \le 1 \\ 0.55 & \text{if } 1 < x \le 2 \\ 0.78 & \text{if } 2 < x \le 3 \end{cases}$$

Graph and find
(a) $\lim_{x \to 1^-} C(x)$, $\lim_{x \to 1^+} C(x)$, and $\lim_{x \to 1} C(x)$.
(b) $\lim_{x \to 1.5^-} C(x)$, $\lim_{x \to 1.5^+} C(x)$, and $\lim_{x \to 1.5} C(x)$.

48. Average Cost. An electric utility company estimates that the cost $C(x)$ of producing x kilowatts of electricity is given by $C(x) = a + bx$, where a represents the fixed costs and

b represents the unit costs. The *average* cost $\overline{C}(x)$ is defined as the total cost of producing x kilowatts divided by x, that is, $\overline{C}(x) = \dfrac{C(x)}{x}$. What happens to the average cost if the amount of kilowatts produced x heads for zero? That is, find $\lim_{x \to 0^+} \overline{C}(x)$. How do you interpret your answer in economic terms?

49. Cost. It is estimated that the cost $C(x)$, in billions of dollars, of maintaining the atmosphere in the United States at an average $x\%$ free of a chemical toxin is given by $C(x) = 10/(100 - x)$. Find the cost of maintaining the atmosphere at levels 90%, 99%, 99.9%, and 99.99% free of this toxin. What happens as $x \to 100^-$? Does your answer make sense?

Enrichment Exercises

50. Determine graphically and numerically if any of the following limits exist or not.
(a) $\lim_{x \to 0^-} (x \cdot 3^{1/x})$ **(b)** $\lim_{x \to 0^+} (x \cdot 3^{1/x})$
(c) $\lim_{x \to 0} (x \cdot 3^{1/x})$

51. Determine graphically and numerically if $\lim_{x \to 0} (2^{-|x|})$ exists or not. If it exists, what is the limit?

52. Rule 5 on limits states that $\lim_{x \to a} (f(x))^n = L^n$, where n is any real number, L^n defined, $L \ne 0$. This exercise explores the possibility that this rule is true if $L = 0$.
(a) Is it true that $\lim_{x \to 0} \sqrt{x^2} = \sqrt{0} = 0$?
(b) Explain why $\lim_{x \to 1^-} \sqrt{x - 1}$ does not exist.
(c) Explain why $\lim_{x \to 0} \sqrt{-x^2}$ does not exist.

53. Let $f(x) = \dfrac{|x|}{x}$ if $x \ne 0$. Is it possible to define $f(0)$ so that $\lim_{x \to 0} f(x)$ exists? Why or why not?

54. Suppose $y = f(x)$ is defined on $(-1, 1)$. Of what importance is knowledge of $f(0)$ to finding $\lim_{x \to 0} f(x)$? Explain.

55. The text mentions two fundamentally different ways in which a limit may not exist. Illustrate these two ways by drawing graphs of two functions.

56. Is it true that if $\lim_{x \to 0} f(x) = 0$ and $\lim_{x \to 0} g(x) = 0$, then $\lim_{x \to 0} [f(x)/g(x)]$ does *not* exist? Explain why this is true, or give an example that shows it is not true.

57. Is it true that if both $\lim_{x \to a} f(x)$ and $\lim_{x \to a} g(x)$ do *not* exist, then $\lim_{x \to a} (f(x) + g(x))$ does *not* exist? Explain why this is true, or give an example that shows it is not true.

58. Finance. The figure shown here appeared in the Business Review for the Federal Reserve Bank of Philadelphia [41].

Explain why you think the figure is drawn so that the curve has a vertical asymptote at z, which is the maximum amount of available space.

59. Average Production. In a study on the output of small farms in Kenya, Carter and Wiebe [42] produced the graph shown here based on their empirical study. Explain why it is reasonable that the graph is drawn asymptotic to $x = 0$.

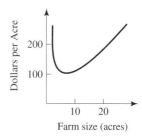

60. Medicine. Two rules have been suggested by Cowling and also by Young to adjust adult drug dosage levels for young children. Let a denote the adult dosage, and let t be the age (in years) of the child. The two rules are given respectively by

$$C = \frac{t + 1}{24} a \qquad \text{and} \qquad Y = \frac{t}{t + 12} a$$

Find the limit as $t \to 0^+$ for both of these. Which seems most realistic for a newborn baby? Why?

SOLUTIONS TO SELF-HELP EXERCISE SET 2.1

1. From the graph

$$\lim_{x \to 1^-} f(x) = 1, \qquad \lim_{x \to 1^+} f(x) = 1, \qquad \lim_{x \to 1} f(x) = 1$$

$$\lim_{x \to 2^-} f(x) \text{ does not exist}, \qquad \lim_{x \to 2^+} f(x) = 1,$$

$$\lim_{x \to 2} f(x) \text{ does not exist} \qquad \lim_{x \to 3^-} f(x) = \frac{1}{2},$$

$$\lim_{x \to 3^+} f(x) = \frac{1}{2}, \qquad \lim_{x \to 3} f(x) = \frac{1}{2} \qquad \lim_{x \to 4^-} f(x) = 0,$$

$$\lim_{x \to 4^+} f(x) = 1, \qquad \lim_{x \to 4} f(x) \text{ does not exist}$$

2. We have

$$\lim_{x \to 1^-} f(x) = \lim_{x \to 1^-} (x^4 - x + 1)$$
$$= 1 - 1 + 1 = 1$$
$$\lim_{x \to 1^+} f(x) = \lim_{x \to 1^+} x^2 = 1$$

Because the limit from the left equals the limit from the right, we then have $\lim_{x \to 1} f(x) = 1$.

3. We have

$$\lim_{x \to 3} \frac{x^2 + x - 12}{x - 3} = \lim_{x \to 3} \frac{(x + 4)(x - 3)}{x - 3}$$
$$= \lim_{x \to 3} (x + 4) = 7$$

4.

$$\lim_{x \to 2} \sqrt{\frac{x^2 - 8}{x^3 - 9}} = \sqrt{\lim_{x \to 2} \frac{x^2 - 8}{x^3 - 9}} \qquad \text{Rule 5}$$

$$= \sqrt{\frac{\lim_{x \to 2} (x^2 - 8)}{\lim_{x \to 2} (x^3 - 9)}} \qquad \text{Rule 4}$$

$$= \sqrt{\frac{-4}{-1}} = 2 \qquad \begin{array}{l}\text{limit of polynomial}\\\text{functions}\end{array}$$

2.2 CONTINUITY, LIMITS AT INFINITY

- Continuity ■ Irrational Powers ■ Limits at Infinity ■ Some Notation
- The Base e and Continuous Compounding ■ The Natural Exponent and Logarithm

APPLICATION: COMPOUND INTEREST

We noted in Section 1.7 that for an account earning compound interest, the more we compound each year, the larger the amount at the end of each year. For example, $1000 deposited into an account that earns 6% annually becomes $1061.36 in one year when compounded quarterly and the larger amount of $1061.68 when compounded monthly. Suppose we compound monthly, then daily, then by the hour, then by the minute, then by the second, and so on indefinitely. What happens to the amount in the account at the end of a year? For the answer see Example 3 and the discussion that follows.

■ **World Population and the Food Supply**

Currently it is estimated that the world population will swell in the next 50 years from 5.3 billion in 1990, to more than 10 billion by 2050. A primary concern is the effect this rate of population growth will have on the world's food supply. Factors that will need to be carefully measured include the rise of urbanization, soil erosion, and global warming. Some environmentalists believe, however, that potential problems with the food supply can be avoided if developing countries take three steps: expanding the amount of cultivated land, extending the growing seasons, and decreasing the estimated 40% of fresh food that is damaged or wasted each year.

Source: Scientific American March 1994, pp. 36–42.

APPLICATION: A POPULATION MODEL FOR THE UNITED STATES

The equation $p(t) = 197.27/(1 + 49.58e^{-0.03134(t - 1790)})$ was used about 150 years ago by the mathematician-biologist P. Verhulst to model the population $p(t)$ in millions of the United States, where t is the calendar year and $e \approx 2.71828$. What does this model predict will happen to the population in the long-term? See Example 5 for the answer.

Continuity

We already noticed that we can calculate limits of polynomials by using substitution. Thus, if $p(x)$ is a polynomial, then $\lim_{x \to a} p(x) = p(a)$. Any function with this property is said to be **continuous** at $x = a$.

Definition of a Function Continuous at a Point

The function $f(x)$ is **continuous** at $x = a$ if $\lim\limits_{x \to a} f(x) = f(a)$.

The rules for continuity are very similar to those for limits. For example, the quotient of two continuous functions is continuous at points at which the denominator is not zero. This follows immediately from the rule for the limit of a quotient. We can use this rule to determine at which points a rational function is continuous. Because a rational function is a quotient of polynomials and because a polynomial is continuous at all points, a rational function is continuous at every point at which the denominator is not zero. We then have the following:

Continuity of Polynomial and Rational Functions

- A polynomial function is continuous everywhere.
- A rational function is continuous at every point at which the denominator is not zero.

It is helpful to notice that the definition of continuity implies that a function is continuous at $x = a$ only if

1. $\lim\limits_{x \to a} f(x)$ exists.
2. $f(a)$ is defined.
3. $\lim\limits_{x \to a} f(x) = f(a)$.

E X A M P L E 1   Continuity

Determine the points at which the function whose graph is shown in Figure 2.13 is not continuous.

S O L U T I O N ■ Since $f(1)$ is not defined, the function is not continuous at $x = 1$.

We notice that $f(2)$ is defined and $f(2) = 3$. Also $\lim\limits_{x \to 2} f(x)$ exists and equals 2. However, $f(x)$ is not continuous at $x = 2$ since $\lim\limits_{x \to 2} f(x) = 2 \neq 3 = f(2)$.

Finally, $f(x)$ is not continuous at $x = 3$ and $x = 4$, because neither $\lim\limits_{x \to 3} f(x)$ nor $\lim\limits_{x \to 4} f(x)$ exist. ■

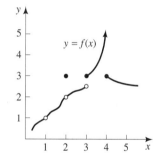

Figure 2.13
$y = f(x)$ is not continuous at $x = 1, 2, 3,$ and 4.

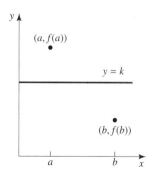

Figure 2.14
If $y = f(x)$ is continuous on $[a, b]$, the graph of $y = f(x)$ must cross the line $y = k$.

Notice from Figure 2.13 that the function is not continuous at precisely the points where the graph has ''holes,'' or ''gaps.'' This is true in general. Thus, a function is continuous at a point on an interval if the graph of the function *has no hole or gap* there.

We now consider a particularly important property of continuous functions that is used extensively in the text.

Suppose $f(x)$ is continuous at every point on an open interval that contains a and b and that $f(a) > k$ and $f(b) < k$ (Fig. 2.14). Can you draw a graph that connects the two points $(a, f(a))$ and $(b, f(b))$ with*out* touching the line $y = k$? If the graph does not touch the line $y = k$, then the graph must ''jump'' over this line and thus create a hole, or gap, in the graph. This cannot happen since the function is continuous at all points between a and b. Thus, an intermediate point $c \in (a, b)$ must exist so that $f(c) = k$. This is called the *intermediate value theorem* and can be used to establish the following theorem:

Constant Sign Theorem

> If $f(x)$ is continuous on (a, b) and $f(x) \neq 0$ for any x in (a, b), then either $f(x) > 0$ for all x in (a, b) or $f(x) < 0$ for all x in (a, b).

If the conclusion were not true, then $f(x)$ would be positive at one point and negative at another. Thus, the graph of $y = f(x)$ is above the x-axis ($y = 0$) at the first point and below the x-axis at the second. From the preceding discussion, we know that the graph of $y = f(x)$ must cross the line $y = 0$, and thus $f(x)$ must be zero at some intermediate point. But this contradicts the hypothesis that $f(x)$ is never zero on (a, b).

This theorem gives us a practical way (called a sign chart) to determine the sign of a continuous function over an interval. We use sign charts in Chapter 4.

Irrational Powers

If a is a positive number and r is a rational number, that is, $r = m/n$, where m and n are integers, then we know that

$$a^r = a^{m/n} = \sqrt[n]{a^m}$$

This simply means the number a is multiplied by itself m times to obtain a number, say y, and then the nth root of y is found. What happens, however, if we wish to take a^x, where x is an irrational number? In this case, it is impossible to write x as the quotient of integers. How can we give a meaning, then, to the expression a^x? For example, it is known that π is an irrational number. What does the expression 2^{π} mean?

To answer this question we first note that every irrational number has a decimal expansion. For example $\pi = 3.14159 \cdots$ and the decimal is infinite.

The sequence of numbers 3, 3.1, 3.14, 3.141, 3.1415, 3.14159, 3.141592, 3.1415927, . . . are all *rational* and have π as a limit. Thus, in the first place, it makes perfectly good sense to raise 2 to any of these powers. Doing this we obtain the sequence 2^3, $2^{3.1}$, $2^{3.14}$, $2^{3.141}$, $2^{3.1415}$, $2^{3.14159}$, $2^{3.141592}$, $2^{3.1415927}$, . . . , or to six decimal places, 8, 8.574188, 8.815241, 8.821353, 8.824411, 8.824962, 8.824974, 8.824978, . . . It can be shown that this sequence has a limit. We then *define* 2^{π} to be this limit. It is in this manner that we can define what is meant by a^x when x is a real and irrational number. Furthermore, this definition assures us that the function a^x, so defined, is **continuous**.

With a^x defined now for all real numbers, it can be shown that all the usual rules for exponents are still satisfied.

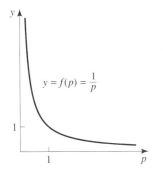

$y = f(p) = \dfrac{1}{p}$

Figure 2.15
As $p \to \infty$, $f(p) \to 0$. Thus,
$\lim\limits_{p \to \infty} f(p) = 0$.

Limits at Infinity

Suppose we are given a demand equation $f(p) = \dfrac{1}{p}$, where p is the price of an item and $f(p)$ is the number of items sold at a price equal to p. Let us see what happens to f as p becomes very large.

From Table 2.5, as p becomes larger, $f(p)$ approaches zero. In such a case we say that the *limit of $f(p)$ as p goes to ∞ is zero* and write $\lim\limits_{p \to \infty} 1/p = 0$.

This makes sense. If we set the price very high, then the number of items sold should be very small. From Table 2.5 we can construct Figure 2.15. Also using a window of dimension $[0, 10]$ by $[-0.2, 1]$ gives Screen 2.6.

We have the following definitions:

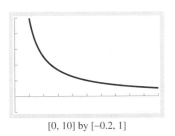

$[0, 10]$ by $[-0.2, 1]$

Screen 2.6
As x becomes larger, the graph of $y = 1/x$ approaches the x-axis.

TABLE 2.5

p	1	10	100	1000	$\to$	∞
$1/p$	1	0.1	0.01	0.001	$\to$	0

Definition of Limits at Infinity

1. If $f(x)$ approaches the number L as x becomes large without bound, then we say that L is the limit of $f(x)$ as x approaches ∞, and we write

$$\lim_{x \to \infty} f(x) = L$$

2. If $f(x)$ approaches the number K as x becomes a large negative number without bound, then we say that K is the limit of $f(x)$ as x approaches $-\infty$, and we write

$$\lim_{x \to -\infty} f(x) = K$$

We refer to the lines $y = L$ and $y = K$ as **horizontal asymptotes**.

Thus, for $y = f(p) = 1/p$, $y = 0$ is a horizontal asymptote.

EXAMPLE 2   Insect Infestation

Asante [43] showed that the proportion y of apple trees infested with the wooly apple aphid was approximated by

$$y = 0.33 \, \frac{x}{x + 41.30}$$

where x is the aphid density measured by the average number of aphids per tree. What proportion of trees are infested as the aphid density increases without bound?

SOLUTION ■ To determine the limit as $x \to \infty$ in a rational expression, we need to *divide the numerator and denominator by x to the highest power present.* In this case, this means dividing numerator and denominator by x. Because limits as $x \to \infty$ satisfy the usual rules for limits, we obtain

$$\lim_{x \to \infty} \left(0.33 \, \frac{x}{x + 41.30} \right) = \lim_{x \to \infty} \left(0.33 \, \frac{x}{x + 41.30} \cdot \frac{1/x}{1/x} \right)$$

$$= 0.33 \lim_{x \to \infty} \left(\frac{1}{1 + \dfrac{41.30}{x}} \right)$$

$$= 0.33 \, \frac{1}{1 + \lim_{x \to \infty} \left(\dfrac{41.30}{x} \right)}$$

$$= 0.33 \, \frac{1}{1 + 0}$$

$$= 0.33$$

Thus, about one-third of the trees become infested as the aphid density increases without bound. ■

EXPLORATION *1*

Graphical Support for Example 2

Graph the function given in Example 2 on your graphing calculator, and use the TRACE feature to support the limit found there.

Some Notation

The values of the function $f(x) = x^2$ become large without bound as x becomes large without bound. We use the notation $\lim_{x \to \infty} x^2 = \infty$ to describe this situation. This notation is not to be construed to suggest that the limit exists. It does not. We can also make corresponding definitions for limits when $x \to -\infty$. For example, with this notation, we have $\lim\limits_{x \to -\infty} x^3 = -\infty$.

In a similar fashion, we use the notation $\lim\limits_{x \to 0^+} \dfrac{1}{x} = \infty$ to describe the fact that $\dfrac{1}{x}$ becomes large without bound when x approaches 0 from the right and $\lim\limits_{x \to 0^-} \dfrac{1}{x} = -\infty$ to describe the fact that $\dfrac{1}{x}$ becomes negatively large without bound when x approaches 0 from the left.

The Base e and Continuous Compounding

We noted in Section 1.7 that for an account earning compound interest, the more we compound each year, the larger the amount at the end of each year. For example, $1000 deposited into an account that earns 6% annually becomes $1061.36 in one year when compounded quarterly and the larger amount of $1061.68 when compounded monthly. Suppose we compound monthly, then daily, then by the hour, then by the minute, then by the second, and so on indefinitely. What happens to the amount in the account at the end of a year? We refer to this process as **continuous compounding**.

Recall from Section 1.7 that if an account with an initial principal P and an annual rate of r, where r is expressed as a decimal, is compounded m times a year, then the future amount in the account after t years is

$$F(t) = P\left(1 + \frac{r}{m}\right)^{mt}$$

EXAMPLE 3   Compounding over Smaller and Smaller Time Intervals

Suppose $1000 is deposited into an account that earns 6% annually. Find the principal at the end of the first year if compounded

(a) annually (b) monthly (c) weekly (d) daily (e) hourly (f) by the minute

SOLUTIONS ▪ In all cases $P = \$1000$, $r = 0.06$, and $t = 1$.

(a) Here $m = 1$. Thus, we have

$$F = P(1 + r) = \$1000(1.06) = \$1060.00.$$

(b) Here $m = 12$. Thus, using a calculator, we have

$$F = P\left(1 + \frac{r}{m}\right)^m = \$1000\left(1 + \frac{0.06}{12}\right)^{12} = \$1061.68$$

(c) Here $m = 52$. Thus,

$$F = P\left(1 + \frac{r}{m}\right)^m = \$1000\left(1 + \frac{0.06}{52}\right)^{52} = \$1061.80$$

(d) Here $m = 365$. Thus,

$$F = P\left(1 + \frac{r}{m}\right)^m = \$1000\left(1 + \frac{0.06}{365}\right)^{365} = \$1061.83$$

(e) Here $m = 8760$. Thus,

$$F = P\left(1 + \frac{r}{m}\right)^m = \$1000\left(1 + \frac{0.06}{8760}\right)^{8760} = \$1061.84$$

TABLE 2.6

Compounded	Amount At End of 1 Year
yearly	$1060.00
monthly	1061.68
weekly	1061.80
daily	1061.83
hourly	1061.84
by the minute	1061.84

(f) Here $m = 525,600$. Thus,

$$F = P\left(1 + \frac{r}{m}\right)^m = \$1000\left(1 + \frac{0.06}{525,600}\right)^{525,600} = \$1061.84$$

All of this is summarized in Table 2.6. ∎

Thus, as the number of compoundings per year increases, the amount of money in the account seems to be heading for a definite limit.

We then need to analyze in general what happens as the number of compoundings increases without bound, that is, as $m \to \infty$.

To do this set $n = m/r$ in the preceding formula for F. This gives $m = rn$ and $r/m = 1/n$. Then

$$F(n) = P\left(1 + \frac{r}{m}\right)^{mt}$$
$$= P\left(1 + \frac{1}{n}\right)^{rnt}$$
$$= P\left[\left(1 + \frac{1}{n}\right)^n\right]^{rt}$$

Since $n = m/r$, as the number m of compoundings increases without bound, that is, $m \to \infty$, then also $n \to \infty$. Thus, we wish to find $\lim_{x\to\infty}\left(1 + \frac{1}{n}\right)^n$. Table 2.7 indicates that this limit is 2.71828 (to five decimal places). This number is of such importance the letter e is set aside for it.

The Number e

The number e is defined to be $e = \lim_{n\to\infty}\left(1 + \frac{1}{n}\right)^n$, and $e = 2.71828\ldots$

Using this in the preceding formula for $F(n)$ then yields

$$\lim_{n\to\infty} F(n) = \lim_{n\to\infty} P\left[\left(1 + \frac{1}{n}\right)^n\right]^{rt}$$
$$= P\left[\lim_{n\to\infty}\left(1 + \frac{1}{n}\right)^n\right]^{rt}$$
$$= P[e]^{rt}$$
$$= Pe^{rt}$$

We refer to this as *continuous* compounding. The compounding is not done just every day, or every hour, or every minute, or every second, but *continuously*.

TABLE 2.7

n	10	100	1000	10,000	100,000	1,000,000
$\left(1 + \frac{1}{n}\right)^n$	2.593742	2.704814	2.716924	2.718146	2.718268	2.718280

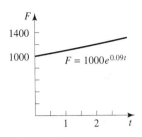

Figure 2.16
$1000 at 9% a year compounded continuously grows according to $F = 1000e^{0.09t}$.

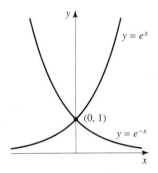

Figure 2.17
Since $e > 1$, $y = e^x$ increases, whereas $y = e^{-x}$ decreases. Both graphs go through the point (0, 1).

EXPLORATION 2

Graphical Support for e

Using your graphing calculator, graph $y = [1 + (1/x)]^x$ on a screen with dimensions [0, 1000] by [2.71, 2.72]. Use TRACE to see what is happening for large x.

EXAMPLE 4   Finding Amounts in a Continuously Compounded Account

Suppose $1000 is invested at an annual rate of 9% compounded continuously. How much is in the account after (a) 1 year (b) 3 years?

SOLUTIONS ■

(a) Here $P = \$1000$, $r = 0.09$, and $n = 1$. Thus, using a calculator, we have

$$F = Pe^{rt} = \$1000e^{0.09} = \$1094.17$$

(b) Here $P = \$1000$, $r = 0.09$, and $n = 3$. Thus,

$$F = Pe^{rt} = \$1000e^{(0.09)(3)} = \$1309.96$$

Refer to Figure 2.16.   ■

We can also determine the effective yield and present value for continuous compounding in the same way as we did in Section 1.7. This is done in the exercises.

Since $e > 1$, we know from Section 1.7 that e^x is an increasing function. Furthermore, since $2 < e < 3$, the graph of e^x lies between the graphs of 2^x and 3^x. The graphs of $y = e^x$ and $y = e^{-x}$ are given in Figure 2.17.

EXPLORATION 3

Solving for Time

Using a graphing calculator, find the time for the amount in the account in Example 2 to reach $1500. Graph on the same screen $y_1 = 1000e^{0.09x}$ and $y_2 = 1500$. Use screen dimensions of [0, 5] by [1000, 2000]. Then use the ZOOM feature to find the point of intersection.

The natural exponential function e^{kx} also arises in many areas of science. We now consider one example.

EXAMPLE 5   Population of the United States in the Long-Term

The equation

$$p(t) = \frac{197.27}{1 + 49.58e^{-0.03134(t - 1790)}}$$

was used about 150 years ago by the mathematician-biologist P. Verhulst to model the population $p(t)$ of the United States in millions, where t is the calendar year. Find the population this model predicts for 1880, 1930, and 1990. Compare with the actual population of 50.2, 123.2, and 250. What does this model predict will happen to the population in the long-term?

SOLUTION ■ We have

$$p(1880) = 49.9$$

$$p(1930) = 122.0$$

$$p(1990) = 180.3$$

Thus, this model gave extraordinarily accurate projections of the population of the United States for over 100 years after being formulated. Only in recent years has the model gone a bit astray. To determine the long-term prediction of the model we need to find $\lim_{t \to \infty} p(t)$. Using the rules of limits, we have

$$\lim_{t \to \infty} p(t) = \lim_{t \to \infty} \frac{197.27}{1 + 49.58e^{-0.03134(t - 1790)}}$$

$$= \frac{197.27}{1 + 49.58 \lim_{t \to \infty} e^{-0.03134(t - 1790)}}$$

$$= \frac{197.27}{1 + 49.58(0)}$$

$$= 197.27$$

This model, formulated in the year 1837, predicted a limiting population of the United States of very nearly 200 million. Screen 2.7 shows a graph of $y = p(x)$ using a window with dimensions [1790, 2190] by [0, 200]. ∎

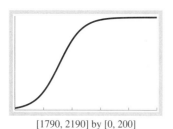

[1790, 2190] by [0, 200]

Screen 2.7
The population $y = 197.27/[1 + 49.58e^{-0.03134(x-1790)}]$ grows fast at first and then levels out and heads for a limiting value.

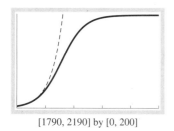

[1790, 2190] by [0, 200]

Screen 2.8
During the earlier years, the solid-line graph of the logistic curve $y_1 = 197.27/[1 + 49.58e^{-0.03134(x-1790)}]$ is about the same as the dotted graph of the exponential function $y_2 = 3.926(1.031)^{x-1790}$

We showed in Section 1.7 that the population of the United States in the years from 1790 to about 1805 was growing exponentially according to $p(t) = 3.926(1.031)^{t - 1790}$. Is this in contradiction with the equation given in Example 5? To check this we graph this last exponential function on the same screen as the logistic equation (Screen 2.8). Notice the close agreement for the first 50 years after 1790. Thus, the logistic curve is *approximately an exponential curve* during the early years.

EXPLORATION 4

Confirming the Long-Term Behavior Graphically

Graph the function $y = p(x)$ in the previous example on larger and larger intervals, and using the TRACE feature, confirm graphically the limit found above.

An equation of the form $L/(1 + ae^{-kt})$, such as given in Example 5, is called a **logistic equation**. Logistic equations are commonly used by biologists to model human and other populations. We will have more to say about this in Chapter 4 and later.

The Natural Exponent and Logarithm

In Section 1.8 we defined the logarithm $\log_a x$ to the base a for any positive number $a \neq 1$. Since e is a positive number with $e \neq 1$, we can define $\log_e x$. This is called the **natural logarithm** and is also denoted by $\ln x$. Thus, we have the following:

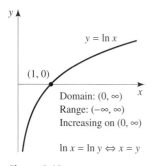

Figure 2.18
The graph of $y = \ln x$ increases and goes through the point (1, 0).

Natural Logarithm

If $x > 0$, the **natural logarithm of x**, denoted by $\ln x$, is defined as follows:

$$y = \ln x \quad \text{if, and only if,} \quad x = e^y$$

Since $e > 1$, we know from Section 1.8 that the graph of $y = \ln x$ is increasing as shown in Figure 2.18. Because the natural logarithm is a logarithm, it satisfies all rules of logarithms given in Section 1.8.

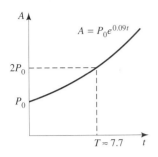

Figure 2.19
An account growing at 9% a year and compounding continuously doubles in value in about 7.7 years.

EXAMPLE 6   Doubling Time

Find the time for the account in Example 4 to double.

SOLUTION ■ The account in Example 4 was growing at 9% a year compounding continuously and had an initial amount of $1000. The amount at the end of t years is $A(t) = 1000e^{0.09t}$. Let T be the time it takes for the account to double. Then

$$2000 = A(T)$$
$$= 1000e^{0.09T}$$
$$2 = e^{0.09T}$$
$$\ln 2 = \ln e^{0.09T}$$
$$= 0.09T$$
$$T = \frac{\ln 2}{0.09} \approx 7.7$$

Thus, it takes about 7.7 years for this account to double in value (Fig. 2.19).   ■

We now use $\ln x$ to write any exponential function $y = a^x$, with $a > 0$, as e^{kx} for some constant k. To see how to do this, notice that

$$a^x = (e^{\ln a})^x = e^{x \ln a}$$

We have shown the following:

Writing Exponential Functions in Terms of the Natural Exponent

Let $a > 0$. If $k = \ln a$, then

$$a^x = e^{kx}$$

EXAMPLE 7   Writing an Exponential Function in Terms of the Natural Exponent

In Section 1.7 we used $P = 4469(1.033)^t$ to model the population of the United States in the early years, where $t = 0$ corresponded to 1794. Write $(1.033)^t$ as e^{kt}, and rewrite the model using the exponent e.

SOLUTION ■ Since $\ln 1.033 \approx 0.0325$, we have

$$P = 4469(1.033)^t$$
$$\approx 4469e^{0.0325t}$$ ■

SELF-HELP EXERCISE SET 2.2

1. Suppose a state levies an income tax of 1% on incomes over $20,000 but less than or equal to $25,000 and 2% on all income if income is over $25,000. At what income levels is the tax discontinuous? Of what significance are these points? (In the parlance of the tax experts, these are referred to as "cliffs.")

2. Find $\displaystyle\lim_{x \to \infty} \frac{2x^3 + x - 1}{x^3 + 1}$

3. Due to an unfortunate nuclear accident, some radioactive iodine-131 is released into the atmosphere and is absorbed by the grass in some fields, which then is ingested by some cows. The milk from the cows has three times the maximum allowable amounts of iodine-131. How much time must elapse until it is safe to drink the milk from these animals? The half-life of iodine-131 is eight days. (Such an accident occurred in Michigan.) The amount of iodine-131 at time t is $A(t) = A_0 e^{-kt}$ for some positive constant k.

EXERCISE SET 2.2

1. Find where the function given in the graph is not continuous.

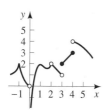

In Exercises 2 through 10 sketch a graph (without using your graphing calculator) to find where each of the given functions is not continuous.

2. $f(x) = |x|$

3. $f(x) = |x - 2|$

4. $f(x) = \begin{cases} x & \text{if } x < 1 \\ 2 & \text{if } x = 1 \\ x - 1 & \text{if } x > 1 \end{cases}$

5. $f(x) = \begin{cases} -x + 1 & \text{if } x < 0 \\ x^2 & \text{if } x \geq 0 \end{cases}$

6. $f(x) = \begin{cases} -x + 1 & \text{if } x < 1 \\ x^2 - 1 & \text{if } x > 1 \end{cases}$

7. $f(x) = \begin{cases} x^2 & \text{if } x < 1 \\ 2 & \text{if } x = 1 \\ x & \text{if } x > 1 \end{cases}$

8. $f(x) = \begin{cases} -x + 2 & \text{if } x < 1 \\ 0 & \text{if } x = 1 \\ x^2 & \text{if } x > 1 \end{cases}$

9. $f(x) = \begin{cases} x & \text{if } x < 0 \\ \dfrac{1}{x} & \text{if } x > 0 \end{cases}$

10. $f(x) = \dfrac{x - 2}{|x - 2|}$

In Exercises 11 through 14 determine where the given functions are continuous.

11. $f(x) = 5x^7 - 3x^2 + 4$

12. $f(x) = \dfrac{x}{x^2 + x}$

13. $f(x) = \dfrac{x}{x^2 - 1}$

14. $f(x) = \dfrac{x - 1}{x^2 - 1}$

15. Suppose $f(x)$ is continuous on $(0, 2)$ and never zero there and $f(1) > 0$. What can you say about the sign of f on $(0, 2)$?

16. Find a function $f(x)$ defined on $[0, 1]$ that changes its sign on $[0, 1]$ but is never zero on $[0, 1]$. Can this function be continuous on $[0, 1]$?

17. Suppose $f(x)$ is continuous on $(0, 2)$ and never zero there and $f(1) < 0$. What can you say about the sign of f on $(0, 2)$?

18. Suppose $f(x)$ is continuous on $(0, 4)$, $f(2) = 0$, and f has no other zeros on $(0, 4)$. If $f(1) = 2$ and $f(3) = -4$, what can you say about the sign of f on $(0, 4)$?

In Exercises 19 through 30 find the limits. Support your answer with a graphing calculator.

19. $\lim\limits_{x \to \infty} \dfrac{2x^2 + 1}{3x^2 - 1}$

20. $\lim\limits_{x \to -\infty} \dfrac{4x^2 - x - 1}{3x^2 + 1}$

21. $\lim\limits_{x \to \infty} \dfrac{3x^2 + x + 1}{2x^4 - 1}$

22. $\lim\limits_{x \to \infty} \dfrac{x^3 - 1}{2x^4 + 1}$

23. $\lim\limits_{x \to -\infty} (x^2 + 1)$

24. $\lim\limits_{x \to -\infty} \dfrac{x^4 - x^2 + x - 1}{x^3 + 1}$

25. $\lim\limits_{x \to \infty} (1 + 2e^{-x})$

26. $\lim\limits_{x \to -\infty} (3 - 2e^x)$

27. $\lim\limits_{x \to \infty} \dfrac{10}{2 + e^{-x}}$

28. $\lim\limits_{x \to \infty} \dfrac{2}{1 + e^x}$

29. $\lim\limits_{x \to \infty} \dfrac{e^x - 1}{e^x + 1}$

30. $\lim\limits_{x \to \infty} \dfrac{e^x - e^{-x}}{e^x + e^{-x}}$

In Exercises 31 through 35 find the limits graphically. Confirm algebraically.

31. $\lim\limits_{x \to \infty} \dfrac{\sqrt{x} + 1}{x + 2}$

32. $\lim\limits_{x \to \infty} \dfrac{x^2 - 1}{\sqrt{x} - 3}$

33. $\lim\limits_{x \to \infty} \sqrt{4 - \dfrac{1}{x}}$

34. $\lim\limits_{x \to -\infty} \sqrt{1 - \dfrac{1}{x^2}}$

35. $\lim\limits_{x \to \infty} (\sqrt{x^2 + 1} - x)$

36. Write 2^x as e^{kx} for some k.

37. Write 10^x as e^{kx} for some k.

Applications

38. Biology. According to Wiess' law, the intensity of an electric current required to excite a living tissue (threshold strength) is given by $i = \dfrac{A}{t} + B$, where t is the duration of the current and A and B are positive constants. What happens as $t \to \infty$? Does your answer seem reasonable?

39. Nutrition. The Morgan-Mercer-Flodid model (see [44] and [45]) $r = \dfrac{ab + cx^k}{b + x^k}$ characterizes the nutritional responses of higher organisms, where r is the weight gain and x is the nutrient intake. What happens to the weight gain as the nutrient intake becomes large without bound? Use your graphing calculator to verify your answer.

40. Biology. Wiedenmann and coworkers [46] showed that the proportion y of a certain host larvae parasitized by *Cotesian flavipes* was approximated by $y = 1 - e^{-1.48x}$, where x is the parasite/host ratio. What proportion of the larvae are parasitized as the parasite/host ratio becomes large without bound? What ratio results in 90% of the host being parasitized?

41. Spread of Technology. When a new useful technology is introduced into a particular industry, it naturally spreads through this industry. A certain model (see Mansfield [47]) of this effect indicates that if t is the number of years after the introduction of the new technology, the percentage of the firms in this industry using the new technology is given by $p(t) = \dfrac{100}{1 + 2e^{-0.15t}}$. To what extent does this new technology spread throughout this industry in the long-term?

42. Taxes. A certain state has a tax on electricity of 1% of the monthly electricity bill with the first $50 of the bill exempt from tax. Let $P(x)$ be the percentage paid in taxes and x the amount of the bill in dollars. Graph this function, and find where it is continuous. What is the significance of the points of discontinuity?

43. Interest. A $100,000 savings certificate has annual interest of 8% compounded quarterly credited to the account only at the end of each quarter. Determine the amount $A(t)$ in the account during the first year, and find points at which the function is discontinuous. What is the significance of these points?

44. Costs. A telephone company charges $1.50 for the first minute for a certain long-distance call and $0.25 for each additional minute. Graph this cost function, and determine where the function is continuous. What is the significance of the points of discontinuity? Are some people affected by these points?

45. Salary. A certain computer salesman earns a base salary of $20,000 plus 5% of the total sales if sales are above $100,000 and 10% of total sales if sales are above $1,000,000. Let $S(x)$ be his salary and x the amount in dollars of computers he sells. Find the discontinuities of $S(x)$. What is the significance of these discontinuities? Do you think that the salesman is affected if his sales are almost at one of the discontinuities?

46. Demand. Sailors [48] use the demand function

$$D(x) = \begin{cases} \dfrac{1}{p^2}, & 0 < p < 10 \\ 0, & p > 10 \end{cases}$$

Determine where this function is continuous.

47. Effective Yield. Suppose an account earns interest at an annual rate of 6% and is continuously compounding. If $1000 is deposited in such an account, find the amount at the end of one year. Use this to determine the effective annual yield.

48. Effective Yield. Suppose an account earns interest at an annual rate of r, expressed as a decimal, and is continuously compounding. Show that the effective annual interest rate r_{eff} is given by $r_{\text{eff}} = e^r - 1$.

49. Effective Yield. Refer to Exercise 48. One bank has an account that pays an annual interest rate of 6.1% compounded annually and a second bank pays an annual interest rate of 6% compounded continuously. In which bank would your money earn more? Why?

50. Present Value. How much money must grandparents set aside at the birth of their grandchild if they wish to have $20,000 by the grandchild's 18th birthday? They can earn 9% compounded continuously.

51. Depreciation. Huang [49] used the mathematical model $\ln D = -3.3195 + 1.0629 \ln K$, where D is depreciation charges and K is equipment assets. Solve for D, and write D as $D = aK^b$.

52. Mathematical Modeling. The equation $y = ax^b$ is often used as a mathematical model. Researchers must decide on the values of a and b. A common way of doing this is to first take logarithms of each side and find a *linear* relationship between $Y = \ln y$ and $X = \ln x$. Find this relationship. What is the Y-intercept? What is the slope?

53. New Technology. In studying the diffusion of adoption of a certain planting technique through the farming community Traxler and Byerlee [50] used the model $\ln \dfrac{S}{c - S} = a + bt$, where S is the percentage of farms using the innovation and t is the time since 1980. Solve for S. If $b > 0$, what happens to S as t becomes large?

54. Irrigation. Chakravorty and Roumasset [51] found that the relationship between the amount y of irrigation water received as a function of the distance x from the source was approximated by $y = Qe^{-0.002x}$, where x is measured in meters, y in cubic meters, and Q is the amount sent. Find the distance at which the water received is half the water sent.

55. Biology. Korzukhin and Porter [52] showed that the population of fire ants was approximated by $P = P_0 e^{-t/13}$, where P_0 was the population at the time of the death of the queen ant and t is the time since her death. Using this model, determine the time it takes for the population to drop to 10% of the size of the population at the time of the death of the queen. Take $P_0 = 100$ and graph.

56. Biology. Yang [53] showed that the expected life y of *Bactrocera dorsalis* in days was approximated by $y = e^{7.0179 - 0.1048T}$, where T is the temperature in degrees Celsius. What is the expected life when $T = 0$? For what temperature is the expected life 1 day? Graph.

Enrichment Exercises

57. Use the constant-sign theorem to show that the polynomial $y = x^5 - 4x^2 - x + 3$ has a root in the interval $(1.25, 2)$.

58. Determine what the graph of $y = p(x) = x^5 - 10x^3 + 9x + 10$ looks like for very large values of $|x|$ by evaluating

$$\lim_{x \to \infty} \frac{p(x)}{x^5}.$$

59. Determine what the graph of $y = p(x) = x^6 - 9x^3 + 8x - 10$ looks like for very large values of $|x|$ by evaluating

$$\lim_{x \to \infty} \frac{p(x)}{x^6}.$$

60. Study the previous two exercises carefully, and show that the graph of a polynomial function $y = p(x) = a_n x^n + a_{n-1} x^{x-1} + \cdots + a_1 x + a_0$, with $a_n \neq 0$, looks like the graph of $y = a_n x^n$ for very large values of $|x|$.

61. Study the results of the previous exercise carefully, and show that for very large values of $|x|$ the graphs of polynomial functions given there fall into only four different cases, accordingly as n is even or odd and a_n is positive or negative. Sketch a typical graph for each of these four cases.

62. Using the results in Exercises 60 and 61 match the given polynomials with their graphs. There can be only one possible match. Explain why you made each choice.

(a) $p_1(x) = x^{14} + g_1(x)$, where $g_1(x)$ is a polynomial of order 13.

(b) $p_2(x) = -x^{18} + g_2(x)$, where $g_2(x)$ is a polynomial of order 17.

(c) $p_3(x) = x^{15} + g_3(x)$, where $g_3(x)$ is a polynomial of order 14.

(d) $p_4(x) = -x^{17} + g_4(x)$, where $g_4(x)$ is a polynomial of order 16.

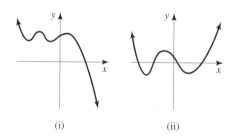

(i)   (ii)

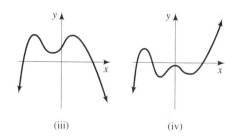

(iii)   (iv)

63. Explain why no polynomial can have a graph like the one shown in the figure.

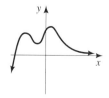

64. Your graphing calculator has a TAN^{-1} key with your calculator in radian mode. Graph $y = f(x) = \text{TAN}^{-1}\, x$. Determine graphically and numerically the two limits $\lim_{x \to \infty} f(x)$ and $\lim_{x \to -\infty} f(x)$.

65. Explain why no polynomial can have a graph like the one shown in the figure.

66. Is it true that if $\lim_{x \to \infty} f(x) = \infty$ and $\lim_{x \to \infty} g(x) = \infty$, then

$$\lim_{x \to \infty} \left(\frac{f(x)}{g(x)} \right)$$ does *not* exist? Explain why this is true, or give an example that shows it is not true.

67. Explain why a polynomial function of odd degree must have at least one zero.

68. Explain the difference between requiring a function $f(x)$ to be continuous at $x = a$ and requiring $\lim_{x \to a} f(x)$ to exist.

69. Show that every real number c has an nth root if n is odd. *Hint:* Consider the polynomial $f(x) = x^n - c$.

70. If f is continuous at a and g is not continuous at a, then $f \cdot g$ may or may not be continuous at a. Give examples to illustrate this.

71. Use the change of base theorem given in Exercise 50, Section 1.8, to find $\log_{10} x$ in terms of natural logarithms.

72. Population. If a population grows according to $P(t) = P_0 e^{kt}$ and if the population at time T is P_1, then show that

$$T = \frac{1}{k} \ln \frac{P_1}{P_0}$$

73. Exponential Decay. Suppose a quantity is decaying according to $y = P_0 e^{-kt}$, where $k > 0$. Let T be the time it takes for the quantity to be reduced to half of its original amount. Show that $\ln 2 = kT$.

74. Exponential Growth. Suppose a quantity is growing according to $y = P_0 e^{kt}$, where $k > 0$. Let T be the time it takes for the quantity to double in size. Show that $\ln 2 = kT$.

75. Forestry. It is not easy to measure the height of a Scots pine, but a simple matter to measure the diameter at breast height. Foresters [54] use the formula $H = f(D) = 30.27 e^{-18.58/D}$, where H is the height in meters of the Scots pine, and D is its diameter at breast height in cm. Using a graphing calculator explore what happens to H as D gets closer and closer to 0. Does your answer make sense?

76. Biology. The accompanying figure can be found in Deshmukh [55] and shows the dry weight of each shoot of 29 species of plants versus the density of shoots. The figure illustrates the $\frac{3}{2}$ **thinning law** in biology. The data points fall more or less in a linear pattern. The line drawn is the line "closest" to the data points. (See Section 7.5 for how this line is obtained.) Explain where the "$\frac{3}{2}$" comes from.

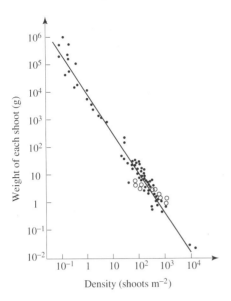

Density (shoots m^{-2})

77. **Agriculture.** Agricultural economists [56] have recently used a new mathematical model to describe crop response to nutrients. The model is $y = m(1 - be^{-kx})$, where x is a measure of the nutrient supplied to the plant and y a measure of the yield. The constants m, k, and b are all positive, with $b < 1$. The number m is called the **asymptotic plateau**.
(a) Show that $y < m$ for $x \geq 0$.
(b) Take $m = 2$, $k = 1$, and $b = 0.5$, and use a graphing calculator to graph $y = 2(1 - 0.5e^{-x})$. Explore what happens as x gets larger and larger. Do you now see why m is called the asymptotic plateau? Explain.

SOLUTIONS TO SELF-HELP EXERCISE SET 2.2

1. If x represents an individual's income and $t(x)$ the tax on that income, then

$$t(x) = \begin{cases} 0 & \text{if } x \leq 20{,}000 \\ (0.01)x & \text{if } 20{,}000 < x \leq 25{,}000 \\ (0.02)x & \text{if } x > 25{,}000 \end{cases}$$

The graph is shown in the figure. The breaks at $x = 20{,}000$ and $x = 25{,}000$ represent discontinuities. If you earn \$20,000 in this state, you pay no taxes, but if you earn \$20,000.01 you pay taxes of \$200. In the parlance of tax experts, your tax bill falls off a "cliff." Naturally you will try your best not to earn that extra one cent beyond the discontinuity at \$20,000. So do such discontinuities affect a person's behavior.

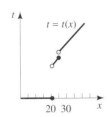

2. Divide numerator and denominator by x to the highest power, which is x^3, and obtain

$$\lim_{x \to \infty} \frac{2x^3 + x - 1}{x^3 + 1} = \lim_{x \to \infty} \frac{2 + \dfrac{1}{x^2} - \dfrac{1}{x^3}}{1 + \dfrac{1}{x^3}}$$

$$= \frac{2 + 0 - 0}{1 + 0} = 2$$

3. Let P_0 be the amount of iodine-131 in the grass immediately after the accident. The amount at time t is then $P(t) = P_0 e^{-kt}$, where t is given in days. We first need to find the decay constant k for iodine-131. Since the half-life is eight days, we must have $0.5A_0 = A(8) = A_0 e^{-k8}$. Then

$$\frac{1}{2} A_0 = A(8) = A_0 e^{-k8}$$

$$\frac{1}{2} = e^{-k8} \qquad \text{cancel } A_0$$

$$\ln\left(\frac{1}{2}\right) = \ln(e^{-k8}) = -k8 \quad \ln e^z = z$$

Since $\ln\left(\frac{1}{2}\right) = -\ln 2$, this yields

$$k = \frac{\ln 2}{8} \approx \frac{0.69315}{8} \approx 0.0866$$

Let T_s be the time when the iodine-131 returns to a safe level. A safe level in terms of P_0 is $\frac{1}{3}P_0$, that is, one-third the initial unsafe amount. Then

$$\frac{1}{3} P_0 = P(T_s) = P_0 e^{-0.0866T_s}$$

$$\frac{1}{3} = e^{-0.0866T_s}$$

$$\ln \frac{1}{3} = \ln(e^{-0.0866T_s}) = -0.0866T_s$$

or since $\ln \frac{1}{3} = -\ln 3$,

$$T_s = \frac{\ln 3}{0.0866} \approx 12.6$$

Thus, 13 days need to elapse before the milk is again safe.

2.3 RATES OF CHANGE AND SLOPE

■ Velocity   ■ Other Rates of Change   ■ Slope of the Tangent Line

APPLICATION: VELOCITY

It is known from physics that if a baseball is dropped from the top of a building, the distance s in feet that it will fall is given by $s = s(t) = 16t^2$, where t is in seconds. What is the velocity at $t = 1$? See Example 1 for the answer.

APPLICATION: INSTANTANEOUS RATE OF CHANGE

A small fertilizer plant sells all the fertilizer it produces. Suppose that the profit, $P(x)$, in dollars of producing x tons of fertilizer is given by

$$P(x) = 100(-x^2 + 12x - 20)$$

How fast are profits changing with respect to changes in x when $x = 4$? See Example 4 for the answer.

APPLICATION: SLOPE OF THE TANGENT LINE

What is the slope of the tangent line to the curve given in the previous example at $x = 4$? See Example 5 for the answer.

Velocity

EXAMPLE 1   Calculating the Instantaneous Velocity

If a baseball is dropped from the side of a building as in Figure 2.20, it can be shown using basic principles of physics that the distance s in feet that the ball falls is given by $s = s(t) = 16t^2$, where t is measured in seconds. Find the velocity $v(1)$ at the instant $t = 1$.

SOLUTION   ■   The formula

$$\text{Velocity} = \frac{\text{distance}}{\text{time}}$$

is familiar. But no time elapses in the "instant" $t = 1$, and no distance is traveled in this instant of time. The formula for velocity then becomes $\frac{0}{0}$ and is meaningless.

Let us consider another approach. We can evaluate the (average) velocity on any interval of time $[1, 1 + h]$, and then take h smaller and smaller and see what happens (Fig. 2.21). We have

$$\text{Average velocity on } [1, 1 + h] = \frac{\text{change in distance}}{\text{change in time}}$$

$$= \frac{s(1 + h) - s(1)}{h}$$

$$= \frac{16(1 + h)^2 - 16}{h}$$

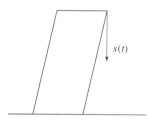

Figure 2.20
Path of a baseball dropped from the top of a building.

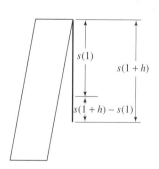

Figure 2.21
$s(1 + h) - s(1)$ is the distance the ball travels during the time interval $[1, 1 + h]$.

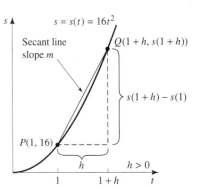

Figure 2.22
The expression $[s(1 + h) - s(1)]/h$ is the average velocity on the time interval $[1, 1 + h]$ and also the slope of the indicated secant line.

Refer to Figure 2.22, where $h > 0$. Figure 2.22 shows the important observation that the average velocity on the interval $[1, 1 + h]$ is also the slope of the secant line from $(1, 16)$ to $(1 + h, s(1 + h))$. Table 2.8 shows the average velocity using selected values of h. As we see, the average velocity seems to be heading for the limiting value of 32.

Let us confirm this analytically. First find the average velocity on any interval $[1, 1 + h]$ with $h > 0$. We obtain

$$\text{Average velocity} = \frac{s(1 + h) - s(1)}{h} \qquad \text{definition}$$

$$= \frac{16(1 + h)^2 - 16}{h} \qquad (1 + h)^2 = 1 + 2h + h^2$$

$$= \frac{16 + 32h + 16h^2 - 16}{h} \qquad \text{simplify}$$

$$= \frac{(32 + 16h)(h)}{h} \qquad \text{cancel } h$$

$$= 32 + 16h$$

where the h can be canceled, since h is never zero. We get the same formula when $h < 0$. Then

$$v(1) = \lim_{h \to 0} (32 + 16h) = 32$$

This confirms our previous work. Thus, the instantaneous velocity when $t = 1$ is 32 ft/s. ∎

TABLE 2.8

h	Interval	Average Velocity
0.1	[1, 1.1]	$\dfrac{16(1.1)^2 - 16}{0.1} = 33.6000$
0.01	[1, 1.01]	$\dfrac{16(1.01)^2 - 16}{0.01} = 32.1600$
0.001	[1, 1.001]	$\dfrac{16(1.001)^2 - 16}{0.001} = 32.0160$
0.0001	[1, 1.0001]	$\dfrac{16(1.0001)^2 - 16}{0.0001} = 32.0016$

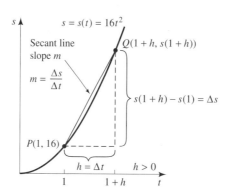

Figure 2.23
The change in s, $s(1 + h) - s(1)$, is denoted by Δs. The change in t is h and is denoted by Δt.

Average and Instantaneous Velocity

Suppose $s = s(t)$ describes the position of an object at time t. The **average velocity** from c to $c + h$ is

$$\text{Average velocity} = \frac{s(c + h) - s(c)}{h}$$

The **instantaneous velocity** (or simply velocity) $v(c)$ at time c is

$$v(c) = \lim_{h \to 0} (\text{average velocity}) = \lim_{h \to 0} \frac{s(c + h) - s(c)}{h}$$

if this limit exists.

EXPLORATION 1

Limit

Verify graphically that $\lim_{h \to 0} [16(1 + h)^2 - 16]/h = 32$. First graph the function $y = \dfrac{16(1 + x)^2 - 16}{x}$ in a window with dimensions $[-0.1, 0.1]$ by $[30, 32]$. Now ZOOM repeatedly to verify that the limit as $x \to 0$ is 32.

We denote the change in the distance on the interval $[1, 1 + h]$ by Δs (pronounced "delta ess"). We denote the change in time h by Δt. Thus, the average velocity on the interval $[1, 1 + h]$ can be written more suggestively as

$$\text{Average velocity on } [1, 1 + h] = \frac{\Delta s}{\Delta t}$$

This is shown in Figure 2.23, where we recall that the average velocity on the interval $[1, 1 + h]$ is just the slope of the secant line from $P(1, 16)$ to $Q(1 + h, s(1 + h))$.

Now we see how to connect the instantaneous velocity with the slope of a certain tangent line. Figure 2.24 shows the tangent line to the curve $s = 16t^2$ at $(1, 16)$. Careful examination of Figure 2.24 indicates that the slope of the secant lines is approaching the slope of the tangent line to the curve at $(1, 16)$. Thus, the instantaneous velocity at $t = 1$ is the slope of the line tangent to the curve $s = 16t^2$ at $t = 1$.

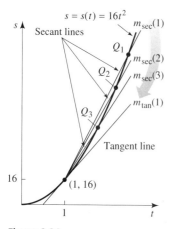

Figure 2.24
As the point slides along the graph and heads for $(1, 16)$, the slopes of the corresponding secant lines head for the slope of the tangent line at $t = 1$.

EXAMPLE 2   Finding the Instantaneous Velocity

Suppose a ball is thrown straight upward with an initial velocity (that is, velocity at the time of release) of 128 ft/s. Suppose, furthermore, that the point at which the ball is released is considered to be at zero height. Then from physics it is known that the

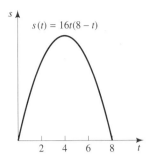

Figure 2.25
The ball rises and then falls as indicated by this graph.

height $s(t)$ in feet of the ball at time t in seconds is given by $s(t) = -16t^2 + 128t$. Find the velocity $v(2)$ at the instant $t = 2$.

SOLUTION ■ Since $s(t) = 16t(8 - t)$, the function is easily graphed, as shown in Figure 2.25. (Do not confuse the *graph* in the figure with the actual *path* of the ball. The ball goes straight up and then straight down.)

Since $s(2) = -16(2)^2 + 128(2) = 192$, the average velocity from 2 to $2 + h$ is

$$\frac{\Delta s}{\Delta t} = \frac{s(2 + h) - s(2)}{h} \qquad \frac{\text{change in } s}{\text{change in } t}$$

$$= \frac{[-16(2 + h)^2 + 128(2 + h)] - [192]}{h}$$

$$= \frac{-64 - 64h - 16h^2 + 256 + 128h - 192}{h}$$

$$= \frac{64h - 16h^2}{h}$$

$$= \frac{h(64 - 16h)}{h}$$

$$= 64 - 16h \quad ■$$

To find the instantaneous velocity at time $t = 2$, we must take the limit of the average velocity as h goes to zero. We then have the following:

$$v(2) = \lim_{h \to 0} \frac{s(2 + h) - s(2)}{h}$$

$$= \lim_{h \to 0} (64 - 16h)$$

$$= 64$$

This means that the ball is moving up at 64 ft/s at the instant when $t = 2$.

Other Rates of Change

We can take the average rate of change and the instantaneous rate of change for any function $y = f(x)$. We now do this and take care to note any geometric interpretations.

Consider the graph of the function $y = f(x)$ shown in Figure 2.26. The term $h = \Delta x$ represents the change in x on the interval $[c, c + h]$, and $f(c + h) - f(c) = \Delta y$ represents the corresponding change in y on this interval. In analogy with average velocity, we have the following:

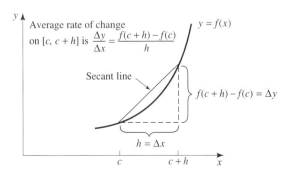

Figure 2.26
The average rate of change is the slope of the secant line.

Average Rate of Change

The average rate of change of $y = f(x)$ with respect to x from c to $c + h$ is the ratio

$$\frac{\text{Change in } y}{\text{Change in } x} = \frac{\Delta y}{\Delta x} = \frac{f(c + h) - f(c)}{h}$$

Taking the limit of this last expression as $h \to 0$ yields the **instantaneous rate of change** at $x = c$. We have the following:

Instantaneous Rate of Change

Given a function $y = f(x)$, the instantaneous rate of change of y with respect to x at $x = c$ is given by

$$\lim_{h \to 0} \frac{f(c + h) - f(c)}{h}$$

if this limit exists.

EXAMPLE 3   Finding the Average Rate of Change

A small fertilizer plant sells all the fertilizer it produces. Suppose that the profit, $P(x)$, in dollars of producing x tons of fertilizer is given by

$$P(x) = 100(-x^2 + 12x - 20)$$

Find the average rate of change of profits on the interval (a) [2, 4], (b) [5, 7], (c) [8, 10].

SOLUTIONS ■ Since $P(x) = -100(x - 2)(x - 10)$, the function is easily graphed as in Figure 2.27.

(a) Here $x = 2$, and $h = 2$. Thus,

$$\frac{\Delta P}{\Delta x} = \frac{P(x + h) - P(x)}{h} = \frac{P(4) - P(2)}{2} = \frac{1200 - 0}{2} = 600$$

(b) Here $x = 5$, and $h = 2$. Thus,

$$\frac{\Delta P}{\Delta x} = \frac{P(x + h) - P(x)}{h} = \frac{P(7) - P(5)}{2} = \frac{1500 - 1500}{2} = 0$$

(c) Here $x = 8$, and $h = 2$. Thus,

$$\frac{\Delta P}{\Delta x} = \frac{P(x + h) - P(x)}{h} = \frac{P(10) - P(8)}{2} = \frac{0 - 1200}{2} = -600$$

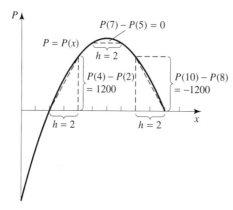

Figure 2.27

The average rate of change on [2, 4] is $\frac{1200}{2} =$ 600, on [5, 7] is 0, and on [8, 10] is $\frac{-1200}{2} =$ −600.

Thus, on the first interval, profits are increasing at the average rate of $600 per ton, on the second interval they are not increasing or decreasing on average, and on the third interval profits are decreasing on average by $600 per ton. ■

EXAMPLE 4   Finding the Instantaneous Rate of Change

Find the instantaneous rate of change of profits at $x = 4$ in Example 3.

SOLUTION ■ Since $P(4) = -1200$, the average rate of change of $P = 100(-x^2 + 12x - 20)$ as x changes from 4 to $4 + h$ is

$$\frac{P(4 + h) - P(4)}{h} = 100 \frac{[-(4 + h)^2 + 12(4 + h) - 20] - [12]}{h}$$

$$= 100 \frac{-16 - 8h - h^2 + 48 + 12h - 20 - 12}{h}$$

$$= 100 \frac{(4 - h)h}{h}$$

$$= 100(4 - h)$$

Taking the limit as $h \to 0$ of the expression for the average rate of change gives the instantaneous rate of change

$$\lim_{h \to 0} \frac{P(4 + h) - P(4)}{h} = \lim_{h \to 0} [100(4 - h)]$$

$$= 100(4) = 400$$

Thus, at $x = 4$, profits are increasing at the instantaneous rate of $400 per ton. ■

Slope of the Tangent Line

Notice from Figure 2.26 that the average rate of change from c to $c + h$ is also the slope m_{sec} of the secant line from the point $P(c, f(c))$ to the point $Q(c + h, f(c + h))$.

Average Rate of Change Is the Slope of the Secant Line

The average rate of change of $y = f(x)$ from c to $c + h$ is the slope of the secant line from $P(c, f(c))$ to the point $Q(c + h, f(c + h))$.

EXPLORATION 2

Tangent Line

■■■ Graph $y_1 = x^2$ and $y_2 = 2x - 1$ on the same screen of dimensions [0, 2] by
■ ■ [−1, 5]. Explain what you see. Use the ZOOM feature once to be sure.
■■■

Figure 2.28 shows what happens to the slopes of the secant lines as h approaches 0. Close examination of this figure indicates that m_{sec} approaches $m_{tan}(c)$, the slope of the tangent line to the curve at the point c. That is

$$\lim_{h \to 0} \frac{f(c + h) - f(c)}{h} = \lim_{h \to 0} m_{sec}$$

$$= m_{tan}(c)$$

provided the limit exists.

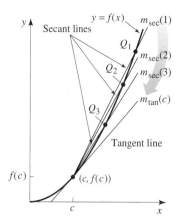

Figure 2.28
As the point slides along the graph and heads for $(c, f(c))$, the slopes of the corresponding secant lines head for the slope of the tangent line at $x = c$.

EXPLORATION 3

Tangent Line

 The program SEC in the Technology Resource Manual graphs $y = f(x)$, the tangent line at the indicated point, and a sequence of secant lines for values of h approaching zero and thus demonstrates Figure 2.28. Use this program on the function $y = x^2$ at $x = 1$ with a window of dimensions [0.8, 2] by [0, 4].

Definition of Tangent Line

The **tangent line** to the graph of $y = f(x)$ at $x = c$ is the line through the point $(c, f(c))$ with slope

$$m_{\tan}(c) = \lim_{h \to 0} \frac{f(c + h) - f(c)}{h}$$

provided this limit exists.

Notice now that the above limit (if it exists) that defines the slope $m_{\tan}(c)$ of the tangent line is the very same limit that defines the instantaneous rate of change at c. We then have the following.

Instantaneous Rate of Change and the Slope of the Tangent Line

If the instantaneous rate of change of $f(x)$ with respect to x exists at a point c, then it is the slope of the tangent line at that point.

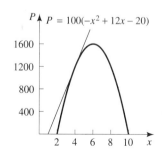

Figure 2.29
The instantaneous rate of change at $x = 4$ is the slope of the tangent line at $x = 4$.

EXAMPLE 5   Slope of Tangent Line

Find the slope of the line tangent to the curve in Example 4 at $x = 4$.

SOLUTION ■ We saw that for the profit function $P(x) = 100(-x^2 + 12x - 20)$ in Example 4, the instantaneous rate of change at $x = 4$ was 400. Thus, $m_{\tan}(4) = 400$, as indicated in Figure 2.29. ■

EXAMPLE 6   Equation of Tangent Line

Find the equation of the line tangent to the curve in Example 4 at $x = 4$. Support the answer using a graphing calculator.

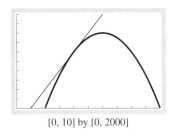

[0, 10] by [0, 2000]

Screen 2.9
$y_2 = 400x - 400$ is tangent to y_1 $= 100(-x^2 + 12x - 20)$ at $x = 4$.

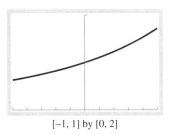

[−1, 1] by [0, 2]

Screen 2.10
As $x \to 0$, it appears that $y = \dfrac{e^x - 1}{x} \to 1$.

SOLUTION ■ From the previous example we know that the slope is $m_{\tan}(4)$ = 400. At $x = 4$, $y = P(4) = 1200$. Thus, the equation of the line tangent to the curve $y = P(x)$ at (4,1200) is

$$y - 1200 = 400(x - 4)$$
$$y = 400x - 1600 + 1200$$
$$= 400x - 400$$

Screen 2.9 shows a graph of $y_1 = 100(-x^2 + 12x - 20)$ and $y_2 = 400x - 400$ on a window of dimension [0, 10] by [0, 2000]. The line is tangent to the curve. ■

We now determine the slope of the tangent line to the graph of the curve $y = e^x$ at (0, 1). In the process, we evaluate an important limit that we will need to use in the next chapter.

EXAMPLE 7   Slope and Equation of a Tangent Line

Find the slope and the equation of the tangent line to the graph of the curve $y = f(x) = e^x$ at (0, 1).

SOLUTION ■ The slope of the tangent line at $x = 0$ is

$$m_{\tan}(0) = \lim_{h \to 0} \frac{f(0 + h) - f(0)}{h}$$
$$= \lim_{h \to 0} \frac{e^{0 + h} - e^0}{h}$$
$$= \lim_{h \to 0} \frac{e^h - 1}{h}$$

To obtain some idea of what this might be we use our graphing calculator to graph $y = \dfrac{e^x - 1}{x}$, using a window of dimensions [−1, 1] by [0, 2] and obtain Screen 2.10. The limit appears to be 1.

We then support this numerically by evaluating $(e^h - 1)/h$ for ever smaller values of h, the results of which are shown in the table.

h	−0.1	−0.01	−0.001	→	0	←	0.001	0.01	0.1
$\dfrac{e^h - 1}{h}$	0.9516	0.9950	0.9995	→	?	←	1.0005	1.0050	1.0517

The table then supports our conjecture that

$$\lim_{h \to 0} \frac{e^h - 1}{h} = 1$$

Of course, none of this actually constitutes a proof. Although too difficult to do here, it can be proven that this limit is indeed 1.

Let $m_{\tan}(0) = m$. The equation of the tangent line is then

$$y - f(0) = m(x - 0)$$
$$y - 1 = 1 \cdot (x - 0)$$
$$y = x + 1$$

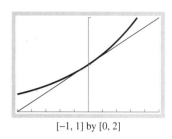

[−1, 1] by [0, 2]

Screen 2.11
$y_2 = x + 1$ is tangent to $y_1 = e^x$ at $x = 0$.

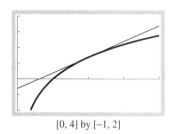

[0, 4] by [−1, 2]

Screen 2.12
$y_2 = 0.5x - 0.307$ is approximately the equation of the tangent line to $y_1 = \ln x$ at $x = 2$.

Screen 2.11 shows a graph of $y_1 = e^x$ and $y_2 = x + 1$ using a window of dimensions $[−1, 1]$ by $[0, 2]$. ∎

EXAMPLE 8   Equation of Tangent Line

Find the approximate equation of the line tangent to the curve $y = f(x) = \ln x$ at $x = 2$.

SOLUTION ∎ We can find an approximate equation for the tangent line by finding an approximation to $m_{\tan}(2)$. We know that by definition

$$m_{\tan}(2) = \lim_{h \to 0} \frac{f(2 + h) - f(2)}{h}$$

$$= \lim_{h \to 0} \frac{\ln(2 + h) - \ln 2}{h}$$

If h is small, say $h = 0.001$, then we have

$$m_{\tan}(2) \approx \frac{\ln(2 + 0.001) - 2}{0.001}$$

Using a calculator we find that the last term is approximately 0.500.
Let $m_{\tan}(2) = m$. Then the tangent line is given by

$$y - \ln 2 = m(x - 2)$$

$$y \approx 0.500(x - 2) + \ln 2$$

$$\approx 0.500x - 1.000 + \ln 2$$

$$\approx 0.500x - 0.307$$

Screen 2.12 shows a graph of $y_1 = \ln x$ and $y_2 = 0.5x - 0.307$ using a window of dimensions $[0, 4]$ by $[−1, 2]$. ∎

SELF-HELP EXERCISE SET 2.3

1. For $s(t) = 16t^2$, find the instantaneous velocity $v(2)$ at $t = 2$.

2. Suppose a ball is thrown straight upward with an initial velocity (that is, velocity at the time of release) of 96 ft/s. Suppose, furthermore, that the point at which the ball is released is considered to be zero height; then from physics it is known that the height $s(t)$ in feet of the ball at time t in seconds is given by $s(t) = -16t^2 + 96t$. Find the instantaneous velocity $v(1)$ at $t = 1$.

3. Find the equation of the line tangent to the curve $y = f(x) = x^2$ at the point $(−1, 1)$.

EXERCISE SET 2.3

1. If a baseball is dropped from the top of a building, the distance s in feet that the ball falls is given by $s(t) = 16t^2$, where t is in seconds. Find the instantaneous velocity of the ball at $t = 3$.

2. For the ball in the previous exercise, find the instantaneous velocity when $t = 4$.

In Exercises 3 through 8 refer to the following. Suppose a ball is thrown straight upward with an initial velocity (that is, velocity at the time of release) of 128 ft/s and that the point at which the ball is released is considered to be zero height. The height $s(t)$ in feet of the ball at time t in seconds is then given by $s(t) = -16t^2 + 128t$. Let $v(t)$ be the instantaneous velocity at time t. Find $v(t)$ for the indicated values of t.

3. $v(1)$
4. $v(3)$
5. $v(4)$
6. $v(5)$
7. $v(6)$
8. $v(7)$

9. Find the equation of the line tangent to the curve $s = -16t^2 + 64t$ at the point $(1, 48)$. Support your answer by graphing the equation and the tangent line on the same screen.

10. Find the equation of the line tangent to the curve $s = -16t^2 + 64t$ at the point $(3, 48)$. Support your answer by graphing the equation and the tangent line on the same screen.

In Exercises 11 through 22 find the instantaneous rates of change of the given functions at the indicated points.

11. $f(x) = 2x + 3$, $c = 2$

12. $f(x) = -3x + 4$, $c = 3$

13. $f(x) = x^2 - 1$, $c = 1$

14. $f(x) = -x^2 + 3$, $c = 2$

15. $f(x) = -2x^2 + 3$, $c = 2$

16. $f(x) = -3x^2 - 1$, $c = 1$

17. $f(x) = x^2 + 2x + 3$, $c = -1$

18. $f(x) = x^2 - 3x + 4$, $c = -2$

19. $f(t) = -2t^2 - 4t - 2$, $c = 3$

20. $f(r) = 3r^2 - 4r + 2$, $c = 3$

21. $f(u) = u^3$, $c = 1$

22. $f(v) = -v^3 + 1$, $c = 1$

23. Between which pair of consecutive points on the curve is the average rate of change positive? Negative? Zero?

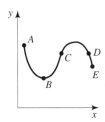

24. Between which pair of consecutive points on the curve is the average rate of change largest? Smallest?

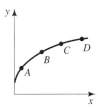

25. Identify the points on the curve for which the slope of the tangent line is positive. Negative. Zero.

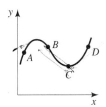

26. Which point on the curve has the tangent line with largest slope? Smallest slope?

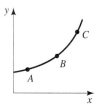

27. The table gives some values (rounded to five decimal places) of $f(x) = \sqrt{x}$ near $x = 1$. From this estimate the slope of the tangent line to $y = \sqrt{x}$ at $x = 1$.

x	0.98	0.99	1	1.01	1.02
$\sqrt{x}$	0.98995	0.99499	1	1.00499	1.00995

28. The table gives some values of $f(x) = \ln x$ near $x = 1$. From this estimate the slope of the tangent line to $y = \ln x$ at $x = 1$.

x	0.980	0.990	1	1.010	1.020
$\ln x$	-0.020	-0.010	0	0.010	0.020

29. Find the equation of the line tangent to the curve $s = 2x^2 - x$ at the point $(2, 6)$. Support your answer by graphing the equation and the tangent line on the same screen.

30. Find the equation of the line tangent to the curve $s = -3x^2 + 2x$ at the point $(2, -8)$. Support your answer by graphing the equation and the tangent line on the same screen.

In Exercises 31 through 34 use the program SEC in the Technology Resource Manual to graph $y = f(x)$, the tangent line at the indicated point, and a sequence of secant lines for values of h approaching zero, thus demonstrating Figure 2.28 of the text. Use the given dimensions of the viewing window.

31. $f(x) = x^3$, $a = 1$, screen dimensions [0.75, 2] by [0.75, 8]

32. $f(x) = x^2$, $a = 2$, screen dimensions [0.8, 2] by [0.8, 4]

33. $f(x) = 1/x$, $a = 0.5$, screen dimensions [0.3, 2] by [0.5, 2]

34. $f(x) = \sqrt{x}$, $a = 1$, screen dimensions [0.75, 2] by [0.85, 1.5]

In Exercises 35 through 40 find the approximate equation of the tangent line to the given curve at the indicated point c by using the approximation $[f(c + 0.001) - f(c)]/0.001$ for $m_{\tan}(c)$. Then graph $y = f(x)$ and the tangent line in an appropriate window of your graphing calculator to check your work.

35. $y = f(x) = e^x$, $c = 2$

36. $y = f(x) = x^4$, $c = 1$

37. $y = f(x) = \ln x$, $c = 4$

38. $y = f(x) = \sqrt[3]{x}$, $c = 8$

39. $y = f(x) = \sqrt{x}$, $c = 4$

40. $y = f(x) = \dfrac{1}{x^2}$, $c = 1$

Applications

41. The table gives the number of law enforcement officers killed in the line of duty in the United States between 1979 and 1988. Find the average rate of change of those killed on the interval **(a)** [1979, 1982], **(b)** [1982, 1984], **(c)** [1985, 1988], **(d)** [1986, 1988].

Year	1979	1980	1981	1982	1983	1984	1985	1986	1987	1988
Killed	164	165	157	164	152	147	148	131	147	155

42. The table gives the total U.S. consumption of petroleum in millions of barrels per day. Find the average rate of change of the consumption of petroleum on the interval **(a)** [1950, 1980], **(b)** [1975, 1985], **(c)** [1980, 1985].

Year	1950	1955	1960	1965	1970	1975	1980	1985
Amount	6.5	8.5	9.8	11.5	14.7	16.3	17.1	15.7

43. The table gives the number of work stoppages in the United States between 1960 and 1985. Find the average rate of change of the number of work stoppages on the intervals **(a)** [1960, 1970], **(b)** [1960, 1975], **(c)** [1970, 1985].

Year	1960	1965	1970	1975	1980	1985
Number	222	268	381	235	187	54

44. Population. The table gives the population in hundreds of thousands of a city over a period of years. Find the average rate of change of the population with respect to time measured in years during **(a)** 1985 to 1987 **(b)** 1985 to 1988 **(c)** 1987 to 1989 **(d)** 1987 to 1991 **(e)** 1985 to 1991.

Year	1985	1986	1987	1988	1989	1990	1991
Number	3.0	2.8	2.6	2.6	2.9	3.1	3.2

45. Profits. The profits in millions of dollars of a truck manufacturer is given by $P = -x^2 + 10x - 16$, where x is the number in thousands of trucks sold. Find the instantaneous rate of change when x is **(a)** 3 **(b)** 5 **(c)** 7. Interpret your answer.

46. Profits. Repeat the previous exercise when x is **(a)** 1 **(b)** 8

47. Revenue. The revenue in hundreds of dollars for a kitchen sink manufacturer is given by $R = 10x - 0.01x^2$, where x is the number of sinks sold. Find the instantaneous rate of change when x is **(a)** 100 **(b)** 500 **(c)** 600. Interpret your answer.

48. Revenue. Repeat the previous exercise when x is **(a)** 200 **(b)** 700.

49. Sales. Suppose that t weeks after the end of an advertising campaign the weekly sales are $100(10t - t^2)$. At what instantaneous rate are sales changing when t equals **(a)** 1 **(b)** 5 **(c)** 6? Interpret your answer.

50. Sales. Suppose that t weeks after the end of an advertising campaign the weekly sales are $100(8t - t^2)$. At what instantaneous rate are sales changing when t equals **(a)** 2 **(b)** 4 **(c)** 6? Interpret your answer.

51. Cost. Suppose that the cost, in thousands of dollars, for a jacket manufacturer is given by $C = 6 + 2x^2$, where x is the number of thousands of jackets produced. Find the instantaneous rate of change of costs when x is **(a)** 0 **(b)** 1 **(c)** 4.

52. Cost. Repeat the previous exercise when x is **(a)** 2 **(b)** 3.

53. Finance. A study of Dutch manufacturers [57] found that the total cost C in thousands of gilders incurred by a company for hiring (or firing) x workers was approximated by $C = 0.0071x^2$. Find the rate of change of costs with respect to workers hired when 100 workers are hired.

Enrichment Exercises

54. State in complete sentences what is meant by average velocity, instantaneous velocity, and the difference between the two.

55. On a large sheet of paper draw a smooth curve. Pick a point on the curve, and with a ruler draw a tangent line to the curve at this point. Now draw a sequence of secant lines as done in Figure 2.28. Demonstrate that the slopes of the secant lines are heading for the slope of the tangent line.

56. Suppose you traveled 100 miles in your car in 2 h. Explain why there must have been a time when your (instantaneous) velocity was precisely 50 mph.

57. On your graphing calculator draw graphs of the three exponential functions $y_1 = 2^x$, $y_2 = e^x$, $y_3 = 3^x$ on the same screen. Decide which graph has a tangent line with the largest slope at $x = 0$ and which has the smallest.

58. In Example 7 we found that the slope of the tangent line to the curve $y = e^x$ at $(0, 1)$ is 1. Do a similar analysis to find the approximate slope of the tangent line to the curve $y = 2^x$ at $(0, 1)$. Is your answer consistent with the result in Exercise 57?

59. Repeat the previous exercise for $y = 3^x$. Is your answer consistent with the result in Exercise 57?

SOLUTIONS TO SELF-HELP EXERCISE SET 2.3

1. First find the average velocity.

$$\text{Average velocity} = \frac{s(2 + h) - s(2)}{h}$$

$$= \frac{16(2 + h)^2 - 64}{h}$$

$$= \frac{64 + 64h + 16h^2 - 64}{h}$$

$$= 64 + 16h$$

Now take the limit of this expression and obtain

$$v(2) = \lim_{h \to 0} (\text{average velocity}) = \lim_{h \to 0} (64 + 16h) = 64$$

2. Since $s(t) = 16t(6 - t)$, the function is easily graphed.

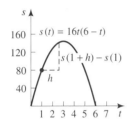

(a) The average velocity on the interval $[1, 1 + h]$ is

$$\frac{s(1 + h) - s(1)}{h} = \frac{[-16(1 + h)^2 + 96(1 + h)] - [80]}{h}$$

$$= \frac{-16 - 32h - 16h^2 + 96 + 96h - 80}{h}$$

$$= \frac{h(64 - 16h)}{h}$$

$$= 64 - 16h$$

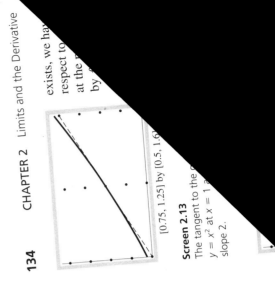

[0.75, 1.25] by [0.5, 1.6]

Screen 2.13
The tangent to the
$y = x^2$ at $x = 1$
slope 2.

... (1, 1) is then $y - 1 =$

$2(x + 1)$.

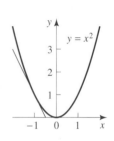

2.4 THE DERIVATIVE

■ The Derivative ■ Nonexistence of the Derivative

APPLICATION: INSTANTANEOUS RATE OF CHANGE

Suppose that $p(x) = \dfrac{1}{x}$ is the price in dollars of a pen, where x is the number of these pens sold per year (in millions). Find the instantaneous rate of change of the price with respect to the number of pens sold per year (in millions) when 2 million pens are sold. See Example 2 for the answer.

The Derivative

Given a function $y = f(x)$ for which

$$\lim_{h \to 0} \frac{f(c + h) - f(c)}{h}$$

e seen that this limit represents the instantaneous rate of change of y with
x at $x = c$ and also the slope of the line tangent to the graph of the function
point $(c, f(c))$. This limit (when it exists) is called the **derivative** and is denoted
$f'(c)$.

Let us see if we can use our graphing calculator to estimate $f'(1)$ graphically, when
$f(x) = x^2$. Using the MODE function, turn the calculator to Grid On. Set $y_1 = x^2$. Set
the dimensions of the window to $[0.75, 1.25]$ by $[0.5, 1.6]$, and set the RANGE variables
$x_{scl} = 0.25$ and $y_{scl} = 0.25$. Graph and obtain Screen 2.13, where the dashed line is an
(imaginary) tangent line. Notice that moving over one horizontal dot represents the
same distance as moving up one dot, since both are equal to 0.25. The slope appears
to be 2. Thus, we suspect $f'(1) = 2$.

We can gain some confidence in our answer if we graph the tangent line. With
$m = f'(1) = 2$, the equation of the tangent line is $y - 1 = 2(x - 1)$, or $y = 2x - 1$.
Inputting $y_2 = 2x - 1$ in the calculator and graphing gives Screen 2.14. Since the line
looks tangent to the curve at $(1, 1)$, it does appear that we have the tangent line and
thus that $m = 2 = f'(1)$.

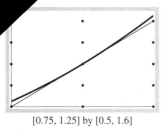

[0.75, 1.25] by [0.5, 1.6]

Screen 2.14
$y_2 = 2x - 1$ appears tangent to
$y_2 = x^2$ at $x = 1$.

EXPLORATION 1

Confirm analytically using the limit definition that $f'(1) = 2$ if $f(x) = x^2$.

As we noted earlier

$$f'(c) = \lim_{h \to 0} \frac{f(c + h) - f(c)}{h}$$

when the limit exists. Replacing the letter c with x we then have the following definition:

Definition of Derivative

If $y = f(x)$, the derivative of $f(x)$, denoted by $f'(x)$, is defined as

$$f'(x) = \lim_{h \to 0} \frac{f(x + h) - f(x)}{h}$$

if this limit exists.

Remark. Other notations for this derivative are y', $\dfrac{dy}{dx}$, and $\dfrac{d}{dx} f(x)$.

EXAMPLE 1   Finding the Derivative of a Function

Let $y = f(x) = 1/x$, $x \neq 0$. Find $f'(x)$.

SOLUTION ■ First find an expression for the average rate of change of y
with respect to x from x to $x + h$ and simplify. This gives

$$\frac{\Delta y}{\Delta x} = \frac{f(x + h) - f(x)}{h} \qquad \frac{\text{change in } y}{\text{change in } x}$$

$$= \frac{[1/(x + h)] - (1/x)}{h}$$

$$= \frac{1}{h}\left[\frac{1}{x + h} - \frac{1}{x} \right]$$

$$= \frac{1}{h}\left[\frac{x - (x + h)}{x(x + h)} \right]$$

$$= \frac{1}{h} \left[\frac{-h}{x(x + h)} \right] \quad \text{cancel the } h$$

$$= -\frac{1}{x(x + h)}$$

Now to find the derivative, take the limit of this expression as $h \to 0$, obtaining

$$f'(x) = \lim_{h \to 0} \frac{f(x + h) - f(x)}{h}$$

$$= \lim_{h \to 0} \frac{-1}{x(x + h)}$$

$$= -\frac{1}{x^2}$$

provided $x \neq 0$. ∎

Notice that the derivative $f'(x)$ is itself a *function*. In the previous example the original function was $f(x) = \frac{1}{x}$, whereas the derivative of this function is the function $f'(x) = -\frac{1}{x^2}$, which for any x gives the slope of the tangent line at x to the graph of $y = f(x)$ (or the instantaneous rate of change at x).

EXAMPLE 2   Finding an Instantaneous Rate of Change

Suppose that $p(x) = \frac{1}{x}$ is the price in dollars of a pen, where x is the number of these pens sold per year (in millions). Find the instantaneous rate of change of the price with respect to the number of pens sold per year (in millions) when 2 million pens are sold. Check your answer by finding the numerical derivative on your graphing calculator.

SOLUTION ∎ Since the instantaneous rate of change is the same as the derivative, the instantaneous rate of change of price with respect to the number of millions of pens sold per year when 2 million pens are sold is $p'(2)$, where $p(x) = \frac{1}{x}$.

From the previous example, we have $p'(2) = -\frac{1}{(2)^2} = -0.25$. Thus, when 2 million pens are sold, the price is dropping at a rate of $0.25 per million sold per year.

Many graphing calculators can find the approximate derivative using a numerical routine. We refer to the numerical derivative of $y = f(x)$ at $x = a$ found in this manner as $n\text{Der } f(a)$. The program in the Teaching Resource Manual to access this function is called NDER. We find that on one graphing calculator, $n\text{Der } p(2) = -0.2500000625$. This is certainly close enough to -0.25 to support our answer. ∎

A project at the end of this chapter gives details on how your graphing calculator finds the numerical derivative.

The general procedure for finding the derivative is summarized as the following three-step process:

To Find the Derivative of $f(x)$.

1. Find $\dfrac{f(x + h) - f(x)}{h}$.

2. Simplify.

3. Take the limit as $h \to 0$ of the simplified expression. That is,

$$f'(x) = \lim_{h \to 0} \frac{f(x + h) - f(x)}{h}$$

EXAMPLE 3   Finding the Derivative of a Function

Find the derivative of $y = f(x) = \sqrt{x}$.

SOLUTION ■ Doing the first two steps gives

$$\frac{f(x + h) - f(x)}{h} = \frac{\sqrt{x + h} - \sqrt{x}}{h}$$

$$= \frac{\sqrt{x + h} - \sqrt{x}}{h} \cdot \frac{\sqrt{x + h} + \sqrt{x}}{\sqrt{x + h} + \sqrt{x}} \quad \text{rationalize the numerator}$$

$$= \frac{x + h - x}{h(\sqrt{x + h} + \sqrt{x})}$$

$$= \frac{h}{h(\sqrt{x + h} + \sqrt{x})} \quad \text{cancel the } h$$

$$= \frac{1}{\sqrt{x + h} + \sqrt{x}}$$

Now take the limit as $h \to 0$ of the last expression, using the rules of limits. This gives

$$\frac{dy}{dx} = \lim_{h \to 0} \frac{f(x + h) - f(x)}{h}$$

$$= \lim_{h \to 0} \frac{1}{\sqrt{x + h} + \sqrt{x}}$$

$$= \frac{1}{2\sqrt{x}}, \quad \text{if } x > 0 \quad ■$$

At this time we do not have the mathematical tools to determine analytically the derivative of $f(x) = \ln x$. But realizing that the derivative is a *function* and using our graphing calculators, we can see what the derivative must be. By the limit definition of derivative we know that

$$f'(x) = \lim_{h \to 0} \frac{f(x + h) - f(x)}{h}$$

$$= \lim_{h \to 0} \frac{\ln(x + h) - \ln(x)}{h}$$

If h is small, say $h = 0.001$, then from our knowledge of limits,

$$f'(x) \approx \frac{\ln(x + 0.001) - \ln x}{0.001}$$

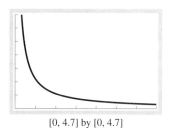

[0, 4.7] by [0, 4.7]

Screen 2.15
Since the derivative of $\ln x$ is approximately $y_1 = [\ln(x + 0.001) - \ln x]/0.001$ and the graph of this function appears to be the same as the graph of $y_2 = 1/x$, we strongly suspect that $(d/dx)(\ln x) = 1/x$.

Screen 2.15 shows a graph of $y_1 = g(x) = \dfrac{\ln(x + 0.001) - \ln x}{0.001}$ using a window of dimension [0, 4.7] by [0, 4.7]. We must have $f'(x) \approx g(x)$. Using the TRACE feature, we create the table shown here.

x	0.25	0.5	1	2	3	4
$g(x)$	3.9920	1.9980	0.9995	0.4999	0.3333	0.2500

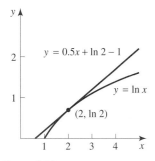

Figure 2.30
Shown is the graph of the tangent line $y = 0.5x + \ln(2) - 1$ at $x = 2$ to the graph of $y = \ln x$.

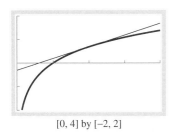

[0, 4] by [−2, 2]

Screen 2.16
$y_2 = 0.5x - 1 + \ln 2$ is tangent to $y_1 = \ln x$ at $x = 2$.

It is rather apparent that $g(x) \approx \dfrac{1}{x}$. If we graph $y_2 = 1/x$ on the same screen, we obtain nothing new. This verifies graphically that $g(x) \approx \dfrac{1}{x}$. Thus, we have ample graphical and numerical evidence to believe that

$$\frac{d}{dx} \ln x = \frac{1}{x}$$

In Section 3.4 we show analytically that this is indeed the case.

EXAMPLE 4   Finding the Equation of the Tangent Line

Find the equation of the tangent line to $y = f(x) = \ln x$ at the point where $x = 2$ (Fig. 2.30). Support your answer using nDer on your graphing calculator and also by graphing $y = \ln x$ and the tangent line on the same screen.

SOLUTION ■ Here $c = 2$ and $f(c) = f(2) = \ln 2$. Also from the previous discussion, $f'(x) = \dfrac{1}{x}$ and $f'(2) = \frac{1}{2}$. Thus, the slope of the tangent line is $m = f'(2) = \frac{1}{2}$. The equation of the tangent line is then

$$y - \ln 2 = \frac{1}{2}(x - 2)$$

On one graphing calculator nDer $f(2) = 0.5000042$, which verifies our answer. Screen 2.16 shows the graphs of $y_1 = \ln x$, and the tangent line $y_2 = 0.5x - 1 + \ln 2$ using a window of dimension [0, 4] by [−2, 2].

We can use nDer on very complicated functions.

EXPLORATION 2

Using nDer

Find the equation of the tangent line to the graph of the curve $y = f(x) = x^5 + x^3 - x + 1$ at $x = -1$ by using nDer to approximate the derivative. Support your answer graphically by graphing the function and the tangent line using a window with dimensions [−1.5, 0] by [−4, 2].

Nonexistence of the Derivative

EXAMPLE 5   A Function Whose Derivative Does Not Exist at One Point

Let $f(x) = |x|$. Does $f'(0)$ exist?

SOLUTION ■ Referring to Figure 2.31, we first note that

Figure 2.31
$\lim\limits_{h \to 0^-} \dfrac{f(0 + h) - f(0)}{h} = -1$, while $\lim\limits_{h \to 0^+}$ $\dfrac{f(0 + h) - f(0)}{h} = 1$. Since these two limits are not equal, $\lim\limits_{h \to 0} \dfrac{f(0 + h) - f(0)}{h}$ and $f'(0)$ do not exist.

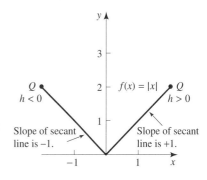

$$\lim_{h \to 0^+} \frac{f(0 + h) - f(0)}{h} = \lim_{h \to 0^+} \frac{h - 0}{h}$$

$$= \lim_{h \to 0^+} 1$$

$$= 1$$

On the other hand,

$$\lim_{h \to 0^-} \frac{f(0 + h) - f(0)}{h} = \lim_{h \to 0^-} \frac{-h - 0}{h}$$

$$= \lim_{h \to 0^-} -1$$

$$= -1$$

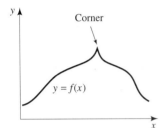

Figure 2.32
The derivative does not exist at a corner.

Since the limit from the left does not equal the limit from the right,

$$\lim_{h \to 0} \frac{f(0 + h) - f(0)}{h}$$

does not exist, and thus $f'(0)$ does not exist.   ∎

Remark. The very same situation occurs whenever a function has a corner. Thus, when the graph of a function has a corner, such as in Figure 2.32, $\lim_{h \to 0^-} \dfrac{f(c + h) - f(c)}{h}$ does not equal $\lim_{h \to 0^+} [f(c + h) - f(c)]/h$. Thus, $\lim_{h \to 0} [f(c + h) - f(c)]/h$, and therefore $f'(c)$ does not exist.

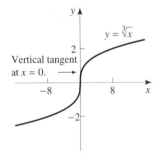

Figure 2.33
The derivative does not exist if the tangent is vertical.

EXAMPLE 6   A Function with a Vertical Tangent

If $f(x) = \sqrt[3]{x}$, determine if $f'(0)$ exists.

SOLUTION ■ A very carefully drawn graph of this function, such as the one shown in Figure 2.33, indicates that the curve *does* have a tangent line at $x = 0$. Because the tangent line is *vertical*, however, the slope does not exist. Therefore $f'(0)$ cannot exist.   ∎

EXPLORATION 3

Vertical Tangent

Verify Figure 2.33 on your graphing calculator.

An important connection between the existence of the derivative and continuity is detailed in the following theorem:

A Differentiable Function Is Continuous

If $y = f(x)$ has a derivative at $x = c$, then $f(x)$ is continuous at $x = c$.

The converse of this theorem is not true. That is, a function may be continuous at points where it is not differentiable. For example, the functions graphed in Figures 2.31, 2.32, and 2.33 are continuous at all points in their domains, but each has one value for which the function is not differentiable.

This theorem does say that if a function is not continuous at a point $x = c$, then $f'(c)$ cannot exist.

We now summarize the circumstances for which the derivative does not exist.

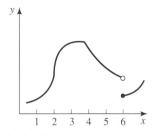

Figure 2.34
The derivative does not exist at the vertical tangent at $x = 2$, at the corner at $x = 4$, and at the point of discontinuity at $x = 6$.

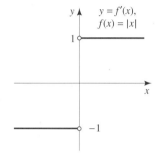

Figure 2.35
We can draw the graph of $y = f'(x)$ by realizing that $f'(x)$ is the slope of the tangent line to the graph of $y = f(x)$.

When the Derivative Fails to Exist

The derivative fails to exist in the following three circumstances:

1. The graph of the function has a corner.
2. The graph of the function has a vertical tangent.
3. The graph of the function has a break (discontinuity).

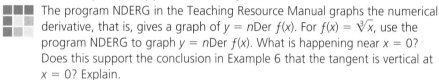

EXAMPLE 7   Determining Graphically When the Derivative Exists

Find the points where the derivative of the function whose graph is shown in Figure 2.34 fails to exist.

SOLUTION ■ The derivative does not exist at $x = 2$, 4, and 6 because at $x = 2$ the tangent is vertical, at $x = 4$ the graph has a corner, and at $x = 6$ the function is not continuous. ■

EXPLORATION 4

Additional Graphical Support for Example 6

The program NDERG in the Teaching Resource Manual graphs the numerical derivative, that is, gives a graph of $y = n\text{Der } f(x)$. For $f(x) = \sqrt[3]{x}$, use the program NDERG to graph $y = n\text{Der } f(x)$. What is happening near $x = 0$? Does this support the conclusion in Example 6 that the tangent is vertical at $x = 0$? Explain.

The following example emphasizes again that the derivative $f'(x)$ is itself a function and therefore can be graphed:

EXAMPLE 8   Drawing a Graph of $y = f'(x)$

Find $f'(x)$, and graph $y = f'(x)$ if $f(x) = |x|$.

SOLUTION ■ From Figure 2.31 we notice that for $x > 0$, $f(x) = |x| = x$ is a line with slope equal to 1. Thus, for $x > 0$, $f'(x) = m_{\tan}(x) = 1$. In a similar fashion we have for $x < 0$, $f'(x) = m_{\tan}(x) = -1$. In the graph shown in Figure 2.35 we notice that no value is given for $x = 0$, since we saw in Example 5 that $f'(0)$ does not exist. ■

EXPLORATION 5

Graphing the Derivative Function

Using the NDERG program, verify Figure 2.35.

SELF-HELP EXERCISE SET 2.4

1. If $y = f(x) = \dfrac{1}{x + 1}$, find $f'(x)$ when $x \neq -1$.

2. Find the equation of the line tangent to the curve $y = \sqrt{x}$ at the point (4, 2).

EXERCISE SET 2.4

In Exercises 1 through 6 find $f'(x)$ using the limit definition of $f'(x)$.

1. $f(x) = 5x - 3$

2. $f(x) = -2x + 3$

3. $f(x) = x^2 + 4$

4. $f(x) = 2x^2 - x$

5. $f(x) = 3x^2 + 3x - 1$

6. $f(x) = -3x^2 - x + 1$

In Exercises 7 through 12 find dy/dx using the limit definition.

7. $y = -2x^3 + x - 3$

8. $y = 3x^3 - 4x + 1$

9. $y = \dfrac{1}{x+2}, \quad x \neq -2$

10. $y = \dfrac{2}{3x+4}, \quad x \neq -\dfrac{4}{3}$

11. $y = \dfrac{1}{2x-1}, \quad x \neq \dfrac{1}{2}$

12. $y = \dfrac{3}{2-3x}, \quad x \neq \dfrac{2}{3}$

In Exercises 13 through 16 find dy/dt using the limit definition.

13. $y = \sqrt{t+1}, \quad t > -1$

14. $y = \sqrt{t-3}, \quad t > 3$

15. $y = \sqrt{2t+5}, \quad t > -\dfrac{5}{2}$

16. $y = \sqrt{1-t}, \quad t < -1$

17. In each of the problems in Exercises 1, 3, and 5 find the equation of the tangent line to the curve at the point $(1, f(1))$. Support your answer by using a graphing calculator to graph the function and the tangent line on the same screen.

18. In each of the problems in Exercises 2, 4, and 6 find the equation of the tangent line to the curve at the point $(-1, f(-1))$. Support your answer by using a graphing calculator to graph the function and the tangent line on the same screen.

19. Find the equation of the tangent line to the curve $y = \ln x$ at the point $(1, 0)$. Recall that $\dfrac{d}{dx} \ln x = \dfrac{1}{x}$. Support your answer by using a graphing calculator to graph the function and the tangent line on the same screen.

20. Find the equation of the tangent line to the curve $y = \sqrt{x}$ at the point $(4, 2)$. Recall that $\dfrac{d}{dx} \sqrt{x} = \dfrac{1}{2\sqrt{x}}$. Support your answer by using a graphing calculator to graph the function and the tangent line on the same screen.

In Exercises 21 through 28 sketch a graph of $y = f(x)$ to find where $m_{\tan}(x)$ and thus $f'(x)$ exist. Use your graphing calculator to help you in Exercises 27 and 28.

21. $f(x) = \begin{cases} 0 & \text{if } x \leq 0 \\ x & \text{if } x > 0 \end{cases}$

22. $f(x) = -|x|$

23. $f(x) = -|x-1|$

24. $f(x) = \begin{cases} x & \text{if } x \leq 1 \\ -x+2 & \text{if } x > 1 \end{cases}$

25. $f(x) = \begin{cases} x & \text{if } x \leq 1 \\ 1 & \text{if } x > 1 \end{cases}$

26. $f(x) = \begin{cases} -x & \text{if } x \leq 0 \\ 0 & \text{if } x > 0 \end{cases}$

27. $f(x) = \begin{cases} \sqrt{1-(x-1)^2} & \text{if } 0 \leq x \leq 1 \\ -\sqrt{1-(x+1)^2} & \text{if } -1 \leq x < 0 \end{cases}$

28. $f(x) = \begin{cases} -\sqrt{1-(x-1)^2} & \text{if } 0 \leq x \leq 1 \\ \sqrt{1-(x+1)^2} & \text{if } -1 \leq x < 0 \end{cases}$

29. In this exercise we estimate $f'(1)$ graphically if $f(x) = x^3$. Set the MODE to Grid On on your graphing calculator and graph $y_1 = x^3$ in a window with dimensions $[0.75, 1.25]$ by $[0.25, 2]$. Use $x_{scl} = 0.25$ and $y_{scl} = 0.25$. Now look at the window and estimate $f'(1)$.

30. In this exercise we estimate $f'(1)$ graphically if $f(x) = 2 - x^2$. Set the MODE to Grid On on your graphing calculator and graph $y_1 = 2 - x^2$ in a window with dimensions

$[0.75, 1.25]$ by $[0.5, 1.6]$. Use $x_{scl} = 0.25$ and $y_{scl} = 0.25$. Now look at the window and estimate $f'(1)$.

In Exercises 31 through 36, draw a graph of $y = f'(x)$ for the graphs shown here.

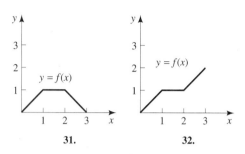

31.   **32.**

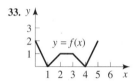

33.

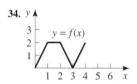

34.

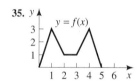

35.

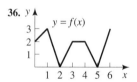

36.

In Exercises 37 through 40 use a graphing calculator to determine if $f'(x)$ exists at the indicated point as follows. **(a)** Graph $y = f(x)$, and try to estimate the tangent at the indicated point. **(b)** Use the NDERG program to graph $y = n\text{Der } f(x)$, and observe the graph near the indicated point.

37. $f(x) = \sqrt[5]{x}, \quad x_0 = 0$

38. $f(x) = \sqrt[3]{x-1}, \quad x_0 = 1$

39. $f(x) = \sqrt[3]{x^2}, \quad x_0 = 0$

40. $f(x) = \sqrt[3]{(x-4)^2}, \quad x_0 = 4$

Applications

41. Profits. The profits in millions of dollars of a firm is given by $P = \sqrt{4x + 1} - 2$, where x is the number in thousands of items sold. Find the instantaneous rate of change when x is **(a)** 2 **(b)** 6. Interpret your answer.

42. Revenue. The revenue function for a firm is given by $R = \sqrt{2x + 5}$. Find the instantaneous rate of change when x is **(a)** 2 **(b)** 10. Interpret your answer.

43. Sales. Suppose that t weeks after the end of an advertising campaign the weekly sales are $s = \dfrac{100}{t + 1}$. At what rate are sales changing when t equals **(a)** 1 **(b)** 4? Interpret your answer.

44. Cost. Suppose that the cost function for a firm is given by $C = x + \sqrt{x}$. Find the instantaneous rate of change of costs when x is **(a)** 1 **(b)** 4 **(c)** 9.

45. Price. Suppose the price of an item as a function of the number sold is given by $p(x) = \dfrac{5}{3x + 1}$. Find the instantaneous rate of change of the price when x is **(a)** 3 **(b)** 8.

46. Finance. Granger [58] in Great Britain found that the relationship between yield y on 20-year government bonds was approximated by $y = -3.42 + (15.23/x)$, where x is demand deposits divided by gross national product. Find the rate of change of y with respect to x.

Enrichment Exercises

In Exercises 47 and 48 a graph of $y = f(x)$ is given together with several tangent lines. Estimate $f'(x)$ at the points where the tangent lines are shown, and use this as an aid in graphing $y = f'(x)$.

47.

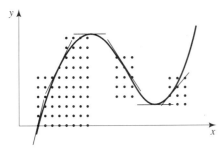

48.

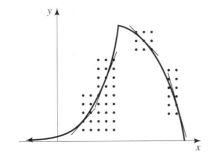

49. In Example 6 and Exploration 3, we indicated graphically why the derivative of $f(x) = \sqrt[3]{x^2}$ does not exist at $x = 0$. Confirm this analytically using the limit definition of the derivative.

50. Use the limit definition of derivative to show that $f'(0)$ does not exist if $f(x) = \sqrt[5]{x^2}$.

51. In this exercise is another way in which your graphing calculator can mislead you. Let $f(x) = \sqrt{x^2 + 0.0001}$. Graph on a window with dimensions $[-10, 10]$ by $[0, 10]$. From the graph can you tell if $f'(0)$ exists? After deciding this question, support your answer graphically and numerically

as follows. Graph $y = \dfrac{\sqrt{h^2 + 0.0001} - \sqrt{0.0001}}{h}$ on a window with dimensions $[-0.1, 0.1]$ by $[-1, 1]$. What do you conclude about the limit of this expression as h goes to zero? Support your answer numerically by evaluating this expression for values of h, such as 0.0001 and 0.00001.

52. Use the limit definition of derivative to show that $f'(0)$ does exist if $f(x)$ is the function given in the previous exercise.

53. Let $f(x) = 0$ if $x < 0$ and x^2 if $x \geq 0$. Explore graphically and numerically if $f'(0)$ exists. Confirm analytically, that is, using the limit definition of derivative.

54. Explain in complete sentences the circumstances under which the derivative of a function does not exist.

55. Your graphing calculator has a SIN key. With your calculator in radian mode, graph $y = f(x) = \sin x$ using a window with dimensions $[-6.14, 6.14]$ by $[-1, 1]$ in order to familiarize yourself with this function. As you can see, this function moves back and forth between -1 and 1. We wish to estimate $f'(0)$. For this purpose graph using a window with dimensions $[-0.5, 0.5]$ by $[-0.5, 0.5]$. From the graph estimate $f'(0)$.

56. Let $y = f(x) = \sin x$. With your calculator in radian mode graph on your graphing calculator. As you can readily see, $\sin(0) = 0$. By definition

$$f'(0) = \lim_{h \to 0} \frac{\sin(0 + h) - \sin(0)}{h} = \lim_{h \to 0} \frac{\sin h}{h}$$

Graph $\dfrac{\sin x}{x}$ on your graphing calculator to estimate the limit as $x \to 0$. Support numerically. Compare your answer to what you found in the previous exercise.

57. Your graphing calculator has a COS key. Graph $y = f(x) = \cos x$ using a window with dimensions $[-6.14, 6.14]$ by $[-1, 1]$ in order to familiarize yourself with this function. As you can see, this function moves back and forth between -1 and 1. We wish to estimate $f'(\pi/2)$, where $\pi/2 \approx 1.57$. For this purpose graph using a window with dimensions $[1.07, 2.07]$ by $[-0.5, 0.5]$. From the graph estimate $f'(1.57)$.

58. Let $y = f(x) = \cos x$. With your calculator in radian mode graph on your graphing calculator. As you can readily see, $\cos(\pi/2) = 0$. By definition

$$f'\left(\frac{\pi}{2}\right) = \lim_{h \to 0} \frac{\cos[(\pi/2) + h] - \cos(\pi/2)}{h}$$

$$= \lim_{h \to 0} \frac{\cos[(\pi/2) + h]}{h}$$

Graph $\dfrac{\cos(\pi/2 + x)}{x}$ on your graphing calculator to estimate the limit as $x \to 0$. Support numerically. Compare your answer to what you found in the previous exercise.

59. With your calculator in radian mode graph $y_1 = \sin x$ and $y_2 = \cos x$ on your graphing calculator, and familiarize yourself with these functions. Now replace $y_1 = \sin x$ with

$$y_1 = \frac{\sin(x + 0.001) - \sin x}{0.001}$$ and graph. This latter function is approximately the derivative of $\sin x$. How does the graph of this latter function compare with the graph of $\cos x$? Does this show that $\dfrac{d}{dx}(\sin x) = \cos x$?

60. With your calculator in radian mode graph $y_1 = \cos x$ and $y_2 = -\sin x$ on your graphing calculator, and familiarize yourself with these functions. Now replace $y_1 = \cos x$ with

$$y_1 = \frac{\cos(x + 0.001) - \cos x}{0.001}$$ and graph. This latter function is approximately the derivative of $\cos x$. How does the graph of this latter function compare with the graph of $-\sin x$? Does this show that $\dfrac{d}{dx}(\cos x) = -\sin x$?

SOLUTIONS TO SELF-HELP EXERCISE SET 2.4

1. First find an expression for the average rate of change of y with respect to x from x to $x + h$, and simplify. This gives

$$\frac{f(x + h) - f(x)}{h} = \frac{\dfrac{1}{x + h + 1} - \dfrac{1}{x + 1}}{h}$$

$$= \frac{1}{h}\left[\frac{1}{x + h + 1} - \frac{1}{x + 1}\right]$$

$$= \frac{1}{h}\left[\frac{x + 1 - (x + h + 1)}{(x + 1)(x + h + 1)}\right]$$

$$= \frac{1}{h}\left[\frac{-h}{(x + 1)(x + h + 1)}\right]$$

$$= -\frac{1}{(x + 1)(x + h + 1)}$$

Now find the derivative by taking the limit of this expression as $h \to 0$, obtaining

$$f'(x) = \lim_{h \to 0} \frac{f(x + h) - f(x)}{h}$$

$$= \lim_{h \to 0} \frac{-1}{(x + 1)(x + h + 1)} = -\frac{1}{(x + 1)^2}$$

provided $x \ne -1$.

2. From Example 3 of the text, we have $\dfrac{d}{dx}(\sqrt{x}) = \dfrac{1}{2\sqrt{x}}$. Thus,

$$m_{\tan}(4) = f'(4) = \frac{1}{2\sqrt{4}} = \frac{1}{4}$$

Thus, the equation of the tangent line at $x = 4$ is

$$y - 2 = \frac{1}{4}(x - 4)$$

2.5 THE TANGENT LINE APPROXIMATION

■ Local Linearity ■ The Tangent Line Approximation

APPLICATION: APPROXIMATING CHANGES IN FUNCTIONS

One month after an experimental drug was used to treat a tumor, the radius of the spherical tumor was measured to have decreased from 2 cm to 1.8 cm. Using the derivative, what linear function can be used to estimate the change in the volume of the spherical tumor? See Example 2 for the answer.

APPLICATION: MAKING APPROXIMATIONS

What is an easy-to-calculate approximation to ln 1.1 without using a calculator or computer? See Example 3 for the answer.

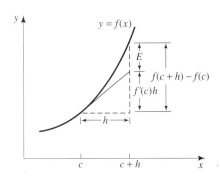

Figure 2.36
If $f'(c)$ exists, $f(c + h) - f(c) \approx f'(c)h$ for small h.

Local Linearity

Given a function $y = f(x)$ differentiable at c, we know that

$$f'(c) = \lim_{h \to 0} \frac{f(c + h) - f(c)}{h}$$

From our knowledge of limits, we know that if h is small, then

$$f'(c) \approx \frac{f(c + h) - f(c)}{h}$$

Multiplying both sides of this expression by h yields

$$f(c + h) - f(c) \approx f'(c)h$$

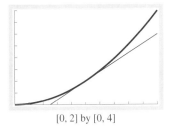

[0, 2] by [0, 4]

Screen 2.17
The graph of $y_1 = x^2$ is nearly the same as the graph of the tangent line $y_2 = 2x - 1$ near $x = 1$.

The term $f(c + h) - f(c)$ is the actual change in y and is shown in Figure 2.36. The term $f'(c)h$ gives the change in y that occurs when moving along the tangent line. According to Figure 2.36, we expect the change in y along the tangent line to be an increasingly better approximation to the actual change in y as h becomes smaller. The term $E = [f(c + h) - f(c)] - f'(c)h$ represents the error in using the term $f'(c)h$ to approximate the actual change in y. We see that as h approaches zero, this error E also approaches zero.

Approximating the Change in y

[0.75, 1.25] by [0.5, 1.5]

Screen 2.18
As we ZOOM in we see that the graph of $y_1 = x^2$ looks more like the graph of the tangent line $y_2 = 2x - 1$ near $x = 1$.

If $y = f(x)$ is a differentiable function at c, then the change in y, given by $f(c + h) - f(c)$, can be approximated by the linear function $f'(c)h$. That is,

$$f(c + h) - f(c) \approx f'(c)h \qquad [1]$$

if h is small.

E X A M P L E 1   The Tangent Line Approximation

Graph $y_1 = x^2$ and the tangent line to the graph of this curve at the point $(1, 1)$. Check to see how close the tangent line is to the curve. Use the ZOOM feature to check further.

[0.94, 1.06] by [0.875, 1.125]

Screen 2.19
As we ZOOM in even further we see that the graph of $y_1 = x^2$ is about the same as the graph of the tangent line $y_2 = 2x - 1$ near $x = 1$.

S O L U T I O N ■ At the beginning of the previous section we found the tangent line to the graph of the curve $y = x^2$ at $(1, 1)$ to be $y = 2x - 1$. Screen 2.17 shows graphs of $y_1 = x^2$ and the tangent line $y_2 = 2x - 1$ using a window with dimensions [0, 2] by [0, 4]. We can already see that the graph of the tangent line nearly merges with the graph of the curve on the interval [0.8, 1.2]. ZOOMing in once at the point $(1, 1)$ by a factor of 4 gives Screen 2.18. ZOOMing one more time by the same factor gives Screen 2.19. Notice that in this last window the graph of $y = x^2$ appears to be a straight line! We see under magnification that the graph of the curve near $x = 1$ is about the same as the graph of the tangent line. ■

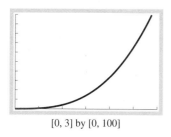

[0, 3] by [0, 100]

Screen 2.20
We see the graph of $y = \frac{4}{3}\pi x^3$ on [0, 3].

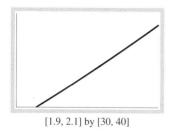

[1.9, 2.1] by [30, 40]

Screen 2.21
Near $x = 2$, the graph of $y = \frac{4}{3}\pi x^3$ is about the same as the graph of a line with slope 50.3.

The following example illustrates a way in which this approximation can be used:

EXAMPLE 2   Approximating the Change in a Cancerous Growth

One month after an experimental drug was used to treat a tumor, the radius of the spherical tumor was measured to have decreased from 2 cm to 1.8 cm. Use Equation [1] to approximate the change in volume.

SOLUTION ■ If r is the radius, then the volume V is given by $V(r) = \frac{4}{3}\pi r^3$. Since r goes from 2 to 1.8, we take $r_0 = 2$ and $h = -0.2$. Screen 2.20 shows a graph of this function in a window of dimensions [0, 3] by [0, 100]. By ZOOMing several times about the point (2, 33.5), where $V(2) \approx 33.5$, we obtain Screen 2.21. This graph is not distinguishable from a straight line of slope about 50.3. Thus, $V'(2) \approx 50.3$. The actual change in volume is given by

$$V(r_0 + h) - V(r_0) = V(1.8) - V(2)$$

Using Equation (1) to approximate this change, we have

$$V(1.8) - V(2) = V(r_0 + h) - V(r_0)$$
$$\approx V'(r_0)h$$
$$\approx 50.3(-0.2)$$
$$= -10.1$$

Compare this number to the actual change in volume.

$$\Delta V = V(1.8) - V(2) = -9.1 \quad ■$$

The Tangent Line Approximation

We now turn our attention to establishing a fundamentally important geometrical fact discovered in Example 1 concerning the behavior of a function near a point where the function is differentiable. Briefly stated, if $f'(c)$ exists, then for values of x near c, the graph of the function $y = f(x)$ is approximately the same as the graph of the tangent line through $(c, f(c))$.

We now see analytically why this is true. Move the $f(c)$ term in Equation (1) to the right-hand side, and obtain

$$f(c + h) \approx f(c) + f'(c)h$$

Think of c as fixed and h as the variable, and let $x = c + h$. Then the last equation becomes

$$f(x) \approx f(c) + f'(c)(x - c)$$

Recall that the right-hand side of this last equation is the equation of the line tangent to the curve at c. Thus, we are saying that the graph of $y = f(x)$ is approximately the same as the graph of the line tangent to the curve at $x = c$ (Fig. 2.37).

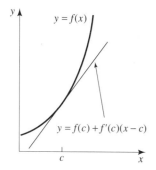

Figure 2.37
If $f'(c)$ exists, the graph of $y = f(x)$ is approximately the same as the graph of the tangent line for x near c.

Tangent Line Approximation

If $y = f(x)$ is differentiable at c, then for values of x near c

$$f(x) \approx f(c) + f'(c)(x - c) \qquad [2]$$

Thus, for values of x near c the graph of the curve $y = f(x)$ is approximately the same as the graph of the tangent line through the point $(c, f(c))$.

Another way of viewing this approximation is to realize that using the tangent line to approximate the function $y = f(x)$ in Figure 2.37 near c is the same as assuming the function changes at a constant rate of change near c given by the instantaneous rate of change at c.

EXAMPLE 3   The Tangent Line Approximation

Use the tangent line approximation in Equation (2) to find an approximation to ln 1.1 without resorting to a calculator.

SOLUTION ■ Let $f(x) = \ln x$, $x = 1.1$, and $c = 1$. Recall from the last section that $f'(x) = 1/x$. Using the approximation in Equation (2), we have

$$\ln 1.1 = f(1.1)$$
$$\approx f(1) + f'(1)(1.1 - 1)$$
$$= \ln 1 + \frac{1}{1}(0.1)$$
$$= 0 + 0.1$$
$$= 0.1 \quad ■$$

Compare this to the actual ln 1.1, which to four decimal places is 0.0953. This approximation is off by only 0.0047. In Figure 2.36, this means that $E \approx -0.0047$.

EXPLORATION 1

Tangent Line Approximation

 On your graphing calculator graph $y_1 = \ln x$ and the tangent line $y_2 = x - 1$ to this graph at $(1, 0)$ on the same screen of dimensions $[0, 3]$ by $[-2, 2]$. Notice already how close the graph of the tangent line is to the graph of the curve. ZOOM just once about the point $(1, 0)$, and observe how close the graph of the curve is to the graph of the tangent line.

EXAMPLE 4   Tangent Line Approximation

Let $y = f(x) = e^x$. In Example 7 of Section 2.3 we showed that $f'(0) = 1$. Therefore, near $x = 0$ we expect that y should increase at the same rate as x. Make a table of values for e^x about $x = 0$ in increments of 0.05 and verify this.

SOLUTION ■ We have

x	-0.10	-0.05	0	0.05	0.10
e^x	0.905	0.951	1	1.051	1.105

Notice that a change 0.05 in x results in a change of about 0.05 in y. ■

EXAMPLE 5   Estimating Population

The table gives the annual population of the United States from 1790 to 1794. Based on the data from 1790 to 1792 and using the ideas in this section, approximate the population in 1793 and 1794.

Year	1790	1791	1792	1793	1794
Population (in thousands)	3926	4056	4194	4332	4469

SOLUTION ■ A graph of the data from 1790 to 1792 is shown in Figure 2.38. Recall that in the tangent line approximation to a curve at c_0, we assume that the rate of change of the function near c_0 is constant and given by the instantaneous

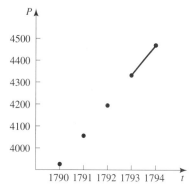

Figure 2.38
We assume that the population increases at the same rate in 1793–1794 as the average rate in 1790–1792.

rate of change at c_0. We have no curve in Figure 2.37. By analogy, however, we can calculate the average rate of change from 1790 to 1792 and assume that the rate of change of population continues at this value for the next two years. This is indicated in Figure 2.38 by the line drawn through the last two data points.

The average rate of change from 1790 to 1792 is

$$\frac{4194 - 3926}{2} = 134$$

If the population is assumed to continue at this rate for the next year, then in thousands

The approximate population in 1793 is $4194 + 134(1) = 4328$

The approximate population in 1794 is $4194 + 134(2) = 4462$ ∎

From Figure 2.36 we should not be surprised if the approximations become less accurate the further x is removed from c_0. In the same way, we should not be surprised if our estimate of the population in 1794 in Example 5 is less accurate than the estimate for 1793. In fact, the population in thousands in 1793 and 1794 was 4332 and 4469, respectively. The error in the first instance was 4000 and in the second 7000.

SELF-HELP EXERCISE SET 2.5

1. The outside of a cube with edges of length 10 in. is painted with a paint 0.10 in. thick. Use the tangent line approximation to estimate the volume of paint used. Use the ZOOM feature on your calculator to estimate any derivative that you need.

EXERCISE SET 2.5

In Exercises 1 through 6 use the linear function $f(c) + f'(c)(x - c)$ to approximate the given quantity. Recall from the previous section that $\frac{d}{dx}(\sqrt{x}) = \frac{1}{2\sqrt{x}}$, $\frac{d}{dx}(\ln x) = \frac{1}{x}$, and $\frac{d}{dx}(x^{-1}) = -x^{-2}$. Use a calculator to find the actual answer, and compare your approximation to this. (Of course the answer on your calculator is also an approximation.) In each case use your graphing calculator to graph the curve and the tangent line at the appropriate point to support your answer.

1. $\sqrt{15.5}$

2. $\sqrt{25.5}$

3. $\ln 0.95$

4. $\ln 1.05$

5. $\dfrac{1}{1.05}$

6. $\dfrac{1}{0.96}$

In Exercises 7 through 12 graph the function on your graphing calculator, and use the ZOOM feature repeatedly about the point $(c, f(c))$ until the graph looks like a straight line. Use the TRACE function on your calculator to help find the slope of this line. What is $f'(c)$?

7. $y = \sqrt[3]{x}$, $c = 1$

8. $y = x^3$, $c = 2$

9. $y = e^x \ln x$, $c = 2$

10. $y = e^x$, $c = 1$

11. $f(x) = 0.5x^4 - 2\sqrt{x} + 1$, $c = 1$

12. $f(x) = xe^x$, $c = 0$

In Exercises 13 through 18 use the linear function $f(c) + f'(c)(x - c)$ to approximate the given quantity. Use the results of Exercises 7 through 12. Use a calculator to find the actual answer, and compare your approximation to this. (Of course the answer on your calculator is also an approximation.) In each case use your graphing calculator to graph the curve and the tangent line at the appropriate point to support your answer.

13. $\sqrt[3]{1.1}$

14. $(1.8)^3$

15. $e^{1.9} \ln 1.9$

16. $e^{1.1}$

17. $0.5(1.1)^4 - 2\sqrt{1.1} + 1$

18. $0.1e^{0.1}$

19. If the side of a square decreases from 4 in. to 3.8 in., use the linear approximation formula to estimate the change in area. Use the ZOOM feature on your calculator to estimate any derivative that you need.

20. If the side of a cube decreases from 10 ft to 9.6 ft, use the linear approximation formula to estimate the change in volume. Use the ZOOM feature on your calculator to estimate any derivative that you need.

Applications

21. Demand. The demand equation for a certain product is given by $p(x) = \dfrac{10}{\sqrt{x+1}}$. Use the linear approximation formula to estimate the change in price p as x changes from 99 to 102. Use the numerical derivative feature on your calculator to estimate any derivative that you need.

22. Revenue. The revenue equation for a certain product is given by $R(x) = 10\sqrt{x}(100 - x)$. Use the linear approximation formula to estimate the change in revenue as x changes from 400 to 402. Use the numerical derivative on your calculator to estimate any derivative that you need.

23. Cost. The cost equation for a certain product is given by $C(x) = \sqrt{x^3}(x + 100) + 1000$. Use the linear approximation formula to estimate the change in costs as x changes from 900 to 904. Use the numerical derivative on your calculator to estimate any derivative that you need.

24. Profit. The profit equation for a certain product is given by $P(x) = \sqrt{x}(1000 - x) - 5000$. Use the linear approximation formula to estimate the change in profits as x changes from 25 to 27. Use the numerical derivative on your calculator to estimate any derivative that you need.

25. Biology. Use the linear approximation formula to estimate the change in volume of a spherical bacterium as the radius increases from 2 microns to 2.1 microns. Use the numerical derivative on your calculator to estimate any derivative that you need.

26. Learning. The time T it takes to memorize a list of n items is approximated by $T(n) = 3n\sqrt{n} - 3$. Find the approximate change in time required from memorizing a list of 12 items to memorizing a list of 14 items. Use the numerical derivative feature on your calculator to estimate any derivative that you need.

27. Pollution. The pollution abatement and control expenditures in billions of dollars made during selected years in this country are given in the table. Based on this data estimate the total expenditures for 1989 and for 1990.

Year	1984	1986	1988
Expenditures (in billions of dollars)	68.9	78.7	86.1

Enrichment Exercises

28. You observe an object moving in a straight line and note that at precisely noon the instantaneous velocity is 45 ft/s. Approximately how far does the object move in the next 5 s? Explain carefully what facts, learned in this section, you are using to obtain your answer.

29. In Exercise 51 of the previous section we examined the graph of $y = f(x) = \sqrt{x^2 + 0.0001}$ to see if $f'(0)$ existed or not. Graph on your graphing calculator using a window of dimensions $[-10, 10]$ by $[-1, 10]$. Notice that the graph appears to have a corner at $x = 0$. If the graph does have a corner at $x = 0$, then $f'(0)$ does not exist. Using the ZOOM feature, zoom repeatedly about the point $\left(0, \sqrt{0.0001}\right)$ until the graph does not change anymore. Explain what you have observed and what relevance these observations have to do with the existence of $f'(0)$.

30. Graph the function $y = f(x) = |x|$ on your graphing calculator, and repeatedly ZOOM about the point $(0, 0)$. Explain what you observe and what relevance your observations have to do with the existence of $f'(0)$.

31. Biology. An anatomist wishes to measure the surface area of a bone assumed to be a cylinder. The length l can be measured with great accuracy and can be assumed to be known exactly. The radius of the bone varies slightly throughout its length, however, making it difficult to decide what the radius should be. Discuss how slight changes in the radius r approximately affect the surface area. Use the numerical derivative feature on your calculator to estimate any derivative that you need.

32. A cone has a height of 6 in. The radius has been measured as 1 in. with a possible error of 0.01 in. Estimate the maximum error in the volume $(\pi r^2 h/3)$ of the cone. Use the numerical derivative on your calculator to estimate any derivative that you need.

33. In Exercise 59 of the previous section you found that $\dfrac{d}{dx}(\sin x) = \cos x$. Notice from your calculator that $\sin 0 = 0$ and $\cos 0 = 1$. Use all of this and the tangent line approximation to approximate $\sin 0.02$. Compare your approximation to the value your calculator gives.

34. In Exercise 60 of the previous section you found that $\frac{d}{dx}(\cos x) = -\sin x$. Notice from your calculator that

$\sin 0 = 0$ and $\cos 0 = 1$. Use all of this and the tangent line approximation to approximate $\cos 0.02$. Compare your approximation to the value your calculator gives.

SOLUTIONS TO SELF-HELP EXERCISE SET 2.5

1. The volume of a cube with edges of length x is given by $V(x) = x^3$. The paint increases the length of each edge by twice 0.10 or by 0.20. Thus, $c = 10$ and $h = 0.20$. Then $V(10.2) - V(10) \approx V'(10)(0.2)$. Graphing $y = x^3$ and

ZOOMing about the point $(10, 1000)$ a few times gives a straight line with slope about 300. Thus,

$$V(10.2) - V(10) \approx V'(10)(0.2) = 300(0.2) = 60$$

Compare this to $V(10.2) - V(10) \approx 61$.

CHAPTER **2**

SUMMARY OUTLINE

■ **Limit of a function.** If as x approaches a, $f(x)$ approaches the number L, then we say that L is the limit of $f(x)$ as x approaches a, and write $\lim_{x \to a} f(x) = L$. p. 99.

■ **Limits from left and right.** If as x approaches a from the left, $f(x)$ approaches the number M, then we write $\lim_{x \to a^-} f(x) = M$. If as x approaches a from the right, $f(x)$ approaches the number N, then we write $\lim_{x \to a^+} f(x) = N$. p. 99.

■ **Existence of a limit.** $\lim_{x \to a} f(x) = L$ if, and only if, both of the limits $\lim_{x \to a^-} f(x)$ and $\lim_{x \to a^+} f(x)$ exist and equal L. p. 100.

■ **Rules for limits.** p. 105. Assume that $\lim_{x \to a} f(x) = L$ and $\lim_{x \to a} g(x) = M$. Then

$$\lim_{x \to a} cf(x) = cL$$

$$\lim_{x \to a} (f(x) \cdot g(x)) = L \cdot M$$

$$\lim_{x \to a} (f(x) \pm g(x)) = L \pm M$$

$$\lim_{x \to a} \frac{f(x)}{g(x)} = \frac{L}{M} \quad \text{if } M \neq 0$$

$$\lim_{x \to a} (f(x))^n = L^n, \quad \text{if } L^n \text{ is defined and } L \neq 0$$

■ **Limit of polynomial function.** If $p(x)$ is a polynomial function, then $\lim_{x \to a} p(x) = p(a)$. p. 104.

■ **Continuity.** The function $f(x)$ is continuous at $x = a$ if $\lim_{x \to a} f(x) = f(a)$. p. 110.

■ A polynomial function is continuous everywhere. p. 110.

■ A rational function is continuous at every point at which the denominator is not zero. p. 110.

■ **Constant-sign theorem.** If $f(x)$ is continuous on (a, b) and $f(x) \neq 0$ on (a, b), then $f(x)$ is either always positive or always negative on (a, b). p. 111.

■ **Average and instantaneous velocity.** These are, respectively, $\frac{s(t + h) - s(t)}{h}$ and $\lim_{h \to 0} \frac{s(t + h) - s(t)}{h}$, where $s(t)$ represents position at time t. p. 124.

■ **Average and instantaneous rate of change** for $f(x)$ are, respectively, $\frac{f(x + h) - f(x)}{h}$ and $\lim_{h \to 0} \frac{f(x + h) - f(x)}{h}$. p. 126.

■ **The slope of the tangent line** to the graph of $y = f(x)$ at $x = c$ is $\lim_{h \to 0} \frac{[f(c + h) - f(c)]}{h}$, if this limit exists. p. 128.

- The instantaneous rate of change of $f(x)$ at c equals the slope of the tangent line to the graph of $f(x)$ at c. p. 128.

- **Definition of derivative.** The derivative of a function $y = f(x)$, denoted by $f'(x)$ or $\dfrac{dy}{dx}$, is defined to be $f'(x) = \lim\limits_{h \to 0} \dfrac{f(x + h) - f(x)}{h}$, if this limit exists. p. 134.

- The **equation of the tangent line** to the graph of the curve $y = f(x)$ at $x = c$ is $y - f(c) = f'(c)(x - c)$. p. 128.

- A differentiable function is continuous. p. 138.

- The number e to five decimals is 2.71828. p. 114.

- An account with an initial amount of P that is growing at an annual rate of r (expressed as a decimal) compounded continuously becomes Pe^{rt} after t years. p. 114.

- If a sum of money is invested at an annual rate of r expressed as a decimal and is compounded continuously, then the **effective yield** is $r_{\text{eff}} = e^r - 1$. p. 119.

- If an account earns an annual rate of r expressed as a decimal and is compounded continuously, then the amount P, called the **present value**, needed presently in this account now so that a future balance of A will be attained in t years is $P = Ae^{-rt}$. p. 115.

- **Natural Logarithm.** If $x > 0$, the natural logarithm of x, denoted by $\ln x$, is defined as $y = \ln x$ if, and only if, $x = e^y$. p. 116.

- **Properties of Natural Logarithms.** p. 82, 83 and 116.

$$e^{\ln x} = x \quad \text{if } x > 0 \qquad \ln e^x = x \quad \text{for all } x$$

$$\ln 1 = 0 \qquad\qquad\qquad \ln xy = \ln x + \ln y$$

$$\ln \frac{x}{y} = \ln x - \ln y \qquad \ln x^a = a \ln x$$

- The last three equations require that $x > 0$, $y > 0$.

- For any $a > 0$, $a^x = e^{kx}$ if $k = \ln a$. p. 117.

- If a quantity grows according to $Q(t) = Q_0 e^{kt}$, where $k > 0$, k is called the **growth constant**. p. 115.

- If a quantity decays according to $Q(t) = Q_0 e^{-kt}$, where $k > 0$, k is called the **decay constant**. p. 115.

- The **half-life** of a radioactive substance is the time that must elapse for half of the material to decay. p. 120.

- **Tangent Line Approximation.** If $y = f(x)$ is a differentiable function at c, then $f(c + h) - f(c) \approx f'(c)h$ if h is small. p. 144.

CHAPTER **2**

REVIEW EXERCISES

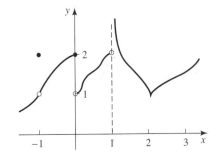

1. For the function in the figure, find $\lim\limits_{x \to a^-} f(x)$, $\lim\limits_{x \to a^+} f(x)$, and $\lim\limits_{x \to a} f(x)$ for $a = -1, 0, 1, 2$.

2. For the function in the previous figure, find all points where $f(x)$ is not continuous.

3. For the function in the figure in Exercise 1, find all points where $f(x)$ does not have a derivative.

In Exercises 4 through 25 find the limits if they exist.

4. $\lim_{x \to 1} 12.7$

5. $\lim_{x \to 1} \pi^3$

6. $\lim_{x \to 1} (x^2 + 2x - 3)$

7. $\lim_{x \to -1} (x^3 - 3x - 5)$

8. $\lim_{x \to 0} (3x^5 + 2x^2 - 4)$

9. $\lim_{x \to 1} \dfrac{2x^2 + 1}{x + 2}$

10. $\lim_{x \to 2} \dfrac{x + 5}{x - 3}$

11. $\lim_{x \to 1} \dfrac{x^2 + x - 2}{x - 1}$

12. $\lim_{x \to 3} \dfrac{x^2 + x - 12}{x - 3}$

13. $\lim_{x \to 2} \dfrac{x^2 + 1}{x - 2}$

14. $\lim_{x \to -2} \dfrac{x + 3}{x + 2}$

15. $\lim_{x \to 2} \sqrt{x^3 - 4}$

16. $\lim_{x \to -1} \sqrt{10 - x^2}$

17. $\lim_{x \to 4} \left(\sqrt{x} - \dfrac{1}{\sqrt{x}} \right)$

18. $\lim_{x \to -2} \sqrt{1 - x^2}$

19. $\lim_{x \to \infty} \dfrac{2x^3 + x - 3}{x^3 + 3}$

20. $\lim_{x \to -\infty} \dfrac{x^3 + x + 2}{x^4 + 1}$

21. $\lim_{x \to \infty} \dfrac{x^4 + x^2 + 1}{x^3 + 1}$

22. $\lim_{x \to \infty} \dfrac{1}{1 + e^x}$

23. $\lim_{x \to \infty} (1 + e^{-2x})$

24. $\lim_{x \to \infty} \dfrac{1}{1 + \ln x}$

25. $\lim_{x \to 0^+} \dfrac{1}{1 + \ln x}$

In Exercises 26 through 31 use the definition of limit to find the derivative.

26. $f(x) = 2x^2 + x - 3$

27. $f(x) = x^2 + 5x + 1$

28. $f(x) = \dfrac{1}{x^2}$

29. $f(x) = \dfrac{1}{2x + 5}$

30. $f(x) = \sqrt{x + 7}$

31. $f(x) = \sqrt{2x - 1}$

In Exercises 32 through 35 find the equation of the tangent line to the graph of each of the functions at the indicated point. Support your answer using a graphing calculator.

32. $y = x^3 + 1$, $c = 2$

33. $y = x^{3/2} + 2$, $c = 4$

34. $y = x^5 + 3x^2 + 2$, $c = 1$

35. $y = \dfrac{1}{\sqrt{x}}$, $c = 9$

In Exercises 36 through 39 ZOOM in on the graph of the curve at the point $(c, f(c))$ until the graph is a straight line. Estimate $f'(c)$ by estimating the slope of the straight line.

36. $f(x) = \sqrt[5]{x}$, $c = 1$

37. $f(x) = \sqrt[3]{x^2}$, $c = 8$

38. $f(x) = \dfrac{1}{x^3}$, $c = 2$

39. $f(x) = \sqrt{x} + \dfrac{1}{\sqrt{x}}$, $c = 4$

In Exercises 40 through 43 use the derivatives found in the previous four exercises and the linear approximation to find approximations of the following quantities:

40. $\sqrt[5]{1.05}$

41. $\sqrt[4]{(7.7)^2}$

42. $\dfrac{1}{(2.08)^3}$

43. $\sqrt{4.016} + \dfrac{1}{\sqrt{4.016}}$

44. Determine by numerical and graphical means, if $f'(0)$ exists with $f(x) = \sqrt[5]{x^2}$.

45. Show analytically, using the limit definition of derivative, that $f'(0)$ does not exist if $f(x) = \sqrt[5]{x^2}$.

46. If $f(2) = 3$ and $f(x) = \dfrac{x^2 - x - 2}{x - 2}$ if $x \neq 2$, is $f(x)$ continuous at $x = 2$? Why or why not?

47. Show analytically, using the limit definition of derivative, that $f'(0)$ exists if $f(x) = 0$ when $x < 0$ and $f(x) = x^3$ when $x \geq 0$.

48. **Average Cost.** A person making a long-distance telephone call to a neighboring town is charged \$1.00 for the first minute or fraction thereof and \$0.20 for each additional minute or fraction thereof. If t is the length of the telephone call and $C(t)$ the cost of the call, then the average cost, $\overline{C}(t)$, is defined to be $\overline{C}(t) = \dfrac{C(t)}{t}$. What happens to the average cost if the length t of the call becomes an unbounded positive number, that is, what is $\lim_{t \to \infty} \overline{C}(t)$? How do you interpret your answer in economic terms?

49. **Biology.** According to Wiess's law, the intensity of an electric current required to excite a living tissue (threshold strength) is given by $i = \dfrac{A}{t} + B$, where t is the duration of the current and A and B are positive constants. What happens as $t \to 0^+$? Does your answer seem reasonable?

50. **Medicine.** Suppose that the concentration c of a drug in the blood stream t hours after it is taken orally is given by $c = \dfrac{5t}{t^2 + 2}$. What happens to the drug concentration as $t \to 0^+$? Does this make sense?

51. **Costs.** A telephone company charges \$1.00 for the first minute or fraction thereof for a certain long-distance call and \$0.50 for each additional minute or fraction thereof. Graph this cost function, and determine where the function is continuous.

52. **Profits.** The daily profits in dollars of a firm is given by $P = 100(-x^2 + 8x - 12)$, where x is the number of items sold. Find the instantaneous rate of change when x is **(a)** 3 **(b)** 4 **(c)** 5. Interpret your answer.

53. **Velocity.** A particle moves according to the law $s(t) = t^2 - 8t + 15$, where s is in feet and t in seconds. Find the instantaneous velocity at $t = 3, 4, 5$. Interpret your answers.

54. **Biology.** Use the numerical derivative feature on your graphing calculator and the linear approximation formula to estimate the change in volume of a cylindrical bacterium as the radius increases from 2 microns to 2.1 microns while the length stays at 10 microns.

55. Stirling's Formula. The approximation $n! \approx \sqrt{2\pi n}(n/e)^n$ is called Stirling's formula. Test this for some values of n.

56. Biology. Gowen [59] determined that the equation $y = y_0 e^{-kr}$ approximated the number of surviving tobacco viruses on tobacco plants, where r is radiation measured by r that the virus was exposed to, and k is a positive constant. What should be the radiation, that is, the amount of r, that will kill 90% of the virus? Your answer should be in terms of k.

57. Chemical Kinetics. The amount $x(t)$ of a certain substance in a chemical reaction depends on the amounts A and B of two other chemicals present and is given by

$$x(t) = A\left(a + \frac{B - A}{A - Be^{k(B-A)t}}\right)$$

where k is a positive constant and $A \neq B$. Determine what happens in the long-term. *Hint:* Consider two cases, $A > B$ and $A < B$.

CHAPTER **2**

PROJECTS

Project 1
A Close Look at *n*Der

We defined the derivative of $f(x)$ at $x = a$ to be

$$f'(a) = \lim_{h \to 0} \frac{f(a + h) - f(a)}{h}$$

The fraction

$$\frac{f(a + h) - f(a)}{h} \qquad [3]$$

is the slope of the secant line through $(a, f(a))$ and $(a + h, f(a + h))$. The slope of this secant line approaches the slope of the line tangent to the curve $y = f(x)$ at the point $(a, f(a))$ as h approaches zero.

Now consider the expression

$$\frac{f(a + h) - f(a - h)}{2h} \qquad [4]$$

It is this expression that your graphing calculator uses to find the numerical derivative we call *n*Der.

(a) This also is the slope of a secant line. On a graph draw this line.

(b) Geometrically, explain why

$$f'(a) = \lim_{h \to 0} \frac{f(a + h) - f(a - h)}{2h}$$

(c) Your graphing calculator uses the expression $[f(a + h) - f(a - h)]/2h$ for small h to find the numerical derivative we call *n*Der. Use *n*Der to find the numerical derivative of $f(x) = |x|$ at $x = 0$. Explain why you obtain the answer 0 when we know from Example 5 in Section 2.5 that $f'(0)$ does not exist.

(d) Consider $f(x) = |x^{1/3}|$. From reviewing Example 6 in Section 2.5 we see that the slope of the tangent line to the curve $y = f(x)$ at $x = 0$ is vertical, and therefore $f'(0)$ does not exist. Now use *n*Der to find $f'(0)$. Why is *n*Der insisting that $f'(0) = 0$?

(e) Find the limit in (b) if $f(x) = x^2$.

Project 2
Drug Kinetics

The figure shows a model, referred to as the *two-compartment model*, frequently used in drug kinetics. The figure indicates that a drug enters the bloodstream, either by injection or orally. The drug flows between the blood and tissue at different rates and is cleared at another rate, all of which rates depend on the particular drug. From certain mathematical theory, not presented here, the concentrations of the drug in the bloodstream and in the tissues is approximated by an equation of the form

$$c(t) = Ae^{-\alpha t} + Be^{-\beta t} \qquad [1]$$

where α and β are positive constants and $\alpha < \beta$. It is quite remarkable that in physiological and pharmaceutical applications such a model gives excellent approximations in describing the kinetics.

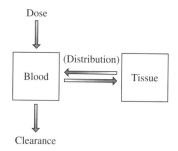

Dose

(Distribution)

Blood

Tissue

Clearance

The practical problem for researchers is to determine the constants A, B, α, and β for a given drug. In this project we consider the *method of exponential peeling* to estimate these constants.

To understand how and why this method works, we need to understand what is happening to $c(t)$ in Equation (1) for large t. We then use these insights to tackle a specific problem.

(a) Using a window with dimensions [0, 20] by [0, 5] graph $y_1 = 2e^{-0.1x} + 3e^{-0.2x}$, $y_2 = 2e^{-0.1x}$, and $y_3 = 3e^{-0.2x}$.

(b) Describe what you see. In particular, which of the two functions y_2 or y_3 is approximately equal to y_1 for large t?

(c) Now give a convincing mathematical explanation for your answer in (b) that uses the exponents -0.1 and -0.2.

(d) Now explain what you think $c(t)$ in Equation (1) approximates for large values of t.

(e) In part (d) you should have concluded that $c(t) \approx Ae^{-\alpha t}$ for large t. The method of "peeling" is based on this observation. Show for large t that we must have $\ln c(t) \approx \ln A - \alpha t$. Notice that the graph of $y = \ln A - \alpha t$ is a *straight line* with slope $-\alpha$ and y-intercept $\ln A$. Thus, we can determine α and A by determining the slope and t-intercept of this line.

(f) Creatinine is a component of urine originally formed from creatinine phosphate, a substance that supplies energy to the muscles. In an experiment carried out by Sepirstein and colleagues [60] 2 g of creatinine were injected into the bloodstream of dogs. Subsequent concentrations in the blood are shown in the table. Using this data and assuming the concentration satisfies an equation of the form of Equation (1), use the method of peeling to determine approximately the values of A and α.

t (min)	10	20	30	40	50	60	70
$c(t)$ (g)	0.14390	0.04273	0.01316	0.00416	0.00134	0.00044	0.00014

(g) Now estimate the constants B and β.

Project 3

Limits

(a) The text mentions two fundamentally different ways in which a limit may not exist. There is a third way. This project explores this third way. Familiarize yourself with the function $f(x) = \sin x$ by graphing this function while in radian mode in a window with dimension $[-10, 10]$ by $[-1, 1]$. (There should be a SIN key.) As you can see, this function moves back and forth between 1 and -1. We wish to determine if $\lim\limits_{x \to \infty} \sin x$ exists or not. Graph this function in a window with dimensions $[0, 50]$ by $[-1, 1]$. What is happening? Graph again in a window with dimensions $[50, 100]$ by $[-1, 1]$. What is happening? Graph again in a window with dimensions $[100, 150]$ by $[-1, 1]$. What is happening? What do you conclude about the limit?

(b) We wish to determine if $\lim\limits_{x \to 0} \sin(1/x)$ exists or not. Graph this function in a window with dimensions $[-0.314, 0.314]$ by $[-1, 1]$. What is happening? Graph again in a window with dimensions $[-0.0314, 0.0314]$ by $[-1, 1]$. What is happening? Graph again in a window with dimensions $[-0.00314, 0.00314]$ by $[-1, 1]$. What is happening? What do you conclude about the limit?

(c) Determine graphically and numerically if $\lim\limits_{x \to \infty} (x \sin x)$ exists or not.

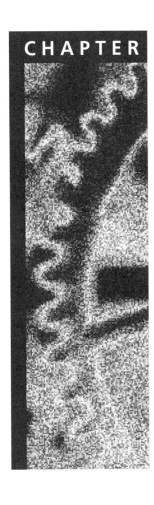

3

Rules for the Derivative

In the previous chapter, we learned how to find the derivatives of functions by taking the limit, that is, by using the definition of derivative. In this chapter we develop some basic rules for derivatives that can then be used to find the derivatives of a wide variety of functions without having to take limits.

3.1 DERIVATIVES OF POWERS, EXPONENTS, AND SUMS

■ Derivatives of Constants ■ Derivatives of Powers ■ Derivative of the Exponential ■ Derivative of a Constant Times a Function ■ Derivatives of Sums and Differences ■ Rates of Change in Business and Economics

---

APPLICATION: A COST FUNCTION FOR A MANUFACTURING PLANT

A plant manufactures small boats. Fixed costs amount to $50,000 and include items such as interest on loans, cost of electricity and heating, and wages for management and for the basic work force. One lot of materials to manufacture one boat costs $2000, but the materials supplier gives a discount of $75 per boat for each lot of materials purchased. Furthermore, management knows from experience that if too many boats are made in this plant, bottlenecks occur that disrupt the production line, an increasing amount of overtime must be worked and paid for, and the workers become inefficient when they put in long hours of overtime. Management estimates these latter costs to be $0.15e^x$, where x is the number of boats made. Naturally, we expect that manufacturing an increasing number of boats is increasingly more cost-efficient, but trying to produce too many at one plant can result in poor cost efficiencies and lead instead to out-of-control costs. Is all this true for this plant? With $C(x)$ the total cost function, answer by making a table of values for $C'(x)$ for x going from 1 to 12, and interpret the meaning of these numbers. See Example 7 for the answer.

Derivatives of Constants

In the previous chapter, we learned how to find the derivatives of functions by taking the limit. In this section we discover some basic rules for derivatives that can then be used to find the derivatives of a wide variety of functions without having to take limits. We begin by finding the derivative of a constant function.

Suppose $f(x)$ is a constant function, that is, $f(x) = C$, where C is a constant. Since the value of the function never changes, the instantaneous rate of change must be zero. Also, the graph of this function is given in Figure 3.1 and is just a horizontal line. The slope of any such horizontal line is zero. Thus, we have two reasons to suspect that the derivative should be zero. Using the definition of derivative, we have

$$f'(x) = \lim_{h \to 0} \frac{f(x + h) - f(x)}{h}$$

$$= \lim_{h \to 0} \frac{C - C}{h}$$

$$= \lim_{h \to 0} \frac{0}{h}$$

$$= \lim_{h \to 0} 0$$

$$= 0$$

We can write this as a basic rule.

Derivative of a Constant

For any constant C

$$\frac{d}{dx}(C) = 0$$

$y = f(x) = C$

Slope is zero.

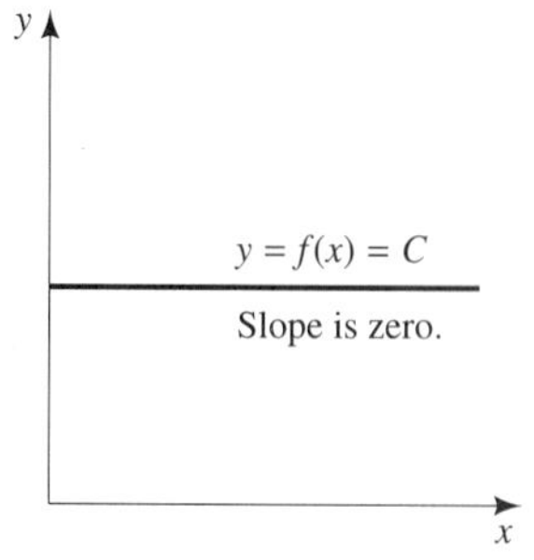

Figure 3.1
The slope of a horizontal line is zero. Thus, $\frac{d}{dx}(C) = 0$.

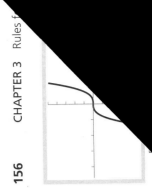

EXAMPLE 1   Finding the Deriva

Find

(a) $\dfrac{d}{dx}(3.4)$ (b) $f'(x)$ if $f(x) = -6$

SOLUTIONS ∎

(a) $\dfrac{d}{dx}(3.4) = 0$ (b) $f'(x) = \dfrac{d}{dx}(-6)$

Derivatives of Powers

In the preceding section we found the deriv

found that $\dfrac{d}{dx}(x^{1/2}) = \frac{1}{2}x^{-1/2}$ and $\dfrac{d}{dx}(x^{-1}) = -x^{-2}$.

Furthermore, since the function $f(x) = x$ represents a straight line with constant slope equal to one (Fig. 3.2), we suspect that $\dfrac{d}{dx}(x) = 1$.

Let us now see if we can determine the derivative of $f(x) = x^2$. We know that by definition

$$f'(x) = \lim_{h \to 0} \frac{(x + h)^2 - x^2}{h}$$

If h is small, say $h = 0.01$, then we should have

$$f'(x) \approx \frac{(x + 0.01)^2 - x^2}{0.01}$$

Therefore, the graph of $y = f'(x)$ should be approximated by the graph of $y = \dfrac{(x + 0.01)^2 - x^2}{0.01}$. The graph of this latter function on a screen of dimensions $[-4.7, 4.7]$ by $[-10, 10]$ is shown in Screen 3.1. We see that the graph is a straight line of slope 2 and goes through $(0, 0)$. Thus, we strongly suspect that $f'(x) = 2x$.

In a similar fashion, if $f(x) = x^3$, then the graph of $y = f'(x)$ should be approximated by the graph of $y_1 = \dfrac{(x + 0.01)^3 - x^3}{0.01}$. Graphing this last function on a screen with the same dimensions as before, we obtain Screen 3.2. The graph looks like the graph of $y = kx^2$. Since the graph goes through the point $(1, 3)$, we try $k = 3$. Taking

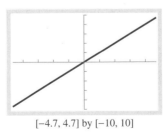

[-4.7, 4.7] by [-10, 10]

Screen 3.1
Since the derivative of x^2 is approximately $y_1 = \dfrac{(x + 0.01)^2 - x^2}{0.01}$, and the graph of this function appears to be the same as the graph of $y_2 = 2x$, we strongly suspect that $\dfrac{d}{dx}(x^2) = 2x$.

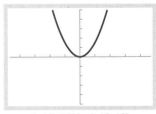

[-4.7, 4.7] by [-10, 10]

Screen 3.2
Since the derivative of x^3 is approximately $y_1 = \dfrac{(x + 0.01)^3 - x^3}{0.01}$, and the graph of this function appears to be the same as the graph of $y_2 = 3x^2$, we strongly suspect that $\dfrac{d}{dx}(x^3) = 3x^2$.

Figure 3.2
The slope of the line $y = x$ is 1.
Thus, $\dfrac{d}{dx}(x) = 1$.

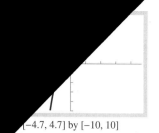

[−4.7, 4.7] by [−10, 10]

Screen 3.3
Since the derivative
of x^4 is approximately
$y_1 = \dfrac{(x + 0.01)^4 - x^4}{0.01}$, and the
graph of this function appears
to be the same as the graph of
$y_2 = 4x^3$, we strongly suspect
that $\dfrac{d}{dx}(x^4) = 4x^3$.

$y_2 = 3x^2$ gives Screen 3.2 again. Since the two graphs are the same, we strongly suspect that $f'(x) = 3x^2$.

At this point you might guess that if $f(x) = x^4$, $f'(x) = 4x^3$. Graphing the two functions $y_1 = \dfrac{(x + 0.01)^4 - x^4}{0.01}$ and $y_2 = 4x^3$ gives Screen 3.3. Since both graphs are approximately the same, we have graphical support for our guess.

Thus, it appears that if $f(x) = x^n$, then $f'(x) = nx^{n-1}$. This is indeed true.

Derivative of a Power

If n is any real number (n may or may not be an integer),

$$\frac{d}{dx}(x^n) = nx^{n-1}$$

We shall give the proof only in the case that $n = 3$, the case for a general positive integer being similar. The case when n is not a positive integer must be done differently and is somewhat difficult.

If $n = 3$, then

$$\frac{f(x + h) - f(x)}{h} = \frac{(x + h)^3 - x^3}{h}$$

$$= \frac{(x^3 + 3x^2h + 3xh^2 + h^3) - x^3}{h}$$

$$= \frac{3x^2h + 3xh^2 + h^3}{h}$$

$$= \frac{h(3x^2 + 3xh + h^2)}{h}$$

$$= 3x^2 + 3xh + h^2$$

Now to find $f'(x)$ we let $h \to 0$ in this last expression to obtain

$$f'(x) = \lim_{h \to 0} \frac{f(x + h) - f(x)}{h}$$

$$= \lim_{h \to 0} \frac{(x + h)^3 - x^3}{h}$$

$$= \lim_{h \to 0} (3x^2 + 3xh + h^2)$$

$$= 3x^2 + 0 + 0$$

$$= 3x^2$$

This completes the proof for the derivative of x^3.

EXAMPLE 2   Finding the Derivatives of Some Powers

Find

(a) $\dfrac{d}{dx}(x^7)$ (b) $f'(t)$ if $f(t) = t^{-4}$ (c) $\dfrac{dy}{dv}$ if $y = v^{4/3}$

SOLUTIONS ∎

(a) $\dfrac{d}{dx}(x^7) = 7x^{7-1} = 7x^6$

(b) $f'(t) = \dfrac{d}{dt}(t^{-4}) = -4t^{-4-1} = -4t^{-5}$

(c) $\dfrac{dy}{dv} = \dfrac{d}{dv}(v^{4/3}) = \dfrac{4}{3}v^{(4/3)-1} = \dfrac{4}{3}v^{1/3}$ ∎

Sometimes a function $f(x)$ is actually x to some power, without this being apparent at first.

EXAMPLE 3 Finding the Derivatives of Some Powers

Find

(a) $\dfrac{dy}{dx}$ if $y = \sqrt[3]{x^2}$ (b) $\dfrac{d}{dx} f(x)$ if $f(x) = 1/x^2$

SOLUTIONS ■

(a) $\dfrac{dy}{dx} = \dfrac{d}{dx}\left(\sqrt[3]{x^2}\right) = \dfrac{d}{dx}(x^{2/3}) = \dfrac{2}{3}x^{-1/3}$

(b) $\dfrac{d}{dx} f(x) = \dfrac{d}{dx}(1/x^2) = \dfrac{d}{dx}(x^{-2}) = -2x^{-3}$ ■

EXPLORATION 1

Finding the Derivative

■■■
■ ■ Support the answer in Example 3(b) by graphing $y_1 = f(x)$ and $y_2 = $
■ ■ nDer $f(x)$ on the same screen.

Derivative of the Exponential

We now consider the derivative of e^x. To determine the derivative of $f(x) = e^x$ we first use the fact that $f'(x) \approx \dfrac{f(x + h) - f(x)}{h}$ if h is small. Taking $h = 0.01$, the graph of $y = f'(x)$ is approximated by the graph of

$$y_1 = \frac{f(x + h) - f(x)}{h}$$

$$= \frac{e^{x+h} - e^x}{h}$$

$$= \frac{e^{x+0.01} - e^x}{0.01}$$

The graph of this last function is shown in Screen 3.4 using a window of dimensions $[-3, 3]$ by $[0, 10]$. The graph looks like that of an exponential function, $y = e^{kx}$, $k > 0$. Trying $k = 1$ and graphing $y_2 = e^x$ on the same screen does not give anything different, indicating that the two graphs are the same. We thus strongly suspect that

$$\frac{d}{dx}(e^x) = e^x$$

We now present further support for this conclusion. With $f(x) = e^x$ use the limit definition of the derivative and obtain

$$f'(x) = \lim_{h \to 0} \frac{f(x + h) - f(x)}{h}$$

$$= \lim_{h \to 0} \frac{e^{x+h} - e^x}{h}$$

$$= \lim_{h \to 0} \frac{e^x e^h - e^x \cdot 1}{h}$$

$$= \lim_{h \to 0} e^x \frac{e^h - 1}{h}$$

$$= e^x \lim_{h \to 0} \frac{e^h - 1}{h}$$

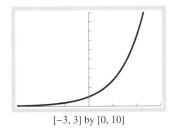

$[-3, 3]$ by $[0, 10]$

Screen 3.4

Since the derivative of e^x is approximately $y_1 = \dfrac{e^{x+0.01} - e^x}{0.01}$, and the graph of this function appears to be the same as the graph of $y_2 = e^x$, we strongly suspect that $\dfrac{d}{dx}(e^x) = e^x$.

To continue, we must evaluate

$$\lim_{h \to 0} \frac{e^h - 1}{h}$$

But recall that we already found this limit to be 1 in Example 7 in Section 2.3.
Using this, we then have

$$f'(x) = e^x \cdot \lim_{h \to 0} \frac{e^h - 1}{h}$$

$$= e^x \cdot 1$$

$$= e^x$$

The Derivative of e^x

The derivative of e^x is again e^x, that is,

$$\frac{d}{dx}(e^x) = e^x$$

Derivative of a Constant Times a Function

We now establish a rule for finding the derivative of a constant times a function.

Derivative of a Constant Times a Function

If $f'(x)$ exists, then

$$\frac{d}{dx}(Cf(x)) = C\frac{d}{dx}f(x)$$

We use the definition of derivative to give a proof of this rule.

$$[Cf(x)]' = \lim_{h \to 0} \frac{Cf(x + h) - Cf(x)}{h}$$

$$= \lim_{h \to 0} C\frac{f(x + h) - f(x)}{h}$$

$$= C\lim_{h \to 0} \frac{f(x + h) - f(x)}{h}$$

$$= Cf'(x)$$

In Section 2.4 we graphically found the derivative of $\ln x$. Since we will now be using this formula, we recall it here.

The Derivative of $\ln x$

If $x > 0$, then

$$\frac{d}{dx}(\ln x) = \frac{1}{x}$$

EXAMPLE 4   Finding the Derivative of a Constant Times a Function

Find the derivatives of

(a) $-5x^3$ (b) $\sqrt{2}e^x$ (c) $\dfrac{3}{x^8}$ (d) $\ln x^4$

SOLUTIONS ■

(a) $\dfrac{d}{dx}(-5x^3) = -5\dfrac{d}{dx}(x^3)$

$\qquad = -5(3x^{3-1})$

$\qquad = -15x^2$

(b) $\dfrac{d}{dx}(\sqrt{2}e^x) = \sqrt{2}\dfrac{d}{dx}(e^x)$

$\qquad = \sqrt{2}e^x$

(c) $\dfrac{d}{dx}\left(\dfrac{3}{x^8}\right) = \dfrac{d}{dx}(3x^{-8})$

$\qquad = 3\dfrac{d}{dx}(x^{-8})$

$\qquad = 3(-8x^{-9})$

$\qquad = -24x^{-9}$

(d) $\dfrac{d}{dx}(\ln x^4) = \dfrac{d}{dx}(4\ln x)$

$\qquad = 4\dfrac{d}{dx}(\ln x)$

$\qquad = 4\cdot\dfrac{1}{x}$ ■

Derivatives of Sums and Differences

The last rule in this section is for derivatives of sums and differences.

Derivative of Sums or Differences

If $f'(x)$ and $g'(x)$ exist, then $\dfrac{d}{dx}[f(x) \pm g(x)]$ exists and

$$\frac{d}{dx}[f(x) \pm g(x)] = \frac{d}{dx}f(x) \pm \frac{d}{dx}g(x)$$

We show only the proof for the derivative of a sum, but the proof for the derivative of a difference is similar. If $p(x) = f(x) + g(x)$, then

$$p'(x) = \lim_{h\to 0}\frac{p(x+h) - p(x)}{h}$$

$$= \lim_{h\to 0}\frac{[f(x+h) + g(x+h)] - [f(x) + g(x)]}{h}$$

$$= \lim_{h\to 0}\frac{[f(x+h) - f(x)] + [g(x+h) - g(x)]}{h}$$

$$= \lim_{h\to 0}\frac{f(x+h) - f(x)}{h} + \lim_{h\to 0}\frac{g(x+h) - g(x)}{h}$$

$$= f'(x) + g'(x)$$

EXAMPLE 5 Derivatives of Sums and Differences

Find the derivatives of

(a) $x^4 + x^3$ (b) $-\dfrac{2}{x^3} - 3\sqrt{x^3} + \pi$

SOLUTIONS ■

(a) $\dfrac{d}{dx}(x^4 + x^3) = \dfrac{d}{dx}(x^4) + \dfrac{d}{dx}(x^3)$

$= 4x^3 + 3x^2$

(b) $\dfrac{d}{dx}\left(-\dfrac{2}{x^3} - 3\sqrt{x^3} + \pi\right) = -2\dfrac{d}{dx}(x^{-3}) - 3\dfrac{d}{dx}(x^{3/2}) + \dfrac{d}{dx}(\pi)$

$= -2(-3x^{-4}) - 3\left(\dfrac{3}{2}x^{1/2}\right) + 0$

$= 6x^{-4} - \dfrac{9}{2}x^{1/2}$ ■

EXAMPLE 6 Finding Horizontal Tangents

Graph $y = x^4 + x^3$ in the standard viewing window.

(a) Roughly estimate the value(s) of x at which the graph of $y = f(x)$ has a horizontal tangent.

(b) Confirm your answer(s) in part (a) using calculus.

SOLUTIONS ■

(a) The graph is shown in Screen 3.5. We see that a horizontal tangent appears to occur at approximately $x = 0$.

(b) The tangent is horizontal when the slope is zero, that is, at points where the derivative is zero. From Example 5b we have

$$0 = f'(x)$$
$$= 4x^3 + 3x^2$$
$$= x^2(4x + 3)$$

Thus, $x = 0$, but also $x = -0.75$. We missed the second value in part (a) because our graph missed some hidden behavior! Screen 3.6 shows the graph of $y = f(x)$ on a screen of dimension $[-1.5, 0.5]$ by $[-0.15, 0.15]$. On this screen we clearly see the behavior that was hidden before. ■

Example 6 demonstrates that we cannot rely completely on a graphing calculator. Only by using calculus can we be certain that we have found *all* the hidden behavior.

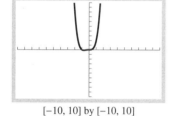

[−10, 10] by [−10, 10]

Screen 3.5
The graph of $y = x^4 + x^3$ seems to have a horizontal tangent at $x = 0$.

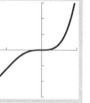

[−1.5, 0.5] by [−0.15, 0.15]

Screen 3.6
A closer look at the graph of $y = x^4 + x^3$ indicates hidden behavior not seen on Screen 3.5.

EXPLORATION 2

Finding Horizontal Tangents

 Graph $y = -\dfrac{2}{x^3} - 3\sqrt{x^3} + \pi$ in the standard viewing window.

(a) Roughly estimate the value(s) of x at which the graph of $y = f(x)$ has a horizontal tangent.

(b) Find the value(s) in part (a) to two decimal places by graphing $y = f'(x)$ and using ZOOM.

(c) Confirm your answer(s) in parts (a) and (b) using calculus. (Refer to Example 5b.)

Rates of Change in Business and Economics

To make intelligent business and economic decisions, one needs to know the effect of changes in production and sales on costs, revenues, and profits. That is, one needs to know the rate of change of cost, revenue, and profits. For example, knowing if the rate of change of profits with respect to an increase in production is positive or negative is very useful information. In economics, the word "marginal" refers to an instantaneous rate of change, that is, to the derivative. Thus, if $C(x)$ is the cost function, we define the marginal cost to be $C'(x)$. If the revenue and profit functions are given by $R(x)$ and $P(x)$, we also define $R'(x)$ and $P'(x)$ to be the marginal revenue and marginal profit, respectively.

EXAMPLE 7   Marginal Cost

A plant manufactures small boats. Fixed costs amount to $50,000 and include items such as interest on loans, costs of electricity and heating, wages for management and for the basic work force. One lot of materials to manufacture one boat costs $2000, but the materials supplier gives a discount per boat of $75 for each lot of materials purchased. Furthermore, management knows from experience that if too many boats are made in this plant, bottlenecks occur that disrupt the production line, an increasing amount of overtime must be worked and paid for, and the workers become inefficient when they put in long hours of overtime. Management estimates these latter costs to be $0.15e^x$, where x is the number of boats made. Naturally, we expect that manufacturing an increasing number of boats is increasingly more cost-efficient, but trying to produce too many at one plant can result in poor cost efficiencies and lead to out-of-control costs. Is all this true for this plant? With $C(x)$ the total cost function answer by doing the following. Make a table of values for $C'(x)$ for x going from 1 to 12. What is the meaning of these numbers? Interpret what is happening to the rate of change of costs.

S O L U T I O N ■ If x is the number of boats made, then the material cost is $(2000 - 75x)$ per boat and the total material cost is $(2000 - 75x)x$. The total cost function is then

$$C(x) = 0.15e^x + (2000 - 75x)x + 50{,}000 = 0.15e^x - 75x^2 + 2000x + 50{,}000$$

Screen 3.7 shows a graph of $y = C(x)$ using a window of dimensions [0, 18.8] by [0, 100,000]. We notice naturally that total costs increase as more boats are made. The marginal cost is

$$C'(x) = 0.15e^x - 150x + 2000$$

Screen 3.8 shows a graph of $y = C'(x)$ using a window of dimensions [0, 18.8] by [0, 5000]. Then using the TRACE feature or by directly calculating, we have

$$C'(10) = 0.15e^{10} - 150(10) + 2000 \approx 3804$$

Thus, cost increases at the rate of $3804 per boat when 10 boats are being produced.

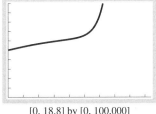

[0, 18.8] by [0, 100,000]

Screen 3.7
The cost $y = 0.15e^x - 75x^2 + 2000x + 50{,}000$ increases slowly and then more dramatically as more boats are made.

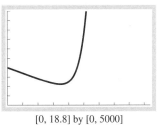

[0, 18.8] by [0, 5000]

Screen 3.8
The marginal cost $y = 0.15e^x - 150x + 2000$ initially decreases but then increases dramatically when a larger number of boats are made.

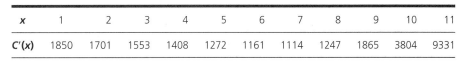

x	1	2	3	4	5	6	7	8	9	10	11
C'(x)	1850	1701	1553	1408	1272	1161	1114	1247	1865	3804	9331

The table gives the other values for $C'(x)$. For example, cost increases at the rate of $1850 per boat when one boat is being produced, at the smaller rate of $1701 per boat when two boats are being produced, and at the enormous rate of $9331 per boat when 11 boats are being produced. The table shows the rate of change of costs decreasing at first, but then begins to increase at an ever faster rate. This is typical of cost functions. Of course, the more boats made, the greater the total cost. Because of economies of scale, however, the rate of increase of costs itself normally decreases initially. That is, the manufacturing process becomes more cost-efficient as more boats are made. But if production is increased too much, the rate of increase of costs increases due to such factors as the less efficient use of equipment and payment of overtime. ■

EXPLORATION **3**

Tangent Line

 On your graphing calculator graph the cost function given in the previous example and the tangent line to this curve at $x = 10$ on the same screen. How is the slope of this tangent line related to the marginal cost $C'(10)$ found in Example 7?

SELF-HELP EXERCISE SET 3.1

1. Find the derivative of each of the following:

   (a) $x^7 + 4x^{2.5} - 3e^x$ **(b)** $\sqrt{t} + \dfrac{1}{\sqrt{t}} + 5 \ln t$

2. Let $f(x) = 0.125x^2 + 0.50$.
   (a) Find $f'(x)$.
   (b) Find the equation of the line tangent to the curve $y = f(x)$ at the point $x = 2$.

3. Suppose the cost function is given by $C(x) = 10 + 0.02x^2$, where x is the number of items produced and $C(x)$ is the cost of producing x items. Find the marginal cost for any x. Find $C'(30)$. Interpret what this means.

4. Lightning strikes a tree in a forest and ignites a fire that spreads in a circular manner. What is the instantaneous rate of change of the burnt area with respect to the radius r of the burnt area when $r = 50$ ft?

EXERCISE SET 3.1

In Exercises 1 through 27 find $\dfrac{d}{dx} f(x)$.

1. $f(x) = 13$ **2.** $f(x) = \sqrt{2}$

3. $f(x) = e + \pi$ **4.** $f(x) = 4\pi^2$

5. $f(x) = x^{23}$ **6.** $f(x) = x^7$

7. $f(x) = x^{1.4}$ **8.** $f(x) = x^{3.1}$

9. $f(x) = x^e$ **10.** $f(x) = x^{0.03}$

11. $f(x) = x^{4/3}$ **12.** $f(x) = x^{2/5}$

13. $f(x) = 3x^{-1.7}$ **14.** $f(x) = \dfrac{1}{3} e^x$

15. $f(x) = 0.24\sqrt{x}$ **16.** $f(x) = 0.24\sqrt[3]{x}$

17. $f(x) = 2e^x - 1$ **18.** $f(x) = 4 - 3 \ln x$

19. $f(x) = \pi x + 2 \ln x$ **20.** $f(x) = 3e^x - \pi^2 x$

21. $f(x) = 4 - x - x^2$ **22.** $f(x) = 3x^2 + x - 1$

23. $f(x) = \dfrac{1}{\pi} x^2 + \pi x - 1$

24. $f(x) = \pi^2 x^2 + \dfrac{1}{\sqrt{2}} x + \sqrt{3} + 2 \ln x$

25. $f(x) = 1 - x - x^2 - x^3 - 10 \ln x$

26. $f(x) = 0.02e^x + 2x^3 - 2x^2 + 3x - 10$

27. $f(x) = x^5 - 5e^x + 1$

In Exercises 28 through 33 find $f'(x)$.

28. $f(x) = 1 - x^4 - x^8$

29. $f(x) = 0.35x^2 - 0.3e^x + 0.45$

30. $f(x) = \ln 3 + e^x - 0.001x^2$

31. $f(x) = e^4 - \ln x - 0.01x^2 - 0.03x^3$

32. $f(x) = 3e^x - 0.003x^4 + 0.01x^3$

33. $f(x) = \dfrac{e^x + x^2 - 2x + 4}{6}$

In Exercises 34 through 39 find $f'(x)$. Support your answer by graphing $y_1 = f'(x)$ and $y_2 = n\text{Der } f(x)$ on the same screen.

34. $f(x) = \dfrac{x^2 - 3x - 6}{3x}$ **35.** $f(x) = 3x^3 + \dfrac{3}{x^3}$

36. $f(x) = \dfrac{1}{x} + x$ **37.** $f(x) = x^2 - \dfrac{1}{x}$

38. $f(x) = x^{-2} + x^{-3} + 3 \ln \left(\dfrac{x}{4} \right)$

39. $f(x) = 4 \ln (5x) - 3x^{-3} + 4x^2$

In Exercises 40 through 45 find $\dfrac{d}{dt} f(t)$.

40. $f(t) = (2t)^5 + e^{t+2}$ **41.** $f(t) = (-3t)^3 - e^{t-3}$

42. $f(t) = \dfrac{1}{t^2} + \dfrac{2}{t^3}$ **43.** $f(t) = \dfrac{1}{t} + \dfrac{1}{t^2} + \dfrac{1}{t^3}$

44. $f(t) = \sqrt{t} - \dfrac{1}{\sqrt{t}}$ **45.** $f(t) = \sqrt[3]{t} - \dfrac{1}{\sqrt[3]{t}}$

In Exercises 46 through 51 find $\dfrac{d}{du} f(u)$.

46. $f(u) = 2u^{3/2} + 4u^{3/4}$ **47.** $f(u) = 3u^{5/3} - 3u^{2/3}$

48. $f(u) = (u - 1)^3$ **49.** $f(u) = (u + 1)(2u + 1)$

50. $f(u) = \dfrac{(u + 1)^2}{u}$ **51.** $f(u) = \dfrac{u + 3/u}{\sqrt{u}}$

In Exercises 52 through 64 find $\dfrac{dy}{dx}$.

52. $y = \dfrac{x^2 + x - 2}{x}$ **53.** $y = \sqrt{4x} - \dfrac{1}{\sqrt{9x}}$

54. $y = x^{1.5} - x^{-1.5} + \ln x^{-2}$

55. $y = \pi x^\pi + \ln x^2$ **56.** $y = \dfrac{1}{\pi} x^{1/\pi}$

57. $y = x^{\sqrt{2}} - \dfrac{1}{x^{\sqrt{2}}}$ **58.** $y = 3x^{-3/5} - 6x^{-5/2}$

59. $y = x^{-3/2} + x^{-4/5} + \ln \sqrt{x}$

60. $y = x^{-2} + 3x^{-4}$ **61.** $y = \sqrt[3]{x^2}$

62. $y = \sqrt{x^3}$ **63.** $y = \dfrac{1}{\sqrt[5]{x^2}}$

64. $y = \dfrac{1}{\sqrt[3]{x^4}}$

In Exercises 65 through 70 find the equation of the tangent line of the given function at the indicated point. Support your answer using a graphing calculator.

65. $y = f(x) = x^2 + e^x + 1, \quad c = 0$

66. $y = f(x) = x^4 + 3x^2 - 1, \quad c = -1$

67. $y = f(x) = x^{-1} + x^{-2} + \ln x, \quad c = 1$

68. $y = f(x) = \sqrt{x} + x, \quad c = 4$

69. $y = f(x) = \dfrac{1}{x} + x^2, \quad c = -1$

70. $y = f(x) = x^{-3/2}, \quad c = 4$

In Exercises 71 through 74 graph the functions on your graphing calculator, and roughly estimate the values where the tangent to the graph of $y = f(x)$ is horizontal. Confirm your answer using calculus.

71. $y = f(x) = x^2 - 1$ **72.** $y = f(x) = x^3 + 3$

73. $y = f(x) = x^3 - 9x + 3$

74. $y = f(x) = 7 + 12x - x^3$

Applications

75. Revenue. If the revenue R in dollars for a product is given by $R(x) = -2x^2 + 24x$, where x is the number of items sold, find the marginal revenue for any x. Find $R'(3)$ and $R'(10)$, and interpret the meanings of these numbers. Graph $y = R(x)$ on $[0, 12]$.

76. Costs. If the cost C in dollars for a product is given by $C(x) = 0.001e^x + 0.2x^2 + 32$, where x is the number of items sold, find the marginal cost for any x. Find $C'(10)$, and interpret the meaning of this number. Graph $y = C(x)$ on $[0, 12]$.

77. Profits. If the demand equation and cost equation for a product are as given in the previous two problems, find the marginal profit for any x. Find $P'(3)$ and $P'(10)$, and interpret the meanings of these numbers. Graph $y = P(x)$ on $[0, 12]$.

78. Demand. Suppose the demand equation for a product is given by $p = 1000/\sqrt{x}$, where x is the number of items sold and p is the price in dollars. Find the instantaneous rate of change of price with respect to the number sold.

79. Velocity. A particle moves according to the law $s(t) = 2t^3 - 4t + 10$. Find $v(t)$.

80. Velocity. A particle moves according to the law $s(t) = -t^3 + 10 + 2/t$. Find $v(t)$.

81. Biology. A population of bacteria is growing according to the law $N(t) = 0.01e^t + 10t + 1000$, where t is measured in hours. What is the rate of change of the population with respect to time when $t = 10$?

82. Medicine. A spherical tumor of radius r cm is growing. What is the rate of change of the volume of the tumor with respect to the radius when $r = 4$? *Hint:* $V = \dfrac{4}{3}\pi r^3$.

83. Biology. Zonneveld and Kooijman [61] determined that the equation $y = 2.81L^2$ approximated the lettuce intake of a pond snail versus the shell length. Find the rate of change of lettuce intake with respect to shell length.

84. Biology. Kaakeh and coworkers [62] artificially infested one-year-old apple trees to a controlled aphid infestation. They showed that the average shoot length y in centimeters was approximated by $y = f(x) = 62.8 + 0.486x - 0.003x^2$, where x is the number of days since the aphid infestation. Find $f'(x)$. Find $f'(60)$ and $f'(110)$, and explain what these numbers mean.

85. Biology. Miller [63] determined that the equation $y = 0.12(L^2 + 0.0026L^3 - 16.8)$ approximated the reproduction rate, measured by the number of eggs per year, of the rock goby versus its length in centimeters. Find the rate of change of reproduction with respect to length.

86. Medicine. Poiseuille's law states that the total resistance R to blood flow in a blood vessel of constant length l and radius r is given by $R = \dfrac{al}{r^4}$, where a is a positive constant. Find the rate of change of resistance with respect to the radius when $r = 3$.

87. Biology. The free water vapor diffusion coefficient D in soft woods is given by

$$D(T) = \frac{167.2}{P}\left(\frac{T}{273}\right)^{1.75}$$

where T is the temperature and P is the constant atmospheric pressure measured in appropriate units. Find the rate of change of D with respect to T.

88. Biology. The chemical d-limonene is extremely toxic to some insects. Karr and Coats [64] showed that the number of days y for a cockroach nymph (*Blatta germanica*) to reach the adult stage when given a diet containing this chemical was approximated by $y = f(x) = 125 - 3.43 \ln x$, where x is the percentage of this chemical in the diet. Find $f'(x)$ for any $x > 0$. Interpret what this means. What does this say about the effectiveness of this chemical as a control of the cockroach nymph? Graph $y = f(x)$.

Enrichment Exercises

In Exercises 89 through 91 graph the function on your graphing calculator. **(a)** Roughly estimate the values of x for which the tangent to the graph of $y = f(x)$ is horizontal. **(b)** Now graph $y = f'(x)$ and use the ZOOM feature to solve $f'(x) = 0$ to two decimal places. **(c)** How do you relate the solutions of $f'(x) = 0$ found in part (b) to the values of x found in part (a)? **(d)** Can you confirm your answers found in part (b) using calculus? Why or why not?

89. $y = f(x) = x^5 + 3x^2 - 9x + 4$

90. $y = f(x) = x^5 - x^2 - x - 1$

91. $y = f(x) = x^7 - x^3 - x - 1$

92. Let L_1 be the line tangent to the graph of $y = 1/x^2$ at $x = 1$, and let L_2 be the line tangent to the graph of $y = -x^2 - 1.5x + 1$ at $x = -1$. Show that the two tangent lines are perpendicular.

93. Let L_1 be the tangent line to the graph of $y = x^4$ at $x = -1$, and let L_2 be the tangent line to the graph of $y = x^4$ at $x = 1$. Show that these two tangent lines intersect on the y-axis.

94. Let $f(x) = g(x) + 4x^3$, and suppose $g'(2)$ does *not* exist. Can $f'(2)$ exist? Explain why or why not.

95. Suppose both $f(x)$ and $g(x)$ are *not* differentiable at $x = a$. Does this imply that $h(x) = f(x) + g(x)$ is *not* differentiable at $x = a$? Explain why this is true or give an example that shows it is not true.

In Exercises 96 through 99 find $f'(x)$. (Recall from Exercises 59 and 60 in Section 2.4 that $\dfrac{d}{dx}(\sin x) = \cos x$ and $\dfrac{d}{dx}(\cos x) = -\sin x$.)

96. $f(x) = x^3 + 4\cos x$

97. $f(x) = x + x^2 - 3\sin x$

98. $f(x) = 2e^x - 3\sin x$

99. $f(x) = 5\ln x + 2\cos x$

100. Biology. Potter and colleagues [65] showed that the mortality rate y (percentage) of a New Zealand thrip was approximated by $y = f(T) = 81.12 + 0.465T - 0.828T^2 + 0.04T^3$, where T is the temperature measured in degrees Celsius. Graph on your graphing calculator using a window of dimensions [0, 20] by [0, 100]. **(a)** Estimate the value of the temperature where the tangent line to the curve $y = f(T)$ is horizontal. **(b)** Check your answer using calculus. (You will need to use the quadratic formula.) **(c)** Did you miss any points in part (a)? **(d)** Is this another example of how your graphing calculator can mislead you? Explain.

SOLUTIONS TO SELF-HELP EXERCISE SET 3.1

1. (a) $\dfrac{d}{dx}(x^7 + 4x^{2.5} - 3e^x) = \dfrac{d}{dx}(x^7) + 4\dfrac{d}{dx}(x^{2.5}) - 3\dfrac{d}{dx}(e^x)$

$= 7x^6 + 10x^{1.5} - 3e^x$

(b) $\dfrac{d}{dt}\left(\sqrt{t} + \dfrac{1}{\sqrt{t}} + 5\ln t\right) = \dfrac{d}{dt}(t^{1/2} + t^{-1/2} + 5\ln t)$

$= \dfrac{1}{2}t^{-1/2} - \dfrac{1}{2}t^{-3/2} + \dfrac{5}{t}$

2. (a) $f'(x) = \dfrac{d}{dx}(0.125x^2 + 0.50) = 0.125\dfrac{d}{dx}(x^2) + \dfrac{d}{dx}(0.50)$

$= 0.25x$

(b) Since $c = 2$, $y_0 = f(c) = f(2) = 1$. The slope of the tangent line at $x = 2$ is $f'(2) = 0.25(2) = 0.50$. Thus, the equation of the line tangent to the curve at $x = 2$ is

$$y - 1 = 0.50(x - 2)$$

3. The marginal cost for any x is

$$C'(x) = \dfrac{d}{dx}(10 + 0.02x^2) = \dfrac{d}{dx}(10) + 0.02\dfrac{d}{dx}(x^2) = 0.04x$$

Then $C'(30) = 0.04(30) = 1.20$. Thus, the cost increases at a rate of \$1.20 per item when 30 items are being produced.

4. The area of a circle of radius r is given by $A = A(r) = \pi r^2$. Then

$$\dfrac{dA}{dr} = A'(r) = 2\pi r$$

and $A'(50) = 2\pi(50) \approx 314$. Thus, the burnt area increases at the rate of approximately 314 ft^2 per foot of increase in r when $r = 50$.

3.2 DERIVATIVES OF PRODUCTS AND QUOTIENTS

■ Product Rule ■ Quotient Rule

APPLICATION: MARGINAL REVENUE

Suppose the demand equation for a certain product is given by $p = e^{-x}$, where x is the number of items sold (in thousands) and p the price in dollars. What is the marginal revenue? For what values of x is the marginal revenue positive? Negative? Zero? Interpret what this means. See Example 4 for the answer.

Product Rule

We continue to develop formulas that permit us to take derivatives without having to resort to using the definition of the derivative. In this section we develop the rules for differentiating the product of two functions and the quotient of two functions.

Since the derivative of the sum is the sum of the derivatives, one might suspect that the derivative of the product is the product of the derivatives. This is *not* true in general. For example, on the one hand

$$\frac{d}{dx}(x^3 \cdot x^2) = \frac{d}{dx}(x^5) = 5x^4$$

but on the other hand

$$\frac{d}{dx}(x^3) \cdot \frac{d}{dx}(x^2) = 3x^2 \cdot 2x = 6x^3$$

The correct rule for taking the derivative of the product of two functions follows:

Product Rule

If both $f'(x)$ and $g'(x)$ exist, then $\frac{d}{dx}(f(x) \cdot g(x))$ exists and

$$\frac{d}{dx}(f(x) \cdot g(x)) = f(x) \cdot \frac{d}{dx}g(x) + g(x) \cdot \frac{d}{dx}f(x)$$

We now consider two examples that use the product rule.

EXAMPLE 1 Using the Product Rule

Find $h'(x)$ if $h(x) = e^x \ln x$

SOLUTION ■ To differentiate $h(x)$, think of $h(x)$ as a product of the two functions: $f(x) = e^x$ and $g(x) = \ln x$. Then

$$h'(x) = f(x) \cdot \frac{d}{dx}g(x) + g(x) \cdot \frac{d}{dx}f(x)$$

$$= e^x \cdot \frac{d}{dx}(\ln x) + (\ln x) \cdot \frac{d}{dx}e^x$$

$$= e^x \frac{1}{x} + (\ln x)e^x$$

$$= e^x\left(\frac{1}{x} + \ln x\right) \quad ■$$

EXAMPLE 2   Using the Product Rule

Find $h'(x)$ if $h(x) = (x^4 + 3x^2 + 2x) \ln x$.

SOLUTION ■ We think of $h(x)$ as $f(x) \cdot g(x)$, where

$$f(x) = (x^4 + 3x^2 + 2x), \quad g(x) = \ln x$$

Then

$$h'(x) = f(x) \cdot \frac{d}{dx} g(x) + g(x) \cdot \frac{d}{dx} f(x)$$

$$= (x^4 + 3x^2 + 2x) \cdot \frac{d}{dx} (\ln x) + (\ln x) \cdot \frac{d}{dx} (x^4 + 3x^2 + 2x)$$

$$= (x^4 + 3x^2 + 2x) \cdot \left(\frac{1}{x}\right) + (\ln x) \cdot (4x^3 + 6x + 2)$$

$$= x^3 + 3x + 2 + (4x^3 + 6x + 2) \ln x \quad ■$$

Quotient Rule

Now that we know that the derivative of the product is not the product of the derivatives, we suspect that the derivative of the quotient is not the quotient of the derivatives. This is indeed the case. The correct formula for the derivative of the quotient is as follows:

Quotient Rule

If $f'(x)$ and $g'(x)$ exist, then $\dfrac{d}{dx}\left(\dfrac{f(x)}{g(x)}\right)$ exists and

$$\frac{d}{dx}\left(\frac{f(x)}{g(x)}\right) = \frac{g(x) \cdot \dfrac{d}{dx} f(x) - f(x) \cdot \dfrac{d}{dx} g(x)}{[g(x)]^2}$$

at points where $g(x) \neq 0$.

EXAMPLE 3   Using the Quotient Rule

Find $h'(x)$ if $h(x) = \dfrac{e^x}{x^2 + 1}$.

SOLUTION ■ Here we think of $h(x)$ as the quotient of two functions $\dfrac{f(x)}{g(x)}$, where

$$f(x) = e^x \qquad g(x) = x^2 + 1$$

Then

$$h'(x) = \frac{g(x) \cdot \dfrac{d}{dx} f(x) - f(x) \cdot \dfrac{d}{dx} g(x)}{[g(x)]^2}$$

$$= \frac{(x^2 + 1) \cdot \dfrac{d}{dx} (e^x) - e^x \cdot \dfrac{d}{dx} (x^2 + 1)}{(x^2 + 1)^2}$$

$$= \frac{(x^2 + 1)e^x - e^x(2x)}{(x^2 + 1)^2}$$

$$= \frac{(x^2 - 2x + 1)e^x}{(x^2 + 1)^2} \quad ■$$

Graphical Support for Example 3

Graph $y_1 = h'(x)$ given in Example 3 and $y_2 = n$Der $h(x)$ on the same screen. Does this support your answer to Example 3? Why?

EXAMPLE 4   Instantaneous Rate of Change of Revenue

Suppose the demand equation for a certain product is given by $p = e^{-x}$, where x is the number of items sold (in thousands) and p the price in dollars. What is the marginal revenue? For what values of x is the marginal revenue positive? Negative? Zero? Interpret what this means.

SOLUTION ■ We have $R(x) = xp(x) = xe^{-x} = \dfrac{x}{e^x}$. We think of $R(x)$ as the

quotient of two functions, $\dfrac{f(x)}{g(x)}$, where $f(x) = x$ and $g(x) = e^x$. Then

$$R'(x) = \frac{g(x) \cdot \dfrac{d}{dx} f(x) - f(x) \cdot \dfrac{d}{dx} g(x)}{[g(x)]^2}$$

$$= \frac{e^x \cdot \dfrac{d}{dx}(x) - (x) \cdot \dfrac{d}{dx} e^x}{(e^x)^2}$$

$$= \frac{e^x(1) - xe^x}{e^{2x}}$$

$$= \frac{(1 - x)e^x}{e^{2x}}$$

$$= (1 - x)e^{-x}$$

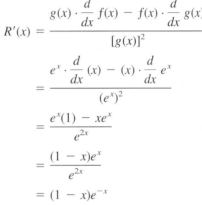

[0, 7] by [0, 0.5]

Screen 3.9
The revenue $y = xe^{-x}$ initially increases, then decreases as the number of items sold increases.

Since e^{-x} is always positive, the sign of $R'(x)$ is the same as the sign of the factor $(1 - x)$. Thus, $R'(x) > 0$ when $0 < x < 1$, $R'(1) = 0$, and $R'(x) < 0$ when $x > 1$. Screen 3.9 shows a graph of $y = R(x)$ using a window of dimensions [0, 7] by [0, 0.5]. Notice the revenue curve is rising on (0, 1) and falling on (1, ∞). ■

We now consider average cost, use the quotient rule to develop a formula used in economics relating marginal average cost with marginal cost, and give a graphical demonstration of this formula.

Average Cost

If $C(x)$ is the cost of producing x items, then the **average cost**, denoted by $\overline{C}(x)$, is the cost of all the items divided by the number of items, that is,

$$\overline{C}(x) = \frac{C(x)}{x}$$

The marginal average cost is given by $\dfrac{d}{dx} \overline{C}(x)$.

EXAMPLE 5   Application of the Quotient Rule

Use the quotient rule to find a formula for the marginal average cost in terms of the cost and the marginal cost.

SOLUTION ■ Using the quotient rule we have

$$\frac{d}{dx}\,\overline{C}(x) = \frac{d}{dx}\left[\frac{C(x)}{x}\right]$$

$$= \frac{x \cdot \dfrac{d}{dx}\,C(x) - C(x) \cdot \dfrac{d}{dx}\,(x)}{x^2}$$

$$= \frac{xC'(x) - C(x)}{x^2}$$

$$= \frac{1}{x}\left[C'(x) - \frac{C(x)}{x}\right]$$

$$= \frac{1}{x}[C'(x) - \overline{C}(x)] \quad ■$$

In words, this formula says that marginal average cost is found by taking marginal cost less average cost and dividing the result by the number of items sold.

EXAMPLE 6   Application of the Quotient Rule

If the cost function is given by $C(x) = 2x^3 - x^2 + 1$, find (a) marginal cost, (b) average cost, (c) marginal average cost.

SOLUTION ■ Marginal cost is given by

$$C'(x) = 6x^2 - 2x$$

Average cost is given by

$$\overline{C}(x) = 2x^2 - x + \frac{1}{x}$$

Marginal average cost is given by

$$\frac{d}{dx}\,\overline{C} = \frac{d}{dx}\left(2x^2 - x + \frac{1}{x}\right) = 4x - 1 - \frac{1}{x^2}$$

To obtain marginal average cost we can also use the formula given in Example 5, which gives

$$\frac{d}{dx}\,\overline{C} = \frac{1}{x}\left[(6x^2 - 2x) - \left(2x^2 - x + \frac{1}{x}\right)\right]$$

$$= \frac{1}{x}\left(4x^2 - x - \frac{1}{x}\right)$$

$$= 4x - 1 - \frac{1}{x^2} \quad ■$$

EXPLORATION 2

An Important Point in Economics

Notice that the result of Example 5 indicates that marginal average cost is zero when the marginal cost equals the average cost. For the cost function given in Example 6 graph the average and marginal cost functions on a screen of dimension [0, 3] by [0, 5]. Notice carefully that the graph of $y = C'(x)$ cuts the graph of $y = \overline{C}(x)$ at the point where the graph of $y = \overline{C}(x)$ has a tangent line with slope zero, which is the point where marginal average cost is zero.

SELF-HELP EXERCISE SET 3.2

Find the derivative of each of the following. Do not simplify.

1. $(x^2 - e^x)(x^3 + x + \ln x)$ **2.** $\dfrac{x^{3/2} + 1}{x^4 + 1}$

EXERCISE SET 3.2

In Exercises 1 through 8 use the product rule to find the derivative.

1. $y = x^3 e^x$ **2.** $y = \sqrt{x} e^x$

3. $y = \sqrt{x} \ln x$ **4.** $y = x^4 \ln x$

5. $y = e^x(x^2 - 3)$ **6.** $y = (2x^3 + 3) \ln x$

7. $y = (4e^x - x^5) \ln x$

8. $y = (x^3 - 3 \ln x)(2e^x + 3x)$

In Exercises 9 through 12 use the product rule to find $f'(x)$.

9. $f(x) = (e^x + 1)(\sqrt{x} + 1)$

10. $f(x) = (x^2 - 3)(\sqrt[3]{x} - \ln x)$

11. $f(x) = (2x - 3 \ln x)\left(x + \dfrac{1}{x}\right)$

12. $f(x) = \left(e^x + \dfrac{1}{x}\right)\left(1 + \dfrac{1}{x^2}\right)$

In Exercises 13 through 24 use the quotient rule to find the derivative. Support your answer by graphing $y_1 = f'(x)$ and $y_2 = n\text{Der } f(x)$ on the same screen.

13. $f(x) = e^{-x} \ln x$ **14.** $f(x) = \dfrac{\ln x}{x^2 + 3}$

15. $f(x) = \dfrac{1}{\ln x}$ **16.** $f(x) = (x + 3)e^{-x}$

17. $f(x) = \dfrac{x}{x + 2}$ **18.** $f(x) = \dfrac{x + 1}{x - 2}$

19. $f(x) = \dfrac{x - 3}{x + 5}$ **20.** $f(x) = \dfrac{3}{x + 3}$

21. $f(x) = \dfrac{e^x}{x - 2}$ **22.** $f(x) = \dfrac{2x - 3}{4x - 1}$

23. $f(x) = \dfrac{2 - 3x}{2x - 1}$ **24.** $f(x) = \dfrac{3 - 2e^x}{1 - 2x}$

In Exercises 25 through 32 use the quotient rule to find $\dfrac{dy}{du}$.

25. $y = \dfrac{3e^u}{u^2 + 1}$ **26.** $y = \dfrac{-4 \ln u}{u^4 + 3}$

27. $y = \dfrac{u + 1}{u^2 + 2}$ **28.** $y = \dfrac{2u - 1}{u^3 - 1}$

29. $y = \dfrac{u^3 - 1}{u^2 + 3}$ **30.** $y = \dfrac{\sqrt{u}}{u^2 + e^u + 1}$

31. $y = \dfrac{\sqrt[3]{u}}{u^3 - e^u - 1}$ **32.** $y = \dfrac{\sqrt[4]{u}}{u^2 + 1}$

Applications

33. Revenue. Suppose the demand equation for a certain product is given by $p = \dfrac{1}{1 + x^2}$, where x is the number of items sold and p the price in dollars. Find the marginal revenue.

34. Cost. It is estimated that the cost $C(x)$ in millions of dollars, of maintaining the toxic emissions of a certain chemical plant $x\%$ free of the toxins is given by $C(x) = \dfrac{4}{100 - x}$. Find the marginal cost.

35. Cost. A large corporation has already cornered 20% of the disposable diaper market and now wishes to undertake a very extensive advertising campaign to increase its share of this market. It is estimated that the advertising cost $C(x)$ in billions of dollars, of attaining an $x\%$ share of this market is given by $C(x) = \dfrac{1}{10 - 0.1x}$. What is the marginal cost?

36. Productivity. The number of items $n(t)$ produced per employee per day over the last 10 years in a certain plant is given by $n(t) = \dfrac{30t}{t + 1}$, where t is the number of years from 10 years ago. What is the rate of change of n with respect to t?

37. Biology. A population N of deer in a state forest follows the law $N(t) = \dfrac{1000e^t}{1 + e^t}$, where t is in years. Find the rate of change of N with respect to t.

38. Spread of Technology. If t is the number of years after the introduction of a certain new technology, the percentage of the firms in this industry using the new technology is given by $p(t) = \dfrac{100}{1 + e^{-t}}$. What is the rate of change of p with respect to t?

39. Medicine. Suppose that the concentration c of a drug in the blood t hours after it is taken orally is given by $c = \dfrac{3t}{2t^2 + 5}$. What is the rate of change of c with respect to t?

40. Chemistry. Salt water with a concentration 0.1 lb of salt per gallon flows into a large tank that initially holds 100 gal of pure water. If 5 gal of salt water per minute flows into the tank, show that the concentration of salt in the tank is given by $c(t) = \dfrac{t}{200 + 10t}$, where t is measured in minutes. What is the rate of change of c with respect to t?

41. Propensity to Save. The amount of money a family saves is a function $S(I)$ of its income I. The *marginal propensity to save* is $\dfrac{dS}{dI}$. If $S(I) = \dfrac{2I^2}{5(I + 50,000)}$, find the marginal propensity to save.

42. Biology. Theoretical studies of photosynthesis assume that the gross rate R of photosynthesis per unit of leaf area is given by $R(E) = \dfrac{aE}{1 + bE}$, where E is the incident radiation

per unit leaf area, and a and b are positive constants. Find the rate of change of R with respect to E.

43. Fishery. The cost function for wild crawfish was estimated by Bell [66] to be $C(x) = \dfrac{7.12}{13.74x - x^2}$, where x is the number of millions of pounds of crawfish caught and C is the cost in millions of dollars. Find the marginal cost.

44. Biology. Pilarska [67] used the equation $y = 15.97\dfrac{x}{1.47 + x}$ to model the ingestion rate y of an individual female rotifer as a function of the density x of green algae it feeds on. Determine the rate of change of the ingestion rate with respect to the density. What is the sign of this rate of change? What is the significance of this sign?

Enrichment Exercises

45. (a) Differentiate e^{2x} by writing $e^{2x} = e^x e^x$.
(b) Now differentiate e^{3x} by writing $e^{3x} = e^{2x}e^x$ and the result in part (a).
(c) Do you see a pattern? What do you think $\dfrac{d}{dx}e^{nx}$ is when n is an integer?

46. (a) Differentiate e^{-x} by writing $e^{-x} = \dfrac{1}{e^x}$.
(b) Now differentiate e^{-2x} by writing $e^{-2x} = e^{-x}e^{-x}$ and using the result of part (a).
(c) Now differentiate e^{-3x} by writing $e^{-3x} = e^{-2x}e^{-x}$ and using the result of part (b).
(d) Do you see a pattern? What do you think $\dfrac{d}{dx}e^{-nx}$ is when n is an integer?

47. Suppose $f(x) = (x - a)^2 g(x)$ and assume $g(x)$ is differentiable at $x = a$. Show that the x-axis is tangent to the graph of $y = f(x)$ at $x = a$.

48. Let $f(x) = |x| \cdot x$. Can you use the product rule of differentiation to find $f'(0)$? Explain why or why not.

In Exercises 49 through 54 find $f'(x)$. (Recall from Exercises 59 and 60 in Section 2.4 that $\dfrac{d}{dx}(\sin x) = \cos x$ and $\dfrac{d}{dx}(\cos x) = -\sin x$.)

49. $f(x) = x \sin x$
50. $f(x) = x^2 \cos x$
51. $f(x) = e^x \cos x$
52. $f(x) = (\ln x)(\sin x)$
53. $f(x) = \dfrac{\sin x}{x}$
54. $f(x) = \dfrac{\cos x}{e^x}$

55. The trigonometric function $\tan x$ is defined as $\tan x = \dfrac{\sin x}{\cos x}$. Find $\dfrac{d}{dx}\tan x$.

56. The trigonometric function $\cot x$ is defined as $\cot x = \dfrac{\cos x}{\sin x}$. Find $\dfrac{d}{dx}\cot x$.

SOLUTIONS TO SELF HELP EXERCISE SET 3.2

1. $\dfrac{d}{dx}[(x^2 - e^x)(x^3 + x + \ln x)] = (x^2 - e^x)\dfrac{d}{dx}(x^3 + x + \ln x) + (x^3 + x + \ln x)\dfrac{d}{dx}(x^2 - e^x)$

$= (x^2 - e^x)\left(3x^2 + 1 + \dfrac{1}{x}\right) + (x^3 + x + \ln x)(2x - e^x)$

2. $\dfrac{d}{dx}\left(\dfrac{x^{3/2} + 1}{x^4 + 1}\right) = \dfrac{(x^4 + 1)\dfrac{d}{dx}(x^{3/2} + 1) - (x^{3/2} + 1)\dfrac{d}{dx}(x^4 + 1)}{(x^4 + 1)^2} = \dfrac{(x^4 + 1)\left(\dfrac{3}{2}x^{1/2}\right) - (x^{3/2} + 1)(4x^3)}{(x^4 + 1)^2}$

3.3 THE CHAIN RULE

■ The Chain Rule ■ The General Power Rule ■ Derivatives of Complex Expressions

---

APPLICATION: SPREAD OF THE BLACK DEATH

Based on some theoretical analysis, Murray [68] determined that the velocity in miles per year of the spread of the Black Death in Europe from 1347 through 1350 was approximated by the equation $V = 200\sqrt{0.008x - 15}$ if $x \geq 1875$ and $V(x) = 0$ if $0 \leq x \leq 1875$, where x is the population density measured in people per square mile. What is $\dfrac{dV}{dx}$ when $x > 1875$, the sign of this quantity when $x > 1875$, and the significance of this sign? For the answer see Example 3.

APPLICATION: MARGINAL COST

What is the marginal cost if $C(x) = \dfrac{e^x}{\sqrt{x^2 + 1}}$? See Example 6 for the answer.

The Chain Rule

Consider the situation in Figure 3.3 in which $y = (f \circ g)(x) = f[g(x)]$ is the composition of the two functions $f(x)$ and $g(x)$. We wish to find a formula for the derivative of $(f \circ g)(x)$ in terms of the derivatives of $f(x)$ and $g(x)$. This formula is perhaps the most important rule of differentiation.

To gain some insight into how this formula can be obtained, let us set $u = g(x)$. Then we have $y = f(u)$, where $u = g(x)$. Since we are looking for the derivative of y with respect to x, we need to determine the relationship between the change in y and the change in x. We can write

$$\frac{\Delta y}{\Delta x} = \frac{\text{change in } y}{\text{change in } x}$$

$$= \frac{\text{change in } y}{\text{change in } u} \cdot \frac{\text{change in } u}{\text{change in } x}$$

$$= \frac{\Delta y}{\Delta u} \cdot \frac{\Delta u}{\Delta x}$$

This formula suggests that

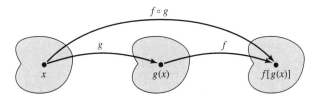

Figure 3.3
The composition of two functions.

$$\frac{dy}{dx} = \frac{dy}{du} \cdot \frac{du}{dx}$$

This can be shown to be true, but a correct proof is too complicated to include here. We now state two forms of the chain rule.

Chain Rule

If $y = f(u)$ and $u = g(x)$, and both of these functions are differentiable, then the composite function $y = f[g(x)]$ is differentiable and

$$\frac{dy}{dx} = \frac{dy}{du} \cdot \frac{du}{dx}$$

or

$$\frac{d}{dx} f[g(x)] = f'[g(x)] \cdot g'(x)$$

EXAMPLE 1 Verifying the Chain Rule

The function $y = 6x - 21 = 3(2x - 7)$ can be written as $y = f[g(x)]$, where $y = f(u) = 3u$ and $u = g(x) = 2x - 7$. Verify the chain rule in this case.

SOLUTION ■ We have

$$\frac{dy}{dx} = 6, \qquad \frac{dy}{du} = 3, \qquad \frac{du}{dx} = 2$$

Since $6 = 3 \cdot 2$,

$$\frac{dy}{dx} = \frac{dy}{du} \cdot \frac{du}{dx} \quad \blacksquare$$

EXAMPLE 2 Using the Chain Rule

Find the derivative of $x = (x^2 + 3)^{78}$.

SOLUTION ■ We could expand this function using the binomial theorem and obtain a polynomial of degree 156 and then differentiate, a very time-consuming process. An alternative way of finding the derivative is to use the chain rule. First recognize that this function is *something* raised to the 78th power, where the something is $(x^2 + 3)$. This suggests that y can be viewed as the composition of two functions, the first function takes x into $x^2 + 3$, and then the second function raises this quantity to the 78th power. In terms of the chain rule, set $y = f(u) = u^{78}$ and $u = g(x) = x^2 + 3$. Then

$$y = f(u) = f[g(x)] = [g(x)]^{78} = (x^2 + 3)^{78}$$

Since $\dfrac{dy}{du} = \dfrac{d}{du}(u^{78}) = 78u^{77}$ and $\dfrac{du}{dx} = \dfrac{d}{dx}(x^2 + 3) = 2x$, the chain rule yields

$$\frac{dy}{dx} = \frac{dy}{du} \cdot \frac{du}{dx}$$
$$= 78u^{77} \cdot (2x)$$
$$= 78(x^2 + 3)^{77}\, 2x$$
$$= 156x(x^2 + 3)^{77} \quad \blacksquare$$

The General Power Rule

Example 2 suggests how to take the derivative of any *function* raised to any power. Now we see how to find the derivative of $[g(x)]^n$, where n is any real number. Here we recognize $y = [g(x)]^n$ as the composition of two functions, the first function takes x to $g(x)$, and the second function raises $g(x)$ to the nth power. If we set $y = f(u) = u^n$ and $u = g(x)$, then

$$y = f[g(x)] = [g(x)]^n$$

Then the chain rule yields

$$\frac{dy}{dx} = \frac{dy}{du} \cdot \frac{du}{dx}$$

$$= nu^{n-1} \cdot \frac{d}{dx} g(x)$$

$$= n[g(x)]^{n-1} \cdot \frac{d}{dx} g(x)$$

We state this result as the general power rule.

General Power Rule

If n is any real number and $g'(x)$ exists, then

$$\frac{d}{dx} [g(x)]^n = n[g(x)]^{n-1} \cdot \frac{d}{dx} g(x)$$

Warning. A common error is to ignore the $\frac{d}{dx} g(x)$ term in the general power formula.

EXAMPLE 3 Using the General Power Rule

Based on some theoretical analysis Murray [68] determined that the velocity in miles per year of the spread of the Black Death in Europe from 1347 through 1350 was approximated by the equation $V = 200\sqrt{0.008x - 15}$ if $x \geq 1875$ and $V(x) = 0$ if $0 \leq x \leq 1875$, where x is the population density measured in people per square mile. Find $\frac{dV}{dx}$ when $x > 1875$. Determine the sign when $x > 1875$ and the significance of this sign. Graph.

SOLUTION ■ We first note that $0.008x - 15 \geq 0$ if $x \geq 1875$. Rewrite V as $V = 200(0.008x - 15)^{1/2}$. Using the general power rule, when $g(x) = 0.008x - 15$ and $n = 1/2$ then for $x > 1875$

$$\frac{dV}{dx} = 200 \frac{d}{dx} (0.008x - 15)^{1/2}$$

$$= 200 \left(\frac{1}{2}\right) (0.008x - 15)^{-1/2} \frac{d}{dx} (0.008x - 15)$$

$$= 200 \left(\frac{1}{2}\right) (0.008x - 15)^{-1/2}(0.008)$$

$$= 0.8(0.008x - 15)^{-1/2}$$

When $x > 1875$ the derivative is positive, indicating that the velocity increases with an increase in density. This is reasonable for an infectious disease. (To transmit this disease fleas must jump from rats to humans and humans must be close enough to

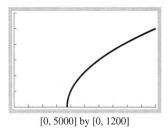

[0, 5000] by [0, 1200]

Screen 3.10

$y = 200\sqrt{0.008x - 15}$ gives the velocity of the spread of the Black Death in Europe in 1347 through 1350 where x is the population density.

infect other humans.) Screen 3.10 shows a graph using a window of dimensions [0, 5000] by [0, 1200]. ∎

EXAMPLE 4   Using the General Power Rule

Find the derivative of $\dfrac{1}{(x^3 + x - 5)^4}$ using the general power rule.

SOLUTION ∎ We must recognize this as some function raised to a power. The power is *minus* 4. Then we can use the general power formula to obtain

$$\frac{d}{dx}\left[\frac{1}{(x^3 + x - 5)^4}\right] = \frac{d}{dx}[(x^3 + x - 5)^{-4}]$$

$$= -4(x^3 + x - 5)^{-5}\frac{d}{dx}(x^3 + x - 5)$$

$$= -4(x^3 + x - 5)^{-5}(3x^2 + 1)$$

$$= -4(3x^2 + 1)(x^3 + x - 5)^{-5} \quad ∎$$

Notice that the general power rule reduces to the ordinary power rule discussed in a previous section. Thus, using the general power rule yields

$$\frac{d}{dx}(x^n) = nx^{n-1}\frac{d}{dx}(x)$$

$$= nx^{n-1}(1)$$

$$= nx^{n-1}$$

Derivatives of Complex Expressions

We now consider some examples that use the general power rule together with other differentiation rules learned earlier.

EXAMPLE 5   Finding the Derivatives of Complex Expressions

Find the derivative of $(x^3 + 5)^{10} \ln x$.

SOLUTION ∎ We first think of this expression as the product of two functions $f(x)g(x)$, where $f(x) = (x^3 + 5)^{10}$ and $g(x) = \ln x$, and then use the product rule of differentiating. We then use the general power rule where necessary. Thus, we obtain

$$\frac{d}{dx}[(x^3 + 5)^{10} \ln x] = (x^3 + 5)^{10}\frac{d}{dx}(\ln x) + (\ln x)\frac{d}{dx}(x^3 + 5)^{10} \qquad \text{product rule}$$

$$= (x^3 + 5)^{10}\frac{1}{x} + (\ln x)10(x^3 + 5)^9\frac{d}{dx}(x^3 + 5) \qquad \begin{matrix}\text{general} \\ \text{power rule}\end{matrix}$$

$$= (x^3 + 5)^{10}\frac{1}{x} + (\ln x)10(x^3 + 5)^9(3x^2)$$

$$= (x^3 + 5)^{10}\frac{1}{x} + 30x^2(x^3 + 5)^9 \ln x \quad ∎$$

EXAMPLE 6   Finding the Derivative of Complex Expressions

Find the marginal cost if $C(x) = \dfrac{e^x}{\sqrt{x^2 + 1}}$.

∎ Black Death

The Black Plague, also known as the Black Death, appeared in Europe in the late 1300s. This pneumonic plague received its name because of the effect it has on its victims; due to respiratory difficulty, black spots appear on the victim's skin shortly after the disease is contracted. The plague reached England in 1348, Germany in 1349, Russia in 1350, and ran rampant in Europe's rat-infested cities. There is limited information available about death rates; however, an example of the mortality rate is Florence, Italy, which lost two thirds of its 100,000 inhabitants in 1348.

A modern-day outbreak of the pneumonic plague occurred in India from August through October 1994. The first cases of the disease were documented in Surat, but the plague was quickly spread to other cities, including New Delhi and Calcutta. This was the first serious epidemic of this plague since early in the twentieth century.

Source: Encyclopedia Americana, 1994 ed.; The New York Times, September 28, 1994.

SOLUTION ■

$$C'(x) = \frac{d}{dx}\left[\frac{e^x}{(x^2+1)^{1/2}}\right]$$

$$= \frac{(x^2+1)^{1/2}\dfrac{d}{dx}(e^x) - e^x\dfrac{d}{dx}[(x^2+1)^{1/2}]}{[(x^2+1)^{1/2}]^2} \qquad \text{quotient rule}$$

$$= \frac{(x^2+1)^{1/2}\,e^x - e^x\dfrac{1}{2}(x^2+1)^{-1/2}\dfrac{d}{dx}(x^2+1)}{(x^2+1)} \qquad \text{general power rule}$$

$$= \frac{(x^2+1)^{1/2}\,e^x - e^x\dfrac{1}{2}(x^2+1)^{-1/2}(2x)}{(x^2+1)}$$

$$= \frac{(x^2+1)^{1/2}\,e^x - xe^x(x^2+1)^{-1/2}}{x^2+1}$$

$$= e^x(x^2+1)^{-1/2}\frac{(x^2+1)-x}{x^2+1}$$

$$= e^x\frac{x^2-x+1}{(x^2+1)^{3/2}} \qquad ■$$

EXPLORATION **1**

Graphical Support for Example 5

Graph $y_1 = C'(x)$ and $y_2 = n\text{Der } C(x)$ on the same screen. Are they the same?

SELF-HELP EXERCISE SET 3.3

1. Find the derivative of $\sqrt{3x^2 + 2e^x}$.

2. Use the general power rule and the product rule to derive the quotient rule. *Hint:* Write

$$\frac{f(x)}{g(x)} = f(x)\cdot[g(x)]^{-1}$$

EXERCISE SET 3.3

In Exercises 1 through 22 find the derivative.

1. $f(x) = (2x+1)^7$

2. $f(x) = (3x-2)^{11}$

3. $f(x) = (x^2+1)^4$

4. $f(x) = (1-x^2)^5$

5. $f(x) = (2e^x-3)^{3/2}$

6. $f(x) = (3-x)^{2/3}$

7. $f(x) = \sqrt{e^x}$

8. $f(x) = e^{4x}$

9. $f(x) = \sqrt{2x+1}$

10. $f(x) = \sqrt{1-3x}$

11. $f(x) = 6\sqrt[3]{x^2+1}$

12. $f(x) = -15\sqrt[5]{2x^3-3}$

13. $f(x) = \dfrac{1}{e^x+1}$

14. $f(x) = \dfrac{4}{x^2+1}$

15. $f(x) = -\dfrac{2}{x^3+1}$

16. $f(x) = \dfrac{1}{\sqrt{x+1}}$

17. $f(x) = \dfrac{\ln x}{\sqrt{x^2+1}}$

18. $f(x) = \dfrac{3}{\sqrt{e^x+1}}$

19. $f(x) = (2x+1)(3x-2)^5$

20. $f(x) = (2x-3)^5(4x+7)$

21. $f(x) = (\ln x)^4$

22. $f(x) = 1/\sqrt{\ln x}$

In Exercises 23 through 34 find $\dfrac{dy}{dx}$.

23. $y = e^x(x^2+1)^8$

24. $y = (1-x^3)^4\ln x$

25. $y = (1-x)^4(2x-1)^5$

26. $y = (7x+3)^3(x^2-4)^6$

27. $y = \sqrt{x}(4x+3)^3$

28. $y = -3\sqrt{x}(1-x^3)^3$

29. $y = e^x(\ln x)^2$

30. $y = e^{2x}\sqrt{\ln x}$

31. $y = \dfrac{3}{(x+1)^4}$

32. $y = \dfrac{-3}{(3e^x+1)^3}$

33. $y = \dfrac{e^x+1}{(x+3)^3}$

34. $y = \dfrac{x^3+x^2-1}{(x+2)^5}$

In Exercises 35 and 36 find $f'(x)$. Support your answer by graphing $y_1 = f'(x)$ and $y_2 = n\text{Der } f(x)$ on the same screen.

35. $f(x) = \sqrt{\sqrt{x}+1}$

36. $f(x) = \sqrt[3]{\sqrt{x}+1}$

Applications

37. Demand. The demand equation for a certain product is given by $p = \dfrac{1}{\sqrt{1 + x^2}}$, where x is the number of items sold and p is the price in dollars. Find the instantaneous rate of change of p with respect to x.

38. Revenue. Find the instantaneous rate of change of revenue with respect to the number sold if the demand equation is given in the previous problem.

39. Compound Interest. If \$1000 is invested at an interest rate of i compounded monthly, the amount in the account at the end of 4 years is given by

$$A = 1000\left(1 + \frac{1}{12}i\right)^{48}$$

Find the rate of change of A with respect to i.

40. Forestry. The volume V in board feet of certain trees was found to be given by $V = 10 + 0.008(d - 4)^3$ for $d > 10$, where d is the diameter. Find the rate of change of V with respect to d when $d = 12$.

41. Fishery. Let n denote the number of prey in a school of fish, r the detection range of the school by a predator, and

c a constant that defines the average spacing of fish within the school. The volume of the region within which a predator can detect the school, called the *visual volume*, has been given by biologists as

$$V = \frac{4\pi r^3}{3}\left(1 + \frac{c}{r}\, n^{1/3}\right)^3$$

With c and r considered as constants, one then has $V = V(n)$. Find the rate of change of V with respect to n.

42. Fishery. In the previous problem suppose c and n are constants. Thus, $V = V(r)$. Find the rate of change of V with respect to r.

43. Psychology. The psychologist L. L. Thurstone suggested that the relationship between the learning time T of memorizing a list of length n is a function of the form $T = an\sqrt{n - b}$, where a and b are positive constants. Find the rate of change of T with respect to n.

44. Biology. Chan and associates [69] showed that the pupal weight y in milligrams of the Mediterranean fruit fly was approximated by $y = 5.5\sqrt{\ln x}$, where x is the percentage of protein in the diet. Find the rate of change of weight with respect to change in protein level.

Enrichment Exercises

45. Find $\dfrac{dy}{dx}$ if $y = \sqrt{\sqrt{\sqrt{x + 1} + 1}}$

46. Suppose $y = f(x)$ is a differentiable function and $f'(4) = 7$. Let $h(x) = f(x^2)$. Find $h'(2)$.

47. Study the solution to Self-Help Exercise 2. Explain how you can do Exercise 33 *without* using the quotient rule. Do Exercise 33 without using the quotient rule.

48. Study the solution to Self-Help Exercise 2. Now do Exercise 34 *without* using the quotient rule.

For Exercises 49 through 52 recall the derivatives of $\sin x$ and $\cos x$ you found in Exercises 59 and 60 in Section 2.4 and find $f'(x)$.

49. $f(x) = (\sin x)^3$

50. $f(x) = (\cos x)^4$

51. $f(x) = \sqrt{\cos x}$

52. $f(x) = \dfrac{1}{\sqrt{\sin x}}$

SOLUTIONS TO SELF-HELP EXERCISE SET 3.3

1. We must recognize this expression as some function raised to a power. In this case the power is $n = \frac{1}{2}$. Using the general power rule, we obtain

$$\frac{d}{dx}\left(\sqrt{3x^2 + 2e^x}\right) = \frac{d}{dx}[(3x^2 + 2e^x)^{1/2}]$$

$$= \frac{1}{2}(3x^2 + 2e^x)^{-1/2}\frac{d}{dx}(3x^2 + 2e^x)$$

$$= \frac{1}{2}(3x^2 + 2e^x)^{-1/2}(6x + 2e^x)$$

$$= (3x + e^x)(3x^2 + 2e^x)^{-1/2}$$

2. First apply the product rule, and then use the general power rule when appropriate to obtain

$$\frac{d}{dx}\left[\frac{f(x)}{g(x)}\right] = \frac{d}{dx}(f(x) \cdot [g(x)]^{-1})$$

$$= [g(x)]^{-1} \cdot \frac{d}{dx}f(x)$$

$$+ f(x) \cdot \frac{d}{dx}([g(x)]^{-1}) \quad \text{product rule}$$

$$= [g(x)]^{-1} \cdot \frac{d}{dx}f(x)$$

$$+ f(x)\left(-[g(x)]^{-2}\frac{d}{dx}g(x)\right) \quad \text{general power rule}$$

$$= \frac{\frac{d}{dx}f(x)}{g(x)} \cdot \frac{g(x)}{g(x)} - \frac{f(x)\frac{d}{dx}g(x)}{g^2(x)}$$

$$= \frac{g(x) \cdot \frac{d}{dx}f(x) - f(x) \cdot \frac{d}{dx}g(x)}{g^2(x)}$$

3.4 DERIVATIVES OF EXPONENTIAL AND LOGARITHMIC FUNCTIONS

■ Derivatives of Exponential Functions ■ Derivatives of Logarithmic Functions

APPLICATION: SURVIVAL CURVES

A survival curve for white males in the United States in the period 1969 through 1971 was determined by Elandt and Johnson [70] to be $y = 0.988e^{-0.0013t - (0.01275t)^2}$, where t is measured in years. What is the rate of change with respect to t and the sign of this rate of change? What is the significance of this sign? See Example 1 for the answer.

APPLICATION: MARGINAL REVENUE

Let the demand equation be given by $p(x) = e^{-0.02x}$. What is the marginal revenue? For what value of x is the marginal revenue zero? See Example 3 for the answer.

Derivatives of Exponential Functions

We already know that the derivative of e^x is again e^x, that is,

$$\frac{d}{dx}(e^x) = e^x$$

We need to be able to differentiate functions of the form $y = e^{f(x)}$. Do this by setting $y = e^u$ where $u = f(x)$, then

$$\frac{dy}{du} = \frac{d}{du}e^u = e^u$$

Now use the chain rule and obtain

$$\frac{d}{dx}(e^{f(x)}) = \frac{dy}{dx}$$

$$= \frac{dy}{du} \cdot \frac{du}{dx}$$

$$= e^u \frac{du}{dx}$$

$$= e^{f(x)} f'(x)$$

The Derivative of $e^{f(x)}$

$$\frac{d}{dx}(e^{f(x)}) = e^{f(x)} f'(x)$$

EXAMPLE 1 The Derivatives of e Raised to a Function

A survival curve for white males in the United States in the period 1969 through 1971 was determined by Elandt and Johnson [70] to be $y = 0.988e^{-0.0013t - (0.01275t)^2}$, where t is measured in years. Determine the rate of change with respect to t and the

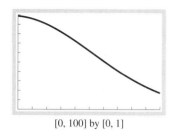

[0, 100] by [0, 1]

Screen 3.11
Notice how the curve $y = 0.988e^{-0.0013x - (0.01275x)^2}$ drops increasingly faster and then levels out.

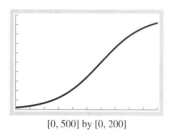

[0, 500] by [0, 200]

Screen 3.12
The population $y = \dfrac{197.3}{1 + 49.6e^{-0.0313x}}$ initially increases fast but then levels out.

sign of this rate of change. What is the significance of this sign? Use your graphing calculator to graph.

SOLUTION ■ Using the above formula with $f(t) = -0.0013t - (0.01275t)^2$ yields

$$\frac{d}{dt}(0.988e^{-0.0013t-(0.01275t)^2}) = 0.988e^{-0.0013t-(0.01275t)^2}\frac{d}{dt}(-0.0013t-(0.01275t)^2)$$

$$= 0.988e^{-0.0013t-(0.01275t)^2}(-0.0013$$
$$-2(0.01275t)(0.01275))$$

$$= -0.988(0.0013 + 2(0.01275)^2t)e^{0.0013t-(0.01275t)^2}$$

The rate of change is always negative. This simply means that the number of survivors is decreasing at all times. Using a window with dimensions [0, 100] by [0, 1] we obtain Screen 3.11. Curves such as these are used by insurance companies. ■

EXAMPLE 2   Derivative of Function Containing Exponential

We noted in Section 2.2 that the logistic equation $p(t) = \dfrac{197.3}{1 + 49.6e^{-0.0313t}}$ was used about 150 years ago by the mathematician-biologist P. Verhulst to model the population $p(t)$ in millions of the United States, where t is the calendar year and $t = 0$ corresponds to 1790. Find the rate of change of this population with respect to time and the sign of this rate of change. What is the significance of this sign? Use your graphing calculator to graph.

SOLUTION ■ Using the quotient rule for derivatives we have

$$\frac{d}{dt}\left(\frac{197.3}{1 + 49.6e^{-0.0313t}}\right) = \frac{(1 + 49e^{-0.0313t})\frac{d}{dt}(197.3) - (197.3)\frac{d}{dt}(1 + 49.6e^{-0.0313t})}{(1 + 49.6e^{-0.0313t})^2}$$

$$= \frac{0 - (197.3)49.6e^{-0.0313t}(-0.0313)}{(1 + 49.6e^{-0.0313t})^2}$$

$$= \frac{(197.3)(49.6)(0.0313)e^{-0.0313t}}{(1 + 49.6e^{-0.0313t})^2}$$

Note that this quantity is positive for all time. This indicates that the population is increasing at all time. The graph in Screen 3.12 uses a window with dimensions [0, 300] by [0, 200]. ■

EXAMPLE 3   Marginal Revenue

Let the demand equation be given $p(x) = e^{-0.02x}$. Find the marginal revenue. Find where the marginal revenue is zero.

SOLUTION ■ We have $R(x) = xp(x) = xe^{-0.02x}$. Thus,

$$R'(x) = \frac{d}{dx}(xe^{-0.02x})$$

$$= e^{-0.02x}\frac{d}{dx}(x) + x\frac{d}{dx}(e^{-0.02x})$$

$$= e^{-0.02x}(1) + x(-0.02e^{-0.02x})$$

$$= e^{-0.02x}(1 - 0.02x)$$

This is the marginal revenue. Now recall that e^z is never zero, no matter what the value of z. Thus, marginal revenue $R'(x) = e^{-0.02x}(1 - 0.02x)$ is zero only when $(1 - 0.02x) = 0$ or $x = 50$. ■

We have already seen how to differentiate the expression $e^{f(x)}$. We now see how to differentiate the expression $a^{f(x)}$ for any positive value of a. First recall that for any $a > 0$, $a^x = e^{x \ln a}$. Differentiating this yields

$$\frac{d}{dx}(a^x) = \frac{d}{dx}(e^{x \ln a})$$

$$= e^{x \ln a} \frac{d}{dx}(x \ln a)$$

$$= e^{x \ln a}(\ln a)$$

$$= a^x \ln a$$

Using the chain rule, we can write the more general form of this as follows:

The Derivative of $a^{f(x)}$

$$\frac{d}{dx}(a^{f(x)}) = a^{f(x)} f'(x) \ln a$$

EXAMPLE 4   Finding the Derivative of $a^{f(x)}$

Find $\dfrac{d}{dx}(2^{x^2 + x + 1})$.

SOLUTION ■ From the previous formula we have

$$\frac{d}{dx}(2^{x^2 + x + 1}) = 2^{x^2 + x + 1} \frac{d}{dx}(x^2 + x + 1) \cdot \ln 2$$

$$= 2^{x^2 + x + 1}(2x + 1) \ln 2 \quad ■$$

Derivatives of Logarithmic Functions

We have already determined graphically that the derivative of $\ln x$ is $1/x$. We can now determine the derivative analytically using the fact that the derivative of e^x is again e^x.

To show this, suppose $x > 0$, then $x = e^{\ln x}$. Now we differentiate this last expression and obtain

$$\frac{d}{dx}(x) = \frac{d}{dx}e^{\ln x}$$

which is just

$$1 = e^{\ln x} \frac{d}{dx}(\ln x)$$

Then

$$\frac{d}{dx}(\ln x) = \frac{1}{e^{\ln x}}$$

$$= \frac{1}{x}$$

We shall now use this formula to find the derivative of an expression such as $y = \ln f(x)$. We first let $u = f(x)$ and notice that

$$\frac{dy}{du} = \frac{d}{du}\ln u = \frac{1}{u}$$

Now the chain rule shows that

$$\frac{d}{dx}(\ln f(x)) = \frac{dy}{dx}$$

$$= \frac{dy}{du} \cdot \frac{du}{dx}$$

$$= \frac{1}{u}\frac{du}{dx}$$

$$= \frac{1}{f(x)}f'(x)$$

Thus,

$$\frac{d}{dx}[\ln f(x)] = \frac{f'(x)}{f(x)}$$

This formula can be used to show that

$$\frac{d}{dx}\ln|x| = \frac{1}{x}$$

To see this, consider $y = \ln(-x)$ where $x < 0$ and set $f(x) = -x$. Then

$$\frac{d}{dx}[\ln(-x)] = \frac{f'(x)}{f(x)}$$

$$= \frac{-1}{-x}$$

$$= \frac{1}{x}$$

This in turn can be used to establish the following more general formula:

Derivative of Logarithmic Function

$$\frac{d}{dx}[\ln|f(x)|] = \frac{f'(x)}{f(x)}$$

Remark. Notice that the derivative of an expression such as ln $|f(x)|$ has *no* logarithm in the answer.

EXPLORATION *1*

Derivatives of Logarithmic Functions

Take the derivative of $f(x) = \ln(kx^n)$, where $k > 0$, $x > 0$, and n is any real number. Where does the k appear? Surprised?

EXAMPLE 5   Derivative of a Logarithmic Function

Find the derivative of $\ln|x^2 + x + 1|$.

SOLUTION ■ Use the preceding formula with $f(x) = x^2 + x + 1$. Then

$$\frac{d}{dx}(\ln|x^2 + x + 1|) = \frac{\frac{d}{dx}(x^2 + x + 1)}{x^2 + x + 1}$$

$$= \frac{2x + 1}{x^2 + x + 1} \quad ■$$

We can make the task of differentiating expressions involving the logarithm easier by exploiting the properties of logarithms. The following example illustrates this:

EXAMPLE 6   Differentiating Using Properties of Logarithms

Find the derivative of $\ln \dfrac{x+1}{x^2+1}$ for $x > -1$.

SOLUTION ■ Use the fact that

$$\ln \frac{x+1}{x^2+1} = \ln(x+1) - \ln(x^2+1)$$

Then

$$\frac{d}{dx}\left(\ln \frac{x+1}{x^2+1}\right) = \frac{d}{dx}[\ln(x+1)] - \frac{d}{dx}[\ln(x^2+1)]$$

$$= \frac{\frac{d}{dx}(x+1)}{x+1} - \frac{\frac{d}{dx}(x^2+1)}{x^2+1}$$

$$= \frac{1}{x+1} - \frac{2x}{x^2+1} \quad ■$$

The last differentiation formula we give in this section is for $\log_a |x|$. To find the derivative of this function, first recall the theorem for change of base

$$\log_a x = \frac{\log_b x}{\log_b a}$$

given in Exercise 50 of Section 1.8. Using this with $b = e$, we have

$$\log_a |x| = \frac{\ln |x|}{\ln a}$$

$$\frac{d}{dx}\log_a |x| = \frac{d}{dx}\left(\frac{\ln |x|}{\ln a}\right)$$

$$= \frac{1}{\ln a}\cdot\frac{d}{dx}\ln |x|$$

$$= \frac{1}{\ln a}\cdot\frac{1}{x}$$

$$= \frac{1}{x \ln a}$$

From the chain rule we then have the following:

$$\boxed{\frac{d}{dx}\log_a |f(x)| = \frac{f'(x)}{f(x)\ln a}}$$

EXAMPLE 7   Derivative of Logarithm to Base *a*

Find $\dfrac{d}{dx}\log x^2$.

S O L U T I O N ■ We use the last formula and obtain

$$\frac{d}{dx} \log x^2 = \frac{d}{dx} \log_{10} x^2$$

$$= \frac{2x}{x^2 \ln 10}$$

$$= \frac{2}{x \ln 10} \quad ■$$

SELF-HELP EXERCISE SET 3.4

1. Find $f'(x)$ if $y = f(x) = e^{-0.50x^2}$. Find where $f'(x)$ is zero.

2. Find $g'(x)$ if $y = g(x) = x \ln x$ for $x > 0$. Find where $g'(x)$ is zero.

EXERCISE SET 3.4

In Exercises 1 through 42 find the derivative.

1. $f(x) = e^{4x}$
2. $f(x) = e^{-0.1x}$
3. $f(x) = e^{-3x}$
4. $f(x) = 4e^{\pi x}$
5. $f(x) = e^{2x^2 + 1}$
6. $f(x) = e^{4x^3 - x - 1}$
7. $f(x) = e^{\sqrt{x}}$
8. $f(x) = e^{\sqrt[3]{x}}$
9. $f(x) = x^2 e^x$
10. $f(x) = \sqrt[3]{x} e^x$
11. $f(x) = x^5 e^x$
12. $f(x) = (x^3 + 1)e^x$
13. $f(x) = \dfrac{x}{e^x}$
14. $f(x) = (x^3 - 2)e^x$
15. $f(x) = x^2 e^{x^2}$
16. $f(x) = \dfrac{e^x}{x^2 + 1}$
17. $f(x) = \sqrt{e^x + 1}$
18. $f(x) = (e^x + 2)^8$
19. $f(x) = \sqrt{x} e^{\sqrt{x}}$
20. $f(x) = \sqrt{x^3} e^{3x}$
21. $f(x) = \dfrac{x}{1 + e^x}$
22. $f(x) = \dfrac{x}{1 + e^{-x}}$
23. $f(x) = \dfrac{e^{2x} - e^{-2x}}{e^{2x} + e^{-2x}}$
24. $f(x) = (e^{3x} + 1)^4$
25. $f(x) = \sqrt{e^{3x} + 2}$
26. $f(x) = \sqrt{2 + e^{-3x}}$
27. $f(x) = 5^x$
28. $f(x) = 3^{3x}$
29. $f(x) = 3^{\sqrt{x}}$
30. $f(x) = 7^{x^7}$
31. $f(x) = x3^x$
32. $f(x) = 3^x \cdot 5^x$
33. $f(x) = \ln |x^3 + x^2 + 1|$
34. $f(x) = \ln(x^4 + x^2 + 3)$
35. $f(x) = x^2 \ln(x^2)$
36. $f(x) = \dfrac{\ln(x^2)}{x + 1}$
37. $f(x) = \sqrt{\ln |x^2 + x + 1|}$
38. $f(x) = \ln(x + 1) \sqrt{x^2 + 1}$
39. $f(x) = \ln \left| \dfrac{x}{x + 1} \right|$
40. $f(x) = \ln \sqrt{\left| \dfrac{x}{x + 1} \right|}$
41. $f(x) = e^{-x^2} \ln(x^2)$
42. $f(x) = \dfrac{\ln(x^2)}{e^x + e^{-x}}$

In Exercises 43 through 52 take $x > 0$ and find the derivatives.

43. $f(x) = x^2 \ln x$
44. $f(x) = (x^3 + 2) \ln x$
45. $f(x) = \dfrac{\ln x}{x^2 + 1}$
46. $f(x) = \sqrt{\ln x}$
47. $f(x) = (\ln x)^{11}$
48. $f(x) = e^{x^2} \ln x$
49. $f(x) = \ln(x^3)$
50. $f(x) = \ln(x^7)$
51. $f(x) = \log(e^x + 1)$
52. $f(x) = \log_2(e^{3x} + x^2)$

In Exercises 53 through 64 find the derivative, and find where the derivative is zero. Assume $x > 0$ in Exercises 61 through 64.

53. $y = xe^{-3x}$
54. $y = xe^x$
55. $y = e^{-x^2}$
56. $y = e^{-x^4}$
57. $y = x^2 e^x$
58. $y = x^4 e^x$
59. $y = x^2 e^{-x}$
60. $y = x^2 e^{-0.5x^2}$
61. $y = x \ln x^2$
62. $y = x - \ln x$
63. $y = \dfrac{\ln x}{x^2}$
64. $y = (\ln x)^3$

Applications

65. **Revenue.** Suppose the price and demand of a commodity is related by $p(x) = e^{-2x}$. Find the marginal revenue $R(x)$, and find where the marginal revenue is zero.

66. **Profits.** Suppose the profit equation is given by $P(x) = xe^{-0.5x^2}$. Find the marginal profit, and find where the marginal profit is zero.

67. Cost. If the cost function is given by $C(x) = \ln(x^2 + 5) + 100$, find the marginal cost. Find where the marginal cost is positive.

68. Biology. Pearl [71] determined that the equation $y = 34.53e^{-0.018x}x^{-0.658}$ approximated the rate of reproduction of fruit flies, where x is the density of the fruit flies in flies per bottle. Find the rate of change of the reproduction rate, and determine the sign of this rate of change. What is the significance of this sign? Use your graphing calculator to graph.

69. Biology. The monomolecular growth assumption states that the rate of change R of growth of a substance at time t is given by $R(t) = \dfrac{A_0}{b-1}(be^{-kt} - 1)$, where $b > 1$ and $k > 0$ are constants, and A_0 is the initial amount of the substance. Show that $R'(t)$ is always negative.

70. Biology. The Gompertz growth curve is given by $P(t) = ae^{-be^{-kt}}$, where a, b, and k are positive constants. Show that $P'(t)$ is always positive.

Enrichment Exercises

In Exercises 71 and 72 find $f'(x)$.

71. $f(x) = e^{e^{e^x}}$

72. $f(x) = \ln[\ln(\ln x)]$

73. Population. In Example 2 the text considered a logistic model of the population of the United States formulated by P. Verhulst about 150 years ago. The following model is an updated version. If $p(t)$ is the population of the United States in millions and t is the time in years with $t = 0$ corresponding to the year 1790, then $p(t) = \dfrac{500}{1 + 124e^{-0.024t}}$. **(a)** Find $\lim\limits_{t \to \infty} p(t)$. **(b)** Find $p'(t)$. **(c)** Determine the sign of $p'(t)$ found in part (b). Interpret the significance of this sign. **(d)** Graph $y = p(t)$ on your graphing calculator using a window with dimensions [0, 400] by [0, 500]. Does your graph verify your answers to parts (a) and (c)? Explain. **(e)** The population of the United States in 1790 was about 4 million and growing during the earlier years at about 2.4% per year. Graph the exponential function $y = 4e^{0.024t}$ on the same screen as that in part (d). How does the graph of this exponential function compare with the graph of the logistic function in part (d)? What do you conclude about the logistic model $y = p(t)$ during the first 50 years as compared to the exponential model given here?

74. We showed in the text that $\dfrac{d}{dx}\log_a |f(x)| = \dfrac{f'(x)}{f(x)\ln a}$. What value of a makes this formula as simple as possible? Explain. What is the formula in this simplest case?

75. You found in Exercises 59 and 60 in Section 2.4 that

$$\frac{d}{dx}(\sin x) = \cos x \quad \text{and} \quad \frac{d}{dx}(\cos x) = -\sin x$$

Use the chain rule to show that

$$\frac{d}{dx}[\sin f(x)] = [\cos f(x)] \cdot f'(x)$$

$$\frac{d}{dx}[\cos f(x)] = -[\sin f(x)] \cdot f'(x)$$

In Exercises 76 through 82 use the formulas found in the Exercise 75 to find the indicated derivatives.

76. $\dfrac{d}{dx}\sin e^x$

77. $\dfrac{d}{dx}\cos x^3$

78. $\dfrac{d}{dx}\cos e^{x^2}$

79. $\dfrac{d}{dx}\sin(\ln x)$

80. $\dfrac{d}{dx}\sin(\cos x)$

81. $\dfrac{d}{dx}\cos(\sin x)$

82. $\dfrac{d}{dx}\ln(\sin x)$

SOLUTIONS TO SELF-HELP EXERCISE SET 3.4

1. We have

$$f'(x) = \frac{d}{dx}(e^{-0.50x^2}) = e^{-0.50x^2}\frac{d}{dx}(-0.50x^2) = -xe^{-0.50x^2}$$

Since $e^{-0.50x^2}$ is positive for all x, $f'(x)$ is only zero when $x = 0$.

2. Since $g(x) = x \ln x$, using the product rule we have

$$g'(x) = x \cdot \frac{d}{dx}(\ln x) + (\ln x) \cdot \frac{d}{dx}(x)$$

$$= x \cdot \frac{1}{x} + (\ln x) \cdot (1)$$

$$= 1 + \ln x$$

This is zero when $\ln x = -1$. According to the definition of the logarithm, this happens if and only if $x = e^{-1}$.

3.5 MARGINAL ANALYSIS IN ECONOMICS

■ Marginal Analysis ■ Elasticity of Demand

APPLICATION: APPROXIMATE COST OF THE NEXT ITEM

Suppose the total cost to produce x barrels of a beverage is given by $C(x) = 5000 + x^2$. Use the derivative to find the approximate cost of the next barrel when $x = 10$. See Example 1 for the answer.

Marginal Analysis

It would be convenient if a manager knew what the additional cost would be of producing one additional unit, or what the change in revenue would be on selling a single additional unit. We shall see that under reasonable conditions, the derivative represents an approximation to the actual change in cost or revenue on production or sale of one additional unit.

To be more specific, suppose a manager knows that the cost of producing x units is given by some function $C(x)$. Suppose the plant has produced x units, and the manager wishes to know to some reasonable approximation what the cost of the next item will be. Naturally, the exact cost is given by $C(x + 1) - C(x)$. We shall now see that under reasonable conditions $C'(x)$ is an approximation to this number.

We use the tangent line approximation (Fig. 3.4). If $\Delta x = [(x + 1) - (x)] = 1$ is small, then

$$
\begin{aligned}
C(x + 1) - C(x) &= \Delta C \\
&\approx C'(x)\Delta x \\
&= C'(x) \cdot 1 \\
&= C'(x)
\end{aligned}
$$

This shows that $C'(x)$ is the **approximate cost of the next unit**.

In the same way we can show for the revenue and profit functions that

$$ R(x + 1) - R(x) \approx R'(x) $$

and

$$ P(x + 1) - P(x) \approx P'(x) $$

This says that the revenue obtained from the sale of the next item is approximately equal to $R'(x)$, and the profit from the sale of the next item is approximately equal to $P'(x)$.

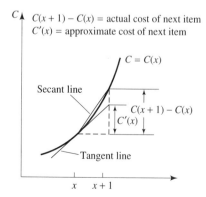

Figure 3.4
The cost of the next item, $C(x + 1) - C(x)$ is approximately $C'(x)$.

Economic Interpretation of Derivative

If $C(x)$, $R(x)$, and $P(x)$ give the cost, revenue, and profit functions, respectively, then under reasonable circumstances $C'(x)$, $R'(x)$, and $P'(x)$ are, respectively, the approximate cost, revenue, and profit associated with the $(x + 1)$st item. The terms $C'(x)$, $R'(x)$, and $P'(x)$ are referred to, respectively, as the **marginal cost**, **marginal revenue**, and **marginal profit**.

EXAMPLE 1   Marginal Cost

Suppose the total cost in dollars to produce x barrels of a beverage is given by $C(x) = 10 + x^2$. Find the marginal cost and the cost of the next barrel when $x = 10$.

SOLUTION ■ Since $C'(x) = 2x$, $C'(10) = 20$, which is the marginal cost when $x = 10$. The actual cost of the next barrel is given by

$$C(11) - C(10) = 21$$

which is remarkably close to the marginal cost. ■

EXPLORATION *1*

Graphical Support for Example 1

 Graph $y = 10 + x^2$ and the tangent line at $x = 10$ to this curve on the screen with dimensions [10, 11] by [110, 131]. Are the two graphs close? Explain.

EXAMPLE 2   Marginal Cost

Suppose a cost function is given by $C(x) = 5e^{0.02x} + 0.001x^3 - 0.3x^2 + 50x + 100$, where x is the number of items produced and $C(x)$ is the cost in dollars to produce x items. Find the marginal cost for any x. Find $C'(60)$, $C'(80)$, and $C'(150)$. Interpret what is happening.

SOLUTION ■ The marginal cost is

$$C'(x) = 0.1e^{0.02x} + 0.003x^2 - 0.6x + 50$$

Then

$$C'(60) = 0.1e^{0.02(60)} + 0.003(60)^2 - 0.6(60) + 50 = 25.13$$

$$C'(80) = 0.1e^{0.02(80)} + 0.003(80)^2 - 0.6(80) + 50 = 21.70$$

$$C'(150) = 0.1e^{0.02(150)} + 0.003(150)^2 - 0.6(150) + 50 = 29.51$$

Thus, cost is increasing at the rate of $25.13 per item when 60 items are being produced, at the lesser rate of $21.70 per item when 80 items are being produced, and at the greater rate of $29.51 per item when 150 items are being produced.

The marginal cost function, $y = C'(x)$, is graphed in Figure 3.5. Notice that the marginal cost decreases initially, and then at some value the marginal cost begins to increase. This is typical of cost functions. Because of economies of scale, unit costs normally decrease initially. But if production is increased too much, unit costs increase due to the use of less efficient equipment and payment of overtime, and so on. ■

Figure 3.5
The marginal cost decreases, then increases.

EXPLORATION *2*

Slope and Rate of Change

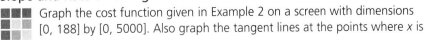 Graph the cost function given in Example 2 on a screen with dimensions [0, 188] by [0, 5000]. Also graph the tangent lines at the points where x is

60, 80, and 150. Observe how the slope of the tangent line changes, and relate this to the previous observations concerning the rates of change.

If a manufacturer sells x units of an item and each item is sold at a price of $\$p$, then the total revenue R is given by

$$R = (\text{number sold})(\text{price of each}) = xp$$

Sometimes the price p is a function $p = p(x)$ of the number sold x. This function is called the **demand equation** and was introduced in Section 1.2.

EXAMPLE 3   Marginal Revenue

Suppose that a manufacturer knows that the demand equation is given by $p(x) = 4.5 - 0.5x$. Find the marginal revenue.

SOLUTION ■ Since $R = xp$ and $p = 4.5 - 0.5x$,

$$R = x(4.5 - 0.5x) = 4.5x - 0.5x^2$$

and therefore marginal revenue is given by

$$R'(x) = 4.5 - x \quad ■$$

Notice that $R'(2) = 2.5$, which is positive. Thus, revenue is increasing at this point. Notice that $R'(8) = -3.5$, indicating that revenue is decreasing at this point.

EXPLORATION 3

Slope and Rate of Change

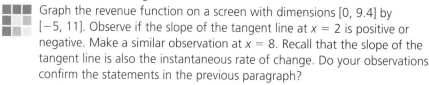 Graph the revenue function on a screen with dimensions [0, 9.4] by [−5, 11]. Observe if the slope of the tangent line at $x = 2$ is positive or negative. Make a similar observation at $x = 8$. Recall that the slope of the tangent line is also the instantaneous rate of change. Do your observations confirm the statements in the previous paragraph?

EXAMPLE 4   Marginal Profit and Break-Even Quantities

Suppose that a manufacturer knows that the cost in thousands of dollars of producing x items is given by $C(x) = 0.5x + 6$ and that the demand equation is known to be $p(x) = 4.5 - 0.5x$. Find the marginal profit and the break-even quantities.

SOLUTION ■ If P is profit, then since $R(x) = xp(x)$

$$\begin{aligned} P(x) &= R(x) - C(x) \\ &= x(4.5 - 0.5x) - (0.5x + 6) \\ &= -0.5x^2 + 4x - 6 \end{aligned}$$

Thus, marginal profit is given by

$$P'(x) = -x + 4$$

Notice that $P'(2) = 2$, implying that profits are increasing at this point. But $P'(6) = -2$, indicating that profits are decreasing at this point.

To find the break-even quantities, we must set profits equal to zero or, equivalently, set revenue equal to cost. Thus, setting $P = 0$, yields

$$\begin{aligned} 0 &= -0.5(x^2 - 8x + 12) \\ &= -0.5(x - 2)(x - 6) \end{aligned}$$

The break-even quantities are therefore $x = 2$ and $x = 6$ (Fig. 3.6). ■

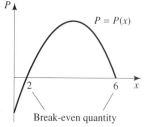

Figure 3.6
As the number of items sold increases, profits first increase, but then decrease, giving two break-even quantities.

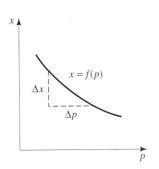

Figure 3.7
A typical demand curve slopes downward.

Slope of Tangent Line

 Using your graphing calculator graph the profit function $y_1 = P(x)$, the tangent line to the curve at $x = 2$, and the tangent line to the curve at $x = 6$ on a screen of dimensions [0, 9.4] by [−8, 4]. Relate the sign of the slope of the tangent line at each point to whether the profits are increasing or decreasing.

Elasticity of Demand

Let us now consider gauging the sensitivity of the demand for a product to changes in the price of the product. Any reasonable demand curve given by $x = f(p)$ is a *decreasing function* (Fig. 3.7). That is, an increase in price will lead to a decrease in demand.

Economists measure this sensitivity of demand to changes in price by the ratio of the percentage change in demand to the percentage change in price. As usual, we let Δp represent the change in price and Δx represent the change in demand. This is shown in Figure 3.7. Notice that if Δp is positive then Δx must be *negative* since the function is decreasing. Also if Δp is negative then Δx must be positive. (That is, a decrease in price leads to greater demand.) We conclude that Δp and Δx must be of opposite signs.

The percentage change in demand is given by the expression $\dfrac{\Delta x}{x}\,100$, and the

percentage change in price is given by the expression $\dfrac{\Delta p}{p}\,100$. The ratio is then

$$\frac{\dfrac{\Delta x}{x}\,100}{\dfrac{\Delta p}{p}\,100} = \frac{p\,\Delta x}{x\,\Delta p}$$

As noted above, Δp and Δx must be of opposite signs, and, therefore, this ratio is *negative*. Since economists prefer to work with positive numbers, they actually work with

$$-\frac{p}{x}\frac{\Delta x}{\Delta p}$$

A large ratio means that a small change in price leads to a relatively large change in demand. A small ratio (say 0.1) means that a large change in price is needed to effect a relatively small change in demand.

Now if $x = f(p)$ is differentiable, then

$$\lim_{\Delta p \to 0} \frac{\Delta x}{\Delta p} = \frac{dx}{dp}.$$

Then we define the term E by

$$E = \lim_{\Delta p \to 0} -\frac{p}{x}\frac{\Delta x}{\Delta p} = -\frac{p}{x}\frac{dx}{dp}$$

This term is called the *elasticity of demand* and measures the instantaneous sensitivity of demand to price.

We say that demand is *inelastic* if $E < 1$, *elastic* if $E > 1$, and of unit elasticity if $E = 1$.

Thus, if the percentage change in demand is less than the percentage change in price, demand is inelastic and $E < 1$. If the percentage change in demand is greater than the percentage change in price, demand is elastic and $E > 1$. Finally, if the percentage change in demand is equal to the percentage change in price, demand is of unit elasticity and $E = 1$.

Elasticity of Demand

If the demand equation is given by $x = f(p)$, then

$$E = -\frac{p}{x}\frac{dx}{dp}$$

is called the **elasticity of demand**.

1. If $E > 1$, then demand is **elastic**.
2. If $E < 1$, then demand is **inelastic**.
3. If $E = 1$, then demand has **unit elasticity**.

EXAMPLE 5   Finding Elasticity of Demand

Suppose

$$x = f(p) = 100 - 2p$$

Find the elasticity of demand when p equals (a) 10 (b) 40 (c) 25.

SOLUTIONS ■ Since $\dfrac{dx}{dp} = -2$,

$$E = -\frac{p}{x}\frac{dx}{dp} = -\frac{p}{100-2p}(-2) = \frac{2p}{100-2p}$$

(a) For $p = 10$,

$$E = \frac{2p}{100-2p} = \frac{20}{80} = 0.25$$

Demand is inelastic. The percentage change in demand is one-fourth the percentage change in price.

(b) For $p = 40$,

$$E = \frac{2p}{100-2p} = \frac{80}{20} = 4$$

Demand is elastic. The percentage change in demand is four times the percentage change in price.

(c) For $p = 25$,

$$E = \frac{2p}{100-2p} = \frac{50}{50} = 1$$

Demand is of unit elasticity. The percentage change in demand is the same as the percentage change in price. ■

There is an interesting connection between the revenue R and elasticity E. We suppose we have $x = f(p)$. Then since $R = px$,

$$\frac{dR}{dp} = x + p\frac{dx}{dp}$$

$$= x + x\frac{p}{x}\frac{dx}{dp}$$

$$= x - xE$$

$$= x(1 - E)$$

If demand is inelastic, $E < 1$, and then $R'(p) > 0$, and therefore $R(p)$ is increasing. If demand is elastic, $E > 1$, and then $R'(p) < 0$, and therefore $R(p)$ is decreasing.

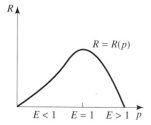

Figure 3.8
The graph shows that if $E < 1$, increasing price increases revenue. But if $E > 1$, increasing price decreases revenue.

If $E < 1$, increasing price therefore increases revenue. If $E > 1$, increasing price decreases revenue. Revenue is optimized when $E = 1$ (Fig. 3.8).

SELF-HELP EXERCISE SET 3.5

1. Suppose the cost function is given by $C(x) = 10 + 0.02x^2$. Find $C'(30)$. Use this to estimate the cost of the 31st item. Compare this estimate with the actual cost of the 31st item.

2. Let $x = 82 - p^2$. Find the elasticity of demand if (a) $p = 1$ (b) $p = 9$.

EXERCISE SET 3.5

In Exercises 1 through 12 estimate the cost of the next item or the revenue or profits from the sale of the next item for the indicated functions and the indicated x.

1. $C(x) = 20 + 0.01x^2, \quad x = 30$

2. $C(x) = 5 + 2x + 0.03x^2, \quad x = 30$

3. $C(x) = \sqrt{x}(x + 12) + 50, \quad x = 36$

4. $C(x) = \sqrt{x^3} + 100, \quad x = 100$

5. $R(x) = x^2 - 10x, \quad x = 20$

6. $R(x) = 0.1x^3 - 8x, \quad x = 20$

7. $R(x) = \sqrt{x}(x - 24), \quad x = 16$

8. $R(x) = \sqrt{x + 10}, \quad x = 90$

9. $P(x) = x^2 - 17x - 60, \quad x = 50$

10. $P(x) = x^3 - 13x - 30, \quad x = 10$

11. $P(x) = \sqrt[3]{x}(x - 54) - 50, \quad x = 27$

12. $P(x) = \sqrt{2x - 1}, \quad x = 41$

Applications

In Exercises 13 through 22 find the elasticity E at the given points and determine if demand is inelastic, elastic, or of unit elasticity.

13. $x = 10 - 2p$, (a) $p = 1$ (b) $p = 2.5$ (c) $p = 4.5$

14. $x = 20 - 4p$, (a) $p = 1$ (b) $p = 2.5$ (c) $p = 4$

15. $x = 24 - 3p$, (a) $p = 1$ (b) $p = 4$ (c) $p = 6$

16. $x = 20 - 5p$, (a) $p = 2$ (b) $p = 3$ (c) $p = 1$

17. $x = \dfrac{1}{10 + p}$, (a) $p = 5$ (b) $p = 10$ (c) $p = 100$

18. $x = 10 + \dfrac{1}{p}$, (a) $p = 0.1$ (b) $p = 1$

19. $x = 10 + \dfrac{1}{p^2}$, (a) $p = 0.01$ (b) $p = 1$

20. $x = \dfrac{10}{p}$, (a) $p = 2$ (b) $p = 3$

21. $x = \dfrac{10}{p^2}$, (a) $p = 1$ (b) $p = 3$

22. $x = \dfrac{10}{\sqrt{p}}$, (a) $p = 1$ (b) $p = 3$

In Exercises 23 through 26 find (a) the value, if any, of x for which $E = 1$ (b) the value(s), if any, of x for which the revenue is maximized.

23. $x = 30 - 10p$

24. $x = 10 - 0.1p$

25. $x = \dfrac{1}{p^3}$

26. $x = \dfrac{1}{1 + p}$

27. **Revenue.** If the demand equation for a product is given by $p = -0.5x + 31$, find the marginal revenue.

28. **Costs.** If the cost equation for a product is given by $C = 0.5x^2 + 20x + 24$, find the marginal cost.

29. **Profits.** If the demand equation and cost equation for a product are as given in the previous two problems, find the marginal profit and the break-even quantities.

Enrichment Exercises

30. **Costs.** Suppose the cost equation for a product is given by $C = 0.1x^3 - 1.5x^2 + 10x + 10$, where x is the number of items produced and $C(x)$ is the cost in dollars to produce x items.
    (a) Find the marginal cost for any x.
    (b) Find $C'(3)$, $C'(5)$, and $C'(12)$. Interpret what is happening.

    (c) Graph the cost function on a screen with dimensions [0, 15] by [0, 200]. Also graph the tangent lines at the points where x is 3, 5, and 12. Observe how the slope of the tangent line changes, and relate this to the observations made above concerning the rates of change.

31. Marginal Cost. Suppose the cost function is given by $C(x) = 0.01x^3 - x^2 + 50x + 100$, where x is the number of items produced and $C(x)$ the cost in dollars to produce x items.

(a) Find the marginal cost for any x.

(b) Find $C'(20)$, $C'(40)$, and $C'(60)$. Interpret what is happening.

(c) Graph the cost function on a screen with dimensions $[0, 90]$ by $[0, 5000]$. Also graph the tangent lines at the points where x is 20, 40, and 60. Observe how the slope of the tangent line changes, and relate this to the observations made above concerning the rates of change.

SOLUTIONS TO SELF-HELP EXERCISE SET 3.5

1. $C'(x) = \dfrac{d}{dx}(10 + 0.02x^2) = \dfrac{d}{dx}(10) + 0.02\dfrac{d}{dx}(x^2) = 0.04x$.

Then $C'(30) = 0.04(30) = 1.20$. Thus, the approximate cost of the 31st item is \$1.20. The actual cost of the 31st item is $C(31) - C(30) = \$1.22$.

2. We have

$$E = -\frac{p}{x}\frac{dx}{dp} = -\frac{p}{82 - p^2}(-2p) = \frac{2p^2}{82 - p^2}$$

(a) When $p = 1$, $E = \dfrac{2}{81} < 1$. Thus, demand is inelastic.

(b) When $p = 9$, $E = 162 > 1$. Thus, demand is elastic.

CHAPTER **3**

SUMMARY OUTLINE

- **Rules for Derivatives.** If $f(x)$ and $g(x)$ are continuous, C a constant, and n a real number, then

$$\frac{d}{dx}[f(x) \cdot g(x)] = f(x) \cdot \frac{d}{dx}g(x) + g(x) \cdot \frac{d}{dx}f(x)$$

$$\frac{d}{dx}\left[\frac{f(x)}{g(x)}\right] = \frac{g(x) \cdot \dfrac{d}{dx}f(x) - f(x) \cdot \dfrac{d}{dx}g(x)}{g^2(x)} \quad \text{if } g(x) \neq 0$$

$$\frac{d}{dx}(C) = 0 \qquad\qquad \frac{d}{dx}(x^n) = nx^{n-1}$$

$$\frac{d}{dx}[Cf(x)] = C\frac{d}{dx}f(x) \qquad\qquad \frac{d}{dx}[f(x) \pm g(x)] = \frac{d}{dx}f(x) \pm \frac{d}{dx}g(x)$$

$$\frac{d}{dx}\{f[g(x)]\} = f'[g(x)] \cdot \frac{d}{dx}g(x) \qquad\qquad \frac{d}{dx}\{[g(x)]^n\} = n[g(x)]^{n-1} \cdot \frac{d}{dx}g(x)$$

$$\frac{d}{dx}(e^x) = e^x \qquad\qquad \frac{d}{dx}(\ln x) = \frac{1}{x}$$

$$\frac{d}{dx}[e^{f(x)}] = e^{f(x)}f'(x) \qquad\qquad \frac{d}{dx}[a^{f(x)}] = a^{f(x)}f'(x)\ln a$$

$$\frac{d}{dx}[\ln f(x)] = \frac{f'(x)}{f(x)}$$

- If $C(x)$, $R(x)$, and $P(x)$ are the cost, revenue, and profit functions, respectively, then $C'(x)$, $R'(x)$, and $P'(x)$ are the **marginal cost**, **marginal revenue**, and **marginal profit**, respectively. p. 161.

- The terms $C'(x)$, $R'(x)$, $P'(x)$, respectively, approximate the cost, revenue, and profit associated with the next item. p. 185.

- **Elasticity of Demand.** If the demand equation is given by $x = f(p)$, then $E = -\dfrac{p}{x}\dfrac{dx}{dp}$ is called the **elasticity of demand**. If $E > 1$, then demand is **elastic**. If $E < 1$, then demand is **inelastic**. If $E = 1$, then demand has **unit elasticity**. p. 188

CHAPTER 3 REVIEW EXERCISES

In Exercises 1 through 34 find the derivative. Support your answers using nDer $f(x)$.

1. $f(x) = x^6 - 3x^2 + 1$

2. $f(x) = 4x^{6/5} + 5x^{2/5} + 3$

3. $f(x) = 2x^{-3} + 3x^{-2/3}$

4. $f(x) = 2x^{1.3} + 3x^{-2.4}$

5. $f(x) = \pi^2 x^2 + 2x - \pi$

6. $f(x) = \pi^3 x^3 + e^2 x^2 - \pi^2$

7. $f(x) = \sqrt[3]{x^2}$

8. $f(x) = \sqrt[3]{x^{-5}}$

9. $f(x) = (x + 3)(x^3 + x + 1)$

10. $f(x) = (x^2 + 5)(x^5 - 3x + 2)$

11. $f(x) = (\sqrt{x})(x^4 - 2x^2 + 3)$

12. $f(x) = (\sqrt[3]{x})(x^3 - 3x + 5)$

13. $f(x) = (x^{3/2})(x^2 + 4x + 7)$

14. $f(x) = (x^{-7/2})(x^3 + 2x + 1)$

15. $f(x) = \dfrac{x - 1}{x + 1}$

16. $f(x) = \dfrac{\sqrt{x}}{2x + 1}$

17. $f(x) = \dfrac{x^3}{x^2 + 1}$

18. $f(x) = \dfrac{x^2 + x + 1}{3x - 2}$

19. $f(x) = \dfrac{x^2 - 1}{x^4 + 1}$

20. $f(x) = (3x + 1)^{10}$

21. $f(x) = (x^2 + 4)^{20}$

22. $f(x) = (\sqrt{x} - 5)^{5/2}$

23. $f(x) = \sqrt{x^2 + x + 1}$

24. $f(x) = x^3(x^2 + x + 2)^5$

25. $f(x) = (x^2 + 3)(x^4 + x + 1)^5$

26. $f(x) = \dfrac{x^2 + 1}{(x^4 + x + 1)^4}$

27. $f(x) = \sqrt{x^{3/2} + 1}$

28. $f(x) = \sqrt[3]{\sqrt{x} + 1}$

29. $f(x) = e^{-7x}$

30. $f(x) = e^{2x^3 + 1}$

31. $f(x) = (x^2 + x + 2)e^x$

32. $f(x) = \ln(x^2 + 1)$

33. $f(x) = e^{x^2} \ln x$

34. $f(x) = \ln(x^2 + e^x)$

In Exercises 35 through 38 find $\dfrac{dy}{dx}$.

35. $y = \sqrt[5]{x}$

36. $y = \sqrt[3]{x^2}$

37. $y = \dfrac{1}{x^3}$

38. $y = \sqrt{x} + \dfrac{1}{\sqrt{x}}$

In Exercises 39 through 42 use the derivatives found in the previous four exercises and the tangent line approximation to find approximations of the following quantities:

39. $\sqrt[5]{1.05}$

40. $\sqrt[3]{(7.7)^2}$

41. $\dfrac{1}{(2.08)^3}$

42. $\sqrt{4.016} + \dfrac{1}{\sqrt{4.016}}$

43. Profits. The daily profits in dollars of a firm is given by $P = -10x^3 + 800x - 10e^x$, where x is the number of items sold. Find the instantaneous rate of change when x is **(a)** 2 **(b)** 3 **(c)** 4 **(d)** 5. Interpret your answers.

44. Velocity. A particle moves according to the law $s(t) = -2t^4 + 64t + 15$, where s is in feet and t in seconds. Find the instantaneous velocity at $t = 1, 2, 3$. Interpret your answers.

45. Biology. Schotzko and Smith [72] showed that the number y of wheat aphids in an experiment was given approximately by $y = f(t) = 5.528 + 1.360t^{2.395}$, where t is the time measured in days. Find $f'(t)$, and explain what this means.

46. Biology. Talekar and coworkers [73] showed that the number y of the eggs of *Ostrinia furnacalis* per mungbean plant was approximated by $y = f(x) = -57.40 + 63.77x - 20.83x^2 + 2.36x^3$, where x is the age of the mungbean plant and $2 \le x \le 7$. Find $f'(x)$, and explain what this means. Use the quadratic formula to show that $f'(x) > 0$ for $2 \le x \le 7$. What does this say about the preference of *O. furnacalis* for laying eggs on older plants?

47. Biology. Buntin and colleagues [74] showed that the net photosynthetic rate y of tomato leaves was approximated by the equation $y = f(x) = 8.094(1.053)^{-x}$, where x is the number of immature sweet potato whiteflies per cubic centimeter on tomato leaflets. Find $f'(x)$, and explain the significance of the sign of the derivative.

48. Costs. The weekly total cost function for a company in dollars is $C(x)$. If $C'(100) = 25$, what is the approximate cost of the 101st item?

49. Elasticity. Given the demand curve $x = 10 - 2p$, determine if the demand is elastic, inelastic, or of unit elasticity if **(a)** $p = 2$ **(b)** $p = 2.5$ **(c)** $p = 3$.

50. Demand. If the demand for a product is given by $p = \dfrac{300}{3x + 1}$, find the marginal demand, where p and x have the usual meanings.

CHAPTER **3**

PROJECTS

Project 1
Loss of Species Due to Deforestation

In one part of Edward Wilson's wonderful book *The Diversity of Life* [75] he attempts to approximate the number of species lost due to the deforestation of the tropics. He starts with the mathematical model $N = aS^b$, that approximates the number N of different species that exist in an area S of tropical forest. The positive constants a and b depend on the specific region. The constant a is not important to this discussion. He notes that empirical studies indicate that b ranges from about 0.15 to 0.35, and that 0.30 is typical. Thus, we take

$$N = f(S) = aS^{0.30}$$

Notice that since N is an increasing function of S, this formula states that the number of species in an area increases with the size of the area. The formula for N attempts to quantify this relationship.

Estimates from both the ground and the air indicate that the deforestation of the tropics is taking place at about 1.8% per year.

(a) From the formula for N estimate the percentage decrease that will occur per year in the number of species.

(b) Now estimate the actual number of species lost per year using the conservative estimate of $b = 0.15$, and make the conservative estimate that 10,000,000 species exist to start with.

Project 2
Polynomial Extrapolation

Often in applications an experiment or survey yields the values of some function f at a finite number of discrete points. Suppose, then, that we know the values $f(x_1), f(x_2), \ldots, f(x_n)$, at the n distinct (nodal) points $x_1, x_2, \ldots, x_n$. We always assume $x_1 < x_2 < \cdots < x_n$. A theorem in mathematics states that a unique polynomial p exists of degree $\leq n - 1$, such that $p(x_i) = f(x_i)$, for $i = 1, 2, \ldots, n$. Such a polynomial is called an **interpolating** polynomial.

We can use graphing calculators or computers to find this interpolating polynomial. We then can use the interpolating polynomial to approximate the values of f at nonnodal points on the interval (x_1, x_n). This process is called **interpolation**. We also can use the interpolating polynomial to approximate the function f outside the interval (x_1, x_2). This process is called **extrapolation**.

Although the interpolating polynomial can sometimes be useful in both interpolating and extrapolating, its use in both interpolating and extrapolating is fraught with dangers. Sometimes, grossly incorrect answers can be obtained!

Consider again the annual population of the United States from 1790 through 1794. In Section 1.7, we noted that the population increased at almost exactly 3.3% a year. We then created a mathematical model based on the exponential function to model the population. We then used this model to estimate the population in 1804. Biologists have observed many populations, both human and nonhuman, following the exponential model, at least when the population is small enough not to be inhibited by a lack of resources. Thus, had we formulated this exponential model in 1794, we would have had some confidence in using it to project the population over the next 10 years. Recall that the model gave a reasonable approximation to the population in 1804.

Now let us use the interpolating polynomial to extrapolate the population to 1804. Let t be the year and $f(t)$ the population in thousands in the year t. Recall we have

$$f(1790) = 3926$$

$$f(1791) = 4056$$

$$f(1792) = 4194$$

$$f(1793) = 4332$$

$$f(1794) = 4469$$

Let the interpolating polynomial be

$$p(t) = a_0 + a_1(t - 1790) + a_2(t - 1790)^2 + a_3(t - 1790)^3 + a_4(t - 1790)^4$$

Notice that $3926 = p(1790) = a_0$. Thus, $a_0 = 3926$. Also notice that

$$4056 = p(1791)$$
$$= a_0 + a_1 + a_2 + a_3 + a_4$$
$$= 3926 + a_1 + a_2 + a_3 + a_4$$
$$a_1 + a_2 + a_3 + a_4 = 130$$

(a) Use the remaining data in a similar fashion, and obtain a total of four linear equations in the four unknowns a_1, a_2, a_3, a_4. Solve on your calculator, and find the interpolating polynomial $p(t)$.

(b) Use calculus to find $p'(t)$ at $t = 1791$, 1792, 1793, and 1794. What do your answers mean in terms of the growth of the population? Do your answers seem reasonable? Why or why not?

(c) Find $p(1804)$. Compare with the actual population in 1804 of 6,065,000.

(d) Graph on [1790, 1804] on your graphing calculator.

(e) Now extrapolate into the *past* and find $p(1780)$ and $p'(1780)$. Is your answer believable?

(f) Graph on [1780, 1804]. Do you think this gives an accurate model of the population from 1780 to 1790? Why or why not?

(g) What do you think went wrong? Was there any *biological* basis to have confidence in this interpolating polynomial to model the population? Was there any biological basis to have confidence in the exponential model used in Section 1.7? If so, what are they?

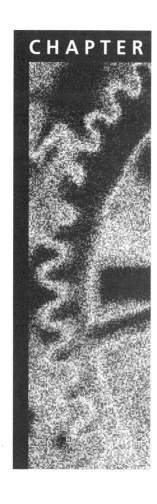

Curve Sketching and Optimization

In the first five sections of this chapter we learn how to sketch the graph of a function and how to find the maximum and minimum of a function on a given interval. To mention just a few applications, we look at the Laffer curve that was used to justify decreasing the personal federal income tax rates during the Reagan administration; we consider the management of renewable natural resources, considering the whaling industry specifically, and we see how to find the quantity of sales that will produce the most profit. In the sixth section we see how to find the derivative when given an equation in two variables, only one of which we can solve for. The last section details a way of solving equations approximately by numerical means when no known method of finding the exact solution exists.

4.1 THE FIRST DERIVATIVE

■ Increasing and Decreasing Functions   ■ Relative Extrema

APPLICATION: FINDING MAXIMUM REVENUE

Let the demand equation be given by $p(x) = e^{-0.02x}$. For what value of x does revenue attain a maximum? For the answer see Example 6.

APPLICATION: TAX REVENUE VERSUS THE TAX RATE

Let x be the percentage of the sum of total income of individuals and total profits of corporations in the United States collected as federal taxes and $R(x)$ the total tax revenue. If x is small, then increasing x should result in an increase in the revenue $R(x)$. If x is large, is it possible that increasing x further decreases revenue? If this is so, can decreasing the rate x of taxation result in an *increase* in the revenue $R(x)$ collected? For the answers given by some of the Reagan administration economists, see Example 1 in Section 4.2

Increasing and Decreasing Functions

In Figure 4.1 we see that the population of the United States has steadily *increased* over the period shown, whereas in Figure 4.2, we see that the percentage of the population that are farmers has steadily *decreased*. In Figure 4.3 we see that the median age of the U.S. population increased from 1860 to 1950, decreased from 1950 to 1970, and then increased again from 1970 to 1990.

In this section we examine how to use the sign of the first derivative to determine whether the function is increasing or decreasing. First we give two definitions.

Definition of Increasing and Decreasing Function

A function $y = f(x)$ is said to be **increasing** (denoted by ↗) on the interval (a, b) if the graph of the function rises while moving left to right or, equivalently, if

$$f(x_1) < f(x_2) \qquad \text{when} \qquad a < x_1 < x_2 < b$$

A function $y = f(x)$ is said to be **decreasing** (denoted by ↘) on the interval (a, b) if the graph of the function falls while moving left to right or, equivalently, if

$$f(x_1) > f(x_2) \qquad \text{when} \qquad a < x_1 < x_2 < b$$

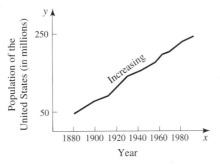

Figure 4.1

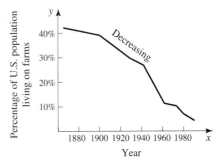

Figure 4.2

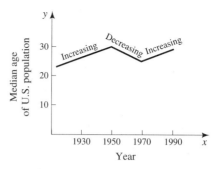

Figure 4.3

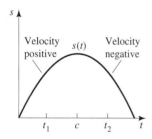

Figure 4.4
Positive velocity means the ball is rising; negative velocity means the ball is falling.

We already know that for straight lines, a positive slope indicates the line is rising and therefore the function is increasing, but if the slope is negative the line is falling and therefore the function is decreasing. A similar situation exists for curves that are not lines.

Suppose a ball is thrown upward and the function $s(t)$ gives the height of the ball in feet, where t is measured in seconds (Fig. 4.4). We already know that the derivative $s'(t)$ gives the velocity $v(t)$ at time t and also the slope of the tangent line to the graph of the curve at t. If the slope or velocity $s'(t_1)$ is positive, we know that the ball is rising at t_1. If the slope or velocity $s'(t_2)$ is negative, we know that the ball is falling at t_2. We can see in Figure 4.4 that for values of t less than c, the slope, or derivative, is positive and the function is increasing, whereas for $t > c$, the slope or derivative is negative and the function is decreasing.

EXPLORATION 1

The Sign of the Derivative

 Graph the function $y_1 = f(x) = x^3 - 6x^2 + 9x + 2$ and its derivative $y_2 = 3x^2 - 12x + 9$ on the same screen with dimensions $[-1, 5]$ by $[-10, 10]$. Locate the two intervals on which $y_1 = f(x)$ is increasing. What is true about the sign of $f'(x)$ on these intervals? Now locate the interval on which $y_1 = f(x)$ is decreasing. What is true about the sign of $f'(x)$ on this interval? Does this make sense to you? Explain.

Hopefully, you just discovered the following theorem.

Test for Increasing or Decreasing Functions

If for all $x \in (a, b)$, $f'(x) > 0$, then $f(x)$ is increasing ($\nearrow$) on (a, b).
If for all $x \in (a, b)$, $f'(x) < 0$, then $f(x)$ is decreasing ($\searrow$) on (a, b).

To find the intervals on which $f(x)$ is either always increasing or always decreasing for a function like the one graphed in Figure 4.5, we need to find the intervals on which

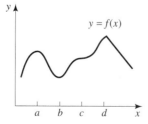

Figure 4.5
The critical values are at $x = a$, b, c, and d.

$f'(x)$ is either always positive or always negative. The simplest way to accomplish this is first to find all values for which the derivative is neither positive nor negative, since normally only a small number of such values exist. Obviously, all values for which the derivative is *zero* fall into this category, but also in this category are all values for which the derivative does *not exist*. These two types of values are so important that they are given a special name, **critical values**.

Definition of Critical Value

A value $x = c$ is a **critical value** if

1. c is in the domain of the function $f(x)$ and
2. $f'(c) = 0$ or $f'(c)$ does not exist.

EXAMPLE 1   Finding Critical Values Graphically

Find the critical values for the function shown in Figure 4.5.

SOLUTION ■   Since the derivative of the function is zero at the values a, b, and c, these three values are critical values. Because the derivative does not exist at the value d, this is also a critical value.   ■

EXAMPLE 2   Finding Critical Points Graphically

Graph $y = f(x) = 1.6x^9 - 8x^5 + 7x^3$ on your graphing calculator using a window of dimensions $[-1.4, 1.4]$ by $[-1.4, 1.4]$, and roughly estimate the critical values. Can you confirm your answer analytically?

SOLUTION ■   The graph is shown in Screen 4.1. We used $x_{scl} = 0.2$. Very roughly, the critical values appear to be at $x = \pm1.15$, $x = \pm0.8$, and $x = 0$, since, at these points, the derivative appears to be zero. We can only confirm our $x = 0$ answer analytically, since we are unable to solve $f'(x) = 14.4x^8 - 40x^4 + 21x^2 = 0$ for the other solutions!   ■

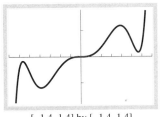

[−1.4, 1.4] by [−1.4, 1.4]

Screen 4.1
Critical values for the function $y = 1.6x^9 - 8x^5 + 7x^3$ appear to be at about $x = \pm1.15$, $x = \pm0.8$ and $x = 0$.

EXPLORATION *2*

Finding Critical Values Graphically

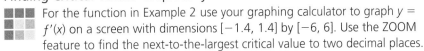 For the function in Example 2 use your graphing calculator to graph $y = f'(x)$ on a screen with dimensions $[-1.4, 1.4]$ by $[-6, 6]$. Use the ZOOM feature to find the next-to-the-largest critical value to two decimal places.

Let us now see how we can use the first derivative to help graph a function.

EXAMPLE 3   Finding Where f Is Increasing or Decreasing

Find the critical values and the intervals where $f(x) = x^3 - 3x^2 + 1$ is increasing or decreasing. Use this information to sketch a graph.

SOLUTION ■

STEP 1. Find the critical values.
   (a) Notice that $f'(x) = 3x^2 - 6x$ exists and is continuous everywhere. Thus, the only critical values are values at which $f'(x) = 0$.
   (b) Setting $f'(x) = 3x^2 - 6x = 3x(x - 2) = 0$ gives $x = 0$ and $x = 2$. Thus, the critical values are $x = 0$ and $x = 2$.

STEP 2. Find the subintervals on which $f'(x)$ has constant sign.

The two critical values divide the real line into the three subintervals: $(-\infty, 0)$, $(0, 2)$, and $(2, \infty)$. The derivative is continuous and never zero on each of these subintervals. By the constant sign theorem of Section 2.2 the sign of $f'(x)$ must therefore be constant on each of these subintervals.

STEP 3. Determine the sign of f' on each subinterval.

Since the sign of $f'(x)$ is constant on the subinterval $(-\infty, 0)$, we need only evaluate f' at one (test) point of this subinterval to determine the sign on the entire subinterval. The same principle applies to the other two subintervals. So take the three test values to be $x_1 = -1$, $x_2 = 1$, and $x_3 = 3$. These three values fall respectively in the three subintervals given in the preceding step (Fig. 4.6). Evaluating f' at these three values gives

$$f'(-1) = +9, \qquad f'(1) = -3, \qquad f'(3) = +9.$$

Thus, $f'(x) > 0$ on the first interval, $f'(x) < 0$ on the second interval, and $f'(x) > 0$ on the third interval.

STEP 4. Use the test for increasing or decreasing functions.

We conclude that $f(x)$ is increasing ($\nearrow$) on the first interval, decreasing ($\searrow$) on the second, and increasing ($\nearrow$) on the third.

Now using this information and the fact that $f(0) = 1$ and $f(2) = -3$, we obtain the graph in Figure 4.7.   ∎

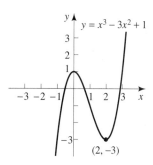

Figure 4.7
The function increases on $(-\infty,0)$ and $(2,\infty)$ and decreases on $(0,2)$.

Warning. In Step 1 it is easy to overlook the solution $x = 0$. A common mistake is to set $f'(x) = 3x^2 - 6x = 0$, cancel x, and obtain $3x - 6 = 0$, or $x = 2$, as the *only* solution.

Procedure for Finding Where a Function Is Increasing or Decreasing

Step 1. Find the critical values. To do this
    a. Locate all values c for which $f'(c) = 0$.
    b. Locate all values c for which $f'(c)$ does not exist but $f(c)$ does.
Step 2. Find the subintervals on which $f'(x)$ has constant sign.
Step 3. Select a convenient test value on each subinterval found in the previous step, and evaluate f' at this value. The sign of f' at this test value is the sign of f' on the whole subinterval.
Step 4. Use the test for increasing or decreasing functions to determine whether f is increasing or decreasing on each subinterval.

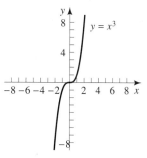

Figure 4.8
This function always increases, but $f'(0) = 0$.

Remark. We have seen that if the derivative is always positive on an interval, then the function increases on that interval, and if the derivative is always negative on an interval, the function always decreases on that interval. A function may increase on an interval, however, without the derivative always being positive on that interval. The graph of the function $f(x) = x^3$ in Figure 4.8 illustrates that this function is increasing throughout the interval $(-\infty, \infty)$. However, $f'(x) = 3x^2$ and is zero at $x = 0$. A similar remark applies to decreasing functions.

Relative Extrema

Now we are interested in studying the peaks and troughs on the graph of a continuous function. For example, the graph in Figure 4.7 has a peak at $(0, 1)$ and a trough at $(2, -3)$. We call the peaks **relative maxima** and the troughs **relative minima**.

Definition of Relative Maximum and Relative Minimum

We say that the quantity $f(c)$ is a relative maximum if $f(x) \leq f(c)$ for all x in some open interval (a, b) that contains c.

We say that the quantity $f(c)$ is a relative minimum if $f(x) \geq f(c)$ for all x in some open interval (a, b) that contains c.

For convenience we also make the following definition:

Definition of Relative Extremum

We say that $f(c)$ is a **relative extremum** if $f(c)$ is a relative maximum or a relative minimum.

Figures 4.6 and 4.7 indicate what happens at relative extrema. Notice that $f'(x)$ is positive to the left of the critical value $x = 0$ and negative to the right. This indicates that f is increasing ($\nearrow$) to the left of $x = 0$ and decreasing ($\searrow$) to the right. This gives $\nearrow \searrow$ and indicates a relative maximum. Notice also that $f'(x)$ is negative to the left of the critical value $x = 2$ and positive to the right. This indicates that f is decreasing ($\searrow$) to the left of $x = 2$ and increasing ($\nearrow$) to the right. This gives $\searrow \nearrow$ and indicates a relative minimum.

This is not the entire story. As the graph of $y = f(x)$ shows in Figure 4.9, $f'(x)$ is negative on both sides of the critical value x_2 and so does not have a relative extremum at x_2. Also, $f'(x)$ is positive on both sides of the critical value x_4 and so does not have a relative extremum at x_4. Thus, only when the derivative *changes sign* at a critical value does a relative extremum occur at this value.

We summarize this discussion as the first derivative test for finding relative extrema.

First Derivative Test

Suppose f is defined on (a, b) and c is a critical value in the interval (a, b).

1. If $f'(x) > 0$ for x near and to the left of c and $f'(x) < 0$ for x near and to the right of c, then we have $\nearrow \searrow$, and $f(c)$ is a relative maximum.

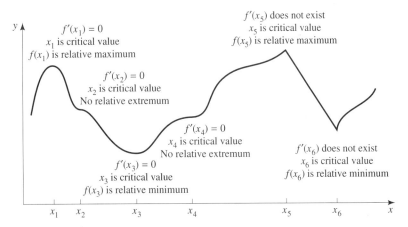

Figure 4.9
Only when the derivative changes sign at a critical point does a relative extremum occur.

2. If $f'(x) < 0$ for x near and to the left of c and $f'(x) > 0$ for x near and to the right of c, then we have $\searrow\nearrow$, and $f(c)$ is a relative minimum.
3. If the sign of $f'(x)$ is the same on both sides of c, then $f(c)$ is not a relative extremum.

Figure 4.10 shows all the possible cases.

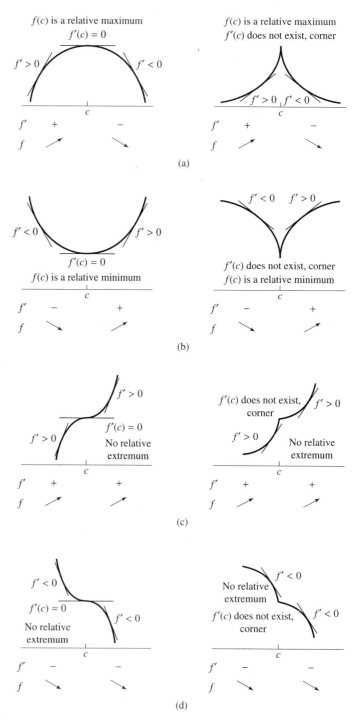

Figure 4.10

EXAMPLE 4 Finding Relative Extrema Graphically

On your graphing calculator graph $y = f(x) = 3x^5 - 20x^3 + 10$, and find the relative extrema.

SOLUTION ■ Using a window with dimensions $[-3, 3]$ by $[-100, 100]$, we obtain a graph that looks very much like Figure 4.11. This could be the correct graph; however, there may be some hidden behavior we are not seeing. No matter how many places we might ZOOM in on the graph, we can never be certain that some relative extrema still remain hidden. But, based on what we see in the window, we believe that the relative extrema occur at about $x = \pm 2$. ■

Only by using calculus can we hope to find all the relative extrema, and then, only if we can find the solutions to $f'(x) = 0$. As we saw in Example 2 we may need to resort to a combination of graphical and numerical procedures to find them. Thus, if we want to solve a variety of problems, we need to know graphical, numerical, and analytical methods.

EXAMPLE 5 Finding the Relative Extrema Using Calculus

Use calculus to confirm the work in Example 4.

SOLUTION ■

STEP 1. Find the critical values.
  (a) The derivative $f'(x) = 15x^4 - 60x^2$ exists and is continuous everywhere. Thus, the only critical values are values at which $f'(x) = 0$.
  (b) Setting $f'(x) = 15x^4 - 60x^2 = 15x^2(x^2 - 4) = 0$ yields $x = 0$ and $x = \pm 2$. Thus, the critical values are at -2, 0, and 2.

STEP 2. These three critical values divide the real line into the four subintervals $(-\infty, -2)$, $(-2, 0)$, $(0, 2)$, and $(2, \infty)$.

STEP 3. Use the test values -3, -1, 1, and 3 (Fig. 4.12). Then notice that

$$f'(-3) > 0, \qquad f'(-1) < 0, \qquad f'(1) < 0, \qquad f'(3) > 0$$

This leads to the sign chart in Figure 4.12.

STEP 4.
  (a) Since $f'(x) > 0$ to the left of $c = -2$ and $f'(x) < 0$ to the right of -2, $f(x)$ is increasing ($\nearrow$) on $(-\infty, -2)$ and decreasing ($\searrow$) on $(-2, 0)$, and thus we have $\nearrow \searrow$ and f has a relative maximum at $x = -2$.

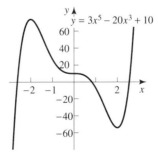

Figure 4.11
We see a relative maximum at $x = -2$, a relative minimum at $x = 2$, but no relative extrema at the critical point at $x = 0$.

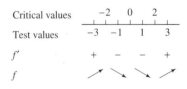

Figure 4.12

(b) Since $f'(x) < 0$ on both sides of $c = 0$, $f(x)$ is decreasing ($\searrow$) on $(-2, 0)$ and $(0, 2)$, and thus f is actually decreasing on the entire interval $(-2, 2)$ and has no relative extremum at $c = 0$.

(c) Since $f'(x) < 0$ to the left of $c = 2$ and $f'(x) > 0$ to the right of 2, f is decreasing ($\searrow$) on $(0, 2)$ and increasing ($\nearrow$) on $(2, \infty)$, and thus we have $\searrow \nearrow$ and f has a relative minimum at $x = 2$.

Notice $f(-2) = 74$, $f(0) = 10$, and $f(2) = -54$. The graph is then given in Figure 4.11. ∎

Remark. Sometimes calculating the *numerical* value of the derivative at the test values can be tedious. Observe, however, that we are only interested in the *sign* of the derivative. Calculating the *sign* is usually less tedious if we have f' factored. Thus, in the previous example, one factor of f' was $15x^2$. This factor is clearly positive at all of the test values, however, and thus the sign of f' at the test values is determined by the sign of the remaining factors $(x + 2)(x - 2)$. Thus, we need only evaluate $(x + 2)(x - 2)$ at the test values. In more complicated situations this alternative procedure can be very useful.

EXAMPLE 6   Finding Relative Extrema

Let the demand equation be given by $p(x) = e^{-0.02x}$. Find where revenue is maximum.

SOLUTION ∎ Revenue is $R(x) = xp(x) = xe^{-0.02x}$. We found $R'(x)$ in Example 3 of Section 3.4, to be $(1 - 0.02x)e^{-0.02x}$. For convenience we repeat the calculation here.

$$\frac{d}{dx}(xe^{-0.02x}) = e^{-0.02x}\frac{d}{dx}(x) + x\frac{d}{dx}(e^{-0.02x})$$
$$= e^{-0.02x}(1) + x(-0.02e^{-0.02x})$$
$$= e^{-0.02x}(1 - 0.02x)$$

From this factorization we see that since $e^{-0.02x}$ is never zero, the only critical point is $x = 50$. Also from this factorization we see that the sign of $R'(x)$ is the same as the sign of the factor $(1 - 0.02x)$. Thus, $R'(x) > 0$ when $0 < x < 50$ and $R'(x) < 0$ when $x > 50$. Therefore, $R(x)$ is increasing on the interval $(0, 50)$ and decreasing on $(50, \infty)$. This implies that the maximum revenue occurs when $x = 50$ (Fig. 4.13). ∎

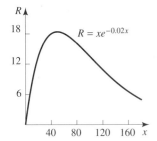

Figure 4.13
A relative maximum at $x = 50$.

EXAMPLE 7   Finding Relative Extrema

Find the critical values, the intervals on which $f(x) = \sqrt[3]{x^2}$ is increasing or decreasing, and the relative extrema.

SOLUTION ∎

STEP 1. Find the critical values.
   (a) Since $f(x) = x^{2/3}$,

$$f'(x) = \frac{2}{3}x^{-1/3} = \frac{2}{3\sqrt[3]{x}}$$

Note that $f'(x)$ exists and is continuous everywhere except at $x = 0$. Since $f'(0)$ does not exist, and $f(0)$ does exist, $x = 0$ is a critical value.
   (b) There are no values for which $f'(x) = 0$, thus $x = 0$ is the only critical value.

STEPS 2 and 3. From the simple form of $f'(x)$, it is apparent that $f'(x) < 0$ when $x < 0$ and $f'(x) > 0$ when $x > 0$. Thus, $f(x)$ is decreasing ($\searrow$) on $(-\infty, 0)$ and increasing ($\nearrow$) on $(0, \infty)$. Then f has a relative minimum at $x = 0$ (Fig. 4.14) ∎

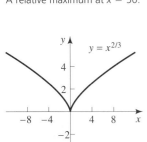

Figure 4.14
A relative minimum at $x = 0$.

EXPLORATION 3

Graphical Support for Example 4

Graph $y = f(x) = x^{2/3}$. Observe what happens near $x = 0$. (To obtain the correct answer on your graphing calculator you may need to write $x^{2/3}$ as $(x^{1/3})^2$ or as $(x^2)^{1/3}$.)

SELF-HELP EXERCISE SET 4.1

1. Suppose a firm has a monopoly on VCRs and x is the number of millions of VCRs sold in a year and $P(x) = -0.1x^2 + 1.1x - 1.0$ is the profit in tens of millions of dollars for selling x million VCRs.
   (a) Find where $P(x)$ is increasing and where it is decreasing. Of what significance is this to the firm?
   (b) Find the break-even values.
   (c) Draw a graph of $P = P(x)$.
   (d) Is this model realistic?

2. Find the critical values and the intervals on which $f(x) = 3x^4 + 8x^3 + 5$ is increasing or decreasing. Draw a graph. Must f' change sign at a critical value?

3. Suppose that $f'(x) > 0$ everywhere. Show that $g(x) = e^{f(x)}$ is increasing everywhere.

EXERCISE SET 4.1

Various graphs are shown here for Exercises 1 through 6. In each case find all values where the function attains a relative maximum. A relative minimum.

1.
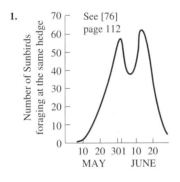
See [76] page 112

2.
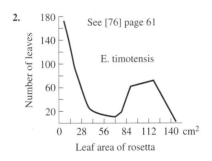
E. timotensis

3.
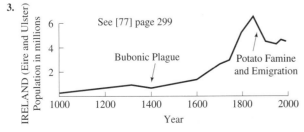
See [77] page 299
Bubonic Plague
Potato Famine and Emigration

4.
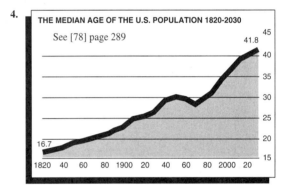
THE MEDIAN AGE OF THE U.S. POPULATION 1820-2030
See [78] page 289

5.
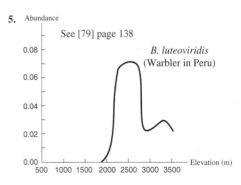
See [79] page 138
B. luteoviridis (Warbler in Peru)

6.
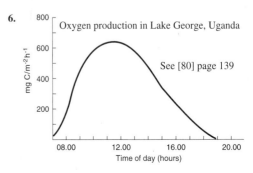
Oxygen production in Lake George, Uganda
See [80] page 139

In Exercises 7 through 18 $f'(x)$ is given in factored form, and $p(x)$ is a function that is always positive. Find all critical values of f, find the largest open intervals on which f is increasing, find the largest open intervals on which f is decreasing, find all relative maxima and all relative minima. Using the additional information given for each function, draw a rough graph of f.

7. $f'(x) = x - 3$, $f(3) = 1$

8. $f'(x) = 4 - x$, $f(4) = 1$

9. $f'(x) = p(x)(x - 2)(x - 4)$, $f(2) = 2$, $f(4) = -2$

10. $f'(x) = -p(x)(x + 2)(x - 1)$, $f(-2) = -15$, $f(1) = 12$

11. $f'(x) = p(x)(x - 1)(x - 3)(x - 5)$, $f(1) = 1$, $f(3) = 4$, $f(5) = 0$

12. $f'(x) = p(x)(x + 1)(x - 2)(x - 5)$, $f(-1) = 3$, $f(2) = 10$, $f(5) = 4$

13. $f'(x) = p(x)(x - 1)(x - 3)(x - 5)^2$, $f(1) = 1$, $f(3) = -2$, $f(5) = 5$

14. $f'(x) = p(x)(x + 2)^2(x - 5)^3(x - 10)$, $f(-2) = -2$, $f(5) = 3$, $f(10) = 0$

15. $f'(x) = p(x)(x + 1)^2(x - 1)^3(x - 2)$, $f(-1) = -2$, $f(1) = 3$, $f(2) = 0$

16. $f'(x) = p(x)(x + 1)^2(x - 1)^4(x - 2)$, $f(-1) = 5$, $f(1) = 3$, $f(2) = 0$

17. $f'(x) = p(x)(x + 1)(x - 1)^3(x - 2)^5$, $f(-1) = 1$, $f(1) = 3$, $f(2) = 0$

18. $f'(x) = p(x)(x + 1)^2(x - 1)^4(x - 2)^4$, $f(-1) = -2$, $f(1) = 3$, $f(2) = 4$

In Exercises 19 through 42 find all critical values, find the largest open intervals on which f is increasing, find the largest open intervals on which f is decreasing, find all relative maxima and all relative minima. Draw a rough graph of f.

19. $f(x) = x^2 - 4x + 1$

20. $f(x) = -2x^2 + 8x - 1$

21. $f(x) = x^3 + 3x + 1$

22. $f(x) = 1 - 5x - x^5$

23. $f(x) = x^4 + 2x^2 + 1$

24. $f(x) = -4x^6 - 6x^4 + 5$

25. $f(x) = 1 - (1 - x)^{2/3}$

26. $f(x) = (x + 3)^{2/3}$

27. $f(x) = x^3 - 3x + 1$

28. $f(x) = x^3 + 3x^2 - 9x + 3$

29. $f(x) = 8x^3 - 24x^2 + 18x + 6$

30. $f(x) = -8x^3 + 6x - 1$

31. $f(x) = x^4 - 8x^2 + 3$

32. $f(x) = 2x^6 - 3x^4 + 1$

33. $f(x) = -x^5 + 5x - 1$

34. $f(x) = 5 + 20x - x^5$

35. $f(x) = 3x^5 - 20x^3 + 5$

36. $f(x) = 3x^5 - 5x^3 + 1$

37. $f(x) = x - \ln x$

38. $f(x) = 1 - \ln x$

39. $f(x) = \dfrac{1}{1 + e^{x^2}}$

40. $f(x) = e^x + e^{-x}$

41. $f(x) = x^2 e^{-x}$

42. $f(x) = e^{-x^2}$

43. For the special cubic equation $f(x) = ax^3 + bx + c$, show that $f(x)$ is always increasing if both a and b are positive and that $f(x)$ is always decreasing if both a and b are negative.

44. Consider the function $\dfrac{x + a}{x + b}$. If $b > a$, show that this function is increasing on $(-\infty, -b)$ and $(-b, \infty)$. If $b < a$, show that this function is decreasing on $(-\infty, -b)$ and $(-b, \infty)$.

In Exercises 45 through 50 use your graphing calculator to graph the function and its derivative on the same screen. Verify that the function increases on intervals where the derivative is positive and decreases on intervals where the derivative is negative.

45. $y = x^2 - 1$

46. $y = 5 - x^2$

47. $y = x^3 + 2x^2 - x - 2$

48. $y = x^3 - 4x$

49. $y = x^4 - 5x^2 + 4$

50. $y = x^5 - 5x^3 + 4x$

Applications

51. Revenue. If the revenue function for a firm is given by $R(x) = 10x - 0.01x^2$, find the value of x for which revenue is maximum.

52. Revenue. If the demand equation for a firm is given by $p = -0.2x + 16$, find the value at which the revenue is maximum.

53. Profit. If the cost function for a firm is given by $C(x) = 5 + 8x$ and the demand equation is the same as in the previous problem, find the value for which the profit is maximum.

54. Revenue. If the revenue function for a firm is given by $R(x) = -x^3 + 36x$, find the value at which revenue is maximized.

55. Profit. If the cost function for a firm is given by $C(x) = 9x + 30$ and the revenue function is the same as in the previous problem, find the value for which the profit is maximized.

56. Average Revenue. The total revenue function for producing x items is given by $R(x) = 100x^2 - 20x^3$. How many items should be produced to maximize $\overline{R}(x) = R(x)/x$, the average revenue per item?

57. Average Revenue. In the previous problem show that average revenue is maximized at the value at which $\overline{R} = R'$.

58. Average Revenue. Suppose the total revenue function for producing x items is given by $R(x)$ and the average revenue per item is given by $\bar{R}(x) = R(x)/x$. If average revenue is at a maximum at some value on $(0, \infty)$, then show that at this value $\bar{R} = R'$.

59. Cost Function. Noreen and Soderstrom [81] suggested that a reasonable cost function for hospitals is $C(x) = px^{0.70}$, where x is in appropriate units of service, p is a positive constant, and $C(x)$ is cost. For what values of x is $C(x)$ increasing? Sketch a possible graph.

60. Firm Size. Philips [82] showed that the number y of research personnel in a firm in one sector of the food, beverage, and tobacco industry was approximated by $y = 2.43 - 2.55x + 2.86x^2$, where x is the number of thousands of total personnel, with x restricted to the interval [0.5, 2.5]. Sketch a graph on the given interval. Is the function increasing on this interval?

61. Profit. In determining the optimal mill price of a manufactured item in a market along a straight line Sailors [83] showed that $P(m) = 2a(b - m)(m - k)/tmk$, was a profit function, where the demand equation is $x = a/p^2$ if $p \leq b$ and zero otherwise; $p = m + tD$, where m is the mill price, t is the freight rate per mile, and D is the distance in miles from the mill; and k is the constant marginal cost. Find the mill price that maximizes profit.

62. Optimum Mill Price. The profit function $P(p) = \frac{1}{6b^2}(p - c)(a - bp)^3 - F$ arises in Beckmann [84] when determining the optimal mill price of a manufactured item in a circular market area. Here p is the mill price, $x = a - bp$ is the demand equation, c is marginal cost, and F is fixed costs. Find the mill price that maximizes profit.

63. Agriculture. Paris [56] showed that plant responses to phosphorus fertilizer were approximated by $y = -0.057 - 0.417x + 0.852\sqrt{x}$, where y is the yield and x the units of nitrogen. Find the number of units of nitrogen that maximizes the yield.

64. Agriculture. Feinerman and colleagues [85] showed that the yield response of corn to nitrogen fertilizer was approximated by $y = -5.119 + 0.099x - 0.004x^{1.5}$, where y is the yield measured in tons per hectare and x is in kilograms per hectare. Find the amount of nitrogen that maximizes the yield.

65. Agriculture. Dai [86] showed that the yield response of corn in Indiana to soil moisture was approximated by $y = -2600 + 2400x - 70x^2 + \frac{2}{3}x^3$, where y is measured in bushels per acre and x is a soil moisture index. Find the value of the soil moisture index that maximizes yield, and find the maximum yield.

66. Medicine. It is well known that when a person coughs, the diameter of the trachea and bronchi shrinks. It has been shown that the flow of air F through the windpipe is given

by $F(r) = k(R - r)r^4$, where r is the radius of the windpipe under the pressure of the air being released from the lungs, R is the radius with no pressure, and k is a positive constant. Find the radius at which the flow is maximum.

67. Autocatalytic Reaction. The rate V of a certain autocatalytic reaction is proportional to the amount x of the product and also proportional to the remaining substance being catalyzed. Thus, if I is the initial amount of the substance, then $V = kx(I - x)$, where k is a positive constant. At what concentration x is the rate maximal?

68. Biology. Lactin and coworkers [87] showed that the feeding rate y of the Colorado potato beetle larvae in square millimeters per hour per larva was approximated by $y = -3.97 + 0.374T - 0.00633T^2$, where T is the temperature in degrees Celsius. Sketch a graph, and find the temperature at which the feeding rate is maximum. Check using your graphing calculator.

69. Biology. Arthur and Zettler [88] showed that the percentage mortality y of the red flour beetle was approximated by $y = f(x) = 9.1 + 86.3e^{-0.64x}$, where x is the number of weeks after a malathion (pesticide) treatment. Use calculus to show that $f(x)$ is a decreasing function. What happens in the long-term? Sketch a graph. Check using your graphing calculator.

70. Biology. Cherry and associates [89] showed that the number y of eggs laid per female of the sugarcane grub was approximated by $y = 14.6e^{-(x-116.6)/60.9}$, where x is a measure of soil moisture. Sketch a graph. Find the value of x that yields the maximum y. Check using your graphing calculator.

71. Biology. Eller and coauthors [90] showed that the percentage y of ovipositing females of a parasite was approximated by $y = -298.1 + 241.1x - 51.9x^2 + 3.4x^3$, where $x = \ln V$ and V the volume of an insecticide in appropriate units. Graph y as a function of V on your graphing calculator. Find the value of x that maximizes y on the interval [10, 400].

72. Biology. The formula $w = w_0 N^{-a}$ is used by plant biologists (see Shainsky and Radosevich [91]), where w is the plant size, w_0 is the maximum plant size in the absence of competitors, N is plant density, and a is a positive constant. According to this formula, for what values of N is plant size decreasing? Sketch a possible graph.

73. Biology. Webb and colleagues [92] showed that the population y of gypsy moths was approximated by $y = f(x) = 10^{3.8811 - 0.1798\sqrt{x}}$, where x was the dose in grams per hectare of a mating disruptant. Find $f'(x)$, and determine if this function is decreasing. Sketch a graph. What happens if the dose x becomes large without bound?

Enrichment Exercises

In Exercises 74 through 77 assume that $f'(x)$ is continuous everywhere and that $f(x)$ has one and only one critical value at $x = 0$. Use the additional information given to determine if $y = f(x)$ attains a relative minimum, relative maximum, or neither at $x = 0$. Explain your reasoning. Sketch a possible graph in each case.

74. $f(-1) = 1$, $f(0) = 3$, $f(2) = 4$

75. $\lim\limits_{x \to \infty} f(x) = \infty$, $\lim\limits_{x \to -\infty} f(x) = -\infty$

76. $f(0) = 1$, $\lim\limits_{x \to \infty} f(x) = 5$, $\lim\limits_{x \to -\infty} f(x) = 3$

77. $f(-1) = 3$, $f(0) = 10$, $f(2) = 4$

78. Suppose $f'(x)$ is continuous everywhere and $f(x)$ has two and only two critical values. Explain why it is not possible for f to have the following values: $f(0) = 10$, $f(1) = 5$, $f(2) = 8$, $f(3) = 6$, $f(4) = 9$.

79. Show that $x^{13} - x^{12} = 10$ has one and only one solution by using an analytical argument.

80. Show that $x^{12} - x^{11} = 10$ has two and only two solutions by using an analytical argument.

81. Using the properties of $f(x) = e^x$, show that $e^x \geq 1 + x$ for all x.

82. Suppose $f(x)$ is positive everywhere. Let $g = 1/f$. If f has a relative maximum at $x = a$, what can you say about g?

83. Suppose $f(x)$ and $g(x)$ are differentiable and negative everywhere. If $f'(x) > 0$ and $g'(x) > 0$ everywhere, show that $h(x) = f(x)g(x)$ is a decreasing function.

84. Suppose $f(x)$ and $g(x)$ are differentiable and positive everywhere. If $f'(x) < 0$ and $g'(x) > 0$ everywhere, show that $h(x) = f(x)/g(x)$ is a decreasing function.

For Exercises 85 through 90 consider the graphs of the following six functions:

85.
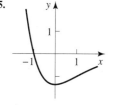

86.
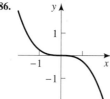

87.
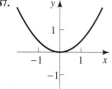

88.
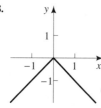

89.

90.
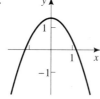

Each of the following figures represent the graph of the derivative of one of the preceding functions, but in a different order. Match the graph of the function to the graph of its derivative.

(a)

(b)
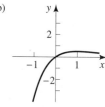

(c)
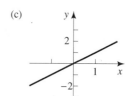

(d)
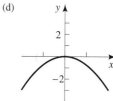

(e)
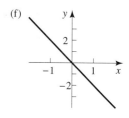

(f)

91. Give conditions on the constants b, c, and d so that the polynomial function $f(x) = x^3 + bx^2 + cx + d$ is increasing on the interval $(-\infty, \infty)$.

92. Use calculus and the properties of cubic polynomials to explain why any polynomial function of the form $f(x) = x^4 + ax^3 + bx^2 + cx + d$ cannot be increasing on all of $(-\infty, \infty)$ or decreasing on all of $(-\infty, \infty)$.

93. With your graphing calculator in radian mode graph $y = \cos x$ using a screen of dimensions $[0, 2\pi]$ by $[-1, 1]$, and determine where on this interval $\cos x$ is positive and where it is negative. Use the fact that $\dfrac{d}{dx}(\sin x) = \cos x$ to then determine where $y = \sin x$ is increasing and where it is decreasing on the interval $(0, 2\pi)$. Check your answer by graphing $y = \sin x$.

94. With your graphing calculator in radian mode graph $y = \sin x$ using a screen of dimensions $[-\pi, \pi]$ by $[-1, 1]$, and determine where on this interval $\sin x$ is positive and where it is negative. Use the fact that $\dfrac{d}{dx}(\cos x) = -\sin x$ to then determine where $y = \cos x$ is increasing and where it is decreasing on the interval $(-\pi, \pi)$. Check your answer by graphing $y = \cos x$.

95. Price Discrimination. The following equation arose in a problem of price discrimination in economics mentioned in Exercise 61.

$$1 - \left(\frac{m}{b}\right)^{x-1} = \left(1 - \frac{k}{m}\right)(x - 1)$$

where $b > k$ and $x > 1$. Show that there is a unique solution for m on the interval (k, b).

96. Accounting. Thornton and Moore [93] used a total audit fee function as $F(w, y) = p(y)q(w)$, where p is the hourly fee charged by the auditor, q the number of hours required to perform the audit, y is the auditor quality, and w is a measure of internal controls. Then the authors state that $p'(y) > 0$ and $q'(w) < 0$. Explain why this is reasonable.

97. Biology. Laudelout [94] used the equation $Q(t) = at + be^{-kt}$ to model the amount Q of carbon dioxide at time t from microbial biomass in soil. Take $a = b = 1$ and $k = 5$, graph on your graphing calculator, and locate the approximate value of x for which the amount of carbon dioxide is the least. Verify using calculus.

98. The Labor Market. Rosenthal [95] indicated that the equation $y = a + b(1 - e^{-ct})$, models the relationship between the proportion y of individuals who have regained employment after being laid off and the time t of unemployment measured in weeks. Sketch a graph. Explain in words what the value a signifies. What happens as $t \to \infty$? Explain in words what it means if $a + b = 1$.

99. Sales of New Products. Dodds [96] determined that the sales of color television sets were approximated by the function $S(t) = 418{,}000 \dfrac{e^{-0.842t}}{(1 + 154e^{-0.842t})^2}$, where t is the number of years after introduction. Determine when $S(t)$ reaches its maximum and graph.

100. Biology. Ring and Benedict [97] studied the yield response of cotton to injury by the bollworm. Normalized yield y (yield with injury divided by yield without injury) was shown to be approximated by $y = 0.1494 \ln x - 5.5931x + 7.12x^2$, where x is the percentage of injured reproductive organs (flower buds plus capsules). Graph on your graphing calculator, and determine the approximate value of x for which y attains a maximum. Confirm using calculus.

SOLUTIONS TO SELF-HELP EXERCISE SET 4.1

1. (a) Taking the derivative gives

$$P'(x) = -0.2x + 1.1$$

Notice that $P'(x)$ is continuous everywhere. Thus, the only critical values are values at which $P'(x) = 0$. Setting $P'(x) = -0.2x + 1.1 = 0$ gives $x = 5.5$ as the only critical value. This divides the interval $(0, \infty)$ into the two intervals $(0, 5.5)$ and $(5.5, \infty)$. Take the test values to be, say, $x_1 = 1$ and $x_2 = 10$. Then

$$P'(1) = -0.2(1) + 1.1 = 0.9 > 0,$$

$$P'(10) = -0.2(10) + 1.1 = -0.9 < 0$$

Thus, $P(x)$ is increasing on $(0, 5.5)$ and decreasing on $(5.5, \infty)$.

Thus, if $0 < x < 5.5$, $P(x)$ is increasing. The firm would be wise to produce and sell more items since profits would increase. If $5.5 < x$, $P'(x) < 0$ and $P(x)$ is decreasing. Producing and selling more items means less profits.

(b) Since

$$P(x) = -0.1(x^2 - 11x + 10) = -0.1(x - 10)(x - 1)$$

$P(1) = 0$ and $P(10) = 0$. This means that profits are zero if the firm sells 1 million or 10 million VCRs in a year. (Apparently, a very substantial price reduction is necessary to sell such a large number of VCRs.)

(c) The graph is given in the figure.

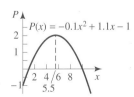

(d) Is this model realistic? It is clearly very unrealistic to think that any firm would know the explicit form of the function $P(x)$; however, experience has shown that the general shape of the graph of the profit function is realistic. Firms are well aware of the fact that flooding the market with their products leads to *decreased* profits and even losses.

2. Taking the derivative gives

$$f'(x) = 12x^3 + 24x^2 = 12x^2(x + 2)$$

The critical values are $x = -2, 0$ where f' is zero. Since the derivative exists everywhere, there are no other critical values.

We divide the real line into three intervals: $(-\infty, -2)$, $(-2, 0)$, $(0, \infty)$.

Pick the three test values -3, -1, and 1 that lie in the intervals given in the preceding step (see the figure). Notice

$$f'(-3) = -108, \qquad f'(-1) = +12, \qquad f'(1) = +36$$

Thus, $f'(x)$ is negative on the first interval and positive on the second and third intervals. Thus, f is decreasing on the first interval and increasing on the second and third intervals. The function has a relative minimum at $x = -2$.

Notice that $f(-2) = -11$ and $f(0) = 5$. The graph is given in the figure.

Note that $f'(0) = 0$, but f' does not change sign at this value.

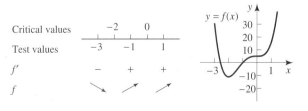

3. We have $g'(x) = f'(x)e^{f(x)}$, but $e^{f(x)}$ is positive no matter what $f(x)$ is. Then $f'(x) > 0$ implies that $g'(x) = f'(x)e^{f(x)} > 0$. Thus, $y = g(x)$ is increasing.

4.2 ADDITIONAL APPLICATIONS

■ Federal Tax Revenue   ■   Management of Renewable Natural Resources

APPLICATION: TAX REVENUE VERSUS THE TAX RATE

Let x be the percentage of the sum of the total income of individuals and total profits of corporations in the United States collected as federal taxes and $R(x)$ the total tax revenue. If x is small, then increasing x should result in an increase in the revenue $R(x)$. If x is large, is it possible that increasing x further will decrease revenue? If this is so, can decreasing the rate x of taxation result in an *increase* in the revenue $R(x)$ collected? For the answers given by some of the Reagan administration's economists, see the following discussion.

APPLICATION: FINDING THE MAXIMUM SUSTAINABLE YIELD

Suppose x is the number of thousands of pounds of a certain species of fish in a small lake and $y = f(x) = x - 0.1x^2$ is the increase in fish biomass over the next year measured in thousands of pounds. What is the maximum number of pounds of fish that can be harvested each year indefinitely? See Example 2 for the answer.

APPLICATION: THE ANTARCTIC BLUE WHALE FISHERY

Let x be the number of thousands of Antarctic blue whales and $f(x)$ the increase in thousands in the number over the next year, where $f(x) = 0.000002(-2x^3 + 303x^2 - 600x)$. What is the maximum sustainable yield? Also what is an estimate of the minimum population level below which the whales inevitably become extinct? What is the carrying capacity? See Example 3 for the answers.

Federal Tax Revenue

In the early 1980s the Reagan administration was struggling with the consequences of lowering the tax rates. Some had the view that if you lowered the tax rate you must lower the total income to the government. Others argued, however, that lowering the tax rates *increase* total tax revenue.

   One of those economists was Arthur Laffer of the University of Southern California. He argued as follows. Let $R = R(x)$, where x is the percentage of income taken by the federal government in the form of taxes and $R(x)$ is the total revenue from taxes from a rate of $x\%$. Obviously, if the rate is set at zero, the government has no revenue. That is, $R(0) = 0$. If the rate x is increased somewhat, surely the revenue $R(x)$ increases somewhat. So $R(x)$ must be an increasing function at least for small values of x. On the other hand, it is reasonable to expect that if $x = 100$, that is, the government takes everything, then probably nobody will wish to earn any money and federal revenue will be zero. Thus, $R(100) = 0$. If x is a little less than 100, then some people will probably earn some money, resulting in some income for the federal government. So one can assume that for x near 100, $R(x)$ is decreasing. At this point the graph of the curve $R = R(x)$ looks like the solid curve in Figure 4.15. Now Laffer made the assumption that the rest of the curve probably looks like the dotted curve in Figure 4.15

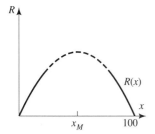

Figure 4.15
As the tax rate increases, tax revenue initially increases but then starts to decrease.

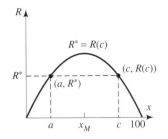

Figure 4.16
According to the Laffer curve, the same revenue can be obtained for two tax rates.

at the intermediate values. (This curve is called a Laffer curve.) Label as x_M the value at which this curve changes direction from increasing to decreasing.

Suppose currently $x = c$. The argument was then made that x_M was *less* than c, as shown in Figure 4.16. Since $R(x)$ is *decreasing* near the value c, increasing x results in a *decrease* of revenue. Decreasing x therefore results in an increase of revenue.

The Reagan administration wished only to lower the tax rate—not to increase federal tax revenue. Drawing a horizontal line at height $R^* = R(c)$ in Figure 4.16 results in this line intersecting the graph of $y = R(x)$ at *two* points, (c, R^*) of course, but also (a, R^*) shown in the figure. Thus, according to this model, the income tax rate could be decreased all the way down to $x = a$ and the government would still have the same revenue.

In a more recent application of the Laffer curve Inman [98] in 1992, writing in the *Business Review of the Federal Reserve Bank of Philadelphia*, drew possible Laffer curves as applied to the city of Philadelphia. The issue was whether raising tax rates would give the extra revenue Philadelphia needed to rescue it from a fiscal crisis or if the increase in tax rates would result in a *decrease* in economic activity leading to a *decrease* in tax revenue.

After a sweeping victory for the Republican Party in the 1994 congressional election, the new congressional leaders were proposing a *decrease* in the capital gains tax rates. They argued that one consequence of lower rates would be increased trading activity and an *increase* in taxes collected from capital gains.

EXAMPLE 1 Finding Where Tax Revenue Increases or Decreases

Suppose in the preceding discussion that $R(x) = \sqrt{x}(100 - x)$. Find where the function is increasing and where it is decreasing.

SOLUTION ■ We are only interested in $0 < x < 100$.

STEP 1.
(a) Since $R(x) = 100x^{1/2} - x^{3/2}$

$$R'(x) = 50x^{-1/2} - \frac{3}{2}x^{1/2}$$

and R' exists and is continuous everywhere on $(0, 100)$. Thus, the only critical values on $(0, 100)$ are values where the derivative is zero.
(b) Setting the derivative equal to zero yields

$$50x^{-1/2} - \frac{3}{2}x^{1/2} = 0 \qquad \text{multiply by } x^{1/2}$$

$$50 = \frac{3}{2}x$$

$$x = \frac{100}{3}$$

Thus, the only critical value on $(0, 100)$ is at $\frac{100}{3}$.

STEP 2. This single critical value then divides $(0, 100)$ into the two intervals $\left(0, \frac{100}{3}\right)$ and $\left(\frac{100}{3}, 100\right)$.

STEP 3. Pick the convenient test values $x_1 = 25$ and $x_2 = 64$, and note that

$$R'(25) = \frac{50}{\sqrt{25}} - \frac{3}{2}\sqrt{25} = \frac{5}{2}, \qquad R'(64) = \frac{50}{\sqrt{64}} - \frac{3}{2}\sqrt{64} = -\frac{23}{4}$$

(See Fig. 4.17.)

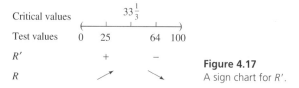

Critical values

Test values   0   25   64   100

R'   +   −

R   ↗   ↘

Figure 4.17
A sign chart for R'.

STEP 4. Thus, $R'(x) > 0$ on $\left(0, \dfrac{100}{3}\right)$, and thus R is increasing ($\nearrow$) there and $R'(x)$ < 0 on $\left(\dfrac{100}{3}, 100\right)$, and thus R is decreasing ($\searrow$) on this second interval. The graph is shown in Figure 4.18.   ∎

It has been pointed out that when the Kennedy administration lowered the tax rate, federal tax revenue *increased*. The Reagan administration, however, had inconclusive results, perhaps due to other factors operating in the economy at that time.

Of course, it is very unrealistic to assume knowledge of an *explicit* formula for $R(x)$. Knowledge of the general *shape* of the graph is very important, however.

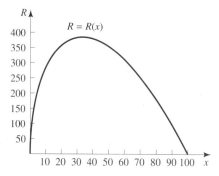

Figure 4.18
Revenue is maximum in this model when the tax rate is 33.3%.

Management of Renewable Natural Resources

As another important application, we now consider a mathematical model of a fishery. We can also apply the ideas presented here to other renewable natural resources. To avoid confusion we emphasize that we are interested in the maximization of productive resources rather than the preservation of natural environments.

A fishery can consist of all or part of an ocean, a lake, a river, a natural or artificial pond, or even an area of water that has been fenced in. To fix a particular situation in our minds, we assume that a corporation engaged in aquaculture (fish farming) has the sole rights to the fish in a fairly small lake. Of course the corporation wishes to harvest the maximum number of fish with the minimum effort and cost. We assume that the corporation does no active management of the fish other than simply harvesting.

Let $y = f(x)$, where x represents the total pounds (**biomass**) of fish at the beginning of a year and y represents the total increase in the biomass of fish over the next year. Thus, the term $x + y$ represents the total biomass one year later.

For convenience we always harvest at the end of the year. We are interested in the *increase* in population since it is this increase that we wish to harvest. We want to predict how much of the resource we can harvest and still allow the resource to replenish itself. The largest possible increase in biomass is called the **maximum sustainable**

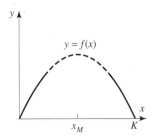

Figure 4.19
As the number of fish increase, the yearly change in the fish population itself increases, but then at some point begins to decrease.

yield. We want to know what the biomass of fish should be to obtain the maximum sustainable yield.

We know that if the biomass is zero, then the increase in the biomass is also zero. Mathematically, this says that $f(0) = 0$. Some fish in the lake result in an increase in the population and biomass, and even more fish result in even more of an increase in the biomass. Thus, f should be increasing, at least initially.

Clearly our small lake has only a finite amount of biological resources to sustain the fish and can only hold so much biomass. This number is referred to as the **carrying capacity**, which we label K. Since K is the saturation level of the species, there can be *no* increase in biomass beyond this level. Thus, $f(K) = 0$. If the biomass x is a little less than K, then we do expect some increase in biomass. Thus, moving from right to left starting at $x = K$, y should increase. Moving from left to right and just to the left of K, the function f should be decreasing.

At this point in the discussion the function looks like the solid curve in Figure 4.19. Biologists now normally make the assumption that the actual curve at the intermediate values is described by the dotted curve in Figure 4.19.

It is therefore to be expected that the increase in the biomass $y = f(x)$ becomes larger and larger (increases) until some value is reached, say $x = x_M$, after which the increase in the biomass begins to get smaller and smaller (decreases). The number $M = f(x_M)$ is then the maximum sustainable yield. For the corporation interested in long-term benefits, the harvesting should be managed so that $x = x_M$. Thus, given these assumptions, if the biomass is at $x = x_M$ at the beginning of the year, the corporation can harvest $f(x_M)$ pounds of fish at the end of the year and see the population renew itself over the subsequent year so that the biomass returns once again a year later to x_M. Furthermore, this can go on indefinitely.

EXAMPLE 2   Finding the Maximum Sustainable Yield

Suppose x is the number of thousands of pounds of a certain species of fish in a small lake described previously and $y = f(x) = x - 0.1x^2$ is the increase in fish biomass over the next year measured in thousands of pounds. What is the carrying capacity and the maximum sustainable yield? What should management do if the fish biomass is at 1000 lb? At 9000 lb?

SOLUTION ■

STEP 1. Find the critical values.
    (a) The derivative

$$f'(x) = 1 - 0.2x = 0.2(5 - x)$$

exists and is continuous everywhere. Thus, the only critical values are values for which $f'(x) = 0$.
    (b) From the factorization of f' it is clear that $f'(x) = 0$ implies that $x = 5$. Thus, $x = 5$ is the only critical value.

STEP 2. This single critical value then divides the nonnegative numbers into two subintervals: $[0, 5)$ and $(5, \infty)$.

STEP 3. It is clear from the factorization of $f'(x)$ that if $0 < x < 5$, $f'(x) > 0$ and if $5 < x$, $f'(x) < 0$.

STEP 4. Thus, f is increasing on $[0, 5)$ and decreasing thereafter, and therefore f attains a relative maximum at $x = 5$. In fact, since f always increases to the left of $x = 5$ and always decreases to the right of $x = 5$, it is clear that the function f attains its largest value at $x = 5$.

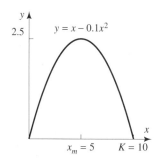

Figure 4.20
A biomass of 5000 lb leads to a maximum increase in fish biomass for the next year.

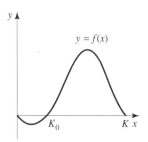

Figure 4.21
If the population becomes less than K_0, then the population decreases.

■ **The Antarctic Blue Whale Fishery**

In the 1960s, the population of the Antarctic blue whale was reduced to levels thought to lead to extinction. Under the protection of the International Whaling Commission the population increased to almost 10,000 in about 10 years and to about 50,000 by 1994.

Fish Catches by Region, 1988
(in metric tons, live weight catches)

Region	Total Catch
World Total	**97,985,400**
Africa	5,310,300
Asia	43,601,200
Europe	12,877,400
North America	9,567,900
Oceania	887,300
South America	14,412,200
(Former) USSR	11,332,100

Furthermore, since

$$f(x) = x - 0.1x^2$$
$$= x(1 - 0.1x)$$

$f(0) = 0$ and $f(10) = 0$. Thus, the carrying capacity is $K = 10$, or 10,000 lb of fish. The maximum sustainable yield is $f(5) = 2.5$, or 2500 lb (Fig. 4.20).

If the fish biomass is 1000 lb, then $x = 1$ and $f(1) = 0.9$ or, 900 lb. Thus, at the end of the year $x + y = 1.9$. Since 1.9 is well below $x_M = 5$ no harvesting should be done. If the biomass is 9000, then $x = 9$. Management can immediately harvest 4000 lb of fish, resulting in a biomass at exactly 5000. ■

Sometimes, an argument is made that the amount of fish (or some other animal species) should be kept at the carrying capacity. One might speculate on the condition of the fish in the fishery if the biomass is at the carrying capacity. Since the resources are such that no additional pound of fish can survive an additional year, the fish must be obtaining the absolute minimal amount of food for survival. This clearly does not represent pleasant living conditions for the fish or sufficient resources for the fish to grow well and reproduce.

We decided that in a relatively small lake a small population of fish would have enough resources to breed and produce more fish. If the lake is relatively large, however, with only a small number of fish, the fish may not be able to find each other in order to breed. (For some species of whales the feeding grounds amount to as much as 10 million square miles.) In this circumstance, instead of the population and biomass *increasing*, they would actually *decrease* due to other natural causes.

Now assume as before that the biomass in pounds of fish is x and the increase in the biomass over the next year is given by $y = f(x)$. The biomass one year later is therefore $x + y$, and we wish to harvest y.

Figure 4.21 indicates that in a relatively large lake the increase in the biomass is *negative* for small values of x, that is, the biomass decreases for small numbers of fish for the reasons indicated earlier. For some larger amount of fish, indicated by K_0, the increase in biomass moves to zero, after which the biomass then increases.

The number K_0, called the **minimum viable biomass level**, is very important. If the biomass of fish in the lake ever goes *below* K_0, then the population will head eventually to zero. Thus, overfishing to the point where the biomass goes below the minimum viable biomass level results in an irrevocable catastrophe (assuming no intervention).

In the following problem, for convenience, we take x to be the *population* of whales.

EXAMPLE 3   Management of a Fishery

Suppose the fishery consisting of the Antarctic blue whale has the function $f(x)$ in the previous discussion, given by

$$f(x) = 0.000002(-2x^3 + 303x^2 - 600x)$$

where x is in thousands of whales. Find the maximum sustainable yield. Also estimate the minimum viable population level and the carrying capacity. Draw a graph of $f(x)$.

SOLUTION ■ We see that

$$f'(x) = 0.000002(-6x^2 + 606x - 600)$$
$$= -0.000012(x^2 - 101x + 100)$$
$$= -0.000012(x - 1)(x - 100)$$

The Coastal Shark Fishery

Rising demand for shark fins in Asia and for shark meat in the United States has caused a sharp rise in the killing of sharks over the last decade, with the result that large coastal sharks are being killed in numbers that outstrip their capacity to reproduce. As a result, the National Marine Fisheries Service of the Federal Government has imposed controls on shark fishing within 200 miles off the Atlantic and Gulf coasts by issuing regulations that place a total quota of 2436 metric tons, dressed weight, on the catch of large coastal sharks by commercial fishermen in 1993 and 2570 metric tons in 1994. Twenty-two shark species are included in this management category, among which are the tiger, lemon, hammerhead, bull, and great white sharks. The total catch of large coastal sharks in the U.S. waters has substantially exceeded 4000 metric tons in each of the five years previous to 1993, according to the fisheries service. The service has calculated that the maximum sustainable yield of these sharks is 3800 metric tons a year, and that the 1993 and 1994 quotas will enable that level to be achieved in 1996.

Thus, the only critical values are $x = 1$ and $x = 100$. Furthermore, from the preceding factorization, f' is negative on $(0, 1)$, positive on $(1, 100)$, and negative again on $(100, \infty)$. Thus, $f(1)$ is a relative minimum and $f(100)$ a relative maximum. Furthermore, we can readily see that $f(100)$ is an absolute maximum on $(0, \infty)$, since $f(0) = 0$, whereas $f(100) = 1.940$.

The maximum sustainable yield is $f(100) = 1.940$, or 1940 whales.

If we now set $f(x) = 0$, we obtain

$$2x^2 - 303x + 600 = 0$$

Using the quadratic formula and a calculator yields $x \approx 2$ and $x \approx 150$. Thus, the minimum viable population level is about 2000 whales, and the carrying capacity is 150,000 whales (Fig. 4.22). ∎

Whaling scientists have estimated these numbers for the carrying capacity and the maximum sustainable yield of the Antarctic blue whale. The minimum viable population level is unknown.

For additional material on mathematical bioeconomics and the optimal management of renewable resources see Clark [99].

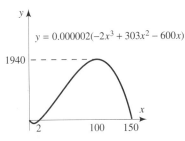

Figure 4.22
A maximum sustainable yield of 1940 occurs when $x = 100$. The minimum viable population occurs when $x = 2$.

EXERCISE SET 4.2

1. **Federal Tax Revenue.** Let x be the percentage of income taken by the federal government in the form of taxes and $R(x) = x(100 - x)$ be the total revenue obtained. Find where $R(x)$ is increasing and find where $R(x)$ is decreasing.

2. **Federal Tax Revenue.** Repeat the previous exercise if $R(x) = x^2(100 - x)$.

3. **Federal Tax Revenue.** Consider again the revenue equation in Exercise 1. Suppose at one time $x = 60$. Find another value of x that gives the same revenue.

4. **Federal Tax Revenue.** Repeat the previous exercise if $x = 70$.

5. **Federal Tax Revenue.** Answer the same questions as in Exercise 1 if $R(x) = \sqrt[3]{x}(100 - x)$.

6. **Federal Tax Revenue.** The analysis leading to the Laffer curve has been criticized by many economists. State some of your own criticisms.

7. **Fishery.** If in Example 2 of the text, $f(x) = 16x^3 -$ x^4, find the carrying capacity and the maximum sustainable yield.

8. **Fishery.** If in Example 2 of the text, $f(x) = 20\sqrt[3]{x} - \sqrt[3]{x^4}$, find the carrying capacity and the maximum sustainable yield.

9. **Fishery.** Dasgupta and Heal [100] studied the blue whale fishery and estimated the function $f(x)$ in Example 2 of the text to be $f(x) = x^{0.8204}(8.356 - x^{0.1996})$, where x is the number of blue whales. Find the carrying capacity and the maximum sustainable yield.

10. **Fishery.** Suppose the function $f(x)$ in the text under the heading of "Management of Renewable Natural Resources" is given by $f(x) = -x^5 + 5x^4 - 4x^3$. Find the maximum sustainable yield. Also find the minimal viable population level and the carrying capacity.

11. **Fishery.** Redo the previous problem if $f(x) = -x^3 + 15x^2 - 27x$.

4.3 THE SECOND DERIVATIVE

■ Concavity ■ The S-Curve ■ Acceleration ■ The Second Derivative Test

---

APPLICATION: ECONOMIES OF SCALE

To determine if any economies of scale exist Haldi and Whitcomb [101] studied over 200 industrial plants and found that a typical cost function was $C(x) = ax^{0.7}$, where x is output in appropriate units, a a positive constant, and $C(x)$ total cost. How can you use calculus to decide if there are economies of scale? See Example 2 for the answer.

APPLICATION: POPULATION STUDY

The model $P(t) = \dfrac{11}{1 + 1.2e^{-0.03t}}$ has been used to estimate the future human population of the earth, where $P(t)$ is population in billions and t is time in years measured from 1990. Determine when the rate of increase in the population is itself increasing (decreasing). How does this rate of increase compare with past rates of increase, and what does the model predict for the future? See Example 5 for the answer.

Concavity

In this section we see that the second derivative measures the rate at which the first derivative is increasing or decreasing. This can then be used in graphing and also used to give another test for relative extrema.

Given a function $y = f(x)$, the second derivative, denoted by $f''(x)$, is defined to be the derivative of the first derivative.

Definition of Second Derivative

Given a function $y = f(x)$, the second derivative, denoted by $f''(x)$, is defined to be the derivative of the first derivative. Thus,

$$f''(x) = \frac{d}{dx}(f'(x))$$

Remark. The second derivative is also denoted by $\dfrac{d^2}{dx^2} f(x)$ or $\dfrac{d^2y}{dx^2}$.

For example, given $y = f(x) = x^4$,

$$\frac{d^2y}{dx^2} = f''(x) = \frac{d}{dx}[f'(x)] = \frac{d}{dx}(4x^3) = 12x^2$$

Remark. Notice that the second derivative is a function, just as the first derivative is a function.

Given a function $y = f(x)$, recall that $n\text{Der } f(c)$ represents a numerical approximation to $f'(c)$ and is executed on your graphing calculator using the program NDER. The program NDERG graphs $y = n\text{Der } f(x)$. In the same way the program $n\text{Der2 } f(c)$

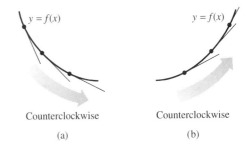

Figure 4.23
When the slope increases, the curve is concave-up.

Counterclockwise   Counterclockwise
(a)   (b)

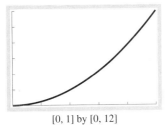

[0, 1] by [0, 12]

Screen 4.2
If we take $x_1 = 1.75$ in Newton's method, we see from the indicated tangent line that x_2 is near the negative root.

represents a numerical approximation to $f''(c)$ and is executed on your graphing calculator using the program NDER2. The program NDER2G graphs $y = n$Der2 $f(x)$. As we saw earlier, if $f(x) = x^4$, then $f''(x) = 12x^2$. On one calculator nDer2 $f(1)$ gave 12.000008 for this function. This is close enough to the exact answer of 12 to lend support. Executing the program NDER2G for $f(x) = x^4$ using a window with dimensions [0, 1] by [0, 12] gives the graph shown in Screen 4.2, a graph that is indistinguishable from that of $y = 12x^2$.

In Figures 4.23a and b, we see two functions that have one characteristic in common. They both have derivatives that are *increasing*. Notice that this means that the slopes are rotating *counterclockwise* as x increases. When this happens we say that the curve is **concave up** (denoted by ⌣). Also notice that in such a case the curve always lies *above* its tangent.

EXPLORATION *1*

Concavity and the Second Derivative

■■■ Graph $y_1 = f(x) = 1.5x^5 - 5x^3 + 5x$ and the *sign* of the second derivative,
■■ $y_2 = sign(30x^3 - 30x)$, on a screen with dimensions [−2, 2] by [−4, 4]. (If
■■   your calculator does not have a sign function, then graph $y_2 = f''x/|f''(x)|$.)
Determine the intervals on which $y = f(x)$ is concave up, and check the sign of the second derivative on these intervals. Now determine the intervals on which $y = f(x)$ is concave down, and check the sign of the second derivative on these intervals. What conclusions do you draw?

We now see that a curve is concave up if $f''(x) > 0$. (Hopefully, this is one of the conclusions that you came to earlier.) Recall that if the derivative of a function is positive, then the function must be increasing. The second derivative is the derivative of the first derivative, however, so if

$$f''(x) = \frac{d}{dx}(f'(x)) > 0$$

then the function $f'(x)$ is increasing. The curve is therefore concave up (⌣).

Similarly, in Figure 4.24a and b, we see two functions, the two derivatives of which are *decreasing*, that is, the slopes are rotating *clockwise* as x increases. When this happens we say that the curve is **concave down** (denoted by ⌢). In such a case the curve always lies *below* its tangent.

Remark. The reader should carefully note that concavity has nothing to do with the function itself increasing or decreasing. In Figures 4.23a and b, the graphs of both functions are concave up, but the first function is decreasing while the second function is increasing. In Figures 4.24a and b, the graphs of both functions are concave down, but the first function is decreasing while the second function is increasing.

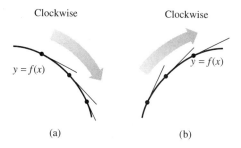

Figure 4.24
When the slope decreases, the curve is concave-down.

We summarize all of this in the following definition and theorem where we assume that $f'(x)$ exists on (a, b).

Definition of Concave Up and Concave Down

1. We say that the graph of f is concave up ($\smile$) on (a, b) if $f'(x)$ is increasing on (a, b).
2. We say that the graph of f is concave down ($\frown$) on (a, b) if $f'(x)$ is decreasing on (a, b).

Test for Concavity

1. If $f''(x) > 0$ on (a, b), then the graph of f is concave up ($\smile$) on (a, b).
2. If $f''(x) < 0$ on (a, b), then the graph of f is concave down ($\frown$) on (a, b).

A point $(c, f(c))$ on the graph of f where the concavity changes is called an **inflection point** and c an **inflection value**. More precisely we have the following definition:

Definition of Inflection Point

A point $(c, f(c))$ on the graph of f is an **inflection point** and c an **inflection value** if $f(c)$ is defined and the concavity of the graph of f changes at $(c, f(c))$.

Figure 4.25 shows four inflection points for four functions. Notice that in each case the graph of the function must cross the tangent line.

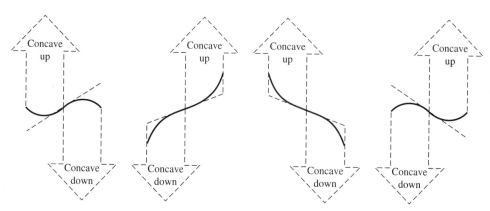

Figure 4.25
At a point where the concavity changes, the curve crosses the tangent line.

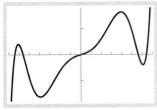

[−1.4, 1.4] by [−1.8, 1.8]

Screen 4.3
The inflection values for the function $y = 1.3x^9 - 8x^5 + 7x^3 + 0.6x$ appear to at about $x = \pm 0.5$, $x = \pm 1.07$, and $x = 0$.

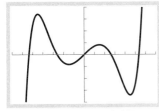

[−1.4, 1.4] by [−40, 40]

Screen 4.4
This is a graph of $y = 93.6x^7 - 160x^3 + 42x$ of the second derivative of the function in Screen 4.3. The second derivative changes sign at the inflection points noted in Screen 4.3.

EXAMPLE 1   Inflection Values

On your graphing calculator find a rough approximation for the inflection values for $f(x) = 1.3x^9 - 8x^5 + 7x^3 + 0.6x$.

SOLUTION ■ Graph $y = f(x)$ on your graphing calculator using a window of dimension [−1.4, 1.4] by [−1.8, 1.8] and obtain Screen 4.3. We set $x_{\text{scl}} = 0.2$. Very roughly the inflection values seem to be $x = \pm 0.5$, $x = \pm 1.07$, and $x = 0$. A graph of $y = f''(x) = 93.6x^7 - 160x^3 + 42x$ shown in Screen 4.4 with a window of dimensions [−1.4, 1.4] by [−40, 40] shows the second derivative changing sign at $x = 0$ and roughly $x = \pm 0.5$ and $x = \pm 1.07$. ■

To find where the graph of f is concave up and where the graph of f is concave down, we need to find the values where $f''(x)$ changes sign. This occurs at values c where $f''(c) = 0$ or where $f''(c)$ does not exist.

EXAMPLE 2   Cost Function

To determine if any economies of scale exist Haldi and Whitcomb [101] studied more than 200 industrial plants and showed that a typical cost function was approximated by $C(x) = ax^{0.7}$, where x is output in appropriate units, a is a positive constant, and $C(x)$ is total cost. The exponent 0.7 is a measure of scale economies. Graph this function, and determine the concavity. What does the concavity tell you about the economy of scale?

SOLUTION ■ We have

$$C(x) = ax^{0.7}$$
$$C'(x) = 0.7ax^{-0.3}$$
$$C''(x) = -0.21ax^{-1.3}$$

We see that $C(x)$ is increasing and concave down, a graph for which is shown in Figure 4.26. Notice that since $C''(x) < 0$, $C'(x)$ is decreasing. That is, marginal cost is decreasing. Thus, for these plants, greater output leads to smaller marginal costs, indicating economies of scale. ■

EXAMPLE 3   Determining Concavity

Discuss the concavity of f if $f(x) = x^3$.

SOLUTION ■ Since $f''(x) = 6x$, $f''(x) < 0$ if $x < 0$, and $f''(x) > 0$ if $x > 0$. Thus, the graph of f is concave down for all values of x on $(-\infty, 0)$ and concave up for all values of x on $(0, \infty)$, and $(0, f(0)) = (0, 0)$ is an inflection point (Fig. 4.27). ■

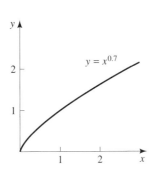

Figure 4.26
Since the curve is concave down, the marginal cost is decreasing.

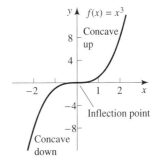

Figure 4.27
This function is concave down on $(-\infty, 0)$ and concave up on $(0, \infty)$, with $(0, 0)$ as an inflection point.

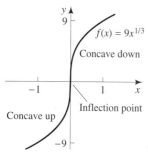

Figure 4.28
This function is concave up on $(-\infty, 0)$ and concave down on $(0, \infty)$, with $(0,0)$ an inflection point.

EXAMPLE 4   Determining Concavity

Discuss the concavity of f if $f(x) = 9x^{1/3}$.

SOLUTION ■ Since $f'(x) = 3x^{-2/3}$, $f''(x) = -2x^{-5/3}$. Thus, $f''(x) > 0$ if $x < 0$, and $f''(x) < 0$ if $x > 0$. So the graph of f is concave up for all values of x on $(-\infty, 0)$ and concave down for all values of x on $(0, \infty)$. Also, $(0, f(0)) = (0, 0)$ is a point of inflection. If we notice that $f'(x)$ is always positive (except at $x = 0$), we can obtain the graph shown in Figure 4.28. ■

The S-Curve

Biologists have noted that populations tend to grow exponentially when the population is small but then level off in time due to the finite amount of available resources. In business, sales of a new product often rise quickly immediately after introduction and then level off later. An equation that models these situations is the logistic equation given by $P(t) = \dfrac{L}{1 + ae^{-kt}}$. We introduced this equation in Section 2.2 and used it to model the population of the United States.

EXPLORATION 2

Exponential Growth in the Long Term

In 1986, the human population of the earth was about 5 billion and growing at 1.6% a year. If we use the exponential model of population growth, then $P(t) = 5e^{0.016t}$, where $P(t)$ is in billions and $t = 0$ corresponds to 1986. Use this model to fill in the table. Given that the surface of the earth has fewer than 2000 trillion square feet with 80% of this water, what do you conclude about using this exponential model over the long-term?

Date	2386	2486	2586	2686	2786
P(t) in trillions					

We now use the logistic equation to model the population of the earth. We will see that concavity plays an important role in understanding the dynamics of the population change.

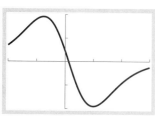

Screen 4.5
A graph of a population model $y = \dfrac{11}{1 + 1.2e^{-0.03x}}$ of the human population of the earth.

Screen 4.6
A graph of $y = n\text{Der2 } p(x)$ of the approximate second derivative of the function given in Screen 4.5.

EXAMPLE 5   Determining the Concavity of a Population Curve

The model $P(t) = \dfrac{11}{1 + 1.2e^{-0.03t}}$ has been used to estimate the future human population of the earth, where $P(t)$ is population in billions and t is time in years measured from 1990. Use your graphing calculator to graph and determine the concavity and what happens in the long-term.

SOLUTION ■ Using a window of dimensions $[-100, 150]$ by $[0, 12]$ we obtain the graph of $y = p(x)$ shown in Screen 4.5. The population is always increasing, and the graph appears to be concave up until about $t = 5$ and then concave down thereafter. It is rather difficult to see on the graph of $y = p(x)$ where the concavity changes. So we need to look at a graph of $y = p''(x)$. The program NDER2G graphs a numerical approximation $n\text{Der2 } f(x)$ to $f''(x)$. Using this, we graph $y_1 = n\text{Der2 } p(x)$ using a window with dimension $[-100, 150]$ by $[-0.001, 0.001]$ and obtain Screen 4.6.

As we see, $p''(x)$ is positive for $x < 6$ and negative for $x > 6$. (You will need to ZOOM to be sure p'' changes sign at 6.) Also we note that $p''(x)$ is approaching zero for large x.

[−100, 150] by [0, 12]

[−100, 150] by [−0.001, 0.001]

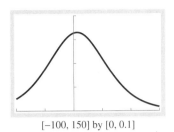

[−100, 150] by [0, 0.1]

Screen 4.7
A graph of $y = n\text{Der } p(x)$ of the approximate first derivative of the function given in Screen 4.5.

We also use NDERG to graph a numerical approximation of $y = p'(t)$. This is shown in Screen 4.7 using a window of dimensions $[-100, 150]$ by $[0, 0.1]$.

We noticed in Screen 4.6 that the second derivative $p''(t)$ is positive on $(-100, 6)$, which should imply that the first derivative $p'(t)$ is increasing there. This is confirmed in Screen 4.7 and also in Screen 4.5, where we saw $y = p(t)$ increasing ever more rapidly on this interval. We also noticed in Screen 4.6 that the second derivative $p''(t)$ is negative on $(6, 150)$. This should imply that the first derivative $p'(t)$ is decreasing there. This is confirmed in Screen 4.7 and in Screen 4.5, where we saw $y = p(t)$ increasing ever more slowly there.

We also noticed that $p''(t)$ is small for large values of t. This should imply that $p'(t)$ is not changing much, and therefore $p(t)$ should not be changing much. This is confirmed in Screen 4.5 where we saw that the graph of $y = p(t)$ is nearly a constant for large values of t. We can see that $p(t)$ is heading for approximately 11 as $t \to \infty$.

In summary, notice that there is an initial period when the curve is concave up, which means that the rate at which the population is increasing is itself increasing. From the graph we see that the slope of the tangent line is increasing. During this period of time the population is often characterized as "exploding." In fact, the concavity of the curve changes at some point and from that point on the curve is concave down. This means that the slope of the tangent line, and thus the rate of increase of $p(t)$, is decreasing. At this point, although the population initially "explodes," the numbers become so large that growth becomes inhibited by the finite amount of food and natural resources available. At this point the growth in the population steadily decreases, and the population heads for some limiting value. ∎

It is interesting to note that the change in concavity in Example 5 is occurring at the present time. Thus, according to this model, the population explosion is over, and the population has just begun to start the leveling off phase in which the finite resource term comes into prominence.

EXPLORATION *3*

Confirming Using Calculus

Confirm the analysis in Example 5 by using calculus. You need to show that

$$p'(t) = \frac{0.396e^{-0.03t}}{(1 + 1.2e^{-0.03t})^2}$$

$$p''(t) = 0.01188e^{-0.03t}\frac{1.2e^{-0.03t} - 1}{(1 + 1.2e^{-0.03t})^3}$$

Also find $\lim_{t \to \infty} p(t)$.

EXPLORATION *4*

Comparing Logistic Growth with Exponential Growth

Graph $y = 5e^{0.03x}$ on the same screen as Screen 4.5. Explain how exponential growth compares with growth given by the logistic equation.

In Figure 4.29, the population of the Northeast of the United States is plotted. Notice how the curve looks like a logistic curve except perhaps for the period from 1930 to 1950. But consider that a major economic depression occurred in the 1930s, a world war in the first half of the 1940s, and a baby boom after the war!

The logistic curve has been used very extensively in modeling. For example, Pearse [102] gave the net stumpage value of an average tree in a typical stand of British Columbia Douglas fir trees (Fig. 4.30). (The trees have no commercial value until they are 30 years old.)

The logistic equation can be used to model the spread of a technological innovation

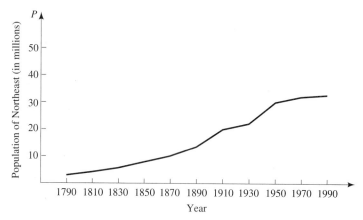

Figure 4.29
The graph of the population of the Northeast looks like a logistic curve.

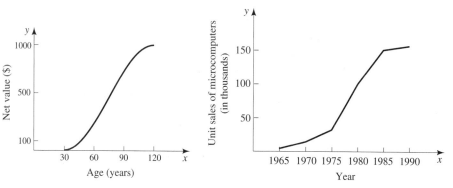

Figure 4.30
The graph of the stumpage value looks like a logistic curve.

Figure 4.31
The graph of the sales of a new innovation looks like a logistic curve.

through an industry. Figure 4.31 shows the number of sales of minicomputers in the United States between 1965 and 1990. This is just one example of how the spread of a new technological innovation tends to explode initially and then levels off.

Acceleration

If $s(t)$ is the distance traveled in time t, then the velocity $v(t)$ is given by $v(t) = s'(t)$. The **acceleration**, denoted by $a(t)$, is defined as the rate of change of velocity. Thus, $a(t) = v'(t) = s''(t)$.

Definition of Acceleration

Acceleration, denoted by $a(t)$, is defined to be the derivative of the velocity, that is,

$$a(t) = v'(t) = s''(t)$$

In some physical situations one can actually "feel" the second derivative. When a person steps heavily on the accelerator of a car, the velocity of the car increases rapidly, and the driver is pressed back against the seat. In this situation the rate of change of velocity is large. That is, $a(t) = s''(t)$ is large. If the driver now slams on the brakes, the velocity decreases rapidly. In this case the acceleration $a(t) = s''(t)$ is negative, and the driver lurches forward in the seat.

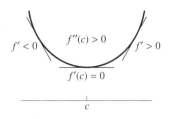

Figure 4.32
$f'(c) = 0$ and $f''(c) > 0$ implies a relative minimum.

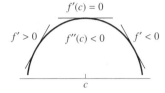

Figure 4.33
$f'(c) = 0$ and $f''(c) < 0$ implies a relative maximum.

The Second Derivative Test

We shall now see that the second derivative can sometimes be used to determine if a point is a relative extremum. Suppose (a, b) is in the domain of definition of f and $c \in (a, b)$. If the second derivative exists and is not zero at a point c where $f'(c) = 0$, then we will see that the sign of $f''(c)$ determines if the value c is a relative minimum or a relative maximum.

Consider, for example, the situation shown in Figure 4.32 in which $f'(c) = 0$ and $f''(c) > 0$. We now show that in this case $f(c)$ is a relative minimum. Notice that since $f''(c) > 0$, so $f'(x)$ is increasing for x near c and therefore $f'(x)$ must be smaller than $f'(c) = 0$ for x just to the left of c and $f'(x)$ must be larger than $f'(c) = 0$ for x just to the right of c. Thus, $f'(x) < 0$ for x to the left of c and $f'(x) > 0$ for x to the right of c. The sign chart of f' is then $- \, 0 \, +$, and we therefore conclude from the first derivative test that $f(c)$ is a relative minimum.

As a second example, consider the situation shown in Figure 4.33 in which we also have $f'(c) = 0$, but this time $f''(c) < 0$. Then, using a similar argument, we see that the sign chart of f' must be $+ \, 0 \, -$, and $f(c)$ must be a relative maximum.

We summarize this as the *second derivative test*.

Second Derivative Test

Suppose f is defined on (a, b), $f'(c) = 0$, and $c \in (a, b)$.

1. If $f''(c) > 0$, then $f(c)$ is a relative minimum.
2. If $f''(c) < 0$, then $f(c)$ is a relative maximum.

EXAMPLE 6   Using the Second Derivative Test

Consider again the function $f(x) = x^3 - 3x^2 + 1$ considered in Example 3 of the Section 4.1. Find where the graph of f is concave up and where it is concave down. Find any inflection points. Use the second derivative test to find the relative extrema. Graph.

SOLUTION ■ We have

$$f'(x) = 3x^2 - 6x = 3x(x - 2)$$

Thus, f' is continuous everywhere and is zero at $x = 0$ and $x = 2$. Also

$$f''(x) = 6x - 6 = 6(x - 1)$$

Since $f''(x) < 0$ on $(-\infty, 1)$ and $f''(x) > 0$ on $(1, \infty)$, the graph of f is concave down for all values of x on $(-\infty, 1)$ and concave up for all values of x on $(1, \infty)$. Since $f(1)$ is defined and the second derivative changes sign at $x = 1$, $(1, f(1)) = (1, -1)$ is an inflection point.

Since $f''(0) = -6 < 0$, the second derivative test indicates that $f(0)$ is a relative maximum. Since $f''(2) = 6 > 0$, the second derivative test indicates that $f(2)$ is a relative minimum.

Recalling that f is increasing on $(-\infty, 0)$ and $(2, \infty)$ and decreasing on $(0, 2)$, we can sketch a graph as shown in Figure 4.34. ■

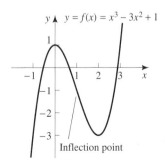

Figure 4.34
The graph is concave down on $(-\infty, 1)$ and concave up on $(1, \infty)$, with $(1, -1)$ as an inflection point.

Remark. The second derivative test can be useful if the second derivative can be easily calculated. A serious problem with the second derivative test is that it can fail to work. For example, if the second derivative is zero or does not exist at a point where the first derivative is zero, then the second derivative fails to supply any information.

EXAMPLE 7   Failure of the Second Derivative Test

Try to use the second derivative test on the following functions. (a) $f(x) = x^4$
(b) $f(x) = -x^4$ (c) $f(x) = x^3$ (d) $f(x) = 9x^{4/3}$.

SOLUTIONS ∎

(a) Here $f'(x) = 4x^3$ and $f''(x) = 12x^2$. Thus, the only critical value is at $x = 0$
and $f''(0) = 0$. Thus, the second derivative test fails, but from the *first* derivative test
it is easily seen that $f(0)$ is a relative minimum (Fig. 4.35).

(b) Here $f'(x) = -4x^3$ and $f''(x) = -12x^2$. Thus, the only critical value is
at $x = 0$ and $f''(0) = 0$. The second derivative test therefore fails, but from the
first derivative test it is easily seen that $f(0)$ is a relative maximum
(Fig. 4.36).

(c) Here $f'(x) = 3x^2$ and $f''(x) = 6x$. Thus, the only critical value is $x = 0$
at which $f''(0) = 0$. The second derivative test therefore fails, but using the first
derivative one can see that $f(0)$ is not a relative minimum or maximum
(Fig. 4.37).

(d) Here $f'(x) = 12x^{1/3}$ and $f''(x) = 4x^{-2/3}$. Thus, the only critical value is
$x = 0$ since $f'(0) = 0$ and $f(0)$ exists. But $f''(0)$ does not exist, and so the second
derivative test cannot be applied. One can see by the first derivative test that $f(0)$ is a
relative minimum (Fig. 4.38). ∎

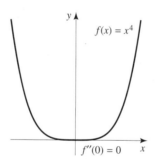

Figure 4.35
The second derivative test cannot
be used since $f''(0) = 0$. The
function has a relative minimum at
$x = 0$.

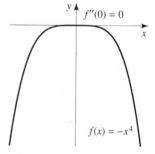

Figure 4.36
The second derivative test cannot
be used since $f''(0) = 0$. The
function has a relative maximum
at $x = 0$.

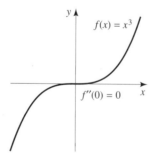

Figure 4.37
The second derivative test cannot
be used since $f''(0) = 0$. The
function has no relative extremum
at $x = 0$.

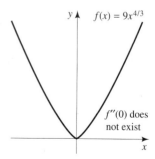

Figure 4.38
The second derivative test cannot
be used since $f''(0)$ does not exist.
The function has a relative
minimum at $x = 0$.

SELF-HELP EXERCISE SET 4.3

1. Suppose a firm believes that the percentage $p(t)$ of the market of its product that it can attain by spending a certain amount on advertising each year for the next six years is given by $p(t) = -\frac{1}{3}t^3 + 3t^2 + 14$, where t is given in years. Find where the graph of the function is concave up and where it is concave down. Discuss the graph of the curve, and interpret.

2. Suppose the second derivative of some functions $y = f(x)$ is given as

$$f''(x) = (x - 1)(x - 2)^2(x - 3)$$

Find the intervals where the graph of the function is concave up and where it is concave down. Find all inflection values.

EXERCISE SET 4.3

In Exercises 1 through 6 find $f''(x)$.

1. $f(x) = 4x^5 + x^2 + x + 3$
2. $f(x) = x^3 + x$
3. $f(x) = \dfrac{1}{x + 1}$
4. $f(x) = \dfrac{1}{x^2 + 1}$
5. $f(x) = \sqrt{2x + 1}$
6. $f(x) = \sqrt{x^2 + 1}$

Define $f^{(3)}(x) = \dfrac{d}{dx}(f''(x))$ and $f^{(4)}(x) = \dfrac{d}{dx}(f^{(3)}(x))$. In Exercises 7 through 12 find $f^{(3)}(x)$ and $f^{(4)}(x)$ for the given functions.

7. $f(x) = 4x^5 + x^2 + x + 3$
8. $f(x) = x^3 + x$
9. $f(x) = \dfrac{1}{x}$
10. $f(x) = \sqrt{x}$
11. $f(x) = \sqrt[3]{x^2}$
12. $f(x) = \dfrac{1}{\sqrt{x}} + \sqrt{x}$

In Exercises 13 through 20 find all inflection values, find the largest open intervals on which the graph of f is concave up, and find the largest open intervals on which the graph of f is concave down for the given functions.

13.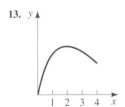

14.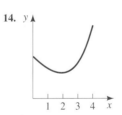

15.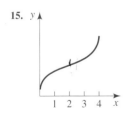

16.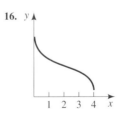

17.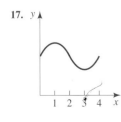

18.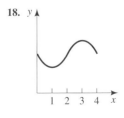

19.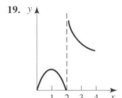

20.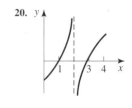

In Exercises 21 through 30, $f''(x)$ is given in factored form. Find all inflections values, find the largest open intervals on which the graph of f is concave up, and find the largest open intervals on which the graph of f is concave down.

21. $f''(x) = x - 2$
22. $f''(x) = (x + 1)(x - 4)$
23. $f''(x) = (x - 1)(x - 3)(x - 5)$
24. $f''(x) = (x - 1)(x - 3)^3(x - 5)$
25. $f''(x) = x(x - 2)^4(x - 4)$
26. $f''(x) = x(x - 2)^2(x - 4)^3$
27. $f''(x) = (x + 2)x^2(x - 2)^2$
28. $f''(x) = (x - 1)^4(x - 3)^2(x - 5)^2$
29. $f''(x) = (x + 2)x^3(x - 2)^5$
30. $f''(x) = (x + 2)^2x^3(x - 2)^4$

In Exercises 31 through 56 find all critical values, inflection values, and the largest open intervals on which the given function is: increasing, decreasing, concave up, and concave down. Find the x-coordinate of each relative minimum and the x-coordinate of each relative maximum. Graph. Support your answers with a graphing calculator.

31. $f(x) = x^5$
32. $f(x) = x^6$
33. $f(x) = x^2 - 8x + 1$
34. $f(x) = 1 + 4x - x^2$
35. $f(x) = x^3 - 3x^2 + 2$
36. $f(x) = 2 - x - x^3$
37. $f(x) = 1 - 9x + 6x^2 - x^3$
38. $f(x) = 1 + 15x + 6x^2 - x^3$
39. $f(x) = 4x^5 - 5x^4 + 1$
40. $f(x) = 6x^7 - 7x^6 + 1$

41. $f(x) = \sqrt{x}, \quad x > 0$

42. $f(x) = \dfrac{1}{\sqrt{x+1}}, \quad x > -1$

43. $f(x) = \sqrt{1-x}, \quad x < 1$

44. $f(x) = \dfrac{1}{\sqrt{1-x}}, \quad x < 1$

45. $f(x) = (x-2)^3$ **46.** $f(x) = (4-x)^3$

47. $f(x) = x^4 + 6x^2 - 2$ **48.** $f(x) = 3 - x^2 - x^4$

49. $f(x) = \sqrt{x^2+1}$ **50.** $f(x) = \dfrac{1}{\sqrt{x^2+1}}$

51. $f(x) = xe^x$ **52.** $f(x) = e^{-x^4}$

53. $f(x) = 5 + 8x^3 - 3x^4$ **54.** $f(x) = 3x^4 - 16x^3 + 3$

55. $f(x) = x^2 e^x$ **56.** $f(x) = e^x - e^{-x}$

57. Show that the graph of the quadratic function $f(x) = ax^2 + bx + c$, $a \neq 0$, is always concave up if $a > 0$ and always concave down if $a < 0$.

58. Show that the quadratic function $f(x) = ax^2 + bx + c$, $a \neq 0$, has an absolute minimum at $x = -\dfrac{b}{2a}$ if $a > 0$ and an absolute maximum at $x = -\dfrac{b}{2a}$ if $a < 0$.

59. Show that the graph of the special fourth-order polynomial function $f(x) = ax^4 + bx^2 + cx + d$ is always concave up if both a and b are positive and always concave down if both a and b are negative.

60. Show that the graph of the cubic function $f(x) = ax^3 + bx^2 + cx + d$, $a \neq 0$, has one and only one inflection point. Find this point.

61. Show that the special cubic function $f(x) = x^3 - 3bx + c$, $b > 0$, has exactly one relative minimum and exactly one relative maximum. Locate and identify them.

In Exercises 62 through 69 you cannot always solve the equations $f'(x) = 0$ exactly to find the critical points. Use the ZOOM feature of your graphing calculator to find the approximate solution to the equations $f'(x) = 0$, and then use the second derivative test to determine the relative extrema. Now sketch a graph of $y = f(x)$. Check your graph by using a graphing calculator.

62. $f(x) = x^4 + x^2 + x + 1$

63. $f(x) = x^4 + x^3 + x + 1$

64. $f(x) = x^5 + x^4 - 3x + 6$

65. $f(x) = 0.20x^5 + 0.50x^2 - x$

66. $f(x) = x^4 - 8x^2 + 2x + 1$

67. $f(x) = x^4 - 4x^3 + 2x + 1$

68. $f(x) = x^4 - 10x^2 - 2x + 1$

69. $f(x) = x^4 - 2x^2 + x + 3$

Applications

70. Revenue. If the revenue function for a firm is given by $R(x) = -x^3 + 144x$, find where the revenue is maximized.

71. Profit. If the cost function for a firm is given by $C(x) = 36x + 5$, and the revenue function is the same as in the previous problem, find where the profit is maximized.

72. Population Dynamics. A general population model similar to that given in Example 5 is $p(t) = \dfrac{ap_0(1+t^2)}{a + bp_0 t^2}$, where p_0 is the initial population, that is, $p(0) = p_0$, and a and b are positive constants. We assume $p_0 < \dfrac{a}{b}$.

(a) Show that $p(t)$ is always increasing for $t > 0$.

(b) Show that the graph of $p(t)$ is concave up at first then turns concave down, and that the inflection point has t-coordinate $\sqrt{a/3bp_0}$.

(c) Draw a graph.

73. Spread of Technology. When a new, useful technology is introduced into a particular industry, it naturally spreads. (See Mansfield [47].) A certain model (very similar to that of Mansfield) of this effect indicates that if t is the number of years after the introduction of the new technology, the percentage of the firms in this industry using the new technology is given by $p(t) = \dfrac{100t^2}{t^2 + 300}$.

(a) Find where $p(t)$ is increasing and where it is decreasing.

(b) Find where the graph of $p(t)$ is concave up and where it is concave down. Find all inflection values.

(c) Draw a graph.

74. Diminishing Returns. Suppose a firm believes that the percentage $p(t)$ of the market it can attain for its product by spending a certain amount on advertising each year for the next three years is given by $p(t) = -2t^3 + 12t^2 + 20$, where t is given in years. Find where the graph of the function is concave up and where it is concave down. Find the inflection point. Where is the point of diminishing returns?

75. Diminishing Returns. Suppose a firm believes that the percentage $p(t)$ of the market it can attain for its product by spending a certain amount on advertising each year for the next 10 years is given by $p(t) = 0.1\dfrac{t^4 + 125}{t^4 + 135}$, where t is given in years. Find where the graph of the function is concave up and where it is concave down. Find the inflection point. Where is the point of diminishing returns?

76. Average Cost. An electric utility company estimates that the cost $C(x)$ of producing x kilowatts of electricity is given by $C(x) = a + bx^2$, where a represents the fixed costs and b is a positive constant. The *average* cost $\overline{C}(x)$ is defined to be the total cost of producing x kilowatts divided by x, that is, $\overline{C}(x) = \dfrac{C(x)}{x}$. Find where the average cost attains a minimum.

77. Biology. The figure from page 93 of Deshmukh [77] shows the result of the classical experiment of Gause using the protozoan *Paramecium caudatum* in the laboratory. On approximately what day did the growth rate of the population change from increasing to decreasing?

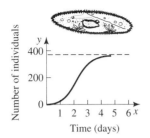

Time (days)

78. Acceleration. A body moves according to the law $s(t) = t^4 - t^2 + 1$. Find the acceleration, and find where it is positive and where it is negative.

79. Spread of Rumors. It has been estimated that a rumor spreads at a rate R that is proportional both to the ratio r of individuals who have heard the rumor and to the ratio $(1 - r)$ who have not. Thus, $R = kr(1 - r)$, where k is a positive constant. For what value of r does the rumor spread fastest?

80. Biology. When predators are not stimulated to search very hard for prey, the rate of change R of the number N of prey attacked per unit of area with respect to the number N of prey is given by $R(x) = kx(ax - N)$, where x is the number of predators, k is the per capita effectiveness of the predators, and a is the maximum per capita effectiveness of the predators over an extended period. Find the number of predators that minimizes R.

81. Learning Curve. Cost engineers have noted that the efficiency of workers in a plant or on a machine increases over time, as the work force gains efficiency and skill in performing the repeated manual tasks. Hirsch [28] reported that the relationship between the cumulative average amount of direct labor input y on a textile machine and the cumulative number of units produced x is given by $y = ax^{-0.175}$, where a is a constant. **(a)** Find the first and second derivative, and sketch a graph. **(b)** If productions doubles, what happens to the average amount of needed labor? What does this mean in terms of learning?

82. Portfolio Management. Christensen and Sorensen [103] showed that the yield i on a 30-year zero-coupon bond was approximated by $i(t) = 0.09 + 0.0025(30 - t) - 0.00005(30 - t)^2$, where t is the time in years. Use calculus to sketch a graph on [0, 30].

Enrichment Exercises

For Exercises 83 through 86 consider the graphs of the following four functions.

83.

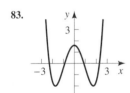

84.

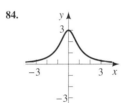

85.

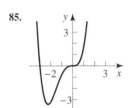

86.

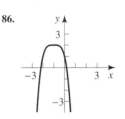

Each of the following figures represents the graph of the second derivative of one of the previous functions, but in a different order. Match the graph of the function to the graph of its second derivative.

(a)

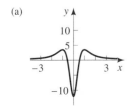

(b)

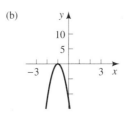

(c)   (d)

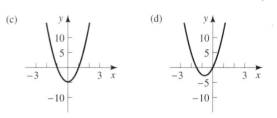

87. If $f''(a)$ exists, explain why $y = f(x)$ must be continuous at $x = a$.

88. Use the concavity of $f(x) = \ln x$ to show that $\ln x \le x - 1$.

89. Assume a polynomial has exactly one relative maximum and two relative minima.
  (a) Sketch a possible graph of $y = f(x)$.
  (b) What is the largest number of zeros that f can have?
  (c) What is the least number of inflection values that f can have?

90. Assume a polynomial has exactly one relative maximum and one relative minimum.
  (a) Sketch a possible graph of $y = f(x)$.
  (b) Decide if the degree of the polynomial is even or odd, and explain how you arrived at your answer.

91. Suppose $y = f(x)$ is a polynomial with two and only two critical values at $x = -2$ and $x = 3$. Also, $f''(-2) > 0$ and $f''(3) < 0$.
  (a) Draw a possible graph.
  (b) Must there be an inflection point? Explain.
  (c) Is the polynomial even or odd? Explain.

92. Suppose $y = f(x)$ is a polynomial with three and only three critical values at $x = -2$, $x = 3$, and $x = 6$. Also, $f''(-2) > 0$, $f''(3) < 0$, and $f''(6) > 0$.

(a) Draw a possible graph.

(b) How many inflection points must there be? Explain.

(c) Is the polynomial even or odd? Explain.

93. Suppose $f''(x)$ is continuous and positive everywhere and $f(x)$ is negative everywhere. Let $g = 1/f$. Show that $g(x)$ is concave down everywhere.

94. Let $f(x) = x^2 p(x)$, where $p(x)$ is a polynomial with $p(0) > 0$. Show that $f(x)$ attains a relative minimum at $x = 0$. What happens if $p(0) < 0$?

95. Let $f(x) = 0$ if $x < 0$ and $f(x) = x^3$ if $x \geq 0$. Use the limit definition to show that $f'(0)$ exists. Then use the limit definition to show that $f''(0)$ exists.

96. Sketch a graph of $y = P'(x)$ below the given graph of $y = P(x)$. Explain carefully what happens to the graph of $y = P'(x)$ when it crosses the two vertical dotted lines.

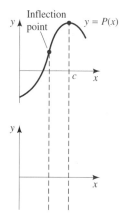

97. Recall $\dfrac{d}{dx} \sin x = \cos x$ and $\dfrac{d}{dx} \cos x = -\sin x$.

(a) Find $\dfrac{d^2}{dx^2} (\sin x)$ and $\dfrac{d^2}{dx^2} (\cos x)$.

(b) With your graphing calculator in radian mode graph $y_1 = \sin x$ and $y_2 = \cos x$ on a screen with dimensions $[0, 2\pi]$ by $[-1, 1]$. Determine where $\sin x$ is positive and where it is negative. Do the same for $\cos x$. Use this information together with the second derivatives found in part (a) to determine where the functions $y_1 = \sin x$ and $y_2 = \cos x$ are concave up and concave down. Verify by closely examining the graphs of these functions.

98. Banking. Keeley and Zimmerman [104] used the logistic model $P(t) = \dfrac{0.21}{1 + e^{2.63 - 1.80t}}$, to describe the growth in the percentage of deposits in money market accounts, $P(t)$, in time t in years. Graph using a window of dimensions $[0, 5]$ by $[0, 0.25]$. Does this look like a logistic curve? Graphically find the inflection point.

99. Biology. Lescano and coworkers [105] showed that the average mortality y at pupation of the Tephritid fruit fly was approximated by $y = \dfrac{e^{3.908 - 0.064t}}{1 + e^{3.908 - 0.64t}}$, where t is the age at irridation. Graph on your graphing calculator. Graphically find the value of t at which the concavity changes.

100. Biology. Roltsch and Mayse [106] showed that the cumulative percentage of emergence y of the moth *H. brillians* was approximated by $y = \dfrac{100}{1 + e^{4.9376 - 0.02351x}}$, where x is the cumulative degree days. Graph on your graphing calculator using a window with dimensions $[0, 1000]$ by $[0, 100]$. Graphically find the value of x at which the concavity changes.

SOLUTIONS TO SELF-HELP EXERCISE SET 4.3

1. Since

$$p'(t) = -t^2 + 6t = t(6 - t)$$

p is always increasing on the interval $(0, 6)$.

Since

$$p''(t) = -2t + 6 = 2(3 - t)$$

$p''(t) > 0$, and thus the graph of p is concave up on $(0, 3)$ and $p''(t) < 0$ and thus the graph of p is concave down on $(3, 6)$. On the interval $(0, 3)$, p is increasing ever faster, while on $(3, 6)$, p is increasing ever slower. We might characterize the interval $(3, 6)$ as the period of diminishing returns.

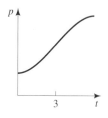

2. According to the given factorization of $f''(x)$, $f''(x) > 0$ on the intervals $(-\infty, 1)$ and $(3, \infty)$, while $f''(x) < 0$ on $(1, 2)$ and $(2, 3)$. Thus, the graph of f is concave up on the intervals $(-\infty, 1)$ and $(3, \infty)$, and concave down on $(1, 3)$.

Since the second derivative changes sign at $x = 1$ and $x = 3$, these values are inflection values.

4.4 MORE CURVE SKETCHING

■ More Curve Sketching   ■ Applications

APPLICATION: FENCE CONSTRUCTION

Three sides of a rectangular fence are to be constructed with a straight river making up the fourth side. The area enclosed must be 2 square miles. What is the perimeter as a function of x if x is in miles, and what is the graph of the resulting function? See Example 3 for the answer.

More Curve Sketching

In the previous three sections we have seen how to use both the first and the second derivative to sketch the graph of a function. We now consider a systematic approach to sketching the graph of a function.

The following checklist is a useful procedure for graphing a function:

Checklist for Graphing a Function

A. Use $f(x)$ to
  1. Determine the domain of the function and the intervals on which the function is continuous.
  2. Determine if the function is symmetrical about the y-axis or the origin.
  3. Find all vertical asymptotes.
  4. Find all horizontal asymptotes.
  5. Find where the function crosses the axes.
B. Use $f'(x)$ to
  6. Find the critical values.
  7. Find intervals where the function is increasing and decreasing.
  8. Find all relative extrema.
C. Use $f''(x)$ to
  9. Find intervals where the graph of the function is concave up and concave down.
  10. Find all inflection values.
D. (Final step.) Use steps A, B, C, and the values of f at the critical values and inflection values to graph.

We will see how to apply this checklist by working through the following examples. Be aware, however, that all the items may not apply when graphing any one particular function.

EXAMPLE 1   Finding the Relative Extrema

Use your graphing calculator to graph $y = f(x) = 6\sqrt[3]{x^5} + 15\sqrt[3]{x^2} = 6x^{5/3} + 15x^{2/3}$. Then use calculus to sketch a more accurate graph.

SOLUTION ■ The graph shown in Screen 4.8 uses a window with dimensions $[-3, 1]$ by $[-10, 10]$. (To obtain the correct graph input $y_1 = 6(x^5)^{(1/3)} + 15(x^2)^{(1/3)}$.) There appears to be a relative maximum at $x = -1$ and a relative minimum at $x = 0$. The curve appears concave down on $(-\infty, 0)$, but the concavity is not discernible on $(0, \infty)$. Of course, any graph drawn by our graphing calculator may have scales too large to show some hidden behavior or may have some interesting behavior occurring just outside the window. The only way to be certain that we have all the interesting behavior is to use calculus.

Let us do just that now.

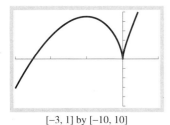

[−3, 1] by [−10, 10]

Screen 4.8
We should use calculus to see if there is any hidden behavior in the graph of $y = 6x^{5/3} + 15x^{2/3}$.

STEP 1. The function is continuous on $(-\infty, \infty)$.

STEP 5. Note that

$$f(x) = x^{2/3}(6x + 15) = 0$$

implies that the graph $y = f(x)$ crosses the x-axis at $x = 0$ and at $x = -2.5$. Since $f(0) = 0$, the graph crosses the y-axis at the origin.

STEP 6. Find the critical values.

(a) The derivative $f'(x) = 10x^{2/3} + 10x^{-1/3}$ exists and is continuous everywhere except at $x = 0$, where f' does not exist but f does. Thus, $x = 0$ is a critical value.

(b) Now setting $f'(x) = 10x^{2/3} + 10x^{-1/3} = 0$ yields

$$10x^{2/3} + 10x^{-1/3} = 0$$

$$x^{2/3} = -\frac{1}{x^{1/3}}$$

$$x^{2/3}(x^{1/3}) = -\frac{1}{x^{1/3}}(x^{1/3})$$

$$x = -1$$

The critical values are then $x = -1$ and $x = 0$.

STEP 7. These two values divide the real line into the three subintervals $(-\infty, -1)$, $(-1, 0)$, and $(0, \infty)$, on each of which f' is continuous and not zero.

Use test values -8, $-\frac{1}{8}$, and 1 (Fig. 4.39). Then

$$f'(-8) = +35, \qquad f'\left(-\frac{1}{8}\right) = -\frac{35}{2}, \qquad f'(1) = +20$$

This leads to the rest of the sign chart given in Figure 4.39. We see that f is increasing on $(-\infty, -1)$ and $(0, \infty)$ and that f is decreasing on $(-1, 0)$.

STEP 8. The first derivative test indicates that $f(-1)$ is a relative maximum and $f(0)$ a relative minimum.

STEP 9. We have

$$f''(x) = \frac{20}{3}x^{-1/3} - \frac{10}{3}x^{-4/3}$$

$$= \frac{10}{3x^{4/3}}(2x - 1)$$

Thus, f is concave down on $(-\infty, 0)$ and $(0, \frac{1}{2})$ and concave up on $(\frac{1}{2}, \infty)$.

STEP 10. Thus, $x = \frac{1}{2}$ is an inflection value.

FINAL STEP. Noting further that $f(-1) = 9$ and $f(0) = 0$, one can put all this together to give the graph shown in Figure 4.40. ∎

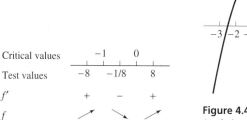

Critical values, Test values chart with f': $+$ $-$ $+$

Figure 4.39

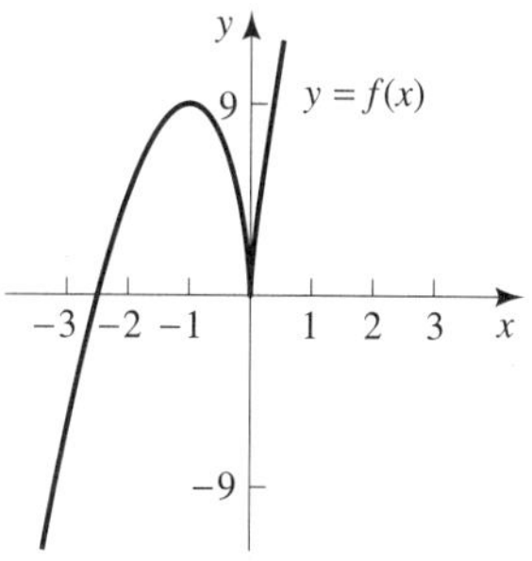

Figure 4.40
A relative maximum at $x = -1$ and a relative minimum at $x = 0$.

EXAMPLE 2   Sketching a Graph

Sketch a graph of the function $f(x) = x^4 - 6x^2 + 4$ using calculus. Support your answer using a graphing calculator.

SOLUTION ■

1. The domain of this polynomial is the entire real line. The function is continuous everywhere.
2. Since $f(-x) = f(x)$, the function is symmetrical about the y-axis. Thus, we need only consider the set of points $x \geq 0$. (For practice and purposes of illustration, however, negative values of x are considered.)
3. There are no vertical asymptotes.
4. We notice that for very large values of $|x|$,

$$f(x) = x^4 \left(1 - \frac{6}{x^2} + \frac{4}{x^4}\right) \approx x^4$$

Thus, as $x \to \infty$ or as $x \to -\infty$, $f(x)$ becomes a large positive number without bound. Thus,

$$\lim_{x \to \infty} f(x) = \infty, \qquad \lim_{x \to -\infty} f(x) = \infty$$

5. The curve crosses the y-axis when $x = 0$ at the point $(0, 4)$. Although normally one cannot hope to find the zeros of a fourth-degree polynomial, in this case it is possible. Setting $z = x^2$ and then setting $f = 0$ gives $z^2 - 6z + 4 = 0$. Thus, by the quadratic formula, $z = 3 \pm \sqrt{5}$ and $x = \pm\sqrt{(3 \pm \sqrt{5})}$. These latter numbers are approximately ± 2.3 and ± 0.9.
6. Since

$$f'(x) = 4x^3 - 12x = 4x(x^2 - 3) = 4x(x + \sqrt{3})(x - \sqrt{3})$$

the three critical values are $-\sqrt{3}$, 0, and $+\sqrt{3}$.

7. The three critical values divide the real line into the four intervals $(-\infty, -\sqrt{3})$, $(-\sqrt{3}, 0)$, $(0, \sqrt{3})$, and $(\sqrt{3}, \infty)$. Figure 4.41 gives the critical values and the test values. At the test values

$$f'(-2) = -8, \qquad f'(-1) = +8, \qquad f'(1) = -8, \qquad f'(2) = +8$$

This then implies the rest of Figure 4.41. This figure indicates that on the successive intervals $(-\infty, -\sqrt{3})$, $(-\sqrt{3}, 0)$, $(0, \sqrt{3})$, and $(\sqrt{3}, \infty)$, $f'(x)$ is negative, positive, negative, and positive, respectively. Thus, on the four intervals, $f(x)$ is successively decreasing ($\searrow$), increasing ($\nearrow$), decreasing ($\searrow$), and increasing ($\nearrow$).

8. By the *first* derivative test, $f(-\sqrt{3})$ and $f(\sqrt{3})$ are relative minima, and $f(0)$ is a relative maximum.

9. Now notice that

$$f''(x) = 12x^2 - 12 = 12(x + 1)(x - 1)$$

The second derivative exists and is continuous everywhere and is zero at the two values -1 and 1. These two values divide the real line into the three subintervals, $(-\infty, -1)$, $(-1, 1)$, and $(1, \infty)$. Figure 4.42 lists the values where the second derivative is zero and the test values for the three intervals.

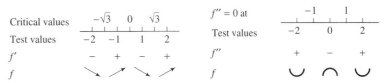

Critical values	$-\sqrt{3}$ 0 $\sqrt{3}$		$f'' = 0$ at	-1 1
Test values	-2 -1 1 2		Test values	-2 0 2
f'	$-$ $+$ $-$ $+$		f''	$+$ $-$ $+$
f	$\searrow \nearrow \searrow \nearrow$		f	$\smile \frown \smile$

Figure 4.41   **Figure 4.42**

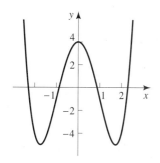

Figure 4.43
Relative minima at $x = \pm\sqrt{3}$ and a relative maximum at $x = 0$.

At the test values,

$$f''(-2) = +36, \qquad f''(0) = -12, \qquad f''(2) = +36$$

From this is constructed a sign chart for f''. From the sign chart of f'' we see that $f''(x)$ is positive on the intervals $(-\infty, -1)$ and $(1, \infty)$, and thus the graph of f is concave up $(\smile)$ on these two intervals. From the sign chart of f'' we see that $f''(x)$ is negative on the interval $(-1, 1)$, and thus the graph of f is concave down $(\frown)$ on this interval.

10. Since $f''(x)$ changes sign at $x = -1$ and $x = +1$, the inflection values are ± 1.

FINAL STEP. The values of f at the critical values $-\sqrt{3}$, 0, and $\sqrt{3}$ are -5, 4, and -5, respectively. The values of f at the inflection values -1 and $+1$ are, respectively, -1 and -1. This information is then put together to give the graph shown in Figure 4.43.

We can support this on our graphing calculator using a window with dimensions $[-3, 3]$ by $[-10, 10]$ and obtain the graph shown in Figure 4.43. ∎

In the preceding example, the *second* derivative test could have been used to find the relative extrema. Since $f''(-\sqrt{3}) > 0$ and $f''(\sqrt{3}) > 0$, the second derivative test indicates that $f(-\sqrt{3})$ and $f(\sqrt{3})$ are relative minima. Since $f''(0) < 0$, the second derivative test indicates that $f(0)$ is a relative maximum.

Applications

Many problems in the applications require finding a minimum value of some positive function f on $(0, \infty)$ when f is very large both for x very small and for x very large. The function should then attain a minimum at some intermediate value of x. The following example is representative of this type.

EXAMPLE 3   Fence Construction

Three sides of a rectangular fence are to be constructed with a straight river making up the fourth side (Fig. 4.44). The area enclosed must be 2 square miles. Write the perimeter as a function of x for x measured in miles. Graph.

Figure 4.44

SOLUTION ∎ According to Figure 4.44, the area is $A = xy$. Since we are given that $A = 2$, we then have that $xy = 2$ and $y = \dfrac{2}{x}$. The perimeter is then given by

$$f = x + 2y = x + 2\frac{2}{x} = x + \frac{4}{x}$$

Since x is the length of one side of the fence, naturally we must have $x > 0$ to make physical sense.

1. The domain of the function is $(0, \infty)$ where it is continuous.
3. As $x \to 0^+$, $f(x)$ increases without bound. That is, $\lim\limits_{x \to 0^+} f(x) = \infty$. Thus, $x = 0$ is a vertical asymptote.
4. For very large values of x,

$$f(x) = x + \frac{4}{x} \approx x$$

5. The curve does not cross the x- or y-axes.
6. Setting $f'(x) = 1 - 4x^{-2} = 0$ yields $x^2 = 4$, which gives the critical value $x = 2$ on the interval $(0, \infty)$.

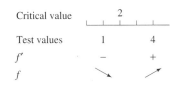

Critical value		2	
Test values	1		4
f'	$-$		$+$
f	↘		↗

Figure 4.45
A sign chart for f'.

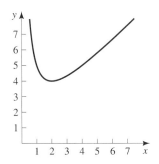

Figure 4.46
An absolute minimum occurs at $x = 2$.

7. This critical value divides $(0, \infty)$ into the two subintervals $(0, 2)$ and $(2, \infty)$. Test values are then chosen for these two subintervals, as indicated in Figure 4.45. We have at the test values

$$f'(1) = -3 < 0, \qquad f'(4) = \frac{3}{4} > 0$$

Thus, f is decreasing on the first subinterval and increasing on the second subinterval.

8. The first derivative test indicates $f(2)$ is a relative minimum.
9. Since $f''(x) = 8x^{-3}$, f'' is positive on $(0, \infty)$. Thus, the graph of f is concave up on $(0, \infty)$.
10. There are no inflection points.

FINAL STEP. $f(2) = 4$. Putting all of this information together then gives the sketch of the graph in Figure 4.46. ∎

EXAMPLE 4   Curve Sketching

Consider $f(x) = \dfrac{\ln x}{x}$, for $x > 0$. Find where f is increasing, where f is decreasing. Find any relative extrema. Find where the graph of f is concave up and where the graph of f is concave down.

SOLUTION ■ First notice that $f(x)$ is continuous on $(0, \infty)$. Also notice that the sign of $f(x)$ on $(0, \infty)$ is the same as the sign of $\ln x$. Thus, $f(x)$ is negative on $(0, 1)$, zero at $x = 1$, and positive on $(1, \infty)$.

Notice that $\lim\limits_{x \to 0^+} \dfrac{\ln x}{x} = -\infty$. Thus, $x = 0$ is a vertical asymptote.

Since $f(100) \approx 0.046052$, $f(1000) = 0.006908$, and $f(10,000) = 0.000921$, $\lim\limits_{x \to \infty} f(x) = 0$ and $y = 0$ is a horizontal asymptote.

$$f'(x) = \frac{d}{dx}\left(x^{-1} \ln x\right)$$

$$= (x^{-1}) \frac{d}{dx}(\ln x) + (\ln x)\frac{d}{dx}(x^{-1})$$

$$= (x^{-1})\left(\frac{1}{x}\right) + (\ln x)(-x^{-2})$$

$$= \frac{1}{x^2}(1 - \ln x)$$

The first factor $1/x^2$ is always positive on $(0, \infty)$. To analyze the factor $(1 - \ln x)$, we recall that $\ln x$ is increasing and $1 - \ln x = 0$ when $x = e$. Thus, $1 - \ln x > 0$ on $(0, e)$ and therefore $f'(x) > 0$ on $(0, e)$. Also $1 - \ln x < 0$ on (e, ∞) and therefore $f'(x) < 0$ on (e, ∞). Since $1 - \ln e = 0$, $f'(e) = 0$. Thus, $f(x)$ is increasing on $(0, e)$ and decreasing on (e, ∞). This implies that $f(x)$ has a relative maximum at $x = e$ and no relative minimum on $(0, \infty)$.

Calculating the second derivative gives

$$f''(x) = (1 - \ln x)\frac{d}{dx}\left(\frac{1}{x^2}\right) + \frac{1}{x^2}\frac{d}{dx}(1 - \ln x)$$

$$= (1 - \ln x)\left(-\frac{2}{x^3}\right) + \frac{1}{x^2}\left(-\frac{1}{x}\right)$$

$$= \frac{2}{x^3}\left(\ln x - \frac{3}{2}\right)$$

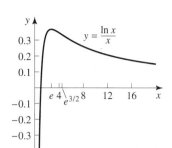

Figure 4.47
An absolute maximum occurs at $x = e$.

The first factor $\frac{2}{x^3}$ is positive for $x > 0$. Notice that $\ln x - \frac{3}{2} = 0$ when $x = e^{3/2}$ and that $\ln x$ is an increasing function. If $0 < x < e^{3/2}$, $\ln x - \frac{3}{2}$ and $f''(x)$ are therefore negative, whereas if $x > e^{3/2}$, $\ln x - \frac{3}{2}$ and $f''(x)$ are positive. Thus, the graph of f is concave down on $(0, e^{3/2})$ and concave up on $(e^{3/2}, \infty)$.

The graph is given in Figure 4.47. ■

We have already seen one example of limited growth in the logistic equation. We now consider another. If a jug of cold milk at 40°F is placed on a table in a room kept at 70°F, the temperature of the milk steadily increases toward the limiting value of 70°F. This is a very familiar example of **limited growth**. An equation that models this is given by $f(x) = L - ae^{-kx}$, where L, a, and k are positive constants. In the next example we see another example of limited growth that uses the same mathematical model.

EXAMPLE 5 Limited Growth

The weekly sales $S(t)$ of a new product satisfy $S(t) = 1000 - 800e^{-0.1t}$, where t is time measured in months from the time of introduction. Graph and indicate what happens in the long-term.

SOLUTION ■ We have

$$S'(t) = -800e^{-0.1t}(-0.1) = 80e^{-0.1t}$$

$$S''(t) = -8e^{-0.1t}$$

Notice that $S'(t) > 0$ and $S''(t) < 0$ for $t \geq 0$. Thus, $S(t)$ is always increasing and concave down. Furthermore,

$$\lim_{t \to \infty} (1000 - 800e^{-0.1t}) = 1000$$

Thus, in the long-term the weekly sales will be 1000, the graph for which is shown in Figure 4.48. ■

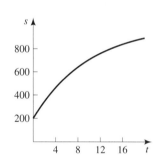

Figure 4.48
The graph is concave down and also heads for a limiting value.

SELF-HELP EXERCISE SET 4.4

1. Sketch a graph of $f(x) = \dfrac{x^2}{x^2 - 1}$.

2. Let $P(t)$ denote the fraction of firms in an industry that are currently using a technological innovation. Assume that $P(t)$ satisfies a logistic curve model and that this innovation eventually spreads throughout the entire industry. Initially 10% of the firms used the innovation, and two years later 20% did. Find the logistic equation that $P(t)$ satisfies.

EXERCISE SET 4.4

In Exercises 1 through 36 go through the checklist and obtain as much information as possible and then sketch a graph of the given function. Support your answer with a graphing calculator.

1. $f(x) = x^4 + 4x^3 + 1$

2. $f(x) = 3x^4 - 16x^3 + 24x^2 - 1$

3. $f(x) = x^4 - 32x^2 + 10$

4. $f(x) = x^4 - 2x^3 + x^2$

5. $f(x) = 3x^5 - 40x^3 + 20$

6. $f(x) = -3x^5 + 10x^3 - 5$

7. $f(x) = \dfrac{1}{x - 1}$

8. $f(x) = \dfrac{x}{x - 1}$

9. $f(x) = \dfrac{x + 1}{x + 2}$

10. $f(x) = \dfrac{x + 1}{x - 1}$

11. $f(x) = \dfrac{1}{x^2 + 1}$

12. $f(x) = \dfrac{1}{x^4 + 1}$

13. $f(x) = x + \dfrac{9}{x}$

14. $f(x) = x + \dfrac{25}{x}$

15. $f(x) = x + \dfrac{1}{3x^3}$

16. $f(x) = x + \dfrac{16}{3x^3}$

17. $f(x) = \dfrac{x^2}{x - 1}$

18. $f(x) = -\dfrac{x^2}{x + 2}$

19. $f(x) = \dfrac{x}{x^2 + 1}$

20. $f(x) = 2 - \dfrac{1}{x^2 + 4}$

21. $f(x) = \dfrac{x - 1}{x^2}$

22. $f(x) = \dfrac{x^2}{1 + x^2}$

23. $f(x) = \dfrac{x^2 + 2x - 4}{x^2}$

24. $f(x) = \dfrac{1}{x^2 - x - 6}$

25. $f(x) = 2\sqrt[3]{x^5} - 5\sqrt[3]{x^2}$

26. $f(x) = 6\sqrt[3]{x^2} - 4x$

27. $f(x) = 5\sqrt[5]{x^4} - 4x$

28. $f(x) = 7\sqrt[7]{x^6} - 6x$

29. $f(x) = xe^{-x^2}$

30. $f(x) = x^2 e^{-0.5x^2}$

31. $f(x) = \dfrac{2}{e^x + e^{-x}}$

32. $f(x) = x \ln x$

33. $f(x) = \dfrac{\ln x}{x^2}$

34. $f(x) = (\ln x)^3$

35. $f(x) = \dfrac{2}{e^x - e^{-x}}$

36. $f(x) = \dfrac{e^x + e^{-x}}{e^x - e^{-x}}$

Applications

37. Cost. It is estimated that the cost $C(x)$ in billions of dollars of maintaining the atmosphere in the United States at an average x percent free of a chemical toxin is given by $C(x) = \dfrac{10}{100 - x}$. Sketch a graph of the function on $(0, 100)$.

38. Cost. A large corporation already has cornered 30% of the toothpaste market and now wishes to undertake a very extensive advertising campaign to increase its share of this market. It is estimated that the advertising cost $C(x)$ in billions of dollars of attaining an x percent share of this market is given by $C(x) = \dfrac{1}{100 - x}$. Sketch a graph of the function on $(0, 100)$.

39. Productivity. A plant manager has worked hard over the past years using various innovative means to increase the number of items produced per employee in his plant and has noticed that the number of items $n(t)$ produced per employee per day over the last 10 years is given by $n(t) = 9 \cdot \dfrac{t}{t + 1}$, where t is the number of years from 10 years ago. Sketch a graph of the function on $(0, \infty)$.

40. Biology. The rate of production P of photosynthesis is related to the light intensity I by the formula $P(I) = \dfrac{aI}{b + I^2}$, where a and b are positive constants. Sketch a graph of this function.

41. Average Cost. Suppose the cost function is given by $C(x) = a + bx^2$, where a and b are positive constants. Sketch a graph of the average costs.

42. Spread of Rumors. In a corporation, rumor spreads of large impending layoffs. We assume that eventually everyone will hear the rumor. The percentage of the work force that has heard the rumor satisfies a logistic equation. If initially 0.1% of the workforce hear the rumor and 12.5% of the workforce hear the rumor after the first day, what percentage has heard the rumor after six days from the start of the rumor?

43. Profits. The profits from the introduction of a new product often explode at first and then level off and thus can sometimes be modeled by a logistic equation. Suppose the annual profits $P(t)$ in millions of dollars t years after the introduction of a certain new product satisfies the logistic equation $P(t) = \dfrac{10}{1 + 20e^{-0.2t}}$. Graph $P(t)$ on your graphing calculator. Find the point where the concavity changes. (This is called the **point of diminishing returns**.) What is the limiting value of the annual profits?

44. Spread of Disease. Jim Smith is a member of a small, completely isolated community of 100 residents. Jim leaves and returns with an infectious disease to which everyone in the community is susceptible. One day after returning Jim gives the disease to one other resident. If the number of persons who get this disease satisfies a logistic equation, find how many of these residents have contracted the disease 10 days after Jim returns.

45. Population. Studies indicate that a small state forest with no deer could hold a maximum of 100 deer. Five deer are transplanted into this forest, and one year later the population has increased to 10. Assuming the deer population satisfies a logistic equation, how long does it take for the deer population to reach 75?

46. Biology. Reed and Semtner [107] showed that the proportional yield loss y of flue-cured tobacco was approximated by $y = f(x) = 0.259(1 - e^{-0.000232x})$, where x is cumulative aphid days in thousands. Use calculus to draw a graph. What happens to the proportional yield as the cumulative aphid days becomes large without bound?

47. Biology. Mayer [108] showed that the number y of screwworm flies a distance x km from animals was approximated by $y = a\left\{\pi b\left[1 + \left(\dfrac{x}{b}\right)^2\right]\right\}^{-1}$, where a and b are positive constants. Use calculus to sketch a graph.

Enrichment Exercises

48. Biology. Overholt and coworkers [109] showed that the proportion y of infested corn plants was approximated by $y = 1 - e^{-x \ln[2.77x^{0.206}/(2.77x^{0.206}-1)]}$, where x is the number of corn borers per corn plant. Use your graphing calculator to graph. Where is the function increasing? Concave down? What happens as $x \to 0^+$?

In Exercises 49 through 52 you are given some conditions that a function must satisfy. In each case graph a function that satisfies the given conditions.

49. $f'(x) > 0$ and $f''(x) > 0$ on $(-\infty, 1)$, $f'(x) < 0$ and $f''(x) > 0$ on $(1, \infty)$, $f(0) = 0$, $\lim\limits_{x \to -\infty} f(x) = -\infty$, $\lim\limits_{x \to 1^-} f(x) = \infty$, $\lim\limits_{x \to 1^+} f(x) = \infty$, $\lim\limits_{s \to \infty} f(x) = 0$

50. $f'(x) > 0$ and $f''(x) > 0$ on $(-\infty, 0)$, $f'(x) < 0$ and $f''(x) < 0$ on $(0, \infty)$, $f''(x) > 0$ on $(-\infty, 0)$, $f''(x) < 0$ on $(0, \infty)$, $f(0) = 0$, $\lim\limits_{x \to -\infty} f(x) = -1$, $\lim\limits_{x \to \infty} f(x) = 1$

51. $f'(x) > 0$ on $(-2, -1)$ and $(-1, 0)$, $f'(x) < 0$ on $(-\infty, -2)$ and $(0, 1)$ and $(1, \infty)$, $f''(x) > 0$ on $(-\infty, -1)$ and $(1, \infty)$, $f''(x) < 0$ on $(-1, 1)$, $f(-2) = 1$, $f(0) = 0$, $\lim\limits_{x \to -\infty} f(x) = \infty$, $\lim\limits_{x \to -1^-} f(x) = \infty$, $\lim\limits_{x \to -1^+} f(x) = -\infty$, $\lim\limits_{x \to 1^-} f(x) = -\infty$, $\lim\limits_{x \to 1^+} f(x) = \infty$, $\lim\limits_{x \to \infty} f(x) = 0$

52. $f'(x) > 0$ on $(-1, 0)$ and $(0, 1)$, $f'(x) < 0$ on $(-\infty, -1)$ and $(1, \infty)$, $f''(x) > 0$ on $(-\infty, 0)$ and $(2, \infty)$, $f''(x) < 0$ on $(0, 2)$, $f(-1) = 1$, $f(1) = 1$, $\lim\limits_{x \to -\infty} f(x) = \infty$, $\lim\limits_{x \to 0^-} f(x) = \infty$, $\lim\limits_{x \to 0^+} f(x) = -\infty$, $\lim\limits_{x \to \infty} f(x) = -1$

For Exercises 53 through 56 consider the graphs of the following four functions:

53.
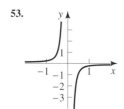

54.
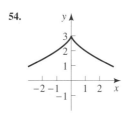

55.
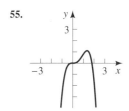

56.
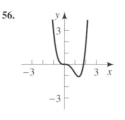

Each of the following figures represents the graph of the derivative of one of the preceding functions, but in a different order. Match the graph of the function to the graph of its derivative.

(a)
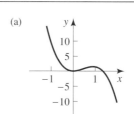

(b)
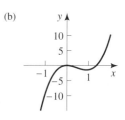

(c)
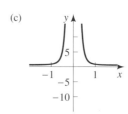

(d)
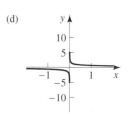

In Exercises 57 through 62 let $P(t) = \dfrac{L}{1 + ae^{-kt}}$. Establish each statement analytically using calculus.

57. $P'(t)$ is positive for $t \geq 0$

58. $\lim\limits_{t \to \infty} P(t) = L$

59. $c = \dfrac{1}{k} \ln a$ is an inflection point

60. $P(t)$ is concave up on $(0, c)$ and concave down on (c, ∞)

61. $P(c) = \dfrac{L}{2}$ where c is given in Exercise 59

62. $a = \dfrac{L}{P(0)} - 1$

63. Technological Innovation. Suppose all firms in an industry eventually institute a technological innovation and the percentage of firms at time t that use the innovation is given by the logistic equation $P(t) = \dfrac{100}{1 + ae^{-kt}}$. If initially, at time $t = 0$, the percentage of firms that use the innovation is $P_0 = P(0)$ and T years later is Q, show that $k = \dfrac{1}{T} \ln \dfrac{Q(1 - P_0)}{P_0(1 - Q)}$.

64. Limited Growth. A jug of milk at a temperature of 40°F is placed in a room held at 70°F. The temperature in degrees Fahrenheit satisfies $T(t) = 70 - 30e^{-0.40t}$, where t is in hours.

(a) Use the model to determine what happens to the temperature of the milk in the long-term. Does your experience confirm your answer?

(b) Find the time it takes for the milk to reach 50°F.

65. Consider $f(t) = L - ae^{-kt}$, where L, a, and k are positive constants. Establish the following. **(a)** $f'(t) > 0$ for $t \geq 0$ **(b)** $\lim\limits_{t \to \infty} f(t) = L$ **(c)** $f''(t) < 0$, $t \geq 0$. Now sketch a graph.

66. Biology. Ali and Gaylor [110] showed that the development rate R of the larvae of the beet armyworm was approximated by $R =$

$$\frac{0.00112te^{24.86-7413.46/t}}{1 + e^{-106.48+30686.56/t} + e^{75.67-23520.93/t}},$$

where t is the temperature in degrees Celsius. Use your graphing calculator to determine the temperature at which the rate R is maximum.

67. Biology. Flamm and colleagues [111] encountered the function $y = 1.27(1 - e^{-6.77x^{0.44}})$ when studying the Southern pine bark beetle. Use calculus to sketch a graph.

SOLUTIONS TO SELF-HELP EXERCISE SET 4.4

1. Use the checklist for graphing a function.

1. The domain of f is all real values except for $x = \pm 1$. The function is continuous at all values except for $x = \pm 1$.
2. Since $f(-x) = f(x)$, the curve is symmetrical about the y-axis.
3. Since

$$f'(x) = \frac{2x(x^2 - 1) - x^2(2x)}{(x^2 - 1)^2} = \frac{-2x}{(x^2 - 1)^2}$$

the only critical value is $x = 0$.

4. Since the denominator in the previous expression for $f'(x)$ is always positive (except for $x = \pm 1$ where f' is not defined), the sign of f', where it exists, is the same as the sign of $-x$. Thus, on the two intervals $(-\infty, -1)$ and $(-1, 0)$, $f'(x) > 0$ and f is increasing. On the two intervals $(0, 1)$ and $(1, \infty)$, $f'(x) < 0$ and f is decreasing.

Critical value		0	
Discontinuity	-1		1
f'	$+$	$+$ $-$	$-$
f	↗	↗ ↘	↘

5. Since f' is positive to the left of 0 and negative to the right, the first derivative test indicates that $f(0) = 0$ is a relative maximum.
6. We have

$$f''(x) = \frac{-2(x^2 - 1)^2 + (2x)2(x^2 - 1)2x}{(x - 1)^4} = \frac{6x^2 + 2}{(x^2 - 1)^3}$$

The numerator of this expression is always positive. Thus, the sign of f'' is the same as the sign of $(x^2 - 1)$. Thus, $f''(x)$ is positive on the two intervals $(-\infty, -1)$ and d1, ∞) and negative on the interval $(-1, 1)$. Thus, the graph of f is concave up on the first two of these intervals and concave down on the last one.

7. There are no points of inflection since $x = +1$ and -1 are not in the domain of definition of f.
8. We have $\lim_{x \to -1^-} f(x) = \infty$ and $\lim_{x \to -1^+} f(x) = -\infty$, whereas $\lim_{x \to 1^-} f(x) = -\infty$ and $\lim_{x \to 1^+} f(x) = \infty$. Certainly $x = \pm 1$ are two vertical asymptotes.
9. Notice that

$$\lim_{x \to \pm\infty} \frac{x^2}{x^2 - 1} = \lim_{x \to \pm\infty} \frac{1}{1 - \frac{1}{x^2}} = \frac{1}{1} = 1$$

Thus, $y = 1$ is a horizontal asymptote.

10. Setting $x = 0$ gives $y = 0$. Setting $y = 0$ gives $x = 0$. Thus, the curve crosses the axes at the point $(0, 0)$.
11. $f(0) = 0$

Putting all this information together gives the following figure.

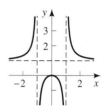

2. Since eventually the entire industry uses the innovation, we must have $L = 1$. Also we have $P_0 = 0.10$. Thus,

$$a = \frac{L}{P_0} - 1 = \frac{L}{0.1L} - 1 = 9$$

Thus,

$$P(t) = \frac{L}{1 + ae^{-kt}} = \frac{1}{1 + 9e^{-kt}}$$

Also $P(2) = 0.20$, and this gives

$$0.20 = P(2) = \frac{1}{1 + 9e^{-2k}}$$

$$1 + 9e^{-2k} = 5$$

$$\frac{4}{9} = e^{-2k}$$

$$\ln \frac{4}{9} = \ln e^{-2k} = -2k$$

$$k = \frac{1}{2} \ln \frac{9}{4} \approx 0.4055$$

Thus,

$$P(t) = \frac{1}{1 + 9e^{-0.4055t}}$$

4.5 OPTIMIZATION

■ Absolute Extrema ■ Optimization ■ Inventory Cost Model

APPLICATION: MINIMIZING REVENUE

A theater owner charges $5 per ticket and sells 250 tickets. By checking other theaters, the owner decides that for every dollar raise in the ticket price, she will lose 10 customers. What should she charge to maximize revenue? See Example 3 for the answer.

APPLICATION: MINIMIZING TOTAL COST

Suppose the start-up costs of each production run for a manufactured item is $50, and suppose it costs $5 to manufacture each item and $0.50 to store each item for one year. Determine the number of items in each run and the number of runs that minimize the total cost if the total number of items to be produced and sold is 5000. See Example 5 for the answer.

Absolute Extrema

In previous sections we studied relative extrema and gave the first and second derivative tests to determine them. In this section we study absolute extrema and find conditions that ensure that a continuous function attains its absolute extrema on a given interval.

Notice that the function in Figure 4.49 attains its smallest value at the left endpoint and its largest value at x_M.

We refer to the smallest value as the **absolute minimum** and to the largest value as the **absolute maximum**. An **absolute extremum** is either an absolute minimum or an absolute maximum. We have the following definitions:

Definition of Absolute Maximum and Minimum

Suppose $f(x)$ is defined on some interval I.

1. We say that $f(x_M)$ is an absolute maximum on the interval I if $x_M \in I$ and

$$f(x) \leq f(x_M) \quad \text{for all } x \in I$$

2. We say that $f(x_m)$ is an absolute minimum on the interval I if $x_m \in I$ and

$$f(x_m) \leq f(x) \quad \text{for all } x \in I$$

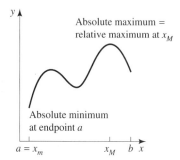

Figure 4.49
The absolute maximum occurs at the critical value x_M, but the absolute minimum occurs at an endpoint.

Continuous functions on a closed and bounded interval have the following very important property:

Extreme Value Theorem

If $f(x)$ is continuous on the closed and bounded interval $[a, b]$, then f attains both its absolute maximum and its absolute minimum on $[a, b]$.

From Figure 4.49, we can see that an absolute extremum for a function can be a relative extremum or *one of the endpoints*. There can be no other possibilities. If $f'(x_0) > 0$ or $f'(x_0) < 0$ and $x_0 \in (a, b)$, then we know from the previous sections that $f(x)$ is increasing in the first case and decreasing in the second. Thus, in neither case can $f(x)$ attain an absolute extremum. The only values that are left are the critical values and the endpoints.

The procedure for finding the absolute extrema on a closed and bounded interval can then be given as follows:

To Find the Absolute Extrema of $f(x)$ on $[a, b]$

1. Locate all critical values.
2. Evaluate $f(x)$ at all the critical values and also at the two values a and b.
3. The absolute maximum of $f(x)$ on $[a, b]$ is the largest number found in Step 2, whereas the absolute minimum of $f(x)$ on $[a, b]$ is the smallest number found in Step 2.

EXAMPLE 1 Finding the Absolute Extrema

Find the absolute extrema of $f(x) = 3x^4 + 4x^3 - 12x^2 + 1$ on $[-1, 1]$ if they exist. On $[-1, 2]$.

SOLUTION ■ Since the function is continuous on each of the intervals, f attains both an absolute minimum and an absolute maximum on each of the intervals.
Since

$$f'(x) = 12x^3 + 12x^2 - 24x = 12x(x^2 + x - 2) = 12x(x - 1)(x + 2)$$

the function is differentiable everywhere, and the critical values are -2, 0, and 1, with only the latter two on the given intervals.

(a) The absolute maximum and the absolute minimum must be among the list of numbers, $f(-1)$, $f(0)$, $f(1)$.

$$f(-1) = -12 \quad \leftarrow \text{absolute minimum}$$
$$f(0) = 1 \quad \leftarrow \text{absolute maximum}$$
$$f(1) = -4$$

Refer to Figure 4.50.

(b) The absolute maximum and the absolute minimum must be among the list of numbers, $f(-1)$, $f(0)$, $f(1)$, $f(2)$.

$$f(-1) = -12 \quad \leftarrow \text{absolute minimum}$$
$$f(0) = 1$$
$$f(1) = -4$$
$$f(2) = 33 \quad \leftarrow \text{absolute maximum}$$

Refer to Figure 4.51. ■

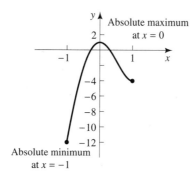

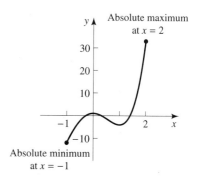

Figure 4.50
The absolute maximum occurs at the critical value $x = 0$, but the absolute minimum occurs at the endpoint $x = -1$.

Figure 4.51
The absolute maximum and absolute minimum occur at endpoints.

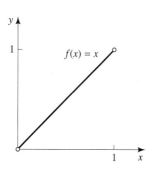

Figure 4.52
A continuous function on an open interval $(0,1)$.

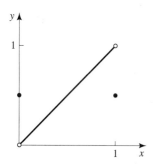

Figure 4.53
A function defined on a closed interval, but which is not continuous there.

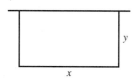

Figure 4.54

EXPLORATION *1*

Extreme Value Theorem

 (a) Consider $f(x) = 1/x$ on $(0, \infty)$. Is $f(x)$ continuous on this interval? Does $f(x)$ have an absolute maximum on this interval? Does $f(x)$ have an absolute minimum on this interval? What does this say about a possible extreme value theorem for continuous functions on an unbounded interval?

(b) Consider the function shown in Figure 4.52. Is $f(x)$ continuous on this interval? Does $f(x)$ have an absolute maximum on this interval? Does $f(x)$ have an absolute minimum on this interval? What does this say about a possible extreme value theorem for continuous functions on an open interval?

(c) Consider the function shown in Figure 4.53. Is $f(x)$ continuous on this interval? Does $f(x)$ have an absolute maximum on this interval? Does $f(x)$ have an absolute minimum on this interval? What does this say about a possible extreme value theorem for functions not necessarily continuous on a closed and bounded interval?

Optimization

EXAMPLE 2 Construction of Fencing

A farmer has 500 yards of fencing with which to fence in three sides of a rectangular pasture. A straight river forms the fourth side. Find the dimensions of the pasture of largest area that the farmer can fence.

SOLUTION ■ The situation is shown in Figure 4.54. Since we do not know either of the dimensions of the rectangle, we label them x and y as shown in the figure. It is these two dimensions that we must determine, where we take the dimensions in yards.

The area A of the pasture is then just $A = xy$. It is A that we wish to maximize. We cannot maximize A in this form. We need to have A as a function of *one* variable not two.

But the two variables x and y are *not* both independent. Rather these two variables are connected by the fact that, on the one hand, we are given that the total amount of fencing is 500 yards, and on the other hand, Figure 4.54 indicates that the total amount of fencing is $x + 2y$. Thus,

$$500 = x + 2y$$

We can now solve this equation for one of the variables, say y, obtaining

$$y = 250 - \frac{1}{2}x$$

Now that we have y in terms of x, we can substitute this value of y in the formula for area, $A = xy$, and obtain

$$A = xy = x\left(250 - \frac{1}{2}x\right) = 250x - \frac{1}{2}x^2$$

Finally, this gives us A as a function of one variable x.

Before proceeding as in the previous sections, we should first determine the domain of the function $A(x)$. Naturally, we must have $x \geq 0$. On the other hand, we must have $x \leq 500$, since $y \geq 0$ and $500 = x + 2y$.

Let us now summarize what we have. We wish to find the maximum value of the *continuously differentiable* function $A = A(x)$ given earlier on the interval $[0, 500]$. This is now just like problems we have worked in previous sections.

We then differentiate, obtaining

$$A'(x) = 250 - x$$

It follows that $A'(x) > 0$ and $A(x)$ is increasing on $(0, 250)$, while $A'(x) < 0$ and $A(x)$ is decreasing on $(250, 500)$. Thus, $A(x)$ attains its absolute maximum at $x = 250$ (Fig. 4.55).

Since $y = 250 - \frac{1}{2}x$, setting $x = 250$ yields

$$y = 250 - \frac{1}{2}(250) = 125$$

Thus, $x = 250$ and $y = 125$.

The maximum area is

$$A = xy = 250 \times 125 = 31{,}250 \quad \blacksquare$$

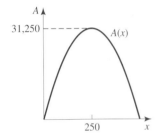

Figure 4.55
The absolute maximum of 31,250 occurs at $x = 250$.

It is not possible to give an explicit procedure for solving the problems we encounter in this section so that the solutions are automatic. The following six-step procedure should be helpful, however:

Six-Step Procedure for Solving Word Problems

1. Identify the unknowns, and denote each with a symbol.
2. Identify the variable that is to be maximized or minimized, and express it as a function of the other variables.
3. Find any relationships between the variables.
4. Express the quantity to be maximized or minimized as a function of only one variable. To do this you may need to solve for one of the variables in one of the equations in Step 3 and substitute this into the formula found in Step 2.
5. Determine the interval over which the function is to be minimized or maximized.
6. Steps 1 through 5 leave you with a function of a single variable on a certain interval to be maximized or minimized. Proceed now as in previous sections.

The six-step procedure applies to Example 2 as follows.

1. We identified the two sides of the rectangle and the area as unknowns.
2. We identified A as the variable to be maximized and wrote $A = xy$.
3. We noted that $500 = x + 2y$.

4. We solved for y in the formula $500 = x + 2y$ and substituted the result into the formula for $A = xy$ to obtain A as a function of x only.
5. We noted that x must lie between 0 and 500.
6. We noted that $A(x) = 250x - \frac{1}{2}x^2$ was to be maximized on the interval $[0, 500]$.

EXAMPLE 3   Maximizing Revenue

A theater owner charges \$5 per ticket and sells 250 tickets. By checking other theaters, she decides that for every dollar raise in ticket price, she will lose 10 customers. What should she charge to maximize revenue?

SOLUTION ■ If p is the price of a ticket, n is the number of tickets sold, and R is the revenue, the formula for revenue is given by

$$\text{Revenue} = (\text{price per ticket}) \times (\text{number of tickets sold})$$

or $$R = pn$$

Thus, her current revenue is

$$5 \times 250 = 1250$$

If she raises the price by \$1, the price of a ticket is $p = 5 + 1$, the number of tickets sold is $n = 250 - 10$, and the revenue is

$$R = pn = (5 + 1)(250 - 10) = 1440$$

If she raises the price by \$2, the price of a ticket is $p = 5 + 2$, the number of tickets sold is $n = 250 - 20$, and the revenue is

$$R = pn = (5 + 2)(250 - 20) = 1610$$

If she raises the price by x dollars, the price of a ticket is $p = 5 + x$, the number of tickets sold is $n = 250 - 10x$, and the revenue is

$$R(x) = pn = (5 + x)(250 - 10x) = 1250 + 200x - 10x^2$$

This function is everywhere continuously differentiable. Since the owner clearly wishes to have the price of each ticket *positive*, we must have $5 + x = p > 0$ or $x > -5$. Also the owner clearly wants to sell at least one ticket, that is, $250 - 10x = n > 0$ or $x < 25$.

We thus wish to maximize the function R given previously on the interval $(-5, 25)$.

Since

$$R'(x) = 200 - 20x = 20(10 - x)$$

there is one critical value at $x = 10$. Also $R'(x) > 0$ on $(-5, 10)$ and $R'(x) < 0$ on $(10, 25)$. Thus, R attains an absolute maximum at $x = 10$ (Fig. 4.56). It is thus clear that the revenue R is maximum when $x = 10$. This means that the price of each ticket should be \$5 + \$10 = \$15.

The maximum revenue is

$$R(10) = (5 + 10)(250 - 10 \times 10) = 2250 \quad ■$$

EXAMPLE 4   Minimizing the Cost of Materials

A manufacturer must produce a sturdy rectangular container with a square base and a volume of 128 cubic feet. The cost of materials making up the top and four sides is \$2 per square foot, and the cost of the materials making up the bottom (which must be reinforced) is \$6 per square foot. Find the dimensions of the box that minimizes the cost of the materials. Use your graphing calculator to estimate the minimum. Confirm analytically.

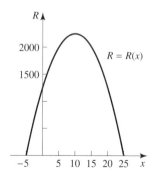

Figure 4.56
The absolute maximum occurs at $x = 10$.

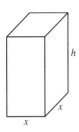

Figure 4.57

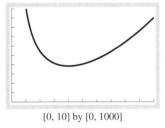

[0, 10] by [0, 1000]

Screen 4.9
A minimum appears to occur at about $x = 4$ for the function $y = 8x^2 + \dfrac{1024}{x}$.

S O L U T I O N ■ In Figure 4.57, each side of the square base is labeled x, and the height is labeled h. We wish to minimize the total cost. The total cost is the cost of the bottom, plus the cost of the top, plus the cost of the four sides.

The area of the bottom is x^2, so the cost of the bottom is $6(x^2) = 6x^2$.

The area of the top is x^2, so the cost of the top is $2(x^2) = 2x^2$. The area of the four sides is $4xh$, so the cost of the four sides is $2(4xh) = 8xh$. Thus, the total cost $C(x)$ is given by

$$C(x) = 6x^2 + 2x^2 + 8xh = 8x^2 + 8xh$$

The volume must be 128, and by Figure 4.57 the volume is also x^2h. Thus,

$$128 = V = x^2h$$

Solving this for h gives

$$h = \frac{128}{x^2}$$

We substitute this into the formula for $C(x)$ and obtain

$$C(x) = 8x^2 + 8x\left(\frac{128}{x^2}\right) = 8x^2 + 1024x^{-1}$$

Now we have C as a function of one variable x, and we wish to find the absolute minimum of C on the interval $(0, \infty)$. A graph is shown in Screen 4.9 on a window of dimension [0, 10] by [0, 1000]. The minimum seems to occur at about $x = 4$.

To confirm analytically, differentiate $C(x)$ and obtain

$$C'(x) = 16x - 1024\,\frac{1}{x^2} = \frac{16}{x^2}\,(x^3 - 64)$$

Setting this equal to zero yields

$$x^3 = 64$$

This gives $x = 4$ as the only critical value. From the preceding factorization of $C'(x)$ we see that $C'(x) < 0$ on $(0, 4)$ and that $C'(x) > 0$ on $(4, \infty)$. Thus, C is decreasing on $(0, 4)$ and increasing on $(4, \infty)$. It is then clear that $C(4)$ is the absolute minimum. This confirms the graph shown in Screen 4.9.

When $x = 4$ feet

$$h = \frac{128}{4^2} = 8 \text{ feet}$$

The minimum cost is

$$\begin{aligned} C &= 8(x^2 + xh) \\ &= 8(4^2 + 4 \times 8) \\ &= \$384 \quad \blacksquare \end{aligned}$$

EXPLORATION 2

Graphing a Sum

The function $y = C(x)$ in Example 4 is the sum of two simpler functions:

$f(x) = 8x^2$ and $g(x) = \dfrac{1024}{x}$. Sketch a rough graph (without using your

graphing calculator) of $y = f(x)$ and $y = g(x)$ on the same graph. The graphs of these two functions are probably familiar. Now sketch a graph of the sum $y = f(x) + g(x) = C(x)$. Notice that since $g(x)$ is extremely large for small x,

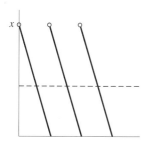

Figure 4.58
Number of items in storage.

the sum must be extremely large. Since $f(x)$ is extremely large for large x, the sum must be extremely large. The minimum must occur at some intermediate value. Confirm all of this on your graphing calculator.

Inventory Cost Model

Suppose a manufacturer intends to produce and sell a large number of items during the next year. All of the items can be manufactured at the start of the year, which would certainly hold down the manufacturing costs (because of the economies of mass production). Doing so would, however, result in a large cost of storing the large number of items until they can be sold. If, on the other hand, a large number of production runs were made and spread out over the year, there would be substantially fewer unsold items requiring storage. Thus, storage costs would be much smaller. Because of the lack of economies inherent in small production runs, however, the manufacturing costs will be large. Presumably there is some middle ground that minimizes the total costs.

We make the simplifying assumption that sales are made at a steady rate throughout the year and the production run occurs when the number of unsold items has been reduced to zero. We assume that each run produces the same number of items, designated by x. Figure 4.58 shows the number of items in storage. From this it is apparent that on average $\frac{1}{2}x$ items are in storage at any time. If s is the cost of storing an item for one year, then the cost of storage is given by the cost of storing one item for one year times the average number in storage or

$$\text{Cost of storage} = s\left(\frac{1}{2}x\right)$$

EXAMPLE 5   Minimizing Total Cost

Suppose the start-up cost of each production run is \$50, it costs \$5 to manufacture each item, and it costs \$0.50 to store each item for one year. Determine the number of items in each run and the number of runs needed to minimize the total cost if the total number of items to be produced and sold is 5000.

SOLUTION ■ As we saw earlier,

$$\text{Cost of storage} = s\left(\frac{1}{2}x\right) = 0.50\left(\frac{1}{2}x\right) = \frac{1}{4}x$$

where x is the number of items produced in each run.

Let the number of runs be given by n. Then the total number of items produced must be given by the number of runs times the number of items in each run or

$$5000 = nx$$

or

$$n = \frac{5000}{x}$$

The total cost of preparing the plant for n runs is the cost of preparing the plant for one run times the number of runs or

$$\text{Total preparation cost} = 50n = 50\frac{5000}{x} = 250{,}000\frac{1}{x}$$

The total cost of manufacturing the 5000 items is given by 5000 times the cost of each item, or

$$\text{Total manufacturing cost} = 5000(5) = 25{,}000$$

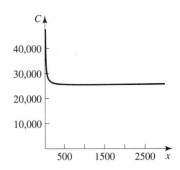

Figure 4.59
An absolute minimum occurs at $x = 1000$. (The function increases slowly after $x = 1000$.)

The total cost $C(x)$ is the sum of the previous three costs, or

$$C(x) = 25,000 + \frac{1}{4}x + 250,000\frac{1}{x}$$

We then wish to find the absolute minimum of this function on the interval [1, 5000]. The function is continuous on this interval.

Differentiating, we obtain

$$C'(x) = \frac{1}{4} - 250,000\frac{1}{x^2} = \frac{1}{4x^2}(x^2 - 1,000,000)$$

This derivative is continuous on [1, 5000]. Setting $C'(x) = 0$, we obtain

$$x^2 = 1,000,000$$

This gives $x = 1000$ as the only critical value on [1, 5000].

From the factorization of $C'(x)$, we see that $C'(x) < 0$ on (0, 1000) and $C'(x) > 0$ on (1000, ∞). Thus C is decreasing on (1, 1000) and increasing on (1000, 5000) (Fig. 4.59). From this it is clear that $C(1000)$ is the absolute minimum of C on [1, 5000].

The number of runs is

$$n = \frac{5000}{x} = \frac{5000}{1000} = 5$$

The minimum cost is

$$C(1000) = 25,000 + \frac{1}{4}(1000) + \frac{250,000}{1000}$$

$$= 25,500 \qquad \blacksquare$$

EXPLORATION 3

Graphing a Sum

The function $y = C(x)$ in Example 5 is the sum of two simpler functions: $f(x) = 25,000 + 0.25x$ and $g(x) = \dfrac{250,000}{x}$. Sketch a rough graph (without using your graphing calculator) of $y = f(x)$ and $y = g(x)$ on the same graph. The graphs of these two functions are probably familiar. Now sketch a graph of the sum $y = f(x) + g(x) = C(x)$. Notice that since $g(x)$ is extremely large for small x, the sum must be extremely large. Since $f(x)$ is extremely large for large x, the sum must be extremely large. The minimum must occur at some intermediate value. Notice how a similar situation occurred in Exploration 2.

SELF-HELP EXERCISE SET 4.5

1. Find the absolute extrema of $f(x) = x^3 - 12x + 1$ on $[-3, 5]$.

2. A package mailed in this country must have length plus girth of no more than 108 inches. Find the dimensions of a rectangular package with square base of greatest volume that can be mailed. (The girth is the length of the perimeter of a rectangular cross section.)

3. The figure shows a lake. A medical team is located at point A and must get to an injured person at point D as quickly as possible. The team can row a boat directly to point D, can row the boat directly across the lake to point B and then run along the shore to D, or can row to some intermediate point C and then run down the shore to D. If the team can row at 6 mph and run at 10 mph, where on the opposite shore should they land the boat?

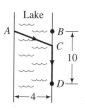

4. A farmer wishes to enclose a rectangular field of area 200 square feet using an existing wall as one of the sides. The cost of the fence for the other three sides is $1 per foot. Find the dimensions of the rectangular field that minimizes the cost of the fence.

EXERCISE SET 4.5

In Exercises 1 through 6 you find various graphs. In each case find all values for which the function attains an absolute maximum. An absolute minimum.

1.

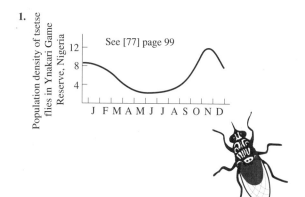

2.

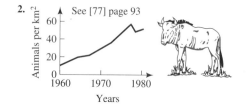

3.

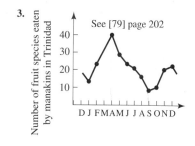

4.

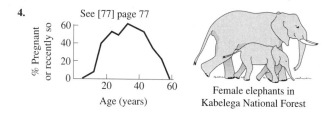

Female elephants in Kabelega National Forest

5.

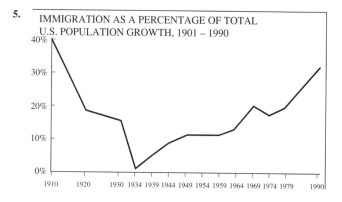

IMMIGRATION AS A PERCENTAGE OF TOTAL U.S. POPULATION GROWTH, 1901–1990

6.

Worms hatched at same time

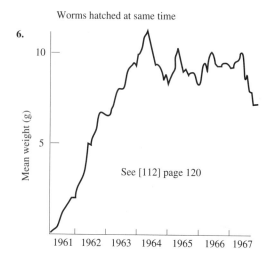

In Exercises 7 and 8 let $f(x) = x^3 - 3x + 1$. Locate the value(s) at which f attains an absolute maximum and the value(s) at which f attains an absolute minimum on each interval.

7. (a) $[2, 4]$ **(b)** $[0, 2]$ **(c)** $[-2, 2]$

8. (a) $[-1, 1]$ **(b)** $[-2, 0]$ **(c)** $[-3, 2]$

In Exercises 9 and 10 let $f(x) = -2x^3 + 3x^2$. Locate the value(s) at which f attains an absolute maximum and the value(s) at which f attains an absolute minimum on each interval.

9. (a) $[-2, 0]$ **(b)** $[0, 2]$ **(c)** $[-2, 2]$

10. (a) $[0, 1]$ **(b)** $[-1, 1]$ **(c)** $[-2, 1]$

In Exercises 11 through 28 locate the value(s) at which each function attains an absolute maximum and the value(s) at which the function attains an absolute minimum, if they exist, of the given function on the given interval.

11. $f(x) = x^2 - 4x + 1$ on $[-1, 1]$

12. $f(x) = x^2 - 2x + 1$ on $[2, 4]$

13. $f(x) = -x^2 + 4x - 2$ on $[0, 3]$

14. $f(x) = x^2 + 2x + 1$ on $[-3, 0]$

15. $f(x) = x^3 - 3x + 1$ on $[-1, 3]$

16. $f(x) = x^3 - 3x^2 + 2$ on $[0, 4]$

17. $f(x) = x^3 + 3x^2 - 2$ on $[-3, 2]$

18. $f(x) = -8x^3 + 6x - 1$ on $[-1, 1]$

19. $f(x) = x^4 - 8x^2 + 3$ on $[-3, 3]$

20. $f(x) = -x^4 + 8x^2 + 1$ on $[-1, 3]$

21. $f(x) = x^2 + 16x^{-2}$ on $[1, 8]$

22. $f(x) = x^2 - 16x^{-2}$ on $[1, 8]$

23. $f(x) = 3 - x + x^2$ on $(-\infty, \infty)$

24. $f(x) = 4 - 2x - x^2$ on $(-\infty, \infty)$

25. $f(x) = x^4 + 6x^2 - 2$ on $(-\infty, \infty)$

26. $f(x) = 3 - x^2 - x^4$ on $(-\infty, \infty)$

27. $f(x) = \sqrt{x^2 + 1}$ on $(-\infty, \infty)$

28. $f(x) = (x - 3)^4$ on $(-\infty, \infty)$

In Exercises 29 and 30 let $f(x) = 3x + x^{-3}$. On each interval locate the value(s) (if they exist) at which f attains an absolute extremum.

29. (a) $[0.5, 2]$ **(b)** $[-1, 3]$
    (c) $(0, 3]$ **(d)** $(-\infty, 0)$

30. (a) $[-2, -0.5]$ **(b)** $[-2, 3]$
    (c) $[-3, 0)$ **(d)** $(0, \infty)$

31. Find two numbers whose sum is 16 and whose product is maximum.

32. Find two numbers whose sum is 24 and whose product is maximum.

33. Find two numbers whose sum is 20 and for which the sum of the squares is a minimum.

34. Find two numbers whose sum is 32 and for which the sum of the squares is a minimum.

35. Find two nonnegative numbers x and y with $2x + y = 30$ for which the term xy^2 is maximized.

36. Find two nonnegative numbers x and y with $x + y = 60$ for which the term x^2y is maximized.

37. Find two numbers x and y with $x - y = 20$ for which the term xy is minimized.

38. Find two numbers x and y with $x - y = 36$ for which the term xy is minimized.

39. What is the area of the largest rectangle that can be enclosed by a circle of radius a?

40. What is the area of the largest rectangle that can be enclosed by a semicircle of radius a?

41. Find the coordinates of the point P that maximizes the area of the rectangle shown within the figure.

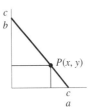

42. A triangle has a base on the x-axis, one side on the y-axis, and the third side goes through the point $(1, 2)$ as shown in the figure. Find the slope of the line through $(1, 2)$ if the area of the triangle is to be a minimum.

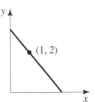

Applications

43. Fencing. Where a river makes a right-angle turn, a farmer with 400 feet of fencing wishes to construct a rectangular fenced-in pasture that uses the river for two sides. What dimensions of the rectangle maximize the area enclosed?

44. Fencing. A farmer with 400 feet of fencing wishes to construct a rectangular pasture with maximum area. What should the dimensions be?

45. Packaging. From a 9-inch × 9-inch piece of cardboard, square corners are cut out so that the sides can be folded up to form a box with no top. What should x be to maximize the volume?

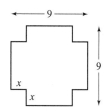

46. Packaging. Find the dimensions of a cylindrical package with circular base of largest volume that can be mailed in this country. Recall that length plus girth cannot be more than 108 in.

47. Cost. A fence is to be built around a 200-square foot rectangular field. Three sides are to be made of wood costing $10 per foot, whereas the other side is made of stone costing $30 per foot. Find the dimensions of the enclosure that minimizes total cost.

48. Cost. A farmer wishes to enclose a rectangular field of area 450 square feet using an existing wall as one of the sides. The cost of the fence for the other three sides is $3 per foot. Find the dimensions of the rectangular field that minimizes the cost of the fence.

49. Cost. What are the dimensions of the rectangular field of 20,000 square feet that minimizes the cost of fencing if one side costs three times as much per unit length as the other three?

50. Fencing. A rectangular corral of 32 square yards is to be fenced off and then divided by a stone fence into two sections. The outer fence costs $10 per yard, and the special dividing stone fence costs $20 per yard. Find the dimensions of the corral that minimizes the cost.

51. Revenue. A bus company charges $10 per person for a sightseeing trip if 30 people travel in a group. If for each person above 30 the company reduces the charge per person by $0.20, how many people are needed to maximize the total revenue for the bus company?

52. Revenue. A hotel has 10 luxury units and during the peak season rents all out at $300 per day. From experience, management knows that one unit becomes vacant for each $50 increase in charge per day. What rent should be charged to maximize revenue?

53. Profit. Suppose in the previous problem, the daily cost of maintaining and cleaning a rented luxury room was $100 per day. What rent should be charged to maximize profit?

54. Profit. An apple orchard produces annual revenue of $50 per tree when planted with 1000 trees. Due to overcrowding, the annual revenue per tree is reduced by 2 cents for each additional tree planted. If the cost of maintaining each tree is $10 per year, how many trees should be planted to maximize total profit from the orchard?

55. Packaging. A company wishes to design a rectangular box with square base and no top that has a volume of 32 cubic inches. What should the dimensions be to yield a minimum surface area? What is the minimum surface area?

56. Packaging. A company wishes to design a rectangular box with square base and no top that has a volume of 27 cubic inches. The cost of the bottom is twice the cost of the sides. What should the dimensions be to minimize the cost? What is the minimum cost?

57. Cost. A closed rectangular box of volume 324 cubic inches is to be made with a square base. If the material for the bottom costs twice as much as the material for the sides and top, find the dimensions of the box that minimize the cost of materials.

58. Construction. Find the most economical proportions for a closed cylindrical can that holds 16 cubic inches.

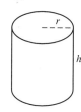

59. Inventory Control. Suppose the start-up cost of each production run is $1000, and it costs $20 to manufacture each item and $2 to store each item for one year. Determine the number of items in each run and the number of runs needed to minimize total cost if the total number of items to be produced and sold is 16,000.

60. Inventory Control. Suppose the start-up costs of each production run is $2500 and that it costs $20 to manufacture each item and $2 to store each item for one year. Determine the number of items in each run and the number of runs needed to minimize total cost if the total number of items to be produced and sold is 10,000.

Enrichment Exercises

61. Construction. Pipe is to be laid connecting A with B in the figure. The cost along the level stretch from A to C is $10 per foot, whereas the cost along the difficult (away from the road) stretch from C to B is $20 per foot. The distance from A to D is 40 feet, and the distance from B to D is 30 feet. At what point C is the cost of laying pipe from A to B to C minimum? Use your graphing calculator to find an approximate solution. Confirm analytically.

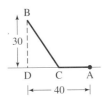

62. Construction. A wire 12 inches long is cut into two pieces with one piece used to construct a square and the other piece used to construct a circle. Where should the wire be cut to minimize the sum of the areas enclosed by the two figures? Use your graphing calculator to find an approximate solution. Confirm analytically.

63. Construction. A wire 12 inches long is cut into two pieces with one piece used to construct a square and the other used to construct an equilateral triangle. Where should the wire be cut to minimize the sum of the areas enclosed by the two figures? (The area of an equilateral triangle is $\frac{\sqrt{3}}{4} r^2$,

where r is the length of a side.) Use your graphing calculator to find an approximate solution. Confirm analytically.

64. Construction. An individual is planning to construct two fenced enclosures, one square and one circular. The cost per yard of constructing the circular enclosure is $8 and of the square enclosure, $4. If the total cost is fixed at $1600, what should be the dimensions of the two figures in order to minimize the area enclosed by the two figures? Use your graphing calculator to find an approximate solution. Confirm analytically.

65. Construction. An individual is planning to construct two fenced enclosures, one square and one an equilateral triangle. The cost per yard of constructing the triangular enclosure is $6 and of the square enclosure is $3. If the total cost is fixed at $1000, what should be the length of each side of the triangular enclosure in order to minimize the area enclosed by the two figures? (The area of an equilateral triangle is $\frac{\sqrt{3}}{4} r^2$, where r is the length of a side.) Use your graphing calculator to find an approximate solution. Confirm analytically.

66. Inventory Control. In the discussion on inventory control in the text, let T be the total number of items to be produced and sold, k is the amount of money needed to prepare each production run, and s the cost of storing one item for an entire year. Show that the number x of items in each run needed to minimize the total cost is $x = \sqrt{2kT/s}$. Notice that the answer does not depend on the cost p of producing each item. (If the total cost is $C(x)$, show that $C''(x) = 2kTx^{-3}$.)

SOLUTIONS TO SELF-HELP EXERCISE SET 4.5

1. We have

$$f'(x) = 3x^2 - 12 = 3(x^2 - 4) = 3(x + 2)(x - 2)$$

Thus, f' is continuous everywhere on the interval $[-3, 5]$, and the extreme value theorem ensures that the function attains both an absolute minimum and an absolute maximum on this interval. Thus, the extrema must be attained at the endpoints, -3 or 5, or the critical values -2 or $+2$. Evaluating f at these four values yields

$$f(-3) = 10$$

$$f(-2) = 17$$

$$f(2) = -15 \quad \leftarrow \text{absolute minimum}$$

$$f(5) = 66 \quad \leftarrow \text{absolute maximum}$$

One can do some further work and construct the following graph:

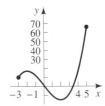

2. See the figure where the length of the package is given as l and each side of the square base is given by s. We wish to maximize the volume V, which is given by $V = s^2 l$. Here V

is a function of *two* variables. We need to make V a function of only *one* variable.

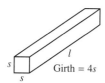

s Girth = 4s

But the two variables s and l are connected by the fact that the length plus the girth can be at most 108 inches. It is clear that, if we are to maximize the volume, then we should take the length plus the girth to be as large as possible, which is 108. Thus, take

$$108 = \text{length} + \text{girth} = l + 4s$$

From this equation we can solve for l and obtain

$$l = 108 - 4s$$

Now substitute this into the formula for V and obtain

$$V = s^2 l = s^2(108 - 4s) = 108s^2 - 4s^3$$

We now have V as a function of one variable.

This function is continuous and differentiable everywhere. Naturally, we must have $s \geq 0$ and also $s \leq 108/4 = 27$ inches.

In summary, we wish to find the absolute maximum of the continuously differentiable function $V = V(s) = 108s^2 - 4s^3$ on the interval $[0, 27]$.

To do this we first differentiate $V(s)$, obtaining

$$V'(s) = 216s - 12s^2 = 12s(18 - s)$$

Since $V'(s) = 0$ implies that $s = 0$ or $s = 18$ inches, the only critical value of $V(s)$ on $(0, 27)$ is at $s = 18$ inches. Furthermore, from the factorization of $V'(s)$, we see that $V'(s) > 0$ on $(0, 18)$ and $V'(s) < 0$ on $(18, 27)$. Thus, V is increasing on $(0, 18)$ and decreasing on $(18, 27)$. From this it is clear that $V(18)$ is an absolute maximum on the interval $[0, 27]$. See the figure. Then

$$l = 108 - 4s = 108 - 4(18) = 36$$

The maximum volume is

$$V = s^2 l = 18^2 \times 36 = 11{,}664 \text{ cubic inches}$$

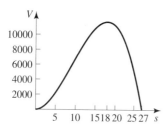

3. Let x be the distance from B to C. Since

$$\text{Time} = \frac{\text{distance}}{\text{rate}}$$

the time spent in the boat is $\dfrac{\sqrt{x^2 + 16}}{6}$, and the time spent running along the shore is $\dfrac{10 - x}{10}$. Thus, the total time of

the trip is given by

$$T(x) = \frac{\sqrt{x^2 + 16}}{6} + \frac{10 - x}{10}$$

We wish to find the minimum of T on the interval $[0, 10]$. We have

$$T'(x) = \frac{x}{6\sqrt{x^2 + 16}} - \frac{1}{10}$$

Then $T'(x) = 0$ if, and only if, $5x = 3\sqrt{x^2 + 16}$, or $25x^2 = 9x^2 + 144$, or $16x^2 = 144$, or $x = 3$. Evaluating $T(x)$ at the critical value $x = 3$ and the endpoints $x = 0$ and $x = 10$ gives

$$T(0) = \frac{5}{3} \approx 1.67 \qquad T(3) = \frac{5}{6} + \frac{7}{10} \approx 1.53,$$

$$T(10) = \frac{\sqrt{116}}{6} \approx 1.80$$

Thus, the minimum occurs at $x = 3$. This means that the medical team should row to a point C located 3 miles below point B and run the remaining 7 miles to the injured person at point D.

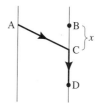

4. Let x and y denote the sides of the rectangular enclosure shown in the figure. Then the area is given by $A = xy = 200$. The cost in dollars is the same as the number of feet of fence. Thus,

$$C(x) = x + 2y = x + 2\frac{200}{x} = x + \frac{400}{x}$$

and

$$C'(x) = 1 - \frac{400}{x^2} = \frac{1}{x^2}(x^2 - 400)$$

Then $C'(x) = 0$ if and only if $x^2 = 400$ or $x = 20$. From the factorization of $C'(x)$ we see that $C'(x) < 0$ on $(0, 20)$ and $C'(x) > 0$ on $(20, \infty)$. Thus, C is decreasing on $(0, 20)$ and increasing on $(20, \infty)$. From this it is clear that $C(20)$ is an absolute minimum.

Solving for y in the equation $xy = 200$ then gives $y = 10$. Thus, the area should be 20 by 10.

4.6 IMPLICIT DIFFERENTIATION AND RELATED RATES

■ Implicit Differentiation ■ Applications ■ Related Rates

APPLICATION: IMPLICITLY DEFINED FUNCTION

The demand function for a certain commodity is given by $p = \dfrac{13}{1 + x^2 + x^3}$.

What is $\dfrac{dx}{dp}$ when $x = 2$? See Example 2 for the answer.

APPLICATION: SPREAD OF AN OIL SLICK

An oil slick is spreading in a circular fashion. The radius of the circular slick is observed to be increasing at 0.1 miles per day. What is the rate of increase of the area of the slick when $r = 2$ miles? See Example 4 for the answer.

Implicit Differentiation

We often have an expression such as $y^6 + y^5 + xy = 0$ in which we are unable to solve for y as an explicit function of x. Nevertheless, we may still need to find $\dfrac{dy}{dx}$ at some specific point $x = x_0$. In this section we shall see how to find $\dfrac{dy}{dx}$ at a point without knowing y as an explicit function of x.

We usually have encountered functions of the form $y = f(x)$, which express y explicitly as a function of x. But some equations we have had to deal with, such as

$$x^2 + y^2 = 1$$

do not give y as an explicit function of x. Still, this last equation can define y as a function of x. If we set x equal to some specific value, then the resulting equation can be solved for one or more values of y. If we specify which value of y to take, we have a function. We then say that y is an **implicit** function of x.

For example, if $x^2 + y^2 = 1$, and we set $x = 0$, then the equation $y^2 = 1$ results, which has solutions $y = -1$ and $y = +1$.

If we are interested in finding the slope to the curve $x^2 + y^2 = 1$ at the point $(1/\sqrt{2}, 1/\sqrt{2})$, we need to solve for y as an *explicit* function of x and then find the derivative. In this case we have

$$y = f(x) = \sqrt{1 - x^2}$$

and

$$f'(x) = -\frac{x}{\sqrt{1 - x^2}}$$

Then

$$f'\left(\frac{1}{\sqrt{2}}\right) = -1$$

(Fig. 4.60).

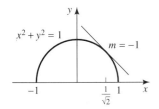

Figure 4.60
The slope of the tangent line at $x = 1/\sqrt{2}$ is -1.

For some equations, however, such as

$$x^4 - 2xy^3 + y^5 = 32$$

is not possible to solve for y explicitly in terms of x. In such cases we still need to know the slope of the curve at a given point. It may seem impossible to ever be able to find $\frac{dy}{dx}$ at a point if we do not know y as a function of x, but the technique of *implicit differentiation* enables us to do precisely this.

Let us now see how we can find the slope of the line tangent to the curve $x^2 + y^2 = 1$ given earlier at the point $(1/\sqrt{2}, 1/\sqrt{2})$ without finding y as an explicit function of x. We first *assume* that it is possible at least theoretically to solve for y explicitly as, say, $y = f(x)$. Then

$$x^2 + [f(x)]^2 = x^2 + y^2 = 1$$

Thus,

$$x^2 + [f(x)]^2 = 1$$

We now differentiate this last equation using the rules we have already learned and obtain

$$\frac{d}{dx} x^2 + \frac{d}{dx} [f(x)]^2 = \frac{d}{dx} 1$$

This gives

$$2x + 2f(x)f'(x) = 0$$

We then solve for $f'(x)$ and obtain

$$\frac{dy}{dx} = f'(x) = -\frac{x}{f(x)} = -\frac{x}{y}$$

At the point under consideration $x = 1/\sqrt{2}$ and also $y = 1/\sqrt{2}$. Using this in the previous equation yields

$$f'\left(\frac{1}{\sqrt{2}}\right) = -\frac{x}{y} = -\frac{1/\sqrt{2}}{1/\sqrt{2}} = -1$$

Notice that we were able to find $f'(1/\sqrt{2})$ without ever knowing what $f(x)$ was explicitly!

EXAMPLE 1   Finding the Derivative of an Implicitly Defined Function

Let $x^4 - 2xy^3 + y^5 = 32$.

(a) Find $\frac{dy}{dx}$.

(b) Find the slope of the line tangent to the curve at $x_0 = 0$.

SOLUTIONS ■

(a) It is extremely difficult, if not impossible, to solve for y *explicitly* as a function of x in this case. We therefore need to differentiate *implicitly*. We assume that $y = f(x)$ for some $f(x)$ and substitute this into the last equation, giving

$$x^4 - 2x[f(x)]^3 + [f(x)]^5 = 32$$

Differentiating, we obtain

$$\frac{d}{dx} x^4 - \frac{d}{dx} 2x[f(x)]^3 + \frac{d}{dx} [f(x)]^5 = \frac{d}{dx} 32$$

Using familiar rules, we obtain

$$4x^3 - 2[f(x)]^3 - (2x)3[f(x)]^2 \, f'(x) + 5[f(x)]^4 \, f'(x) = 0$$

We now solve this equation for $f'(x)$ and obtain

$$(-6x[f(x)]^2 + 5[f(x)]^4) \, f'(x) = 2[f(x)]^3 - 4x^3$$

and finally

$$f'(x) = \frac{2[f(x)]^3 - 4x^3}{-6x[f(x)]^2 + 5[f(x)]^4}$$

or

$$\frac{dy}{dx} = f'(x) = \frac{2y^3 - 4x^3}{-6xy^2 + 5y^4}$$

(b) We wish to find $f'(0)$. But when $x = 0$, the original equation becomes

$$0^4 - 2(0)y^3 + y^5 = 32$$

or $y = 2$. Thus, we substitute $x = 0$ and $y = 2$ in the answer to part (a) and obtain

$$f'(0) = \frac{2(2)^3 - 0}{0 + 5(2)^4} = \frac{1}{5} \quad \blacksquare$$

Remark. In the preceding analysis we *assumed* that we could at least theoretically solve for $y = f(x)$ for some $f(x)$. A theorem called the *implicit function theorem* gives specific conditions under which this assumption is guaranteed. We do not go into these conditions in this text.

Remark. Notice in the previous work that the equation involving $f'(x)$ was *linear* in f' and therefore it was easy to solve for f'. *This is always the case.*

Applications

EXAMPLE 2   Finding the Derivative of an Implicitly Defined Function

The demand function for a certain commodity is given by $p = \dfrac{13}{1 + x^2 + x^3}$. Find the instantaneous rate of change of the number sold with respect to the price when $x = 2$.

SOLUTION ■ In this case it is extremely difficult, if not impossible, to solve for x explicitly as a function of p in order to find $\dfrac{dx}{dp}$. Thus, we try implicit differentiation.

We assume that $x = f(p)$ for some function $f(p)$. Then

$$p = \frac{13}{1 + [f(p)]^2 + [f(p)]^3}$$

Clearing the fraction, we obtain

$$p + p[f(p)]^2 + p[f(p)]^3 = 13$$

Now differentiate this with respect to p and obtain

$$\frac{d}{dp} p + \frac{d}{dp} p[f(p)]^2 + \frac{d}{dp} p[f(p)]^3 = \frac{d}{dp} 13$$

Now use the differentiation techniques from previous sections and obtain

$$1 + [f(p)]^2 + 2pf(p) f'(p) + [f(p)]^3 + 3p[f(p)]^2 f'(p) = 0$$

This is a *linear* equation in f'. Solve for f' and obtain

$$f'(p) = -\frac{1 + [f(p)]^2 + [f(p)]^3}{2pf(p) + 3p[f(p)]^2}$$

or

$$\frac{dx}{dp} = f'(p) = -\frac{1 + x^2 + x^3}{2px + 3px^2}$$

Now when $x = 2$,

$$p = \frac{13}{1 + 2^2 + 2^3} = 1$$

Then

$$f'(1) = -\frac{1 + 2^2 + 2^3}{2 \cdot 1 \cdot 2 + 3 \cdot 1 \cdot 2^2} = -\frac{13}{16} \quad \blacksquare$$

Related Rates

In this sub-section the rate of change of one quantity is related to the rate of change of another quantity by the chain rule.

It is often the case when given a function, such as a cost function $C = C(x)$ that the number of items sold is a function of time. For example, a company may estimate that their sales may increase on a yearly basis at the same rate as the increase in population. Thus, the number sold, x, is actually a function of the time t, or $x = f(t)$.

If now we wish to find the rate of change of costs with respect to time, we are interested in the quantity $\dfrac{dC}{dt}$. But since $C = C(f(t))$, the chain rule then gives

$$\frac{dC}{dt} = \frac{dC}{dx}\frac{dx}{dt}$$

EXAMPLE 3   Finding a Related Rate

Suppose the cost function in thousands of dollars is given by $C(x) = 100 + x^2$, where x is in thousands of units. The company estimates that its rate of change of sales with respect to time will be 0.04. What is the rate of increase of costs with respect to time if the company produces and sells 10,000 items?

SOLUTION ■ Since the rate of change of sales with respect to time is 0.40,

$$\frac{dx}{dt} = 0.04$$

Then

$$\frac{dC}{dt} = \frac{dC}{dx}\frac{dx}{dt} = 2x(0.04)$$

When $x = 10$, this yields

$$\frac{dC}{dt} = 2(10)(0.04) = 0.8$$

That is, costs increase at an annual rate of $800 when 10,000 items are being produced. ■

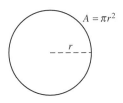

Figure 4.61
Area of a circle of radius r.

EXAMPLE 4   Finding a Related Rate of the Spread of an Oil Slick

An oil slick is spreading in a circular fashion. The radius of the circular slick is observed to be increasing at 0.1 mile per day. What is the rate of increase of the area of the slick when $r = 2$ miles?

SOLUTION ■ If r is the radius of the circular slick in miles, then the area, as shown in Figure 4.61, is given by $A = \pi r^2$. Also $\dfrac{dr}{dt} = 0.1$. Thus,

$$\frac{dA}{dt} = \frac{dA}{dr}\frac{dr}{dt} = 2\pi r(0.1)$$

When $r = 2$, this becomes

$$\frac{dA}{dt} = 2\pi 2(0.1) \approx 1.26$$

Thus, when the radius of the slick is 2 miles, the area of the slick is increasing at a rate of 1.26 square miles per day. ■

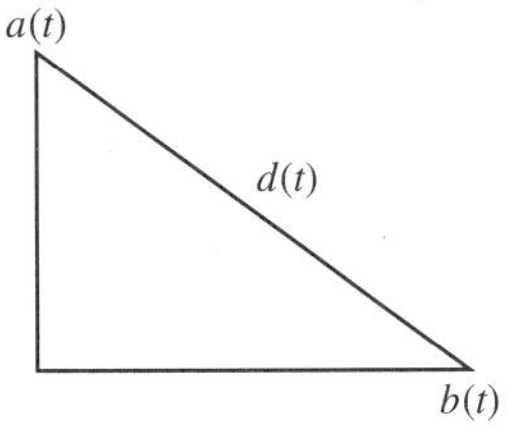

$a(t)$

$d(t)$

$b(t)$

Figure 4.62
The distance between the ships is $d(t)$.

EXAMPLE 5   Change of Distance Between Two Ships

Two ships leave the same port at 1:00 PM. The first ship heads due north and at 2:00 PM (one hour later) is observed to be 3 miles due north of port and going at 6 miles per hour at that instant. The second ship leaves port at the same time as the first ship and heads due east and one hour later is observed to be 4 miles due east of port and going at 7 miles per hour. At what rate is the distance between the two ships changing at 2:00 PM?

SOLUTION ■ Let $a(t)$ and $b(t)$ be the respective distances that the first and second ships are from port at time t after leaving (Fig. 4.62).

If $d(t)$ is the distance between the ships, then

$$d^2(t) = a^2(t) + b^2(t)$$

Differentiating this expression gives

$$2d(t)\,d'(t) = 2a(t)a'(t) + 2b(t)b'(t)$$

Thus,

$$d'(t) = \frac{a(t)a'(t) + b(t)b'(t)}{d(t)}$$

Now we are given that $a(1) = 3$, $b(1) = 4$, $a'(1) = 6$, and $b'(1) = 7$. It follows that

$$d(1) = \sqrt{3^2 + 4^2} = 5$$

Therefore,

$$d'(1) = \frac{3 \cdot 6 + 4 \cdot 7}{5} = 9.2$$

The ships are moving apart at 9.2 miles per hour at 2:00 PM. ■

Notice how we were able to do this problem without knowing explicitly the two functions $a(t)$ and $b(t)$.

1. If $2x^2 \sqrt{y} + y^3 + x^4 = 27$, find $\dfrac{dy}{dx}$.

2. Find the slope of the line tangent to the curve in the previous example at the point $x_0 = 0$.

3. After using an experimental drug, the radius of a spherical tumor is observed to be decreasing at the rate of 0.1 cm per week when the radius is 2 cm. At what rate is the volume of the tumor decreasing?

EXERCISE SET 4.6

In Exercises 1 through 14 differentiate implicitly and find the slope of the curve at the indicated point.

1. $xy + 2x + y = 6$, $(1, 2)$

2. $xy + x + y = -1$, $(-2, -1)$

3. $xy + x + y^2 = 7$, $(1, 2)$

4. $x^2y + x - y^3 = -1$, $(2, -1)$

5. $3x^2 - y^2 = 23$, $(3, 2)$

6. $x^3 - 2y^3 = -15$, $(1, 2)$

7. $x^2y - 2x + y^3 = 1$, $(2, 1)$

8. $\sqrt{xy} + x^2 - y^4 = 17$, $(4, 1)$

9. $x^2 + y^2 + xy^3 = 4$, $(0, 2)$

10. $\sqrt{x} - \sqrt{y} = 1$, $(9, 4)$

11. $x^2y^2 + xy^4 = 2$, $(1, 1)$

12. $xy^2 + xy + y^3 = 5$, $(2, 1)$

13. $x\sqrt{y} - y^2 + x^2 = -13$, $(1, 4)$

14. $\sqrt{xy} + x^3y^2 = 84$, $(1, 9)$

Applications of Implicit Differentiation

In Exercises 15 through 22 the price of a commodity is given as a function of the demand x. Use implicit differentiation to find $\dfrac{dx}{dp}$ for the indicated x.

15. $p = -2x + 15$, $x = 3$

16. $p = -3x + 20$, $x = 5$

17. $p = \dfrac{3}{1 + x}$, $x = 2$

18. $p = \dfrac{5}{1 + x^2}$, $x = 2$

19. $p = \dfrac{15}{3 + x + x^2}$, $x = 3$

20. $p = \dfrac{5}{3 + x + x^3}$, $x = 1$

21. $p = \sqrt{25 - x^2}$, $x = 4$

22. $p = \sqrt{25 - x^4}$, $x = 2$

23. Demand. Suppose the demand x and the price p are related by the equation $16x^2 + 100p^2 = 500$. Find dx/dp at the point where $p = 1$.

24. Demand. Suppose the demand x and the price p are related by the equation $x^2 + 100p^2 = 20{,}000$. Find dp/dx at the point where $x = 100$.

25. Medicine. Poiseuille's law states that the total resistance R to blood flow in a blood vessel of constant length l and radius r is given by $R = \dfrac{al}{r^4}$, where a is a positive constant. Use implicit differentiation to find $\dfrac{dr}{dR}$, where $R = 1$.

26. Medicine. It is well known that when a person coughs, the diameter of the trachea and bronchi shrinks. It has been shown that the flow of air F through the windpipe is given by $F = k(R - r)r^4$, where r is the radius of the windpipe

under the pressure of the air being released from the lungs, R is the radius with no pressure, and k is a positive constant. Suppose $k = 1$ and $R = 2$. Use implicit differentiation to find $\dfrac{dr}{dF}$ when $F = 1$.

27. Biology. Biologists have proposed that the rate of production P of photosynthesis is related to the light intensity I by the formula $P = \dfrac{aI}{b + I^2}$, where a and b are positive constants. Suppose $a = 6$ and $b = 8$. Use implicit differentiation to find $\dfrac{dI}{dP}$ when $P = 1$ and $I = 2$.

28. Biology. The free water vapor diffusion coefficient D in soft woods is given by $D = \dfrac{167.2}{P}\left(\dfrac{T}{273}\right)^{1.75}$, where T is the temperature and P is the constant atmospheric pressure measured in appropriate units. Suppose $P = 1$. Use implicit differentiation to find $\dfrac{dT}{dD}$ when $D = 1$.

Exercises 29 through 40 involve related rates. In these exercises find $\dfrac{dy}{dt}$ given in the indicated information.

29. $y = 3 + 2x$, $\dfrac{dx}{dt} = 3$

30. $y = 3x - 4$, $\dfrac{dx}{dt} = 2$

31. $y = 2 + x^2$, $\dfrac{dx}{dt} = 4$, $x = 2$

32. $y = 3 - 2x^2$, $\dfrac{dx}{dt} = 2$, $x = 3$

33. $y = 3 - 2x^3$, $\dfrac{dx}{dt} = -2$, $x = 1$

34. $y = 1 - x^4$, $\dfrac{dx}{dt} = -1$, $x = 2$

35. $y = \dfrac{1 - x}{1 + x}$, $\dfrac{dx}{dt} = 2$, $x = 1$

36. $y = \dfrac{1 - x}{1 + x^2}$, $\dfrac{dx}{dt} = -2$, $x = -1$

37. $x^2 + y^2 = 5$, $\dfrac{dx}{dt} = 2$, $x = 1$, $y = 2$

38. $x^2 - y^2 = 3$, $\dfrac{dx}{dt} = -2$, $x = 2$, $y = -1$

39. $x^3 + xy + y^2 = 7$, $\dfrac{dx}{dt} = 2$, $x = 1$, $y = 2$

40. $x + x^2y - x^2 = -1$, $\dfrac{dx}{dt} = 3$, $x = -1$, $y = 1$

Applications of Related Rates

41. Demand. If the demand equation is given by $x = 10 - 0.1p$, and the number of items manufactured (and sold) is increasing at the rate of 20 per week, find the rate of change of p with respect to time.

42. Cost. If the cost equation is given by $C(x) = 8 + 0.2x^2$, and the number of items manufactured is increasing at the rate of 20 per week, find the rate of the change of C with respect to time when $x = 4$.

43. Revenue. If the demand equation is given by $x = 10 - 0.1p$, and the number of items manufactured (and sold) is increasing at the rate of 20 per week, find the rate of change of revenue with respect to time when $x = 4$.

44. Profits. If the demand and cost equations are the same as in the two previous problems, find the rate of change of profits with respect to time when $x = 4$.

45. Distance. A ship is observed to be 4 miles due north of port and traveling due north at 5 miles per hour. At the same time another ship is observed to be 3 miles due west of port and traveling due *east* on its way back to port at 4 miles per hour. What is the rate at which the distance between the ships is changing?

46. Distance. Answer the same question as in the previous problem if the first ship is traveling due *south* at 5 miles per hour, everything else remaining the same.

47. Physics. A 10-foot ladder leans against a wall and slides down with the foot of the ladder observed to be moving away from the wall at 4 feet per second when 6 feet from

the wall. At what speed is the top of the ladder moving downward?

48. Medicine. After using an experimental drug, the volume of a spherical tumor is observed to be decreasing at the rate of 2 cm^3 per month when the radius is 5 cm. At what rate is the radius of the tumor decreasing?

49. Physics. A water tank in the shape of a circular cone (see figure) has a radius of 4 yards and a height of 10 yards. If water is being poured into the tank at the rate of 5 cubic yards per minute, find the rate at which the water level is rising when the water level is at 2 yards. *Hint:* $V = \dfrac{1}{3}\pi r^2 h$. Use similar triangles to find r as a function of h, and substitute this into the previous equation. Then differentiate the latter expression.

50. Geometry. If the volume of a cube is increasing at 6 cubic feet per sec, what is the rate at which the sides are increasing when the sides are 10 feet long?

51. Geometry. If the sides of a square are increasing at 2 feet per second, what is the rate at which the area is changing when the sides are 10 feet long?

Enrichment Exercises

52. In Section 3.1 we indicated how to use the limit definition of the derivative to find the derivative of x^n when n is a positive integer, and said that a proof for general n was somewhat difficult. You can now construct a proof in the case that r is a rational number. Let $y = f(x) = x^{p/q}$, where p and q are positive integers. Then $y^q = x^p$. Differentiate and show that

$$\frac{dy}{dx} = \frac{p}{q} x^{(p/q)-1}$$

53. Suppose two functions $f(x)$ and $g(x)$ satisfy $f(g(x)) = x$, and are both differentiable. An example of two such functions are e^x and $\ln x$. If $y_0 = g(c)$ and $g'(c) \neq 0$, show that

$$f'(y_0) = \frac{1}{g'(c)}$$

54. Verify the formula in Exercise 53 if $g(x) = x^2$ and $f(x) = \sqrt{x}$. Take $c = 2$.

55. Verify the formula in Exercise 53 if $g(x) = x^3$ and $f(x) = \sqrt[3]{x}$. Take $c = 2$.

1. Assume that we have $y = f(x)$, then

$$2x^2 \sqrt{f(x)} + [f(x)]^3 + x^4 = 27$$

Now differentiate the preceding expression and obtain

$$4x[f(x)]^{1/2} + x^2[f(x)]^{-1/2} f'(x) + 3[f(x)]^2 f'(x) + 4x^3 = 0$$

Solving for $f'(x)$ gives

$$f'(x) = -\frac{4x[f(x)]^{1/2} + 4x^3}{x^2[f(x)]^{-1/2} + 3[f(x)]^2}$$

$$= -4 \frac{x\sqrt{y} + x^3}{\dfrac{x^2}{\sqrt{y}} + 3y^2}$$

$$= -4 \frac{xy + x^3 \sqrt{y}}{x^2 + 3y^{5/2}}$$

2. Setting $x_0 = 0$ in the original equation gives

$$0 + y^3 + 0 = 27$$

or $y = 3$. Substituting $x = 0$ and $y = 3$ in the answer to the previous exercise yields

$$f'(0) = -4 \frac{(0)(3) + (0)(\sqrt{3})}{0 + (3)3^{5/2}} = 0$$

3. The relationship between the volume V in cubic centimeters and the radius r in centimeters of the tumor is given by $V = \frac{4}{3}\pi r^3$. Thus,

$$\frac{dV}{dt} = 4\pi r^2 \frac{dr}{dt}$$

where t is measured in weeks. We are given that $\frac{dr}{dt} = -0.1$ when $r = 2$. Putting this information into the last displayed line gives

$$\frac{dV}{dt} = 4\pi(2)^2(-0.1) \approx -5.03$$

cubic centimeters per week.

4.7 NEWTON'S METHOD

■ Newton's Method ■ Applications

APPLICATION: BREAK-EVEN QUANTITY

If a profit function for a firm is given by $P(x) = -x^3 + 3x^2 - 1$, what are the approximate break-even quantities? For the answer see Example 4.

Newton's Method

We have often found it necessary to solve an equation of the form $f(x) = 0$. For example, this is necessary in finding the y-intercepts of the equation $y = f(x)$. To find the critical points of a function we need to solve $f'(x) = 0$, which again, is a function set equal to zero. Actually any equation in one variable can be placed in the form $f(x) = 0$ simply by bringing all terms to the left side of the equation. For example $x^5 + x = 5$ can be written as $f(x) = 0$, where $f(x) = x^5 + x - 5$.

In this section we consider the problem of finding the solution(s) of an equation in one variable. As we have just seen this is the same as solving an equation of the form $f(x) = 0$. If $f(x)$ is a polynomial of second degree, then we can find the solutions of $f(x) = 0$ by using the quadratic formula. If $f(x)$ is a higher degree polynomial, we can sometimes solve $f(x) = 0$ by factoring. For example, to solve $x^3 - 4x^2 + 3x = 0$, we factor and obtain

$$x^3 - 4x^2 + 3x = x(x^2 - 4x + 3) = x(x - 1)(x - 3)$$

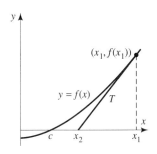

Figure 4.63
With x_1 as the initial guess, x_2 is obtained from Newton's method.

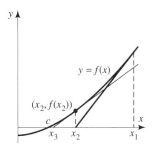

Figure 4.64
The second step of Newton's method gives x_3.

With this factorization, it is apparent that the solutions of $x^3 - 4x^2 + 3x = 0$ are $x = 0$, $x = 1$, and $x = 3$.

Unfortunately, there are no known formulas for finding the roots of a polynomial of degree five or higher, unless the polynomial is in a special form. Our only recourse when we cannot find the solution exactly is to use some method of approximating the solution. In this section we consider Newton's method for finding an approximate solution to $f(x) = 0$. In a wide variety of circumstances this method produces an approximate solution to any desired accuracy.

The idea behind Newton's method is indicated in Figure 4.63. As we see in this figure $f(c) = 0$. Our first step is to make an initial guess, which we label x_1, to the solution. We then draw the line T tangent to the graph of $y = f(x)$ at the point $(x_1, f(x_1))$. We then take the x-intercept x_2 of the tangent line to be the second approximation. The equation of T is

$$y - f(x_1) = f'(x_1)(x - x_1)$$

To find the x-intercept x_2 we set $y = 0$ and $x = x_2$ and solve for x_2. Doing this, we obtain

$$0 - f(x_1) = f'(x_1)(x_2 - x_1)$$

$$f'(x_1)x_2 = f'(x_1)x_1 - f(x_1)$$

If $f'(x_1) \neq 0$, then we can solve for x_2 and obtain

$$x_2 = x_1 - \frac{f(x_1)}{f'(x_1)}$$

We then repeat this procedure, with x_2 replacing x_1, using the line tangent at $(x_2, f(x_2))$, as shown in Figure 4.64. If $f'(x_2) \neq 0$, this gives the third approximation

$$x_3 = x_2 - \frac{f(x_2)}{f'(x_2)}$$

We continue to repeat this procedure obtaining approximations $x_1, x_2, x_3, \ldots, x_n$. Then, if $f'(x_n) \neq 0$, the next approximation x_{n+1} is given by

$$x_{n+1} = x_n - \frac{f(x_n)}{f'(x_n)}$$

In a wide variety of circumstances the numbers x_n approaches the zero c.

Newton's Method

1. To find an approximate solution to $f(x) = 0$, make an initial guess x_1.
2. The subsequent approximations are given by

$$x_{n+1} = x_n - \frac{f(x_n)}{f'(x_n)}$$

The first example we give is the example that Newton used to illustrate the method.

EXAMPLE 1   Using Newton's Method

First sketch a graph of $y = f(x) = x^3 - 2x - 5$, and then, using Newton's method, find the third approximation x_3 to the root of the equation $x^3 - 2x - 5 = 0$.

SOLUTION ■ Since $f'(x) = 3x^2 - 2$, the critical points are $x = \pm\sqrt{2/3}$. Since $f''(x) = 12x$, we readily see that $x = -\sqrt{2/3}$ is a relative maximum and $x = \sqrt{2/3}$ is a relative minimum. If we note further that $f(-\sqrt{2/3}) \approx -3.9$, $f(\sqrt{2/3}) \approx$

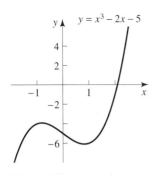

Figure 4.65
There should be a zero for x slightly larger than 2.

-6.1, $f(2) = -1$, and $f(3) = 16$, we can easily sketch the graph shown in Figure 4.65. Since $f(2)$ is negative and $f(3)$ is positive, the graph must cross the x-axis on the interval $(2, 3)$. This means $f(x) = 0$ must have a solution on this interval. The graph indicates that $x = 2$ is close to the zero, so, in Newton's method we take as an initial guess $x_1 = 2$. Then we have the situation shown in Figures 4.63 and 4.64.

For Newton's method we use the formula

$$x_{n+1} = x_n - \frac{f(x_n)}{f'(x_n)} = x_n - \frac{x_n^3 - 2x_n - 5}{3x_n^2 - 2}$$

Then

$$x_2 = x_1 - \frac{x_1^3 - 2x_1 - 5}{3x_1^2 - 2} = 2 - \frac{(2)^3 - 2(2) - 5}{3(2)^2 - 2} = 2.1$$

Finally,

$$x_3 = x_2 - \frac{x_2^3 - 2x_2 - 5}{3x_2^2 - 2} = 2.1 - \frac{(2.1)^3 - 2(2.1) - 5}{3(2.1)^2 - 2} \approx 2.094568 \quad \blacksquare$$

The third approximation x_3 in the preceding example is already accurate to four decimal places.

EXAMPLE 2 Using Newton's Method to Find Square Roots

Find x_4 in Newton's method to find an approximation to $\sqrt{2}$.

SOLUTION ■ To find $\sqrt{2}$, we need to solve the equation $f(x) = x^2 - 2 = 0$ (Fig. 4.66). In Newton's method we have

$$x_{n+1} = x_n - \frac{f(x_n)}{f'(x_n)} = x_n - \frac{x_n^2 - 2}{2x_n}$$

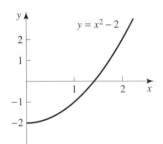

Figure 4.66
The positive zero of $x^2 - 2$ yields the positive square root $\sqrt{2}$.

Taking the initial guess to be $x_1 = 2$, gives

$$x_2 = 2 - \frac{(2)^2 - 2}{2(2)} = 1.5$$

$$x_3 = 1.5 - \frac{(1.5)^2 - 2}{2(1.5)} \approx 1.416667$$

$$x_4 = 1.416667 - \frac{(1.416667)^2 - 2}{2(1.416667)} \approx 1.414216 \quad \blacksquare$$

EXAMPLE 3 Newton's Method to Four-Decimal-Place Accuracy

Find $\sqrt{2}$ correct to four decimal places.

SOLUTION ■ We already found x_4, using Newton's method in Example 2. We now find x_5.

$$x_5 = x_4 - \frac{x_4^2 - 2}{2x_4}$$

$$= 1.414216 - \frac{(1.414216)^2 - 2}{2(1.414216)} \approx 1.414214$$

Since x_4 and x_5 agree to four decimal places (and nearly to five decimal places), we conclude that $\sqrt{2}$, correct to four decimal places, is 1.4142. ■

It is important to realize that if you ask your calculator or computer for $\sqrt{2}$, the answer you obtain, no matter how many decimal places, is still an *approximation*.

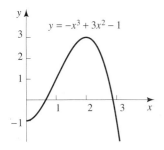

Figure 4.67
There are two break-even quantities, one less than $x = 1$ and one almost $x = 3$.

Applications

EXAMPLE 4 Break-Even Quantity

If a profit function for a firm is given by $P(x) = -x^3 + 3x^2 - 1$, find the approximate break-even quantities by finding x_3 in Newton's method.

SOLUTION ■ First notice that $P'(x) = -3x^2 + 6x = -3x(x - 2)$. Thus, $P(x)$ is increasing on $(0, 2)$ and decreasing on $(2, \infty)$. Notice that $P(0) = -1$, $P(1) = 1$, $P(2) = 3$, and $P(3) = -1$. Thus, $P(x)$ has a zero on the interval $(0, 1)$ and also on the interval $(2, 3)$ (Fig. 4.67). Newton's method gives

$$x_{n+1} = x_n - \frac{-x_n^3 + 3x_n^2 - 1}{-3x_n^2 + 6x_n}$$

To find the first zero take $x_1 = 1$. Then

$$x_2 = x_1 - \frac{-x_1^3 + 3x_1^2 - 1}{-3x_1^2 - 6x_1}$$

$$= 1 - \frac{-(1)^3 + 3(1)^2 - 1}{-3(1)^2 + 6(1)} = \frac{2}{3}$$

Then

$$x_3 = x_2 - \frac{-x_2^3 + 3x_2^2 - 1}{-3x_2^2 + 6x_2}$$

$$= \frac{2}{3} - \frac{-(2/3)^3 + 3(2/3)^2 - 1}{-3(2/3)^2 + 6(2/3)} \approx 0.653$$

To find the second zero take $x_1 = 3$. Then

$$x_2 = x_1 - \frac{-x_1^3 + 3x_1^2 - 1}{-3x_1^2 + 6x_1}$$

$$= 3 - \frac{-(3)^3 + 3(3)^2 - 1}{-3(3)^2 + 6(3)} = \frac{26}{9}$$

Then

$$x_3 = x_2 - \frac{-x_2^3 + 3x_2^2 - 1}{-3x_2^2 + 6x_2}$$

$$= \frac{26}{9} - \frac{-(26/9)^3 + 3(26/9)^2 - 1}{-3(26/9)^2 + 6(26/9)} \approx 2.879 \quad ■$$

$x_1 = 1$	$x_1 = 3$
x_1 1.000000000	3.000000000
x_2 0.666666667	2.888888889
x_3 0.652777778	2.879451567
x_4 0.652703647	2.879385245
x_5 0.652703645	2.879385242

We can see just how fast the numbers x_n are approaching the zeros in the table.

Notice that x_4 is already accurate to about eight decimal places. This is not unusual in Newton's method.

Figure 4.68 illustrates two of the ways in which Newton's method can go wrong. Suppose we are seeking the zero c_2. If we take an initial guess x_1 that is too large, then the values $x_2, x_3, \ldots$, do not head for the root c_2 or for any other number. If we take the initial guess to be $\bar{x}_1$ shown in Figure 4.68, then the values $\bar{x}_2, \bar{x}_3, \ldots$, head for the root c_1. These possibilities illustrate that care must be exercised when using Newton's method.

The following explicit example illustrates one of the problems using Newton's method. Other problems with Newton's method are illustrated in the exercise set. Screen 4.10 shows a graph of $y = -x^3 + 3x^2 - 1$ using dimensions $[-1, 4]$ by $[-5, 5]$. Notice that there are three zeros. Suppose we wanted the middle zero. Use the program NEWT in the Technology Resource Manual with $x_0 = 1.5$, and we see that Newton's

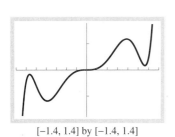

[−1.4, 1.4] by [−1.4, 1.4]

Screen 4.10
A graph of $y = -x^3 + 3x^2 - 1$.

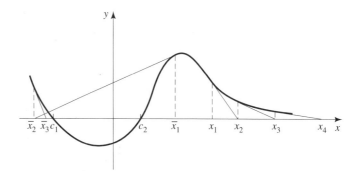

Figure 4.68
Some ways that Newton's method can go astray if the initial guess is not close enough to the zero being sought.

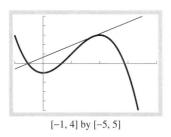

[−1, 4] by [−5, 5]

Screen 4.11
A graph of $y = nDer2\ f(x)$ when $f(x) = x^4$ is indistinguishable from the graph of $y = f''(x) = 12x^2$.

method gives the zero 0.6252703645 found in the previous example. But taking $x_0 = 1.75$ and Newton's method gives the root -0.5320888862. The line tangent to the curve at $x = 1.75$ intersects the x-axis near the negative zero, and Newton's method then converges to this zero. You can readily verify this by using your graphing calculator to draw a tangent line at about $x = 1.75$, obtaining Screen 4.11.

SELF-HELP EXERCISE SET 4.7

1. Let $f(x) = x^4 + 4x^2 + x - 3$. Show that $f(x)$ has two zeros. Determine x_3 in Newton's method in order to find an approximation to the largest zero.

2. Determine x_3 in Newton's method in order to find an approximation to the smallest zero of the function in the previous exercise.

3. Use Newton's method to find the largest zero of the function $f(x)$ in Exercise 1 to five decimal places.

EXERCISE SET 4.7

In Exercises 1 through 8 you are given a function $y = f(x)$, an interval, and an initial value x_1. First show that the function takes on opposite signs at the endpoints of the interval. Then show that the derivative is never zero on this interval. Then use Newton's method to find x_3.

1. $f(x) = x^3 + x - 3$, $[1, 2]$, $x_1 = 2$

2. $f(x) = x^3 + x - 1$, $[0, 1]$, $x_1 = 1$

3. $f(x) = x^4 + x - 3$, $[1, 2]$, $x_1 = 1$

4. $f(x) = x^4 + x - 1$, $[0, 1]$, $x_1 = 1$

5. $f(x) = 8 - x - x^3$, $[1, 2]$, $x_1 = 2$

6. $f(x) = -4x^3 + 3x - 2$, $[-2, -1]$, $x_1 = -2$

7. $f(x) = 3x^4 - 4x^3 - 2$, $[1, 2]$, $x_1 = 2$

8. $f(x) = x^3 + 3x + 1$, $[-1, 0]$, $x_1 = -1$

In Exercises 9 through 14 use Newton's method to find the indicated root $\sqrt[n]{b}$ to four decimal places. Take $x_1 = b$.

9. $\sqrt{3}$

10. $\sqrt{5}$

11. $\sqrt[3]{11}$

12. $\sqrt[3]{5}$

13. $\sqrt[6]{2}$

14. $\sqrt[5]{3}$

In Exercises 15 through 22 use Newton's method to find all the real solutions, to four decimal places, of the given equation.

15. $x^4 + 6x^2 = 2$

16. $x^3 = 2x^2 - 1$

17. $1 - 9x + 6x^2 = x^3$

18. $4x^5 = 5x - 1$

19. $e^{-x} = x$

20. $2 - x = \ln x$

21. $e^{-x} = \ln x$

22. $x^2 = -\ln x$

In Exercises 23 and 24 you cannot solve the equation $f'(x) = 0$ exactly to find the critical value. Find the third approximation in Newton's method using the indicated x_1.

23. $f(x) = x^4 + x^2 + x + 1$, $x_1 = 0$

24. $f(x) = x^4 + x^3 + 0.50x + 1$, $x_1 = -1$

In Exercises 25 and 26 you cannot solve the equation $f'(x) = 0$ exactly to find the two critical values. Find the third approximation in Newton's method using for x_1 the positive or negative integer closest to each zero of $f'(x)$.

25. $f(x) = x^5 + x^4 - 3x + 6$

26. $f(x) = 0.20x^5 + 0.50x^2 - x$

Applications

27. Break-Even Quantity. If a profit function for a firm is given by $P(x) = -x^3 + 3x - 1$, find the approximate break-even quantities to six decimal places by using Newton's method.

28. Break-Even Quantity. If a profit function for a firm is given by $P(x) = -x^5 + 5x - 1$, find the approximate break-even quantities to six decimal places by using Newton's method.

29. Supply and Demand. Suppose a supply equation for a commodity is given by $p = 12 - x$ and a demand equation is given by $p = 1 + x^3$. Use the program NEWT to find the break-even quantity to six decimal places.

30. Supply and Demand. Suppose a supply equation for a commodity is given by $p = 15 - x^2$ and a demand equation is given by $p = x^3$. Use the program NEWT to find the break-even quantity to six decimal places.

31. Use the program NEWT to find the zeros of the functions given in Exercises 15 through 18 to six decimal places.

Enrichment Exercises

32. The function $f(x) = 2x^{1/3} + x$ has a zero at $x = 0$. Try to find this zero using the program NEWT by taking $x_1 = 1$. What is happening? Why do you think this behavior is occurring?

33. Obviously the function $f(x) = \dfrac{x + 1}{x + 2}$ has a zero at $x = -1$.

Try to find this zero using Newton's method with $x_1 = 1$. Explain what is happening by sketching a graph.

34. Obviously the function $f(x) = \dfrac{x - 1}{x^2}$ has a zero at $x = 1$.

Try to find this zero using Newton's method with $x_1 = 2$. Explain what is happening by sketching a graph.

35. The function $f(x) = x^4 + 4x^3 + 1$ has two zeros, one near $x = -4$ and one on the interval $[-1, 0]$. Try to find the zero on the interval $[-1, 0]$ by taking $x_1 = -0.10$. Explain what is happening on a graph.

36. The function $f(x) = 3x^5 - 20x^3 + 5$ has two positive zeros, one less than $x = 1$ and one greater than $x = 2$. Try to find the smaller zero by using Newton's method with $x_1 = 0.10$. Explain what is happening on a graph.

37. Use Newton's method to find the solution to $\cos x = x$ to four decimal places.

38. Use Newton's method to find the smallest positive solution to $\sin x = e^{-x}$ to four decimal places.

SOLUTIONS TO SELF-HELP EXERCISE SET 4.7

1. We have $f'(x) = 4x^3 + 8x + 1$. We can easily see that $f'(x)$ has only one zero since $f''(x) = 12x^2 + 8$ is always positive implies that $f'(x)$ is an increasing function. Since $f'(0) = 1$ and $f'(x)$ is always increasing, $f'(c) = 0$ for negative c. Furthermore, since f' is always increasing, $f'(x) < 0$ for $x < c$ and $f'(x) > 0$ for $x > c$. Thus, $x = c$ is a relative (and absolute) minimum. If we then notice that $f(-1) = 1$, $f(0) = -3$, and $f(1) = 3$, we can sketch the graph shown in the figure. We see that $f(x)$ has a zero on the interval $(-1, 0)$ and on the interval $(0, 1)$. Newton's method in this case gives

$$x_{n+1} = x_n - \frac{x_n^4 + 4x_n^2 + x_n - 3}{4x_n^3 + 8x_n + 1}$$

Taking $x_1 = 1$ yields

$$x_2 = x_1 - \frac{x_1^4 + 4x_1^2 + x_1 - 3}{4x_1^3 + 8x_1 + 1}$$

$$= 1 - \frac{(1)^4 + 4(1)^2 + (1) - 3}{4(1)^3 + 8(1) + 1}$$

$$= 1 - \frac{3}{13} = \frac{10}{13} \approx 0.769231$$

Then

$$x_3 = 0.76923 - \frac{(0.76923)^4 + 4(0.76923)^2 + (0.76923) - 3}{4(0.76923)^3 + 8(0.76923) + 1}$$

$$\approx 0.715053$$

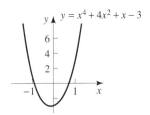

2. Taking $x_1 = -1$ gives

$$x_2 = x_1 - \frac{x_1^4 + 4x_1^2 + x_1 - 3}{4x_1^3 + 8x_1 + 1}$$

$$= -1 - \frac{(-1)^4 + 4(-1)^2 + (-1) - 3}{4(-1)^3 + 8(-1) + 1}$$

$$= -1 - \frac{1}{-11} \approx -0.909091$$

Then

$x_3 = -0.90909$

$- \dfrac{(-0.90909)^4 + 4(-0.90909)^2 + (-0.90909) - 3}{4(-0.90909)^3 + 8(-0.90909) + 1}$

≈ -0.900500

3. We find $x_4 = 0.712403$ and $x_5 = 0.712397$. Since both of these are 0.71240 to five decimal places, we accept this last number to be the zero to five decimal places.

CHAPTER 4

SUMMARY OUTLINE

- A function is said to be **increasing** on (a, b) if $f(x_1) < f(x_2)$ when $x_1 < x_2$ and **decreasing** if $f(x_1) > f(x_2)$. p. 195.

- The number c is a **critical value** for the function $f(x)$ if $f'(c) = 0$ or if $f'(c)$ does not exist but $f(c)$ is defined. p. 197.

- $f(x_M)$ is a **relative maximum** for the function $f(x)$ if $f(x) \leq f(x_M)$ for all x on some interval (a, b) that contains x_M and $f(x_m)$ is a **relative minimum** if $f(x) \geq f(x_m)$ for all x on some interval (a, b) that contains x_m. p. 199.

- **First Derivative Test:** The sign charts for f' on either side of a critical value c have the following consequences: $+ \ -$ implies a relative maximum, $- \ +$ implies a relative minimum, and $+ \ +$ and $- \ -$ both indicate no extremum. p. 199–200.

- The **second derivative** $f''(x)$ of $y = f(x)$ is the derivative of the first derivative, that is, $f''(x) = \dfrac{d}{dx}[f'(x)]$. We also use the notation $f''(x) = \dfrac{d^2y}{dx^2}$. p. 214.

- The graph of a function $f(x)$ is **concave up** ($\smile$) if the derivative $f'(x)$ is an increasing function. This happens if $f''(x) > 0$. The graph of a function $f(x)$ is **concave down** ($\frown$) if the derivative $f'(x)$ is a decreasing function. This happens if $f''(x) < 0$. p. 215.

- An **inflection point** is a point on the graph of f where the concavity changes sign. p. 216.

- **Acceleration** $a(t) = v'(t)$, where $v(t)$ is the velocity. p. 220.

- **Second Derivative Test:** Let $f'(c) = 0$. If $f''(c) > 0$, then f assumes a relative minimum at c. If $f''(c) < 0$, then f assumes a relative maximum at c. p. 221.

- $f(x_M)$ is the **absolute maximum** of the function $f(x)$ on the interval I if $f(x) \leq f(x_M)$ for all $x \in I$, and an **absolute minimum** on the interval I if $f(x) \geq f(x_M)$ for all $x \in I$. p. 236.

- **Extreme Value Theorem:** A continuous function on a closed and bounded interval attains a maximum and a minimum. p. 237.

- A **logistic equation** is any equation of the form $P(t) = \dfrac{L}{1 + ae^{-kt}}$. p. 218

- **Newton's Method.** p. 257. To find an approximate solution to $f(x) = 0$, make an initial guess x_1. Subsequent approximations are given by

$$x_{n+1} = x_n - \frac{f(x_n)}{f'(x_n)}$$

CHAPTER **4**

REVIEW EXERCISES

1. For the function whose graph is shown in the accompanying figure, find all critical values and the largest open intervals on which the function is increasing and the largest open intervals on which the function is decreasing.

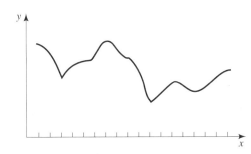

2. Suppose the derivative of a function $y = f(x)$ is given by $f'(x) = (x + 2)(x - 1)$. Find all the largest open intervals on which the function is increasing, the largest open intervals on which the function is decreasing, all critical values, and all relative extrema.

3. Repeat the previous exercise if $f'(x) = (x + 4)(x - 2)^2(x - 5)$.

In Exercises 4 through 25 find the intervals on which the function is increasing, the intervals on which the function is decreasing, the critical values, the intervals on which the graph of the function is concave up, the intervals on which the graph of the function is concave down, the inflection points, the vertical asymptotes, and the horizontal asymptotes. Then sketch a graph.

4. $y = f(x) = 2x^2 + 3x - 4$

5. $y = f(x) = -x^2 + 2x + 5$

6. $y = f(x) = -3x^2 - 6x + 1$

7. $y = f(x) = 3x^2 + 12x - 10$

8. $y = f(x) = \frac{1}{3}x^3 - x + 1$

9. $y = f(x) = -x^3 + 3x + 2$

10. $y = f(x) = x^4 - 4x^3 + 4$

11. $y = f(x) = x^5 - 15x^4 + 5$

12. $y = f(x) = x^4 - 18x^2 + 15$

13. $y = f(x) = -x^5 + 15x^3 - 10$

14. $y = f(x) = x + \frac{4}{x^2}$ 15. $y = f(x) = x^2 + \frac{1}{x^2}$

16. $y = f(x) = x - \frac{1}{x^2}$ 17. $y = f(x) = x^2 - \frac{27}{x^2}$

18. $y = f(x) = \frac{x}{x - 5}$ 19. $y = f(x) = \frac{x - 1}{x - 5}$

20. $y = f(x) = \sqrt{x^6 + 1}$ 21. $y = f(x) = (9 - x)^{1/5}$

22. $y = e^{-x^2 - 1}$ 23. $y = 1 - e^{-x^2}$

24. $y = \ln(x^2 + 1)$ 25. $y = \ln(1 + e^x)$

In Exercises 26 through 35 find the location of all absolute maxima and absolute minima on the given intervals.

26. $f(x) = \frac{x}{3 + x^4}$, $(-\infty, \infty)$

27. $f(x) = 1 - \frac{x}{48 + x^4}$, $(-\infty, \infty)$

28. $f(x) = x - \frac{1}{x}$, $[1, \infty)$

29. $f(x) = \frac{1}{x^3} - x$, $[1, \infty)$

30. $f(x) = 2x^3 - 3x^2 + 1$, $[-1, 2]$

31. $f(x) = 2x^3 - 9x^2 + 12x + 1$, $[0, 3]$

32. $f(x) = (x - 2)^2$, $[0, 3]$ 33. $f(x) = x + \frac{4}{x^2}$, $[1, 3]$

34. $f(x) = x^2 + \frac{16}{x^2}$, $[1, 3]$ 35. $f(x) = x - \frac{1}{x^2}$, $[1, 4]$

In Exercises 36 through 39 find $\frac{dy}{dx}$ at any point (x, y), and then find the slope of the line tangent at the given point.

36. $x^3y^4 - xy = 6$, $(2, 1)$

37. $\sqrt{x} + x\sqrt{y} + y^2 = 7$, $(4, 1)$

38. $\frac{x + y}{x - y} = 3y^2$, $(2, 1)$

39. $\sqrt[3]{y^2 + 7} - x^2y^3 = 2$, $(0, 1)$

40. Show that the graph of the special fourth-order polynomial function $f(x) = ax^4 + bx^2 + cx + d$ is always concave up if both a and b are positive and always concave down if both a and b are negative.

41. If $b > a$ and c is any constant, show that the polynomial function $f(x) = -2x^3 + 3(a + b)x^2 - 6abx + c$ has exactly one relative minimum and exactly one relative maximum. Locate and identify them.

42. Suppose $f''(x) > 0$ everywhere. Show that $y = g(x) = e^{f(x)}$ is concave-up everywhere.

43. Suppose that $f(x) > 0$ and $f''(x) < 0$ everywhere. Show that $y = g(x) = \ln f(x)$ is concave-down.

44. **Average Cost.** An electric utility company estimates that the cost $C(x)$ of producing x kilowatts of electricity is given by $C(x) = a + bx^2$, where a represents the fixed costs and b is the unit cost. The *average* cost $\overline{C}(x)$ is defined to be the total cost of producing x kilowatts divided by x, that is, $\overline{C}(x) = \frac{C(x)}{x}$. Find where the average cost increases and where it decreases.

45. Average Cost. The total cost function for producing x items is given by $C(x) = 1000 + 10x^2$. How many items should be produced to minimize $\overline{C}(x) = C(x)/x$, the average cost per item?

46. Average Cost. In the previous problem show that average cost is minimized at the value at which $\overline{C} = C'$.

47. Average Cost. Suppose the total cost function for producing x items is given by $C(x)$ and the average cost per item is given by $\overline{C}(x) = C(x)/x$. If average cost is at a minimum at some value on $(0, \infty)$, then show that at this value $\overline{C} = C'$.

48. Biology. In the attack equation the rate of change R of the number attacked with respect to the number N vulnerable to attack is given by $R(x) = x^{1-a}(kx - N)$, where x is the number of attackers, a and k are positive constants with $a < 1$. Find the number of attackers that minimizes R.

49. Average Cost. An electric utility company estimates that the cost $C(x)$ of producing x kilowatts of electricity is given by $C(x) = a + bx^3$, where a represents the fixed costs and b is a positive constant. The *average* cost $\overline{C}(x)$ is defined to be the total cost of producing x kilowatts divided by x, that is, $\overline{C}(x) = \dfrac{C(x)}{x}$. Find where the average cost attains a minimum.

50. Cost. An individual is planning to construct two fenced enclosures, one square and one an equilateral triangle. The cost per yard of constructing the triangle is twice that of the square enclosure. If the total cost is fixed at T dollars, what should be the dimensions of the two figures in order to maximize the area enclosed by the two figures? (The area of an equilateral triangle is $\sqrt{3}r^2/4$, where r is the length of a side).

51. Revenue. A bus company charged $60 per person for a sightseeing trip and obtained 40 people for the trip. The company's data indicate that for the same trip, each $2 increase in the price above $60 results in the loss of one customer. What should the company charge to maximize revenue?

52. Profits. The bus company in the previous exercise has fixed costs of $3000 for each trip and additional costs of $4 per customer. What should the company charge to maximize profits?

53. Packaging. A manufacturer wishes to make a cylindrical can of 108 cubic inches. The top and bottom cost twice as much per square inch as the sides. What should the dimensions of the can be to minimize the cost?

54. Inventory Control. A company produces 1000 cases of perfume a year. The start-up cost of each production run is $1000, it costs $50 to manufacture each case, and $1 to store each case for a year. Determine the number of cases to manufacture in each run and the number of runs needed to minimize the total cost.

55. Cost. Assume a firm has a cost function $C = b + ax$, where x is the number of items made and sold. Suppose $p = p(x)$, where p is the price of each item, and assume $p'(x) < 0$. Then profit is given by $P(x) = xp(x) - b - ax$. By taking $P'(x)$ and setting this equal to zero, show mathematically that at equilibrium the adage "relatively high-cost firms produce relatively little" is true. (See Kimmel [113].)

56. Population. An isolated lake was stocked with 50 lake trout. Two years later it was estimated that there were 100 of these trout in this lake. It is estimated that this lake can support 500 of these trout. Assuming that the trout population satisfies a logistic equation, find how many lake trout will be in this lake in 4 years.

57. Physics. A 30-foot ladder leans against a wall and slides down with the top of the ladder observed to be moving down the wall at 4 feet per second when 6 feet above the ground. At what speed is the base of the ladder moving away from the wall?

58. Forest Fires. McAlpine and Wakimoto [114] showed that $R = f(t) = 70.13e^{-2.1645/t}$, where R is head fire rate of spread at time t, and t is time from ignition. Find $\lim_{t \to 0^+} f(t)$ and $\lim_{t \to \infty} f(t)$. Also find where $f(t)$ is increasing. Graph. Does the graph make sense?

59. Biology. Boyce and coworkers [115] showed that the needle temperature $T(u)$ in degrees Celsius of red spruce was approximated by $T(u) = T_{\text{air}} + 0.0061e^{-2.0087u}$, where T_{air} is air temperature, and u is wind velocity in meters per second. Find where $T(u)$ is decreasing. Does this agree with your own experience, at least in cold weather?

60. Biology. Biologists have used the mathematical model $Q(t) = at + be^{-kt}$, where a, b and k are positive and $Q(t)$ is the CO_2 evolved at time t from microbial biomass in the soil. (See Laudelout [116].) What condition must the constants a, b, and k satisfy so that $Q(t)$ is always an increasing function?

In Exercises 61 through 66 match the equation with the given graphs in (a) through (f).

61. $y = 2 - e^{-t}$ **62.** $y = e^{-t}$

63. $y = \dfrac{1}{1 + e^{-t}}$ **64.** $y = t^{0.75}$

65. $y = \dfrac{1}{t^2}$ **66.** $y = e^t$

67. The figure shows a graph of $y = f'(x)$. What can you say about a possible relative extremum of $f(x)$ at $x = 1$? Justify your answer.

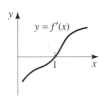

68. The figure shows a graph of $y = f'(x)$. What can you say about a possible relative extremum of $f(x)$ at $x = 1$? Justify your answer.

69. Suppose the function $y = f'(x)$ is decreasing on the interval $(0, 2)$ and $f'(1) = 0$. What can you say about a possible relative extremum of $f(x)$ at $x = 1$? Justify your answer.

70. Suppose the function $y = f'(x)$ is increasing on the interval $(0, 2)$ and $f'(1) = 0$. What can you say about a possible relative extremum of $f(x)$ at $x = 1$? Justify your answer.

71. The figure shows a graph of $y = f''(x)$. Suppose $f'(1) = 0$. What can you say about a possible relative extremum of $f(x)$ at $x = 1$? Justify your answer.

72. The figure shows a graph of $y = f''(x)$. Suppose $f'(1) = 0$. What can you say about a possible relative extremum of $f(x)$ at $x = 1$? Justify your answer.

CHAPTER **4**

PROJECTS

Project 1
A Problem in Price Discrimination

In this project we consider a problem in price discrimination that economists have pondered over the years.

In the type of price discrimination we investigate here we assume that the seller can divide customers into two groups, each of which has its own demand function.

Suppose in Market A the demand equation is given by $x = f(p)$ and in Market B by $x = g(p)$, where x is a measure of the units of the commodity and p is the price in dollars of each unit. The seller wants to ensure that the last unit of output sold in Market A adds the same amount to total revenue as the last unit sold in Market B. To accomplish this, we must sum the two demand curves to obtain the demand curve for the combined Market C. Thus, the combined demand equation is $x = h(p) = f(p) + g(p)$.

The cost function in all markets is assumed to be $C(x) = cx$. Then, for example, the profit function in Market A is

$$P_A(p) = px - cx = pf(p) - cf(p)$$

Then the profit functions in Markets B and C are

$$P_B(p) = pg(p) - cg(p), \qquad P_C(p) = ph(p) - ch(p)$$

respectively.

(a) A basic result that remained unquestioned for many years in the economic literature is that when given the two groups of buyers mentioned here, discrimination raises the price for one group and lowers it for the other. Establish this result mathematically when the profit curves are all concave down. See the figure. Notice that if the seller discriminates, then the seller simply sets the price p_1 in Market A to maximize $P_A(p)$ and sets the price p_2 in Market B to maximize $P_B(p)$. If the seller does not discriminate, however, and sets the same price in both markets, then establish the result by showing that the price p^* that maximizes $P_C(p)$ must be between p_1 and p_2 when all the profit curves are concave down.

(b) This part of the project requires a graphing calculator or computer.

It has been recently reported by Nahata [117] that the situation described here need not be true if the profit functions are no longer assumed to be concave down.

The following counterexample was given:

$$f(p) = -0.25p^3 + 2.0001p^2 - 5.5p + 10$$

$$g(p) = -0.2561p^3 + 2.7p^2 - 9.5p + 12$$

Show in this case that $p^* > p_1 > p_2$ by using your graphing calculator or a computer to find the (single) zero of each of the marginal profit functions.

Project 2
Drug Treatment for Irregular Heartbeat

Ventricular arrhythmias (irregular heartbeat) is frequently treated by the drug *lidocaine*. To be effective in the treatment of arrhythmias the drug must have a concentration above 1.5 mg/L in the bloodstream. A concentration above 6 mg/L in the bloodstream, however, can produce serious side effects and even death.

Let t be the time in minutes from injection of the drug lidocaine and $c(t)$ the concentration at time t in the bloodstream.

(A) If a steady infusion rate of 2.52 mg/L of the drug into the bloodstream is maintained, then the approximate concentration is given by the formula

$$c(t) = 3.5 - 0.5095e^{-0.12043t} - 2.9905e^{-0.0075t}$$

(a) Use calculus to show that the concentration is always increasing.
(b) Use calculus to find $\lim_{t \to \infty} c(t)$.

(c) Verify parts (a) and (b) by graphing on your graphing calculator.
(d) Use Newton's method to determine the time for which the concentration reaches the effective concentration of 1.5 mg/L.
(e) Verify your answer in part (d) by graphing $y_2 = 1.5$ on the same screen as your graph in part (c) and use the ZOOM feature to find the intersection of the two curves.
(B) If in addition to the infusion rate of 2.5 mg/L there is a load dose of 100 mg, then the concentration is given by

$$c(t) = 3.5 + 1.92513e^{-0.12043t} - 2.0918e^{-0.00757t}$$

(a) Graph this function on your graphing calculator. Does the concentration stay above the minimum therapeutic level and at a safe level?
(b) Graph $y = c'(t)$ on your calculator and find where $y = c'(t) = 0$. Use this information to find the minimum level of the concentration.

Project 3
Tracking Radioactive Fallout in a Tropical Rain Forest

The figure shows a model for the cycling of radioactive strontium-90 (from large-scale nuclear testing) in a rain forest. The arrows pointing to the left and right indicate the losses due to natural radioactive decay. The transfers indicated by arrows pointing up or down must be estimated by researchers. For details of how this was done see Jordan and colleagues [118]. The litter consists of leaf, twig, and fruit-fall.

The following four functions approximate the percentage of strontium-90 in the canopy, litter, soil, and wood, respectively, at time t measured in months.

$$p_1(t) = -2.35(0.0011)e^{-0.2314t} + 32.65(0.0885)e^{-0.0029t} + 8.69(-0.5243)e^{-0.0150t}$$
$$-20.22(1.0179)e^{-0.0936t} + 100(0.2226)e^{-0.0508t}$$

$$p_2(t) = -2.35(-0.9978)e^{-0.2314t} + 32.65(0.0273)e^{-0.0029t} + 8.69(-0.1711)e^{-0.0150t}$$
$$-20.22(0.5217)e^{-0.0936t} + 100(0.088)e^{-0.0508t}$$

$$p_3(t) = -2.35(1.0205)e^{-0.2314t} + 32.65(1.8858)e^{-0.0029t} + 8.69(6.4448)e^{-0.0150t}$$
$$-20.22(-1.6392)e^{-0.0936t} + 100(-1.4836)e^{-0.0508t}$$

$$p_4(t) = -2.35(-0.0177)e^{-0.2314t} + 32.65(0.9894)e^{-0.0029t} + 8.69(-5.0822)e^{-0.0150t}$$
$$-20.22(0.0754)e^{-0.0936t} + 100(0.1336)e^{-0.0508t}$$

(a) Graph all of these functions on your graphing calculator, and determine approximately where each have relative maxima and minima. Also determine approximately what the maxima and minima are. (You will have some difficulty in determining the dimensions of an appropriate window, since you initially have little idea of the values of t at which the maxima and minima occur, nor the values of the maxima and minima.)

(b) What can you say about all of these functions for the first couple of years? What is true about all of these functions after about 15 years?

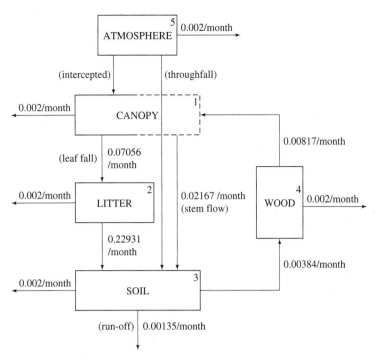

Strontium cycling in a tropical forest.

(c) Make an attempt to use only calculus to find the maxima and minima. What difficulties do you encounter?

Project 4
Relationship Between Wing Length and Altitude

(This project requires a solver routine on a calculator or computer.) By considering the per gram hovering costs of a hummingbird and the air density dependence on elevation, Wolf and Hill [119] showed that the relationship between the length l in centimeters of the wing of a hummingbird and the altitude z in meters that the hummingbird lives is given by

$$l + 0.404 l^{0.6} = 1.404 \left(\frac{1 - 0.0065 z}{288} \right)^{-2.128}$$

This equation gives l as an implicit function of z. Algebraically we cannot solve for l. We must use some graphical or numerical method. Input this equation into your graphing calculator or computer. Take z equal to 500, 1000, 1500, and so on, and use the solver routine to determine l. Graph l as a function of z. Approximately what kind of curve do you obtain? Is $l(z)$ an increasing function? Is your answer reasonable? Why?

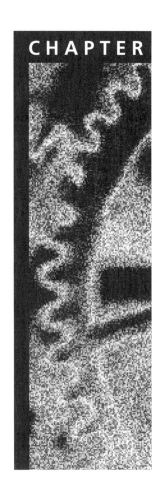

Integration

In this second part of the study of calculus we introduce the integral and see in what sense it is the opposite of the derivative. The integral enables us to the find the area between curves, which as we will see, has many concrete applications. We also discuss how to recover the cost, revenue, and profit functions when given the rates of change of these quantities; how to find the population when given a measurement of the density; how to find important economic quantities, such as consumers' and producers' surplus; and even how to measure to what extent an economic system is distributing income equitably.

5.1 ANTIDERIVATIVES

■ Antiderivatives ■ Rules of Integration ■ Applications

APPLICATION: FINDING COST GIVEN MARGINAL COST

What is the cost function $C(x)$ if the marginal cost is $10e^x + 40x - 100$ and fixed costs are 1000? See Example 8 for the answer.

Antiderivatives

Given a function $f(x) = x^2$, can we find some function $F(x)$ whose derivative is x^2? We know that in differentiating x to a power, the power is reduced by one. Thus, we might tentatively try the function x^3. But $\dfrac{d}{dx}(x^3) = 3x^2$. Thus, we have *three* times our function x^2. Dividing by this number, we obtain $\dfrac{d}{dx}\left(\dfrac{1}{3}x^3\right) = x^2$. This is one answer. But are there others? Recalling that the derivative of a constant is zero, we readily see that

$$\frac{d}{dx}\left(\frac{1}{3}x^3 + C\right) = x^2$$

for any constant C. This gives us the family of answers

$$F(x) = \frac{1}{3}x^3 + C$$

(Fig. 5.1).

We must now address the following general problem. Given a function $f(x)$ we need to find a function $F(x)$ such that $F'(x) = f(x)$. We call $F(x)$ an **antiderivative** of $f(x)$. We use the term *anti*derivative to refer to the fact that finding $F(x)$ is the *opposite* of differentiating.

Definition of Antiderivative

If $F'(x) = f(x)$, then $F(x)$ is called an antiderivative of $f(x)$.

We saw that $\frac{1}{3}x^3 + C$, where C is any constant, are all antiderivatives of x^2. Are there still others? The following theorem indicates that the answer is no:

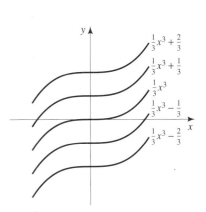

Figure 5.1
Shown are some members of the family $\frac{1}{3}x^3 + C$, all of which are antiderivatives of x^2.

Theorem 1

If $F'(x) = G'(x)$ on (a, b), then on (a, b)

$$F(x) = G(x) + C$$

for some constant C.

That is, if two functions have the same derivative on (a, b), they differ by a constant on (a, b).

EXAMPLE 1 Finding All the Antiderivatives

Find all antiderivatives of x^2.

SOLUTION ■ From the above theorem and work, we know that all antiderivatives of x^2 must be of the form

$$\frac{1}{3} x^3 + C$$

where C is any constant. ■

When we include the arbitrary constant C, we refer to the function $\frac{1}{3}x^3 + C$ as the *general form* of the antiderivative. By this we mean that the collection of all antiderivatives of x^2 can be obtained by adding all possible constants to $\frac{1}{3}x^3$.

The collection of all antiderivatives of a function $f(x)$ is called the **indefinite integral** and is denoted by $\int f(x)\, dx$. This is read as *the integral of $f(x)$*, or more carefully, as *the integral of $f(x)$ with respect to x*. We refer to the symbol $\int$ as the **integral sign** and $f(x)$ as the **integrand**. The symbol $\int$ looks like an S, and in fact stands for *sum*. We shall later see the important connection between the integral and a certain sum.

If we know of one function $F(x)$ for which $F'(x) = f(x)$, then Theorem 1 indicates that the collection of all antiderivatives of $f(x)$ must be of the form $F(x) + C$, where C is a constant. Thus, we have the following definition:

The Indefinite Integral

The collection of all antiderivatives of a function $f(x)$ is called the **indefinite integral** and is denoted by $\int f(x)\, dx$.

If we know one function $F(x)$ for which $F'(x) = f(x)$, then

$$\int f(x)\, dx = F(x) + C$$

where C is an arbitrary constant and called the **constant of integration**.

From Example 1 we see that

$$\int x^2\, dx = \frac{1}{3} x^3 + C$$

The following is also true

$$\int x^3\, dx = \frac{1}{4} x^4 + C$$

This can be readily verified by differentiating the right-hand side to obtain the integrand. Thus,

$$\frac{d}{dx}\left(\frac{1}{4}x^4 + C\right) = x^3$$

Rules of Integration

It would be convenient if we had some rules to find indefinite integrals. The following is the first such rule:

Power Rule

If $n \neq -1$, then

$$\int x^n \, dx = \frac{1}{n+1} x^{n+1} + C$$

In words, increase the power by 1 and divide by the resulting number, then add an arbitrary constant.

To prove the power rule formula we need only differentiate the right-hand side and note that we obtain the integrand. Thus,

$$\frac{d}{dx}\left(\frac{1}{n+1} x^{n+1} + C\right) = x^n$$

EXAMPLE 2   Using the Power Rule

Find (a) $\int x^5 \, dx$ and (b) $\int \frac{1}{\sqrt{x}} \, dx$.

SOLUTIONS ■
(a) Here we set $n = 5$ in the power rule formula and obtain

$$\int x^5 \, dx = \frac{1}{5+1} x^{5+1} + C = \frac{1}{6} x^6 + C$$

(b) Here we first note that $\frac{1}{\sqrt{x}} = x^{-1/2}$ and set $n = -\frac{1}{2}$ in the power rule formula and obtain

$$\int \frac{1}{\sqrt{x}} \, dx = \int x^{-1/2} \, dx = \frac{1}{-\frac{1}{2}+1} x^{-1/2+1} + C = 2x^{1/2} + C \quad ■$$

Remark. As in the previous example, *always remove radical signs before integrating.*

Since $\frac{d}{dx}[kF(x)] = k\frac{d}{dx}F(x)$, we have the following rule:

Constant Times Function Rule

For any constant k

$$\int kf(x) \, dx = k\int f(x) \, dx$$

In words, the integral of a constant times a function is the constant times the integral of the function.

EXAMPLE 3   Using the Constant-Times-Function Rule

Find (a) $\int 3t^9 \, dt$ (b) $\int \dfrac{\sqrt{x}}{\pi} \, dx$

SOLUTIONS ■

(a) Since 3 is a constant

$$\int 3t^9 \, dt = 3 \int t^9 \, dt = \frac{3}{10} t^{10} + C$$

(b) Since $\dfrac{1}{\pi}$ is a constant

$$\int \frac{\sqrt{x}}{\pi} \, dx = \int \frac{1}{\pi} \sqrt{x} \, dx = \frac{1}{\pi} \int \sqrt{x} \, dx = \frac{1}{\pi} \int x^{1/2} \, dx$$

$$= \frac{1}{\pi} \frac{1}{\frac{1}{2} + 1} x^{1/2+1} + C$$

$$= \frac{2}{3\pi} x^{3/2} + C \quad ■$$

Since $\dfrac{d}{dx} [F(x) \pm G(x)] = \dfrac{d}{dx} F(x) \pm \dfrac{d}{dx} G(x)$, we have the following rule:

Sum or Difference Rule

$$\int [f(x) \pm g(x)] \, dx = \int f(x) \, dx \pm \int g(x) \, dx$$

In words, the integral of the sum (or difference) is the sum (or difference) of the integrals.

EXAMPLE 4   Using the Sum and Difference Rules

Find (a) $\int (u^2 + u^3) \, du$ and (b) $\int (2x^4 - \frac{1}{2}x^2) \, dx$.

SOLUTIONS ■

(a) $\displaystyle\int (u^2 + u^3) \, du \quad = \int u^2 \, du + \int u^3 \, du$

$$= \frac{1}{3} u^3 + \frac{1}{4} u^4 + C$$

(b) $\displaystyle\int \left(2x^4 - \frac{1}{2} x^2 \right) dx = \int 2x^4 \, dx - \int \frac{1}{2} x^2 \, dx$

$$= 2 \int x^4 \, dx - \frac{1}{2} \int x^2 \, dx$$

$$= \frac{2}{5} x^5 - \frac{1}{6} x^3 + C \quad ■$$

As we have seen so far, for every rule on derivatives, there is a corresponding rule on integrals. For example

$$\frac{d}{dx} (e^x) = e^x \qquad \text{implies} \qquad \int e^x \, dx = e^x + C$$

We then have the following rule for integration of the exponential function:

Indefinite Integral of the Exponential Function

$$\int e^x \, dx = e^x + C$$

EXAMPLE 5   Integrating Exponential Functions

Find $\int (3e^x - 4x) \, dx$.

SOLUTION ∎

$$\int (3e^x - 4x) \, dx = 3 \int e^x \, dx - 4 \int x \, dx$$
$$= 3e^x - 4 \cdot \tfrac{1}{2}x^2 + C$$
$$= 3e^x - 2x^2 + C \quad \blacksquare$$

Since

$$\frac{d}{dx} (\ln |x|) = \frac{1}{x} \quad \text{implies} \quad \int \frac{1}{x} \, dx = \ln |x| + C$$

we have the following rule.

Indefinite Integral of $x^{-1} = \dfrac{1}{x}$

$$\int \frac{1}{x} \, dx = \ln |x| + C$$

Recall that in the power rule for integrating x^n we restricted $n \neq -1$. (If we try to use this formula with $n = -1$, we obtain a zero in the denominator.) With this last rule we can now find the indefinite integral of x^n for *any* number n.

EXAMPLE 6   Integrating x^{-1}

Find $\int \dfrac{2}{x} \, dx$.

SOLUTION ∎

$$\int \frac{2}{x} \, dx = 2 \int \frac{1}{x} \, dx$$
$$= 2 \ln |x| + C \quad \blacksquare$$

Applications

EXAMPLE 7   Finding Revenue, Given Marginal Revenue

Find the revenue function $R(x)$ if the marginal revenue is $2x - 10$.

SOLUTION ∎   We are given that $R'(x) = 2x - 10$. Thus,

$$R(x) = \int (2x - 10) \, dx$$
$$= x^2 - 10x + C$$

where C is some constant not yet determined.

We know, however, that no sales results in no revenue, that is $R(0) = 0$. Using this gives

$$0 = R(0) = (0)^2 - 10(0) + C = C$$

Thus,

$$R(x) = x^2 - 10x \quad \blacksquare$$

EXAMPLE 8   Finding Cost, Given Marginal Cost

Find the cost function $C(x)$ if the marginal cost is $10e^x + 40x - 100$ and fixed costs are 1000.

SOLUTION ■ We are given that $C'(x) = 10e^x + 40x - 100$. Thus,

$$C(x) = \int (10e^x + 40x - 100)\, dx$$
$$= 10e^x + 20x^2 - 100x + C^*$$

where C^* is some constant not yet determined.
If fixed costs are 1000, this translates into $C(0) = 1000$. Then

$$1000 = C(0) = 10e^0 + 20(0) - 100(0) + C^* = 10 + C^*$$

Thus, $C^* = 990$ and

$$C(x) = 10e^x + 20x^2 - 100x + 990 \quad \blacksquare$$

SELF-HELP EXERCISE SET 5.1

1. Find $\int \left(1 + \sqrt[3]{x} + \dfrac{3}{x} - 2e^x\right) dx$.

2. If a ball is thrown upward with an initial velocity of 20 feet per second, then from physics it can be shown that the velocity (ignoring air resistance) is given by $v(t) = -32t + 20$. Find $s(t)$ if the ball is thrown from 6 feet above ground level and $s(t)$ measures the height of the ball in feet.

EXERCISE SET 5.1

In Exercises 1 through 38 find the antiderivatives.

1. $\int x^{99}\, dx$
2. $\int x^{0.01}\, dx$
3. $\int x^{-99}\, dx$
4. $\int x^{-0.01}\, dx$
5. $\int 5\, dx$
6. $\int dx$
7. $\int \dfrac{5}{y^3}\, dy$
8. $\int \dfrac{1}{2y^5}\, dy$
9. $\int \dfrac{y^{3/2}}{\sqrt{2}}\, dy$
10. $\int \pi y^{2/3}\, dy$
11. $\int \sqrt[3]{u^2}\, du$
12. $\int \sqrt[5]{u^3}\, du$
13. $\int (6x^2 + 4x)\, dx$
14. $\int (8x^3 + 1)\, dx$
15. $\int (x^2 + x + 1)\, dx$
16. $\int (3u^2 + 4u^3)\, du$
17. $\int (\sqrt{2}u^{0.1} - 0.1u^{1.1})\, du$
18. $\int \left(\dfrac{3}{t^2} - 6t^2\right) dt$
19. $\int (t^2 - 1)\, dt$
20. $\int (t^{-2} + 3)\, dt$
21. $\int (\sqrt{t} - \sqrt[3]{t^5})\, dt$
22. $\int (t^3 + 4t^2 + 5)\, dt$
23. $\int (6t^5 - 4t^3 + 1)\, dt$
24. $\int \left(x + \dfrac{1}{x}\right) dx$
25. $\int \left(1 + \dfrac{3}{x}\right) dx$
26. $\int \dfrac{2}{\pi x}\, dx$
27. $\int \left(\pi + \dfrac{1}{x}\right) dx$
28. $\int \dfrac{t^4 + 3}{t^2}\, dt$
29. $\int \dfrac{t + 1}{\sqrt{t}}\, dt$
30. $\int \sqrt{t}(t + 1)\, dt$
31. $\int (e^x - 3x)\, dx$
32. $\int \left(e^x - \dfrac{1}{x}\right) dx$
33. $5 \int e^x\, dx$
34. $\int (6 - e^x)\, dx$
35. $\int (5e^x - 4)\, dx$
36. $\int (6x - e^x)\, dx$
37. $\int \dfrac{x + 1}{x}\, dx$
38. $\int \dfrac{e^{-x} + 1}{e^{-x}}\, dx$

Applications

39. Revenue. Find the revenue function for a knife manufacturer if the marginal revenue, in dollars, is given by $30 - 0.5x$, where x is the number of knives sold.

40. Revenue. Find the revenue function for a paper plate manufacturer if the marginal revenue, in dollars, is given by $200 - 0.26x$, where x is the number of plates sold.

41. Demand. Find the demand function for a salt shaker manufacturer if marginal demand, in dollars, is given by $p'(x) = -x^{-3/2}$, where x is the number of salt shakers sold and $p(1) = 3$.

42. Demand. Find the demand function for a table cloth manufacturer if marginal demand, in dollars, is given by $p'(x) = -x^{-3/2}$, where x is the number of thousands of table cloths sold. Assume $p(1) = 100$.

43. Cost. Find the cost function for an envelope manufacturer if the marginal cost, in dollars, is given by $100 - 0.1e^x$, where x is the number of thousands of envelopes produced, and fixed costs are 1000.

44. Cost. Find the cost function for an adhesive tape manufacturer if the marginal cost, in dollars, is given by $150 - 0.01e^x$, where x is the number of cases of tape produced. Assume $C(0) = 100$.

45. Cost. Find the cost function for a computer disk manufacturer if the marginal cost, in dollars, is given by $30\sqrt{x} - 6x^2$, where x is the number of thousands of disks sold. Assume no fixed costs.

46. Cost. Find the cost function for a spark plug manufacturer if the marginal cost, in dollars, is given by $30x - 4e^x$, where x is the number of thousands of plugs sold. Assume no fixed costs.

47. Velocity. A ball is thrown upward with an initial velocity of 30 feet per second. From physics it can be shown that the velocity (ignoring air resistance) is given by $v(t) = -32t + 30$. Find $s(t)$ if the ball is thrown from 15 feet above ground level and $s(t)$ measures the height of the ball in feet.

48. Acceleration. If a ball is thrown upward, then from physics it can be shown that the acceleration is given by $a(t) = -32$ feet per second. If a ball is thrown upward with an initial velocity of 10 feet per second, find $v(t)$.

49. Velocity. If in the previous problem, the ball is at an initial height of 6 feet, find $s(t)$, the height of the ball after t seconds.

50. Population. A certain insect population is increasing at the rate given by $P'(t) = 2000e^t$, where t is time in years measured from the beginning of 1994 when the population was 200,000. Find the population at the end of 1996.

51. Appreciation. A painting by one of the masters is purchased by a museum for $1,000,000 and increases in value at a rate given by $V'(t) = 100e^t$, where t is measured in years from the time of purchase. What will the painting be worth in 10 years?

52. Marginal Propensity to Consume. Let $C(x)$ represent national consumption in trillions of dollars, where x is disposable national income in trillions of dollars. Suppose that the marginal propensity to consume $C'(x)$ is given by $C'(x) = 1 + \frac{2}{\sqrt{x}}$, and that consumption is $0.1 trillion when disposable income is zero. Find the national consumption function.

53. Marginal Propensity to Save. Let S represent national savings in trillions of dollars. We can assume that disposable national income equals national consumption plus national savings, that is, $x = C(x) + S(x)$. Thus, $C'(x) = 1 - S'(x)$, where $S'(x)$ is the marginal propensity to save. If the marginal propensity to save is given by $0.5x$ and consumption is $0.2 trillion dollars when disposable income is zero, find the consumption function.

54. Sum-of-Years-Digit Depreciation. The sum-of-years-digit depreciation method depreciates a piece of equipment more in the early years of its life and less in the later years. In this method the value decreases at a rate given by $\frac{d}{dt}V(t) = -k(T - t)$, where T is the useful life in years of the equipment, t is the time in years since purchase, and k is a positive constant. (Notice that as t gets closer to T, $V'(t)$ becomes smaller and thus the value decreases less.) If a piece of machinery is purchased for I dollars and has no salvage value, show that $k = \frac{2I}{T^2}$, and write an expression for $V(t)$ in terms of t, I, and T.

55. Agriculture. Carter and Wiebe [120] studied small farms in the African country of Kenya. Their data indicated that the relationship $L'(x) = 0.28 + 0.05x - 0.0001x^2$, held approximately, where x was the size in acres of the farm and L the amount of labor measured in appropriate units. Find $L(x)$.

Enrichment Exercises

56. Explain in complete sentences the difference between an antiderivative of $f(x)$ and the indefinite integral $\int f(x)\, dx$.

57. Find the antiderivatives of e^{2x}. *Hint:* Start by finding the derivative of e^{2x}.

58. In Exercises 59 and 60 of Section 2.4 you found that

$$\frac{d}{dx}(\sin x) = \cos x \quad \text{and} \quad \frac{d}{dx}(\cos x) = -\sin x$$

Using this, find
(a) $\int \sin x\, dx$ **(b)** $\int \cos x\, dx$

59. Find the antiderivatives of $\cos 3x$. *Hint:* Start by finding the derivative of $\sin 3x$ by recalling from Exercise 75 of Section 3.4 that $\frac{d}{dx}[\sin f(x)] = [\cos f(x)] \cdot f'(x)$

60. Find the antiderivatives of $\sin 3x$. *Hint:* Start by finding the derivative of $\cos 3x$ by recalling from Exercise 75 of Section 3.4 that $\frac{d}{dx}[\cos f(x)] = -[\sin f(x)] \cdot f'(x)$

SOLUTIONS TO SELF-HELP EXERCISE SET 5.1

1. First remove the radical, and obtain

$$\int \left(1 + x^{1/3} + \frac{3}{x} - 2e^x\right) dx = \int 1 \, dx + \int x^{1/3} \, dx$$

$$+ 3 \int \frac{1}{x} \, dx - 2 \int e^x \, dx$$

$$= x + \frac{1}{\frac{4}{3}} x^{4/3} + 3 \ln |x| - 2e^x + C$$

$$= x + \frac{3}{4} x^{4/3} + 3 \ln |x| - 2e^x + C$$

2. Since $\frac{ds}{dt}(t) = v(t)$ and $s(0) = 6$, we have

$$s(t) = \int (-32t + 20) \, dt = -32 \int t \, dt + 20 \int 1 \, dt$$

$$= -16t^2 + 20t + C$$

Then since

$$6 = s(0) = -16(0)^2 + 20(0) + C$$

$C = 6$ and

$$s(t) = -16t^2 + 20t + 6$$

5.2 SUBSTITUTION

- The Method of Substitution ■ Applications

APPLICATIONS: PRODUCTION OF OIL

The rate at which a natural resource is extracted tends to increase initially and then fall off later after the easily accessible material has been extracted. Suppose the rate in hundreds of thousands of barrels per year at which oil is being extracted from a field during the early and intermediate stages of the life of the field is given by $P'(t) = \dfrac{t}{(t^2 + 1)^2} + 1$, where t is in years. Find the amount of oil extracted during the first 3 years. See Example 6 for the answer.

The Method of Substitution

The methods in the previous section are inadequate to find all the indefinite integrals that we need. In this section we develop the method of *substitution*. The method of substitution is based on the chain rule. Let us consider the simplest case. Suppose n is any real number with $n \neq -1$ and

$$F(x) = \frac{1}{n + 1} [f(x)]^{n+1}$$

Then using the chain rule, we know that $F'(x) = [f(x)]^n f'(x)$. Thus, $F(x)$ is an antiderivative of $[f(x)]^n f'(x)$. Then

$$\int [f(x)]^n f'(x) \, dx = F(x) + C = \frac{1}{n + 1} [f(x)]^{n+1} + C$$

This gives

General Power Rule

$$\int [f(x)]^n f'(x) \, dx = \frac{1}{n + 1} [f(x)]^{n+1} + C \quad n \neq -1$$

We note that the preceding expression is the correct answer since differentiating the right-hand side gives the integrand on the left-hand side.

The way we use this in practice is as follows. We let $u = f(x)$ and notice that $\frac{du}{dx} = f'(x)$. Now treat du and dx as if they were separate quantities, and rewrite the last equation as $du = f'(x)\,dx$. We then formally *substitute* all of this into the integral, obtaining

$$\int [f(x)]^n\, f'(x)\,dx = \int u^n\,du = \frac{1}{n+1}\,u^{n+1} + C$$

Now we recall that $u = f(x)$ and the previous expression becomes

$$\frac{1}{n+1}\,[f(x)]^{n+1} + C$$

which is the correct answer. Our only justification of this formal procedure is that it always gives the correct answer.

E X A M P L E 1   Using the Method of Substitution

Find $\int 2x(x^2 + 1)^{99}\,dx$.

S O L U T I O N ◾ We could actually solve this problem using the techniques of the previous section by expanding the term $(x^2 + 1)^{99}$ by the binomial theorem and then multiply the result by $2x$. This method of solution provides an expression with 100 terms to integrate.

Instead we try the method of substitution. One should clearly understand that this method is in practice a trial-and-error method.

Suppose we try the substitution $u = x^2 + 1$. Then $du = 2x\,dx$. Substitute all of this into the integral and obtain

$$\int 2x(x^2 + 1)^{99}\,dx = \int (x^2 + 1)^{99}\, 2x\,dx$$
$$= \int u^{99}\,du$$
$$= \frac{1}{100}\,u^{100} + C$$
$$= \frac{1}{100}\,(x^2 + 1)^{100} + C \quad ◾$$

How can we be certain we have the correct answer? Simply differentiate the alleged answer to see if we obtain the integrand. Since

$$\frac{d}{dx}\left[\frac{1}{100}\,(x^2 + 1)^{100} + C\right] = \frac{1}{100}\,(100)(x^2 + 1)^{99}(2x) + 0 = 2x(x^2 + 1)^{99}$$

our answer is correct.

How can one possibly know what to substitute? The preceding process indicates that if you encounter an integrand that is of the form of a function raised to a power, then set u equal to the function as a first try. It just might work! (Fortunately the previous integrand has the factor $2x$ in it, otherwise we would be stuck and would have to resort to some other, as yet unlearned, technique.)

E X A M P L E 2   Using the General Power Rule

Find $\displaystyle\int \frac{3x^2}{\sqrt{x^3 + 3}}\,dx$.

S O L U T I O N ◾ Rewrite the integral as $\int 3x^2(x^3 + 3)^{-1/2}\,dx$. We notice a function $(x^3 + 3)$ raised to the power $-\frac{1}{2}$. Thus, we try $u = x^3 + 3$ and hope for the best. With u defined in this way, we then have $du = 3x^2\,dx$. Substitute all of this into the integral, and obtain

$$\int 3x^2(x^3 + 3)^{-1/2}\, dx = \int (x^3 + 3)^{-1/2} 3x^2\, dx$$

$$= \int u^{-1/2}\, du$$

$$= 2u^{1/2} + C$$

$$= 2(x^3 + 3)^{1/2} + C \quad \blacksquare$$

To check this answer we simply differentiate and notice that it does indeed equal the integrand. That is,

$$\frac{d}{dx}\,[2(x^3 + 3)^{1/2} + C] = 3x^2(x^3 + 3)^{-1/2}$$

EXAMPLE 3   Using the General Power Rule

Find $\displaystyle\int \frac{4\ln x}{x}\, dx.$

SOLUTION ■ If $u = \ln x$, then $du = \dfrac{1}{x}\, dx$. Then

$$\int \frac{4\ln x}{x}\, dx = 4\int \ln x \cdot \frac{1}{x}\, dx$$

$$= 4\int u\, du$$

$$= 2u^2 + C$$

$$= 2(\ln x)^2 + C \quad \blacksquare$$

We now consider the case $n = 1$ that was avoided in the power rule earlier. For the second substitution formula we let $F(x) = \ln|f(x)|$. Then $F'(x) = \dfrac{f'(x)}{f(x)}$, and $F(x)$ is an antiderivative of $f'(x)/f(x)$. Thus,

$$\int \frac{f'(x)}{f(x)}\, dx = F(x) + C = \ln|f(x)| + C$$

This gives

General Logarthmic Rule

$$\int \frac{f'(x)}{f(x)}\, dx = \ln|f(x)| + C$$

In practice, if we let $u = f(x)$, then $du = f'(x)\, dx$. Formally substituting this into the integral yields

$$\int \frac{f'(x)}{f(x)}\, dx = \int \frac{1}{f(x)}\, f'(x)\, dx$$

$$= \int \frac{1}{u}\, du$$

$$= \ln|u| + C$$

$$= \ln|f(x)| + C$$

which is the correct answer.

EXAMPLE 4   Using the Method of Substitution

Find $\displaystyle\int \frac{x + 1}{x^2 + 2x}\, dx.$

SOLUTION ▪ We let $u = x^2 + 2x$, then $du = (2x + 2) \, dx = 2(x + 1) \, dx$. Now formally substitute all of this into the integral and obtain

$$\int \frac{x + 1}{x^2 + 2x} \, dx = \int \frac{1}{x^2 + 2x} (x + 1) \, dx$$

$$= \int \frac{1}{u} \frac{1}{2} \, du \qquad\qquad (x + 1) \, dx = \frac{1}{2} \, du$$

$$= \frac{1}{2} \int \frac{du}{u}$$

$$= \frac{1}{2} \ln |u| + C$$

$$= \frac{1}{2} \ln |x^2 + 2x| + C \quad ▪$$

As the last case of the use of substitution, now let $F(x) = e^{f(x)}$. Then $F'(x) = e^{f(x)} f'(x)$, and so $F(x)$ is an antiderivative of $e^{f(x)} f'(x)$. Thus,

$$\int e^{f(x)} f'(x) \, dx = F(x) + C = e^{f(x)} + C$$

This gives

General Exponential Rule

$$\int e^{f(x)} f'(x) \, dx = e^{f(x)} + C$$

We know this must be the correct answer since differentiating the answer on the right-hand side gives the integrand on the left-hand side.

In practice we let $u = f(x)$ and then $du = f'(x) \, dx$. Formally substituting all of this into the integral gives

$$\int e^{f(x)} f'(x) \, dx = \int e^u \, du$$

$$= e^u + C$$

$$= e^{f(x)} + C$$

which is the correct answer.

EXAMPLE 5 The General Exponential Rule

Find $\int (x^2 + 1) e^{x^3 + 3x + 1} \, dx$.

SOLUTION ▪ Let $u = x^3 + 3x + 1$. Then $du = (3x^2 + 3) \, dx = 3(x^2 + 1) \, dx$. Then

$$\int (x^2 + 1) e^{x^3 + 3x + 1} \, dx = \int e^{x^3 + 3x + 1} (x^2 + 1) \, dx$$

$$= \int e^u \frac{1}{3} \, du \qquad\qquad (x^2 + 1) \, dx = \frac{1}{3} \, du$$

$$= \frac{1}{3} \int e^u \, du$$

$$= \frac{1}{3} e^u + C$$

$$= \frac{1}{3} e^{x^3 + 3x + 1} + C$$

This is the correct answer since

$$\frac{d}{dx}\left(\frac{1}{3}e^{x^3+3x+1}+C\right)=(x^2+1)e^{x^3+3x+1}$$

which is the integrand.

■ Historic Oil Well

The first development oil well in Alaska's Prudhoe Bay, Number DS 1-1, produced over 19 million barrels of oil during its 19-year life. It went into production in December, 1970.

Applications

EXAMPLE 6   Production of Oil

The rate at which a natural resource is extracted tends to increase initially and then fall off later after the easily accessible material has been extracted. (See Dorner [121].) Suppose the rate in hundreds of thousands of barrels per year at which oil is being extracted during the early and intermediate stages of the life of the field is given by

$$P'(t)=\frac{t}{(t^2+1)^2}+1$$

where t is in years (Fig. 5.2). Find the amount of oil extracted during the first three years.

SOLUTION ■ We must have

$$P(t)=\int\left[\frac{t}{(t^2+1)^2}+1\right]dt=\int(t^2+1)^{-2}t\,dt+\int dt$$

The second integral is $t+C$. The first integral requires the substitution $u=t^2+1$. Then $du=2t\,dt$ and

$$\int(t^2+1)^{-2}t\,dt+\int dt=\int u^{-2}\frac{1}{2}\,du+t+C$$

$$=-\frac{1}{2}u^{-1}+t+C$$

$$=-\frac{1}{2}\cdot\frac{1}{t^2+1}+t+C$$

There can be no production of oil in no time, thus,

$$0=P(0)=-\frac{1}{2}+C$$

This implies that $C=\frac{1}{2}$. Thus,

$$P(3)=-\frac{1}{2}\cdot\frac{1}{3^2+1}+3+\frac{1}{2}=3.45$$

Thus, 345,000 barrels of oil are extracted in the first three years. ■

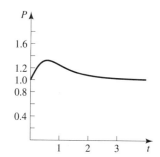

Figure 5.2
Graph of the rate of extraction of oil from a field.

SELF-HELP EXERCISE SET 5.2

1. Find $\displaystyle\int\frac{2x}{\sqrt[3]{x^2+4}}\,dx$ **2.** Find $\displaystyle\int\frac{e^{2x}}{e^{2x}+1}\,dx$ **3.** Find $\displaystyle\int\frac{e^{1/x}}{x^2}\,dx$

EXERCISE SET 5.2

In Exercises 1 through 34 find the indefinite integral.

1. $\displaystyle\int 6(3x+1)^{10}\,dx$ **2.** $\displaystyle\int 8(3-2x)^5\,dx$

3. $\displaystyle\int x(3-x^2)^7\,dx$ **4.** $\displaystyle\int 10x(2x^2+1)^3\,dx$

5. $\displaystyle\int 8(x+1)(2x^2+4x-1)^{3/2}\,dx$

6. $\displaystyle\int (x^3 + 2)(x^4 + 8x + 3)^{1/3}\, dx$

7. $\displaystyle\int (x^3 + x^2 + x + 1)(3x^4 + 4x^3 + 6x^2 + 12x + 1)^3\, dx$

8. $\displaystyle\int (x^4 + x^2 + 2)(3x^5 + 5x^3 + 30x + 1)^5\, dx$

9. $\displaystyle\int 2\sqrt{x + 1}\, dx$

10. $\displaystyle\int \sqrt[3]{x + 1}\, dx$

11. $\displaystyle\int x\sqrt{x^2 + 1}\, dx$

12. $\displaystyle\int x\sqrt{3x^2 + 1}\, dx$

13. $\displaystyle\int \frac{x}{\sqrt[3]{x^2 + 1}}\, dx$

14. $\displaystyle\int \frac{x}{\sqrt[5]{2x^2 + 5}}\, dx$

15. $\displaystyle\int \frac{\sqrt{x^{1/3} + 1}}{x^{2/3}}\, dx$

16. $\displaystyle\int \frac{\ln 2x}{x}\, dx$

17. $\displaystyle\int \frac{\sqrt{\ln x}}{x}\, dx$

18. $\displaystyle\int e^{2x+1}\, dx$

19. $\displaystyle\int e^{1-x}\, dx$

20. $\displaystyle\int 4xe^{x^2-1}\, dx$

21. $\displaystyle\int xe^{1-x^2}\, dx$

22. $\displaystyle\int \frac{e^{1/x^2}}{x^3}\, dx$

23. $\displaystyle\int \frac{e^{\sqrt{x}}}{\sqrt{x}}\, dx$

24. $\displaystyle\int \frac{1}{2x + 1}\, dx$

25. $\displaystyle\int \frac{1}{3x + 5}\, dx$

26. $\displaystyle\int \frac{x^3}{(x^4 + 4)^2}\, dx$

27. $\displaystyle\int \frac{x}{(x^2 + 3)^4}\, dx$

28. $\displaystyle\int \frac{e^x}{e^x + 1}\, dx$

29. $\displaystyle\int \frac{e^{-x}}{e^{-x} + 1}\, dx$

30. $\displaystyle\int \frac{e^x - e^{-x}}{e^x + e^{-x}}\, dx$

31. $\displaystyle\int \frac{e^{2x} - e^{-2x}}{e^{2x} + e^{-2x}}\, dx$

32. $\displaystyle\int \frac{1}{x \ln x}\, dx$

33. $\displaystyle\int \frac{1}{x \ln x^2}\, dx$

34. $\displaystyle\int \frac{1}{x \ln \sqrt{x}}\, dx$

Applications

35. Demand. Find the demand function for a clock manufacturer if marginal demand, in dollars, is given by $p'(x) = -24x/(3x^2 + 1)^2$ and $p(1) = 10$, where x is the number of thousands of clocks sold.

36. Demand. Find the demand function for a cardboard box manufacturer if marginal demand, in dollars, is given by $p'(x) = -4xe^{-x^2}$ and $p(1) = 10$, where x is the number of thousands of boxes sold.

37. Revenue. Find the revenue function for a shoulder bag manufacturer if the marginal revenue, in dollars, is given by $4x(10 - x^2)$, where x is the number of hundreds of bags sold.

38. Revenue. Find the revenue function for a purse manufacturer if the marginal revenue, in dollars, is given by $(1 - x)e^{2x-x^2}$, where x is the number of thousands of purses sold.

39. Cost. Find the cost function for a lipstick manufacturer if the marginal cost, in dollars, is given by $4x/(x^2 + 1)$, where x is the number of cases of lipstick produced, and fixed costs are 1000.

40. Cost. Find the cost function for a hair brush manufacturer if the marginal cost, in dollars, is given by $x^3(100 + x^4)$, where x is the number of cases of brushes produced, and fixed costs are 1000.

41. Production. Suppose the rate in tons per year at which copper is being extracted from a mine during the early and intermediate stages of the life of the mine is given by $P'(t) = \dfrac{-54t^2}{(t^3 + 1)^2} + 11$. Find the amount of copper extracted during the first two years.

42. Depreciation. The value of a piece of machinery purchased for \$10,000 is decreasing at the rate given by $V'(t) =$ $-1000te^{-0.25t^2}$, where t is measured in years from purchase. How much will the machine be worth in two years?

43. Population. The population of a certain city for the next several years is increasing at the rate given by $P'(t) = 2000(t + 10)e^{0.05t^2+t}$, where t is time in years measured from the beginning of 1994 when the population was 100,000. Find the population at the end of 1995.

44. Sales. After the introduction of a new camera, the rate of sales in thousands per month is given by $N'(t) = 10(1 + 2t)^{-5/2}$, where t is in months. Find the number sold in the first four months.

45. Productivity. Suppose the number of items produced on a certain piece of machinery by an average employee is increasing at a rate given by $N'(t) = \dfrac{1}{2t + 1}$, where t is measured in hours since the employee was placed on the machine for the first time. How many items are produced by the average employee in the first three hours?

46. Memorization. Suppose the rate at which an average person can memorize a list of items is given by $N'(t) = \dfrac{15}{\sqrt{3t + 1}}$, where t is the number of hours spent memorizing. How many items does the average person memorize in one hour?

47. Depreciation. The value of a piece of machinery purchased for \$20,000 is decreasing at the rate given by $V'(t) = -1000e^{-0.1t}$, where t is measured in years from purchase. How much will the machine be worth in four years?

48. Radioactive Decay. A certain radioactive substance initially weighing 100 g decays at a rate given by $A'(t) = -20e^{-0.5t}$, where t is measured in years. How much of the substance is there after 10 years?

Enrichment Exercises

49. Try to find $\int (x^5 + x^4 + 1)^2(5x^3 + 4x^2)\, dx$ by using the method of substitution with $u = x^5 + x^4 + 1$. Show that the method fails since there is no way of solving the equation $u = x^5 + x^4 + 1$ for x.

50. Evaluate the integral in the previous exercise. Do *not* try any form of substitution.

In Exercises 51 through 54 use the fact that $\dfrac{d}{dx}(\sin u) = (\cos u)\dfrac{du}{dx}$ and $\dfrac{d}{dx}(\cos u) = (-\sin u)\dfrac{du}{dx}$.

51. $\displaystyle\int 2x \sin x^2\, dx$

52. $\displaystyle\int e^x \cos e^x\, dx$

53. $\displaystyle\int x^2 \cos x^3\, dx$

54. $\displaystyle\int (8x + 2)\sin(4x^2 + 2x)\, dx$

SOLUTIONS TO SELF-HELP EXERCISE SET 5.2

1. Write the integral as $\int (x^2 + 4)^{-1/3}2x\, dx$ and use the substitution $u = x^2 + 4$. Then $du = 2x\, dx$ and

$$\int (x^2 + 4)^{-1/3}2x\, dx = \int u^{-1/3}\, du$$

$$= \frac{3}{2}u^{2/3} + C$$

$$= \frac{3}{2}(x^2 + 4)^{2/3} + C$$

2. Write the integral as $\int (e^{2x} + 1)^{-1}e^{2x}\, dx$. This is a function raised to the power $n = -1$. Let $u = e^{2x} + 1$, then $du = 2e^{2x}\, dx$. Then

$$\int \frac{e^{2x}}{e^{2x} + 1}\, dx = \int \frac{\frac{1}{2}\, du}{u} = \frac{1}{2}\int \frac{du}{u}$$

$$= \frac{1}{2}\ln|u| + C$$

$$= \frac{1}{2}\ln(e^{2x} + 1) + C$$

3. First write the integral as $\int e^{1/x}x^{-2}\, dx$. Let $u = 1/x = x^{-1}$. Then $du = -x^{-2}\, dx$ and

$$\int \frac{e^{1/x}}{x^2}\, dx = \int e^{1/x}x^{-2}\, dx$$

$$= \int e^u(-\,du) = -\int e^u\, du$$

$$= -e^u + C = -e^{1/x} + C$$

5.3 DISTANCE TRAVELED

■ Distance Traveled ■ Accuracy of Approximation

APPLICATION: DISTANCE TRAVELED

An object travels with velocity $v = t^2$, where v is in feet per second and t is in seconds. Find estimates, to any desired accuracy, for the distance the object travels during the time-interval $[0, 1]$. For the answer see Example 1 and the discussion that follows.

Method of Exhaustion

Around 370 BC, the Greek geometer Eudoxus of Cnidus made the method of exhaustion a rigorous procedure. To find the area of a circle, he first inscribed an equilateral triangle P_0 shown in Figure 5.3a and labeled E the area outside the triangle but inside the circle. He then placed a regular hexagon P_1, as shown in Figure 5.3b, and noted that from simple geometrical principles the area outside P_1 and inside the circle was less than $\frac{1}{2}E$. Continuing in this manner, as shown in Figure

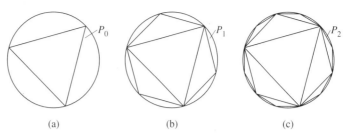

Figure 5.3

5.3c, the area outside the regular 12-sided polygon P_2 and inside the circle is less than $(\frac{1}{2})^2 E$. In such a manner, regular $3 \cdot 2^n$-sided polygons P_n could be inscribed in a circle with the area outside the polygon P_n and inside the circle less than $(\frac{1}{2})^n E$. Since $(\frac{1}{2})^n E \to 0$ as $n \to \infty$, the areas of P_n must approach the area of the circle. Since he knew the areas of P_n, he could find the area of the circle to any desired accuracy.

Bernhard Riemann   1826–1866

Bernhard Riemann clarified the concept of the integral by developing the Riemann sums, which led later to the more important *Lebesgue* integral. He also developed a non-Euclidean geometry called Riemannian geometry that was used by Albert Einstein as a basis for his relativity theory. He also is known for the Riemann zeta function and the associated Riemann hypothesis, a celebrated unproven conjecture. His achievements were all the more remarkable, since he died from tuberculosis at the age of 39.

(The Bettman Archive)

Distance Traveled

As our first goal we wish to estimate the distance an object has traveled in 1 second if we know that the velocity function is given by $v = f(t) = t^2$ in feet per second on the t-interval $[0, 1]$. We are all familiar with monitoring the velocity of an object, since when driving an automobile we look at the speedometer from time to time. As we shall soon see this is a problem that models basic problems in business, economics, biology, and other areas. Thus, a clear understanding of this velocity problem leads to a deeper understanding of basic problems in other areas.

If our car travels at a constant velocity of 55 miles per hour for 2 hours, then, since distance is the product of velocity with time ($d = v \cdot t$), we have traveled $55 \cdot 2 = 110$ miles. It is useful to represent the distance traveled as a certain area. In Figure 5.4 the velocity is constant. Notice, in this case, that the distance traveled in 2 hours, $55 \cdot 2$, is also the area of the rectangle with height equal to the constant velocity of 55 and base equal to the length, 2, of the interval $[0, 2]$. Thus, in this case, distance traveled is the area of the region between the graph of $v = 55$ and the t-axis on the interval $[0, 2]$. But what happens if velocity varies from instant to instant? How can we determine the distance traveled? This is a fundamental problem that we must address.

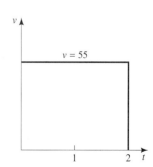

Figure 5.4
The area of the rectangle is base $\times$ height $= 2 \cdot 55 = 110$. This area is also the distance the car travels in 2 hours.

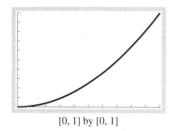

[0, 1] by [0, 1]

Screen 5.1
A graph of $y = x^2$.

The object mentioned earlier has a variable velocity $v = f(t) = t^2$ on the t-interval [0, 1] (Screen 5.1). We first note that the velocity is positive on the interval (0, 1], and therefore, the object always moves in the positive direction. We also note that the velocity is *increasing* on this interval. For starters, let us divide the interval [0, 1] into $n = 5$ subintervals of equal length and create the table of velocities shown here.

Time (seconds)	0	0.2	0.4	0.6	0.8	1
Velocity (feet per second)	0	$(0.2)^2$	$(0.4)^2$	$(0.6)^2$	$(0.8)^2$	$(1)^2$
Velocity	0	0.04	0.16	0.36	0.64	1

Let us first determine an upper estimate of the distance traveled. Since the velocity of our object is always *increasing*, the velocity at any time during the subinterval [0, 0.2] is less than the velocity at the right-hand endpoint $t = 0.2$. The velocity at $t = 0.2$ is $(0.2)^2 = 0.04$. Now imagine another object, which we call the fast object, that travels with a constant velocity of 0.04 during the same fifth of a second. A graph of this velocity is shown in Figure 5.5. Since the velocity of our object is always less than or equal to that of the fast object, the distance our object travels during the first fifth of a second is *less than or equal to* the distance the fast object travels. This latter distance is $v \cdot t = (0.04)(0.2) = 0.008$ feet, and is also the area of the rectangle shown in Figure 5.5.

Likewise, the velocity of our object at any time during the second fifth of a second is less than at the time $t = 0.4$, which is located at the right-hand endpoint of the second interval. The velocity at $t = 0.4$ is $(0.4)^2 = 0.16$. Therefore, during the second fifth of a second the distance our object travels is less than or equal to that of a fast object traveling with a constant velocity of 0.16 indicated in Figure 5.6. This latter distance is $v \cdot t = (0.16)(0.2) = 0.032$ feet and is also the area of the second rectangle indicated in Figure 5.6.

Continuing in this way we obtain Figure 5.7. An upper estimate of the distance traveled on the interval [0, 1] is then

$$\text{Upper estimate} = (0.04)(0.2) + (0.16)(0.2) + (0.36)(0.2)$$
$$+ (0.64)(0.2) + (1)(0.2) = 0.44$$

You can obtain Figure 5.7 on your graphing calculator by executing the program RECT in the Technology Resource Manual. First input $Y_1 = x^2$ and use a window with

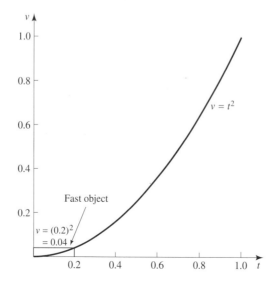

Figure 5.5
The area of the rectangle is base $\times$ height $= (0.2) \cdot (0.2)^2 = 0.008$. This area is also the distance the fast object travels during the first fifth of a second.

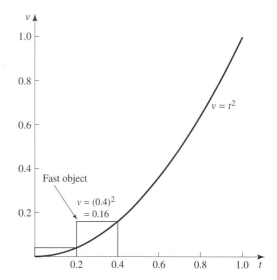

Figure 5.6
The area of the rectangle is $=$ $(0.2) \cdot (0.4)^2 = 0.032$. This area is also the distance the fast object travels during the second fifth of a second.

dimensions $[0, 1]$ by $[0, 1]$. When the program requests N, input 5. By pressing the ENTER (or EXE) key again, the program gives you the numerical value of the upper estimate.

Let us now determine a lower estimate of the distance traveled. Since the velocity of our object is always greater than or equal to zero at any time during the time interval $[0, 0.2]$, the distance our object travels during the first fifth of a second is greater than or equal to 0. For future reference, we notice that our object has 0 velocity at the *left-hand* endpoint of the first interval, and we can visualize the distance 0 as the area of a rectangle with height 0 and width 0.2.

Since the velocity of our object is always *increasing*, the velocity during the second fifth of a second is greater than at time $t = 0.2$, which is located at the left-hand endpoint of the second interval. The velocity at $t = 0.2$ is $(0.2)^2 = 0.04$. Therefore, during the second fifth of a second the distance our object travels is greater than or equal to that of a slow object traveling with a constant velocity of 0.04 indicated in Figure 5.8. This latter distance is $v \cdot t = (0.04)(0.2) = 0.008$ feet and is also the area of the rectangle indicated in Figure 5.8. (The first rectangle on the first subinterval has height equal to zero and therefore does not show in the figure.)

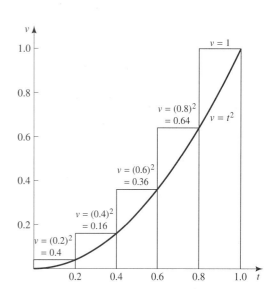

Figure 5.7
The distance the fast object travels is given by the sum of the areas of the five rectangles. This must be greater than the distance our object travels on $[0, 1]$.

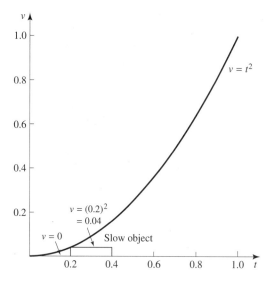

Figure 5.8
The area of the rectangle is $=$ $(0.2) \cdot (0.2)^2 = 0.008$. This area is also the distance the slow object travels during the second fifth of a second.

Continuing in this way we obtain Figure 5.9. A lower estimate of the distance traveled on the interval [0, 1] is then

$$\text{Lower estimate} = (0)(0.2) + (0.04)(0.2) + (0.16)(0.2)$$
$$+ (0.36)(0.2) + (0.64)(0.2) = 0.24$$

You can obtain Figure 5.9 on your graphing calculator by continuing the program RECT by pressing the ENTER (or EXE) key again. Press this key once more, and the program gives you the numerical value of the lower estimate.

In summary, we have

$$0.24 \leq \text{distance traveled} \leq 0.44$$

$$\text{Upper estimate} - \text{lower estimate} = 0.44 - 0.24 = 0.2$$

Figure 5.10 overlays Figure 5.9 with Figure 5.7. The total area of the five shaded rectangles is precisely the lower estimate of the distance traveled by our object, whereas the total area of the five tall rectangles is precisely the upper estimate of the distance traveled by the object. The total area of the five unshaded small rectangles is the difference between the upper and lower estimates. If we slide these unshaded rectangles

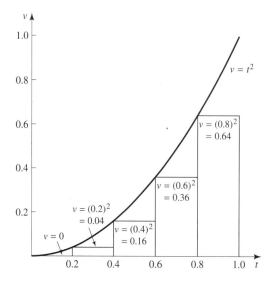

Figure 5.9
The distance the slow object travels is given by the sum of the areas of the five rectangles. (The first rectangle has height 0.) This must be less than the distance our object travels on [0, 1].

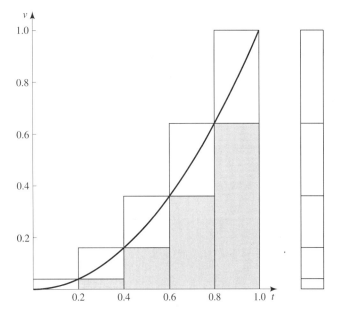

Figure 5.10
The total area of the shaded rectangles equals the left-hand sum, and the total area of the bigger rectangles equals the right-hand sum. The area of the rectangle on the right is the difference between the right- and left-hand sums.

off to the right as indicated in Figure 5.10, we see that the total area of these rectangles must be

$$[f(1) - f(0)](0.2) = [1 - 0](0.2) = 0.2$$

This is the value we obtained earlier.

EXAMPLE 1   Estimating Distance Traveled

Now divide the t-interval $[0, 1]$ into 10 subintervals, and repeat the preceding process.

SOLUTION   ■   Let $t_0 = 0$, $t_1 = 0.1$, $t_2 = 0.2, \ldots, t_{10} = 1$. Since the velocity at any time during the first tenth of a second must be less than the velocity at time $t = 0.1$, the distance traveled during the first tenth of a second must be less than $f(t_1) \cdot (0.1) = (0.1)^2(0.1) = 0.001$, and so on. An upper estimate of the distance traveled is then

$$\text{Upper estimate} = f(t_1) \cdot (0.1) + f(t_2) \cdot (0.1) + \cdots + f(t_{10}) \cdot (0.1)$$

In a similar fashion

$$\text{Lower estimate} = f(t_0) \cdot (0.1) + f(t_1) \cdot (0.1) + \cdots + f(t_9) \cdot (0.1)$$

Instead of calculating these by hand, we use the program RECT again, but this time with $n = 10$. The two screens obtained are overlaid and shown in Figure 5.11. The numerical values obtained are

$$\text{Lower estimate} = 0.285$$

$$\text{Upper estimate} = 0.385$$

Therefore, we have

$$0.285 \leq \text{distance traveled} \leq 0.385$$

$$\text{Upper estimate} - \text{lower estimate} = 0.385 - 0.285 = 0.1 \quad ■$$

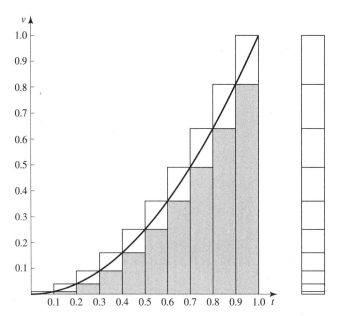

Figure 5.11
Notice that doubling the number of rectangles from that in Figure 5.10 halves the difference between the right- and left-hand sums.

Notice that by doubling the number of subintervals, we have halved the difference between the upper estimate and the lower estimate. The difference is the total area of the 10 unshaded rectangles in Figure 5.11. By taking even more subdivisions, the rectangles shown in Figure 5.11 become thinner, and the difference between the upper and lower estimate, the total area of the unshaded small rectangles, gets even smaller. It thus appears from examination of Figures 5.10 and 5.11 that letting the number of subdivisions n become large without bound results in the upper estimates decreasing and the lower estimates increasing and both estimates heading for the area of the region between the graph of $y = t^2$ and the t-axis on the interval $[0, 1]$. Thus, the *exact* distance traveled by the particle is precisely this latter area.

EXPLORATION *1*

Letting $n \to \infty$

 You can graphically verify the comments made in the previous paragraph by using the program RECT. Input $y = x^2$, and use a window with dimensions $[0, 1]$ by $[0, 1]$. The RECT program requests n. Take n successively equal to 20, 47, and 94. Each time you provide a new n the rectangles associated with the left-hand sums and the right-hand sums are graphed. Does it appear that the area under the rectangles is heading for the area under the curve as $n \to \infty$? (We verify this numerically in the next section.) Check to see if the upper estimates are decreasing and the lower estimates are increasing. Try this program on some other functions.

In general, suppose we have an object traveling at a velocity given by a nonnegative continuous function $v = f(t)$ on the interval $[a, b]$. We divide the interval into n subintervals of equal length $\Delta t = \dfrac{b - a}{n}$. The endpoints of the subintervals are $a = t_0$, $t_1, t_2, \ldots, t_n = b$. We then define

$$\text{Right-hand sum} = f(t_1)\,\Delta t + f(t_2)\,\Delta t + \cdots + f(t_n)\,\Delta t$$

$$\text{Left-hand sum} = f(t_0)\,\Delta t + f(t_1)\,\Delta t + \cdots + f(t_{n-1})\,\Delta t$$

We use the term **right-hand sum** since we use the times at the right-hand ends of the subintervals, and use the term **left-hand sum** since we use the times at the left-hand ends of the subintervals.

If n is large, both the left- and right-hand sums should approximate the total distance traveled by the object on the time-interval $[a, b]$.

Total Distance Traveled

Suppose an object travels at a velocity given by the nonnegative continuous function $v = f(t)$ on the interval $[a, b]$. Divide the interval into n subintervals of equal length $\Delta t = \dfrac{b - a}{n}$. The endpoints of the subintervals are $a = t_0, t_1, t_2, \ldots, t_n = b$. We then define

$$\text{Right-hand sum} = f(t_1)\,\Delta t + f(t_2)\,\Delta t + \cdots + f(t_n)\,\Delta t$$

$$\text{Left-hand sum} = f(t_0)\,\Delta t + f(t_1)\,\Delta t + \cdots + f(t_{n-1})\,\Delta t$$

Furthermore, the limit as $n \to \infty$ of the left-hand sum equals the limit as $n \to \infty$ of the right-hand sum. This common limit is the total distance traveled by the object on the interval $[a, b]$. This distance is also the area of the region between the graph of $v = f(t)$ and the t-axis on the interval $[a, b]$.

In the next section we explore in more detail how to find the exact distance traveled by taking the limits of the left- and right-hand sums. We know that this distance equals the area of the region between the graph of $v = f(t)$ and the t-axis on the interval $[a, b]$. In certain instances we can find this area by using known geometric formulas. The following example is such a case.

EXAMPLE 2 Finding Distance Exactly

Find the distance traveled by the object with a velocity of $v = 2t$ on the interval $[1, 3]$ by finding the area of an appropriate geometric region.

SOLUTION ■ A graph of $v = 2t$ is shown in Figure 5.12. The distance traveled by the object on the interval $[1, 3]$ is the area of the region between the graph of the line $v = 2t$ and the t-axis on the interval $[1, 3]$. The region is a triangle sitting on a rectangle of dimensions 2×2. The area of the rectangle is 4 and of the triangle is 4. Thus, the total area and the distance traveled by the object is 8. ■

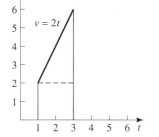

Figure 5.12
The distance traveled on the interval [1, 3] is the area under the graph.

Accuracy of Approximation

If the function $v = f(t)$ is always increasing or always decreasing on the interval $[a, b]$, we say that $v = f(t)$ is **monotonic** on $[a, b]$. From earlier analysis we have the following:

Error When the Function Is Monotonic

If the function $v = f(t)$ is monotonic, then

$$\left|\text{Left-hand sum} - \text{right-hand sum}\right| = \left|f(b) - f(a)\right|\,\Delta t$$

We use absolute values to handle both increasing and decreasing functions in one formula.

EXAMPLE 3 Accuracy of Approximation

Consider the object in Example 1. What is the difference between the upper and lower estimates for the object in Example 1 if we divide the interval into 100 subintervals? Into 1000?

SOLUTION ■ Since the function $v = f(t)$ is increasing, the right-hand sum minus the left-hand sum must equal

$$[f(1) - f(0)] \, \Delta t = [1 - 0] \frac{1}{n} = \frac{1}{n}$$

Thus, if $n = 100$, the difference is 0.01, and if $n = 1000$, the difference is 0.001. ■

When the velocity function $v = f(t)$ is either always increasing or always decreasing on the interval of interest, the left- and right-hand sums always bound the area, giving rise to a very convenient formula for the maximum error of our estimates. This is not necessarily the case for all functions, as we shall see in the next section. Nonetheless, if the function $v = f(t)$ is continuous on the finite interval $[a, b]$, then the left- and right-hand limits still converge to a common limit as $n \to \infty$. We do not, however, have a convenient error estimate formula.

SELF-HELP EXERCISE SET 5.3

1. An object decelerates with a velocity $v = f(t) = \frac{1}{t}$ on the interval $[1, 2]$. Give an upper and a lower estimate for the distance traveled by dividing the interval $[1, 2]$ into five equal subintervals, and calculate by hand the left- and right-hand sums. Sketch a graph similar to Figure 5.4 or 5.5 illustrating the left- and right-hand sums.

2. Identify the region whose area is exactly the distance traveled by the object in Exercise 1 on the interval $[1, 2]$.

EXERCISE SET 5.3

In Exercises 1 through 10 we consider approximations to the distance traveled by an object with velocity $v = f(t)$ on the given interval $[a, b]$.
(a) Make a sketch that illustrates the left- and right-hand sums showing clearly the five rectangles and $t_0, t_1, t_2, t_3, t_4, t_5$.
(b) For $n = 5$. Write out left- and right-hand sums by hand, and calculate each without using the program SUM. Also calculate the difference between the upper and lower estimates.
(c) Finally, use the program SUM found in the Technology Resource Manual to find the left- and right-hand sums when $n = 10$. Also calculate the difference between the upper and lower estimates and compare with part (b).

1. $v = f(t) = 5 - 2t$, $[0, 2]$
2. $v = f(t) = 2t + 1$, $[0, 2]$
3. $v = f(t) = t^2$, $[0, 2]$
4. $v = f(t) = \frac{1}{t}$, $[1, 3]$
5. $v = f(t) = t^2$, $[1, 3]$
6. $v = f(t) = \frac{1}{t}$, $[1, 1.5]$

7. $v = f(t) = t^2 + 1$, $[-2, 3]$
8. $v = f(t) = 10t^2$, $[-3, 2]$
9. $v = f(t) = \sqrt{t}$, $[0, 5]$
10. $v = f(t) = t^3$, $[-2, 3]$

In Exercises 11 through 20 find the distance traveled by the object on the given interval by finding the areas of the appropriate geometric region.

11. $v = f(t) = 4$, $[1, 5]$
12. $v = f(t) = 3$, $[2, 7]$
13. $v = f(t) = t$, $[0, 1]$
14. $v = f(t) = t$, $[0, 2]$
15. $v = f(t) = 2t + 1$, $[0, 2]$
16. $v = f(t) = 3t - 1$, $[1, 2]$
17. $v = f(t) = 6 - 2t$, $[1, 3]$
18. $v = f(t) = 10 - 3t$, $[1, 3]$
19. $v = f(t) = |1 - t|$, $[0, 2]$
20. $v = f(t) = |2 - t|$, $[1, 4]$

Applications

21. An insect travels with a velocity function given by $v = 2t$, where t is measured in seconds and v in feet per second. Find a formula that gives the exact distance this insect travels during the first t seconds. *Hint:* Consider the area of an appropriate geometric region.

22. A bird travels with a velocity function given by $v = 3t + 1$, where t is measured in seconds and v in feet per second. Find a formula that gives the exact distance this bird travels during the first t seconds. *Hint:* Consider the area of an appropriate geometric region.

23. An object travels with a velocity function given by the figure. Use the grid to obtain an upper and lower estimate of the distance traveled by the object. Explain what you are doing.

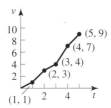

24. An object travels with a velocity function given by the figure. Use the grid to obtain an upper and lower estimate of the distance traveled by the object. Explain what you are doing.

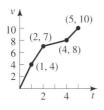

25. The table gives the velocity at 1-second intervals of an accelerating automobile. Give an upper and lower estimate for the distance the car travels during the 4-second interval. Explain what you are doing.

Time (seconds)	0	1	2	3	4
Velocity (feet per second)	10	20	25	40	50

26. A person slams on the brakes of an automobile. The table gives the velocity at 1-second intervals. Give an upper and lower estimate for the distance the car travels during the 4-second interval. Explain what you are doing.

Time (seconds)	0	1	2	3	4
Velocity (feet per second)	50	40	25	10	0

27. An object travels with a velocity function given by $v = \sqrt{t}$, where t is measured in hours and v in miles per second. What is the difference between the upper and lower estimates for the object if we divide the interval $[0, 1]$ into 100 subintervals? Into 1000?

28. An object travels with a velocity function given by $v = 4 - \sqrt{t}$, where t is measured in hours and v in miles per second. What is the difference between the upper and lower estimates for the object if we divide the interval $[0, 1]$ into 100 subintervals? Into 1000?

Enrichment Exercises

29. Refer to Exercise 27. How large must you take n to ensure that the difference between the upper and lower estimates in less than 0.1? 0.01?

30. Figure 5.3 outlines the method of exhaustion formulated by Eudoxus of Cnidus. Let E be the area of the region inside the circle but outside the triangle P_0 shown in Figure 5.3a. Explain why the area inside the circle but outside the hexagon P_1 shown in Figure 5.3b is less than one-half of E.

31. The figure shows the velocity of two runners racing against each other. (They start together at the same time and run the same track.)
   (a) Which runner is ahead after the first minute? Explain your answer.
   (b) Which runner is ahead after the first 5 minutes? Explain your answer.

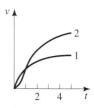

SOLUTIONS TO SELF-HELP EXERCISE SET 5.3

1. We divide the interval [1, 2] into five subintervals of equal length. Then $\Delta t = 0.20$ and $t_0 = 1$, $t_1 = 1.2$, $t_2 = 1.4$, $t_3 = 1.6$, $t_4 = 1.8$, and $t_5 = 2$. From the figure we see that the function $v = 1/t$ is *decreasing* and thus the right-hand sum is a *lower* estimate and the left-hand sum an *upper* estimate of the area we are seeking. Then

$$\text{Right-hand sum} = f(t_1)\,\Delta t + f(t_2)\,\Delta t + f(t_3)\,\Delta t$$
$$+ f(t_4)\,\Delta t + f(t_5)\,\Delta t$$
$$= f(1.2)(0.20) + f(1.4)(0.20)$$
$$+ f(1.6)(0.20) + f(1.8)(0.20)$$
$$+ f(2)(0.20)$$
$$= 0.20\left(\frac{1}{1.2} + \frac{1}{1.4} + \frac{1}{1.6} + \frac{1}{1.8} + \frac{1}{2}\right)$$
$$\approx 0.64563$$

This is the area of the shaded rectangles in the figure.

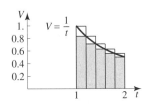

$$\text{Left-hand sum} = f(t_0)\,\Delta t + f(t_1)\,\Delta t + f(t_2)\,\Delta t$$
$$+ f(t_3)\,\Delta t + f(t_4)\,\Delta t$$
$$= f(1)(0.20) + f(1.2)(0.20)$$
$$+ f(1.4)(0.20) + f(1.6)(0.20)$$
$$+ f(1.8)(0.20)$$
$$= 0.20\left(\frac{1}{1} + \frac{1}{1.2} + \frac{1}{1.4} + \frac{1}{1.6} + \frac{1}{1.8}\right)$$
$$\approx 0.74563$$

This is the area of the tall rectangles in the figures. We thus have

$$0.64563 \leq \text{distance traveled} \leq 0.74563$$

Upper estimate − lower estimate
$$= 0.74563 - 0.64563 = 0.10$$

2. The exact distance traveled by the object is the area of the region between the graph of $v = 1/t$ and the t-axis on the interval [1, 2].

5.4 THE DEFINITE INTEGRAL

■ Taking the Limit as $n \to \infty$ ■ The Definite Integral ■ More General Riemann Sums (Optional)

APPLICATION: FINDING DISTANCE TRAVELED

An object travels with a velocity in feet per second given by $v = f(t) = t^2$, where t is in seconds. What distance does the object travel in the first second? See Example 1 for the answer.

APPLICATION: THE BOMBAY PLAGUE OF 1905

Kermack and McKendrick [122] found that the number $D(t)$ of deaths per week during the Bombay plague of 1905–1906 was approximated by $D(t) = 3960(e^{0.2t-3.4} + e^{-0.2t+3.4})^{-2}$, where t is in weeks from the outbreak. What were the approximate number of deaths during the first 30 weeks of this plague? For the answer see Example 3.

Taking the Limit as $n \to \infty$

If the nonnegative function $v = f(t)$ gives the velocity of an object on the interval $[a, b]$, we saw in the previous section how the left- and right-hand sums approximate the distance traveled by the object on the time interval $[a, b]$. We saw that this distance was exactly the area of the region between the graph of $v = f(t)$ and the t-axis on $[a, b]$. We also saw that the absolute value of the difference of the left- and right-hand sums for monotonic functions is bounded by $\left| f(b) - f(a) \right| \Delta t$, making for a convenient estimate for error in this important class of functions.

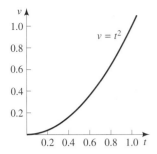

Figure 5.13
The distance traveled is the area under the curve. The only way to find this area is to find the limiting value of the left- and right-hand sums.

In this section we evaluate the distance traveled by the object by determining the limiting value of the left- and right-hand sums when $t \to \infty$.

EXAMPLE 1   Finding Distance Exactly

Suppose an object travels with a velocity in feet per second given by $v = f(t) = t^2$ on the t-interval [0, 1], where t is in seconds. Determine the exact distance traveled.

SOLUTION ■ Figure 5.13 is the graph of the function $v = t^2$ on the interval [0, 1]. We know that the exact distance traveled is the area of the region between the graph of $v = t^2$ and the t-axis on the interval [0, 1]. Unfortunately, there are no formulas from geometry that can give us this area. We are left with no alternative other than determining the limiting value of the left- and right-hand sums.

Since the function $v = t^2$ is increasing, we know that the left-hand sums are all less than the distance traveled and increase with increasing n, whereas the right-hand sums are all greater than the distance traveled and decrease with increasing n. Both the left- and right-hand sums should approach the distance traveled as $t \to \infty$. We have trapped our area between the left- and right-hand sums, and furthermore, we can use the program SUM to calculate any of these sums. Table 5.1 gives some results. (Your calculator may need to run the program SUM for 15 minutes or more to find the last entry in the table.)

Table 5.1 confirms our geometric insight. Both the left- and right-hand sums seem to be approaching a limit. The limit appears to be $\frac{1}{3}$. Thus, the area under the curve $v = t^2$ on the interval [0, 1], and therefore, the distance traveled by the object, appears to be $\frac{1}{3}$.

Furthermore, notice in Table 5.1 that the left-hand sums are increasing, whereas the right-hand sums are decreasing, and, the limit of $\frac{1}{3}$ is between the two sums.

Also notice from the third column in Table 5.1 that the difference between the sums is always $\frac{1}{n}$. We saw in the preceding section that this must be the case. ■

TABLE 5.1

n	Left-Hand Sum	Right-Hand Sum	Difference of Sums
4	0.21875	0.46875	0.25000
8	0.27344	0.39844	0.12500
100	0.32835	0.33835	0.01000
1,000	0.33283	0.33383	0.00100
10,000	0.33328	0.33338	0.00010

EXAMPLE 2   Finding Distance Exactly

Suppose an object travels with a velocity in feet per second given by $v = f(t) = \frac{1}{t}$ on the t-interval [1, 2], where t is in seconds. Determine the exact distance traveled.

SOLUTION ■ See Figure 5.14 for a graph of $v = \frac{1}{t}$. Using the SUM program, we obtain Table 5.2.

Both the left- and right-hand sums are heading for 0.6931 to four decimal places. You just might recognize this number! When we looked in Section 1.8 at the time it takes for an account that is compounding continuously to double, we encountered the number $\ln 2 \approx 0.6931$. Apparently, the area under $y = 1/x$ on the interval [1, 2] is converging to $\ln 2$. We will be able to confirm this in the next section. ■

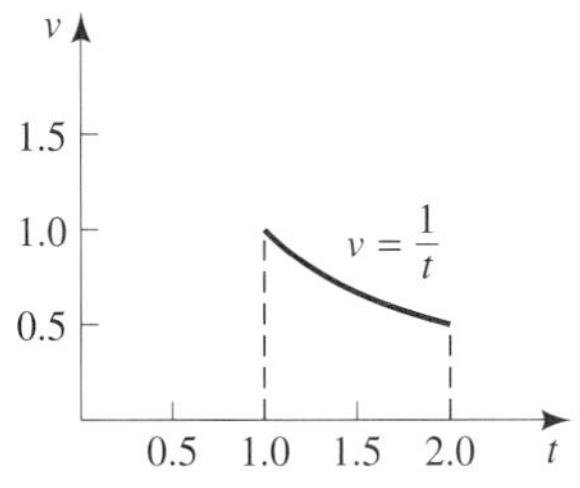

Figure 5.14
The distance traveled is the area under the curve. The only way to find this area is to find the limiting value of the left- and right-hand sums.

TABLE 5.2

n	Left-Hand Sum	Right-Hand Sum	Difference of Sums
5	0.74563	0.64563	0.10000
10	0.71877	0.66877	0.05000
100	0.69565	0.69065	0.00500
1,000	0.69340	0.69290	0.00050
10,000	0.69317	0.69312	0.00005

Notice in Table 5.2 that, since $v = f(t) = 1/t$ is a decreasing function, the left-hand sums decrease, whereas the right-hand sums increase with increasing n. Also, recall from the previous section that the difference between the sums should be bounded by

$$[f(1) - f(2)] \, \Delta x = \left(1 - \frac{1}{2}\right)\left(\frac{1}{n}\right) = \frac{1}{2n}$$

This is also confirmed in Table 5.2.

We are now in a position to see how to find an estimate of the number π. The equation $x^2 + y^2 = 4$ is the equation of a circle of radius 2 centered at $(0, 0)$. The area of this circle is $(2)^2 \pi = 4\pi$. The upper half of this circle is the graph of the function $y = f(x) = \sqrt{4 - x^2}$. The area under the graph of this function in the first quadrant is then just $\frac{1}{4}(4\pi) = \pi$ (Fig. 5.15).

EXPLORATION 1

Estimating π

 Find an estimate of π by estimating the area of the region between the graph of $y = \sqrt{4 - x^2}$ and the x-axis on the interval $[0, 2]$. Do this by reproducing Table 5.2 for this function. Note that $\Delta x = \dfrac{2 - 0}{n} = \dfrac{2}{n}$.

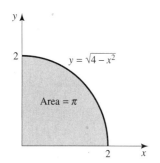

Figure 5.15
The area under the curve $y = \sqrt{4 - x^2}$ on $[0, 2]$ is π.

The Definite Integral

We have seen that we can assign an important *physical* interpretation to the area of the region between the graph of a nonnegative function and the horizontal axis on some interval. But we can also assign important economic, business, and biological interpretations to such areas. We therefore now consider finding the area between the graph of a nonnegative function $y = f(x)$ and the x-axis on the interval $[a, b]$ (Fig. 5.16).

The examples we considered so far were either increasing or decreasing on the interval of interest. Thus, the left- and right-hand sums always bounded the area, giving a very convenient estimate of the maximum error of our estimates. This is not necessarily the case for all functions. Nonetheless, if the function $y = f(x)$ is continuous on the finite interval $[a, b]$, then the limits of the left- and right-hand sums converge to a common limit as $n \to \infty$. If $f(x)$ is nonnegative on $[a, b]$, then it is natural for us, given the geometry, to interpret this common limit as the area of the region between the graph of the function and the x-axis on $[a, b]$.

Before continuing, let us introduce some very suggestive Σ (sigma or sum) notation for the left- and right-hand sums. We divide the x-interval $[a, b]$ into n subintervals of equal length $\Delta x = \dfrac{b - a}{n}$. Let the endpoints of these subintervals be $a = x_0, x_1, x_2,$ $\ldots, x_n = b$. We have the following:

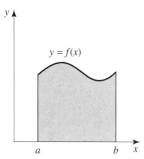

Figure 5.16
We want to find the area between the graph of $y = f(x)$ and the x-axis on $[a, b]$.

$$\text{Right-hand sum} = f(x_1) \, \Delta x + f(x_2) \, \Delta x + \cdots + f(x_n) \, \Delta x$$

$$= \sum_{k=1}^{n} f(x_k) \, \Delta x$$

$$\text{Left-hand sum} = f(x_0)\,\Delta x + f(x_1)\,\Delta x + \cdots + f(x_{n-1})\,\Delta x$$

$$= \sum_{k=0}^{n-1} f(x_k)\,\Delta x$$

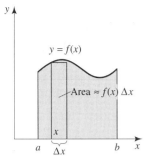

Figure 5.17
The symbol $\int$ suggests summation and the term $f(x)\,dx$ suggests $f(x)\Delta x$, which is the area of a thin rectangle. Thus, $\int_a^b f(x)\,dx$ suggests summing up the areas of the thin rectangles from a to b.

The Definite Integral

Suppose $f(x)$ is a continuous function on the finite interval $[a, b]$. Let $\Delta x = \dfrac{b-a}{n}$. Then the right-hand sum $\sum_{k=1}^{n} f(x_k)\,\Delta x$ and the left-hand sum $\sum_{k=0}^{n-1} f(x_k)\,\Delta x$ satisfy

$$\lim_{n\to\infty} \sum_{k=1}^{n} f(x_k)\,\Delta x = \lim_{n\to\infty} \sum_{k=0}^{n-1} f(x_k)\,\Delta x$$

We refer to this common limit as the **definite integral** of f from a to b and write it as

$$\int_a^b f(x)\,dx$$

We refer to a as the **lower limit** of integration and to b as the **upper limit** of integration and the interval $[a, b]$ as the **interval of integration**. If $f(x)$ is nonnegative on the interval $[a, b]$, we interpret the definite integral $\int_a^b f(x)\,dx$ as the area of the region between the graph of $y = f(x)$ and the x-axis on the interval $[a, b]$. The sums $\sum_{k=1}^{n} f(x_k)\,\Delta x$ and $\sum_{k=0}^{n-1} f(x_k)\,\Delta x$ are called **Riemann sums**.

The notation for the definite integral, $\int_a^b f(x)\,dx$, is very suggestive of a Riemann sum. One can think of the integral sign $\int$ as an elongated S (or Σ) that stands for "sum." Also, the term $f(x)\,dx$ suggests $f(x)\,\Delta x$, which is an area of a rectangle (Fig. 5.17). The notation for the definite integral thus suggests summing the area of such rectangles from a to b. This interpretation is widely used in the applications and can yield significant insights.

EXAMPLE 3 The Bombay Plague of 1905

■ **AIDS Deaths**
The loss of life that resulted from the spread of the Bombay Plague can be compared to modern day deaths from AIDS. In 1994, the World Bank released a report saying 17 million people, including 1 million in the United States, are infected with HIV, and 2 million will die of AIDS each year through the end of the decade. In the United States alone, from 1985 to 1994 the death toll was 197,727. The highest death toll from this disease occurred in 1992 with 34,228 lives lost. Today, HIV/AIDS is clearly the world's greatest plague, bar none.
Source: 1995 World Almanac, p. 971, 840–41; Joyce Price, "Extended Survival, New Drugs Give Hope on World AIDS Day," *The Washington Times*, December 1, 1994.

Kermack and McKendrick [122] found that the number $D(t)$ of deaths per week during the Bombay plague of 1905–1906 was approximated by $D(t) = 3960(e^{0.2t-3.4} + e^{-0.2t+3.4})^{-2}$, where t is in weeks from the outbreak. Find the approximate number of deaths during the first 30 weeks of this plague.

SOLUTION ■ If $N(t)$ is the number of deaths t weeks after the outbreak, then $D(t) = \dfrac{d}{dt}\,N(t)$. The term $D(t)$ is the rate of change of the number of deaths with respect to time. Thus, $D(t)$ is analogous to the velocity $v(t)$ of an object, since $v(t)$ is the rate of change of distance with respect to time.

We can then find the number of deaths in a way very similar to what we used to find the distance traveled when given the velocity. If we divide the t interval $[0, 30]$ into n equal subintervals with $0 = t_0, t_1, \ldots, t_n = 30$ denoting the endpoints, then $D(t_1)\,\Delta t$ approximates the number of deaths during the first interval of time $[t_0, t_1]$, $D(t_2)\,\Delta t$ approximates the number of deaths during the second interval of time $[t_1, t_2]$, and so on. (For example, if $\Delta t = 1$, then $D(t_1)\,\Delta t$ approximates the number of deaths in the first week.) Thus, the right-hand sum

$$\sum_{k=1}^{n} D(t_k)\,\Delta t = D(t_1)\,\Delta t + D(t_2)\,\Delta t + \cdots + D(t_n)\,\Delta t$$

approximates the number of deaths between the time $t = 0$ and $t = 30$. Then, according to the model, $\int_0^{30} D(t)\,dt$ gives the exact number of deaths from the time $t = 0$ to

TABLE 5.3

n	Left-Hand Sum	Right-Hand Sum
2	12,772.2	13,030.3
4	9,688.8	9,817.9
8	9,789.8	9,854.3
30	9,825.2	9,842.4
100	9,832.0	9,837.2
200	9,833.4	9,835.9
2000	9,834.5	9,834.8

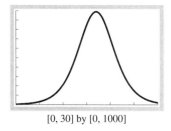

[0, 30] by [0, 1000]

Screen 5.2
A graph of $y = 3960(e^{0.2t-3.4} + e^{-0.2t+3.4})^{-2}$ giving the deaths per week during the Bombay plague

$t = 30$. We assume that no deaths occurred at the very instant of the outbreak. There-fore, the number of deaths during the first 30 weeks of the outbreak is given by the definite integral

$$\int_0^{30} 3960(e^{0.2t-3.4} + e^{-0.2t+3.4})^{-2} \, dt$$

To determine this integral we first use the program SUM to construct Table 5.3. The graph is shown in Screen 5.2 using a window with dimensions [0, 30] by [0, 1000]. The function is not monotonic on the interval [0, 30]. We see from Table 5.3 that both the left- and right-hand sums are converging to 9835, rounded to the nearest integer. Thus $\int_0^{30} 3960(e^{0.2t-3.4} + e^{-0.2t+3.4})^{-2} \, dt \approx 9835$. Thus, according to this model there were about 9835 deaths due to this plague during the first 30 weeks. Since $D(t)$ is nonnegative on [0, 30], this is also the area between the graph of this function and the t-axis on [0, 30]. Notice that neither the left-hand sums nor the right-hand sums always increase or always decrease. Notice that the area 9835 is not bracketed by each of the left- and right-hand sums. ■

EXAMPLE 4　　The Indefinite Integral and Area

Find $\int_1^3 (6 - 2x) \, dx$.

SOLUTION ■ Figure 5.18 shows the graph of $y = 6 - 2x$. The definite in-tegral $\int_1^3 (6 - 2x) \, dx$ is the area between the line $y = 6 - 2x$ and the x-axis on [1, 3]. But this is just a triangle. Since the area of a triangle is one-half the base times the height

$$\int_1^3 (6 - 2x) \, dx = \frac{1}{2}(2)(4) = 4 \quad ■$$

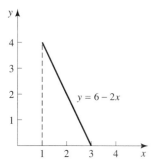

Figure 5.18
$\int_1^3 (6 - 2x) \, dx$ is the area between the line and the x-axis on the interval [1, 3].

EXPLORATION 2

Verifying Example 4 Numerically

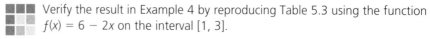 Verify the result in Example 4 by reproducing Table 5.3 using the function $f(x) = 6 - 2x$ on the interval [1, 3].

More General Riemann Sums

The left- and right-hand Riemann sums that we have been studying are actually specific examples of more general Riemann sums that we now consider.

Suppose we have a function $f(x)$ continuous on an interval $[a, b]$. As usual, divide the x-interval $[a, b]$ into n subintervals of equal length $\Delta x = \dfrac{b - a}{n}$. Let the endpoints of these subintervals be $a = x_0, x_1, x_2, \ldots, x_n = b$. Denote I_1 as the first subinterval

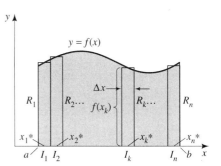

Figure 5.19
A general Riemann sum sums up the areas of the rectangles where x_k^* is any point on the *k*th subinterval.

$[x_0, x_1]$, denote I_2 as the second subinterval $[x_1, x_2]$, and in general denote I_k as the *k*th subinterval $[x_{k-1}, x_k]$. Now for each interval I_k pick a point $x_k^* \in I_k$. The sum

$$\sum_{k=1}^{n} f(x_k^*) \, \Delta x = f(x_1^*) \, \Delta x + f(x_2^*) \, \Delta x + \cdots + f(x_n^*) \, \Delta x$$

is called a **Riemann sum**. The term $f(x_k^*) \, \Delta x$ is the area of a rectangle with base Δx and height $f(x_k^*)$ (Fig. 5.19).

If for each *k* we pick x_k^* as the right-hand point of the subinterval $I_k = [x_{k-1}, x_k]$, then $x_k^* = x_k$ and the Riemann sum $\Sigma_{k=1}^n f(x_k^*) \, \Delta x$ becomes equal to

$$\sum_{k=1}^{n} f(x_k) \, \Delta x = f(x_1) \, \Delta x + f(x_2) \, \Delta x + \cdots + f(x_n) \, \Delta x$$

This is the right-hand sum we have already considered.

If for each *k* we pick x_k^* as the left-hand point of the subinterval $I_k = [x_{k-1}, x_k]$, then $x_k^* = x_{k-1}$, and the Riemann sum $\Sigma_{k=1}^n f(x_k^*) \, \Delta x$ becomes equal to

$$\sum_{k=0}^{n-1} f(x_{k-1}) \, \Delta x = f(x_0) \, \Delta x + f(x_1) \, \Delta x + \cdots + f(x_{n-1}) \, \Delta x$$

This is the left-hand sum we have already considered.

An important theorem on Riemann sums states that if $f(x)$ is continuous on the interval $[a, b]$ and x_k^* is any point in the subinterval I_k, then $\lim_{n \to \infty} \Sigma_{k=1}^n f(x_k^*) \, \Delta x$ exists and is the same limit, independent of how the points x_k^* are chosen in the *k*th subinterval. Thus, for continuous functions

$$\lim_{n \to \infty} \sum_{k=1}^{n} f(x_k^*) \, \Delta x = \int_{a}^{b} f(x) \, dx$$

no matter how the points x_k^* are picked in the subinterval I_k.

We have already seen that picking x_k^* as the left endpoint of the subinterval I_k gives the left-hand limit, and picking x_k^* as the right endpoint of the subinterval I_k gives the right-hand limit. A third important way to obtain a Riemann sum is to pick x_k^* as the midpoint of the interval I_k.

For positive functions that are either always increasing or always decreasing on the interval $[a, b]$, we saw that the left- and right-hand sums bracketed the area $\int_a^b f(x) \, dx$. In such circumstances we naturally expect a better approximation of the area by using rectangles with heights given by the values of the function at the midpoints of the subintervals. The program in the Technology Resource Manual that uses this is called MSUM.

EXAMPLE 5 A Riemann Sum Using the Midpoint

Use MSUM for $n = 1000$ to approximate $\int_1^2 \frac{1}{x} \, dx$. Compare your answer to the correct answer ln 2 and to the left- and right-hand sums found in Table 5.2.

SOLUTION ■ Using the program MSUM for $n = 1000$, we obtain $\int_1^2 \frac{1}{x} \, dx$ ≈ 0.69315. Referring to Table 5.2 we see that the left-hand sum is 0.69340 and the right-hand sum is 0.69290. The integral is exactly ln 2, which to five decimal places is 0.69315. Thus, MSUM for $n = 1000$ gives the correct answer to five decimal places. ■

SELF-HELP EXERCISE SET 5.4

1. Find the definite integral $\int_0^1 2x \, dx$ by finding the area of an appropriate geometric region.

2. Confirm your answer in Exercise 1 by finding the left- and right-hand sums for $n = 10, 100, 1000$, and by determining the limits.

EXERCISE SET 5.4

In Exercises 1 through 12 estimate the given definite integrals by finding the left- and right-hand sums for $n = 10, 100, 1000$. (Notice that all of the integrands are monotonic on the interval of integration.)

1. $\int_0^1 \sqrt{x} \, dx$

2. $\int_0^1 (x^2 + x + 1) \, dx$

3. $\int_0^1 x^3 \, dx$

4. $\int_0^1 x^4 \, dx$

5. $\int_{-2}^0 (4 - x^2) \, dx$

6. $\int_{-1}^0 (1 - x^2) \, dx$

7. $\int_1^3 \frac{1}{x} \, dx$

8. $\int_1^4 \frac{1}{x} \, dx$

9. $\int_0^1 e^x \, dx$

10. $\int_0^1 e^{-x} \, dx$

11. $\int_1^2 \ln x \, dx$

12. $\int_1^2 x \ln x \, dx$

In Exercises 13 through 18 determine n and an upper and lower estimate of the given definite integral so that the difference of the estimates is at most 0.1.

13. $\int_0^1 \sqrt[3]{x} \, dx$

14. $\int_0^1 (x^2 + 2x + 3) \, dx$

15. $\int_0^2 x^3 \, dx$

16. $\int_0^2 x^4 \, dx$

17. $\int_{-2}^0 (4 - x^2) \, dx$

18. $\int_{-1}^0 (1 - x^2) \, dx$

In Exercises 19 through 22 estimate the given definite integrals by finding the left- and right-hand sums for $n = 10, 100, 1000$. (Notice that none of the integrands are monotonic on the interval of integration.)

19. $\int_{-2}^2 (4 - x^2) \, dx$

20. $\int_{-1}^1 (1 - x^2) \, dx$

21. $\int_{-1}^2 \sqrt[3]{|x|} \, dx$

22. $\int_{-1}^2 |x^3| \, dx$

In Exercises 23 through 32 find the given definite integrals by finding the areas of the appropriate geometric region.

23. $\int_{-1}^3 5 \, dx$

24. $\int_{-10}^{-2} 4 \, dx$

25. $\int_0^2 3x \, dx$

26. $\int_1^3 2x \, dx$

27. $\int_1^4 (4 - x) \, dx$

28. $\int_{-1}^4 (4 - x) \, dx$

29. $\int_{-1}^1 \sqrt{1 - x^2} \, dx$

30. $\int_0^3 \sqrt{9 - x^2} \, dx$

31. $\int_0^2 \sqrt{1 - (x - 1)^2} \, dx$

32. $\int_1^2 \sqrt{1 - (x - 1)^2} \, dx$

The *only* way of evaluating the definite integrals of functions, such as e^{-x^2} or $\frac{1}{\ln x}$, is by some approximation technique such as Riemann sums. The definite integral of e^{-x^2} is of critical importance in the applications of probability.

33. Use the left- and right-hand sums for $n = 1000$ to estimate $\int_0^1 e^{-x^2} \, dx$. Using a graph of e^{-x^2} show which approximation must be less than the integral and which greater.

34. Use the left- and right-hand sums for $n = 1000$ to estimate $\int_2^3 \frac{1}{\ln x} \, dx$. Using a graph of $1/\ln x$ show which approximation must be less than the integral and which greater.

Enrichment Exercises

35. Explain carefully what you think is the difference (if any) between the definite integrals $\int_1^4 10x^2(4-x)\,dx$ and $\int_1^4 10t^2(4-t)\,dt$.

36. In Example 1 we found the area of the region between the graph of the curve $y = x^2$ and the x-axis on the interval $[0, 1]$ by first finding the left-hand and right-hand sums using the program SUM. It appeared to us that the sums had the common limit $\frac{1}{3}$. In this exercise you are asked to verify analytically that the right-hand sums converge to $\frac{1}{3}$. To do this divide the interval $[0, 1]$ into n subintervals with right endpoints at $1/n, 2/n, 3/n, \ldots, n/n = 1$.

(a) Then show that the right-hand sum S_n^r is

$$S_n^r = \frac{1}{n}\left[\frac{(1)^2}{n^2} + \frac{(2)^2}{n^2} + \frac{(3)^2}{n^2} + \cdots \frac{(n)^2}{n^2}\right]$$

$$= \frac{1}{n^3}(1 + 2^2 + 3^2 + \cdots + n^2)$$

(b) Now use the formula

$$1 + 2^2 + 3^2 + \cdots n^2 = \frac{n(n+1)(2n+1)}{6}$$

and show that

$$S_n^r = \frac{1}{3} + \frac{1}{2n} + \frac{1}{6n^2}$$

(c) Now take the limit of this expression as $n \to \infty$.

37. Use MSUM for $n = 1000$ to approximate $\int_1^3 \frac{1}{x}\,dx$. Compare your answer with the correct answer $\ln 3$ and to the left- and right-hand sums.

38. In this section we noted that $\int_0^4 \sqrt{1-x^2}\,dx = \pi$. Use MSUM for the values of n found in Table 5.2. Compare your answers with the actual value of π and to the approximations found in Exploration 1.

39. In Example 3 we found that, according to the given model, about 9835 deaths occurred during the first 30 weeks of the Bombay plague of 1905–1906. From Table 5.3 we needed to take $n = 2000$ before the left- and right-hand sums rounded to the same integer. Naturally, we expect greater accuracy using MSUM. Use MSUM on the integral found in Example 3, and find the smallest value of n for which MSUM rounds to 9835. Compare this value of n with the value $n = 2000$.

40. In Exercise 37 you were asked to compare your answer to the correct answer $\ln 3$. What does your calculator give as $\ln 3$? Is this the exact answer? Explain.

SOLUTIONS TO SELF-HELP EXERCISE SET 5.4

1. The definite integral $\int_0^1 2x\,dx$ is the area between the line $y = 2x$ and the x-axis on $[0, 1]$. See the figure. This is a triangle with area one-half the height times the base or $\frac{1}{2}(2)(1) = 1$.

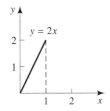

2. The table shows the results of using the program SUM with $f(x) = 2x$ on $[0, 1]$. As we can see, both the left- and right-sums appear to approach 1.

n	Left-Hand Sum	Right-Hand Sum
10	0.9000	1.1000
100	0.9900	1.0100
1000	0.9990	1.0010

5.5 THE FUNDAMENTAL THEOREM OF CALCULUS

- Properties of the Definite Integral
- Average Value
- The Fundamental Theorem of Calculus

APPLICATION: EXTRACTING NATURAL GAS FROM A FIELD

The rate at which a natural resource is extracted often increases at first until the easily accessible part of the resource is exhausted. Then the rate of extraction tends to decline. Suppose natural gas is being extracted from a new field at a rate of $\dfrac{4t}{t^2 + 1}$ billion cubic feet a year, where t is in years. If this field has an estimated 10 billion cubic feet, how long will it take to exhaust this field at the given rate of extraction? See Example 4 for the answer.

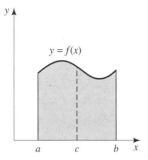

Figure 5.20
The area under the curve from a to b is the area under the curve from a to c plus the area under the curve from c to b. Thus, $\int_a^b f(x)\, dx$ is $\int_a^c f(x)\, dx + \int_c^b f(x)\, dx$.

Properties of the Definite Integral

We state five properties for the definite integral that we defined in the previous section. We assume that $f(x)$ and $g(x)$ are continuous on $[a, b]$.

Properties of Definite Integrals

1. $\displaystyle\int_a^a f(x)\, dx = 0$

2. $\displaystyle\int_a^b kf(x)\, dx = k \int_a^b f(x)\, dx, \quad k \text{ constant}$

3. $\displaystyle\int_a^b [f(x) \pm g(x)]\, dx = \int_a^b f(x)\, dx \pm \int_a^b g(x)\, dx$

4. $\displaystyle\int_a^b f(x)\, dx = \int_a^c f(x)\, dx + \int_c^b f(x)\, dx, \quad a < c < b$

5. $\displaystyle\int_b^a f(x)\, dx = -\int_a^b f(x)\, dx, \quad a < b$

When both $f(x)$ and $g(x)$ are positive, the definite integrals are areas under curves. Thus, Property 1 says that the area under a single point is zero. Property 2 says that the area under kf is k times the area under f. One part of Property 3 says that the area under $f + g$ is the area under f plus the area under g. Property 4 says that, according to Figure 5.20, the area under the curve from a to b is the area under the curve from a to c plus the area under the curve from c to b. The actual proofs require the use of Riemann sums and can be shown to be valid without the restriction that f be nonnegative. The proofs are not given here. Property 5 is actually a *definition*.

The Fundamental Theorem of Calculus

Given an object with nonnegative velocity function $v = f(t)$ we have seen that the definite integral $\int_a^b f(t)\, dt$ is the distance traveled by the object during the time interval $[a, b]$. We know that velocity is the instantaneous rate of change of distance with respect to time. Let $v = s'(t)$, where $s(t)$ is the position of the object at time t. With this notation the distance traveled, or total change of s, during the time interval $[a, b]$ is just $s(b) - s(a)$. Since $s'(t) = f(t)$, we conclude that

$$s(b) - s(a) = \int_a^b s'(t)\, dt$$

By analogy, we can see that a similar formula must exist for any rate of change. Suppose $F'(x)$ is nonnegative and the rate of change of $y = F(x)$ with respect to x. Then, in analogy with velocity, the total change of F, $F(b) - F(a)$ must be the definite integral $\int_a^b F'(x)\, dx$. Thus,

$$F(b) - F(a) = \int_a^b F'(x)\, dx$$

When the velocity function $v = s'(t)$ is nonnegative, the object always moves in the positive direction. As a consequence the total distance traveled by the object on the time interval $[a, b]$ is the same as the total change in s given by $s(b) - s(a)$. Suppose, however, you throw a ball straight up 5 feet and catch it a couple of seconds later when it falls back. The ball has traveled a total distance of 10 feet. But since the ball ends where it began, there is *no* change in its position. Thus, in this case, the total distance traveled is different from the change in position. In this example, the first term $f(t_1)\, \Delta t$ in the right-hand sum

$$f(t_1)\, \Delta t + f(t_2)\, \Delta t + \cdots + f(t_n)\, \Delta t$$

is positive since $f(t_1)$ is positive (as the ball heads up), whereas the last term $f(t_n)\, \Delta t$ in the sum is negative since $f(t_n)$ is negative (as the ball returns). Thus, $f(t_1)\, \Delta t$ corresponds to a *positive* change in position, whereas $f(t_n) \, \Delta t$ corresponds to a *negative* change in position. Thus, we see that the right-hand sum, in this case, does not correspond to the distance traveled by the object but rather with the total change in position.

Thus, whether $s'(t)$ changes sign on the interval $[a, b]$ or not, $s(b) - s(a) = \int_a^b s'(t) \, dt$ represents the total change in s, or the total change in position. In the same way, $F(b) - F(a) = \int_a^b F'(x) \, dx$ represents the total change in F on the interval $[a, b]$. This is called the fundamental theorem of calculus.

Fundamental Theorem of Calculus

Suppose $f(x)$ is continuous on $[a, b]$. If $F(x)$ is an antiderivative of $f(x)$, that is, $f(x) = \dfrac{dF(x)}{dx}$, then

$$\int_a^b f(x) \, dx = F(b) - F(a)$$

For convenience we write

$$F(x)\big|_a^b = F(b) - F(a)$$

We know that if $F(x)$ is an antiderivative of $f(x)$, then any other antiderivative must be of the form $F(x) + C$ for some constant C. Let $G(x) = F(x) + C$. Then $G(x)$ is an antiderivative of $f(x)$. Let us then use $G(x)$ in the fundamental theorem of calculus to evaluate the definite integral. We obtain

$$\int_a^b f(x) \, dx = G(b) - G(a)$$
$$= [F(b) + C] - [F(a) + C]$$
$$= F(b) - F(a)$$

Thus, *any* antiderivative will do.

Using the limit of Riemann sums, we showed numerically in the previous section that

$$\int_0^1 x^2 \, dx = \frac{1}{3}$$

The fundamental theorem of calculus gives us a much easier way of obtaining this answer.

EXAMPLE 1   Using the Fundamental Theorem

Use the fundamental theorem of calculus to find $\int_0^1 x^2 \, dx$.

SOLUTION ■   Since $F(x) = \frac{1}{3}x^3$ is an antiderivative of x^2, according to the fundamental theorem of calculus

$$\int_0^1 x^2 \, dx = \frac{1}{3}x^3 \bigg|_0^1 = \frac{1}{3}(1)^3 - \frac{1}{3}(0)^3 = \frac{1}{3} \quad ■$$

EXPLORATION 1

The Fundamental Theorem

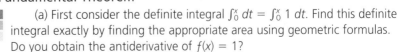
(a) First consider the definite integral $\int_0^x dt = \int_0^x 1 \, dt$. Find this definite integral exactly by finding the appropriate area using geometric formulas. Do you obtain the antiderivative of $f(x) = 1$?

(b) Next consider the definite integral $\int_0^x t \, dt$. Find this definite integral exactly by finding the appropriate area using geometric formulas. Do you obtain the antiderivative of $f(x) = x$?

(c) The program NINT calculates a numerical approximation to $\int_a^b f(t)\, dt$, and the program NINTG graphs the function $\int_a^x f(t)\, dt$. Use NINTG to graph an approximation to $\int_0^x t^2\, dt$ and also on the same screen graph $y = \frac{1}{3}x^3$. Are they the same? What does this have to do with the fundamental theorem of calculus? Explain.

EXAMPLE 2   Finding a Definite Integral

Find $\int_1^2 (1 + 6x^2)\, dx$.

SOLUTION ■ Using Properties 2 and 3,

$$\int_1^2 (1 + 6x^2)\, dx = \int_1^2 dx + 6 \int_1^2 x^2\, dx$$

$$= x\big|_1^2 + 2x^3\big|_1^2$$

$$= (2 - 1) + 2[(2)^3 - (1)^3] = 15 \quad ■$$

EXAMPLE 3   Finding the Area Under a Curve

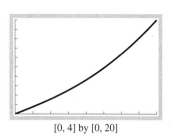

[0, 4] by [0, 20]

Screen 5.3
A graph of a marginal cost function $y = x\sqrt{9 + x^2}$.

Suppose a company has a marginal cost function $y = C'(x) = x\sqrt{9 + x^2}$, where x is the number of thousands of items sold and the cost C is in thousands of dollars. Graph the function $y = C'(x)$ on your graphing calculator. If fixed costs are \$10,000, find the total cost of manufacturing the first 4000 items.

SOLUTION ■ The graph of $y = C'(x)$ is shown in Screen 5.3 using a window with dimensions [0, 4] by [0, 20]. Since cost is measured in thousands of dollars, we have $C(0) = 10$. We are seeking $C(4)$. By the fundamental theorem of calculus

$$C(4) - C(0) = \int_0^4 C'(x)\, dx$$

$$C(4) = C(0) + \int_0^4 C'(x)\, dx$$

$$C(4) = 10 + \int_0^4 x\sqrt{9 + x^2}\, dx$$

Use the method of substitution. Let $u = 9 + x^2$. Then $du = 2x\, dx$. When the integration is with respect to x, then x varies from 0 to 4. If we are to integrate with respect to u, however, we need to use limits of integration that correspond to the change in u. We have $u(0) = 9$ and $u(4) = 25$. Thus,

$$C(4) = 10 + \int_0^4 x\sqrt{9 + x^2}\, dx$$

$$= 10 + \int_0^4 (9 + x^2)^{1/2}\, x\, dx$$

$$= 10 + \int_9^{25} u^{1/2}\, \frac{1}{2}\, du$$

$$= 10 + \frac{1}{2} \cdot \frac{2}{3} u^{3/2}\big|_9^{25}$$

$$= 10 + \frac{1}{3}(25^{3/2} - 9^{3/2})$$

$$= 10 + \frac{1}{3}(125 - 27)$$

$$= \frac{128}{3}$$

or \$42,666.67. ■

■ Natural Gas

Natural gas is a fossil fuel that gained popularity during the Industrial Revolution of the 18th century. Although scientists believe large supplies of natural gas are still available, these reserves are difficult and expensive to access. A 1983 study estimated that at the present rate of consumption, natural gas reserves will be available in the United States until 2259 and in the world until 2604. Nevertheless, if demand continues to increase as it has been, scientists believe the natural gas supply will disappear from the United States in 2003 and from the world in 2028. In response to this data, experts continue to emphasize conservation and the implementation of alternative energy sources, such as solar power and nuclear fusion.

Source: Melvin D. Joesten, David O. Johnston, John T. Netterville, et al., *World of Chemistry* Saunders College Publishing, Philadelphia, PA, 1991.

EXAMPLE 4   Extracting Natural Gas from a Field

The rate at which a natural resource is extracted often increases at first until the easily accessible part of the resource is exhausted. Then the rate of extraction tends to decline. Suppose natural gas is being extracted from a new field at a rate of $A'(t) = \dfrac{4t}{t^2+1}$ billion cubic feet a year, where t is in years since the field was opened. Graph this function on your graphing calculator. If this field has an estimated 10 billion cubic feet of gas, how long will it take to exhaust this field at the given rate of extraction?

SOLUTION ■ The graph of $y = A'(t)$ is shown in Screen 5.4 using a window of dimensions [0, 10] by [0, 3]. Since the field opens at $t = 0$, $A(0) = 0$. Using the fundamental theorem of calculus the amount extracted during the first T years is

$$A(T) = A(0) + \int_0^T \frac{4t}{t^2+1}\,dt = \int_0^T \frac{4t}{t^2+1}\,dt$$

We are seeking the time T for which this definite integral equals 10. Let $u = t^2 + 1$. Then $du = 2t\,dt$. As t goes from 0 to T, u goes from $u(0) = 1$ to $u(T) = T^2 + 1$. Then

$$
\begin{aligned}
10 &= \int_0^T \frac{4t}{t^2+1}\,dt \\
&= 4\int_1^{T^2+1} \frac{\frac{1}{2}du}{u} \\
&= 2\int_1^{T^2+1} \frac{du}{u} \\
&= 2\ln|u|\Big|_1^{T^2+1} \\
&= 2[\ln(T^2+1) - \ln 1] \\
&= 2\ln(T^2+1) \\
5 &= \ln(T^2+1) \\
T^2 + 1 &= e^5 \\
T &= \sqrt{e^5 - 1} \\
&\approx 12.1
\end{aligned}
$$

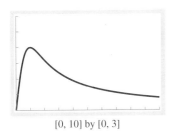

[0, 10] by [0, 3]

Screen 5.4

A graph of $y = \dfrac{4x}{x^2+1}$ giving the rate of extraction of natural gas from a new field.

or about 12.1 years. ■

Average Value

As we shall see, the indefinite integral can represent many different quantities in business, economics, and science. In order to recognize these quantities as definite integrals, we must be able to write them as a limit of Riemann sums. We first consider the average of a function.

The average of a set of n numbers $y_1, y_2, \ldots, y_n$, is

$$\frac{y_1 + y_2 + \cdots + y_n}{n}$$

But what is the average value of a continuous function such as seen in Figure 5.21 on the *interval* $[a, b]$? Here there are an *infinite* number of values of x.

To answer this question let us start with an average such as

$$\frac{f(x_1) + f(x_2) + \cdots + f(x_n)}{n}$$

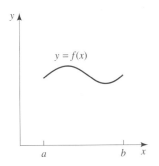

Figure 5.21
We want to define the average value of f on $[a, b]$.

where n is very large and the numbers $x_1, x_2, \ldots, x_n$ are distributed evenly throughout $[a, b]$. In fact let us divide the interval $[a, b]$ into n equal subintervals and denote the right-hand endpoints as $x_1, x_2, \ldots, x_n$. Then if n is very large

$$\frac{f(x_1) + f(x_2) + \cdots + f(x_n)}{n}$$

is an average of a very large number of values of $f(x)$ at points evenly distributed throughout $[a, b]$. Our strategy is to let $n \to \infty$ and see what happens to this average.

First, however, we multiply and divide each term in this last expression by $b - a$ and obtain

$$f(x_1) \frac{b-a}{n} \cdot \frac{1}{b-a} + f(x_2) \frac{b-a}{n} \cdot \frac{1}{b-a} + \cdots + f(x_n) \frac{b-a}{n} \cdot \frac{1}{b-a}$$

Factoring out the term $\dfrac{1}{b-a}$ then gives

$$\frac{1}{b-a} \left[f(x_1) \frac{b-a}{n} + f(x_2) \frac{b-a}{n} + \cdots + f(x_n) \frac{b-a}{n} \right]$$

Replacing $(b-a)/n$ with Δx gives

$$\frac{1}{b-a} \left[f(x_1) \Delta x + f(x_2) \Delta x + \cdots + f(x_n) \Delta x \right]$$

which can be rewritten in the shorter form as

$$\frac{1}{b-a} \sum_{i=1}^{n} f(x_i) \Delta x$$

We recognize this as $1/(b - a)$ times a Riemann sum of $f(x)$ over $[a, b]$. In fact, this is a right-handed sum (Fig. 5.22).

The right-handed Riemann sum approximates the area under the curve $f(x)$ on $[a, b]$. If we let $n \to \infty$ this becomes

$$\frac{1}{b-a} \left[\lim_{n \to \infty} \sum_{i=1}^{n} f(x_i) \Delta x \right]$$

But since the term inside the square brackets is precisely the definition of the definite integral $\int_a^b f(x) \, dx$, this last term is

$$\frac{1}{b-a} \int_a^b f(x) \, dx$$

This is what we define as the average of $f(x)$ over $[a, b]$.

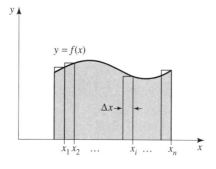

Figure 5.22
The term $\sum_{i=1}^{n} f(x_i) \Delta x$ is a Riemann sum of $f(x)$ over $[a, b]$, which is the sum of the area of the rectangles.

The Average Value of $f(x)$ over $[a, b]$

If $f(x)$ is continuous on $[a, b]$, we define the average value of $f(x)$ on $[a, b]$ to be

$$\frac{1}{b - a} \int_a^b f(x) \, dx$$

Notice that we do not require that $f(x)$ be a nonnegative function.

EXAMPLE 5   Average Amount in an Account

An account with an initial amount of $1000 is compounding continuously at an annual rate of 10% per year. What is the average amount of money in the account over the first 10 years?

SOLUTION ■ The average amount on $[0, 10]$ is given by

$$\frac{1}{b - a} \int_a^b f(t) \, dt = \frac{1}{10 - 0} \int_0^{10} 1000e^{0.1t} \, dt$$

$$= 100 \int_0^{10} e^{0.1t} \, dt$$

$$= 1000e^{0.1t}\big|_0^{10}$$

$$= 1000[e - 1]$$

$$\approx 1718$$

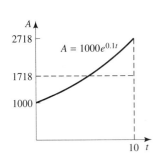

Figure 5.23
The average value of $1000e^{0.1t}$ on $[0, 10]$ is $\frac{1}{10} \int_0^{10} 1000e^{0.1t} \, dt \approx$ 1718.

or $1718. Refer to Figure 5.23. ■

Geometric Interpretation of Average Value

When the function $f(x)$ is nonnegative on the interval $[a, b]$, one can give a simple geometric interpretation of the average value. Let us denote the average value of $f(x)$ on $[a, b]$ as $\bar{y}$. Then

$$\bar{y} = \frac{1}{b - a} \int_a^b f(x) \, dx$$

and thus

$$\bar{y}(b - a) = \int_a^b f(x) \, dx$$

Since $f(x)$ is nonnegative, the definite integral $\int_a^b f(x) \, dx$ is the area under the curve $y = f(x)$ on $[a, b]$. Thus, if we construct a rectangle with base on the interval $[a, b]$ and height $\bar{y}$, the rectangle has area equal to $\bar{y}(b - a)$, which by the preceding formula is also the area under the curve $y = f(x)$ on $[a, b]$ (Fig. 5.24).

Figure 5.24
The area between the graph of $y = f(x)$ and the x-axis from a to b is $\bar{y}(b - a)$ where $\bar{y}$ is the average value of f on $[a, b]$.

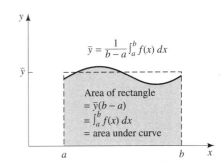

SELF-HELP EXERCISE SET 5.5

1. Evaluate $\int_{-1}^{0} 4xe^{x^2+2}\, dx$.

2. Suppose the rate of sales of a certain brand of bicycle by a retailer in thousands of dollars per week is given by

$$\frac{d}{dt} S(t) = 10t - 0.30t^2, \quad 0 \le t \le 5$$

where t is the number of weeks after an advertising campaign has begun. Find the amount of sales during the third and fourth weeks.

EXERCISE SET 5.5

In Exercises 1 through 24 evaluate the definite integrals.

1. $\displaystyle\int_{1}^{2} 4x^3\, dx$

2. $\displaystyle\int_{-2}^{-1} 6x^2\, dx$

3. $\displaystyle\int_{-2}^{-2} 3x^4\, dx$

4. $\displaystyle\int_{-2}^{2} 2x^5\, dx$

5. $\displaystyle\int_{-1}^{0} (9x^2 - 1)\, dx$

6. $\displaystyle\int_{-1}^{1} (1 - x^4)\, dx$

7. $\displaystyle\int_{0}^{2} (4x^3 - 2x + 1)\, dx$

8. $\displaystyle\int_{1}^{4} \left(\frac{3}{\sqrt{x}} - \frac{6}{\sqrt{x}}\right) dx$

9. $\displaystyle\int_{1}^{2} (x^{-2} + 3x^{-4})\, dx$

10. $\displaystyle\int_{-2}^{-1} (x^{-5} + 1)\, dx$

11. $\displaystyle\int_{-2}^{-1} e^{2x}\, dx$

12. $\displaystyle\int_{-1}^{0} e^{-x}\, dx$

13. $\displaystyle\int_{1}^{3} \frac{1}{2x}\, dx$

14. $\displaystyle\int_{2}^{4} \frac{3}{x}\, dx$

15. $\displaystyle\int_{0}^{1} (2x - 1)^9\, dx$

16. $\displaystyle\int_{-1}^{0} (1 + 2x)^5\, dx$

17. $\displaystyle\int_{0}^{1} x(x^2 - 1)^7\, dx$

18. $\displaystyle\int_{0}^{1} x\sqrt[3]{8 + x^2}\, dx$

19. $\displaystyle\int_{0}^{4} \sqrt{2x + 1}\, dx$

20. $\displaystyle\int_{0}^{4} \frac{1}{\sqrt{2x + 1}}\, dx$

21. $\displaystyle\int_{-1}^{1} xe^{x^2+1}\, dx$

22. $\displaystyle\int_{1}^{4} \frac{e^{\sqrt{x}}}{\sqrt{x}}\, dx$

23. $\displaystyle\int_{1}^{2} \frac{1}{2x + 1}\, dx$

24. $\displaystyle\int_{-2}^{0} \frac{x}{x^2 + 1}\, dx$

In Exercises 25 through 34 find the average value of each of the given functions on the given interval.

25. $f(x) = 6$ on $[0, 10]$

26. $f(x) = -3$ on $[-10, -2]$

27. $f(x) = x$ on $[0, 10]$ 28. $f(x) = x$ on $[-4, 4]$

29. $f(x) = 2x$ on $[-2, 2]$ 30. $f(x) = x^2$ on $[0, 3]$

31. $f(x) = x^3$ on $[-1, 1]$ 32. $f(x) = e^x$ on $[0, \ln 2]$

33. $f(x) = x(x - 1)$ on $[0, 2]$

34. $f(x) = x(x - 1)$ on $[-1, 2]$

35. Refer to the figure. If $F(0) = 1$, what is $F(2)$? $F(4)$?

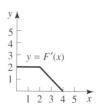

36. Refer to the figure. Suppose $F(1) = 3$ and the area of the shaded region is 10. Find $F(4)$.

Applications

37. **Revenue.** In the following figure the rate $R'(t)$ of revenue in millions of dollars per week is given, where t is in weeks. Find an upper and a lower estimate for the revenue received during this four-week period. Explain what you are doing.

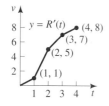

38. **Sales.** In the figure the rate $R'(t)$ of sales for an entire industry in billions of dollars is given, where t is in weeks. Find an upper and a lower estimate for the sales received during this five-week period. Explain what you are doing.

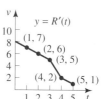

39. Population. In the table the rate of growth $M'(t)$ of a bacteria has been estimated by measuring the rate of increase in mass, where t is in hours. Find two estimates for the change in population during this 20-hour period. Explain what you are doing.

t_i	0	2	4	6	8	10	12	14	16	18	20
$M'(t_i)$	10	14	20	30	50	100	100	80	50	20	0

40. Velocity. In the table the velocity $v(t) = s'(t)$ is given for discrete, equally spaced points $0 = t_0, t_1, \dots, t_5 = 1$, where v is in feet per second and t in seconds. Find two estimates for the change in position during this period. Explain what you are doing.

t_i	0	0.2	0.4	0.6	0.8	1.0
$v(t_i)$	1.0	1.6	2.0	1.8	1.5	1.0

41. Production. Suppose oil is being extracted from a field at a rate given by $\dfrac{d}{dt} P(t) = 0.2e^{0.1t}$, where $P(t)$ is measured in millions of barrels and t in years. At this rate how much oil will be extracted during the second 10-year period?

42. Production. Suppose copper is being extracted from a certain mine at a rate given by $\dfrac{d}{dt} P(t) = 100e^{-0.2t}$, where $P(t)$ is measured in tons of copper and t in years. At this rate how much copper will be extracted during the third year?

43. Sales. Suppose the rate of sales of an item is given by $\dfrac{d}{dt} S(t) = -3t^2 + 36t$, where t is the number of weeks after an advertising campaign has begun. How many items were sold during the third week?

44. Sales. A firm estimates that its sales will increase at a rate given by $\dfrac{d}{dt} S(t) = 10\sqrt{t}$, where t is measured in years. Find the total sales during the second and third years.

45. Natural Resource Depletion. Suppose a new oil reserve has been discovered with an estimated capacity of 10 billion barrels. Suppose production is given in billions of barrels and proceeds at a rate given by $\dfrac{d}{dt} P(t) = 1.5e^{-0.1t}$, where t is measured in years. At this rate, how long will it take to exhaust this resource?

46. Natural Resource Depletion. Suppose a resource, such as oil, gas, or some mineral, has total reserves equal to R. Suppose the resource is being extracted at a rate given by $\dfrac{d}{dt} P(t) = ae^{-kt}$, where a and k are positive constants and t is time. Show that at this rate the resource will be exhausted in time $T = -\dfrac{\ln (1 - kR/a)}{k}$.

47. Chemical Spill. A tank holding 10,000 gallons of a polluting chemical breaks at the bottom and spills out at the rate given by $f'(t) = 400e^{-0.01t}$, where t is measured in hours. How much spills during the third day?

48. Chemical Spill. In the previous problem how long before the tank is emptied?

49. Sum-of-Years-Digit Depreciation. The sum-of-years-digit depreciation method depreciates a piece of equipment more in the early years of its life and less in the later years. In this method the value changes at a rate given by $\dfrac{d}{dt} V(t) = -k(T - t)$, where T is the useful life in years of the equipment, t is the time in years since purchase, and k is a positive constant.
  (a) If the initial value of the equipment is I, show that the scrap value of the equipment is given by $I - kT^2/2$.
  (b) Show that the number of years for the equipment to be worth half of its value is given by $T - \sqrt{T^2 - I/k}$.

50. Sum-of-Years-Digit Depreciation. Suppose in the previous problem a certain computer with an initial value of $51,000 has $T = 10$ and $k = 1000$.
  (a) Find the scrap value of this computer.
  (b) Find the number of years it takes for the computer to be worth half its initial value using this method of depreciation.

51. Sales. From the time a firm introduces a new product, sales in thousands of this product increase at a rate given by $S'(t) = 2 + t$, where t is the time since the introduction of the new product. At the beginning of the third year, an advertising campaign is introduced and sales then increase at a rate given by $S_2'(t) = t^2$, where $t \geq 2$. Find the total sales during the first four years.

52. Production. Suppose copper is being extracted from a certain mine at a rate given by $\dfrac{d}{dt} P(t) = 100e^{-0.2t}$, where $P(t)$ is measured in tons of copper and t in years from the opening of the mine. At the beginning of the sixth year, a mining innovation is introduced to boost production to a rate given by $Q'(t) = 500/t$. Find the total production during the first 10 years.

53. Average Amount of Money. An account with an initial amount of $1000 is compounding continuously at an annual rate of 10% per year. What is the average amount of money in the account over the first 20 years?

54. Average Price. Given the demand equation $p = 100e^{-0.2x}$ find the average price over the demand interval $[10, 20]$.

55. Average Profit. The profit in millions of dollars of a certain firm is given by $P(t) = 6(t - 1)(t - 2)$, where t is measured in years. Find the average profits per year over the period of time $[0, 4]$.

56. Average Revenue. The revenue in thousands of dollars is given by $R(x) = x^2 - 2x$, where x is the number of thousands of items sold. Find the average revenue for the first 10,000 items sold.

57. Average Air Pollution. A study has indicated that the level $L(x)$ of air pollution a distance of x miles from a certain factory is given by $L(x) = e^{-0.1x} + 10$. Find the average level of pollution between 10 and 20 miles from the factory.

58. **Average Amount of a Drug.** The amount of a drug in the body at time t is given by $A(t) = 5e^{-0.2t}$ where t is in hours and A is in milligrams. What is the average amount of drug in the body over the next 5-hour period?

59. **Average Blood Pressure.** The blood pressure in an artery of a healthy individual can change substantially over only a few seconds. The average pressure is used in some studies of blood pressure. Suppose the function $p(t) = 40t^4 - 160t^3 + 160t^2 + 80$ gives the blood pressure in millimeters of mercury in an artery over an interval [0, 2] of time measured in seconds. Find the average blood pressure in this artery over the 2-second interval [0, 2].

Enrichment Exercises

60. **Revenue.** Two competing retail stores open in the same mall at time $t = 0$ and have the marginal revenue functions given in the figure. The marginal revenue is measured in thousands of dollars per day. Estimate the point in time, other than $t = 0$, for which the total sales of the two stores are equal. Justify your answer. *Hint:* Use the fundamental theorem of calculus, and roughly compare the areas of two regions.

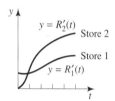

61. **Revenue.** Consider the two competing retail stores in Exercise 60.

    (a) Which store had the most revenue during the first year? Explain carefully why you think so.

    (b) Which store had the most revenue during the first three years? Explain carefully why you think so. *Hint:* Use the fundamental theorem of calculus, and roughly compare the areas of two regions.

62. Suppose for any t that $v(t)$ is the velocity in feet per second of an object at time t measured in seconds. Suppose also that on the t-interval [0, 1], $v(t)$ is a continuous function that takes both negative and positive values. Explain, using Riemann sums, why $\int |v(t)|\, dt$ is the total distance traveled by the object on the t-interval [0, 1].

63. Explain in complete sentences the difference between the definite integral $\int_a^b f(x)\, dx$ and the indefinite integral $\int f(x)\, dx$. Take a specific function $f(x)$ and specific limits of integration, and illustrate.

64. **Transportation Costs.** A plant has a constant market share s, serving a circular market area of radius R with uniform demand density D per square mile and freight rate of T per mile. The transport cost per unit is then T_r. Determine the demand arising in the area at a radius between r and $r + \Delta r$, and sum all these demands up in a radius R. (See Hay and Morris [123].) Use this Riemann sum to find a definite integral that gives the transportation cost over the circular market of radius R.

65. **Demand.** The price p of an item located a distance D in miles from a mill is $p = m + tD$, where m is the mill price and t is the transportation cost per mile. (See Sailors [124].) The demand curve is $x = a/p^2 = a/(m + tD)^2$, where a is a positive constant. The aggregated demand X over an assumed lined (one-dimensional) market, where $q(p)$ is uniformly distributed, is then $X = \int_0^b a/(m + tD)^2\, dD$, where b measures the extent of the market. Find this integral.

66. **Optimum Mill Price.** The price p of an item located a distance D in miles from a mill is $p = m + tD$, where m is the mill price and t is the transportation cost per mile. (See Hsu [125].) The demand curve is $x = e^{-ap} = e^{-a(m+tD)}\, dD$, where a is a positive constant. Then the profit is $P(m) = \int_0^b (m - c)e^{-a(m+tD)}\, dD - F$, where c is (constant) marginal cost, F is fixed cost, and b measures the extent of the market. Find the value of m that maximizes P.

67. Given the graph of $y = f'(x)$, sketch a graph of $y = f(x)$ if $f(0) = 10$ and if the areas of the regions R_1, R_2, R_3, and R_4 are respectively 10, 12, 3, and 2. Give the x- and y-coordinates of critical points and inflection points.

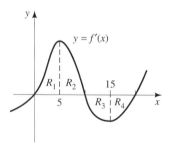

68. (a) Using Riemann sums, outline a proof that if $f(x)$ is continuous on [a, b], $\int_a^b cf(x)\, dx = c \int_a^b f(x)\, dx$. *Hint:* Note that $\sum_{k=1}^n cf(x_k)\, \Delta x = c \sum_{k=1}^n f(x_k)\, \Delta x$.

    (b) Using Riemann sums, outline a proof that if $f(x)$ and $g(x)$ are continuous on [a, b], $\int_a^b [f(x) + g(x)]\, dx = \int_a^b f(x)\, dx + \int_a^b g(x)\, dx$. *Hint:* Note that $\sum_{k=1}^n [f(x_k) + g(x_k)]\, \Delta x = \sum_{k=1}^n f(x_k)\, \Delta x + \sum_{k=1}^n g(x_k)\, \Delta x$.

SOLUTIONS TO SELF-HELP EXERCISE SET 5.5

1. Let $u = x^2 + 2$. Then $du = 2x \, dx$. As x goes from -1 to 0, u goes from $u(-1) = 3$ to $u(0) = 2$. Then

$$\int_{-1}^{0} 4xe^{x^2+2} \, dx = 2 \int_{-1}^{0} e^{x^2+2} 2x \, dx$$

$$= 2 \int_{3}^{2} e^u \, du$$

$$= 2(e^2 - e^3)$$

2. We are seeking the definite integral $\int_{2}^{4} (10t - 0.30t^2) \, dt$. Then

$$\int_{2}^{4} (10t - 0.30t^2) \, dt = (5t^2 - 0.10t^3)\Big|_{2}^{4}$$

$$= (80 - 6.4) - (20 - 0.80) = 54.4$$

or $54,400.

5.6 AREA BETWEEN TWO CURVES

■ Area Between Two Curves   ■ Lorentz Curves

APPLICATION: LORENTZ CURVES

For a given population let $f(x)$ be the proportion of total income that is received by the lowest paid $100x\%$ of income recipients. What definite integral is a measure of the inequality of income distribution for this population? For the answer see the subsection on Lorentz curves.

Area Between Two Curves

Suppose we have two continuous functions $f(x)$ and $g(x)$ on the interval $[a, b]$ with $f(x) \geq g(x) \geq 0$ (Fig. 5.25). We wish to find the area between the two curves.

Clearly the area between the two curves in Figure 5.25 is the area under f less the area under g. That is, the area between the two curves is given by

$$\int_{a}^{b} f(x) \, dx - \int_{a}^{b} g(x) \, dx = \int_{a}^{b} [f(x) - g(x)] \, dx$$

Consider now the more general situation shown in Figure 5.26. We still have $f(x) \geq g(x)$, but both functions can change sign.

It is apparent (see Fig. 5.26) that there exists a positive constant large enough so that the graph of $y = g(x) + k$ remains *above* the x-axis. We also add k to f and let $F(x) = f(x) + k$ and $G(x) = g(x) + k$. The graphs of these functions are shown in Figure 5.27. Adding the very same constant to each function, however, does not change the area between the resulting functions. Thus, it is apparent that the area between F and G is the same as the area between f and g.

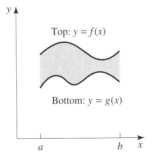

Figure 5.25
The area between the graphs of the two curves from a to b is $\int_{a}^{b} [f(x) - g(x)] \, dx$.

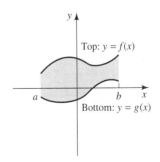

Figure 5.26
The area between the graphs of the two curves from a to b is $\int_{a}^{b} [f(x) - g(x)] \, dx$.

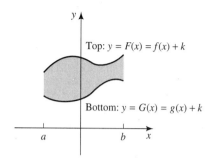

Figure 5.27

But now $F(x) \geq G(x) > 0$. Thus, the area between F and G is just

$$\int_a^b [F(x) - G(x)]\, dx = \int_a^b [(f(x) + k) - (g(x) + k)]\, dx$$

$$= \int_a^b [f(x) - g(x)]\, dx$$

which is exactly the same formula as found before.

Area Between Two Curves

Let $y = f(x)$ and $y = g(x)$ be two continuous functions with $f(x) \geq g(x)$ on $[a, b]$. Then the area between the graphs of the two curves on $[a, b]$ is given by the definite integral

$$\int_a^b [f(x) - g(x)]\, dx$$

We can think of this as the area between the *top* curve and the *bottom* curve or as

$$\int_a^b [\text{top} - \text{bottom}]\, dx$$

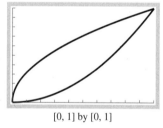

[0, 1] by [0, 1]

Screen 5.5
The region has a top $y_1 = \sqrt{x}$ and a bottom $y_2 = x^2$.

EXAMPLE 1   Finding the Area Between Two Curves

Find the area under the curve $y = \sqrt{x}$ and above the curve $y = x^2$ between $x = 0$ and $x = 1$.

SOLUTION ■ Graphs of the two curves are shown in Screen 5.5 using a window with dimensions [0, 1] by [0, 1]. The graph indicates that the top of the region is given by the function $y = \sqrt{x}$, and the bottom is given by $y = x^2$ (Fig. 5.28). According to the previous discussion, we are seeking

$$\int_a^b [\text{top} - \text{bottom}]\, dx = \int_0^1 [x^{1/2} - x^2]\, dx$$

$$= \left(\frac{2}{3} x^{3/2} - \frac{1}{3} x^3 \right) \Big|_0^1$$

$$= \frac{2}{3} - \frac{1}{3} = \frac{1}{3} \quad \blacksquare$$

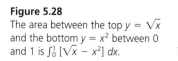

Figure 5.28
The area between the top $y = \sqrt{x}$ and the bottom $y = x^2$ between 0 and 1 is $\int_0^1 [\sqrt{x} - x^2]\, dx$.

Notice that a graph is critical to determine exactly what is the top and what is the bottom curve and also to discern what is the interval of integration.

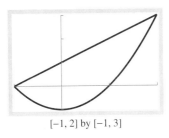

[−1, 2] by [−1, 3]

Screen 5.6
The region has a top $y_1 = x + 1$
and a bottom $y_2 = x^2 - 1$.

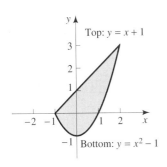

Figure 5.29
The area between the top $y = x + 1$ and the bottom $y = x^2 - 1$ between −1 and 2 is $\int_{-1}^{2} [(x + 1) - (x^2 - 1)]\, dx$.

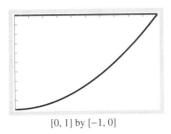

[0, 1] by [−1, 0]

Screen 5.7
The region has a top $y_1 = 0$ and a bottom $y_2 = x^2 - 1$.

EXAMPLE 2   Finding the Area Between Two Curves

Find the area between the two curves $y = x + 1$ and $y = x^2 - 1$.

SOLUTION ■ It is critical that we draw a graph. This is indicated in Screen 5.6 using a window with dimensions [−1, 2] by [−1, 3]. The curve $y = x + 1$ is the top, and the curve $y = x^2 - 1$ is the bottom (Fig. 5.29). To determine where the two curves intersect, set $x^2 - 1 = x + 1$. This gives $0 = x^2 - x - 2 = (x + 1)(x - 2)$. Thus, $x = -1$ and $x = 2$ are the x-coordinates of the points of intersection. Then

$$\int_a^b [\text{top} - \text{bottom}]\, dx = \int_{-1}^{2} [(x + 1) - (x^2 - 1)]\, dx$$
$$= \int_{-1}^{2} [2 + x - x^2]\, dx$$
$$= \left(2x + \frac{1}{2}x^2 - \frac{1}{3}x^3\right)\Big|_{-1}^{2}$$
$$= \left(4 + 2 - \frac{8}{3}\right) - \left(-2 + \frac{1}{2} + \frac{1}{3}\right) = \frac{9}{2} \quad ■$$

EXAMPLE 3   Finding the Area of a Curve Below the x-axis

Find the area of that part of the curve $y = x^2 - 1$ that is below the x-axis and in the fourth quadrant.

SOLUTION ■ We graph $y = x^2 - 1$ using a window with dimensions [0, 1] by [−1, 0]. Refer to Screen 5.7. The curve $y = x^2 - 1$ is the bottom. The top is given by $y = 0$ (Fig. 5.30). The region of integration is easily seen to be [0, 1]. Then

$$\int_a^b [\text{top} - \text{bottom}]\, dx = \int_0^1 [(0) - (x^2 - 1)]\, dx$$
$$= \int_0^1 (1 - x^2)\, dx$$
$$= \left(x - \frac{1}{3}x^3\right)\Big|_0^1$$
$$= \frac{2}{3} - 0 = \frac{2}{3} \quad ■$$

The previous example illustrates how we can geometrically interpret $\int_a^b f(x)\, dx$ when $f(x) \leq 0$ on $[a, b]$. In Figure 5.31, we are given the graph of such a function.

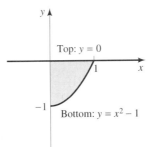

Figure 5.30
The area between the top $y = 0$ and the bottom $y = x^2 - 1$ between 0 and 1 is $\int_0^1 [(0) - (x^2 - 1)]\, dx$.

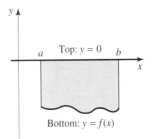

Figure 5.31
If $f(x) \leq 0$ on $[a, b]$, then the area between the x-axis and the graph of the curve $y = f(x)$ between a and b is $-\int_a^b f(x)\, dx$.

The area of the region between the x-axis and the curve $y = f(x)$ from a to b has as the top $y = 0$ and has as the bottom $y = f(x)$. Thus, the area A is just

$$A = \int_a^b [\text{top} - \text{bottom}] \, dx$$

$$= \int_a^b [0 - f(x)] \, dx$$

$$= -\int_a^b f(x) \, dx$$

We therefore have the following:

Area of a Curve Under the x-axis

If a graph of $y = f(x)$ is below the x-axis on $[a, b]$, then the area below the x-axis and above the graph of $y = f(x)$ on $[a, b]$ is

$$\text{Area} = -\int_a^b f(x) \, dx$$

From another point of view, when $f(x) \leq 0$ on $[a, b]$, $\int_a^b f(x) \, dx$ represents the *negative* of the area between the x-axis and the curve $y = f(x)$ from a to b.

Sometimes the region of integration must be divided into subregions.

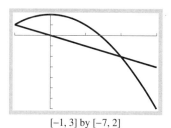

[−1, 3] by [−7, 2]

Screen 5.8
The equation $y_1 = -x$ is the bottom of the left region but the top of the right region. The equation $y_2 = 2 - x^2$ is the top of the left region but the bottom of the right region.

EXAMPLE 4   Graphs with Multiple Points of Intersection

Find the area enclosed between the curves $y = -x$ and $y = 2 - x^2$ on $[-1, 3]$.

SOLUTION ■ Graphs of the two functions are shown in Screen 5.8, using a window with dimensions $[-1, 3]$ by $[-7, 2]$. Notice that the top of the region is *not* given by one single equation. (In fact neither is the bottom.) Refer to Figure 5.32. For example, the top of the region R_1 is given by $y = 2 - x^2$, and the top of the region R_2 is given by $y = -x$. In such a "bow tie" region, we must divide the region into two subregions in order to use our formulas.

To do this we must find the point where the two functions cross. We then set $2 - x^2 = y = -x$ or $0 = x^2 - x - 2 = (x + 1)(x - 2)$. Thus, the two curves cross when $x = -1$ and $x = 2$. So the two intervals of integration are $[-1, 2]$ and $[2, 3]$. Let A be the area we are seeking, A_1 the area of R_1, and A_2 the area of R_2, then

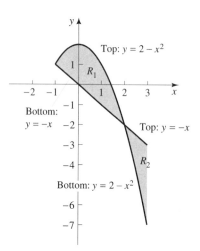

Figure 5.32
The area between the graphs of the two curves from −1 to 3 is the area of R_1 plus the area of R_2.

$$A = A_1 + A_2$$

$$= \int_{-1}^{2} [\text{top of } R_1 - \text{bottom of } R_1] \, dx + \int_{2}^{3} [\text{top of } R_2 - \text{bottom of } R_2] \, dx$$

$$= \int_{-1}^{2} [(2 - x^2) - (-x)] \, dx + \int_{2}^{3} [(-x) - (2 - x^2)] \, dx$$

$$= \int_{-1}^{2} [2 + x - x^2] \, dx + \int_{2}^{3} [x^2 - x - 2] \, dx$$

$$= \left(2x + \frac{1}{2}x^2 - \frac{1}{3}x^3 \right)\Big|_{-1}^{2} + \left(\frac{1}{3}x^3 - \frac{1}{2}x^2 - 2x \right)\Big|_{2}^{3}$$

$$= \left(4 + 2 - \frac{8}{3} \right) - \left(-2 + \frac{1}{2} + \frac{1}{3} \right) + \left(9 - \frac{9}{2} - 6 \right) - \left(\frac{8}{3} - 2 - 4 \right)$$

$$= \frac{19}{3} \quad \blacksquare$$

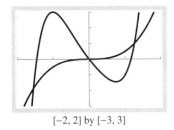

[−2, 2] by [−3, 3]

Screen 5.9
We need to ZOOM to find the x-coordinates of two of the points of intersection of the graphs of $y_1 = x^5 + x^2 - 3x$ and $y_2 = 0.5x^3$. We obtain $x_1 \approx -1.55$ and $x_2 \approx 1.26$.

EXAMPLE 5   Graphs with Multiple Points of Intersection

Find an approximation to the area enclosed between the curves $y = x^5 + x^2 - 3x$ and $y = 0.5x^3$.

SOLUTION ■   Screen 5.9 shows a graph of the two functions $y_1 = x^5 + x^2 - 3x$ and $y_2 = 0.5x^3$ using a window of dimensions [−2, 2] by [−3, 3]. We see that our area is a bow tie region. To find the points of intersection, we set $x^5 + x^2 - 3x = 0.5x^3$. One solution is $x = 0$. We cannot analytically find the others, since the degree of the polynomial is too high. But using the ZOOM feature we can locate the x-coordinates of the two points of intersection as $x = -1.55$ and $x = 1.26$ to two decimal places. Using the program SUM and $n = 64$, we find that

$$A = A_1 + A_2$$

$$= \int_{-1.55}^{0} [\text{top of } R_1 - \text{bottom of } R_1] \, dx + \int_{0}^{1.26} [\text{top of } R_2 - \text{bottom of } R_2] \, dx$$

$$= \int_{-1.55}^{0} [(x^5 + x^2 - 3x) - (0.5x^3)] \, dx + \int_{0}^{1.26} [(0.5x^3) - (x^5 + x^2 - 3x)] \, dx$$

$$\approx 3.25 + 1.36 = 4.61 \quad \blacksquare$$

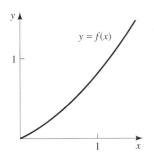

Figure 5.33
A typical Lorentz curve.

Lorentz Curves

For a given population let $f(x)$ be the proportion of total income that is received by the lowest paid $100x\%$ of income recipients. Thus, $f(0.3) = 0.2$ means that the lowest paid 30% of income recipients receive 20% of the total income. Or if $f(0.6) = 0.25$, this means that the lowest paid 60% receive 25% of the total income. We must have $0 \le x \le 1$ and $0 \le f(x) \le 1$.

We only include income recipients, so $f(0) = 0$. Since all income is received by 100% of the recipients, $f(1) = 1$. Also $f(x) \le x$ since the lowest $100x\%$ of the income recipients cannot receive more than $100x\%$ of the total income. The graph of such an income distribution is called a **Lorentz curve**. Figure 5.33 is a typical Lorentz curve.

Suppose a Lorentz curve is given by $f(x) = x^2$. Since $f(0.5) = 0.25$, the lowest paid 50% receive 25% of the total income. Since $f(0.1) = 0.01$, the lowest paid 10% receive 1% of the total income.

A perfect equality of income distribution is represented by the curve $f(x) = x$. This is true since in this case the lowest paid 1% receive 1% of total income, the lowest paid 10% receive 10% of total income, and so on. Any deviation from this perfect equality represents an inequality of income distribution (Fig. 5.34).

A convenient and intuitive measure of this deviation is the *area* between the curve $y = x$ and $y = f(x)$. Economists then define the **coefficient of inequality** to be the ratio

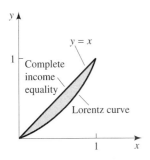

Figure 5.34
The coefficient of inequality is the ratio of the area under the graph of $y = f(x)$ on [0, 1] with the area under $y = x$ on [0, 1]. This equals $2 \int_0^1 [x - f(x)] \, dx$.

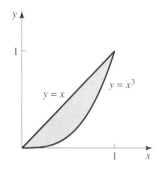

Figure 5.35
The coefficient of inequality is
twice the area between the two
curves, or $2 \int_0^1 (x - x^3)\, dx$.

of this area to the area under the curve $y = x$ that represents perfect equality of income distribution. Since of course the area under $y = x$ on [0, 1] is $\frac{1}{2}$, the ratio is two times the area between $y = x$ and $y = f(x)$. We have the following:

Coefficient of Inequality

The coefficient of inequality L of a Lorentz curve $y = f(x)$ is given by

$$L = 2 \int_0^1 [x - f(x)]\, dx$$

EXAMPLE 6   Calculating the Coefficient of Inequality

Find the coefficient of inequality of the Lorentz curve $y = x^3$ (Fig. 5.35).

SOLUTION ■

$$L = 2 \int_0^1 [x - f(x)]\, dx$$

$$= 2 \int_0^1 (x - x^3)\, dx$$

$$= 2 \left(\frac{1}{2} x^2 - \frac{1}{4} x^4 \right) \Big|_0^1$$

$$= \frac{1}{2} \quad ■$$

SELF-HELP EXERCISE SET 5.6

1. Find the area between the two curves $y = x + 2$ and $y = x^2$.

2. Find the area enclosed by the graphs of the two curves $y = x$ and $y = f(x) = \sqrt[3]{x}$.

EXERCISE SET 5.6

In Exercises 1 through 36 find the area enclosed by the given curves.

1. $y = x^3$, $y = 0$, $x = 1$, $x = 2$

2. $y = x^2$, $y = 0$, $x = -1$, $x = 2$

3. $y = e^{0.1x}$, $y = 1$, $x = 0$, $x = 10$

4. $y = e^{-x}$, $y = 3$, $x = -1$, $x = 0$

5. $y = x^2$, $y = x$, $x = 0$, $x = 1$

6. $y = \sqrt{x}$, $y = x$, $x = 0$, $x = 1$

7. $y = x$, $y = \dfrac{1}{x}$, $x = e$, $x = e^2$

8. $y = 4$, $y = x^2$

9. $y = x^2 - 2x + 1$, $y = x + 1$

10. $y = x^2$, $y = 8 - x^2$

11. $y = x^3$, $y = -x^3$, $x = 0$, $x = 1$

12. $y = x$, $y = -\dfrac{1}{x}$, $x = \dfrac{1}{e}$, $x = e$

13. $y = -x^2$, $y = -3x - 1$, $x = 0$, $x = 2$

14. $y = x^2 - 2x$, $y = x - 4$, $x = 0$, $x = 1$

15. $y = x^2 + 1$, $y = x$, $x = -1$, $x = 1$

16. $y = x + 2$, $y = \sqrt[3]{x}$, $x = -1$, $x = 1$

17. $y = x^2 + 4x + 4$, $y = 4 - x$

18. $y = x^2 - 2x + 1$, $y = 1 + 2x - x^2$

19. $y = x(x - 4)$, $y = x$

20. $y = x(x - 2)$, $y = x$

21. $y = e^{|x|}$, $y = 0$, $x = -1$, $x = 2$

22. $y = e^{-|x|}$, $y = 0$, $x = -1$, $x = 2$

23. $y = x^2$, $y = 4$, $x = 0$, $x = 4$

24. $y = x^3$, $y = 2$, $x = 0$, $x = 2$

25. $y = x^3$, $y = -x$, $x = -1$, $x = 1$

26. $y = e^x$, $y = 2$, $x = 0$, $x = 1$

27. $y = -\sqrt[3]{x}$, $y = -x$, $x = -1$, $x = 1$

28. $y = \sqrt[3]{x}$, $y = -x$, $x = -1$, $x = 1$

29. $y = e^x$, $y = e^{-x}$, $x = -\ln 2$, $x = \ln 2$

30. $y = e^{2x}$, $y = e^{-2x}$, $x = -1$, $x = 2$

31. $y = x^3 - 3x, y = x$

32. $y = x^3 - 3x, y = 2x^2$

33. $y = |x|, y = 3, x = -1, x = 6$ (triple region)

34. $y = |x|, y = \frac{1}{2}x + 2, x = -2, x = 4$ (triple region)

35. $y = \sqrt[3]{x}, y = x, x = -8, x = 1$ (triple region)

36. $y = -\sqrt[3]{x}, y = -x, x = -1, x = 8$ (triple region)

In Exercises 37 through 40 graph the given pair of curves in the same viewing window of your graphing calculator, and use the zoom feature to find the points of intersection to two decimal places. Then use the program SUM with $n = 100$ to estimate the area enclosed by the given pairs of curves.

37. $y = x^5 + x^2 - 3x, y = x$ **38.** $y = x^5 + x^4 - 3x, y = x$

39. $y = x^5 + x^4 - 3x, y = 0.50x^3$

40. $y = x^5 + x^4 - 3x, y = 3x - x^2 - x^5$

Applications

In Exercises 41 through 48 show that the curves are Lorentz curves. (You need to find $f''(x)$.) Then find the coefficient of inequality.

41. $f(x) = x^2$ **42.** $f(x) = x^5$

43. $f(x) = \frac{1}{4}x^2 + \frac{3}{4}x$ **44.** $f(x) = \frac{3}{4}x^2 + \frac{1}{4}x$

45. $f(x) = 0.3x^2 + 0.7x$ **46.** $f(x) = 0.7x^2 + 0.3x$

47. $f(x) = 0.01x^2 + 0.99x$ **48.** $f(x) = 0.99x^2 + 0.01x$

49. Costs. A firm has found that its costs in billions of dollars have been increasing (due primarily to inflation) at a rate given by $C_1'(t) = 1.2t$, where t is the time measured in years since the item began being produced. At the beginning of the second year, an innovation in the production process resulted in costs increasing at a rate given by $C_2'(t) = 0.9\sqrt{t}$. Find the total costs during the first four years. Assume $C_1(0) = 0$.

50. Revenue. A firm has found that its revenue in billions of dollars has been increasing at a rate given by $R'(t) = 3t^2$, where t is the time measured in years since the item was introduced. At the beginning of the third year, new com-

petition resulted in revenue increasing only at a rate given by $R_1'(t) = 12t$. Find the total revenue during the first four years.

51. Sales. From the time a firm introduces a new product sales in thousands of this product have increased at a rate given by $S_1'(t) = 2 + t$, where t is the time since the introduction of the new product. At the beginning of the third year, an advertising campaign was introduced, and sales then increased at a rate given by $S_2'(t) = t^2$. Find the increase in sales due to the advertising campaign during the first two years of this campaign over what the sales would have been without the campaign.

52. Production. Suppose copper is being extracted from a certain mine at a rate given by $\frac{d}{dt}P(t) = 100e^{-0.2t}$, where $P(t)$ is measured in tons of copper and t in years from opening the mine. At the beginning of the sixth year, a new mining innovation is introduced to boost production to a rate given by $Q'(t) = 500/t$. Find the increase in production of copper due to this innovation during the first five years of its use over what copper production would have been without its use.

Enrichment Exercises

53. Find the area between the graph of the curve $y = \sin x$ and the x-axis on the interval $[0, \pi]$.

54. Find the area between the graph of the curve $y = \cos x$ and the x-axis on the interval $[-\pi/2, \pi/2]$.

55. Wage Distribution. The following graph appeared recently in a paper in the *Economic Review Federal Reserve Bank of Dallas* by Phillips [126] and gave a Lorentz curve for the wage distribution for year-round full-time workers in the Texas goods sector for the years 1978 and 1989. What conclusions can you come to about wages in these two years?

Wage Distribution for Year-round, Full-time Workers in the Texas Goods Sector (Lorentz Curve)

Cumulative percent of workers

56. We learned in this section that the area of the shaded region in the accompanying figure is

$$\int_a^c [f(x) - g(x)]\, dx + \int_c^b [g(x) - f(x)]\, dx$$

Explain carefully why the area of the shaded region is also given by $\int_a^b |f(x) - g(x)|\, dx$.

57. Refer to the figure. Suppose $F(2) = 5$ and the area of the shaded region is 10. Find the maximum value attained by F.

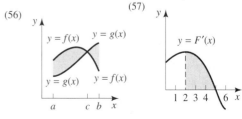

58. On your graphing calculator draw graphs of $y_1 = f(x) = \sin x$ and $y_2 = 0.5$ using a window with dimensions $[0, 3.14]$ by $[0, 1]$. By looking at the graphs and doing no calculations, explain why the average value of f on the interval $[0, 3.14]$ must be between 0.5 and 1.

SOLUTIONS TO SELF-HELP EXERCISE SET 5.6

1. In order to determine the interval of integration we need to determine the values of x for which the two curves in the figure intersect. This happens when $x + 2 = y = x^2$ or $0 = x^2 - x - 2 = (x + 1)(x - 2)$. Thus, $x = -1$ and $x = 2$. Then

$$\int_a^b [\text{top} - \text{bottom}] \, dx = \int_{-1}^{2} [(x + 2) - (x^2)] \, dx$$

$$= \left(\frac{1}{2} x^2 + 2x - \frac{1}{3} x^3 \right) \Big|_{-1}^{2}$$

$$= \left(2 + 4 - \frac{8}{3} \right) - \left(\frac{1}{2} - 2 + \frac{1}{3} \right)$$

$$= \frac{9}{2}$$

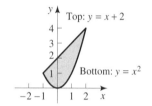

2. The region between the graphs of the two curves in the figure forms a "bow tie" region, so we must divide the region into two subregions in order to use our formulas. If we let A be the area we are seeking and A_1 the area of R_1 and A_2 the area of R_2, then

$$A = A_1 + A_2$$

$$= \int_{-1}^{0} [\text{top of } R_1 - \text{bottom of } R_1] \, dx$$

$$\quad + \int_{0}^{1} [\text{top of } R_2 - \text{bottom of } R_2] \, dx$$

$$= \int_{-1}^{0} [(x) - (x^{1/3})] \, dx + \int_{0}^{1} [(x^{1/3}) - (x)] \, dx$$

$$= \left(\frac{1}{2} x^2 - \frac{3}{4} x^{4/3} \right) \Big|_{-1}^{0} + \left(\frac{3}{4} x^{4/3} - \frac{1}{2} x^2 \right) \Big|_{0}^{1}$$

$$= -\left(\frac{1}{2} - \frac{3}{4} \right) + \left(\frac{3}{4} - \frac{1}{2} \right) = \frac{1}{2}$$

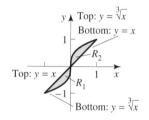

5.7 ADDITIONAL APPLICATIONS OF THE INTEGRAL

- Continuous Money Flow ■ Present Value of a Continuous Income Flow
- Continuous Reinvestment of Income (Optional) ■ Consumers' and Producers' Surplus ■ Density (Optional)

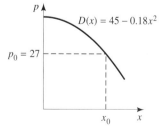

Figure 5.36
Here the current demand is x_0, and the current price is 27.

APPLICATION: PRICE OF AN OIL WELL

Suppose the rate of change in thousands of dollars of total income from an oil well is projected to be $f(t) = 100e^{-0.1t}$, where t is measured in years. If the oil well is being offered for sale for $500,000 and has a useful life of 10 years, and current annual interest rates are at 10%, should you buy it? See Example 4 for the answer.

APPLICATION: CONSUMERS' SURPLUS

Let $p = D(x) = 45 - 0.18x^2$ be the demand equation, where p is the unit price of a commodity and x is the quantity demanded by the consumers at that price. Suppose the current price is $p_0 = 27$. As can be seen in Figure 5.36, $D(x) > p_0$ when $x < x_0$. This means that some consumers are willing to pay a *higher* price for the commodity than p_0. These consumers are then actually experiencing a *savings*. What is the total amount of these savings? See Example 6 for the answer.

APPLICATION: CALCULATING THE QUANTITY OF POLLUTANTS

An industrial plant is located at the center of town, which extends two miles on either side of the plant. Let x be the distance from the plant. A straight highway goes through town, and the concentration (density) of pollutants along this highway is given by $\delta(x) = 1000(24 - 3x^2)$ in numbers of particles per mile per day at the point x. What is the total number of particles of pollutant per day in town along this highway? See Example 7 for the answer.

Continuous Money Flow

We are already familiar with the idea of *continuous* compounding. When money is being compounded continuously, we are not suggesting that the bank is actually continuously placing money into our account. Rather we calculate what is in our account as if this were in fact happening. We then can think of our account as a **continuous flow of money**. This is not unlike the electric utility company that has a meter on our house continuously totaling the amount of electricity we are using. Since a price is set for the electricity, this can be thought of as a continuous flow of income for the utility company.

We then assume that we have a continuous flow of money and that the positive function $f(t)$ represents the *rate of change* of this flow. (We could think of $f(t)$ as the velocity of the flow.) If we let $A(t)$ be the total amount of income obtained from this flow, we are saying that $A'(t) = f(t)$. We always assume that the amount of money from this flow at time zero is zero, that is, we always assume that $A(0) = 0$. Then the fundamental theorem of calculus tells us that

$$A(t) = \int_0^t f(x) \, dx$$

This is the area under the curve $f(x)$ on the interval $[0, t]$ (Fig. 5.37).

EXAMPLE 1 Finding Income from an Income Flow

Suppose an investment brings in an income of $1000 per year. How much income is obtained in three years?

SOLUTION ■ Obviously the answer is $3000. But this can also be viewed as an income flow given by $f(t) = \$1000$ per year on the interval $[0, 3]$. The area under this curve is the definite integral

$$\int_0^3 1000 \, dx = 3000$$

(Fig. 5.38). ■

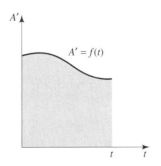

Figure 5.37
A graph of the rate of change of a typical income flow.

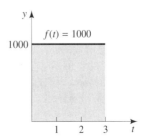

Figure 5.38
The area between the graph of $y = 1000$ and the t-axis from 0 to 3, $\int_0^3 1000 \, dt = 3000$, is the total income obtained.

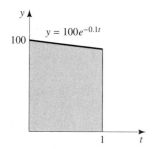

Figure 5.39
The area between the graph of $y = 100e^{-0.1t}$ and the t-axis from 0 to 1, $\int_0^1 100e^{-0.1t}\, dt$, is the total income obtained.

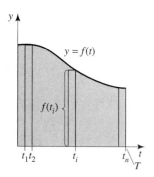

Figure 5.40
Divide the t-interval into n equal subintervals, and in each of these subintervals pick t_i to be the right-hand endpoint.

EXAMPLE 2   Finding Income from an Income Flow

Suppose the rate of change of income in thousands of dollars per year from an oil well is projected to be $f(t) = 100e^{-0.1t}$, where t is measured in years. Find the total amount of money from this well during the first year (Fig. 5.39).

SOLUTION ■ We have

$$A(1) = \int_0^1 f(x)\, dx$$

$$= \int_0^1 100e^{-0.1x}\, dx$$

$$= 100(-10)e^{-0.1t}\big|_0^1$$

$$= 1000\,(1 - e^{-0.1}) \approx 95$$

or approximately \$95,000. ■

Present Value of a Continuous Income Flow

We now take up the subject of **present value** of an income flow and find a formula that yields the present value. This permits us to compare different income flows.

We again assume that we have some continuous flow of income, perhaps from oil or gas wells, accounts with continuous compounding of interest, or other accounts assumed to be compounding continuously. We assume that $f(t)$ gives the rate of change of the total flow of funds, with t measured in some appropriate units of time that, for convenience, we take to be years.

For any $T > 0$ we wish to define a number that we call the present value of this income flow on the interval $[0, T]$. Intuitively we want this number to be the *present amount of money needed to be able to generate the same income flow over the time interval $[0, T]$ as is generated by the given flow.* To do this we must assume that the present money can be invested at a certain given rate (the current available rate) over that time interval, say a rate equal to r compounded continuously.

Recall that if an amount P is invested at an annual rate of r (given as a decimal) and compounded continuously, then the amount becomes equal to $A = Pe^{rt}$ after t years. Solving for P gives the present value $P = Ae^{-rt}$. This is the present amount needed to attain the amount A after t years of continuous compounding at the annual rate of r.

To proceed, divide the interval $[0, T]$ into n equal subintervals of length Δt as in Figure 5.40. In a typical interval I_i we pick a point t_i, say the right-hand endpoint. The area under $f(t)$ on the interval I_i is precisely the income generated by the given flow over the time interval I_i. This area A_i can be approximated by the area of the rectangle given by $f(t_i)\, \Delta t$. The present value of this income A_i is

$$P_i = A_i e^{-rt_i} \approx f(t_i)e^{-rt_i}\, \Delta t$$

Summing all of these present values gives the total present value, P. This is also approximately the sum of all the areas of all n rectangles, or

$$P = \sum_{i=1}^n P_i \approx \sum_{i=1}^n f(t_i)e^{-rt_i}\, \Delta t$$

which is a Riemann sum of $f(t)e^{-rt}$ on the interval $[0, T]$. If we take n larger and larger, the approximations become better and better. Then letting $n \to \infty$, we should obtain the exact value. But this is

$$\lim_{n \to \infty} \sum_{i=1}^n f(t_i)e^{-rt_i}\, \Delta t$$

which is precisely the definition of the definite integral

$$\int_0^T f(t)e^{-rt}\, dt$$

We thus have the following:

Present Value of a Continuous Income Flow

Let $f(t)$ be the continuous rate of change of total income on the interval $[0, T]$, and suppose that we can invest current money at an interest rate of r, where r is a decimal, compounded continuously over the time interval $[0, T]$. Then the present value $P_V(T)$ of this continuous income flow over $[0, T]$ is given by

$$P_V(T) = \int_0^T f(t)e^{-rt}\, dt$$

EXAMPLE 3   Finding the Present Value of a Continuous Income Flow

Suppose that the rate of change of an income flow is given by the constant (and continuous) function $f(t) = \$1000$ per year. If the rate r is given by $r = 0.10$, find $P_V(3)$.

SOLUTION ■

$$\begin{aligned}
P_V(3) &= \int_0^T f(t)e^{-rt}\, dt \\
&= \int_0^3 1000e^{-0.1t}\, dt \\
&= 1000(-10)e^{-0.1t}\big|_0^3 \\
&= 10{,}000(1 - e^{-0.3}) \approx 2592 \quad ■
\end{aligned}$$

EXAMPLE 4   Determining the Price of an Oil Well

If the oil well in Example 2 is being offered for sale for \$500,000 and has a useful life of 10 years, should you buy it if (a) $r = 0.1$ (b) $r = 0.05$?

SOLUTIONS ■
    (a) In this case

$$\begin{aligned}
P_V(10) &= \int_0^T f(t)e^{-rt}\, dt \\
&= \int_0^{10} 100e^{-0.1t}e^{-0.1t}\, dt \\
&= 100 \int_0^{10} e^{-0.2t}\, dt \\
&= 100(-5)e^{-0.2t}\big|_0^{10} \\
&= 500(1 - e^{-2}) \approx 432
\end{aligned}$$

or \$432,000. This is not a good deal since you can generate more income from \$500,000 from an investment at current interest rates.
    (b) In this case

$$\begin{aligned}
P_V(10) &= \int_0^T f(t)e^{-rt}\, dt \\
&= \int_0^{10} 100e^{-0.10t}e^{-0.05t}\, dt \\
&= 100 \int_0^{10} e^{-0.15t}\, dt
\end{aligned}$$

$$= -\frac{100}{0.15} e^{-0.15t} \Big|_0^{10}$$

$$= \frac{100}{0.15} [1 - e^{-1.5}] \approx 518$$

or \$518,000. This is a good deal since you can generate more income from this oil well than you can from \$500,000 from an investment at current interest rates. ■

Remark. Naturally an important consideration in any investment is its riskiness. This has not been taken into account in the decision to buy in the previous example.

Continuous Reinvestment of Income

As we noted before, the formula $A(t) = Pe^{rt}$ relates the amount $A(t)$ in an account after t years with initial amount P and earning an annual rate r compounded continuously. We now suppose that the continuous income stream from the flow is continuously invested, and furthermore we assume that throughout the interval under consideration this income is always invested at the very same annual rate of r compounded continuously. If we let $A^*(T)$ be the amount of money after T years if all the income has been continuously invested at an annual rate of r compounded continuously, then

$$A^*(T) = e^{rT}P_V(T)$$

To see why this is true, recall from the setup in Figure 5.40 that the amount of income obtained during the time interval I_i was $A_i \approx f(t_i) \, \Delta t$. If this money is then invested at an annual rate of r compounded continuously for the next $T - t_i$ years, this money then becomes

$$A_i^*(T) = A_i e^{r(T-t_i)} \approx f(t_i)e^{r(T-t_i)} \, \Delta t$$

or

$$A_i^*(T) \approx e^{rT}f(t_i)e^{-rt_i} \, \Delta t$$

Summing all of these up gives the amount at time T, designated as $A^*(T)$, which according to this last formula, is also approximately equal to

$$\sum_{i=1}^{n} e^{rT}f(t_i)e^{-rt_i} \, \Delta t$$

which can be written as

$$e^{rT} \sum_{i=1}^{n} f(t_i)e^{-rt_i} \, \Delta t$$

Letting $n \to \infty$, the latter becomes

$$e^{rT} \int_0^T f(t)e^{-rt} \, dt = e^{rT}P_V(T)$$

Continuous Investment of Money

Let $A^*(T)$ be the amount of money after T years if all the income has been continuously invested at an annual rate of r compounded continuously, then

$$A^*(T) = e^{rT}P_V(T) = e^{rT} \int_0^T f(t)e^{-rt} \, dt$$

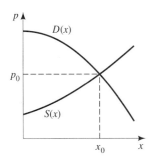

Figure 5.41
The demand and supply curves intersect at (x_0, p_0).

EXAMPLE 5   Finding the Total Amount When Continuously Reinvested

Suppose in Example 3 that the income is continuously invested at the rate r given there. How much money is there after three years?

SOLUTION ■ Using the formula for $A^*(T)$ and the result found in Example 3, we obtain

$$A^*(3) = e^{(0.1)(3)}P_V(3) \approx e^{0.3}(2592) \approx 3499$$

that is, approximately \$3499. ■

Naturally the amount in Example 5 is greater than the amount in Example 1 since in the former case the money is being continuously reinvested as it comes in.

Consumers' and Producers' Surplus

Let $p = D(x)$ be the demand equation, where p is the unit price of a commodity and x is the quantity demanded by the consumers at that price. As we know, this function must be decreasing (Fig. 5.41).

Let $p = S(x)$ be the supply equation, where p is the unit price of a commodity and x is the quantity made available by the producers at that price. As we know, this function must be increasing (see Fig. 5.41). We assume that the demand curve and the supply curve intersect at $x = x_0$. Thus, $D(x_0) = p_0 = S(x_0)$. With (x_0, p_0) the point of intersection p_0 is the equilibrium price. At the equilibrium price consumers purchase the same number of the commodity as the producers supply.

As can be seen in Figure 5.41, $D(x) > p_0$ when $x < x_0$. This means that some consumers are willing to pay a *higher* price for the commodity than p_0. These consumers are then actually experiencing a *savings*. The total amount of such "savings" is called the **consumers' surplus**.

We shall now show that the consumers' surplus is given by the area between the curve $p = D(x)$ and the straight line $p = p_0$ on the interval $[0, x_0]$ which is the definite integral

$$\int_0^{x_0} [D(x) - p_0] \, dx$$

To see this we divide the interval $[0, x_0]$ into n subintervals of equal length Δx. Thus, Δx is the number of consumers in each subinterval. In a typical interval I_i we pick a point x_i, say the right-hand endpoint. Thus, x_i represents a "typical" consumer in the subinterval I_i. The number $p_i - p_0 = D(x_i) - p_0$ then represents the "typical" savings experienced by consumers on this interval. The total savings of consumers on this interval is then approximately the "typical" savings times the number in the interval, or

$$[D(x_i) - p_0] \, \Delta x$$

which is also the area of the rectangle in Figure 5.42.

To find an approximation of the total consumers' surplus, we sum n such terms and obtain

$$\sum_{i=1}^{n} [D(x_i) - p_0] \, \Delta x$$

This is also a right-handed Riemann sum of $[D(x) - p_0]$ over $[a, b]$. For n larger and larger this approximation becomes better and better. Taking the limit as $n \to \infty$, that is,

$$\lim_{n \to \infty} \sum_{i=1}^{n} [D(x_i) - p_0] \, \Delta x$$

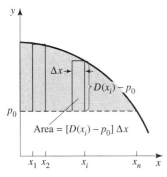

Figure 5.42
The total consumer savings on the ith interval is approximately the area of the indicated rectangle.

we obtain the exact consumers' surplus. By the definition of the definite integral, however, this limit is

$$\int_0^{x_0} [D(x) - p_0] \, dx$$

We thus have the following:

Consumers' Surplus

If $p = D(x)$ is the demand equation, p_0 is the current price of the commodity, and x_0 is the current demand, then the consumers' surplus is given by

$$\int_0^{x_0} [D(x) - p_0] \, dx$$

We also see from Figure 5.41 that $p = S(x) < p_0$ if $x < x_0$, which means that some producers are willing to sell the commodity for *less* than the going price. For these producers the current price of p_0 represents a *savings*. The total of all such savings is called the **producers' surplus**.

In exact analogy with consumers' surplus, the producers' surplus is given by the area between the straight line $p = p_0$ and the curve $p = S(x)$ on the interval $[0, x_0]$, which is the definite integral $\int_0^{x_0} [p_0 - S(x)] \, dx$. We then have the following:

Producers' Surplus

If $p = S(x)$ is the supply equation, p_0 is the current price of the commodity, and x_0 is the current demand, then the producers' surplus is given by

$$\int_0^{x_0} [p_0 - S(x)] \, dx$$

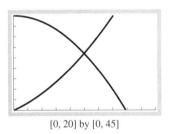

[0, 20] by [0, 45]

Screen 5.10
The supply and demand curves $y_1 = 0.12x^2 + 1.5x$ and $y_2 = 45 - 0.18x^2$ intersect at one point.

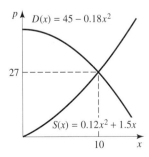

Figure 5.43
The demand and supply curves intersect at (10,27).

EXAMPLE 6   Finding the Consumers' Surplus

If the demand equation is given by $p = D(x) = 45 - 0.18x^2$ and the supply equation is given by $p = S(x) = 0.12x^2 + 1.5x$, find the consumers' surplus.

SOLUTION ■ Graphs of the demand and supply curves are shown in Screen 5.10 using a window of dimensions [0, 20] by [0, 45]. First we need to find the equilibrium quantity. To do this set $S(x) = D(x)$ and obtain

$$0.12x^2 + 1.5x = 45 - 0.18x^2$$

$$0 = 0.30x^2 + 1.5x - 45$$

$$= 0.30[x^2 + 5x - 150]$$

$$= 0.30(x + 15)(x - 10)$$

Since the equilibrium quantity cannot be negative, we ignore the solution $x_0 = -15$. Thus, the equilibrium quantity is $x_0 = 10$. For this value we have $p_0 = D(x_0) = D(10) = 27$ (Fig. 5.43).

The consumers' surplus is then given by

$$\int_0^{x_0} [D(x) - p_0] \, dx = \int_0^{10} [45 - 0.18x^2 - 27] \, dx$$

$$= [18x - 0.06x^3]\Big|_0^{10}$$

$$= 120 \quad ■$$

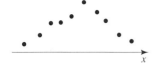

Figure 5.44
The density at discrete points.

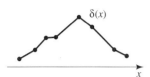

Figure 5.45
A continuous curve through the points in Figure 5.44.

Density

Suppose an industrial plant emits a certain amount of pollutants. We wish to determine the amount of pollutants falling to the ground along some straight line from the plant. At various distances from the plant along this line we place instruments that count the number of pollutants per day falling on a 1-foot stretch. We then accumulate data that give the number of pollutants per foot per day, or the **density**, at various points from the plant, as indicated in Figure 5.44.

We then determine a continuous function $\delta(x)$ whose graph goes through the data points as in Figure 5.45. The function $\delta(x)$ could be obtained just by drawing straight lines between the data points or by a more sophisticated curve-fitting scheme that fits a higher degree polynomial to the data. Or we might have some theoretical means at our disposal that gives us $\delta(x)$. The function $\delta(x)$ is a **density function** and gives the number of pollutants per square foot (per day).

To find the total number of pollutants or the population on the interval $[a, b]$, proceed as follows. For any integer n, divide the interval $[a, b]$ into n subintervals I_1, $I_2, \ldots, I_n$, each of length $\Delta x = \dfrac{b - a}{n}$. In a typical interval I_i, pick a point $x_i \in I_i$, say the right endpoint, to be specific.

As Figure 5.46 indicates, if Δx is small then, since $\delta(x)$ is continuous, $\delta(x)$ is approximately equal to $\delta(x_i)$ on the interval I_i. Thus, the term $P_i = \delta(x_i)\,\Delta x$ is the number per square foot times the length of I_i, or the total number of pollutants on I_i, that is, the population on I_i. Of course, $P_i = \delta(x_i)\,\Delta x$ is also the area of the rectangle shown in Figure 5.46.

The total number of pollutants on $[a, b]$ is given by

$$P = P_1 + P_2 + \cdots + P_n = \sum_{i=1}^{n} P_i \approx \sum_{i=1}^{n} \delta(x_i)\,\Delta x$$

As $n \to \infty$, $\Delta x \to 0$, and this last approximation becomes more and more accurate. Also we notice that $\sum_{i=1}^{n} \delta(x_i)\,\Delta x$ is a right-handed Riemann sum S_n of $\delta(x)$ on $[a, b]$, and by the very definition of the definite integral

$$\lim_{n \to \infty} \sum_{i=1}^{n} \delta(x_i)\,\Delta x = \int_{a}^{b} \delta(x)\,dx$$

Thus, we have the following:

Population Given the Density

If $\delta(x)$ is a continuous function that gives the density in number per unit length, then the total population in the interval $[a, b]$ is

$$P = \int_{a}^{b} \delta(x)\,dx$$

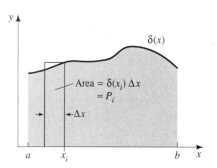

Figure 5.46
The total number of pollutants on the *i*th interval is approximately the area of the rectangle.

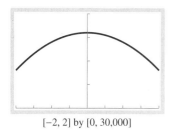

[−2, 2] by [0, 30,000]

Screen 5.11
The graph of $y = 1000(24 - 3x^2)$ giving the density of pollutants along a highway.

EXAMPLE 7 Finding the Population Given the Density

An industrial plant is located at the center of town, which extends two miles on either side of the plant. Let x be the distance from the plant. A straight highway goes through town and the concentration (density) of pollutants along this highway is given by

$$\delta(x) = 1000(24 - 3x^2)$$

in numbers of particles per mile per day at the point x. Screen 5.11 shows a graph of $y = \delta(x)$ using a window with dimensions $[-2, 2]$ by $[0, 30,000]$. Find the total number of particles of pollutant per day in town along this highway.

SOLUTION ■ We have

$$P = \int_a^b \delta(x)\, dx$$

$$= \int_{-2}^2 1000(24 - 3x^2)\, dx$$

$$= 1000(24x - x^3)\big|_{-2}^2$$

$$= 1000[(48 - 8) - (-48 + 8)]$$

$$= 80{,}000 \quad ■$$

SELF-HELP EXERCISE SET 5.7

1. For the oil well in Example 2 find the total amount $A(t)$ of income over 10 years.

2. If the oil well in Example 2 of the text is offered for sale for $500,000 and has a useful life of 8 years and $r = 0.05$, should you buy it?

3. If the demand equation is given by $p = D(x) = 45 - 0.18x^2$ and the supply equation is given by $p = S(x) = 0.12x^2 + 1.5x$, find the producers' surplus. (See Example 6.)

4. A study has indicated that the level $L(x)$ of air pollution a distance of x miles from a certain factory is given by $L(x) = e^{-0.1x} + 1/x^2$. Find the average level of pollution between 10 and 20 miles from the factory.

EXERCISE SET 5.7

Applications

In Exercises 1 through 10 $f(t)$ is the rate of change of total income per unit time. Find **(a)** the total amount of income at the given time T, **(b)** the present value $P_V(T)$ at the given time T for the given interest rate r that is compounding continuously, and **(c)** the amount at the given time T if all income is reinvested continuously at the given annual rate of r compounded continuously. In Exercises 7 through 10, use the program SUM and $n = 100$ to estimate the integrals.

1. $f(t) = 20$, $T = 5$, $r = 10\%$

2. $f(t) = 30$, $T = 10$, $r = 10\%$

3. $f(t) = 100e^{0.05t}$, $T = 20$, $r = 5\%$

4. $f(t) = 100e^{-0.05t}$, $T = 40$, $r = 5\%$

5. $f(t) = 20e^{-0.05t}$, $T = 10$, $r = 10\%$

6. $f(t) = 20e^{0.05t}$, $T = 10$, $r = 10\%$

7. $f(t) = t$, $T = 10$, $r = 10\%$

8. $f(t) = t + 1$, $T = 10$, $r = 5\%$

9. $f(t) = t^2$, $T = 10$, $r = 10\%$

10. $f(t) = t^2 + 1$, $T = 10$, $r = 8\%$

11. **Investment Decision.** An investment that has a continuous return of $150,000 per year for 10 years is being offered for sale for $850,000. Should you buy if the current interest rate r that can be compounded continuously over the next 10 years is **(a)** 0.10? **(b)** 0.15?

12. **Investment Decision.** An investment that has a continuous return of $240,000 per year for 10 years is being offered for sale for $1,500,000. Should you buy if the current interest rate r that can be compounded continuously over the next 10 years is **(a)** 0.08? **(b)** 0.12?

13. **Oil Well.** An oil well is being offered for sale for $310,000 and has a rate of change of income per year given by $f(t) = 100e^{-0.1t}$, in thousands of dollars, and a useful life of five years. Should you buy if the current interest rate r that can be compounded continuously over the next five years is **(a)** 0.10? **(b)** 0.12?

14. Oil Well. An oil well is being offered for sale for $420,000 and has a rate of change of income per year given by $f(t) = 100e^{-0.1t}$, in thousands of dollars, and a useful life of 10 years. Should you buy if the current interest rate r that can be compounded continuously over the next 10 years is **(a)** 0.10? **(b)** 0.12?

In Exercises 15 through 18 find the consumers' surplus, using the given demand equations and the market price p_0.

15. $D(x) = 20 - x^2, p_0 = 4$ **16.** $D(x) = 30 - x^2, p_0 = 5$

17. $D(x) = e^{-x}, p_0 = 0.1$ **18.** $D(x) = 20 - x, p_0 = 10$

In Exercises 19 through 22 find the producers' surplus, using the given supply equations and the market price p_0.

19. $S(x) = 100x, p_0 = 10$ **20.** $S(x) = x^3, p_0 = 8$

21. $S(x) = 9x^2, p_0 = 1$ **22.** $S(x) = e^x, p_0 = e^2$

In Exercises 23 through 26 you are given the demand and supply equation. Find the equilibrium point, and then calculate both the consumers' surplus and the producers' surplus.

23. $D(x) = 12 - x; S(x) = 2x$

24. $D(x) = 50 - x^2; S(x) = 5x$

25. $D(x) = 20 - x^2; S(x) = x$

26. $D(x) = 64 - x^2; S(x) = 3x^2$

27. Density of Pollutants. Find the number of pollutants between 1 and 2 miles from the plant in Example 7 of the text.

28. Density of Pollutants. Find the number of pollutants within 1 mile of the plant given in Example 7 of the text.

29. Density of a Population. The density function in people per mile for the population of a small coastal town is given by $\delta(y) = \dfrac{32{,}000}{(2 + y)^3}$, where y is the distance in miles from the ocean. If the town extends for 2 miles from the ocean, what is the total population of this town?

30. Density of a Population. Find the total population in the previous problem within 1 mile of shore.

SOLUTIONS TO SELF-HELP EXERCISE SET 5.7

1.
$$A(t) = \int_0^{10} f(t)\, dt$$
$$= \int_0^{10} 100e^{-0.1t}\, dt$$
$$= 100(-10)e^{-0.1t}\big|_0^{10}$$
$$= 1000(1 - e^{-1}) \approx 632$$

or $632,000.

2. In this case
$$P_V(8) = \int_0^T f(t)e^{-rt}\, dt$$
$$= \int_0^8 100e^{-0.05t}e^{-0.1t}\, dt$$
$$= 100\int_0^8 e^{-0.15t}\, dt$$
$$= -\frac{100}{0.15}e^{-0.15t}\big|_0^8$$
$$= \frac{100}{0.15}[1 - e^{-1.2}] \approx 466$$

or $466,000. This is not a good deal since this oil well generates less income than investment of $500,000 at current interest rates.

3. The equilibrium demand was already found to be $x_0 = 10$ with an equilibrium price of $p_0 = 27$ in Example 6. The producers' surplus is then given by

$$\int_0^{x_0} [p_0 - S(x)]\, dx = \int_0^{10} (27 - 0.12x^2 - 1.5x)\, dx$$
$$= (27x - 0.04x^3 - 0.75x^2)\big|_0^{10}$$
$$= 155$$

4. The average on the interval $[10, 20]$ is

$$\frac{1}{b - a}\int_a^b f(x)\, dx = \frac{1}{20 - 10}\int_{10}^{20}\left(e^{-0.10x} + \frac{1}{x^2}\right) dx$$
$$= \frac{1}{10}\left(-10e^{-0.10x} - \frac{1}{x}\right)\bigg|_{10}^{20}$$
$$= \frac{1}{10}\left[\left(-10e^{-2} - \frac{1}{20}\right)\right.$$
$$\left. -\left(-10e^{-1} - \frac{1}{10}\right)\right] \approx 0.238$$

- If $F'(x) = f(x)$, then $F(x)$ is called the **antiderivative** of $f(x)$. p. 269.

- If two functions have the same derivative on (a, b), then they differ by a constant on (a, b). p. 270.

- If $F'(x) = f(x)$, then the **indefinite integral** is $\int f(x)\,dx = F(x) + C$, where C is an arbitrary constant called the **constant of integration**. p. 270.

- **Rules of Integration**

$$\int x^n\,dx = \frac{1}{n+1}x^{n+1} + C, \quad n \neq -1$$

$$\int \frac{1}{x}\,dx = \ln|x| + C$$

$$\int [f(x)]^n f'(x)\,dx$$
$$= \frac{1}{n+1}[f(x)]^{n+1} + C, \; n \neq -1$$

$$\int \frac{f'(x)}{f(x)}\,dx = \ln|f(x)| + C$$

$$\int e^{f(x)} f'(x)\,dx = e^{f(x)} + C$$

- Divide the x-interval $[a, b]$ into n subintervals of equal length $\Delta x = \dfrac{b-a}{n}$. Let the endpoints of these subintervals be $a = x_0, x_1, x_2, \ldots, x_n = b$. Then

$$\text{Right-hand sum} = f(x_1)\,\Delta x + f(x_2)\,\Delta x + \cdots + f(x_n)\,\Delta x$$

$$= \sum_{k=1}^{n} f(x_k)\,\Delta x$$

$$\text{Left-hand sum} = f(x_0)\,\Delta x + f(x_1)\,\Delta x + \cdots + f(x_{n-1})\,\Delta x$$

$$= \sum_{k=0}^{n-1} f(x_k)\,\Delta x$$

- Suppose $f(x)$ is a continuous function on the finite interval $[a, b]$. Then the right-hand sum $\sum_{k=1}^{n} f(x_k)\,\Delta x$ and the left-hand sum $\sum_{k=0}^{n-1} f(x_k)\,\Delta x$ satisfy

$$\lim_{n \to \infty} \sum_{k=1}^{n} f(x_k)\,\Delta x = \lim_{n \to \infty} \sum_{k=0}^{n-1} f(x_k)\,\Delta x$$

We refer to this common limit as the **definite integral** of f from a to b and write it as $\int_a^b f(x)\,dx$. We refer to a as the **lower limit** of integration, to b as the **upper limit** of integration, and the interval $[a, b]$ as the **interval of integration**. p. 295.

- If $f(x)$ is nonnegative on the interval $[a, b]$, we interpret the definite integral $\int_a^b f(x)\,dx$ as the area of the region between the graph of $y = f(x)$ and the x-axis on the interval $[a, b]$. p. 295.

- Divide the interval $[a, b]$ into n subintervals of equal length Δx and let x_i be any point in the ith subinterval. Then $S_n = \sum_{i=1}^{n} f(x_i)\,\Delta x$ is called a general **Riemann sum**. p. 297.

- **Common Limit of Riemann Sums.** Suppose that $f(x)$ is continuous on the interval $[a, b]$. If this interval is divided into n subintervals of equal length Δx and x_i is any point in the ith subinterval, then $\lim_{n \to \infty} \sum_{i=1}^{n} f(x_i)\,\Delta x$ exists and is the same limit independent of how the points x_i are chosen in the ith interval. This limit is the definite integral $\int_a^b f(x)\,dx$. p. 297.

■ **Area Under the Curve.** If $f(x)$ is nonnegative and continuous on $[a, b]$, we say the **area under the curve** $y = f(x)$ **from** a **to** b is given by the indefinite integral $\int_a^b f(x)\, dx$. p. 294.

■ **Properties of Definite Integrals**

$$\int_a^a f(x)\, dx = 0$$

$$\int_a^b kf(x)\, dx = k \int_a^b f(x)\, dx, \quad k \text{ constant}$$

$$\int_a^b [f(x) \pm g(x)]\, dx = \int_a^b f(x)\, dx \pm \int_a^b g(x)\, dx$$

$$\int_a^b f(x)\, dx = \int_a^c f(x)\, dx + \int_c^b f(x)\, dx, \quad a < c < b$$

$$\int_b^a f(x)\, dx = -\int_a^b f(x)\, dx, \quad a < b$$

■ **Fundamental Theorem of Calculus.** Suppose $f(x)$ is continuous on $[a, b]$. If $F(x)$ is an antiderivative of $f(x)$, then $\int_a^b f(x)\, dx = F(b) - F(a)$. p. 301.

■ If $f(x)$ is continuous on $[a, b]$, we define the **average value of** $f(x)$ **on** $[a, b]$ to be $\dfrac{1}{b - a} \int_a^b f(x)\, dx$. p. 305.

■ **Population Given the Density.** If $\delta(x)$ is a continuous function that gives the density in number per unit length, then the total population in the interval $[a, b]$ is $P = \int_a^b \delta(x)\, dx$. p. 323.

■ Let $y = f(x)$ and $y = g(x)$ be two continuous functions with $f(x) \geq g(x)$ on $[a, b]$. Then the **area between the graphs of the two curves** on $[a, b]$ is given by the definite integral $\int_a^b [f(x) - g(x)]\, dx$. p. 310.

■ If $p = D(x)$ is the demand equation, p_0 is the current price of the commodity, and x_0 is the current demand, then the **consumers' surplus** is given by $\int_0^{x_0} [D(x) - p_0]\, dx$. p. 322.

■ If $p = S(x)$ is the supply equation, p_0 is the current price of the commodity, and x_0 is the current demand, then the **producers' surplus** is given by $\int_0^{x_0} [p_0 - S(x)]\, dx$. p. 322.

■ A **Lorentz curve** is the graph of any function $y = f(x)$ defined on $[0, 1]$ for which $f(x) \leq x$, $f(0) = 0$, and $f(1) = 1$. p. 313.

■ The **coefficient of inequality** L of a Lorentz curve $y = f(x)$ is given by $L = 2 \int_0^1 [x - f(x)]\, dx$. p. 314.

■ Let $f(t)$ be the continuous rate of change of income on the interval $[0, T]$, and suppose that one can invest current money at an interest rate of r, where r is a decimal, compounded continuously over the time interval $[0, T]$. Then the **present value** $P_V(T)$ of this continuous income flow over $[0, T]$ is given by

$$P_V(T) = \int_0^T f(t)e^{-rt}\, dt \qquad \text{p. 319.}$$

■ Let $A^*(T)$ be the amount of money after T years if all the income has been continuously invested at an annual rate of r compounded continuously, then

$$A^*(T) = e^{rT}P_V(T) = e^{rT} \int_0^T f(t)e^{-rt}\, dt \qquad \text{p. 320.}$$

CHAPTER 5 REVIEW EXERCISES

In Exercises 1 through 9 find the antiderivatives.

1. $\displaystyle\int x^9\, dx$

2. $\displaystyle\int x^{0.50}\, dx$

3. $\displaystyle\int \frac{2}{y^2}\, dy$

4. $\displaystyle\int \sqrt{3}y^2\, dy$

5. $\displaystyle\int (6x^2 + 8x)\, dx$

6. $\displaystyle\int \left(2x - \frac{3}{x}\right)\, dx$

7. $\displaystyle\int \frac{t^2 + 1}{\sqrt{t}}\, dt$

8. $\displaystyle\int e^{3x}\, dx$

9. $\displaystyle\int \frac{e^{2x} + e^{-2x}}{e^{3x}}\, dx$

In Exercises 10 through 18 use the method of substitution to evaluate the integrals.

10. $\displaystyle\int 4(2x + 1)^{20}\, dx$

11. $\displaystyle\int 60x^2(2x^3 + 1)^9\, dx$

12. $\displaystyle\int \frac{x^2}{\sqrt{x^3 + 1}}\, dx$

13. $\displaystyle\int \frac{x + 1}{x^2 + 2x}\, dx$

14. $\displaystyle\int xe^{x^2+5}\, dx$

15. $\displaystyle\int \frac{10x}{(x^2 + 1)^8}\, dx$

16. $\displaystyle\int \frac{1}{\sqrt{xe^{\sqrt{x}}}}\, dx$

17. $\displaystyle\int \frac{e^x}{e^x + 1}\, dx$

18. $\displaystyle\int \frac{(\ln x)^3}{x}\, dx$

19. Find the left- and right-hand sums for $f(x) = 2x^2 + 1$ on the interval $[0, 1]$ for $n = 10, 100, 1000$. Determine $\lim_{n\to\infty} \Sigma_{k=1}^{n} f(x_k)\,\Delta x$, and relate this to some definite integral.

20. Find the left- and right-hand sums for $f(x) = x^3 + 1$ on the interval $[0, 1]$ for $n = 10, 100, 1000$. Determine $\lim_{n\to\infty} \Sigma_{k=1}^{n} f(x_k)\,\Delta x$, and relate this to some definite integral.

21. Find $\displaystyle\int_0^3 \sqrt{9 - x^2}\, dx$ by finding the area of an appropriate geometric figure.

22. Since $\displaystyle\int_1^3 \frac{1}{x}\, dx = \ln 3$, use the left-hand Riemann sum with $n = 4$ to approximate $\ln 3$. Using a graph of $y = \frac{1}{x}$, show why your answer must be *greater* than $\ln 3$.

23. Use the definite integral in Exercise 22 to approximate $\ln 3$ using the right-hand Riemann sum with $n = 4$. Using a graph of $y = \frac{1}{x}$, show why your answer must be *less* than $\ln 3$.

In Exercises 24 through 31 evaluate the definite integrals.

24. $\displaystyle\int_1^3 6x^2\, dx$

25. $\displaystyle\int_{-3}^{-1} 2x\, dx$

26. $\displaystyle\int_{-1}^1 (9x^2 - 2x + 1)\, dx$

27. $\displaystyle\int_1^2 \left(e^{2x} + \frac{1}{x}\right)\, dx$

28. $\displaystyle\int_0^1 4x(x^2 + 1)^9\, dx$

29. $\displaystyle\int_0^2 x^2 e^{x^3+1}\, dx$

30. $\displaystyle\int_0^2 \frac{1}{2x + 1}\, dx$

31. $\displaystyle\int_0^3 \frac{1}{\sqrt{2x + 3}}\, dx$

32. Write an expression for the average value of $y = e^{t^4}$ on the interval $[0, 2]$. Do not attempt to evaluate.

33. Find the average value of $f(x) = x^3$ on $[-2, 2]$.

34. Find the average value of $f(x) = \frac{1}{x}$ on $[1, 3]$.

In Exercises 35 through 46 find the area enclosed by the given curves.

35. $y = x^2$, $y = 0$, $x = 0$, $x = 3$

36. $y = x^2 + 1$, $y = x$, $x = 0$, $x = 2$

37. $y = 2\sqrt{x}$, $y = x$, $x = 0$, $x = 4$

38. $y = \sqrt{x}$, $y = -x^2$, $x = 0$, $x = 1$

39. $y = x^2 + 1$, $y = 1 - 2x$, $x = 0$, $x = 2$

40. $y = x^2$, $y = 18 - x^2$

41. $y = x^2 - 4x + 3$, $y = 2x - 2$

42. $y = x^2 - 5x + 4$, $y = -x^2 + 5x - 4$

43. $y = \sqrt[5]{x}$, $y = x$

44. $y = x^2 - 4x + 3$, $y = 2x - 2$, $x = 0$, $x = 5$

45. $y = \frac{2}{x}$, $y = 1$, $x = 1$, $x = 3$

46. $y = x^2 - 5x + 4$, $y = -x^2 + 5x - 4$, $x = 0$, $x = 4$

47. Revenue. If marginal revenue is given by $2x - 10$, find the revenue function.

48. Cost. If marginal cost is given by $3x^2 + 1$ and fixed costs are 1000, find the cost function.

49. Depreciation. The value of a piece of machinery purchased for \$100,000 is decreasing at a rate given by $V'(t) = -500e^{-0.20t}$, where t is measured in years from date of purchase. How much will the machine be worth in five years?

50. Sales. After an advertising campaign ends, the rate of change of sales (in thousands) of a product is given by $S'(t) = 4te^{-t^2}$, where t is in years. How many will be sold in the next two years?

51. Sales. In the table the rate $R'(t)$ of sales for a firm in millions of dollars per month is increasing and is given in monthly intervals. Assuming $R(0) = 0$, give upper and lower estimates for the amount of sales during the five-month period.

t_i	0	1	2	3	4	5
$R'(t_i)$	2	3	5	6	10	8

52. Population. The population of a certain city is increasing at the rate given by $P'(t) = 2000e^{0.04t}$, where t is time in years measured from the beginning of 1992. Find the change in the population from the beginning of 1994 to the beginning of 1997.

53. **Medicine.** A tumor is increasing in volume at a rate given by $V'(t) = 4e^{0.1t}$, where t is measured in months from the time of diagnosis. How much will the tumor increase in volume during the third and fourth months after diagnosis?

54. **Average Population.** The population of a certain city is given $10{,}000e^{0.1t}$, where t is given in years. Find the average population during the next 10 years.

In Exercises 55 through 57 suppose the demand equation is given by $p = D(x) = 10 - x^2$ and the supply equation is given by $p = S(x) = 3x$.

55. **Supply and Demand.** Find the equilibrium point.

56. **Consumers' Surplus.** Find the consumers' surplus.

57. **Producers' Surplus.** Find the producers' surplus.

58. **Lorentz Curve.** Show that $f(x) = x^4$ is a Lorentz curve, and find the coefficient of inequality.

In Exercises 59 through 61 assume $f(t) = 10e^{-0.04t}$ is the rate of change of total income per year and the interest rate is 8% per year compounded continuously.

59. Find the total amount of income over the first 10 years.

60. Find the present value $P_V(10)$.

61. Find the amount in the account after 10 years if all income is continuously reinvested at the given interest rate.

62. A mine produces a rate of change of total income in thousands of dollars per year given by $f(t) = 100e^{-0.10t}$ and has a useful life of five years. Market interest rates are 8% compounded continuously. What is a reasonable price for this mine, discounting any risk involved?

CHAPTER 5

PROJECTS

Project 1
Lorentz Curve

Recently Chotikapanich [127] suggested the following function as a Lorentz curve:

$$y = f(x) = \frac{e^{kx} - 1}{e^{k} - 1}$$

Show that this curve is indeed a Lorentz curve by showing the following:
(a) $f(0) = 0$
(b) $f(1) = 1$
(c) $0 < f(x) < x < 1$
(d) $f'(x) \geq 0$
(e) $f''(x) > 0$

Project 2
Determining the Coefficient of Inequality

The table gives the accumulated U.S. income by population deciles for 1966. (See Suits [128].)

Population Decile	Cumulated Adjusted Family Income
1	1.21
2	3.88
3	8.13
4	13.92
5	21.16
6	30.22
7	40.02
8	52.29
9	67.45
10	100.00

Use the table to obtain upper and lower estimates of the coefficient of inequality defined in Section 5.6.

Project 3
Theory of Social Security

The figure shows graphs of the earnings $y = W(t)$ and consumption $y = C(t)$ functions for an individual, where R is the time of retirement and D the time of death. We assume a constant rate of interest r throughout this life cycle.

(a) Explain what is the significance of the graph of the consumption curve always being under the graph of the earnings curve. *Hint:* What is $W(t) - C(t)$?

(b) Show that the following budget constraint must hold:

$$\int_0^D C(t)e^{-rt}\, dt = \int_0^R W(t)e^{-rt}\, dt$$

Hint: Start by noticing that $\int_0^D = \int_0^R + \int_R^D$.

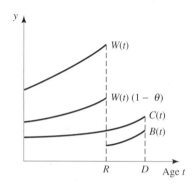

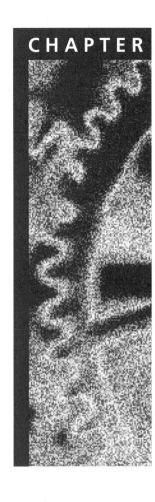

6

Additional Topics
in Integration

This chapter contains a collection of important additional topics in integration. In the first section we learn a very important technique of integration, in the second section we detail how to use a table of integrals, and in the third section we discuss how to evaluate definite integrals using numerical techniques. In the fourth, and final, section we extend the notion of integration to unbounded intervals.

6.1 INTEGRATION BY PARTS

■   Integration by Parts   ■   Application

APPLICATION: SALES OF A NEW PRODUCT

The rate of sales of a new product tend to increase rapidly initially and then decrease. Suppose the rate of sales of a new product is given by $S(t) = 1000t^2e^{-t}$ items per week, where t is the number of weeks from the introduction of the product. How many items are sold in the first three weeks? The answer can be found in Example 4.

Integration by Parts

If $f(x)$ and $g(x)$ are differentiable functions, then the product rule of differentiation says that

$$\frac{d}{dx} f(x)g(x) = f(x) \frac{d}{dx} g(x) + g(x) \frac{d}{dx} f(x)$$

Rearranging the terms yields

$$f(x)g'(x) = [f(x)g(x)]' - g(x)f'(x)$$

Taking the indefinite integral gives

$$\int f(x)g'(x) \, dx = f(x)g(x) - \int g(x)f'(x) \, dx$$

In practice this is written an alternative way. If we let $u = f(x)$ and $v = g(x)$, then $du = f'(x) \, dx$ and $dv = g'(x) \, dx$. The preceding formula then becomes $\int u \, dv = uv - \int v \, du$. This technique is called **integration by parts**.

Integration by Parts

$$\int u \, dv = uv - \int v \, du$$

At first glance it is difficult to believe that anything has been accomplished. The goal, however, is to define u and v so that $\int v \, du$ is easier to find than the original integral, $\int u \, dv$. The following examples illustrate this. Keep in mind that your first choice of u and v may not work.

EXAMPLE 1   Using Integration by Parts

Find $\int xe^x \, dx$.

SOLUTION   ■   If we set $u = x$, and $dv = e^x \, dx$, then $du = dx$, and $v = e^x$, and

$$\int xe^x \, dx = \int u \, dv$$

$$= uv - \int v \, du$$

$$= xe^x - \int e^x \, dx$$

$$= xe^x - e^x + C = e^x(x - 1) + C \quad ■$$

Notice that $\int v \, du = \int e^x \, dx$ was simpler than the original integral. Had we made the substitution $u = e^x$ and $dv = x \, dx$, then we would have $\int v \, du = \int \frac{1}{2}x^2 e^x \, dx$, which is *more* complicated than the original integral.

EXAMPLE 2   Using Integration by Parts

Find $\int x\sqrt{1 + x} \, dx$.

SOLUTION ■ We might first think of letting $u = x$ since du is simple. Let us try this. Let $u = x$, and $dv = \sqrt{1 + x} \, dx$. Of course $du = dx$, and $v = \int \sqrt{1 + x} \, dx$. This last integral is similar to others we have done before and can be readily done by substitution. If we let $w = 1 + x$, then $dw = dx$ and

$$v = \int \sqrt{1 + x} \, dx = \int w^{1/2} \, dw = \frac{2}{3} w^{3/2} = \frac{2}{3}(1 + x)^{3/2}$$

Then

$$\int x\sqrt{1 + x} \, dx = \int u \, dv$$
$$= uv - \int v \, du$$
$$= x\frac{2}{3}(1 + x)^{3/2} - \int \frac{2}{3}(1 + x)^{3/2} \, dx$$

Now we can evaluate the last integral in just the way we found v. Thus,

$$\int x\sqrt{1 + x} \, dx = x\frac{2}{3}(1 + x)^{3/2} - \int \frac{2}{3}(1 + x)^{3/2} \, dx$$
$$= x\frac{2}{3}(1 + x)^{3/2} - \frac{2}{3}\cdot\frac{2}{5}(1 + x)^{5/2} + C$$
$$= \frac{2}{3}x(1 + x)^{3/2} - \frac{4}{15}(1 + x)^{5/2} + C \quad ■$$

EXAMPLE 3   Using Integration by Parts

Find (a) $\int x^3\sqrt{1 + x^2} \, dx$ (b) $\int_0^{\sqrt{3}} x^3\sqrt{1 + x^2} \, dx$.

SOLUTIONS ■
(a) We might be tempted to set $u = x^3$ and $dv = \sqrt{1 + x^2}$, but then we could *not* find v since we cannot find $\int \sqrt{1 + x^2} \, dx$. We can integrate $\int x\sqrt{1 + x^2} \, dx$, however. Thus, we set $u = x^2$, and $dv = x\sqrt{1 + x^2}$. Then $du = 2x \, dx$, and $v = \frac{1}{3}(1 + x^2)^{3/2}$. (The reader should verify the correctness of v.) Thus,

$$\int x^3\sqrt{1 + x^2} \, dx = \int u \, dv$$
$$= uv - \int v \, du$$
$$= x^2\frac{1}{3}(1 + x^2)^{3/2} - \int \frac{1}{3}(1 + x^2)^{3/2} \, 2x \, dx$$
$$= \frac{1}{3}x^2(1 + x^2)^{3/2} - \frac{2}{15}(1 + x^2)^{5/2} + C$$

(b) To evaluate this definite integral we use the antiderivative found in part (a).

$$\int_0^{\sqrt{3}} x^3\sqrt{1 + x^2}\, dx = \left[\frac{1}{3} x^2(1 + x^2)^{3/2} - \frac{2}{15}(1 + x^2)^{5/2}\right]\Bigg|_0^{\sqrt{3}}$$

$$= \left(4^{3/2} - \frac{2}{15}\cdot 4^{5/2}\right) - \left(0 - \frac{2}{15}\right)$$

$$= \frac{58}{15} \quad \blacksquare$$

Sometimes one must use integration by parts *more than once*. Sometimes $\int v\, du$ is definitely simpler but requires another integration by parts. The following is such an example.

Application

EXAMPLE 4   Repeated Integration by Parts

Suppose the rate of sales of a new product is given by $S'(x) = 1000x^2 e^{-x}$ items per week, where x is the number of weeks from the introduction of the product. How many items are sold in the first three weeks?

SOLUTION ■ Screen 6.1 shows a graph of $y = 1000x^2 e^{-x}$ using a window with dimensions [0, 10] by [0, 1000]. We are seeking $1000 \int_0^3 x^2 e^{-x}\, dx$. We first find $\int x^2 e^{-x}\, dx$. We set $u = x^2$, and $dv = e^{-x}\, dx$. Then $du = 2x\, dx$, and $v = -e^{-x}$. Then

$$\int x^2 e^{-x}\, dx = \int u\, dv$$

$$= uv - \int v\, du$$

$$= x^2\left(-e^{-x}\right) - \int -e^{-x}2x\, dx$$

$$= -x^2 e^{-x} + 2\int xe^{-x}\, dx$$

The integral we are left with, $\int xe^{-x}\, dx$, is definitely simpler, but to evaluate it requires another integration by parts. To do this we let $u = x$, and $dv = e^{-x}\, dx$. Then $du = dx$, and $v = -e^{-x}$, and

$$\int xe^{-x}\, dx = \int u\, dv$$

$$= uv - \int v\, du$$

$$= x(-e^{-x}) - \int -e^{-x}\, dx$$

$$= -xe^{-x} + \int e^{-x}\, dx$$

$$= -xe^{-x} - e^{-x} + C^*$$

Substituting this back in the above equation yields

$$\int x^2 e^{-x}\, dx = -x^2 e^{-x} + 2\int xe^{-x}\, dx$$

$$= -x^2 e^{-x} + 2[-xe^{-x} - e^{-x} + C^*]$$

$$= -e^{-x}[x^2 + 2x + 2] + C$$

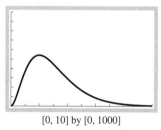

[0, 10] by [0, 1000]

Screen 6.1
A graph of $y = 1000x^2 e^{-x}$ giving the rate of sales of a new product.

where we have set $C = 2C^*$ for convenience. Then

$$1000 \int_0^3 x^2 e^{-x} \, dx = -1000 e^{-x}(x^2 + 2x + 2) \Big|_0^3$$

$$= -1000 e^{-3}(17) + 1000 e^0(2)$$

$$= 1000(2 - 17e^{-3})$$

$$\approx 1153.6 \quad \blacksquare$$

SELF-HELP EXERCISE SET 6.1

1. Find **(a)** $\int \ln x \, dx$ and **(b)** $\int_1^e \ln x \, dx$. *Hint:* Notice that $\ln x = 1 \cdot \ln x$.

2. Find $\int \dfrac{xe^x}{(x + 1)^2} \, dx$. *Hint:* You may need to try several choices of u and dv before one works.

EXERCISE SET 6.1

In Exercises 1 through 20 evaluate.

1. $\int 2xe^{2x} \, dx$

2. $\int xe^{-2x} \, dx$

3. $\int_1^{e^2} x \ln x \, dx$

4. $\int_e^{e^2} x^2 \ln x \, dx$

5. $\int_1^e \ln x^2 \, dx$

6. $\int_1^e \ln x^3 \, dx$

7. $\int \dfrac{x}{\sqrt{1 + x}} \, dx$

8. $\int x(1 + x)^{3/2} \, dx$

9. $\int x(1 + x)^{10} \, dx$

10. $\int x(1 + x)^5 \, dx$

11. $\int x(x + 2)^{-2} \, dx$

12. $\int x(x + 2)^{-3} \, dx$

13. $\int x^5 \sqrt{1 + x^3} \, dx$

14. $\int (x + 1)^{2/3} x \, dx$

15. $\int x^3(1 + x^2)^{10} \, dx$

16. $\int \dfrac{x^3}{\sqrt{1 + x^2}} \, dx$

17. $\int x^2 e^{2x} \, dx$

18. $\int x^2 e^{-2x} \, dx$

19. $\int_0^2 x^3 e^{x^2} \, dx$

20. $\int_1^2 (\ln x)^2 \, dx$

Applications

21. **Sales.** The rate of sales of a new product tends to increase rapidly initially and then fall off. Suppose the rate of sales of a new product is given by $S'(t) = 1000te^{-3t}$ items per week, where t is in weeks from the introduction of the product. How many items are sold in the first four weeks? Assume $S(0) = 0$.

22. **Sales.** Suppose the rate of sales of a new product is given by $S'(t) = 1000t^2 e^{-4t}$ items per month, where t is in months from the introduction of the product. How many items are sold in the first year? Assume $S(0) = 0$.

23. **Profits.** The marginal profits of an electric can opener manufacturer are given by $P'(x) = 0.1x\sqrt{x + 2}$, where x is measured in thousands of can openers and $P(x)$ is measured in thousands of dollars. Find the total profits generated by increasing the number of can openers from 7000 to 14,000.

24. **Consumers' Surplus.** Find the consumers' surplus if $p = D(x) = 25 - x^3\sqrt{5 + x^2}$ is the demand equation, with x

measured in thousands of units and $p_0 = 1$ is the current price of the commodity. *Hint:* $D(2) = 1$.

25. **Costs.** The cost function, in thousands of dollars, of a coat manufacturer is given by $x(1 + x)^4$, where x is in thousands of coats. Find the average cost for the first 2000 coats.

26. **Price.** The demand equation for a bath towel manufacturer is given by $p = p(x) = x(x + 1)^{-2}$, where p is in dollars and x is measured in thousands of towels. Find the average price over the interval $[1, 8]$.

27. **Population.** The population of a new town is given by $P(t) = 1000 \ln t$, where $t \geq 1$ is measured in years. What is the average population over the time from $t = 1$ to $t = e$?

28. **Income Distribution.** Find the coefficient of inequality for the Lorentz curve $y = xe^{x-1}$.

Enrichment Exercises

29. Establish the following reduction formula:

$$\int (\ln x)^n \, dx = x(\ln x)^n - n \int (\ln x)^{n-1} \, dx$$

30. If $f(0) = g(0) = 0$, show that

$$\int_0^b f(x)g''(x) \, dx = f(a)g'(a) - f'(a)g(a) + \int_0^a f''(x)g(x) \, dx$$

31. Use integration by parts to show that

$$\int f(x) \, dx = xf(x) - \int xf'(x) \, dx$$

SOLUTIONS TO SELF-HELP EXERCISE SET 6.1

1. (a) Our choices for u are very limited. Let $u = \ln x$, and $dv = dx$. Then $du = \dfrac{1}{x} \, dx$, $v = x$, and

$$\int \ln x \, dx = \int u \, dv$$

$$= uv - \int v \, du$$

$$= (\ln x)x - \int x \frac{1}{x} \, dx$$

$$= x \ln x - \int dx$$

$$= x \ln x - x + C$$

(b)

$$\int_1^e \ln x \, dx = (x \ln x - x) \Big|_1^e$$

$$= (e \ln e - e) - (1 \cdot \ln 1 - 1)$$

$$= (0) - (-1) = 1$$

2. Let $u = xe^x$, and $dv = \dfrac{1}{(1 + x)^2} \, dx$. Then $du = e^x(x + 1) \, dx$, and $v = \dfrac{-1}{x + 1}$.

$$\int \frac{xe^x}{(1 + x)^2} \, dx = \int u \, dv$$

$$= uv - \int v \, du$$

$$= xe^x \left(\frac{-1}{x + 1} \right) - \int \left(-\frac{1}{x + 1} \right) e^x(x + 1) \, dx$$

$$= -\frac{xe^x}{x + 1} + \int e^x \, dx$$

$$= -\frac{xe^x}{x + 1} + e^x + C$$

$$= \frac{e^x}{x + 1} + C$$

6.2 INTEGRATION USING TABLES

■ Using Tables of Integrals ■ Application

APPLICATION: SALES

The rate of sales in millions of dollars per week of a toy is given by $S'(x) = \dfrac{e^x}{e^{2x} - 1}$, where x is the number of weeks after the end of an advertising campaign. Find the total dollar sales during the second and third week. See Example 4 for the answer.

Using Tables of Integrals

In previous sections we learned how to integrate many expressions, including $\int x^n \, dx$, $\int e^{kx} \, dx$, and so on. We were also introduced to some techniques of integration, such as substitution and integration by parts. A substantial number of different *techniques* of integration exist, many of which require considerable skill and ingenuity. Unfortunately, no hard and fast rules determine when to use each technique. In the face of these difficulties the most convenient technique is to refer to a table of integrals. We can always check an answer in any table by merely differentiating the answer to see if we obtain the integrand.

EXAMPLE 1   Using a Table of Integrals

Find $\int \dfrac{1}{x(2x + 3)}\, dx$.

SOLUTION ■ By looking through the table of integrals found in Appendix B.2, we notice that the integrand has the form of the formula in the sixth entry if we set $a = 2$ and $b = 3$. Thus, we have

$$\int \frac{1}{x(2x + 3)}\, dx = \frac{1}{3} \ln \left| \frac{x}{2x + 3} \right| + C \quad ■$$

Remark. Notice that we added the arbitrary constant C. The table does not include this constant so as to save space.

EXAMPLE 2   Using a Table of Integrals

Find $\int \dfrac{1}{\sqrt{9x^2 + 1}}\, dx$.

SOLUTION ■ If we look through the table we do not actually find this integral, but it appears to look somewhat like entry 22. Suppose we factor out the 9 from the integrand. Then we obtain

$$\frac{1}{3} \int \frac{1}{\sqrt{x^2 + \frac{1}{9}}}\, dx$$

This integral is of the form of entry 22 with $a = \frac{1}{3}$. Thus,

$$\int \frac{1}{\sqrt{9x^2 + 1}}\, dx = \frac{1}{3} \int \frac{1}{\sqrt{x^2 + \frac{1}{9}}}\, dx$$

$$= \frac{1}{3} \ln \left| x + \sqrt{x^2 + \frac{1}{9}} \right| + C \quad ■$$

EXAMPLE 3   Repeated Use of a Table of Integrals

Find $\int x^2 e^{-x}\, dx$.

SOLUTION ■ Although this integral can be done using integration by parts, one can also do it by using entry 38 in the table by taking $n = 2$ and $a = -1$. Doing this, we obtain

$$\int x^2 e^{-x}\, dx = \frac{1}{-1} x^2 e^{-x} - \frac{2}{-1} \int x e^{-x}\, dx$$

$$= -x^2 e^{-x} + 2 \int x e^{-x}\, dx$$

Now we are left to evaluate $\int xe^{-x}\,dx$, which we can do by using the very same entry, but this time take $n=1$ and $a=-1$. For this latter integral we then obtain

$$\int xe^{-x}\,dx = \frac{1}{-1}\,xe^{-x} - \frac{1}{-1}\int x^0 e^{-x}\,dx$$

$$= -xe^{-x} + \int e^{-x}\,dx$$

$$= -xe^{-x} - e^{-x} + C^*$$

Using this yields

$$\int x^2 e^{-x}\,dx = -x^2 e^{-x} + 2\int xe^{-x}\,dx$$

$$= -x^2 e^{-x} + 2(-xe^{-x} - e^{-x} + C^*)$$

$$= -e^{-x}(x^2 + 2x + 2) + C$$

where we set $C = 2C^*$ for convenience.   ∎

Application

■ Power Rangers

The hottest toy trend of the early 1990s, both in the United States and abroad, has been the Power Rangers action figures. These Japanese exports have earned close to $1 billion in sales for 1994 alone. In comparison with other children favorites, the Power Rangers have fared well. Cabbage Patch dolls brought in $550 million during their best year and the Teen-Age Mutant Ninja Turtles action figures $450 million. The only competitor in the Power Rangers financial league is the American favorite, Barbie which grossed $1 billion in sales for both 1993 and 1994. With a stronghold on 40% of the action figure market, the popularity of the Power Rangers has lead to shortages of the toys and is now an item that can be found on the black market.

Source: The New York Times,
December 5, 1994, pp. A1, D3.

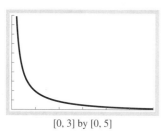

[0, 3] by [0, 5]

Screen 6.2

A graph of $y = \dfrac{e^x}{e^{2x} - 1}$ giving the rate of sales per week of a toy.

E X A M P L E 4   Using Substitution in a Table of Integrals

The rate of sales in millions of dollars per week of a toy is given by $S'(x) = \dfrac{e^x}{e^{2x} - 1}$, where x is the number of weeks after the end of an advertising campaign. Find the total dollar sales during the second and third week.

S O L U T I O N ■ The graph of $y = \dfrac{e^x}{e^{2x} - 1}$ is shown in Screen 6.2 using a window with dimensions [0, 3] by [0, 5]. We are seeking $\displaystyle\int_1^3 \frac{e^x}{e^{2x} - 1}\,dx$. The integrand involves exponentials, yet no formula in the table under the exponential heading applies. If we make the substitution $u = e^x$, then $du = e^x\,dx$, and the integral becomes

$$\int \frac{e^x}{e^{2x} - 1}\,dx = -\int \frac{e^x}{1 - e^{2x}}\,dx = -\int \frac{1}{1 - u^2}\,du$$

This last integral can be found in entry 13 in the table with $a = 1$. Thus,

$$\int \frac{e^x}{e^{2x} - 1}\,dx = -\int \frac{1}{1 - u^2}\,du$$

$$= -\frac{1}{2}\ln\left|\frac{u + 1}{u - 1}\right| + C$$

$$= \frac{1}{2}\ln\left|\frac{u - 1}{u + 1}\right| + C$$

$$= \frac{1}{2}\ln\left|\frac{e^x - 1}{e^x + 1}\right| + C$$

Then

$$\int_1^3 \frac{e^x}{e^{2x} - 1}\,dx = \frac{1}{2}\ln\left|\frac{e^x - 1}{e^x + 1}\right|\,\Bigg|_1^3 \approx 0.336$$

or about $336,000.   ∎

1. Find $\displaystyle\int \frac{4x}{\sqrt{x^4 + 4}}\, dx$

EXERCISE SET 6.2

In Exercises 1 through 26 evaluate using the tables of integrals in Appendix B2.

1. $\displaystyle\int x\sqrt{3x + 2}\, dx$

2. $\displaystyle\int \frac{1}{x\sqrt{4 - x^2}}\, dx$

3. $\displaystyle\int \sqrt{x^2 - 9}\, dx$

4. $\displaystyle\int \frac{x}{2x - 1}\, dx$

5. $\displaystyle\int \frac{1}{x\sqrt{1 - x}}\, dx$

6. $\displaystyle\int \frac{\sqrt{4 - x^2}}{x}\, dx$

7. $\displaystyle\int \frac{1}{2x^2 + 5x + 2}\, dx$

8. $\displaystyle\int \frac{1}{x^2\sqrt{x^2 + 9}}\, dx$

9. $\displaystyle\int x\sqrt{x^2 - 9}\, dx$

10. $\displaystyle\int \frac{x^2}{(2x + 3)^2}\, dx$

11. $\displaystyle\int \frac{1}{x^2(x - 3)}\, dx$

12. $\displaystyle\int \frac{x}{(9 - x^2)^{3/2}}\, dx$

13. $\displaystyle\int \frac{x^2}{(x^2 + 16)^{3/2}}\, dx$

14. $\displaystyle\int \frac{\sqrt{x^2 + 16}}{x}\, dx$

15. $\displaystyle\int \frac{1}{x(2x^3 + 1)}\, dx$

16. $\displaystyle\int \frac{1}{2 + 3e^{4x}}\, dx$

17. $\displaystyle\int \frac{1}{x^2\sqrt{1 - 4x^2}}\, dx$

18. $\displaystyle\int \frac{1}{(1 - 4x^2)^{3/2}}\, dx$

19. $\displaystyle\int \frac{x^2}{\sqrt{9x^2 - 1}}\, dx$

20. $\displaystyle\int \frac{1}{(4x^2 + 1)^{3/2}}\, dx$

21. $\displaystyle\int (\ln x)^3\, dx$

22. $\displaystyle\int x^3 e^{2x}\, dx$

23. $\displaystyle\int \frac{e^{3x}}{e^x + 1}\, dx$

24. $\displaystyle\int \frac{e^{2x}}{(e^x + 1)^2}\, dx$

25. $\displaystyle\int x^2 e^{-2x}\, dx$

26. $\displaystyle\int [\ln(x + 1)]^3\, dx$

Applications

27. Revenue. Find the revenue function of a toaster manufacturer if marginal revenue, in hundreds of dollars, is given by $x\sqrt{x + 1}$, where x is in hundreds of toasters.

28. Sales. The rate of change of the number of sales of refrigerators by a firm is given by $\dfrac{t}{(t^2 + 9)^{3/2}}$, where t is in weeks. How many refrigerators are sold in the first four weeks?

29. Profit. If the rate of change of profit in thousands of dollars per week is given by $P'(t) = \dfrac{10}{(t^2 + 9)^{3/2}}$, where t is measured in weeks and $P(0) = 0$, find $P = P(t)$.

30. Costs. The marginal cost, in thousands of dollars, of a brick manufacturer is given by $\dfrac{x}{\sqrt{x^2 + 9}}$, where x is in thousands of bricks. If fixed costs are $10,000, find $C = C(x)$.

31. Consumers' Surplus. Find the consumers' surplus if $p = D(x) = \dfrac{4e}{e + e^{0.10x}}$, is the demand equation, with x mea-

sured in thousands of units and $p_0 = 2$ is the current price of the commodity.

32. Producers' Surplus. Find the producers' surplus if $p = S(x) = x\sqrt{x^2 + 5}$ is the supply equation, with x measured in thousands of units and $p_0 = 6$ is the current price of the commodity. *Hint:* $S(2) = 6$.

33. Pollutants. The concentration (density) of pollutants at a distance of x miles east of an industrial plant measured in thousands of particles per mile per day is given by $\delta(x) = \dfrac{1}{x^2 + 5x + 4}$. Find the amount of pollutants between $x = 0$ and $x = 5$ miles.

34. Biology. The rate of change of a certain population of plant in a small region is given by $P'(t) = \dfrac{1000}{(9t^2 + 1)^{3/2}}$, where t is in months. If the initial population is zero, find the population at any time. What happens as $t \to \infty$?

Enrichment Exercises

35. Often when using tables or otherwise, you can obtain two very different looking answers. Consider the integral $\int \dfrac{x}{(x+1)^2}\, dx$. Integrate by parts and obtain $-\dfrac{x}{x+1} + \ln|x+1| + C$. Integrate by substitution and obtain $\dfrac{1}{x+1} + \ln|x+1| + C$. Does this mean that

$$-\frac{x}{x+1} + \ln|x+1| = \frac{1}{x+1} + \ln|x+1|$$

Let y_1 be the left-hand side and y_2 the right-hand side. On your graphing calculator graph y_1, y_2, and $y_3 = y_1 - y_2$. Explain what you have discovered.

36. Check the answer to Exercise 13. *Hint:* Let $F(x)$ be the answer in the text. Then show that

$$F'(x) = \frac{-16}{(x^2 + 16)^{3/2}} + \frac{\sqrt{x^2 + 16} + x}{x\sqrt{x^2 + 16} + x^2 + 16}$$

Denote the second fraction as $g(x)$. Since $F'(x) = \dfrac{x^2}{(x^2+16)^{3/2}}$, we must have $g(x) = \dfrac{x^2 + 16}{(x^2 + 16)^{3/2}}$. You probably cannot establish this algebraically. Establish using your graphing calculator with a window of dimensions $[-20, 20]$ by $[0, 0.3]$. Use TRACE to move back and forth between the two functions.

SOLUTIONS TO SELF-HELP EXERCISE SET 6.2

1. Use entry 22 in the table of integrals after using the substitution $u = x^2$. Then $du = 2x\, dx$ and

$$\int \frac{4x}{\sqrt{x^4 + 4}}\, dx = 2 \int \frac{2x}{\sqrt{x^4 + 4}}\, dx$$

$$= 2 \int \frac{1}{\sqrt{u^2 + 4}}\, du$$

$$= 2\ln|u + \sqrt{u^2 + 4}| + C$$

$$= 2\ln|x^2 + \sqrt{x^4 + 4}| + C$$

6.3 NUMERICAL INTEGRATION

■ Introduction ■ The Trapezoidal Rule ■ Simpson's Rule ■ Errors

APPLICATION: FINDING APPROXIMATIONS OF DEFINITE INTEGRALS

Find approximations for $\displaystyle\int_0^2 \sqrt{4 - x^2}\, dx = \pi$ and $\displaystyle\int_1^2 \frac{1}{x}\, dx = \ln 2$. See Examples 3 and 5.

Introduction

In Chapter 5, we introduced the Riemann sum and used left- and right-hand sums to estimate definite integrals. Otherwise the only method we have for evaluating the definite integral $\int_a^b f(x)\, dx$ requires us to find an antiderivative $F(x)$ of $f(x)$, evaluate F at a and b, and use the formula

$$\int_a^b f(x)\, dx = F(b) - F(a)$$

If we cannot do this, our method fails and we must return to numerical methods, such as those touched on in Chapter 5. This section introduces other methods of estimating definite integrals and develops ways to calculate the magnitude of the errors associated with these estimates.

It turns out however that the method for determining definite integrals by finding antiderivatives *does* fail for many important and rather simple-looking integrals, such as $\int_0^1 e^{-x^2}\, dx$ and $\int_2^3 \dfrac{1}{\ln x}\, dx$. The first of these integrals plays a vital role in continuous probability theory.

Of course $\int_0^1 e^{-x^2}\, dx$ exists since e^{-x^2} is continuous on $[0, 1]$. It has been *proven*, however, that e^{-x^2} has no elementary antiderivative, that is, one that can be expressed as finite combinations of algebraic, exponential, logarithmic, and trigonometric functions. In effect we cannot write down the antiderivative in a form that permits us to evaluate the antiderivative at the two points 0 and 1.

This difficulty is not as disastrous as it might first appear, for two reasons. The first is the existence of *numerical* methods, which are the subject of this section, that permit us to find definite integrals such as $\int_0^1 e^{-x^2}\, dx$ to (for all practical purposes) any desired accuracy. Thus, knowing this integral to an accuracy of for example 5 or 10 decimal places is sufficient for any practical purpose.

The second reason is that many of the functions encountered in practice come from data collected in various ways. In a typical situation one does not know what the function $f(x)$ is at *every* point x but rather only at the discrete points $x_1, x_2, \ldots, x_n$, where f has been measured (Fig. 6.1). The numerical methods given in this section are well suited to producing reasonable approximations to the definite integral $\int_a^b f(x)\, dx$ under such conditions.

If $f(x)$ is continuous on the interval $[a, b]$, we *defined* the definite integral $\int_a^b f(x)\, dx$ as the common limit of the left- and right-hand Riemann sums. That is, we divided the interval $[a, b]$ into n subintervals of equal length $\Delta x = \dfrac{b - a}{n}$, with endpoints at $a = x_0, x_1, \ldots, x_n = b$. If $f(x)$ is continuous, the limit of the left-hand sum $\lim_{n \to \infty} \sum_{k=0}^{n-1} f(x_k)\, \Delta x$ equals the limit of the right-hand sum $\lim_{n \to \infty} \sum_{k=1}^{n} f(x_k)\, \Delta x$, which equals the definite integral $\int_a^b f(x)\, dx$.

If n is large, we expect both the left- and right-hand Riemann sums $\sum_{k=0}^{n-1} f(x_k)\, \Delta x$ and $\sum_{k=1}^{n} f(x_k)\, \Delta x$ to approximate the definite integral $\int_a^b f(x)\, dx$. This represents a *numerical* method of approximating the definite integral by using certain thin rectangles. Refer to Figure 6.2 where the left-hand sum is indicated. We will not dwell on left- and right-hand Riemann sums here since they were covered in the previous chapter.

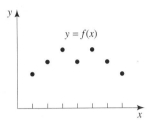

Figure 6.1
Typically, we only know $f(x)$ at discrete values of x.

The Trapezoidal Rule

The left- and right-hand Riemann sums approximate $\int_a^b f(x)\, dx$ by rectangles with heights given by the values of f evaluated at the endpoints of the subintervals. In general, a better approximation can be obtained by using trapezoids as shown in Figure 6.3. Once again the interval $[a, b]$ has been divided into n subintervals each of length $\Delta x = \dfrac{b - a}{n}$. Also $x_0 = a, x_1 = a + \Delta x, x_2 = a + 2\Delta x, \ldots, x_n = a + n\Delta x$.

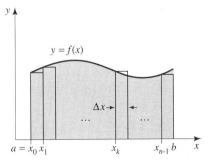

Figure 6.2
The sum of the areas of the rectangles is the left-hand sum $\sum_{k=0}^{n-1} f(x_k)\, \Delta x$.

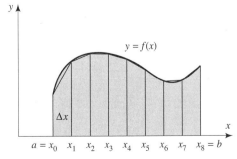

Figure 6.3
Approximate the area with trapezoids.

Recall that the area of a trapezoid is given by multiplying the length of the base times the average of the heights of the sides. Thus, the area of the n trapezoids is given by

$$T_n = \frac{f(x_0) + f(x_1)}{2} \Delta x + \frac{f(x_1) + f(x_2)}{2} \Delta x + \frac{f(x_2) + f(x_3)}{2} \Delta x$$

$$+ \cdots + \frac{f(x_{n-2}) + f(x_{n-1})}{2} \Delta x + \frac{f(x_{n-1}) + f(x_n)}{2} \Delta x$$

$$= \Delta x \left[\frac{1}{2} f(x_0) + f(x_1) + f(x_2) + \cdots + f(x_{n-1}) + \frac{1}{2} f(x_n) \right]$$

We then have

Trapezoidal Rule

For any integer n divide the interval $[a, b]$ into n subintervals of equal length with endpoints $x_0, x_1, \ldots, x_n$. If

$$T_n = \frac{b - a}{n} \left[\frac{1}{2} f(x_0) + f(x_1) + f(x_2) + \cdots + f(x_{n-1}) + \frac{1}{2} f(x_n) \right]$$

then

$$\int_a^b f(x) \, dx \approx T_n$$

In general, the trapezoidal rule gives much closer approximations than using the left- or right-hand Riemann sums. We will return to this point later.

EXAMPLE 1 Using the Trapezoidal Rule

Approximate $\int_0^2 \sqrt{4 - x^2} \, dx = \pi$ by using T_4.

SOLUTION ■ See Figure 6.4 in which we see that T_4 must be smaller than the area that is π. We have $\Delta x = \dfrac{2 - 0}{4} = 0.50$ and the points x_0, x_1, x_2, x_3, and x_4, are, respectively, 0, 0.50, 1.00, 1.50, and 2. From the preceding formula we have that

$$T_4 = \frac{2 - 0}{4} \left[\frac{1}{2} f(0) + f(0.5) + f(1) + f(1.5) + \frac{1}{2} f(2) \right]$$

$$= \frac{1}{2} \left[\frac{1}{2} \cdot 2 + \sqrt{4 - 0.25} + \sqrt{4 - 1} + \sqrt{4 - 2.25} + \frac{1}{2} \cdot 0 \right]$$

$$\approx 2.996 \quad ■$$

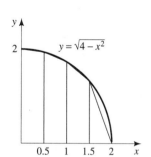

Figure 6.4
Using the trapezoid rule to approximate π.

Now notice that

$$\int_1^2 \frac{1}{x} \, dx = \ln x \Big|_1^2 = \ln 2$$

Thus, we can approximate $\ln 2$ by approximating the integral $\int_1^2 \frac{1}{x} \, dx$.

EXAMPLE 2 Using the Trapezoidal Rule

Use T_4 to find an approximation to $\int_1^2 \frac{1}{x} \, dx = \ln 2$.

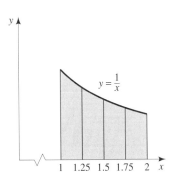

Figure 6.5
Using the trapezoid rule to approximate ln 2.

S O L U T I O N ■ In Figure 6.5 we can see that T_4 must be larger than the area, that is, $T_4 > \ln 2$. We have $\Delta x = \dfrac{2-1}{4} = 0.25$ and the points x_0, x_1, x_2, x_3, and x_4, are, respectively, 1, 1.25, 1.5, 1.75, 2. Thus,

$$T_4 = \frac{2-1}{4}\left[\frac{1}{2}f(1) + f(1.25) + f(1.5) + f(1.75) + \frac{1}{2}f(2)\right]$$

$$= \frac{1}{4}\left(\frac{1}{2} + \frac{4}{5} + \frac{2}{3} + \frac{4}{7} + \frac{1}{4}\right)$$

$$\approx 0.6970$$

Compare this with $\ln 2 = 0.6931$ to four decimal places. ■

Simpson's Rule

James Gregory 1638–1675

It was actually the English mathematician James Gregory who in 1668 discovered Simpson's rule. Thomas Simpson (1710–1761), after whom Simpson's rule is named, wrote a very popular calculus textbook.

(The Granger Collection)

When approximating the area under a curve using left- or right-hand Riemann sums, we use thin "slices" consisting of thin rectangles, that is, a slice with the top a horizontal line (polynomial of degree **zero**). When approximating using the trapezoidal rule, we use thin slices with the top a line that connects the curve at two points (polynomial of degree **one**).

Intuitively we might expect trapezoids to be the better approximation, in general, since the straight line forming the top of a trapezoid hits the curve at *both* endpoints of the subinterval, whereas the top of a rectangle used in a left- or right-hand sum need only hit the curve at *one* endpoint. In general it is true that trapezoids yield better approximations than rectangles from left- or right-hand sums.

It is easy to criticize the use of trapezoids, however, since the typical curve actually "bends," whereas the top of a trapezoid is a straight line (polynomial of degree **one**). We might then think of using for the top a convenient curve that "bends." The obvious next choice is a polynomial of degree **two**, that is, a parabola. For technical reasons we need to divide the interval $[a, b]$ into an *even* number n of subintervals and, as usual, of equal length.

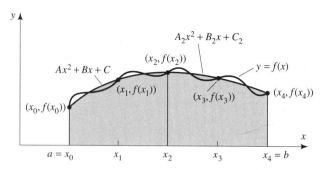

Figure 6.6
Simpson's rule uses different quadratics for the tops of pairs of subintervals.

In Figure 6.6 the interval has been divided into four subintervals. We then wish to find a second-degree polynomial $Ax^2 + Bx + C$ that goes through the three points $(x_0, f(x_0))$, $(x_1, f(x_1))$, and $(x_2, f(x_2))$, and use this as a top over the interval $[x_0, x_2]$.

In the same way we take the unique second-degree polynomial $A_2x^2 + B_2x + C_2$ that goes through the three points $(x_2, f(x_2))$, $(x_3, f(x_3))$, and $(x_4\ f(x_4))$ as the top over the subinterval $[x_2, x_4]$. We refer to the sum of the areas under these two polynomials as S_4, the 4 referring to the original four subintervals.

It turns out (no proof given) that the area under the first parabola from x_0 to x_2 is

$$\frac{\Delta x}{3} [f(x_0) + 4f(x_1) + f(x_2)]$$

where as usual $\Delta x = (b - a)/n$. In the same way, the area under the second parabola from x_2 to x_4 is

$$\frac{\Delta x}{3} [f(x_2) + 4f(x_3) + f(x_4)]$$

Thus, the sum of these two is

$$S_4 = \frac{\Delta x}{3} [f(x_0) + 4f(x_1) + 2f(x_2) + 4f(x_3) + f(x_4)]$$

In general, we have the following:

Simpson's Rule

For any even integer n, divide the interval $[a, b]$ into n equal subintervals with endpoints $a = x_0, x_1, \ldots, x_n = b$, and let

$$S_n = \frac{b - a}{3n} \left[f(x_0) + 4f(x_1) + 2f(x_2) + 4f(x_3) \right.$$

$$\left. + \cdots + 2f(x_{n-2}) + 4f(x_{n-1}) + f(x_n) \right]$$

Then

$$S_n \approx \int_a^b f(x)\, dx$$

Notice in this formula that, with the exceptions of the first and last terms, there is a 4 in front of a term $f(x_k)$ if k is *odd* and a 2 in front of a term $f(x_k)$ if k is *even*.

EXAMPLE 3   Using Simpson's Rule

Use Simpson's rule for S_4 to approximate $\int_0^2 \sqrt{4 - x^2}\, dx = \pi$.

SOLUTION ■   The interval $[0, 2]$ is divided into four subintervals with $x_0 = 0$, $x_1 = 0.50$, $x_2 = 1$, $x_3 = 1.50$, $x_4 = 2$. Using Simpson's rule, we then have

$$S_4 = \frac{2 - 0}{3 \cdot 4}\left[f(0) + 4f(0.5) + 2f(1) + 4f(1.5) + f(2) \right]$$

$$= \frac{1}{6}\left(2 + 4\sqrt{4 - 0.25} + 2\sqrt{4 - 1} + 4\sqrt{4 - 2.25} + 0 \right)$$

$$\approx 3.084 \quad ■$$

This compares with $T_4 = 2.996$ for the trapezoidal rule, and we see that S_4 is much closer to π than T_4 is.

EXAMPLE 4   Comparing the Two Rules

Use the programs TRAP and SIMP found in the Technology Resource Manual to find the trapezoidal and Simpson approximation for $\int_0^2 \sqrt{4 - x^2}\, dx = \pi$ for $n = 100, 200, 400$. Compare the results.

SOLUTION ■   The table gives the results. The errors listed are $|T_n - \pi|$ and $|S_n - \pi|$, respectively. Notice how the approximations improve as n becomes larger. Also notice that Simpson's rule is substantially better than the trapezoidal rule for each n. ■

n	T_n	Error	S_n	Error
100	3.14042	0.00117	3.14113	0.00046
200	3.14118	0.00041	3.14143	0.00016
400	3.14145	0.00014	3.14154	0.00005

EXAMPLE 5   Using Simpson's Rule

Use S_4 to find an approximation to $\int_1^2 \frac{1}{x}\, dx = \ln 2$.

SOLUTION ■   Here $x_0 = 1$, $x_1 = 1.25$, $x_2 = 1.50$, $x_3 = 1.75$, $x_4 = 2$ and

$$S_4 = \frac{2 - 1}{3 \cdot 4}\left[f(1) + 4f(1.25) + 2f(1.5) + 4f(1.75) + f(2) \right]$$

$$= \frac{1}{12}\left(1 + 4 \cdot \frac{4}{5} + 2 \cdot \frac{2}{3} + 4 \cdot \frac{4}{7} + \frac{1}{2} \right)$$

$$\approx 0.6933$$

Compare this with $T_4 = 0.6970$ for the trapezoidal rule, and we see that S_4 is much closer to $\ln 2$ than T_4 is, since $\ln 2 = 0.6931$ to four decimal places. ■

EXAMPLE 6   Comparing the Two Rules

Use the programs TRAP and SIMP to find the trapezoidal and Simpson approximation for $\int_1^2 \frac{1}{x}\, dx = \ln 2$ for $n = 8, 16, 32$. Compare the results.

SOLUTION ■ The table gives the results. The errors listed are $|T_n - \ln 2|$ and $|S_n - \ln 2|$, respectively. Notice how the approximations improve as n becomes larger. Also notice that Simpson's rule is substantially better than the trapezoidal rule for each n. ■

n	T_n	Error	S_n	Error
8	0.69412185	0.00097467	0.69315453	0.00000735
16	0.69339120	0.00024402	0.69314765	0.00000047
32	0.69320821	0.00006103	0.69314721	0.00000003

Errors

An approximation is worth little if we do not know how close it is to the correct answer. It is of course of vital importance to know just how accurate our approximations really are.

We now state two important results that give error estimates.

Error in the Trapezoidal Rule

The maximum possible error incurred in using the trapezoidal rule is

$$\frac{M(b - a)^3}{12n^2}$$

where $|f''(x)| \le M$ on $[a, b]$.

Naturally we need to assume that $f''(x)$ exists and is bounded on $[a, b]$.

Error in Simpson's Rule

The maximum possible error incurred in using Simpson's rule is

$$\frac{M(b - a)^5}{180n^4}$$

where $|f^{(4)}(x)| \le M$ on $[a, b]$.

Naturally we need to assume that $f^{(4)}(x)$ exists and is bounded on $[a, b]$.

Remark. Since these error formulas are correct, no matter what function f is used, they are extremely conservative error bounds. That is, most likely the error in any particular problem is much less.

EXAMPLE 7 Finding Error Bounds

Use the preceding error-bound formulas to find the largest possible errors incurred in the previous calculations of T_4 and S_4 for the definite integral $\int_1^2 \frac{1}{x}\, dx$.

SOLUTIONS ■

(a) *Trapezoidal Rule Error.* We first need to find a bound of $f''(x)$ on the interval $[1, 2]$. We have $f(x) = x^{-1}$, $f'(x) = -x^{-2}$, and $f''(x) = 2x^{-3}$. So we need to find an upper bound on $2x^{-3}$ on [1, 2]. Since the function $2x^{-3}$ is decreasing on this interval, it attains its maximum at the left endpoint $x = 1$ where we see that $f''(1) = 2$. Thus, we take $M = 2$ in the error-bound formula given earlier. Then

$$\frac{M(b - a)^3}{12n^2} = \frac{2(2 - 1)^3}{12(4)^2} \approx 0.01$$

As we noted earlier the actual error was $T_4 - \ln 2 \approx 0.004$.

(b) *Simpson's Rule Error.* We first need to find a bound of $f^{(4)}(x)$ on the interval [1, 2]. We have $f''(x) = 2x^{-3}$, so $f^{(3)}(x) = -6x^{-4}$ and $f^{(4)}(x) = 24x^{-5}$. So we need to find an upper bound on $24x^{-5}$ on [1, 2]. Since the function $24x^{-5}$ is decreasing on this interval, it attains its maximum at the left endpoint $x = 1$, where we see that $f^{(4)}(1) = 24$. Thus, we take $M = 24$ in the error-bound formula given earlier. Then

$$\frac{M(b - a)^5}{180n^4} = \frac{24(2 - 1)^5}{180(4)^4} \approx 0.0005$$

As we noted earlier, the actual error was $S_4 - \ln 2 \approx 0.0002$. ∎

EXPLORATION *1*

Controlling the Error

 Suppose you wish to find $\int_1^2 \frac{1}{x}\, dx = \ln 2$ with an error no greater than

0.0001. (a) If you use the trapezoidal rule how large must you take *n*? (b) If you use Simpson's rule how large must you take *n*?

If the original function $f(x)$ was a *first*-degree polynomial, that is, $f(x) = Ax + B$, then the error-bound formula for the trapezoidal rule indicates that there is **no** error, since in this case $f''(x) = 0$ and therefore $M = 0$. This is naturally what we expect since the tops of the approximating trapezoids are themselves *first*-degree polynomials.

Very surprisingly, however, the Simpson's rule error-bound formula indicates that Simpson's rule is exact not just for *second*-degree polynomials, as you might suspect, but also for **third**-degree polynomials. This is true since the fourth derivative of a third-degree polynomial is identically zero, and thus $M = 0$ in the preceding formula. For this reason Simpson's rule gives much better approximations than one might otherwise suspect.

$f(x)$ on [0, 2]	x^2	x^3	$1/(x + 1)$	$\sqrt{1 + x^2}$
Trapezoidal ($n = 2$)	3.000	5.000	1.167	3.032
Simpson's ($n = 2$)	2.667	4.000	1.111	2.964
Exact (to three decimals)	2.667	4.000	1.099	2.958

SELF-HELP EXERCISE SET 6.3

1. Use T_8 to approximate $\int_1^2 \frac{1}{x}\, dx$.

2. Use S_8 to approximate $\int_1^2 \frac{1}{x}\, dx$.

3. Using the error-bound formulas, find the maximum error in the previous two exercises.

EXERCISE SET 6.3

In Exercises 1 through 8 find T_2 and S_2, and for Exercises 1 through 6 compare your approximations to the correct answer.

1. $\int_0^2 x^3 \, dx$

2. $\int_0^2 x^2 \, dx$

3. $\int_1^3 \frac{1}{x} \, dx$

4. $\int_0^2 x^4 \, dx$

5. $\int_0^1 \sqrt{1 - x^2} \, dx$

6. $\int_0^2 \frac{1}{x + 1} \, dx$

7. $\int_0^1 e^{x^2} \, dx$

8. $\int_2^3 \frac{1}{\ln x} \, dx$

In Exercises 9 through 16 find T_4 and S_4, and for Exercises 9 through 14 compare your approximations to the correct answer and to the approximations obtained in Exercises 1 through 6.

9. $\int_0^2 x^3 \, dx$

10. $\int_0^2 x^2 \, dx$

11. $\int_1^3 \frac{1}{x} \, dx$

12. $\int_0^2 x^4 \, dx$

13. $\int_0^1 \sqrt{1 - x^2} \, dx$

14. $\int_0^2 \frac{1}{x + 1} \, dx$

15. $\int_0^1 e^{x^2} \, dx$

16. $\int_2^3 \frac{1}{\ln x} \, dx$

In Exercises 17 through 24 use the programs TRAP and SIMP to find T_{100} and S_{100}, and for Exercises 17 through 22 compare your approximations to the correct answer and to the approximations obtained in Exercises 1 through 6.

17. $\int_0^2 x^3 \, dx$

18. $\int_0^2 x^2 \, dx$

19. $\int_1^3 \frac{1}{x} \, dx$

20. $\int_0^2 x^4 \, dx$

21. $\int_0^1 \sqrt{1 - x^2} \, dx$

22. $\int_0^2 \frac{1}{x + 1} \, dx$

23. $\int_0^1 e^{x^2} \, dx$

24. $\int_2^3 \frac{1}{\ln x} \, dx$

Applications

25. Revenue. In the table the rate $R'(t)$ of revenue in millions of dollars per week is estimated in some manner for discrete, equally spaced points $0 = t_0, t_1, \ldots, t_4 = 8$, where t is in weeks. Find **(a)** T_4 and **(b)** S_4. The answers are approximations to $\int_0^8 R'(t) \, dt = R(8) - R(0)$, which is the total revenue for the eight-week period.

t_i	0	2	4	6	8
$R'(t_i)$	1.0	0.8	0.5	0.7	0.8

26. Sales. In the table the rate $R'(t)$ of sales for an entire industry in billions of dollars is given for discrete, equally spaced points $0 = t_0, t_1, \ldots, t_{10} = 10$, where t is in weeks. Find **(a)** T_{10} and **(b)** S_{10}. The answers are approximations to $\int_0^{10} R'(t) \, dt = R(10) - R(0)$, which is the total sales for the 10-week period.

t_i	0	1	2	3	4	5	6	7	8	9	10
$R'(t_i)$	10	12	15	16	20	18	16	15	12	10	6

27. Production. In the table the rate $P'(t)$ of production of steel in thousands of tons is estimated in some manner for discrete, equally spaced points $0 = t_0, t_1, \ldots, t_4 = 16$, where t is in weeks. Find **(a)** T_4 and **(b)** S_4. The answers are approximations to $\int_0^{16} P'(t) \, dt = P(16) - P(0)$, which is the total production of steel for the eight-week period.

t_i	0	4	8	12	16
$P'(t_i)$	1	2	3	5	4

28. Population. In the table the rate of growth $M'(t)$ of a bacteria has been estimated by measuring the rate of increase in mass at discrete, equally spaced points $0 = t_0, t_1, \ldots, t_{10} = 20$, where t is in hours. Find **(a)** T_{10} and **(b)** S_{10}. The answers are approximations to $\int_0^{20} M'(t) \, dt = M(20) - M(0)$, which is the change in mass for the 20-hour period.

t_i	0	2	4	6	8	10	12	14	16	18	20
$M'(t_i)$	10	14	20	30	50	100	100	80	50	20	0

29. Velocity. In the table the velocity $v(t) = s'(t)$ is given for discrete, equally spaced points $0 = t_0, t_1, \ldots, t_6 = 3$. Find **(a)** T_6 and **(b)** S_6. The answers are approximations to $\int_0^3 v(t) \, dt = s(3) - s(0)$, which is the total distance traveled.

t_i	0	0.5	1	1.5	2	2.5	3
$v(t_i)$	1.0	1.6	2.0	1.8	1.5	1.0	0.5

In Exercises 30 through 34 find the maximum error for T_{10} and S_{10}:

30. $\int_0^2 x^3 \, dx$

31. $\int_0^2 x^2 \, dx$

32. $\int_1^3 \frac{1}{x} \, dx$

33. $\int_0^2 x^4 \, dx$

34. $\int_0^2 \frac{1}{x + 1} \, dx$

35. A tract of city land bounded by four roads is shown in the figure. The figure shows measurements taken every 100

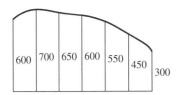

600 700 650 600 550 450 300

feet. If property in this location goes for $10,000 an acre, approximately what is this tract worth? *Hint:* Use Simp-

son's rule to estimate the area and note that there are 43,560 square feet to an acre.

36. If x is large, the logarithmic integral $\text{li}(x) = \int_2^x \dfrac{dt}{\ln t}$ approximates the number of prime numbers less than x. Find an approximation to $\text{li}(200)$. (There are 46 primes less than 200.)

Enrichment Exercises

37. In Example 1 the curve was concave down and the trapezoidal rule gave an underestimate. In Example 2 the curve was concave up and the trapezoidal rule gave an overestimate. Does the trapezoidal rule always give an underestimate when the graph of the function is concave down and an overestimate when the graph of the function is concave up? Explain.

38. Can you use the trapezoidal rule error-bound formula to find an error bound for T_{10} for $\int_0^1 \sqrt{x}\, dx$? Why or why not? Explain using complete sentences.

39. For any $b > 0$ show directly (without using the Simpson's rule error-bound formula) that S_2 gives the exact value for $\int_0^b x^3\, dx$.

40. In Examples 4 and 6 we used the trapezoidal and Simpson rules to estimate $\int_1^2 \dfrac{1}{x}\, dx$ and $\int_0^2 \sqrt{4 - x^2}\, dx$. Recall that the approximation for the first integral was more accurate with $n = 32$ than the second with $n = 400$. By studying the error rule formulas and the integrands give a reasonable explanation of how this could happen.

41. Show that you can obtain the trapezoidal rule by taking the average of the left- and right-hand sums.

42. Section 5.4 contains a discussion of MSUM that uses the rectangles with heights equal to the values of f at the midpoint of the subintervals.
 (a) In Example 4 we used the trapezoidal rule to approximate $\int_0^2 \sqrt{4 - x^2}\, dx = \pi$ using $n = 100, 200$, and 400. Use the midpoint rule MSUM to approximate this same integral for the same values of n. Determine the errors. Compare the errors used in the midpoint rule with the errors in the trapezoidal rule.
 (b) In Example 6 we used the trapezoidal rule to approximate $\int_1^2 \dfrac{1}{x}\, dx = \ln 2$ using $n = 8, 16$, and 32. Use the midpoint rule MSUM to approximate this same integral for the same values of n. Determine the errors. Compare the errors used in the midpoint rule with the errors in the trapezoidal rule.
 (c) Based on your results in parts (a) and (b), speculate on how accurate the midpoint rule is compared with the trapezoidal rule.

SOLUTIONS TO SELF-HELP EXERCISE SET 6.3

1. We have $\Delta x = \dfrac{2 - 1}{8} = 0.125$, and the points $x_0, x_1, x_2,$ $\ldots, x_8$, are, respectively, 1, 1.125, 1.25, $\ldots$, 1.875, and 2. Thus,

$$T_8 = \frac{1 - 0}{8}\left[\frac{1}{2} f(1) + f(1.125) + f(1.25)\right.$$
$$+ f(1.375) + f(1.50) + f(1.625) + f(1.75)$$
$$\left. + f(1.875) + \frac{1}{2} f(2)\right]$$
$$= \frac{1}{8}\left(\frac{1}{2} + \frac{1}{1.125} + \frac{1}{1.25} + \frac{1}{1.375} + \frac{1}{1.5}\right.$$
$$\left. + \frac{1}{1.625} + \frac{1}{1.75} + \frac{1}{1.875} + \frac{1}{4}\right)$$
$$\approx 0.6941219$$

2. The points $x_0, x_1, \ldots, x_8$, are the same as in the previous exercise. Then

$$S_8 = \frac{2 - 1}{3 \cdot 8}\left[f(1) + 4f(1.125) + 2f(1.25)\right.$$

$$+ 4f(1.375) + 2f(1.5) + 4f(1.625) + 2f(1.75)$$
$$\left. + 4f(1.875) + f(2)\right]$$
$$= \frac{1}{24}\left(1 + 4\,\frac{1}{1.125} + 2\,\frac{1}{1.25} + 4\,\frac{1}{1.375}\right.$$
$$\left. + 2\,\frac{1}{1.5} + 4\,\frac{1}{1.625} + 2\,\frac{1}{1.75} + 4\,\frac{1}{1.875} + \frac{1}{2}\right)$$
$$\approx 0.6931545$$

3. Just as in Example 5 of the text, $|f''(x)| \le 2$, $|f^{(4)}(x)| \le 24$, and $b - a = 2 - 1 = 1$. For this exercise $n = 8$. Thus, a bound on the trapezoid error is

$$\frac{M(b - a)^3}{12n^2} = \frac{2(2 - 1)^3}{12(8)^2} \approx 0.0026$$

Compare this with the actual error of 0.001. A bound on the Simpson's error is

$$\frac{M(b - a)^5}{180n^4} = \frac{24(2 - 1)^5}{180(8)^4} \approx .00003$$

Compare this with the actual error of 0.000007.

6.4 IMPROPER INTEGRALS

■ Improper Integrals   ■ Applications

APPLICATIONS: SALES OF AN ENGINE PART OVER THE LONG-TERM

To sell a large number of its most recent engines to the aircraft manufacturers of the world, an aircraft engine manufacturer had to guarantee to supply all spare parts for this engine in perpetuity. One year from now the company plans to stop manufacturing this engine and wishes in the next year to make all the spare parts they will ever need. They estimate that the rate at which engine manifolds will be needed per year is given by $r(t) = 200e^{-0.1t}$, where t is in years from now. How many manifolds will be needed? See Example 3 for the answer.

Improper Integrals

Up to now we have considered definite integrals $\int_a^b f(x)\,dx$, where the interval $[a, b]$ of integration was *bounded*, that is, both a and b are finite numbers. But in some important applications one must consider *infinite* intervals of integration.

Consider, for example, $\int_1^\infty \dfrac{1}{x^2}\,dx$. We define this as follows. First, calculate $\int_1^b \dfrac{1}{x^2}\,dx$, which is the area of the region between the graph of $y = \dfrac{1}{x^2}$ and the x-axis on the interval $[1, b]$ (Fig. 6.7). We obtain

$$\int_1^b \frac{1}{x^2}\,dx = -\frac{1}{x}\Big|_1^b$$

$$= -\frac{1}{b} + 1 = 1 - \frac{1}{b}$$

We then take the limit of this expression as $b \to \infty$ (Fig. 6.8). We obtain

$$\int_1^\infty \frac{1}{x^2}\,dx = \lim_{b\to\infty}\int_1^b \frac{1}{x^2}\,dx$$

$$= \lim_{b\to\infty}\left(1 - \frac{1}{b}\right)$$

$$= 1$$

Thus, we declare that $\int_1^\infty \dfrac{1}{x^2}\,dx = 1$. We also declare that the area of the unbounded region between the graph of $y = \dfrac{1}{x^2}$ and the x-axis on the interval $[1, \infty)$ is 1.

Of course, for some functions, $y = f(x)$, the limit $\lim_{b\to\infty}\int_1^b f(x)\,dx$ does not exist. In such a case we say that the integral **diverges**.

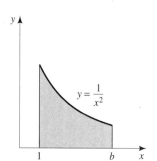

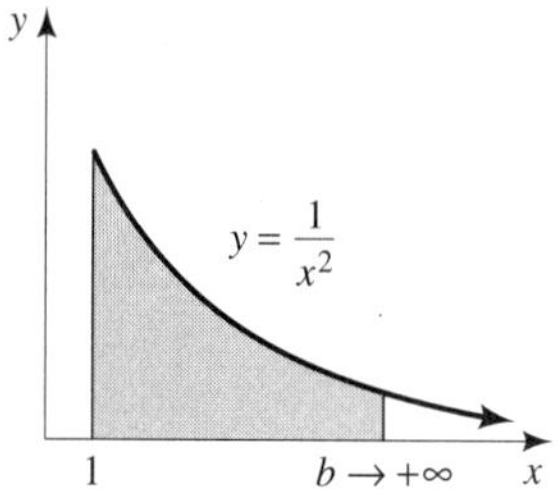

Figure 6.7
The area between the graph of the curve $y = \dfrac{1}{x^2}$ and the x-axis from 1 to b is $\int_1^b x^{-2}\,dx$.

Figure 6.8
We define
$\int_1^\infty x^{-2}\,dx$ as $\lim_{b\to\infty}\int_1^b x^{-2}\,dx$.

Definition of $\int_a^\infty f(x)\,dx$

If $f(x)$ is continuous on $[a, \infty)$, we define

$$\int_a^\infty f(x)\,dx = \lim_{b\to\infty}\int_a^b f(x)\,dx$$

if this limit exists as a number. Otherwise we say that the integral **diverges**. If the limit exists as a number, we say that the integral **converges** to that number.

Remark. When the integral is over an infinite interval we call it an **improper integral**.

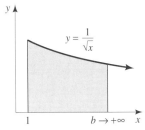

Figure 6.9
$\int_1^\infty \frac{1}{\sqrt{x}}\, dx = \lim_{b\to\infty} \int_1^b \frac{1}{\sqrt{x}}\, dx.$

EXAMPLE 1   Finding an Improper Integral

Find $\displaystyle\int_1^\infty \frac{1}{\sqrt{x}}\, dx$.

SOLUTION ■ According to our definition we first need to evaluate $\int_1^b \frac{1}{\sqrt{x}}\, dx$, for any $b > 1$. Of course this just represents the area under the curve on the interval $[1, b]$ (Fig. 6.9). We have

$$\int_1^b x^{-1/2}\, dx = 2x^{1/2}\Big|_1^b$$
$$= 2(\sqrt{b} - 1)$$

Now we need to take the limit of this expression as $b \to \infty$ (see Fig. 6.9). Doing this, we obtain

$$\int_1^\infty x^{-1/2}\, dx = \lim_{b\to\infty} \int_1^b x^{-1/2}\, dx$$
$$= \lim_{b\to\infty} 2(\sqrt{b} - 1)$$

which is unbounded and thus does not exist. This integral then diverges.   ■

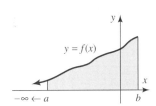

Figure 6.10
$\int_{-\infty}^b f(x)\, dx = \lim_{a\to-\infty} \int_a^b f(x)\, dx.$

Just as we defined an improper integral on the interval $[a, \infty)$, so also we can define an improper integral on the interval $(-\infty, b]$. We now state this definition (Fig. 6.10).

Definition of $\int_{-\infty}^b f(x)\, dx$

If $f(x)$ is continuous on $(-\infty, b]$, we define

$$\int_{-\infty}^b f(x)\, dx = \lim_{a\to-\infty} \int_a^b f(x)\, dx$$

if this limit exists. If the limit does not exist, we say that the integral **diverges**.

Finally, we define an improper integral of the form $\int_{-\infty}^\infty f(x)\, dx$ (Fig. 6.11).

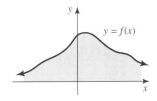

Figure 6.11
$\int_{-\infty}^\infty f(x)\, dx = \int_{-\infty}^0 f(x)\, dx +$
$\int_0^\infty f(x)\, dx$ if both of the latter two integrals converge.

Definition of $\int_{-\infty}^\infty f(x)\, dx$

If $f(x)$ is continuous on $(-\infty, \infty)$ we define

$$\int_{-\infty}^\infty f(x)\, dx = \int_{-\infty}^0 f(x)\, dx + \int_0^\infty f(x)\, dx$$

if both integrals on the right converge. If one or both of the integrals on the right diverges, then we say that $\int_{-\infty}^\infty f(x)\, dx$ diverges.

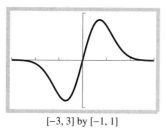

[−3, 3] by [−1, 1]

Screen 6.3
A graph of $y = 2xe^{-x^2}$.

EXAMPLE 2   Finding an Improper Integral

Find $\int_{-\infty}^{\infty} 2xe^{-x^2}\, dx$.

SOLUTION ■ A graph of the function is shown in Screen 6.3 using a window with dimensions [−3, 3] by [−1, 1]. According to the previous definition we need to find both $\int_{-\infty}^{0} 2xe^{-x^2}\, dx$ and $\int_{0}^{\infty} 2xe^{-x^2}\, dx$. To find the latter integral we need to first find $\int_{0}^{b} 2xe^{-x^2}\, dx$ for every $b > 0$. Then

$$\int_{0}^{b} 2xe^{-x^2}\, dx = -e^{-x^2}\Big|_{0}^{b}$$
$$= 1 - e^{-b^2}$$
$$= 1 - \frac{1}{e^{b^2}}$$

Now taking the limit of this last expression as $b \to \infty$, we obtain

$$\lim_{b \to \infty} \int_{0}^{b} 2xe^{-x^2}\, dx = \lim_{b \to \infty}\left(1 - \frac{1}{e^{b^2}}\right)$$
$$= 1$$

For the other integral we have for any $a < 0$,

$$\int_{a}^{0} 2xe^{-x^2}\, dx = -e^{-x^2}\Big|_{a}^{0}$$
$$= -1 + e^{-a^2}$$
$$= \frac{1}{e^{a^2}} - 1$$

Now taking the limit of this last expression as $a \to -\infty$, we obtain

$$\lim_{a \to -\infty} \int_{a}^{0} 2xe^{-x^2}\, dx = \lim_{a \to -\infty}\left(\frac{1}{e^{a^2}} - 1\right)$$
$$= -1$$

Thus,

$$\int_{-\infty}^{\infty} 2xe^{-x^2}\, dx = \int_{-\infty}^{0} 2xe^{-x^2}\, dx + \int_{0}^{\infty} 2xe^{-x^2}\, dx$$
$$= -1 + 1 = 0 \quad ■$$

Applications

The large aircraft manufacturers of the world do not produce the engines for their aircraft. Rather they purchase them from several large engine manufacturers. In this country General Electric and United Technologies dominate this fiercely competitive industry.

EXAMPLE 3   Spare Parts Production

To sell a large number of its most recent engines to the aircraft manufacturers of the world, an aircraft engine manufacturer had to guarantee to supply all spare parts for this engine in perpetuity. One year from now the company plans to stop manufacturing this engine and wishes in the next year to make all the spare parts they will ever need. They estimate that the rate at which engine manifolds will be needed per year

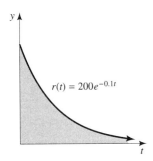

Figure 6.12
Graph of the rate of required
engine manifolds.

is given by $r(t) = 200e^{-0.1t}$, where t is in years from now (Fig. 6.12). How many
manifolds will be needed?

S O L U T I O N ■ Since the integral $\int_0^b r(t)\,dt$ is the amount needed from now
to b years from now, $\int_0^\infty r(t)\,dt$ is the amount needed for all time. Then

$$\int_0^b r(t)\,dt = 200\int_0^b e^{-0.10t}\,dt$$

$$= -2000e^{-0.10t}\Big|_0^b$$

$$= -2000(e^{-0.10b} - 1)$$

$$\int_0^\infty r(t)\,dt = \lim_{b\to\infty}\int_0^b r(t)\,dt$$

$$= \lim_{b\to\infty} -2000(e^{-0.10b} - 1)$$

$$= 2000$$

Thus, the company needs to make 2000 engine manifolds. ■

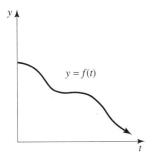

Figure 6.13
A graph of a typical rate of change
of a continuous income flow.

We now assume that we have a continuous income flow of money and that the
positive function $f(t)$ represents the rate of change of this flow. If the flow of money
continues perpetually, then the function $f(t)$ must be given on the interval $[0, \infty)$ (Fig.
6.13). The present value of all future income is denoted by $P_V(\infty)$. We have the following:

Present Value of a Continuous Income Flow on $[0, \infty)$

Let $f(t)$ be the continuous rate of change of total income on the interval $[0, \infty)$, and
suppose that one can invest current money at an interest rate of r in perpetuity, where
r is a decimal, compounded continuously forever. Then the present value $P_V(\infty)$ of
this continuous income flow over $[0, \infty)$ is given by

$$P_V(\infty) = \int_0^\infty f(t)e^{-rt}\,dt$$

$$= \lim_{T\to\infty}\int_0^T f(t)e^{-rt}\,dt$$

E X A M P L E 4   Present Value on an Infinite Interval

The rate of change of income in thousands of dollars per year from an oil well is
projected to be $f(t) = 100e^{-0.1t}$, where t is measured in years. What is a fair initial
price for the oil well as a function of r?

S O L U T I O N ■

$$P_V(\infty) = \int_0^\infty f(t)e^{-rt}\,dt$$

$$= \lim_{T\to\infty}\int_0^T f(t)e^{-rt}\,dt$$

$$= \lim_{T\to\infty}\int_0^T 100e^{-0.1t}e^{-rt}\,dt$$

$$= 100\lim_{T\to\infty}\int_0^T e^{-(r+0.1)t}\,dt$$

$$= 100\lim_{T\to\infty}\frac{-1}{r+0.1}e^{-(r+0.1)t}\Big|_0^T$$

$$= \frac{100}{r + 0.1} \lim_{T \to \infty} (1 - e^{-(r+0.1)T})$$

$$= \frac{100}{r + 0.1} \quad \blacksquare$$

Remark. Notice that as r increases the value of the oil well goes down. Thus, if long-term interest rates go up, the value of the well drops.

SELF-HELP EXERCISE SET 6.4

1. Repeat Example 3 if $r(t) = 1000(t + 1)^{-3/2}$.

2. Find $\int_{-\infty}^{\infty} f(x) \, dx$, where $f(x) = 1$ if $|x| < 1$ and $f(x) = x^{-2}$ if $|x| \geq 1$.

EXERCISE SET 6.4

In Exercises 1 through 20 evaluate.

1. $\int_1^{\infty} \frac{1}{x^3} \, dx$

2. $\int_1^{\infty} \frac{1}{\sqrt[3]{x}} \, dx$

3. $\int_1^{\infty} \frac{1}{\sqrt[4]{x}} \, dx$

4. $\int_1^{\infty} \frac{1}{x^4} \, dx$

5. $\int_1^{\infty} \frac{1}{x^{1.01}} \, dx$

6. $\int_1^{\infty} \frac{1}{x^{0.99}} \, dx$

7. $\int_1^{\infty} \frac{1}{x} \, dx$

8. $\int_{-\infty}^{-1} \frac{1}{x^2} \, dx$

9. $\int_{-\infty}^0 e^x \, dx$

10. $\int_0^{\infty} e^{-x} \, dx$

11. $\int_{-\infty}^{-1} \frac{1}{x^3} \, dx$

12. $\int_{-\infty}^{-1} \frac{1}{\sqrt[3]{x}} \, dx$

13. $\int_0^{\infty} x^2 e^{-x^3} \, dx$

14. $\int_{-\infty}^0 x^2 e^{x^3} \, dx$

15. $\int_{-\infty}^0 \frac{x}{(x^2 + 1)^2} \, dx$

16. $\int_{-\infty}^{\infty} e^{-|x|} \, dx$

17. $\int_{-\infty}^{\infty} \frac{x}{(x^2 + 1)^4} \, dx$

18. $\int_{-\infty}^{\infty} \frac{x}{x^2 + 1} \, dx$

19. $\int_{-\infty}^{\infty} x^3 e^{-x^4} \, dx$

20. $\int_{-\infty}^{\infty} \frac{x^3}{(x^4 + 3)^2} \, dx$

Applications

21. Production. Repeat Example 3 if $r(t) = 1000(t + 1)^{-5/2}$.

22. Mining. It is estimated that platinum is being extracted from a certain mine at the rate of $r(t) = 10 \dfrac{t}{(t^2 + 1)^2}$ tons per year, where t is in years. How much will be extracted from this mine in the long-term?

23. Sales. A company estimates that the rate of increase, in millions of dollars, of sales for a new product is given by $r(t) = 40e^{-2t}$, where t is in years. If this rate continues forever, what will be the eventual sales?

24. Profits. A company estimates that the rate of increase in millions of dollars of profits from a new product is given by $(t + 1)^{-4/3}$, where t is in years. If this rate continues forever, what will be the eventual profits?

25. Radioactive Waste. Suppose radioactive waste from a closed dump site is entering the atmosphere over an area at a rate given by $r(t) = 100e^{-kt}$ tons per year, where t is measured in years. Assuming that this rate continues forever and $k = 0.05$, find the total amount of waste that enters the atmosphere.

26. Radioactive Waste. Repeat the previous exercise if $k = 0.10$.

27. Disease. A contagious disease is spreading into the world's human population at the rate of $p'(t) = 10e^{-0.05t}$, where $p(t)$ is measured in millions and t in years. If this rate continues forever, what is the total number of people that will be infected with this disease?

28. Biology. A population of mice in an area is growing at the rate of $p'(t) = \dfrac{1}{(t + 1)^{5/2}}$, where p is measured in thousands and t in months. If this rate continues forever, what will be the total number of mice in this area?

29. Pollutants. Pollutants are leaking from a closed dumpsite into a small lake nearby. The number of thousands of particles of pollutants that is accumulating in the lake is given by $N(t) = \dfrac{1}{(t + 1)^2}$, where t is measured in years. Assuming this equation holds forever, how many particles of pollutants eventually enter this lake from the dump?

30. By making the change of variable $x = -\ln u$, show that the improper integral $\int_0^{\infty} f(x) \, dx$ becomes $\int_0^1 u^{-1} f(-\ln u) \, du$. Use this and Simpson's rule for $n = 128$ to approximate $\int_0^{\infty} \dfrac{e^{-x}}{1 + x^2} \, dx$.

In Exercises 31 through 34 $f(t)$ is the rate of change of total income per year, and r is the annual interest rate compounding continuously. Find $P_V(\infty)$.

31. $f(t) = 100$, $r = 8\%$ **32.** $f(t) = 1000$, $r = 12\%$

33. $f(t) = 100e^{-0.1t}$, $r = 8\%$

34. $f(t) = 1000e^{-0.2t}$, $r = 10\%$

35. Continuous Income Flow. An investment with a continuous income flow of \$10,000 per year forever is being sold. Assume that the current annual interest rate of 10% compounded continuously will continue forever. Should you buy if the price is **(a)** \$110,000? **(b)** \$90,000?

36. Continuous Income Flow. An investment with a continuous income flow of \$16,000 per year forever is being sold.

Assume that the current annual interest rate of 8% compounded continuously will continue forever. Should you buy if the price is **(a)** \$250,000? **(b)** \$180,000?

37. Investment Decision. An investment with a continuous income flow of $24,000e^{0.04t}$ in dollars per year forever is being sold. Assume that the current annual interest rate of 8% compounded continuously will continue forever. Should you buy if the price is **(a)** \$700,000? **(b)** \$550,000?

38. Investment Decision. An investment with a continuous income flow of $10,000e^{0.05t}$ in dollars per year forever is being sold. Assume that the current annual interest rate of 10% compounded continuously will continue forever. Should you buy if the price is **(a)** \$300,000? **(b)** \$250,000?

Enrichment Exercises

39. A student argues that $\int_{-\infty}^{\infty} x \, dx = 0$ since $\int_{-a}^{a} x \, dx = 0$ for every a. Explain why this is or is not correct using the definitions given in this section.

40. Suppose $0 \le f(x) \le \dfrac{1}{x^2}$ on the interval $[1, \infty)$. What can you say about $\int_1^{\infty} f(x) \, dx$? Explain carefully.

41. Suppose $f(x) \ge \dfrac{1}{\sqrt{x}}$ on the interval $[1, \infty)$. What can you say about $\int_1^{\infty} f(x) \, dx$? Explain carefully.

42. The integral $\displaystyle\int_0^1 \dfrac{1}{\sqrt{x}} \, dx$ is an improper integral (of a type not considered in this section) since the function $f(x) = \dfrac{1}{\sqrt{x}}$ is unbounded on the interval $(0, 1]$. Give a reasonable

definition of this improper integral, and then evaluate using your definition.

43. If $f(x)$ is continuous, $\int_{-\infty}^{\infty} f(x) \, dx$ is convergent, and c is any real number, show that

$$\int_{-\infty}^{\infty} f(x) \, dx = \int_{-\infty}^{c} f(x) \, dx + \int_{c}^{\infty} f(x) \, dx$$

44. Law Enforcement. Shavell [129] let $x(h)$ be the enforcement effort specific to apprehending those who commit acts causing harm h and considered the term $\int_0^{\infty} x(h) \, dh$. Explain what you think this integral stands for and explain why.

SOLUTIONS TO SELF-HELP EXERCISE SET 6.4

1. In the integral $\int_0^b (t + 1)^{-3/2} \, dt$ make the change of variable $u = t + 1$. Then $du = dt$ and

$$1000 \int_0^b (t + 1)^{-3/2} \, dt = 1000 \int_1^{b+1} u^{-3/2} \, du$$

$$= -2000 u^{-1/2}\big|_1^{b+1}$$

$$= 2000 \left(1 - \frac{1}{\sqrt{b + 1}}\right)$$

$$1000 \int_0^{\infty} (t + 1)^{-3/2} \, dt = 1000 \lim_{b \to \infty} \int_0^b (t + 1)^{-3/2} \, dt$$

$$= \lim_{b \to \infty} 2000 \left(1 - \frac{1}{\sqrt{b + 1}}\right) = 2000$$

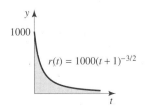

2. By symmetry, notice from the figure that $\int_{-\infty}^0 f(x) \, dx = \int_0^{\infty} f(x) \, dx$. Then

$$\int_{-\infty}^{\infty} f(x) \, dx = \int_{-\infty}^0 f(x) \, dx + \int_0^{\infty} f(x) \, dx$$

$$= 2 \int_0^{\infty} f(x) \, dx$$

$$= 2 \left(\int_0^1 dx + \int_1^{\infty} x^{-2} \, dx\right)$$

$$= 2 + 2 \lim_{b \to \infty} \int_1^b x^{-2} \, dx$$

$$= 2 + 2 \lim_{b \to \infty} (-x^{-1}\big|_1^b)$$

$$= 2 + 2 \lim_{b \to \infty} \left(1 - \frac{1}{b}\right) = 4$$

- **Integration by Parts:** $\int u\,dv = uv - \int v\,du$. p. 332.
- **Trapezoidal Rule.** p. 344. For any integer n, divide the interval $[a, b]$ into n subintervals of equal length with endpoints $a = x_0, x_1, \ldots, x_n = b$. If

$$T_n = \frac{b-a}{n}\left[\frac{1}{2}f(x_0) + f(x_1) + f(x_2) + \cdots + f(x_{n-1}) + \frac{1}{2}f(x_n)\right]$$

then

$$\int_a^b f(x)\,dx \approx T_n2$$

- **Simpson's Rule.** p. 344. For any even integer n divide the interval $[a, b]$ into n equal subintervals with endpoints $a = x_0, x_2, \ldots, x_n = b$, and let

$$S_n = \frac{b-a}{3n}\left[f(x_0) + 4f(x_1) + 2f(x_2) + 4f(x_3)\right.$$

$$\left. + \cdots + 2f(x_{n-2}) + 4f(x_{n-1}) + f(x_n)\right]$$

Then

$$S_n \approx \int_a^b f(x)\,dx2$$

- **Error in the Trapezoidal Rule.** The maximum possible error incurred in using the trapezoidal rule is $\dfrac{M(b-a)^3}{12n^2}$, where $|f''(x)| \le M$ on $[a, b]$. p. 346.
- **Error in Simpson's Rule.** The maximum possible error incurred in using Simpson's rule is $\dfrac{M(b-a)^5}{180n^4}$, where $|f^{(4)}(x)| \le M$ on $[a, b]$. p. 346.
- If $f(x)$ is continuous on $[a, \infty)$, define

$$\int_a^\infty f(x)\,dx = \lim_{b \to \infty} \int_a^b f(x)\,dx2$$

if this limit exists as a finite number. Otherwise we say that the integral **diverges**. If the limit exists as a finite number we say that the integral **converges** to that number. p. 350.

- If $f(x)$ is continuous on $(-\infty, b]$, define

$$\int_{-\infty}^b f(x)\,dx = \lim_{a \to -\infty} \int_a^b f(x)\,dx$$

if this limit exists. If the limit does not exist we say that the integral **diverges**. p. 351.

- If $f(x)$ is continuous on $(-\infty, \infty)$ define

$$\int_{-\infty}^\infty f(x)\,dx = \int_{-\infty}^0 f(x)\,dx + \int_0^\infty f(x)\,dx$$

if both integrals on the right exist (as finite numbers). If one or both of the integrals on the right diverges then we say that $\int_{-\infty}^\infty f(x)\,dx$ **diverges**. p. 351.

- Let $f(t)$ be the continuous rate of change of income on the interval $[0, \infty)$ and suppose that one can invest current money at an interest rate of r in perpetuity, where r is a decimal, compounded continuously forever. Then the **present value** $P_V(\infty)$ of this continuous income flow over $[0, \infty)$ is given by

$$P_V(\infty) = \int_0^\infty f(t)e^{-rt}\,dt = \lim_{T \to \infty} \int_0^T f(t)e^{-rt}\,dt.\text{ p. 353.}$$

CHAPTER **6**

REVIEW EXERCISES

In Exercises 1 through 6 integrate by parts.

1. $\displaystyle\int xe^{5x}\,dx$

2. $\displaystyle\int \frac{x}{\sqrt{5+3x}}\,dx$

3. $\displaystyle\int x^5 \ln x\,dx$

4. $\displaystyle\int \frac{x}{(x+1)^2}\,dx$

5. $\displaystyle\int x^3\sqrt{9-x^2}\,dx$

6. $\displaystyle\int \frac{\ln x}{\sqrt{x}}\,dx$

In Exercises 7 through 12 evaluate using the table of integrals.

7. $\displaystyle\int \frac{x}{2x+3}\,dx$

8. $\displaystyle\int \frac{1}{x(x+4)^2}\,dx$

9. $\displaystyle\int \frac{1}{x^2\sqrt{4-x^2}}\,dx$

10. $\displaystyle\int \frac{\sqrt{9-x^2}}{x}\,dx$

11. $\displaystyle\int \sqrt{x^2+1}\,dx$

12. $\displaystyle\int \frac{1}{1+e^x}\,dx$

In Exercises 13 through 16 evaluate the improper integrals.

13. $\displaystyle\int_1^\infty x^{-1.1}\,dx$

14. $\displaystyle\int_1^\infty x^{-0.9}\,dx$

15. $\displaystyle\int_{-\infty}^\infty x^2 e^{-|x^3|}\,dx$

16. $\displaystyle\int_{-\infty}^\infty e^{-x}\,dx$

In Exercises 17 through 18 approximate using the trapezoidal rule and Simpson's rule for $n=2$, and compare your answers to the correct answer.

17. $\displaystyle\int_0^2 5x^4\,dx$

18. $\displaystyle\int_1^4 \frac{1}{x}\,dx$

19. Repeat Exercise 17 for $n=4$.

20. Repeat Exercise 18 for $n=4$.

21. A mine produces a rate of change of total income in thousands of dollars per year given by $f(t)=100e^{-0.10t}$ in perpetuity. Market interest rates are at 8% compounded continuously and will stay the same forever. What is a reasonable price for this mine discounting any risk involved?

CHAPTER **6**

PROJECTS

Project 1
Coefficient of Inequality

The table gives the accumulated U.S. income by population deciles for 1966. (See Suits [128]). Use the table to calculate the coefficient of inequality. Do this by using the trapezoidal rule to approximate the needed definite integral.

Population Decile	Cumulated Adjusted Family Income
1	1.21
2	3.88
3	8.13
4	13.92
5	21.16
6	30.22
7	40.02
8	52.29
9	67.45
10	100.00

Project 2
Measuring Cardiac Output

An important medical procedure is to measure cardiac output. Insufficient output indicates a problem with the heart.

To measure cardiac output of a patient, a measured amount of dye is placed in a vein close to where the blood enters the right atrium of the heart. The dye then mixes with the blood in this chamber. Samples of the blood are regularly taken in a peripheral artery near the aorta where the blood leaves the heart. An instrument called a *densitometer* can make the withdrawals and dye concentration estimates as frequently as five times a second. The figure shows a model of the situation, in which the pulsing of the heart acts to mix the dye and the blood.

A dye must be selected that is rapidly removed from the circulatory system, so as to minimize the effects of recirculation. One such dye is indocyanine which is removed by the liver.

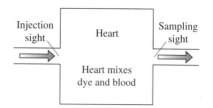

Injection sight | Heart | Sampling sight

Heart mixes dye and blood

Let F measure cardiac output in liters per second and $c(t)$ in milligrams per liter be the concentration of dye in the blood (measured by the densitometer). Then the rate at which dye leaves the heart is given by $r(t) = Fc(t)$. If the recirculation problem has been minimized, then the amount A of dye injected is also the total amount of dye leaving the heart. Then $A = \int_0^\infty Fc(t)\,dt$. We then solve for F as follows.

$$A = \int_0^\infty Fc(t)\,dt$$

$$= F\int_0^\infty c(t)\,dt$$

$$F = \frac{A}{\int_0^\infty c(t)\,dt}$$

In a matter of only 10 or 15 seconds, the concentration $c(t)$ of dye is essentially zero. Thus, the integral $\int_0^\infty Fc(t)\,dt$ can be approximated by $\int_0^b Fc(t)\,dt$, where b is about 10 or 15. This latter integral can then be approximated by using Simpson's rule.

(a) The table gives the concentration over a 12-second period. Use the procedure outlined earlier to approximate the patient's cardiac output. (An output over 7 liters per minute is considered healthy.)

(b) Use the data in part (a) to approximate the *central blood volume*.

$$V = \frac{F\int_0^\infty tc(t)\,dt}{\int_0^\infty c(t)\,dt}$$

t (s)	0	1	2	3	4	5	6	7	8	9	10	11	12
$c(t)$ (mg/L)	0	0	1.7	5.6	9.2	8.4	5.2	3.8	2.1	1.0	0.5	0.2	0

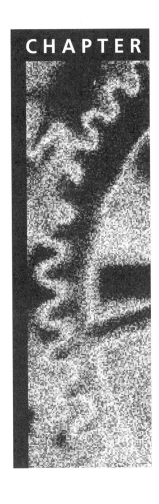

7

Functions of Several Variables

We now consider functions of more than one variable. In the first section we see that the graph of a function $z = f(x, y)$ of two variables represents a surface in three dimensions. In the second section we will see that the partial derivatives measure the rates of change of a function when moving in certain directions, just as the rate of change of height (from the ground) when moving in different directions from a point on the side of a mountain differs depending in which direction you move. In the third and fourth sections we develop ways of finding the relative and absolute extrema of functions of several variables. In the fifth section we give an important application to curve fitting. In the sixth and seventh sections we extend the notions of differential and integral to functions of several variables.

7.1 FUNCTIONS OF SEVERAL VARIABLES

■ Functions of Several Variables   ■ Cartesian Coordinate System   ■ Level Curves   ■ Continuity

---

APPLICATION: REVENUE THAT DEPENDS ON TWO VARIABLES

Determine the revenue of a firm that sells x thousands of boxes of regular corn flakes at a price p dollars per box and y thousands of boxes of crunchy corn flakes at a price q dollars per box if $p = 4 - 2x + y$ and $q = 6 + x - 2y$. For the answer see Example 3.

Functions of Several Variables

We have become familiar with functions of *one* variable. For example, the area A of a circle is a function of its radius r and can be written as $A = A(r) = \pi r^2$. Describing the area of a rectangle of width w and length l, however, requires the function of *two* variables given by $A = A(w, l) = wl$. The notation $A(2, 3)$ means the area of a rectangle with width $w = 2$ and length $l = 3$, which is obtained by replacing w with 2 and replacing l with 3 in the formula for area; thus, $A(2, 3) = 2 \cdot 3 = 6$. In the same way $A(7, 4) = 7 \cdot 4 = 28$ is the area of a rectangle with width $w = 7$ and length $l = 4$.

If a company sells pens at \$2 each and sells x pens, then the revenue is a function of *one* variable and is given by $R = R(x) = 2x$ dollars. If, in addition, the company sells y pencils at 5 cents each, then the revenue, in dollars, now becomes a function of *two* variables and is given by $R = R(x, y) = 2x + 0.05y$. Thus, if 3000 pens and 4000 pencils are sold, the revenue is

$$R(3000, 4000) = 2(3000) + 0.05(4000) = \$6200$$

Functions of Two Variables

A **function of two variables**, denoted by $z = f(x, y)$, is a rule that associates to every point (x, y) in some set D called the **domain** a unique number denoted by $z = f(x, y)$.

The variables x and y are the **independent variables**, and z is the **dependent variable**.

If a domain is not specified, we assume the domain is the largest possible set for which $f(x, y)$ makes sense.

EXAMPLE 1   Evaluating Functions

Let $f(x, y) = 2x - 3y$, $g(x, y) = \dfrac{1}{x + y}$, $h(x, y) = \sqrt{9 - x^2 - y^2}$. Find $f(5, 1)$, $g(2, 3)$, $h(1, 2)$.

SOLUTION ■

$$f(5, 1) = 2(5) - 3(1) = 7$$

$$g(2, 3) = \frac{1}{2 + 3} = \frac{1}{5}$$

$$h(1, 2) = \sqrt{9 - (1)^2 - (2)^2} = 2 \qquad ■$$

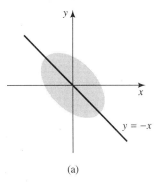

(a)

Figure 7.1

(a) The domain of $g(x,y) = \dfrac{1}{x + y}$ excludes the line $y = -x$.

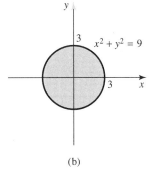

(b)

Figure 7.1
(b) The domain of $h(x,y) = \sqrt{9 - x^2 - y^2}$ is the region inside and on the circle $x^2 + y^2 = 9$.

EXAMPLE 2　Finding the Domain

Find the domain of each of the following functions: (a) $f(x, y) = 2x - 3y$
(b) $g(x, y) = \dfrac{1}{x + y}$ (c) $h(x, y) = \sqrt{9 - x^2 - y^2}$.

SOLUTIONS ■

(a) One can calculate $2x - 3y$ for any x and any y, so the domain is the entire xy-plane.

(b) One can only calculate $\dfrac{1}{x + y}$ if the denominator is not zero. This happens when $y \neq -x$. Thus, the domain of g is all points in the xy-plane except for points on the line $y = -x$ (Fig. 7.1a).

(c) One can only form the square root of a nonnegative number. Thus, the domain of h is all points (x, y) for which $9 - x^2 - y^2 \geq 0$, which is the same as $x^2 + y^2 \leq 9$. This is a circle of radius 3 centered at the origin (see Fig. 7.1b). ■

EXAMPLE 3　Revenue, Cost, and Profit Functions

Suppose a firm produces two types of corn flakes, regular and crunchy. Naturally if the price of one of them increases, the demand for that one decreases. Because these two types compete with each other, however, if the price of one of them increases, the demand for the other *increases*. Let the number of thousands of boxes of regular and crunchy corn flakes sold by the manufacturer per week be x and y, respectively, and the price per box be p and q, respectively. Suppose

$$p = 4 - 2x + y$$
$$q = 6 + x - 2y$$

(a) Find the weekly revenue function $R(x, y)$, and find the revenue when a box of regular corn flakes sells for \$1 and a box of crunchy corn flakes for \$2.

(b) If the weekly cost function of the firm is $C(x, y) = 1000(10 + 2x + 3y)$, find the weekly profit function $P(x, y)$ and $P(1, 2)$.

SOLUTIONS ■

(a) Since the revenue on the sale of regular corn flakes is

$$\text{(Number sold)} \times \text{(price)} = (1000x)(p) = 1000x(4 - 2x + y)$$

and the revenue on the sale of crunchy corn flakes is

$$\text{(Number sold)} \times \text{(price)} = (1000y)(q) = 1000y(6 + x - 2y)$$

the total revenue $R(x, y)$ is the sum of these two. Thus,

$$R(x, y) = 1000x(4 - 2x + y) + 1000y(6 + x - 2y)$$

Then

$$R(1, 2) = 1000(1)[4 - 2(1) + (2)] + 1000(2)[6 + (1) - 2(2)] = 10,000$$

The weekly revenue is therefore \$10,000.

(b) Since profit equals revenue less cost,

$$\begin{aligned} P(x, y) &= R(x, y) - C(x, y) \\ &= 1000x(4 - 2x + y) + 1000y(6 + x - 2y) - 1000(10 + 2x + 3y) \\ &= 1000(2x - 2x^2 + 3y - 2y^2 + 2xy - 10) \end{aligned}$$

Then

$$P(1, 2) = 1000[2(1) - 2(1)^2 + 3(2) - 2(2)^2 + 2(1)(2) - 10]$$
$$= -8000$$

This represents a loss of \$8000 per week. ∎

Next we consider production functions, which play an important role in business. The total output of a firm P, measured by the total number of items produced per year, certainly depends on labor costs L and capital investment K, with labor costs measured by dollars per year spent on wages and capital measured by dollars of capital investment per year. Assume that we explicitly know this relationship as $P = f(K, L)$, Then we call $f(K, L)$ the **production function** of the firm.

One important class of production functions are the **Cobb-Douglas production functions**. They are

$$f(K, L) = AL^b K^{1-b}$$

where A and b are positive constants with $0 < b < 1$.

EXAMPLE 4   Evaluating a Cobb-Douglas Production Function

Suppose the production function of a firm is given by the Cobb-Douglas production function

$$P(K, L) = 5L^{1/3}K^{2/3}$$

where P is the number of thousands of items produced each year, K is the capital investment per year in millions of dollars, and L is the labor cost per year in millions of dollars. Find the number of items produced if \$8 million is invested in capital and \$27 million in labor.

SOLUTION ∎ Since

$$P(8, 27) = 5(27)^{1/3}(8)^{2/3} = 5(3)(4) = 60$$

60,000 are produced. ∎

Cartesian Coordinate System

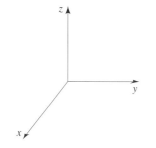

Figure 7.2
The three-dimensional coordinate system.

We are already familiar with a two-dimensional (Cartesian) coordinate system. This is necessary if we are to graph a function $y = f(x)$ of *one* variable since we need to plot the pair of points $(x, y) = (x, f(x))$ for all x in the domain of f. To graph a function $z = f(x, y)$ of *two* variables we need a three-dimensional (Cartesian) coordinate system.

A point in three-dimensional space can be uniquely represented by an ordered triplet of numbers (x, y, z), and every ordered triplet (x, y, z) can uniquely represent a point in three-dimensional space. The coordinate system is shown in Figure 7.2. The three axes shown in Figure 7.2, the x-axis, the y-axis, and the z-axis, are all perpendicular to one another.

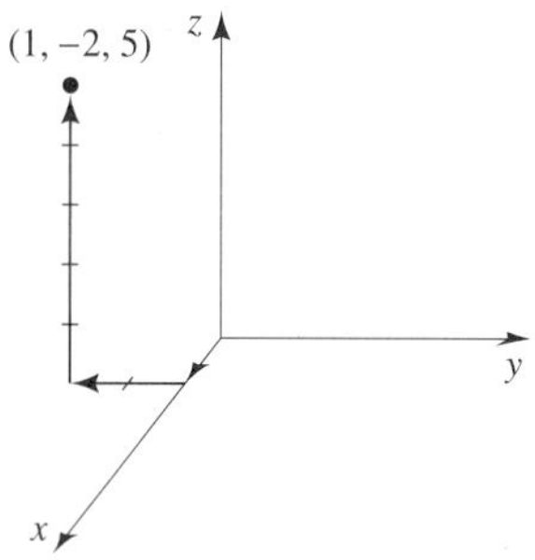

Figure 7.3
To get to $(1, -2, 5)$ go 1 unit in the x-direction, then 2 units in the negative y-direction, and then 5 units in the z-direction.

EXAMPLE 5   Using the Cartesian Coordinate System

Locate $(1, -2, 5)$ in a three-dimensional Cartesian coordinate system.

SOLUTION ∎ To find the point $(1, -2, 5)$ start at the origin $(0, 0, 0)$ and move 1 unit in the positive x-direction along the x-axis, then move 2 units in the direction of the negative y-axis, then move 5 units vertically in the positive z-axis (Fig. 7.3). ∎

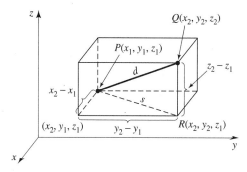

Figure 7.4
The line segment *PQ* is the diagonal of the box.

We would like to find a formula for the distance d between the two points $P(x_1, y_1, z_1)$ and $Q(x_2, y_2, z_2)$ shown in Figure 7.4. We construct a rectangular box so that the line from P to Q forms its diagonal, as shown in Figure 7.4. The bottom of the box lies in a plane, and so the distance s can be found from the Pythagorean theorem to be

$$s = \sqrt{(x_2 - x_1)^2 + (y_2 - y_1)^2}$$

Now the triangle through the three points R, P, and Q forms a right triangle, and again using the Pythagorean theorem, we have that

$$d^2 = s^2 + (z_2 - z_1)^2$$
$$= (x_2 - x_1)^2 + (y_2 - y_1)^2 + (z_2 - z_1)^2$$

Thus, we have the following:

The Distance Formula

The distance between the two points (x_1, y_1, z_1) and (x_2, y_2, z_2) is
$$d = \sqrt{(x_2 - x_1)^2 + (y_2 - y_1)^2 + (z_2 - z_1)^2}$$

EXAMPLE 6   Finding Distance Between Two Points

Find the distance between the two points $(2, 3, -2)$ and $(1, -4, -1)$ shown in Figure 7.5.

SOLUTION ■ Setting $(2, 3, -2) = (x_1, y_1, z_1)$ and $(1, -4, -1) = (x_2, y_2, z_2)$ yields

$$d = \sqrt{(x_2 - x_1)^2 + (y_2 - y_1)^2 + (z_2 - z_1)^2}$$
$$= \sqrt{[(1) - (2)]^2 + [(-4) - (3)]^2 + [(-1) - (-2)]^2}$$
$$= \sqrt{1 + 49 + 1} = \sqrt{51} \quad ■$$

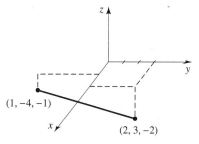

Figure 7.5
The distance from $(1, -4, -1)$ to $(2, 3, -2)$ is $\sqrt{51}$.

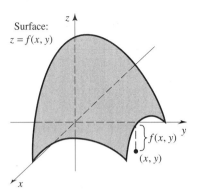

Figure 7.6
The graph of $z = f(x,y)$ is a surface in three dimensions.

We now define the **graph** of a function $z = f(x, y)$. A graph of a function enables us to see a picture of the function (Fig. 7.6).

Graph of a Function of Two Variables

Given a function $z = f(x, y)$ with domain D, the **graph** of f is the set of points (x, y, z) with $z = f(x, y)$ and $(x, y) \in D$. Since this is described by *two* variables, the result is a two-dimensional object called a **surface**.

For example, if (x, y) represents a point in the state of Vermont, and $z = f(x, y)$ the height above sea level, then the graph of $z = f(x, y)$ is the surface of Vermont, including the mountains, valleys, and plains.

EXAMPLE 7   Graphing a Function

Graph the function

$$z = f(x, y) = \sqrt{9 - x^2 - y^2}$$

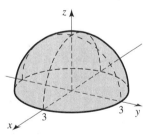

Figure 7.7
A graph of $z = f(x,y) = \sqrt{9 - x^2 - y^2}$.

SOLUTION ■ In Example 2 the domain was found to be all points inside and on the circle in the xy-plane with center at $(0, 0)$ and radius 3. To graph the function $f(x, y)$, square each side of $z = \sqrt{9 - x^2 - y^2}$ and obtain

$$z^2 = 9 - x^2 - y^2$$

which can be written as

$$x^2 + y^2 + z^2 = 9 \qquad [1]$$

The left-hand side is the square of the distance from the point (x, y, z) to $(0, 0, 0)$. Thus, a point (x, y, z) satisfies Equation 1 if, and only if, this point is on the sphere with center at $(0, 0, 0)$ and radius 3. But since z must be *positive*, the graph of $f(x, y)$ is the surface given by the top half of this sphere (Fig. 7.7). ■

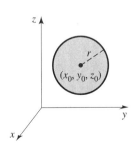

Figure 7.8
A sphere of radius r and center (x_0, y_0, z_0).

We can use the distance formula to find the equation of the sphere centered at the point (x_0, y_0, z_0) with radius r shown in Figure 7.8. The point (x, y, z) is on this sphere if, and only if, the distance from the point (x, y, z) to the point (x_0, y_0, z_0) is r, that is,

$$\sqrt{(x - x_0)^2 + (y - y_0)^2 + (z - z_0)^2} = r$$

or

$$(x - x_0)^2 + (y - y_0)^2 + (z - z_0)^2 = r^2$$

Equation of Sphere

The equation of the sphere centered at (x_0, y_0, z_0) with radius r is given by

$$(x - x_0)^2 + (y - y_0)^2 + (z - z_0)^2 = r^2$$

The graph of an equation of the form $Ax + By + Cz = D$ is a plane.

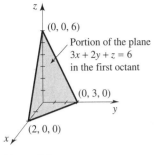

Figure 7.9

EXAMPLE 8   Graphing a Plane

Graph $3x + 2y + z = 6$.

SOLUTION ■ Since we know the graph is a plane, we can determine the plane by locating the intercepts. Setting $y = z = 0$ yields $x = 2$. Thus, the point $(2, 0, 0)$ is on the plane. Also setting $x = z = 0$ yields $y = 3$. Thus, $(0, 3, 0)$ is on the plane. Finally, setting $x = y = 0$ yields $z = 6$. Thus, $(0, 0, 6)$ is on the plane. From this we can graph the plane (Fig. 7.9). ■

Level Curves

A topographic map conveys a description of a *three*-dimensional portion of the surface of the earth by using a *two*-dimension map. This is done by using **level curves**. If we intersect a surface given by the graph of $z = f(x, y)$ with a plane given by $z = z_0$, we obtain the equation $z_0 = f(x, y)$. The graph of this equation in the xy-plane is called a level curve for f (Fig. 7.10).

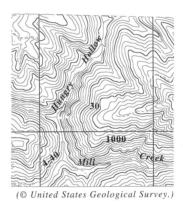

(© United States Geological Survey.)

Thus, in Figure 7.11, the curve with a 3000 next to it means that the surface above this curve has the constant height 3000. The next curve shown inside this one has the number 3500 next to it. This conveys the information that the surface is *rising* as we

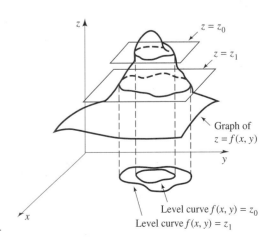

Figure 7.10
Some level curves for the given $z = f(x,y)$.

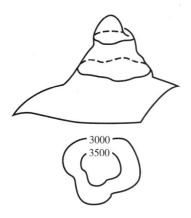

Figure 7.11
The curve with 3000 indicates the points at which the surface is 3000 units above the *xy*-plane.

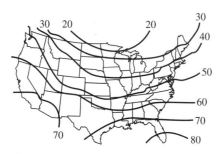

Figure 7.12
Isobars give level curves on which the temperature is the same.

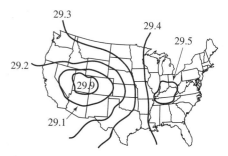

Figure 7.13
Levels curves on which the barometric pressure is the same.

move from the level curve $z_0 = 3000$ to the level curve $z_1 = 3500$. This indicates that we are moving up a hill, as shown in Figure 7.11.

Figure 7.12 gives a typical weather chart. Here the temperature is $T = T(x, y)$. These level curves are also referred to as **isobars**, that is, curves on which the temperature is the same.

Figure 7.13 shows a typical chart of the barometric pressure. On each level curve the barometric pressure is the same.

The following example shows how we can exploit level curves to determine how a surface looks.

EXAMPLE 9   Graphing Using Level Curves

Graph $z = 4 - x^2 - y^2$ by using level curves.

SOLUTION ■ To find the level curves we find the intersection of each horizontal plane $z = z_0$ with the surface and obtain

$$z_0 = 4 - x^2 - y^2$$
$$x^2 + y^2 = 4 - z_0$$

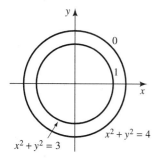

Figure 7.14
By setting $z = f(x,y) = 4 - x^2 - y^2 = 1$ we obtain the level curve $x^2 + y^2 = 3$. By setting $z = f(x,y) = 4 - x^2 - y^2 = 0$ we obtain the level curve $x^2 + y^2 = 4$.

Clearly we obtain the empty set if $z_0 > 4$. This conveys the information that no part of the surface extends above the plane $z = 4$. If $z_0 = 4$, we obtain $x^2 + y^2 = 0$, which is true only if $x = 0$ and $y = 0$. Thus, the level curve for $z_0 = 4$ consists of the single point $(0, 0)$. If now $z_0 < 4$, we obtain a circle $x^2 + y^2 = 4 - z_0$ in the *xy*-plane with center at $(0, 0)$ and radius $\sqrt{4 - z_0}$. For example, if $z_0 = 1$ we have the level curve shown in Figure 7.14 with a "1" next to it. This indicates that along this level curve the surface is always 1 unit above the *xy*-plane. The level curve is a circle centered at $(0, 0)$ with radius $\sqrt{3}$. The level curve for $z_0 = 0$ is also shown in

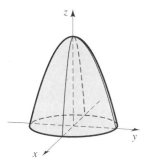

Figure 7.15
A graph of $z = 4 - x^2 - y^2$.

Figure 7.14 and is a circle centered at $(0, 0)$ with radius 2. This conveys the information that moving outward from $(0, 0)$ in any direction results in the height of the surface decreasing. Figure 7.15 then gives a three-dimensional picture of the graph. ∎

Continuity

We say that $f(x)$ is *continuous* at $x = a$ if $f(x)$ approaches $f(a)$ no matter how x approaches a. A similar definition is used for a function of two variables.

Continuity

We say that $f(x, y)$ is **continuous** at (a, b) if $f(x, y)$ approaches $f(a, b)$ no matter how (x, y) approaches (a, b).

Geometrically, a function $f(x)$ of a single variable is continuous on some interval if the curve that represents the graph of the function has no breaks, gaps, or holes. In the same way a function $f(x, y)$ of two variables is continuous on some region if the surface that represents the graph of the function has no breaks, gaps, or holes. Thus, the function whose graph is shown in Figure 7.16 is continuous everywhere in D, whereas the graph of the function shown in Figure 7.17 has a break and so is not continuous throughout D.

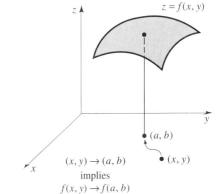

Figure 7.16
The function $f(x,y)$ is continuous at (a,b) if $f(x,y) \to f(a,b)$ no matter how $(x,y) \to (a,b)$.

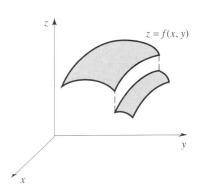

Figure 7.17
This function is not continuous because the graph has a break.

SELF-HELP EXERCISE SET 7.1

1. The box shown in the figure has a square base and open top. Find the surface area $S(x, y)$ as a function of x and y. Then find $S(5, 8)$.

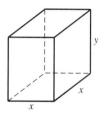

2. Plot the two points $(1, 2, -5)$ and $(0, 4, -3)$, and find the distance between them.

3. Use level curves to graph $z = x^2 + y^2 - 3$.

EXERCISE SET 7.1

In Exercises 1 through 10 evaluate the function at the indicated points.

1. $f(x, y) = 3x + 4y + 2$, $(1, 2), (0, 2), (-2, 4)$

2. $f(x, y) = 2x - 5y$, $(0, 0), (-2, 4), (2, -3)$

3. $f(x, y) = 2x^2 + y^2 - 4$, $(1, 2), (2, -3), (-1, -2)$

4. $f(x, y) = x^2 + x - y^2$, $(2, 3), (-1, 4), (-2, -3)$

5. $f(x, y) = \dfrac{x}{x + y}$, $(1, 2), (-1, 2), (-2, -4)$

6. $f(x, y) = \dfrac{y}{x + y}$, $(2, 1), (-1, 2), (-2, -1)$

7. $f(x, y) = \sqrt{8 - x^2 - y}$, $(1, 1), (1, -2), (-2, 0)$

8. $f(x, y) = \sqrt{9 + x^2 + 2y^2}$, $(0, 0), (1, 2), (-1, -3)$

9. $f(x, y) = \dfrac{1}{\sqrt{x - y}}$, $(2, 1), (5, 1), (-1, -5)$

10. $f(x, y) = \sqrt[3]{xy}$, $(8, 1), (9, 3), (-1, -1)$

In Exercises 11 through 16 find the domain of the given function.

11. $f(x, y) = x + 3y$ **12.** $f(x, y) = x - 3y$

13. $f(x, y) = \dfrac{x}{x + y}$ **14.** $f(x, y) = \dfrac{y}{x - 2y}$

15. $f(x, y) = \sqrt{16 - x^2 - y^2}$

16. $f(x, y) = \dfrac{1}{\sqrt{16 - x^2 - y^2}}$

In Exercises 17 and 18 find the distance between the given two points.

17. $(1, 2, 3)$ and $(0, 4, 5)$

18. $(3, -1, -2)$ and $(2, -3, -5)$

In Exercises 19 through 26 describe the surface.

19. $x + 2y + 3z = 12$ **20.** $3x - 2y + z = 6$

21. $2x + 5y + 10z = 20$ **22.** $y = 4$

23. $z = 3$ **24.** $x^2 + y^2 + z^2 = 25$

25. $x^2 + y^2 + z^2 = 36$

26. $(x - 1)^2 + (y - 2)^2 + z^2 = 4$

Exercises 27 through 32 give graphs of functions. Match the functions with the level curves shown in the figures labeled with letters from (a) to (f).

27. $f(x, y) = \dfrac{1}{1 + x^2 + y^2}$

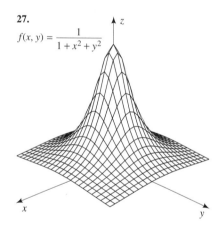

28.

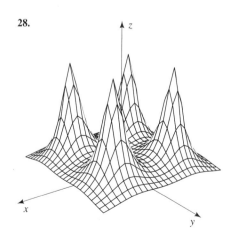

29.
$$f(x, y) = \frac{xy(x^2 - y^2)}{x^2 + y^2}$$

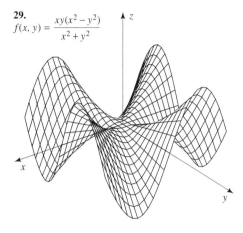

30.
$$f(x, y) = e^{-x^2} + e^{-y^2}$$

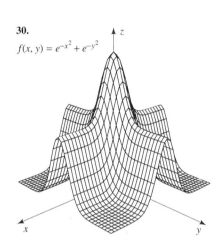

31.
$$f(x, y) = e^{-y}\cos x$$

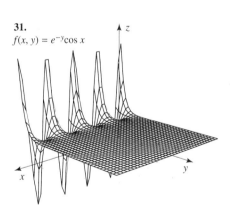

32.
$$f(x, y) = 25x^2 - x^4 - y^4$$

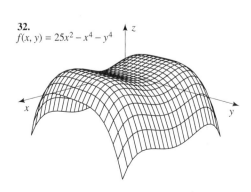

(a)

(b)

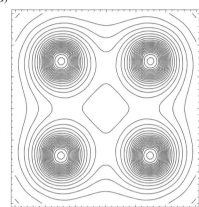

(c)

(d)

(e)

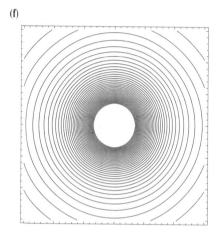

(f)

In Exercises 33 through 36 find the given level curves for the indicated functions, and describe the surface.

33. $z = f(x, y) = x^2 + y^2$, $z_0 = 0, z_0 = 1, z_0 = 4, z_0 = 9$

34. $z = f(x, y) = 1 + x^2 + y^2$, $z_0 = 1, z_0 = 2, z_0 = 5$, $z_0 = 10$

35. $z = f(x, y) = 1 - x^2 - y^2$, $z_0 = 1, z_0 = 0, z_0 = -3$, $z_0 = -8$

36. $z = f(x, y) = 2 - x^2 - y^2$, $z_0 = 2, z_0 = 1, z_0 = -2$, $z_0 = -7$

37. Find the surface area $S(x, y)$ of the box with square base and lid as a function of x and y. Find $S(3, 5)$.

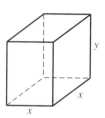

38. Find the surface area $S(r, h)$ of the cylinder as a function of r and h. Find $S(5, 10)$.

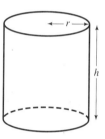

Applications

39. Demand. A company sells cameras and film to be used in the camera. Let x and y be the number of cameras sold per day and the number of rolls of film sold per day, respectively, and p and q the respective prices. Suppose $p = 1400 - 12x - y$ and $q = 802 - 3x - 0.5y$. Find the revenue function $R(x, y)$ and $R(100, 200)$.

40. Cost. Suppose the daily cost function $C(x, y)$ for the company in the previous exercise is $C(x, y) = 15,000 + 50x + 0.5y$. Find $C(100, 200)$.

41. Profits. Find the profit function $P(x, y)$ for the company discussed in the previous two exercises. Find $P(100, 200)$.

42. Interest. If $1000 is compounded continuously at an annual rate of r and for t years, write the amount $A(r, t)$ as a function of r and t. Find $A(0.10, 5)$.

43. Interest. If $1000 is compounding m times a year at an annual rate of 8% for t years, write the amount $A(m, t)$ as a function of m and t. Find $A(12, 5)$.

44. Cobb-Douglas Production Function. A production function is given by $f(K, L) = 10K^{0.25}L^{0.75}$. Find $f(16, 81)$.

45. Medicine. In the Fick method for directly measuring cardiac output, the cardiac output C is given by $C = 100x/y$. Here x is the carbon dioxide, in cubic centimeters per minute, released by the lungs, and y represents the change in carbon dioxide content of the blood leaving the lungs from when it entered, measured in cubic centimeters of carbon dioxide per 100 cm³ of blood per minute. **(a)** What is the domain of this function? **(b)** Find $C(6, 3)$. **(c)** Find $C(1, 2)$.

46. Biology. The transfer of energy by convection from an animal results from a temperature difference between the animal's surface temperature and the surrounding air temperature. The convection coefficient is given by $h(V, D) = \dfrac{kV^{1/3}}{D^{2/3}}$, where k is a constant, V is the wind velocity in centimeters per second, and D is the diameter of the animal's body in centimeters. **(a)** What is the domain of this function? **(b)** Find $h(8, 27)$. **(c)** Find $h(64, 8)$.

47. Medicine. The flow, Q in cubic centimeters per second, of blood from a large vessel to a small capillary has been described by $Q(d, P) = 0.25C\pi d^2\sqrt{P}$, where C is a constant, d is the diameter of the capillary, and P is the difference in pressure from the large vessel to the capillary. **(a)** What is the domain of this function? **(b)** Find $Q(4, 9)$ if $C = 2$. **(c)** Find $Q(2, 16)$ if $C = 2$.

48. Biology. The following formula relates the surface area A in square meters of a human to the weight w in kilograms and height h in meters: $A(w, h) = 2.02w^{0.425}h^{0.725}$. What is the domain of this function? Estimate your surface area.

Enrichment Exercises

49. Continuity. Let $f(x, y) = \dfrac{x^2 y}{x^4 + y^2}$ if $(x, y) \neq (0, 0)$ and $f(0, 0) = 0$. Let (x, y) approach $(0, 0)$ along the straight line $y = mx$, and see what happens to $f(x, y)$. In fact, show $\lim\limits_{x \to 0} f(x, mx) = 0$, no matter what m is. Can you conclude that f is continuous at $(0, 0)$? If you answered yes, be sure to do the next exercise.

50. Continuity. Let f be the function in Exercise 49. Now let (x, y) approach $(0, 0)$ along the curves $y = mx^2$ and see what happens to $f(x, y)$. That is, find $\lim\limits_{x \to 0} f(x, mx^2)$. Do you obtain the same limit for every value of m? What do you now conclude about the continuity of f at $(0, 0)$?

SOLUTIONS TO SELF-HELP EXERCISE SET 7.1

1. The area of the base is x^2, whereas the area of each of the four sides is xy. Thus, the total surface area is $S(x, y) = x^2 + 4xy$. Then $S(5, 8) = 5^2 + 4(5)(8) = 185$.

2. The two points $(1, 2, -5)$ and $(0, 4, -3)$ are plotted in the figure.

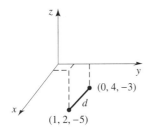

The distance between the 2 points is given by

$$d = \sqrt{(x_2 - x_1)^2 + (y_2 - y_1)^2 + (z_2 - z_1)^2}$$
$$= \sqrt{(0 - 1)^2 + (4 - 2)^2 + [(-3) - (-5)]^2}$$
$$= \sqrt{1 + 4 + 4} = 3$$

3. To find the level curves to $z = x^2 + y^2 - 3$, set $z = z_0$, $z_0 = x^2 + y^2 - 3$ or $x^2 + y^2 = z_0 + 3$. If $z_0 < -3$, this yields the empty set. If $z_0 = -3$, this gives one single point $(0, 0)$. If $z_0 > -3$, this gives a circle centered at $(0, 0)$ with radius $\sqrt{z_0 + 3}$. As z_0 increases, the radius increases. Thus, for example, if $z_0 = 1$, the level curve is a circle with radius 2 shown in the figure. If $z_0 = 6$, the level curve is a circle with radius 3 shown in the figure. Moving away from $(0, 0)$ results in z increasing.

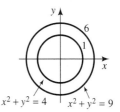

7.2 PARTIAL DERIVATIVES

- Partial Derivatives ■ Competitive and Complementary Demand Relations
- Second Partial Derivatives

APPLICATION: COMPETING COMMODITIES

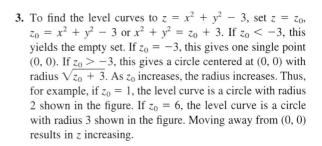

Suppose two commodities A and B have respective unit prices p and q and respective demand equations

$$x_A = 25 + \frac{q^3}{p^2} \quad \text{and} \quad x_B = 150 + \frac{2p^2}{q^3}$$

Can you determine if these two commodities compete with each other? For the answer see Example 4.

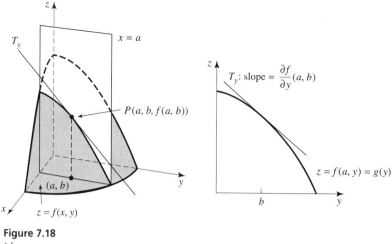

Figure 7.18
$$\frac{\partial f}{\partial y}(a,b) = g'(b).$$

Partial Derivatives

Suppose we are given a function $z = f(x, y)$ with domain D. Given a point $(a, b) \in D$, the plane $x = a$ that passes through this point also intersects the surface as indicated in Figure 7.18a. The intersection of the plane with the surface forms a curve $z = f(a, y) = g(y)$ shown in Figure 7.18a. In Figure 7.18b the plane $x = a$ has been removed for better examination. Notice that with $x = a$ this curve is a function of *one* variable, y. The slope of the tangent line to this curve is then given simply by

$$g'(b) = \lim_{\Delta y \to 0} \frac{f(a, b + \Delta y) - f(a, b)}{\Delta y}$$

We refer to this as the **partial derivative of f with respect to y** and denote it by $\frac{\partial f}{\partial y}(a, b)$. The important practical point is that this derivative is calculated as an ordinary derivative with the x-variable treated as a *constant*.

In the same way the plane $y = b$ intersects the surface in a curve shown in Figure

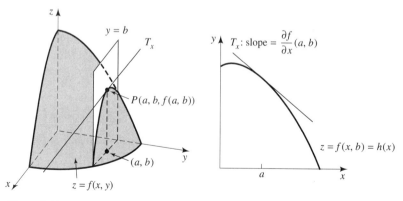

Figure 7.19
$$\frac{\partial f}{\partial x}(a,b) = h'(a).$$

7.19a. This curve is given by $z = f(x, b) = h(x)$, and the slope of the tangent line to this curve is given by

$$h'(a) = \lim_{\Delta x \to 0} \frac{f(a + \Delta x, b) - f(a, b)}{\Delta x}$$

We refer to this as the **partial derivative of f with respect to x** and denote it by $\frac{\partial f}{\partial x}(a, b)$. It is calculated as an ordinary derivative with the y-variable treated as a *constant*.

We then have the following definitions:

Partial Derivatives

If $z = f(x, y)$, then the **partial derivative of f with respect to x** is

$$\frac{\partial f}{\partial x}(x, y) = \lim_{\Delta x \to 0} \frac{f(x + \Delta x, y) - f(x, y)}{\Delta x}$$

and the **partial derivative of f with respect to y** is

$$\frac{\partial f}{\partial y}(x, y) = \lim_{\Delta y \to 0} \frac{f(x, y + \Delta y) - f(x, y)}{\Delta y}$$

if the limits exist.

Remark. If $z = f(x, y)$ we also denote $\frac{\partial f}{\partial x}(x, y)$ by $\frac{\partial z}{\partial x}$, f_x, or $f_x(x, y)$; and $\frac{\partial f}{\partial y}(x, y)$ by $\frac{\partial z}{\partial y}$, f_y or $f_y(x, y)$.

If we move in the positive x-direction from the point (x, y), the partial derivative with respect to x, $\frac{\partial f}{\partial x}(x, y)$, is the instantaneous rate of change of f with respect to x. If we move in the positive y-direction from the point (x, y), the partial derivative with respect to y, $\frac{\partial f}{\partial y}(x, y)$, is the instantaneous rate of change of f with respect to y.

If we think of the surface of our function as the surface of a mountain and the point $(x, y, f(x, y))$ as a point on the surface on which we are currently standing, then as we move in the positive x-direction, $\frac{\partial f}{\partial x}(x, y)$ measures our rate of ascent if this partial derivative is positive or the rate of descent if the partial derivative is negative (Fig. 7.20). Similarly, as we move in the positive y-direction, $\frac{\partial f}{\partial y}(x, y)$ measures our rate of ascent if this partial derivative is positive or the rate of descent if the partial derivative is negative (Fig. 7.21).

EXAMPLE 1   Finding Partial Derivatives

Let $f(x, y) = x^2 + 2x^3y^4 + \frac{1}{3}x^3 - y$. Find (a) $\partial f/\partial x$ at $(1, 0)$ and (b) $\partial f/\partial y$ at $(1, 0)$. Interpret these numbers.

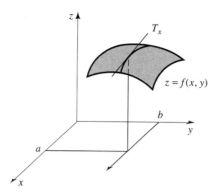

Figure 7.20

The slope of the tangent line T_x is $\dfrac{\partial f}{\partial x}$ (a,b).

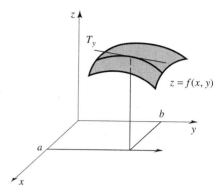

Figure 7.21

The slope of the tangent line T_y is $\dfrac{\partial f}{\partial y}$ (a,b).

S O L U T I O N S ■

(a) To find $\partial f/\partial x$ treat y as a *constant* and differentiate with respect to x and obtain

$$\frac{\partial f}{\partial x}(x,\ y) = \frac{\partial}{\partial x}(x^2) + \frac{\partial}{\partial x}(2y^4x^3) + \frac{\partial}{\partial x}\left(\frac{1}{3}x^3\right) - \frac{\partial}{\partial x}(y)$$

$$= \frac{\partial}{\partial x}(x^2) + 2y^4\frac{\partial}{\partial x}(x^3) + \frac{1}{3}\frac{\partial}{\partial x}(x^3) - \frac{\partial}{\partial x}(y)$$

$$= 2x + 2y^43x^2 + x^2 + 0$$

$$= 2x + 6x^2y^4 + x^2$$

Then

$$\frac{\partial f}{\partial x}(1,\ 0) = 2(1) + 6(1)^2(0)^4 + (1) = 3$$

With y kept at $y = 0$, the instantaneous rate of change of f with respect to x when $x = 1$ is 3. So as a point moves from $(1, 0)$ in the positive x-direction the function is increasing at 3 z-units per each x-unit (Fig. 7.22).

(b) To find $\partial f/\partial y$, treat x as a *constant* and differentiate with respect to y to obtain

$$\frac{\partial f}{\partial y}(x,\ y) = \frac{\partial}{\partial y}(x^2) + \frac{\partial}{\partial y}(2x^3y^4) + \frac{\partial}{\partial y}\left(\frac{1}{3}x^3\right) - \frac{\partial}{\partial y}(y)$$

$$= \frac{\partial}{\partial y}(x^2) + 2x^3\frac{\partial}{\partial y}(y^4) + \frac{\partial}{\partial y}\left(\frac{1}{3}x^3\right) - \frac{\partial}{\partial y}(y)$$

$$= 0 + 2x^34y^3 + 0 - 1$$

$$= 8x^3y^3 - 1$$

Then

$$\frac{\partial f}{\partial y}(1,\ 0) = -1$$

With x kept at $x = 1$, the instantaneous rate of change of f with respect to y when $y = 0$ is -1. So as a point moves from $(1, 0)$ in the positive y-direction, the function is decreasing at 1 z-unit for each y-unit (see Fig. 7.22). ■

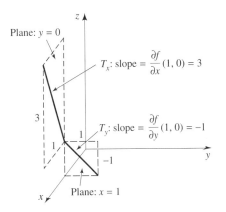

Figure 7.22

$\frac{\partial f}{\partial x}(1,0) = 3$ means that the instantaneous rate of change of f with respect to x is 3 as we move in

the positive x-direction from the point $(1,0)$. $\frac{\partial f}{\partial y}(1,0) = -1$ means that the instantaneous rate of

change of f with respect to y is -1 as we move in the positive y-direction from the point $(1,0)$.

EXPLORATION *1*

Partial Derivative at a Point

 Show that the following method is a correct alternative way of doing Example 1. It is helpful to keep Figures 7.18 and 7.19 in mind.

We have $f(x, y) = x^2 + 2x^3y^4 + \frac{1}{3}x^3 - y$. We first wish to find $\frac{\partial f}{\partial x}(1, 0)$.

To do this, first set $y = 0$, and obtain

$$f(x, 0) = x^2 + \frac{1}{3}x^3$$

$$\frac{d}{dx}f(x, 0) = 2x + x^2$$

Then setting $x = 1$ in the last expression gives $\frac{\partial f}{\partial x}(1, 0) = 2(1) + (1)^2 = 3$.

To find $\frac{\partial f}{\partial y}(1, 0)$ first set $x = 1$ and consider $f(1, y) = (1)^2 + 2(1)^3y^4 + \frac{1}{3}(1)^3 - y$. Then

$$f(1, y) = \frac{4}{3} + 2y^4 - y$$

$$\frac{d}{dy}f(1, y) = 8y^3 - 1$$

Then setting $y = 0$ in this last expression gives $\frac{\partial f}{\partial y}(1, 0) = 8(0)^3 - (1) = -1$.

Because of the result of Exploration 1 we can use our graphing calculator to find numerical derivatives. Although the method works on the most complex functions, we illustrate the method on the relatively easy function found in Example 1. If we want $\frac{\partial f}{\partial x}(1, 0)$, then we first find $h(x) = f(x, 0) = x^2 + \frac{1}{3}x^3$. We use the calculator to find the numerical derivative of $h(x) = f(x, 0)$ at $x = 1$. On one calculator we find nDer $h(1) = 3.00003333$, which is close enough to 3 to verify the answer.

To find $\dfrac{\partial f}{\partial y}$ $(1, 0)$ on our calculator first find $g(y) = f(1, y) = 2y^4 - y + \frac{4}{3}$. We then use the calculator to find the numerical derivative of $g(y) = f(1, y)$ at $y = 0$. On one calculator we find nDer $g(0) = -1$, exactly.

In the previous section Cobb-Douglas production functions $f(K, L) = AL^b K^{1-b}$ were introduced. Recall that $P = f(K, L)$ is the total output of a firm, L the cost of labor, and K is capital investment.

The partial derivative $\partial f/\partial K$ is called the **marginal productivity of capital**, and the partial derivative $\partial f/\partial L$ is called the **marginal productivity of labor**.

EXAMPLE 2   Finding Marginal Productivities

Suppose the production function of a firm is given by the Cobb-Douglas production function.

$$P = f(K, L) = 27L^{1/3}K^{2/3}$$

where P is the number of thousands of items produced each year, K is the capital investment per year in millions of dollars, and L is the labor cost per year in millions of dollars. Find the marginal productivities when $K = 8$ and $L = 27$, and interpret the result.

S O L U T I O N ■ The marginal productivity of capital and the marginal productivity of labor are given respectively by

$$\frac{\partial f}{\partial K} (K, L) = 18L^{1/3}K^{-1/3}$$

and

$$\frac{\partial f}{\partial L} (K, L) = 9L^{-2/3}K^{2/3}$$

Substituting $K = 8$ and $L = 27$ yields

$$\frac{\partial f}{\partial K} (8, 27) = 18(27)^{1/3}(8)^{-1/3} = 18(3)\left(\frac{1}{2}\right) = 27$$

and

$$\frac{\partial f}{\partial L} (8, 27) = 9(27)^{-2/3}(8)^{2/3} = 9\left(\frac{1}{9}\right)(4) = 4$$

Thus, with $L = 27$, the rate of change of production with respect to K at $K = 8$ is 27. With $K = 8$, the rate of change of production with respect to L at $L = 27$ is 4. We notice that production increases much more rapidly with an increase in capital spending, as compared with an increase in labor costs. Putting all of a (small) sum of money into capital investment therefore leads to a much greater increase in production than putting all of this money into increasing labor from this point. ■

EXPLORATION *2*

Partial Derivative at a Point

Verify the partial derivatives found in Example 2 by using the numerical derivative illustrated in the two paragraphs following Exploration 1.

Competitive and Complementary Demand Relations

We are already familiar with a demand relation that connects the demand for a product to the price of that product. But the demand for a commodity such as home heating oil depends not only on the price of oil but also on the price of natural gas, and vice versa.

Suppose then that there are two commodities A and B such that the price of one affects the price of the other. Let the unit price of the first commodity be p and the unit price of the second commodity be q. We assume that their respective demands x_A and x_B are functions of p and q. Thus,

$$x_A = f(p, q) \qquad \text{and} \qquad x_B = g(p, q)$$

We then have the following definitions that generalize the idea of marginal demand from one dimension:

$$\frac{\partial x_A}{\partial p} = \text{marginal demand for } A \text{ with respect to } p$$

$$\frac{\partial x_A}{\partial q} = \text{marginal demand for } A \text{ with respect to } q$$

$$\frac{\partial x_B}{\partial p} = \text{marginal demand for } B \text{ with respect to } p$$

$$\frac{\partial x_B}{\partial q} = \text{marginal demand for } B \text{ with respect to } q$$

If the price of the second commodity is held constant, then the demand x_A should decrease if p increases. But this implies* that

$$\frac{\partial x_A}{\partial p} < 0$$

In the same way if the price of the first commodity is held constant, then the demand x_B should decrease if q increases. But this implies* that

$$\frac{\partial x_B}{\partial q} < 0$$

Now suppose that the two commodities A and B are, respectively, home heating oil and natural gas. Then an increase in the price q of natural gas naturally leads to an increase in demand x_A for home heating oil and an increase in the price p of home heating oil leads to an increase in demand x_B for natural gas. This implies that

$$\frac{\partial x_A}{\partial q} > 0 \qquad \text{and} \qquad \frac{\partial x_B}{\partial p} > 0$$

In this case we say that the two commodities are **competitive**.

On the other hand suppose that the two commodities A and B are automobiles and gasoline. Then an increase in the price of one leads to a decrease in the demand for the other. This implies that

$$\frac{\partial x_A}{\partial q} < 0 \qquad \text{and} \qquad \frac{\partial x_B}{\partial p} < 0$$

In this case we say that the two commodities are **complementary**.

Figure 7.23 shows the graphs of typical demand functions when the two commodities are competitive and when they are complementary.

---

*Mathematically we could have the partial derivative equal zero, but in practice this rarely happens.

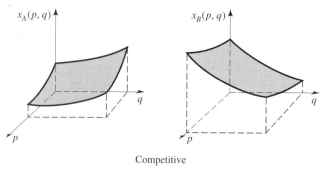

Figure 7.23

Competitive and Complementary Demand Relations

Suppose we have two commodities A and B with respective unit prices p and q and respective demand equations $x_A = f(p, q)$ and $x_B = g(p, q)$.
If

$$\frac{\partial x_A}{\partial q} > 0 \quad \text{and} \quad \frac{\partial x_B}{\partial p} > 0$$

then we say that the two commodities are **competitive**.
If

$$\frac{\partial x_A}{\partial q} < 0 \quad \text{and} \quad \frac{\partial x_B}{\partial p} < 0$$

then we say that the two commodities are **complementary**.

EXAMPLE 3   Competitive and Complementary Commodities

If

$$x_A = 50 - 2p^3 - q^2 \quad \text{and} \quad x_B = 75 - p^2 - 5q^3$$

determine whether A and B are competitive, complementary, or neither.

SOLUTION ■ First notice that these functions are reasonable demand functions since

$$\frac{\partial x_A}{\partial p} = -6p^2 < 0 \quad \text{and} \quad \frac{\partial x_B}{\partial q} = -15q^2 < 0$$

Since

$$\frac{\partial x_A}{\partial q} = -2q < 0 \quad \text{and} \quad \frac{\partial x_B}{\partial p} = -2p < 0$$

we see that the two commodities are complementary. ■

EXAMPLE 4 Competitive and Complementary Commodities

If

$$x_A = 25 + \frac{q^3}{p^2} \quad \text{and} \quad x_B = 150 + \frac{2p^2}{q^3}$$

determine whether A and B are competitive, complementary, or neither.

SOLUTION ■ First notice that these functions are reasonable demand functions since

$$\frac{\partial x_A}{\partial p} = -2p^{-3}q^3 < 0 \quad \text{and} \quad \frac{\partial x_B}{\partial q} = -6q^{-4}p^2 < 0$$

Since

$$\frac{\partial x_A}{\partial q} = \frac{3q^2}{p^2} > 0 \quad \text{and} \quad \frac{\partial x_B}{\partial p} = \frac{4p}{q^3} > 0$$

we see that the two commodities are competitive. ■

Second Partial Derivatives

We know that the second derivative of a function of one variable is defined to be the derivative of the derivative. We do something very similar for partial derivatives.

Second Partial Derivatives

Given a function $z = f(x, y)$, we define the following four second partial derivatives.

1. $\dfrac{\partial^2 f}{\partial x^2} = \dfrac{\partial}{\partial x}\left(\dfrac{\partial f}{\partial x}\right)$

2. $\dfrac{\partial^2 f}{\partial y\, \partial x} = \dfrac{\partial}{\partial y}\left(\dfrac{\partial f}{\partial x}\right)$

3. $\dfrac{\partial^2 f}{\partial y^2} = \dfrac{\partial}{\partial y}\left(\dfrac{\partial f}{\partial y}\right)$

4. $\dfrac{\partial^2 f}{\partial x\, \partial y} = \dfrac{\partial}{\partial x}\left(\dfrac{\partial f}{\partial y}\right)$

Remark. We also use the notation

$$\frac{\partial^2 f}{\partial x^2} = f_{xx} \qquad \frac{\partial^2 f}{\partial y^2} = f_{yy}$$

$$\frac{\partial^2 f}{\partial y\, \partial x} = (f_x)_y = f_{xy} \qquad \frac{\partial^2 f}{\partial x\, \partial y} = (f_y)_x = f_{yx}$$

EXAMPLE 5 Finding Second Partial Derivatives

Find all four second partial derivatives of $f(x, y) = e^{xy}$.

SOLUTION ■ We first find the two first partial derivatives.

$$f_x = ye^{xy} \qquad f_y = xe^{xy}$$

Then

$$f_{xx} = \frac{\partial}{\partial x}\left(f_x\right) = \frac{\partial}{\partial x}\left(ye^{xy}\right) = y^2 e^{xy}$$

$$f_{xy} = \frac{\partial}{\partial y}\left(f_x\right) = \frac{\partial}{\partial y}\left(ye^{xy}\right)$$

$$= y\frac{\partial}{\partial y}e^{xy} + e^{xy}\frac{\partial}{\partial y}y$$

$$= xye^{xy} + e^{xy}$$

$$f_{yy} = \frac{\partial}{\partial y}\left(f_y\right) = \frac{\partial}{\partial y}\left(xe^{xy}\right) = x^2 e^{xy}$$

$$f_{yx} = \frac{\partial}{\partial x}\left(f_y\right) = \frac{\partial}{\partial x}\left(xe^{xy}\right)$$

$$= x\frac{\partial}{\partial x}e^{xy} + e^{xy}\frac{\partial}{\partial x}x$$

$$= xye^{xy} + e^{xy} \quad \blacksquare$$

Remark. Notice that in the previous example $f_{xy} = f_{yx}$. This is no accident. This *always* happens if all second partial derivatives are continuous.

So far in this chapter we have only considered functions of *two* variables. We can have functions of any number of variables. For example, $w = f(x, y, z)$ is a function of three variables. The partial derivative of f with respect to one of the variables is obtained by taking the derivative with respect to that variable while treating the other variables as constants.

EXAMPLE 6 Finding Partial Derivatives of a Function of Three Variables

Find all three first partial derivatives of $w = f(x, y, z) = x^2 y^3 z^4$.

SOLUTION ■ We have

$$f_x = 2xy^3 z^4 \qquad f_y = 3x^2 y^2 z^4 \qquad f_z = 4x^2 y^3 z^3 \quad \blacksquare$$

SELF-HELP EXERCISE SET 7.2

1. If $f(x, y) = x^2 y^3$, find the two first partial derivatives, evaluate them at $(1, 1)$, and interpret this result.

2. Find all second partial derivatives of the function in the previous exercise.

EXERCISE SET 7.2

In Exercises 1 through 20 find both first partial derivatives. Then evaluate each partial derivative at the indicated point. Verify your answer to each partial derivative at the indicated point by using *n*Der and the method found in the two paragraphs following Exploration 1.

1. $f(x, y) = x^2 + y^2$, $(1, 3)$

2. $f(x, y) = x^2 - y^2$, $(1, 2)$

3. $f(x, y) = x^2 y - x^3 y^2 + 10$, $(1, 2)$

4. $f(x, y) = xy - x^2 y^3 + y^4$, $(1, -1)$

5. $f(x, y) = \sqrt{xy}$, $(1, 1)$

6. $f(x, y) = \sqrt{1 + xy}$, $(0, 1)$

7. $f(x, y) = \sqrt{1 + x^2 y^2}$, $(1, 0)$

8. $f(x, y) = \sqrt{1 + x^4 y^3}$, $(1, 1)$

9. $f(x, y) = e^{2x+3y}$, $(1, 1)$ 10. $f(x, y) = e^{x-y^2}$, $(0, 3)$

11. $f(x, y) = xye^{xy}$, $(1, 1)$ 12. $f(x, y) = xye^{-xy}$, $(1, 1)$

13. $f(x, y) = \ln(x + 2y)$, $(1, 0)$

14. $f(x, y) = \ln(x^2 - y)$, $(1, e)$

15. $f(x, y) = e^{xy} \ln x, \quad (1, 0)$

16. $f(x, y) = e^{xy} \ln y, \quad (0, 1)$

17. $f(x, y) = \dfrac{1}{xy}, \quad (1, 2)$ **18.** $f(x, y) = \dfrac{y}{x}, \quad (2, 1)$

19. $f(x, y) = \dfrac{x - y}{x^2 + y^2}, \quad (2, 1)$

20. $f(x, y) = \dfrac{x^2 - y^2}{x^2 + y^2}, \quad (2, 3)$

In Exercises 21 through 30 find all four of the second partial derivatives. In each case check to see if $f_{xy} = f_{yx}$.

21. $f(x, y) = x^2 y^4$ **22.** $f(x, y) = \dfrac{x^2}{y^3}$

23. $f(x, y) = e^{2x - 3y}$ **24.** $f(x, y) = e^{x - y}$

25. $f(x, y) = \sqrt{xy}$ **26.** $f(x, y) = \ln(x + 2y)$

27. $f(x, y) = x^2 e^y$ **28.** $f(x, y) = x \ln y$

29. $f(x, y) = \sqrt{x^2 + y^2}$ **30.** $f(x, y) = xye^{xy}$

In Exercises 31 through 38 find all three first partial derivatives.

31. $f(x, y, z) = xyz$ **32.** $f(x, y, z) = e^{xyz}$

33. $f(x, y, z) = \sqrt{x^2 + y^2 + z^2}$

34. $f(x, y, z) = \dfrac{x^2}{x^2 + y^2}$

35. $f(x, y, z) = e^{x + 2y + 3z}$

36. $f(x, y, z) = xe^y + ye^z$

37. $f(x, y, z) = \ln(x + 2y + 5z)$

38. $f(x, y, z) = xy \ln(y + 2z)$

In Exercises 39 through 44 determine if the demand relations are competitive or complementary.

39. $x_A = 100 - p - q, x_B = 200 - 2p - 4q$

40. $x_A = 100 - p + 2q, x_B = 50 + 2p - 4q$

41. $x_A = 20 - p^3 + q^3, x_B = 10 + p - q^3$

42. $x_A = e^{-pq}, x_B = e^{-p - q}$

43. $x_A = 10 - \dfrac{p}{q}, x_B = 10 + p - q$

44. $x_A = 30 - \dfrac{p}{q}, x_B = 40 + \dfrac{p}{q}$

Applications

45. Compounding. If $1000 is invested at an annual rate of r and compounded monthly, then the amount after t years is given by

$$A(r, t) = 1000\left(1 + \frac{r}{12}\right)^{12t}$$

Find $\partial A(r, t)/\partial r$. Interpret your answer.

46. Continuous Compounding. If $1000 is invested at an annual rate of r and compounded continuously, the amount after t years is given by $A(r, t) = 1000e^{rt}$. Find $\partial A(r, t)/\partial r$. Interpret your answer.

47. Fishery. Let N denote the number of prey in a school of fish, r the detection range of the school by a predator, and c a constant that defines the average spacing of fish within the school. The volume of the region within which a predator can detect the school, called the *visual volume*, has been given by biologists as

$$V(r, N) = \frac{4\pi r^3}{3}\left(1 + \frac{c}{r}N^{1/3}\right)^3$$

Find **(a)** $\partial V(r, N)/\partial r$. **(b)** $\partial V(r, N)/\partial N$. Interpret your answers.

48. Biology. The transfer of energy by convection from an animal results from a temperature difference between the animal's surface temperature and the surrounding air temperature. The convection coefficient is given by

$$h(V, D) = \frac{kV^{1/3}}{D^{2/3}}$$

where k is a constant, V is the wind velocity in centimeters

per second, and D is the diameter of the animal's body in centimeters. Find the two first partial derivatives, and interpret your answers.

49. Biology. If p represents the total density of leaves in a tree per unit of ground area, the leaves are uniformly distributed among n layers, and r is the circular radius of the leaf, then the amount L of light penetrating all n layers is given approximately by

$$L(r, n, p) = \left(1 - \frac{p\pi r^2}{n}\right)^n$$

Find the first partial derivatives of L with respect to r and with respect to p, and state what each of these means.

50. Biology. The concentration C of oxygen at the surface of the roots of plants is given by

$$C(Q, r, R) = A + \frac{Qr^2}{2D}\ln\frac{r}{R}$$

where A and D are constants, Q is the oxygen consumption of the roots, r is the radius of the root, and R is the radius of the root plus the moisture film. Find the three first partial derivatives, and interpret your answers.

51. Forestry. Smith and colleagues [130] estimated that the leaf area of the lodgepole pine is

$$A = A(S, D) = 0.079S^{1.43}D^{-0.73}$$

where A is the leaf area, S the cross-sectional area at breast height, and D the distance from breast height to the center of the live crown. Find the two first partial derivatives of $A(S, D)$.

52. Engineering Production Function. An engineering production function in the natural gas transmission industry was given by Cullen [131] as

$$Q = Q(H, d, L) = 0.33 \frac{H^{0.27} d^{1.8}}{L^{0.36}}$$

where Q is the output in cubic feet of natural gas, H is the station horsepower, d is the inside diameter of the transmission line in inches, and L is the length of the pipe line. Find the three first partial derivatives of $Q(H, d, L)$.

Enrichment Exercises

53. Advertising. The economists Dorfman and Steiner [132] used the function $x = x(A, P)$ to describe the quantity demanded when advertising at a level of A and P is the price of a unit of advertising. The assumption is made that

$$\frac{\partial x}{\partial A} > 0 \quad \text{and} \quad \frac{\partial x}{\partial P} < 0$$

Explain why you think these two conditions are reasonable.

54. Product Liability. The economists Viscusi and Moore [133] used the function $C(s, L)$ to describe unit liability cost, where s is a measure of product safety and L measures

the percentage of the total liability to be paid. The standard assumption is that

$$\frac{\partial C}{\partial s} < 0 \quad \text{and} \quad \frac{\partial C}{\partial L} > 0$$

Explain why you think these two conditions are reasonable.

55. Let $z = f(x, y)$.

(a) Suppose $\frac{\partial f}{\partial x}(1, 2) = 3$. Interpret this partial derivative as an instantaneous rate of change.

(b) Suppose $\frac{\partial f}{\partial y}(1, 2) = -5$. Interpret this partial derivative as an instantaneous rate of change.

SOLUTIONS TO SELF-HELP EXERCISE SET 7.2

1. (a) $\frac{\partial f}{\partial x}(x, y) = 2xy^3$

$\frac{\partial f}{\partial x}(1, 1) = 2(1)(1)^3 = 2$

(b) $\frac{\partial f}{\partial y}(x, y) = 3x^2y^2$

$\frac{\partial f}{\partial y}(1, 1) = 3(1)^2(1)^2 = 3$

With y kept at $y = 1$, the instantaneous rate of change of f with respect to x when $x = 1$ is 2. Thus, as a point (x, y) moves in the positive x-direction, the function f is increasing at 2 z-units per each x-unit. Also with x kept at $x = 1$, the

instantaneous rate of change of f with respect to y when $y = 1$ is 3. Thus, as a point (x, y) moves in the positive y-direction, the function f is increasing at 3 z-units per each y-unit.

2. $\frac{\partial^2 f}{\partial x^2} = \frac{\partial}{\partial x}\left(\frac{\partial f}{\partial x}\right) = \frac{\partial}{\partial x}(2xy^3) = 2y^3$

$\frac{\partial^2 f}{\partial y \partial x} = \frac{\partial}{\partial y}\left(\frac{\partial f}{\partial x}\right) = \frac{\partial}{\partial y}(2xy^3) = 6xy^2$

$\frac{\partial^2 f}{\partial y^2} = \frac{\partial}{\partial y}\left(\frac{\partial f}{\partial y}\right) = \frac{\partial}{\partial y}(3x^2y^2) = 6x^2y$

$\frac{\partial^2 f}{\partial x \partial y} = \frac{\partial}{\partial x}\left(\frac{\partial f}{\partial y}\right) = \frac{\partial}{\partial x}(3x^2y^2) = 6xy^2$

7.3 EXTREMA OF FUNCTIONS OF TWO VARIABLES

■ Relative Extrema ■ Absolute Extrema

APPLICATION: FINDING AN ABSOLUTE MAXIMUM

A package mailed in this country must have length plus girth of no more than 108 inches. What are the dimensions of a rectangular package of greatest volume that can be mailed? See Example 4 for the answer.

Relative Extrema

We have already defined relative and absolute extrema for a function of *one* variable. In the search for these extrema, we noted the need to find all critical points and found both the first and second derivative useful in determining the extrema. In this section we give the corresponding definitions of relative extrema, absolute extrema, and critical point for a function of *two* variables and also give an appropriate second derivative test.

Relative Maximum and Relative Minimum

Suppose $z = f(x, y)$ is a function defined on some domain D. We say that f has a **relative maximum** at $(a, b) \in D$ if there exists a circle centered at (a, b) and entirely in D, such that

$$f(x, y) \leq f(a, b)$$

for all points (x, y) inside this circle.

   We say that f has a **relative minimum** at $(a, b) \in D$ if there exists a circle centered at (a, b) and entirely in D, such that

$$f(x, y) \geq f(a, b)$$

for all points (x, y) inside this circle.

   We say that the function $z = f(x, y)$ has a **relative extremum** at $(a, b) \in D$ if f has either a relative maximum or a relative minimum at (a, b).

   See Figure 7.24 for examples.

   For a typical domain D shown in Figure 7.25, we see that a point $(a, b) \in D$ has a circle centered at (a, b) and entirely in D if, and only if, the point (a, b) is not on the boundary of D. This is true in general and simply means that a relative extremum cannot occur at a boundary point of D. This is analogous to the one-variable case, in which

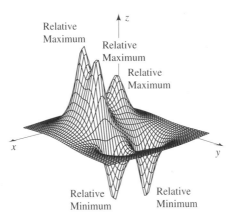

Figure 7.24
The relative extrema of $z = f(x,y)$.

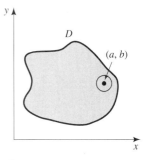

Figure 7.25
A point (a,b) is interior to D if there exists a circle centered at (a,b) and entirely inside D.

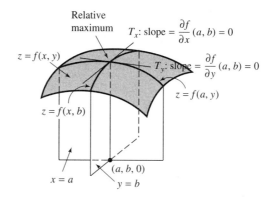

Figure 7.26

If $z = f(x,y)$ attains a relative maximum at (a,b), then $\dfrac{\partial f}{\partial x}$ $(a,b) = 0$ and $\dfrac{\partial f}{\partial y}$ $(a,b) = 0$.

the relative extrema were by definition excluded from occurring at the endpoints (boundary) of the interval $[a, b]$.

For the remainder of this chapter, we *always assume* that the given function $z = f(x, y)$ has *continuous* first and second partial derivatives in its domain of definition.

Suppose we then have such a function $z = f(x, y)$ and that this function has a relative maximum at (a, b). If we intersect the surface given by the graph of $z = f(x, y)$ with the plane $y = b$ as in Figure 7.26, then $f(a, b)$ is also a relative maximum of the curve $z = g(x) = f(x, b)$ of intersection. This implies from our knowledge of *one* variable that the slope m_x of the tangent line at this point must be *zero*. But the slope of this tangent line is the partial derivative with respect to x. Thus, $f_x(a, b) = 0$. In the same way we see from Figure 7.26 that at (a, b) the slope of the line tangent to the curve of intersection of the surface with the plane $x = a$ is also zero and is just $f_y(a, b)$. Thus, $f_y(a, b) = 0$. Similar remarks apply for the case of a relative minimum. We thus have the following important result:

Necessary Condition for Relative Extrema

If $z = f(x, y)$ is defined and both first partial derivatives exist for all values of (x, y) inside some circle about (a, b) and f assumes a relative extremum at (a, b) then

$$\frac{\partial f}{\partial x}(a, b) = 0 \qquad \text{and} \qquad \frac{\partial f}{\partial y}(a, b) = 0$$

A point not on the boundary of the domain of f where *both* first partial derivatives are zero is called a **critical point**. (Recall that we are only dealing with functions whose first partial derivatives exist in their domains of definition.)

Critical Point

Assume that both first partial derivatives of $z = f(x, y)$ exist in the domain of definition of f. **Critical points** are points (a, b), not on the boundary of the domain of f, for which

$$\frac{\partial f}{\partial x}(a, b) = 0 \qquad \text{and} \qquad \frac{\partial f}{\partial y}(a, b) = 0$$

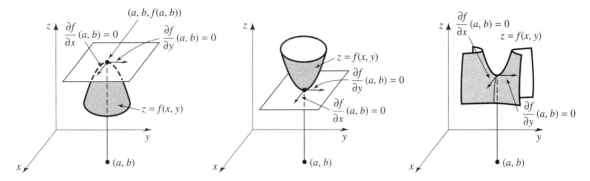

Figure 7.27
Each function has a critical point at (a,b). Function (a) has a relative maximum at (a,b), function (b) a relative minimum, and function (c) neither a relative maximum nor a relative minimum.

In Figure 7.27 we see the three possibilities that can happen at a critical point. We can have (a) a relative maximum, (b) a relative minimum, or (c) a **saddle point**. As Figure 7.27 indicates, at a saddle point both partial derivatives are zero, but the function does not attain a relative maximum or a relative minimum.

When dealing with a function of one variable, we know from the second derivative test that if the first derivative is zero at a point and the second derivative is positive at the same point, then the function has a relative minimum, with a corresponding result if the second derivative is negative. One might then suspect that if at a critical point for a function of *two* variables *all* the second partial derivatives were positive the function would have a relative minimum. *This is not true, in general.* The following example illustrates this:

EXAMPLE 1   Function with Positive Second Derivatives but No Extrema

For the function $f(x, y) = x^2 + 3xy + y^2$ show that

$$f_x(0, 0) = 0, \quad f_y(0, 0) = 0, \quad f_{xx}(0, 0) > 0, \quad f_{yy}(0, 0) > 0, \quad f_{xy}(0, 0) > 0$$

but f assumes no relative extrema at $(0, 0)$.

SOLUTION   ■   Since $f_x(x, y) = 2x + 3y$, $f_x(0, 0) = 0$, and since $f_y(x, y) = 3x + 2y$, $f_y(0, 0) = 0$. Also $f_{xx}(x, y) = 2 > 0$, $f_{yy}(x, y) = 2 > 0$, and $f_{xy}(x, y) = 3 > 0$. If now we move along the line $y = x$, we see that

$$f(x, x) = x^2 + 3xx + x^2 = 5x^2$$

so that f increases as we move away from $(0, 0)$ along this line. But along the line $y = -x$,

$$f(x, -x) = x^2 - 3xx + x^2 = -x^2$$

so that f decreases as we move away from $(0, 0)$ along this line.

In summary, moving away from the origin in one direction we ascend, but moving away from the origin in a different direction we descend. Thus, we have neither a minimum nor a maximum.   ■

The second derivative test for functions of more than one variable is therefore more complicated. We now state this test.

Second Derivative Test for Functions of Two Variables

For $z = f(x, y)$ assume that f_{xx}, f_{yy}, and f_{xy} all exist for every point inside a circle centered at (a, b) and that (a, b) is a critical point, that is,

$$f_x(a, b) = 0 \qquad \text{and} \qquad f_y(a, b) = 0$$

Define the number $\Delta(a, b)$ by

$$\Delta(a, b) = f_{xx}(a, b)f_{yy}(a, b) - [f_{xy}(a, b)]^2$$

Then

1. $\Delta(a, b) > 0$ and $f_{xx}(a, b) < 0$ implies that f has a relative maximum at (a, b).
2. $\Delta(a, b) > 0$ and $f_{xx}(a, b) > 0$ implies that f has a relative minimum at (a, b).
3. $\Delta(a, b) < 0$ implies that (a, b) is a saddle point, that is, f has neither a relative minimum nor a relative maximum at (a, b).
4. If $\Delta(a, b) = 0$, f can have a relative maximum, a relative minimum, or a saddle point.

Figure 7.28 summarizes this test.

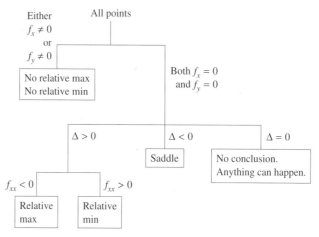

Figure 7.28

EXAMPLE 2   Using the Second Derivative Test

Find the relative extrema of

$$f(x, y) = -x^2 - y^2 + 2x + 4y + 5$$

SOLUTION ■ We first locate all critical points by finding both first partial derivatives and setting them both equal to zero. Thus,

$$f_x = -2x + 2 = 0 \qquad f_y = -2y + 4 = 0$$

The first equation implies that $x = 1$, and the second equation implies that $y = 2$; these are the only solutions, and thus $(1, 2)$ is the only critical point. We note that

$$f_{xx} = -2, \qquad f_{yy} = -2, \qquad f_{xy} = 0$$

Thus,

$$\Delta(1, 2) = (-2)(-2) - (0)^2 = 4$$

Since $\Delta(1, 2) > 0$ and $f_{xx}(1, 2) < 0$, the second derivative test indicates that f has a relative maximum at $(1, 2)$. ■

EXAMPLE 3 Using the Second Derivative Test

Find the relative extrema of

$$f(x, y) = x^2 - 2xy + \frac{1}{3}y^3 - 3y$$

SOLUTION ■ We first locate all critical points by finding both first partial derivatives and setting them both equal to zero. Thus,

$$f_x = 2x - 2y = 0 \qquad f_y = -2x + y^2 - 3 = 0$$

The first equation implies that $y = x$. Substituting this into the second equation yields $-2x + x^2 - 3 = 0$, or

$$0 = x^2 - 2x - 3 = (x - 3)(x + 1)$$

This has two solutions $x = -1$ and $x = 3$. Since $x = y$, this then indicates that the critical points are $(-1, -1)$ and $(3, 3)$. Notice that

$$f_{xx} = 2, \qquad f_{yy} = 2y, \qquad f_{xy} = -2$$

Thus, for the first critical point $(-1, -1)$, $f_{yy}(-1, -1) = -2$ and

$$\Delta(-1, -1) = (2)(-2) - (-2)^2 = -8$$

Since $\Delta(-1, -1) < 0$, the second derivative test indicates that f has a saddle point at $(-1, -1)$.

For the second critical point $(3, 3)$, $f_{yy}(3, 3) = 6$ and

$$\Delta(3, 3) = (2)(6) - (-2)^2 = 8$$

Since $\Delta(3, 3) > 0$ and $f_{xx}(3, 3) > 0$, the second derivative test indicates that f has a relative minimum at $(3, 3)$. ■

Absolute Extrema

We have the following definitions for functions of two variables:

Absolute Maximum and Absolute Minimum

We say that f has an **absolute maximum** on D at $(a, b) \in D$ if for all $(x, y) \in D$

$$f(x, y) \le f(a, b)$$

We say that f has an **absolute minimum** on D at $(a, b) \in D$ if for all $(x, y) \in D$

$$f(x, y) \ge f(a, b)$$

Also we say that f has an **absolute extremum** on D at $(a, b) \in D$ if f has either an absolute maximum or an absolute minimum on D at (a, b).

When we have a differentiable function of one variable $y = f(x)$, we know that a point at which the function attains an absolute extremum on a *bounded and closed* interval $[a, b]$ can occur at the critical points on the open interval (a, b) or *can occur at the endpoints* where the derivative need not be zero. A similar situation exists for functions of two variables.

We call the interval $[a, b]$ *closed* since it includes its endpoints. In the same way we say that a region D in the xy-plane is **closed** if it includes its boundary.

We know that a continuous function of one variable on a closed and bounded interval $[a, b]$ assumes its absolute maximum and its absolute minimum on the interval. We have a similar theorem for functions of two variables.

Existence of Absolute Extrema on a Closed and Bounded Region

> If $z = f(x, y)$ is continuous on the closed and bounded region D, then f assumes its absolute extrema in D, that is, there exist two points $(a, b) \in D$ and $(A, B) \in D$ such that
>
> $$f(a, b) \leq f(x, y) \leq f(A, B) \quad \text{for all } (x, y) \in D$$

If (x_c, y_c) is a critical point of $f(x, y)$, then it is a candidate for a relative extremum and therefore a candidate for an absolute extremum. But also *every point* on the boundary of D is a candidate for an absolute extremum.

EXAMPLE 4   Finding an Absolute Maximum

A package mailed in this country must have length plus girth of no more than 108 inches. Find the dimensions of a rectangular package of greatest volume that can be mailed.

SOLUTION ■ Let x, y, and z represent the dimensions of the package, with z denoting the length (Fig. 7.29). We wish to find the maximum volume that is given by $V = xyz$. As Figure 7.29 indicates, the girth is $2x + 2y$ so that the condition that the girth plus the length be no more than 108 inches can be written as

$$\text{Girth} + \text{length} = 2x + 2y + z \leq 108$$

Naturally, $x \geq 0$, $y \geq 0$, and $z \geq 0$. Since we wish the largest possible volume, we can assume that

$$2x + 2y + z = 108$$

Solving this equation for z and substituting into the equation for V yields

$$V = xy(108 - 2x - 2y)$$
$$= 108xy - 2x^2y - 2xy^2$$

Since $2x + 2y + z = 108$ and $z \geq 0$, we must have

$$2x + 2y \leq 108$$

Thus, we are searching for the absolute maximum of $V(x, y) = 108xy - 2x^2y - 2xy^2$ on the region D shown in Figure 7.30.

We first search for critical points inside D. Setting the two partial derivatives equal to zero, we obtain

$$0 = V_x = 108y - 4xy - 2y^2 = y(108 - 4x - 2y)$$
$$0 = V_y = 108x - 2x^2 - 4xy = x(108 - 2x - 4y)$$

The first equation implies that $y = 0$, or $108 - 4x - 2y = 0$. Since $y = 0$ is on the boundary of D, we discard this solution. The second equation implies that $x = 0$, or $108 - 2x - 4y = 0$. Since $x = 0$ is on the boundary of D, we discard this solution. Thus, the only critical point in the interior of D is given by the solution to

$$108 - 4x - 2y = 0 \quad \text{and} \quad 108 - 2x - 4y = 0$$

Subtracting the second equation from twice the first yields $108 - 6x = 0$, or $x = 18$. Substituting this back into one of the two equations then yields $y = 18$. Thus, $(18, 18)$ is the only critical point of $V(x, y)$ inside the region D, and so it is certainly a candidate for the absolute maximum.

But all points of the boundary of D are also candidates for an absolute maximum. We notice, however, that on the boundary of D either $x = 0$ or $y = 0$ or $z = 0$, and all of these give $V = 0$, that is, $V = 0$ on the boundary of D. Thus, V does not assume its absolute maximum on the boundary of D.

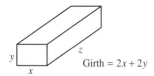

Girth = $2x + 2y$

Figure 7.29

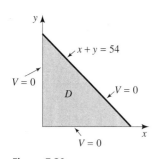

$x + y = 54$

$V = 0$

D

$V = 0$

$V = 0$

Figure 7.30

We conclude that V assumes its absolute maximum on D at (18, 18) *without even checking* to see if $V(18, 18)$ is a relative maximum, since the critical point (18, 18) is now the only possible place in D where V could assume its absolute maximum. Finally, when $x = 18$ and $y = 18$ are substituted into the equation $2x + 2y + z = 108$, we obtain $z = 36$ for the final dimension of the box.

Let us double-check our work to see if $V(18, 18)$ is indeed a relative maximum. We have

$$V_{xx} = -4y, \qquad V_{yy} = -4x, \qquad V_{xy} = 108 - 4x - 4y$$

and

$$V_{xx}(18, 18) = -72, \qquad V_{yy}(18, 18) = -72, \qquad V_{xy}(18, 18) = -36$$

Thus,

$$\Delta(18, 18) = (-72)(-72) - (-36)^2 > 0$$

and $V_{xx}(18, 18) < 0$, and so by the second derivative test V has a relative maximum at this point. ∎

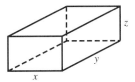

Figure 7.31

EXAMPLE 5   Finding an Absolute Minimum

Find the dimensions of the rectangular box of volume 8 ft^3 that minimizes the surface area.

SOLUTION ∎ If as in Figure 7.31 the dimensions of the box are given by x, y, and z, the volume is $V = xyz$, and also, according to the problem, $V = 8$. Thus, $8 = xyz$. The surface area S is given by $S = 2xy + 2xz + 2yz$. Since $8 = xyz$, we have $z = 8/xy$ and

$$S = 2xy + 2x\,\frac{8}{xy} + 2y\,\frac{8}{xy}$$

$$= 2xy + \frac{16}{y} + \frac{16}{x}$$

We thus wish to minimize S over the region D given by $x > 0$ and $y > 0$.

We find the critical points by taking both first partial derivatives and setting them equal to zero, obtaining

$$0 = S_x = 2y - \frac{16}{x^2} = \frac{2}{x^2}(x^2y - 8)$$

$$0 = S_y = 2x - \frac{16}{y^2} = \frac{2}{y^2}(xy^2 - 8)$$

The first of these equations gives $x^2y = 8$, and the second gives $xy^2 = 8$. Then $x^2y = xy^2$, and since $x > 0$ and $y > 0$, this implies that $x = y$. Then $8 = x^2y = x^3$, and so $x = 2$ and also $y = 2$. Thus, (2, 2) is the only critical point of $S(x, y)$ in the region D.

If we now look carefully at the formula $S(x, y) = 2xy + 16/y + 16/x$, we see that since $x > 0$ and $y > 0$, $S(x, y)$ becomes unbounded if either x or y become large without bound or if either x or y go to zero. Thus, we can visualize the surface $S(x, y)$ as extremely high near either axis and also for large x and y (Fig. 7.32). We conclude from this that $S(x, y)$ cannot have a minimum for any small value of either x or y or that $S(x, y)$ cannot have a minimum for any large value of either x or y.

Thus, the single critical point (2, 2) is the *only* candidate for a point at which S assumes an absolute minimum, and, therefore, S *must* assume its absolute minimum at this point. Substituting $x = 2$ and $y = 2$ into the equation $8 = xyz$ yields $z = 2$. Thus, the box must be a cube with each side equal to 2 feet.

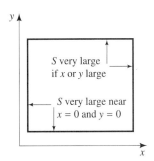

Figure 7.32

Let us double-check our work to see if $(2, 2)$ is indeed a relative minimum. We have

$$S_{xx} = \frac{32}{x^3}, \qquad S_{yy} = \frac{32}{y^3}, \qquad S_{xy} = 2$$

and

$$S_{xx}(2, 2) = 4, \qquad S_{yy}(2, 2) = 4, \qquad S_{xy}(2, 2) = 2$$

Thus,

$$\Delta(2, 2) = (4)(4) - (2)^2 = 12 > 0$$

and $S_{xx}(2, 2) > 0$, and so by the second derivative test V has a relative minimum at this point. ■

SELF-HELP EXERCISE SET 7.3

1. Determine the relative extrema, if any, for the function $f(x) = -x^3 - y^3 - 3xy$.

2. Suppose a firm has two products X and Y that compete with each other. Let the unit prices for the products X and Y be, respectively, p and q and the demand equations, respectively,

$$f(p, q) = 4 - p + q$$
$$h(p, q) = 8 + p - 2q$$

where $f(p, q)$ and $h(p, q)$ are the number of items of X and Y, respectively, produced and sold. Find where the revenue attains a relative maximum.

EXERCISE SET 7.3

In Exercises 1 through 20 find all critical points, and determine whether each point is a relative minimum, relative maximum, or a saddle point.

1. $f(x, y) = x^2 + xy + 2y^2 - 8x + 3y$
2. $f(x, y) = x^2 + 2y^2 - xy + 3y$
3. $f(x, y) = -x^2 + xy - y^2 + 3x + 8$
4. $f(x, y) = -5x^2 - xy - y^2 - 4x - 8y$
5. $f(x, y) = -x^2 + 2xy + 3y^2 - 8y$
6. $f(x, y) = 3x^2 - 2xy + y^2 + x$
7. $f(x, y) = -3x^2 + xy - y^2 - 4x - 3y$
8. $f(x, y) = -x^2 - xy - 3y^2 + 4x + 2y$
9. $f(x, y) = 2x^2 + xy + y^2 + 4$
10. $f(x, y) = 2x^2 - xy + y^2 - 8x + 7$
11. $f(x, y) = -x^2 + 2xy + 2y^2 - 4x - 8y$
12. $f(x, y) = -2x^2 + 2xy - y^2 + 2x + 4y$
13. $f(x, y) = x^3 + y^3 - 3xy - 1$
14. $f(x, y) = 2x^3 - 3x^2 - 12x + y^2 - 2y$
15. $f(x, y) = xy - x^3 - y^2$
16. $f(x, y) = -x^2 + y^3 + 6x - 12y$
17. $f(x, y) = \dfrac{3}{xy} - \dfrac{1}{x^2 y} - \dfrac{1}{xy^2}$

18. $f(x, y) = e^{xy}$
19. $f(x, y) = e^{x^2 - y^2}$
20. $f(x, y) = xy + \dfrac{x}{y^2}$

21. Show that $f(x, y) = x^4 y^4$ has a relative minimum at $(0, 0)$ and that $\Delta(0, 0) = 0$.

22. Show that $f(x, y) = -x^4 y^4$ has a relative maximum at $(0, 0)$ and that $\Delta(0, 0) = 0$.

23. Show that $f(x, y) = x^3 y^3$ has a saddle point at $(0, 0)$ and that $\Delta(0, 0) = 0$.

24. Show that if $\Delta(a, b) > 0$ and $f_{xx}(a, b) > 0$, then $f_{yy}(a, b) > 0$.

In Exercises 25 through 28 determine whether the given function is increasing or decreasing at the point (a) $(1, 0)$ as y increases (b) $(0, 1)$ as x increases (c) $(1, 5)$ as y increases (d) $(1, 5)$ as x increases.

25. $f(x, y) = x^2 + y^2 + xy + x + y$
26. $f(x, y) = -x^2 - y^2 + xy + x + y$
27. $f(x, y) = x^2 + y^2 - xy - x - y$
28. $f(x, y) = -x^2 - y^2 - xy - x - y$

Applications

In Exercises 29 through 30 do the following:

(a) Find the single critical point in the interior of the first quadrant.

(b) Show that the region over which the function is to be maximized is a triangular region and that the function is zero on the boundary of this triangle. Conclude from this alone that the function assumes its maximum at the critical point found in part (a).

(c) Check your work by using the second derivative test to verify that the critical point found in part (a) is a point at which the function attains a relative maximum.

29. **Construction.** An architect is attempting to fit a rectangular closet of maximal volume into a chopped-off corner of a building. She determines that the available space is as shown in the figure, where the constricting plane is given by $4x + 2y + z = 12$ with the dimensions given in feet. What are the dimensions of the closet of maximum volume?

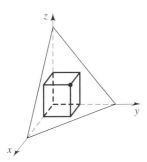

30. **Number Theory.** Find three positive numbers whose sum is 48 and whose product is as large as possible.

In Exercises 31 and 32 do the following:

(a) Find the single critical point in the first quadrant.

(b) Show that the function to be minimized is large for large x and y and also for small x and y. Conclude from this alone that the function assumes its minimum at the critical point found in part (a).

(c) Check your work by using the second derivative test to verify that the critical point found in part (a) is a point at which the function attains a relative minimum.

31. **Cost.** Find the dimensions of the cheapest rectangular box with open top and volume of 96 ft^3 with the cost per square feet of the base three times that of the cost of each side.

32. **Number Theory.** Find three positive numbers whose product is 64 and whose sum is as large as possible.

In Exercises 33 through 40 (except for Exercise 36), there is only one critical point in the domain under consideration. Locate this critical point and use the second derivative test. Assume that the relative extrema are the absolute extrema.

33. **Cost.** A firm has two separate plants producing the same item. Let x and y be the amounts produced in the respective plants with respective cost functions given by $C_1 = 100 - 12x + 3x^2/100$ and $C_2 = 100 - 10y + y^3/3000$. What allocation of production in the two plants minimizes the firm's cost?

34. **Profit.** A firm manufactures and sells two products, X and Y, that sell for \$15 and \$10 each, respectively. The cost of producing x units of X and y units of Y is

$$C(x, y) = 400 + 7x + 4y + 0.01(3x^2 + xy + 3y^2)$$

Find the values of x and y that maximize the firm's profit.

35. **Competitive Pricing.** Suppose a firm has two products X and Y that compete with each other. Let the unit price for X and Y be, respectively, p and q and the demand equations respectively $p(x, y) = 4 - x + 3y$ and $q(x, y) = 8 + x - 5y$, where x and y are the number of items of X and Y, respectively, produced and sold. Find where the revenue attains an absolute maximum for $x \geq 0$ and $y \geq 0$.

36. **Competitive Pricing.** Refer to Exercise 35. If the demand equations are now given by $p(x, y) = 4 - x + 2y$ and $q(x, y) = 8 + x - 2y$, show that the revenue does **not** attain an absolute maximum for $x \geq 0$ and $y \geq 0$. Do this by showing that along the line $y = \frac{3}{4}x$

$$R(x, y) = R\left(x, \frac{3}{4}x\right) = 10x + \frac{1}{8}x^2$$

which is not bounded.

37. **Competitive Pricing.** If the demand equations are the same as in Exercise 35 and the cost function is given by $C(x, y) = 2x + 4y + 10,000$, find the point at which the profit attains a maximum for $x \geq 0$ and $y \geq 0$.

38. **Geometry.** A rectangular box without a top is to have a surface area of 48 ft.^{2} What dimensions yield the maximum volume?

39. **Cost.** A firm has three separate plants producing the same product and has a contract to sell 1000 units of its product. Each plant has a different cost function. Let x, y, and z be the number of items produced at the three plants, and assume the respective cost functions are

$$C_1(x) = 100 + 0.03x^2, \qquad C_2(y) = 100 + 2y + 0.05y^2,$$

$$C_3(z) = 100 + 12z$$

Find the allocation of production in the three plants that minimizes the firm's total cost.

40. **Facility Location.** A company has three stores. The first is located 4 miles east and 2 miles north of the center of town, the second is located 2 miles east and 1 mile north of the center of town, and the third store is located 1 mile east and 6 miles south of the center of town. A warehouse is to be located convenient to these three stores. Find the location of the warehouse if the sum of the squares of the distances from each store to the warehouse is to be minimized.

Enrichment Exercises

41. Profit. A firm sells its product in the United States and in Japan. The cost function given in dollars is $C(z) = 100,000 + 80z + 0.05z^2$, where z is the amount produced. Let x be the amount sold in the United States and y the amount sold in Japan. Then $x \geq 0$ and $y \geq 0$. Considering the two different cultures, we assume the following two different demand equations (given in dollars)

$$p_U(x) = 500 - 0.05x \quad \text{and} \quad p_J(y) = 400 - 0.03y$$

Suppose that at most 3000 units of the product can be sold in each country. What are the values of x and y that maximize the firm's profit $P(x, y)$?

42. Profit. Suppose in Exercise 41 the production facilities are located in the United States and the transportation cost of moving the product to Japan is \$110 per item shipped. With this additional cost now find the values of x and y that maximize the firm's profit.

43. Profit. Suppose in Exercise 42 the transportation cost of moving the product to Japan is \$330 per item shipped. Now find the values of x and y that maximize the firm's profit.

44. Profit. Suppose in Exercise 41 the production facilities limit the total number of items produced to no more than 1500. Now find the values of x and y that maximize the firm's profit.

45. Agriculture. Paris [134] showed that an approximate relationship between yield y of corn and the amounts of nitrogen N and phosphorus P fertilizer used is given by

$$y = -0.075 + 0.584N + 0.664P$$
$$- 0.158N^2 - 0.18P^2 + 0.081PN$$

where y, N, and P are in appropriate units. Find the number of units of nitrogen and phosphorous that maximize the corn yield according to this model. Use your graphing calculator to find the critical point.

46. Agriculture. The same study as mentioned in the previous exercise indicated that another approximate relationship between yield y of corn and the amounts of nitrogen N and phosphorus P fertilizer used is also given by

$$y = -0.057 - 0.316N - 0.417P + 0.635N^{1/2}$$
$$+ 0.852P^{1/2} + 0.341(PN)^{1/2}$$

where y, N, and P are in appropriate units. Find the number of units of nitrogen and phosphorus that maximize the corn yield according to this model. Use your graphing calculator to find the critical point.

47. The Production Function. Suppose a production function is in the Cobb-Douglas form: $P = ax^b y^{1-b}$ and a cost function is $C = p_1 x + p_2 y$, where x is the number of items X made at a cost of p_1 and y is the number of items Y made at a cost of p_2. Find the cost as a function of P, that is, find $C = f(P)$. *Hint:* Solve for x in the production function, substitute this x in the cost function. The cost function then becomes a function of y. Differentiate with respect to y, set equal to zero, and solve for y. Use this to find x, substitute into the cost equation, and finally obtain

$$C = \left(\frac{1}{b} - 1\right)^b \left(\frac{p_1}{p_2}\right)^b \left(\frac{p_2}{1 - b}\right) \frac{P}{a}$$

SOLUTIONS TO SELF-HELP EXERCISE SET 7.3

1. Taking the first partial derivatives and setting them equal to zero gives

$$f_x = -3x^2 - 3y = 0 \qquad f_y = -3y^2 - 3x = 0$$

From the first of these, we obtain $y = -x^2$. Substituting this into the second then gives $3x^4 + 3x = 3x(x^3 + 1)$. Therefore, $x = 0$ or $x = -1$. The critical points are thus $(0, 0)$ and $(-1, -1)$. Also

$$f_{xx} = -6x, \qquad f_{yy} = -6y, \qquad f_{xy} = -3$$

At $(0, 0)$

$$\Delta(0, 0) = f_{xx}(0,0)f_{yy}(0, 0) - (f_{xy}(0, 0))^2 = -9 < 0$$

and therefore $(0, 0)$ is a saddle point.
At $(-1, -1)$

$$\Delta(-1, -1) = f_{xx}(-1, -1)f_{yy}(-1, -1) - (f_{xy}(-1, -1))^2$$
$$= (6)(6) - (-3)^2 = 27$$

and $f_{xx}(-1, -1) = 6 > 0$. Therefore, $f(-1, -1)$ is a relative minimum.

2. The revenue $R(p, q)$ is

$$R(p, q) = pf(p, q) + qh(p, q)$$
$$= 4p + 8q + 2pq - p^2 - 2q^2$$

Take

$$D = \{(p, q) : p \geq 0, q \geq 0\}$$

We then are seeking to maximize $R(p, q)$ over the region D. We first look for critical points interior to D by setting the first partial derivatives equal to zero.

$$0 = R_p = 4 + 2q - 2p$$
$$0 = R_q = 8 + 2p - 4q$$

This yields the two linear equations

$$p - q = 2$$
$$p - 2q = -4$$

Subtracting the two equations yields $q = 6$. This gives $p = 8$.

Using the second derivative test to classify this critical point, we have

$$R_{pp} = -2. \qquad R_{qq} = -4, \qquad R_{pq} = 2$$

Thus,

$$\Delta(8, 6) = (-2)(-4) - (2)^2 = 4 > 0$$

and $R_{pp}(8, 6) < 0$, which implies that the $R(p, q)$ has a relative maximum at $(8, 6)$. (It can be shown that this is also an absolute maximum.)

7.4 LAGRANGE MULTIPLIERS

■ The Method of Lagrange Multipliers   ■ Applications

(The Bettmann Archive)

Count Joseph Louis Lagrange 1736–1813

After Leonhard Euler, Lagrange is considered the greatest mathematician of the 18th century. Lagrange had wide interests and made fundamental contributions to the theory of equations, differential equations, and number theory. He was the first great mathematician to recognize the unsatisfactory state of the foundations of analysis and calculus and to introduce more rigor into these subjects.

APPLICATION: MAXIMIZING PRODUCTION

Suppose we have a Cobb-Douglas production function $f(x, y) = 15x^{1/3}y^{2/3}$, where x is the number of units of labor, y is the number of units of capital, and f is the number of units of a certain product that is produced. If each unit of labor costs \$200, and each unit of capital costs \$100, and the total expense for both is limited to \$7,500,000, what is the number of units of labor and capital needed to maximize production? The answer can be found in Example 3.

The Method of Lagrange Multipliers

There is no maximum volume of a rectangular box of dimensions x, y, and z, since $V(x, y, z) = xyz$ is obviously unbounded in the first octant. In Example 4 of the previous section, however, we were asked to maximize the volume of such a box subject to the **constraint** that the girth plus the length be 108, that is, subject to $g(x, y, z) = 2x + 2y + z - 108 = 0$. Subject to the given constraint, a maximum existed and we found it. Consider a second example. The minimum surface area of a rectangular box is clearly zero. In the preceding section, however, we were asked in Example 5 to minimize the surface area $S(x, y, z) = 2xy + 2xz + 2yz$ of a box subject to the *constraint* that the volume be 8, that is $g(x, y, z) = xyz - 8 = 0$. Subject to the given constraint, a *positive* minimum existed and we found it. Consider now a third example. The maximum of $f(x, y) = 4 - x^2 - y^2$ is clearly 4. The maximum of f subject to the constraint that $g(x, y) = y - 1 = 0$, however, is 3 and therefore different (Fig. 7.33). Notice that the maximum or minimum of a function can be very different from the maximum or minimum subject to a *constraint*.

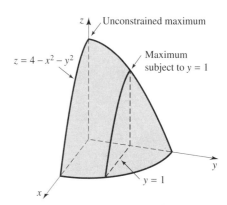

Figure 7.33
The maximum is different if subjected to the constraint $y = 1$.

The preceding are examples of a situation that arises often. We wish to maximize or minimize some function $f(x, y, z)$ subject to a constraint $g(x, y, z) = 0$, or we wish to maximize or minimize some function of two variables $f(x, y)$ subject to a constraint $g(x, y) = 0$. When solving such problems in the previous section, we merely solved the constraint equation $g(x, y, z) = 0$ for z and substituted the result in $f(x, y, z)$, obtaining a function of only *two* variables and then found the extrema. In general, this may be extremely difficult, if not impossible, to do. But even if this can be done, the resulting equation can sometimes be unnecessarily complicated. An alternative method, called the *method of Lagrange multipliers*, avoids these difficulties. We state the method for three independent variables.

Method of Lagrange Multipliers

Candidates (x_c, y_c, z_c) for a relative extrema of $f(x, y, z)$ subject to the constraint of $g(x, y, z) = 0$ can be found among the critical points (x_c, y_c, z_c, λ) of the auxiliary function

$$F(x, y, z, \lambda) = f(x, y, z) + \lambda g(x, y, z)$$

that is, among the solutions of the equations

$$F_x = 0, \qquad F_y = 0, \qquad F_z = 0, \qquad g = 0 \qquad [1]$$

It is important to realize that this method only yields candidates for the extrema. The test to determine whether the relative extrema are minima or maxima is too complicated to be given here. We noted in the previous section, however, that there is often only one candidate and that we can argue in some way that this candidate is the one we are seeking.

Applications

We first revisit a "fence problem" from an earlier chapter to gain some geometric insight into Lagrange multipliers.

EXAMPLE 1   Using Lagrange Multipliers

What are the dimensions of the rectangular pasture of largest area that can be enclosed by a fence of fixed perimeter P?

SOLUTION ■ Let x and y be the dimensions of the rectangular pasture. Then we are seeking to maximize the area $A(x, y) = xy$ subject to the constraint $2x + 2y = P$. Let us see if we can solve this problem geometrically. Figure 7.34 shows several level curves for the function $A(x, y)$. Notice from the figure that the maximum cannot occur on the level curve $A(x, y) = A_1$, since there are other level curves (such as $A(x, y) = A_2$) on which $A(x, y)$ is larger and that also intersect the constraint equation given by $g(x, y) = 2x + 2y - P = 0$. Notice that the maximum must occur at a point where a level curve is *tangent to the constraint curve*. Because of the symmetry of the curves, this must be where $x = y$.

Now let us solve using Lagrange multipliers. Form the function

$$F(x, y) = A(x, y) + \lambda g(x, y) = xy + \lambda(2x + 2y - P)$$

and solve the equations

$$F_x = y + 2\lambda = 0$$

$$F_y = x + 2\lambda = 0$$

$$g = 2x + 2y - P = 0$$

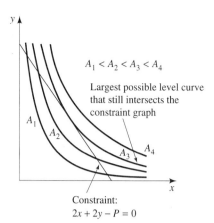

Figure 7.34
Level curves for $A(x,y) = xy$.

Constraint:
$2x + 2y - P = 0$

The first two equations yield $x = y$. Substituting this into the constraint equation then yields $x = y = P/4$. ■

It turns out in general (though not shown here) that the *critical points* of the auxiliary function F in the Lagrange multipliers method occur at points where *a level curve of f is tangent to the constraint curve g.* Because of the difficulty of visualizing the graphs involved, this is much less useful when the dimension of the problem is higher.

Let us redo Example 4 of the previous section using this new method.

EXAMPLE 2   Using Lagrange Multipliers

A package mailed in this country must have length plus girth of no more than 108 inches. Find the dimensions of a rectangular package of greatest volume that can be mailed.

SOLUTION ■ Let x, y, and z represent the dimensions of the package, with z denoting the length (Fig. 7.35). We wish to maximize the volume $V = xyz$. The condition that the girth plus the length be no more than 108 inches can be written as

$$\text{Girth} + \text{length} = 2x + 2y + z \leq 108$$

Naturally $x \geq 0$, $y \geq 0$, and $z \geq 0$. Since we wish the largest possible volume, we can assume that

$$2x + 2y + z = 108$$

Thus, our constraint equation is given by $g(x, y, z) = 2x + 2y + z - 108 = 0$.
We then form the auxiliary function $F(x, y, z, \lambda)$ given by

$$F(x, y, z, \lambda) = xyz + \lambda(2x + 2y + z - 108)$$

and solve the equations

$$F_x = yz + 2\lambda = 0$$

$$F_y = xz + 2\lambda = 0$$

$$F_z = xy + \lambda = 0$$

$$g(x, y, z) = 2x + 2y + z - 108 = 0$$

Solving the first three equations for $-\lambda$ and setting them equal yields

$$-\lambda = \frac{yz}{2} = \frac{xz}{2} = xy$$

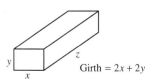

Girth $= 2x + 2y$

Figure 7.35

This immediately yields $x = y$ and then $z = 2y = 2x$. Substituting this into the constraint equation then yields

$$2x + 2x + 2x = 108$$

or $x = 18$, and then $y = x = 18$ and $z = 2x = 36$. We then must argue as we did in the previous section that this point must be where the function attains its maximum volume. ■

EXAMPLE 3   Using Lagrange Multipliers

Suppose we have a Cobb-Douglas production function

$$f(x, y) = 15x^{1/3}y^{2/3}$$

where x is the number of units of labor, y is the number of units of capital, and f is the number of units of a certain product that is produced. If each unit of labor costs $200, and each unit of capital costs $100, and the total expense for both is limited to $7,500,000, find the number of units of labor and capital needed to maximize production.

SOLUTION ■ We have that $200x + 100y = 7,500,000$. Thus, the constraint equation is

$$g(x, y) = 200x + 100y - 7,500,000 = 0$$

Form the auxiliary function $F(x, y, \lambda)$ given by

$$F(x, y, \lambda) = 15x^{1/3}y^{2/3} + \lambda(200x + 100y - 7,500,000)$$

and solve the equations

$$F_x = 5x^{-2/3}y^{2/3} + 200\lambda = 0$$
$$F_y = 10x^{1/3}y^{-1/3} + 100\lambda = 0$$
$$g(x, y) = 200x + 100y - 7,500,000 = 0$$

Solving the first two equations for $-\lambda$ and setting them equal yields

$$-\lambda = \frac{x^{-2/3}y^{2/3}}{40} = \frac{x^{1/3}y^{-1/3}}{10}$$

Multiplying each side of this equation by $x^{2/3}y^{1/3}$ yields $y = 4x$. Substituting this into the constraint equation yields

$$200x + 100(4x) = 7,500,000$$
$$x = 12,500$$

and $y = 50,000$. The relative maximum production is then

$$f(12,500, \ 50,000) = 15(12,500)^{1/3}(50,000)^{2/3} \approx 472,500$$

units. It turns out that this is also the maximum production. ■

Every plane has an equation of the form $ax + by + cz = d$. There are a number of ways of finding the perpendicular distance from a point to a plane. The following way uses Lagrange multipliers.

EXAMPLE 4   Using Lagrange Multipliers

Find the shortest distance from the point $(0, 0, 0)$ to the plane $2x + 2y + z = 9$.

S O L U T I O N ■ We minimize the square of the distance d to avoid working with square roots. Thus, we wish to minimize

$$d^2 = f(x, y, z) = x^2 + y^2 + z^2$$

subject to the constraint

$$g(x, y, z) = 2x + 2y + z - 9 = 0$$

We form the auxiliary function $F(x, y, z, \lambda)$ given by

$$F(x, y, z, \lambda) = x^2 + y^2 + z^2 + \lambda(2x + 2y + z - 9)$$

and solve the equations

$$F_x = 2x + 2\lambda = 0$$
$$F_y = 2y + 2\lambda = 0$$
$$F_z = 2z + \lambda = 0$$
$$g(x, y, z) = 2x + 2y + z - 9 = 0$$

Solving the first three equations for $-\lambda$ and setting them equal yields

$$-\lambda = x = y = 2z$$

This yields $y = x$ and $z = \frac{1}{2}x$. Now substituting back into the constraint equation yields

$$9 = 2x + 2y + z$$
$$= 2x + 2x + \frac{1}{2}x$$
$$= \frac{9}{2}x$$
$$x = 2$$

Then $y = 2$ and $z = 1$. Substituting this into f yields

$$d^2 = 2^2 + 2^2 + 1^2$$
$$= 9$$

Thus, $d = 3$. From geometric considerations, we know this must be the absolute minimum. ■

SELF-HELP EXERCISE SET 7.4

1. Find the dimensions of the rectangular box with lid of volume 8 ft³ that minimizes the surface area using Lagrange multipliers. (This was done in Example 5 of the previous section without using Lagrange multipliers.)

EXERCISE SET 7.4

In Exercises 1 through 32 solve using Lagrange multipliers.

1. Minimize $f(x, y) = 3x^2 + y^2$ subject to the constraint $x + y - 4 = 0$.

2. Minimize $f(x, y) = x^2 + 3y^2$ subject to the constraint $x - y + 4 = 0$.

3. Maximize $f(x, y) = -x^2 + xy - 4y^2$ subject to the constraint $x + y + 4 = 0$.

4. Maximize $f(x, y) = -5x^2 - xy - y^2$ subject to the constraint $x + y + 20 = 0$.

5. Minimize $f(x, y) = 2x^2 + xy + y^2 + x$ subject to the constraint $2x + y - 3 = 0$.

6. Minimize $f(x, y) = 2x^2 - xy + y^2 + 7y$ subject to the constraint $x - 2y + 1 = 0$.

7. Maximize $f(x, y) = -2x^2 + xy - y^2 + 3x + y$ subject to the constraint $2x + 3y + 11 = 0$.

8. Maximize $f(x, y) = -x^2 - xy - 3y^2 + x - y$ subject to the constraint $-x + y + 2 = 0$.

9. Maximize $f(x, y) = xy$ in the first quadrant subject to the constraint $x^2 + y^2 - 8 = 0$.

10. Minimize $f(x, y) = xy^2$ in the first quadrant subject to the constraint $x^2 + y^2 - 8 = 0$.

11. Maximize $f(x, y) = xy$ in the first quadrant subject to the constraint $x + y - 8 = 0$.

12. Maximize $f(x, y) = x^2 y$ in the first quadrant subject to the constraint $x + y - 9 = 0$.

13. Minimize $f(x, y) = x^2 + 2y^2 + 4z^2$ subject to the constraint $x + y + z - 7 = 0$.

14. Minimize $f(x, y) = x + 2y - 3z$ subject to the constraint $x^2 + 4y^2 - z = 0$.

15. Maximize $f(x, y) = xyz$ in the first quadrant subject to the constraint $x^2 + y^2 + 4z^2 - 12 = 0$.

16. Maximize $f(x, y) = xy + yz$ subject to the constraint $x^2 + y^2 + z^2 - 16 = 0$.

17. Find two positive numbers whose sum is 20 and whose product is as large as possible.

18. Find two positive numbers whose sum is 40 such that the sum of their squares is as small as possible.

19. Find three positive numbers whose sum is 36 and whose product is as large as possible.

20. Find three positive numbers whose product is 64 and whose sum is as small as possible.

Applications

21. **Construction.** Find the most economical proportions for a closed cylindrical can (cola can) that will hold 16π in.^{3}, if the cost of the top, bottom, and side are the same.

22. **Construction.** A manufacturer wishes to construct an open rectangular box by removing squares from the corners of a square piece of cardboard and bending up sides as shown in the figure. What are the dimensions of the box of maximum volume?

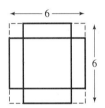

23. **Fencing.** A person wishes to enclose a rectangular parking lot, using a building as one boundary and adding fencing for the other boundaries. If 400 feet of fencing is available, find the dimensions of the largest parking lot that can be enclosed.

24. **Cost.** What are the dimensions of the rectangular field of 20,000 ft^2 that minimizes the cost of fencing if one side costs three times as much as the other three?

25. **Cost.** A closed rectangular box of volume 324 in.^{3} is to be made with a square base. If the material for the bottom costs twice as much per square foot as the material for the sides and top, find the dimensions of the box that minimize the cost of materials.

26. **Production.** Suppose a Cobb-Douglas production function is given by $f(x, y) = 100x^{0.75}y^{0.25}$, where x is the number of units of labor, y is the number of units of capital, and f is the number of units of a certain product that is produced. If each unit of labor costs \$100, and each unit of capital costs \$200, and the total expense for both is limited to \$1,000,000, find the number of units of labor and capital needed to maximize production.

27. **Production.** Suppose a Cobb-Douglas production function is given by $f(x, y) = 100x^{0.20}y^{0.80}$, where x is the number

of units of labor, y is the number of units of capital, and f is the number of units of a certain product that is produced. If each unit of labor costs \$100, and each unit of capital costs \$200, and the total expense for both is limited to \$1,000,000, find the number of units of labor and capital needed to maximize production.

28. **Geometry.** A rectangular box without top is to have a surface area of 48 ft^2. What dimensions yield the maximum volume?

29. **Construction.** An architect is attempting to fit a rectangular closet of maximal volume into a chopped-off corner of a building. She determines that the available space is as shown in the figure in which the constricting plane is given by $4x + 2y + z = 12$ with the dimensions given in feet. What are the dimensions and volume of the closet of maximum volume?

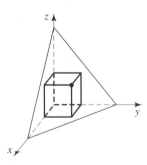

30. **Cost.** Find the dimensions of the cheapest rectangular box with open top and volume of 96 ft^3 with the cost per square foot of the base three times that of the cost of each side.

31. **Cost.** A firm has three separate plants producing the same product and has a contract to sell 1000 units of its product. Each plant has a different cost function. Let x, y, and z be the number of items produced at the three plants, and assume the respective cost functions are

$$C_1(x) = 100 + 0.03x^2, \qquad C_2(y) = 100 + 2y + 0.05y^2,$$
$$C_3(z) = 100 + 12z$$

Find the allocation of production in the three plants that minimizes the firm's total cost.

Enrichment Exercises

32. Show that the shortest distance from the point (x_0, y_0, z_0) to the plane $ax + by + cz = E$ is

$$d = \frac{|E - ax_0 - by_0 - cz_0|}{\sqrt{a^2 + b^2 + c^2}}$$

33. Suppose a production function is given by $P = f(x, y)$ and a cost function by $C = p_1 x + p_2 y$, where x is the number of items X made at a cost of p_1 and y is the number of items Y made at a cost of p_2. Use Lagrange multipliers to establish the important economic observation that if the firm wishes to produce P_1 items while minimizing cost, then

$$\frac{f_x}{p_1} = \frac{f_y}{p_2}$$

34. Suppose the production function in the previous exercise was a Cobb-Douglas production function of the form $P = ax^b y^{1-b}$ and the cost function was the same. Use the result in that exercise to show that $p_2 y = \dfrac{1 - b}{b} p_1 x$.

SOLUTIONS TO SELF-HELP EXERCISE SET 7.4

1. Let the dimensions of a box be given by the three numbers x, y, and z. Then the volume is $V = xyz$, and also, according to the problem, $V = 8$. Thus, $8 = xyz$. From this we obtain the constraint equation $g(x, y, z) = xyz - 8 = 0$. The surface area S is given by $S(x, y, z) = 2xy + 2xz + 2yz$. It is this surface area that we wish to minimize.

We then form the auxiliary function $F(x, y, z, \lambda)$ given by

$$F(x, y, z, \lambda) = 2xy + 2xz + 2yz + \lambda(xyz - 8)$$

and solve the equations

$$F_x = 2y + 2z + \lambda yz = 0$$
$$F_y = 2x + 2z + \lambda xz = 0$$
$$F_z = 2x + 2y + \lambda xy = 0$$
$$g(x, y, z) = xyz - 8 = 0$$

Solving the first three equations for $-\lambda$ and setting them equal yields

$$-\lambda = 2\frac{y + z}{yz} = 2\frac{x + z}{xz} = 2\frac{x + y}{xy}$$

Setting the first two fractions equal yields $xy + xz = xy + yz$, giving $x = y$, whereas the other equation yields $xy + yz = xz + yz$, which implies that $y = z$. Thus, $x = y = z$. Substituting this into the constraint equation yields $x^3 - 8 = 0$ or $x = 2$. Then $y = 2$ and $z = 2$. We then must argue as we did in the previous section that this point must be where the function attains its minimum surface.

7.5 METHOD OF LEAST SQUARES

■ The Method of Least Squares ■ Correlation

(David Eugene Smith Collection/Rare Book and Manuscript Library/ Columbia University)

Karl Gauss, the Missing Planet, and the Method of Least Squares

Karl Gauss (1777–1855), one of the greatest mathematicians of all time, first formulated the method of least squares. He did this in 1794 at the age of 16! In 1781, Sir William Herschel discovered the planet Uranus, bringing the number of planets known then to seven. At that time astronomers were certain that an eighth planet existed between Mars and Jupiter, but the search for the missing planet had proved fruitless. On January 1, 1801, Giuseppe Piazzi, director of the Palermo observatory, announced the discovery of a new planet Ceres in this location. (It was subsequently realized that Ceres was actually a large asteroid.) But unfortunately, a few weeks later he lost sight of it. Gauss took up the challenge of determining the orbit from the few recorded observations. He had in his possession one remarkable mathematical tool—the method of least squares, which he had not bothered to publish. Using this method, he predicted the orbit. Astronomers around the world were astonished to find Ceres exactly where he said it would be. This incident catapulted him to fame.

The Method of Least Squares

Let x be the number of items produced and sold and p the price of each item. Suppose x_1 items were sold at a price of p_1, and x_2 sold at a price of p_2. If we then *assume* that the demand equation is linear, then of course there is only *one* straight line through these two points, and we can easily calculate the equation $y = ax + b$ of this line.

But suppose that we have more than two data points. Suppose, as in Table 7.1, we have five points available. The p_i are the prices in dollars for a certain product, and the x_i are the corresponding demands for the product in the number of thousands of items sold per day.

TABLE 7.1

x_i	1	2	3	5	9
p_i	10	9	8	7	5

These are plotted in Figure 7.36, which is called a **scatter diagram**. If we examine the scatter diagram, we see clearly that the points do not lie on any straight line but seem to be scattered in a more or less linear fashion. Under such circumstances we might be justified in assuming that the demand equation was more or less a straight line. But what straight line? Any line that we draw will miss most of the points. We might then think to draw a line that is somehow closest to the data points. To actually follow such a procedure, we need to state exactly how we are to measure this closeness. In this section we measure this closeness in a manner that leads us to the **method of least squares**.

First notice that to be given a nonvertical straight line is the same as to be given two numbers a and b with the equation of the straight line given as $y = ax + b$. Suppose now we are given n data points $(x_1, y_1), (x_2, y_2), \ldots, (x_n, y_n)$, and a line $y = ax + b$. We then define $d_1 = y_1 - (ax_1 + b)$, and note from Figure 7.37 that $|d_1|$ is just the vertical distance from the first data point (x_1, y_1) to the line $y = ax + b$. Doing the same for all the data points, we then have

$$d_1 = y_1 - (ax_1 + b)$$

$$d_2 = y_2 - (ax_2 + b)$$

$$\vdots$$

$$d_n = y_n - (ax_n + b)$$

where $|d_2|$ is the vertical distance from the second data point (x_2, y_2) to the line $y = ax + b$, and so on.

Now if all the data points were on the line $y = ax + b$, then all the distances $|d_1|$, $|d_2|, \ldots, |d_n|$, would be zero. Unfortunately, this is rarely the case. We then use the sum of the squares of these distances

$$d = d_1^2 + d_2^2 + \cdots + d_n^2$$

as a measure of how close the set of data points is to the line $y = ax + b$. Notice that this number d is different for different straight lines—large if the straight line is far removed from the data points and small if the straight line passes close to all the data

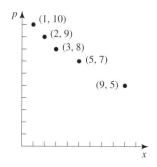

Figure 7.36
Discrete points on the demand curve.

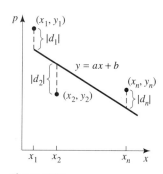

Figure 7.37
We measure the distance from each point to the line.

points. We then seek the line or, what is the same thing, the two numbers a and b, that make this sum of squares the least, hence, the name *least squares*.

Let a and b be independent variables and define the function of two variables f as

$$f(a, b) = d_1^2 + d_2^2 + \cdots + d_n^2$$
$$= (ax_1 + b - y_1)^2 + (ax_2 + b - y_2)^2 + \cdots + (ax_n + b - y_n)^2$$

If we are to minimize $f(a, b)$ we need to find the critical points. Then

$$\frac{\partial f(a, b)}{\partial a} = 2(ax_1 + b - y_1)x_1 + 2(ax_2 + b - y_2)x_2 + \cdots + 2(ax_n + b - y_n)x_n$$

$$= 2(x_1^2 + x_2^2 + \cdots + x_n^2)a + 2(x_1 + x_2 + \cdots + x_n)b$$
$$- 2(x_1 y_1 + x_2 y_2 + \cdots + x_n y_n)$$

Also

$$\frac{\partial f(a, b)}{\partial b} = 2(ax_1 + b - y_1) + 2(ax_2 + b - y_2) + \cdots + 2(ax_n + b - y_n)$$

$$= 2(x_1 + x_2 + \cdots + x_n)a + 2nb - 2(y_1 + y_2 + \cdots + y_n)$$

To find the critical points of $f(a, b)$ set these two first partial derivatives equal to zero and obtain

$$(x_1^2 + x_2^2 + \cdots x_n^2)a + (x_1 + x_2 + \cdots + x_n)b = x_1 y_1 + x_2 y_2 + \cdots + x_n y_n$$

$$(x_1 + x_2 + \cdots + x_n)a + nb = y_1 + y_2 + \cdots + y_n$$

Or in more compact form

$$\left(\sum_{i=1}^{n} x_i^2\right) a + \left(\sum_{i=1}^{n} x_i\right) b = \sum_{i=1}^{n} x_i y_i$$

$$\left(\sum_{i=1}^{n} x_i\right) a + nb = \sum_{i=1}^{n} y_i$$

If $x_1, x_2, \ldots, x_n$ are distinct, it can be shown that this system of two linear equations in a and b has one and only one solution and at this point the function $f(a, b)$ has a relative and absolute minimum.

Method of Least Squares

The line $y = ax + b$ closest to the n data points $(x_1, y_1), (x_2, y_2), \ldots, (x_n, y_n)$ can be found by solving the following two linear equations for a and b.

$$\left(\sum_{i=1}^{n} x_i^2\right) a + \left(\sum_{i=1}^{n} x_i\right) b = \sum_{i=1}^{n} x_i y_i$$

$$\left(\sum_{i=1}^{n} x_i\right) a + nb = \sum_{i=1}^{n} y_i$$

E X A M P L E 1   Using the Method of Least Squares to Find a Demand Function

(a) Find the best fitting line through the data points in Table 7.1 and thus find a linear demand function.

(b) Estimate the price if the demand is 6000.

S O L U T I O N   ■

(a) Create Table 7.2.

Our equations are

$$120a + 20b = 132$$
$$20a + 5b = 39$$

TABLE 7.2

x_i	y_i	x_i^2	$x_i y_i$
1	10	1	10
2	9	4	18
3	8	9	24
5	7	25	35
9	5	81	45
Sum 20	39	120	132

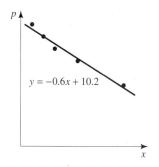

Figure 7.38
The line closest to the data points.

These two equations can be solved by eliminating one of the variables. For example, to eliminate the variable b, multiply the second equation by -4, and add the result to the first, giving

First equation	$120a + 20b = 132$
Multiply -4 times second equation	$-80a - 20b = -156$
Add	$40a = -24$

This yields $a = -0.6$. Since $a = -0.6$ and $20a + 5b = 39$,

$$b = \frac{39 - 20a}{5} = 10.2$$

Thus, the equation of the best fitting straight line that we are seeking is

$$y = -0.6x + 10.2$$

The graph is shown in Figure 7.38.

(b) The answer to the second question is

$$p = y = (-0.6)(6) + 10.2 = 6.6$$

that is, if 6000 items are to be sold, then the price should be $6.60. ■

Correlation

We have just seen how to determine a functional relationship between two variables. This process is called **regression analysis**. We now wish to determine the strength or degree of association between two variables, referred to as **correlation analysis**. The strength of association is measured by the **correlation coefficient** defined as

$$r = \frac{n \sum x_i y_i - \sum x_i \sum y_i}{\sqrt{[n \sum x_i^2 - (\sum x_i)^2][n \sum y_i^2 - (\sum y_i)^2]}}.$$

The value of this correlation coefficient ranges from $+1$ for two variables with perfect positive correlation to -1 for two variables with perfect negative correlation. See Figure 7.39 for examples.

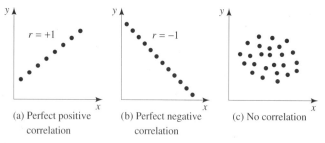

(a) Perfect positive correlation

(b) Perfect negative correlation

(c) No correlation

Figure 7.39

EXAMPLE 2   Correlation

Find the correlation coefficient for the data in Table 7.1.

SOLUTION ■ From Table 7.2 we have

$$r = \frac{n \sum x_i y_i - \sum x_i \sum y_i}{\sqrt{[n \sum x_i^2 - (\sum x_i)^2][n \sum y_i^2 - (\sum y_i)^2]}}$$

$$= \frac{5(132) - (20)(39)}{\sqrt{[5(120) - (20)^2][5(319) - (39)^2]}}$$

$$\approx -0.99$$

This indicates a high negative correlation, which can easily be seen by observing the original data in Figure 7.36. We conclude that there is a strong negative correlation between price and demand in this instance. Thus, we expect increases in prices to lead to decreases in demand. ∎

SELF-HELP EXERCISE SET 7.5

1. Using the method of least squares, find and graph the best fitting line through the three points (0, 0), (2, 2), (3, 2). Find the correlation coefficient.

EXERCISE SET 7.5

In Exercises 1 through 8 find the best fitting straight line to the given set of data, using the method of least squares. Graph this straight line on a scatter diagram. Find the correlation coefficient.

1. (0, 0), (1, 2), (2, 1)

2. (0, 1), (1, 2), (2, 2)

3. (0, 0), (1, 1), (2, 3), (3, 3)

4. (0, 0), (1, 2), (2, 2), (3, 0)

5. (1, 4), (2, 2), (3, 2), (4, 1)

6. (0, 0), (1, 1), (2, 2), (3, 4)

7. (0, 4), (1, 2), (2, 2), (3, 1), (4, 1)

8. (0, 0), (1, 2), (2, 1), (3, 2), (4, 4)

Applications

9. **Sales.** A firm has just introduced a new product and the sales during the first five weeks are given by the table. Find the best fitting straight line, using the method of least squares. What do you think sales will be next week?

Week Number (x_i)	1	2	3	4	5
Units Sold (y_i)	10	16	20	26	30

10. **Costs.** In the last four years a firm has produced varying amounts of an item at a certain plant and incurred varying total costs given by the table, in which the number of items is in thousands per year and the total cost is in millions of dollars per year. Find the best fitting straight line, using the method of least squares. What do you think costs per year would be next year if production was set at 2500 units?

Number Produced (x_i)	2	3	1	3
Total Cost (y_i)	2	4	2	3

11. **Sales.** A firm has undertaken four different radio advertising campaigns, and the sales in millions of dollars versus the amount spent on advertising in millions of dollars is given in the table. Find the best fitting straight line, using the method of least squares. What do you think sales would be in the next campaign if $3 million was spent on radio advertising?

Cost of Advertising (x_i)	1	2.5	3.5	4
Sales in Dollars (y_i)	3.5	5	6	6.5

12. **Insect Temperature.** An insect cannot control its body temperature; the body temperature depends on the temperature of the surrounding air. The data in the table were collected for a particular insect species. Find the best fitting straight line, using the method of least squares. Find the correlation coefficient. What does this imply about the strength of the relationship between the two variables?

Air Temperature (x_i)	18.2	23.0	25.6	30.4
Insect Temperature (y_i)	19.7	24.5	26.2	31.5

13. **Median Age of U.S. Population.** The table gives the median age of the U.S. population. Find the best fitting straight line, using the method of least squares. Find the correlation coefficient. What does this imply about the strength of the relationship between the two variables?

Year (x_i)	1820	1880	1950	1990
Median Age (y_i)	17	20	30	33

14. **Demand.** A firm has set different prices for its product in different cities and has obtained the data in the table relating the price of the product to the number of thousands sold per week at that price. Find the best fitting straight line, using the method of least squares. Graph this function. What do you think sales per week would be if the price were set at $7?

Price in Dollars (x_i)	8	9	10	11	12	13	14	15	16	17
Sales Volume in Thousands (y_i)	10	8	7	6	5	4.5	4	3.6	3	2

Enrichment Exercises

15. The selling expense and sales data from a number of stores for one product marketed by a company are shown in the table. The managers want to know if there is any diminishing marginal returns relationship between the selling expenses and the resulting sales generated by those expenses.

Store	Selling Expense (in thousands of dollars)	Sales (in millions of units)
A	35	4.5
B	5	1.0
C	15	2.0
D	25	4.0
E	60	5.5
F	20	3.0
G	45	5.0
H	10	1.5
I	40	5.0
J	55	5.5

Help them as follows.

(a) Using the table, find the least squares fit $y = mx + b$, where y is sales and x is selling expenses. Also find the correlation coefficient.

(b) Now consider the possibility that, with y and x defined as in part (a), $y = ax^c$. By taking logarithms, obtain

$\ln y = \ln a + c \ln x$. Let $Y = \ln y$ and $X = \ln x$. Then, $y = mx + b$ is equivalent to $Y = cX + \ln a$. Take the logarithms of the data in the table and find the least square fit for $Y = cX + \ln a$. Find this correlation coefficient.

(c) Based on the correlation coefficients found in parts (a) and (b), which model appears to fit the data better?

(d) How does this answer the manager's question concerning a diminishing marginal relationship?

(e) By looking carefully at the graph of the data, can you think of another possible mathematical model that may also fit the data? *Hint:* Review Section 4.4.

---

1. Create the table shown here. This leads to the following two linear equations:

$$13a + 5b = 10$$
$$5a + 3b = 4$$

	x_i	y_i	x_i^2	$x_i y_i$
	0	0	0	0
	2	2	4	4
	3	2	9	6
Sum	5	4	13	10

Multiplying the first equation by 3 and the second by -5 and adding then gives

Multiply first equation by 3	$39a + 15b = 30$
Multiply second equation by -5	$-25a - 15b = -20$
Add	$14a = 10$

or $a = \frac{5}{7}$. Then since $5a + 3b = 4$,

$$b = \frac{4}{3} - \frac{5}{3}a = \frac{4}{3} - \frac{5}{3}\frac{5}{7} = \frac{1}{7}$$

The best fitting line is then $y = \frac{5}{7}x + \frac{1}{7}$.

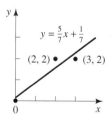

The correlation coefficient is

$$r = \frac{n \, \Sigma \, x_i y_i - \Sigma \, x_i \, \Sigma \, y_i}{\sqrt{[n \, \Sigma \, x_i^2 - (\Sigma \, x_i)^2][n \, \Sigma \, y_i^2 - (\Sigma \, y_i)^2]}}$$

$$= \frac{3(10) - (5)(4)}{\sqrt{[3(13) - (5)^2][3(8) - (4)^2]}}$$

$$\approx 0.94$$

This indicates a strong correlation.

7.6 TOTAL DIFFERENTIALS AND APPROXIMATIONS

■ Total Differentials and Approximations ■ Applications

APPLICATION: FINDING EASY-TO-CALCULATE APPROXIMATIONS

Suppose we have a Cobb-Douglas production function $f(x, y) = 81x^{1/3}y^{2/3}$, where x is the number of units of labor, y is the number of units of capital, and f is the number of units of a certain product that is produced. Find the approximation for the change in production given changes in labor Δx and changes in capital Δy as a *linear* function of Δx and Δy. For the answer see Example 2 and the discussion that follows.

Total Differentials and Approximations

Recall that for a function of *one* variable, $y = f(x)$, the actual change in y as x changes from a to $a + \Delta x$ is given by

$$\Delta y = f(a + \Delta x) - f(a)$$

Recall also that if Δx is small, this actual change in y can be approximated by the change along the tangent line, that is,

$$\Delta y \approx f'(a) \cdot \Delta x$$

Thus, for small Δx, the graph of $y = f(x)$ is approximately the same as the graph of the tangent line (Fig. 7.40). We shall now extend these ideas to functions of *two* variables.

We have already seen that $\dfrac{\partial f}{\partial x}(a, b)$ gives the slope of the line T_x tangent to the curve in which the plane $y = b$ intersects the surface $z = f(x, y)$ and that $\dfrac{\partial f}{\partial y}(a, b)$ gives the slope of the line T_y tangent to the curve in which the plane $x = a$ intersects the surface $z = f(x, y)$. The lines T_x and T_y determine a plane. This plane is called the **tangent plane** to the surface at the point $P = (a, b, z_0)$, where $z_0 = f(a, b)$ (Fig. 7.41).

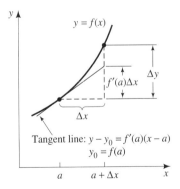

Figure 7.40
If $f'(a)$ exists, the graph of $y = f(x)$ is nearly linear near $x = a$.

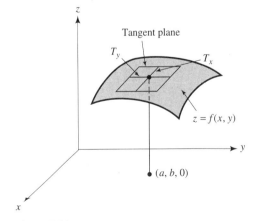

Figure 7.41
The two tangent lines T_x and T_y form the tangent plane.

It can be shown, but we do not do so here, that this tangent plane is given by the equation

$$z - z_0 = f_x(a, b)(x - a) + f_y(a, b)(y - b)$$

Notice carefully in the previous equation that the point (a, b, z_0) is on the *surface*, whereas the point (x, y, z) is on the *tangent plane*. If we let

$$\Delta x = x - a \quad \text{and} \quad \Delta y = y - b$$

and denote the corresponding change in z as $\Delta z = z - z_0$, then the equation of the tangent plane can be written as

$$\Delta z = f_x(a, b)\, \Delta x + f_y(a, b)\, \Delta y$$

This indicates how the changes Δx and Δy produce the change Δz *on the tangent plant*. Since the tangent plane *is tangent to* the surface, we fully expect from the geometry of the situation (see Fig. 7.41) that if the changes in Δx and Δy are small, the actual change in z, which is $f(a + \Delta x, b + \Delta y) - f(a, b) = \Delta z$, is approximately the change on the tangent plane. Thus, we have the following:

Tangent Plane Approximations

Suppose $f_x(a, b)$ and $f_y(a, b)$ exist.
If Δx and Δy are small, then

$$\Delta z = f(a + \Delta x, b + \Delta y) - f(a, b) \approx f_x(a, b) \cdot \Delta x + f_y(a, b) \cdot \Delta y$$

EXAMPLE 1   Using Differentials for an Approximation

Use differentials to approximate $\sqrt[3]{(2.06)^2 + (1.97)^2}$.

SOLUTION ■ Since 2.06 and 1.97 are very close to 2 and we can easily take the cube root of $2^2 + 2^2 = 8$, we set $a = 2 = b$ and $\Delta x = 0.06$ and $\Delta y = -0.03$. We take

$$f(x, y) = \sqrt[3]{x^2 + y^2}$$

Then $f(a, b) = f(2, 2) = \sqrt[3]{8} = 2$ and

$$f_x(x, y) = \frac{2x}{3(x^2 + y^2)^{2/3}} \qquad f_y(x, y) = \frac{2y}{3(x^2 + y^2)^{2/3}}$$

and

$$f_x(2, 2) = \frac{1}{3} \qquad f_y(2, 2) = \frac{1}{3}$$

Thus,

$$\begin{aligned}
\sqrt[3]{(2.06)^2 + (1.97)^2} &= f(a + \Delta x, b + \Delta y) \\
&\approx f(a, b) + \Delta z \\
&= f(a, b) + f_x(a, b) \cdot \Delta x + f_y(a, b) \cdot \Delta y \\
&= 2 + \frac{1}{3}\Delta x + \frac{1}{3}\Delta y \\
&= 2 + \frac{1}{3}(0.06) + \frac{1}{3}(-0.03) \\
&= 2.01
\end{aligned}$$

The actual value of $\sqrt[3]{(2.06)^2 + (1.97)^2}$ is 2.0103 to four decimal places. Thus, in this case, the error using the tangent plane approximation is very small. ■

Remark. Notice that for $x_0 = 2$ and $y_0 = 2$, we have the formula

$$\Delta z \approx \frac{1}{3} \cdot \Delta x + \frac{1}{3} \cdot \Delta y$$

which can be used to give quick estimates of Δz for a variety of values of Δx and Δy.

More generally for any (x, y) we have

$$\Delta z \approx \frac{2x}{3(x^2 + y^2)^{2/3}} \cdot \Delta x + \frac{2y}{3(x^2 + y^2)^{2/3}} \cdot \Delta y$$

where $\Delta z = f(x + \Delta x, y + \Delta y) - f(x, y)$.

Applications

We now give a number of applications.

EXAMPLE 2   Using Differentials for an Approximation

Suppose we have a Cobb-Douglas production function

$$f(x, y) = 81x^{1/3}y^{2/3}$$

where x is the number of units of labor, y is the number of units of capital, and f is the number of units of a certain product that is produced. Using the tangent plane, find an approximation for the change in production as x changes from 27 to 29 and as y changes from 125 to 122.

SOLUTION ■ We take $a = 27$ and $\Delta x = 2$. Also $b = 125$ and $\Delta y = -3$. We first need to calculate the first partial derivatives. Thus

$$f_x(x, y) = 27x^{-2/3}y^{2/3} \quad \text{and} \quad f_y(x, y) = 54x^{1/3}y^{-1/3}$$

and

$$f_x(27, 125) = 27(27)^{-2/3}(125)^{2/3} = 27(9)^{-1}(25) = 75$$

and

$$f_y(27, 125) = 54(27)^{1/3}(125)^{-1/3} = 54(3)(5)^{-1} = 32.4$$

Then

$$\begin{aligned}
\Delta z &= f(27 + 2, 125 - 3) - f(27, 125) \\
&= f(a + \Delta x, b + \Delta y) - f(a, b) \\
&\approx f_x(a, b) \cdot \Delta x + f_y(a, b) \cdot \Delta y \\
&= f_x(27, 125)(2) + f_y(27, 125)(-3) \\
&= (75)(2) + (32.4)(-3) \\
&= 52.8
\end{aligned}$$

The actual change in z is

$$\Delta z = f(29, 122) - f(27, 125) \approx 46.5 \quad ■$$

Remark. Notice that for $(a, b) = (27, 125)$ we have the formula

$$\Delta z \approx (75) \cdot \Delta x + (32.4) \cdot \Delta y$$

This formula can be used to give quick estimates of the change in production given various changes in x and y. More generally, for any (x, y) we have

$$\Delta z \approx 27x^{-2/3}y^{2/3} \cdot \Delta x + 54(x)^{1/3}(y)^{-1/3} \cdot \Delta y$$

where $\Delta z = f(x + \Delta x, y + \Delta y) - f(x, y)$.

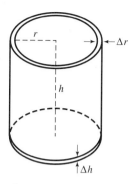

Figure 7.42
The side has thickness Δr and the top and bottom has thickness Δh.

EXAMPLE 3 Using Differentials for an Approximation

(a) Use the tangent line approximation to estimate the change in volume of a cylindrical can of inside radius r and height h for any small changes in r and h when $r = 2$ and $h = 5$ (Fig. 7.42).

(b) Suppose the side of the can is 0.02 inches thick and the top and bottom arc 0.03 inches thick. Use the formula in part (a) to estimate the volume of the metal in the can.

SOLUTIONS ■

(a) The volume $V(r, h)$ is given by

$$V(r, h) = \pi r^2 h$$

The first partial derivatives are

$$V_r(r, h) = 2\pi rh \qquad V_h(r, h) = \pi r^2$$

Then

$$\begin{aligned}
\Delta V &= V(r + \Delta r, h + \Delta h) - V(r, h) \\
&\approx V_r(r, h) \cdot \Delta r + V_h(r, h) \cdot \Delta h \\
&= 2\pi rh \cdot \Delta r + \pi r^2 \cdot \Delta h \\
&= 2\pi(2)(5) \cdot \Delta r + \pi(2)^2 \cdot \Delta h \\
&= 20\pi \cdot \Delta r + 4\pi \cdot \Delta h
\end{aligned}$$

(b) Take $\Delta r = 0.02$. Since the top and bottom are each 0.03 inches thick, $\Delta h = 2(0.03)$. Then we have

$$\Delta V \approx 20\pi \cdot \Delta r + 4\pi \cdot \Delta h = 20\pi(0.02) + 4\pi(0.06) \approx 2.01$$

or 2.01 in.^{3} ■

SELF-HELP EXERCISE SET 7.6

1. A firm has a cost function given by

$$C(x, y) = x^3 + 3xy + y^2$$

Currently, $x = 20$ and $y = 100$. Use the tangent line approximation to approximate the change in costs ΔC as a function of the change in x and the change in y.

2. Take $\Delta x = \Delta y = 1$ in the answer to the previous problem to estimate the change in costs of producing one more of each item. Compare your approximation to the actual answer.

EXERCISE SET 7.6

In Exercises 1 through 16 use the tangent plane approximation to estimate

$$\Delta z = f(a + \Delta x, b + \Delta y) - f(a, b)$$

for the given function at the given point and for the given values of Δx and Δy.

1. $f(x, y) = x^2 + y^3$, $(a, b) = (2, 1)$, $\Delta x = 0.1$, $\Delta y = 0.2$

2. $f(x, y) = 2x^3 + xy$, $(a, b) = (1, 2)$, $\Delta x = 0.1$, $\Delta y = 0.3$

3. $f(x, y) = x^2 y^3$, $(a, b) = (3, 2)$, $\Delta x = 0.1$, $\Delta y = -0.1$

4. $f(x, y) = x^3 y^4$, $(a, b) = (-1, 1)$, $\Delta x = 0.1$, $\Delta y = -0.2$

5. $f(x, y) = \dfrac{x - y}{x + y}$, $(a, b) = (1, 2)$, $\Delta x = 0.2$, $\Delta y = 0.1$

6. $f(x, y) = \dfrac{x + y}{x - y}$, $(a, b) = (3, 4)$, $\Delta x = -0.1$, $\Delta y = 0.2$

7. $f(x, y) = \dfrac{x}{x^2 + y^2}$, $(a, b) = (3, 4)$, $\Delta x = -0.1$, $\Delta y = 0.2$

8. $f(x, y) = \dfrac{y}{x^2 + y^2}$, $(a, b) = (3, 4)$, $\Delta x = 0.02$, $\Delta y = -0.03$

9. $f(x, y) = \ln(x + 2y)$, $(a, b) = (2, 4)$, $\Delta x = 0.1$, $\Delta y = 0.2$

10. $f(x, y) = \ln(x^2 + y^2)$, $(a, b) = (1, 2)$, $\Delta x = 0.5$, $\Delta y = -0.5$

11. $f(x, y) = \sqrt{x + y}$, $(a, b) = (4, 5)$, $\Delta x = -0.1$, $\Delta y = -0.2$

12. $f(x, y) = \sqrt{x^2 + y}$, $(a, b) = (2, 5)$, $\Delta x = \dfrac{1}{30}$, $\Delta y = \dfrac{1}{15}$

13. $f(x, y) = e^{x+y}$, $(a, b) = (1, -1)$, $\Delta x = 0.02$, $\Delta y = 0.03$

14. $f(x, y) = e^{xy}$, $(a, b) = (2, 0.5)$, $\Delta x = 0.02$, $\Delta y = -0.01$

15. $f(x, y) = xe^y$, $(a, b) = (0, 1)$, $\Delta x = 0.1$, $\Delta y = 0.2$

16. $f(x, y) = xye^{xy}$, $(a, b) = (1, 2)$, $\Delta x = 0.02$, $\Delta y = 0.01$

In Exercises 17 through 28 use the tangent line approximation to estimate the given quantities.

17. $\sqrt{4.04}\,\sqrt{9.06}$

18. $\sqrt{4.08}\,\sqrt{8.94}$

19. $\sqrt{(3.02)^2 + (4.03)^2}$

20. $\sqrt{(3.01)^2 + (3.97)^2}$

21. $\sqrt{(2.06)^3 + (0.97)^2}$

22. $\sqrt{(1.94)^3 - (1.03)^2}$

23. $(1.02)^7 + (1.98)^3$

24. $(1.02)^7(1.98)^3$

25. $(0.99)^7(2.98)^3$

26. $(0.97)^7 + (1.98)^3$

27. $e^{(1.02)^2 - (0.98)^2}$

28. $e^{(2.01)^3 - 7.97}$

Applications

29. Cost. A cost function in thousands of dollars for a firm producing two products X and Y is given by $C(x, y) = 1000 + x^3 + xy + y^2$, where x and y are the respective number in thousands of X and Y produced. If 1000 of X and 2000 of Y are currently being produced, use the tangent plane approximation to estimate the change in cost if the firm produces an additional 100 of the X items and 300 of the Y items.

30. Revenue. Suppose the revenue function for the firm in the previous problem is given by $R(x, y) = \sqrt{xy}\,(10 - xy)$, and the firm is currently selling all they produce. Use the tangent plane approximation to estimate the change in revenue if the firm produces and sells an additional 100 of the X items and 300 of the Y items.

31. Production. Suppose we have a Cobb-Douglas production function $f(x, y) = 100x^{1/4}y^{3/4}$, where x is the number of units of labor, y is the number of units of capital, and f is the number of units of a certain product that is produced. Using the tangent plane, find an approximation for the change in production as x changes from 16 to 18 and as y changes from 81 to 80.

32. Manufacturing. Use the tangent plane approximation to estimate the volume of metal in a closed cylindrical can of inner radius 3 inches and inner height 6 inches if the metal is 0.03 inches thick.

33. Manufacturing. Use the tangent plane approximation to estimate the volume of metal in a closed rectangular box with square bottom and top with each side of inner length 1 foot and inner height 3 feet if the metal is 0.05 feet thick.

34. Biology. In measuring the volume of a rod-shaped bacterium (right circular cylinder), suppose an error of 2% is made measuring its radius and a 1% error in measuring its length. Using the tangent line approximation, estimate the greatest relative error ($|\Delta V|/V$) in calculating its volume when $r = 1$ unit and $h = 10$ units.

35. Biology. The formula shown here relates the surface area A in square meters of a human to the weight w in kilograms and height h in meters: $A(w, h) = 2.02w^{0.425}h^{0.725}$. Use the tangent plane approximation to estimate the difference in surface area of two humans, one weighing 100 kg and 2 m tall and the other weighing 101 kg and 2.1 m tall.

36. Medicine. Poiseuille's law states that the resistance of a blood vessel of length L and radius r is given by $R = \dfrac{kL}{r^4}$, where k is a constant. Use the tangent plane approximation to show how small changes in L and r affect the resistance R.

Enrichment Exercises

37. Find an approximation for $f(1.01, 0.02)$, where $f(x, y) = \dfrac{x^3 \ln(x + 1) + (x - 1)e^{yx^2 - 0.4y^2}}{\sqrt{x^3 + \ln(y + 1)}}$. The function $f(x, y)$ is rather complicated. Calculating the first partial derivatives is very time-consuming and prone to error. Instead use your calculator to find the numerical derivatives. (See the two paragraphs following Exploration 1 in Section 7.2.)

38. Find a square about the point $(5, 7)$ in which the volume of the cylinder $V = \pi r^2 h$ given in Example 3 does not vary by more than approximately ± 0.1.

SOLUTIONS TO SELF-HELP EXERCISE SET 7.6

1. We have $C_x(x, y) = 3x^2 + 3y$ and $C_y(x, y) = 3x + 2y$. Thus, $C_x(20, 100) = 1500$, $C_y(20, 100) = 260$, and

$$\Delta C \approx 1500\Delta x + 260\Delta y$$

2. Setting $\Delta x = \Delta y = 1$ in the previous formula yields

$$\Delta C \approx 1500(1) + 260(1) = 1760$$

The actual change in C is given by

$$C(21, 101) - C(20, 100) = 25,825 - 24,000 = 1825$$

7.7 DOUBLE INTEGRALS

■ Riemann Sums and Integrals   ■ Iterated Integrals   ■ Applications

APPLICATION: CONCENTRATION OF POLLUTANTS

An industrial plant is located at the precise center of a town shaped as a square with each side of length 4 miles. If the plant is placed at the point $(0, 0)$ of the xy-plane, then certain pollutants are dispersed is such a manner that the concentration at any point (x, y) in town is given by $C(x, y) = 1000(24 - 3x^2 - 3y^2)$, where $C(x, y)$ is the number of particles of pollutants per square mile of surface per day at a point (x, y) in town. What is the average concentration of these pollutants in this town? For the answer see Example 5.

Riemann Sums and Integrals

In this section we are first interested in finding the volume V between a surface given by $z = f(x, y)$ and a region D in the xy-plane. We then see that this can be used for many applications. We assume that $f(x, y)$ is nonnegative on D (Fig. 7.43).

The domain D is shown in the xy-plane in Figure 7.44. In this figure the domain D has been encased with a rectangle R as shown, where

$$R(x, y) = \{(x, y) : a \le x \le b, c \le y \le d\}$$

For any integer n we divide both intervals $[a, b]$ and $[c, d]$ into n equal subintervals and define

$$\Delta x = \frac{b - a}{n} \qquad \Delta y = \frac{d - c}{n}$$

This gives us a collection of subrectangles shown in Figure 7.45. We label all subrectangles that lie *entirely inside* D in some manner by going from left to right and top to bottom as indicated in Figure 7.45. Let the total number of subrectangles be $N = N(n)$, and label them as $R_1, R_2, \ldots, R_N$. (If the region D were the rectangle R, then $N = N(n) = n^2$.) In each R_i pick some point $(x_i, y_i) \in R_i$. Now for each $i = 1, 2, \ldots, N$, construct a rectangular box as in Figure 7.46 of height $f(x_i, y_i)$ and base R_i. If n is large, the volume

$$V_i = f(x_i, y_i) \, \Delta x \, \Delta y$$

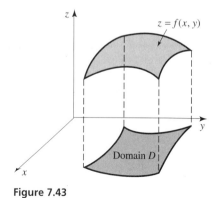

Figure 7.43

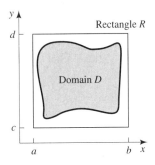

Figure 7.44
Encase the domain D in a rectangle.

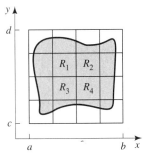

Figure 7.45
Divide the interval $[a, b]$ and $[c, d]$ into n equal subintervals giving n^2 subrectangles.

of this rectangular box approximates the area under the surface $z = f(x, y)$ and over R_i. The sum of all such volumes then approximates the volume we are seeking. This sum is called a **Riemann sum of f on D** and is denoted by S_n. We thus have

$$V \approx S_n = V_1 + V_2 + \cdots + V_N$$
$$= f(x_1, y_1)\, \Delta x\, \Delta y + f(x_2, y_2)\, \Delta x\, \Delta y + \cdots + f(x_N, y_N)\, \Delta x\, \Delta y$$
$$= \sum_{i=1}^{N} f(x_i, y_i)\, \Delta x\, \Delta y$$

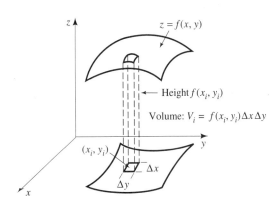

Figure 7.46
The Riemann sum is the sum of the volumes of boxes like the one shown.

EXAMPLE 1   Finding a Riemann Sum

Find the Riemann sum of $f(x, y) = 6xy^2$ on the rectangle

$$R = \{(x, y) : 0 \le x \le 2, 0 \le y \le 1\}$$

for $n = 2$ by taking the points $(x_i\ y_i)$ at the top right of each subrectangle.

SOLUTION ■ Each interval $[0, 2]$ and $[0, 1]$ is divided into two parts and

$$\Delta x = \frac{2 - 0}{2} = 1 \qquad \Delta y = \frac{1 - 0}{2} = \frac{1}{2}$$

The four subrectangles are obtained as shown in Figure 7.47, where the four points (x_1, y_1), (x_2, y_2), (x_3, y_3), (x_4, y_4) are, respectively, $(1, 1)$, $(2, 1)$, $(1, 0.50)$, $(2, 0.50)$. Note that the function increases as either x or y increases. Thus, on any subrectangle

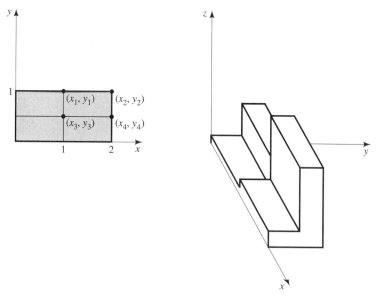

Figure 7.47
The Riemann sum S_2 is the sum of the volumes of the four boxes on the right.

R_i, the function is largest at the top right point (x_i, y_i). Thus, the top of each of the rectangular boxes is above the surface, and the Riemann sum is larger than the volume. Then

$$S_2 = f(x_1, y_1) \, \Delta x \, \Delta y + f(x_2, y_2) \, \Delta x \, \Delta y + f(x_3, y_3) \, \Delta x \, \Delta y + f(x_4, y_4) \, \Delta x \, \Delta y$$
$$= f(1, 1) \, \Delta x \, \Delta y + f(2, 1) \, \Delta x \, \Delta y + f(1, 0.50) \, \Delta x \, \Delta y + f(2, 0.50) \, \Delta x \, \Delta y$$
$$= (6) \frac{1}{2} + (12) \frac{1}{2} + (1.50) \frac{1}{2} + (3) \frac{1}{2}$$
$$= 11.25 \quad \blacksquare$$

If we take the points (x_i, y_i) at the lower left of R_i, then the tops of the rectangular boxes are under the surface, and the corresponding Riemann sum S_n^* is less than the volume.

The table shows the results of calculating the Riemann sums taking (x_i, y_i) at the lower left and the upper right. You can obtain the table using a graphing calculator and the program VOL1 found in the Technology Resource Manual. (Taking $n = 100$ will already require the calculator to run for many minutes.)

If, in the previous discussion, we then let n become larger and larger, we expect the Riemann sum to become a better and better approximation to the volume V. By letting $n \to \infty$, we intuitively see that we should obtain the volume V. According to the table, the volume appears to be 4. (In Example 2, we shall see that this is the case.)

For a function $f(x, y)$ that is continuous on D, the limit $\lim\limits_{n \to \infty} S_n$ exists and is the same limit no matter how the points (x_i, y_i) in R_i are chosen. The result also holds whether or not $f(x, y)$ is nonnegative. We have the following theorem:

n	n^2	S_n^*	S_n
4	16	1.969	7.031
8	64	2.871	5.379
100	10,000	3.901	4.100
500	250,000	3.980	4.020

Definite Integral

Let $f(x, y)$ be continuous on the bounded region D. Then the **definite integral of f over D** is

$$\int \int_D f(x, y) \, dA = \lim_{n \to \infty} \sum_{i=1}^{N} f(x_i, y_i) \, \Delta x \, \Delta y$$

where the limit exists and is the same no matter how the points (x_i, y_i) are chosen in R_i.

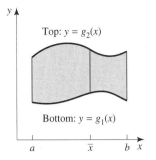

Figure 7.48
This domain has a top and bottom
each given by one function.

When $f(x, y) \geq 0$, then the definite integral is intuitively the volume under the surface and over the domain D. We then *define* the volume to be this Riemann integral.

Volume as a Definite Integral

Let $f(x, y)$ be continuous on the bounded region D and nonnegative there. Then the volume under the surface $z = f(x, y)$ and over the domain D is the definite integral of f over D or

$$V = \int \int_D f(x, y) \, dA$$

Iterated Integrals

Suppose the domain D of $z = f(x, y)$ has a "top" and a "bottom" given, respectively, by $y = g_2(x)$ and $y = g_1(x)$ as in Figure 7.48. For any fixed $\bar{x} \in [a, b]$, draw a straight line connecting the top and the bottom of the domain D, as shown in Figure 7.48. As shown in Figure 7.49, the plane $x = \bar{x}$ cuts out a plane region with top $z = z(\bar{x}, y)$. We can readily find the area $A(\bar{x})$ on the plane $x = \bar{x}$ under the curve $z = f(\bar{x}, y)$ as

$$A(\bar{x}) = \int_{g_1(\bar{x})}^{g_2(\bar{x})} f(x, y) \, dy$$

This is the cross-sectional area of a slice of the volume.

For any integer n, we now divide the interval $[a, b]$ into n equal subintervals each with base of length $\Delta x = \dfrac{b - a}{n}$. The ith such interval I_i is shown in Figure 7.50. Each interval then slices out a thin volume under the surface as shown in Figure 7.51. For each i, we pick any point $x_i \in I_i$. By virtue of continuity and the smallness of Δx, the

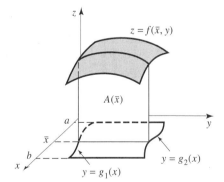

Figure 7.49
The slice at $\bar{x}$ cuts out an area between the curve $z = f(\bar{x},y)$ and the xy-plane from $g_1(\bar{x})$ to $g_2(\bar{x})$.

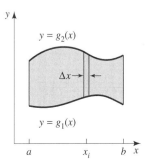

Figure 7.50
We take a vertical slice of thickness Δx.

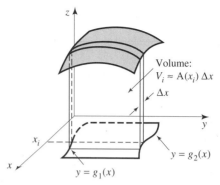

Figure 7.51
The slice in Figure 7.50 then cuts out the volume shown.

volume V_i of this slice is approximately the thickness, Δx, times the surface area of either side. The surface area of either side is approximately $A(x_i)$, given before. Thus,

$$V_i \approx A(x_i) \, \Delta x$$

If we sum all n of these we obtain, approximately, the volume V under the surface $z = f(x, y)$ over the domain D. We have

$$
\begin{aligned}
V &= V_1 + V_2 + \cdots + V_n \\
&\approx A(x_1) \, \Delta x + A(x_2) \, \Delta x + \cdots + A(x_n) \, \Delta x \\
&= \sum_{i=1}^{n} A(x_i) \, \Delta x
\end{aligned}
$$

But we recognize this as a Riemann sum of $A(x)$ on the interval $[a, b]$. Thus, as $n \to \infty$,

$$\sum_{i=1}^{n} A(x_i) \, \Delta x \to \int_a^b A(x) \, dx$$

Thus, we have

$$V = \int_a^b A(x) \, dx$$

where

$$A(x) = \int_{g_1(x)}^{g_2(x)} f(x, y) \, dy$$

We must understand that in this last integral for $A(x)$, x *is held constant* during the integration. Putting this together, we have

$$V = \int_a^b \left(\int_{g_1(x)}^{g_2(x)} f(x, y) \, dy \right) dx$$

We thus have the following result:

Iterated Integral

If the domain D is given as in Figure 7.48, and $z = f(x, y) \geq 0$ is continuous on D, then the volume V between the surface $z = f(x, y)$ and the domain D is

$$\int \int_D f(x, y) \, dA = \int_a^b \left(\int_{g_1(x)}^{g_2(x)} f(x, y) \, dy \right) dx$$

Remarks. The following two remarks are critical:

1. In the preceding iteration, $\int_{g_1(x)}^{g_2(x)} f(x, y) \, dy$ is to be done *first*, that is, we integrate with respect to y first.
2. When calculating $\int_{g_1(x)}^{g_2(x)} f(x, y) \, dy$, treat x as a *constant*. Thus, we integrate first with respect to y, while holding x constant.

Since there are two basic types of regions to master, we must be able to distinguish between the two. Toward this end we should get in the habit when taking a double integral over a region such as shown in Figures 7.48 or 7.49 of taking the following steps:

Procedure for Setting up the Iterated Integral

1. Clearly label the "top" and "bottom" of the domain D.
2. Draw the slice shown in Figure 7.48 so that it is obvious that x is being held constant while the y is varying from the "bottom" to the "top." This gives the limits of integration for the *inner* integral.
3. Now "sum" these slices "on x," that is, x goes from a to b. This gives the limits of integration on the *outer* integral.

EXAMPLE 2   An Iterated Integral over a Rectangle

Find the volume under the surface given by $z = 6xy^2$ over the domain given by the rectangle

$$D = \{(x, y) : 0 \le x \le 2, 0 \le y \le 1\}$$

that is, find

$$\int \int_D 6xy^2 \, dA$$

SOLUTION ■ The region D has a bottom given by $y = g_1(x) = 0$ and a top given by $y = g_2(x) = 1$ (Fig. 7.52). The slice shown in Figures 7.52 or 7.53 indicates that while holding x constant, we integrate with respect to y from the bottom to the top. Thus, the limits of integration on the inner integral go from $y = 0$ to $y = 1$. Now the slices are "summed" on x from 0 to 2. This gives the limits of integration on the outer integral. Thus,

$$\int \int_D f(x, y) \, dA = \int_a^b \left[\int_{g_1(x)}^{g_2(x)} f(x, y) \, dy \right] dx$$

$$= \int_0^2 \left(\int_0^1 6xy^2 \, dy \right) dx$$

$$= \int_0^2 \left(2xy^3 \Big|_{y=0}^{y=1} \right) dx$$

$$= \int_0^2 2x \, dx$$

$$= x^2 \Big|_0^2$$

$$= 4 \quad ■$$

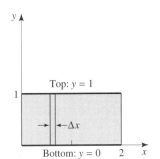

Figure 7.52
We take a vertical slice of thickness Δx.

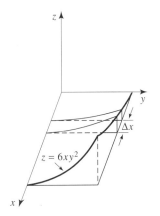

Figure 7.53
The slice shown in Figure 7.52 cuts out the volume shown.

EXAMPLE 3   Finding an Iterated Integral

Find the volume under the surface given by $z = 48xy$ over the domain given by the region D in the first quadrant of the xy-plane bounded by $y = x$ and $y = x^2$.

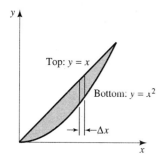

Figure 7.54
We take a vertical slice of thickness Δx.

S O L U T I O N ■   The region D has a bottom given by $y = g_1(x) = x^2$ and a top given by $y = g_2(x) = x$ (Fig. 7.54). The slice shown in Figures 7.54 and 7.55 indicates that while holding x constant, we integrate with respect to y from the bottom to the top. Thus, the limits of integration on the inner integral go from $y = x^2$ to $y = x$. Now the slices are ''summed'' on x from 0 to 1. This gives the limits of integration on the outer integral.

Thus

$$
\begin{aligned}
\int\!\!\int_D f(x,\,y)\,dA &= \int_a^b \left[\int_{g_1(x)}^{g_2(x)} f(x,\,y)\,dy \right] dx \\
&= \int_0^1 \left(\int_{x^2}^{x} 48xy\,dy \right) dx \\
&= \int_0^1 \left(24xy^2 \Big|_{y=x^2}^{y=x} \right) dx \\
&= \int_0^1 (24x^3 - 24x^5)\,dx \\
&= (6x^4 - 4x^6)\big|_0^1 \\
&= 2 \quad ■
\end{aligned}
$$

Now let us consider the case in which the domain D of $z = f(x, y)$ has a ''left'' and a ''right'' given, respectively, by $x = g_1(y)$ and $x = g_2(y)$ as in Figure 7.56. We proceed in a manner similar to what we did before, this time summing thin slices of volumes shown in Figure 7.57. This leads to the following iterated integral:

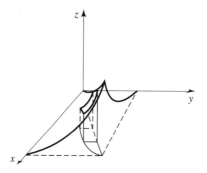

Figure 7.55
The slice shown in Figure 7.54 cuts out the volume shown.

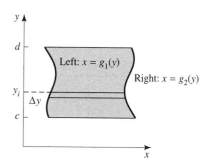

Figure 7.56
We take a horizontal slice of thickness Δy.

Iterated Integral

If the domain D is given as in Figure 7.56, and $z = f(x, y) \geq 0$ is continuous on D, then the volume V between the surface $z = f(x, y)$ and the domain D is

$$
\int\!\!\int_D f(x,\,y)\,dA = \int_c^d \left[\int_{g_1(y)}^{g_2(y)} f(x,\,y)\,dx \right] dy
$$

In this iterated integral we say we are *integrating with respect to x first*, whereas in the previous iterated integral we say that we are *integrating with respect to y first*.

Remarks. The following two remarks are critical:

1. Notice that $\int_{g_1(y)}^{g_2(y)} f(x, y)\,dx$ is to be done *first*.
2. When calculating $\int_{g_1(y)}^{g_2(y)} f(x, y)\,dx$, treat y as a constant.

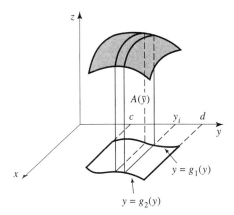

Figure 7.57
The slice shown in Figure 7.56 cuts out the thin slice of volume shown.

When taking a double integral over a region such as shown in Figure 7.56, get in the habit of following the procedures listed here.

Procedure for Setting up an Iterated Integral

1. Clearly label the "left" and the "right" of the domain D.
2. Draw the slice shown in Figure 7.56 so that it is obvious that y is being held constant while x varies from the "left" to the "right." This gives the limits of integration for the *inner* integral.
3. Now sum these slices "on y," that is, the slices go from c to d. This gives the limits of integration on the *outer* integral.

EXAMPLE 4   Integrating with Respect to x First

Redo Example 3 by integrating first with respect to x.

SOLUTION ■ This region has a left side and a right side. The left side is given by $x = g_1(y) = y$, and the right side is given by $x = g_2(y) = \sqrt{y}$ (Fig. 7.58). The slice shown in the figure indicates that we integrate from the left to the right while holding y constant. Thus, the limits of integration on the inner integral go from $x = y$ to $x = \sqrt{y}$. Now the slices are summed on y from 0 to 1. This gives the limits of integration on the outer integral. Then

$$\int\int_D f(x, y)\, dA = \int_c^d \left[\int_{g_1(y)}^{g_2(y)} f(x, y)\, dx \right] dy$$

$$= \int_0^1 \left(\int_y^{\sqrt{y}} 48xy\, dx \right) dy$$

$$= \int_0^1 \left[(24x^2 y)\big|_{x=y}^{x=\sqrt{y}} \right] dy$$

$$= \int_0^1 (24y^2 - 24y^3)\, dy$$

$$= (8y^3 - 6y^4)\big|_0^1$$

$$= 2 \quad ■$$

Applications

We have already defined the average value of a continuous function $f(x)$ of *one* variable on the interval $[a, b]$ to be

$$\text{Average value} = \frac{1}{b - a} \int_a^b f(x)\, dx$$

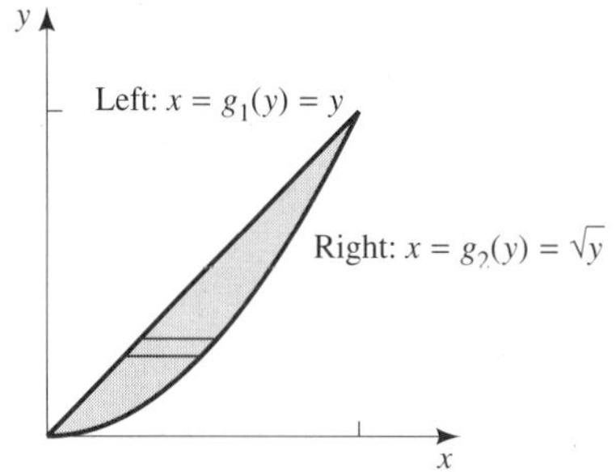

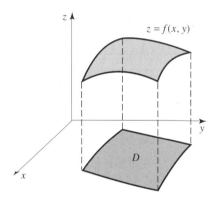

Figure 7.58
We take a horizontal slice of thickness Δy.

Figure 7.59
We want to define the average value of $z = f(x,y)$ over the region D.

For a nonnegative function f, this is the same as the area under the function $y = f(x)$ on the interval $[a, b]$, divided by the length of this interval.

There is a similar definition for the **average value** of a function $z = f(x, y)$ of *two* variables on a domain D in the xy-plane such as shown in Figure 7.59.

Average Value

If $f(x, y)$ is a continuous function on a domain D, the average value of f on D is

$$\frac{1}{A(D)} \int \int_D f(x, y) \, dA$$

where $A(D)$ is the area of the domain D.

Notice that for a nonnegative function f, this is the same as the volume under the surface $z = f(x, y)$ and above the domain D divided by the area of the domain.

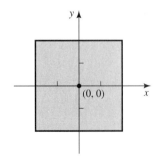

Figure 7.60
The industrial plant is located at the center of the town shaped as a square.

EXAMPLE 5 Average Concentration of Pollutants

An industrial plant is located at the precise center of a town shaped as a square with each side of length 4 miles. If the plant is placed at the point $(0, 0)$ of the xy-plane (Fig. 7.60), then certain pollutants are dispersed in such a manner that the concentration at any point (x, y) in town is given by

$$C(x, y) = 1000(24 - 3x^2 - 3y^2)$$

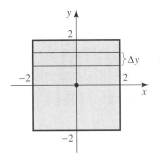

Figure 7.61
We take a horizontal slice of thickness Δy.

where $C(x, y)$ is the number of particles of pollutants per square mile of surface per day at a point (x, y) in town. What is the average concentration of these pollutants in this town?

S O L U T I O N ■ This concentration varies over the area of the town, being largest naturally at the center of town where the plant is located and smallest at the four corners of town. In fact $C(0, 0) = 24{,}000$, whereas $C(2, 2) = C(2, -2) = C(-2, -2) = C(-2, 2) = 0$. The concentration function $C(x, y)$ can be obtained by taking samples at various points around the town and then "fitting a surface to the data" in a fashion analogous to the way we fitted a straight line to data points by the method of least squares in Section 7.5 when dealing with functions of one variable. Each sample can be a pollutant count over, say, a single square foot of surface. Multiplying this number by the number of square feet in a square mile then gives the number of particles per square mile. This last number is then $C(x, y)$ at the center of the 1-foot-by-1-foot square.

The average value is the double integral $\int \int_D C(x, y)\, dA$, divided by the area of D, where D is the square shown in Figure 7.60. If the domain D were some irregularly shaped region, we would have to resort to Riemann sums to approximate this double integral. But the region D has a "bottom" and "top" and also a "left" and "right." Thus, we can evaluate this double integral as an iterated integral and can also integrate first with respect to x or first with respect to y. We integrate first with respect to x, as indicated by the slice in Figure 7.61. The left is given by $x = -2$ and the right by $x = 2$. The slices are summed from $y = -2$ to $y = 2$. Thus,

$$
\begin{aligned}
\int \int_D C(x, y)\, dA &= 1000 \int_{-2}^{2} \left[\int_{-2}^{2} (24 - 3x^2 - 3y^2)\, dx \right] dy \\
&= 1000 \int_{-2}^{2} [(24x - x^3 - 3xy^2)|_{x=-2}^{x=2}]\, dy \\
&= 1000 \int_{-2}^{2} [(48 - 8 - 6y^2) - (-48 + 8 + 6y^2)]\, dy \\
&= 1000 \int_{-2}^{2} (80 - 12y^2)\, dy \\
&= 1000(80y - 4y^3)|_{-2}^{2} \\
&= 1000[(160 - 32) - (-160 + 32)] \\
&= 256{,}000
\end{aligned}
$$

■ **Air Pollution**

What does the U.S. Environmental Protection Agency consider one of the nation's worst environmental hazards? The air you breath, indoors! Indoor pollution causes more deaths and costs companies in the U.S. an estimated $60 billion a year in health care and lost productivity. It is at least two times as polluted as outdoor air, government experts say. Yet there are no state or federal rules, only suggested guidelines.

Source: Bruce Finley, "Indoor Air More Hazardous Yet No State, Federal Health Regulations Exist," *The Denver Post*, November 7, 1994.

The area of D is the area of the square: 16. Thus, the average concentration of pollutants is $256{,}000/16 = 16{,}000$ particles per square mile, or about 0.0006 particles per square foot per day. ■

SELF-HELP EXERCISE SET 7.7

1. Find the volume under the surface given by $z = 10 - x^2 - y^2$ and over the domain given by the rectangle

$$D = \{(x, y) : 0 \le x \le 1, 0 \le y \le 3\}$$

that is, find

$$\int \int_D (10 - x^2 - y^2)\, dA$$

2. Find the volume under the surface given by $z = 8xy$ over the domain given by the triangular region D in the xy-plane bounded by $y = x$, $y = 0$, and $x = 1$ by integrating first with respect to y.

3. Redo the previous exercise by integrating first with respect to x.

EXERCISE SET 7.7

In Exercises 1 through 4 find the Riemann sum S_n for the given n for $f(x, y) = 6xy^2$ on the rectangle $R = \{(x, y) : 0 \le x \le 2, 0 \le y \le 1\}$, by taking the points (x_i, y_i) at the indicated points of each subrectangle. (Compare with Example 1 of the text.)

1. S_2, (x_i, y_i) at the center of each subrectangle

2. S_3, (x_i, y_i) at the center of each subrectangle

3. S_2, (x_i, y_i) at the lower left of each subrectangle

4. S_4, (x_i, y_i) at the center of each subrectangle

5. Find S_2 for $z = x^2y$ for $D = \{(x, y) : 0 \le x \le 4, 0 \le y \le 2\}$, where (x_i, y_i) are taken at the lower left of each subrectangle.

6. Repeat the previous exercise for S_4.

In Exercises 7 through 28 find the double integral over the indicated region D in two ways. (a) Integrate first with respect to x. (b) Integrate first with respect to y.

7. $\displaystyle\int\int_D (2x + 3y)\, dA$, $D = \{(x, y) : 0 \le x \le 1, 0 \le y \le 1\}$

8. $\displaystyle\int\int_D (x^2 + y)\, dA$, $D = \{(x, y) : 0 \le x \le 1, 0 \le y \le 1\}$

9. $\displaystyle\int\int_D xy\, dA$, $D = \{(x, y) : 0 \le x \le 2, 0 \le y \le 1\}$

10. $\displaystyle\int\int_D xy\, dA$, $D = \{(x, y) : 0 \le x \le 1, 0 \le y \le 3\}$

11. $\displaystyle\int\int_D 6x^2y\, dA$, $D = \{(x, y) : 1 \le x \le 2, 2 \le y \le 3\}$

12. $\displaystyle\int\int_D 6xy^2\, dA$, $D = \{(x, y) : 1 \le x \le 2, 2 \le y \le 3\}$

13. $\displaystyle\int\int_D y\sqrt{y^2 + x}\, dA$, $D = \{(x, y) : 0 \le x \le 1, 0 \le y \le 1\}$

14. $\displaystyle\int\int_D x\sqrt{x^2 + y}\, dA$, $D = \{(x, y) : 0 \le x \le 1, 0 \le y \le 4\}$

15. $\displaystyle\int\int_D \frac{1}{xy}\, dA$, $D = \{(x, y) : 1 \le x \le 2, 1 \le y \le 3\}$

16. $\displaystyle\int\int_D \left(1 + \frac{x}{y}\right)dA$, $D = \{(x, y) : 1 \le x \le 2, 1 \le y \le 3\}$

17. $\displaystyle\int\int_D e^{x+y}\, dA$, $D = \{(x, y) : 0 \le x \le 2, 0 \le y \le 3\}$

18. $\displaystyle\int\int_D xye^{0.5(x^2+y^2)}\, dA$, $D = \{(x, y) : 0 \le x \le 2, 0 \le y \le 3\}$

19. $\displaystyle\int\int_D \frac{1}{(1 + x + y)^3}\, dA$, $D = \{(x, y) : 0 \le x \le 2, 0 \le y \le 3\}$

20. $\displaystyle\int\int_D \frac{1}{(1 + x + y)^2}\, dA$, $D = \{(x, y) : 1 \le x \le 2, 1 \le y \le 3\}$

21. $\displaystyle\int\int_D xe^{x^2}\, dA$, $D = \{(x, y) : 0 \le x \le 2, 0 \le y \le 1\}$

22. $\displaystyle\int\int_D ye^{y^2}\, dA$, $D = \{(x, y) : 0 \le x \le 2, 0 \le y \le 1\}$

23. $\displaystyle\int\int_D (x + y)\, dA$, D is the triangular region with vertices at $(0, 0), (1, 0), (1, 1)$

24. $\displaystyle\int\int_D (x + y)\, dA$, D is the triangular region with vertices at $(0, 0), (0, 1), (1, 1)$

25. $\displaystyle\int\int_D (x^2 + y^2)\, dA$, D is the triangular region with vertices at $(0, 0), (1, 0), (1, 1)$

26. $\displaystyle\int\int_D (x^2 + y^2)\, dA$, D is the triangular region with vertices at $(0, 0), (0, 1), (1, 1)$

27. $\displaystyle\int\int_D xy\, dA$, D is the triangular region with vertices at $(0, 0), (1, 0), (1, 1)$

28. $\displaystyle\int\int_D x^2y\, dA$, D is the triangular region with vertices at $(1, 0), (2, 0), (2, 1)$

In Exercises 29 through 34 find the volume under the surface of the given function and over the indicated region.

29. $f(x, y) = 1$, D is the region in the first quadrant bounded by the curves $y = x$ and $y = x^3$.

30. $f(x, y) = 1$, D is the region in the first quadrant bounded by the curves $y^2 = x^3$, $y = x$.

31. $f(x, y) = xy^2$, D is the region in the first quadrant bounded by the curves $y = x$ and $y = x^2$.

32. $f(x, y) = \dfrac{x}{y^2}$, D is the region in the first quadrant bounded by the curves $x = 0$, $x = y^{3/2}$, $y = 1$, $y = 2$.

33. $f(x, y) = xy$, D is the region in the first quadrant bounded by the curves $y = x$, $x = y^2$.

34. $f(x, y) = xe^y$, D is the region in the first quadrant bounded by the curves $y = 0$, $y = x^2$, $x = 0$, $x = 1$.

In Exercises 35 through 42 integrating first with respect to one of x or y is difficult while integrating first with respect to the other is not. Find $\int\int_D f(x, y)\, dA$ the easier way.

35. $f(x, y) = x$, D is the region bounded by the curves $y = x^2 + 1$, $y = x$, $x = -1$, $x = 1$.

36. $f(x, y) = x$, D is the region in the fourth quadrant bounded by the curves $y = -x^2$, $y = -3x - 1$, $x = 0$, $x = 2$.

37. $f(x, y) = 1$, D is the region in the first and fourth quadrants bounded by the curves $x = 1 - y$, $y = x^2 - 1$, $x = 0$.

38. $f(x, y) = 1$, D is the region in the first quadrant bounded by the curves $x = y - 2$, $x = y^3$, $x = 0$, $x = 1$.

39. $f(x, y) = 1$, D is the region in the first quadrant bounded by the curves $xy = 1$, $y = x$, $y = 2$.

40. $f(x, y) = y$, D is the region in the first quadrant bounded by the curves $x = 1$, $x = y^2$.

41. $f(x, y) = y$, D is the region in the first quadrant bounded by the curves $y = 2x$, $y = x$, $x = 2$.

42. $f(x, y) = y$, D is the region in the first quadrant bounded by the curves $y = \frac{1}{2}x$, $y = x$, $y = 2$.

Applications

43. Concentration of Pollutants. In Example 5 of the text, what is the average concentration of pollutants in the region

$$D_1 = \{(x, y) : 0 \le x \le 1, 0 \le y \le 2\}$$

44. Concentration of Pollutants. If D is the region given in Example 5 and D_1 is the region given in the previous problem, find the average concentration of pollutants over the region consisting of D less D_1.

45. Cobb-Douglas Production Function. What is the average

value of the Cobb-Douglas production function $f(x, y) = x^{1/3}y^{2/3}$ for the range of x and y given by

$$D = \{(x, y) : 8 \le x \le 27, 1 \le y \le 8\}$$

46. Cobb-Douglas Production Function. What is the average value of the Cobb-Douglas production function $f(x, y) = x^{2/3}y^{1/3}$ for the range of x and y given by

$$D = \{(x, y) : 8 \le x \le 27, 1 \le y \le 8\}$$

Enrichment Exercises

47. The iterated integral

$$\int_0^1 \left(\int_y^1 2e^{x^2} \, dx \right) dy$$

is impossible to do, if the integration must be done first with respect to x. Change the order of integration and find the double integral.

48. Density. The concentration $C(x, y)$ given in Example 5 is an example of a **density** function that we usually denote in general by $\delta(x, y)$. The density function gives the number per unit area. In the case of the pollutants we counted the number of particles per square mile. In general, if $\delta(x, y)$ is a continuous function that gives the density in number per unit of area, then the total number in D is given by

$$\int \int_D \delta(x, y) \, dA$$

Find the total number of particles of pollutant in town per day in Example 5.

49. Population. The density function for a coastal town is given by

$$\delta(x, y) = \frac{120{,}000}{(2 + x + y)^3}$$

in people per square mile, where x and y are in miles and where the town is the square

$$D = \{(x, y) : 0 \le x \le 2, 0 \le y \le 2\}$$

(The shore runs along the x-axis.) Find the total population of this town.

50. Population. Find the total population living in the town given in the previous problem that are within 1 mile of shore.

51. For the integral in Exercise 9 use your graphing calculator and the program VOL for $n = 20$, 40, and 100 to find two Riemann sums for each n. The sum denoted as LLSUM in this program is formed by taking (x_i, y_i) at the lower left of the subrectangles, and the sum denoted as URSUM is formed by taking (x_i, y_i) at the upper right of the subrectangles. By examining the integrand show that LLSUM must be a lower bound and URSUM must be an upper bound.

52. For the integral in Exercise 15 use your graphing calculator and the program VOL for $n = 20$, 40, and 100 to find two Riemann sums for each n. The sum denoted as LLSUM in this program is formed by taking (x_i, y_i) at the lower left of the subrectangles, and the sum denoted as URSUM is formed by taking (x_i, y_i) at the upper right of the subrectangles. By examining the integrand show that URSUM must be a lower bound and LLSUM must be an upper bound.

SOLUTIONS TO SELF-HELP EXERCISE SET 7.7

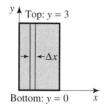

1. The region D has a bottom given by $y = g_1(x) = 0$ and a top given by $y = g_2(x) = 3$. The slice shown in the figure indicates that while holding x constant, we integrate with respect to y from the bottom to the top. Thus, the limits of integration on the inner integral go from $y = 0$ to $y = 3$. Now the slices are summed on x from 0 to 1. This gives the limits of integration on the outer integral. Thus,

$$\int\int_D f(x, y)\, dA = \int_a^b \left[\int_{g_1(x)}^{g_2(x)} f(x, y)\, dy \right] dx$$

$$= \int_0^1 \left[\int_0^3 (10 - x^2 - y^2)\, dy \right] dx$$

$$= \int_0^1 \left[\left(10y - x^2 y - \frac{1}{3} y^3 \right) \Big|_{y=0}^{y=3} \right] dx$$

$$= \int_0^1 (30 - 3x^2 - 9)\, dx$$

$$= \int_0^1 (21 - 3x^2)\, dx$$

$$= (21x - x^3)\big|_0^1 = 20$$

2. The region D has a bottom given by $y = g_1(x) = 0$ and a top given by $y = g_2(x) = x$. The slice shown in the figure indicates that while holding x constant, we integrate with respect to y from the bottom to the top. Thus, the limits of integration on the inner integral go from $y = 0$ to $y = x$.

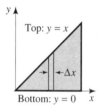

Now the slices are summed on x from 0 to 1. This gives the limits of integration on the outer integral. Thus,

$$\int\int_D f(x, y)\, dA = \int_a^b \left[\int_{g_1(x)}^{g_2(x)} f(x, y)\, dy \right] dx$$

$$= \int_0^1 \left(\int_0^x 8xy\, dy \right) dx$$

$$= \int_0^1 [(4xy^2)|_{y=0}^{y=x}]\, dx$$

$$= \int_0^1 4xx^2\, dx$$

$$= x^4\big|_0^1 = 1$$

3. We notice that this region does have a left side and a right side. The left side is given by $x = g_1(y) = y$, and the right side is given by $x = g_2(y) = 1$. The slice shown in the figure indicates we integrate from the left to the right while holding y constant. Thus, the limits of integration on the inner integral go from $x = y$ to $x = 1$. Now the slices are summed on y from 0 to 1. This gives the limits of integration on the outer integral. Then

$$\int\int_D f(x, y)\, dA = \int_c^d \left[\int_{g_1(y)}^{g_2(y)} f(x, y)\, dx \right] dy$$

$$= \int_0^1 \left(\int_y^1 8xy\, dx \right) dy$$

$$= \int_0^1 (4x^2 y|_{x=y}^{x=1})\, dy$$

$$= \int_0^1 (4y - 4y^3)\, dy$$

$$= (2y^2 - y^4)\big|_0^1 = 1$$

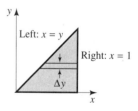

C H A P T E R **7**

SUMMARY OUTLINE

- A **function of two variables**, denoted by $z = f(x, y)$, is a rule that associates to every point (x, y) in some set D (called the **domain**) a unique point denoted by $z = f(x, y)$. p. 360.

- The **distance between the two points** (x_1, y_1, z_1) and (x_2, y_2, z_2) is   p. 363.

$$d = \sqrt{(x_2 - x_1)^2 + (y_2 - y_1)^2 + (z_2 - z_1)^2}$$

- Given a function $z = f(x, y)$ with domain D, the **graph** is the set of points (x, y, z) with $z = f(x, y)$ and $(x, y) \in D$. Since this is described by *two* variables, the result is a two-dimensional object called a **surface**. p. 364.

- The equation of the **sphere** centered at (x_0, y_0, z_0) with radius r is given by p. 365.

$$(x - x_0)^2 + (y - y_0)^2 + (z - z_0)^2 = r^2$$

- The graph of an equation of the form $Ax + By + Cz = D$ is a **plane**. p. 365.

- The graph of $f(x, y) = z_0$ in the xy-plane is called a **level curve**.

- We say that $f(x, y)$ is **continuous** at (a, b), if $f(x, y)$ approaches $f(a, b)$ no matter how (x, y) approaches (a, b). p. 365.

- If $z = f(x, y)$, then the **partial derivative of f with respect to x** is

$$\frac{\partial f}{\partial x}(x, y) = \lim_{\Delta x \to 0} \frac{f(x + \Delta x, y) - f(x, y)}{\Delta x}$$

and the **partial derivative of f with respect to y** is

$$\frac{\partial f}{\partial y}(x, y) = \lim_{\Delta y \to 0} \frac{f(x, y + \Delta y) - f(x, y)}{\Delta y}$$

if the limits exist. p. 373.

- If

$$\frac{\partial x_A}{\partial q} > 0 \qquad \text{and} \qquad \frac{\partial x_B}{\partial p} > 0$$

then we say that the two commodities are **competitive**. If

$$\frac{\partial x_A}{\partial q} < 0 \qquad \text{and} \qquad \frac{\partial x_B}{\partial p} < 0$$

then we say that the two commodities are **complementary**. p. 378.

- Given a function $z = f(x, y)$, we define the following four, second partial derivatives.

1. $\dfrac{\partial^2 f}{\partial x^2} = \dfrac{\partial}{\partial x}\left(\dfrac{\partial f}{\partial x}\right)$ 3. $\dfrac{\partial^2 f}{\partial y^2} = \dfrac{\partial}{\partial y}\left(\dfrac{\partial f}{\partial y}\right)$

2. $\dfrac{\partial^2 f}{\partial y\, \partial x} = \dfrac{\partial}{\partial y}\left(\dfrac{\partial f}{\partial x}\right)$ 4. $\dfrac{\partial^2 f}{\partial x\, \partial y} = \dfrac{\partial}{\partial x}\left(\dfrac{\partial f}{\partial y}\right)$

- We say that f has a **relative maximum** at $(a, b) \in D$ if there exists a circle centered at (a, b) and entirely in D, such that

$$f(x, y) \le f(a, b)$$

for all points (x, y) inside this circle. p. 380.

     We say that f has a **relative minimum** at $(a, b) \in D$ if there exists a circle centered at (a, b) and entirely within D, such that

$$f(x, y) \ge f(a, b)$$

for all points (x, y) inside this circle.

     We say that the function $z = f(x, y)$ has a **relative extremum** at $(a, b) \in D$ if f has either a relative maximum or a relative minimum at (a, b). p. 380.

- **Necessary Condition for Relative Extrema.** If $z = f(x, y)$ is defined, and both first partial derivatives exist for all values of (x, y) inside some circle about (a, b), and f assumes a relative extremum at (a, b), then p. 384.

$$\frac{\partial f}{\partial x}(a, b) = 0 \qquad \text{and} \qquad \frac{\partial f}{\partial y}(a, b) = 0$$

- Suppose that both partial derivatives of $z = f(x, y)$ exist in its domain of definition. **Critical points** are points (a, b), not on the boundary of the domain of f, for which

$$\frac{\partial f}{\partial x}(a, b) = 0 \quad \text{and} \quad \frac{\partial f}{\partial y}(a, b) = 0$$

- **Second Derivative Test for Functions of Two Variables.** For $z = f(x, y)$ assume that f_{xx}, f_{yy}, and f_{xy} all exist for every point near (a, b). Suppose (a, b) is a critical point; that is, assume that $f_x(a, b) = 0$ and $f_y(a, b) = 0$. Define

$$\Delta(a, b) = f_{xx}(a, b)f_{yy}(a, b) - [f_{xy}(a, b)]^2$$

Then p. 386.

1. $\Delta(a, b) > 0$ and $f_{xx}(a, b) < 0$ implies that $f(a, b)$ is a relative maximum,
2. $\Delta(a, b) > 0$ and $f_{xx}(a, b) > 0$ implies that $f(a, b)$ is a relative minimum,
3. $\Delta(a, b) < 0$ implies (a, b) is a saddle point, that is, (a, b) is neither a relative minimum nor a relative maximum,
4. $\Delta(a, b) = 0$ implies that the test fails, that is, when $\Delta(a, b) = 0$, there can be a relative maximum, a relative minimum, or a saddle.

- We say that f has an **absolute maximum** on D at $(a, b) \in D$ if for all $(x, y) \in D$, $f(x, y) \leq f(a, b)$. p. 387.

    We say that f has an **absolute minimum** on D at $(a, b) \in D$ if for all $(x, y) \in D$, $f(x, y) \geq f(a, b)$.

    Also we say that f has an **absolute extremum** on D at $(a, b) \in D$ if f has either an absolute maximum or an absolute minimum on D at (a, b).

- **Existence of Absolute Extrema on a Closed and Bounded Region.** If $z = f(x, y)$ is continuous on the closed and bounded region D, then f assumes its absolute extrema in D, that is, there exist two points $(a, b) \in D$ and $(A, B) \in D$ such that

$$f(a, b) \leq f(x, y) \leq f(A, B) \quad \text{for all } (x, y) \in D \qquad \text{p. 388.}$$

- **Method of Lagrange Multipliers.** Candidates (x_c, y_c, z_c) for a relative extremum of $f(x, y, z)$, subject to the constraint $g(x, y, z) = 0$, can be found among the critical points (x_c, y_c, z_c, λ) of the auxiliary function p. 390.

$$F(x, y, z, \lambda) = f(x, y, z) + \lambda g(x, y, z)$$

that is, among the solutions of the equations

$$F_x = 0, \quad F_y = 0, \quad F_z = 0, \quad g = 0$$

- **Method of Least Squares.** The best fitting straight line $y = ax + b$ through the n data points $(x_1, y_1), (x_2, y_2), \ldots, (x_n, y_n)$ can be found by solving the following two linear equations for a and b. p. 401.

$$\left(\sum_{i=1}^{n} x_i^2\right) a + \left(\sum_{i=1}^{n} x_i\right) b = \sum_{i=1}^{n} x_i y_i$$

$$\left(\sum_{i=1}^{n} x_i\right) a + \qquad nb = \sum_{i=1}^{n} y_i$$

- **Equation of Tangent Plane.** The plane that is tangent to the surface $z = f(x, y)$ at the point $P = (a, b, z_0)$ is given by the equation

$$z - z_0 = f_x(a, b)(x - a) + f_y(a, b)(y - b)$$

where $z_0 = f(a, b)$. p. 406.

- **Tangent Plane Approximation.** If the first partial derivatives of $f(x, y)$ are continuous at (a, b) and if Δx and Δy are small then p. 406.

$$f(a + \Delta x, b + \Delta y) - f(a, b) \approx f_x(a, b)\,\Delta x + f_y(a, b)\,\Delta y$$

■ **Riemann Sum.** For any integer n divide both intervals $[a, b]$ and $[c, d]$ into n equal subintervals and define $\Delta x = \dfrac{b - a}{n}$ and $\Delta y = \dfrac{d - c}{n}$. This gives a collection of subrectangles. Label all subrectangles that lie *entirely inside* D (the domain of $f(x, y)$) in some manner. Let the total number of subrectangles be N. In each subrectangle R_i pick some point $(x_i, y_i) \in R_i$. The Riemann sum is defined as p. 411.

$$\sum_{i=1}^{N} f(x_i, y_i)\, \Delta x\, \Delta y$$

■ Let $f(x, y)$ be continuous on the bounded region D. Then the **definite integral of f over D** is

$$\int\int_D f(x, y)\, dA = \lim_{n \to \infty} \sum_{i=1}^{N} f(x_i, y_i)\, \Delta x\, \Delta y$$

where the limit exists and is the same no matter how the points (x_i, y_i) are chosen in R_i.

■ Let $f(x, y)$ be continuous on the bounded region D and nonnegative there. Then the volume under the surface $z = f(x, y)$ and over the domain D is the definite integral of f over D or

$$V = \int\int_D f(x, y)\, dA$$

■ If the domain D is given as in the figure and $z = f(x, y) \geq 0$ is continuous on D, then the volume V between the surface $z = f(x, y)$ and the domain D is

$$\int\int_D f(x, y)\, dA = \int_a^b \left[\int_{g_1(x)}^{g_2(x)} f(x, y)\, dy \right] dx$$

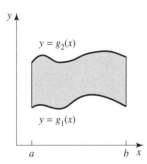

■ If the domain D is given as in the figure and $z = f(x, y) \geq 0$ is continuous on D, then the volume V between the surface $z = f(x, y)$ and the domain D is

$$\int\int_D f(x, y)\, dA = \int_c^d \left[\int_{g_1(y)}^{g_2(y)} f(x, y)\, dx \right] dy$$

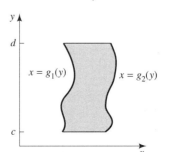

CHAPTER **7**

REVIEW EXERCISES

1. Evaluate the function $z = f(x, y) = 2x^2 + 3y - 1$ at the points $(0, 0), (1, 2), (2, -1)$.

2. Evaluate the function $z = f(x, y) = \dfrac{2x^2}{5y}$ at the points $(0, 1)$, $(1, 2), (2, -1)$.

3. Determine the domains of the functions in the previous two exercises.

4. Find the distance between the two points $(1, -2, 5)$ and $(3, -4, -1)$.

5. For the function $z = f(x, y) = 16 - x^2 - y^2$, find the level surfaces corresponding to $z_0 = 16, z_0 = 12, z_0 = 7$. Describe the surface $z = f(x, y)$.

6. Find both first partial derivatives to the function $z = f(x, y) = x + x^2y^5 + 10$.

7. Evaluate the two first partial derivatives found in the previous exercise at the point $(2, 1)$.

8. Find both first partial derivatives to the function $z = f(x, y) = e^{-xy} + \ln xy$.

9. Find all three second partial derivatives of the function in Exercise 6.

10. Find all three second partial derivatives of the function in Exercise 8.

11. Find all three first partial derivatives to the function $z = f(x, y, z) = e^{-xyz} + xy^2z^3$.

In Exercises 12 through 19 find all the critical points and determine whether each critical point is a relative minimum, relative maximum, or a saddle point.

12. $f(x, y) = x^2 - xy + 4y^2$

13. $f(x, y) = -x^2 + 2xy - 8y^2$

14. $f(x, y) = -x^2 + xy - y^2 - 3x$

15. $f(x, y) = (x - 2)^2 + (y + 3)^2$

16. $f(x, y) = x^3 + y^3 + 3xy$

17. $f(x, y) = 3x^2 + y^3 - 6xy - 9y$

18. $f(x, y) = y^3 - 3y^2 - x^2 - 9y + 10x$

19. $f(x, y) = 12xy - x^3 - 36y^2$

20. Let $z = f(x, y) = ax^2 + 2bxy + cy^2$, where a, b, and c are constants. Show that $(0, 0)$ is a critical point. Show that the point $(0, 0)$ is a relative minimum if $ac - b^2 > 0$ and $a > 0$, and a relative maximum if $ac - b^2 > 0$ and $a < 0$.

21. Using Lagrange multipliers, minimize $z = 4x^2 + 2xy - 3y^2$ subject to the constraint $x + y = 1$.

22. Using Lagrange multipliers, maximize $z = -2x^2 - xy - y^2 - x$ subject to the constraint $2x + y - 1 = 0$.

23. Use the tangent line approximation to estimate
$$\Delta z = f(x_0 + \Delta x, y_0 + \Delta y) - f(x_0, y_0)$$
if $f(x, y) = 2x^2 + 3y^2, (x_0, y_0) = (1, 2), \Delta x = 0.1, \Delta y = 0.2$.

24. Repeat the previous exercise if $f(x, y) = x^2y^4, (x_0, y_0) = (1, 2), \Delta x = 0.2, \Delta y = -0.1$.

25. Use the tangent line approximation to estimate $\sqrt{3.01^2 + 3.98^2}$.

In Exercises 26 through 31 find the double integral over the indicated region D in two ways. **(a)** *Integrate first with respect to* **x**; **(b)** *integrate first with respect to* **y**.

26. $\displaystyle\int\int_D (3x + 6y)\, dA, \quad D = \{(x, y): 0 \le x \le 2, 1 \le y \le 2\}$.

27. $\displaystyle\int\int_D (12x^2y^3)\, dA, \quad D = \{(x, y): 0 \le x \le 1, 1 \le y \le 3\}$.

28. $\displaystyle\int\int_D (x^2 + y)\, dA$, where D is the triangular region with vertices at $(0, 0), (1, 0), (1, 1)$.

29. $\displaystyle\int\int_D (x + xy)\, dA$, where D is the triangular region with vertices at $(0, 0), (0, 1), (1, 1)$.

30. $\displaystyle\int\int_D (x + y)\, dA$, where D is the region between the two curves $y = 1$ and $y = x^2$.

31. $\displaystyle\int\int_D (x + y)\, dA$, where D is the region between the two curves $y = x$ and $y = \sqrt{x}$.

In Exercises 32 through 35 change the order of integration and then evaluate.

32. $\displaystyle\int_0^1 \int_0^x (x + y)\, dy\, dx$

33. $\displaystyle\int_0^1 \int_0^{\sqrt{x}} (x + y)\, dy\, dx$

34. $\displaystyle\int_0^1 \int_0^{\sqrt{y}} (x + y)\, dx\, dy$

35. $\displaystyle\int_0^1 \int_y^{\sqrt{y}} dx\, dy$

36. Find the average value of $f(x, y) = 8xy^3$ on the region
$$D = \{(x, y): 0 \le x \le 2, 1 \le y \le 3\}$$

37. **Cobb-Douglas Production Function.** A production function is given by $f(K, L) = 10K^{0.20}L^{0.80}$. Find $f(243, 32)$.

38. **Cobb-Douglas Production Function.** Find the first partial derivatives of the production function given in the previous exercise.

39. **Cobb-Douglas Production Function.** Find all the second partial derivatives of the production function given in the previous exercise.

40. **Cost.** A cost function is given by $C(x, y) = 1000 + x^2 + y^2 - xy$. Find $C(10, 1)$.

41. **Cost.** Find the first partial derivatives of the cost function given in the previous exercise.

42. **Cost.** Find all the second partial derivatives of the cost function given in the previous exercise.

43. Continuous Compounding. If P is invested at an annual rate of r and compounded continuously, the amount after t years is $A(P, r, t) = Pe^{rt}$. Find the three first partial derivatives of this function.

44. Packaging. Find the dimensions of the cheapest rectangular box of volume 16 ft^3 for which the cost per square foot of the top and bottom is twice that of the sides.

45. Minimizing Production Costs. The daily cost function of a firm that produces x units of a product X and y units of another product Y is given by $C(x, y) = 244 + x^2 - 20x + y^2 - 24y$. Find the number of units of each of the products that minimize the cost.

46. Maximizing Profit. The firm in the previous exercise can sell each unit of product X for \$2 and each unit of product Y for \$4. Find the levels of production that maximize the profits.

47. Competitive Pricing. A firm makes x thousand units of regular toothpaste and y thousand units of mint toothpaste. Suppose the prices are p and q, respectively, and the demand equations are given by

$$p = 4 - x + y$$
$$q = 14 + x - 2y$$

Find the number of units of each that maximize revenue.

48. Fishery. A small lake is to be stocked with two types of fish. There are x thousand of the first fish and y thousand of the second, the average weight of the first is $(3 - 2x - y)$ pounds and of the second is $(4 - x - 3y)$. Find the numbers of each fish that maximize the total weight of these fish.

49. Measurements. A rectangular cardboard box of dimensions 10 inches by 10 inches by 20 inches is being produced. Use the tangent plane approximation to approximate the volume of cardboard if the sides are 0.15 inches thick and the top and bottom are 0.25 inches thick.

50. Fishery. Karpoff [135] used the function $h = h(e, N, X)$, where h = individual firm harvest rate, e = individual firm effort rate, N = number of participating firms (or vessels), and X = resource stock size. He then assumes

$$\frac{\partial h}{\partial e} > 0, \qquad \frac{\partial h}{\partial N} < 0, \qquad \frac{\partial h}{\partial X} > 0$$

Explain why you think these three conditions are reasonable.

CHAPTER **7**

PROJECTS

Project 1
**The Theory
of Economic Growth**

This discussion is based on a paper by Robert M. Solow (Nobel Prize winner in economics) [136].

Let the total rate of production for a country at time t be designated as $Y(t)$. Part of the output at any instant is consumed, and the rest is saved. We assume that the fraction s of output that is saved is a constant, so that the rate of savings is $sY(t)$. We let $K(t)$ denote the stock of capital, and then net investment is the rate of increase of this capital stock, or $\frac{dK}{dt}$. Thus, we must have

$$\frac{dK}{dt} = sY$$

We assume a Cobb-Douglas production function of the form

$$Y = K^a L^{1-a}$$

where L is the rate of labor input. Thus,

$$\frac{dK}{dt} = sY = sK^a L^{1-a}$$

We now assume that the labor force increases at a constant rate n. Thus,

$$L(t) = L_0 e^{nt}$$

where L_0 is the initial labor force. Substituting this into the last equation then gives

$$\frac{dK}{dt} = sK^a (L_0 e^{nt})^{1-a}$$

(**a**) Integrate this last equation and find $K(t)$, assuming that K_0 is the initial capital stock. Now define $k = K/L$ as the level of output per unit of labor, and find an expression for $k(t)$.

(**b**) From part (a) determine approximately what $k(t)$ is for large t. (*Note:* The term $\left(\dfrac{s}{n}\right)^{1/b}$ is referred to as the equilibrium value of the capital/labor ratio.

(**c**) What happens to the long-term value of $k(t)$ if the savings rate s is larger? If the rate of labor growth n is larger? Does this model indicate that a country with higher savings rate is richer or poorer? What about a higher labor growth?

By looking at a large set of economic data on many countries Mankiw and colleagues [137] showed that the Solow model given here is correct to a first approximation. This paper then modifies the Solow model and shows that the modifications are even better predictors of the direction and magnitudes of the effects of savings and labor. The improvement comes from including another term that accounts for human capital. Specifically, this model assumes a production function of the form

$$Y(t) = K(t)^a H(t)^b L(t)^{1-a-b}$$

where H is the stock of human capital and the other variables are the same as in the Solow model.

8

Differential Equations

This chapter presents only the barest introduction to differential equations. The field of differential equations is an extremely active area of ongoing research. Differential equations appear not only throughout physics and engineering but also in many areas of biology, economics, and the social sciences.

The first section of this chapter sets down some basics, the second section gives an important technique for solving certain differential equations with a number of applications, and the third section presents a method for finding an approximate solution to a differential equation. The last two sections give a large variety of applications to population and related problems.

8.1 DIFFERENTIAL EQUATIONS

■ Solutions of Differential Equations ■ Models Involving Differential Equations

APPLICATION: A DIFFERENTIAL EQUATION

From physics it is known that the rate of change of the temperature T with respect to time t of a hot object placed in a medium held at a constant temperature C is negatively proportional to the difference of the temperature of the object and the surrounding medium. Write $\dfrac{dT}{dt}$ as a function of T. The answer is found in Example 8.

Solutions of Differential Equations

A **differential equation** is an equation that involves an unknown function together with one or more of its derivatives. A **first order differential equation** is an equation that involves an unknown function and its first derivative. Examples of first order differential equations are

$$\frac{dy}{dx} = 2y, \qquad \frac{dy}{dx} = y + 2x, \qquad \frac{dy}{dx} = 8x^3y^2$$

In this text we only consider first order differential equations that can be written in the form

$$\frac{dy}{dx} = f(x, y) \qquad\qquad [1]$$

By a **solution** of Equation 1, we mean the following:

Solution of a Differential Equation

We say that $y = y(x)$ is a **solution** of Equation 1 on the interval (a, b) if $y(x)$ is differentiable on (a, b) and for all $x \in (a, b)$

$$\frac{d}{dx} y(x) = f[x, y(x)]$$

Remark. Notice that a solution to a differential equation is a *function.*

EXAMPLE 1   Showing a Function Is a Solution of a Differential Equation

Show that $y = y(x) = -\dfrac{1}{2x^4 + 1}$ is a solution of the differential equation

$$\frac{dy}{dx} = 8x^3y^2$$

on the interval $(-\infty, \infty)$.

SOLUTION ■ First notice that the denominator of $\overline{y}(x)$ is never zero and that $\overline{y}(x)$ is differentiable everywhere. Furthermore, for all x

$$\frac{d}{dx} y(x) = \frac{d}{dx}\left(-\frac{1}{2x^4 + 1} \right)$$

$$= \frac{8x^3}{(2x^4 + 1)^2}$$

$$= 8x^3\left(-\frac{1}{2x^4 + 1}\right)^2$$

$$= 8x^3\bar{y}^2$$

Thus, $y(x)$ is a solution of the differential equation on the interval $(-\infty, \infty)$. ■

EXAMPLE 2   Showing a Function Is a Solution of a Differential Equation

Show that $y = y(x) = 5e^{2x}$ is a solution of the differential equation

$$\frac{dy}{dx} = 2y \qquad\qquad [2]$$

on the interval $(-\infty, \infty)$.

SOLUTION ■ First notice that $y(x)$ is differentiable everywhere. Furthermore, for all x

$$\frac{d}{dx}y(x) = \frac{d}{dx}5e^{2x} = 10e^{2x} = 2(5e^{2x}) = 2y(x)$$

Thus, the function $y = y(x) = 5e^{2x}$ is a solution of Equation 2 on the interval $(-\infty, \infty)$. ■

EXAMPLE 3   Finding the Solution Intervals

Find the largest interval(s) on which the function $y = -\dfrac{1}{x}$ is a solution of the differential equation

$$\frac{dy}{dx} = y^2$$

SOLUTION ■ First notice that $y(x)$ is not differentiable at $x = 0$ but is differentiable on the intervals $(-\infty, 0)$ and $(0, \infty)$. Furthermore, for any x on these two intervals

$$\frac{d}{dx}y(x) = \frac{d}{dx}\left(-\frac{1}{x}\right) = \frac{1}{x^2} = y^2(x)$$

Therefore, $y(x)$ is a solution to the given differential equation on the two intervals $(-\infty, 0)$ and $(0, \infty)$. ■

EXAMPLE 4   Showing a Function Is a Solution of a Differential Equation

Show that for any constant C, $y = y(x) = Ce^{2x}$ is a solution of Equation 2 on the interval $(-\infty, \infty)$.

SOLUTION ■ First notice that for any constant C the function $y = y(x) = Ce^{2x}$ is differentiable on the interval $(-\infty, \infty)$. Furthermore, for any constant C and any x

$$\frac{d}{dx}y(x) = \frac{d}{dx}Ce^{2x} = 2Ce^{2x} = 2(Ce^{2x}) = 2y(x)$$

Thus, for any constant C the function $y = y(x) = Ce^{2x}$ is a solution of Equation 2 on the interval $(-\infty, \infty)$. ■

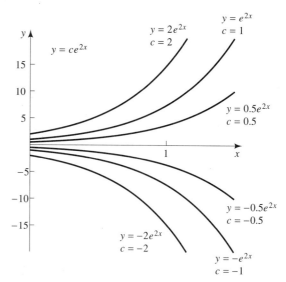

Figure 8.1
Several members of the family $y = Ce^{2x}$, all of which are solutions of $\frac{dy}{dx} = 2y$.

This means that Equation 2 has an *infinite* number of solutions. A different solution is obtained by taking a different value for C. Figure 8.1 then shows a *family* of solutions to Equation 2.

We are already familiar with solving differential equations of a very simple type. Finding the solutions of the differential equation $\frac{dy}{dx} = g(x)$ is the same as finding the antiderivatives of $g(x)$.

EXAMPLE 5   Finding the Solutions of a Differential Equation

Find all solutions of the differential equation

$$\frac{dy}{dx} = 6x^2 \qquad [3]$$

SOLUTION ■ We know from Chapter 5 that a solution of this particular differential equation must be an antiderivative of $6x^2$. We also know that *all* antiderivatives and thus all solutions of the given differential equation are of the form $y = 2x^3 + C$ for some constant C. Furthermore, since these functions are all differentiable everywhere, they are all solutions on the interval $(-\infty, \infty)$. ■

Since *every* solution of Equation 3 must be of the form $y = 2x^3 + C$, we say that $y = 2x^3 + C$ is the **general solution** of Equation 3. In solving Equation 3 we integrated. This process of integration gave an arbitrary constant of integration. In the next section we will see that to solve a first order differential equation, we must, at some point, integrate. This integration then produces an arbitrary constant of integration, which then yields the general solution. In another example in the next section we will see that $y = Ce^{2x}$ is the general solution of $\frac{dy}{dx} = 2y$.

By assigning a particular number to the constant C, a **particular solution** is obtained. The usual way a particular value is assigned to the constant C is by specifying an **initial condition** $y(x_0) = y_0$. A differential equation given together with an initial condition is called an **initial value problem**.

Initial Value Problem

We call the problem of finding a solution to

$$\frac{dy}{dx} = f(x, y) \qquad y(x_0) = y_0$$

on some interval containing x_0 an **initial value problem**.

EXAMPLE 6 Finding a Solution to an Initial Value Problem

Using the fact that $y = Ce^{2x}$ is the general solution to Equation 2, find the (particular) solution $y(x)$ to the initial value problem

$$\frac{dy}{dx} = 2y \qquad y(0) = 10$$

SOLUTION ■ We know that $y(x) = Ce^{2x}$ for some C. Using the initial condition yields

$$10 = y(0) = Ce^{2(0)} = C$$

Thus, the particular solution of the initial value problem is $y(x) = 10e^{2x}$. ■

Models Involving Differential Equations

We now see how to determine differential equations that arise in applications. We will wait until the next section, however, to find the solutions of these differential equations.

EXAMPLE 7 Finding a Differential Equation

Biologists argue that when a population is small compared with the available resources, the population grows at a rate that is proportional to the population. Find the differential equation that the population must satisfy.

SOLUTION ■ If $P(t)$ is the population at time t, the rate of change of the population is $\frac{dP}{dt}$. If this is proportional to the population P, then the population satisfies the differential equation

$$\frac{dP}{dt} = kP$$

for some positive constant of proportionality k. ■

EXAMPLE 8 Finding a Differential Equation

From physics it is known that the rate of change of the temperature T with respect to time t of a hot object placed in a medium held at a constant temperature C, is negatively proportional to the difference of the temperature of the object and the surrounding medium. Find the differential equation that the temperature of the object satisfies.

SOLUTION ■ It t is time, we are given that $\frac{dT}{dt}$ is negatively proportional to $(T - C)$. Thus,

$$\frac{dT}{dt} = -k(T - C)$$

for some positive constant k. ■

EXAMPLE 9 Finding a Differential Equation

When a new technological innovation is introduced into an industry, there is some evidence to support the conclusion that the rate of change of the percentage of products produced using this innovation is proportional to the product of the percentage using this innovation with the percentage not using it. Find the differential equation that the percentage $P(t)$ satisfies.

SOLUTION ■ The percentage not using the innovation is $(100 - P)$. Thus, the differential equation must be

$$\frac{dP}{dt} = kP(100 - P)$$

for some positive constant of proportionality k. ■

We shall see how to find the general solution of these last three differential equations in the next section.

SELF-HELP EXERCISE SET 8.1

1. Show that $y = y(x) = e^x(x^2 + 1)$ is a solution of the differential equation

$$\frac{dy}{dx} = y + 2xe^x$$

on the interval $(-\infty, \infty)$.

2. Find the general solution to the differential equation

$$\frac{dy}{dx} = 4x + 5$$

3. Using the results of the previous exercise, solve the initial value problem

$$\frac{dy}{dx} = 4x + 5 \qquad y(1) = 10$$

4. Suppose $P(t)$ is the number of tons of copper per year extracted from a mine and that the rate of change of P is negatively proportional to P. Find the differential equation that P must satisfy.

EXERCISE SET 8.1

In Exercises 1 through 10 verify that the given $\bar{y}(x)$ is a solution on the indicated interval of the given differential equation for any constant C.

1. $y(x) = x^3 + C$, $(-\infty, \infty)$, $y' = 3x^2$

2. $y(x) = x + C$, $(-\infty, \infty)$, $y' = 1$

3. $y(x) = e^x + C$, $(-\infty, \infty)$, $y' = e^x$

4. $y(x) = \frac{1}{3}e^{3x} + C$, $(-\infty, \infty)$, $y' = e^{3x}$

5. $y(x) = Ce^{x^2}$, $(-\infty, \infty)$, $y' = 2xy$

6. $y(x) = Ce^{x^3}$, $(-\infty, \infty)$, $y' = 3x^2y$

7. $y(x) = Ce^{x^3+x^2}$, $(-\infty, \infty)$, $y' = (3x^2 + 2x)y$

8. $y(x) = Cx$, $(0, \infty)$, $y' = \frac{y}{x}$

9. $y(x) = Ce^{-1/x}$, $(0, \infty)$, $y' = \frac{y}{x^2}$

10. $y(x) = Ce^{-1/x}$, $(-\infty, 0)$, $y' = \frac{y}{x^2}$

In Exercises 11 through 20 find the solution to the initial value problem using your answers from the previous 10 problems.

11. $y(x) = x^3 + C$, $(-\infty, \infty)$, $y' = 3x^2$, $y(1) = 1$

12. $y(x) = x + C$, $(-\infty, \infty)$, $y' = 1$, $y(2) = 10$

13. $y(x) = e^x + C$, $(-\infty, \infty)$, $y' = e^x$, $y(0) = 0$

14. $y(x) = \frac{1}{3}e^{3x} + C$, $(-\infty, \infty)$, $y' = e^{3x}$, $y(0) = 4$

15. $y(x) = Ce^{x^2}$, $(-\infty, \infty)$, $y' = 2xy$, $y(0) = 10$

16. $y(x) = Ce^{x^3}$, $(-\infty, \infty)$, $y' = 3x^2y$, $y(1) = e$

17. $y(x) = Ce^{x^3+x^2}$, $(-\infty, \infty)$, $y' = (3x^2 + 2x)y$, $y(0) = 100$

18. $y(x) = Cx$, $(0, \infty)$, $y' = \frac{y}{x}$, $y(2) = 4$

19. $y(x) = Ce^{-1/x}$, $(0, \infty)$, $y' = \frac{y}{x^2}$, $y(1) = 1$

20. $y(x) = Ce^{-1/x}$, $(-\infty, 0)$, $y' = \frac{y}{x^2}$, $y(-1) = 1$

In Exercises 21 through 26 solve the given differential equation.

21. $y' = 4x^3$, $y(0) = 3$ **22.** $y' = x + 3$, $y(0) = 1$

23. $y' = e^{2x}$, $y(0) = 3$

24. $y' = 6x^2 + 4x + 5$, $y(0) = 1$

25. $y' = 2xe^{x^2}$, $y(0) = 1$ **26.** $y' = \dfrac{2}{x + 1}$, $y(0) = 0$

Applications

27. Revenue. Suppose the rate of change of revenue R with respect to number of sales x satisfies

$$\frac{dR}{dx} = xe^{-x^2}, \quad R(0) = 0$$

Find $R(x)$.

28. Cost. Suppose the rate of change of cost C with respect to number of sales x satisfies

$$\frac{dC}{dx} = 12x^3 + 2x, \quad C(0) = 10{,}000$$

Find $C(x)$.

29. Psychology. The Webner-Fechner law in the quantitative theory of psychology asserts that the rate of change dR/dS of a reaction R with respect to a stimulus S is inversely proportional to the stimulus. Set up the differential equation.

30. Medicine. If a glucose solution is given intravenously, the concentration in the blood tends to increase at a constant rate R less an amount that is proportional to the present concentration of glucose. Find the differential equation governing the glucose concentration C.

31. Chemistry. Some chemical reactions involve a reactant Y that reacts with itself. Let y be the concentration of Y. Often the rate of change of this reaction is proportional to the negative of the square of the concentration. Find the differential equation that y satisfies.

32. Chemistry. Answer the same questions as in Exercise 31 if the reaction is third order, that is, if the rate of change of this reaction is proportional to the negative of the cube of the concentration.

33. Chemistry. Two chemicals A and B react to produce a third Y. If a is the concentration of A and b is the concentration of B, then for some reactions, the rate of change of the concentration y of Y is proportional to the difference of the concentration of A and the concentration of Y times the difference of the concentration of B and the concentration of Y. Find the differential equation that y satisfies.

34. Biology. In a certain area a mouse population has been happily obeying the differential equation

$$\frac{dP}{dt} = 0.04P - 0.01P^2$$

where t is in hours. At time $t = 0$ a family moves into the area with a housecat who proceeds to terrorize the mouse population. Frightened by this cat, the mice migrate out of the area at the rate of 0.03 mice per hour. Modify the differential equation to account for the migration.

35. Mixing. A tank initially contains 10 pounds of salt dissolved in 200 gallons of water. Water containing 1 pound of salt per gallon starts entering the tank at a rate of 5 gallons per minute. The solution is well stirred and leaves at the same rate. Find a differential equation involving the amount $A(t)$ of salt in the tank at time t.

Enrichment Exercises

36. Define $y = f(x)$ as follows:

$$f(x) = \begin{cases} -1 & x \le 0 \\ 1 & x > 0 \end{cases}$$

   (a) Explain carefully and in complete sentences why $y = f(x)$ is *not* a solution of the differential equation $\dfrac{dy}{dx} = 0$ on the interval $(-\infty, \infty)$.

   (b) Explain carefully why $y = f(x)$ *is* a solution of the differential equation $\dfrac{dy}{dx} = 0$ on the interval $(0, \infty)$.

37. Explain in complete sentences the difference between a particular solution of a differential equation and the general solution. Give specific examples.

38. Explain the difference between the general solution of the differential equation $\dfrac{dy}{dx} = f(x, y)$ and the solution of the initial value problem

$$\frac{dy}{dx} = f(x, y), \quad y(x_0) = y_0$$

SOLUTIONS TO SELF-HELP EXERCISE SET 8.1

1. First notice that the function $y = \bar{y}(x) = e^x(x^2 + 1)$ is differentiable on the interval $(-\infty, \infty)$. Furthermore, for all x

$$\frac{d\bar{y}}{dx} = \frac{d}{dx} e^x(x^2 + 1)$$

$$= e^x(2x) + e^x(x^2 + 1)$$
$$= e^x(x^2 + 1) + 2xe^x$$
$$= \bar{y}(x) + 2xe^x$$

Thus, the function $y = \bar{y}(x) = e^x(x^2 + 1)$ is a solution of the differential equation on $(-\infty, \infty)$.

2. The general solution to the differential equation

$$\frac{dy}{dx} = 4x + 5$$

is the indefinite integral $\int(4x + 5)\,dx = 2x^2 + 5x + C$.

3. Using the results of the previous exercise

$$10 = y(1) = 2(1)^2 + 5(1) + C = 7 + C$$

Thus, $C = 3$ and the particular solution of the initial value problem is $\bar{y}(x) = 2x^2 + 5x + 3$.

4. Since the rate of change of P is negatively proportional to P, the differential equation must be

$$\frac{dP}{dt} = -kP$$

where k is a positive constant.

8.2 SEPARATION OF VARIABLES

■ Method of Separation of Variables ■ Applications

APPLICATION: POPULATION GROWTH

A population satisfies the initial value problem

$$\frac{dP}{dt} = kP - bP^2, \quad P(0) = P_0$$

where k and b are positive constants. What is $P(t)$? See Example 4 for the answer.

APPLICATION: POPULATION OF THE WORLD

What is a reasonable estimate of the future human population of the earth? See Example 5 for an answer.

Method of Separation of Variables

In this section we solve differential equations of the type

$$\frac{dy}{dx} = f(x)g(y) \tag{1}$$

by the **method of separation of variables**. We will see that many differential equations in the applications are in the form of Equation 1.

To solve this differential equation, proceed as follows:

Solving Equation 1 by Separation of Variables

1. Place all the y-terms on the left and all the x-terms on the right (separating the variables).

$$\frac{dy}{g(y)} = f(x)\,dx$$

2. Then integrate

$$\int \frac{dy}{g(y)} = \int f(x)\,dx$$

3. Solve for y if possible.

Remark. The problem of solving Equation 1 has been reduced to *two* problems in integration. We need only use *one* constant of integration.

Applications

We saw in Example 7 of the previous section that when populations are small compared with the available resources, the rate of growth of the population is proportional to the population. This implies that the populations satisfied a differential equation of the form $\frac{dP}{dt} = kP$, where k is a constant. We now find the solutions of this differential equation.

EXAMPLE 1   Solving by Separation of Variables

Find the general solution of the differential equation

$$\frac{dP}{dt} = kP \qquad\qquad [2]$$

where k is a constant.

S O L U T I O N ■ In this particular differential equation, we can place all P-terms on one side and all t-terms on the other side to obtain

$$\frac{dP}{P} = k\,dt$$

Now integrate both sides and obtain

$$\int \frac{1}{P}\,dP = \int k\,dt$$

$$\ln |P| = kt + C^*$$

where C^* is an arbitrary constant. This is equivalent to

$$|P| = e^{kt+C^*} = e^{C^*}e^{kt}$$

Thus,

$$P = e^{C^*}e^{kt} \qquad \text{or} \qquad P = -e^{C^*}e^{kt}$$

We then replace the constants e^{C^*} and $-e^{C^*}$ with the single constant C to obtain

$$P = Ce^{kt} \quad ■$$

This is the general solution since *all* solutions must be in this form. The general solution *always* contains an arbitrary constant of integration.

A positive constant k describes the situation when the population is growing. Figure 8.2 indicates solutions for various values of C when $k = 2$.

EXAMPLE 2   Finding the Solution to an Initial Value Problem

Suppose a population satisfies Equation 2 and the population at time t_0 is P_0. Then the population satisfies

$$\frac{dP}{dt} = kP, \quad P(t_0) = P_0$$

Solve this initial value problem.

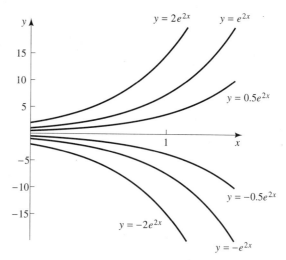

Figure 8.2
Several members of the family $y = Ce^{2x}$, all of which are solutions of $\dfrac{dy}{dx} = 2y$.

S O L U T I O N ■ From Example 1 we know that $P(t)$ must be given by $P(t) = Ce^{kt}$. The initial condition implies that

$$P_0 = P(t_0) = Ce^{kt_0}$$

thus, $C = P_0 e^{-kt}$, and the solution of the initial value problem is

$$
\begin{aligned}
P(t) &= Ce^{kt} \\
&= P_0 e^{-kt_0} e^{kt} \\
&= P_0 e^{k(t - t_0)} \quad \blacksquare
\end{aligned}
$$

Remark. In the very important special case that $t_0 = 0$, this becomes

$$P(t) = P_0 e^{kt}$$

E X A M P L E 3   An Initial Value Problem

If a cold object is placed in a surrounding medium kept at a constant temperature C, then the rate of change of the temperature $T(t)$ with respect to time t is proportional to the difference of the temperature of the surrounding medium and the object and thus satisfies

$$\frac{dT}{dt} = k(C - T), \quad T(0) = T_0$$

Assume that $T_0 < C$, that is, that the initial temperature is less than the temperature of the surrounding medium. Assume also that $k > 0$. Find the solution of the initial value problem.

S O L U T I O N ■ To solve the differential equation we place all the T-terms on the left and all the t-terms on the right and integrate. Thus,

$$\frac{dT}{C - T} = k\,dt$$

$$\int \frac{dT}{C - T} = \int k\,dt$$

$$-\ln |C - T| = kt + C^*$$

$$\ln(C - T) = -kt - C^*$$

where C^* is an arbitrary constant of integration and where we have $\ln|C - T| = \ln(C - T)$ since we know that the temperature T is always less than C. The initial condition then implies that

$$\ln(C - T_0) = k \cdot 0 - C^*$$

Thus,

$$\ln(C - T) = -kt + \ln(C - T_0)$$

$$\ln \frac{C - T}{C - T_0} = -kt$$

$$\frac{C - T}{C - T_0} = e^{-kt}$$

$$C - T = (C - T_0)e^{-kt}$$

$$T = C - (C - T_0)e^{-kt} \quad \blacksquare$$

We encountered this equation before as an example of limited growth. For example, if the cold object was a glass of cold milk on a table in a room kept at a constant temperature of 70°F, then the temperature of the milk increases and heads for the limiting value of 70°F.

We can verify this analytically. First notice that since $k > 0$,

$$\lim_{t \to \infty} T(t) = C$$

Also notice that $T'(t) = (C - T_0)e^{-kt}$. Thus, when $T_0 < C$, $T'(t) > 0$, and the temperature $T(t)$ always increases. Finally, $T''(t) = -(C - T_0)e^{-kt}$. Therefore, when $T_0 < C$, $T''(t) < 0$, and the curve $y = T(t)$ is concave down. The graph is shown in Figure 8.3.

Biologists have noted that populations tend to grow exponentially when the population is small but then level off in time due to the finite amount of available resources. If the population is larger than that supportable by the available resources, then the population has been observed to decrease to some limiting value. The Dutch mathematician-biologist P. Verhulst in 1838 replaced the growth model given in Equation 2 with a more realistic model that incorporates this finite resource factor.

Notice that Equation 2 says that the increase in the population becomes larger as the population becomes larger. This is reasonable when the population is small. But as the population becomes large enough to be affected by the limited available resources, this increase should slow. Thus, the equation should be modified to reflect this fact.

Verhulst suggested subtracting a positive term from the right-hand side of Equation 2 that would become larger if the population became larger. The simplest such term is bP^2, where b is a positive constant. Doing this gives

$$P'(t) = kP - bP^2 \qquad [3]$$

The term b is assumed to be very small compared with k. In fact, as we will see, for the population of the earth in billions $b \approx 0.003$. Thus, if the population is less than a billion, the term bP^2 in Equation 3 is much smaller than the term kP. In this case the right-hand side of Equation 3 is approximately kP, which just gives Equation 2 again. As the population of P increases, however, the term bP^2 becomes larger and slows up the change in population.

It is instructive to factor the right-hand side of Equation 3, obtaining

$$P'(t) = k\left(1 - \frac{b}{k}P\right)P$$

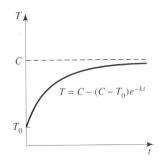

Figure 8.3
The graph indicates that the temperature T increases toward the limiting value of C.

It is furthermore useful to let $L = k/b$ and finally obtain

$$P'(t) = k\left(1 - \frac{P}{L}\right)P \qquad [4]$$

This equation is called the **logistic equation**. From this equation one can see that if $P < L$, then the right-hand side is *positive* and the population increases. Furthermore, as the population increases, the term in front of P in Equation 4 becomes smaller, that is, the growth rate slows. Thus, it is instructive to view the term

$$r = k\left(1 - \frac{P}{L}\right)$$

as a changing growth rate. As P increases toward L, this growth rate r slows to zero. Should P ever exceed L, then this growth rate becomes *negative*, and the population decreases.

Also notice from Equation 4 that *two* constant functions satisfy this equation. The first is just $P = 0$. The second is $P = L$. These constant functions are also referred to as **equilibrium points**, or **fixed points**.

We now find the solution of the logistic equation.

EXAMPLE 4   Solving an Initial Value Problem

A population satisfies the initial value problem

$$\frac{dP}{dt} = kP - bP^2, \quad P(0) = P_0$$

where k and b are positive constants. Find $P(t)$.

SOLUTION ■ Once again we place all the P-terms on the left and all the t-terms on the right obtaining

$$\frac{dP}{P(k - bP)} = dt \qquad [5]$$

Before we can integrate the left, we need to put the fraction on the left in a more convenient form. We try to find constants A and B such that

$$\frac{1}{P(k - bP)} = \frac{A}{P} + \frac{B}{k - bP}$$

By multiplying by $P(k - bP)$, we see that this is the same as

$$1 = A(k - bP) + BP$$

A very convenient way to find A and B is to notice that since this last equation must be true for *all* P, then it certainly must be true for $P = 0$ and for $P = k/b$. If $P = 0$, the equation becomes simply $1 = A(k)$, or $A = 1/k$. Also if $P = k/b$, then the equation becomes $1 = B\dfrac{k}{b}$, or $B = b/k$. Thus, we have

$$\frac{1}{P(k - bP)} = \frac{1}{k}\frac{1}{P} + \frac{1}{k}\frac{b}{k - bP}$$

Putting this back into Equation 4 and integrating, we obtain

$$\int \frac{1}{k}\frac{1}{P} + \int \frac{1}{k}\frac{b}{k-bP} = \int dt$$

$$\frac{1}{k}\ln|P| - \frac{1}{k}\ln|k-bP| = t + C^*$$

where C^* is an arbitrary constant of integration. Continuing, we obtain

$$\ln\left|\frac{P}{k-bP}\right| = kt + kC^*$$

$$\left|\frac{P}{k-bP}\right| = e^{kt}e^{kC^*}$$

$$\frac{P}{k-bP} = \pm e^{kt}e^{kC^*} = Ce^{kt}$$

where $C = \pm e^{kC^*}$. To solve explicitly for P we have

$$P = C(k-bP)e^{kt}$$

$$P + CbPe^{kt} = Cke^{kt}$$

$$P = \frac{kCe^{kt}}{1 + Cbe^{kt}}$$

Using the initial condition, we obtain

$$P_0 = P(0) = \frac{kC}{1 + Cb}$$

$$(1 + Cb)P_0 = kC$$

$$P_0 = C(k - bP_0)$$

$$C = \frac{P_0}{k - bP_0}$$

if $P_0 \neq k/b$. Putting this back into the equation for P, we obtain

$$P = \frac{\dfrac{P_0}{k-bP_0}ke^{kt}}{1 + \dfrac{P_0 b}{k-bP_0}e^{kt}}$$

$$= \frac{P_0 ke^{kt}}{k - bP_0 + bP_0 e^{kt}}$$

$$= \frac{kP_0}{bP_0 + (k - bP_0)e^{-kt}}$$

If we divide the numerator and denominator of the right-hand side of this last equation by b and recall that $L = k/b$, we then obtain

$$P(t) = \frac{LP_0}{P_0 + (L - P_0)e^{-kt}}$$

This is the form of the solution that is most convenient. ∎

We already noticed from Equation 4 that, if $P(t) < L$, then $P'(t) > 0$. Thus, in this case, the population increases. Notice also that

$$\lim_{t\to\infty} P(t) = \lim_{t\to\infty} \frac{LP_0}{P_0 + (L - P_0)e^{-kt}} = L$$

Therefore, when the initial population is less than L, the population steadily increases to the limiting value of L. Finally, we check the concavity. We could calculate $P''(t)$,

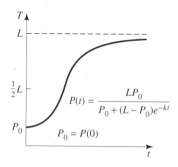

Figure 8.4
When $P_0 < L$, the logistic curve increases toward the limiting value of L. The curve is concave up when $P(t) < 0.5L$ and concave down when $P(t) > 0.5L$.

but this is tedious. Since we are only interested, for now, in the *sign* of $P''(t)$, we can do the following. From Equation 4 we have

$$P' = k\left(1 - \frac{P}{L}\right)P$$

$$= kP - \frac{k}{L}P^2$$

$$P'' = kP' - \frac{2k}{L}PP'$$

$$= k\left(1 - \frac{2}{L}P\right)P'$$

We noted that when $P(t) < L$, $P'(t) > 0$. Thus, in this situation, $P''(t)$ has the same sign as the factor $\left(1 - \frac{2}{L}P\right)$. Since this factor is positive when $P < \frac{L}{2}$ and negative when $P > \frac{L}{2}$, we see that $P''(t)$ is positive when $P < \frac{L}{2}$ and negative when $P > \frac{L}{2}$. Thus, when the population starts out small, the population curve is concave up until the population reaches $\frac{L}{2}$. Then the curve turns concave down. Figure 8.4 shows a graph.

We now summarize the significance of the three constants P_0, k, and L. The initial population (at time $t = 0$) is P_0, k is the growth rate of the population when the population is small compared with the available resources, and L is the limiting value of the population.

Many world organizations are very concerned about the growth of the human population of the earth, particularly in regard to our ability to produce enough food to feed a growing population. A necessary first problem is to make some estimate of the future limiting human population of the earth. In the next example, we do this.

EXAMPLE 5   Estimating the Future Population of the World

Using historical records to estimate the past human population of the earth, biologists estimate the correct value of k in the logistic equation to be $k = 0.03$. In 1990, the population of the earth was about $5\frac{1}{3}$ billion growing at about 1.6% per year. Estimate the constant b in the logistic equation, and obtain an estimate for the limiting value of the human population of the earth.

SOLUTION ■ We rewrite Equation 3 as

$$\frac{1}{P}\frac{dP}{dt} = k - bP$$

If $\Delta t = 1$ year, then

$$\frac{1}{P}\frac{\Delta P}{\Delta t} = \frac{\Delta P}{P}$$

is the fractional change in the population, which is given as 1.6/100. But we can make the estimate

$$\frac{1}{P}\frac{dP}{dt} \approx \frac{1}{P}\frac{\Delta P}{\Delta t} \approx 0.016$$

Putting this information in the previous rewritten logistic equation gives

$$0.016 \approx \frac{1}{P}\frac{dP}{dt} = k - bP \approx 0.03 - b \cdot \frac{16}{3}$$

This yields

$$b = (0.014)\frac{3}{16} \approx 0.0026$$

The limiting value of the population of the earth is

$$L = \frac{k}{b} = \frac{0.03}{0.0026} = 11.5$$

or about 11.5 billion. ∎

It is interesting to note that the change in concavity occurs at half this value, which is approximately the population at this time. Thus, according to this model, the population "explosion" is over, and the population has just begun to start the leveling-off phase in which the finite resource term comes into prominence.

Recall that the population $P(t)$ satisfies

$$P(t) = \frac{LP_0}{P_0 + (L - P_0)e^{-kt}}$$

Let $t = 0$ correspond to 1990. Then, as we noted, the population of the earth is presently about one-half of L. Also recall that $k = 0.03$. Thus,

$$P(t) = \frac{11.5}{1 + e^{-0.03t}}$$

Screen 8.1 shows a graph of this function using a window with dimensions $[-200, 200]$ by $[0, 12]$. Notice, according to this model, how the population rose rapidly during the last 200 years and how the population growth will slow in the future.

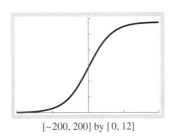

[−200, 200] by [0, 12]

Screen 8.1

The logistic curve $y = \dfrac{11.5}{1 + e^{-0.03t}}$ models the human population of the earth in billions. Notice that the population increased rapidly in the past; but, according to this model, the growth of the population will slow with the population heading for the limiting value of 11.5 billion.

SELF-HELP EXERCISE SET 8.2

1. Solve the initial value problem

$$\frac{dy}{dx} = \frac{y}{x+1}, \quad y(1) = 4$$

EXERCISE SET 8.2

In Exercises 1 through 14 solve the given differential equations.

1. $y' = 4x^3 y, \quad y(0) = 3$

2. $y' = xy, \quad y(0) = 1$

3. $y' = xy^2, \quad y(1) = 2$

4. $y' = y^2, \quad y(2) = 1$

5. $y' = e^y, \quad y(0) = 1$

6. $y' = e^{-y}, \quad y(0) = 1$

7. $y' = e^{x-y}, \quad y(0) = 0$

8. $y' = x/y, \quad y(1) = 1$

9. $y' = \frac{x+1}{y^2}, \quad y(0) = 1$

10. $y' = \frac{y^2}{x+1}, \quad y(0) = 1$

11. $y' = \frac{y}{x}, \quad y(1) = 1$

12. $y' = \sqrt{\frac{y}{x}}, \quad y(1) = 2$

13. $y' = \frac{x+1}{y^2+1}, \quad y(0) = 3$

14. $y' = \frac{xy^2 + x}{y}, \quad y(0) = 0$

Applications

15. Geometry. Find a family of curves with the slope of their tangent line at any point $P(x, y)$ given by $-\frac{x}{y}$. Graph these curves.

16. Limited Growth. In Example 3 of the text, the case $T_0 < C$ was solved. Solve the initial value problem when $T_0 \ge C$. *Hint:* Consider two cases. First $T_0 > C$ and then $T_0 = C$.

17. Psychology. The Webner-Fechner law in the quantitative theory of psychology asserts that the rate of change dR/dS of a reaction R with respect to a stimulus S is inversely proportional to the stimulus. Thus,

$$\frac{dR}{dS} = \frac{k}{S}$$

(a) Find the general solution. **(b)** Find the solution such that $R(S_0) = 0$ and graph.

18. Medicine. If a glucose solution is given intravenously, the concentration in the blood tends to increase at a constant rate R except that the glucose is continuously converted and excreted at a rate that is proportional to the present concentration of glucose. Thus, the differential equation governing the glucose concentration C is given by

$$\frac{dC}{dt} = R - kC, \quad C(0) = c_0$$

where k is a positive constant. Find $C(t)$.

19. Chemistry. Suppose a chemical reaction involves a reactant Y that reacts with itself. Let y be the concentration of Y. Often the rate of change of this reaction is proportional to the negative of the square of the concentration. (This is called a *second-order reaction*). This yields the differential equation

$$\frac{dy}{dt} = -ky^2, \quad y(0) = y_0$$

where k is a positive constant. Find the solution. Also find the time when half of the reactant is left.

20. Chemistry. Answer the same questions as in the previous problem if the reaction is third order, that is, if

$$\frac{dy}{dt} = -ky^3, \quad y(0) = y_0$$

where k is a positive constant.

21. Chemistry. Two chemicals A and B react to produce a third Y. If a is the concentration of A and b is the concentration of B, then for some reactions, the rate of change of the concentration y of Y is proportional to $(a - y)(b - y)$. Thus, y satisfies the differential equation

$$\frac{dy}{dt} = k(a - y)(b - y)$$

where k is a positive constant. Solve this differential equation assuming that $y(0) = 0$ and that $a - y > 0$ and $b - y > 0$. *Hint:*

$$\frac{1}{(a-y)(b-y)} = -\frac{1}{(b-a)(y-a)} + \frac{1}{(b-a)(y-b)}$$

22 Gompertz Growth. This law assumes that the population drops rapidly at even small times. This law is governed by the initial value problem

$$\frac{dP}{dt} = bPe^{-kt}, \quad P(0) = P_0$$

where k and b are positive constants. Solve, and graph the solution.

23. Population. Estimate the limiting population of the United States if $k = 0.025$ in the logistic equation. The population in 1990 was 250 million and growing at 1% a year.

24. Population. In 1845, Verhulst used the logistic model to predict the population of the United States. He took $k =$

0.03134, $b = (1.5887)10^{-10}$, and $P_0 = (3.9)10^6$, where $t = 0$ corresponds to 1790. Using this model find Verhulst's projected population of the United States in 1890 and 1930. Compare your answer with the actual population of 62 and 124 million, respectively.

25. **Spread of Infectious Disease.** A farmer has a herd of 100 cows. One cow contracts an infectious disease that spreads to four cows by the next day. The rate of change of the number of cows with the disease is proportional to the product of the number that have the disease and the number that do not. The veterinarian identifies the disease and sends for a vaccine. How many cows will become infected if the vaccine arrives in two days? Three days?

26. **Biology.** In certain predator/prey situations, the "effectiveness" E of a predator is given by

$$\frac{dE}{dN} = -k\frac{E}{N}$$

where N is the number of predators, and k is a positive constant. Solve for the "effectiveness."

27. **Population.** A confined and protected wildlife area is stocked with two timber wolves that grow according to the logistic equation. Five years later 10 wolves were counted in this area. The area is estimated to be able to hold 30 wolves. How many wolves will there be in this area 10 years after their introduction?

28. **Biology.** In a certain area a mouse population has been happily obeying the logistic equation

$$\frac{dP}{dt} = 0.04P - 0.01P^2$$

where t is in hours. At time $t = 0$ a family moves into the area with a housecat who proceeds to terrorize the mouse population. Frightened by this cat, the mice migrate out of the area at the rate of 0.03 mice per hour. To account for the migration, the differential equation becomes

$$\frac{dP}{dt} = 0.04P - 0.01P^2 - 0.03$$

Solve this new differential equation and find what happens as $t \to \infty$.

29. **Allometry.** Although some body parts of children grow slower than other body parts, the *allometric law*, which is supported by empirical data, states that the specific growth rates of two different body parts are proportional. That is

$$\frac{1}{y}\frac{dy}{dt} = k\frac{1}{x}\frac{dx}{dt}$$

where $x(t)$ and $y(t)$ are the sizes of two different body parts of the same person at time t and k is a positive constant. Cancel dt, and solve.

Enrichment Exercises

30. **Time of Death.** At noon the body of a murdered man is found in a room that has been maintained at a constant temperature of 70°F. The coroner arrives at 1:00 PM and notes that the body temperature is 90°F. An hour later the temperature of the body is noted to be 87°F. If the body temperature is 98.6°F at the time of death, find the time of death.

31. **Mixing.** A tank initially contains 10 pounds of salt dissolved in 200 gallons of water. Water containing 1 pound of salt per gallon starts entering the tank at a rate of 5 gallons per minute. The solution is well stirred and leaves at the same rate. Find a differential equation involving the amount $A(t)$ of salt in the tank at time t, and solve.

SOLUTIONS TO SELF-HELP EXERCISE SET 8.2

1. We separate variables and integrate to obtain

$$\frac{dy}{y} = \frac{dx}{x + 1}$$

$$\int \frac{dy}{y} = \int \frac{dx}{x + 1}$$

$$\ln|y| = \ln|x + 1| + C^*$$

where C^* is a constant of integration. Then

$$\ln\left|\frac{y}{x + 1}\right| = C^*$$

$$\left|\frac{y}{x + 1}\right| = e^{C^*}$$

$$\frac{y}{x + 1} = \pm e^{C^*} = C$$

$$y = C(x + 1)$$

This is the general solution. The figure shows the graphs of some of them.

To find the particular solution we set $y(1) = 4$ in this last equation. We obtain

$$4 = y(1) = C(1 + 1) = 2C$$

Thus, $C = 2$, and the solution of the initial value problem is

$$y = 2(x + 1)$$

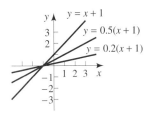

8.3 APPROXIMATE SOLUTIONS
TO DIFFERENTIAL EQUATIONS

- Introduction   ■   Direction Fields   ■   Euler's Method

(Northwind Picture Archives)

Leonhard Euler   1707–1783

Euler is widely considered the greatest mathematician of the 18th century. He was also the most prolific mathematician of all time. His collected works take up 75 large volumes. He also had a very wide range of interests in mathematics and science. Seventeen years before his death he went blind but did not permit his blindness to diminish his mathematical output. Commenting on his loss of sight, Euler observed, "I'll have fewer distractions."

APPLICATION: APPROXIMATE SOLUTIONS

In applications explicit solutions of differential equations can rarely be found. Find a technique that yields an approximate solution to

$$\frac{dy}{dx} = -y + x + 1, \quad y(0) = 1$$

See Example 3 for the technique.

Introduction

The previous section developed the technique of separation of variables to find explicit solutions to certain specific differential equations. There are many other techniques for solving a variety of other types of differential equations. We do not pursue these techniques any further in this text.

Despite the large number of methods available to solve various differential equations, one routinely in practice encounters differential equations for which no *explicit* solution is available. In such a situation, in which the exact solution is not known, we have two possible ways to proceed. We can either try to establish theorems that indicate the *qualitative* behavior of the solution or develop methods that give an *approximate* solution. These latter methods require a great deal of repetitive computations to be useful. But with the current easy access to powerful computers, these latter *numerical* methods have become very practical.

Suppose that we are interested in the solution to the initial value problem

$$\frac{dy}{dx} = f(x, y), \quad y(x_0) = y_0 \qquad [1]$$

We assume that there is in fact a unique (exact) solution that we designate as $y = \phi(x)$ on the interval I under consideration. That is, we have

$$\frac{d}{dx}\,\phi(x) = f[x, \phi(x)] \quad x \in I$$

$$\phi(x_0) = y_0$$

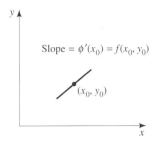

Figure 8.5
Since the slope of the solution curve through the point (x_0, y_0) must have slope $f(x_0, y_0)$, the graph of the solution curve is approximately the same as the given line segment for x near x_0.

Direction Fields

We first notice some geometric facts concerning Equation 1. The solution $y = \phi(x)$ must satisfy $\phi(x_0) = y_0$. This means that this solution passes through the point (x_0, y_0). Furthermore, since

$$\frac{d}{dx}\,\phi(x_0) = f[x_0,\, \phi(x_0)]$$
$$= f(x_0, y_0)$$

the slope of the line tangent to the solution curve $y = \phi(x)$ at (x_0, y_0) is $f(x_0, y_0)$.

Thus, a small line segment drawn through the point (x_0, y_0) with slope $f(x_0, y_0)$ approximates the solution (Fig. 8.5).

By plotting a lot of such line segments, we can create a **direction field** and obtain some insights into how the solutions behave without knowing the solutions explicitly.

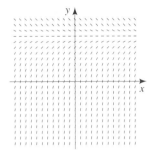

Figure 8.6
The direction field for $\dfrac{dy}{dx} = 4 - y$.

EXAMPLE 1   Finding a Direction Field

Find the direction field for the differential equation

$$\frac{dy}{dx} = 4 - y$$

SOLUTION ■ We find curves, called **isoclines**, on which dy/dx takes particular values. This helps to create the direction field. Notice, for example, that when $y = 5$, dy/dx is always equal to -1. This is indicated in Figure 8.6, where small line segments with slope -1 are drawn along the line $y = 5$. When $y = 3$, dy/dx equals 1. When $y = 2$, dy/dx equals 2, and so on. ■

We get the overwhelming feeling from looking at the direction field in Figure 8.6 that the graphs of the solutions must approach $y = 4$ asymptotically. This is indeed the case, as this is a particular example of the equation $\dfrac{dT}{dt} = k(C - T)$ that we considered in Example 3 of the previous section.

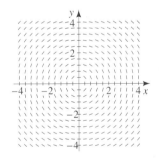

Figure 8.7
The direction field for $\dfrac{dy}{dx} = -\dfrac{x}{y}$.

EXAMPLE 2   Finding a Direction Field

Find the direction field for the differential equation

$$\frac{dy}{dx} = -\frac{x}{y}$$

SOLUTION ■ We find the isoclines. Notice, for example, that when $y = x$, dy/dx is always equal to -1. This is indicated in Figure 8.7, where small line segments with slope -1 are drawn along the line $y = x$. When $y = 2x$, dy/dx equals $-\frac{1}{2}$. When $y = x/2$, dy/dx equals -2. When $x = 0$ ($y \neq 0$), then dy/dx equals zero, and so on. This is shown in Figure 8.7. ■

We get the overwhelming feeling from looking at the direction field in Figure 8.7 that the graphs of the solutions must be circles or perhaps ellipses surrounding $(0, 0)$. This is in fact the case. The reader can easily solve the differential equation in Example 2 by the techniques of the previous section and see this.

Euler's Method

We now see how to construct an approximate solution when no exact solution is known. We take some small value of h and let $x_1 = x_0 + h$, $x_2 = x_1 + h$, $\ldots$, $x_{k+1} = x_k + $

h. Thus, we have a uniform spacing between all the points x_i with the uniform distance between two successive ones being h. The *actual* solution at x_k is $\phi(x_k)$. We will develop a method that gives us an *approximate* value for $\phi(x_k)$. We designate this approximate value as y_k. Thus,

$$y_k \approx \phi(x_k)$$

Notice that we do *not* have y_k and $\phi(x_k)$ equal but rather only *approximately* equal.

The procedure that we use is the tangent line method, or Euler's method. We begin by realizing that the solution to Equation 1 goes through the point (x_0, y_0) and has slope equal to $f(x_0, y_0)$ at this point. The equation for the tangent line through the point (x_0, y_0) with slope $f(x_0, y_0)$ is

$$y - y_0 = f(x_0, y_0)(x - x_0)$$

Since h is small, we expect the actual solution $\phi(x)$ to remain close to this tangent line on the (small) interval $[x_0, x_1]$ of length h. Thus, we expect the value of y on this tangent line at the point x_1, denoted as y_1, to be approximately equal to $\phi(x_1)$. We obtain y_1 by replacing x with x_1 in the equation for the tangent line and solving for y. Thus,

$$\begin{aligned} y_1 &= y_0 + f(x_0, y_0)(x_1 - x_0) \\ &= y_0 + hf(x_0, y_0) \end{aligned}$$

Figure 8.8 indicates what we have done.

We do not know the exact solution $\phi(x_1)$ at the point x_1. The closest thing we have available is the approximation y_1. We then take another straight line through the point (x_1, y_1) with slope $f(x_1, y_1)$. Since the exact solution through this point must have this as slope, it seems that this is the best we can do. The equation for this straight line is

$$y - y_1 = f(x_1, y_1)(x - x_1)$$

We expect the exact solution to be approximately this straight line on the interval $[x_1, x_2]$ and take the value y_2 on this straight line at the point x_2 to be the approximation to the value of the exact solution $\phi(x_2)$. We have

$$\begin{aligned} y_2 &= y_1 + f(x_1, y_1)(x_2 - x_1) \\ &= y_1 + hf(x_1, y_1) \end{aligned}$$

We continue this process so that in general we have

$$y_k = y_{k-1} + hf(x_{k-1}, y_{k-1})$$

and we expect

$$y_k \approx \phi(x_k)$$

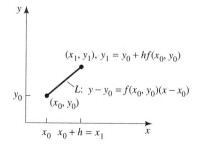

Figure 8.8
In Euler's method we approximate the solution through (x_0, y_0) on $[x_0, x_0 + h]$ by the indicated line L, which is the line tangent to the actual solution curve through (x_0, y_0).

In practice we often are seeking the approximate value of the solution at some point b and are not interested in the approximations at the intermediate points. In this case, we take n to be some integer and define

$$h = \frac{b - x_0}{n}$$

Then

$$x_n = x_0 + nh = x_0 + n\frac{b - x_0}{n} = b$$

Thus,

$$y_n \approx \phi(b)$$

Euler's Method

Given the initial value problem

$$\frac{dy}{dx} = f(x, y), \quad y(x_0) = y_0$$

and some point $b > x_0$. For any integer n define $h = \dfrac{b - x_0}{n}$, and set $x_1 = x_0 + h$, $x_2 = x_1 + h, \ldots, x_n = x_{n-1} + h = b$. Calculate the numbers

$$y_1 = y_0 + hf(x_0, y_0)$$
$$y_2 = y_1 + hf(x_1, y_1)$$
$$\cdots$$
$$y_n = y_{n-1} + hf(x_{n-1}, y_{n-1})$$

Then

$$y_k \approx \phi(x_k), \quad k = 1, 2, \ldots, n$$

where $\phi(x)$ is the exact solution of the initial value problem. In particular

$$y_n \approx \phi(b)$$

Refer to Figure 8.9.

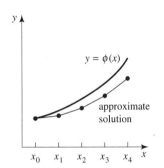

Figure 8.9
The graph of the approximate solution from Euler's method is made of line segments determined by the method.

EXAMPLE 3   Using Euler's Method

Use Euler's method to find the approximations to the initial value problem

$$\frac{dy}{dx} = -y + x + 1, \quad y(0) = 1$$

on the interval $[0, 1]$ by taking $n = 5$.

SOLUTION ■ We take $b = 1$. Then since $n = 5$, we have $h = 0.2$ and $x_k = 0.2k$. Since $f(x, y) = -y + x + 1$, we then have

$$y_0 = 1$$
$$y_k = y_{k-1} + hf[x_{k-1}, y_{k-1}]$$
$$= y_{k-1} + 0.2[-y_{k-1} + 0.2(k - 1) + 1]$$
$$= 0.8y_{k-1} + 0.04(k - 1) + 0.2$$

for $k = 1, 2, \ldots, 5$. Thus,

$$y_1 = 0.8y_0 + 0.04(0) + 0.2 = 0.8 + 0.2 = 1$$
$$y_2 = 0.8y_1 + 0.04(1) + 0.2 = 0.8 + 0.4 + 0.2 = 1.04$$

$$y_3 = 0.8y_2 + 0.04(2) + 0.2 = 0.8(1.04) + 0.08 + 0.2 = 1.112$$

$$y_4 = 0.8y_3 + 0.04(3) + 0.2 = 0.8(1.112) + 0.12 + 0.2 = 1.2096$$

$$y_5 = 0.8y_4 + 0.04(4) + 0.2 = 0.8(1.2096) + 0.16 + 0.2 = 1.32768$$

x_k	y_k	$\phi(x_k)$	e_k
0.0	1.0000	1.0000	0.0000
0.2	1.0000	1.0187	0.0187
0.4	1.0400	1.0703	0.0303
0.6	1.1120	1.1488	0.0368
0.8	1.2096	1.2493	0.0397
1.0	1.3277	1.3679	0.0402

This gives the table at left. The actual solution is $\phi(x) = x + e^{-x}$ as can be readily checked. We then can measure the error, which is $e_k = |y_k - \phi(x_k)|$. We notice how the errors increase. We expect our approximations to get worse as we move away from the initial point. ■

EXAMPLE 4 Using Euler's Method

Use Euler's method in the previous example to find the approximations on the interval $[0, 1]$ by taking $n = 10$.

SOLUTION ■ Again $b = 1$. Then since $n = 10$, we have $h = 0.1$ and $x_k = 0.1k$. Since $f(x, y) = -y + x + 1$, we then have

$$y_0 = 1$$

$$\begin{aligned} y_k &= y_{k-1} + hf[x_{k-1}, y_{k-1}] \\ &= y_{k-1} + 0.1[-y_{k-1} + 0.1(k - 1) + 1] \\ &= 0.9y_{k-1} + 0.01(k - 1) + 0.1 \end{aligned}$$

for $k = 1, 2, \ldots, 10$. Thus,

$$y_1 = 0.9y_0 + 0.01(0) + 0.1 = 0.9 + 0.1 = 1$$

$$y_2 = 0.9y_1 + 0.01(1) + 0.1 = 0.9 + 0.1 + 0.1 = 1.01$$

$$y_3 = 0.9y_2 + 0.01(2) + 0.1 = 0.9(1.01) + 0.2 + 0.1 = 1.029$$

x_k	y_k	$\phi(x_k)$	e_k
0.0	1.0000	1.0000	0.0000
0.1	1.0000	1.0048	0.0048
0.2	1.0100	1.0187	0.0087
0.3	1.0290	1.0408	0.0118
0.4	1.0561	1.0703	0.0142
0.5	1.0905	1.1065	0.0160
0.6	1.1314	1.1488	0.0174
0.7	1.1783	1.1966	0.0183
0.8	1.2305	1.2493	0.0189
0.9	1.2874	1.3066	0.0191
1.0	1.3487	1.3679	0.0192

Continuing, we obtain the table at left. (You can use the program EULER found in the Technology Resource Manual.) Recall that the actual solution is $\phi(x) = x + e^{-x}$. We then can measure the error which is $e_k = |y_k - \phi(x_k)|$.

Comparing this table with the previous one, we see that the approximate solutions are more accurate with smaller h. This is what is to be expected in general. We see more than that. We notice that halving h resulted in halving the error. This is also true in general. ■

EXPLORATION 1

Support for Example 4

Use the program EULERG on the initial value problem in Example 4. Use a window with dimensions $[0, 1]$ by $[0, 2]$.

It turns out that there exists a constant M such that the errors $e_n = |y_n - \phi(b)|$ satisfy

$$e_n \le Mh$$

Although in general M is difficult to find, this formula at least tells us how changing h should affect the error. In particular taking h half as small should reduce the error about in half. And taking h one-tenth as small should reduce the error to one-tenth of the previous error.

Finally, we point out that there are methods much better than the Euler method. But we only present here a barest introduction to finding approximate solutions.

SELF-HELP EXERCISE SET 8.3

1. Use Euler's method to find the approximations to the initial value problem

$$\frac{dy}{dx} = -2xy^2, \quad y(0) = 1$$

on the interval $[0, 1]$ by taking $n = 5$.
Compare the approximate solution with the actual solution

$$y = \phi(x) = \frac{1}{x^2 + 1} \text{ at } x = 1.$$

EXERCISE SET 8.3

In Exercises 1 through 6 match the differential equations with the direction fields given in (a) through (f).

1. $y' = x$

2. $y' = x + y$

3. $y' = 1 + y^2$

4. $y' = \dfrac{x}{y}$

5. $y' = x^2 - y^2$

6. $y' = x - y$

(a)

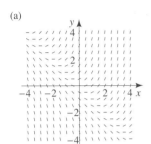

(b)

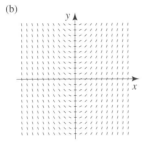

(c)

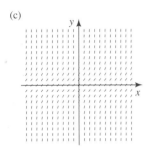

(d)

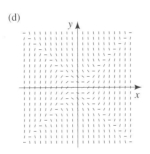

(e) (f)

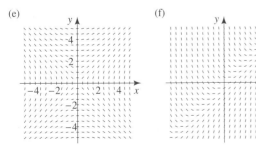

In Exercises 7 through 16 use Euler's method for the given differential equations for the indicated values. Compare the approximate answers with the actual solution that is given. Compare the answers of the even-numbered problems with the answer to the previous problem.

7. $y' = y$, $y(0) = 1$. Find $y_5 \approx \phi(1)$ with $h = 0.2$.
($\phi(x) = e^x$).

8. $y' = y$, $y(0) = 1$. Find $y_{10} \approx \phi(1)$ with $h = 0.1$.
($\phi(x) = e^x$).

9. $y' = x + y$, $y(0) = 0$. Find $y_5 \approx \phi(1)$ with $h = 0.2$.
($\phi(x) = e^x - x - 1$).

10. $y' = x + y$, $y(0) = 0$. Find $y_{10} \approx \phi(1)$ with $h = 0.1$.
($\phi(x) = e^x - x - 1$).

11. $y' = \dfrac{y}{x} + 1$, $y(1) = 0$. Find $y_5 \approx \phi(2)$ with $h = 0.2$.
($\phi(x) = x \ln x$).

12. $y' = \dfrac{y}{x} + 1$, $y(1) = 0$. Find $y_{10} \approx \phi(2)$ with $h = 0.1$.
($\phi(x) = x \ln x$).

13. $y' = \dfrac{y}{1 + x}$, $y(0) = 1$. Find $y_5 \approx \phi(1)$ with $h = 0.2$.
($\phi(x) = 1 + x$).

14. $y' = \dfrac{y}{1 + x}$, $y(1) = 0$. Find $y_{10} \approx \phi(1)$ with $h = 0.1$.
($\phi(x) = 1 + x$).

15. $y' = -xy^2$, $y(2) = 1$. Find $y_5 \approx \phi(3)$ with $h = 0.2$.
$\left(\phi(x) = \dfrac{2}{x^2 - 2}\right)$.

16. $y' = -xy^2$, $y(2) = 1$. Find $y_{10} \approx \phi(3)$ with $h = 0.1$.
$\left(\phi(x) = \dfrac{2}{x^2 - 2}\right)$.

In Exercises 17 through 22 use the program DFIELD found in the Technology Resource Manual to draw a direction field for the given differential equation.

17. $y' = -\dfrac{x}{y}$

18. $y' = \dfrac{y}{x}$

19. $y' = x^2 + y^2$

20. $y' = 2x + y$

21. $y' = 5 - y$

22. $y' = y(1 - 0.2y)$

In Exercises 23 through 26 use the program EULERG to plot a graph of the approximate solution given by Euler's's method and also the graph of the indicated exact solution.

23. $y' = x + y$, $y(0) = 0$. Find $y_{10} \approx \phi(1)$ with $h = 0.1$.
($\phi(x) = e^x - x - 1$).

24. $y' = \dfrac{y}{x} + 1$, $y(1) = 0$. Find $y_{10} \approx \phi(2)$ with $h = 0.1$.
($\phi(x) = x \ln x$).

25. $y' = \dfrac{y}{1 + x}$, $y(0) = 1$. Find $y_{10} \approx \phi(1)$ with $h = 0.1$.
($\phi(x) = 1 + x$).

26. $y' = -xy^2$, $y(2) = 1$. Find $y_{10} \approx \phi(3)$ with $h = 0.1$.
$\left(\phi(x) = \dfrac{2}{x^2 - 2}\right)$.

Enrichment Exercises

27. Can the solution to the differential equation $y' = 5 - y$ that begins at the point $(1, 7)$ ever get below the line $y = 5$? Why or why not?

28. (a) Use Euler's method to investigate what happens to the solution of

$$\frac{dy}{dx} = x^2 + y^2, \quad y(0) = 1$$

on the interval $[0, 2]$. In particular, does the solution have a vertical asymptote? Approximately where?

(b) Now show that the solution of $\frac{dy}{dx} = x^2 + y^2$, $y(0) = 1$, increases faster on the interval $[0, 1)$ than does the solution of $\frac{dy}{dx} = y^2$, $y(0) = 1$. Solve the latter initial value

problem and thus show that the solution of the first problem must become infinite at some point on the interval $[0, 1)$.

In Exercises 29 through 34 consider a solution curve for each one of the direction fields given in figures (a) through (f) in Exercises 1 through 6. Write a sentence or two describing the qualitative behavior of the solution over the long-term. Your answer may be different depending on which solution curve you have chosen.

29. Slope field (a)   **30.** Slope field (b)

31. Slope field (c)   **32.** Slope field (d)

33. Slope field (e)   **34.** Slope field (f)

SOLUTIONS TO SELF-HELP EXERCISE SET 8.3

1. We have $h = \frac{1}{5} = 0.20$, $y_0 = 1$, and $x_k = 0.2k$. Thus,

$$
\begin{aligned}
y_n &= y_{n-1} + hf(x_{n-1}, y_{n-1}) \\
&= y_{n-1} + 0.20[-2x_{n-1}y_{n-1}^2] \\
&= y_{n-1} - 0.40x_{n-1}y_{n-1}^2 \\
y_1 &= y_0 - 0.40x_0y_0^2 = (1) - 0.40(0)(1)^2 = 1
\end{aligned}
$$

$y_2 = y_1 - 0.40x_1y_1^2 = (1) - 0.40(0.20)(1)^2 = 0.92$

$y_3 = y_2 - 0.40x_2y_2^2 = (0.92) - 0.40(0.40)(0.92)^2 = 0.7846$

$y_4 = y_3 - 0.40x_3y_3^2 = (0.7846) - 0.40(0.60)(0.7846)^2 = 0.6368$

$y_5 = y_4 - 0.04x_4y_4^2(0.6368) - 0.40(0.80)(0.6368)^2 = 0.5071$

Thus, $y_5 = 0.5071$, while $\phi(x_5) = \phi(1) = 0.5000$.

8.4 QUALITATIVE ANALYSIS

■ Qualitative Analysis   ■ Logistic Growth with a Threshold

APPLICATION: LOGISTIC GROWTH WITH A THRESHOLD

For many species when the population becomes too small, the members are unable to find each other to breed. In this situation a small population results in a decrease in the population. The differential equation

$$\frac{dP}{dt} = -kP\left(1 - \frac{P}{T}\right)\left(1 - \frac{P}{L}\right)$$

models this effect. Can you use your knowledge of mathematics to determine the qualitative behavior of solutions without finding an explicit solution? See Example 4 for the answer.

Qualitative Analysis

For most differential equations that arise in applications, explicit solutions are not known. In this situation we have two choices: we can seek numerical solutions or we can attempt to determine the *qualitative* behavior of the solutions. In the previous section we determined how to find numerical solutions. In this section we see how to use our knowledge of mathematics to determine the qualitative behavior of solutions in situations in which we are unable to determine the explicit solution.

In determining qualitative behavior, we seek to determine if our solution is increasing, decreasing, heading for some limiting value, or displaying some other useful quality. First we revisit some differential equations for which we already know the explicit solutions. We then determine some qualitative behavior of these differential equations without using our knowledge of the explicit solutions. We do this to gain skills that will then be used in situations for which we do not have explicit solutions. In one application we consider differential equations that model a fishery and determine the effect of different harvesting policies on the fish population.

We begin by looking at the differential equation

$$\frac{dQ}{dt} = kQ \qquad [1]$$

For the moment, we ignore our knowledge of the explicit solution of Equation 1 and only consider what information we can gain about the solutions by just looking at the form of Equation 1 and our knowledge of mathematics. We first consider the case that $k > 0$. This is the situation when, for example, $Q(t)$ represents the quantity of a population and the rate of change of the population is proportional to the population. Since this is an applied problem, we only consider the case when $Q \geq 0$.

EXAMPLE 1   Qualitative Behavior

Determine the qualitative behavior of Equation 1 in the case $k > 0$ and $Q(t) \geq 0$.

SOLUTION ■ We shall find it very useful to find constant solutions of our differential equation. Since the derivative of a constant is identically zero, in order for a constant to be a solution to Equation 1, the left-hand side of Equation 1 must be identically zero. Thus, to find a constant solution we must set the right-hand side of Equation 1 equal to zero and solve. This gives $kQ = 0$ or $Q = 0$. It is clear that $Q(t) = 0$ is a solution to Equation 1. This is also a physically apparent solution, since a population of zero should imply a population that is always zero.

Now we determine the qualitative behavior of the other solutions. In the case under consideration, $k > 0$, and thus Equation 1 immediately implies that $Q'(t) > 0$ whenever $Q(t) > 0$. This is indicated in Figure 8.10. Furthermore, from Equation 1 we see that

$$Q''(t) = \frac{d}{dt}(Q'(t)) = \frac{d}{dt}(kQ(t)) = kQ'(t) = k^2Q(t)$$

This shows that if $Q(t) > 0$, then $Q''(t) > 0$. This is also indicated in Figure 8.10. From Figure 8.10 we can imply a great deal about the qualitative behavior of solutions. From Figure 8.10 we see that for $t \geq 0$ and $Q > 0$, both $Q'(t) > 0$ and $Q''(t) > 0$. Thus, any solution in this region must be increasing and concave up. This then implies the graph shown in Figure 8.11. ■

We already know that the solution of Equation 1 is given by $Q(t) = Q_0 e^{kt}$, where $Q(0) = Q_0$. The graph of this function agrees with the graph found in Figure 8.11.

Since the graphs of the nonconstant solutions move away from the graph of the constant solution $Q = 0$, we say that the constant $P = 0$ is **unstable**.

We now consider two differential equations that model limited growth. In the first, we consider Newton's law of cooling and heating. According to this law, if a substance is placed in a medium that is held at constant temperature C, then the rate of change of the temperature of the substance with respect to time is proportional to the difference

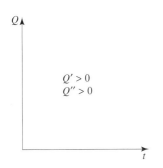

Figure 8.10

If $k > 0$, the solutions of $\dfrac{dQ}{dt} = kQ$

must have Q' and Q'' positive in the first quadrant.

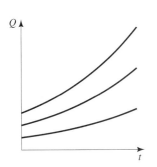

Figure 8.11

These are some solutions with positive first and second derivatives in the first quadrant.

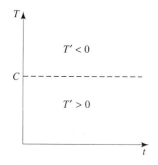

Figure 8.12
This indicates that solutions must be decreasing when $T > C$ and increasing when $T < C$.

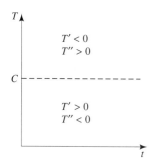

Figure 8.13
This indicates that when $T > C$ solutions are decreasing and concave up. Also, when $T < C$ solutions are increasing and concave down.

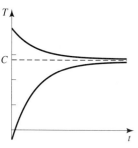

Figure 8.14
The top curve is decreasing and concave up, whereas the bottom curve is increasing and concave down.

of the temperature of the substance and C. Thus, if $T(t)$ is the temperature of the substance at time t, we have

$$\frac{dT}{dt} = k(C - T) \qquad [2]$$

for $k > 0$.

EXAMPLE 2   Qualitative Behavior

For Equation 2 determine the qualitative behavior without referring to the explicit solution found in Section 8.2.

SOLUTION ■ We first find any constant solutions by setting the right-hand side of Equation 2 equal to zero. This gives $k(C - T) = 0$, and thus $T = C$. Thus $T(t) = C$ is the only constant solution. This solution is physically plausible since an object that starts out at temperature C and is placed in a medium of constant temperature C naturally always remains at temperature C.

Now we look at the other solutions. Notice from Equation 2 that since $k > 0$, $T'(t) > 0$ when $T(t) < C$, $T'(t) < 0$ when $T(t) > C$, and $T'(t) = 0$ when $T(t) = C$. Using this information, we construct Figure 8.12. Differentiating Equation 2 yields

$$T''(t) = \frac{d}{dt} T(t) = \frac{d}{dt} k(C - T) = -kT'(t) = -k[k(C - T)] = -k^2(C - T)$$

This shows that $T''(t) > 0$ when $T > C$, $T''(t) < 0$ when $T < C$, and $T''(t) = 0$ when $T = C$. Putting this information together with that found in Figure 8.12 then gives Figure 8.13. From Figure 8.13 we see that if $T(t) > C$, $T(t)$ is decreasing and concave up, whereas if $T'(t) < C$, $T(t)$ is increasing and concave down. This is shown in Figure 8.14. ■

Since the graphs of the nonconstant solutions move asymptotically toward the graph of the constant solution $P = C$, we say that the constant solution is **asymptotically stable**.

EXPLORATION 1

Support for Figure 8.14

Support the previous analysis using the program EULERG on the differential equation $y' = 70 - y$. Set the dimensions of the window as [0, 5] by [0, 100]. When requested, set $x_0 = 0$ and $y_0 = 10$ for the first solution. For the second, set $x_0 = 0$ and $y_0 = 100$. Your two graphs should look just like those in Figure 8.14, only dotted.

We next consider the logistic equation. We did a little qualitative analysis on this equation in Section 8.2.

EXAMPLE 3   Qualitative Behavior of the Logistic Equation

Determine the qualitative behavior of the logistic equation

$$\frac{dP}{dt} = kP - bP^2 \qquad [3]$$

where k and b are positive constants, without referring to the explicit solution found in Section 8.2. Only consider the physically meaningful case that $P(t) \geq 0$.

SOLUTION ■ It is helpful to put Equation 3 into another form. Write the right-hand side as

$$kP - bP^2 = kP\left(1 - \frac{b}{k}P\right) = kP\left(1 - \frac{P}{L}\right) \qquad [4]$$

where $L = \dfrac{k}{b}$.

To find the constant solutions, set the right-hand side of Equation 4 equal to zero and obtain

$$kP\left(1 - \frac{P}{L}\right) = 0$$

This gives two constant solutions: $P = 0$ and $P = L$.

To determine the qualitative behavior of the other solutions first notice that

$$P' = kP\left(1 - \frac{P}{L}\right)$$

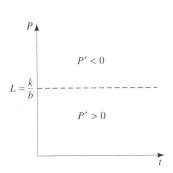

Figure 8.15
This indicates that solutions are decreasing when $P > k/b$ and increasing when $P < k/b$.

If $P(t) > 0$, then $P'(t) > 0$ when $P < L$ and $P'(t) < 0$ when $P > L$. This gives Figure 8.15. Also from Equation 4

$$P''(t) = kP'(t) - 2\frac{k}{L}P(t)P'(t) = \frac{2k}{L}P'(t)\left[\frac{L}{2} - P(t)\right]$$

We now consider two cases, $P(t) > L$ and $P(t) < L$. In the first case when $P(t) > L$, we notice from Figure 8.15 that $P'(t) < 0$. But for these values of $P(t)$, the factor $\left[\dfrac{L}{2} - P(t)\right]$ is negative. Thus, $P'' = \dfrac{2k}{L}P'\left(\dfrac{L}{2} - P\right) > 0$. This is shown in Figure 8.16. In the case that $P(t) < L$, Figure 8.15 shows that $P'(t) > 0$. Since $P'' = \dfrac{2k}{L}P'\left(\dfrac{L}{2} - P\right)$ and in this case $P' > 0$, P'' has the same sign as the factor $\left(\dfrac{L}{2} - P\right)$. Thus, the factor $\left[\dfrac{L}{2} - P(t)\right]$ and also $P''(t)$ are positive when $0 < P(t) < \dfrac{L}{2}$, and this factor and $P''(t)$ are negative when $\dfrac{L}{2} < P(t) < L$. This is then shown in Figure 8.16. Figure 8.16 has the first quadrant split into three regions. In the region $P > L$ we see that $P' < 0$ and $P'' > 0$. Thus, in this region, $P(t)$ is decreasing and concave up as shown in Figure 8.17. In the region $\dfrac{L}{2} < P < L$ we see that $P' > 0$ and $P'' < 0$. Thus, in this region the solutions are increasing and concave down as indicated in Figure 8.17. Finally, in the region $0 < P < \dfrac{L}{2}$ we see that $P' > 0$ and $P'' > 0$. Thus, in this region $P(t)$ is increasing and concave up. ■

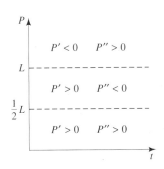

Figure 8.16
Using the second derivative information indicates that the solutions are concave up when $0 < P < 0.5L$ and $P > L$ and concave down when $0.5L < P < L$.

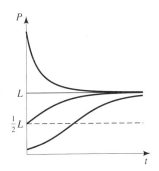

Figure 8.17
The top curve decreases and is concave up. The next curve increases and is concave down. The bottom curve increases and is concave up when $P < 0.5L$, and then concave down when $P > 0.5L$.

We see from Figure 8.17 that the constant solution $P = 0$ is unstable, but the constant solution $P = L$ is asymptotically stable.

Notice that we are able to obtain significant information about the solutions *without ever knowing the explicit solution* found in Section 8.2. We were able to do this by using our knowledge of mathematics.

EXPLORATION 2

Support for Figure 8.17

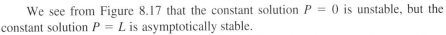 Support Figure 8.17 by using the program EULERG on the differential equation $y' = 0.2y - 0.02y^2$. Use a window with dimensions $[0, 40]$ by $[0, 15]$.

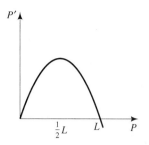

Figure 8.18
We see from this figure that if
$P < L$, $P' > 0$ and the population
increases. If $P > L$, $P' < 0$ and the
population decreases.

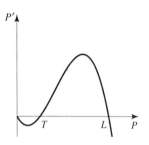

Figure 8.19
We see from this figure that if $0 <$
$P < T$, $P' < 0$ and the population
decreases. If $T < P < L$, $P' > 0$
and the population increases. If
$P > L$, $P' < 0$ and the population
decreases.

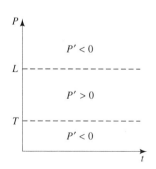

Figure 8.20
We see from this figure that the
population must decrease when
$P > L$ and when $P < T$, whereas
the population increases when
$T < P < L$.

Always take $x_0 = 0$ when requested. Take $y_0 = 0.2$, 6, and 14, respectively, and obtain graphs just like the ones shown in Figure 8.17, except that yours should be dotted.

Logistic Growth with a Threshold

In the logistic equation

$$P' = kP\left(1 - \frac{P}{L}\right)$$

it is instructive to graph P' versus the right-hand side (Fig. 8.18). As Figure 8.18 indicates, the rate of change of population P' itself increases until the population reaches $\frac{L}{2}$. After this, P' decreases and becomes zero at L and negative for any larger population.

The limiting population L is called the **carrying capacity**. As we see in either Figure 8.17 or Figure 8.18 a population above the carrying capacity results in a *decrease* in population.

Notice that according to this model, no matter how small the population, the population *increases*. This is not always the case in applications such as a fishery. For many species when the population becomes too small, the fish are unable to find each other to breed. In this situation small numbers of fish result in a *decrease* in population. To account for this possibility, the right-hand side of the logistic equation needs to be modified so that the graph of P' versus the new right-hand side looks like that shown in Figure 8.19. According to Figure 8.19 notice that for a small population P' is negative, indicating that for small populations the population decreases. At some threshold value $P = T$, the function is zero, and for larger populations it turns positive. The value T is called the **minimum viable population level**. If the population level ever goes below this level, then it heads for zero and extinction.

Perhaps the simplest differential equation that models this situation is given by

$$\frac{dP}{dt} = -kP\left(1 - \frac{P}{T}\right)\left(1 - \frac{P}{L}\right) \quad [5]$$

where $T < L$. The graph P' versus the right-hand side of Equation 5 looks the same as Figure 8.19.

EXAMPLE 4   Qualitative Behavior of Logistic Growth with Threshold

Determine the qualitative behavior of Equation 5.

SOLUTION ■ Equation 5 has three constant solutions: $P = 0$, $P = T$, and $P = L$. Referring to either the right-hand side of Equation 5 or to Figure 8.19, we see that $P' < 0$ when $0 < P < T$, $P' > 0$ when $T < P < L$, and $P' < 0$ when $P > L$. This gives Figure 8.20. To find P'' we could differentiate Equation 5 and factor. Instead we demonstrate an alternative method for finding the sign of $P''(t)$ that is more insightful. Set $f(P)$ equal to the right-hand side of Equation 5, so that $P' = f(P)$. By the chain rule we have

$$P''(t) = \frac{d}{dt} P'(t) = \frac{d}{dt} f(P) = \frac{df(P)}{dP} \cdot \frac{dP(t)}{dt} = f'(P) \cdot P'(t)$$

Now we can determine the sign of P' from Figure 8.20 and the sign of $f'(P)$ from determining where the function $P' = f(P)$ graphed in Figure 8.21 is increasing and where it is decreasing.

For example, when $0 < P(t) < T$, Figure 8.20 indicates that $P'(t) < 0$. Figure 8.21 indicates that $f(P)$ is a decreasing function of P on $(0, I_1)$ and an increasing

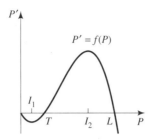

Figure 8.21
The function $f(P)$ is decreasing on $(0, I_1)$ and (I_2, ∞) and increasing on (I_1, I_2).

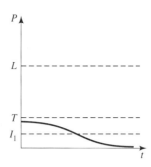

Figure 8.22
This figure shows a solution with $P(t) < T$, decreasing, concave down when $I_1 < P(t) < T$, and concave up when $P(t) < I_1$.

function of P on (I_1, T). Thus, $f'(P)$ is negative on $(0, I_1)$ and positive on (I_1, T). Since $P''(t) = f'(P) \cdot P'(t)$, we see that $P''(t)$ is positive on $(0, I_1)$ and negative on (I_1, T). We therefore conclude that for any solution $P(t)$ with $0 < P(t) < T$ must be decreasing, concave down when $I_1 < P(t) < T$ and concave up when $0 < P(t) < I_1$. An example of such a solution is shown in Figure 8.22. You will be asked to finish this problem in Self-Help Exercise 2 and in Exercise 1. ■

EXPLORATION *3*

Support for Figure 8.22

 Support Figure 8.22 by using the program EULERG on the differential equation $y' = -y(1 - y)(1 - 0.2y)$. Take a window with dimensions $[0, 5]$ by $[0, 1]$. Set $x_0 = 0$ and $y_0 = 0.9$.

SELF-HELP EXERCISE SET 8.4

1. Determine the qualitative behavior of Equation 1 in the case where $k < 0$ and $Q(t) \geq 0$. Find the constant solution, and determine its stability type.

2. Consider Equation 5. Determine what the graphs of solutions look like in the strip $T < P(t) < L$. Using your graphs and the results of Example 4 determine the stability of the constant solution $P(t) = T$. Verify your analysis using the program EULERG on the differential equation $y' = -y(1 - y)(1 - 0.2y)$. Use a window with dimensions $[0, 5]$ by $[0, 5]$. Take $x_0 = 0$ and $y_0 = 1.1$.

EXERCISE SET 8.4

1. Consider Equation 5. Determine what the graphs of solutions look like in the strip $L < P(t)$. Using your graphs and the results of Self-Help Exercise 2 determine the stability of the constant solution $P(t) = L$.

In Exercises 2 through 18 **(a)** find the constant solutions, **(b)** determine the qualitative behavior and the stability of the constant solutions, and **(c)** graph the solutions.

2. $P' = P^4$

3. $P' = P^2$

4. $P' = P + P^2$

5. $P' = -P^2$

6. $P' = -P - P^2$

7. $P' = (1 - P)^3$

8. $P' = (1 - P)^5$

9. $P' = (1 - P)^2$

10. $P' = (1 - P)^4$

11. $P' = P - 1$

12. $P' = (P - 1)^3$

13. $P' = P - 2\sqrt{P}$

14. $P' = P^2(P - 1)$

15. $P' = (P - 1)(P - 2)$

16. $P' = (1 - P)(2 - P)$

17. $P' = P(P - 1)(P - 2)$

18. $P' = P(1 - P)(2 - P)$

In Exercises 19 through 22 use the program DFIELD to draw a direction field for the given differential equation.

19. $P' = 4 - P$

20. $P' = P - 5$

21. $P' = P(1 - 0.20P)$

22. $P' = -P(1 - 0.5P)(1 - 0.2P)$

SOLUTIONS TO SELF-HELP EXERCISE SET 8.4

1. As in Example 1 the only constant solution is $Q(t) = 0$. Since $Q'(t) = kQ(t)$ and $k < 0$, we see that $Q'(t) < 0$ when $Q(t) > 0$. We already showed above that $Q''(t) = k^2Q(t)$. Thus, $Q''(t) > 0$ when $Q(t) > 0$. Thus, when $t \geq 0$ and $Q(t) > 0$, $Q'(t) < 0$ and $Q''(t) > 0$. This information is shown in part (a) of the figure.

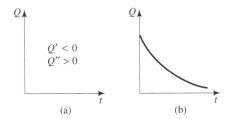

(a)   (b)

From this figure we then conclude that in the region $t \geq 0$ and $Q > 0$, $Q(t)$ is decreasing and concave up. This gives part (b) of the figure.

Since the graphs of the nonconstant solutions move asymptotically toward the graph of the constant solution $Q = 0$, the constant $P = 0$ is asymptotically stable.

We already know that the solution of Equation 1 is given by $Q(t) = Q_0e^{kt}$. The graph of this function, when $k < 0$, agrees with the graph found in part (b) of the figure above. Recall that an example of this situation is radioactive decay.

2. When $T < P(t) < L$, Figure 8.20 indicates that $P'(t) > 0$. Figure 8.21 indicates that $f(P)$ is an increasing function of P on (T, I_2) and a decreasing function of P on (I_2, L). Thus, $f'(P)$ is positive on (T, I_2) and negative on (I_2, L). Since $P''(t) = f'(P) \cdot P'(t)$, we see that $P''(t)$ is positive on (T, I_2) and negative on (I_2, L). We therefore conclude that for any solution $P(t)$ with $T < P(t) < L$ must be increasing, concave up when $T < P(t) < I_2$ and concave down when $I_2 < P(t) < L$. An example of such a solution is shown in the figure. This figure and Figure 8.22 show that the constant solution $P = T$ is unstable.

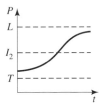

Using the program EULERG on the given differential equation gives a graph like the one shown in the figure, except dotted.

8.5 HARVESTING A RENEWABLE RESOURCE

■ Constant Harvesting ■ Proportional Harvesting

APPLICATION: HARVESTING

Suppose a species satisfies a logistic equation but is now subject to harvesting at a constant rate of h. Then the resulting differential equation is

$$\frac{dP}{dt} = kP\left(1 - \frac{P}{L}\right) - h. \qquad [1]$$

Can you determine the qualitative behavior? For an answer see Example 1.

Constant Harvesting

EXAMPLE 1 Qualitative Behavior with Constant Harvesting

Determine the qualitative behavior of Equation 1 in the case $h < \frac{1}{4}kL$.

SOLUTION ■ First let $f(P) = kP\left(1 - \frac{P}{L}\right)$. The graph is a parabola and is shown in Figure 8.23. We wish to find the maximum of this function. Since $f'(P) = k\left(1 - \frac{2}{L}P\right)$, the maximum occurs at $P = \frac{L}{2}$. The maximum is $f\left(\frac{L}{2}\right) = \frac{1}{4}kL$. Thus,

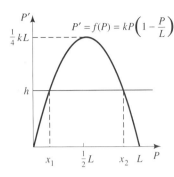

Figure 8.23
The function $f(P)$ is the increase in the population, and h is the amount harvested. Here we have $h < 0.25kL$.

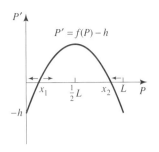

Figure 8.24
When $x_1 < P < x_2$, $P' > 0$, and the population increases. When $0 < P < x_1$ or $P > x_2$, $P' < 0$, and the population decreases.

if h is positive and $h < \frac{1}{4}kL$, the horizontal line of height h intersects the parabola at two points with x-coordinates x_1 and x_2 shown in Figure 8.23. The graph of $P' = f(P) - h$ is then shown in Figure 8.24. Notice that the zeros of the function $f(P) - h$ are x_1 and x_2, indicating that the two constant solutions of Equation 1 are $P = x_1$ and $P = x_2$. (Notice that $P = 0$ is *not* a constant solution.)

From Figure 8.24 we see that if the population $P(t)$ is larger than x_2, then $P'(t)$ is negative and the population heads for x_2. If the population is between x_1 and x_2, then $P'(t)$ is positive, and the population again heads for x_2. But if the population satisfies $P(t) < x_1$, then $P'(t)$ is negative, and the population decreases and approaches 0. But in this case the approach is *not* asymptotic (since $P = 0$ is not a constant solution), but rather the solution $P(t)$ reaches the value 0 in *finite* time. See Example 2. This, of course, means the population becomes extinct in finite time.

From the preceding analysis and the arrows in Figure 8.24 we see that the constant solution $P = x_2$ is asymptotically stable, whereas the solution $P = x_1$ is unstable. ∎

EXPLORATION 1

Support for Example 1

Support some of the results found in Example 1 by using the program EULERG on the differential equation $\frac{dy}{dx} = y\left(1 - \frac{y}{16}\right)$. Use a window with dimensions [0, 20] by [0, 20]. Take $x_0 = 0$ and $y_0 = 5$, and verify that the solution heads monotonically to $y = 16$. Also take $x_0 = 0$ and $y_0 = 20$, and verify that the solution heads monotonically to $y = 16$. (The case when $y_0 < x_1$ is left to Exploration 2.)

We describe the constant solution $P = x_1$ as a **repeller** since all solutions near to this one move away from it.

Notice that as long as $h < \frac{1}{4}kL$ and the population is larger than x_1, the population of the fish heads for the asymptotically stable point x_2. If $h > \frac{1}{4}kL$, then no matter what the population, $P' < 0$ and the population heads for zero and extinction in finite time. Thus, whether the harvesting rate is greater or less than the maximum $\frac{1}{4}kL$ of $f(P)$ is of utmost importance.

The passenger pigeon is a historically important example of overharvesting. This pigeon was hunted to extinction from an original population estimated to be 7 billion, a fact that ignited an upsurge of interest in conservation.

Many other factors need to be considered in a model of a fishery. We do not go into them here. For more detail the reader is referred to the outstanding text by Clark [99].

E X A M P L E 2   Extinction

Suppose the fishery satisfies the differential equation

$$\frac{dP}{dt} = P\left(1 - \frac{P}{16}\right) - 3$$

where t is in years and $P(t)$ is in thousands. (a) Determine the constant solutions. (b) If $P(0) = 3$, show that the population becomes extinct in finite time, and find the time.

S O L U T I O N S ■

(a) We have

$$P\left(1 - \frac{P}{16}\right) - 3 = -\frac{1}{16}\,(P^2 - 16P + 48) = -\frac{1}{16}\,(P - 4)(P - 12)$$

Thus, the constant solutions are $P = x_1 = 4$ and $P = x_2 = 12$, corresponding to populations of 4000 and 12,000.

(b) As we showed in part (a) the differential equation is

$$\frac{dP}{dt} = -\frac{1}{16}\,(P - 4)(P - 12)$$

Separate the variables, and obtain

$$\frac{dP}{(P - 4)(P - 12)} = -\frac{1}{16}\,dt \qquad\qquad [2]$$

To integrate the left side of this equation we need to find constants A and B so that

$$\frac{1}{(P - 4)(P - 12)} = \frac{A}{P - 4} + \frac{B}{P - 12}$$

This is true if, and only if,

$$1 = A(P - 12) + B(P - 4)$$

We need to solve this for A and B. An easy way to do so is to note that since it must be satisfied for all P, it must be satisfied for $P = 4$ and $P = 12$. When $P = 4$, this equation becomes $1 = A(-8)$, which gives $A = -\frac{1}{8}$. When $P = 12$, this equation becomes $1 = B(8)$, which gives $B = \frac{1}{8}$. Thus,

$$\frac{1}{(P - 4)(P - 12)} = \frac{1}{8}\left(\frac{1}{P - 12} - \frac{1}{P - 4}\right)$$

Substituting this into Equation 2 gives

$$\frac{1}{8}\left(\frac{1}{P - 12} - \frac{1}{P - 4}\right)dP = -\frac{1}{16}\,dt$$

Multiplying by 8 and integrating gives

$$\ln(P - 12) - \ln(P - 4) = -0.50t + C$$

or

$$\ln\frac{P(t) - 12}{P(t) - 4} = -0.50t + C$$

In this equation use the fact that $P(0) = 3$ and obtain $\ln 9 = C$. This gives

$$\ln\frac{P(t) - 12}{P(t) - 4} = -0.50t + \ln 9$$

$$\ln \frac{1}{9} \frac{P(t) - 12}{P(t) - 4} = -0.50t$$

$$\frac{P - 12}{P - 4} = 9e^{-0.50t}$$

$$P - 12 = 9(P - 4)e^{-0.50t}$$

$$P = \frac{12 - 36e^{-0.50t}}{1 - 9e^{-0.50t}}$$

Let t^* be the time for the population to become extinct. Then substitute $P(t^*) = 0$ into this last equation and obtain

$$0 = 12 - 36e^{-0.50t^*}$$

$$e^{-0.50t^*} = \frac{1}{3}$$

$$-0.50t^* = \ln \frac{1}{3}$$

$$t^* = 2 \ln 3 \approx 2.2$$

or 2.2 years.   ∎

■ The Coastal Shark Fishery
Rising demand for shark fins in Asia and for shark meat in the United States has caused a sharp rise in the killing of sharks over the last decade, with the result that large coastal sharks are being killed in numbers that outstrip their capacity to reproduce. As a result the National Marine Fisheries Service of the Federal Government has imposed controls on shark fishing within 200 miles of the Atlantic and Gulf coasts by issuing regulations that placed a total quota of 2436 metric tons, dressed weight, on the catch of large coastal sharks by commercial fishermen in 1993 and 2570 metric tons in 1994. Twenty-two shark species are in this management category, including the tiger, lemon, hammerhead, bull, and great white. The total catch of large coastal sharks in the United States waters has substantially exceeded 4000 metric tons in each of the last five years, according to the fisheries service. The service has calculated that the maximum sustainable yield of these sharks is 3800 metric tons a year, and that the 1993 and 1994 quotas should enable that level to be achieved in 1996.

EXPLORATION 2

Support for Extinction

Support the extinction in 2.2 years found in Example 2 by using the program EULERG on the differential equation given in Example 2. Use a window with dimensions [0, 3] by [0, 3].

Proportional Harvesting

In the preceding examples we noticed the dangers inherent in a constant harvesting policy. We now consider a harvest that is proportional to the population and given by $h = EP$, where E measures "effort" in units appropriate to the particular fishery. Thus, the differential equation becomes

$$\frac{dP}{dt} = kP\left(1 - \frac{P}{L}\right) - EP \qquad [3]$$

If E is not too large (the consequences of a large effort are explored in Exercise 1), then a typical graph is shown in Figure 8.25.

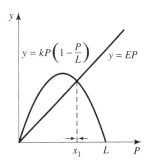

Figure 8.25
When harvesting is proportional to P with E not too large, notice that the population always increases when $P < x_1$. This avoids extinction due to overharvesting.

EXAMPLE 3 Qualitative Behavior with Proportional Harvesting

Determine the qualitative behavior of Equation 3 for the case shown in Figure 8.25.

SOLUTION ■ Figure 8.25 shows one zero x_1 of the function $kP\left(1 - \dfrac{P}{L}\right) - EP$. Thus, $P = x_1$ is the only constant solution of Equation 3. We see from Figure 8.25 that if the population is between 0 and x_1, then $kP\left(1 - \dfrac{P}{L}\right) - EP$ is positive and the population increases and heads for x_1. If the population is greater than x_1, then $kP\left(1 - \dfrac{P}{L}\right) - EP$ is negative and the population decreases and also heads for x_1. Thus, the constant solution $P = x_1$ is asymptotically stable. ■

Thus, in proportional harvesting (with E not too large) there are theoretically no populations that result in eventual extinction. Despite this clear advantage over constant harvesting, proportional harvesting is extremely difficult for a government to institute. If the population of the resource declines, for whatever reason, the fishing industry is required to decrease its catch. This normally means the retiring of boats and equipment, layoffs of employees, and less profit, all consequences likely to be vigorously resisted.

EXPLORATION 3

Support for Proportional Harvesting

Support the result found in proportional harvesting that x_1 is stable by using the program EULERG on the differential equation $y' = y(1 - y/16) - 0.1y$. Use a window with dimensions [0, 20] by [0, 20]. Take $x_0 = 0$ and $y_0 = 1$ and also $x_0 = 0$ and $y_0 = 20$.

SELF-HELP EXERCISE SET 8.5

1. Consider the situation of proportional harvesting given by

$$\frac{dP}{dt} = P(1 - P) - 2P$$

where the effort is $E = 2$. Without solving explicitly, show that the population must head for zero. Now find the explicit solution in the case that $P(0) = 1$, and verify this. Support your answers using the program EULERG on the given differential equation. Take a window with dimensions [0, 5] by [0, 1].

EXERCISE SET 8.5

1. Determine the values of E in Equation 3 that always result in the population becoming extinct.

2. In Equation 1 take $h = \frac{1}{4}kL$, and determine the qualitative behavior.

3. If $E = k$ in Equation 3, determine the qualitative behavior.

In Exercises 4 through 12 differential equations are given, including a constant harvesting term. Without explicitly finding the constant solution(s), determine their number and their sta-bility. Support your answers using the program EULERG on the given differential equation.

4. $P' = 2\sqrt{P} - P - 0.50$ 5. $P' = 2\sqrt{P} - P - 1$

6. $P' = 2\sqrt{P} - P - 2$ 7. $P' = P^2(3 - P) - 2$

8. $P' = P^2(3 - P) - 4$ 9. $P' = P^2(3 - P) - 6$

10. $P' = \sqrt{P}(16 - P) - 10$ 11. $P' = \sqrt{P}(16 - P) - 16$

12. $P' = \sqrt{P}(16 - P) - 20$

Applications

13. **Proportional Harvesting.** The differential equation $P' = Pe^{-P} - EP$ describes a model of a fishery with proportional harvesting. **(a)** Determine the values of E that give one positive constant solution. **(b)** Without finding this constant solution explicitly, determine its stability. Support your answers using the program EULERG on the differential equation.

14. **Proportional Harvesting.** Repeat the previous exercise using the differential equation $P' = P^3(3 - P) - EP$. Support your answers using the program EULERG on the differential equation.

15. **Proportional Harvesting.** The differential equation $P' = 2\sqrt{P} - P - EP$ describes a model of a fishery with proportional harvesting. **(a)** Show that there is exactly one positive constant solution no matter what the value of E. **(b)** Show that this solution is always stable. Support your answers using the program EULERG on the differential equation.

16. **Proportional Harvesting.** Repeat the previous exercise using the differential equation $P' = \sqrt{P}(12 - P)$. Support your answers using the program EULERG on the differential equation.

17. **Depensation.** The differential equation $P' = P^2(3 - P) - EP$ describes a model of a fishery with proportional harvesting that exhibits a characteristic known as **depensation**, that is, a curve that is initially concave up. **(a)** Graphically determine that there is a value E^* so that if $0 < E < E^*$, the differential equation has two positive constant solutions. **(b)** Without finding these constant solutions explicitly, determine their stability. Support your answers using the program EULERG on the differential equation.

18. **Depensation.** Repeat the previous exercise using the differential equation $P' = P^3(3 - P) - EP$. Support your answers using the program EULERG on the differential equation.

19. For the differential equation in Exercise 17 find the values of E that give two positive constant solutions.

20. Find the values of E that give two positive constant solutions for the differential equation in Exercise 18.

SOLUTIONS TO SELF-HELP EXERCISE SET 8.5

1. The differential equation is

$$P' = P(1 - P) - 2P = -P(1 + P)$$

Separating variables gives

$$-\frac{dP}{P(1 + P)} = dt$$

We want to find constants A and B so that

$$-\frac{1}{P(1 - P)} = \frac{A}{P} + \frac{B}{1 + P}$$

This is true if, and only if,

$$-1 = A(1 + P) + BP$$

Setting $P = 0$ yields $A = -1$, whereas setting $P = -1$ yields $B = 1$. Thus,

$$-\frac{1}{P(1 - P)} = \frac{1}{P + 1} - \frac{1}{P}$$

Then we have

$$\left(\frac{1}{P + 1} - \frac{1}{P}\right) dP = dt$$

$$\ln(P + 1) - \ln(P) = t + C$$

$$\ln\frac{P + 1}{P} = t + C$$

Since $P(0) = 1$, this gives $\ln 2 = C$ and

$$\ln\frac{1}{2}\frac{P + 1}{P} = t$$

$$\frac{1}{2}\frac{1 + P}{P} = e^t$$

$$P + 1 = 2Pe^t$$

$$P = \frac{1}{2e^t - 1}$$

From this we see that $\lim\limits_{t \to \infty} P(t) = 0$.

CHAPTER **8**

SUMMARY OUTLINE

- We say that $y = y(x)$ is a **solution** of the differential equation $\dfrac{dy}{dx} = f(x, y)$ on the interval (a, b) if $y(x)$ is differentiable on (a, b) and for all $x \in (a, b)$ $\dfrac{d}{dx} y(x) = f[x, y(x)]$. p. 430.

- We call the problem of finding a solution to $\dfrac{dy}{dx} = f(x, y)$, $y(x_0) = y_0$ on some interval containing x_0 an **initial value problem**. p. 433.

- To solve $\dfrac{dy}{dx} = f(x)g(y)$ by separation of variables place all the y-terms on the left and all the x-terms on the right (separating the variables) and obtain $\dfrac{dy}{g(y)} = f(x)\, dx$. Then integrate and obtain $\displaystyle\int \dfrac{dy}{g(y)} = \int f(x)\, dx$. Solve for y if possible. p. 436.

- **Euler's Method.** Given the initial value problem

$$\frac{dy}{dx} = f(x, y), \quad y(x_0) = y_0$$

and some point $b > x_0$. For any integer n define $h = \dfrac{b - x_0}{n}$, and set $x_1 = x_0 + h, x_2 = x_1 + h, \ldots, x_n = x_{n-1} + h = b$. Calculate the numbers

$$y_1 = y_0 + hf(x_0, y_0)$$
$$y_2 = y_1 + hf(x_1, y_1)$$
$$\cdots$$
$$y_n = y_{n-1} + hf(x_{n-1}, y_{n-1})$$

Then $y_k \approx \phi(x_k)$, $k = 1, 2, \ldots, n$, where $\phi(x)$ is the exact solution of the initial value problem. In particular $y_n \approx \phi(b)$. p. 449.

- The constant solution $y = y_0$ of $\dfrac{dy}{dx} = f(x, y)$ is said to be **asymptotically stable** if the graphs of solutions near $y = y_0$ are asymptotic to $y = y_0$. p. 454.

- The constant solution $y = y_0$ of $\dfrac{dy}{dx} = f(x, y)$ is said to be **unstable** if the graphs of solutions near $y = y_0$ move away from the graph of $y = y_0$. p. 453.

CHAPTER **8**

REVIEW EXERCISES

1. Show that $y = y(x) = 2x^3 e^{-3x}$ is a solution of the differential equation

$$\frac{dy}{dx} = -3y + 6x^2 e^{-3x}$$

on the interval $(-\infty, \infty)$.

2. Show that $y = y(x) = 2x^3 e^{-3x} + Ce^{-3x}$ is a solution on the interval $(-\infty, \infty)$ to the differential equation in the previous exercise for any constant C.

3. The function given in the previous exercise is the general solution to the differential equation given in Exercise 1.

Using this fact, find the solution to

$$\frac{dy}{dx} = -3y + 6x^2 e^{-3x}, \quad y(0) = 5$$

In Exercises 4 through 9 find the general solution.

4. $\dfrac{dy}{dx} = 10x^4$

5. $\dfrac{dy}{dx} = 6x^2 + 4x$

6. $\dfrac{dy}{dx} = 10x^4 y$

7. $\dfrac{dy}{dx} = 5\dfrac{x^4}{y}$

8. $\dfrac{dy}{dx} = \dfrac{y^4}{3x^2}$

9. $\dfrac{dy}{dx} = xe^{x^2-y}$

10. Construct a direction field for

$$\frac{dy}{dx} = 3 - y$$

and suggest from this what the graphs of the solutions look like.

11. Construct a direction field for

$$\frac{dy}{dx} = 2xy + x$$

and suggest from this what the graphs of the solutions look like.

12. Use Euler's method to find the approximations to

$$\frac{dy}{dx} = x^2 + 3y, \quad y(0) = 1$$

on the interval $[0, 1]$ by taking $n = 5$.

13. Repeat the previous exercise using $n = 10$.

14. Biology. A population $P(t)$ of fish is growing according to

$$\frac{dP}{dt} = kP - h$$

where k is the natural growth rate and h is the amount of harvesting per unit time. If the population at $t = 0$ is P_0, solve this differential equation and find $P(t)$.

15. Biology. Biologists have argued that some populations satisfy the differential equation

$$\frac{dP}{dt} = -kP \ln P$$

Show that $P(t) = e^{-be^{-kt}}$ is a solution.

In Exercises 16 through 19 **(a)** find the constant solutions, **(b)** determine the qualitative behavior and the stability of the constant solutions, and **(c)** graph the solutions.

16. $P' = P^4$ **17.** $P' = -P^4$

18. $P' = (P - 1)(P - 3)$ **19.** $P' = P(P - 2)(P - 5)$

20. Constant Harvesting. The differential equation $P' = 2\sqrt{P} - P - h$ models a fishery with a constant rate of harvesting. Determine the values of h that give two constant solutions, and, without solving, determine the stability of each.

21. Proportional Harvesting. The differential equation $P' = P(2 - P) - EP$ models a fishery with a proportional rate of harvesting. Determine the values of E that give one positive constant solution, and, without solving, determine the stability of this solution.

CHAPTER **8**

PROJECTS

Project 1
Water Pollution in the Great Lakes*

Depending on the pollutant, a polluted river can often cleanse itself rather rapidly once the pollution is stopped. The cleansing is effected by the flow of water, which carries the pollutant away. A similar rapid cleansing can also occur for lakes. For example, it is estimated that Lake Erie could be drained in only 2.6 years if the inflow were stopped. It would take 189 years, however, to drain Lake Superior if the inflow were stopped. The table shows the number of years required for each of the Great Lakes to drain if inflow stops.

Lake	Years to Drain
Superior	189.0
Michigan	30.8
Ontario	7.8
Erie	2.6

We now formulate a model of the time needed to cleanse a lake using the natural cleansing effect of the flow of water through the lake. The model makes the following three assumptions:

1. The average inflow and outflow rates are approximately equal and denoted by r.
2. The pollutants are uniformly distributed in the water, with L denoting the concentration in the lake and I the concentration in the inflow.
3. The pollutants are removed only by the flow of water out of the lake.

*Adapted from Martin Eisen, *Mathematical Methods and Models in the Biological Sciences*, Prentice Hall, New Jersey, 1988, p. 97–98.

During any time interval Δt, we must have

Change in total pollutants = total inflow of pollutants − total outflow of pollutants

If V is the volume of the lake, then VL is the total amount of pollutants, and the last equation implies that

$$\Delta(VL) = Ir\,\Delta t - Lr\,\Delta t$$

$$\frac{\Delta(VL)}{\Delta t} = Ir - Lr$$

$$\frac{\Delta L}{\Delta t} = \frac{r}{V}I - \frac{r}{V}L$$

This yields the differential equation

$$\frac{dL}{dt} = \frac{r}{V}I - \frac{r}{V}L$$

If T is the number of years required to drain the lake if the inflow stops, then $T = V/r$. Thus, we can write the last differential equation as

$$\frac{dL}{dt} = \frac{1}{T}I - \frac{1}{T}L$$

or

$$\frac{dL}{dt} + \frac{1}{T}L = \frac{1}{T}I \tag{1}$$

(a) Solve Equation 1, and obtain

$$L(t) = e^{-t/T}\left[L(0) + T^{-1}\int_0^t I(s)e^{s/T}\,ds\right] \tag{2}$$

Hint: Multiply each side of Equation 1 by $e^{t/T}$, and notice that the left side can be written as $\frac{d}{dt}[e^{t/T}L(t)]$.

(b) If the inflow of pollution stops ($I = 0$), then show from Equation 2 that

$$t = T\ln\left[\frac{L(0)}{L(t)}\right] \tag{3}$$

(c) Suppose the inflow of pollution to Lake Erie stops. Using Equation 3 find the number of years for the pollution to be reduced by one-half.

(d) The preceding results do not apply to Lake Ontario, since approximately 84% of its inflow comes from Lake Erie, which is a source of pollution not controlled directly. You are now asked to make modifications of the model to account for this. Let the subscripts e and o refer to Erie and Ontario, respectively, and let I be the inflow into Lake Ontario not from Lake Erie. Show that the differential equation for the concentration of pollutants in Lake Ontario is

$$\frac{d}{dt}L_o = V_o^{-1}[L_e r_e + I(r_o - r_e) - L_o r_o]$$

(e) If about five-sixths of the inflow to Lake Ontario is the outflow from Lake Erie, show that

$$L_o(t) = e^{-t/T}\left\{L_o(0) + \frac{1}{6T}\int_0^t [5L_e(s) + I(s)]e^{s/T}\,ds\right\}$$

(f) If all the inflow of pollution to Lake Erie and Lake Ontario stops except for the flow from Lake Erie to Lake Ontario, find the time for the pollution in Lake Ontario to be reduced by half. Your answer should be in terms of $L_e(0)/L_o(0)$.

Project 2
Predator and Prey

An insect pest population has been damaging agricultural crops in a certain farming community over a number of years. The pest population periodically gets large and does a lot of damage. The farmer naturally is considering using a pesticide, which he knows will reduce but not eliminate the pest population. The farmer is also aware that another insect, called the predator, that is not harmful to his crops preys on the pest insect population and keeps the pest somewhat in control. The pesticide used to control the pest (or prey) also kills the predator. Should he use the pesticide or not?

To answer this question we first construct a mathematical model of the situation. Let $P(t)$ be the prey population and $Q(t)$ the predator population in the farming community at large, where t is the time measured in years and the populations are measured in millions. We assume that the prey population grows exponentially in the absence of the predator since the prey have ample crops to consume. Conversations with farmers in the area indicate that the growth constant for these prolific pests is $k = 1$. In the absence of the predator we therefore assume that the prey population $P(t)$ satisfies

$$\frac{dP}{dt} = P$$

Since the predator insect only feeds on the prey and nothing else, we assume that the predator population decays exponentially in the absence of the prey. Research on the predator indicates a decay constant of 0.5. In the absence of the prey we therefore assume that the predator population $Q(t)$ satisfies

$$\frac{dQ}{dt} = -0.5Q .$$

Because an increase in the predator population should inhibit the growth of the prey population, we subtract the term aPQ, $a > 0$, from the first equation. Since an increase in the prey population should increase the growth of the predator population, we add the term bPQ, $b > 0$, to the second equation. We tentatively "guesstimate" $a = 0.5$ and $b = 0.05$.

The model that results for this situation is the system of two differential equations:

$$\left.\begin{array}{l} \dfrac{dP}{dt} = P - 0.5PQ = P(1 - 0.5Q) \\[2mm] \dfrac{dQ}{dt} = -0.5Q + 0.05PQ = Q(-0.5 + 0.05P) \end{array}\right\} \qquad \text{[1]†}$$

For such a system we seek two functions, $P(t)$ and $Q(t)$, that satisfy the system in Equation 1, that is, satisfy

$$\frac{dP(t)}{dt} = P(t) - 0.5P(t)Q(t)$$

$$\frac{dQ(t)}{dt} = -0.5Q(t) + 0.05P(t)Q(t)$$

Our goal is to determine the qualitative behavior of the populations.

Constant solutions now play important roles. To find the constant solutions, set the right-hand sides of each part in Equation 1 equal to zero. This yields the two constant solutions $P = 0$, $Q = 0$, and $P = 10$, $Q = 2$. In the PQ-plane the graphs of these two solutions are the two points $(0, 0)$ and $(10, 2)$.

(a) Use the graphing calculator and the program EULERS to find approximate solutions to Equation 1. First remove the second line of the program that reads ClrDraw. (This permits all our graphed curves to remain on the screen.) Use a window with dimensions [0, 20] by [0, 4]. When requested, select $X0 = 12$, $Y0 = 2$, $H = 0.1$ and $T = 100$. Graph three more solutions by changing $X0$ to 15, 18, and 20, keeping the other numbers the same. Describe what you see on the screen. What does this imply about the predator and prey populations over the years? Is your description consistent with the observations of the farmers concerning the pest population over the years? Explain.

(b) Not wishing to use insecticides indiscriminately, let us decide to use the insecticides in our model in proportion to the pest population. In our model let us tentatively assume an insecticide application that results in a rate of pest (prey) kill given by $0.5P$ and a rate of predator kill

---

†Equation 1 is called the Lotka-Volterra equations, after the work of Alfred J. Lotka in 1925 and Vito Volterra in 1926.

given by 0.5Q. Modify Equation 1 to include these kill rates. Find the constant solutions. Use the program EULERS to graph a number of solutions. Describe what happens. Does the predator population decrease? Does the pest population decrease? Does our model indicate that we should proceed with the insecticide use as tentatively indicated?

A similar phenomenon was first observed by Umberto D'Ancona in about 1925. Puzzled by observing this phenomenon, he took his data to his colleague, Vito Volterra, mentioned earlier, who then worked out the mathematical model presented here.

Project 3
Average Number of Predator and Prey

In the previous project you should have seen that the number of prey $P(t)$ and the number of predators $Q(t)$ is periodic, or at least approximately so. Suppose for the system in Equation 1 that the period is T. Then, $P(T) = P(0)$ and $Q(T) = Q(0)$. The average number of predators is

$$\frac{1}{T - 0} \int_0^T Q(t)\, dt$$

Find this quantity. *Hint:* First notice from Equation 1 that $\dfrac{dP}{dt} = P(1 - 0.50Q)$. Therefore,

$\dfrac{d}{dt}(\ln P) = \dfrac{P'(t)}{P(t)} = (1 - 0.50Q)$. Also find the average number of prey. What have you discovered?

Project 4
Competing Species

Suppose two populations, $P(t)$ and $Q(t)$, are supported by a common food supply. Suppose that each population satisfies a logistic equation in the absence of the other and that each population interferes with the other population's use of the food supply.

Therefore, in the absence of the Q-population, assume that $P(t)$ satisfies the specific differential equation

$$\frac{dP}{dt} = P\left(1 - \frac{P}{3}\right)$$

and in the absence of the P-population $Q(t)$ satisfies

$$\frac{dQ}{dt} = 2Q\left(1 - \frac{Q}{4}\right)$$

Since the populations interfere with each other in competing for the common food supply, an increase in one population should result in a slower rate of growth for the other. To account for this, we subtract the specific term $PQ/9$ from the first logistic equation and $PQ/4$ from the second and obtain the **system** of differential equations

$$\left.\begin{aligned}
\frac{dP}{dt} &= P\left(1 - \frac{P}{3}\right) - \frac{1}{9}PQ \\[2mm]
\frac{dQ}{dt} &= 2Q\left(1 - \frac{Q}{4}\right) - \frac{1}{4}PQ
\end{aligned}\right\}
\qquad [1]$$

This system of differential equations then models the dynamics of these two competing populations. Since explicit solutions are very difficult, if not impossible, to find, we seek to determine the qualitative behavior of the populations. For example, we ask if the two competing populations can coexist or if one population eventually drives the other to extinction.

Answer this question by using the program EULERS using a window with dimensions [0, 10] by [0, 10]. Always select $H = 0.1$ and $T = 100$. Take a variety of initial points, but first remove the second line of the program that reads ClrDraw. (This permits all graphed curves to remain on the screen.) Explain what you see. How does this answer the question?

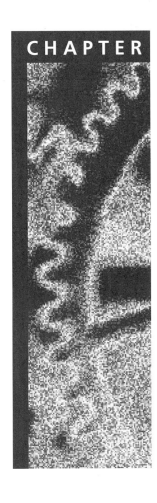

CHAPTER

9

The Trigonometric Functions

9.1 ANGLES

■ Angles ■ Degree Measure ■ Radian Measure ■ Applications

APPLICATION: MODEL OF A NATIONAL ECONOMY

The sine and cosine functions, sin x and cos x, form the basis of mathematical models that indicate oscillations in the total economic activity of national economies, prices of commodities, international trade, inventory accumulation, and investments, as well as providing fundamental models in many areas of biology and physics. One such model was given by Phillips [138]. He assumed that $T(t) = 20 - 1.58e^{0.2t} \cos(0.98t + 0.89)$, where $T(t)$ denotes the total national income of a country at time t over the preceding year. Phillips referred to this as "explosive oscillation." How is the function cos x defined? How can you use graphing principles to obtain a rough idea of the graph? What is happening to this economy? What should be done? For the answers to these questions read this section and see Example 6 of the next section.

APPLICATION: MEASURING THE DISTANCE TRAVELED

Each tire on a certain automobile has a radius of 1 foot. During a trip one of the wheels turned 50,000 revolutions. How far has the car gone? See Example 3 for the answer.

Angles

$A \bullet\!\!\longrightarrow$

Figure 9.1
The half line is called a ray. Point A is the initial point.

A half line, shown in Figure 9.1, is a **ray**. Point A is called the **initial point** of the ray. An **angle** is formed by rotating a ray about its initial point. The initial position of the ray is called the **initial side** of the angle, the position of the ray at the end of its rotation is called the **terminal side** of the angle, and the initial point of the ray is called the **vertex** of the angle. If, as in Figure 9.2a, the ray rotates in the counterclockwise direction, we say that the angle is **positive**, whereas, if as in Figure 9.2b, the rotation is in the clockwise direction, we say the angle is **negative**.

An angle, such as shown in Figure 9.2a, can be named by its vertex A. If we label points B and C, as shown in Figure 9.3, then we can also refer to the angle as BAC. In this notation the vertex is always the middle letter.

In forming an angle the rotation of the ray about its initial point A is not restricted in any way. Thus, for example, the ray may make more than one rotation about its initial point A, as shown in Figure 9.4. Notice that this implies that many angles may have the same initial and terminal sides. Such angles are called **coterminal**.

If we introduce a rectangular coordinate system, then the **standard position** of an angle is obtained by placing the vertex at the origin and the initial side along the positive x-axis, as shown in Figure 9.5. We normally use the Greek letter θ to label an angle,

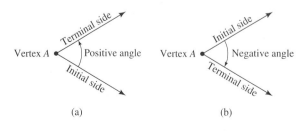

(a)   (b)

Figure 9.2
(a) The angle is positive since the ray rotates in a counterclockwise direction. (b) The angle is negative since the ray rotates in a clockwise direction.

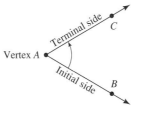

Figure 9.3
Angle *BAC*.

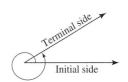

Figure 9.4
Many angles can have the same initial and terminal sides.

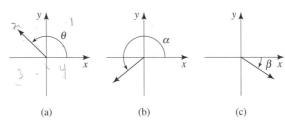

    (a)   (b)   (c)   (d)

Figure 9.5
Shown are four angles in standard position, each in a different quadrant.

but we also use α, β, and γ. An angle in standard position is said to be in the quadrant of its terminal side. Thus, the positive angle in Figure 9.5a is in Quadrant II, whereas the positive angle in Figure 9.5b is in Quadrant III. The negative angle in Figure 9.5c is in Quadrant IV, and the negative angle in Figure 9.5d is in Quadrant I.

Degree Measure

There are two common ways of measuring angles: by degrees or by radians. We first consider measuring angles by degrees, a measure that was first devised by the Babylonians about 4000 years ago and still widely used today. The angle obtained by rotating a ray one complete revolution is designated as having a measure of 360 degrees (Fig. 9.6). Thus, **one degree**, written 1°, is the measure of the angle formed by 1/360 of a rotation.

    There is, of course, a distinction between an angle and its measure. The distinction, however, is cumbersome and rarely made. Thus, for example, we replace the precise statement "an angle θ whose measure is 360°," with the less precise statement "$\theta = 360°$."

    An angle of measure 180° is then an angle formed by $180/360 = \frac{1}{2}$ of a revolution. This is shown in Figure 9.7a, where we note that such an angle is also called a **straight angle**. As Figure 9.7b indicates, an angle of measure 90° is then $\frac{1}{4}$ of a rotation, and is also referred to as a **right angle**. An angle θ is **acute** if $0° < \theta < 90°$, and **obtuse** if $90° < \theta < 180°$ (Fig. 9.8).

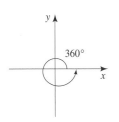

Figure 9.6
Rotating a ray one complete revolution gives 360°.

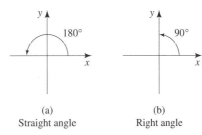

      (a)   (b)
  Straight angle   Right angle

Figure 9.7

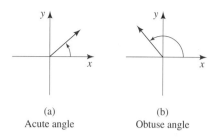

(a)
Acute angle

(b)
Obtuse angle

Figure 9.8

EXAMPLE 1   Angles

Let $\theta = 60°$. (a) Describe this angle in terms of the amount of rotation about the origin this angle represents. (b) Find two angles, one positive and one negative, that are coterminal with $\theta = 60°$.

SOLUTIONS ∎

(a) Since $60/360 = \frac{1}{6}$, $60°$ represents an angle formed by $\frac{1}{6}$ of a rotation. This is shown in Figure 9.9a.

(b) Another positive angle that has the same terminal side is given by

$$\alpha = 60° + 360° = 420°$$

as shown in Figure 9.9b and obtained by rotating the initial ray counterclockwise through $\frac{1}{6}$ plus 1 rotation. A negative angle that has the same terminal side is

$$\beta = 60° - 360° = -300°$$

as shown in Figure 9.9c and is obtained by rotating the initial ray clockwise through $\frac{5}{6}$ of a rotation. ∎

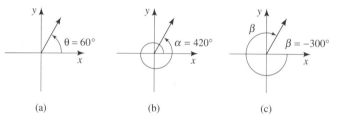

(a)   (b)   (c)

Figure 9.9
The angles α and β are coterminal with $\theta = 60°$.

Radian Measure

We now consider the radian measure of angles. The radian measure of an angle is used in calculus, since the resulting differentiation and integration formulas are simpler than those with degrees.

Consider the circle with radius 1 unit shown in Figure 9.10. The angle determined by an arc of length 1 unit is said to have a measure of *one radian*. Since the circumference of the unit circle is $2\pi(1) = 2\pi$ units, there are 2π radians in one full revolution of the circle. From Figure 9.11 we also see that there are $360°$ in one full revolution of the circle. Thus, we have found a connection between the two measures:

$$360° = 2\pi \text{ radians}$$

or

$$\boxed{180° = \pi \text{ radians}}$$

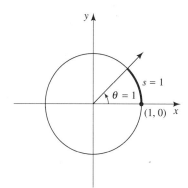

Figure 9.10
The angle determined by an arc length of 1 unit is said to have a measure of 1 radian.

Therefore,

$$1° = \frac{\pi}{180} \text{ radians}$$

Figure 9.11
One full revolution gives 360° and also 2π radians.

and

$$1 \text{ radian} = \left(\frac{180}{\pi}\right)°$$

When we speak of an angle of 2 radians, we drop the term radians. If θ has radian measure 2, we simply write $\theta = 2$. There should be no confusion as to whether radian or degree measure is being used, since if θ is an angle with degree measure 2, we write $\theta = 2°$.

EXAMPLE 2   Degree and Radian Measure

(a) Find the radian measure of θ if $\theta = 150°$ and if $\theta = -300°$.

(b) Find the degree measure of θ if $\theta = \frac{9\pi}{4}$ and if $\theta = -\frac{\pi}{12}$.

SOLUTIONS ■ (a) Since $1° = \frac{\pi}{180}$, to obtain 150°, we simply multiply each side of this equation by 150 and obtain

$$150° = 150\left(\frac{\pi}{180}\right) = \frac{5\pi}{6}$$

In the same way

$$-300° = -300\left(\frac{\pi}{180}\right) = -\frac{5\pi}{3}$$

(b) Since 1 radian $= \left(\frac{180}{\pi}\right)°$,

$$\frac{9\pi}{4} = \left(\frac{9\pi}{4} \cdot \frac{180}{\pi}\right)° = 405°$$

In the same way

$$-\frac{\pi}{12} = \left(-\frac{\pi}{12} \cdot \frac{180}{\pi}\right)° = -15° \quad ■$$

The table shows the relationship between the radian and degree measures of a number of common angles.

Several of these angles are shown in standard position in Figure 9.12.

Radians	0	$\dfrac{\pi}{6}$	$\dfrac{\pi}{4}$	$\dfrac{\pi}{3}$	$\dfrac{\pi}{2}$	$\dfrac{2\pi}{3}$	$\dfrac{3\pi}{4}$	$\dfrac{5\pi}{6}$	π
Degrees	0°	30°	45°	60°	90°	120°	135°	150°	180°

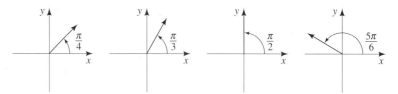

Figure 9.12

Applications

Figure 9.13 shows a circle of radius r_1, a circle of radius r, and an angle θ that cuts an arc of length s_1 on the circle of radius r_1 and an arc of length s on the circle of radius r. Since the sectors formed are similar

$$\frac{s_1}{r_1} = \frac{s}{r}$$

If $r_1 = 1$, then we know that the radian measure of θ is just s_1. But then

$$\theta = s_1 = \frac{s_1}{1} = \frac{s}{r}$$

Thus, the radian measure of an angle θ is the length s of the arc of a circle cut by the angle divided by the radius r of the circle, no matter what the radius is (Fig. 9.14). If θ is measured in radians, then

$$\boxed{\theta = \frac{s}{r}}$$

It is useful to solve this equation for s and obtain

$$\boxed{s = r\theta}$$

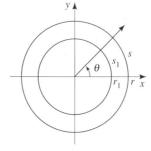

Figure 9.13

We must have $\dfrac{s_1}{r_1} = \dfrac{s}{r}$.

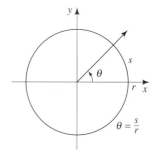

Figure 9.14

If θ is measured in radians, then

$$\theta = \frac{s}{r}.$$

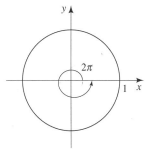

Figure 9.15
One revolution gives 2π radians.

EXAMPLE 3 Measuring the Distance Traveled

Each tire on a certain automobile has a radius of 1 foot. During a trip a measuring device records that one of the wheels has turned 50,000 revolutions. How far has the car gone?

SOLUTION ■ Each revolution of the wheel results in an angle of 2π radians (Fig. 9.15). Thus, 50,000 revolutions gives an angle

$$\theta = 50,000(2\pi) = 100,000\pi$$

Since $s = r\theta$ and $r = 1$ foot, the distance traveled in feet is

$$s = r\theta = (1)(100,000\pi) = 100,000\pi \approx 314,159$$

Since there are 5280 feet in a mile, this is also $314,159/5280 \approx 59.5$ miles.

SELF-HELP EXERCISE SET 9.1

1. Convert 225° to radian measure.

2. Convert $-\dfrac{2\pi}{3}$ to degree measure.

3. Make a sketch of the angle 3π.

4. Suppose in Example 3 the tire has lost $\frac{1}{2}$ inch of thread. How far has the car gone if the wheels turn 50,000 revolutions?

EXERCISE SET 9.1

In Exercises 1 through 6 convert each of the given angles to radian measure.

1. 15°
2. 50°
3. 240°
4. 270°
5. −420°
6. −720°

In Exercises 7 through 12 convert each of the given angles to degree measure.

7. $\dfrac{5\pi}{12}$
8. $\dfrac{13\pi}{12}$
9. $\dfrac{7\pi}{3}$
10. $\dfrac{5\pi}{6}$
11. -4π
12. -5π

13. Identify the quadrant in which each of the angles in Exercises 1, 3, and 5 lies.

14. Identify the quadrant in which each of the angles in Exercises 7, 9, and 10 lies.

In Exercises 15 through 18 express each of the angles shown in radian measure.

15.
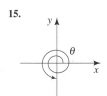

16.
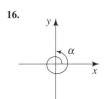

17.
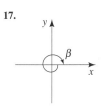

18.
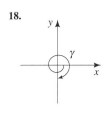

In Exercises 19 through 24 sketch each of the given angles in standard position.

19. 315°
20. −315°
21. $\dfrac{5\pi}{4}$
22. $-\dfrac{5\pi}{4}$
23. $\dfrac{5\pi}{2}$
24. $\dfrac{9\pi}{4}$

In Exercises 25 through 30 place the given angle in standard position and find two other angles, one positive and one negative, that are coterminal.

25. 135°
26. 270°
27. $\dfrac{4\pi}{3}$
28. $\dfrac{3\pi}{2}$
29. $-\dfrac{4\pi}{3}$
30. $-\dfrac{3\pi}{2}$

In Exercises 31 through 34 an angle θ cuts an arc of given length s on a circle of given radius r. Find the measure of θ in (a) radians (b) degrees.

31. $s = 5, r = 2$
32. $s = 10, r = 4$
33. $s = 20, r = 10$
34. $s = 9, r = 3$

In Exercises 35 through 38 the given angle θ cuts an arc of length s on a circle of given radius r. Find s.

35. $\theta = 1, r = 2$
36. $\theta = 2, r = 3$
37. $\theta = 45°, r = 2$
38. $\theta = 30°, r = 5$

Applications

39. Distances on Earth. The distance between two points A and B on the surface of the earth is actually the length of the arc of a circle from A to B having center C at the center of the earth. Assuming the diameter of the earth is 8000 miles and assuming the earth's surface is smooth, find the distance between A and B if the angle ACB is **(a)** $1°$ **(b)** $45°$.

40. Nautical Mile. Refer to Exercise 39. If the angle ACB is $\left(\frac{1}{60}\right)°$, then the distance from A to B is defined as one nautical mile. Find the number of ordinary (statute) miles in a nautical mile.

41. Distances on Earth. Refer to Exercise 39. If two points A and B are 1000 miles apart, find the angle ACB in **(a)** degrees **(b)** radians.

42. Angular Speed. Suppose the car in Example 3 of the text traveled at constant speed and made the trip in 1 hour. Find the angular velocity of the wheel, that is, the average number of radians turned per second.

43. Distance Traveled. A car is driven one-half revolution in a counterclockwise direction about a perfectly circular beltway that surrounds a city, with the body of the driver exactly 1 mile from the center of the circle. **(a)** How far does the driver travel? **(b)** How far does a passenger in the front seat located 3 feet from the driver travel?

44. Ferris Wheel. A person sits in a seat of a ferris wheel that is located 25 feet from the center. What total distance does the person travel in 30 revolutions of the ferris wheel?

Enrichment Exercises

45. Race Track. A race track is constructed in the form of a circle with two lanes laid out. The radius of the inside of the inner lane is 210 feet and the lanes are 3 feet across. Since each runner must stay within their respective lanes, the runner in the outer lane must be given a starting block located ahead of the other runner. How far ahead should this starting block be located if each runner is to have the distance traveled along the inside of their lanes equal while the runner in the inside lane runs four revolutions of the track?

46. Distance of a Ride. Two children take a ride on a merry-go-round. The first child is located 10 feet from the center and the second child 14 feet. In 20 revolutions of the merry-go-round how far does each child travel?

47. Engineering. A large winch of diameter 3 feet lifts cargo out of the hull of a ship (see figure). **(a)** How far is the cargo lifted if the winch rotates three times? **(b)** How many radians must the winch rotate to lift the cargo 10 feet?

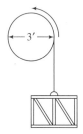

48. Gears. In the figure above right the larger gear has radius r_1 and the smaller r_2. If the larger gear rotates through an angle of radian measure θ_2, what is the angle of rotation, in radians, of the smaller gear in terms of θ_2?

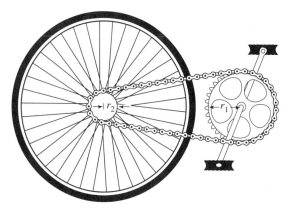

49. Draw a straight angle, a right angle, an acute angle, and an obtuse angle.

50. Given any straight angle, right angle, acute angle, and obtuse angle, rank them according to the size of the measure of their angles.

51. The figure shows two rays starting at the same point. Draw six different angles, each of which has one of the rays as the initial side and the other as the terminal side.

52. Do you think the equation $2 = 2°$ is true? Why or why not?

SOLUTIONS TO SELF-HELP EXERCISE SET 9.1

1. Since $1° = \dfrac{\pi}{180}$,

$$225° = 225\left(\dfrac{\pi}{180}\right) = \dfrac{5\pi}{4}$$

2. Since 1 radian $= \left(\dfrac{180}{\pi}\right)°$,

$$-\dfrac{2\pi}{3} = \left(-\dfrac{2\pi}{3}\dfrac{180}{\pi}\right)° = -120°$$

3.

4. We now have $r = 1 - \dfrac{0.5}{12} \approx \dfrac{11.5}{12}$ feet. Then

$$s = r\theta = \dfrac{11.5}{12}(100{,}000\pi) \approx 301{,}069$$

feet, or $301{,}069/5280 \approx 57.0$ miles.

9.2 THE SINE AND THE COSINE

- Definition of the Sine and Cosine ■ Some Basic Trigonometric Identities
- Right Triangle Interpretation ■ Graphing the Sine and Cosine Functions
- Applications to Business and Economics

APPLICATION: MODEL OF A NATIONAL ECONOMY

Paul Samuelson [139], 1970 Nobel Prize winner in economics, formulated mathematical models that described the tendency for economic activity to rise and fall periodically over time. Suppose $T(t) = 3 + 0.2 \sin(1.6t + 1.2)$ denotes the total national income in trillions of dollars of a country at time t over the preceding year. How is the function $y = \sin x$ defined? What is the graph of $y = T(t)$? What is the significance for the economy? For the answers to these questions, see immediately below and Example 5.

Definition of the Sine and Cosine

Let θ be the angle shown in Figure 9.16 whose terminal side cuts the unit circle at the point $P(x, y)$. We define the **sine** of the angle θ, written $\sin \theta$, to be the y-coordinate of P. Similarly, we define the **cosine** of the angle θ, written $\cos \theta$, to be the x-coordinate of P.

Definition of the Sine and the Cosine

Let θ be the angle shown in Figure 9.16 whose terminal side cuts the unit circle at the point $P(x, y)$. We then define

$$\sin \theta = y \qquad \cos \theta = x$$

It is useful to know the sine and cosine values of a number of angles. For example, if $\theta = 0$, Figure 9.17 indicates that $P(x, y) = (1, 0)$ and $x = 1$ and $y = 0$. Thus,

$$\sin 0 = y = 0 \qquad \cos 0 = x = 1$$

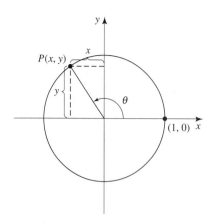

Figure 9.16
We define sin $\theta = y$ and cos $\theta = x$.

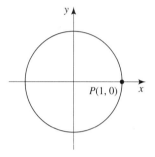

Figure 9.17
For $\theta = P(x, y) = (1, 0)$. Thus, sin 0 $= y = 0$ and cos 0 $= x = 1$.

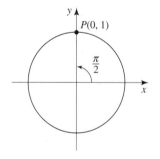

Figure 9.18
For $\theta = \pi/2$, $P(x, y) = (0, 1)$. Thus, $\sin(\pi/2) = y = 1$ and cos $(\pi/2) = x = 0$.

EXAMPLE 1 Finding Trigonometric Values

Evaluate the sine and cosine functions for $\theta = \pi/2$.

SOLUTION ■ According to Figure 9.18, $P(x, y) = (0, 1)$ and $x = 0$ and $y = 1$. Thus,

$$\sin \frac{\pi}{2} = y = 1 \qquad \cos \frac{\pi}{2} = x = 0 \qquad ■$$

EXPLORATION 1

Evaluating Sine and Cosine

Evaluate the sine and cosine functions at $\theta = \pi$ and $\theta = 3\pi/2$.

EXAMPLE 2 Finding Trigonometric Values

Evaluate the sine and cosine functions for (a) $\theta = \pi/4$ (b) $\theta = 3\pi/4$.

SOLUTIONS ■

(a) Since the angle $\pi/4$ is one-half of the angle $\pi/2$ formed by the positive x- and positive y-axis, the point $P(x, y)$ lies on the line that bisects this latter angle. Thus, $y = x$ (Fig. 9.19). Since $x^2 + y^2 = 1$ and $y = x$, we have $2x^2 = 1$, or $x = \sqrt{2}/2$. (We take the positive square root since our point $P(x, y)$ lies in the first quadrant.) Therefore,

$$x = y = \frac{1}{\sqrt{2}} = \frac{\sqrt{2}}{2}$$

and

$$\sin \frac{\pi}{4} = y = \frac{\sqrt{2}}{2} \qquad \cos \frac{\pi}{4} = x = \frac{\sqrt{2}}{2}$$

(b) Since the angle $3\pi/4$ is midway between the angle $\pi/2$, formed by the positive y-axis, and the angle π, formed by the negative x, the point $P(x, y)$ lies on the line $y = -x$ (Fig. 9.20). Since $x^2 + y^2 = 1$ and $y = -x$, we have $2x^2 = 1$, or $x = -\sqrt{2}/2$. (We take the negative square root since our point $P(x, y)$ lies in the second quadrant.) Therefore,

$$x = \frac{-\sqrt{2}}{2} \qquad y = \frac{\sqrt{2}}{2}$$

and

$$\sin \frac{3\pi}{4} = y = \frac{\sqrt{2}}{2} \qquad \cos \frac{3\pi}{4} = x = -\frac{\sqrt{2}}{2} \qquad ■$$

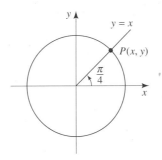

Figure 9.19
The angle $\pi/4$ has a terminal side given by the line $y = x$.

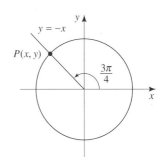

Figure 9.20
The angle $3\pi/4$ has a terminal side given by the line $y = -x$.

EXPLORATION *2*

Evaluating Sine and Cosine

Evaluate the sine and cosine functions at $\theta = 5\pi/4$ and $\theta = 7\pi/4$.

Table 9.1 gives the values of the sine and cosine functions for some important values of θ. (Exercise 58 indicates how to establish the values of sine and cosine for $\theta = \pi/3$.)

As Figure 9.21 indicates, the angle θ is coterminal with the angle $\theta + 2\pi$. Thus,

$$\sin(\theta + 2\pi) = \sin\theta \qquad \text{and} \qquad \cos(\theta + 2\pi) = \cos\theta$$

Such a function is called **periodic**. We now state the following definitions:

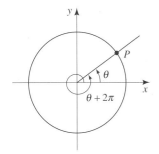

Figure 9.21
The angle θ is coterminal with the angle $\theta + 2\pi$.

Definition of Periodic Function and Period

A function $f(x)$ with the property that

$$f(x + p) = f(x)$$

for all x is called a **periodic function**. The smallest such p is called the **period** p.

The sine and cosine functions, then, are both periodic with a period of 2π or less. An examination of Table 9.1, however, indicates that the periods of these two functions are not less than 2π. Thus, the sine and cosine functions both have periods of 2π.

Actually, the angle θ is coterminal with any angle $\theta + 2n\pi$, where n is any positive or negative integer. Thus,

$$\sin(\theta + 2n\pi) = \sin\theta \qquad \cos(\theta + 2n\pi) = \cos\theta$$

TABLE 9.1

θ	0	$\dfrac{\pi}{6}$	$\dfrac{\pi}{4}$	$\dfrac{\pi}{3}$	$\dfrac{\pi}{2}$	$\dfrac{2\pi}{3}$	$\dfrac{3\pi}{4}$	$\dfrac{5\pi}{6}$	π	$\dfrac{7\pi}{6}$	$\dfrac{5\pi}{4}$	$\dfrac{4\pi}{3}$	$\dfrac{3\pi}{2}$	$\dfrac{5\pi}{3}$	$\dfrac{7\pi}{4}$	$\dfrac{11\pi}{6}$
$\sin\theta$	0	$\dfrac{1}{2}$	$\dfrac{\sqrt{2}}{2}$	$\dfrac{\sqrt{3}}{2}$	1	$\dfrac{\sqrt{3}}{2}$	$\dfrac{\sqrt{2}}{2}$	$\dfrac{1}{2}$	0	$-\dfrac{1}{2}$	$-\dfrac{\sqrt{2}}{2}$	$-\dfrac{\sqrt{3}}{2}$	-1	$-\dfrac{\sqrt{3}}{2}$	$-\dfrac{\sqrt{2}}{2}$	$-\dfrac{1}{2}$
$\cos\theta$	1	$\dfrac{\sqrt{3}}{2}$	$\dfrac{\sqrt{2}}{2}$	$\dfrac{1}{2}$	0	$-\dfrac{1}{2}$	$-\dfrac{\sqrt{2}}{2}$	$-\dfrac{\sqrt{3}}{2}$	-1	$-\dfrac{\sqrt{3}}{2}$	$-\dfrac{\sqrt{2}}{2}$	$-\dfrac{1}{2}$	0	$\dfrac{1}{2}$	$\dfrac{\sqrt{2}}{2}$	$\dfrac{\sqrt{3}}{2}$

EXAMPLE 3 Finding Trigonometric Values

Find $\sin \dfrac{25\pi}{4}$.

SOLUTION ■ Since

$$\frac{25\pi}{4} = \frac{\pi}{4} + 2(3)\pi$$

$$\sin \frac{25\pi}{4} = \sin\left(\frac{\pi}{4} + 2(3)\pi\right) = \sin \frac{\pi}{4} = \frac{\sqrt{2}}{2}$$

where the last equality was obtained in Example 2. ■

Some Basic Trigonometric Identities

We have already seen the value of algebraic identities. For example, it is not obvious how to integrate $\int \dfrac{dx}{x^2 + x}$. But if we use the identity

$$\frac{1}{x^2 + x} = \frac{1}{x} - \frac{1}{x + 1}$$

then we easily see that

$$\int \frac{1}{x^2 + x}\, dx = \int \frac{1}{x}\, dx - \int \frac{1}{x + 1}\, dx = \ln |x| - \ln |x + 1| + C$$

In a similar way, we will find trigonometric identities useful. We now derive some of them.

We see from Figure 9.22 that if the angle θ cuts the unit circle at the point $P(x, y)$, then the angle $-\theta$ cuts the unit circle at $P(x, -y)$. Thus,

$$\sin(-\theta) = -y = -\sin \theta$$

$$\cos(-\theta) = x = \cos \theta$$

These are referred to as the identities for negatives.

Another useful basic identity is the **Pythagorean identity**. For the angle θ in Figure 9.23, $P(x, y)$ is on the unit circle. This implies that $x^2 + y^2 = 1$. But $x = \cos \theta$ and $y = \sin \theta$. From the Pythagorean theorem we then must have

$$\sin^2 \theta + \cos^2 \theta = 1$$

Here we use the standard convention and write, for example, $\sin^2 \theta$ to mean $(\sin \theta)^2$.

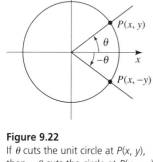

Figure 9.22
If θ cuts the unit circle at $P(x, y)$, then $-\theta$ cuts the circle at $P(x, -y)$.

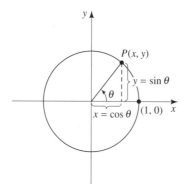

Figure 9.23
Since $P(x, y)$ is on the unit circle, $1 = x^2 + y^2 = \sin^2 x + \cos^2 x$.

We summarize these basic identities.

Some Basic Trigonometric Identities

Identities for Negatives: $\sin(-\theta) = -\sin\theta$ $\cos(-\theta) = \cos\theta$
Pythagorean Identity: $\sin^2\theta + \cos^2\theta = 1$

Some additional identities are considered in the exercise sets.

Right Triangle Interpretation

An alternative definition of the sine and cosine functions is useful in applications, particularly when right triangles are involved. Figure 9.24 shows an angle θ that cuts the unit circle at the point $P(\bar{x}, \bar{y})$ and also cuts a circle of radius r at the point $Q(x, y)$. The two triangles OAP and OBQ are similar, and therefore from geometry we know that the ratios of corresponding sides of these two triangles are equal. Thus, for example, the ratio of the length of PA to the length of OP equals the ratio of the length of QB to the length of OQ. This gives

$$\sin\theta = \bar{y} = \frac{\bar{y}}{1} = \frac{y}{r}$$

In a similar manner, we have

$$\cos\theta = \bar{x} = \frac{\bar{x}}{1} = \frac{x}{r}$$

These derivations do not depend on the angle θ being in the first quadrant. The same derivation applies for θ in any quadrant. Thus, we have the following alternative definition of the sine and cosine functions.

Alternative Definition of the Sine and Cosine Functions

Let θ be the angle shown in Figure 9.25 whose terminal side cuts the circle of radius r at the point $P(x, y)$. We then define

$$\sin\theta = \frac{y}{r} \qquad \cos\theta = \frac{x}{r}$$

This alternative definition is particularly useful when working with right triangles. Figure 9.26 shows a right triangle and an acute angle θ. We see that $\sin\theta$ is just the

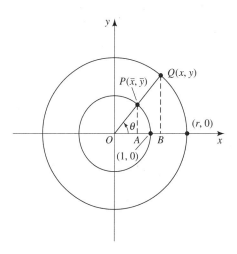

Figure 9.24
The two triangles OAP and OBQ are similar.

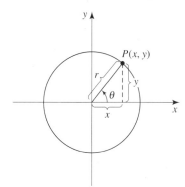

Figure 9.25

In general, $\sin \theta = \dfrac{y}{r}$ and $\cos \theta = \dfrac{x}{r}$.

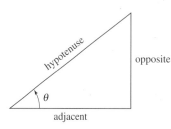

Figure 9.26

For right triangles, $\sin \theta = \dfrac{\text{opposite}}{\text{hypotenuse}}$, and $\cos \theta = \dfrac{\text{adjacent}}{\text{hypotenuse}}$.

length of the opposite side divided by the length of the hypotenuse. The following definition gives the sine and cosine functions of the acute angle θ in the triangle of Figure 9.26.

Right Triangle Interpretation of the Sine and Cosine Functions

Let θ be the acute angle shown in the right triangle in Figure 9.26. Then

$$\sin \theta = \frac{\text{opposite}}{\text{hypotenuse}} \qquad \cos \theta = \frac{\text{adjacent}}{\text{hypothenuse}}$$

This right triangle interpretation is particularly useful in determining one side of a right triangle when the length of another side is known and an instrument measures one angle of the triangle.

EXAMPLE 4   Measuring Heights

An instrument located 100 feet from a tree as shown in Figure 9.27 measures the indicated angle in radians as 0.81. Find the height of the tree.

SOLUTION ■ If h is the height of the tree and c is the distance from the instrument to the top of the tree, then

$$\frac{\sin 0.81}{\cos 0.81} = \frac{h/c}{100/c} = \frac{h}{100}$$

$$h = 100 \, \frac{\sin 0.81}{\cos 0.81}$$

$$\approx 100 \times \frac{0.724287}{0.689498}$$

$$\approx 105.0$$

or about 105 feet.   ■

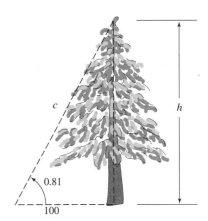

Figure 9.27
What is the height of the tree?

Graphing the Sine and Cosine Functions

Until now we labeled angles with the Greek letters, θ, α, β, or λ. Henceforward, we only measure angles by using radian measure and label our angles with the letter x.

You can easily find the graphs of the sine and cosine functions by using your graphing calculator. We will not do this just yet, but rather we see what the graphs must look like by using our mathematical knowledge and the definitions of the sine and cosine functions. In the process we obtain skills in graphing and a deeper understanding of these functions that will prove helpful later. We use our graphing calculators to confirm our analysis.

The Graph of $y = \sin x$

We begin by graphing $y = \sin x$. Refer to Figure 9.28, where the angle x intersects the unit circle at the point $P(a, b)$. Then $y = \sin x = b$. We trace the point $P(a, b)$ around the unit circle and make careful note of what is happening to the y-coordinate. This gives us the graph of $y = \sin x = b$.

Figure 9.28a indicates that as x goes from 0 to $\pi/2$, $y = \sin x = b$ goes from 0 to 1. This is indicated in Figure 9.29 on the interval $[0, \pi/2]$. Continuing, the rest of Figure 9.28 indicates that as x goes from $\pi/2$ to π to $3\pi/2$ to 2π, $x = \sin x = b$ goes from 1 to 0 to -1 and back to 0. This gives Figure 9.29.

We can readily confirm this on our graphing calculator. Set the mode functions to radian, parametric, and simultaneous. Set the range variables to $T_{min} = 0$, $T_{max} = 6.28$, $T_{step} = 0.01$. Use a viewing window with dimensions $[-1, 6.26]$ by $[-2, 2]$. Set $x_1 = \cos t$, $y_1 = \sin t$ (giving the unit circle), and $x_2 = t$, $y_2 = \sin t$, and graph. As the graph slowly unfolds, note carefully the y-coordinates. You obtain Screen 9.1. Repeat, if needed, and use $T_{step} = 0.03$ to speed up the graphing. Thus, we have the graph of $y = \sin x$ shown in Figure 9.29.

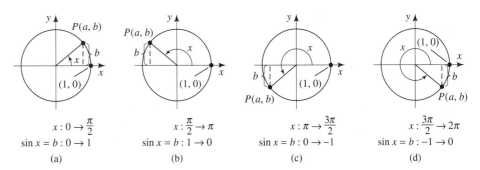

$$x : 0 \to \frac{\pi}{2}$$
$$\sin x = b : 0 \to 1$$
(a)

$$x : \frac{\pi}{2} \to \pi$$
$$\sin x = b : 1 \to 0$$
(b)

$$x : \pi \to \frac{3\pi}{2}$$
$$\sin x = b : 0 \to -1$$
(c)

$$x : \frac{3\pi}{2} \to 2\pi$$
$$\sin x = b : -1 \to 0$$
(d)

Figure 9.28
As the point $P(a, b)$ moves around the circle tracing out the angle x, $\sin x$ is given by the y-coordinate b.

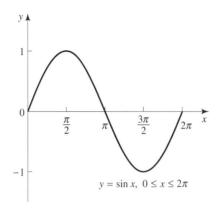

$$y = \sin x, \ 0 \le x \le 2\pi$$

Figure 9.29
The graph of $y = \sin x$ on $[0, 2\pi]$.

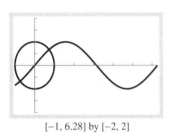

[−1, 6.28] by [−2, 2]

Screen 9.1
As your graphing calculator slowly traces out the unit circle, the y-coordinate on the unit circle gives the values $y = \sin x$.

Recall, however, that the sine function is periodic of period 2π. Thus, to complete the graph of $y = \sin x$, we repeat the graph found on the interval $[0, 2\pi]$ on the intervals $[2\pi, 4\pi]$, $[4\pi, 6\pi]$, . . . , and also repeat on the intervals $[-2\pi, 0]$, $[-4\pi, -2\pi]$, . . . , and obtain the graph shown in Figure 9.30. Notice that the domain of the function $y = \sin x$ is $(-\infty, \infty)$ and the range is $[-1, 1]$.

You can readily verify the graph shown in Figure 9.30 by graphing $y = \sin x$ on your graphing calculator. Your calculator has a SIN key. (Be sure your calculator is in radian mode.)

The graph shown in Figure 9.30 indicates that $y = \sin x$ is concave down on $(0, \pi)$, $(2\pi, 3\pi)$, . . . , and concave up on $(\pi, 2\pi)$, $(3\pi, 4\pi)$, ... We shall establish this in the next section when we find the second derivative.

The Graph of $y = \cos x$

To graph $y = \cos x$ we refer to Figure 9.31, where we notice that $a = \cos x$. Now from Figure 9.31a we see that as x varies from 0 to $\dfrac{\pi}{2}$, $y = \cos x = a$ varies from 1 to 0.

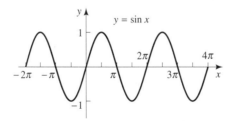

Figure 9.30
The graph of $y = \sin x$.

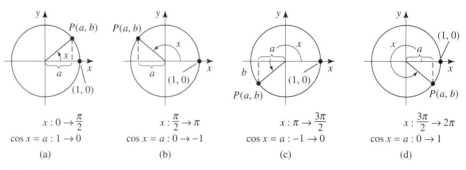

$$x : 0 \to \frac{\pi}{2}$$
$$\cos x = a : 1 \to 0$$
(a)

$$x : \frac{\pi}{2} \to \pi$$
$$\cos x = a : 0 \to -1$$
(b)

$$x : \pi \to \frac{3\pi}{2}$$
$$\cos x = a : -1 \to 0$$
(c)

$$x : \frac{3\pi}{2} \to 2\pi$$
$$\cos x = a : 0 \to 1$$
(d)

Figure 9.31
As the point $P(a, b)$ moves around the circle tracing out the angle x, $\cos x$ is given by the x-coordinate a.

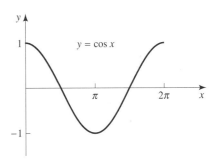

Figure 9.32
The graph of $y = \cos x$ on $[0, 2\pi]$.

This is indicated in Figure 9.32 on the interval $[0, \pi/2]$. From the rest of Figure 9.31, we see that as x varies from $\dfrac{\pi}{2}$ to π to $3\pi/2$ to 2π, $a = \cos x$ varies from 0 to -1 to 0 to 1. We then have the graph of $y = \cos x$ shown in Figure 9.32. Thus, the graph of $y = \cos x$ decreases on the interval $(0, \pi)$ from 1 to -1, and increases on the interval $(\pi, 2\pi)$ from -1 to 1.

EXPLORATION *3*

Graph of cos *x*

 Verify Figure 9.32 on your graphing calculator as follows. Set the mode functions as radian, parametric, and simultaneous. Set the range variables to $T_{min} = 0$, $T_{max} = 6.28$, $T_{step} = 0.03$. Use a viewing window with dimensions $[-1, 6.28]$ by $[-2, 2]$. Set $x_1 = \cos t$, $y_1 = \sin t$, and $x_2 = t$, $y_2 = \cos t$, and graph. Repeat, if needed, and change T_{step} to speed or slow the dynamics.

In the next section we show that $y = \cos x$ is concave down on $(-\pi/2, \pi/2)$, $(3\pi/2, 5\pi/2), \ldots$, and concave up on $(\pi/2, 3\pi/2)$, $(5\pi/2, 7\pi/2), \ldots$.

The graph of $y = \cos x$ is periodic of period 2π. We then obtain the graph shown in Figure 9.33. You can verify this on your graphing calculator.

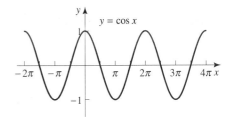

Figure 9.33
The graph of $y = \cos x$.

Applications to Business and Economics

Paul Samuelson [139], the 1970 Nobel Prize winner in economics, formulated mathematical models that described the tendency for economic activity to rise and fall periodically over time. Periodicity is also observed in the prices of commodities, international trade, inventory accumulation, and investments. The sine and cosine functions form the basis of all of these models. Furthermore, as we see in the exercise sets, these functions also describe periodic phenomena in biology.

We now consider a variation of the Paul Samuelson accelerator-multiplier model of a national economy that he gave in 1939 that uses the sine function. (You are asked to provide more details in Project 1 at the end of this chapter.) Let $T(t)$ denote the total national income of a country at time t over the preceding year. This model then predicts that the national income will approximately satisfy an equation of the form $T(t) = A + B\sin(at + b)$. We then wish to obtain some insights into the graphs of such functions.

EXAMPLE 5   Graph of the Income of a National Economy

Suppose $T(t) = 3 + 0.2 \sin(1.6t + 1.2)$ denotes the total national income in trillions of dollars of a country at time t over the preceding year. Graph this function. Confirm with your graphing calculator. Explain the significance for the economy.

SOLUTION ■ We rewrite the function $T(t)$ as follows:

$$T(t) = 3 + 0.2 \sin(1.6t + 1.2)$$
$$= 3 + 0.2 \sin 1.6(t + 0.75)$$

To determine the graph of this function we need to recall some graphing techniques from Section 1.5. As we now see, the key is to find the graph of $y = \sin 1.6t$. We know that $\sin x$ is periodic with period 2π. Then, $\sin 1.6t$ traces out one period as $1.6t$ goes from 0 to 2π, or as t goes from 0 to $\dfrac{2\pi}{1.6} \approx 3.9$. This is shown in Figure 9.34, where we see that $y = \sin 1.6t$ is a periodic function with period approximately 3.9 and moving back and forth between $y = 1$ and $y = -1$.

From the basic graphing principles the graph of $y = \sin 1.6(t + 0.75)$ is the graph of $y = \sin 1.6t$ shifted to the left by 0.75 units (Fig. 9.35). Then the graph of $y = 0.2 \sin 1.6(t + 0.75)$ is the graph of $y = \sin 1.6(t + 0.75)$ contracted by the factor 0.2 (Fig. 9.36). Finally, the graph of $y = 3 + 0.2 \sin 1.6(t + 0.75)$ is the graph of $y = 0.2 \sin 1.6(t + 0.75)$ shifted up by 3 units (Fig. 9.37). This graph can readily be reproduced using your graphing calculator. From the graph we see that economic activity, as measured by $T(t)$, periodically rises and falls, oscillating between 3.2 and 2.8 trillion dollars, and completing one full period in about 3.9 years. ■

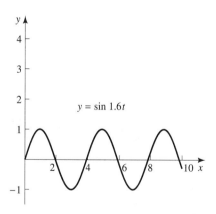

Figure 9.34
A graph of $y = \sin 1.6t$.

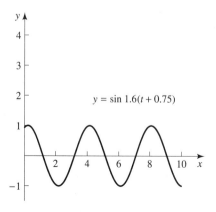

Figure 9.35
The graph of $y = \sin 1.6(t + 0.75)$ is the graph in Figure 9.34 shifted 0.75 units to the left.

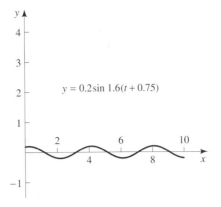

Figure 9.36
The graph of $y = 0.2 \sin 1.6(t + 0.75)$ is the graph in Figure 9.35 contracted by 0.2.

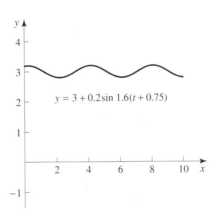

Figure 9.37
The graph of $y = 3 + 0.2 \sin 1.6(t + 0.75)$ is the graph in Figure 9.36 shifted up by 3.

Phillips [138] also formulated models of national economies based on the sine and cosine functions. In the next example we consider a slight modification of one of his models.

EXAMPLE 6   Another Model of a National Economy

Using the same notation as in Example 5, let $T(t) = 20 - 1.58e^{0.2t} \cos(0.98t + 0.89)$. Phillips [138] referred to this as "explosive oscillation." Use graphing principles to obtain a rough idea of the graph. Confirm using your graphing calculator. What is happening to this economy? What should be done?

SOLUTION ■ First write $\cos(0.98t + 0.89)$ as $\cos 0.98(t + 0.91)$, where $0.91 \approx 0.89/0.98$. We first consider the function $y = \cos 0.98t$. Using the same reasoning found in the previous example, this function is periodic with period $\dfrac{2\pi}{0.98} \approx$ 6.4 and moves back and forth between $y = 1$ and $y = -1$. Shifting this graph to the left by 0.91 units then gives the graph of $y = \cos 0.98(t + 0.91)$. Since the graph of $y = 1.58e^{0.2t}$ is strictly increasing, the graph of $y = 1.58e^{0.2t} \cos 0.98(t + 0.91)$ should move back and forth between ever-increasing values. Adding 20 shifts the graph up 20 units. The graph of $y = 20 - 1.58e^{0.2t} \cos(0.98t + 0.89)$ is shown in Screen 9.2 using a window with dimensions [0, 18] by [0, 60]. Notice this economy is moving into ever more dramatic boom and bust periods, as measured by $T(t)$. The national government must take some action against such an "explosively oscillating" economy. Some possibilities include increasing taxes and interest rates during appropriate times to inhibit the explosive boom periods and decreasing taxes and interest rates to boost economic activity at appropriate times to avoid the sharp recessionary periods. ■

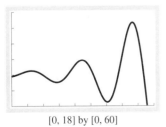

[0, 18] by [0, 60]

Screen 9.2
Total national income is graphed and is oscillating wildly.

SELF-HELP EXERCISE SET 9.2

1. Find the values of the sine and cosine functions for the indicated angle shown in the figure.

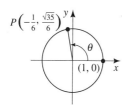

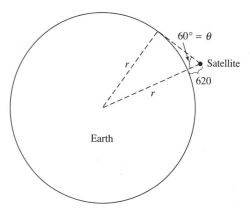

2. An earth satellite is 620 miles above the surface of the earth and measures the angle indicated in the figure. Use this information to find the radius of the earth.

3. Graph $y = -3\cos(\pi x - \pi)$.

EXERCISE SET 9.2

In Exercises 1 through 4 find the values of the sine and cosine functions for the indicated angles shown in the figure.

1. α 2. β
3. γ 4. θ

1.

2.
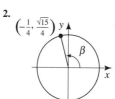

3.
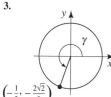

4.

In Exercises 5 through 14 find the exact values of the sine and cosine functions of the given angle.

5. $\dfrac{13\pi}{6}$ 6. $\dfrac{9\pi}{4}$

7. $\dfrac{7\pi}{3}$ 8. $\dfrac{5\pi}{2}$

9. $\dfrac{19\pi}{6}$ 10. $\dfrac{13\pi}{4}$

11. $\dfrac{10\pi}{3}$ 12. $\dfrac{7\pi}{2}$

13. $180°$ 14. $270°$

In Exercises 15 through 18 find the values of the sine and cosine functions for the indicated angle shown in the figure.

15. α 16. β
17. γ 18. θ

15.

16.

17.   **18.**
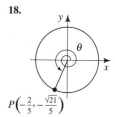

In Exercises 19 through 24 evaluate the given functions.

19. $\cos 31\pi$ 20. $\cos 5\pi$
21. $\sin 6\pi$ 22. $\sin 810°$
23. $\cos 540°$ 24. $\sin 630°$

In Exercises 25 through 28 find the lengths of the remaining two sides of the given triangle.

25.   **26.**
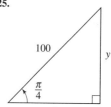   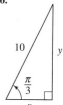

27.   **28.**
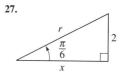   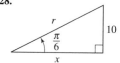

In Exercises 29 through 32 find the values of the sine and cosine functions of the given angle.

29. α **30.** β

31. γ **32.** θ

In Exercises 33 through 42 draw a graph.

33. $y = 5 \sin x$ **34.** $y = -4 \cos x$

35. $y = 2 + \cos x$ **36.** $y = \sin x - 1$

37. $y = 3 \sin 2x$ **38.** $y = -2 \cos 3x$

39. $y = 2 \cos\left(\dfrac{x}{2}\right)$ **40.** $y = 3 \cos(2x + \pi)$

41. $y = -2 \sin\left(\dfrac{x}{2} - \pi\right)$ **42.** $y = 3 \sin(2\pi x + \pi)$

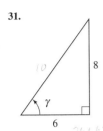

29.   **30.**

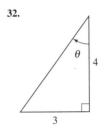

31.   **32.**

Applications

43. Measuring Heights. The shadow from a building is 20 feet long and makes an angle of 30° with the ground. How high is the building?

44. Measuring Heights. A 15-foot ladder leans against a building and makes an angle of 45° with the ground. At what height on the wall does the top of the ladder rest?

45. Measuring Distances. A person stands on a vertical cliff at the edge of the ocean and observes a boat at an angle of depression of 60°. If this person is 100 feet above sea level, how far away is the boat from the base of the cliff?

46. Measuring Heights. A ramp 50 feet long makes an angle of 60° with the ground. How high does the ramp rise?

47. Measuring Distances. A gun at ground level is raised to an angle of 45° and fires a bullet that travels in a straight line at a constant speed of 500 feet per second. How far above ground level is the bullet 4 seconds later?

48. Measuring Heights. A kite is flying at one end of 100 feet of a taut line with the other end tied at ground level. If the angle the straight line makes with the ground is 30°, how high is the kite?

49. Demand. A firm that supplies home heating oil notices that its demand for heating oil, in thousands of gallons per month, satisfies the equation $A(t) = 500 + 500 \cos \dfrac{\pi}{6} t$, where t is in months and $t = 0$ corresponds to January 1. Plot this function. Find the month for which demand is maximum and the month for which demand is minimum.

50. Length of Daylight. The number of hours of daylight in a certain city is given by $N(t) = 12 + 3 \sin \dfrac{2\pi}{365} t$, where t is the number of days from March 20. Plot this function.

51. Temperature. The average temperature in a certain northern city is given by

$$T(t) = 40 + 50 \sin\left[\frac{2\pi}{365}(t - 111)\right]$$

where t is the number of days from January 1. Plot this function. What is the range of average temperatures? Approximately what day is the coldest? Hottest?

52. Temperature. The average temperature in a certain southern city is given by

$$T(t) = 65 + 25 \cos\left[\frac{2\pi}{365}(t - 20)\right]$$

where t is the number of days from July 1. Plot this function. What is the range of average temperatures? Approximately what day is the coldest?

53. Predator and Prey. Suppose the population of a predator is given by $P(t) = 300 + 200 \sin 4\pi t$, where t is in years. A prey population is given by $Q(t) = 2000 + 1000 \cos 4\pi t$. Graph these two functions on the same coordinate system. Notice the lag in the predator population. Why do you think this happens?

54. Biology. Samples of insect pests are commonly taken using nets, and these samples are then used to provide estimates of insect density for determination of control actions. Schotzko and O'Keefe [140] showed that there is a relationship between the time of sampling and insect density. This information is important in determining the time at which sampling is most accurate. They showed that the mean number y of pea aphids was approximated by $y = 17.76 + 13.22 \sin x$, where x is the time of day with $x = 0$ corresponding to 9:00 A.M. Graph using a screen with dimensions [0, 6.26] by [0, 35] and set $x_{scl} = \pi/24 \approx 0.262$. Approximately what time of day is the number the largest? Smallest?

55. **Biology.** Lyons and Liehold [141] showed that the equation

$$ y = 3.5749\left(36.724 + \sin\frac{2\pi x - 270.89}{360}\right) $$

held approximately, where y is the mean hatch time in days of egg masses of gypsy moths and x is the aspect in degrees of an egg on a tree, where 130° indicates due south. Graph on your graphing calculator, and determine the degree x for which mean hatch time is most. Least. Does this coincide with your knowledge of the direction that the sun makes? Explain.

Enrichment Exercises

56. **(a)** Use your graphing calculator to find the period of $y = f(x) = 3 \sin x + 2 \cos 3x$.
    (b) What is the period of $\sin x$? What is the period of $\cos 3x$?
    (c) Use your answer to part (b) to explain your answer to part (a).

57. Do you think $\sin 2 = \sin 2°$? Why or why not?

58. Evaluate the sine and cosine functions for $\theta = \pi/3$. *Hint:* Refer to the figure. Notice that the point Q has been placed on the unit circle so that the angle QOA is $2\pi/3$. Thus, the y-axis bisects the angle QOP, and therefore, the x-coordinate of Q must be $-x$. This implies that the length of the line segment QP is $2x$. Since the angle QOP is $\pi/3$, the two triangles QOP and POA are identical, except that the first has been rotated by $\pi/3$ radians. This then implies that the length of the two line segments QP and PA are equal. Use this fact to solve for x.

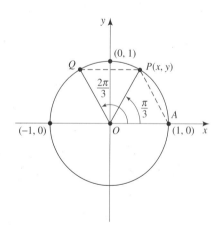

59.

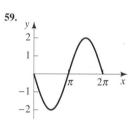

60.

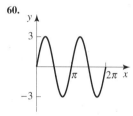

61.

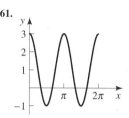

62. **Biology.** The gupy (*Lebistes teticulatus*) uses gravitation and incident light to adjust its position in the water. It has been hypothesized (see [142], pp. 106–107) that the gupy is influenced by two forces, a vertical gravitational force F and a force L caused by the light sensation and parallel to the incident rays. (See the figure.) These two forces combine to give a resultant force R that determines the position of the gupy. The angles α and β can be measured. Show that

$$ \frac{\sin \alpha}{\sin \beta} = \frac{L}{F} $$

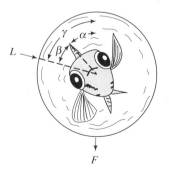

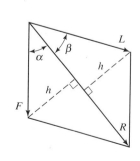

In Exercises 59 through 61 graphs of some trigonometric functions are given. Find a possible function in each case.

SOLUTIONS TO SELF-HELP EXERCISE SET 9.2

1. $\sin \theta = y = \dfrac{\sqrt{35}}{6}$, $\cos \theta = x = -\dfrac{1}{6}$

2. With r in miles we have from the figure that

$$\sin 60° = \frac{r}{r + 620}$$

$$\frac{\sqrt{3}}{2} = \frac{r}{r + 620}$$

$$\frac{\sqrt{3}}{2}(r + 620) = r$$

$$\frac{\sqrt{3}}{2} \, 620 = \left(1 - \frac{\sqrt{3}}{2}\right) r$$

$$r \approx 4008$$

3. We first note that $y = -3 \cos(\pi x - \pi) = -3 \cos \pi(x - 1)$. The graph of $y = \cos \pi x$ completes one period when πx

goes from 0 to 2π, or x goes from 0 to 2. The graph of $y = \cos \pi(x - 1)$ is obtained by shifting the graph of $y = \cos \pi x$ to the right by 1 unit. The graph of $3 \cos \pi(x - 1)$ is obtained by vertically expanding the graph of $\cos \pi(x - 1)$ by a factor of 3. Finally, the graph of $y = -3 \cos \pi(x - 1)$ is obtained by reflecting the graph of $y = 3 \cos \pi(x - 1)$ across the y-axis.

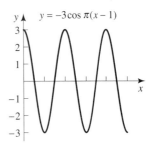

$y = -3\cos \pi(x - 1)$

9.3 DIFFERENTIATION OF THE SINE AND COSINE FUNCTIONS

■ Derivatives of Sine and Cosine Functions ■ Applications

APPLICATION: MAXIMIZING WATER FLOW

A rain gutter is to be constructed from a sheet of metal 12 inches wide by bending the sheet at 4 inches from each side through an angle θ. What value of θ yields the maximum amount of water that the gutter can carry?

Derivatives of Sine and Cosine Functions

First we need to find the derivative of $y = f(x) = \sin x$. Using the definition of derivative, we know that

$$f'(x) = \lim_{h \to 0} \frac{f(x + h) - f(x)}{h}$$

$$= \lim_{h \to 0} \frac{\sin(x + h) - \sin(x)}{h}$$

Now, if h is small, say $h = 0.001$, then by the definition of limit, we have

$$f'(x) \approx \frac{\sin(x + h) - \sin x}{h}$$

$$= \frac{\sin(x + 0.001) - \sin x}{0.001}$$

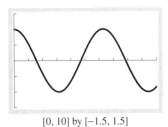

[0, 10] by [−1.5, 1.5]

Screen 9.3

If $f(x) = \sin x$, then $g(x) = \dfrac{\sin(x + 0.001) - \sin x}{0.001} \approx f'(x)$.
The graph of $y = g(x)$ is shown and looks like the graph of $y = \cos x$. Thus, we suspect that $f'(x) = \cos x$.

Let $g(x) = \dfrac{\sin(x + 0.001) - \sin x}{0.001}$. Then, $f'(x) \approx g(x)$. Graph $y = g(x)$ on your graphing calculator using a window with dimensions [0, 10] by [−1.5, 1.5] and obtain Screen 9.3. The graph looks remarkably like the graph of $y = \cos x$. Graphing $y =$

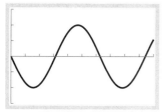

[0, 10] by [−1.5, 1.5]

Screen 9.4
If $f(x) = \cos x$, then $g(x) = \dfrac{\cos(x + 0.001) - \cos x}{0.001} \approx f'(x)$.
The graph of $y = g(x)$ is shown and looks like the graph of $y = -\sin x$. Thus, we suspect that $f'(x) = -\sin x$.

$\cos x$ in the same window gives nothing else, indicating that $g(x) \approx \cos x$. We then conclude that

$$\frac{d}{dx} \sin x = \cos x$$

This can be confirmed mathematically. (See Exercises 36 through 39 for suggestions.)
In the same way, if $y = f(x) = \cos x$, then $f'(x)$ is approximated by $h(x) = \dfrac{\cos(x + 0.001) - \cos x}{0.001}$. We graph $y = h(x)$ using a window with dimensions [0, 10] by [−1.5, 1.5] and obtain Screen 9.4. The graph looks like that of $y = -\sin x$. Graphing $y = -\sin x$ on the same screen gives nothing new, indicating the $f'(x) \approx -\sin x$. We then conclude that

$$\frac{d}{dx} \cos x = -\sin x$$

This can be confirmed mathematically. (See Exercises 40 and 41 for suggestions.)
Using the chain rule, we then have the following two formulas:

Derivatives of Sine and Cosine Functions

$$\frac{d}{dx} \sin f(x) = [\cos f(x)] \cdot f'(x)$$

$$\frac{d}{dx} \cos f(x) = -[\sin f(x)] \cdot f'(x)$$

EXAMPLE 1   Derivatives Involving Sine and Cosine

Find (a) $\dfrac{d}{dx} (\sin x^3)$, (b) $\dfrac{d}{dx} \sqrt{\cos x}$ (c) $\dfrac{d}{dx} e^{\cos 4x}$

SOLUTIONS ■
(a) Here we have $\sin f(x)$ with $f(x) = x^3$.

$$\frac{d}{dx} (\sin x^3) = (\cos x^3) \frac{d}{dx} (x^3)$$

$$= (\cos x^3)(3x^2)$$

$$= 3x^2 \cos x^3$$

(b)
$$\frac{d}{dx} \sqrt{\cos x} = \frac{d}{dx} (\cos x)^{1/2}$$

$$= \frac{1}{2} (\cos x)^{-1/2} \frac{d}{dx} (\cos x)$$

$$= \frac{1}{2} (\cos x)^{-1/2} (-\sin x)$$

$$= -\frac{\sin x}{2\sqrt{\cos x}}$$

(c)
$$\frac{d}{dx} e^{\cos 4x} = e^{\cos 4x} \frac{d}{dx} (\cos 4x)$$

$$= e^{\cos 4x} (-\sin 4x) \frac{d}{dx} (4x)$$

$$= e^{\cos 4x} (-\sin 4x)4$$

$$= -4e^{\cos 4x} \sin 4x \quad ■$$

EXAMPLE 2 The Graph of $y = \sin x$

Let $y = f(x) = \sin x$. (a) Determine where this function is concave up and concave down on $(0, 2\pi)$. (b) Find the critical points on this interval. (c) Determine the value on this interval where the function attains a relative maximum and where it attains a relative minimum.

SOLUTIONS ■

(a) We have

$$\frac{d^2}{dx^2} \sin x = \frac{d}{dx}\left(\frac{d}{dx} \sin x\right)$$

$$= \frac{d}{dx}(\cos x)$$

$$= -\sin x$$

Since $-\sin x$ is negative on $(0, \pi)$ and positive on $(\pi, 2\pi)$, $\sin x$ is concave down on $(0, \pi)$ and concave up on $(\pi, 2\pi)$.

(b) Since $\dfrac{d}{dx}(\sin x) = \cos x$, the critical values of $y = \sin x$ are the zeros of

$\cos x$. On the interval $(0, 2\pi)$ these are $x = \dfrac{\pi}{2}$ and $x = \dfrac{3\pi}{2}$.

(c) Since

$$f''\left(\frac{\pi}{2}\right) = -\sin \frac{\pi}{2} = -1$$

by the second derivative test, $y = \sin x$ attains a relative maximum at $x = \dfrac{\pi}{2}$. Since

$$f''\left(\frac{3\pi}{2}\right) = -\sin \frac{3\pi}{2} = 1$$

by the second derivative test, $y = \sin x$ attains a relative minimum at $x = \dfrac{3\pi}{2}$ (Fig. 9.38). ■

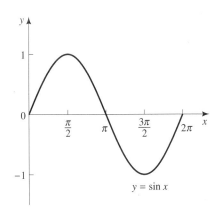

Figure 9.38
The graph of $y = \sin x$.

Applications

EXAMPLE 3 Optimization

A rain gutter is to be constructed from a sheet of metal 12 inches wide by bending the sheet at 4 inches from each side through an angle θ. What value of θ yields the maximum amount of water that the gutter can carry? Refer to Figure 9.39.

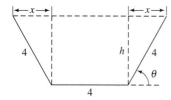

Figure 9.39
A rain gutter.

SOLUTION ■ The volume of water flowing in the gutter is maximized when the cross-sectional area is maximized. Thus, we must find the maximum area of the trapezoid shown in Figure 9.39. The area of a trapezoid is the average of the lengths of the parallel sides times the height. Since the lengths of the sides are $L_1 = 4$ and $L_2 = 4 + 2x$,

$$A(\theta) = \frac{1}{2}(L_1 + L_2)h$$

$$= \frac{1}{2}[4 + (4 + 2x)]h$$

$$= \frac{1}{2}[4 + (4 + 8\cos\theta)]4\sin\theta \quad \left(\text{since } \sin\theta = \frac{h}{4}\right)$$

$$= 16(1 + \cos\theta)\sin\theta$$

We wish to find the absolute maximum of this continuous function over the interval $[0, \pi/2]$. The function is differentiable on this interval so the critical points are points where the derivative is zero. We have

$$A'(\theta) = 16\left[(1 + \cos\theta)\frac{d}{d\theta}(\sin\theta) + \sin\theta\frac{d}{d\theta}(1 + \cos\theta)\right]$$

$$= 16[(1 + \cos\theta)(\cos\theta) + \sin\theta(-\sin\theta)]$$

$$= 16(\cos\theta + \cos^2\theta - \sin^2\theta)$$

$$= 16(\cos\theta + \cos^2\theta + \cos^2\theta - 1)$$

$$= 16(2\cos^2\theta + \cos\theta - 1)$$

$$= 16(2\cos\theta - 1)(\cos\theta + 1)$$

Setting $A'(\theta) = 0$ yields $\cos\theta = \frac{1}{2}$ or $\cos\theta = -1$. Since $\cos\theta \neq -1$ on the interval $[0, \pi/2]$, we must have $\cos\theta = \frac{1}{2}$. This implies $\theta = \pi/3$. We then check the values of A at this critical point and at the two endpoints of the interval $[0, \pi/2]$.

$$A(0) = 16(1 + 1)(0) = 0$$

$$A\left(\frac{\pi}{3}\right) = 16\left(1 + \frac{1}{2}\right)\left(\frac{\sqrt{3}}{2}\right) = 12\sqrt{3} \approx 20.8$$

$$A\left(\frac{\pi}{2}\right) = 16(1 + 0)(1) = 16$$

Thus, we see that the maximum occurs when $\theta = \frac{\pi}{3}$ or 60°. Screen 9.5 shows a graph of $y = A(x)$ using a window of dimensions $[0, 1.57]$ by $[0, 25]$, with $x_{scl} = 0.25$ and $y_{scl} = 5$. ■

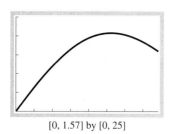

[0, 1.57] by [0, 25]

Screen 9.5
A graph of $y = 16(1 + \cos x)\sin x$.

EXAMPLE 4 Measuring the Speed of a Rocket

A rocket is launched vertically in a level area and is tracked by a rotating radar screen located 2000 feet away from the launch point. When the radar measures an angle of elevation of 45°, the angle is increasing at the rate of 0.12 radians per second. How fast is the rocket rising at that moment? Refer to Figure 9.40.

SOLUTION ■ We are given that $\frac{d\theta}{dt} = 0.12$ radians per second when $\theta = \frac{\pi}{4}$. We want $\frac{dy}{dt}$ when $\theta = \frac{\pi}{4}$. We notice from Figure 9.40 that

$$\frac{\sin\theta}{\cos\theta} = \frac{y/c}{2000/c} = \frac{y}{2000}$$

or $y = 2000\dfrac{\sin\theta}{\cos\theta}$.

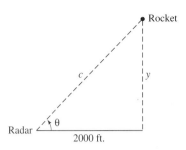

Figure 9.40
The radar follows the rocket.

Differentiating with respect to t gives

$$\frac{dy}{dt} = 2000 \frac{(\cos \theta) \dfrac{d}{dt} \sin \theta - (\sin \theta) \dfrac{d}{dt} \cos \theta}{\cos^2 \theta}$$

$$= 2000 \frac{(\cos \theta)(\cos \theta) \dfrac{d\theta}{dt} - (\sin \theta)(-\sin \theta) \dfrac{d\theta}{dt}}{\cos^2 \theta}$$

$$= 2000 \frac{\cos^2 \theta + \sin^2 \theta}{\cos^2 \theta} \frac{d\theta}{dt}$$

$$= 2000 \frac{1}{\cos^2 \theta} \frac{d\theta}{dt}$$

where we used the Pythagorean identity $\cos^2 \theta + \sin^2 \theta = 1$. Since $\cos \dfrac{\pi}{4} = 1/\sqrt{2}$
we have

$$\left. \frac{dy}{dt} \right|_{\theta = \pi/4} = 2000(\sqrt{2})^2(0.12) = 480$$

or 480 feet per second. ∎

SELF-HELP EXERCISE SET 9.3

1. Show that $\dfrac{d}{dx}\left(\dfrac{\cos x}{\sin x}\right) = -\dfrac{1}{\sin^2 x}$.

2. Let $y = f(x) = \cos x$. **(a)** Determine where this function is concave up and concave down on $(-\pi/2, 3\pi/2)$. **(b)** Find the critical points on this interval. **(c)** Determine the value on this interval where the function attains a relative maximum and where it attains a relative minimum.

EXERCISE SET 9.3

In Exercises 1 through 14 find the derivative of the given function.

1. $f(x) = \sin 5x$

2. $f(x) = \cos 4x$

3. $f(x) = \cos(x^2 + x + 1)$

4. $f(x) = \sin(x^3 - 2x + 1)$

5. $f(x) = \sin e^{3x}$

6. $f(x) = \cos e^{x^2}$

7. $f(x) = \ln(\cos x^2)$

8. $f(x) = \ln(\sin x^2)$

9. $f(x) = \sin^3 x \cos^4 x$

10. $f(x) = \dfrac{\cos 3x}{\cos x}$

11. $f(x) = \dfrac{\sin 3x}{\sin x}$

12. $f(x) = \cos(\sin x)$

13. $f(x) = \sin(\cos x)$

14. $f(x) = \sin(\sin x)$

15. Find the equation of the line tangent to the graph of $y = \sin 3x$ at $x = \pi/12$.

16. Find the equation of the line tangent to the graph of $y = \cos 3x$ at $x = \pi/12$.

In Exercises 17 through 20 find the maximum and minimum of the given function on the indicated interval.

17. $f(x) = \sin x + \cos x$ $[0, 2\pi]$

18. $f(x) = \sin x - \cos x$, $[-\pi, \pi]$

19. $f(x) = x + \sin x$, $[0, 2\pi]$

20. $f(x) = e^x \sin x$, $[0, 2\pi]$

In Exercises 21 through 24 use the first and second derivatives of each function to sketch its graph.

21. $f(x) = \sin x + \cos x$, $[0, 2\pi]$

22. $f(x) = \sin x - \cos x$, $[-\pi, \pi]$

23. $f(x) = x - \sin x$, $[0, 2\pi]$ **24.** $f(x) = e^x \cos x$, $[0, 2\pi]$

Applications

25. Measuring Velocity. A ship moves along a straight path directly away from a helicopter hovering 4 miles above the water. The radar on the helicopter is kept focused on the ship for several seconds. The radar screen rotates at 0.03 radians per minute when the angle that the radar beam makes with the horizontal is 30°. How fast is the ship going at this time?

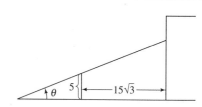

26. Measuring Speed of Rotation. A radar, located 200 feet off a straight highway, tracks a car traveling at 55 miles per hour. What is the speed of rotation of the radar beam at the time the car is closest to the radar?

27. Length of Ladder. A ladder is being carried down a hallway 9 feet wide. At the end of the hall there is a right angle turn into a hallway 9 feet wide. What is the length of the longest ladder that can be carried horizontally around the corner?

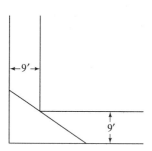

28. Length of Ladder. The 5-foot fence in the figure at top right is $15\sqrt{3}$ feet from the building. What is the length of the shortest ladder that will reach the building from the ground? *Hint:* Write θ as a function of the length of the ladder.

29. Predator and Prey. Suppose a predator population is given by $P(t) = 300 + 200 \sin 4\pi t$, where t is in years, and a prey population is given by $Q(t) = 2000 + 1000 \cos 4\pi t$. Find the maximum and minimum populations of the two species.

30. Demand. A firm that supplies home heating oil notices that its demand for heating oil, in thousands of gallons, satisfies the equation $A(t) = 500 + 500 \cos \dfrac{\pi}{6} t$, where t is in months and $t = 0$ corresponds to January 1. Find the maximum and minimum of this function.

31. Temperature. The temperature in a certain northern city is given by

$$T(t) = 40 + 50 \sin\left[\frac{2\pi}{365}(t - 111)\right]$$

where t is the number of days from January 1. Find the maximum and minimum temperatures and the exact days when these extrema are attained.

32. Temperature. The temperature in a certain southern city is given by

$$T(t) = 65 + 25 \cos\left[\frac{2\pi}{365}(t - 20)\right]$$

where t is the number of days from July 1. Find the maximum and minimum temperatures and the exact days when these extrema are attained.

Enrichment Exercises

33. Let $f(t) = \sin g(t)$. Find $f'(1)$ if $g(1) = \pi/3$ and $g'(1) = 8$.

34. Suppose $y = f(x)$ is a differentiable function and periodic with period p. Show that $y = f'(x)$ is also periodic.

35. Show that $\dfrac{d}{d\theta}(\sin \theta°) = \dfrac{\pi}{180} \cos \theta°$.

Exercises 36 through 41 give a mathematical derivation of the derivative formulas for $\sin x$ and $\cos x$.

36. Show numerically that $\lim\limits_{h \to 0} \dfrac{\sin h}{h} = 1$. Confirm graphically.

37. Find numerically $\lim\limits_{h \to 0} \dfrac{\cos h - 1}{h}$. Confirm graphically.

38. Establish the limit in the previous exercise analytically. *Hint:* Multiply by $\dfrac{\cos h + 1}{\cos h + 1}$, use the substitution $1 - \cos_x^2 = \sin^2 x$, and use the result of Exercise 36.

39. Use the limit definition of derivative to show that $\dfrac{d}{dx} \sin x = \cos x$. *Hint:* Use the identity $\sin(x + y) = \sin x \cos y + \cos x \sin y$, a derivation of which can be found in any book on trigonometry. Then use the limits in Exercises 36 and 37.

40. Use your graphing calculator to establish the identity

$\sin\left(\dfrac{\pi}{2} - x\right) = \cos x$. Do this by setting $y_1 = \sin\left(\dfrac{\pi}{2} - x\right)$ and $y_2 = \cos x$ and by graphing both functions in the same window. In a similar fashion establish the identity $\cos\left(\dfrac{\pi}{2} - x\right) = \sin x$.

41. Use the results of the previous exercise to show that $\dfrac{d}{dx}(\cos x) = -\sin x$.

SOLUTIONS TO SELF-HELP EXERCISE SET 9.3

1. $\dfrac{d}{dx}\left(\dfrac{\cos x}{\sin x}\right) = \dfrac{\sin x \dfrac{d}{dx}(\cos x) - (\cos x)\dfrac{d}{dx}(\sin x)}{\sin^2 x}$

$= \dfrac{(\sin x)(-\sin x) - (\cos x)(\cos x)}{\sin^2 x}$

$= \dfrac{-\sin^2 x - \cos^2 x}{\sin^2 x}$

$= \dfrac{-1}{\sin^2 x}$

2. (a) We have

$$\dfrac{d^2}{dx^2} \cos x = \dfrac{d}{dx}\left(\dfrac{d}{dx} \cos x\right)$$

$= \dfrac{d}{dx}(-\sin x)$

$= -\cos x$

Since $-\cos x$ is negative on $(-\pi/2, \pi/2)$ and positive on $(\pi/2, 3\pi/2)$, $\cos x$ is concave down on $(-\pi/2, \pi/2)$ and concave up on $(\pi/2, 3\pi/2)$.

(b) Since $\dfrac{d}{dx}(\cos x) = -\sin x$, the critical values of $y = \cos x$ are the zeros of $\sin x$. On the interval $(-\pi/2, 3\pi/2)$ these are $x = 0$ and $x = \pi$.

(c) Since $f''(0) = -\cos 0 = -1$, by the second derivative test, $y = \cos x$ attains a relative maximum at $x = 0$. Since $f''(\pi) = -\cos \pi = 1$, by the second derivative test, $y = \cos x$ attains a relative minimum at $x = \pi$.

9.4 INTEGRALS OF THE SINE AND COSINE FUNCTIONS

■ Integrating the Sine and Cosine Functions   ■ Applications

APPLICATION: AVERAGE INSECT POPULATION

An insect population of $P(t)$ is given by

$$P(t) = 41{,}000 + 40{,}000 \sin \dfrac{\pi t}{180}$$

where t is in days and $t = 0$ corresponds to April 1. What is the average population for the period from April 1 through May 30? For the answer see Example 5.

Integrating the Sine and Cosine Functions

For every differentiation formula there is a corresponding integration formula. Since $\dfrac{d}{dx}(-\cos x) = \sin x$,

$$\int \sin x \, dx = -\cos x + C$$

Since $\dfrac{d}{dx}(\sin x) = \cos x$,

$$\int \cos x\, dx = \sin x + C$$

Integration of Sine and Cosine

$$\int \sin x\, dx = -\cos x + C$$

$$\int \cos x\, dx = \sin x + C$$

EXAMPLE 1   Substitution

Find $\int 4x \sin x^2\, dx$.

SOLUTION ■ We let $u = x^2$. Then $du = 2x\, dx$ and

$$\int 4x \sin x^2\, dx = \int 2(\sin x^2)\, 2x\, dx$$
$$= 2 \int \sin u\, du$$
$$= -2 \cos u + C$$
$$= -2 \cos x^2 + C \quad ■$$

EXAMPLE 2   Substitution

Find $\int 12 \sin^3 x \cos x\, dx$.

SOLUTION ■ Let $u = \sin x$. Then $du = \cos x\, dx$ and

$$\int 12 \sin^3 x \cos x\, dx = 12 \int u^3\, du$$
$$= 3u^4 + C$$
$$= 3 \sin^4 x + C \quad ■$$

EXAMPLE 3   Area

Find the area between $y = \sin x$ and the x-axis on $[0,\ \pi]$.

SOLUTION ■ From Figure 9.41 we see that we are seeking $\int_0^\pi \sin x\, dx$. Then

$$\int_0^\pi \sin x\, dx = -\cos x \Big|_0^\pi = -\cos \pi + \cos 0 = -(-1) + 1 = 2 \quad ■$$

EXAMPLE 4   Integration by Parts

Find $\int x \sin x\, dx$.

SOLUTION ■ Let $u = x$ and $dv = \sin x\, dx$. Then $du = dx$ and $v = -\cos x$. Then integrating by parts gives

$$\int u\, dv = uv - \int v\, du$$
$$= -x \cos x - \int (-\cos x)\, dx$$

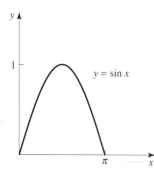

Figure 9.41
The area between the graph of $y = \sin x$ and the y-axis on $[0,\ \pi]$ is $\int_0^\pi \sin x\, dx$.

$$= -x \cos x + \int \cos x \, dx$$
$$= -x \cos x + \sin x + C \quad \blacksquare$$

Applications

EXAMPLE 5   Average Population

An insect population $P(t)$ is given by

$$P(t) = 41,000 + 40,000 \sin \frac{\pi t}{180}$$

where t is in days and $t = 0$ corresponds to April 1. Find the average population for the period from April 1 through May 30.

SOLUTION ■   Screen 9.6 shows a graph of $y = P(x)$ using a window of dimensions $[0, 360]$ by $[0, 85,000]$. The average daily population for the period from April 1 through May 30 is

$$\frac{1}{60} \int_0^{60} \left(41,000 + 40,000 \sin \frac{\pi t}{180} \right) dt$$

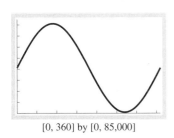

$[0, 360]$ by $[0, 85,000]$

Screen 9.6
A graph of $y = 41,000 + 40,000$ $\sin(\pi x/180)$.

Now let $u = \dfrac{\pi t}{180}$. Then $du = \dfrac{\pi}{180} dt$ and

$$\frac{1}{60} \int_0^{60} \left(41,000 + 40,000 \sin \frac{\pi t}{180} \right) dt = \frac{1}{60} \int_0^{\pi/3} (41,000 + 40,000 \sin u) \frac{180}{\pi} du$$

$$= \frac{3}{\pi} (41,000u - 40,000 \cos u) \Big|_0^{\pi/3}$$

$$= \frac{3}{\pi} \left(41,000 \cdot \frac{\pi}{3} - 40,000 \cdot \cos \frac{\pi}{3} \right)$$

$$- \frac{3}{\pi} (0 - 40,000 \cos 0)$$

$$= \frac{3}{\pi} \left(41,000 \cdot \frac{\pi}{3} - 40,000 \cdot \frac{1}{2} \right)$$

$$- \frac{3}{\pi} (-40,000)$$

$$= 41,000 - \frac{3}{\pi} (20,000) + \frac{1}{\pi} (120,000)$$

$$\approx 60,098.6 \quad \blacksquare$$

SELF-HELP EXERCISE SET 9.4

1. Find the area between $y = \cos x$ and the x-axis between $-\pi/2$ and $\pi/2$.

2. Find $\displaystyle\int \frac{\sin x}{\sqrt{\cos x}} dx$

EXERCISE SET 9.4

In Exercises 1 through 16 evaluate each of the integrals.

1. $\displaystyle\int \sin 5x \, dx$

2. $\displaystyle\int \cos 7x \, dx$

3. $\displaystyle\int 10x^4 \cos x^5 \, dx$

4. $\displaystyle\int 12x^2 \sin x^3 \, dx$

5. $\displaystyle\int \sqrt{\sin x} \cos x \, dx$

6. $\displaystyle\int \cos^5 x \sin x \, dx$

7. $\displaystyle\int \frac{\sin x}{2 + \cos x} dx$

8. $\displaystyle\int \frac{\cos x}{2 + \sin x} dx$

9. $\int \dfrac{\sin(\ln x)}{x}\, dx$

10. $\int \dfrac{\cos(\sqrt{x})}{\sqrt{x}}\, dx$

11. $\int x \cos x\, dx$

12. $\int 2x \sin 2x\, dx$

13. $\int x^2 \sin x\, dx$

14. $\int x^2 \cos x\, dx$

15. $\int_0^{\pi/3} \sin x\, dx$

16. $\int_{\pi/2}^{3\pi/2} \cos x\, dx$

In Exercises 17 through 20 find the area enclosed by the given curves.

17. $y = \cos x, \ y = \sin x, \ x = 1, \ x = \pi/4$

18. $y = 2 \sin x, \ y = \sin 2x, \ x = 0, \ x = \pi$

19. $y = \sin x, \ y = \cos x, \ x = \pi/2, \ x = 5\pi/4$

20. $y = 2 \cos x, \ y = \cos 2x, \ x = -\pi/2, \ x = \pi/2$

In Exercises 21 and 22 find the average value of the given function on the indicated interval.

21. $y = a \sin \dfrac{2\pi}{b} x, \quad [0, b/2]$

22. $y = a \cos \dfrac{2\pi}{b} x, \quad [-b/4, b/4]$

Applications

23. Predator and Prey. Suppose the population of a predator is given by $P(t) = 300 + 200 \sin \dfrac{2\pi}{12} t$, where t is in months. Find the average number of predators in a year.

24. Demand. A firm that supplies home heating oil notices that its demand for heating oil, in thousands of gallons, satisfies the equation $A(t) = 500 + 500 \cos \dfrac{\pi}{6} t$, where t is in months and $t = 0$ corresponds to January 1. Find the average demand for heating oil in a year.

25. Temperature. The temperature in a certain northern city is given by

$$T(t) = 40 + 50 \sin \left[\dfrac{2\pi}{365}(t - 111) \right]$$

where t is the number of days from January 1. Find the average temperature in a year.

26. Temperature. The temperature in a certain southern city is given by

$$T(t) = 65 + 25 \cos \left[\dfrac{2\pi}{365}(t - 20) \right]$$

where t is the number of days from July 1. Find the average temperature in a year.

Enrichment Exercises

27. Establish the identity $\sin^2 x = \frac{1}{2}(1 - \cos 2x)$ using your graphing calculator.

28. Use the result of the previous exercise to find $\int \sin^2 x\, dx$.

29. Establish the identity $\cos^2 x = \frac{1}{2}(1 + \cos 2x)$ using your graphing calculator.

30. Use the result of the previous exercise to find $\int \cos^2 x\, dx$.

31. Let $f(x) = a_1 \sin x + a_2 \sin 2x + \cdots + a_n \sin nx$. Show

$$a_k = \dfrac{1}{\pi} \int_{-\pi}^{\pi} f(x) \sin kx\, dx, \text{ for } k = 1, 2, \cdots, n$$

Hint: Use the identity $\sin x \sin y = \frac{1}{2}[\cos(x - y) - \cos(x + y)]$.

32. Show that $\int_{-\pi}^{\pi} \sin mx \cos nx\, dx = 0$ for any integers m and n. *Hint:* Use the identity

$$\sin x \cos y = \dfrac{1}{2}[\sin(x - y) + \sin(x + y)]$$

SOLUTIONS TO SELF-HELP EXERCISE SET 9.4

1. The requested area is given by

$$\int_{-\pi/2}^{\pi/2} \cos x\, dx = \sin x \bigg|_{-\pi/2}^{\pi/2}$$

$$= \sin \dfrac{\pi}{2} - \sin \left(-\dfrac{\pi}{2}\right) = 1 - (-1) = 2$$

2. Let $u = \cos x$. Then $du = -\sin x\, dx$ and

$$\int \dfrac{\sin x}{\sqrt{\cos x}}\, dx = \int \dfrac{-du}{\sqrt{u}} = -\int u^{-1/2}\, du$$

$$= -2u^{1/2} + C = -2\sqrt{\cos x} + C$$

9.5 OTHER TRIGONOMETRIC FUNCTIONS

- Other Trigonometric Functions ■ Derivatives of Trigonometric Functions
- Graphs ■ Integration of Trigonometric Functions

APPLICATION: MEASURING THE SPEED OF A BOAT USING RADAR

The radar on a hovering helicopter 4000 feet above sea level measures the angle of depression of a speed boat heading in the direction of the helicopter to be 32° and one minute later to be 40°. What is the average speed of the boat during that minute? See Example 1 for the answer.

Other Trigonometric Functions

On several occasions in this chapter we had need to consider the expression $\dfrac{\sin \theta}{\cos \theta}$. This function occurs so often in applications that we give it the name *tangent* and denote $\dfrac{\sin \theta}{\cos \theta}$ as $\tan \theta$. Three additional functions of $\sin \theta$ and $\cos \theta$ arise often and so are also given special names. These are *cotangent* (cot), *secant* (sec), and *cosecant* (csc). We now give the definitions of these functions.

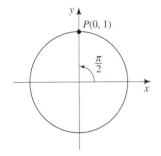

Figure 9.42

$\tan \pi/2 = \dfrac{\sin \pi/2}{\cos \pi/2} = 1/0$ is undefined.

Definition of Tangent, Cotangent, Secant, Cosecant

$$\tan \theta = \frac{\sin \theta}{\cos \theta} \qquad \cot \theta = \frac{\cos \theta}{\sin \theta}$$

$$\sec \theta = \frac{1}{\cos \theta} \qquad \csc \theta = \frac{1}{\sin \theta}$$

Remark. The previous functions are not defined for values of θ for which the denominators are zero. For example, from Figure 9.42, $\tan \pi/2 = \dfrac{1}{0}$ is undefined.

The six functions $\sin \theta$, $\cos \theta$, $\tan \theta$, $\cot \theta$, $\sec \theta$, $\csc \theta$ are called the *trigonometric functions*.

Consider Figure 9.43. Recall that $\sin \theta = \dfrac{y}{r}$ and $\cos \theta = \dfrac{x}{r}$. Thus,

$$\tan \theta = \frac{\sin \theta}{\cos \theta} = \frac{y/r}{x/r} = \frac{y}{x}$$

$$\cot \theta = \frac{\cos \theta}{\sin \theta} = \frac{x/r}{y/r} = \frac{x}{y}$$

Referring to Figure 9.44 we have, as a consequence of these formulas, the following right triangle interpretations

$$\tan \theta = \frac{\text{opposite side}}{\text{adjacent side}}$$

$$\cot \theta = \frac{\text{adjacent side}}{\text{opposite side}}$$

Similar formulas can be written for $\sec \theta$ and $\csc \theta$.

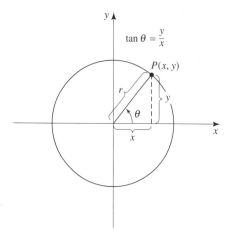

$$\tan \theta = \frac{y}{x}$$

Figure 9.43

$$\tan \theta = \frac{y}{x}$$

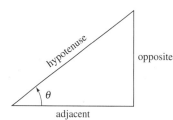

Figure 9.44

$$\tan \theta = \frac{\text{opposite}}{\text{adjacent}}$$

The following is an example of how the right triangle interpretations can be used in an application.

EXAMPLE 1   Measuring the Speed of a Boat Using Radar

The radar on a helicopter hovering 4000 feet above sea level measures the angle of depression of a speed boat heading in the direction of the helicopter to be 32° and 1 minute later to be 40°. Find the average speed of the boat during that minute.

SOLUTION ■ In Figure 9.45 the helicopter is located at point O, and the speed boat is initially at point A and 1 minute later at point B. We first need to find the distance x that the boat traveled during the given minute. From geometry we know that $\beta = 40°$ and $\alpha = 32°$. From triangle OCB we have

$$\tan \beta = \tan 40° = \frac{4000}{b}$$

or solving for b and using a calculator operating in degree mode,

$$b = \frac{4000}{\tan 40°} \approx 4767$$

Using triangle OCA, we have

$$\tan \alpha = \tan 32° = \frac{4000}{b + x}$$

$$b + x = \frac{4000}{\tan 32°}$$

$$4767 + x = \frac{4000}{\tan 32°}$$

$$x \approx 1634$$

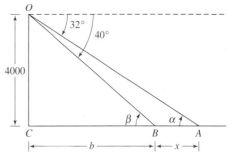

Figure 9.45
The stationary helicopter at O spots a boat at A and 1 minute later at B.

feet. Thus, the average speed is 1634 feet per minute or, since there are 5280 feet in a mile, the average speed is also

$$\frac{1634}{5280} \times 60 \approx 18.6$$

miles per hour. ∎

Derivatives of Trigonometric Functions

We now find the derivatives of $\tan x$ and $\sec x$.

EXAMPLE 2   Derivative of Tangent

Find the derivative of $\tan x$.

SOLUTION ∎

$$\frac{d}{dx}(\tan x) = \frac{d}{dx}\left(\frac{\sin x}{\cos x}\right)$$

$$= \frac{\cos x \dfrac{d}{dx}(\sin x) - (\sin x)\dfrac{d}{dx}(\cos x)}{\cos^2 x}$$

$$= \frac{(\cos x)(\cos x) - (\sin x)(-\sin x)}{\cos^2 x}$$

$$= \frac{\cos^2 x + \sin^2 x}{\cos^2 x}$$

$$= \frac{1}{\cos^2 x}$$

$$= \sec^2 x \quad ∎$$

EXAMPLE 3   Derivative of the Secant

Find the derivative of $\sec x$.

SOLUTION ∎

$$\frac{d}{dx}(\sec x) = \frac{d}{dx}[(\cos x)^{-1}]$$

$$= -(\cos x)^{-2}\frac{d}{dx}(\cos x)$$

$$= -(\cos x)^{-2}(-\sin x)$$

$$= \frac{1}{\cos x} \frac{\sin x}{\cos x}$$

$$= \sec x \tan x \quad \blacksquare$$

We can use the chain rule to find the more general differentiation formulas. We now list these together with the differentiation formulas for the remaining two trigonometric functions. The differentiation formulas for $\cot x$ and $\csc x$ are derived in the same way we derived the formulas for $\tan x$ and $\sec x$.

Derivatives of the Remaining Four Trigonometric Functions

$$\frac{d}{dx}[\tan f(x)] = [\sec^2 f(x)]f'(x)$$

$$\frac{d}{dx}[\sec f(x)] = [\sec f(x)][\tan f(x)]f'(x)$$

$$\frac{d}{dx}[\cot f(x)] = -[\csc^2 f(x)]f'(x)$$

$$\frac{d}{dx}[\csc f(x)] = -[\csc f(x)][\cot f(x)]f'(x)$$

Graphs

We now use some of the preceding derivative formulas to sketch a graph of $y = \tan x$. As will become apparent later, we first need to sketch the graph on $(-\pi/2, \pi/2)$. On this interval $\cos x$ is positive, thus, $\tan x = \dfrac{\sin x}{\cos x}$ has the same sign as $\sin x$. Therefore, $\tan x$ is negative on $(-\pi/2, 0)$ and positive on $(0, \pi/2)$. Also $\tan 0 = 0$. Since $\dfrac{d}{dx} \tan x = \sec^2 x$, $\tan x$ is always increasing (no matter what the interval). As $x \rightarrow \pi/2$, $\sin x \rightarrow 1$ and $\cos x$ approaches 0 through positive values. Thus $\lim\limits_{x \rightarrow (\pi/2)^-} \tan x = \infty$. In a similar fashion we see that $\lim\limits_{x \rightarrow (-\pi/2)^+} \tan x = -\infty$.

To determine the concavity, we need to find the second derivative. We have

$$\frac{d^2}{dx^2} \tan x = \frac{d}{dx}\left(\frac{d}{dx} \tan x\right)$$

$$= \frac{d}{dx} \sec^2 x$$

$$= 2(\sec x) \cdot \frac{d}{dx} \sec x$$

$$= 2 \sec x \sec x \tan x$$

$$= 2 \sec^2 x \tan x$$

Thus, the sign of the second derivative of $\tan x$ is the same as the sign of $\tan x$ (no matter what the interval). Therefore, $\tan x$ is concave down on $(-\pi/2, 0)$ and concave up $(0, \pi/2)$. We then obtain the part of the graph shown in Figure 9.46 on the interval $(-\pi/2, \pi/2)$. Using certain trigonometric identities, it is possible to show that $\tan x$ is periodic with period π. Using this fact, we can obtain the graph shown in Figure 9.46.

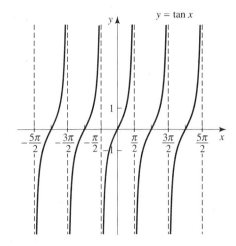

Figure 9.46
A graph of $y = \tan x$.

We can obtain the graphs of $y = \cot x$, $y = \sec x$, and $y = \csc x$ by using the reciprocal identities

$$\cot x = \frac{1}{\tan x} \qquad \sec x = \frac{1}{\cos x} \qquad \csc x = \frac{1}{\sin x}$$

You are invited to sketch these graphs in the exercises.

Integration of Trigonometric Functions

As we know, to every differentiation formula there corresponds an integration formula. Therefore, since $\dfrac{d}{dx} \tan x = \sec^2 x$, we must have

$$\int \sec^2 x \, dx = \tan x + C$$

Other such formulas are

$$\int \csc^2 x \, dx = -\cot x + C$$

$$\int \sec x \tan x \, dx = \sec x + C$$

$$\int \csc x \cot x \, dx = -\csc x + C$$

We continue with another example.

EXAMPLE 4   Integral of Tangent

Find the integral of $\tan x$.

SOLUTION ■ Since $\tan x = \dfrac{\sin x}{\cos x}$, use the substitution $u = \cos x$. Then $du = -\sin x$. Then

$$\int \tan x \, dx = \int \frac{\sin x}{\cos x} \, dx$$

$$= \int \frac{-du}{u}$$

$$= -\ln |u| + C$$

$$= -\ln |\cos x| + C \quad ■$$

SELF-HELP EXERCISE SET 9.5

1. Show that $\dfrac{d}{dx}(\cot x) = -\csc^2 x$.

2. Find $\int \sec^5 x \tan x \, dx$.

EXERCISE SET 9.5

In Exercises 1 through 8 find **(a)** $\tan \theta$ **(b)** $\sec \theta$ **(c)** $\cot \theta$ **(d)** $\csc \theta$ for the given θ.

1. $\theta = 0$

2. $\theta = \dfrac{\pi}{2}$

3. $\theta = \dfrac{3\pi}{2}$

4. $\theta = \pi$

5. $\theta = \dfrac{\pi}{4}$

6. $\theta = \dfrac{3\pi}{4}$

7. $\theta = \dfrac{\pi}{3}$

8. $\theta = \dfrac{\pi}{6}$

In Exercises 9 through 16 find **(a)** $\tan \theta$ **(b)** $\sec \theta$ **(c)** $\cot \theta$ **(d)** $\csc \theta$, where θ is the radian measure of the angle shown.

9.
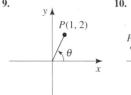

10.

11.
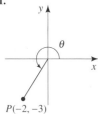

12.
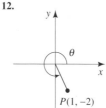

13.
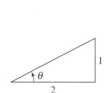

14.
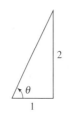

15.
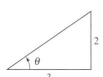

16.
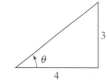

In Exercises 17 through 30 find the derivative of the given function.

17. $\tan^5 3x$

18. $\cot^3 x^2$

19. $x \sec^2 x$

20. $x \csc^4 x$

21. $\dfrac{\cot x^2}{x}$

22. $\dfrac{\sec^2 x}{x}$

23. $\csc(\ln 5x^2)$

24. $\tan[\ln(x^2 + 1)]$

25. $e^{\tan x^2}$

26. $e^{\cot x^2}$

27. $\tan(\sec x)$

28. $\cot(\csc x)$

29. $\sin(\cos x + \tan x)$

30. $\cos(\sin x + \cot x)$

In Exercises 31 through 42 evaluate each of the integrals.

31. $\displaystyle\int 9 \sec^2 3x \, dx$

32. $\displaystyle\int 12x^2 \sec x^3 \tan x^3 \, dx$

33. $\displaystyle\int e^{\tan x} \sec^2 x \, dx$

34. $\displaystyle\int e^x \sec^2 e^x \, dx$

35. $\displaystyle\int \tan^2 x \sec^2 x \, dx$

36. $\displaystyle\int \tan x \sec^2 x \, dx$

37. $\displaystyle\int \tan 5x \, dx$

38. $\displaystyle\int \cot 3x \, dx$

39. $\displaystyle\int 8x \cot x^2 \, dx$

40. $\displaystyle\int 6x \tan x^2 \, dx$

41. $\displaystyle\int 2x \sec x^2 \, dx$

42. $\displaystyle\int 2x \csc x^2 \, dx$

Applications

43. Measuring Heights. The angle of inclination from ground level to the top of a mountain is 30°. On level ground 2 miles closer to the mountain the angle of inclination is 45°. How high is the mountain?

44. Distance Traveled. Dawn swims toward a vertical cliff at the edge of the ocean and knows that the cliff is 200 feet above sea level. At one point Dawn estimates that the angle of inclination from herself to the top of the cliff is 30° and 1 minute later is 60°. How far has she traveled in this minute?

45. A dune buggy is located 2 miles south of a highway that runs east and west. The buggy heads in a straight line in a northeast direction at an angle of 60° from north. If the buggy travels at a steady 10 miles per hour across the level sand, how long will it take before the buggy intersects the highway?

46. Estimating Speed. An airplane passes 5000 feet over a tower, and 15 seconds later the angle of depression from the plane to the tower is 60°. What was the average speed of the plane during this time?

Enrichment Exercises

47. Show $\tan(-\theta) = -\tan \theta$.

48. Show $1 + \tan^2 \theta = \sec^2 \theta$.

49. Graph $y = \sec x$.

50. Graph **(a)** $y = \cot x$ **(b)** $y = \csc x$.

51. Medicine. The figure is a schematic of a human femur bone. A bone goes from G to E (the femoral shaft), E is a joint, and another bone goes from E to B. The angles α and

β play an important role in orthopedics. For example, improper sizes for the angles α and β can be responsible for hip-joint problems. For further details see Morscher [143]. The formula $\dfrac{\tan \beta_1}{\tan \beta} = \cos \alpha$ is useful. Establish this.

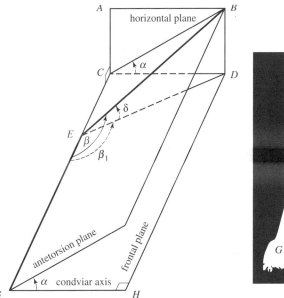
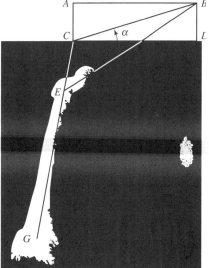

SOLUTIONS TO SELF-HELP EXERCISE SET 9.5

1.
$$\frac{d}{dx}(\cot x) = \frac{d}{dx}\left(\frac{\cos x}{\sin x}\right)$$
$$= \frac{\sin x \frac{d}{dx}(\cos x) - (\cos x)\frac{d}{dx}(\sin x)}{\sin^2 x}$$
$$= \frac{(\sin x)(-\sin x) - (\cos x)(\cos x)}{\sin^2 x}$$
$$= \frac{-\sin^2 x - \cos^2 x}{\sin^2 x}$$
$$= \frac{-1}{\sin^2 x}$$
$$= -\csc^2 x$$

2. Let $u = \sec x$. Then $du = \sec x \tan x \, dx$ and
$$\int \sec^5 x \tan x \, dx = \int \sec^4 x \sec x \tan x \, dx$$
$$= \int u^4 \, du = \frac{1}{5} u^5 + C = \frac{1}{5} \sec^5 x + C$$

CHAPTER **9**

SUMMARY OUTLINE

- An **angle** is formed by rotating a ray about its initial point. p. 470.

- The initial position of the ray is called the **initial side** of the angle. p. 470.

- The position of the ray at the end of its rotation is called the **terminal side** of the angle. p. 470.

- The initial point of the ray is called the **vertex** of the angle. p. 470.

- If the rotation of the ray is in the counterclockwise direction, we say that the angle is **positive**, whereas, if the rotation is in the clockwise direction, we say the angle is **negative**. p. 470.

- The **standard position** of an angle is obtained by placing the vertex at the origin and the initial side along the positive x-axis. p. 470.

- An angle in standard position is said to be in the quadrant of its terminal side. p. 471.

- **One degree**, written $1°$, is the measure of the angle formed by 1/360 of a rotation. p. 471.

- An angle of measure $180°$ is a **straight angle**. p. 471.

- An angle of measure $90°$ is a **right angle**. p. 471.

- An angle θ is **acute** if $0° < \theta < 90°$, and **obtuse** if $90° < \theta < 180°$. p. 471.

- The radian measure of θ is the length s of the arc of the unit circle cut by θ. p. 471.

- $1° = \dfrac{\pi}{180}$ radians $\quad$ 1 radian $= \left(\dfrac{180}{\pi}\right)°$ p. 473.

- The Trigonometric Functions.
  Let θ be the angle whose terminal side cuts the unit circle at the point $P(x, y)$. We then define

$$\sin \theta = y \qquad \cos \theta = x$$

$$\tan \theta = \frac{\sin \theta}{\cos \theta} \qquad \cot \theta = \frac{\cos \theta}{\sin \theta}$$

$$\csc \theta = \frac{1}{\sin \theta} \qquad \sec \theta = \frac{1}{\cos \theta}$$

- Some Trigonometric Identities. p. 480.

$$\sin(-\theta) = -\sin \theta \qquad \cos(-\theta) = \cos \theta$$

$$\sin^2 \theta + \cos^2 \theta = 1$$

- Alternative Definition of the Trigonometric Functions. p. 481 and 501.
  Let θ be the angle whose terminal side cuts the circle of radius r at the point $P(x, y)$. We then define

$$\sin \theta = \frac{y}{r} \qquad\qquad \cos \theta = \frac{x}{r}$$

$$\tan \theta = \frac{y}{x}, \quad x \neq 0 \qquad \cot \theta = \frac{x}{y}, \quad y \neq 0$$

$$\sec \theta = \frac{r}{x}, \quad x \neq 0 \qquad \csc \theta = \frac{r}{y}, \quad y \neq 0$$

- Right Triangle Interpretation of the Trigonometric Functions. p. 482 and 501.
  Let θ be the acute angle shown in the right triangle in Figure 9.26. Then

$$\sin \theta = \frac{\text{opposite}}{\text{hypotenuse}} \qquad \cos \theta = \frac{\text{adjacent}}{\text{hypotenuse}} \qquad \tan \theta = \frac{\text{opposite}}{\text{adjacent}}$$

$$\cot \theta = \frac{\text{adjacent}}{\text{opposite}} \qquad \sec \theta = \frac{\text{hypotenuse}}{\text{adjacent}} \qquad \csc \theta = \frac{\text{hypotenuse}}{\text{opposite}}$$

- Derivatives of the Trigonometric Functions p. 492 and 504.

$$\frac{d}{dx}[\sin f(x)] = [\cos f(x)]f'(x) \qquad \frac{d}{dx}[\cos f(x)] = -[\sin f(x)]f'(x)$$

$$\frac{d}{dx}[\tan f(x)] = [\sec^2 f(x)]f'(x) \qquad \frac{d}{dx}[\sec f(x)] = [\sec f(x)][\tan f(x)]f'(x)$$

$$\frac{d}{dx}[\cot f(x)] = -[\csc^2 f(x)]f'(x) \qquad \frac{d}{dx}[\csc f(x)] = -[\csc f(x)][\cot f(x)]f'(x)$$

- Trigonometric Integration Formulas p. 498 and 505.

$$\int \sin x \, dx = -\cos x + C \qquad \int \cos x \, dx = \sin x + C$$

$$\int \sec^2 x \, dx = \tan x + C \qquad \int \csc^2 x \, dx = -\cot x + C$$

$$\int \sec x \tan x \, dx = \sec x + C \qquad \int \csc x \cot x \, dx = -\csc x + C$$

CHAPTER 9

REVIEW EXERCISES

In Exercises 1 through 6 convert each degree measure to radian measure.

1. $90°$

2. $120°$

3. $270°$

4. $450°$

5. $-225°$

6. $75°$

In Exercises 7 through 12 convert each radian measure to degree measure.

7. $\dfrac{11\pi}{12}$

8. $\dfrac{7\pi}{4}$

9. $\dfrac{-5\pi}{2}$

10. 5π

11. $\dfrac{11\pi}{20}$

12. $\dfrac{\pi}{6}$

In Exercises 13 through 16 find **(a)** $\sin\theta$ **(b)** $\cos\theta$ **(c)** $\tan\theta$ **(d)** $\sec\theta$ **(e)** $\cot\theta$ **(f)** $\csc\theta$, where θ is the radian measure of the angle shown.

13.

14.

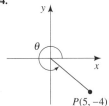

15.

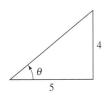

16.

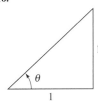

In Exercises 17 through 20, graph the given function.

17. $y = 2 \sin 6x$

18. $y = -3 \cos 8x$

19. $y = 2 \sin(2x - 2)$

20. $y = 3 \cos \pi(x + 1)$

In Exercises 21 through 35 find the derivative.

21. $y = 3 \sin \pi x$

22. $y = 4 \cos \dfrac{x}{3}$

23. $y = 2 \tan(x^2 + 1)$

24. $y = \cot(x + 1)$

25. $y = \sec^2 x$

26. $y = \sec x \csc x$

27. $y = 3 \sin^3 x^3$

28. $y = \sqrt{\cos x + 1}$

29. $y = x^2 \tan x^2$

30. $y = \sec e^x$

31. $y = e^{\sin x + \cos x}$

32. $y = \cot(\ln |x|)$

33. $y = \dfrac{\sin x}{x}$

34. $y = \dfrac{\sin x}{2 + \cos x}$

35. $y = \tan(\sin x)$

In Exercises 36 through 47 find the integral.

36. $\displaystyle\int \sin 8x \, dx$

37. $\displaystyle\int \cos \dfrac{x}{2} \, dx$

38. $\displaystyle\int \sec^2 \dfrac{x}{2} \, dx$

39. $\displaystyle\int \sec 3x \tan 3x \, dx$

40. $\displaystyle\int 6 \csc^2 x \, dx$

41. $\displaystyle\int 6 \tan 3x \, dx$

42. $\displaystyle\int e^x \cos e^x \, dx$

43. $\displaystyle\int e^{\cos x} \sin x \, dx$

44. $\displaystyle\int \dfrac{\cos(\ln x)}{x} \, dx$

45. $\displaystyle\int \dfrac{\cos x}{3 + \sin x} \, dx$

46. $\displaystyle\int \dfrac{\sin x}{(3 + \cos x)^2} \, dx$

47. $\displaystyle\int \sin(\ln x) \, dx$

48. Medicine. Many infectious diseases progress in a periodic fashion. Bliss and Blevins [144] showed that the incidence of measles over a 31-year period in Baltimore approximately satisfied

$$y = 1.56 - 0.7 \cos 0.5t + 0.146 \sin 0.5t$$
$$- 0.26 \cos t + 0.156 \sin t$$

where y is the logarithm to the base 10 of the number of cases of measles and t is measured in months, with $t = 0$ corresponding to October. In a screen of dimensions [0, 16] by [−1, 3] graph the functions $y_1 = -0.7 \cos 0.5x + 0.146 \sin 0.5x$, $y_2 = -0.26 \cos x + 0.156 \sin x$, and $y_3 = 1.56 + y_1 + y_2$. The time is extended beyond one year to see the periodic nature of the graph. Find the approximate time when the incidence is maximum. Minimum. Check your estimates by graphing the numerical derivative.

49. Forestry. Beer's law [145] states that $I = I_0 e^{-kL/\cos \theta}$, where I is the incident radiation at the base of a forest canopy, I_0 is the incident radiation at the top of the canopy, θ is the solar zenith angle, and L is a leaf area index. Find $\dfrac{dI}{d\theta}$.

CHAPTER **9**

PROJECTS

Project **1**
A Model of a National Economy

Paul Samuelson [139], 1970 Nobel Prize winner in economics, formulated mathematical models that described the tendency for economic activity to rise and fall periodically over time. Periodicity is also observed in the prices of commodities, international trade, inventory accumulation, and investments.

We now construct a mathematical model based on Samuelson's analysis. To construct our mathematical model of a national economy, we make four assumptions. First, we assume that total national income is the sum of consumer expenditure, private investment, and government expenditures. Let $T(n)$, $C(n)$, $I(n)$, and $G(n)$, denote total national income, consumer expenditures, private investment (which includes buying new machinery and building new factories and so on), and government expenditures, respectively, during the nth time period. Then

$$T(n) = C(n) + I(n) + G(n) \qquad [1]$$

Secondly, we assume that the total savings of individuals, companies, and government in any time period is a fixed proportion of their incomes in the previous time period. Thus, there is a positive constant a, called the **marginal propensity to consume**, such that

$$C(n + 1) = aT(n) \qquad [2]$$

Thirdly, we assume that private investment in any period is proportional to the change in consumption over the past period. For example, if consumption increases, then more machinery and factories are needed to produce the goods being consumed. If consumption decreases, then existing machinery and factories can be retired. Thus, there exists a positive constant b, called the **constant of adjustment** so that

$$I(n + 1) = b[C(n + 1) - C(n)] \qquad [3]$$

Finally, we assume that government expenditures are constant. By taking appropriate units of money we take the constant to be 1. Thus,

$$G(n) = 1 \qquad [4]$$

The last equation is more realistic if we measure money by discounting for inflation. Then, Equation 4 says the *value* of government expenditures is constant.

Equations 1, 2, 3, and 4 then represent a mathematical model of a national economy. We shall now manipulate these four equations and obtain one single equation that gives total income. We have

$$T(n + 2) = C(n + 2) + I(n + 2)$$
$$+ G(n + 2) \qquad \text{Replace } n \text{ with } n + 2 \text{ in Equation 1}$$
$$= C(n + 2) + I(n + 2) + 1 \qquad \text{From Equation 2}$$
$$= C(n + 2) + b[C(n + 2)$$
$$- C(n + 1)] + 1 \qquad \text{Replace } n \text{ with } n + 1 \text{ in Equation 3}$$
$$= (1 + b)C(n + 2) - bC(n + 1) + 1 \qquad \text{Simplify}$$
$$= (1 + b)aT(n + 1) - baT(n) + 1 \qquad \text{Use Equation 2}$$

To further simplify and for later convenience, we let $(1 + b)a = p$ and $ab = q^2$ and obtain

$$T(n + 2) = pT(n + 1) - q^2 T(n) + 1 \qquad [5]$$

as the equation giving the total income that we now work with. Equation 5 is an example of a **dynamical system**, or **difference equation**, and is a variation of the accelerator-multiplier model that Samuelson gave in 1939.

(a) If $1 - p - q^2 \neq 0$, show that a **solution** of Equation 5 is

$$T(n) = q^n(A \cos n\theta + B \sin n\theta) + k \qquad [6]$$

where A and B are any constants and $\tan \theta = \sqrt{4q^2 - p^2}/p$ and $k = 1/(1 - p + q^2)$. (Actually, Equation 6 gives the general solution to Equation 5, that is, any solution of Equation 5 must be in the form of Equation 6. *Hint:* Use trigonometric identities $\sin(x + y) = \sin x \cos y + \cos x \sin y$ and $\cos(x + y) = \cos x \cos y - \sin x \sin y$ and show that $\cos \theta = p/(2q)$.

(b) Solve

$$T(n + 2) = T(n + 1) - T(n) + 1, \quad T(0) = 1, T(1) = \sqrt{3}$$

Hint: First find θ. Use the condition $T(0) = 1$ to determine A, and use $T(1) = \sqrt{3}$ to determine B. Graph $T(n)$ on your graphing calculator. Explain what is happening. Is the solution periodic? What is the period? What does this mean for this particular economy?

(c) Solve

$$T(n + 2) = T(n + 1) - 0.5T(n) + 4, \quad T(0) = 5, T(1) = 4$$

Hint: First find θ. Use the condition $T(0) = 5$ to determine A, and use $T(1) = 4$ to determine B. Graph $T(n)$ on your graphing calculator. Explain what is happening. Is the solution periodic? Is the solution oscillating? Explain. What is happening to the peaks and troughs in the long term? What does this mean for this particular economy?

P r o j e c t   2
Oscillations

Equation 5 in Project 1 is a discrete **second-order** dynamical system (difference equation). We now look at a continuous analog

$$\frac{d^2 y}{dt^2} + b\frac{dy}{dt} + cy = d \qquad [1]$$

which is a second-order differential equation. This differential equation is fundamental in oscillation theory in mechanics and electricity. It is also used to describe oscillatory economic and business activity. (See Phillips [138].)

(a) Consider

$$\frac{d^2 y}{dt^2} + a^2 y = d \qquad [2]$$

Show that $y = A \cos at + B \sin at + \dfrac{d}{a^2}$ satisfies Equation 2 for any constants A and B.

(b) Find C and θ such that

$$A \cos at + B \sin at = C \sin(at - \theta)$$

Sketch a graph when the constants C, a, and θ are all positive. *Hint:* Use the trigonometric identity for the sine of the sum of two angles and take θ so that $\tan \theta = \dfrac{B}{A}$.

(c) If $b^2 - 4c < 0$, show that the solution of Equation 1 is

$$e^{-(b/2)t}(A \cos \omega t + B \sin \omega t) + \frac{d}{c}$$

where $\omega = 0.5\sqrt{4c - b^2}$. Place this expression in an alternative form by using the result in part (a), and sketch a graph when $b > 0$. What does the graph look like if $b < 0$?

P r o j e c t   3
Resonance

In Project 2 we noted that the second-order differential equation $y'' + a^2 y = d$ can model total national income under certain circumstances. The constant d is government expenditures, assumed to be constant. Let us now assume that government expenditures is cyclical and of the form $\sin \omega t$. Then the differential equation that models such an economy is

$$y'' + a^2 y = \sin \omega t \qquad [1]$$

(a) Show that if $\omega \neq a$ there is a solution to this differential equation of the form $y = A \cos \omega t + B \sin \omega t$. Do this by differentiating this quantity twice and substituting this quantity and its second derivative into Equation 1. Determine what A and B must be.

(b) What happens to the solution you found in part (a) as ω approaches a? How do you describe what is happening to the economy in this case?

(c) Specifically, let $a = 2$ and take $\omega = 1.5, 1.9, 1.99$ and graph on your graphing calculator. Does this confirm your answer to part (b)?

Project 4
Vascular Branching

The transportation of blood from the heart to all areas of the body has naturally evolved to be as efficient as possible. In this project we look at one aspect of this efficiency: *vascular branching*. The figure shows a schematic of a main blood vessel and a branch vessel. We wish to find the angle θ_0 that minimizes the total resistance of the blood along the path ADC.

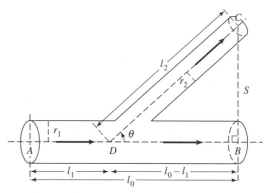

Poiseuille's law states that the resistance R of the blood is given by $R = k\dfrac{L}{r^4}$, where r is the diameter of the vessel, L the length of the vessel, and k a constant determined by the viscosity of the blood.

(a) Let $AB = L_0$, $AD = L_1$, and $DC = L_2$. Show that

$$L_2 = \frac{s}{\sin \theta} \qquad L_0 - L_1 = s \cdot \cot \theta \qquad\qquad [1]$$

(b) Since the total resistance R along the path ADC is the resistance R_1 along AD plus the resistance R_2 along DC, we can use Poiseuille's law to obtain

$$R = R_1 + R_2 = k\frac{L_1}{r_1^4} + k\frac{L_2}{r_2^4}$$

In this last equation substitute L_1 and L_2 from Equation 1 and obtain a function $R = R(\theta)$. Use calculus to find the value of θ, in terms of r_1 and r_2, that maximizes R. What is this θ if $r_2/r_1 = 0.5$?

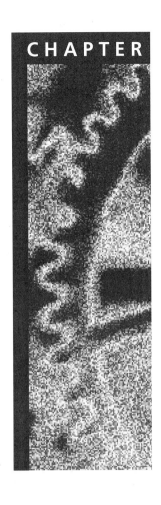

CHAPTER

10

Taylor Polynomials and Infinite Series

This chapter is written for considerable flexibility. Section 10.7 has been written to permit the reader to go from Section 10.2 directly to Section 10.7. Thus Sections 10.1, 10.2, and 10.7 constitute a subchapter on Taylor polynomials and Taylor series.

10.1 TAYLOR POLYNOMIALS

- Approximating $f(x) = e^x$ ■ The nth-Order Taylor Polynomial
- Applications

APPLICATION: NUMBER OF SUSPENDED PARTICULATES

The number of total suspended particulates per mile from an industrial plant is given by $\delta(x) = 100{,}000e^{-x^2}$, where x is measured in miles from the plant. Approximate the total number of particulates within 0.1 miles of the plant. See Example 6 for the answer.

How can we evaluate logarithmic, exponential, and trigonometric functions? We can, of course, use a calculator. But how can the calculator be programmed to give us an answer? Of more importance, how can we evaluate new functions that are not found on a calculator or computer? As we shall see in this and the next section a Taylor polynomial can be constructed that approximates many important functions to any desired accuracy. Furthermore, a Taylor polynomial exists that approximates the function to any desired accuracy over an entire interval. It is often advantageous to replace the original function with the approximating polynomial since the polynomial is easier to evaluate and differentiate, and often, substantially easier to integrate.

We will see that taking a higher degree Taylor polynomial results in a higher degree of accuracy in approximating a suitable function. In the last section of this chapter we will see that allowing the degree of the polynomial to increase without bound results in the approximation becoming exact and allowing us to represent the function as an infinite series.

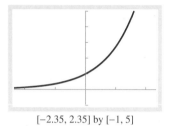

[−2.35, 2.35] by [−1, 5]

Screen 10.1
A graph of $y = e^x$.

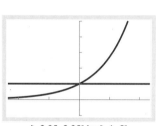

[−2.35, 2.35] by [−1, 5]

Screen 10.2
The constant that best approximates $y = f(x) = e^x$ near $x = 0$ is $y = p_0(x) = 1$. Notice that $p_0 = f(0)$.

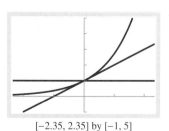

[−2.35, 2.35] by [−1, 5]

Screen 10.3
A better approximation to $y = f(x) = e^x$ near $x = 0$ than $y = 1$ is the tangent line, given by $y = p_1(x) = 1 + x$. Notice that $p_1(0) = f(0)$ and $p_1'(0) = f'(0)$.

Approximating $f(x) = e^x$

Screen 10.1 shows the graph of $f(x) = e^x$ using a window with dimensions $[-2.35, 2.35]$ by $[-1, 5]$. What constant $p_0(x) = a_0$ best approximates $f(x) = e^x$ near $x = 0$? Obviously the polynomial

$$p_0(x) = 1$$

since $f(0) = e^0 = 1$. This is seen on Screen 10.2, which uses the same dimensions as Screen 10.1. Now what polynomial $p_1(x) = a_0 + a_1x$ best approximates e^x near $x = 0$? It is the polynomial $p_1(x) = f(0) + f'(0)x$, since this is the equation of the line tangent to the graph of $y = e^x$ at $x = 0$, that is, $p_1(x)$ satisfies $p_1(0) = f(0)$ and $p_1'(0) = f'(0)$. Since $f'(x) = e^x$, $=f'(0) = e^0 = 1$, and

$$p_1(x) = 1 + x$$

See Screen 10.3.

What polynomial

$$p_2(x) = a_0 + a_1x + a_2x^2$$

best approximates $f(x) = e^x$ near $x = 0$? Clearly one that is tangent to the graph of the curve $y = e^x$ at $x = 0$ and has the same second derivative at $x = 0$, that is, one for which $p_2(0) = f(0)$, $p_2'(0) = f'(0)$, and $p_2''(0) = f''(0)$. We have

$$f(x) = e^x, \quad p_2(x) = a_0 + a_1x + a_2x^2$$

$$f'(x) = e^x, \quad p_2'(x) = a_1 + 2a_2x$$

$$f''(x) = e^x, \quad p_2''(x) = 2a_2$$

Then we require that

$$1 = e^0 = f(0) = p_2(0) = a_0 \quad \text{implies } a_0 = 1$$

$$1 = e^0 = f'(0) = p_2'(0) = a_1 \quad \text{implies } a_1 = 1$$

$$1 = e^0 = f''(0) = p_2''(0) = 2a_2 \quad \text{implies } a_2 = \frac{1}{2}$$

Thus,

$$p_2(x) = 1 + x + \frac{1}{2}x^2$$

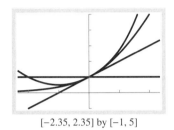

[−2.35, 2.35] by [−1, 5]

Screen 10.4

An even better approximation to $y = f(x) = e^x$ near $x = 0$ than $y = p_0(x)$ or $y = p_1(x)$ is the quadratic given by $y = p_2(x) = 1 + x + 0.5x^2$. Notice that $p_2(0) = f(0)$, $p_2'(0) = f'(0)$, and $p''(0) = f''(0)$.

See Screen 10.4.

We can continue this process indefinitely. For example, the polynomial

$$p_3(x) = a_0 + a_1x + a_2x^2 + a_3x^3$$

that best approximates $f(x) = e^x$ near $x = 0$, should satisfy the conditions

$$p_3(0) = f(0) = e^0 = 1$$

$$p_3'(0) = f'(0) = e^0 = 1$$

$$p_3''(0) = f''(0) = e^0 = 1$$

$$p_3^{(3)}(0) = f^{(3)}(0) = e^0 = 1$$

Since

$$p_3'(x) = a_1 + 2a_2x + 3a_3x^2$$

$$p_3''(x) = 2a_2 + 6a_3x$$

$$p_3^{(3)}(x) = 6a_3$$

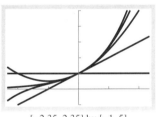

[−2.35, 2.35] by [−1, 5]

Screen 10.5

A still better approximation to $y = f(x) = e^x$ near $x = 0$ than $y = p_0(x)$, $y = p_1(x)$, or $y = p_2(x)$ is the cubic polynomial given by $y = p_3(x) = 1 + x + 0.5x^2 + x^3/6$. Notice that $p_3(0) = f(0)$, $p_3'(0) = f'(0)$, $p_3''(0) = f''(0)$, and $p_3^{(3)}(0) = f^{(3)}(0)$.

we have

$$1 = p_3(0) = a_0 \quad \text{implies } a_0 = 1$$

$$1 = p_3'(0) = a_1 \quad \text{implies } a_1 = 1$$

$$1 = p_3''(0) = 2a_2 \quad \text{implies } a_2 = \frac{1}{2}$$

$$1 = p_3^{(3)}(0) = 6a_3 \quad \text{implies } a_3 = \frac{1}{6}$$

Thus

$$p_3(x) = 1 + x + \frac{1}{2}x^2 + \frac{1}{6}x^3$$

See Screen 10.5.

Using the TRACE function on Screen 10.5 we can readily create Table 10.1. This table and Screen 10.5 illustrate that the higher the degree of the polynomial constructed, the better the approximations to the function $f(x) = e^x$.

Another way to see this principle graphically is to graph $y_1 = |e^x - p_0(x)| = |e^x - 1|$, $y_2 = |e^x - p_1(x)| = |e^x - 1 - x|$, $y_3 = |e^x - p_2(x)| = |e^x - 1 - x - 0.5x^2|$, and $y_4 = |e^x - p_3(x)| = |e^x - 1 - x - 0.5x^2 - x^3/6|$ on the same screen using a window with dimensions $[-2.35, 2.35]$ by $[0, 3]$. We thereby obtain Screen 10.6, which shows that the higher the degree of the polynomial constructed the smaller the error for x near 0. Screen 10.6 indicates this is certainly true on the interval $[-1.5, 2.35]$. Using TRACE on Screen 10.6, we can obtain Table 10.2.

From Table 10.2 notice that $p_3(0.2)$ approximates $e^{0.2}$ to three decimal places and that $p_3(0.1)$ approximates $e^{0.1}$ to five decimal places. Also notice from the screens and

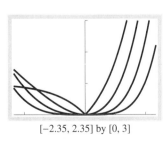

[−2.35, 2.35] by [0, 3]

Screen 10.6

Notice how the error in the approximation of $y = e^x$ with $p_n(x)$ decreases as n increases from 0 to 1 to 2 to 3, for x near 0.

TABLE 10.1

x	−0.2	−0.1	0	0.1	0.2	1
$p_0(x)$	1.00000	1.000000	1	1.000000	1.00000	1.000
$p_1(x)$	0.80000	0.900000	1	1.100000	1.20000	2.000
$p_2(x)$	0.82000	0.905000	1	1.105000	1.22000	2.500
$p_3(x)$	0.81867	0.904833	1	1.105167	1.22133	2.667
e^x	0.81873	0.904837	1	1.105171	1.22140	2.718

TABLE 10.2

x	−0.2	−0.1	0	0.1	0.2	1
$|e^x - p_0(x)|$	0.18127	0.095163	0	0.105171	0.22140	1.718
$|e^x - p_1(x)|$	0.01873	0.004837	0	0.005171	0.02140	0.718
$|e^x - p_2(x)|$	0.00127	0.000163	0	0.000171	0.00140	0.218
$|e^x - p_3(x)|$	0.00006	0.000004	0	0.000004	0.00007	0.052

the tables that the approximations seem to get worse the farther we go from $x = 0$. This is to be expected.

EXAMPLE 1 Approximating $\sqrt{e}$

Use the previously mentioned polynomial $p_3(x)$ to approximate $\sqrt{e}$.

SOLUTION ■ We have that $\sqrt{e} = e^{0.5}$. Then

$$p_3(x) = 1 + x + \frac{1}{2}x^2 + \frac{1}{6}x^3$$

$$p_3(0.5) = 1 + (0.5) + \frac{1}{2}(0.5)^2 + \frac{1}{6}(0.5)^3$$

$$= 1 + \frac{1}{2} + \frac{1}{8} + \frac{1}{48} \approx 1.6458 \quad ■$$

The actual value of $e^{0.5}$ to four decimals is 1.6487.

The *n*th-Order Taylor Polynomial

We can continue the process that gave us p_1, p_2, and p_3 to any polynomial $p_n(x)$. We can also follow the same construction using a wide variety of functions at any point $x = a$.

Given a function $f(x)$ with n derivatives at $x = a$, we want to find a polynomial

$$p_n(x) = a_0 + a_1(x - a) + a_2(x - a)^2 + \cdots + a_n(x - a)^n$$

that satisfies

$$f^{(k)}(a) = p_n^{(k)}(a), \quad \text{for } k = 0, 1, 2, \ldots, n$$

Notice that

$$p_n'(x) = a_1 + 2a_2(x - a) + 3a_3(x - a)^2 + \cdots na_n(x - a)^{n-1}$$

$$p_n''(x) = 2a_2 + (3)(2)a_3(x - a) + \cdots + n(n - 1)a_n(x - a)^{n-2}$$

$$p_n^{(3)}(x) = (3)(2)(1)a_3 + \cdots + n(n - 1)(n - 2)a_n(x - a)^{n-3}$$

$$\vdots$$

$$p_n^{(n)}(x) = n(n - 1)(n - 2) \cdots (1)a_n = n!a_n$$

Thus,

$$f(a) = p_n(a) = a_0 \qquad \text{implies } a_0 = f(a)$$

$$f'(a) = p'_n(a) = a_1 \qquad \text{implies } a_1 = f'(a)$$

$$f''(a) = p''_n(a) = 2a_2 \qquad \text{implies } a_2 = \frac{f''(a)}{2}$$

$$f^{(3)}(a) = p_n^{(3)}(a) = 3!a_3 \quad \text{implies } a_3 = \frac{f^{(3)}(a)}{3!}$$

$$\vdots$$

$$f^{(n)}(a) = p_n^{(n)}(a) = n!a_n \quad \text{implies } a_n = \frac{f^n(a)}{n!}$$

Thus, we have

$$a_k = \frac{f^{(k)}(a)}{k!}, \quad k = 0, 1, 2, 3, \ldots, n$$

The polynomial $p_n(x)$ is called the **nth-order Taylor polynomial for the function $f(x)$ at $x = a$.**

Taylor Polynomial at $x = a$

The **nth-order Taylor polynomial for the function $f(x)$ at $x = a$** is

$$p_n(x) = f(a) + f'(a)(x - a) + \frac{f''(a)}{2!}(x - a)^2 + \cdots + \frac{f^{(n)}(a)}{n!}(x - a)^n$$

provided f has n derivatives at $x = a$.

Remark. We speak of a Taylor polynomial of *order n* rather than *degree n* because $f^{(n)}(a)$ may be zero.

EXAMPLE 2   Finding an nth-Order Taylor Polynomial

Find the nth-order Taylor polynomial for $f(x) = e^x$ at $x = 0$.

SOLUTION ■ We have $f^{(k)}(x) = e^x$ for $k = 0, 1, 2, \ldots, n$, and thus,

$$f^{(k)}(0) = e^0 = 1, \quad k = 0, 1, 2, \ldots, n$$

Therefore,

$$a_k = \frac{f^{(k)}(0)}{k!} = \frac{1}{k!}$$

and

$$p_n(x) = a_0 + a_1(x - 0) + a_2(x - 0)^2 + \cdots + a_n(x - 0)^n$$

$$= 1 + x + \frac{1}{2!}x^2 + \cdots + \frac{1}{n!}x^n \quad ■$$

EXPLORATION 1

Taylor Polynomials

Find the third-order Taylor polynomial for $y = e^x$ at $x = 0$, and graph this polynomial together with $y = e^x$ using your graphing calculator in a window with dimensions $[-5, 5]$ by $[-20, 50]$. Compare how close the two graphs are. Repeat with the fourth-, fifth-, and sixth-order Taylor polynomials. Are the approximations getting better with larger n? Take even larger values of n, increase the dimension of the window, and repeat.

EXAMPLE 3   Finding a Fourth-Order Taylor Polynomial

Find the fourth-order Taylor polynomial for $f(x) = \sin x$ at $x = 0$. Graph $y_1 = \sin x$ and $y_2 = p_4(x)$ using a window of dimensions $[-2, 2]$ by $[-1, 1]$.

S O L U T I O N ■ We have

$$f(x) = \sin x, \quad \text{and } f(0) = \sin(0) = 0$$

$$f'(x) = \cos x, \quad \text{and } f'(0) = \cos(0) = 1$$

$$f''(x) = -\sin x, \quad \text{and } f''(0) = -\sin(0) = 0$$

$$f^{(3)}(x) = -\cos x, \quad \text{and } f^{(3)}(0) = -\cos(0) = -1$$

$$f^{(4)}(x) = \sin x, \quad \text{and } f^{(4)}(0) = \sin(0) = 0$$

Thus, for $p_4(x) = a_0 + a_1 x + a_2 x^2 + a_3 x^3 + a_4 x^4$,

$$a_0 = f(0) = 0$$

$$a_1 = f'(0) = 1$$

$$a_2 = \frac{f''(0)}{2} = 0$$

$$a_3 = \frac{f^{(3)}(0)}{3!} = -\frac{1}{6}$$

$$a_4 = \frac{f^{(4)}(0)}{4!} = 0$$

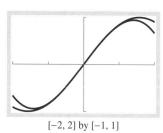

[−2, 2] by [−1, 1]

Screen 10.7
We see how well $y = \sin x$ is approximated by the Taylor polynomial $p_3 = x - x^3/6$ for x near 0.

and

$$p_4(x) = x - \frac{1}{6} x^3$$

See Screen 10.7. ■

Remark. Notice that the *fourth*-order Taylor polynomial p_4 found in Example 3 is a polynomial of degree *three* since $f^{(4)}(0) = 0$.

EXAMPLE 4   Finding Approximations of sin x

Use the result of Example 3 to find an approximation to $\sin(0.2)$.

S O L U T I O N ■ We have, to six decimal places,

$$p_4(0.2) = (0.2) - \frac{1}{6}(0.2)^3 = 0.198667 \quad ■$$

Remark. To six decimal places $\sin(0.2) = 0.198669$, so that $p_4(0.2)$ in Example 4 has an error of only 0.000003.

EXAMPLE 5   Finding a Third-Order Taylor Polynomial

Find the third-order Taylor polynomial for $f(x) = \ln x$ at $x = 1$. Graph $y_1 = \ln x$ and $y_2 = p_3(x)$ using a window of dimensions $[0, 3]$ by $[-3, 3]$.

S O L U T I O N ■ We have

$$f(x) = \ln x, \quad \text{and } f(1) = 0$$

$$f'(x) = \frac{1}{x}, \quad \text{and } f'(1) = 1$$

$$f''(x) = -x^{-2}, \quad \text{and } f''(1) = -1$$
$$f^{(3)}(x) = 2x^{-3} \quad \text{and } f^{(3)}(1) = 2$$

Then for

$$p_3(x) = a_0 + a_1(x - 1) + a_2(x - 1)^2 + a_3(x - 1)^3$$

we have

$$a_0 = f(1) = 0$$
$$a_1 = f'(1) = 1$$
$$a_2 = \frac{f''(1)}{2} = \frac{-1}{2}$$
$$a_3 = \frac{f^{(3)}(1)}{3!} = \frac{2}{6} = \frac{1}{3}$$

Therefore, the third-order Taylor polynomial at $x = 1$ is

$$p_3(x) = (x - 1) - \frac{1}{2}(x - 1)^2 + \frac{1}{3}(x - 1)^3 \quad \blacksquare$$

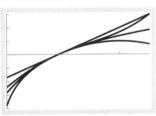

[0.15, 2.5] by [−2, 1.5]

Screen 10.8
We see the graphs of $y = f(x) = \ln x$ and the Taylor polynomials about $x = 1$: $p_1(x) = x - 1$, $p_2(x) = x - 1 - 0.5(x - 1)^2$, $p_3(x) = x - 1 - 0.5(x - 1)^2 + (x - 1)^3/3$.

You can verify that the first- and second-order Taylor polynomials for $\ln x$ at $x = 1$ are

$$p_1(x) = x - 1 \qquad p_2(x) = (x - 1) - \frac{1}{2}(x - 1)^2$$

Screen 10.8 has dimensions [0.15, 2.5] by [−2, 1.5] and shows the graphs of $y = \ln x$, $y = p_1(x)$, $y = p_2(x)$, and $y = p_3(x)$. Using TRACE on Screen 10.8, we can obtain Table 10.3. Again we see that the higher the order of the Taylor polynomial the more accurate the approximation, and that the accuracy decreases as we move away from $x = 1$.

Screen 10.9 has a window of dimensions [0.15, 2.5] by [0, 1] and shows the graphs of $y = |\ln x - p_1(x)| = |\ln x - x + 1|$, $y = |\ln x - p_2(x)| = |\ln x - x + 1 + 0.5(x - 1)^2|$, and $y = |\ln x - p_3(x)| = |\ln x - x + 1 + 0.5(x - 1)^2 - (x - 1)^3/3|$. Using TRACE on Screen 10.9, we can create Table 10.4. Again, we see graphically and numerically how the error decreases with increasing degree of the approximating Taylor polynomial.

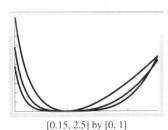

[0.15, 2.5] by [0, 1]

Screen 10.9
Notice how the error in the approximation of $y = \ln x$ with $p_n(x)$ decreases as n increases from 0 to 1 to 2 to 3, for x near 1.

TABLE 10.3

x	1.0	1.1	1.2	1.3
$p_1(x)$	0	0.10000	0.20000	0.30000
$p_2(x)$	0	0.09500	0.18000	0.25500
$p_3(x)$	0	0.09533	0.18267	0.26400
$\ln x$	0	0.09531	0.18232	0.26236

TABLE 10.4

x	1.0	1.1	1.2	1.3
$|\ln x - p_1(x)|$	0	0.00469	0.01768	0.03764
$|\ln x - p_2(x)|$	0	0.00031	0.00232	0.00736
$|\ln x - p_3(x)|$	0	0.00002	0.00035	0.00164

Applications

EXAMPLE 6

The density, measured in numbers per mile, of total suspended particulates from an industrial plant is given by

$$\delta(x) = 100{,}000e^{-x^2}$$

where x is measured in miles from the plant. Approximate the total number of particulates within 0.1 mile of the plant using the second-order Taylor polynomial for $f(x) = e^{x^2}$ at $x = 0$.

SOLUTION ■ Recalling the discussion of density in Section 5.7, the total number of particulates within 0.1 mile of the plant is given by the integral

$$100{,}000 \int_{-0.1}^{0.1} e^{-x^2}\, dx = 200{,}000 \int_{0}^{0.1} e^{-x^2}\, dx$$

Recall that it is impossible to express the antiderivative of e^{-x^2} as an elementary function. Thus, this definite integral must be found using an approximation technique. We first find the second-order Taylor polynomial for e^{-x^2} at $x = 0$. We use the Taylor polynomial at $x = 0$ since $x = 0$ is in the middle of the interval of integration.

$$f(x) = e^{-x^2}, \qquad\qquad \text{and } f(0) = 1$$
$$f'(x) = -2xe^{-x^2}, \qquad \text{and } f'(0) = 0$$
$$f''(x) = (4x^2 - 2)e^{-x^2}, \quad \text{and } f''(0) = -2$$

Thus,

$$a_0 = f(0) = 1$$
$$a_1 = f'(0) = 0$$
$$a_2 = \frac{f''(0)}{2!} = \frac{-2}{2} = -1$$

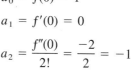

[−1, 1] by [0, 1]

Screen 10.10
Shown are graphs of $y = f(x) = e^{-x^2}$ and $p_2(x) = 1 - x^2$.

and

$$p_2(x) = a_0 + a_1 x + a_2 x^2$$
$$= 1 - x^2$$

Screen 10.10 has a window with dimensions [−1, 1] by [0, 1] and shows graphs of $y = e^{-x^2}$ and $y = p_2(x) = 1 - x^2$. Then

$$\int_{0}^{0.1} e^{-x^2}\, dx \approx \int_{0}^{0.1} p_2(x)\, dx$$
$$= \int_{0}^{0.1} (1 - x^2)\, dx$$
$$= \left(x - \frac{1}{3} x^3\right)\Big|_{0}^{0.1}$$
$$= [(0.1) - \frac{1}{3}(0.1)^3]$$
$$\approx 0.099667$$

Thus, the total number of particulates within 0.1 mile of the plant is

$$200{,}000 \int_{0}^{0.1} e^{-x^2}\, dx \approx 200{,}000(0.099667) \approx 19{,}933 \quad ■$$

SELF-HELP EXERCISE SET 10.1

1. Find the third-order Taylor polynomial for $f(x) = \ln(x + 1)$ at $x = 0$.

2. Find the third-order Taylor polynomial for $f(x) = \sqrt[4]{x}$ at $x = 1$, and then use this polynomial to find an approximation of $\sqrt[4]{1.1}$.

EXERCISE SET 10.1

In Exercises 1 through 10 find the Taylor polynomial $p(x)$ of the indicated function $f(x)$ at $x = 0$. On your graphing calculator, graph $y = f(x)$ and $y = p(x)$ on the same screen. Also graph $y = |f(x) - p(x)|$. Is $p(x)$ close to $f(x)$ for x near 0?

1. $f(x) = e^{-x}$, $p_4(x)$
2. $f(x) = e^{2x}$, $p_4(x)$

3. $f(x) = \sqrt{x + 2}$, $p_4(x)$
4. $f(x) = \sqrt[3]{x - 1}$, $p_4(x)$

5. $f(x) = \dfrac{1}{x + 1}$, $p_3(x)$
6. $f(x) = \sqrt{x + 1}$, $p_3(x)$

7. $f(x) = \ln(x + 2)$, $p_4(x)$
8. $f(x) = e^{1/(x+1)}$, $p_2(x)$

9. $f(x) = \cos x$, $p_5(x)$
10. $f(x) = \tan x$, $p_2(x)$

In Exercises 11 through 18 find the Taylor polynomial $p(x)$ of the indicated function $f(x)$ at the given point. On your graphing calculator, graph $y = f(x)$ and $y = p(x)$ on the same screen. Also graph $y = |f(x) - p(x)|$. Is $p(x)$ close to $f(x)$ for x near the indicated point?

11. $f(x) = x^2$, $p_2(x)$ at $x = 1$

12. $f(x) = x^3$, $p_3(x)$ at $x = 2$

13. $f(x) = \dfrac{1}{x}$, $p_4(x)$ at $x = 1$

14. $f(x) = \sqrt{x}$, $p_4(x)$ at $x = 4$

15. $f(x) = e^{2x}$, $p_3(x)$ at $x = 1$

16. $f(x) = \ln(4x)$, $p_3(x)$ at $x = 1$

17. $f(x) = \sin x$, $p_4(x)$ at $x = \dfrac{\pi}{2}$

18. $f(x) = \cos x$, $p_5(x)$ at $x = \dfrac{\pi}{2}$

In Exercises 19 through 28 find the nth-order Taylor polynomial at $x = 0$.

19. x^4
20. x^5

21. $x^2 + 2x + 3$
22. $x^3 + x + 4$

23. e^{-x}
24. e^{2x}

25. $\dfrac{1}{1 - x}$
26. $\ln(2 + x)$

27. $\sin x$
28. $\cos x$

In Exercises 29 through 32 find the nth-order Taylor polynomial at any point $x = a$.

29. x^4
30. $\ln x$, $a > 0$

31. e^x
32. $\dfrac{1}{x}$, $a \neq 0$

33. Use the fourth-order Taylor polynomial for $f(x) = e^{-x}$ at $x = 0$ to approximate $e^{-0.20}$.

34. Use the fourth-order Taylor polynomial for $f(x) = e^{-x}$ at $x = 0$ to approximate $e^{-0.25}$.

35. Use the second-order Taylor polynomial for $f(x) = \sqrt{x}$ at $x = 4$ to approximate $\sqrt{5}$.

36. Use the second-order Taylor polynomial for $f(x) = \sqrt{x}$ at $x = 9$ to approximate $\sqrt{10}$.

37. Use the third-order Taylor polynomial for $f(x) = \sqrt{x}$ at $x = 16$ to approximate $\sqrt{15}$.

38. Use the third-order Taylor polynomial for $f(x) = \sqrt[3]{x}$ at $x = 8$ to approximate $\sqrt[3]{9}$.

Using your graphing calculator in Exercises 39 through 44, graph the given function and the indicated Taylor polynomial on the given interval in the same viewing rectangle. Does the Taylor polynomial appear to approximate the function?

39. $f(x) = \sin x$, $p_3(x)$ at $x = 0$, $[-1, 1]$.
40. $f(x) = \cos x$, $p_4(x)$ at $x = 0$, $[-1, 1]$.
41. $f(x) = e^{-x}$, $p_5(x)$ at $x = 0$, $[-1, 2]$.
42. $f(x) = \sqrt{4 - x^2}$, $p_2(x)$ at $x = 0$, $[-2, 2]$.
43. $f(x) = e^{-x^2/2}$, $p_2(x)$ at $x = 0$, $[-2, 2]$.
44. $f(x) = e^{1/(x+1)}$, $p_2(x)$ at $x = 0$, $[0, 1]$.

Applications

45. Demand Equation. Given the demand equation $p = D(x) = e^{1/(x+1)}$ use the second-order Taylor polynomial for $D(x) = e^{1/(x+1)}$ at $x = 0$ to approximate the average price over the demand interval $[0, 0.5]$.

46. Supply Equation. Given the supply equation $p = S(x) = \ln(x + 1)$, use the third-order Taylor polynomial for $S(x) = \ln(x + 1)$ at $x = 0$ to approximate the average price over the supply interval $[0, 0.5]$.

47. Demand Equation. Given the demand equation $p = D(x) = e^{1/(x+1)}$, use the second-order Taylor polynomial for $D(x) = e^{1/(x+1)}$ at $x = 1$ to approximate the average price over the demand interval $[1, 1.5]$.

48. Supply Equation. Given the supply equation $p = S(x) = \ln(x + 1)$, use the third-order Taylor polynomial for $S(x) = \ln(x + 1)$ at $x = 1$ to approximate the average price over the supply interval $[1, 1.5]$.

49. Oil Production. Suppose the rate in millions of barrels per year at which oil is being extracted from a field is given by $R(t) = e^{-0.10t^2}$, where t is in years. Use the second-order Taylor polynomial for $R(t)$ at $t = 0$ to approximate the total production during the first two years.

50. Copper Production. Suppose the rate in thousands of tons per year at which copper is being extracted from a mine is given by $R(t) = e^{-0.20t^2}$, where t is in years. Use the second-order Taylor polynomial for $R(t)$ at $t = 0$ to approximate the total production during the first two years.

51. Memorization. Suppose the rate at which an average person can memorize a list of items is given by $R(t) = \dfrac{15}{\sqrt{3t^2 + 1}}$, where t is the number of hours spent memorizing. Use the second-order Taylor polynomial for $R(t)$ at $t = 0$ to approximate the number memorized in 30 minutes.

52. Average Amount of a Drug. The amount of a drug in the body at time t is given by $A(t) = \dfrac{1}{t^2 + 1}$, where t is in hours and A in milligrams. Use the second-order Taylor polynomial for $A(t)$ at $t = 0$ to approximate the average amount of the drug in the body over the first 0.5 hour.

53. Memorization. Suppose the rate at which an average person can memorize a list of items is given by $R(t) = \dfrac{15}{\sqrt{3t^2 + 1}}$, where t is the number of hours spent memorizing. Use the second-order Taylor polynomial for $R(t)$ at $t = 1$ to approximate the number memorized in the time interval $[1, 1.5]$.

54. Average Amount of a Drug. The amount of a drug in the body at time t is given by $A(t) = \dfrac{1}{t^2 + 1}$, where t is in hours and A in milligrams. Use the second-order Taylor polynomial for $A(t)$ at $t = 1$ to approximate the average amount of the drug in the body over the time interval $[1, 1.5]$.

55. Air Pollution. A study has indicated that the level of air pollution at a distance x miles from a certain factory is given by $L(x) = \sqrt{4 - x^2}$. Use the second-order Taylor polynomial for $L(x)$ at $x = 0$ to approximate the average amount of air pollution within 1 mile of the factory.

56. Air Pollution. With $L(x)$ given in the previous exercise, use the second-order Taylor polynomial for $L(x)$ at $x = 1$ to approximate the average amount of air pollution between 1 and 2 miles of the factory.

Enrichment Exercises

In Exercises 57 through 60 you are shown the graphs of four functions. In each case let $p_2(x) = a_2x^2 + a_1x + a_0$ be the second-order Taylor polynomial for the indicated function at $x = 0$. In each case indicate what the signs of the coefficients a_2, a_1, and a_0 must be.

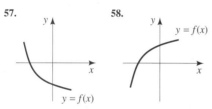

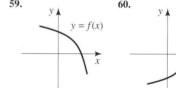

61. Find the second-order Taylor polynomial $p_2(x)$ of $f(x) = 2x^2 - 3x + 4$ at $x = 1$. Write $p_2(x)$ in the form $Ax^2 + Bx + C$. What important observation can you make?

62. Find the third-order Taylor polynomial of $f(x) = 4x^3 + 3x^2 - x + 2$ at $x = 0$. What important observation can you make?

63. Based on the observations in Exercises 61 and 62 how do you think the Taylor polynomial of order n of a polynomial $P(x)$ of degree n is related to the polynomial $P(x)$?

64. In Example 6 we took the Taylor polynomial at $x = 0$, which is at the center of the interval of integration $[-1, 1]$. Explain why you think this is a reasonable choice.

SOLUTIONS TO SELF-HELP EXERCISE SET 10.1

1. We have

$$f(x) = \ln(1 + x), \quad \text{and} \quad f(0) = 0$$

$$f'(x) = \frac{1}{x + 1}, \quad \text{and} \quad f'(0) = 1$$

$$f''(x) = -(x + 1)^{-2}, \quad \text{and} \quad f''(0) = -1$$

$$f^{(3)}(x) = 2(x + 1)^{-3}, \quad \text{and} \quad f^{(3)}(0) = 2$$

Then for

$$p_3(x) = a_0 + a_1x + a_2x^2 + a_3x^3$$

we have

$$a_0 = f(0) = 0$$

$$a_1 = f'(0) = 1$$

$$a_2 = \frac{f''(0)}{2} = \frac{-1}{2}$$

$$a_3 = \frac{f^{(3)}(0)}{3!} = \frac{2}{6} = \frac{1}{3}$$

Therefore the third-order Taylor polynomial is

$$p_3(x) = x - \frac{1}{2}x^2 + \frac{1}{3}x^3$$

2. We have

$$f(x) = x^{1/4}, \quad \text{and} \quad f(1) = 1$$

$$f'(x) = \frac{1}{4}x^{-3/4}, \quad \text{and} \quad f'(1) = \frac{1}{4}$$

$$f''(x) = -\frac{3}{16}x^{-7/4}, \quad \text{and} \quad f''(1) = -\frac{3}{16}$$

$$f^{(3)}(x) = \frac{21}{64}x^{-11/4}, \quad \text{and} \quad f^{(3)}(1) = \frac{21}{64}$$

Then for

$$p_3(x) = a_0 + a_1(x - 1) + a_2(x - 1)^2 + a_3(x - 1)^3$$

we have

$$a_0 = f(1) = 1$$

$$a_1 = f'(1) = \frac{1}{4}$$

$$a_2 = \frac{f''(1)}{2!} = \frac{-3/16}{2} = -\frac{3}{32}$$

$$a_3 = \frac{f^{(3)}(1)}{3!} = \frac{21/64}{6} = \frac{7}{128}$$

Therefore, the third-order Taylor polynomial is

$$p_3(x) = 1 + \frac{1}{4}(x - 1) - \frac{3}{32}(x - 1)^2 + \frac{7}{128}(x - 1)^3$$

An approximation to $\sqrt[4]{1.1}$ is

$$p_3(1.1) = 1 + \frac{1}{4}(1.1 - 1) - \frac{3}{32}(1.1 - 1)^2$$

$$+ \frac{7}{128}(1.1 - 1)^3 \approx 1.024117$$

It is interesting to note that to six decimal places $\sqrt[4]{1.1} = 1.024114$, so that $p_3(1.1)$ is in error by only 0.000003.

10.2 ERRORS IN TAYLOR POLYNOMIAL APPROXIMATIONS

■ Taylor's Theorem ■ Applications

APPLICATION: NUMBER OF SUSPENDED PARTICULATES

The number of total suspended particulates per mile from an industrial plant is given by $\delta(x) = 100{,}000e^{-x^2}$, where x is measured in miles from the plant. In the previous section we approximated the total number of particulates within 0.1 mile of the plant by using the second-order Taylor polynomial for e^{-x^2} at $x = 0$. Estimate the error in this approximation. See Example 5 for the answer.

Taylor's Theorem

In the previous section we saw that a function can be approximated by a Taylor polynomial. In this section we use Taylor's theorem to see just how good the approximation is.

If $f(x)$ is the given function and p_n is the nth-order Taylor polynomial for $f(x)$ at $x = a$ that approximates $f(x)$, we define the error or the remainder $R_n(x)$ to be $f(x) - p_n(x)$. Thus,

$$f(x) = p_n(x) + R_n(x)$$

Taylor's theorem gives us a valuable estimate on the size of $|R_n(x)|$.

Taylor's Theorem

Suppose that the function f and its first $n + 1$ derivatives are continuous on the interval $[c, d]$ that contains $x = a$. Then for all x in $[c, d]$,

$$f(x) = f(a) + f'(a)(x - a) + \frac{f''(a)}{2!}(x - a)^2 + \cdots + \frac{f^{(n)}(a)}{n!}(x - a)^n + R_n(x)$$

with

$$|R_n(x)| \le \frac{M}{(n+1)!} |x - a|^{n+1}$$

where

$$|f^{(n+1)}(t)| \le M \quad \text{for } t \text{ between } a \text{ and } x$$

EXAMPLE 1 Using Taylor's Theorem

Find the third-order Taylor polynomial for $f(x) = e^x$ at $x = 0$, and use it to find an approximation to $e^{0.5} = \sqrt{e}$. Then use Taylor's theorem to give an upper bound on the error.

SOLUTION ■ At the beginning of the previous section we found that the third-order Taylor polynomial at $x = 0$ for $f(x) = e^x$ is

$$p_3(x) = 1 + x + \frac{1}{2} x^2 + \frac{1}{6} x^3$$

and we also found that $p_3(0.5)$ to four decimal places is 1.6458 and approximates $e^{0.5}$.

Now let us apply Taylor's theorem to find an estimate for the error in this approximation. First notice that $f(x) = e^x$ and all its derivatives are continuous everywhere. We have $n = 3$, so to find an estimate of the error, we need to first find a bound on $f^{(4)}(x) = e^x$ on the interval $[0, 0.5]$. Since e^x is an increasing function, e^x attains its maximum at $x = 0.5$. The maximum is $e^{0.5}$. This is, of course, precisely the number we are seeking. Since $e < 3$, we can obtain a rough estimate as follows:

$$e^{0.5} = \sqrt{e} < \sqrt{3} < \sqrt{4} = 2$$

Therefore we have

$$|R_n(x)| \le \frac{M}{(n+1)!} |x - a|^{n+1}$$
$$\le \frac{2}{(3+1)!} |0.5 - 0|^{3+1}$$
$$= \frac{2}{24} (0.5)^4$$
$$\approx 0.0052$$

Thus, our estimate that $\sqrt{e} \approx 1.6458$ is accurate at least to within 0.0052. ■

Remark. To four decimal places $\sqrt{e} = 1.6487$, so that $p_3(0.5)$ has an error of 0.0029. This is smaller than our bound in Example 1, which is to be expected. The bound given in Taylor's theorem applies to *all* functions and thus, in general, is conservative.

EXAMPLE 2 Using Taylor's Theorem

Find the fourth-order Taylor polynomial for $f(x) = \sin x$ at $x = 0$ and use it to find an approximation to $\sin(0.2)$. Then use Taylor's theorem to give an upper bound on the error.

SOLUTION ■ In Examples 3 and 4 of the previous section we found that the fourth-order Taylor polynomial for $\sin x$ at $x = 0$ is

$$p_4(x) = x - \frac{1}{6} x^3$$

and we also found that to six decimal places $p_4(0.2) = 0.198667$. In Taylor's theorem we have $n = 4$ and must seek a bound on $f^{(5)}(x)$ on the interval $[0, 0.2]$. But $f^{(5)}(x) = \cos x$ and therefore

$$\left| f^{(5)}(x) \right| = |\cos x| \le 1$$

Thus,

$$|R_n(x)| \le \frac{M}{(n+1)!} |x - a|^{n+1}$$

$$\le \frac{1}{(4+1)!} |0.2 - 0|^{4+1}$$

$$= \frac{1}{120} (0.2)^5$$

$$\approx 0.000003$$

Thus, our estimate that $\sin(0.2) \approx 0.198667$ is accurate to within 0.000003. ∎

EXAMPLE 3　Using Taylor's Theorem

Find the third-order Taylor polynomial for $f(x) = \ln x$ at $x = 1$, and use it to find an approximation to $\ln 1.1$. Then use Taylor's theorem to give an upper bound on the error.

SOLUTION ∎ In Example 5 of the previous section we found that the third-order Taylor polynomial for $\ln x$ at $x = 1$ is

$$p_3(x) = (x - 1) - \frac{1}{2} (x - 1)^2 + \frac{1}{3} (x - 1)^3$$

We also found that to five decimal places $p_3(1.1) = 0.09533$. In Taylor's theorem we have $n = 3$ and must seek a bound on $f^{(4)}(x)$ on the interval $[1, 1.1]$. But $f^{(4)}(x) = -6x^{-4}$. Since $\left| f^{(4)}(x) \right| = 6x^{-4}$ is a decreasing function, the maximum occurs at the left endpoint of the interval $[1, 1.1]$, or at $x = 1$. Thus, on this interval,

$$\left| f^{(4)}(x) \right| = 6x^{-4} \le 6$$

Therefore

$$|R_n(x)| \le \frac{M}{(n+1)!} |x - a|^{n+1}$$

$$\le \frac{6}{(3+1)!} |1.1 - 1|^{3+1}$$

$$= \frac{6}{24} (0.1)^4$$

$$\approx 0.000025$$

Thus, our estimate that $\ln 1.1 \approx 0.09531$ is accurate to within 0.000025. ∎

EXAMPLE 4　An Accuracy Requirement

Find n so that the nth-order Taylor polynomial for $\sin x$ at $x = 0$ approximates $\sin x$ on the interval $[0, \pi/4]$ to within three decimal places.

SOLUTION ∎ Since our approximation must be accurate to within three decimal places, we must have an error of at most 0.0005. Since $f(x) = \sin x$, the derivatives are $\cos x$, $-\sin x$, $-\cos x$, or $\sin x$. Therefore,

$$\left| f^{(n+1)}(x) \right| \le 1$$

for all n and x. Applying Taylor's theorem, we have

$$|R_n(x)| \le \frac{M}{(n+1)!}|x-a|^{n+1}$$

$$\le \frac{1}{(n+1)!}\left|\frac{\pi}{4}-0\right|^{n+1}$$

$$\le \frac{1}{(n+1)!}$$

where we use the rough estimate that $\pi/4 \le 1$. We then find our n if

$$\frac{1}{(n+1)!} \le |R_n(x)| \le 0.0005$$

If we try $n = 5$, we obtain

$$\frac{1}{(5+1)!} \approx 0.0014$$

This does not work. If we then try $n = 6$, we obtain

$$\frac{1}{(6+1)!} \approx 0.0002$$

This is less than 0.0005. Thus, $p_6(x)$ approximates $\sin x$ on the interval $[0, \pi/4]$ to within three decimal places. ■

Applications

To do the next example we need the following property of definite integrals. We assume that $f(x)$ is continuous on $[a, b]$.

$$\left|\int_a^b f(x)\,dx\right| \le \int_a^b |f(x)|\,dx \tag{1}$$

EXAMPLE 5   Number of Suspended Particulates

The density, in numbers per mile, of total suspended particulates from an industrial plant is given by $\delta(x) = 100{,}000e^{-x^2}$, where x is measured in miles from the plant. In the preceding section we approximated the total number of particulates within 0.1 mile of the plant by using the second-order Taylor polynomial approximation of $f(x) = e^{-x^2}$ at $x = 0$. Find an upper bound on the error in this approximation.

SOLUTION ■ Since we approximated with a Taylor polynomial of order 2, we have $n = 2$ in Taylor's theorem. We first seek a bound on $|f^{(3)}(x)|$ on the interval $[0, 0.1]$. We have

$$f'(x) = -2xe^{-x^2}$$
$$f''(x) = (4x^2 - 2)e^{-x^2}$$
$$f^{(3)}(x) = [-2x(4x^2-2)+8x]e^{-x^2} = (12x - 8x^3)e^{-x^2}$$

First note that $e^{-x^2} \le 1$ for $x \in [0, 0.1]$. Then on the interval $[0, 0.1]$

$$|f^{(3)}(x)| = |(12x-8x^3)e^{-x^2}| = |12x-8x^3|\cdot|e^{-x^2}|$$
$$\le |12x-8x^3| = |x|\cdot|12-8x^2|$$

$$\le |x|(12)$$
$$\le (0.1)(12) = 1.2$$

Thus,

$$|R_2(x)| \le \frac{M}{(2+1)!} |x-a|^{2+1}$$
$$\le \frac{1.2}{6} |0.1 - 0|^3$$
$$\approx 0.0002$$

The expression for the exact number of particulates within 0.1 mile of the plant is

$$100,000 \int_{-0.1}^{0.1} e^{-x^2} \, dx = 200,000 \int_{0}^{0.1} e^{-x^2} \, dx$$

We approximated this in the last section by evaluating

$$200,000 \int_{0}^{0.1} p_2(x) \, dx$$

where $p_2(x)$ is the second-order Taylor polynomial for e^{-x^2} at $x = 0$. For this last integral we obtained 19,933 when rounded to the nearest whole number. The error is

$$E = \left| 200,000 \int_{0}^{0.1} e^{-x^2} \, dx - 200,000 \int_{0}^{0.1} p_2(x) \, dx \right|$$

Using Equation (1), we have that

$$E = 200,000 \left| \int_{0}^{0.1} e^{-x^2} \, dx - \int_{0}^{0.1} p_2(x) \, dx \right|$$
$$= 200,000 \left| \int_{0}^{0.1} [e^{-x^2} - p_2(x)] \, dx \right|$$
$$\le 200,000 \int_{0}^{0.1} |e^{-x^2} - p_2(x)| \, dx$$
$$= 200,000 \int_{0}^{0.1} |R_3(x)| \, dx$$
$$\le 200,000 \int_{0}^{0.1} (0.0002) \, dx$$
$$= 200,000(0.1)(0.0002) = 4$$

Therefore, rounded to the nearest whole number, a bound on the error is 4. ∎

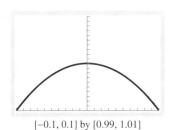

[−0.1, 0.1] by [0.99, 1.01]

Screen 10.11
The graphs of $y = f(x) = e^{-x^2}$ and $p_2(x) = 1 - x^2$ are indistinguishable on [−0.1, 0.1].

Screen 10.11 uses a window with dimensions [−0.1, 0.1] by [0.99, 1.01] and shows the graphs of $y = e^{-x^2}$ and $y = p_2(x) = 1 - x^2$. Notice that the graphs are so close that we cannot distinguish between them on the interval (−0.1, 0.1). Thus, we should not be surprised if the areas under them differ by a small amount.

SELF-HELP EXERCISE SET 10.2

1. Find a bound on the error using the second-order Taylor polynomial for $\sqrt[4]{x}$ at $x = 1$ to approximate $\sqrt[4]{4.1}$.

2. Find n so that the nth-order Taylor polynomial for $\ln x$ at $x = 1$ approximates $\ln x$ on the interval [1, 1.5] with an error of at most 0.001.

In Exercises 1 through 10 find a bound on the error using the indicated Taylor polynomial at $x = 0$ over the given interval.

1. e^{-x}, $p_4(x)$, $[0, 0.5]$ **2.** e^{2x}, $p_4(x)$, $[0, 0.4]$

3. $\sqrt{x + 4}$, $p_4(x)$, $[0, 0.25]$

4. $\sqrt[3]{x + 8}$, $p_4(x)$, $[0, 0.125]$

5. $\dfrac{1}{x + 1}$, $p_3(x)$, $[0, 0.2]$ **6.** $\sqrt{x + 1}$, $p_3(x)$, $[0, 0.5]$

7. $\ln(x + 2)$, $p_4(x)$, $[0, 0.5]$ **8.** $\ln(x + 1)$, $p_4(x)$, $[0, 1]$

9. $\cos x$, $p_5(x)$, $[0, 0.4]$ **10.** $\tan x$, $p_2(x)$, $[0, 0.4]$

In Exercises 11 through 18 find a bound on the error using the indicated Taylor polynomial at the given point over the given interval.

11. x^2, $p_2(x)$ at $x = 1$, $[1, 1.5]$

12. x^3, $p_3(x)$ at $x = 2$, $[2, 2.2]$

13. $\dfrac{1}{x}$, $p_4(x)$ at $x = 1$, $[1, 1.2]$

14. $\sqrt{x}$, $p_4(x)$ at $x = 4$, $[4, 4.5]$

15. e^{2x}, $p_3(x)$ at $x = 1$, $[1, 1.25]$

16. $\ln(4x)$, $p_3(x)$ at $x = 1$, $[1, 1.5]$

17. $\sin x$, $p_4(x)$ at $x = \dfrac{\pi}{2}$, $\left[\dfrac{\pi}{2}, \pi\right]$

18. $\cos x$, $p_5(x)$ at $x = \dfrac{\pi}{2}$, $\left[\dfrac{\pi}{2}, \pi\right]$

In Exercises 19 through 26 find the n so that the nth-order Taylor polynomial at the indicated point and over the indicated interval has an error of at most 0.001.

19. x^5, $x = 0$, $[0, 1.5]$ **20.** x^6, $x = 0$, $[1, 1.5]$

21. $\ln(x + 1)$, $x = 0$, $[0, 0.5]$ **22.** e^x, $x = 0$, $[0, 0.4]$

23. e^{-x}, $x = 0$, $[0, 0.5]$ **24.** $\dfrac{1}{x}$, $x = 1$, $[1, 1.5]$

25. $\cos x$, $x = 0$, $\left[0, \dfrac{\pi}{4}\right]$ **26.** $\sin x$, $x = 0$, $\left[0, \dfrac{\pi}{4}\right]$

Applications

27. Demand Equation. Given the demand equation $p = D(x) = 1/(x + 1)$, find a bound on the error in using the second-order Taylor polynomial for $D(x) = 1/(x + 1)$ at $x = 0$ to approximate the average price over the demand interval $[0, 0.5]$.

28. Supply Equation. Given the supply equation $p = S(x) = \ln(x + 1)$, find a bound on the error in using the third-order Taylor polynomial for $S(x) = \ln(x + 1)$ at $x = 0$ to approximate the average price over the supply interval $[0, 0.5]$.

29. Demand Equation. Given the demand equation $p = D(x) = 1/(x + 1)$, find a bound on the error in using the second-order Taylor polynomial for $D(x) = 1/(x + 1)$ at $x = 1$ to approximate the average price over the demand interval $[1, 1.5]$.

30. Supply Equation. Given the supply equation $p = S(x) = \ln(x + 1)$, find a bound on the error in using the third-order Taylor polynomial for $S(x) = \ln(x + 1)$ at $x = 1$ to approximate the average price over the supply interval $[1, 1.5]$.

31. Supply Equation. Given the supply equation $p = S(x) = \ln(x + 1)$, find the smallest n for which the error in using the third-order Taylor polynomial for $S(x)$ at $x = 0$ to approximate the average price over the supply interval $[0, 0.5]$ is at most 0.01.

32. Supply Equation. Repeat the previous exercise if the error is at most 0.001.

33. Oil Production. Suppose the rate in millions of barrels per year at which oil is being extracted from a field is given by $R(t) = e^{-0.10t^2}$, where t is in years. Find a bound on the

error in using the second-order Taylor polynomial for $R(t)$ at $t = 0$ to approximate the total production during the first two years.

34. Copper Production. Suppose the rate in thousands of tons per year at which copper is being extracted from a mine is given by $R(t) = e^{-0.20t^2}$, where t is in years. Find a bound on the error in using the second-order Taylor polynomial for $R(t)$ at $t = 0$ to approximate the total production during the first year.

35. Memorization. Suppose the rate at which an average person can memorize a list of items is given by $R(t) = \dfrac{15}{\sqrt{3t^2 + 1}}$, where t is the number of hours spent memorizing. Find a bound on the error in using the second-order Taylor polynomial for $R(t)$ at $t = 0$ to approximate the number of items memorized in the first 30 minutes.

36. Average Amount of a Drug. The amount of a drug in the body at time t is given by $A(t) = \dfrac{1}{t^2 + 1}$, where t is in hours and A in milligrams. Find a bound on the error in using the second-order Taylor polynomial for $A(t)$ at $t = 0$ to approximate the average amount of the drug in the body over the first 0.5 hour.

37. Memorization. Suppose the rate at which an average person can memorize a list of items is given by $R(t) = \dfrac{15}{\sqrt{3t^2 + 1}}$, where t is the number of hours spent memorizing. Find a bound on the error in using the second-order Taylor polynomial for $R(t)$ at $t = 1$ to approximate the number memorized in the time interval $[1, 1.5]$.

38. **Average Amount of a Drug.** The amount of a drug in the body at time t is given by $A(t) = \dfrac{1}{t^2 + 1}$, where t is in hours and A in milligrams. Find a bound on the error in using the second-order Taylor polynomial for $A(t)$ at $t = 1$ to approximate the average amount of the drug in the body over the time interval $[1, 1.5]$.

39. **Air Pollution.** A study has indicated that the level of air pollution a distance x miles from a certain factory is given

by $L(x) = \sqrt{4 - x^2}$. Find a bound on the error in using the second-order Taylor polynomial for $L(x)$ at $x = 0$ to approximate the average amount of air pollution within 1 mile of the factory.

40. **Air Pollution.** With $L(x)$ given in the previous exercise, find a bound on the error in using the second-order Taylor polynomial for $L(x)$ at $x = 1$ to approximate the average amount of air pollution between 1 and 1.5 miles of the factory.

Enrichment Exercises

41. In Example 3 of the previous section we saw that the third-order Taylor polynomial for $f(x) = \sin x$ at $x = 0$ is $p_3(x)$
$= x - \dfrac{1}{6}x^3$. Use this to show $\lim\limits_{x \to 0} \dfrac{\sin x}{x} = 1$.

42. As we noted before, our graphing calculator uses the expression $\dfrac{f(x + h) - f(x - h)}{2h}$ to approximate $f'(x)$. Geometric considerations seem to indicate that this expression should be a better approximation, in general, than the expression $\dfrac{f(x + h) - f(x)}{h}$. Use Taylor polynomials to show that this is indeed the case. *Hint:* From Taylor's theorem we have

$$f(x) = f(a) + f'(a)(x - a) + \frac{f''(a)}{2}(x - a)^2 + R_2(x)$$

Using this formula, replace x with $a + h$ and again with $a - h$. Now determine a bound on the expression

$$\left| \frac{f(a + h) - f(a - h)}{2h} - f'(a) \right|$$

using the bound on $R_2(x)$. For small values of h how does this bound compare with a bound on

$$\left| \frac{f(a + h) - f(a)}{h} - f'(a) \right|$$

What do you conclude concerning the error of each of these expressions in approximating $f'(a)$?

43. To determine the derivative of e^x in Section 3.1 we needed the result $\lim\limits_{h \to 0} \dfrac{e^h - 1}{h} = 1$. Show this by using an appropriate Taylor polynomial.

1. We first need to find a bound on $\left| f^{(3)}(x) \right|$ on $[1, 1.1]$. We have

$$f(x) = x^{1/4}$$

$$f'(x) = \frac{1}{4}x^{-3/4}$$

$$f''(x) = -\frac{3}{16}x^{-7/4}$$

$$f^{(3)}(x) = \frac{21}{64}x^{-11/4}$$

We see that $f^{(3)}(x)$ is a decreasing function and thus assumes its maximum on the interval $[1, 1.1]$ at the left endpoint $x = 1$. The maximum is 21/64. Then

$$\left| R_2(x) \right| \le \frac{M}{(2 + 1)!}\left| x - a \right|^{2+1}$$

$$\le \frac{21/64}{6}\left| 1.1 - 1 \right|^3$$

$$\approx 0.00005$$

2. We first need to find a bound on $\left| f^{(n)}(x) \right|$. Then

$$f(x) = \ln x$$

$$f'(x) = x^{-1}$$

$$f''(x) = -x^{-2}$$

$$f^{(3)}(x) = 2x^{-3}$$

$$f^{(4)}(x) = -3!x^{-4}$$

From this pattern we see that

$$f^{(n+1)}(x) = (-1)^n(n!)x^{-n-1}$$

and

$$\left| f^{(n+1)}(x) \right| = (n!)\left| x \right|^{-n-1}$$

This is a decreasing function on the interval $[1, 1.5]$ and thus assumes its maximum at the left endpoint $x = 1$. The maximum is $n!$. Then

$$\left| R_n(x) \right| \le \frac{M}{(n + 1)!}\left| x - a \right|^{n+1}$$

$$\le \frac{n!}{(n + 1)!}\left| 1.5 - 1 \right|^{n+1}$$

$$= \frac{1}{n + 1}(0.5)^{n+1}$$

We want this to be at most 0.001. Thus, we want

$$\frac{1}{n + 1}(0.5)^{n+1} \le 0.001$$

If we try $n = 6$, we obtain

$$\frac{1}{6 + 1} (0.5)^{6+1} \approx 0.0011$$

This does not work. Now try $n = 7$ and obtain

$$\frac{1}{7 + 1} (0.5)^{7+1} \approx 0.00049$$

This works. So $n = 7$ is the smallest integer for which the error is at most 0.001 on the interval $[1, 1.5]$.

10.3 INFINITE SEQUENCES

- Infinite Sequences ■ Limit of an Infinite Sequence ■ Applications

APPLICATION: DEPRECIATION

A stamping machine, which originally cost $100,000, depreciates in value by 5% each year. Find the value after n years. What is happening to this value in the long-term? For the answer see Example 6.

Infinite Sequences

An infinite sequence is an ordered list of numbers

$$a_1, a_2, a_3, \ldots, a_n, \ldots$$

The number a_1 is called the **first term**, a_2 the **second term**, and a_n the **nth term**. Since for every positive integer we have a number, we can regard a sequence as a function whose domain is the set of positive integers. If f is this function, then $f(n) = a_n$ for $n = 1, 2, \ldots$.

Infinite Sequence

An **infinite sequence** of numbers is a function whose domain is the set of positive integers. If f is the defining function, we write $f(n) = a_n$ and call a_n the **nth term**, or **general term**. The sequence is denoted by $\{a_1, a_2, a_3, \ldots\}$ or $\{a_n\}$.

EXAMPLE 1 Writing Infinite Sequences

Write the first four terms of the following infinite sequences:

(a) $\{2n\}$ (b) $\{(-1)^n\}$ (c) $\left\{1 - \dfrac{1}{n}\right\}$

SOLUTIONS ■ In each case we take $n = 1, 2, 3,$ and 4, in succession.

(a) $a_1 = 2(1) = 2, a_2 = 2(2) = 4, a_3 = 2(3) = 6, a_4 = 2(4) = 8$
The sequence is $\{2, 4, 6, 8, \ldots\}$.

(b) $a_1 = (-1)^1 = -1, a_2 = (-1)^2 = 1, a_3 = (-1)^3 = -1, a_4 = (-1)^4 = 1$
The sequence is $\{-1, 1, -1, 1, \ldots\}$.

(c) $a_1 = 1 - \dfrac{1}{1} = 0, a_2 = 1 - \dfrac{1}{2} = \dfrac{1}{2}, a_3 = 1 - \dfrac{1}{3} = \dfrac{2}{3}, a_4 = 1 - \dfrac{1}{4} = \dfrac{3}{4}$
The sequence is $\{0, \frac{1}{2}, \frac{2}{3}, \frac{3}{4}, \ldots\}$. ■

Since an infinite sequence is a function, we can graph the function by graphing the points $(n, f(n)) = (n, a_n)$.

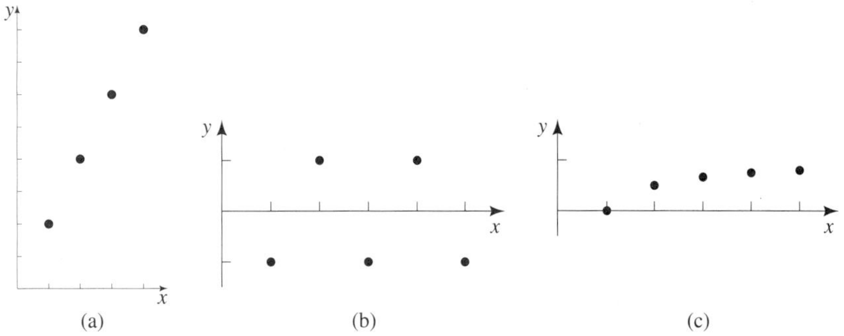

Figure 10.1
(a) The sequence $y_n = 2n$ is unbounded and has no limit. (b) The sequence $y_n = (-1)^n$ bounces between 1 and -1 and has no limit. (c) The sequence $y_n = 1 - 1/n$ heads for 1.

EXAMPLE 2 Graphing Infinite Sequences

Graph the infinite sequences in Example 1.
SOLUTION ■ The graphs are shown in Figure 10.1. ■

EXAMPLE 3 Finding the General Term

Find the general term for the sequence $\{1, 4, 9, 16, \ldots\}$.
SOLUTION ■ We observe that

$$1 = (1)^2, \qquad 4 = (2)^2, \qquad 9 = (3)^2, \qquad 16 = (4)^2$$

and conclude that $a_n = n^2$ is the nth, or general, term. ■

Limit of an Infinite Sequence

The graphs of the three sequences shown in Figure 10.1 behave differently. The first becomes large without bound. The second bounces forever back and forth between two different numbers. The third graph is asymptotic to the horizontal line $y = 1$. This is true since $a_n = 1 - \dfrac{1}{n}$ approaches the number 1 as n becomes large without bound. In this case we say that $L = 1$ is the **limit** of $a_n = 1 - \dfrac{1}{n}$ as n becomes large without bound. We also say that $a_n = 1 - \dfrac{1}{n}$ **converges** to $L = 1$. The first two sequences in Figure 10.1 do not approach any one number and therefore have no limit. In this case we say that the sequences **diverges**.

Limit of an Infinite Sequence

We say that a sequence $\{a_n\}$ has **limit** L, and write

$$\lim_{n \to \infty} a_n = L$$

if a_n approaches L as n becomes large without bound. In this case we also say that the sequence **converges** to L. If the limit does not exist, we say that the sequence **diverges**.

EXAMPLE 4 Limits of Sequences

Graphically decide whether each of the following sequences converges or diverges. Confirm analytically.

(a) $\left\{\dfrac{2n + 1}{n}\right\}$ (b) $\{(-1)^n n\}$

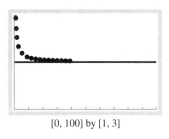

[0, 100] by [1, 3]

Screen 10.12
The graph of $y_n = 2 + 1/n$ is asymptotic to the graph of $y = 2$.

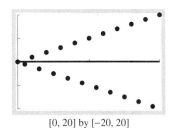

[0, 20] by [−20, 20]

Screen 10.13
The graph indicates that the sequence $y_n = (-1)^n n$ does not have a limit.

SOLUTION ■

(a) Set your graphing calculator on parametric and dot mode. Set $x_1 = t$, $y_1 = (2t + 1)/t$. For the range variables set $T_{min} = 0$, $T_{max} = 100$, $T_{step} = 1$, $x_{min} = 0$, $x_{max} = 100$, $y_{min} = 1$, $y_{max} = 3$. Also set $x_2 = t$, $y_2 = 2$. Graph to obtain Screen 10.12. Using the TRACE function, we see that as t becomes large, $y_1 = (2t + 1)/t$ seems to be heading for 2.

Analytically, $\lim_{n \to \infty} \left(\frac{2n + 1}{n} \right) = \lim_{n \to \infty} \left(2 + \frac{1}{n} \right) = 2$, since $\frac{1}{n}$ approaches 0 as n becomes large without bound. Therefore this sequence converges to 2.

(b) Set your graphing calculator on parameter and dot mode. Set $x_1 = t$, $y_1 = t(-1)^t$. For the range variables set $T_{min} = 0$, $T_{max} = 20$, $T_{step} = 1$, $x_{min} = 0$, $x_{max} = 20$, $y_{min} = -20$, $y_{max} = 20$. Also set $x_2 = t$, $y_2 = 0$. Graph to obtain Screen 10.13. The dots do not approach any number, so the limit does not exist.

Analytically, we write out the first number of terms and obtain

$$-1, 2, -3, 4, -5, 6, \ldots$$

We see that these numbers are not approaching any number. Thus, $\lim_{n \to \infty} (-1)^n n$ does not exist, and the sequence diverges. ■

EXPLORATION 1

Alternative Graphs of Sequences

We can graph sequences in an alternative manner. Again set the mode on parameter and dot. Set $x_1 = (2t + 1)/t$, $y_1 = 1$. For the range variables set $T_{min} = 1$, $T_{max} = 100$, $T_{step} = 1$, $x_{min} = 1.9$, $x_{max} = 2.5$, $y_{min} = 0$, $y_{max} = 2$. Explain what you see and how this is related to Screen 10.13.

Since sequences are functions, we have limit properties of sequences that parallel limit properties of continuous functions.

Limit Properties of Sequences

Assume that c is any number and

$$\lim_{n \to \infty} a_n = A \qquad \text{and} \qquad \lim_{n \to \infty} b_n = B$$

Then

Rule 1. $\lim_{n \to \infty} ca_n = c \lim_{n \to \infty} a_n = cA$

Rule 2. $\lim_{n \to \infty} (a_n \pm b_n) = \lim_{n \to \infty} a_n \pm \lim_{n \to \infty} b_n = A \pm B$

Rule 3. $\lim_{n \to \infty} (a_n \cdot b_n) = \left(\lim_{n \to \infty} a_n \right) \cdot \left(\lim_{n \to \infty} b_n \right) = A \cdot B$

Rule 4. $\lim_{n \to \infty} \frac{a_n}{b_n} = \frac{\lim_{n \to \infty} a_n}{\lim_{n \to \infty} b_n} = \frac{A}{B}$, if $\lim_{n \to \infty} b_n = B \neq 0$

Rule 5. $\lim_{n \to \infty} (a_n)^k = A^k$, if k is any real number and A^k is defined, $A \neq 0$.

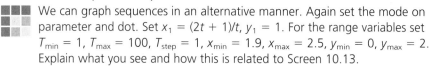 **EXAMPLE 5** Using the Properties of Limits

Find $\lim_{n \to \infty} \dfrac{5n^3 + 7n^2 - 1}{3n^3 - n + 4}$.

SOLUTION ■ Divide numerator and denominator by n^3, and obtain

$$\lim_{n\to\infty} \frac{5n^3 + 7n^2 - 1}{3n^3 - n + 4} = \lim_{n\to\infty} \frac{5 + \dfrac{7}{n} - \dfrac{1}{n^3}}{3 - \dfrac{1}{n^2} + \dfrac{4}{n^3}}$$

$$= \frac{\lim\limits_{n\to\infty}\left(5 + \dfrac{7}{n} - \dfrac{1}{n^3}\right)}{\lim\limits_{n\to\infty}\left(3 - \dfrac{1}{n^2} + \dfrac{4}{n^3}\right)}$$

$$= \frac{5 + 0 + 0}{3 + 0 + 0} = \frac{5}{3} \qquad ■$$

Applications

EXAMPLE 6   Depreciation

A stamping machine, which originally cost \$100,000, depreciates in value by 5% each year. Find the value after n years, and determine what is happening to this value in the long term.

SOLUTION ■ Each year the value is 95% of the value the previous year. Thus, the value for the first several years is

$$\$100{,}000(0.95), \qquad \$100{,}000(0.95)^2, \qquad \$100{,}000(0.95)^3, \qquad \$100{,}000(0.95)^4$$

and we see that the value after n years is $\$100{,}000(0.95)^n$.

We first graph the sequence $(0.95)^n$, $x_1 = t$, $y_1 = (0.95)^t$. For the range variables set $T_{\min} = 0$, $T_{\max} = 200$, $T_{\text{step}} = 1$, $x_{\min} = 0$, $x_{\max} = 200$, $y_{\min} = 0$, $y_{\max} = 1$. Graph to obtain Screen 10.14. You can use the TRACE function to check the values of $(0.95)^t$. From this we conclude that

$$\lim_{n\to\infty}(0.95)^n = 0$$

Table 10.5 gives the values of $(0.95)^n$ for selected values of n. Thus, we have that

$$\lim_{n\to\infty} 100{,}000(0.95)^n = 100{,}000 \lim_{n\to\infty}(0.95)^n = 100{,}000(0) = 0$$

and the sequence $\{100{,}000(0.95)^n\}$ converges to 0. ■

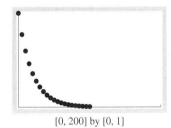

[0, 200] by [0, 1]

Screen 10.14
The graph indicates that the sequence $y_n = (0.95)^n$ has 0 as a limit.

EXPLORATION 2

r^n When $0 < r < 1$

Have your graphing calculator retain all the settings from the preceding example. Change $(0.95)^t$ to $(0.97)^t$, and graph. What happens? Repeat for $(0.98)^t$. Repeat for $(0.99)^t$, except change $T_{\max}$ and $x_{\max}$ to 400. What do you conclude about the convergence of the sequence r^n for $0 < r < 1$?

TABLE 10.5

n	$(0.95)^n$	n	$(0.95)^n$
1	0.95000	40	0.12851
2	0.90250	60	0.04607
3	0.85738	80	0.01652
4	0.81451	100	0.00592
5	0.77378	150	0.00046
10	0.59874	200	0.00004
20	0.35849		

Exploration 2 suggests that if $0 < r < 1$, then r^n approaches 0 as n becomes large without bound. This is indeed true.

EXPLORATION 3

r^n When $-1 < r < 0$

 Have your graphing calculator retain all the settings from the previous exploration. Change $(0.95)^t$ to $(-0.95)^t$, and graph. What happens? Repeat for $(-0.96)^t$. Repeat for $(-0.98)^t$. What do you conclude about the convergence of the sequence r^n for $-1 < r < 0$?

In Exploration 3 you noticed that $(-0.95)^n$, although alternating between positive and negative numbers, approaches 0 as n becomes large without bound. In a similar manner this happens for any sequence $\{r^n\}$ with $-1 < r < 0$. Thus, we have

$$\lim_{n \to \infty} r^n = 0 \quad \text{if } |r| < 1$$

EXAMPLE 7 Appreciation

A savings account with $1000 grows at 5% annually. Find the value of this account after n years, and determine what is happening to this value in the long-term.

SOLUTION ■ Since the account grows by 5% each year, the value, in dollars, for the next several years is

$$1000(1.05) = 1050.00, \qquad 1000(1.05)^2 = 1102.50, \qquad 1000(1.05)^3 = 1157.63$$

The value of the account after n years is $1000(1.05)^n$. This value becomes large without bound as n becomes large without bound. Thus, the sequence $\{1000(1.05)^n\}$ diverges. ■

Since the sequence $\{1000(1.05)^n\}$ diverges, the sequence $\{(1.05)^n\}$ also diverges. (Why is this so?) The same reasoning applies to any sequence $\{r^n\}$ with $r > 1$. This being the case it is then true that any sequence $\{r^n\}$ with $r < -1$ also diverges. We then have

$$\{r^n\} \text{ diverges if } |r| > 1$$

EXPLORATION 4

r^n When $|r| > 1$

Verify graphically that r^n diverges if $|r| > 1$ by taking specific values of r, with $|r| > 1$.

SELF-HELP EXERCISE SET 10.3

1. Find $\lim\limits_{n \to \infty} \dfrac{n^2 - 1}{n^3 + n + 1}$.

2. Determine if the sequence $\{(-1.001)^n\}$ converges or diverges.

EXERCISE SET 10.3

In Exercises 1 through 10 write out the first five terms of the given sequence.

1. $\{3n + 2\}$

2. $\left\{\dfrac{2n}{n+1}\right\}$

3. $\left\{\dfrac{n-1}{n+1}\right\}$

4. $\{2^{n-1}\}$

5. $\left\{\dfrac{2^n}{n}\right\}$

6. $\left\{\dfrac{2^n}{n!}\right\}$

7. $\left\{\left(-\dfrac{1}{2}\right)^n\right\}$

8. $\left\{\dfrac{(-1)^{n-1}}{n^2}\right\}$

9. $\{\cos n\pi\}$

10. $\left\{\sin\left(\dfrac{n\pi}{2}\right)\right\}$

In Exercises 11 through 18 find the general term of each of the given sequences.

11. $5, 7, 9, 11, \ldots$

12. $8, 11, 14, 17, \ldots$

13. $3, 9, 27, 81, 243, \ldots$

14. $3, 5, 9, 17, 33, \ldots$

15. $2, -4, 8, -16, 32, \ldots$

16. $1, -1, 1, -1, 1, \ldots$

17. $\dfrac{1}{2}, \dfrac{2}{3}, \dfrac{3}{4}, \dfrac{4}{5}, \dfrac{5}{6}, \ldots$

18. $1, \dfrac{1}{2}, \dfrac{1}{6}, \dfrac{1}{24}, \dfrac{1}{120}, \ldots$

In Exercises 19 through 24 sketch a graph of each of the given sequences.

19. $\left\{\dfrac{(-1)^n}{n}\right\}$

20. $\left\{(-1)^n \dfrac{n}{n+1}\right\}$

21. $\left\{1 + \dfrac{1}{2^n}\right\}$

22. $\left\{-1 + \dfrac{1}{n+1}\right\}$

23. $\left\{\dfrac{n^2}{n+1}\right\}$

24. $\{n^2 - n\}$

In Exercises 25 through 48 determine whether the given sequence converges or diverges. If the sequence converges, find its limit.

25. $\left\{\dfrac{3n}{n+1}\right\}$

26. $\left\{\dfrac{n}{2n+1}\right\}$

27. $\left\{\dfrac{2n^2-1}{n^2+1}\right\}$

28. $\left\{\dfrac{2n^3-1}{n^3+1}\right\}$

29. $\left\{\dfrac{n^2-1}{n^3+n+1}\right\}$

30. $\left\{\dfrac{n-1}{n^2+1}\right\}$

31. $\left\{\dfrac{n^2}{n+1}\right\}$

32. $\left\{\dfrac{n^3}{n^2+1}\right\}$

33. $\left\{\left(\dfrac{10}{11}\right)^n\right\}$

34. $\left\{\left(\dfrac{99}{100}\right)^n\right\}$

35. $\{(1.001)^n\}$

36. $\{(2.4)^n\}$

37. $\{(-0.99)^n\}$

38. $\{(-1.5)^n\}$

39. $\left\{\dfrac{1}{\sqrt{n}}\right\}$

40. $\{\sqrt{n}\}$

41. $\left\{\dfrac{1}{n} - \dfrac{1}{n+1}\right\}$

42. $\left\{\dfrac{1}{n+1} - \dfrac{1}{n+2}\right\}$

43. $\{\ln(n+1) - \ln(n)\}$

44. $\{e^{-n}\}$

45. $\left\{\sqrt{\dfrac{4n-1}{n+1}}\right\}$

46. $\left\{\sqrt[3]{\dfrac{3-8n}{n+1}}\right\}$

47. $\left\{\dfrac{e^n}{e^n+1}\right\}$

48. $\left\{\dfrac{2^n}{3^n+5^n}\right\}$

Applications

49. Medicine. Suppose that the concentration C_n, in appropriate units, of a drug in the bloodstream n hours after injection is given by $C_n = \dfrac{8n}{n^2+1}$. What happens to this concentration in the long-term?

50. Spread of Technological Innovation. The percentage P_n of firms in an industry that use a technological innovation n years after being introduced is given by $P_n = \dfrac{200n^3}{2n^3+1}$. What percentage of the firms eventually use this innovation?

51. Profit. In the nth year after introducing a new product a company makes a profit, P_n, in millions of dollars, where $P_n = 10 - 9e^{-0.10n}$. What is happening to these profits in the long-term?

52. Population Growth. In the nth year after placing five deer in a forest where there were none before, the population is estimated to be $P_n = \dfrac{500}{5 + 95e^{-0.25n}}$. What happens to this population in the long-term?

Enrichment Exercises

53. What is $\lim\limits_{n\to\infty}\left(1 + \dfrac{1}{n}\right)^n$?

54. What is $\lim\limits_{n\to\infty}\left(1 + \dfrac{x}{n}\right)^n$?

55. Show that the sequence $\sqrt{3}, \sqrt{\sqrt{3}}, \sqrt{\sqrt{\sqrt{3}}}, \ldots$, converges to 1. *Hint:* The sequence can be written as $3^{1/2}$, $3^{1/4}, 3^{1/8}, \ldots$. First decide what is the limit of the second sequence $3^{1/2}, 3^{1/3}, 3^{1/4}, 3^{1/5}, 3^{1/6}, 3^{1/7}, \ldots$. Now use this to determine the limit of the original sequence.

56. (a) Show $\lim\limits_{n\to\infty} \dfrac{(6.1)^n}{n!} = 0$.

   Hint: Show $a_{13} < 0.5a_{12}$, $a_{14} < (0.5)^2 a_{12}$, and so on.

   (b) Now show $\lim\limits_{n\to\infty} \dfrac{(24.6)^n}{n!} = 0$.

   (c) Based on the results in parts (a) and (b), for what values of x do you suspect that $\lim\limits_{n\to\infty} \dfrac{x^n}{n!} = 0$?

SOLUTIONS TO SELF-HELP EXERCISE SET 10.3

1. Divide numerator and denominator by n^3, and obtain

$$\lim_{n\to\infty} \frac{n^2 - 1}{n^3 + n + 1} = \lim_{n\to\infty} \frac{\dfrac{1}{n} - \dfrac{1}{n^3}}{1 + \dfrac{1}{n^2} + \dfrac{1}{n^3}}$$

$$= \frac{\lim\limits_{n\to\infty}\left(\dfrac{1}{n} - \dfrac{1}{n^3}\right)}{\lim\limits_{n\to\infty}\left(1 + \dfrac{1}{n^2} + \dfrac{1}{n^3}\right)}$$

$$= \frac{0 + 0}{1 + 0 + 0} = \frac{0}{1} = 0$$

2. This is a sequence $\{x^n\}$, where $x = -1.001 < -1$. Therefore, this sequence diverges.

10.4 INFINITE SERIES

- Infinite Series   ■ Geometric Series   ■ Properties of Infinite Series
- Applications

APPLICATION: DETERMINATION OF EFFECTIVE AND SAFE DRUG DOSAGES

A constant dose of a drug raises the concentration $C(t)$ of the drug in the bloodstream by D and is injected intravenously at fixed time intervals T. Clinical evidence suggests that the decrease in concentration is exponential and satisfies $C(t) = C(t_0)e^{-kt}$, where t_0 is the time of injection. (The constant k is called the *elimination constant*.) Since the drug is administered intravenously, we can assume that the concentration of the drug *immediately* after injection is increased by D. What happens to the concentration over an extended period? For the answer see Example 11.

Infinite Series

If we have an infinite sequence $\{a_k\}$ and try to add the terms, we obtain an expression of the form

$$a_1 + a_2 + a_3 + \cdots + a_k + \cdots$$

We call this expression an **infinite series** and also write it using the sigma notation as

$$a_1 + a_2 + a_3 + \cdots + a_k + \cdots = \sum_{k=1}^{\infty} a_k$$

However, an immediate question arises as to the meaning of this expression. How do we sum an *infinite* number of terms?

Our approach to first define an infinite sequence of **partial sums** $\{S_n\}$ as follows.

$$S_1 = \sum_{k=1}^{1} a_k = a_1$$

$$S_2 = \sum_{k=1}^{2} a_k = a_1 + a_2$$

$$S_3 = \sum_{k=1}^{3} a_k = a_1 + a_2 + a_3$$

$$\vdots$$

$$S_n = \sum_{k=1}^{n} a_k = a_1 + a_2 + a_3 + \cdots + a_n$$

Notice that each of these expressions is a sum of a *finite* number of terms and therefore makes perfectly good sense. If the *sequence* of partial sums $\{S_n\}$ converges to S, then we say that the infinite *series* $\sum_{k=1}^{\infty} a_k$ **converges** to the sum S. If the sequence of partial sums does not converge, then we say that the infinite series **diverges**.

Sum of an Infinite Series

Given an infinite series

$$\sum_{k=1}^{\infty} a_k = a_1 + a_2 + a_3 + \cdots + a_k + \cdots$$

we define the **partial sums** S_n by

$$S_n = \sum_{k=1}^{n} a_k = a_1 + a_2 + a_3 + \cdots + a_n$$

If the sequence of partial sums $\{S_n\}$ converges to S, that is, if

$$\lim_{n \to \infty} S_n = S$$

then we say that the infinite series $\sum_{k=1}^{\infty} a_k$ **converges** to the sum S and write

$$\sum_{k=1}^{\infty} a_k = S$$

Otherwise we say that the infinite series **diverges**.

EXPLORATION 1

Showing Convergence of Infinite Series Graphically

By graphing the partial sums on your graphing calculator give support to the conjecture that the series

$$1 + (0.5) + (0.5)^2 + (0.5)^3 + \cdots$$

converges to the sum 2. Support numerically.

Only in a few special cases can we determine an explicit formula for the partial sums S_n, which makes it difficult for us to determine whether an infinite series converges or diverges. To overcome this difficulty we develop a variety of tests in this section and in several to follow that indirectly determine convergence or divergence.

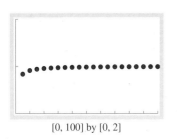

[0, 100] by [0, 2]

Screen 10.15
The graph indicates that the sequence $y_n = \sum_{k=1}^{n} 1/k(k + 1)$ has 1 as a limit.

EXAMPLE 1 Convergence of Infinite Series Graphically and Numerically, Confirm Analytically

Show that the infinite series

$$\sum_{k=1}^{\infty} \frac{1}{k(k + 1)} = \frac{1}{1 \cdot 2} + \frac{1}{2 \cdot 3} + \frac{1}{3 \cdot 4} + \cdots$$

converges.

SOLUTION ■ Using our graphing calculators we graph $\sum_{k=1}^{n} \frac{1}{k(k + 1)}$ using a window with dimensions [0, 100] by [0, 2] and obtain Screen 10.15. Notice that the sequence of partial sums appears to be converging to 1. Using the TRACE feature, we obtain the information in the table. Notice that the sequence of partial sums appears to be converging to 1.

n	10	20	30	40	50	60	70	80	90	100
S_n	0.9091	0.9524	0.9677	0.9756	0.9804	0.9836	0.9859	0.9877	0.9890	0.9901

Now let us do the problem analytically. First notice that

$$\frac{1}{k(k + 1)} = \frac{1}{k} - \frac{1}{k + 1}$$

Then

$$\sum_{k=1}^{n} \frac{1}{k(k + 1)} = \sum_{k=1}^{n} \left(\frac{1}{k} - \frac{1}{k + 1} \right)$$

$$= \left(1 - \frac{1}{2} \right) + \left(\frac{1}{2} - \frac{1}{3} \right) + \left(\frac{1}{3} - \frac{1}{4} \right) + \cdots + \left(\frac{1}{n} - \frac{1}{n + 1} \right)$$

$$= 1 + \left(-\frac{1}{2} + \frac{1}{2} \right) + \left(-\frac{1}{3} + \frac{1}{3} \right) + \cdots \left(-\frac{1}{n} + \frac{1}{n} \right) - \frac{1}{n + 1}$$

$$= 1 - \frac{1}{n + 1}$$

Since

$$\lim_{n \to \infty} S_n = \lim_{n \to \infty} \left(1 - \frac{1}{n + 1} \right) = 1$$

the infinite series converges to the sum $S = 1$. ■

EXAMPLE 2 Divergence of Infinite Series

Show that the infinite series

$$\sum_{k=1}^{\infty} (-1)^{k-1} = 1 - 1 + 1 - 1 + 1 - 1 + \cdots$$

diverges.

SOLUTION ■ The partial sums are

$$S_1 = 1, S_2 = 1 - 1 = 0, S_3 = 1 - 1 + 1 = 1, S_4 = 1 - 1 + 1 - 1 = 0, \ldots$$

Since this sequence alternates between 1 and 0, the sequence of partial sums does not converge, and thus the infinite series diverges. ■

Geometric Series

An important series is the **geometric series**. This is a series of the form

$$\sum_{k=1}^{\infty} ar^{k-1} = a + ar + ar^2 + ar^3 + \cdots + ar^{k-1} + \cdots$$

The nth partial is

$$S_n = \sum_{k=1}^{n} ar^{k-1} = a + ar + ar^2 + ar^3 + \cdots + ar^{n-1} \qquad [1]$$

EXPLORATION **2**

Convergence of Geometric Series

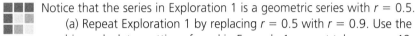 Notice that the series in Exploration 1 is a geometric series with $r = 0.5$.

(a) Repeat Exploration 1 by replacing $r = 0.5$ with $r = 0.9$. Use the graphing calculator settings found in Example 1, except take $y_{max} = 12$.

(b) Now take $r = 1.05$. Set $t_{min} = 0$, $t_{max} = 20$, $t_{step} = 1$, $x_{min} = 0$, $x_{max} = 20$, $y_{min} = 0$, $y_{max} = 50$. Do you think the series converges? Why or why not?

To see if the sequence of partial sums $\{S_n\}$ converges we need to put the S_n in another form. To do this multiply Equation (1) by r and obtain

$$rS_n = ar + ar^2 + ar^3 + \cdots + ar^{n-1} + ar^n \qquad [2]$$

Subtract Equation (2) from Equation (1), and if $r \neq 1$ obtain

$$S_n - rS_n = a + ar + ar^2 + ar^3 + \cdots + ar^{n-1}$$
$$\qquad\qquad -ar - ar^2 - ar^3 - \cdots - ar^{n-1} - ar^n$$
$$= a - ar^n$$
$$(1 - r)S_n = a - ar^n$$
$$S_n = \frac{a - ar^n}{1 - r}$$

Now recall from the previous section that

$$\lim_{n \to \infty} r^n = 0$$

if $|r| < 1$, and does not exist if $|r| > 1$. Thus, if $|r| > 1$ (and $a \neq 0$), the sequence of partial sums diverges and thus the geometric series diverges. If $|r| < 1$, then

$$\lim_{n \to \infty} S_n = \lim_{n \to \infty} \frac{a - ar^n}{1 - r} = \frac{a}{1 - r}$$

Thus when $|r| < 1$ the geometric series converges to the sum $\dfrac{a}{1 - r}$. There are two remaining cases, $r = -1$ and $r = 1$. When $r = -1$, the sequence of partial sums is $\{a, 0, a, 0, \ldots\}$. This sequence diverges if $a \neq 0$. For $r = 1$, the nth partial sum is $S_n = a + a + \cdots + a = na$, and this sequence diverges (when $a \neq 0$). Thus in this case the geometric series diverges. We summarize these facts in the following definition:

Geometric Series

The geometric series

$$\sum_{k=1}^{\infty} ar^{k-1} = a + ar + ar^2 + ar^3 + \cdots + ar^{k-1} + \cdots$$

is convergent if $|r| < 1$ with sum

$$\sum_{k=1}^{\infty} ar^{k-1} = \frac{a}{1-r}$$

If $|r| \geq 1$ and $a \neq 0$, then the geometric series diverges.

EXAMPLE 3 Geometric Series

Show that the infinite series

$$\sum_{k=1}^{\infty} \frac{4}{3^{k-1}}$$

is a geometric series, and find the sum if it converges.

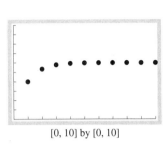

[0, 10] by [0, 10]

Screen 10.16
The graph indicates that the sequence $y_n = \Sigma_{k=1}^n 4 \cdot 3^{-k+1}$ has 6 as a limit.

SOLUTION ■ We first determine an answer graphically. Screen 10.16 shows a graph of $\sum_{k=1}^{n} \frac{4}{3^{k-1}}$ using a window with dimensions [0, 10] by [0, 10]. The geometric series appears to be converging to 6. Using the TRACE function we can obtain the information in the table, which also suggests that the geometric series is converging to 6.

n	2	4	6	8	10
S_n	5.33333	5.92593	5.99177	5.99909	5.99990

Analytically, since $\frac{4}{3^{k-1}} = 4\left(\frac{1}{3}\right)^{k-1}$, we see that

$$\sum_{k=1}^{\infty} \frac{4}{3^{k-1}} = \sum_{k=1}^{\infty} 4\left(\frac{1}{3}\right)^{k-1}$$

Thus, the series is a geometric series with $r = \frac{1}{3}$ and $a = 4$ and sum

$$\frac{a}{1-r} = \frac{4}{1-\frac{1}{3}} = 4 \cdot \frac{3}{2} = 6$$

■

EXAMPLE 4 Geometric Series

Determine if the infinite series

$$\sum_{k=1}^{\infty} (1.04)^k$$

converges or diverges.

[0, 20] by [0, 50]

Screen 10.17
The graph indicates that the sequence $y_n = \Sigma_{k=1}^n (1.04)^k$ becomes unbounded.

SOLUTION ■ We first determine an answer graphically. Screen 10.17 shows a graph of $\Sigma_{k=1}^n (1.04)^k$ using a window with dimensions [0, 20] by [0, 50]. The graph indicates that the series diverges.

Analytically, this is a geometric series with $r = 1.04$. Since $|r| = 1.04 > 1$, this geometric series diverges. ■

EXAMPLE 5 Geometric Series

Find the sum of the infinite series

$$\frac{1}{2} + \left(\frac{1}{2}\right)^2 + \left(\frac{1}{2}\right)^3 + \cdots + \left(\frac{1}{2}\right)^k + \cdots$$

SOLUTION ■ We first determine an answer graphically. Screen 10.18 shows a graph of $\Sigma_{k=1}^{n}(\frac{1}{2})^k$ using a window with dimensions [0, 20] by [0, 2]. The graph indicates that the series converges to 1. The TRACE function provides numerical support for this conclusion.

Analytically, factor out $\frac{1}{2}$ from each of the terms and obtain

$$\frac{1}{2}(1) + \frac{1}{2}\left(\frac{1}{2}\right)^1 + \frac{1}{2}\left(\frac{1}{2}\right)^2 + \cdots + \frac{1}{2}\left(\frac{1}{2}\right)^{k-1} + \cdots$$

This is a geometric series with $r = \frac{1}{2}$ and $a = \frac{1}{2}$. Thus, this geometric series converges to

$$\frac{a}{1-r} = \frac{\frac{1}{2}}{1-\frac{1}{2}} = 1 \quad ■$$

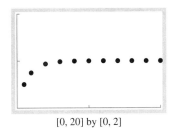

[0, 20] by [0, 2]

Screen 10.18
The graph indicates that the sequence $y_n = \Sigma_{k=1}^{n}(0.5)^k$ has 1 as a limit.

EXAMPLE 6 Infinite Decimals

Find the rational number that has the repeated decimal representation 0.4444. . . .

SOLUTION ■ By definition of the decimal representation we have

$$0.4444\ldots = \frac{4}{10} + \frac{4}{100} + \frac{4}{1000} + \cdots$$

$$= \frac{4}{10}(1) + \frac{4}{10}\left(\frac{1}{10}\right) + \frac{4}{10}\left(\frac{1}{10}\right)^2 + \cdots$$

This is a geometric series with $a = \dfrac{4}{10}$ and $r = \dfrac{1}{10}$. The sum is

$$\frac{a}{1-r} = \frac{4/10}{1-(1/10)} = \frac{4}{10} \cdot \frac{10}{9} = \frac{4}{9} \quad ■$$

Properties of Infinite Series

Since the sum of a convergent series is the limit of its partial sums, convergent series have properties that parallel the properties of limits. We now give two such important properties.

Properties of Infinite Series

Let $\Sigma_{k=1}^{\infty} a_k$ and $\Sigma_{k=1}^{\infty} b_k$ be convergent infinite series and c be any real number. Then

1. $\displaystyle\sum_{k=1}^{\infty} ca_k = c\sum_{k=1}^{\infty} a_k.$

2. $\displaystyle\sum_{k=1}^{\infty}(a_k \pm b_k) = \sum_{k=1}^{\infty} a_k \pm \sum_{k=1}^{\infty} b_k$

EXAMPLE 7 Using the Properties of Infinite Series

Find the sum of

$$\sum_{k=1}^{\infty}\left[\frac{4}{k(k+1)} + \left(\frac{2}{3}\right)^{k-1}\right]$$

SOLUTION ■ Recall from Example 1 that

$$\sum_{k=1}^{\infty} \frac{1}{k(k+1)} = 1$$

Also notice that the series $\sum_{k=1}^{\infty}(\frac{2}{3})^{k-1}$ is a geometric series with $a = 1$ and $r = \frac{2}{3}$ and the sum

$$\frac{a}{1 - r} = \frac{1}{1 - \frac{2}{3}} = 3$$

Now we are prepared to use the two properties of convergent series given previously.

$$\sum_{k=1}^{\infty}\left[\frac{4}{k(k + 1)} + \left(\frac{2}{3}\right)^{k-1}\right] = \sum_{k=1}^{\infty}\frac{4}{k(k + 1)} + \sum_{k=1}^{\infty}\left(\frac{2}{3}\right)^{k-1} \qquad \text{Property 2}$$

$$= 4\sum_{k=1}^{\infty}\frac{1}{k(k + 1)} + \sum_{k=1}^{\infty}\left(\frac{2}{3}\right)^{k-1} \qquad \text{Property 1}$$

$$= 4(1) + 3 = 7 \quad \blacksquare$$

We now give a third property of infinite series that leads to a test for divergence. We first notice that

$$S_n = a_1 + a_2 + \cdots + a_{n-1} + a_n$$
$$= S_{n-1} + a_n$$

Thus

$$a_n = S_n - S_{n-1}$$

If the series converges to S, then

$$\lim_{n\to\infty} a_n = \lim_{n\to\infty}(S_n - S_{n-1}) = \lim_{n\to\infty} S_n - \lim_{n\to\infty} S_{n-1} = S - S = 0$$

In summary, if $\sum_{k=1}^{\infty} a_k$ converges to S, then $\lim_{k\to\infty} a_k = 0$. As a consequence we have the following test for divergence.

Divergence Test for Infinite Series

If $\lim_{k\to\infty} a_k \neq 0$, then the infinite series $\sum_{k=1}^{\infty} a_k$ diverges.

EXAMPLE 8   Using the Test for Divergence

Determine whether the series

$$\sum_{k=1}^{\infty} k$$

converges or diverges.

SOLUTION ■ Since $\lim_{k\to\infty} k$ does not exist, then the limit does not equal zero, and the series $\sum_{k=1}^{\infty} k$ diverges. ■

It is important to realize that there are *divergent* series for which $\lim_{k\to\infty} a_k = 0$. The following is such a series.

EXPLORATION 3
Divergent Series

 Graphically show that the series $\sum_{k=1}^{\infty} \frac{1}{k}$ diverges.

In the next section, we show analytically that this series diverges. Thus, $\lim_{k\to\infty} a_k = 0$ does *not* imply that the series converges.

We can always add a finite number of terms to a series or delete a finite number of terms from a series without affecting its convergence or divergence. For example, if $\sum_{k=1}^{\infty} a_k$ converges, then $\sum_{k=i}^{\infty} a_k$ converges for any $i > 1$ and

$$\sum_{k=1}^{\infty} a_k = a_1 + a_2 + \cdots + a_{i-1} + \sum_{k=i}^{\infty} a_k$$

On the other hand, if $\sum_{k=i}^{\infty} a_k$ converges for some $i > 1$, then $\sum_{k=1}^{\infty} a_k$ converges.

For example,

$$\left(\frac{1}{2}\right)^2 + \left(\frac{1}{2}\right)^3 + \left(\frac{1}{2}\right)^4 + \cdots = \left(-1 - \frac{1}{2}\right) + \left(1 + \frac{1}{2}\right) + \left(\frac{1}{2}\right)^2 + \left(\frac{1}{2}\right)^3 + \cdots$$

$$= -1 - \frac{1}{2} + \sum_{k=1}^{\infty} \left(\frac{1}{2}\right)^{k-1}$$

$$= -1 - \frac{1}{2} + \frac{1}{1 - \frac{1}{2}} = -1 - \frac{1}{2} + 2 = \frac{1}{2}$$

Applications

EXAMPLE 9 Perpetuity

A corporation wishes to endow a university with a sum of money large enough to ensure that an annual interest of \$100,000 can be obtained at the end of the next year and at the end of every year forever. If interest rates are 7% compounded continuously, what should this sum be?

SOLUTION ■ The amount needed now to ensure \$100,000 at the end of the first year is

$$P_1 = 100{,}000e^{-0.07(1)}$$

The amount needed now to ensure \$100,000 at the end of the second year is

$$P_2 = 100{,}000e^{-0.07(2)}$$

We then see that the amount needed now to ensure \$100,000 at the end of the kth year is

$$P_k = 100{,}000e^{-0.07(k)}$$

Therefore the amount of the endowment must be

$$100{,}000e^{-0.07(1)} + 100{,}000e^{-0.07(2)} + 100{,}000e^{-0.07(3)}$$
$$+ \cdots + 100{,}000e^{-0.07(k)} + \cdots$$

We can write this as

$$100{,}000e^{-0.07}[1 + e^{-0.07(1)} + e^{-0.07(2)} + \cdots + e^{-0.07(k-1)} + \cdots]$$

This is a geometric series with $a = 100{,}000e^{-0.07}$ and $r = e^{-0.07}$. Thus, the sum is

$$\frac{a}{1-r} = \frac{100{,}000e^{-0.07}}{1 - e^{-0.07}} \approx 1{,}379{,}155$$

or \$1,379,155. ■

One way the Federal Reserve exercises control over the banking system and on monetary growth is by setting reserve requirements. Typically, the Federal Reserve may require banks to have on reserve at least 5% of any amount of money that is lent out. (The Federal Reserve changes this percentage from time to time, depending on economic conditions.)

EXAMPLE 10 Multiplier Effect

Suppose the Federal Reserve injects $2 billion into the monetary system, and this money is deposited in some banks. These banks loan out $2(0.95) billion and still meet the reserve requirements. The money $2(0.95) billion is then deposited in other banks. These other banks then loan out 95% of this, or $2(0.95)^2$ billion. How much money enters the system if this continues (a) for five turnovers? (b) Indefinitely?

SOLUTIONS ■

(a) The amount of money, in billions, in the system after five such moves is

$$2 + 2(0.95) + 2(0.95)^2 + 2(0.95)^3 + 2(0.95)^4$$

But this is just the sum S_5 of the first five terms of a geometric series with $a = 2$ and $r = 0.95$. This is

$$S_n = a\frac{1 - r^n}{1 - r} = 2\frac{1 - (0.95)^5}{0.05} \approx 9$$

or $9 billion.

(b) If the number of turnovers continues indefinitely, the amount of money entering the system would be

$$2 + 2(0.95) + 2(0.95)^2 + 2(0.95)^3 + \cdots$$

The sum is

$$S = \frac{a}{1 - r} = \frac{2}{1 - 0.95} = 40$$

or $40 billion. ■

The fraction $\dfrac{1}{1 - r} = 20$ is called the "multiplier effect." This means that for each of the original dollars injected into the system, very nearly 20 dollars can be loaned out if the money went through many transactions as indicated above. In normal times one would only expect five or six turnovers, so that reaching the upper limit of 20 times the original amount is not likely.

EXAMPLE 11 Determination of Effective and Safe Drug Dosages

A constant dose of a drug raises the concentration $C(t)$ of the drug in the bloodstream by D and is injected intravenously at fixed time intervals T. Clinical evidence suggests that the decrease in concentration is exponential and satisfies $C(t) = C(0)e^{-bt}$, where $t = 0$ is the time of injection. (The constant b is called the *elimination constant*.) Since the drug is administered intravenously, we can assume that the concentration of the drug *immediately* after injection is increased by D. Figure 10.2 shows graphically the effect of the repeated equal dosages of the drug. Determine what happens to the sequence $\{C_n\}$.

SOLUTION ■ After the initial injection, the concentration immediately before the second injection at time T is $R_1 = De^{-bT}$ and immediately after the injection is

$$C_1 = D + De^{-bT}$$

according to our assumptions. Then the concentration immediately before $2T$ is $R_2 = C_1e^{-bT}$ and immediately after is

$$\begin{aligned} C_2 &= D + C_1e^{-bT} \\ &= D + [D + De^{-bT}]e^{-bT} \\ &= D + De^{-bT} + De^{-2bT} \end{aligned}$$

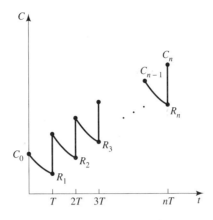

Figure 10.2
The graph gives the concentration C of a drug in the bloodstream being injected at fixed time intervals T.

We then see that

$$C_n = D + De^{-bT} + De^{-2bT} + \cdots + De^{-nbT}$$

If this continues indefinitely, then we need to sum the series

$$D + De^{-bT} + De^{-2bT} + \cdots + De^{-kbT} + \cdots$$

where C_n is the nth partial sum. Notice that

$$e^{-kbT} = (e^{-bT})^k$$

Thus, this series is

$$D + De^{-bT} + D(e^{-bT})^2 + \cdots + D(e^{-bT})^k + \cdots$$

But this is a geometric series with $a = D$ and $r = e^{-bT}$. The sum is then

$$\frac{a}{1 - r} = \frac{D}{1 - e^{-bT}} \quad \blacksquare$$

Often a drug is ineffective below a concentration C_L and harmful above some concentration C_H. An important clinical problem is to determine values of T and D so that eventually the concentration remains above C_L but below C_H. This is considered in the exercise set.

SELF-HELP EXERCISE SET 10.4

In Exercises 1 through 2 determine if the infinite series is convergent or divergent. If convergent, find its sum.

1. $\displaystyle\sum_{k=1}^{\infty} 4(0.9)^{k-1}$ **2.** $\displaystyle\sum_{k=1}^{\infty} 4(0.9)^{k}$

EXERCISE SET 10.4

In Exercises 1 through 4, find the limit of the nth partial sum of the infinite series to determine whether the series converges or diverges. If the series converges, find the sum.

1. $\displaystyle\sum_{k=1}^{\infty} 2$ **2.** $\displaystyle\sum_{k=1}^{\infty} (-2)$

3. $\displaystyle\sum_{k=1}^{\infty} \left(\frac{1}{k + 1} - \frac{1}{k + 2} \right)$

4. $\displaystyle\sum_{k=1}^{\infty} \left[\frac{1}{\ln(k + 1)} - \frac{1}{\ln(k + 2)} \right]$

In Exercises 5 through 26 determine if the infinite series converges or diverges. If the series converges, find its sum.

5. $\displaystyle\sum_{k=1}^{\infty} \frac{1}{5^{k-1}}$ **6.** $\displaystyle\sum_{k=1}^{\infty} \frac{1}{(2.1)^{k-1}}$

7. $\displaystyle\sum_{k=1}^{\infty} \frac{2^{k-1}}{3^{k-1}}$ **8.** $\displaystyle\sum_{k=1}^{\infty} \frac{3^{k-1}}{5^{k-1}}$

9. $\displaystyle\sum_{k=1}^{\infty} \frac{5^{k-1}}{3^{k-1}}$ **10.** $\displaystyle\sum_{k=1}^{\infty} \frac{7^{k-1}}{5^{k-1}}$

11. $\sum_{k=1}^{\infty} (0.9999)^{k-1}$

12. $\sum_{k=1}^{\infty} (1.0001)^{k-1}$

13. $\sum_{k=1}^{\infty} \frac{2^k}{3^{k-1}}$

14. $\sum_{k=1}^{\infty} \frac{2^{k-1}}{3^k}$

15. $\sum_{k=1}^{\infty} \frac{2^k - 3^k}{5^k}$

16. $\sum_{k=1}^{\infty} \frac{1 + 2^k}{3^k}$

17. $\sum_{k=1}^{\infty} \frac{k}{k+1}$

18. $\sum_{k=1}^{\infty} \frac{k+1}{k+2}$

19. $\sum_{k=1}^{\infty} \frac{k^2}{k+1}$

20. $\sum_{k=1}^{\infty} \frac{k^2 + 1}{k+2}$

21. $\sum_{k=1}^{\infty} \frac{2 \cdot 3^{k-1} - 3 \cdot 5^{k-1}}{7^{k-1}}$

22. $\sum_{k=1}^{\infty} e^{-0.3k}$

23. $\sum_{k=2}^{\infty} \frac{1}{3^{k-1}}$

24. $\sum_{k=2}^{\infty} \frac{1}{4^{k-1}}$

25. $\sum_{k=2}^{\infty} \frac{1}{5^k}$

26. $\sum_{k=2}^{\infty} \frac{1}{4^k}$

In Exercises 27 through 32 write the decimal as a rational number.

27. $0.7777\ldots$

28. $0.5555\ldots$

29. $0.272727\ldots$

30. $0.234234234\ldots$

31. $3.4272727\ldots$

32. $1.5234234234\ldots$

Applications

33. Multiplier Effect. A city is attempting to attract a convention to use its civic center. It believes that 80% of every dollar spent by the people attending the convention that ends up in the hands of local businesses is spent locally by these local businesses, and 80% of this locally spent money in turn is spent locally, and so on. Find the effect that one dollar will have if this goes on indefinitely.

34. Bouncing Ball. A ball is dropped from a height of 6 feet and bounces up and down indefinitely. If the ball rebounds every time to two-thirds of its height on the previous bounce, how far will the ball travel?

35. Perpetuity. You wish to donate to your town museum a sum of money large enough to ensure that an annual interest of $1000 can be obtained at the end of the next year and at the end of every year forever. If interest rates are 8% compounded continuously, what should this sum be?

36. Spread of Infectious Disease. Twenty people return from a trip to their hometown carrying an infectious disease. For every five people who have this disease, four are infected the following day. How many eventually become infected in town?

Enrichment Exercises

37. Give a proof of Property 1 of infinite series.

38. Give a proof of Property 2 of infinite series.

39. Suppose that $a_k \neq 0$ for all k and that the infinite series $\sum_{k=1}^{\infty} a_k$ converges. Prove that the infinite series $\sum_{k=1}^{\infty} \frac{1}{a_k}$ diverges.

40. Find two divergent infinite series $\sum_{k=1}^{\infty} a_k$ and $\sum_{k=1}^{\infty} b_k$ for which the infinite series $\sum_{k=1}^{\infty} (a_k + b_k)$ converges.

41. If $|x| < 1$, show that the infinite series $\sum_{k=1}^{\infty} x^k$ converges to $\frac{x}{1-x}$.

42. If $|x| < 1$, show that the infinite series

$$1 + x^2 + x^4 + x^6 + \cdots$$

converges to $\frac{1}{1 - x^2}$.

43. Determination of Effective and Safe Drug Dosages. What happens to R_n in Figure 10.2 in Example 11 in the long-term?

44. Determination of Effective and Safe Drug Dosages. Suppose the drug in Example 11 is ineffective below a concentration C_L and harmful above a concentration of C_H. Find values for T and D so that eventually the concentration of the drug is at least C_L and at most C_H.

SOLUTIONS TO SELF-HELP EXERCISE SET 10.4

1. This is the infinite series

$$4 + 4(0.9) + 4(0.9)^2 + 4(0.9)^3 + \cdots$$

This is a geometric series with $a = 4$ and $r = 0.9$. Since $|r| = 0.9 < 1$, the series converges to

$$\frac{a}{1-r} = \frac{4}{1 - 0.9} = 40$$

2. Let $b_n = 4(0.9)^k$ and $a_k = 4(0.9)^{k-1}$. Then notice that

$$b_k = 4(0.9)^k = (0.9)[4(0.9)^{k-1}] = (0.9)a_k$$

Thus, the terms of the infinite series in this example are all (0.9) times the corresponding terms of the infinite series in the previous example. Therefore

$$\sum_{k=1}^{\infty} b_n = \sum_{k=1}^{\infty} (0.9)a_n = (0.9)\sum_{k=1}^{\infty} a_n = (0.9)(40) = 36$$

10.5 THE INTEGRAL AND COMPARISON TESTS

■ The Integral Test ■ The Comparison Test ■ The Limit Comparison Test
■ Applications

APPLICATION: MULTIPLIER EFFECT

Suppose the Federal Reserve injects $2 billion into the monetary system and this money is deposited in some banks. Suppose these banks loan out between 90% and 95% of this money. Suppose 90% to 95% of the loaned money is again deposited in other banks. If this continues indefinitely, what is an upper and lower bound on the amount of money that enters the system? See Example 8 for the answer.

The Integral Test

In this section we present three tests for determining whether a series $\sum_{k=1}^{\infty} a_k$ is convergent. We first consider the integral test.

The integral test requires that a_k be positive and decreasing, that is,

1. $a_k > 0$, for $k = 1, 2, 3, \ldots$
2. $a_1 > a_2 > a_3 > \cdots$

The first step is to define a function $y = f(x)$ that is positive on $[0, \infty)$ and decreasing there and for which

$$f(k) = a_k \quad \text{for all } k$$

In Figure 10.3 we also see drawn rectangles with heights of length a_k and bases of length 1. Thus, the area of the first rectangle is a_1, the area of the second rectangle is a_2, and so on. Then the finite series

$$a_1 + a_2 + a_3 + \cdots$$

can be thought of as the total area of all the rectangles. Since the tops of the rectangles shown in Figure 10.3 lie above the graph of $y = f(x)$, the area of any rectangle is greater than the area under the curve on the same interval. As a consequence

$$S_n = \sum_{k=1}^{n} a_k = a_1 + a_2 + \cdots + a_n \geq \int_{1}^{n+1} f(x)\, dx$$

Now as $n \to \infty$, the integral $\int_{1}^{n+1} f(x)\, dx$ becomes the improper integral $\int_{1}^{\infty} f(x)\, dx$. This suggests that if this improper integral is infinite, then the total area of all the

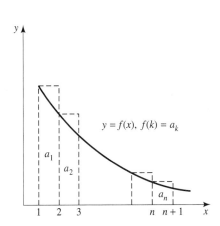

Figure 10.3
The area under the curve is less than the area under the rectangles.

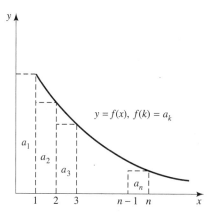

Figure 10.4
The area under the curve is greater than the area under the rectangles.

rectangles, which is also $\lim_{n\to\infty} S_n$, is infinite, and thus the series diverges. This is the first part of the **integral test**. If the improper integral $\int_1^\infty f(x)\,dx$ diverges, then the infinite series $\Sigma_{k=1}^\infty a_k$ diverges.

Now for the second part of the integral test look at Figure 10.4, where the rectangles are now inscribed. From this figure we see that

$$a_2 + a_3 + \cdots + a_n \le \int_1^n f(x)\,dx$$

Thus, by adding a_1 to each side we obtain

$$S_n = \sum_{k=1}^n a_k = a_1 + a_2 + \cdots + a_n \le a_1 + \int_1^n f(x)\,dx$$

Now as $n \to \infty$, the right-hand side becomes $a_1 + \int_1^\infty f(x)\,dx$. If the improper integral $\int_1^\infty f(x)\,dx$ converges (to some finite value), then this suggests that $\lim_{n\to\infty} S_n$ converges. This is the second part of the integral test. If the improper integral $\int_1^\infty f(x)\,dx$ converges, then the infinite series $\Sigma_{k=1}^\infty a_k$ converges.

Integral Test

Let a_k be a positive decreasing sequence and let $f(x)$ be positive, continuous, and decreasing on the interval $[1, \infty)$ with $f(k) = a_k$. The infinite series $\Sigma_{k=1}^\infty a_k$ and the improper integral $\int_1^\infty f(x)\,dx$ both converge or both diverge. That is

1. If $\int_1^\infty f(x)\,dx$ is convergent, then $\Sigma_{k=1}^\infty a_k$ is convergent.
2. If $\int_1^\infty f(x)\,dx$ is divergent, then $\Sigma_{k=1}^\infty a_k$ is divergent.

Remark. Do not make the mistake of thinking that if the improper integral $\int_1^\infty f(x)\,dx$ is convergent, then the sum of the infinite series $\Sigma_{k=1}^\infty a_k$ equals the value of the improper integral. All we know, in this case, is that the sum of the infinite series is *less* than $a_1 + \int_1^\infty f(x)\,dx$.

Normally the function f in the integral test is obtained by replacing the k in a_k with x, that is, $f(x) = a_x$.

EXAMPLE 1 Integral Test

Show that the series $\sum_{k=1}^\infty \dfrac{1}{k}$ diverges.

S O L U T I O N ■ The terms $a_k = \dfrac{1}{k}$ are positive and decreasing. Take $f(x) = \dfrac{1}{x}$. Then $f(k) = \dfrac{1}{k}$. We then have

$$\int_1^\infty f(x) \, dx = \int_1^\infty \frac{1}{x} \, dx$$

$$= \lim_{b \to 0} \int_1^b \frac{1}{x} \, dx$$

$$= \lim_{b \to 0} \ln x \big|_1^b$$

$$= \lim_{b \to 0} \ln b$$

which is infinite. Thus, the improper integral $\displaystyle\int_1^\infty \frac{1}{x} \, dx$ and also the infinite series $\displaystyle\sum_{k=1}^\infty \frac{1}{k}$ diverge. ■

The series $\displaystyle\sum_{k=1}^\infty \frac{1}{k}$ is called the **harmonic series**.

The **p-series** is a series of the form

$$\sum_{k=1}^\infty \frac{1}{k^p}$$

where p is a positive constant. The harmonic series is a p-series with $p = 1$.

E X A M P L E 2   The p-Series

Find the values of p for which the p-series converges and the values of p for which it diverges.

S O L U T I O N ■ Because for $p = 1$ the p-series is the harmonic series, we already know that the p-series diverges when $p = 1$. We now consider the other cases and define $f(x) = \dfrac{1}{x^p}$. Since

$$f'(x) = -px^{-p-1} = -\frac{p}{x^{p+1}}$$

is negative for all $x \ge 0$, $f(x)$ is a decreasing function on $(0, \infty)$. The integral test then applies. We then have

$$\int_1^\infty \frac{1}{x^p} \, dx = \lim_{b \to \infty} \int_1^b x^{-p} \, dx$$

$$= \lim_{b \to \infty} \frac{1}{1 - p} x^{-p+1} \bigg|_1^b$$

$$= \lim_{b \to \infty} \left(\frac{b^{-p+1}}{1 - p} - \frac{1}{1 - p} \right)$$

$$= \lim_{b \to \infty} \frac{b^{-p+1}}{1 - p} - \frac{1}{1 - p}$$

This limit exists if and only if $-p + 1 < 0$, that is, if and only if, $p > 1$. Thus, the improper integral converges if $p > 1$ and diverges if $p < 1$. Therefore by the integral test, the p-series converges if $p > 1$ and diverges if $p \le 1$. ■

The p-Series

The *p*-series

$$\sum_{k=1}^{\infty} \frac{1}{k^p} = 1 + \frac{1}{2^p} + \frac{1}{3^p} + \frac{1}{4^p} + \cdots$$

converges if $p > 1$ and diverges if $p \leq 1$.

EXAMPLE 3   The *p*-Series

Determine if the following infinite series converge or diverge:

(a) $\displaystyle\sum_{k=1}^{\infty} \frac{1}{k^2}$ (b) $\displaystyle\sum_{k=1}^{\infty} \frac{1}{\sqrt{k}}$ (c) $\displaystyle\sum_{k=1}^{\infty} \frac{1}{k^{0.999}}$ (d) $\displaystyle\sum_{k=1}^{\infty} \frac{1}{k^{1.001}}$

SOLUTION ■ All these series are *p*-series.
(a) Here $p = 2 > 1$; therefore, this series converges.
(b) Here $p = 0.5 < 1$; therefore, this series diverges.
(c) Here $p = 0.999 < 1$; therefore, this series diverges.
(d) Here $p = 1.001 > 1$; therefore, this series converges. ■

The Comparison Test

The comparison test is also based on comparing areas. Suppose we have two infinite series, $\sum_{k=1}^{\infty} a_k$ and $\sum_{k=1}^{\infty} b_k$, with $0 < a_k \leq b_k$. Then Figure 10.5 illustrates the situation. The series $\sum_{k=1}^{\infty} a_k$ represents the total area under the smaller rectangles. This must be smaller than the total area under the larger rectangles, which represents the series $\sum_{k=1}^{\infty} b_k$.

Thus, if the series $\sum_{k=1}^{\infty} b_k$ converges, the series $\sum_{k=1}^{\infty} a_k$ with the smaller terms converges. If the series $\sum_{k=1}^{\infty} a_k$ diverges, then the series $\sum_{k=1}^{\infty} b_k$ with the larger terms also diverges.

The Comparison Test

Suppose $\sum_{k=1}^{\infty} a_k$ and $\sum_{k=1}^{\infty} b_k$ are two infinite series with positive terms and that $0 < a_k \leq b_k$ for all k. Then
(a) If $\sum_{k=1}^{\infty} b_k$ converges, then $\sum_{k=1}^{\infty} a_k$ converges and $\sum_{k=1}^{\infty} a_k \leq \sum_{k=1}^{\infty} b_k$.
(b) If $\sum_{k=1}^{\infty} a_k$ diverges, then $\sum_{k=1}^{\infty} b_k$ diverges.

EXAMPLE 4   The Comparison Test

Determine whether the series $\displaystyle\sum_{k=1}^{\infty} \frac{3}{5k^2 + 2k + 3}$ converges or diverges.

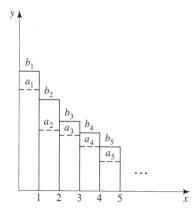

Figure 10.5
The area under the smaller rectangles is $\sum_{k=1}^{\infty} a_k$ and is less than the area under the larger rectangles, which is $\sum_{k=1}^{\infty} b_k$.

SOLUTION ■ Since for all positive k, $5k^2 + 2k + 3 > 5k^2$,

$$a_k = \frac{3}{5k^2 + 2k + 3} \le \frac{3}{5k^2} = b_k$$

The series $\sum_{k=1}^{\infty} \frac{1}{k^2}$ converges since it is a p-series with $p = 2 > 1$. Thus, the series $\sum_{k=1}^{\infty} \frac{3}{5k^2} = \frac{3}{5}\sum_{k=1}^{\infty} \frac{1}{k^2}$ converges. Then by the comparison test, the series $\sum_{k=1}^{\infty} \frac{3}{5k^2 + 2k + 3}$ converges. ■

EXAMPLE 5 The Comparison Test

Determine whether the series $\sum_{k=1}^{\infty} \frac{1}{k - 0.9}$ converges or diverges.

SOLUTION ■ For all positive k

$$a_k = \frac{1}{k} < \frac{1}{k - 0.9} = b_k$$

Since the harmonic series $\sum_{k=1}^{\infty} \frac{1}{k}$ diverges, by the comparison test, the series $\sum_{k=1}^{\infty} \frac{1}{k - 0.9}$ diverges. ■

The Limit Comparison Test

Suppose we wish to determine if the series $\sum_{k=1}^{\infty} \frac{1}{2^k + 1}$ converges or diverges. We might think to compare this series with the convergent geometric series $\sum_{k=1}^{\infty} \frac{1}{2^k}$. Since

$$\frac{1}{2^k + 1} < \frac{1}{2^k}$$

the comparison test indicates that $\sum_{k=1}^{\infty} \frac{1}{2^k + 1}$ converges.

Now suppose we wish to determine if the series $\sum_{k=1}^{\infty} \frac{1}{2^k - 1}$ converges or diverges. We might again think to compare this series with the convergent geometric series $\sum_{k=1}^{\infty} \frac{1}{2^k}$. But

$$\frac{1}{2^k - 1} > \frac{1}{2^k}$$

does not help in the comparison test. Still, for large k, the kth terms of the two series are nearly equal. Thus, we strongly suspect that this series also converges. In such a case the following test can be used.

The Limit Comparison Test

Let $\Sigma_{k=1}^{\infty} a_k$ and $\Sigma_{k=1}^{\infty} b_k$ be two series with positive terms. If

$$\lim_{k \to \infty} \frac{a_k}{b_k} = \rho$$

exists and $\rho \ne 0$, then either both series converge or both series diverge.

The following is a very rough sketch of the proof. If

$$\lim_{k \to \infty} \frac{a_k}{b_k} = \rho$$

then for large k, $a_k \approx \rho b_k$, and the terms of one series approximate a nonzero constant times the corresponding terms of the other series. Thus, they should both converge or both diverge together.

EXAMPLE 6 Limit Comparison Test

Determine whether the series $\displaystyle\sum_{k=1}^{\infty} \frac{1}{2^k - 1}$ converges or diverges.

SOLUTION ■ For large k the kth term $\dfrac{1}{2^k - 1}$ is approximately $\dfrac{1}{2^k}$. We then use the limit comparison test comparing the given series to the convergent geometric series $\displaystyle\sum_{k=1}^{\infty} \frac{1}{2^k}$. We have

$$\lim_{k \to \infty} \left(\frac{\frac{1}{2^k}}{\frac{1}{2^k - 1}} \right) = \lim_{k \to \infty} \left(\frac{2^k - 1}{2^k} \right)$$

$$= \lim_{k \to \infty} \left(1 - \frac{1}{2^k} \right) = 1$$

Since this limit is finite and is not zero, the given series converges. ■

EXAMPLE 7 Limit Comparison Test

Determine whether the series $\displaystyle\sum_{k=1}^{\infty} \frac{2}{3\sqrt{k} - 2}$ converges or diverges.

SOLUTION ■ For large k the kth term of the series is approximately $\dfrac{2}{3\sqrt{k}} = \dfrac{2}{3k^{1/2}}$. This is $\dfrac{2}{3}$ times $\dfrac{1}{k^{1/2}}$, and so we compare our series to the series $\displaystyle\sum_{k=1}^{\infty} \frac{1}{k^{1/2}}$, which is a p-series with $p = \frac{1}{2} < 1$ and therefore divergent. We then have

$$\lim_{k \to \infty} \left(\frac{\frac{1}{k^{1/2}}}{\frac{2}{3k^{1/2} - 2}} \right) = \lim_{k \to \infty} \left(\frac{3k^{1/2} - 2}{2k^{1/2}} \right)$$

$$= \lim_{k \to \infty} \left(\frac{3}{2} - \frac{1}{k^{1/2}} \right) = \frac{3}{2}$$

Since this limit is finite and is not zero, the given series diverges. ■

Applications

EXAMPLE 8 Multiplier Effect

Suppose the Federal Reserve injects $2 billion into the monetary system and this money is deposited in some banks. Suppose these banks loan out between 90% and 95% of this money. Suppose 90% to 95% of the loaned money is again deposited in other banks. If this continues indefinitely, find an upper and lower bound on the amount of money that enters the system.

S O L U T I O N ■ Let a_1 be 2, a_2 be the amount in billions that enters the system during the first turnover, a_3 be the amount in billions that enters the system during the next turnover, and so on. Then

$$2(0.90) \leq a_2 \leq 2(0.95)$$

$$2(0.90)^2 \leq a_3 \leq 2(0.95)^2$$

and so on. The amount that we are seeking is $\Sigma_{k=1}^{\infty} a_k$. A lower bound on the amount of money entering the system is

$$\sum_{k=1}^{\infty} 2(0.09)^{k-1} = 2 + 2(0.90) + 2(0.90)^2 + 2(0.90)^3 + \cdots$$

and an upper bound is

$$\sum_{k=1}^{\infty} 2(0.95)^{k-1} = 2 + 2(0.95) + 2(0.95)^2 + 2(0.95)^3 + \cdots$$

These are two convergent geometric series. The sum of the first is

$$S = \frac{a}{1 - r} = \frac{2}{1 - 0.90} = 20$$

and the sum of the second is

$$S = \frac{a}{1 - r} = \frac{2}{1 - 0.05} = 40$$

Thus, by the comparison test, we conclude that between $20 and $40 billion eventually enters the system. ■

SELF-HELP EXERCISE SET 10.5

1. Consider the infinite series $\displaystyle\sum_{k=1}^{\infty} \frac{k}{2k^2 - 1}$. Use the following tests to show that the series diverges. **(a)** Integral test **(b)** Comparison test **(c)** Limit comparison test

EXERCISE SET 10.5

In Exercises 1 through 10 use the integral test to determine whether the infinite series is convergent or divergent. If the integral test fails, use another test.

In Exercises 11 through 20 use the comparison test to determine whether the infinite series is convergent or divergent.

1. $\displaystyle\sum_{k=1}^{\infty} \frac{1}{k + 3}$

2. $\displaystyle\sum_{k=1}^{\infty} \frac{2}{3k + 1}$

3. $\displaystyle\sum_{k=1}^{\infty} \frac{4}{\sqrt[3]{k}}$

4. $\displaystyle\sum_{k=1}^{\infty} \frac{2}{k^{3/2}}$

5. $\displaystyle\sum_{k=1}^{\infty} ke^{-k^2}$

6. $\displaystyle\sum_{k=1}^{\infty} k^2 e^{k^3}$

7. $\displaystyle\sum_{k=1}^{\infty} \frac{k}{\sqrt{k^2 + 1}}$

8. $\displaystyle\sum_{k=1}^{\infty} \frac{k}{(k^2 + 1)^3}$

9. $\displaystyle\sum_{k=1}^{\infty} \frac{1}{k \ln(k + 1)}$

10. $\displaystyle\sum_{k=1}^{\infty} ke^{-k}$

11. $\displaystyle\sum_{k=1}^{\infty} \frac{5}{k}$

12. $\displaystyle\sum_{k=1}^{\infty} \frac{1}{2k^2}$

13. $\displaystyle\sum_{k=1}^{\infty} \frac{1}{k^2 + k + 1}$

14. $\displaystyle\sum_{k=1}^{\infty} \frac{1}{k^3 + k^2 + 1}$

15. $\displaystyle\sum_{k=1}^{\infty} \frac{1}{4\sqrt[3]{k} - \pi}$

16. $\displaystyle\sum_{k=1}^{\infty} \frac{1}{4\sqrt{k} - \pi}$

17. $\displaystyle\sum_{k=1}^{\infty} \frac{k^{1/2}}{k^{5/2} + 1}$

18. $\displaystyle\sum_{k=1}^{\infty} \frac{1}{\sqrt{k^4 + k + 1}}$

19. $\displaystyle\sum_{k=1}^{\infty} \frac{1}{3k^2 - k - 1}$

20. $\displaystyle\sum_{k=1}^{\infty} \frac{2}{k + 1}$

In Exercises 21 through 42 determine whether the infinite series is convergent or divergent.

21. $\displaystyle\sum_{k=1}^{\infty} \frac{1}{1 + \sqrt{k}}$

22. $\displaystyle\sum_{k=1}^{\infty} \frac{3^k - 1}{2^k}$

23. $\displaystyle\sum_{k=1}^{\infty} \frac{k + 1}{2k^3 - 1}$

24. $\displaystyle\sum_{k=1}^{\infty} \frac{k + 1}{\sqrt{2k^6 - 1}}$

25. $\displaystyle\sum_{k=1}^{\infty} \frac{2k + 1}{2k - 1}$

26. $\displaystyle\sum_{k=1}^{\infty} \frac{\sqrt{k^2 + 1}}{k}$

27. $\displaystyle\sum_{k=1}^{\infty} \frac{1 + 2^k}{1 + 3^k}$

28. $\displaystyle\sum_{k=1}^{\infty} \frac{k^2 + 1}{k^4 + k + 1}$

29. $\displaystyle\sum_{k=1}^{\infty} \frac{k^3 + k + 1}{k^6 + k^3 + 1}$

30. $\displaystyle\sum_{k=1}^{\infty} \frac{1}{\sqrt{2k - 1}}$

31. $\displaystyle\sum_{k=1}^{\infty} \frac{\sqrt{k}}{2k + 1}$

32. $\displaystyle\sum_{k=1}^{\infty} \frac{3^k - 1}{2^k - 1}$

33. $\displaystyle\sum_{k=1}^{\infty} \frac{k + 1}{k2^k}$

34. $\displaystyle\sum_{k=1}^{\infty} \frac{e^{\sqrt{k}}}{\sqrt{k}}$

35. $\displaystyle\sum_{k=1}^{\infty} e^{-k}$

36. $\displaystyle\sum_{k=1}^{\infty} \sin\frac{\pi}{k}$

37. $\displaystyle\sum_{k=1}^{\infty} \frac{1 + \sqrt{k}}{k^2 + 1}$

38. $\displaystyle\sum_{k=1}^{\infty} \frac{k}{\sqrt{k^6 + 1}}$

39. $\displaystyle\sum_{k=1}^{\infty} \frac{1 + \sqrt{k}}{1 + k}$

40. $\displaystyle\sum_{k=1}^{\infty} \frac{1 + k}{1 + \sqrt{k}}$

41. $\displaystyle\sum_{k=1}^{\infty} \frac{\sin^2 k}{k^2}$

42. $\displaystyle\sum_{k=1}^{\infty} \frac{\cos^2 k}{k^2}$

Applications

43. Multiplier Effect. A city is attempting to attract a convention to use its civic center. It believes that between 60% and 80% of every dollar spent by the people attending the convention that ends up in the hands of local businesses is spent by these local businesses locally, and between 60% and 80% of this locally spent money is in turn spent locally, and so on. Find an upper and lower bound for the effect that one dollar will have if this goes on indefinitely.

44. Bouncing Ball. A ball is dropped from a height of 6 feet and bounces up and down indefinitely. If the ball rebounds every time to between two-thirds and three-quarters of its height on the previous bounce, find an upper and lower bound on how far the ball travels.

45. Perpetuity. You wish to donate a sum of money to your town museum large enough to ensure that an annual interest of $1000 can be obtained at the end of next year and at the end of every year forever. If interest rates are between 7% and 8% compounded continuously, find an upper and lower bound for what this sum should be.

46. Spread of Infectious Disease. Twenty people return from a trip to their hometown carrying an infectious disease. For every five people who have this disease, between three and four are infected the following day. Find an upper and lower bound on how many eventually become infected in town.

Enrichment Exercises

47. If S_n is the nth partial sum of the harmonic series, show that $S_n \le 1 + \ln n$.

48. We know that the harmonic series diverges. The inequality in the previous exercise indicates just how slowly. Show that the sum of the first million terms is less than 15 and the sum of the first billion terms is less than 22.

49. Show that if $a_k \ge 0$ and $\Sigma\, a_k$ converges then $\Sigma\, a_k^2$ converges.

50. Show that if $a_k > 0$ and $\Sigma\, a_k$ converges, then $\Sigma \ln(1 + a_k)$ converges. *Hint:* Compare $\ln(1 + x)$ with x.

51. Show that the series $\Sigma_{k=1}^{\infty} e^{-k^2}$ converges. *Hint:* Compare the series with one that you can use the integral test on.

SOLUTIONS TO SELF-HELP EXERCISE SET 10.5

1. (a) Let $f(x) = \dfrac{x}{2x^2 - 1}$. Then

$$f'(x) = \frac{(2x^2 - 1)(1) - (x)(4x)}{(2x^2 - 1)^2}$$

$$= -\frac{2x^2 + 1}{(2x^2 - 1)^2}$$

is negative on $[1, \infty)$, and thus $f(x)$ is decreasing on this interval. Then

$$\int_1^{\infty} \frac{x}{2x^2 - 1}\, dx = \lim_{b \to \infty} \int_1^b \frac{x}{2x^2 - 1}\, dx \qquad \text{Let } u = 2x^2 - 1$$

$$= \lim_{b \to \infty} \frac{1}{4} \ln(2x^2 - 1)\Big|_1^b$$

$$= \lim_{b \to \infty} \frac{1}{4} \ln(2b^2 - 1) = \infty$$

Since the improper integral is divergent, so also is the series.

(b) Since $\dfrac{k}{2k^2 - 1} \geq \dfrac{k}{2k^2} = \dfrac{1}{2k}$, we compare the series with

the test series $\displaystyle\sum_{k=1}^{\infty} \dfrac{1}{2k}$. This is $\dfrac{1}{2}\displaystyle\sum_{k=1}^{\infty}\dfrac{1}{k}$, which is a nonzero constant times the divergent harmonic series, and, therefore, divergent. Thus, the original series is divergent.

(c) For the limit comparison, use the harmonic series $\displaystyle\sum_{k=1}^{\infty}$

$\dfrac{1}{k}$. Since $b_k = \dfrac{1}{k}$, we have

$$\lim_{k\to\infty}\left(\frac{\frac{1}{k}}{\frac{k}{2k^2-1}}\right) = \lim_{k\to\infty}\left(\frac{2k^2-1}{k^2}\right)$$

$$= \lim_{k\to\infty}\left(2 - \frac{1}{k^2}\right) = 2$$

Since this limit is finite and is not zero, the given series diverges.

10.6 THE RATIO TEST AND ABSOLUTE CONVERGENCE

■ The Ratio Test ■ Alternating Series ■ Absolute Convergence

The Ratio Test

We now consider another test for the convergence or divergence of infinite series, the ratio test. We assume we have an infinite series $\Sigma_{k=1}^{\infty} a_k$ of nonzero terms and that the limit of the ratio

$$\lim_{k\to\infty}\frac{|a_{k+1}|}{|a_k|}$$

exists and equals r. As a consequence, for large k, $\dfrac{|a_{k+1}|}{|a_k|} \approx r$. Then if j is such a large k, $|a_{j+1}| \approx r|a_j|$. Then $|a_{j+2}| \approx r|a_{j+1}| \approx r^2|a_j|$. In general, we then see that for large k

$$|a_{j+k}| \approx r^k|a_j|$$

In other words the infinite series for large k is approximately a geometric series with ratio r. Thus, we expect the series $\Sigma_{k=1}^{\infty} a_k$ to converge if $0 \leq r < 1$ and diverge if $r > 1$. This gives the ratio test.

The Ratio Test

Let $\Sigma_{k=1}^{\infty} a_k$ be an infinite series of nonzero terms and suppose that

$$\lim_{k\to\infty}\frac{|a_{k+1}|}{|a_k|} = r$$

1. If $0 \leq r < 1$, then the series converges.
2. If $r > 1$, then the series diverges.
3. If $r = 1$, the test provides no information. (The series may converge or may diverge.)

EXAMPLE 1 Using the Ratio Test

Determine whether the series $\displaystyle\sum_{k=1}^{\infty}\dfrac{k}{3^k}$ is convergent or divergent.

SOLUTION ■ Here $a_k = \dfrac{k}{3^k}$. To obtain a_{k+1} we replace k with $k + 1$ in the expression for a_k. This gives

$$a_{k+1} = \frac{k+1}{3^{k+1}}$$

Then

$$\frac{|a_{k+1}|}{|a_k|} = \frac{\left|\dfrac{k+1}{3^{k+1}}\right|}{\left|\dfrac{k}{3^k}\right|} = \frac{\dfrac{k+1}{3^{k+1}}}{\dfrac{k}{3^k}}$$

$$= \frac{k+1}{k} \cdot \frac{3^k}{3^{k+1}}$$

$$= \frac{1}{3} \cdot \frac{k+1}{k} = \frac{1}{3} \cdot \left(1 + \frac{1}{k}\right)$$

We then take the limit of this ratio and obtain

$$r = \lim_{k \to \infty} \frac{|a_{k+1}|}{|a_k|} = \lim_{k \to \infty} \frac{1}{3}\left(1 + \frac{1}{k}\right) = \frac{1}{3}$$

Since $r = \frac{1}{3} < 1$, the series converges by the ratio test. ■

To do the next example we need to notice that $(k+1)! = (k+1)k!$ and to recall from Section 1.7 the following important limit:

$$\lim_{k \to \infty}\left(1 + \frac{1}{k}\right)^k = e$$

EXAMPLE 2   Using the Ratio Test

Determine whether the series $\displaystyle\sum_{k=1}^{\infty} \frac{k^k}{k!}$ is convergent or divergent.

SOLUTION ■ We have $a_k = \dfrac{k^k}{k!}$, and therefore,

$$a_{k+1} = \frac{(k+1)^{k+1}}{(k+1)!}$$

Then

$$\frac{|a_{k+1}|}{|a_k|} = \frac{\dfrac{(k+1)^{k+1}}{(k+1)!}}{\dfrac{k^k}{k!}}$$

$$= \frac{(k+1)^{k+1}}{(k+1)!} \cdot \frac{k!}{k^k}$$

$$= \frac{k!}{(k+1)!} \cdot \frac{(k+1)^{k+1}}{k^k}$$

$$= \frac{k!}{(k+1)k!} \cdot \frac{(k+1)(k+1)^k}{k^k}$$

$$= \frac{1}{(k+1)} \cdot \frac{(k+1)(k+1)^k}{k^k}$$

$$= \left(\frac{k+1}{k}\right)^k$$

Now take the limit of this ratio, and obtain

$$r = \lim_{k \to \infty} \frac{|a_{k+1}|}{|a_k|} = \lim_{k \to \infty}\left(1 + \frac{1}{k}\right)^k = e$$

Since $r = e \approx 2.71828 > 1$, the series diverges by the ratio test.   ■

E X A M P L E 3   Using the Ratio Test

Using the ratio test, try to determine whether each of the following series is convergent or divergent:

(a) $\displaystyle\sum_{k=1}^{\infty} \frac{1}{k}$ (b) $\displaystyle\sum_{k=1}^{\infty} \frac{1}{k^2}$

S O L U T I O N S   ■

(a) Here $a_k = \dfrac{1}{k}$ and

$$a_{k+1} = \frac{1}{k+1}$$

Then

$$\frac{|a_{k+1}|}{|a_k|} = \frac{\dfrac{1}{k+1}}{\dfrac{1}{k}} = \frac{k}{k+1}$$

Taking the limit of this ratio we obtain

$$r = \lim_{k \to \infty} \frac{|a_{k+1}|}{|a_k|}$$

$$= \lim_{k \to \infty} \frac{k}{k+1}$$

$$= \lim_{k \to \infty} \frac{1}{1 + \dfrac{1}{k}}$$

$$= \frac{1}{1+0} = 1$$

Since $r = 1$, the ratio test yields no conclusion.

(b) Here $a_k = \dfrac{1}{k^2}$ and

$$a_{k+1} = \frac{1}{(k+1)^2}$$

Then

$$\frac{|a_{k+1}|}{|a_k|} = \frac{\dfrac{1}{(k+1)^2}}{\dfrac{1}{k^2}} = \frac{k^2}{(k+1)^2} = \left(\frac{k}{k+1}\right)^2$$

Taking the limit of this ratio we obtain

$$r = \lim_{k \to \infty} \frac{|a_{k+1}|}{|a_k|}$$

$$= \lim_{k \to \infty} \left(\frac{k}{k+1}\right)^2$$

$$= \lim_{k \to \infty} \left(\frac{1}{1+\frac{1}{k}}\right)^2$$

$$= \left(\frac{1}{1+0}\right)^2 = 1$$

Since $r = 1$, the ratio test yields no conclusion. ∎

The first series in Example 3 is the divergent harmonic series. The second series in Example 3 is the convergent p-series with $p = 2$. Thus, the two series in Example 3 illustrate the fact that when the ratio r in the ratio test is $r = 1$, the series may converge or may diverge. Therefore, in this case, we can come to no conclusion.

Alternating Series

An **alternating series** is an infinite series whose terms are alternately positive and negative. Here are three examples.

$$\sum_{k=1}^{\infty} \frac{(-1)^{k-1}}{k} = 1 - \frac{1}{2} + \frac{1}{3} - \frac{1}{4} + \frac{1}{5} - \frac{1}{6} + \cdots -$$

$$\sum_{k=1}^{\infty} \frac{(-1)^{k-1}}{2^{k-1}} = 1 - \frac{1}{2} + \frac{1}{4} - \frac{1}{8} + \frac{1}{16} - \frac{1}{32} + \cdots -$$

$$\sum_{k=1}^{\infty} (-1)^k \frac{k}{k+1} = -\frac{1}{2} + \frac{2}{3} - \frac{3}{4} + \frac{4}{5} - \frac{5}{6} + \cdots -$$

As we will see shortly, the first series converges. The second series is a geometric series with $r = -\frac{1}{2}$, and so it converges. The third series diverges since the nth term does not approach zero.

If the absolute values of the terms of an alternating series decrease and converge to zero, then the series converges. This is called the **alternating series test**.

Alternating Series Test

The alternating series

$$\sum_{k=1}^{\infty} (-1)^{k-1} a_k = a_1 - a_2 + a_3 - a_4 + \cdots$$

converges if

1. $a_k > 0$ for all k.
2. $a_{k+1} \le a_k$ for all k.
3. $\lim_{k \to \infty} a_k = 0$.

A proof is beyond the scope of this text.

EXAMPLE 4   Alternating Series

Determine whether the alternating harmonic series

$$\sum_{k=1}^{\infty} \frac{(-1)^{k-1}}{k} = 1 - \frac{1}{2} + \frac{1}{3} - \frac{1}{4} + \frac{1}{5} - \frac{1}{6} + \cdots$$

converges.

SOLUTION ■ The series is an alternating series $\sum_{k=1}^{\infty} (-1)^{k-1} a_k$, where $a_k = \frac{1}{k}$. Since $\frac{1}{k} > 0$ for all k, $\left\{ \frac{1}{k} \right\}$ is a decreasing sequence and $\lim_{k \to \infty} \frac{1}{k} = 0$, the series converges by the alternating series test.   ■

Absolute Convergence

Given any series $\sum_{k=1}^{\infty} a_k$, we can also consider the series

$$\sum_{k=1}^{\infty} |a_k| = |a_1| + |a_2| + |a_3| + \cdots$$

whose kth term is the absolute value of the kth term of the original series.

Absolute Convergence

We say that the series $\sum_{k=1}^{\infty} a_k$ is **absolutely convergent** if the corresponding series of absolute values

$$\sum_{k=1}^{\infty} |a_k| = |a_1| + |a_2| + |a_3| + \cdots$$

converges.

Remark. If $\sum_{k=1}^{\infty} a_k$ is a series with positive terms, then $|a_k| = a_k$, and absolute convergence is the same as convergence.

EXAMPLE 5   Absolute Convergence

Determine if the series

$$\sum_{k=1}^{\infty} \frac{(-1)^{k-1}}{k^2} = 1 - \frac{1}{4} + \frac{1}{9} - \frac{1}{16} + \cdots$$

is absolutely convergent.

SOLUTION ■ The series

$$\sum_{k=1}^{\infty} \left| \frac{(-1)^{k-1}}{k^2} \right| = \sum_{k=1}^{\infty} \frac{1}{k^2}$$

is a p-series with $p = 2$ and is convergent. Therefore the original series is absolutely convergent.   ■

EXAMPLE 6   Absolute Convergence

Determine if the convergent alternating harmonic series

$$\sum_{k=1}^{\infty} \frac{(-1)^{k-1}}{k} = 1 - \frac{1}{2} + \frac{1}{3} - \frac{1}{4} + \cdots$$

is absolutely convergent.

SOLUTION ■ The series

$$\sum_{k=1}^{\infty} \left| \frac{(-1)^{k-1}}{k} \right| = \sum_{k=1}^{\infty} \frac{1}{k}$$

is the harmonic series and is divergent. Therefore the original series is not absolutely convergent. ■

The alternating harmonic series is convergent but not absolutely convergent. We call such a series **conditionally convergent**.

Conditional Convergence

We say that an infinite series $\sum_{k=1}^{\infty} a_k$ is **conditionally convergent** if the series is convergent but not absolutely convergent.

Although a convergent series need not be absolutely convergent, an absolutely convergent series must be convergent.

Absolute Convergence Implies Convergence

If a series $\sum_{k=1}^{\infty} a_k$ is absolutely convergent, then it is convergent.

A proof is not given.

If the series $\sum_{k=1}^{\infty} a_k$ contains negative terms, we test the series $\sum_{k=1}^{\infty} |a_k|$ of positive terms for convergence. If $\sum_{k=1}^{\infty} |a_k|$ converges, then $\sum_{k=1}^{\infty} a_k$ converges. If $\sum_{k=1}^{\infty} |a_k|$ diverges, however, the series $\sum_{k=1}^{\infty} a_k$ may or may not converge.

EXAMPLE 7 Conditional Convergence

Determine if the alternating series

$$\sum_{k=1}^{\infty} \frac{(-1)^{k-1}}{k^{1/2}} = 1 - \frac{1}{\sqrt{2}} + \frac{1}{\sqrt{3}} - \frac{1}{2} + \cdots$$

is conditionally convergent.

SOLUTION ■ The series is alternating with $a_k = \frac{1}{k^{1/2}}$. Since a_k is positive and decreasing, and has limit zero, this series converges by the alternating series test.

The series

$$\sum_{k=1}^{\infty} \left| \frac{(-1)^1}{k^{1/2}} \right| = \sum_{k=1}^{\infty} \frac{1}{k^{1/2}}$$

is a p-series with $p = \frac{1}{2} < 1$ and is divergent. The original series is conditionally convergent since it is convergent but not absolutely convergent. ■

A **rearrangement** of an infinite series $\sum_{k=1}^{\infty} a_k$ is a series obtained by changing the order of the terms. For example, one rearrangement of $\sum_{k=1}^{\infty} a_k$ is

$$a_2 + a_1 + a_4 + a_3 + \cdots$$

If a series is absolutely convergent with sum s, then it turns out that any rearrangement is convergent with sum s. Thus, an absolutely convergent series behaves like a finite sum.

SELF-HELP EXERCISE SET 10.6

1. Use the ratio test to determine whether the infinite series $\displaystyle\sum_{k=1}^{\infty} \frac{k}{e^k}$ is convergent or divergent.

2. Determine whether the alternating series $\displaystyle\sum_{k=1}^{\infty} (-1)^{k-1} \frac{k}{e^k}$ is convergent or divergent.

3. Determine whether the series $\displaystyle\sum_{k=1}^{\infty} (-1)^{k-1} \frac{k}{e^k}$ converges absolutely. If the series does not converge absolutely, then determine whether it converges conditionally or diverges.

EXERCISE SET 10.6

In Exercises 1 through 20 use the ratio test, where applicable, to determine whether the infinite series is convergent or divergent.

1. $\displaystyle\sum_{k=1}^{\infty} \frac{k}{2^k}$

2. $\displaystyle\sum_{k=1}^{\infty} \frac{k}{5^k}$

3. $\displaystyle\sum_{k=1}^{\infty} \frac{3^k}{k}$

4. $\displaystyle\sum_{k=1}^{\infty} \frac{3^k}{k+3}$

5. $\displaystyle\sum_{k=1}^{\infty} \frac{3^k}{k^2}$

6. $\displaystyle\sum_{k=1}^{\infty} \frac{k!}{k^k}$

7. $\displaystyle\sum_{k=1}^{\infty} \frac{k^k}{3^k k!}$

8. $\displaystyle\sum_{k=1}^{\infty} \frac{k}{e^k}$

9. $\displaystyle\sum_{k=1}^{\infty} \frac{2^k}{k!}$

10. $\displaystyle\sum_{k=1}^{\infty} \frac{2^k}{e^k}$

11. $\displaystyle\sum_{k=1}^{\infty} \frac{k^2+1}{2^k}$

12. $\displaystyle\sum_{k=1}^{\infty} k^2 e^{-k}$

13. $\displaystyle\sum_{k=1}^{\infty} \frac{e^{3k}}{k!}$

14. $\displaystyle\sum_{k=1}^{\infty} \frac{\sqrt{k}}{k!}$

15. $\displaystyle\sum_{k=1}^{\infty} \frac{k!}{10^{5k}}$

16. $\displaystyle\sum_{k=1}^{\infty} \frac{(k+1)^2}{e^k}$

17. $\displaystyle\sum_{k=1}^{\infty} \frac{k^3}{3^k}$

18. $\displaystyle\sum_{k=1}^{\infty} \frac{k^4}{4^k}$

19. $\displaystyle\sum_{k=1}^{\infty} \frac{k! k!}{(2k)!}$

20. $\displaystyle\sum_{k=1}^{\infty} \frac{2^k+5}{3^k}$

In Exercises 21 through 30 determine whether the alternating series is convergent or divergent.

21. $\displaystyle\sum_{k=1}^{\infty} \frac{(-1)^{k-1}}{k+1}$

22. $\displaystyle\sum_{k=1}^{\infty} \frac{(-1)^{k-1}}{k^3}$

23. $\displaystyle\sum_{k=1}^{\infty} \frac{(-1)^{k-1}}{\sqrt{k}}$

24. $\displaystyle\sum_{k=1}^{\infty} \frac{\cos \pi k}{k}$

25. $\displaystyle\sum_{k=1}^{\infty} (-1)^{k-1} e^{-k}$

26. $\displaystyle\sum_{k=1}^{\infty} \frac{(-1)^{k-1}}{\ln(k+2)}$

27. $\displaystyle\sum_{k=1}^{\infty} (-1)^{k-1} \frac{k}{k+1}$

28. $\displaystyle\sum_{k=1}^{\infty} (-1)^{k-1} \frac{k}{\sqrt{k^2+1}}$

29. $\displaystyle\sum_{k=1}^{\infty} (-1)^{k-1} \frac{k^2}{k+1}$

30. $\displaystyle\sum_{k=1}^{\infty} (-1)^{k-1} \frac{k}{\sqrt{k+1}}$

In Exercises 31 through 40 determine whether the series converges absolutely. If the series does not converge absolutely, then determine whether it converges conditionally or diverges.

31. $\displaystyle\sum_{k=1}^{\infty} \frac{(-1)^{k-1}}{k^3}$

32. $\displaystyle\sum_{k=1}^{\infty} \frac{(-1)^{k-1}}{2^k}$

33. $\displaystyle\sum_{k=1}^{\infty} \frac{(-1)^{k-1}}{k!}$

34. $\displaystyle\sum_{k=1}^{\infty} \frac{(-1)^{k-1} 2^k}{k!}$

35. $\displaystyle\sum_{k=1}^{\infty} \frac{(-1)^{k-1}}{\sqrt[3]{k}}$

36. $\displaystyle\sum_{k=1}^{\infty} \frac{(-1)^{k-1}}{\sqrt[5]{k}}$

37. $\displaystyle\sum_{k=1}^{\infty} \frac{(-1)^{k-1} k^2}{(k+1)!}$

38. $\displaystyle\sum_{k=1}^{\infty} \frac{\cos \pi k}{2^k}$

39. $\displaystyle\sum_{k=1}^{\infty} \frac{(-1)^{k-1}}{1+\sqrt{k}}$

40. $\displaystyle\sum_{k=1}^{\infty} \frac{(-1)^{k-1} k!}{3^k}$

Enrichment Exercises

41. (a) Show that if $\sum_{k=1}^{\infty} |a_k|$ converges and $b_k = a_k$ if $a_k \geq 0$ and $b_k = 0$ if $a_k < 0$, then $\sum_{k=1}^{\infty} b_k$ converges.
    (b) Show that if $\sum_{k=1}^{\infty} |a_k|$ converges and $c_k = 0$ if $a_k \geq 0$ and $c_k = a_k$ if $a_k < 0$, then $\sum_{k=1}^{\infty} c_k$ converges.

42. Use the result in Exercise 41 to show that if $\sum_{k=1}^{\infty} a_k$ converges absolutely, then

$$\sum_{k=1}^{\infty} a_k = \sum_{k=1}^{\infty} b_k + \sum_{k=1}^{\infty} c_k$$

where b_k and c_k are defined in Exercise 41.

43. Show that if $\Sigma\, a_k$ is absolutely convergent, then

$$\left| \sum_{k=1}^{\infty} a_k \right| \leq \sum_{k=1}^{\infty} |a_k|$$

44. If $|x| \leq 1$, show that $\displaystyle\lim_{n \to \infty} \frac{x^n}{n} = 0$ for all x. *Hint:* Consider the series $\displaystyle\sum_{n=1}^{\infty} \frac{x^n}{n}$.

45. Suppose $a_k \geq 0$ and $\Sigma_{k=1}^{\infty} a_k$ converges. What can you say about the convergence or divergence of the series $\Sigma_{k=1}^{\infty} a_k \cos n$?

1. Here $a_k = \dfrac{k}{e^k}$. To obtain a_{k+1} we replace $k + 1$ with k in the expression for a_k. This gives

$$a_{k+1} = \frac{k + 1}{e^{k+1}}$$

Then

$$\frac{|a_{k+1}|}{|a_k|} = \frac{\left| \dfrac{k + 1}{e^{k+1}} \right|}{\left| \dfrac{k}{e^k} \right|} = \frac{\dfrac{k + 1}{e^{k+1}}}{\dfrac{k}{e^k}}$$

$$= \cdot \frac{k + 1}{k} \cdot \frac{e^k}{e^{k+1}}$$

$$= \frac{1}{e} \cdot \frac{k + 1}{k} = \frac{1}{e}\left(1 + \frac{1}{k} \right)$$

We then take the limit of this ratio, and obtain

$$r = \lim_{k \to \infty} \frac{|a_{k+1}|}{|a_k|} = \lim_{k \to \infty} \frac{1}{e}\left(1 + \frac{1}{k} \right) = \frac{1}{e}$$

Since $r = \dfrac{1}{e} < 1$, the series converges by the ratio test.

2. The series is an alternating series with $a_k = \dfrac{k}{e^k}$. Notice that $a_k > 0$ for all $k = 1, 2, \dots$. Let $f(x) = xe^{-x}$. Then $f(k) = a_k$ and

$$f'(x) = e^{-x} - xe^{-x} = e^{-x}(1 - x) < 0$$

for $x > 1$. Therefore, $f(x)$ is a decreasing function on $[1, +\infty)$ and a_k is also decreasing. Furthermore, you can check on your graphing calculator that

$$\lim_{x \to \infty} \frac{x}{e^x} = 0$$

Thus,

$$\lim_{k \to \infty} \frac{k}{e^k} = 0$$

and the series converges by the alternating series test.

3. Exercise 1 indicates that this series converges absolutely.

10.7 TAYLOR SERIES

■ Taylor Series ■ Examples

Taylor Series

Given a function $f(x)$ with derivatives up to order n existing at $x = a$, we saw in Section 10.1 how to find a polynomial

$$p_n(x) = a_0 + a_1(x - a) + a_2(x - a)^2 + \cdots + a_n(x - a)^n$$

called the nth-order Taylor polynomial, that satisfies the conditions

$$f^{(k)}(a) = p_n^{(k)}(a), \quad k = 0, 1, 2, \dots, n \tag{1}$$

As we saw, Equation (1) implies that

$$a_k = \frac{f^{(k)}(a)}{k!}, \quad k = 0, 1, 2, \dots, n$$

Thus, the nth-order Taylor polynomial for $f(x)$ at $x = a$ is

$$p_n(x) = f(a) + f'(a)(x - a) + \frac{f''(a)}{2!}(x - a)^2 + \cdots + \frac{f^{(n)}(a)}{n!}(x - a)^n$$

For example, we saw that the nth-order Taylor polynomial for $f(x) = e^x$ at $x = 0$ is

$$p_n(x) = 1 + x + \frac{1}{2!}x^2 + \frac{1}{3!}x^3 + \cdots + \frac{1}{n!}x^n$$

If $f(x)$ is a function with all derivatives existing at $x = a$, then the nth-order Taylor polynomial for $f(x)$ at $x = a$ exists for every n. For the examples we considered in Sections 10.1 and 10.2, increasing the order of the Taylor polynomial $p_n(x)$ increases the accuracy of the approximation of $p_n(x)$ to $f(x)$. It is natural to ask if n increases without bound, does $p_n(x)$ approach $f(x)$? That is, when does $\lim_{n\to\infty} p_n(x) = f(x)$?

In the case that all derivatives of $f(x)$ exist at $x = a$, we use the notation

$$p_n(x) = a_0 + a_1(x - a) + a_2(x - a)^2 + \cdots + a_n(x - a)^n$$

$$= \sum_{k=0}^{\infty} a_k(x - a)^k$$

where

$$a_k = \frac{f^{(k)}(a)}{k!}, \quad k = 0, 1, 2, \ldots, n \tag{2}$$

With this notation, we then ask for what values of x does the limit

$$\lim_{n\to\infty} p_n(x) = \lim_{n\to\infty} \sum_{k=0}^{n} a_k(x - a)^k = f(x)$$

We also write this last limit as

$$\lim_{n\to\infty} \sum_{k=0}^{n} a_k(x - a)^k = \sum_{k=0}^{\infty} a_k(x - a)^k$$

and refer to the expression $\sum_{k=0}^{\infty} a_k(x - a)^k$ as a **Taylor series*** for f at $x = a$, and also write

$$\sum_{k=0}^{\infty} a_k(x - a)^k = a_0 + a_1(x - a) + a_2(x - a)^2 + \cdots + a_k(x - a)^k + \cdots$$

If $\lim_{n\to\infty} p_n(x) = \lim_{n\to\infty} \sum_{k=0}^{n} a_k(x - a)^k = f(x)$ exists, we say the Taylor series **converges** to $f(x)$ at x. If the limit does not exist, we say the series **diverges**.

Taylor Series

If f is a function with derivatives of all orders at $x = a$, then the **Taylor series for f at $x = a$** is

$$\sum_{k=0}^{\infty} a_k(x - a)^k = a_0 + a_1(x - a) + a_2(x - a)^2 + \cdots + a_k(x - a)^k + \cdots$$

where

$$a_k = \frac{f^{(k)}(a)}{k!}, \quad k = 0, 1, 2, \ldots, n$$

The Taylor series **converges at x to $f(x)$** if

$$\lim_{n\to\infty} p_n(x) = \lim_{n\to\infty} \sum_{k=0}^{n} a_k(x - a)^k = f(x)$$

If the limit does not exist, we say the series **diverges**.

If $\lim_{n\to\infty} \sum_{k=1}^{n} a_k(x - a)^k = f(x)$, we say that $\sum_{k=0}^{\infty} a_k(x - a)^k = f(x)$.

Recall from Section 10.2 that we wrote

$$f(x) = p_n(x) + R_n(x)$$

---

*Readers of the sections on infinite series will recognize $p_n(x)$ as the $(n + 1)$st partial sum of the infinite series $\sum_{k=0}^{\infty} a_k(x - a)^k$

We see that $\lim_{n \to \infty} p_n(x) = f(x)$ if and only if $\lim_{n \to \infty} R_n(x) = 0$.

Recall from Section 10.2 that Taylor's theorem gave a bound on $R_n(x)$.

Taylor's Theorem

Suppose that the function f and all its derivatives exist on the interval $[c, d]$ that contains $x = a$. Then for all x in $[c, d]$,

$$f(x) = p_n(x) + R_n(x)$$

with

$$|R_n(x)| \leq M \frac{|x - a|^{n+1}}{(n + 1)!}$$

where

$$\left|f^{(n+1)}(t)\right| \leq M \quad \text{for } t \text{ between } a \text{ and } x$$

In applying Taylor's theorem in many of the examples that follow, we need the fact that

$$\lim_{n \to \infty} \frac{x^n}{n!} = 0 \quad \text{for any real number } x$$

To see why this is true, consider a specific x, say $x = 6.1$. Let $a_n = \dfrac{(6.1)^n}{n!}$. Take n to be some number larger than twice 6.1, say 13. Then

$$
\begin{aligned}
a_{13} &= \frac{(6.1)^{13}}{13!} \\
&= \frac{6.1}{13} \cdot \frac{(6.1)^{12}}{12!} \\
&< (0.5)a_{12}
\end{aligned}
$$

Then

$$
\begin{aligned}
a_{14} &= \frac{(6.1)^{14}}{14!} \\
&= \frac{6.1}{13} \cdot \frac{(6.1)^{13}}{13!} \\
&< (0.5)a_{13} \\
&< (0.5)^2 a_{12}
\end{aligned}
$$

We then see that $a_{15} < (0.5)^3 a_{12}$, and so on. Thus, we see that $a_n \to 0$ as $n \to \infty$.

Examples

EXAMPLE 1   Taylor Series Representation

Find the Taylor series for $f(x) = \cos x$ at $x = 0$ and show this Taylor series converges to $\cos x$ for all x.

SOLUTION ■ We have

$$f(x) = \cos x, \quad \text{and } f(0) = 1$$

$$f'(x) = -\sin x, \quad \text{and } f'(0) = 0$$

$$f''(x) = -\cos x, \quad \text{and} \quad f''(0) = -1$$

$$f^{(3)}(x) = \sin x, \quad \text{and} \quad f^{(3)}(0) = 0$$

$$f^{(4)}(x) = \cos x, \quad \text{and} \quad f^{(4)}(0) = 1$$

Since $f^{(4)}(x) = f(x)$, the numbers $f^{(k)}(0)$ repeat themselves. We see that $f(x) = \cos x$ has derivatives of all orders at $x = 0$ and the Taylor series for $f(x) = \cos x$ at $x = 0$ is

$$1 - \frac{1}{2}x^2 + \frac{1}{4!}x^4 - \frac{1}{6!}x^6 + \cdots = \sum_{k=0}^{\infty} \frac{(-1)^k}{(2k)!}x^{2k}$$

Notice the derivatives of $\cos x$ are $-\sin x$, $-\cos x$, $\sin x$, or $\cos x$. But the absolute values of all of these functions are bounded by 1. Thus, $|f^{(n+1)}(t)| \leq 1$ for all t, and we can take $M = 1$ in Taylor's theorem. Then

$$|R_n(x)| \leq M \frac{|x - a|^{n+1}}{(n + 1)!}$$

$$\leq \frac{|x|^{n+1}}{(n + 1)!}$$

But since $\lim\limits_{n \to \infty} \frac{|x|^n}{n!} = 0$, $\lim\limits_{n \to \infty} R_n(x) = 0$ and $\lim\limits_{n \to \infty} p_n(x) = f(x)$. This last limit is the

same as saying $\cos x = \sum_{k=0}^{\infty} \frac{(-1)^k}{(2k)!}x^{2k}$. This holds for any x. ∎

EXPLORATION *1*

Convergence of Taylor Series

 Graph $y_1 = \cos x$ and $\sum_{k=1}^{n} \frac{(-1)^k}{(2k)!}x^{2k}$, for various values of n on a screen with dimensions $[-10, 10]$ by $[-2, 2]$. Does it appear that the Taylor polynomials are converging to $\cos x$?

E X A M P L E 2   Convergence of Taylor Series

Find the Taylor series for $f(x) = e^x$ at $x = 0$ and show this Taylor series converges to e^x for all x.

S O L U T I O N ∎ We noted at the beginning of this section that the nth-order Taylor polynomial for $f(x) = e^x$ at $x = 0$ is

$$p_n(x) = 1 + x + \frac{1}{2!}x^2 + \frac{1}{3!}x^3 + \cdots + \frac{1}{n!}x^n$$

Since e^x has derivatives of all orders at $x = 0$, we have that the Taylor series for e^x at $x = 0$ is

$$1 + x + \frac{1}{2!}x^2 + \frac{1}{3!}x^3 + \cdots + \frac{1}{k!}x^k + \cdots = \sum_{k=0}^{\infty} \frac{1}{k!}x^k$$

Here $a_k = \dfrac{1}{k!}$.

Let x be any number. To show that $e^x = \sum_{k=0}^{\infty} \frac{1}{k!}x^k$ we must show that $\lim\limits_{n \to \infty} R_n(x)$ $= 0$. All the derivatives of e^t are again e^t. Thus, $f^{(n+1)}(t) = e^t$. Since e^t is an increasing function, $|f^{(n+1)}(t)| = e^t \leq e^{|x|}$ for all t between 0 and x. Thus, in Taylor's theorem, we have $M = e^{|x|}$, and

$$|R_n(x)| \le M \frac{|x - a|^{n+1}}{(n + 1)!}$$

$$\le e^{|x|} \frac{|x|^{n+1}}{(n + 1)!}$$

But since $\lim\limits_{n \to \infty} \dfrac{|x|^n}{n!} = 0$, $\lim\limits_{n \to \infty} R_n(x) = 0$ and $\lim\limits_{n \to \infty} p_n(x) = f(x)$. This last limit is the

same as saying $e^x = \sum\limits_{k=0}^{\infty} \dfrac{1}{k!} x^k$. This holds for any x. ■

EXPLORATION 2

Convergence of Taylor Series

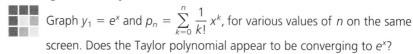 Graph $y_1 = e^x$ and $p_n = \sum\limits_{k=0}^{n} \dfrac{1}{k!} x^k$, for various values of n on the same screen. Does the Taylor polynomial appear to be converging to e^x?

EXAMPLE 3   Taylor Series Representation

Find the Taylor series for $f(x) = \sin x$ at $x = \pi/4$ and show this Taylor series converges to $\sin x$ for all x.

SOLUTION ■ We have

$$f(x) = \sin x, \quad \text{and} \quad f\left(\frac{\pi}{4}\right) = \frac{\sqrt{2}}{2}$$

$$f'(x) = \cos x, \quad \text{and} \quad f'\left(\frac{\pi}{4}\right) = \frac{\sqrt{2}}{2}$$

$$f''(x) = -\sin x, \quad \text{and} \quad f''\left(\frac{\pi}{4}\right) = \frac{-\sqrt{2}}{2}$$

$$f^{(3)}(x) = -\cos x, \quad \text{and} \quad f^{(3)}\left(\frac{\pi}{4}\right) = \frac{-\sqrt{2}}{2}$$

$$f^{(4)}(x) = \sin x, \quad \text{and} \quad f^{(4)}\left(\frac{\pi}{4}\right) = \frac{\sqrt{2}}{2}$$

Since $f^{(4)}(x) = f(x)$, the numbers $f^{(k)}(\pi/4)$ repeat themselves. We see that $f(x) = \sin x$ has derivatives of all orders at $x = \pi/4$ and the Taylor series for $f(x) = \sin x$ at $x = \pi/4$ is

$$\frac{\sqrt{2}}{2} + \frac{\sqrt{2}}{2}\left(x - \frac{\pi}{4}\right) - \frac{\sqrt{2}}{2 \cdot 2}\left(x - \frac{\pi}{4}\right)^2 - \frac{\sqrt{2}}{2 \cdot 3!}\left(x - \frac{\pi}{4}\right)^3$$

$$+ \frac{\sqrt{2}}{2 \cdot 4!}\left(x - \frac{\pi}{4}\right)^4 + \cdots = \sum_{k=0}^{\infty} \frac{\sqrt{2}}{2 \cdot k!}(-1)^{k(k-1)/2}\left(x - \frac{\pi}{4}\right)^k$$

Notice the derivatives of $\sin x$ are either $\cos x$, $-\sin x$, $-\cos x$, or $\sin x$. But the absolute value of all of these functions are bounded by 1. Thus, $|f^{(n+1)}(t)| \le 1$ for all t, and we can take $M = 1$ in Taylor's theorem. Then

$$|R_n(x)| \le M \frac{|x - a|^{n+1}}{(n + 1)!}$$

$$\le \frac{|x|^{n+1}}{(n + 1)!}$$

But since $\lim\limits_{n \to \infty} \dfrac{|x|^n}{n!} = 0$, $\lim\limits_{n \to \infty} R_n(x) = 0$ and $\lim\limits_{n \to \infty} p_n(x) = f(x)$. This last limit is the

same as saying $\sin x = \sum\limits_{k=0}^{\infty} \dfrac{\sqrt{2}}{2 \cdot k!}(-1)^{k(k-1)/2}\left(x - \frac{\pi}{4}\right)^k$. This holds for any x. ■

EXPLORATION *3*

Convergence of Taylor Series

Graph $y_1 = \sin x$ and $\displaystyle\sum_{k=0}^{n} \frac{\sqrt{2}}{2 \cdot k!} (-1)^{k(k-1)/2} \left(x - \frac{\pi}{4}\right)^k$ for various values of n on a screen with dimensions $[-10, 10]$ by $[-2, 2]$. Does it appear that the Taylor polynomials are converging to $\sin x$?

EXAMPLE 4   Convergence of Taylor Series

Find the Taylor series for $f(x) = \dfrac{1}{1 - x}$ at $x = 0$ and show this Taylor series converges to $\dfrac{1}{1 - x}$ for all $|x| < 0.5$.

S O L U T I O N S ■ We first need to find the derivatives of $f(x)$ at $x = 0$. We have

$$f(x) = (1 - x)^{-1}, \qquad \text{and} \quad f(0) = 1$$
$$f'(x) = (1 - x)^{-2}, \qquad \text{and} \quad f'(0) = 1$$
$$f''(x) = 2(1 - x)^{-3}, \qquad \text{and} \quad f''(0) = 2$$
$$f^{(3)}(x) = 3!(1 - x)^{-4}, \qquad \text{and} \quad f^{(3)}(0) = 3!$$
$$\vdots$$
$$f^{(k)}(x) = k!(1 - x)^{-(k+1)}, \quad \text{and} \quad f^{(k)}(0) = k!$$

Thus, f has derivatives of all orders at $x = 0$ and

$$a_k = \frac{f^{(k)}(0)}{k!} = \frac{k!}{k!} = 1$$

The Taylor series is

$$1 + x + x^2 + x^3 + \cdots + x^k + \cdots = \sum_{k=0}^{\infty} x^k$$

As usual we must find a bound on $f^{(n+1)}(t)$, which is $(n + 1)!(1 - t)^{-(n+2)}$. Let x be any point in $(-0.5, 0.5)$. Notice that $f^{(n+1)}(t)$ is positive and increasing on $(-0.5, 0.5)$. Therefore, the maximum of $f^{(n+1)}(t)$ for t between 0 and x is at most $f^{(n+1)}(|x|) = (n + 1)!(1 - |x|)^{-(n+2)}$. We use this bound in Taylor's theorem and obtain

$$|R_n(x)| \leq M \frac{|x - a|^{n+1}}{(n + 1)!}$$

$$\leq \frac{(n + 1)!}{(1 - |x|)^{n+2}} \cdot \frac{|x|^{n+1}}{(n + 1)!}$$

$$\leq \frac{1}{1 - |x|} \cdot \left(\frac{|x|}{1 - |x|}\right)^{n+1}$$

$$\leq \frac{1}{1 - |x|} \cdot \left(\frac{1}{\frac{1}{|x|} - 1}\right)^{n+1}$$

But since $|x| < 0.5$, $\left(\dfrac{1}{\frac{1}{|x|} - 1}\right)$ is less than 1. Thus, $\left(\dfrac{1}{\frac{1}{|x|} - 1}\right)^{n+1} \to 0$ as $n \to \infty$,

and also $R_n(x) \to 0$. Therefore, for $|x| < 0.5$, $\dfrac{1}{1 - x} = \displaystyle\sum_{k=0}^{\infty} x^k$. ■

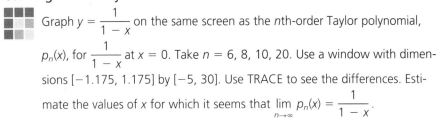

EXPLORATION **4**

Convergence of Taylor Series

Graph $y = \dfrac{1}{1 - x}$ on the same screen as the nth-order Taylor polynomial,

$p_n(x)$, for $\dfrac{1}{1 - x}$ at $x = 0$. Take $n = 6, 8, 10, 20$. Use a window with dimensions $[-1.175, 1.175]$ by $[-5, 30]$. Use TRACE to see the differences. Estimate the values of x for which it seems that $\displaystyle\lim_{n \to \infty} p_n(x) = \dfrac{1}{1 - x}$.

If you do Exploration 4 you will conclude that the Taylor polynomials $p_n(x)$ appear to be converging to $\dfrac{1}{1 - x}$ on the interval $(-1, 1)$. This is in fact the case. This can be established using a more general form of the error $R_n(x)$ than indicated in Taylor's theorem. We do not pursue this here. Readers of Section 10.4 will recognize the series $\sum_{k=0}^{\infty} x^k$ as a geometric series that converges to $\dfrac{1}{1 - x}$ for $|x| < 1$.

SELF-HELP EXERCISE SET 10.7

1. Find the Taylor series for $f(x) = \dfrac{1}{2 - x}$ at $x = 0$, and show that this Taylor series converges to $\dfrac{1}{2 - x}$ for all $|x| < 1$.

Using your graphing calculator to graph Taylor polynomials, estimate the values of x for which the Taylor series converges to $f(x)$.

EXERCISES 10.7

In Exercises 1 through 10 find the Taylor series for $f(x)$ at $x = a$, and show that the Taylor series for $f(x)$ at $x = a$ converges to $f(x)$ for all real numbers x.

1. $f(x) = e^{3x}, \quad a = 0$

2. $f(x) = e^{-x}, \quad a = 0$

3. $f(x) = \sin x, \quad a = 0$

4. $f(x) = \cos 2x, \quad a = 0$

5. $f(x) = \cos 3x, \quad a = 0$

6. $f(x) = \sin 3x, \quad a = 0$

7. $f(x) = \sin x, \quad a = \dfrac{\pi}{2}$

8. $f(x) = \cos x, \quad a = \dfrac{\pi}{2}$

9. $f(x) = \cos x, \quad a = \pi$

10. $f(x) = \sin x, \quad a = \pi$

In Exercises 11 through 20 find the Taylor series for $f(x)$ at $x = a$. Also use your graphing calculator to graph the Taylor polynomials $p_n(x)$ for various values of n, and estimate an interval $(a - c, a + c)$ on which the Taylor series for $f(x)$ at $x = a$ converges to $f(x)$.

11. $f(x) = \dfrac{1}{4 - x}, \quad x = 0$

12. $f(x) = \dfrac{1}{2 + 3x}, \quad x = 0$

13. $f(x) = \ln(1 + 2x), \quad x = 0$

14. $f(x) = \ln(2 + 3x), \quad x = 0$

15. $f(x) = \dfrac{1}{x}, \quad x = 1$

16. $f(x) = \dfrac{1}{x}, \quad x = 2$

17. $f(x) = \ln x, \quad x = 1$

18. $f(x) = \ln x, \quad x = 2$

19. $f(x) = \ln(x + 1), \quad x = 1$

20. $f(x) = \ln(x + 1), \quad x = 2$

In Exercises 21 through 24, find the Taylor series for $f(x)$ at $x = 0$. Use Taylor's theorem to determine some interval on which the Taylor series for $f(x)$ at $x = 0$ converges to $f(x)$.

21. $f(x) = \dfrac{1}{6 - x}$

22. $f(x) = \dfrac{1}{1 + x}$

23. $f(x) = \ln(1 - x)$

24. $f(x) = \ln(1 + x)$

Enrichment Exercises

25. Suppose $f(x)$ has the Taylor series at $x = 0$ given by

$$x + \frac{1}{3^2} x^2 + \frac{1}{3^3} x^3 + \frac{1}{3^4} x^4 + \cdots + \frac{1}{3^n} x^n + \cdots$$

Find $f'(0), f''(0), f^{(5)}(0)$.

26. The Taylor series for $f(x) = x^2 \sin x^2$ at $x = 0$ is

$$x^4 - \frac{1}{3!} x^8 + \frac{1}{5x} x^{12} + \cdots$$

Find $f''(0), f^{(8)}(0)$.

27. Is it true that the graph of every Taylor polynomial at $x = 0$ of $\sin x$ must eventually move away from the graph of $y = \sin x$ as $|x|$ becomes large? You may first wish to use your graphing calculator to suggest an answer. Explain why or why not.

28. **(a)** Take the derivative, term by term, of the Taylor series for $\cos x$ at $x = 0$ found in Example 1 of the text. Compare your answer to the Taylor series found in Exercise 3 for $\sin x$.

(b) Take the derivative, term by term, of the Taylor series for $\sin x$ at $x = 0$ found in Exercise 3. Compare your answer to the Taylor series found in Example 1 of the text for $\cos x$.

(c) Take the derivative, term by term, of the Taylor series

for e^x at $x = 0$ found in Example 2 of the text. What do you obtain?

(d) In view of parts (a), (b), and (c) what do you speculate about how a Taylor series can be differentiated?

29. **(a)** The Taylor series for $f(x) = e^x$ at $x = 0$ was found in Example 2. Substitute $3x$ for x in this series, and compare the result with the Taylor series for $f(x) = e^{3x}$ that you found in Exercise 1.

(b) The Taylor series for $f(x) = \cos x$ at $x = 0$ was found in Example 1. Substitute $3x$ for x in this series, and compare the result with the Taylor series for $f(x) = \cos 3x$ that you found in Exercise 5.

(c) In view of parts (a) and (b) what do you speculate of how a Taylor series might be found by substitution?

SOLUTIONS TO SELF-HELP EXERCISE SET 10.7

We first need to find the derivatives of $f(x)$ at $x = 0$. We have

$$f(x) = (2 - x)^{-1}, \qquad \text{and} \quad f(0) = \frac{1}{2}$$

$$f'(x) = (2 - x)^{-2}, \qquad \text{and} \quad f'(0) = \frac{1}{2^2}$$

$$f''(x) = 2(2 - x)^{-3}, \qquad \text{and} \quad f''(0) = \frac{2}{2^3}$$

$$f^{(3)}(x) = 3!(2 - x)^{-4}, \qquad \text{and} \quad f^{(3)}(0) = \frac{3!}{2^4}$$

$$\vdots$$

$$f^{(k)}(x) = k!(2 - x)^{-(k+1)}, \quad \text{and} \quad f^{(k)}(0) = \frac{k!}{2^{k+1}}$$

Thus, f has derivatives of all orders at $x = 0$ and

$$a_k = \frac{f^{(k)}(0)}{(k!)} = \frac{k!}{(k!)2^{k+1}} = \frac{1}{2^{k+1}}$$

The Taylor series is

$$\frac{1}{2} + \frac{1}{4}x + \frac{1}{8}x^2 + \cdots + \frac{1}{2^{k+1}}x^k + \cdots = \sum_{k=0}^{\infty} 2^{-(k+1)}x^k$$

As usual we must find a bound on $f^{(n+1)}(t)$, which is $(n + 1)!(2 - t)^{-(n+2)}$. Let x be any point in $(-1, 1)$. Notice that $f^{(n+1)}(t)$ is positive and increasing on $(-1, 1)$. Therefore, the

maximum of $f^{(n+1)}(t)$ for t bewteen 0 and x is at most $f^{(n+1)}(|x|)$ $= (n + 1)!(2 - |x|)^{-(n+2)}$. We use this bound in Taylor's theorem and obtain

$$|R_n(x)| \le M\frac{|x - a|^{n+1}}{(n + 1)!}$$

$$\le \frac{(n + 1)!}{(2 - |x|)^{n+2}} \cdot \frac{|x|^{n+1}}{(n + 1)!}$$

$$\le \frac{1}{2 - |x|} \cdot \left(\frac{|x|}{2 - |x|}\right)^{n+1}$$

$$\le \frac{1}{2 - |x|} \cdot \left(\frac{1}{\frac{2}{|x|} - 1}\right)^{n+1}$$

But since $|x| < 1$, $\left(\dfrac{1}{\frac{2}{|x|} - 1}\right)$ is less than 1. Thus,

$\left(\dfrac{1}{\frac{2}{|x|} - 1}\right)^{n+1} \to 0$ as $n \to \infty$, and also $R_n(x) \to 0$. Therefore,

for $|x| < 1$, $\dfrac{1}{2 - x} = \displaystyle\sum_{k=0}^{\infty} 2^{-(k+1)}x^k$.

This Taylor series actually converges to $f(x)$ on the wider interval $(-2, 2)$, as experimenting on a graphing calculator indicates, but the proof requires additional analysis of the remainder $R_n(x)$ and will not be pursued here.

CHAPTER 10

SUMMARY OUTLINE

■ The **nth-order Taylor polynomial for the function $f(x)$ at $x = a$** is

$$p_n(x) = f(a) + f'(a)(x - a) + \frac{f''(a)}{2}(x - a)^2 + \cdots + \frac{f^{(n)}(a)}{n!}(x - a)^n$$

provided f has n derivatives at $x = a$. p. 517.

■ **Taylor's Theorem.** Suppose that the function f and its first $n + 1$ derivatives are continuous on the interval $[c, d]$ that contains $x = a$. Then for all x in $[c, d]$,

$$f(x) = f(a) + f'(a)(x - a) + \frac{f''(a)}{2!}(x - a)^2 + \cdots + \frac{f^{(n)}(a)}{n!}(x - a)^n + R_n(x)$$

with $|R_n(x)| \le \dfrac{M}{(n + 1)!}|x - a|^{n+1}$, where $|f^{(n+1)}(t)| \le M$ for t between a and x.

p. 523.

- An **infinite sequence** of numbers is a function whose domain is the set of positive integers. If f is the defining function, we write $f(n) = a_n$ and call a_n the **nth term**, or **general term**. The sequence is denoted by $\{a_1, a_2, a_3, \ldots\}$ or $\{a_n\}$. p. 530.

- **Limit of a Sequence.** We say that a sequence $\{a_n\}$ has **limit** L, and write $\lim_{n\to\infty} a_n = L$, if a_n approaches L as n becomes large without bound. In this case we also say that the sequence **converges** to L. If the limit does not exist, we say that the sequence **diverges**. p. 531.

- **Limit Properties of Sequences** p. 532
  Assume that c is any number, $\lim_{n\to\infty} a_n = A$, and $\lim_{n\to\infty} b_n = B$. Then

  Rule 1. $\lim_{n\to\infty} ca_n = c \lim_{n\to\infty} a_n = cA$

  Rule 2. $\lim_{n\to\infty} (a_n \pm b_n) = \lim_{n\to\infty} a_n \pm \lim_{n\to\infty} b_n = A \pm B$

  Rule 3. $\lim_{n\to\infty}(a_n \cdot b_n) = \left(\lim_{n\to\infty} a_n\right) \cdot \left(\lim_{n\to\infty} b_n\right) = A \cdot B$

  Rule 4. $\lim_{n\to\infty} \dfrac{a_n}{b_n} = \dfrac{\lim_{n\to\infty} a_n}{\lim_{n\to\infty} b_n} = \dfrac{A}{B}$ if $\lim_{n\to\infty} b_n = B \neq 0$.

  Rule 5. $\lim_{n\to\infty}(a_n)^k = A^k$, if k is any real number and A^k is defined and $A \neq 0$.

- $\lim_{n\to\infty} r^n = 0$ if $|r| < 1$ p. 534

- $\{r^n\}$ diverges if $|r| > 1$ p. 534

- An expression of the form $a_1 + a_2 + a_3 + \cdots + a_k + \cdots$ is called an **infinite series** and is also written as $\sum_{k=1}^{\infty} a_k$. p. 536.

- **Sum of an Infinite Series**
  Given an infinite series $\sum_{k=1}^{\infty} a_k = a_1 + a_2 + a_3 + \cdots + a_k + \cdots$ we define the **partial sums** S_n by $S_n = \sum_{k=1}^{n} a_k = a_1 + a_2 + a_3 + \cdots + a_n$. If the sequence of partial sums $\{S_n\}$ converges to S, that is, if $\lim_{n\to\infty} S_n = S$, then we say that the infinite series $\sum_{k=1}^{\infty} a_k$ **converges** to the sum S and write $\sum_{k=1}^{\infty} a_k = S$. Otherwise we say that the infinite series **diverges**. p. 537.

- A geometric series is an infinite series of the form

$$\sum_{k=1}^{\infty} ar^{k-1} = a + ar + ar^2 + ar^3 + \cdots + ar^{k-1} + \cdots$$

It is convergent if $|r| < 1$ with sum $\sum_{k=1}^{\infty} ar^{k-1} = \dfrac{a}{1-r}$. If $|r| \geq 1$ and $a \neq 0$, then the geometric series diverges. p. 539.

- **Properties of Infinite Series**
  Let $\sum_{k=1}^{\infty} a_k$ and $\sum_{k=1}^{\infty} b_k$ be convergent infinite series and c any real number. Then

  1. $\sum_{k=1}^{\infty} ca_k = c \sum_{k=1}^{\infty} a_k$
  2. $\sum_{k=1}^{\infty}(a_k \pm b_k) = \sum_{k=1}^{\infty} a_k \pm \sum_{k=1}^{\infty} b_k$ p. 541.

- **Divergence Test for Infinite Series.** If $\lim_{k\to\infty} a_k \neq 0$, then the infinite series $\sum_{k=1}^{\infty} a_k$ diverges. p. 542.

- **Integral Test.** Let a_k be a positive, decreasing sequence and let $f(x)$ be positive, continuous, and decreasing on the interval $[1, \infty)$ with $f(k) = a_k$. Then the infinite series $\sum_{k=1}^{\infty} a_k$ and the improper integral $\int_1^{\infty} f(x)\, dx$ both converge or both diverge. That is

  (a) If $\int_1^{\infty} f(x)\, dx$ is convergent, then $\sum_{k=1}^{\infty} a_k$ is convergent.
  (b) If $\int_1^{\infty} f(x)\, dx$ is divergent, then $\sum_{k=1}^{\infty} a_k$ is divergent. p. 548.

- The series $\displaystyle\sum_{k=1}^{\infty} \frac{1}{k}$, is called the **harmonic series.** p. 549.

- The **p-series** is a series of the form $\displaystyle\sum_{k=1}^{\infty} \frac{1}{k^p}$, where p is a positive constant. The harmonic series is a p-series with $p = 1$. p. 549.

- **Convergence of p-Series.** The p-series $\displaystyle\sum_{k=1}^{\infty} \frac{1}{k^p} = 1 + \frac{1}{2^p} + \frac{1}{3^p} + \frac{1}{4^p} + \cdots$ converges if $p > 1$ and diverges if $p \leq 1$. p. 550.

- **The Comparison Test.** Suppose $\Sigma_{k=1}^{\infty}\, a_k$ and $\Sigma_{k=1}^{\infty}\, b_k$ are two infinite series with positive terms and that $0 < a_k \leq b_k$ for all k. Then

  (a) If $\Sigma_{k=1}^{\infty}\, b_k$ converges, $\Sigma_{k=1}^{\infty}\, a_k$ converges and $\Sigma_{k=1}^{\infty}\, a_k \leq \Sigma_{k=1}^{\infty}\, b_k$.
  (b) If $\Sigma_{k=1}^{\infty}\, a_k$ diverges, $\Sigma_{k=1}^{\infty}\, b_k$ diverges. p. 550.

- **The Limit Comparison Test.** Let $\Sigma_{k=1}^{\infty}\, a_k$ and $\Sigma_{k=1}^{\infty}\, b_k$ be two series with positive terms. If $\displaystyle\lim_{k\to\infty} \frac{a_k}{b_k} = \rho$ exists and $\rho \neq 0$, then either both series converge or both series diverge. p. 551.

- **The Ratio Test.** Let $\Sigma_{k=1}^{\infty}\, a_k$ be an infinite series of nonzero terms, and suppose that $\displaystyle\lim_{k\to\infty} \frac{|a_{k+1}|}{|a_k|} = r$.

  1. If $0 \leq r < 1$, then the series converges.
  2. If $r > 1$, then the series diverges.
  3. If $r = 1$, the test provides no information. (The series may converge or may diverge.) p. 555.

- An **alternating series** is an infinite series whose terms are alternately positive and negative. p. 558.

- **Alternating Series Test.** The alternating series

$$\sum_{k=1}^{\infty} (-1)^{k-1} a_k = a_1 - a_2 + a_3 - a_4 + \cdots$$

converges if $a_k > 0$ for all k, $a_{k+1} \leq a_k$ for all k, and $\displaystyle\lim_{k\to\infty} a_k = 0$. p. 558.

- We say that the series $\Sigma_{k=1}^{\infty}\, a_k$ is **absolutely convergent** if the corresponding series of absolute values $\Sigma_{k=1}^{\infty}\, |a_k| = |a_1| + |a_2| + |a_3| + \cdots$ converges. p. 559.

- **Conditional Convergence.** We say that an infinite series $\Sigma_{k=1}^{\infty}\, a_k$ is **conditionally convergent** if the series is convergent but not absolutely convergent. p. 560.

- If a series $\Sigma_{k=1}^{\infty}$ is absolutely convergent, then it is convergent. p. 560.

- If f is a function with derivatives of all orders at $x = a$, then the **Taylor series for f at $x = a$** is

$$\sum_{k=0}^{\infty} a_k(x - a)^k = a_0 + a_1(x - a) + a_2(x - a)^2 + \cdots + a_k(x - a)^k + \cdots$$

where $a_k = \dfrac{f^{(k)}(a)}{k!}, \quad k = 0, 1, 2, \ldots, n$.

  The Taylor series **converges at x** to $f(x)$ if $\displaystyle\lim_{n\to\infty} p_n(x) = \lim_{n\to\infty} \Sigma_{k=0}^{n}\, a_k(x - a)^k = f(x)$ and **diverges at $x = a$** if this limit does not exist. p. 563.

CHAPTER **10**

REVIEW EXERCISES

1. Find the fourth-order Taylor polynomial for $f(x) = e^{-2x}$ at $x = 0$.

2. Find the third-order Taylor polynomial for $f(x) = \sqrt[4]{x}$ at $x = 1$.

3. Find the fourth-order Taylor polynomial for $f(x) = \ln 5x$ at $x = 1$.

4. Find the third-order Taylor polynomial for $f(x) = 1/x^2$ at $x = 1$.

5. Find the nth-order Taylor polynomial for $f(x) = e^{-3x}$ at $x = 0$.

6. Use the fourth-order Taylor polynomial for $f(x) = \sqrt[4]{x}$ at $x = 1$ to approximate $\sqrt[4]{1.2}$.

7. Use the fourth-order Taylor polynomial for $f(x) = \ln x$ at $x = 1$ to approximate $\ln 1.2$.

8. Find a bound on the error using the Taylor polynomial $p_2(x)$ for $f(x) = \sqrt{x + 4}$ at $x = 0$ on the interval $[0, 0.5]$.

9. Find a bound on the error using the Taylor polynomial $p_4(x)$ for $f(x) = \ln(x + 1)$ at $x = 0$ on the interval $[0, 0.4]$.

10. Find n so that the nth-order Taylor polynomial for $f(x) = 1/x^2$ at $x = 1$ has an error of at most 0.001 on the interval $[1, 1.5]$.

11. Write out the first five terms of the sequence $\{2^n\}$.

12. Find the general term of the sequence $\{2, -4, 8, -16, \ldots\}$

13. Sketch a graph of the sequence $\left\{1 + \dfrac{(-1)^n}{n}\right\}$.

In Exercises 14 through 16 determine whether the sequence converges or diverges. If the sequence converges, find its limit.

14. $\left\{\dfrac{4n^3 - 3}{2n^3 + n^2 + 1}\right\}$ 15. $\{(0.998)^n\}$

16. $\{(-1.003)^n\}$

In Exercises 17 through 22 determine if the infinite series converges or diverges. If the series converges, find its sum.

17. $\displaystyle\sum_{k=1}^{\infty} \dfrac{1}{(0.993)^{k-1}}$ 18. $\displaystyle\sum_{k=1}^{\infty} \dfrac{1}{(1.002)^{k-1}}$

19. $\displaystyle\sum_{k=1}^{\infty} \dfrac{2^{k+1}}{3^{k-1}}$ 20. $\displaystyle\sum_{k=1}^{\infty} \dfrac{\sqrt{k^2 + 4}}{k}$

21. $\displaystyle\sum_{k=1}^{\infty} \dfrac{2 + 3^{k-1}}{5^{k-1}}$ 22. $\displaystyle\sum_{k=2}^{\infty} \dfrac{1}{7^{k-1}}$

23. Use the integral test to determine whether the series $\displaystyle\sum_{k=1}^{\infty} \dfrac{2}{k^{4/3}}$ converges or diverges.

24. Use the integral test to determine whether the series $\displaystyle\sum_{k=1}^{\infty} \dfrac{2}{k^{2/3}}$ converges or diverges.

25. Use the comparison test to determine whether the series $\displaystyle\sum_{k=2}^{\infty} \dfrac{1}{\sqrt{k^3 + 1}}$ converges or diverges.

26. Use the comparison test to determine whether the series $\displaystyle\sum_{k=2}^{\infty} \dfrac{1}{\sqrt[3]{k^2 - 1}}$ converges or diverges.

In Exercises 27 through 30 determine whether the infinite series converges or diverges.

27. $\displaystyle\sum_{k=2}^{\infty} \dfrac{1}{\sqrt{k^3 - 2}}$ 28. $\displaystyle\sum_{k=1}^{\infty} \dfrac{1}{\sqrt[3]{k^2 + 3}}$

29. $\displaystyle\sum_{k=1}^{\infty} \cos\dfrac{1}{k}$ 30. $\displaystyle\sum_{k=1}^{\infty} \dfrac{k + 2}{k3^k}$

In Exercises 31 through 34 use the ratio test to determine whether the infinite series converges or diverges.

31. $\displaystyle\sum_{k=1}^{\infty} \dfrac{k}{9^k}$ 32. $\displaystyle\sum_{k=1}^{\infty} \dfrac{5^k}{k^2}$

33. $\displaystyle\sum_{k=1}^{\infty} \dfrac{5^k}{k!}$ 34. $\displaystyle\sum_{k=1}^{\infty} (k!)e^{-k}$

35. Determine whether the alternating series $\displaystyle\sum_{k=1}^{\infty} \dfrac{(-1)^k}{e^k}$ converges or diverges.

36. Determine whether the alternating series $\displaystyle\sum_{k=1}^{\infty} \dfrac{(-1)^k k}{k + 1}$ converges or diverges.

37. Determine whether the series $\displaystyle\sum_{k=1}^{\infty} \dfrac{(-1)^k k}{k^2 + 1}$ converges absolutely. If the series does not converge absolutely, then determine whether it converges conditionally or diverges.

38. Determine whether the series $\displaystyle\sum_{k=1}^{\infty} \dfrac{(-1)^k k}{k^3 + 1}$ converges absolutely. If the series does not converge absolutely, then determine whether it converges conditionally or diverges.

In Exercises 39 through 42 find the Taylor series for the given function at the indicated value of x.

39. $f(x) = e^{-5x}, \quad x = 0$ 40. $f(x) = \ln 2x, \quad x = 1$

41. $f(x) = \dfrac{1}{5 - x}, \quad x = 0$ 42. $f(x) = \dfrac{1}{5 - 2x}, \quad x = 2$

43. Show that the Taylor series found in Exercise 39 converges to e^{-5x} for all real numbers x.

44. Show that the Taylor series found in Exercise 41 converges to $\dfrac{1}{5 - x}$ for $|x| < 2.5$

45. **Population Growth.** In the nth year after placing 10 moose in a forest where there were none before, the population is

estimated to be $P_n = \dfrac{500}{5 + 45e^{-0.50n}}$. What happens to this population in the long-term?

46. Multiplier Effect. A city is attempting to attract a convention to use its civic center. It believes that for every dollar spent by the people attending the convention and that ends up in the hands of local businesses, these local businesses spend between 70% and 90% locally, and between 70% and 90% of this locally spent money is in turn spent locally, and so on. Find an upper and lower bound for the effect that one dollar has if this goes on indefinitely.

CHAPTER **10**

PROJECTS

Project 1
Taylor Series and Differential Equations

A very important application of Taylor series is found in differential equations. For example, to find an infinite series solution to the second-order differential equation

$$\frac{d^2y}{dx^2} - y = 0, \quad y(0) = 1, \, y'(0) = 1 \qquad [1]$$

we assume a solution in the form of a Taylor series

$$y = a_0 + a_1x + a_2x^2 + a_3x^3 + a_4x^4 + \cdots$$

Recall that for a Taylor series we must have $y(0) = a_0$ and $y'(0) = a_1$. Thus, from Equation (1), we have $a_0 = 1 = a_1$. Therefore,

$$y = 1 + x + a_2x^2 + a_3x^3 + a_4x^4 + \cdots$$

Differentiating, we obtain

$$y' = 1 + 2a_2x + 3a_3x^2 + 4a_4x^3 + \cdots$$
$$y'' = 2a_2 + 6a_3x + 12a_4x^2 + \cdots$$

Then Equation (1) becomes

$$\begin{aligned} 0 &= y'' - y \\ &= (2a_2 + 6a_3x + 12a_4x^2 + \cdots) - (1 + x + a_2x^2 + a_3x^3 + a_4x^4 + \cdots) \\ &= (2a_2 - 1) + (6a_3 - 1)x + (12a_4 - a_2)x^2 + \cdots \end{aligned}$$

Since the coefficient of each x^k on the right must equal the coefficient of x^k on the left (all of these are zero), we have

$$2a_2 - 1 = 0$$
$$6a_3 - 1 = 0$$
$$12a_4 - a_2 = 0$$
$$\vdots$$

This gives

$$a_2 = \frac{1}{2}$$

$$a_3 = \frac{1}{6} = \frac{1}{3!}$$

$$a_4 = \frac{a_2}{12} = \frac{1}{24} = \frac{1}{4!}$$
$$\vdots$$

We then have the solution

$$y = 1 + x + \frac{1}{2}x^2 + \frac{1}{3!}x^3 + \frac{1}{4!}x^4 + \cdots$$

Although the series obtained is not always recognizable, in this case, we see that the series is the Taylor series of e^x at $x = 0$. Thus, we conclude that $y = e^x$ is a solution to Equation (1). This can easily be checked.

(a) Find a series solution to the second-order differential equation

$$y'' + y = 0, \quad y(0) = 0, \, y'(0) = 1$$

Do you recognize what function this Taylor series represents?

(b) Find a series solution to the second-order differential equation

$$y'' + y = 0, \quad y(0) = 1, \, y'(0) = 0$$

Do you recognize what function this Taylor series represents?

P r o j e c t   2
Chaos

We have already noted that the logistic equation

$$\frac{dP}{dt} = k\left(1 - \frac{P}{L}\right)P \qquad [1]$$

has been successfully used to model the growth of populations, especially when the populations become large enough to be inhibited by the finite resources available to the population. See the figure.

We already noted in Chapter 8 and can readily see from examining Equation (1) that when the population satisfies $P < L$, P' is positive and the population increases. When $P > L$, P' is negative and the population decreases. Also $P(t) = L$ is a constant solution. Thus, in the logistic Equation (1), *it is not possible for populations to oscillate*. Biologists, however, are well aware that many populations do oscillate, and that these oscillations can play important roles in the environment.

We need to examine why Equation (1) is not accurately modeling this observed oscillating behavior. Notice that the population model in Equation (1) assumes that the population is *continuously* adjusting its growth rate based on the current population. We know that this is not realistic since the reproductive process of any species of animals or insects is not instantaneous but instead requires a certain amount of time to adjust.

Thus, in this case, a more realistic model is the discrete version of Equation (1). One easy way to proceed from the continuous model given by Equation (1) to a discrete model is to note that

$$\frac{dP(t)}{dt} \approx \frac{P(t + \Delta t) - P(t)}{\Delta t}$$

by definition of the derivative. Taking $\Delta t = 1$ unit of time and labeling t_n as the nth time period, the right-hand side of the last equation becomes $P(t_{n+1}) - P(t_n)$, which is just the change in population during the $(n + 1)$st period. Labeling $P(t_n) = P_n$, the discrete version of Equation (1) is

$$P_{n+1} - P_n = k\left(1 - \frac{P_n}{L}\right)P_n$$

We can choose the size of the units that measure population to be in thousands, millions, or anything we wish. For computational purposes it is helpful to choose units so that $L = \dfrac{k}{1+k}$.

Then the logistic equation becomes

$$P_{n+1} = (k + 1)P_n - (k + 1)P_n^2$$

Setting $r = k + 1$, then finally gives

$$P_{n+1} = rP_n(1 - P_n) \qquad [2]$$

When the ecologist R. M. May first studied logistic Equation (2), he was very surprised to find that the solutions behaved very differently when r was greater than 3. He carried out a program of intense numerical exploration into the behavior of the solutions to this seemingly simple nonlinear equation. He was surprised at the complexity of the behavior of the solutions and naturally began to think what implications this had in ecology.

As you will see, when $r < 3$, the solutions of Equation (2) behave much as the solutions of Equation (1) do. But when $r > 3$, the solutions of Equation (2) oscillate. The following exercises are designed to give a feeling for what can happen.

(a) Take $r = 3.3$ in Equation (2), and use the program LOGISTIC to graph the solution. (Take $P_0 = 0.4$.) Describe what you observe.

(b) Take $r = 3.5$ in Equation (2), and use the program LOGISTIC to graph the solution.

(Take $P_0 = 0.4$.) Again describe what you observe. You may need to write down all P_n (to three decimal places) up to $n = 15$ or 20 to better see what is happening.

It can be shown that if $r > 3$ but less than $1 + \sqrt{6} \approx 3.449$, then the dynamical system in Equation (2) has a two-cycle (period 2) that is an attractor. Such an example was given in part (a). For values of r slightly larger than $1 + \sqrt{6} \approx 3.449$, the solutions approach an attracting four-cycle, that is, a solution that repeats itself every four time periods. Such an example was given in part (b).

It can also be shown that when r is between the two numbers 3.449 and 3.544 (rounded to three decimal places), there always exists an attracting four-cycle.

Thus, when the parameter r passes through the number 3 or 3.449, the solutions are said to **bifurcate** into period-doubling solutions.

The story does not end here. There are numbers $c_1, c_2, c_3, \ldots$, all less than the number 3.57 (to two decimal places), for which

$$c_1 < c_2 < c_3 < \cdots$$

that have the following properties. These numbers approach 3.57 and when $c_n < r < c_{n+1}$, the dynamical system in Equation (2) has an attracting 2^n-cycle. In other words, no matter how large a value of n you pick you can find a number r close and to the left of 3.57 so that the dynamical system has an attracting 2^n-cycle. We have already noted that $c_1 = 3$, $c_2 = 3.449$, and $c_3 = 3.544$.

This model may explain the cyclical fluctuations in populations with large growth rates such as occur for certain insects and small mammals[†]. Similar behavior has been observed in mechanics, astronomy, meteorology, and chemistry.

If we then imagine an insect population satisfying the logistics equation and the growth rate k and thus $r = k + 1$ increases, then as r passes through 3, the insect population would refuse to settle down but instead oscillates between two values in alternate time periods. Now as r passes through the value 3.449, the population settles down into eventually repeating every four years. As r passes through the value 3.544, the cycle doubles again and again is stable. Now as r increases, the period-doubling bifurcations appear ever faster.

(c) Now take $r = 3.7$ in Equation (2), and find the solution with $P_0 = 2.000$. Consider $n \leq 40$. Repeat with $P_0 = 2.001$. Do these two solutions that started out close together stay together? This is referred to as **chaos**. Why do you think this might be called chaos?

Project 3
Fourier Series

One very unsettling aspect of Taylor series expansions of *periodic* functions is that the Taylor polynomial approximations are *not* periodic functions. Thus, for example, we know that

$$\sin x = x - \frac{1}{3!} x^3 + \frac{1}{5!} x^5 + \cdots$$

and that this series converges for all x. Despite this, the Taylor polynomials, such as $p_1(x) = x$, $p_3(x) = x - \frac{1}{3!} x^3$, and so on, are not periodic.

Therefore, if we have a periodic function, whether a trigonometric function or not, a Taylor polynomial approximation lacks the important property of periodicity. A continuous periodic function repeats itself in endless fashion and is bounded, whereas the Taylor polynomial approximation, being a polynomial, is not periodic and furthermore eventually becomes unbounded as $|x|$ becomes large.

Thus, for periodic functions, it might make considerably more sense to expand in terms of other *periodic* functions. This we consider now. Look at the function $y = f(x)$ shown in the figure. To an electrical engineer this might represent a "square wave" that models the flow of electricity in a circuit in which a switch is repeatedly turned on and off.

(a) Suppose the function $y = f(x)$ shown in the figure can be written as the infinite sum of trigonometric functions given by

$$f(x) = a_0 + a_1 \cos x + a_2 \cos 2x + a_3 \cos 3x + \cdots$$
$$+ b_1 \sin x + b_2 \sin 2x + b_3 \sin 3x + \cdots$$
$$= a_0 + \sum_{k=1}^{\infty} (a_k \cos kx + b_k \sin kx)$$

[†]R. M. May, "Biological Populations Obeying Difference Equations: Stable Points, Stable Cycles, and Chaos," *Journal of Theoretical Biology* 51(1975), pp. 511–524.

Assuming the series on the right can be integrated term by term, show that we must have

$$a_0 = \frac{1}{2\pi} \int_{-\pi}^{\pi} f(x)\, dx$$

$$a_k = \frac{1}{\pi} \int_{-\pi}^{\pi} f(x) \cos kx\, dx \quad k \geq 0$$

$$b_k = \frac{1}{\pi} \int_{-\pi}^{\pi} f(x) \sin kx\, dx \quad k \geq 0$$

for all k. *Hint:* Use the identities

$$\sin x \sin y = \frac{1}{2}[\cos(x - y) - \cos(x + y)]$$

$$\sin x \cos y = \frac{1}{2}[\sin(x - y) + \sin(x + y)]$$

$$\cos x \cos y = \frac{1}{2}[\cos(x + y) + \cos(x - y)]$$

(b) Now evaluate a_k and b_k for the specific function shown in the figure.
(c) In part (b) you should have obtained $b_k = 0$ for all k. Thus,

$$f(x) = a_0 + \sum_{k=1}^{\infty} a_k \cos kx$$

Using your graphing calculator, graph $y = f(x)$ and then $y = a_0$, and $y = S_n(x) = \sum_{k=1}^{n} a_k \cos kx$ for $n = 1, 2, 3, 4, 5, 6, 7$, or even larger values of n. Notice how the approximations become better and better!

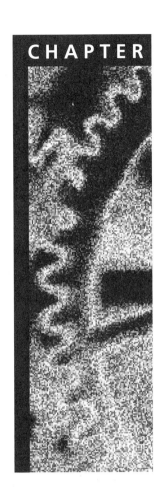

11

Probability and Calculus

In this chapter we apply the integral calculus to continuous probability distributions. If you are already familiar with discrete probability, the first section can be skipped.

11.1 DISCRETE PROBABILITY

■ Events ■ Probability ■ Random Variables ■ Probability Distributions ■ Expected Value ■ Variance and Standard Deviation

North Wind Archives Picture

The Beginnings of Mathematical Probability

In 1654 the famous mathematician Blaise Pascal (1623–1662) had a friend, Chevalier de Mere, a member of the French nobility and a gambler, who wanted to adjust gambling stakes so that he would be assured of winning if he played long enough. This gambler raised some of the following questions with Pascal: In eight throws of a die a player is to attempt to throw a 1, but after three unsuccessful trials the game is interrupted. How should he be indemnified? Pascal wrote to the leading mathematician of that day, Pierre de Fermat (1601–1665), about these problems, and their resulting correspondence represents the beginnings for the modern theory of mathematical probability.

APPLICATION: FINDING THE PROBABILITY OF OBTAINING A CONTRACT

Three companies are competing for a contract. It is known that the first two companies have an equal chance of obtaining the contract and that the third company has only one-half the chance of the first company. What is the likelihood that the first company will get the contract? See Example 3 for the answer.

Events

We first lay the foundations to probability theory by studying sample spaces, events, and related ideas. We begin with the following definitions of experiments and outcomes.

Experiments and Outcomes

An **experiment** is an activity that has observable results. The results of the experiment are called **outcomes**.

The following are some examples of experiments. Flip a coin, and observe whether it falls "heads" or "tails." Throw a die (a small cube marked on each face with from one to six dots), and observe the number of dots on the top face.

We continue by defining some additional needed terms.

Sample Spaces and Trials

A **sample space** of an experiment is the set of all possible outcomes of the experiment. Each repetition of an experiment is called a **trial**.

In the experiment of flipping a coin and observing whether it falls heads or tails, the sample space is $S = \{\text{heads, tails}\}$. For the experiment of throwing a die and observing the number of dots on the top face the sample space is the set

$$S = \{1, 2, 3, 4, 5, 6\}.$$

Events

Given a sample space S for an experiment, an **event** is any subset E of S.

<u>**EXAMPLE 1**</u>   Determining an Event

For the experiment of throwing a die and observing the number of dots on the top face the sample space is $S = \{1, 2, 3, 4, 5, 6\}$. Determine the event E: "An even number of dots shows on the top face."

SOLUTION ■ We have $E = \{2, 4, 6\}$. ■

<u>**EXAMPLE 2**</u>   Determining a Sample Space

Two dice, identical except that one is white and the other is red, are tossed and the number of dots on the top face of each is observed. What is the sample space for this experiment?

SOLUTION ■ The outcomes can be considered ordered pairs. For example (2, 3) means two dots on the top face of the white die and three dots on the top face of the red die. The sample space S is

$$S = \{(1, 1), (1, 2), (1, 3), (1, 4), (1, 5), (1, 6),$$
$$(2, 1), (2, 2), (2, 3), (2, 4), (2, 5), (2, 6),$$
$$(3, 1), (3, 2), (3, 3), (3, 4), (3, 5), (3, 6),$$
$$(4, 1), (4, 2), (4, 3), (4, 4), (4, 5), (4, 6),$$
$$(5, 1), (5, 2), (5, 3), (5, 4), (5, 5), (5, 6),$$
$$(6, 1), (6, 2), (6, 3), (6, 4), (6, 5), (6, 6)\} \quad ■$$

We can use union, intersection, and complement to describe events.

Union of Two Events

If E and F are two events, then $E \cup F$ is the union of the two events and consists of the set of outcomes that are in E or F or both.

Thus, the event $E \cup F$ is the event that "E or F occurs."

Intersection of Two Events

If E and F are two events, then $E \cap F$ is the intersection of the two events and consists of the set of outcomes that are in both E and F.

Thus, the event $E \cap F$ is the event that both "E and F occur."

Complement of an Event

If E is an event, then E^c is the complement of E and consists of the set of outcomes that are not in E.

Thus, the event E^c is the event that "E does not occur."

The Impossible Event

The empty set, ϕ, is called the **impossible event**.

For example, if H is the event that a head shows on flipping a coin and T is the event that a tail shows, then $H \cap T = \phi$. The event $H \cap T$ means that both heads and tails show, which is impossible.

Since $S \subset S$, S is itself an event. We call S the **certainty event** since any outcome of the experiment must be in S.

The Certainty Event

Let S be a sample space. The event S is called the **certainty event**.

Probability

We are now in a position to consider the idea of probability.

To gain some insight into what probability is consider a very important type of problem that arises every day in business and science: find a *practical* way to estimate the likelihood of certain events occurring. For example, a food company may seek a practical method of estimating the likelihood that a new type of candy will be enjoyed by consumers. The most obvious procedure for the company to follow is to select a consumer, have the consumer taste the candy, and then record the result. This should be repeated many times and the final totals tabulated to give the fraction of consumers who enjoy the candy. This fraction is then a practical estimate of the likelihood that consumers will enjoy this candy.

In mathematical terms, let L be the event that the consumer enjoys the candy and N the number of consumers that have been tested. If $f(L)$ is the number or frequency of times a consumer in this group enjoys the candy, then the fraction of consumers who enjoy the candy is given by $\dfrac{f(L)}{N}$. We call this fraction the relative frequency.

For a general event E in a sample space S we have the following:

Relative Frequency

If an experiment is repeated N times and an event E occurs $f(E)$ times, then the fraction

$$\frac{f(E)}{N}$$

is called the **relative frequency** of the event.

The food company takes N as large as time and money permits. The resulting relative frequency $\dfrac{f(L)}{N}$ is then a *practical* estimate of the likelihood that a randomly chosen consumer will enjoy this candy.

It is very important to notice that the relative frequency of an event $A \cup B$ for which $A \cap B = \phi$ is simply obtained as the sum of the relative frequencies of A and B. That is,

$$f(A \cup B) = f(A) + f(B), \quad \text{if } A \cap B = \phi$$

This follows simply because, in repeated trials of an experiment, the total number of times an event $A \cup B$ occurs when $A \cap B = \phi$ just equals the sum of the number of occurrences of each of the outcomes A and B.

Thus we have the following:

Finding the Relative Frequency of an Event

Suppose an experiment has been repeated N times and A and B are two events in some sample space S. Then

$$f(A \cup B) = f(A) + f(B), \quad \text{if } A \cap B = \phi \qquad [1]$$

■ **Frederick Mosteller and the Dice Experiment**

Frederick Mosteller has been president of the American Association for the Advancement of Science, the Institute of Mathematical Statistics, and the American Statistical Association. He once decided that "it would be nice to see if the actual outcome of a real person tossing real dice would match up with the theory." He then engaged Willard H. Longcor to buy some dice, toss them, and keep careful records of the outcomes. Mr. Longcor then tossed the dice on his floor at home so that the dice bounced on the floor and then up against the wall and then landed back on the floor. After doing this several thousand times, his wife became troubled by the noise. He then placed a rug on the floor and on the wall, and then proceeded to quietly toss his dice *millions* of times, keeping careful records of the outcomes. In fact, he was so careful and responsible about his task, that he threw away his data on the first 100,000 tosses, since he had a nagging worry that he might have made some mistake keeping perfect track.

We have just given a practical way to assign a number to an event that indicates the approximate likelihood that the event will occur. We now see how to assign a number to an event that gives the *exact* likelihood that the event will occur. This number is assigned on the basis of the percentage of times the event E occurs over the long-term when the experiment is repeated over and over again. More specifically, the number is assigned on the basis of the long-term behavior of the relative frequency $\frac{f(E)}{N}$.

Consider a *theoretical* example. We flip a fair coin, that is, a coin that shows no preference for heads or tails. We naturally expect heads to show as often as tails. Thus, we expect heads to show one-half of the time, and if H is the event that a head shows, we expect that as the number N of times the coin is flipped becomes larger and larger, the relative frequencies $\frac{f(H)}{N}$ should become closer and closer to exactly 0.50. The number 0.50 is called the **probability** of H, and is denoted by $p(H)$.

EXPLORATION *1*

Relative Frequency

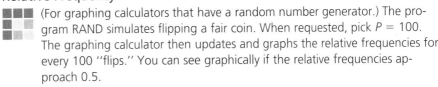 (For graphing calculators that have a random number generator.) The program RAND simulates flipping a fair coin. When requested, pick $P = 100$. The graphing calculator then updates and graphs the relative frequencies for every 100 "flips." You can see graphically if the relative frequencies approach 0.5.

For a general event E in a sample space S we have the following (intuitive) definition:

The Probability of an Event E

Let S be the sample space of an experiment, let E be an event in S, and let N be the number of times the experiment has been repeated. The **probability of E**, denoted by $p(E)$, is the number that the relative frequency $\frac{f(E)}{N}$ approaches as N becomes large without bound.

We now consider some basic properties of probability. The definition immediately implies several things about the numbers $p(E)$. First, if E is an event in a sample space S and the experiment is repeated N times, then the number of occurrences $f(E)$ of the event E can be at most N and of course at least 0. That is, $0 \le f(E) \le N$. This implies that

$$0 \le \frac{f(E)}{N} \le 1$$

Since for large N, $\frac{f(E)}{N} \approx p(E)$, we then must have $0 \le p(E) \le 1$. Second, since $f(S) = N$, $\frac{f(S)}{N} = \frac{N}{N} = 1$. Thus, we must have $p(S) = 1$. Finally, since $f(\phi) = 0$, $\frac{f(\phi)}{N} = \frac{0}{N} = 0$. Thus, we must have $p(\phi) = 0$.

We now summarize what we just learned in the previous two paragraphs as the first three properties of probability.

Properties of Probability

Let E be an event in a sample space S. Then

Property 1: $0 \le p(E) \le 1$
Property 2: $p(S) = 1$
Property 3: $p(\phi) = 0$

We are now prepared to consider another important property of probability. We now see that Equation (1) is not only satisfied by relative frequency but also by probability in general. We then have the following:

Property 4

Let A and B be two events in some sample space S. Then

$$p(A \cup B) = p(A) + p(B), \quad \text{if } A \cap B = \phi$$

To see why this is true, suppose the experiment is repeated N times. Then since $A \cap B = \phi$

$$f(A \cup B) = f(A) + f(B)$$

and therefore

$$\frac{f(A \cup B)}{N} = \frac{f(A)}{N} + \frac{f(B)}{N} \tag{2}$$

Now for large N, we have $\dfrac{f(A \cup B)}{N} \approx p(A \cup B)$, $\dfrac{f(A)}{N} \approx p(A)$, and $\dfrac{f(B)}{N} \approx p(B)$. Then Equation (2) implies that

$$p(A \cup B) = p(A) + p(B)$$

We have so far taken the point of view that the nature of probability is part of the study of natural and human phenomena in the same way that physics or economics are. Probability has only been discussed in the context of events in business or science. Subjects such as "randomness" and the "long-term" behavior of the relative frequency of an event were deemed intuitive and within the experience of the reader.

We now take an alternative view of probability by regarding probability as a part of *mathematics*. As such, we need to set down a set of abstract definitions and construct a set of axioms that govern the theory, so as to place the theory on a firm foundation. Using the logic of mathematics we can then derive other properties that probability must satisfy. Such a mathematical theory is said to be a **model** of the phenomena of probability. One important advantage of this approach is that all terms are clearly defined, and "experience" is not a prerequisite for understanding the development. The axioms that are laid down to describe the phenomena of probability must be very carefully chosen so that the resulting theory actually describes the phenomena we are studying. In other words, the theory must not contradict our intuitive notions of probability already gained. We choose as axioms the four properties of probability given earlier. These four, in effect, capture the "essence" of what we intuitively think of as probability.

We then have the following definition:

Definition of Probability

Let S be a sample space. Assign to each event in S a number $p(E)$ so that the following four properties are satisfied.

Property 1: $0 \le p(E) \le 1$ for any event E
Property 2: $p(S) = 1$

Property 3: $p(\phi) = 0$

Property 4: Let A and B be two events in S. Then

$$p(A \cup B) = p(A) + p(B), \quad \text{if } A \cap B = \phi$$

Just how to assign the probabilities to each outcome must be determined by the information on hand.

EXAMPLE 3   Finding the Probability of an Event

Three companies are competing for a contract. It is known that the first two companies have an equal probability of obtaining the contract and that the third company has only one-half the probability of the first company. What is the probability that the first company will get the contract?

SOLUTION ■ Let s_1 represent the outcome that the first company receives the contract, s_2 the second, and s_3 the third. Then the sample space is $S = \{s_1, s_2, s_3\}$. Let $p(s_1)$, $p(s_2)$, and $p(s_3)$ denote the respective probabilities that s_1, s_2, and s_3 will occur. From the statement of the problem $p(s_1) = p(s_2)$ and $p(s_3) = 0.5p(s_1)$. From Properties 2 and 4 we have

$$\begin{aligned} 1 &= p(S) \\ &= p(\{s_1, s_2, s_3\}) \\ &= p(s_1) + p(s_2) + p(s_3) \\ &= p(s_1) + p(s_1) + 0.5p(s_1) \\ &= 2.5p(s_1) \end{aligned}$$

$$p(s_1) = \frac{1}{2.5} = 0.40 \quad ■$$

The following properties of probability follow from the first four properties and are left as exercises.

Additional Properties of Probability

Property 5: Given any event $E = \{e_1, e_2, \ldots, e_r\}$, where $e_1, e_2, \ldots, e_r$ are all distinct, then

$$p(\{e_1, e_2, \ldots, e_r\}) = p(e_1) + p(e_2) + \cdots + p(e_r)$$

Property 6: Let $S = \{s_1, s_2, \ldots, s_n\}$ and let $p(s_1) = p_1, p(s_2) = p_2, \ldots, p(s_n) = p_n$. Then

$$p_1 + p_2 + \cdots + p_n = 1$$

Property 7: If E is any event, then

$$p(E^c) = 1 - p(E)$$

Property 8: If E and F are any two events, then

$$p(E \cup F) = p(E) - p(F) - p(E \cap F)$$

Not all the probabilities were the same in Example 3. We now consider the important case when all the probabilities are the same. Let S be the finite sample space $S = \{s_1, s_2, \ldots, s_n\}$, and assume that the probability of each of the outcomes is the same. We refer to such a sample space as a **uniform sample space**. Thus,

$$p(s_1) = p(s_2) = \cdots = p(s_n)$$

If we let p be this common probability, then from Property 6 we must have that

$$p(S) = p(s_1) + p(s_2) + \cdots + p(s_n) = np$$

But $p(S) = 1$. Thus, $p = \dfrac{1}{n}$.

Now, given any event $E = \{e_1, e_2, \ldots, e_r\}$ in S, we have, using Property 5, that

$$
\begin{aligned}
p(E) &= p(\{e_1, e_2, \ldots, e_r\}) \\
&= p(e_1) + p(e_2) + \cdots + p(e_r) \\
&= r\,\frac{1}{n} \\
&= \frac{n(E)}{n(S)}
\end{aligned}
$$

Thus, we have the following:

Probability of an Event in the Uniform Sample Space

If S is a finite uniform sample space and E is any event, then

$$p(E) = \frac{\text{number of elements in } E}{\text{number of elements in } S} = \frac{n(E)}{n(S)}$$

EXAMPLE 4   Using Counting

Two fair dice are tossed. Assuming that the probability of every outcome shown in Example 2 is equally likely, find the probability that a sum shows that is equal to (a) 3 (b) 7.

SOLUTIONS ■ We first comment that the assumption that any outcome is just as likely as any other seems reasonable since it does not seem possible to argue based on the physical set-up how one outcome, such as $(2, 4)$, can be any more or less likely than any other outcome, such as $(3, 6)$.

The total number of outcomes is $n(S) = 36$.

(a) The event "the sum is 3" is given by $E = \{(1, 2), (2, 1)\}$. Thus,

$$p(E) = \frac{n(E)}{n(S)} = \frac{2}{36} = \frac{1}{18}$$

(b) The event "the sum is 7" is given by

$$F = \{(1, 6), (2, 5), (3, 4), (4, 3), (5, 2), (6, 1)\}$$

Thus,

$$p(F) = \frac{n(F)}{n(S)} = \frac{6}{36} = \frac{1}{6} \quad ■$$

Random Variables

Outcomes of experiments are not always real numbers. The outcomes of experiments can be "heads," "above average," or "defective"—none of which are real numbers. It is often useful to assign a real number to each outcome of an experiment. For example, the outcome of the experiment "your grade in math" is a *letter*. But for purposes that you are well aware of, each letter is assigned a real number. Usually 4 is assigned to an "A," 3 to a "B," 2 to a "C," 1 to a "D," and 0 to an "F." Usually there is some rational basis on which the assignment of a real number to an outcome is made. For example, since a grade of ''A'' is ''better'' than a grade of ''B,'' it makes sense to assign a *higher* number to the grade ''A'' than to ''B.'' Also the assignment of the numbers to each letter is done in such a manner that the numerical difference between two successive letters is equal. Assigning 0 to ''F'' makes sense since no credit is given for a grade of ''F.'' Finally, using the numbers 0, 1, 2, 3, and 4 makes it easy to calculate the ''average'' grade and for this number to have a readily understood meaning.

When *numbers* are assigned to the outcomes of experiments according to some rule, the rule is referred to as a random variable. As we have seen in assigning numbers to letter grades, the assignments of numbers to the outcomes of experiments is normally done in a manner that is reasonable and most importantly in a manner that permits these numbers to be used for interpretation and comparison.

Random Variables

A **random variable** is a rule that assigns precisely one real number to each outcome of an experiment.

Thus, a random variable is a real valued *function* with the domain equal to the set of outcomes of an experiment.

Remark. Unless otherwise specified, when the outcomes of an experiment are themselves numbers, the random variable is the function that simply assigns each number to itself.

There are three types of random variables as we now indicate.

Finite Discrete, Infinite Discrete, and Continuous Random Variables

1. A random variable is **finite discrete** is it assumes only a finite number of values.
2. A random variable is **infinite discrete** if it takes on an infinite number of values that can be listed in a sequence, so that there is a first one, a second one, a third one, and so on.
3. A random variable is said to be **continuous** if it can take any of the infinite number of values in some interval of real numbers.

If a random variable denotes the number of dots on the top face of a tossed die, then the random variable can take the finite number of values in the set $\{1, 2, \ldots, 6\}$ and thus is finite discrete. If the random variable denotes the number of flips it takes to obtain a head, then the random variable can take any of the infinitely many values in the set $\{1, 2, \ldots\}$ and thus is infinite discrete. If the random variable denotes the life of a light bulb in hours, then the random variable can take on any value in the interval $[0, \infty)$ and thus is continuous.

Probability Distributions

Suppose now that the outcomes of an experiment are the finite set of real numbers $\{x_1, x_2, \ldots, x_n\}$ and that these numbers are all the different real numbers assigned to the outcomes of an experiment. We have usually denoted the probability of the outcome x_k by $p(x_k)$. We now denote this same probability by

$$P(X = x_k)$$

We use this latter notation since it is widely used in statistics and has some advantages as we shall soon see. The **probability distribution of the random variable X** is a listing of all the probabilities associated with all possible values of the random variable. Such a listing is often given in a table.

Probability Distribution of the Random Variable X

Suppose the random variable X can take the values $x_1, \ldots, x_n$. The probability distribution of the random variable X is a listing of all the probabilities associated with all possible values of the random variable, that is, $p_1, \ldots, p_n$, where $p_1 = p(x_1), \ldots, p_n = p(x_n)$.

Suppose a pair of fair dice are tossed. Let X denote the random variable that gives the sum of the top faces. The probability distribution is given in the table.

x	2	3	4	5	6	7	8	9	10	11	12
$P(X = x)$	$\frac{1}{36}$	$\frac{2}{36}$	$\frac{3}{36}$	$\frac{4}{36}$	$\frac{5}{36}$	$\frac{6}{36}$	$\frac{5}{36}$	$\frac{4}{36}$	$\frac{3}{36}$	$\frac{2}{36}$	$\frac{1}{36}$

We can gain more insight into this probability distribution by presenting the data in the table in graphical form. The numbers $2, 3, \ldots, 12$ represent the possible outcomes. These numbers are located on the horizontal axis shown in Figure 11.1. Above each number is drawn a rectangle with base equal to one unit and height equal to the probability of that number. For example, above the number 4 is a rectangle with height 3/36. Such graphs are called histograms and give a vivid description of how the probability is distributed. (You can reproduce Figure 11.1 using the histogram feature on your graphing calculator.) As a consequence, comparison can be more easily made between two histograms than between two tables of probability.

We now use histograms to find the probability of events by finding the areas under appropriate rectangles.

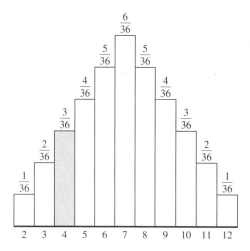

Figure 11.1
Since the base of each rectangle is one unit, the area is the same as the height. Thus, the rectangle above 4 has area equal to 3/36, which is the probability of obtaining a 4.

EXAMPLE 5   Finding Probability Using Histograms

Use the histogram in Figure 11.1 to find the probability that the sum of the number of dots on the top faces of two fair dice when tossed will be at least 9 and less than 12.

SOLUTION ■ We are seeking

$$P(X = 9) + P(X = 10) + P(X = 11)$$

But the probability that $X = 9$, $P(X = 9)$ is the area of the rectangle above 9,

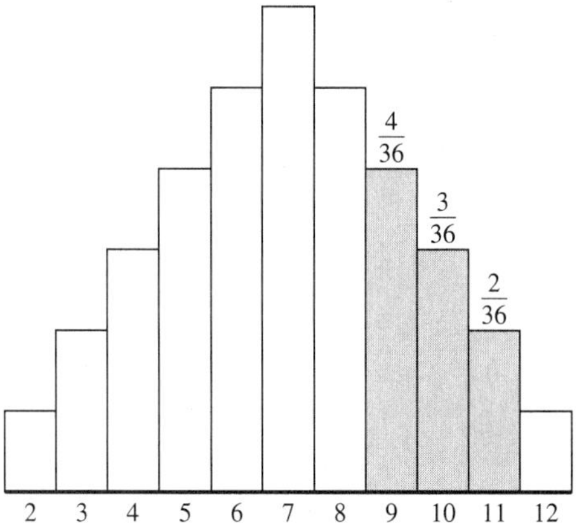

Figure 11.2
The area of the three rectangles is the
probability of obtaining a 9, 10, or 11.

$P(X = 10)$ is the area of the rectangle above 10, and so on. Thus, we are seeking the area of the shaded region in Figure 11.2. This is

$$\frac{4}{36} + \frac{3}{36} + \frac{2}{36} = \frac{1}{4} \quad \blacksquare$$

A shorthand notation for the probability found in this example is $P(9 \le X < 12)$.

Expected Value

Let $x_1, x_2, \ldots, x_n$ be the numerical outcomes of an experiment. We now define a number that measures the ''center'' of the data.

We begin with a notion of *average*, which is familiar to most. If you have three test grades of 95, 85, and 93, the average is the sum of all the grades divided by the total number of grades, or

$$\frac{95 + 85 + 93}{3} = \frac{273}{3} = 91$$

In general we have the following definition of *average*.

Average

The **average** of the n numbers $x_1, x_2, \ldots, x_n$ is

$$\text{Average} = \frac{x_1 + x_2 + \cdots + x_n}{n}$$

Suppose the sample space S has n outcomes $x_1, x_2, \ldots, x_n$ with respective probabilities $p_1, p_2, \ldots, p_n$. Now suppose the experiment has been repeated a large number of times N so that x_1 is observed to occur a frequency of f_1 times, x_2 is observed f_2 times, and so on. (We must have $f_1 + f_2 + \cdots + f_n = N$). Then the average is

$$\text{Average} = \frac{\overbrace{x_1 + \cdots + x_1}^{f_1 \text{ times}} + \overbrace{x_2 + \cdots + x_2}^{f_2 \text{ times}} + \cdots + \overbrace{x_n + \cdots + x_n}^{f_n \text{ times}}}{N}$$

$$= \frac{x_1 f_1 + x_2 f_2 + \cdots + x_n f_n}{N}$$

$$= x_1 \frac{f_1}{N} + x_2 \frac{f_2}{N} + \cdots + x_n \frac{f_n}{N}$$

But as N gets larger we expect the relative frequency $\dfrac{f_1}{N}$ to approach p_1. Similarly for the other relative frequencies. Thus, the right-hand side of the last displayed line approaches

$$x_1 p_1 + x_2 p_2 + \cdots + x_n p_n$$

We call this the **expected value**, or **mean**. In the language of random variables we then have the following:

Expected Value, or Mean

Let X denote the random variable that takes on the values $x_1, x_2, \ldots, x_n$, and let the associated probabilities be $p_1, p_2, \ldots, p_n$. The expected value, or **mean**, of the random variable X, denoted by $E(X)$ or by the Greek letter μ, is defined to be

$$E(X) = x_1 p_1 + x_2 p_2 + \cdots + x_n p_n$$

The expected value is an average in some sense. As we saw above, if the random phenomenon is repeated a large number of times, the average of the observed outcomes approaches the expected value.

Often we do not know the probabilities, but an experiment has been performed N times. Then the probabilities $p_1, p_2, \ldots, p_n$ are the relative frequencies (empirical probabilities). This is illustrated in the next example.

EXAMPLE 6   Calculating the Mean

You are thinking of buying a small manufacturer of quality yachts and have obtained the data in the table from the owners summarizing how often various numbers of yacht sales occur per quarter. Find the average or mean number sold per quarter.

Number Sold in One Quarter	1	2	3	4	5
Frequency of Occurrence	10	15	16	7	2

Number Sold in One Quarter	1	2	3	4	5
Probability of Occurrence	$\dfrac{10}{50} = 0.20$	$\dfrac{15}{50} = 0.30$	$\dfrac{16}{50} = 0.32$	$\dfrac{7}{50} = 0.14$	$\dfrac{2}{50} = 0.04$

SOLUTION ■ We first notice that

$$N = 10 + 15 + 16 + 7 + 2 = 50$$

Using the relative frequencies as the (empirical) probability yields the table. Then

$$E(X) = 1(0.20) + 2(0.30) + 3(0.32) + 4(0.14) + 5(0.04) = 2.52 \quad ■$$

You can obtain the mean found in Example 6 using the statistical functions on your graphing calculator.

$E(X)$ is what we ''expect'' over the long term. But we must realize that $E(X)$ need not be an actual outcome. In Example 6, the expected value of sales per quarter was 2.52.

In the case that all the n outcomes are equally likely, the probabilities are just $\dfrac{1}{n}$ and then

$$E(X) = x_1\frac{1}{n} + x_2\frac{1}{n} + \cdots + x_n\frac{1}{n}$$

$$= \frac{x_1 + x_2 + \cdots + x_n}{n}$$

which is the average.

Variance and Standard Deviation

We have seen that the mean, or the expected value, is one measure of the "center" of the data. We now define numbers called the **variance** and the **standard deviation** that measure the dispersion of data about the expected value. If the data are clustered close to the expected value, then the variance and the standard deviation are small. If the data are widely dispersed away from the expected value, then the variance and standard deviation are large.

Knowing the variance or dispersion is often of critical interest. Suppose the manufacturer of precision ball bearings requires their 1-cm bearings to have a diameter of 1 cm with an error of no more than 0.001 cm. If one production line produces these bearings with an average diameter of exactly 1 cm, but not one single bearing with a diameter that falls within the interval (0.999, 1.001), then all the bearings would be rejected. On the other hand, if another production line produces these bearings with an average diameter of 1.0008, a bit off from exactly 1, and all the diameters fall within the interval (0.999, 1.001), then all of these bearings are acceptable. Thus, the first production line produces, on average, a perfect bearing, yet all are rejected. The second production line produces, on average, a much less perfect bearing, yet all are accepted. The dispersion, in this case, plays a critical role.

Suppose an experiment has the n outcomes $x_1, x_2, \ldots, x_n$, which we assume for the present are all equally likely, and μ is the mean. Then it is tempting to measure the dispersion by taking the average of all the differences

$$x_1 - \mu, x_2 - \mu, \ldots, x_n - \mu$$

The average of these differences must be zero, however. (Why?) To avoid this we take the average of the *square* of the above differences and call this the **variance**. Thus,

$$\mathrm{Var} = \frac{(x_1 - \mu)^2 + (x_2 - \mu)^2 + \cdots + (x_n - \mu)^2}{n}$$

In general suppose there are the n outcomes $x_1, x_2, \ldots, x_n$, with respective probabilities $p_1, p_2, \ldots, p_n$. Now suppose the experiment has been repeated a large number of times N so that x_1 is observed to occur a frequency of f_1 times, x_2 is observed f_2 times, and so on. (We must have $f_1 + f_2 + \cdots + f_n = N$). Then if we set $\mu = E(X)$, the variance is

$$\mathrm{Var} = \frac{\overbrace{(x_1 - \mu)^2 + \cdots + (x_1 - \mu)^2}^{f_1 \text{ times}} + \cdots + \overbrace{(x_n - \mu)^2 + \cdots + (x_n - \mu)^2}^{f_n \text{ times}}}{N}$$

$$= \frac{(x_1 - \mu)^2 f_1 + (x_2 - \mu)^2 f_2 + \cdots + (x_n - \mu)^2 f_n}{N}$$

$$= (x_1 - \mu)^2 \frac{f_1}{N} + (x_2 - \mu)^2 \frac{f_2}{N} + \cdots + (x_n - \mu)^2 \frac{f_n}{N}$$

But as N gets larger, we expect the relative frequency $\dfrac{f_1}{N}$ to approach p_1 and similarly for the other relative frequencies. Thus, the right-hand side of the last displayed line approaches

$$(x_1 - \mu)^2 p_1 + (x_2 - \mu)^2 p_2 + \cdots + (x_n - \mu)^2 p_n$$

We call this term the **variance**. In the language of random variables we then have the following:

Variance and Standard Deviation

Let X denote the random variable that takes on the values $x_1, x_2, \ldots, x_n$, and let the associated probabilities be $p_1, p_2, \ldots, p_n$. Then if $\mu = E(X)$, the **variance** of the random variable X, denoted by $\mathrm{Var}(X)$, is

$$\mathrm{Var}(X) = (x_1 - \mu)^2 p_1 + (x_2 - \mu)^2 p_2 + \cdots + (x_n - \mu)^2 p_n = \sum_{k=1}^{n} (x_k - \mu)^2 p_k$$

The **standard deviation**, denoted by $\sigma(X)$ is

$$\sigma(X) = \sqrt{\mathrm{Var}(X)}$$

EXAMPLE 7   Comparing Variances of Two Probability Distributions

Figure 11.3 shows the histograms of two probability distributions both with means equal to 1. Find the variance of each, and compare the two.

SOLUTIONS ■ The variance of the first probability distribution shown in Figure 11.3 is

$$(0 - 1)^2(0.10) + (1 - 1)^2(0.80) + (2 - 1)^2(0.10) = 0.20$$

The variance of the second probability distribution shown in Figure 11.3 is

$$(0 - 1)^2 \frac{1}{3} + (1 - 1)^2 \frac{1}{3} + (2 - 1)^2 \frac{1}{3} = \frac{2}{3}$$

The variance of the second probability distribution is much larger than the variance of the first. This just reflects the fact, seen from Figure 11.3, that the second probability distribution is more dispersed from the mean than the first one is. ■

Using the statistical features of your graphing calculator you can obtain the variances found in Example 7.

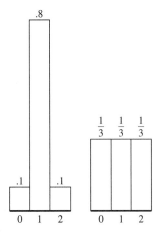

Figure 11.3
Histograms of two probability distributions both with means equal to 1.

SELF-HELP EXERCISE SET 11.1

1. Consider the following version of the game of chance known as roulette: the roulette wheel has 18 red numbers, 18 black numbers, and 1 (green) zero. Suppose you bet on a color, say red. If red occurs, you win back twice your bet. If black occurs, you lose your bet. If 0 occurs, the wheel is spun again until 0 does not occur. If red occurs, then you obtain your original bet back. If black comes up, you lose your original bet. Find the expected "return."

2. The table gives two probability distributions associated with the same outcomes. Draw a histogram of each probability distribution. By inspection of these histograms (and making

no calculations), estimate the means and estimate which one has the largest variance. Now calculate the means, variances, and standard deviations.

Outcome	1	2	3	4	5
(a) Probability	0.1	0.2	0.4	0.2	0.1
(b) Probability	0.4	0.1	0	0.1	0.4

EXERCISE SET 11.1

1. Two tetrahedrons (four-sided), each with equal sides with from one to four dots, are identical except that one is red and the other white. The two tetrahedrons are tossed, and the total number of dots on the bottom faces is observed.
   (a) What is the sample space for this experiment?
   (b) Let X represent the random variable that gives the total number of dots on the bottom two faces. Find the probability distribution for X.
   (c) Find the probability of obtaining a 4.
   (d) Find the probability of obtaining a 5 or higher.
   (e) Find the expected value of X.
   (f) Find the standard deviation of X.

2. An urn has five identical balls except that they are numbered from 1 to 5. A person randomly picks a ball, replaces the ball back in the urn, and then randomly picks a ball again and then records the sum of the numbers on the two selected balls.
   (a) What is the sample space for this experiment?
   (b) Let X represent the random variable that gives the sum of the numbers on the two selected balls. Find the probability distribution for X.
   (c) Find the probability of obtaining a 6.
   (d) Find the probability of obtaining a 9 or higher.
   (e) Find the expected value of X.
   (f) Find the standard deviation of X.

3. The probability distribution of the random variable X is given in the table. What must z be?

Random Variable x	0	1	2	3	4
P(X = x)	0.2	0.1	0.1	z	0.3

4. (a) A two-year investment yields an average annual return of 100% over the two-year period and a first-year return of 300%. What was the return during the second year?
   (b) If you invested $1000 at the beginning of this two-year period, how much would you have at the end?

(c) What does this say about a percentage loss compared with a percentage gain? For example, if you lost 50%, what percentage gain is needed to bring the investment back to even?

5. A company is giving away $10,000 as an advertising promotion. To enter you must submit a letter with your address on the entry form. The winning entry will then be randomly selected with your chance of winning the same as anyone else's who enters. If 100,000 people enter this contest and your only cost is a 32-cent stamp, what is your expected return (to the nearest cent)?

6. In the daily numbers game, a person picks any of the 1000 numbers from 0 to 999. A $1 wager on one number returns $600 if that number hits; otherwise nothing is returned. If the probability of any one number is the same as any other, find the expected return on this wager.

7. One version of roulette has a wheel with 38 numbers (1 through 36 plus 0 and 00) in 38 equally spaced slots. Half of the numbers from 1 to 36 are red and the other half are black. The numbers 0 and 00 are green. Consider a $1 bet on red. If red comes up, you are returned $2, otherwise you lose the $1 wagered. Find the expected return for this bet.

8. Using the version of roulette in the previous exercise, consider a $1 bet on a number from 1 to 36. If your number hits, you have $36 returned to you; otherwise you lose your $1 bet. Find the expected return for this bet.

9. Using the version of roulette in Exercise 7, consider a $1 bet that can be made on two adjacent numbers from 1 through 36. If one of the numbers comes up, the player wins back $18; otherwise nothing is won. What is the expected value of this bet?

10. A fair die is tossed. If X denotes the random variable giving the number of the top face of the die, find $E(X)$.

11. A lottery has a grand prize of $500,000, two runner-up prizes of $100,000 each, and 100 consolation prizes of $1000 each. If 1 million tickets are sold for $1 each and the probability of any ticket winning is the same as any other, find the expected return on a $1 ticket.

12. A lottery has a grand prize of $50,000, 5 runner-up prizes of $5000 each, 10 third-place prizes of $1000, and 100 consolation prizes of $10 each. If 100,000 tickets are sold for $1 each and the probability of any ticket winning is the same as any other, find the expected return on a $1 ticket.

13. Two fair dice are tossed. If you roll a total of 7, you win $6; otherwise you lose $1. What is the expected return of this game?

14. Two coins are taken at random (without replacement) from a bag containing five nickels, four dimes, and one quarter. Let X denote the random variable given by the total value of the two coins. Find $E(X)$.

In Exercises 15 through 18 find the mean, variance, and standard deviation of the random variable having the probability distribution given in the table.

15.

Random Variable x	−3	0	1
P(X = x)	0.2	0.5	0.3

16.

Random Variable x	−2	1	3
P(X = x)	0.1	0.3	0.6

17.

Random Variable x	−2	0	1	2	4
P(X = x)	0.1	0.3	0.1	0.2	0.3

18.

Random Variable x	−2	−1	1	2	5
P(X = x)	0.2	0.1	0.25	0.25	0.2

Applications

19. Sales. A salesman makes two stops when in Pittsburgh. The first stop yields a sale 10% of the time, the second stop 15% of the time, and a sale at both stops 4% of the time. What proportion of the time does a trip to Pittsburgh result in no sale?

20. Bidding on Contracts. An aerospace firm has three bids on government contracts. They know that the contracts are most likely to be divided up among a number of companies. They decide that the probability of obtaining exactly one contract is 0.6, of exactly two contracts is 0.15, and of exactly three contracts is 0.04. What is the probability they will obtain at least one contract? No contracts?

21. Quality Control. An inspection of computers manufactured at a plant reveals that 2% of the monitors are defective, 3% of the keyboards are defective, and 1% have both defective. **(a)** Find the probability that a computer at this plant has at least one of these defects. **(b)** Find the probability that a computer at this plant has none of these defects.

22. Medicine. A new medication produces headaches in 5% of the users, upset stomach in 15%, and both in 2%. **(a)** Find the probability that at least one of these side effects is produced. **(b)** Find the probability that neither of these side effects is produced.

23. Investment Returns. The tables give all the possible returns and the associated probabilities of two investments: A and B. Find the expected value of each investment and compare the two.

Outcome of A	$1000	$2000	$1000
Probability	0.1	0.8	0.1

Outcome of B	−$1000	$0	$9000
Probability	0.8	0.1	0.1

24. Investment Advisor. An investment advisor informs you that his average annual return for the last 3 years is 100%. Furthermore, he says his annual return during each of the first 2 years of this 3-year period was 200%. Find how the people did who followed his advice by finding the return for the third year of this 3-year period.

25. Quality Control. Electrical switches are manufactured with the probability of 5% that any one is defective. If 50 are chosen at random, what is the expected number of defective switches in this batch?

26. Medicine. For a certain heart operation, the probability of survival is 0.95. If 10 of these operations are performed every week, what is the expected number of deaths due to this operation?

27. Comparing Investments. Two car dealerships are up for sale. The table gives the number of cars sold per day together with the associated probabilities. The average profits per car at the first dealership is $400 and at the second is $300. Which dealership yields the highest daily profit?

	First			
Number Sold in One Day	0	1	2	3
Probability of Occurrence	0.50	0.30	0.15	0.05

	Second				
Number Sold in One Day	0	1	2	3	4
Probability of Occurrence	0.60	0.20	0.05	0.05	0.10

28. Comparing Investments. Two motels are up for sale. The two tables at top of page 593 give the number of rooms rented per day together with the associated probabilities. The average profits per room rented at the first motel is $20 and at the second is $21. Which motel yields the highest daily profit?

Motel 1

Number Rented in One Day	5	6	7	8	9	10
Probability of Occurrence	0.10	0.30	0.40	0.10	0.05	0.05

Motel 2

Number Rented in One Day	3	4	5	6	7
Probability of Occurrence	0.05	0.05	0.10	0.20	0.60

29. Insurance. An insurance company sells a 65-year-old man a 1-year life insurance policy. The policy, which cost $100, pays $5000 in the event the man dies during the next year. If there is a 1% chance that the man will die in the next year, find the probability distribution for this financial transaction for the insurance company, and find their expected return. What meaning does this expected value have to the insurance company?

30. Insurance. An insurance company sells a $10,000, 5-year term life insurance policy to an individual for $700. Find the expected return for the company if the probability that the individual will live for the next 5 years is 0.95.

31. Sales. The number of sales per week and the associated probabilities of two car salesmen, A and B, are given in the table. Find the mean, variance, and standard deviation for the sales per week of each. Which one sells the most cars? Which one is the most consistent?

Number of Sales per Week	0	1	2	3	4
Probability A	0.50	0.30	0.10	0.07	0.03
Probability B	0.40	0.20	0.20	0.10	0.10

32. Employee Absences. The table gives the probabilities that two employees, A and B, will have the given number of absences from work per month. Find the mean, variance, and standard deviation for each. Which attendance is the best? Which attendance is the most consistent?

Number of Absences per Month	0	1	2	3	4
Probability A	0.90	0.04	0.03	0.02	0.01
Probability B	0.85	0.05	0.05	0.05	0.00

33. Scholastic Aptitude Test Scores. The table gives the verbal SAT scores for recent years. Find the average, variance, and standard deviation of each. Are the scores of the males or females varying the most?

Year	1987	1988	1989	1990	1991
Males	435	435	434	429	426
Females	425	422	421	419	418

34. Drug Use. The table gives the percentage of current users of alcohol and cigarettes in the 18-to-25-year-old age group in the United States. Find the average, variance, and standard deviation of each. Which group is varying the most?

Year	1974	1979	1982	1985	1988	1991
Alcohol	69	76	71	71	65	64
Cigarettes	49	43	40	37	35	32

Enrichment Exercises

35. Give a proof of Property 5 of probability.

36. Give a proof of Property 6 of probability.

37. Give a proof of Property 7 of probability.

38. Give a proof of Property 8 of probability.

SOLUTIONS TO SELF-HELP EXERCISE SET 11.1

1. A tree diagram is shown in the figure at the top of page 595. On a $1 bet there is a probability of $\frac{18}{37}$ on the first spin of red occurring and of winning $1 ($2 less $1 bet). There is a probability of $\frac{18}{37}$ of black occurring and of losing $1. Then there is a probability of $\frac{1}{2} \cdot \frac{1}{37}$ of 0 and then red, which

results in breaking even and a probability of $\frac{1}{2} \cdot \frac{1}{37}$ of 0 and then black with a subsequent loss of $1. The expected value is

$$1 \cdot \frac{18}{37} - 1 \cdot \frac{18}{37} + 0 \cdot \frac{1}{2} \cdot \frac{1}{37} - 1 \cdot \frac{1}{2} \cdot \frac{1}{37} \approx -0.0135$$

On this bet you expect to lose about 1.35% per bet in the long run.

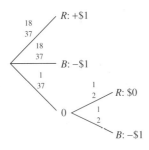

From the histograms one sees that the probability is evenly distributed about 3 in each case. Thus the mean should be 3. In the first case the probability is clustered about the mean, whereas in the second case the probability is more dispersed away from the mean. Thus the variance in the second case should be larger.

2. The following figure shows the histograms.

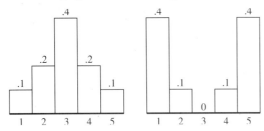

(a) $\mu = x_1p_1 + x_2p_2 + x_3p_3 + x_4p_4 + x_5p_5$

$= 1(0.1) + 2(0.2) + 3(0.4) + 4(0.2) + 5(0.1)$

$= 3$

$\text{var} = (x_1 - \mu)^2p_1 + (x_2 - \mu)^2 p_2 + (x_3 - \mu)^2 p_3 + (x_4 - \mu)^2 p_4 + (x_5 - \mu)^2 p_5$

$= (1 - 3)^2 (0.1) + (2 - 3)^2 (0.2) + (3 - 3)^2 (0.4) + (4 - 3)^2 (0.2) + (5 - 3)^2 (0.1)$

$= 1.2$

$\sigma = \sqrt{1.2} \approx 1.095$

(b) $\mu = 1(0.4) + 2(0.1) + 3(0.0) + 4(0.1) + 5(0.4)$

$= 3$

$\text{var} = (1 - 3)^2 (0.4) + (2 - 3)^2 (0.1) + (3 - 3)^2 (0.0) + (4 - 3)^2 (0.1) + (5 - 3)^2 (0.4)$

$= 3.4$

$\sigma = \sqrt{3.4} \approx 1.844$

11.2 CONTINUOUS PROBABILITY DENSITY FUNCTIONS

■ Continuous Probability Density Functions   ■ Applications

APPLICATION: THE LIFE SPAN OF A LIGHT BULB

How can you determine the probability that a light bulb will last a certain number of hours? See Example 6 for the answer.

Continuous Probability Density Functions

We have seen for a discrete random variable in a finite sample space how to associate the probability of an event with the area of rectangles. For a continuous random variable that can take any value in some interval (a, b), we can associate with the probability of an event an area under a curve. This curve is called a probability density function.

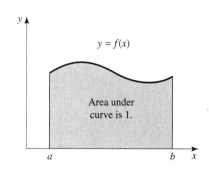

Figure 11.4
If $f(x)$ is to be a probability density function on (a, b), then $f(x) \geq 0$ and the area between the graph of $y = f(x)$ and the x-axis on (a, b) must be 1.

Probability Density Function

A function $f(x)$ defined on (a, b) is called a **probability density function** on (a, b) if the following are satisfied:

1. For all $x \in (a, b)$, $f(x) \geq 0$.
2. The area between the graph of $y = f(x)$ and the x-axis from a to b is one, that is,

$$\int_a^b f(x) \, dx = 1$$

For the interval (a, b), a may be $-\infty$, b may be ∞, or both (Fig. 11.4).

We can now define the probability associated with a continuous random variable.

Probability of a Continuous Random Variable

Suppose $f(x)$ is a continuous probability density function on (a, b) associated with the random variable X. The probability that the random variable X assumes a value in the interval $(c, d) \subset (a, b)$, written $P(c < X < d)$, is defined to be the area between the graph of $y = f(x)$ and the x-axis from c to d. That is,

$$P(c \leq X \leq d) = \int_c^d f(x) \, dx$$

(Fig. 11.5).

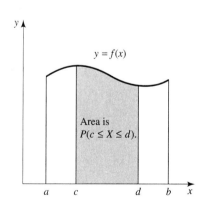

Figure 11.5
$P(c \leq X \leq d) = \int_c^d f(x) \, dx$

Remark. Since the area under a single point is zero,

$$P(X = c) = P(c \leq X \leq c) = \int_c^c f(x) \, dx = 0$$

and

$$P(c < X < d) = P(c \le X \le d)$$

Sometimes we can obtain the probability density function by theoretical analysis. Other times we can obtain the probability density function (or a close approximation) from an experiment or from the collection and analysis of data. We now give an example of each instance.

EXAMPLE 1   Determining a Probability Density Function Empirically

Table 11.1 gives the percentage distribution of the male population in the United States between the ages of 18 and 24 by height.

For example, the probability that a male between the ages of 18 and 24 will have a height between 69 and 70 inches is 0.1477. On your graphing calculator produce a histogram of this data using rectangles with base 1 unit like that shown in Figure 11.6. For example, the rectangle above 69.5 has base 1 unit and height 0.1477. Since the base of this rectangle is 1 unit, the area of the rectangle equals the height 0.1477, which is also the probability that the height of the males will be between 69 and 70 inches. Let $P(c \le X \le d)$ be the probability that one of the males considered here has a height between c and d inches. Find a function $f(x)$ such that

$$P(c \le X \le d) = \int_c^d f(x)\, dx$$

at least approximately.

SOLUTION ■ On your graphing calculator graph a line plot of the data over the histogram. You should obtain a graph like that shown in Figure 11.7. The func-

TABLE 11.1

Height (inches)	Percent	Height (inches)	Percent
<65	3.85	70–71	14.59
65–66	4.39	71–72	11.89
66–67	7.94	72–73	8.40
67–68	10.70	73–74	4.20
68–69	12.21	74–75	3.43
69–70	14.77	>75	3.83

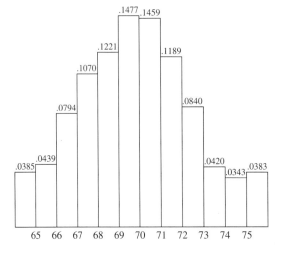

Figure 11.6
The area of each rectangle is the probability that the heights are between the numbers at the base of the rectangle.

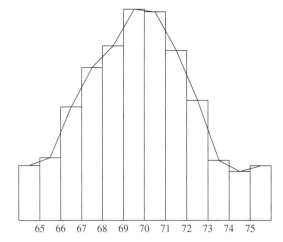

Figure 11.7
The straight line connecting the midpoints of the tops of the rectangles is approximately a probability density function.

65 66 67 68 69 70 71 72 73 74 75

tion $y = f(x)$ that gives the line plot is the function we are seeking. This is true, since according to Figure 11.7, the area under f on any subinterval $[c, d]$ is approximately the area of the rectangles enclosed by c and d. For example, the area under the curve $y = f(x)$ on the interval $[69, 70]$ is $\int_{69}^{70} f(x)\, dx$. This is also approximately the area of the rectangle above 69.5, which is the probability that the males under consideration will have heights between 69 and 70 inches. Thus,

$$P(69 \leq X \leq 70) \approx \int_{69}^{70} f(x)\, dx$$

and in general we have that

$$P(c \leq X \leq d) \approx \int_{c}^{d} f(x)\, dx$$

(If this were not accurate enough we could subdivide the 1-inch intervals further.) ∎

EXAMPLE 2   Determining a Probability Density Function by Theoretical Means

Suppose we have a pointer attached to the center of a circle of circumference 10 inches and free to spin in a completely random manner, as shown in Figure 11.8. Let X be the random variable given by the point on the interval $[0, 10]$ at which the pointer ends. (We assume that the pointer is perfectly sharp and does point at one single point.) Find the continuous function $f(x)$ that is defined on $[0, 10]$, such that

$$P(c \leq X \leq d) = \int_{c}^{d} f(x)\, dx$$

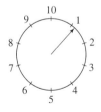

Figure 11.8
A random spinner.

SOLUTION ∎ From our knowledge of the physical setup we know that the probability that the pointer will give a value in any one interval is the same probability as any other interval of the *same* length. We now show that the probability density function must be a constant.

From the physical setup we know that for any $s, t \in [0, 10]$ and for any $h > 0$,

$$P(s \leq X \leq s + h) = P(t \leq X \leq t + h)$$

since the intervals $[s, s + h]$ and $[t, t + h]$ have the same length h. If we furthermore assume that $f(x)$ is *continuous*, then for very small h, $f(x)$ varies very little on

$[s, s + h]$, and the average value of $f(x)$ on $[s, s + h]$ is approximately equal to $f(s)$. Thus, we have

$$f(s) \approx \text{average value of } f(x) \text{ on } [s, s + h]$$

$$= \frac{1}{h} \int_s^{s+h} f(x)\, dx$$

$$= \frac{1}{h} P(x \le X \le s + h)$$

$$= \frac{1}{h} P(t \le X \le t + h)$$

$$= \frac{1}{h} \int_t^{t+h} f(x)\, dx$$

$$= \text{average value of } f(x) \text{ on } [t, t + h]$$

$$\approx f(t)$$

Since this approximation improves as h becomes smaller, we conclude that $f(s) = f(t)$. Since s and t were *any* points in $[0, 10]$, this means that $f(x)$ is a constant, say C.

Since $f(x) = C$, then

$$1 = P(0 \le X \le 10) = \int_0^{10} f(x)\, dx = \int_0^{10} C\, dx = 10C$$

and $f(x) = C = 1/10$. If an event E is some interval $E = [c, d]$, then

$$P(E) = P(c \le X \le d) = \int_c^d f(x)\, dx = \int_c^d \frac{1}{10}\, dx = \frac{d - c}{10} = \frac{\text{size of } E}{\text{size of } S}$$

This is intuitively what we expect, since this is analogous to a finite sample space with equally likely single events. ■

EXAMPLE 3   Finding a Probability Density Function

Find the only probability density function on $[a, b]$ that is a constant.

SOLUTION ■ Set $f(x) = k$. Then the first condition in the definition of a probability density function requires that $k \ge 0$. The second condition requires that

$$1 = \int_a^b f(x)\, dx = \int_a^b k\, dx = k(b - a)$$

Thus, $f(x) = k = 1/(b - a)$ (Fig. 11.9). ■

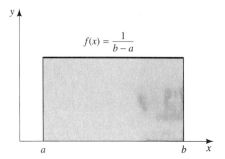

Figure 11.9
A uniform probability density function on $[a, b]$.

When the probability density function is constant, as in Figure 11.9, we say it is a **uniform** probability density function. Notice that in this case

$$P(c \le X \le d) = \int_c^d \frac{1}{b-a} \, dx = \frac{d-c}{b-a} = \frac{\text{length of } [c, d]}{\text{length of } [a, b]}$$

This is an extension of the notion in a *finite* sample space S with equally likely single events, since in this case the probability $P(E)$ of an event E is given by

$$P(E) = \frac{\text{size of } E}{\text{size of } S}$$

The uniform probability function arises in any situation in which *any two intervals of equal length give rise to the same probability.*

Uniform Probability Density Function

1. The probability distribution on $[a, b]$ with the constant probability density function $f(x) = \dfrac{1}{b-a}$ is called the **uniform probability function**.

2. For the uniform probability density function on $[a, b]$

$$P(c \le X \le d) = \frac{d-c}{b-a}$$

EXAMPLE 4   Finding a Probability Density Function

Find k so that the function $f(x) = kx$ is a probability density function on $[0, 1]$.

SOLUTION ■ The first condition requires that $k \ge 0$. The second condition requires that

$$1 = \int_a^b f(x) \, dx = \int_0^1 kx \, dx = k \left. \frac{1}{2} x^2 \right|_0^1 = k \frac{1}{2}$$

Thus, $k = 2$ and $f(x) = 2x$ is a probability density function on $[0, 1]$.   ■

We have assumed that the interval $[a, b]$ is *finite*. But there are many situations in which we need to consider infinite intervals. For example, we may wish to consider the probability that a light bulb will last for a certain amount of time. It is then convenient to take the interval under consideration to be $[0, \infty)$ if we have no idea of how long the bulb may last. We naturally continue to require that $f(x) \ge 0$ on the interval I under consideration and that the definite integral over I still be one, where we interpret the integral as an improper integral when I is infinite.

EXAMPLE 5   Finding a Probability Density Function
on an Infinite Interval

Given a positive constant k, find C so that the function $f(x) = Ce^{-kx}$ is a probability density function on $[0, \infty)$

SOLUTION ■ Again the first condition in the definition of a probability density function requires that $C \ge 0$. The second condition requires that

$$1 = \int_0^\infty f(x) \, dx$$

$$= \lim_{b \to \infty} \int_0^b Ce^{-kx} \, dx$$

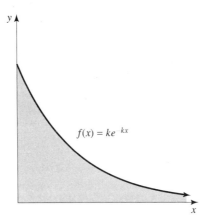

$f(x) = ke^{-kx}$

Figure 11.10
An exponential probability density function.

$$= \lim_{b \to \infty} \left. -\frac{C}{k} e^{-kx} \right|_0^b$$

$$= \lim_{b \to \infty} \frac{C}{k} (1 - e^{-kb})$$

$$= \frac{C}{k}$$

Thus, $C = k$ and $f(x) = ke^{-kx}$ is a probability density function on $[0, \infty)$. We refer to this as an **exponential probability density function** (Fig. 11.10). ■

Applications

The probability of the life of products often can be approximated by a probability with an exponential probability density function.

E X A M P L E 6   The Life Span of a Light Bulb

Suppose that the life span of a certain light bulb has a probability density function given by $f(t) = 0.01e^{-0.01t}$, where t is given in hours. Find the probability that the light bulb will last (a) at most 100 hours, (b) at least 100 hours, (c) between 100 and 200 hours.

S O L U T I O N ■ First notice from the previous example that $f(t) = 0.01e^{-0.01t}$ is in fact a probability density function, where $k = 0.01$.
(a) Here the interval is $[0, 100]$. Thus (see Fig. 11.11),

$$P(0 \le X \le 100) = \int_0^{100} f(x)\, dx = \int_0^{100} 0.01e^{-0.01t}\, dt$$
$$= \left. -e^{-0.01t} \right|_0^{100} = 1 - e^{-1} \approx 0.63$$

(b) Here the interval is $[100, \infty)$. Therefore we need to find

$$P(100 \le X) = \int_{100}^{\infty} f(x)\, dx$$

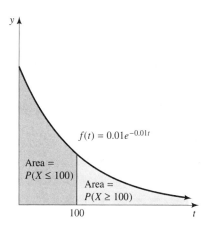

Figure 11.11
Notice that $P(X \geq 100) = 1 - P(X \leq 100)$.

where $f(x) = 0.01e^{-0.01t}$. This involves an improper integral. We can calculate the requested probability by avoiding this improper integral as follows:

$$P(100 \leq X) = 1 - P(X \leq 100)$$

$$= 1 - \int_0^{100} f(x)\, dx$$

$$\approx 1 - 0.63 \qquad \text{From (a)}$$

$$= 0.37$$

See Figure 11.11.

(c) Here the interval is [100, 200]. Thus (Fig. 11.12),

$$P(100 \leq X \leq 200) = \int_{100}^{200} f(x)\, dx$$

$$= \int_{100}^{200} 0.01e^{-0.01t}\, dt$$

$$= -e^{-0.01t}\big|_{100}^{200}$$

$$= e^{-1} - e^{-2} \approx 0.23 \quad \blacksquare$$

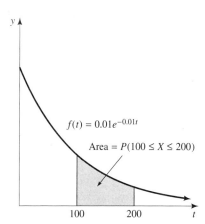

Figure 11.12
$P(100 \leq X \leq 200) = \int_{100}^{200} f(x)\, dx$

SELF-HELP EXERCISE SET 11.2

1. What must be the value of k if $f(x) = kx^2$ is to be a probability density function on [1, 4]?

2. The life in years of a certain automobile has a probability density function given by $f(t) = 0.1e^{-0.1t}$ on $[0, \infty)$. Find the probability that the life of the vehicle is (**a**) at most 5 years, (**b**) at least 5 years, (**c**) between 5 and 10 years.

EXERCISES 11.2

In Exercises 1 through 12 find k such that the given function is a probability density function on the given interval.

1. $f(x) = kx$, $[1, 2]$ **2.** $f(x) = kx$, $[2, 3]$

3. $f(x) = kx^2$, $[0, 1]$ **4.** $f(x) = kx^2$, $[1, 2]$

5. $f(x) = kx^3$, $[0, 2]$ **6.** $f(x) = kx^4$, $[-1, 0]$

7. $f(x) = \dfrac{k}{x}$, $[1, e]$ **8.** $f(x) = \dfrac{k}{x}$, $[1, e^2]$

9. $f(x) = \dfrac{k}{x^2}$, $[1, \infty)$ **10.** $f(x) = \dfrac{k}{x^3}$, $[1, \infty)$

11. $f(x) = \dfrac{kx^2}{(1 + x^3)^2}$, $[0, \infty)$

12. $f(x) = \dfrac{kx^2}{(1 + x^3)^3}$, $[0, \infty)$

In Exercises 13 through 22 determine why each of the functions is *not* a probability density function on the given interval.

13. $f(x) = \dfrac{3}{2}x - 1$, $[0, 2]$ **14.** $f(x) = 3 - x^2$, $[0, 3]$

15. $f(x) = \dfrac{2}{3}(x^2 - 3x + 2)$, $[0, 3]$

16. $f(x) = \dfrac{1}{9}(x^2 - 4x)$, $[0, 3]$

17. $f(x) = \dfrac{1}{(x + 2)^2}$, $[0, \infty)$

18. $f(x) = \dfrac{1}{(x + 3)^2}$, $[0, \infty)$

19. $f(x) = \dfrac{1}{\sqrt{x}}$, $[1, \infty)$ **20.** $f(x) = \dfrac{1}{x}$, $[1, \infty)$

21. $f(x) = e^{-x}$, $(-\infty, \infty)$ **22.** $f(x) = e^x$, $(-\infty, \infty)$

23. Show that the function $f(x) = x/8$ is a probability density function on $[0, 4]$. Find **(a)** $P(0 \le X \le 1)$, **(b)** $P(1 \le X \le 2)$, **(c)** $P(0 \le X \le 4)$.

24. Show that the function $f(x) = \dfrac{2x}{15}$ is a probability density function on $[1, 4]$. Find **(a)** $P(1 \le X \le 2)$, **(b)** $P(2 \le X \le 3)$, **(c)** $P(X = 3)$.

25. Show that the function $f(x) = 6x(1 - x)$ is a probability density function on $[0, 1]$. Find **(a)** $P(0 \le X \le 0.5)$, **(b)** $P(0.25 \le X \le 0.5)$, **(c)** $P(0.5 \le X \le 1)$.

26. Show that the function $f(x) = x^3/20$ is a probability density function on $[1, 3]$. Find **(a)** $P(1 \le X \le 2)$, **(b)** $P(2 \le X \le 3)$, **(c)** $P(X = 3)$.

Applications

27. Commuting Time. The time, measured in minutes, required by a certain individual to commute to work is a random phenomenon obeying a uniform probability law over the interval $[20, 25]$. If this person leaves home promptly at 7:06 AM, what is the probability that he will arrive at work by 7:29 AM?

28. Roulette. The circumference of a wheel is divided into 37 equal lengths (as in roulette) with numbers 0 through 36 marking the dividing lines. The wheel is twirled and after coming to rest, the point on the wheel located opposite a certain fixed marker is noted. Assuming the uniform probability, what is the probability that the point thus chosen will lie **(a)** between 0 and 10, **(b)** between two successive integers the first of which is odd, **(c)** on the number 7?

29. Waiting Time. A bus arrives at a certain stop every 30 minutes. A person arrives at this bus stop at random. What is the probability that this person **(a)** must wait no more than 5 minutes before the bus arrives, **(b)** must wait at least 12 minutes before the bus arrives, **(c)** must wait between 6 and 10 minutes, **(d)** must wait at least 30 minutes, **(e)** must wait at most 30 minutes?

30. Waiting Time. A geyser erupts every 60 minutes without fail. If a person arrives randomly to see this event not knowing the schedule of the geyser, what is the probability that this person will have to wait **(a)** at least 15 minutes, **(b)** at most 35 minutes, **(c)** between 30 and 45 minutes, **(d)** at most 1 hour?

31. Water Consumption. The daily demand for water, in thousands of gallons, in a city is a continuous random variable with probability density function $f(x) = 1 - 0.50x$, on the interval $[0, 2]$. **(a)** What is the probability that this city will use less than 1500 gallons of water in one day? **(b)** What is the probability that this city will use more than 1500 gallons of water in one day? **(c)** What is the probability that this city will use between 500 and 1500 gallons of water in one day?

32. Electricity Demand. The daily demand for electricity, in thousands of kilowatt-hours at an industrial plant is a continuous random variable with probability density function $f(x) = \dfrac{1}{21}(10 - x^2)$, on the interval $[0, 3]$. **(a)** What is the probability that this plant will use less than 2000 kilowatt-hours in one day? **(b)** What is the probability that this plant will use more than 1000 kilowatt-hours in one day? **(c)** What is the probability that this plant will use between 1000 and 2000 kilowatt-hours in one day?

33. Battery Life. The life in hours of a certain battery has a probability density function given by $f(t) = 0.15e^{-0.15t}$ on $[0, \infty)$. Find the probability that the life of the battery is **(a)** at most 10 hours, **(b)** at least 10 hours, **(c)** between 20 and 40 hours.

34. Life of Part. The life in days of a critical electronics component has a probability density function given by $f(t) = 0.001e^{-0.001t}$ on $[0, \infty)$. Find the probability that the life of this component is **(a)** at most 1000 days, **(b)** at least 1000 days, **(c)** between 1000 and 2000 days.

35. VCR Life. The life in years of a certain VCR is given by the probability density function $f(t) = 1/(t + 1)^2$ on $[0, \infty)$. Show that $f(t)$ is indeed a probability density function on $[0, \infty)$. Find the probability that the life of the VCR is **(a)** at most 2 years, **(b)** at least 2 years, **(c)** between 1 and 4 years.

36. Waiting Time. The waiting time in hours at an auto muffler shop is a random variable with a probability density function given by $f(x) = \frac{4}{15}(x + 1)^3$ on $[0, 1]$. Find the probability that you will have to wait **(a)** at most 30 minutes, **(b)** at least 30 minutes, **(c)** between 15 and 45 minutes.

37. Medicine. The length of time in days for recovery from a certain disease is a random variable with probability density function $f(t) = 0.02e^{-0.02t}$ on the interval $[0, \infty)$. Find the probability that a person with this disease will recover in **(a)** less than 5 days, **(b)** more than 5 days, **(c)** between 5 and 10 days.

38. Learning. The number of hours for a new employee to learn a certain task is a random variable with probability density function $f(x) = \dfrac{1}{(x + 1)^2}$, on the interval $[0, \infty)$. Find the probability that a new employee will learn this task in **(a)** less than 2 hours, **(b)** more than 2 hours, **(c)** between 1 and 2 hours.

Enrichment Exercises

39. Show that $f(x) = \dfrac{x^2}{\sqrt{2\pi}} e^{-x^2/2}$ is a probability density function on $(-\infty, \infty)$ by evaluating the integral $2\displaystyle\int_0^b \dfrac{x^2}{\sqrt{2\pi}} e^{-x^2/2}\, dx$ numerically for $b = 5$, 6, and 10.

40. Show that $f(x) = \dfrac{9}{\pi(1 + 9x^2)}$ is a probability density function on $\left(0, \dfrac{\sqrt{3}}{3}\right)$ by evaluating the integral $\displaystyle\int_0^{\sqrt{3}/3} \dfrac{9}{\pi(1 + 9x^2)}\, dx$ numerically.

41. Show that $f(x) = \dfrac{4}{\pi(x^2 - 2x + 2)}$ is a probability density function on the interval $(1, 2)$ by evaluating the integral $\displaystyle\int_1^2 \dfrac{4}{\pi(x^2 - 2x + 2)}\, dx$ numerically.

42. Show that $f(x) = \dfrac{1}{(\ln(2 + \sqrt{3}))(x \cos \ln x)}$ is a probability density function on the interval $(1, e^{\pi/3})$ by evaluating the integral $\displaystyle\int_1^{e^{\pi/3}} \dfrac{1}{(\ln(2 + \sqrt{3}))(x \cos \ln x)}\, dx$ numerically.

SOLUTIONS TO SELF-HELP EXERCISE SET 11.2

1. We must have
$$1 = \int_1^4 kx^2\, dx = \frac{k}{3} x^3 \Big|_1^4 = \frac{k}{3}(64 - 1) = 21k$$
Thus, $k = 1/21$.

2. (a) We are seeking $P(X \le 5)$. Then
$$P(X \le 5) = \int_0^5 0.1e^{-0.1t}\, dt =$$
$$-e^{-0.1t}\Big|_0^5 = 1 - e^{-0.5} \approx 0.39$$

(b) We are seeking $P(X \ge 5)$. Then
$$P(X \ge 5) = 1 - P(X \le 5) \approx 1 - 0.39 = 0.61$$

(c) We are seeking $P(5 \le X \le 10)$. Then
$$P(5 \le X \le 10) = \int_5^{10} 0.1e^{-0.1t}\, dt$$
$$= -e^{-0.1t}\Big|_5^{10}$$
$$= e^{-0.5} - e^{-1} \approx 0.24$$

11.3 EXPECTED VALUE AND VARIANCE

■ Mean, Variance, Standard Deviation ■ Median

APPLICATION: MEAN, VARIANCE, AND STANDARD DEVIATION OF THE LIFE OF A LIGHT BULB

Suppose that the life span of a certain light bulb is a random variable with a probability density function given by $f(t) = 0.01e^{-0.01t}$, where t is given in hours. What is the mean and variance? See Example 3 for the answer.

Mean, Variance, Standard Deviation

Recall from discrete probability that if $x_1, x_2, \ldots, x_n$ are a *finite* number n of possible values of a random variable X with probabilities $p_1, p_2, \ldots, p_n$, respectively, the *mean*, or *expected value*, is defined as

$$E(X) = x_1 p_1 + x_2 p_2 + \cdots + x_n p_n$$

We now want to extend the notion of expected value to the case of a continuous probability distribution $f(x)$ on the interval (a, b). But we can no longer talk about the probability that the random variable takes on *one single* value but rather must consider the probability that the random variable takes on values in some *interval*.

If the interval (a, b) is finite, we divide the interval $[a, b]$ into n subintervals, I_1, $I_2, \ldots, I_n$, each of length $\Delta x = \dfrac{b - a}{n}$. In each subinterval I_i pick a point $x_i \in I_i$, say at the midpoint as shown in Figure 11.13. The probability that the random variable X lies in I_i is precisely the area under the curve $y = f(x)$ on the interval I_i. This area (see Fig. 11.13) is approximately equal to the area of the rectangle $f(x_i)\, \Delta x$. Thus, we have

$$
\begin{aligned}
E(X) &= x_1 P(x \in I_1) + x_2 P(x \in I_2) + \cdots + x_n P(x \in I_n) \\
&\approx x_1 f(x_1)\, \Delta x + x_2 f(x_2)\, \Delta x + \cdots + x_n f(x_n)\, \Delta x \\
&= \sum_{i=1}^{n} x_i f(x_i)\, \Delta x
\end{aligned}
$$

which is a Riemann sum for $xf(x)$ on $[a, b]$. The larger n is the better the approximation. Now letting $n \to \infty$ we obtain by definition the definite integral $\displaystyle\int_a^b xf(x)\, dx$. We have then motivated the following definition.

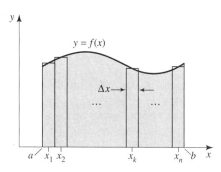

Figure 11.13
The probability that the random variable lies in I_k is the area under the curve $y = f(x)$ on the interval I_k.

Expected Value, or Mean, of a Continuous Random Variable

The **expected value** $E(X)$ (or **mean** μ) of a random variable X with a continuous probability density function $f(x)$ on the interval (a, b), is defined to be

$$E(X) = \mu = \int_a^b xf(x)\, dx$$

Recall from discrete probability that if $x_1, x_2, \ldots, x_n$ are a *finite* number n of possible values of a random variable X with probabilities $p_1, p_2, \ldots, p_n$, respectively, the *variance* measures the spread of these values and is defined as

$$
\begin{aligned}
\mathrm{Var}(X) &= (x_1 - \mu)^2 p_1 + (x_2 - \mu)^2 p_2 + \cdots + (x_n - \mu)^2 p_n \\
&= \sum_{i=1}^{n} (x_i - \mu)^2 p_i
\end{aligned}
$$

If we are to define a corresponding number in the case of a continuous random variable with continuous probability density function $f(x)$ on the interval (a, b), we proceed as we did in defining the expected value. Then

$$\text{Var}(X) = \sum_{i=1}^{n} (x_i - \mu)^2 P(x_i \in I_i)$$

$$\approx \sum_{i=1}^{n} (x_i - \mu)^2 f(x_i) \, \Delta x$$

which is a Riemann sum for $(x - \mu)^2 f(x)$ over $[a, b]$. Letting $n \to \infty$ we obtain the definite integral $\int_a^b (x - \mu)^2 f(x) \, dx$. This motivates the following definition:

Variance of a Continuous Random Variable

The **variance** of a random variable X with a continuous probability density function $f(x)$ on the interval (a, b), is defined to be

$$\text{Var}(X) = \int_a^b (x - \mu)^2 f(x) \, dx$$

This formula for variance can be simplified. First recall that

$$\mu = \int_a^b x f(x) \, dx \qquad \text{and} \qquad \int_a^b f(x) \, dx = 1$$

Then

$$\text{Var}(X) = \int_a^b (x - \mu)^2 f(x) \, dx$$

$$= \int_a^b (x^2 - 2x\mu + \mu^2) \, f(x) \, dx$$

$$= \int_a^b x^2 f(x) \, dx - 2\mu \int_a^b x f(x) \, dx + \mu^2 \int_a^b f(x) \, dx$$

$$= \int_a^b x^2 f(x) \, dx - 2\mu\mu + \mu^2$$

$$= \int_a^b x^2 f(x) \, dx - \mu^2$$

Alternative Form of Variance

If $f(x)$ is a continuous probability density function on the interval (a, b) with mean μ, an alternative formula for the variance $\text{Var}(X)$ is given by

$$\text{Var}(X) = \int_a^b x^2 f(x) \, dx - \mu^2$$

We also define the standard deviation σ as the square root of the variance.

Standard Deviation

The **standard deviation**, denoted by σ, is the square root of the variance, that is,

$$\sigma = \sqrt{\text{Var}(X)}$$

EXAMPLE 1   Finding the Mean and the Variance

Find the mean and variance of the random variable with the constant or uniform
probability density function $f(x) = 1/b$ on $[0, b]$.

SOLUTIONS ■   (a) For the mean we have

$$\mu = \int_0^b xf(x)\,dx$$

$$= \frac{1}{b}\int_0^b x\,dx$$

$$= \frac{1}{b}\cdot\frac{1}{2}x^2\Big|_0^b$$

$$= \frac{1}{b}\cdot\frac{1}{2}(b^2 - 0)$$

$$= \frac{b}{2}$$

(b) We use the simplified formula for Var(X) and first find $\int_0^b x^2 f(x)\,dx$. Thus,

$$\int_0^b x^2 f(x)\,dx = \frac{1}{b}\int_0^b x^2\,dx$$

$$= \frac{1}{b}\cdot\frac{1}{3}x^3\Big|_0^b$$

$$= \frac{1}{b}\cdot\frac{1}{3}(b^3 - 0)$$

$$= \frac{1}{3}b^2$$

Now from (a) we have $\mu^2 = \frac{1}{4}b^2$, thus,

$$\text{Var}(X) = \int_0^b x^2 f(x)\,dx - \mu^2$$

$$= \frac{1}{3}b^2 - \frac{1}{4}b^2$$

$$= \frac{1}{12}b^2 \quad ■$$

We can do similar calculations to obtain the following:

Mean, Variance, and Standard Deviation of a Random Variable with a Constant Probability Density Function

Let X be a random variable with constant probability density function $f(x) = 1/(b - a)$ on $[a, b]$. Then

$$\mu = \frac{1}{2}(b + a)$$

$$\text{Var}(X) = \frac{1}{12}(b - a)^2$$

$$\sigma = \frac{1}{2\sqrt{3}}(b - a)$$

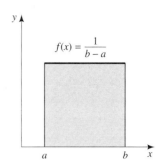

Figure 11.14
A uniform probability density function on $[a, b]$ has a variance of $(b - a)^2/12$.

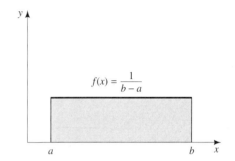

Figure 11.15
Notice that since b is larger than that in Figure 11.14, the variance is larger. The probability density function is spread out more.

Remark. According to Figures 11.14 and 11.15, the probability density function spreads out as b increases. But since $\text{Var}(X) = \dfrac{(b - a)^2}{12}$, increasing b is the same as increasing $\text{Var}(X)$. This shows how the variance measures the spread of a probability density function. A larger variance means a more spread-out function.

On unbounded intervals the formulas are the same but require improper integrals.

EXAMPLE 2   Finding Mean and Variance on an Infinite Interval

Find the mean and the variance of the random variable with probability density function given by $f(x) = 3x^{-4}$ on the unbounded interval $[1, \infty)$. Refer to Figure 11.16.

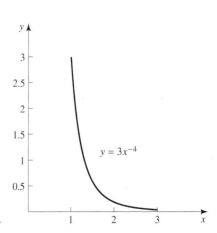

Figure 11.16
$f(x) = 3x^{-4}$ is a probability density function on $[1, \infty)$.

SOLUTIONS ■

(a) We have

$$\mu = \int_1^{\infty} xf(x)\,dx$$

$$= \lim_{b\to\infty} \int_1^b x3x^{-4}\,dx$$

$$= \lim_{b\to\infty} \int_1^b 3x^{-3}\,dx$$

$$= \lim_{b\to\infty} \left(-\frac{3}{2}x^{-2}\Big|_1^b \right)$$

$$= \lim_{b\to\infty} \frac{3}{2}(1 - b^{-2})$$

$$= \frac{3}{2}$$

(b) We have

$$\text{Var}(X) = \int_1^{\infty} x^2 f(x)\,dx - \mu^2$$

$$= \lim_{b\to\infty} \int_1^b 3x^{-2}\,dx - \left(\frac{3}{2}\right)^2$$

$$= \lim_{b\to\infty} \left(-3x^{-1}\Big|_1^b \right) - \frac{9}{4}$$

$$= \lim_{b\to\infty} 3(1 - b^{-1}) - \frac{9}{4}$$

$$= \frac{3}{4}$$

(c) $\sigma = \sqrt{\text{Var}(X)} = \sqrt{3}/2$ ■

The following result for the exponential probability density function is left as Exercise 34.

Mean, Variance, and Standard Deviation of a Random Variable with an Exponential Probability Density Function

Let X be a random variable with an exponential probability density function $f(x) = ke^{-kx}$, $k > 0$, on $[0, \infty)$. Then

$$\mu = \frac{1}{k}$$

$$\text{Var}(X) = \frac{1}{k^2}$$

$$\sigma = \frac{1}{k}$$

EXAMPLE 3   Finding the Mean and Variance on an Infinite Interval

Suppose that the life span of a certain light bulb is a random variable with a probability density function given by $f(t) = 0.01e^{-0.01t}$, where t is given in hours (Fig. 11.17). Find the mean, variance, and standard deviation.

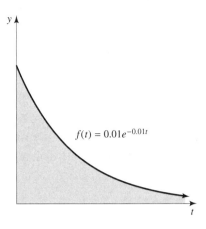

Figure 11.17
$f(x) = 0.01e^{-0.01t}$ is a probability density function on $[0, \infty)$.

S O L U T I O N ■ Substituting $k = 0.01$ in the previous example yields

$$\mu = \frac{1}{k} = \frac{1}{0.01} = 100$$

$$\mathrm{Var}(X) = \frac{1}{k^2} = \frac{1}{(0.01)^2} = 10{,}000$$

$$\sigma = \sqrt{\mathrm{Var}(X)} = 100 \quad ■$$

We then expect a typical light bulb from this group to last 100 hours, but the lifetimes will vary greatly by virtue of the large variance.

Median

For any continuous random variable, the median outcome x_m is that outcome for which $P(X \le x_m) = \frac{1}{2}$.

The Median of a Continuous Random Variable

If $f(x)$ is the probability density function on (a, b) associated with this continuous random variable, then the **median**, denoted by x_m, is that number for which

$$\int_a^{x_m} f(x)\, dx = \frac{1}{2}$$

Remark. As noted in Figure 11.18 the area under $f(x)$ on $[a, x_m]$ is exactly $\frac{1}{2}$, as is the area under $f(x)$ on $[x_m, b]$.

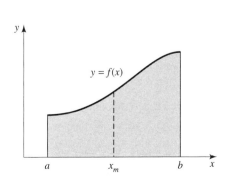

Figure 11.18
At the median x_m, the area to the left and the area to the right are both equal to 0.5.

EXAMPLE 4 Finding the Median

Find the median of the random variable with probability density function given by

$$f(x) = ke^{-kx} \quad k > 0$$

on the unbounded interval $[0, \infty)$ (Fig. 11.19).

SOLUTION ■

$$\frac{1}{2} = \int_0^{x_m} ke^{-kx} \, dx$$
$$= -e^{-kx}\big|_0^{x_m}$$
$$= 1 - e^{-kx_m}$$

Thus,

$$e^{-kx_m} = \frac{1}{2}$$

$$-kx_m = \ln\frac{1}{2} = -\ln 2$$

$$x_m = \frac{1}{k}\ln 2 \approx 0.693\,\frac{1}{k} \quad ■$$

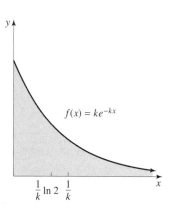

$f(x) = ke^{-kx}$

$\frac{1}{k}\ln 2 \quad \frac{1}{k}$

Figure 11.19

The exponential probability density function has mean $\frac{1}{k}$ and median $\frac{1}{k}\ln 2$.

EXAMPLE 5 Finding the Median

Find the median of the random variable with probability density function given by

$$f(x) = \frac{2}{\sqrt{2\pi}}e^{-0.5x^2}$$

on the unbounded interval $[0, \infty)$ (Fig. 11.20).

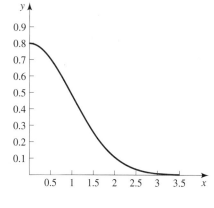

Figure 11.20

$f(x) = \dfrac{2}{\sqrt{2\pi}}e^{-0.5x^2}$ is a probability density function on $[0, \infty)$.

SOLUTION ■ We need to solve the equation

$$\frac{1}{2} = \int_0^x \frac{2}{\sqrt{2\pi}} e^{-0.5t^2}\, dt$$

for x to obtain the median. Unfortunately, there is no known way of finding this integral. We can, however, find an approximate solution using a graphing calculator. Define $F(x)$ to be the integral. Then use the program NINTG to draw a graph. Use a window with dimensions [0, 1] by [0, 1] and obtain Screen 11.1, the intersection of $y = F(x)$ and $y = \frac{1}{2}$ occurs when $x \approx 0.67$. Thus, the median is approximately 0.67. ■

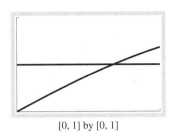

[0, 1] by [0, 1]

Screen 11.1

The graphs of $y = \frac{2}{\sqrt{2\pi}} \int_0^x e^{-0.5x^2}\, dx$ and $y = 0.5$ intersect at about 0.67.

SELF-HELP EXERCISE SET 11.3

1. Find the mean, variance, and standard deviation of the random variable with probability density function $f(x) = 2x$ on [0, 1].

2. Find the mean, variance, and standard deviation of the random variable with exponential probability density function $f(x) = 0.5e^{-0.5x}$ on [0, ∞).

EXERCISES 11.3

In Exercises 1 through 10 find the **(a)** mean μ, **(b)** variance, **(c)** standard deviation σ, and **(d)** median of the given probability density functions on the given interval.

1. $f(x) = \frac{1}{4}$, [1, 5]   **2.** $f(x) = \frac{1}{8}$, [2, 10]

3. $f(x) = \frac{x}{8}$, [0, 4]   **4.** $f(x) = \frac{2x}{15}$, [1, 4]

5. $f(x) = 6x(1 - x)$, [0, 1]

6. $f(x) = \frac{x^3}{20}$, [1, 3]

7. $f(x) = 0.001e^{-0.001x}$, [0, ∞)

8. $f(x) = 0.15e^{-0.15x}$, [0, ∞)

9. $f(x) = \frac{4}{x^5}$, [1, ∞)   **10.** $f(x) = \frac{5}{x^6}$, [1, ∞)

11. Find the formula for the median of the uniform probability density function $f(x) = 1/(b - a)$ on [a, b].

12. For the uniform probability density function $f(x) = 1/(b - a)$, on [a, b], show that

$$P(\mu - \sigma \le X \le \mu + \sigma) = \frac{1}{\sqrt{3}}$$

where μ and σ are, respectively, the mean and standard deviation of the uniform distribution.

13. For the probability density function $f(x) = ke^{-kx}$, $k > 0$, on [0, ∞), show that

$$P(\mu - \sigma \le X \le \mu + \sigma) = 1 - e^{-2}$$

where μ and σ are, respectively, the mean and standard deviation of the distribution.

14. For the probability density function $f(x) = ke^{-kx}$, $k > 0$, on [0, ∞), show that

$$P(X \le \mu) = 1 - e^{-1}$$

where μ is the mean.

15. For the probability density function $f(x) = ke^{-kx}$, $k > 0$, on [0, ∞), show that

$$P(\mu \le X \le \infty) = e^{-1}$$

where μ is the mean of the distribution.

16. Let $f(x) = xe^{-x}$. **(a)** Show $f(x)$ is a probability density function on [0, ∞); **(b)** find the mean; **(c)** find the variance.

Applications

17. Waiting Time. The waiting time, in minutes, in the speedy checkout line in a certain grocery store is a continuous random variable with a probability density function $f(x) = 0.20 - 0.02x$, on the interval [0, 10]. Find the expected waiting time, the variance, and the standard deviation.

18. Product Life. The time in years before a product wears out is a continuous random variable with a probability density function $f(x) = 0.15 - 0.0015x^2$ on the interval [0, 10]. Find the expected life of the product, the variance, and the standard deviation.

19. **Heating Oil Consumption.** The daily consumption for heating oil, in thousands of gallons, in a town is a continuous random variable with probability density function $f(x) = 0.06x - 0.006x^2$ on the interval $[0, 10]$. Find the expected consumption, the variance, and the standard deviation.

20. **Demand.** The daily demand for electricity in thousands of kilowatt-hours in a city is a continuous random variable with probability density function $f(x) = 0.012x^2 - 0.0012x^3$ on the interval $[0, 10]$. Find the expected demand, the variance, and the standard deviation.

21. **Water Consumption.** The daily demand for water, in thousands of gallons, in a city is a continuous random variable with probability density function $f(x) = 1 - 0.50x$ on the interval $[0, 2]$. Find the expected demand, the variance, and the standard deviation.

22. **Electricity Demand.** The daily demand for electricity, in thousands of kilowatt-hours, at an industrial plant is a continuous random variable with probability density function $f(x) = \frac{1}{21}(10 - x^2)$, on the interval $[0, 3]$. Find the expected demand, the variance, and the standard deviation.

23. **Revenue.** The daily revenue, in thousands of dollars, of a certain company is a continuous random variable with a probability density function $f(x) = \frac{3}{(x+1)^4}$, on the interval $[0, \infty)$. Estimate the expected revenue, the variance, and the standard deviation using Simpson's rule with $n = 200$ on the interval $[0, 50]$. (See Exercise 35 for additional details.)

24. **Profit.** The daily profit, in thousands of dollars, of a certain company is a continuous random variable with a probability density function $f(x) = \frac{4}{(x+1)^5}$, on the interval $[0, \infty)$. Estimate the expected profit, the variance, and the standard deviation using Simpson's rule with $n = 200$ on the interval $[0, 50]$.

25. **Shelf Life.** The shelf life, in months, of a certain cereal is a continuous random variable with a probability density function $f(x) = \frac{3}{14}\sqrt{x+1}$, on $[0, 3]$. Find the expected shelf life, the variance, and the standard deviation. Use Simpson's rule with $n = 50$.

26. **Waiting Time.** The waiting time in hours at an auto muffler shop is a random variable with a probability density function given by $f(x) = \frac{4}{15}(x+1)^3$ on $[0, 1]$. What is the expected waiting time, the variance, and the standard deviation? Use Simpson's rule with $n = 50$.

27. **Medicine.** The length of time in days for recovery from a certain disease is a random variable with probability density function $f(t) = 0.02e^{-0.02t}$ on the interval $[0, \infty)$. Find the expected recovery time, the variance, and the standard deviation.

28. **Learning.** The number of hours for a new employee to learn a certain task is a random variable with probability density function $f(x) = \frac{6}{x^7}$, on the interval $[1, \infty)$. Find the expected learning time, the variance, and the standard deviation.

Enrichment Exercises

29. Obtain the answers to Exercise 23 by evaluating the integrals exactly, that is, do not use numerical methods.

30. Obtain the answers to Exercise 24 by evaluating the integrals exactly, that is, do not use numerical methods.

31. Obtain the answers to Exercise 25 by evaluating the integrals exactly, that is, do not use numerical methods.

32. Obtain the answers to Exercise 26 by evaluating the integrals exactly, that is, do not use numerical methods.

33. Let $f(x) = (\beta + 1)(\beta + 2)x^\beta(1 - x)$. Then $f(x)$ is a probability density function on $[0, 1]$ for any $\beta \geq 0$. (a) Take $\beta = 2$, and find the median numerically as was done in Example 5. (b) Can you find the median exactly by integrating and solving the appropriate equation? Why or why not?

34. Find the mean and the variance of the random variable with probability density function given by $f(x) = ke^{-kx}$, $k > 0$, on the unbounded interval $[0, \infty)$.

35.* The integrals in Exercise 23 do not converge rapidly enough to deduce the correct answers using Simpson's rule with $n = 200$.

(a) Check this by graphing $y = 3x(x+1)^{-4}$ on a screen with dimensions $[0, 4.7]$ by $[0, 0.5]$.
(b) To overcome the difficulty of slow convergence let $x = e^{6t}$ and show that $\mu = \int_{-\infty}^{\infty} \frac{18e^{12t}}{(e^{6t}+1)^4}dt$
(c) Graph $y = \frac{18e^{12t}}{(e^{6t}+1)^4}$ on a screen with dimensions $[-1.175, 1.175]$ by $[0, 1.5]$. Notice how rapidly this function approaches zero.
(d) Now evaluate the integral in part (b) on the interval $[-2, 2]$ using Simpson's rule with $n = 100$.

36.* (Continuation of Exercise 35.)
(a) With $x = e^{6t}$ show that
$$\text{Var} = \int_{-\infty}^{\infty} \frac{18e^{18t}}{(e^{6t}+1)^4}dt - \mu^2$$
(b) Graph $y = 18e^{18t}(e^{6t}+1)^{-4}$ on a screen with dimensions $[-4, 4]$ by $[0, 1.5]$ and notice how rapidly this function approaches zero.
(c) Evaluate the integral in part (a) on the interval $[-4, 4]$ using Simpson's rule with $n = 100$, $n = 200$.

*This exercise is due to George Hukle of the University of Kansas.

SOLUTIONS TO SELF-HELP EXERCISE SET 11.3

1.

$$\mu = \int_a^b xf(x)\,dx = \int_0^1 x(2x)\,dx = \frac{2}{3}x^3 \Big|_0^1 = \frac{2}{3}$$

$$\begin{aligned} \text{Var}(X) &= \int_a^b x^2 f(x)\,dx - \mu^2 \\ &= \int_0^1 x^2(2x)\,dx - \left(\frac{2}{3}\right)^2 \\ &= \frac{1}{2}x^4 \Big|_0^1 - \frac{4}{9} = \frac{1}{18} \end{aligned}$$

$$\sigma = \sqrt{\text{Var}(X)} = \sqrt{\frac{1}{18}} = \frac{1}{3\sqrt{2}}$$

2. According to the formulas developed in the text:

$$\mu = \frac{1}{k} = \frac{1}{0.5} = 2$$

$$\text{Var}(X) = \frac{1}{k^2} = \frac{1}{(0.5)^2} = 4$$

$$\sigma = \frac{1}{k} = 2$$

11.4 THE NORMAL DISTRIBUTION

■ The Standard Normal Distribution   ■   The Normal Distribution

---

APPLICATION: FINDING NORMALLY DISTRIBUTED PROBABILITIES

Assume the weights of adult males in this country have a mean of 180 pounds and a standard deviation of 30 pounds. What is the probability that the weight of an adult male in this country is between 160 and 200 pounds? See Example 2 for the answer.

The Standard Normal Distribution

The normal probability distribution is arguably the most important of the probability distributions. The reason is that many natural phenomena obey a normal distribution, and also, under various conditions, the normal distribution approximates many other probability laws and is easy to work with.

We first must consider the *standard* normal distribution.

Standard Normal Distribution

The random variable Z with continuous probability density function given by

$$f(z) = \frac{1}{\sqrt{2\pi}}\, e^{-z^2/2}$$

on the interval $(-\infty, \infty)$ is said to satisfy the **standard normal distribution** with mean $\mu = 0$ and standard deviation $\sigma = 1$.

Our convention is to reserve the letter Z for a random variable that has a standard normal distribution.

The graph of this function is shown in Figure 11.21. Notice its bell-shaped character and its symmetry about $x = 0$. We can readily see that it *is* a probability density function, that is,

$$\int_{-\infty}^{\infty} \frac{1}{\sqrt{2\pi}}\, e^{-z^2/2}\, dz = 1$$

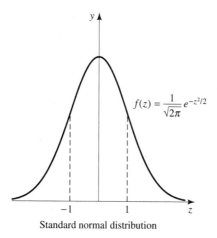

$$f(z) = \frac{1}{\sqrt{2\pi}} e^{-z^2/2}$$

Standard normal distribution

Figure 11.21

A graph of the standard normal distribution.

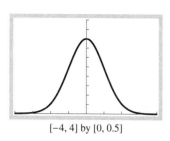

[−4, 4] by [0, 0.5]

Screen 11.2

A graph of $y = \dfrac{1}{\sqrt{2\pi}} e^{-0.5x^2}$

TABLE 11.2

Interval	Area
[−1, 1]	0.682689493
[−2, 2]	0.954499733
[−3, 3]	0.997300192
[−4, 4]	0.999936654
[−5, 5]	0.999999427
[−6, 6]	0.999999998

First graph $y = \dfrac{1}{\sqrt{2\pi}} e^{-z^2/2}$ using a screen with dimensions $[-3, 3]$ by $[0, 0.5]$, and obtain Screen 11.2. Notice how quickly the function drops to near zero. Now integrate $2 \displaystyle\int_0^L \dfrac{1}{\sqrt{2\pi}} e^{-x^2/2}\, dx$ over some intervals using Simpson's rule with $n = 50$ and obtain Table 11.2.

From the table we not only see that the area under the graph of the function on $(-\infty, \infty)$ is 1 but also just how little area is under the "tails."

Since this distribution is evenly spread about the origin, one suspects that $\mu = 0$, which is the case. It can be shown that the standard deviation σ is 1. (See Exercise 39.)

According to Table 11.2, about two-thirds of the area under the curve is found over the interval $[-1, 1]$, whereas slightly more than 95% of the area under the curve is found over the interval $[-2, 2]$. Slightly more than 99.7% of the area under the curve is found over the interval $[-3, 3]$ (Fig. 11.22). These numbers should be kept in mind as rough checks on calculations.

Since

$$P(c \le Z \le d) = \int_c^d \frac{1}{\sqrt{2\pi}} e^{-z^2/2}\, dz$$

naturally, we need to calculate this definite integral. However, recall that the function $e^{-x^2/2}$ is a function that can only be integrated by numerical means! But the importance

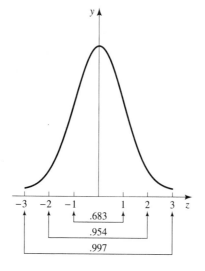

.683

.954

.997

Figure 11.22

Areas under the graph of the standard normal distribution over various intervals.

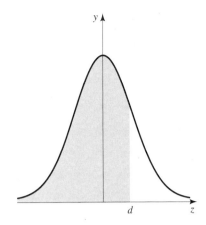

Figure 11.23
The area under the curve for the standard normal distribution to the left of d, shown as the shaded region, is designated as $\mathcal{A}(d)$.

of the normal distribution is so great that tables have been created that can be used to find such integrals. The tables give the values of

$$\int_{-\infty}^{d} \frac{1}{\sqrt{2\pi}} \, e^{-z^2/2} \, dz$$

for various discrete values of d which is the area under the curve to the left of d, shown as the shaded region in Figure 11.23. We designate this area as $\mathcal{A}(d)$. Use the table in Appendix B.3 to find this number.

If we are asked for $P(Z \le d)$, this is just $\mathcal{A}(d)$. Thus,

$$P(Z \le d) = \mathcal{A}(d)$$

where one looks up $\mathcal{A}(d)$ in Appendix B.3.

If we are asked for $P(d \le Z)$, this is

$$\int_{d}^{\infty} \frac{1}{\sqrt{2\pi}} \, e^{-z^2/2} \, dz$$

which is also the unshaded area under the curve in Figure 11.23, that is, the area to the right of d. This area also is the entire area under the curve (which is one) less the area under the curve to the left of d. Thus,

$$P(d \le Z) = 1 - \mathcal{A}(d)$$

Finally if we are asked for $P(c \le Z \le d)$ this is

$$\int_{c}^{d} \frac{1}{\sqrt{2\pi}} \, e^{-z^2/2} \, dz$$

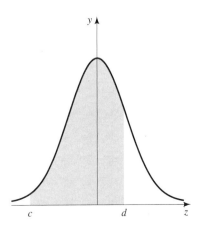

Figure 11.24
The area under the curve for the standard normal distribution between c and d, shown as the shaded region, is $\mathcal{A}(d) - \mathcal{A}(c)$.

which is the area under the curve from c to d shown in Figure 11.24. This can be obtained as the area to the left of d, $\mathcal{A}(d)$, less the area to the left of c, $\mathcal{A}(c)$. Thus,

$$P(c \le Z \le d) = \mathcal{A}(d) - \mathcal{A}(c)$$

where both the numbers $\mathcal{A}(d)$ and $\mathcal{A}(c)$ are obtained in Appendix B.3.

Finding the Probabilities of a Random Variable with a Standard Normal Distribution

If Z is a random variable with a standard normal probability density function, then

$$P(Z \le d) = \int_{-\infty}^{d} \frac{1}{\sqrt{2\pi}} e^{-z^2/2}\, dz = \mathcal{A}(d)$$

$$P(d \le Z) = \int_{d}^{\infty} \frac{1}{\sqrt{2\pi}} e^{-z^2/2}\, dz = 1 - \mathcal{A}(d)$$

$$P(c \le Z \le d) = \int_{c}^{d} \frac{1}{\sqrt{2\pi}} e^{-z^2/2}\, dz = \mathcal{A}(d) - \mathcal{A}(c)$$

EXAMPLE 1   Finding Probabilities of a Normally Distributed Random Variable

Using the table of the normal distribution found in Appendix B.3 find (a) $P(Z \le 1.12)$, (b) $P(Z \ge 1.12)$, (c) $P(-0.50 \le Z \le 0.71)$.

SOLUTIONS ■ (a) Here

$$P(Z \le 1.12) = \mathcal{A}(1.12)$$

which, according to the table in Appendix B.3, is 0.8686.
  (b) Here

$$P(Z \ge 1.12) = 1 - \mathcal{A}(1.12) = 1 - 0.8686 = 0.1314$$

  (c) Here

$$P(-0.50 \le Z \le 0.71) = \mathcal{A}(0.71) - \mathcal{A}(-0.50)$$

which, according to the table in Appendix B.3, is

$$0.7611 - 0.3085 = 0.4526 \quad ■$$

The Normal Distribution

We are now prepared to consider the normal distribution.

The Normal Distribution

The continuous random variable X is normally distributed with mean μ and standard deviation σ if its probability density function is given by

$$f(x) = \frac{1}{\sigma\sqrt{2\pi}} e^{-0.5[(x-\mu)/\sigma]^2} \quad \text{on } (-\infty, \infty)$$

(Fig. 11.25).

Showing that this function *is* a probability function, that is,

$$\int_{-\infty}^{\infty} \frac{1}{\sigma\sqrt{2\pi}} e^{-0.5[(x-\mu)/\sigma]^2}\, dx = 1$$

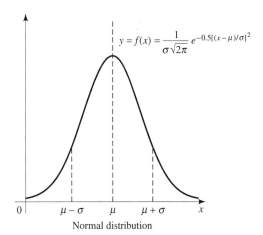

Figure 11.25
A typical graph of a normal distribution with mean μ and standard deviation σ.

Normal distribution

and that its mean is μ, and that its standard deviation is σ is beyond the scope of this text.

Notice that the standard normal distribution is obtained by setting $\mu = 0$ and $\sigma = 1$.

Figure 11.26 shows some normal density functions for $\sigma = 1$ and different values of μ. The curves for different values of μ and the same σ are simply translations of each other. Figure 11.27 shows several normal density functions all with $\mu = 0$ and different values of σ. The curves are all centered on the same point $\mu = 0$ but are more spread out for larger σ.

A normal probability density function has many properties that make it easy to manipulate. Thus, for mathematical convenience we sometimes assume that a random phenomenon has a normal density function if its actual density function has a shape similar to the bell shape of the normal density function. An example is the probability distribution function associated with the weight of a human being. In doing this one must, however, exercise some care since the "tails" of the normal density function have small but finite areas. Thus, we might conclude that there is *some* probability that a human being could have *negative* weight.

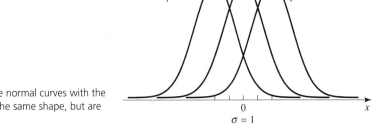

Figure 11.26
Notice that the normal curves with the same σ have the same shape, but are centered at μ.

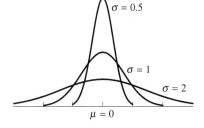

Figure 11.27
Since $\mu = 0$ all these normal curves are centered at $x = 0$. The larger the standard deviation σ, the more spread out the curves.

Also it may happen that a random phenomenon does not have a normal density function but that a simple numerical transformation of the measurement does.

Naturally, given specific values for μ and σ, we are interested in calculating

$$P(c \leq X \leq d) = \int_c^d \frac{1}{\sigma\sqrt{2\pi}} e^{-0.5[(x-\mu)/\sigma]^2} \, dx$$

for a whole variety of numbers c and d. Do we need a complete table of values for every pair of numbers μ and σ? Fortunately not.

A simple substitution changes this definite integral to one that has the *standard* probability density function as the integrand. Let

$$z = \frac{x - \mu}{\sigma}$$

Then $dz = dx/\sigma$. Also when $x = c$, $z = (c - \mu)/\sigma$ and when $x = d$, $z = (d - \mu)/\sigma$. Thus,

$$P(c \leq X \leq d) = \int_c^d \frac{1}{\sigma\sqrt{2\pi}} e^{-0.5[(x-\mu)/\sigma]^2} \, dx$$
$$= \int_{(c-\mu)/\sigma}^{(d-\mu)/\sigma} \frac{1}{\sigma\sqrt{2\pi}} e^{-0.5z^2} \, \sigma \, dz$$
$$= \int_{(c-\mu)/\sigma}^{(d-\mu)/\sigma} \frac{1}{\sqrt{2\pi}} e^{-0.5z^2} \, dz$$

This is the area under the *standard* normal density function on the interval $[(c - \mu)/\sigma, (d - \mu)/\sigma]$, which is $\mathcal{A}[(d - \mu)/\sigma] - \mathcal{A}[(c - \mu)/\sigma]$. This area can be found in the table in Appendix B.3 as before.

Finding Normally Distributed Probabilities

If X is a normally distributed random variable with mean μ and standard deviation σ, then

$$P(X \leq d) = \mathcal{A}\left(\frac{d - \mu}{\sigma}\right)$$

$$P(d \leq X) = 1 - \mathcal{A}\left(\frac{d - \mu}{\sigma}\right)$$

$$P(c \leq X \leq d) = \mathcal{A}\left(\frac{d - \mu}{\sigma}\right) - \mathcal{A}\left(\frac{c - \mu}{\sigma}\right)$$

EXAMPLE 2 Finding Normally Distributed Probabilities

Assume the weights of adult males in this country have a mean of 180 pounds and a standard deviation of 30 pounds. Assuming a normal distribution, find the probability that the weight of an adult male in this country is (a) less than 160 pounds (b) greater than 160 pounds (c) between 160 and 200 pounds.

SOLUTIONS ■

(a)
$$P(X \leq d) = \mathcal{A}\left(\frac{d - \mu}{\sigma}\right)$$
$$= \mathcal{A}\left(\frac{160 - 180}{30}\right)$$
$$\approx \mathcal{A}(-0.67)$$
$$= 0.2514 \qquad \text{From Appendix B.3}$$

(Fig. 11.28).

(b) This is just one less the probability found in (a), or about 0.75.

(c)

$$P(c \leq X \leq d = \mathscr{A}\left(\frac{d - \mu}{\sigma}\right) - \mathscr{A}\left(\frac{c - \mu}{\sigma}\right)$$

$$= \mathscr{A}\left(\frac{200 - 180}{30}\right) - \mathscr{A}\left(\frac{160 - 180}{30}\right)$$

$$\approx \mathscr{A}(0.67) - \mathscr{A}(-0.67)$$

$$= 0.7486 - 0.2514 \qquad \text{From (a) and Appendix B.3}$$

$$\approx 0.50$$

(Fig. 11.29). ∎

Remark. Notice that part (c) indicated that about 50% of the area under the standard normal distribution lies on the interval $[-0.67, 0.67]$.

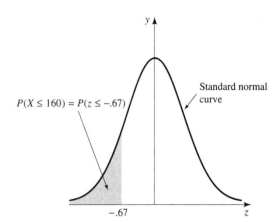

Figure 11.28
The area of the shaded region is $P(X \leq 160) = P(Z \leq -0.67) = \mathscr{A}(-0.67)$

$P(X \leq 160) = P(z \leq -.67]$.

Standard normal curve

$-.67$ z

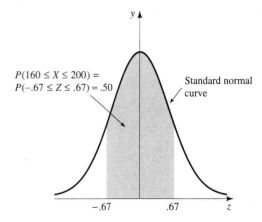

Figure 11.29
The area of the shaded region is $P(160 \leq X \leq 200) = P(-0.67 \leq Z \leq 0.67)$

$P(160 \leq X \leq 200) = P(-.67 \leq Z \leq .67) \approx .50$

Standard normal curve

$-.67$ $.67$ z

SELF-HELP EXERCISE SET 11.4

In each of the following assume that X is a normally distributed random variable with $\mu = 7$ and $\sigma = 10$. Find

1. $P(X \leq 5)$ **2.** $P(X \geq 5)$

3. $P(5 \leq X \leq 10)$

In Exercises 1 through 12 find the indicated probabilities given that Z is a random variable with a standard normal distribution.

1. $P(Z \le 0.5)$ **2.** $P(Z \le 0.75)$

3. $P(Z \le -0.3)$ **4.** $P(Z \le -1.5)$

5. $P(Z \ge 1.5)$ **6.** $P(Z \ge 0.6)$

7. $P(Z \ge -0.8)$ **8.** $P(Z \ge -1.42)$

9. $P(1.0 \le Z \le 1.5)$ **10.** $P(0 \le Z \le 1.5)$

11. $P(-1.0 \le Z \le -0.5)$ **12.** $P(-0.5 \le Z \le 1.0)$

In Exercises 13 through 20 find the indicated probabilities assuming that X is a random variable with a normal distribution with the given mean and standard deviation.

13. $P(X \le 50)$, $\mu = 38, \sigma = 8$

14. $P(X \le 200)$, $\mu = 100, \sigma = 50$

15. $P(X \ge 0.01)$, $\mu = 0.006, \sigma = 0.002$

16. $P(X \ge -10)$, $\mu = 20, \sigma = 30$

17. $P(100 \le X \le 150)$, $\mu = 20, \sigma = 100$

18. $P(10 \le X \le 20)$, $\mu = 5, \sigma = 10$

19. $P(0.01 \le X \le 0.02)$, $\mu = 0.005, \sigma = 0.01$

20. $P(-3 \le X \le 10)$, $\mu = 4, \sigma = 10$

21. Find the area under the standard normal density function on the interval $[-2, 2]$.

22. Find the area under the standard normal density function on the interval $[-3, 3]$.

23. Show that the function $f(x) = e^{-0.5x^2}$ is increasing on $(-\infty, 0)$, decreasing on $(0, \infty)$, and has a maximum at $x = 0$.

24. For any $\mu > 0$ and $\sigma > 0$, show that the function $f(x) = e^{-0.5[(x-\mu)/\sigma]^2}$ is increasing on $(-\infty, \mu)$, decreasing on (μ, ∞), and has a maximum at $x = \mu$.

25. Show that the function $f(x) = e^{-0.5x^2}$ is concave up on the intervals $(-\infty, -1)$ and $(1, \infty)$, whereas it is concave down on $(-1, 1)$.

26. Show that the function $f(x) = e^{-0.5[(x-\mu)/\sigma]^2}$ is concave up on the intervals $(-\infty, \mu - \sigma)$ and $(\mu + \sigma, +\infty)$, whereas it is concave down on $(\mu - \sigma, \mu + \sigma)$.

Applications

27. Quality Control. A machine produces ball bearings with diameters normally distributed. The mean diameter is 3.50 cm, and the standard deviation is 0.02 cm. Quality requirements demand a ball bearing to be rejected if the diameter is more than 0.05 cm different from the mean. What percentage of bearings will be rejected?

28. Quality Control. The length, in feet, of a certain structural steel beam is normally distributed with the mean 10 feet, and the standard deviation 1 inch. Quality requirements demand a beam to be rejected if the length is more than 2 inches different from the mean. What percentage of beams will be rejected?

29. Revenue. The weekly revenue, in thousands of dollars, of a certain retail store is normally distributed with mean 1000 and standard deviation 600. What percentage of weeks will revenue **(a)** exceed 1600, **(b)** be less than 600, **(c)** be between 800 and 1200?

30. Sales. The number of a particular item sold in a week is normally distributed with mean 1000 and standard deviation 400. What percentage of weeks will sales **(a)** exceed 1500, **(b)** be less than 600, **(c)** be between 800 and 1200?

31. Manufacturing. The number of ounces of detergent in an advertised 1-pound box is normally distributed with mean 16.1 ounces and standard deviation 0.3. What percentage of boxes have at least 1 pound of detergent in them?

32. Manufacturing. A company regularly orders 2000 copies of a newsletter from a printer. The number of copies delivered is normally distributed with mean 2000 and standard deviation 100. The delivery is not acceptable if the number delivered is less than 90% of the original order. What percentage of deliveries is not acceptable?

33. Manufacturing. The amount of tape on a roll is normally distributed with a mean of 30 feet with a standard deviation of 2 feet. What percentage of rolls will have less than 27 feet?

34. Manufacturing. The amount of soda in a 16-ounce can is normally distributed with a mean of 16 ounces, and the standard deviation is 0.50 ounce. What percentage of these cans will have less than 15 ounces?

35. Medicine. The time, in hours, to perform a certain operation is normally distributed with a mean of 3 hours and a standard deviation of 1 hour. What percentage of times will this operation be **(a)** less than 1 hour, **(b)** more than 3 hours, **(c)** between 1 and 3 hours?

36. Biology. The weight, in pounds, of a certain type of adult squirrel is normally distributed with a mean of 3 pounds and a standard deviation of 0.50 pound. What percentage of these squirrels have weight **(a)** less than 2 pounds, **(b)** greater than 4 pounds, **(c)** between 2 and 4 pounds?

37. Learning Time. The learning time for a particular task on an assembly line is normally distributed with a mean of 5 hours and a standard deviation of 2 hours. If an employee is given an 8-hour shift in which to learn the task, what is the probability that the task will be learned before the shift ends?

38. Testing. A corporation notes that scores on an intelligence test for newly hired employees is normally distributed with a mean of 100 and a standard deviation of 20. What percentage of new employees should score **(a)** above 110, **(b)** below 80, **(c)** between 95 and 105?

Enrichment Exercises

39. Create a table like Table 11.2 to show that $\int_{-\infty}^{\infty} \frac{z^2}{\sqrt{2\pi}} e^{-z^2/2} dz = 1$ and thus that the standard deviation of the standard normal distribution is 1.

40. Estimate the value of $\int_0^{\infty} e^{-0.5x^2} dx$ by noting that

$$\int_0^{\infty} e^{-0.5x^2} dx = \int_0^6 e^{-0.5x^2} dx + \int_6^{\infty} e^{-0.5x^2} dx$$

Estimate the first integral on the right-hand side of the last equation by using Simpson's rule with $n = 50$. Show that the second integral is smaller than $\int_6^{\infty} e^{-3x} dx$, which is smaller than 0.00000001.

SOLUTIONS TO SELF-HELP EXERCISE SET 11.4

1.

$$P(X \leq d) = \mathscr{A}\left(\frac{d - \mu}{\sigma}\right)$$

$$= \mathscr{A}\left(\frac{5 - 7}{10}\right) = \mathscr{A}(-0.20)$$

$$= 0.4207 \qquad \text{From Appendix B.3}$$

2. This is just one less the probability found in Exercise 1 or about 0.5793.

3.

$$P(c \leq X \leq d) = \mathscr{A}\left(\frac{d - \mu}{\sigma}\right) - \mathscr{A}\left(\frac{c - \mu}{\sigma}\right)$$

$$= \mathscr{A}\left(\frac{10 - 7}{10}\right) - \mathscr{A}\left(\frac{5 - 7}{10}\right)$$

$$= \mathscr{A}(0.30) - \mathscr{A}(-0.20)$$

$$= 0.6179 - 0.4207 \qquad \text{From (a) and Appendix B.3}$$

$$= 0.1972$$

CHAPTER **11**

SUMMARY OUTLINE

■ A function $f(x)$ defined on (a, b) is called a **probability density function** on (a, b) if the following are satisfied.

1. For all $x \in (a, b)$, $f(x) \geq 0$.
2. The area between the graph of $y = f(x)$ and the x-axis from a to b is one, that is

$$\int_a^b f(x) \, dx = 1.$$

For the interval (a, b), a may be $-\infty$, b may be ∞, or both. p. 595

■ Suppose $f(x)$ is a continuous probability density function on (a, b) for a continuous random variable X. The **probability that the random variable X assumes a value in the interval** $(c, d) \subset (a, b)$, written $P(c < X < d)$, is defined to be the area between the graph of $y = f(x)$ and the x-axis from c to d. That is, p. 595

$$P(c < X < d) = \int_c^d f(x) \, dx$$

■ The probability distribution on $[a, b]$ with the constant probability density function $f(x) = \frac{1}{b - a}$ is called the **uniform probability function**. For the uniform probability function $P(c \leq X \leq d) = \frac{d - c}{b - a}$. p. 599

■ The probability density function given by $f(x) = ke^{-kx}$ on $[0, \infty)$ is called the **exponential probability density function**. p. 600

■ The **expected value** $E(X)$ (or **mean** μ) of a random variable X with a continuous probability density function $f(x)$ on the interval (a, b), is defined to be $E(X) = \mu = \int_a^b xf(x) \, dx$. p. 604

■ The **variance** of a random variable X with a continuous probability density function $f(x)$ on the interval (a, b), is defined to be $\text{Var}(X) = \int_a^b (x - \mu)^2 f(x) \, dx$. p. 605

- **Alternative Form of Variance.** If $f(x)$ is a continuous probability density function on the interval (a, b) with mean μ, the variance $\text{Var}(X)$ is also given by $\int_a^b x^2 f(x)\, dx - \mu^2$. p. 605

- The **standard deviation**, denoted by σ, is the square root of the variance, that is, $\sigma = \sqrt{\text{Var}(X)}$. p. 605

- The mean, variance, and standard deviation of a random variable with a constant probability density function $f(x) = 1/(b - a)$ on $[a, b]$ is $\mu = \dfrac{1}{2}(b + a)$, $\text{Var}(X) = \dfrac{1}{12}(b - a)^2$, $\sigma = \dfrac{1}{2\sqrt{3}}(b - a)$. p. 606

- The mean, variance, and standard deviation of a random variable with an exponential probability density function $f(x) = ke^{-kx}$, $k > 0$, on $[0, \infty)$ is $\mu = \dfrac{1}{k}$, $\text{Var}(X) = \dfrac{1}{k^2}$, $\sigma = \dfrac{1}{k}$. p. 608

- For any continuous random variable, the **median** outcome x_m is that outcome for which $P(X \le x_m) = \frac{1}{2}$. p. 609

- The random variable Z with continuous probability distribution given by the function $f(z) = \dfrac{1}{\sqrt{2\pi}} e^{-z^2/2}$, on the interval $(-\infty, \infty)$ is said to satisfy the **standard normal distribution** with mean $\mu = 0$ and standard deviation $\sigma = 1$. p. 613

- If Z is a random variable with a standard normal probability density function, then

$$P(Z \le d) = \int_{-\infty}^{d} \frac{1}{\sqrt{2\pi}} e^{-z^2/2}\, dz = \mathscr{A}(d)$$

$$P(d \le Z) = \int_{d}^{\infty} \frac{1}{\sqrt{2\pi}} e^{-z^2/2}\, dz = 1 - \mathscr{A}(d)$$

$$P(c \le Z \le d) = \int_{c}^{d} \frac{1}{\sqrt{2\pi}} e^{-z^2/2}\, dz = \mathscr{A}(d) - \mathscr{A}(c)$$

- The continuous random variable X is normally distributed with a mean μ and a standard deviation σ if its probability density function is given by

$$f(x) = \frac{1}{\sigma\sqrt{2\pi}} e^{-0.5[(x-\mu)/\sigma]^2} \quad \text{on } (-\infty, \infty)$$

- If X is a normally distributed random variable with a mean μ and a standard deviation σ, then

$$P(X \le d) = \mathscr{A}\!\left(\frac{d - \mu}{\sigma}\right)$$

$$P(d \le X) = 1 - \mathscr{A}\!\left(\frac{d - \mu}{\sigma}\right)$$

$$P(c \le X \le d) = \mathscr{A}\!\left(\frac{d - \mu}{\sigma}\right) - \mathscr{A}\!\left(\frac{c - \mu}{\sigma}\right)$$

CHAPTER **11**

REVIEW EXERCISES

1. During a recent four-round golf tournament the frequencies of scores shown in the table were recorded on a par-5 hole.

Score	3	4	5	6	7	8
Frequency	4	62	157	22	4	1

(a) Find the probability that each of the scores was made by a random player.

(b) Find the probability that a score of par or lower was recorded.

(c) Find the probability that a score of less than par was recorded.

2. An urn has 10 white, 5 red, and 15 blue balls. A ball is drawn at random. What is the probability that the ball will be **(a)** red? **(b)** red or white? **(c)** not white?

3. If E and F are mutually disjoint sets in a sample space S with $p(E) = 0.25$ and $p(F) = 0.35$, find

$$p(E \cup F) \qquad p(E \cap F) \qquad p(E^c)$$

4. A furniture manufacturer notes that 6% of its reclining chairs have a defect in the upholstery, 4% a defect in the reclining mechanism, and 1% have both defects. **(a)** Find the probability that a recliner has at least one of these defects. **(b)** Find the probability that a recliner has none of these defects.

5. The probability distribution of the random variable X is given in the table.

Random Variable x	0	1	2	3	4	5
P(X = x)	0.20	0.10	0.05	0.15	0.18	0.32

(a) Draw a histogram. Find **(b)** $P(X = 0)$ **(c)** $P(X \leq 2)$ **(d)** $P(0 < X \leq 3)$ **(e)** $P(X \geq 2)$. Identify each of the probabilities as an area.

6. Find the expected value of the random variable given in the previous exercise.

7. Lottery. A lottery has a grand prize of $1,000,000, a second prize of $100,000, and 100 consolation prizes of $2000 each. If 1 million tickets are sold and the probability of any ticket winning is the same as any other, find the expected return on a $1 ticket.

8. Life Insurance. An insurance company sells a $10,000, 5-year term life insurance policy to an individual for $800. Find the expected return for the company if the probability that the individual will live for the next 5 years is 0.96.

9. The pitcher Cy Young holds the all-time record for the most wins in a lifetime with 511, a record that is unlikely to be matched. The table gives his win and loss record for the 5-year period beginning with 1900. Find the mean, variance, and standard deviation for his wins and also for his losses for this 5-year period. Which is varying the most?

Year	1900	1901	1902	1903	1904
Number of Wins	20	33	32	28	27
Number of Losses	18	10	10	9	16

10. Find the variance and the standard deviation of the random variable given in Exercise 5.

11. Find k such that kx is a probability density function on $[1, 3]$.

12. Find k such that $ke^{-|x|}$ is a probability density function on $(-\infty, \infty)$.

13. Find k such that $\dfrac{2}{x^2}$ is a probability density function on $[1, k]$.

14. Show that $\frac{3}{8}x^2$ is a probability density function on $[0, 2]$. Find **(a)** $P(0 \leq X \leq 1)$, **(b)** $P(X > 1)$, **(c)** $P(X = 1)$.

15. Find the mean of the random variable with probability density function $f(x) = \frac{1}{3}$ on $[0, 3]$.

16. Find the variance and standard deviation of the random variable with the probability density function given in the previous exercise.

17. Find the median of the random variable with the probability density function given in Exercise 15.

18. Find the mean of the random variable with probability density function $f(x) = \dfrac{1}{18}x$ on $[0, 6]$.

19. Find the variance and standard deviation of the random variable with the probability density function given in the previous exercise.

20. Find the median of the random variable with the probability density function given in Exercise 18.

21. Suppose Z is a random variable with a standard normal probability density function. Find **(a)** $P(Z \leq 1.25)$, **(b)** $P(Z \geq 1.25)$, **(c)** $P(1 \leq Z \leq 1.25)$.

22. Suppose X is a random variable with a normal probability density function with a mean $\mu = 2$ and a standard deviation $\sigma = 5$. Find **(a)** $P(X \leq 1)$, **(b)** $P(X \geq 1)$, **(c)** $P(1 \leq X \leq 1.5)$.

23. Manufacturing. The life in years of a certain automobile has a probability density function given by $f(t) = 0.20e^{-0.20t}$. Find the probability that the life of this car is **(a)** at most 10 years, **(b)** at least 10 years, **(c)** between 5 and 10 years.

24. Shelf Life. The shelf life, in months, of a certain cereal is a continuous random variable with a probability density function $f(x) = \dfrac{3}{98}x\sqrt{9 + x^2}$ on $[0, 4]$. Find the probability that the shelf life is **(a)** at most 1 month, **(b)** at least 1 month, **(c)** between 1 and 2 months.

25. Manufacturing. The length, in feet, of wire produced on a certain machine is normally distributed with a mean of 10 feet and a standard deviation of 0.30 feet. What percentage of these will be cut less than 9.4 feet? More than 9.4 feet? Between 9.4 and 10.6 feet?

PROJECTS

P r o j e c t   1
**Health Care Unit
Location**

A state government is trying to locate a new health care unit to serve the needs of the city of Adams and the four smaller cities indicated in the figure. The table gives the population and coordinates of the centers of the five cities. For computational purposes we make the assumption that *all* residents of a city are located precisely at the center of each city and that there are no residents between cities. The health care unit should be located at the population ''center'' of the five cities. The probability that a randomly selected individual in this community is in North Adams is just the population of North Adams divided by the total population of this community of five cities. Similarly for the other four cities. If X and Y are the random variables that take on the values of x_i and y_i, respectively, then the population ''center'' and therefore the location of the health unit should be located at the coordinate $(E(X), E(Y))$. Find this point.

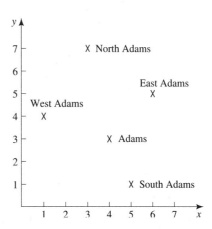

		Population Coordinates	
City	**Population**	x_i	y_i
North Adams	11,000	3	7
West Adams	12,000	1	4
Adams	42,000	4	3
East Adams	17,000	6	5
South Adams	18,000	5	1

P r o j e c t   2
Diffusion

We consider the diffusion (dispersion) of a population in a one-dimensional region over time. If $u(x, t)$ is the ''concentration'' (see below) of the population at a point x and at time t, then u must be a solution of the *partial differential equation*

$$\frac{\partial u}{\partial t} = \alpha \frac{\partial^2 u}{\partial x^2} \tag{1}$$

Equation (1) is also called the *diffusion equation*.

(a) Show that the function $u(x, t) = \dfrac{e^{-x^2/4\alpha t}}{\sqrt{4\pi\alpha t}}$ is a solution to Equation (1) by taking the appropriate partial derivatives and placing into Equation (1).

(b) Show that for any fixed t, $u(x, t) = \dfrac{e^{-x^2/4\alpha t}}{\sqrt{4\pi\alpha t}}$ is a normal distribution on $(-\infty, \infty)$. Do this by determining the standard deviation σ.

(c) From (b) the probability of finding an individual at time t in the interval (a, b) is $\int_a^b u(x, t)\, dx$. Take $\alpha = 1$ and graph $u = u(x, t)$ on your graphing calculator when $t = 1, 2, 4$. Explain how what you see is related to the dispersion of the population. What seems to be happening over time?

Project 3
Binomial Distribution and Its Approximation by a Normal Distribution

Experiments in which there are just two outcomes are called **Bernoulli trials**. Some examples follow. Flip a coin and see if heads or tails turns up. Test a transistor to see if it is defective or not. Examine a patient to see if a particular disease is present or not. Take a free throw in basketball and make the basket or not.

We commonly refer to the two outcomes of a Bernoulli trial as "success" (S) or "failure" (F). We agree to always write p for the probability of "success" and q for the probability of "failure." Naturally $q = 1 - p$.

We are actually interested in a **repeated Bernoulli trial**, that is, repeating many times an experiment with two outcomes. We make a very fundamental assumption: The successive Bernoulli trials are independent of one another. Thus, for example, flipping a coin 10 times is a repeated binomial trial. Tossing a die 20 times and seeing if an even number or an odd number occurs each time is another example.

Given a Bernoulli trial repeated n times, we are interested in determining the probability that a specific number of successes occurs. If k is the number of successes, then we denote $b(k\colon n, p)$ the probability of exactly k successes in n repeated Bernoulli trials in which the probability of success is p. It turns out that

$$b(k\colon n, p) = \frac{n!}{k!(n - k)!}\, p^k q^{n-k}$$

A derivation can be found in a book on probability.

Under certain conditions a binomial distribution can be approximated by a normal distribution. Let X be the random variable that gives the number of successes in a Bernoulli trial repeated n times. Then the histogram for the probability distribution of X is approximated by the normal curve with mean $\mu = np$ and standard deviation $\sigma = \sqrt{npq}$. Demonstrate this using your graphing calculator.

(a) Create histograms when $p = 0.7$ and $n = 3, 5$, and 10.

(b) Do these histograms for increasing n appear more like a normal curve?

(c) For $n = 10$ graph over your histogram the appropriate approximating normal distribution.

REFERENCES

[1] Duane Chapman. 1987. Computation techniques for intertemporal allocation of natural resources. *Amer. J. Agr. Econ.* 69:134–142.

[2] Wilhelmina A. Leigh. 1988. The social preference for fair housing during the civil rights movement and since. *Amer. Econ. Rev.* 78(2):156–162.

[3] Katherine C. Ewel and Henry L. Gholz. 1991. A simulation model of the role of belowground dynamics in a Florida pine plantation. *Forest Sci.* 37(2):397–438.

[4] S. R. Aiken and C. H. Leigh. 1984. A second national park for Peninsular Malaysia? *Biol. Conser.* 29:253–276.

[5] Duane J. Clow and N. Scott Urquhart. 1984. *Mathematics in Biology.* New York: Ardsley House.

[6] E. R. Pianka. 1986. *Ecology and Natural History of Desert Lizards.* Princeton, N.J.: Princeton University Press.

[7] G. G. Simpson, A. Roe, and R. C. Lewontin. 1960. *Quantitative Zoology.* New York: Harcourt, Brace.

[8] Geoffrey Clarke and Leslie McKenzie. 1992. Fluctuating asymmetry as a quality control indicator for insect mass rearing process. *J. Econ. Entomol.* 85:2045–2050.

[9] Lawrence Hribar, Daniel LePrince, and Lane Foil. 1992. Feeding sites of some Louisiana tabanidae on fenvalevate-treated and control cattle. *J. Econ. Entomol.* 85:2279–2285.

[10] Gary Pierzynski and A. Paul Schwab. 1993. Bioavailability of zinc, cadmium, and lead in a metal contaminated alluvial soil. *J. Environ. Qual.* 22:247–254.

[11] Thomas Gilligan. 1992. Imperfect competition and basing-point pricing. *Amer. Econ. Rev.* 82:1106–1119.

[12] John Shea. 1993. Do supply curves slope up? *Quart. J. Econ.* cviii:1–32.

[13] Andrew Schmitz, Dale Sigurdson, and Otto Doering. 1986. Domestic farm policy and the gains from trade. *Amer. J. Agr. Econ.* 68:820–827.

[14] Carl F. Christ. 1985. Early progress in estimating quantitative economic relationships in America. *Amer. Econ. Rev.* 75:39–52.

[15] Herb Taylor, 1988. Experimental economics. *Bus. Rev. Fed. Res. Bank of Philadelphia.* Mar–Apr:15–25.

[16] Keith Knapp. 1987. Dynamic equilibrium in markets for perennial crops. *Amer. Agr. J. Econ.* 69:97–105.

[17] Sunwoong Kim and Hamid Mohtadi. 1992. Labor specialization endogenous growth. *Amer. Econ. Rev.* 82:404–408.

[18] Roger Guesnerie. 1992. An exploration of the eductive justification of the rational-expectations hypothesis. *Amer. Econ. Rev.* 82:1254–1278.

[19] *The Wall Street Journal.* August 17, 1994.

[20] D. G. Senanayake, S. F. Pernal, and N. J. Holliday. 1993. Yield response of potatoes to defoliation by the potato flea beetle in Manitoba. *J. Econ. Entomol.* 86(5):1527–1533.

[21] J. M. Hardman. 1989. Model simulating the use of miticides to control European red mite in Nova Scotia apple orchards. *J. Econ. Entomol.* 82:1411–1422.

[22] Ujjayant Chakravorty and James Roumasset. 1994. Incorporating economic analysis in irrigation design and management. *J. Water Res. Plan. Man.* 120:819–835.

[23] C. J. Sutton. 1974. Advertising, concentration, and competition. *Econ. J.* 84:56–69.

[24] Milton H. Spencer. 1968. *Managerial Economics.* Homewood, Il.: Richard D. Irwin, Inc.

[25] Margriet Caswell, Erik Lichtenberg, and David Zilberman. 1990. The effects of pricing policies on water conservation and drainage. *Amer. J. Agr. Econ.* 72:883–890.

[26] Kevin B. Grier, Michael C. Munger, and Brian E. Roberts. 1991. The industrial organization of corporate political participation. *Southern Econ. J.* 57:727–738.

[27] W. Kaakeh, D. G. Pfeiffer, and R. P. Marini. 1992. Combined effects of spirea aphid and nitrogen fertilization on net photosynthesis, total chlorophyll content, and greenness of apple leaves. *J. Econ. Entomol.* 85:939–946.

[28] Werner Hirsch. 1956. Firm progress ratios. *Econometrics.* 24:137–138.

[29] S. Richman. 1958. The transformation of energy by *Daphnia pulex. Ecol. monogr.* 28:273–291.

[30] Robert C. McNeil, Russ Lea, Russell Ballard, et al. 1988. Predicting fertilizer response of Loblolly pine using foliar and needle-fall nutrients samples in different seasons. *Forest Sci.* 34:698–707.

[31] Darrell J. Bosch and Vernon R. Eidman. 1987. Valuing information when risk preferences are neutral. *Amer. J. Agr. Econ.* 69:658–668.

[32] M. B. Usher, A. C. Brown, and S. E. Bedford. 1992. Plant species richness in farm woodlands. *Forestry.* 65:1–13.

[33] K. D. Klopzig, K. F. Raffa, and E. B. Smalley. 1991. Association of an insect-fungal complex with red pine decline in Wisconsin. *Forest Sci.* 37:1119–1139.

[34] R. Harmer. 1992. Relationship between shoot length, bud number, and branch production in *Quercus petraea*. *Forestry.* 65:1–13.

[35] G. S. Tolley and V. S. Hastings. 1960. Optimal water allocation: The North Platte River. *Quart. J. of Econ.* 74:279–295.

[36] Russell Tronstad and Russell Gum. 1994. Cow culling decisions adapted for management with CART. *Amer. J. Agr. Econ.* 76:237–249.

[37] S. SriRamaratnam, David A. Bessler, M. Edward Rister, et al. 1987. Fertilization under uncertainty: An analysis based on producer yield expectations. *Amer. J. Agr. Econ.* 69:349–357.

[38] Blodwyn M. McIntyre, Martha A. Scholl, and John T. Sigmon. 1990. A quantitative description of a deciduous forest canopy using a photographic technique. *Forest Sci.* 36:381–393.

[39] Joseph Buongiorno, Sally Dahir, Hsien-Chih Lu, et al. 1994. Tree size diversity and economic returns in uneven-aged forest stands. *Forest Sci.* 40:5–17.

[40] Robert MacArthur and Edward O. Wilson. 1967. *The Theory of Island Biogeography.* Princeton, N.J.: Princeton University Press.

[41] Theodore M. Crone. 1989. Office vacancy rates: How should we interpret them? *Bus. Rev. Fed. Res. Bank of Philadelphia.* May–June:3–12.

[42] Michael R. Carter and Keith D. Wiebe. 1990. Access to capital and its impact on agrarian structure and productivity in Kenya. *Amer. J. Agr. Econ.* 72:1146–1150.

[43] Stephen Asante. 1993. Spatial and temporal distribution patterns of *Eriosoma lanigerum* on apple. *Environ. Entomol.* 22:1060–1065.

[44] T. P. Parkin, D. R. Shelton, and J. A. Robinson. 1991. Evaluation of methods for characterizing carbofuran hydrolysis in soil. *J. Environ. Qual.* 20:763–769.

[45] Paul Morgan, L. Preston Mercer, and Nestor Flodin. 1975. General model for nutritional responses of higher organisms. *Proc. Nat. Acad. Sci. U.S.A.* 72:4327–4331.

[46] Robert Wiedenmann, J. W. Smith, Jr. and Patricia Darrell. 1992. Laboratory rearing and biology of the parasite *Cotesian flavipes*. *Environ. Entomol.* 21:1160–1167.

[47] E. Mansfield. 1961. Technical change and the rate of imitation. *Econometrica.* 29(4):741–766.

[48] Joel Sailors, M. L. Greenhut, and H. Ohta. 1985. Reverse dumping: A form of spacial price discrimination. *J. Indust. Econ.* 34:167–181.

[49] Kuo S. Huang. 1991. Factor demands in the U.S. food-manufacturing industry. *Amer. J. Agr. Econ.* 73:615–620.

[50] Greg Traxler and Derek Byerlee. 1992. Economic returns to crop management research in a post-green revolution setting. *Amer. J. Agr. Econ.* 74:573–582.

[51] Ujjayant Chakravorty and James Roumasset. 1991. Efficient spatial allocation of irrigation water. *Amer. J. Agr. Econ.* 166:165–173.

[52] Michael Korzukhin and Sanford Porter. 1994. Spatial model of territorial competition and population dynamics in fire ants. *Environ. Entomol.* 23:912–922.

[53] Pinyun Yang. 1994. Temperature influences on the development and demograph of *Bactrocera dorsalis* in China. *Environ. Entomol.* 23:971–974.

[54] Juha Lappi. 1991. Calibration of height and volume equations with random parameters. *Forest Sci.* 37:781–801.

[55] Ian Deshmukh. 1986. *Ecology and Tropical Biology.* Boston: Blackwell Scientific Publishers.

[56] Quirino Paris. 1992. The von Liebig hypothesis. *Amer. Agr. Econ.* 74:1019–1028.

[57] Gerard A. Pfann and Bart Verspagen. 1989. The structure of adjustment costs for labour in the Dutch manufacturing sector. *Econ. Lett.* 29:365–371.

[58] Clive W. J. Granger. 1993. What are we learning about the long run? *Econ. J.* 103:307–317.

[59] J. W. Gowen. 1964. Effects of x-rays of different wave lengths on viruses. In: *Statistics and Mathematics in Biology.* Kempthorne, Bancroft, Gowen, et al. (eds.). New York-London: Hafner.

[60] L. A. Sepirstein, D. G. Vidt, M. J. Mandel. 1955. Volumes of distributions and clearances of intravenously injected creatinine in the dog. *Amer. J. Phys.* 181:330–336.

[61] C. Zonneveld and S. Kooijman. 1989. The application of a dynamic energy budget model to *Lymnaea stagnalis*. *Funct. Ecol.* 3:269–278.

[62] Walid Kaakeh, Douglas Pfeiffer, and Richard Marini. 1992. Combined effects of spirea aphid and nitrogen fertilization on shoot growth in young apple trees. *J. Econ. Entomol.* 85:496–505.

[63] P. J. Miller. 1961. Age, growth, and reproduction of the rock goby. *J. Mar. Biol. Ass. U. K.* 41:737–769.

[64] L. L. Karr and J. R. Coats. 1992. Effects of four monoterpenoids on growth and reproduction of the German cockroach. *J. Econ. Entomol.* 85:424–429.

[65] Murray Potter, Alan Carpenter, Adrienne Stocker, et al. 1994. Controlled atmospheres for the post-harvest disinfestation of *Thrips obscuratus*. *J. Econ. Entomol.* 87:1251–1255.

[66] Frederick W. Bell. 1986. Competition from fish farming in influencing rent dissipation: The crawfish fishery. *Amer. J. Agr. Econ.* 68:95–101.

[67] J. Pilarska. 1977. Eco-physiological studies on *Brachionus rubens*. *Pol. Arch. Hydrobiol.* 24:319–328.

[68] J. D. Murray. 1989. *Mathematical Biology.* New York: Springer-Verlag.

[69] Harvey Chan, James Hansen, and Stephen Tam. 1990. Larval diets from different protein sources for Mediterranean fruit flies. *J. Econ. Entomol.* 83:1954–1958.

[70] R. C. Elandt-Johnson and N. L. Johnson. 1980. Survival Models and Data Analysis. New York: J. Wiley & Sons.

[71] R. Pearl, J. R. Miner, and S. L. Parker. 1927. Experimental studies on the duration of life. *Am. Nat.* 61:289–318.

[72] D. J. Schotzko and C. M. Smith. 1991. Effects of host plant on the between-plant spatial distribution of the Russian wheat aphid. *J. Econ. Entomol.* 84:1725–1734.

[73] Narayan Talekar, Chih Pin Lin, Yii Fei Yin, et al. 1991. Characteristics of infestation by *Ostrinia furnacalis* in mungbean. *J. Econ. Entomol.* 84:1499–1502.

[74] G. David Buntin, David Gilbertz, and Ronald Oetting. 1993. Chlorophyll loss and gas exchange in tomato leaves after feeding injury by *Bemisia tabace. J. Econ. Entomol.* 86:517–522.

[75] Edward O. Wilson. 1992. *The Diversity of Life.* New York: W. W. Norton & Co.

[76] F. Vuilleumier and M. Monasterio. 1986. *High Altitude Tropical Biogeography.* New York: Oxford University Press.

[77] Ian Deshmukh. 1986. *Ecology and Tropical Biology.* Boston: Blackwell Scientific Publications.

[78] John W. Wright. 1992. *The Universal Almanac 1993.* New York: Universal Press.

[79] Robert MacArthur. 1972. *Geographical Ecology.* New York: Harper & Row.

[80] A. I. Payne. 1986. *The Ecology of Tropical Lakes and Rivers.* New York: John Wiley & Sons.

[81] Eric Noreen and Naomi Soderstrom. 1994. Are overhead costs strictly proportional to activity? *J. Accounting Econ.* 17:255–278.

[82] Louis Phlips. 1971. *Effects of Industrial Concentration.* Amsterdam-London: North Holland.

[83] Joel Sailors, M. L. Greenhut, and H. Ohta. 1985. Reverse dumping: A form of spacial price discrimination. *J. Indust. Econ.* 34(2):167–181.

[84] Martin Beckmann. 1968. *Location Theory.* New York: Random House.

[85] E. Feinerman, E. Choi, and S. Johnson. 1990. Uncertainty and split nitrogen application in corn production. *Amer. J. Agr. Econ.* 72:975–984.

[86] Q. Dai, J. Fletcher, and J. Lee. 1993. Incorporating stochastic variables in crop response models. *Amer. J. Agr. Econ.* 75:377–386.

[87] Derek Lactin. N. J. Holliday, and L. L. Laman. 1993. Temperature dependence of feeding by Colorado potato beetle larvae. *Environ. Entomol.* 22:784–790.

[88] Frank Arthur and J. Larry Zettler. 1991. Malathion resistance in *Tribolium castancum. J. Econ. Entomol.* 84:721–726.

[89] R. H. Cherry, F. J. Coale, and P. S. Porter. 1990. Oviposition and survivorship of sugarcane grubs at different soil moisture. *J. Econ. Entomol.* 83:1355–1359.

[90] F. J. Eller, R. R. Heath, and S. M. Ferkovich. 1990. Factors affecting oviposition by the parasitoid *Microplitis croceipes. J. Econ. Entomol.* 83:398–404.

[91] Lauri J. Shainsky and Stevin R. Radosevich. 1991. Analysis of yield-density relationships in experimental stands of Douglass fir and red alder seedlings. *Forest Sci.* 37(2):574–592.

[92] R. E. Webb, B. A. Leonhardt, J. R. Plimmer, et al. 1990. Effect of racemic disparlure released from grids of plastic ropes on mating success of gypsy moth as influenced by dose and by population density. *J. Econ. Entomol.* 83(3):910–916.

[93] Daniel B. Thornton and Giora Moore. 1993. Auditor choice and audit fee determinants. *J. Bus. Finance Accounting* 20(3):333–349.

[94] H. Laudelout. 1993. Chemical and microbiological effects of soil liming in a broad-leaved forest ecosystem. *Forest Ecol. Manage.* 61:247–261.

[95] Leslie Rosenthal. 1990. Time to re-establishment of equilibrium for a group of redundant workers. *Applied Econ.* 22:83–95.

[96] Wellesley Dodds. 1973. An application of the Bass model in long-term new product forecasting. *J. Marketing Res.* 10:308–315.

[97] Dennis Ring and John Bennedict. 1993. Comparison of insect injury-cotton yield response functions and economic injury levels for *Helicoverpa zea* and *Heliothis virescens* in the lower Gulf coast of Texas. *J. Econ. Entomol.* 86(4):1228–1235.

[98] Robert Inman. 1992. Can Philadelphia escape its fiscal crisis with another tax increase? *Bus. Rev. Fed. Res. Bank of Philadelphia.* Sept–Oct:5–20.

[99] Colin Clark. 1976. *Mathematical Bioeconomics.* New York: John Wiley & Sons.

[100] P. S. Dasgupta and G. M. Heal. 1979. *Economic Theory and Exhaustible Resources.* Cambridge, England: Cambridge University Press.

[101] J. Haldi and D. Whitcomb. 1967. Economies of scale in industrial plants. *J. Pol. Econ.* 75:373–385.

[102] P. Pearse. 1967. The optimal forest rotation. *Forest Chronicle* 43:178–195.

[103] Peter Oue Christensen and Gjarne G. Sorensen. 1994. Duration, convexity, and time value. *J. Portfolio Manage.* 20(2):51–60.

[104] Michael C. Keeley and Gary C. Zimmerman. 1985. Competition for money market deposit accounts. *Econ. Rev. Fed. Res. Bank of San Francisco.* Spring(2):5–27.

[105] H. G. Lescano, B. C. Congdon, and N. W. Heather. 1994. *J. Econ. Entomol.* 87(5):1256–1261.

[106] William Roltsch and Mark Mayse. 1993. Simulation phenology model for the western grapeleaf skeletonizer. *Env. Entomol.* 22(3):577–586.

[107] David Reed and Paul Semtner. 1992. Effects of tobacco aphid populations on flue-cured tobacco production. *J. Econ. Entomol.* 85(5):1963–1971.

[108] David G. Mayer. 1993. Estimation of dispersal distances. *J. Econ. Entomol.* 22(2):368–380.

[109] W. A. Overholt et al. 1990. Distribution and sampling of southwestern corn borer in preharvest corn. *J. Econ. Entomol.* 83(4):1370–1375.

[110] Abbas Ali and M. J. Gaylor. 1992. Effects of temperature and larval diet on development of the beet armyworm. *Env. Entomol.* 21(4):780–786.

[111] Richard Flamm, Paul Pulley, and Robert Coulson. 1993. Colonization of disturbed trees by the southern pine bark beetle guild. *Env. Entomol.* 22(1):62–70.

[112] J. Phillipson. 1971. Methods of study in quantitative soil ecology: Population, production, and energy flow. Oxford, England: Blackwell Scientific Publications.

[113] Sheldon Kimmel. 1992. Effects of cost changes on oligopolists' profits. *J. Industrial Econ.* XL(4):442–449.

[114] R. S. McAlpine and R. H. Wakimoto. 1991. The acceleration of fire from point source to equilibrium spread. *Forest Sci.* 37(5):1314–1337.

[115] Richard L. Boyce, Andrew J. Friedland, Elizabeth T. Webb, et al. 1991. Modeling the effect of winter climate on high-elevation red spruce shoot water contents. *Forest Sci.* 37(6):1567–1580.

[116] H. Laudelout. 1993. Chemical and microbiological effects of soil liming in a broad-leaved forest ecosystem. *Forest Eco. Manage.* 61(3):247–261.

[117] Babu Nahuta, Krzysztof Ostaszewski, and P. K. Sahoo. 1990. Direction of price changes in third-degree price discrimination. *Amer. Econ. Rev.* 80:1254–1258.

[118] Carl F. Jordan, Jerry R. Kline, and Donald S. Sasser. 1973. A simple model of strontium and manganese dynamics in a tropical rain forest. *Health Physics.* 24:477–489.

[119] Larry Wolf and Frank Hill. 1986. Physiological and ecological adaptations of high montane sunbirds and hummingbirds. In: *High Altitude Tropical Biogeography.* Oxford, England: University Press.

[120] Michael R. Carter and Keith D. Wiebe. 1990. Access to capital and its impact on agrarian structure and productivity in Kenya. *Amer. J. Agr. Econ.* 72:1146–1150.

[121] Peter Dorner and Mahmoud El-Shafie. 1980. *Resources and Development.* Madison, Wisconsin: The University of Wisconsin Press.

[122] W. O. Kermack and A. G. McKendrick. 1933. Contributions to the mathematical theory of epidemics. *Proc. Roy. Soc. A.* 141:94–122.

[123] Donald A. Hay and Derek J. Morris. 1979. *Industrial Economics.* New York: Oxford University Press.

[124] Joel Sailors, M. L. Greenhut, and H. Ohta. 1985. Reverse dumping: A form of spatial price discrimination. *J. Indust. Econ.* 34(2):167–181.

[125] Soon-Ken Hsu. 1979. Monopoly output under alternative spatial pricing techniques: Comment. *Amer. Econ. Rev.* 69:678–679.

[126] Keith R. Phillips. 1991. The effects of the growing service sector on wages in Texas. *Econ. Rev. Fed. Res. Bank of Dallas.* Nov:15–28.

[127] Duangkamon Chotikapanich. 1993. A comparison of alternative functional forms for the Lorentz curve. *Econ. Lett.* 41:129–138.

[128] Danial B. Suits. 1977. Measurement of tax progressivity. *Amer. Econ. Rev.* 67(4):747–752.

[129] Steven Shavell. 1991. Law enforcement: Specific versus general enforcement of law. *J. Pol. Econ.* 99(5):1088–1108.

[130] Frederick W. Smith, D. Arthur Sampson, James N. Long. 1991. Comparison of leaf area index estimates from tree allometrics and measured light interception. *Forest Sci.* 37:1682–1688.

[131] Jeffrey L. Cullen. 1978. Production, efficiency, and welfare in the natural gas transmission industry. *Amer. Econ. Rev.* 68(3):311–323.

[132] R. Dorfman and P. O. Steiner. 1954. Optimal advertising and optimal quality. *Amer. Econ. Rev.* 44:826–836.

[133] Skip Viscusi and Michael J. Moore. 1993. Product liability, research and development, and innovation. *J. Pol. Econ.* 101(1):161–184.

[134] Quirino Paris. 1992. The von Liebig hypothesis. *Amer. J. Agr. Econ.* 74(4):1017–1028.

[135] Jonathan M. Karpoff. 1987 Suboptimal controls in common resource management: The case of the fishery. *J. Polit. Econ.* 95(1):179–194.

[136] Robert M. Solow. 1956. A contribution to the theory of economic growth. *Quart. J. Econ.* LXX:65–94.

[137] N. Gregory Mankiw, David Romer, and David N. Weil. 1992. A contribution to the empirics of economic growth. *Quart. J. Econ.* CVII(2):407–437.

[138] A. W. Phillips. 1957. Stabilization policy in a closed economy. *Economic J.* 67:265–77.

[139] Paul Samuelson. 1949. Foundations of Economic Analysis (Chapter XI). Boston: Harvard University Press.

[140] D. J. Schotzko and L. O'Keefe. 1989. Comparison of sweep net, d-vac, and absolute sampling. *J. Econ. Entomol.* 82:491–506.

[141] D. Barry Lyons and Andrew Liebold. 1992. Spatial distribution and hatch times of egg masses of gypsy moth. *Environ. Entomol.* 21:354–358.

[142] Edward Batschelet. 1971. *Introduction to Mathematics for Life Sciences.* Springer-Verlag, New York.

[143] E. Morscher. 1967. Development and clinical significance of the anteversion of the femoral neck. *Wiederherstellungschir. u. Traum.* 9:107–125.

[144] C. I. Bliss and D. L. Blevins. 1959. The analysis of seasonal variation in measles. *Amer. J. Hyg.* 70:328–334.

[145] N. J. Smith. 1991. Predicting radiation attenuation in stands of Douglas-fir. *Forest Sci.* 37:1213–1223.

APPENDIX

REVIEW

A.1 EXPONENTS AND ROOTS

■ Exponents ■ Roots ■ Rational Exponents ■ Rationalizing

Exponents

The symbol x^2 denotes $x \cdot x$, the symbol x^3 denotes $x \cdot x \cdot x$, and, in general, we have the following.

Exponents

If n is a positive integer

$$x^n = \underbrace{x \cdot x \cdot \cdots \cdot x.}_{n \text{ factors}}$$

The number n is called the **exponent**, and x is called the **base**.

Examples

$$2^4 = 2 \cdot 2 \cdot 2 \cdot 2 = 16$$

$$\left(-\frac{1}{2}\right)^3 = \left(-\frac{1}{2}\right)\left(-\frac{1}{2}\right)\left(-\frac{1}{2}\right) = -\frac{1}{8}$$

Zero Exponent

If $x \neq 0$, then

$$x^0 = 1.$$

The expression 0^0 is not defined.

Examples

$$2^0 = 1 \qquad \pi^0 = 1 \qquad (a + b)^0 = 1$$

Negative Exponent

If n is a positive integer, then

$$x^{-n} = \frac{1}{x^n}.$$

Examples

$$2^{-1} = \frac{1}{2} \qquad x^{-3} = \frac{1}{x^3} \qquad \left(\frac{2}{3}\right)^{-2} = \frac{1}{(\frac{2}{3})^2} = \frac{1}{\frac{4}{9}} = \frac{9}{4}$$

Properties of Exponents

If m and n are integers, the following properties are true.

Property	Example
1. $x^m x^n = x^{m+n}$	$x^2 \cdot x^3 = x^{2+3} = x^5$
2. $(x^m)^n = x^{mn}$	$(x^2)^3 = x^{2\cdot3} = x^6$
3. $(xy)^n = x^n y^n$	$(2 \cdot 4)^3 = 2^3 \cdot 4^3 = 8 \cdot 64 = 512$
4. $\left(\dfrac{x}{y}\right)^n = \dfrac{x^n}{y^n}$ if $y \neq 0$	$\left(\dfrac{2}{3}\right)^2 = \dfrac{2^2}{3^2} = \dfrac{4}{9}$
5. $\dfrac{x^m}{x^n} = x^{m-n}$ if $x \neq 0$	$\dfrac{x^5}{x^2} = x^{5-2} = x^3$
$\dfrac{x^m}{x^n} = \dfrac{1}{x^{n-m}}$ if $x \neq 0$	$\dfrac{x^2}{x^5} = \dfrac{1}{x^{5-2}} = \dfrac{1}{x^3}$

Examples

$$(2x^2y^3)(3x^3y^4) = (2)(3)(x^2x^3)(y^3y^4) = 6x^5y^7$$

$$(2x^2y^3z)^4 = 2^4(x^2)^4(y^3)^4(z)^4 = 16x^8y^{12}z^4$$

$$\left(\frac{2x^2}{3y^3}\right)^3 = \frac{(2x^2)^3}{(3y^3)^3} = \frac{8x^6}{27y^9}$$

$$(a^4b^{-2})^{-3} = (a^4)^{-3}(b^{-2})^{-3} = (a^{-12})(b^6) = \frac{b^6}{a^{12}}$$

$$(a^2b^3)^2(a^4b)^{-1} = (a^2b^3)^2\frac{1}{a^4b} = \frac{a^4b^6}{a^4b} = b^5.$$

We can also do the last example as follows.

$$(a^2b^3)^2(a^4b)^{-1} = (a^4b^6)(a^{-4}b^{-1}) = a^{4-4}b^{6-1} = b^5.$$

Roots

Roots

- If n is a positive integer and $y^n = x$, then we say that y is an nth root of x.
- If n is odd, there is exactly one nth root, which we denote by $y = \sqrt[n]{x} = x^{1/n}$.
- If n is even and x is negative, there is no (real) root.
- If n is even and x is positive, we use the symbols $y = \sqrt[n]{x} = x^{1/n}$ to denote the positive nth root.
- If $n = 2$, we write $\sqrt[2]{x} = \sqrt{x}$.

The symbol $\sqrt{}$ is called a **radical sign**.

Examples

$$\sqrt{9} = 3 \qquad \sqrt[3]{-64} = -4 \qquad \sqrt[4]{\frac{1}{16}} = \frac{1}{2}$$

Properties of Roots

Radical Formulation	Exponent Formulation
1. $\sqrt[n]{xy} = \sqrt[n]{x}\,\sqrt[n]{y}$	$(xy)^{1/n} = x^{1/n}y^{1/n}$
2. $\sqrt[n]{\dfrac{x}{y}} = \dfrac{\sqrt[n]{x}}{\sqrt[n]{y}}$	$\left(\dfrac{x}{y}\right)^{1/n} = \dfrac{x^{1/n}}{y^{1/n}},\ y \neq 0$
3. $\sqrt[m]{\sqrt[n]{x}} = \sqrt[mn]{x}$	$(x^{1/n})^{1/m} = x^{1/mn}$

Examples

$$\sqrt{40} = \sqrt{4\cdot 10} = \sqrt{4}\sqrt{10} = 2\sqrt{10}$$
$$\sqrt[3]{-24} = \sqrt[3]{(-8)(3)} = \sqrt[3]{-8}\sqrt[3]{3} = -2\sqrt[3]{3}$$
$$\sqrt{\sqrt[3]{64}} = \sqrt[6]{64} = 2$$
$$\sqrt[3]{8x^6y^{10}} = \sqrt[3]{8}\sqrt[3]{x^6}\sqrt[3]{y^{10}} = 2\sqrt[3]{(x^2)^3}\sqrt[3]{(y^3)^3y} = 2x^2\sqrt[3]{(y^3)^3}\sqrt[3]{y} = 2x^2y^3\sqrt[3]{y}$$

Rational Exponents

Rational Exponents

If m and n are integers, $x \neq 0$, and $x^{1/n}$ exists, then
$$x^{m/n} = (x^{1/n})^m = (\sqrt[n]{x})^m$$
or, equivalently,
$$x^{m/n} = (x^m)^{1/n} = \sqrt[n]{x^m}.$$

Examples

$$4^{3/2} = (\sqrt{4})^3 = 2^3 = 8 \qquad \text{or} \qquad 4^{3/2} = \sqrt{4^3} = \sqrt{64} = 8$$
$$(-8)^{2/3} = (\sqrt[3]{-8})^2 = (-2)^2 = 4 \quad \text{or}\ (-8)^{2/3} = \sqrt[3]{(-8)^2} = \sqrt[3]{64} = 4$$

The properties of exponents are also true for rational exponents.

Example

$$x^{3/2}y^{2/3}(\sqrt[5]{4x}^{1/3}y^{1/2})^{-5/2} = 4x^{3/2}y^{2/3}(4^{1/5})^{-5/2}(x^{1/3})^{-5/2}(y^{1/2})^{-5/2}$$
$$= 4x^{3/2}y^{2/3}4^{-1/2}x^{-5/6}y^{-5/4}$$
$$= 4x^{3/2-5/6}\frac{1}{2}y^{2/3-5/4}$$
$$= 2x^{4/6}y^{-7/12} = 2\frac{x^{2/3}}{y^{7/12}}$$

Rationalizing

To remove a radical from the denominator of a quotient, multiply both the numerator and denominator of the quotient by an appropriate expression.

Examples

$$\frac{1}{\sqrt{3}} = \frac{1}{\sqrt{3}}\frac{\sqrt{3}}{\sqrt{3}} = \frac{\sqrt{3}}{(\sqrt{3})^2} = \frac{\sqrt{3}}{3}$$

$$\sqrt[5]{\frac{x}{y^2}} = \sqrt[5]{\frac{x}{y^2}\frac{y^3}{y^3}} = \frac{\sqrt[5]{xy^3}}{\sqrt[5]{y^5}} = \frac{\sqrt[5]{xy^3}}{y}$$

We can also rationalize the numerator.

Example

$$2 + \sqrt{3} = \frac{2 + \sqrt{3}}{1}\frac{2 - \sqrt{3}}{2 - \sqrt{3}} = \frac{(2)^2 - (\sqrt{3})^2}{2 - \sqrt{3}} = \frac{1}{2 - \sqrt{3}}$$

EXERCISE SET A.1

In Exercises 1 through 64 simplify. Write the answers without using negative exponents.

1. x^2x^8
2. y^3y^{11}
3. $(x^3)^5$
4. $(y^4)^3$
5. $(x+2)^4(x+2)$
6. $(y+3)^5(y+3)^2$
7. $\frac{x^9}{x^4}$
8. $\frac{y^3}{y^7}$
9. $\frac{(x+1)^5}{(x+1)^2}$
10. $\frac{(x^2y)^2}{(xy^3)^3}$
11. 8^0
12. $(2^0+2)^0$
13. 2^{-1}
14. 4^{-2}
15. $\left(\frac{1}{10}\right)^{-1}$
16. $\left(\frac{4}{9}\right)^{-2}$
17. $\left(\frac{xy^2}{z}\right)^0$
18. $(a^3bc^0)^{-2}$
19. $(a^2b)^2(ab^3)^{-1}$
20. $(ab^2c^{-1})^{-2}$
21. $\left(\frac{x^{-1}y^{-1}}{z}\right)^{-1}$
22. $\left(\frac{x^2y^{-2}z}{x^3yz^{-2}}\right)^{-2}$
23. $\left(\frac{x^2y^{-5}}{x^{-2}y^{-3}}\right)^{-3}$
24. $\frac{(x+1)^3(x+2)^4}{[(x+1)(x+2)]^3}$

25. $\sqrt[3]{-64}$
26. $\sqrt[5]{32}$
27. $\sqrt[5]{-32}$
28. $\sqrt[3]{-\frac{1}{9}}$
29. $\sqrt[4]{(-7)^4}$
30. $\sqrt[4]{7^4}$
31. $\sqrt[3]{(-3)^3}$
32. $\sqrt[3]{3^3}$
33. $\sqrt{\frac{25}{4}}$
34. $\sqrt[3]{-\frac{8}{27}}$
35. $\sqrt[3]{-\frac{8}{125}}$
36. $\sqrt[5]{\frac{243}{32}}$
37. $\sqrt{\sqrt{16}}$
38. $\sqrt[3]{\sqrt{64}}$
39. $8^{2/3}$
40. $64^{2/3}$
41. $(0.001)^{1/3}$
42. $\left(-\frac{1}{125}\right)^{1/3}$
43. $36^{-3/2}$
44. $125^{2/3}$
45. $32^{4/5}$
46. $32^{-4/5}$
47. $64^{4/3}$
48. $\left(\frac{9}{16}\right)^{-5/2}$
49. $\left(\frac{1000}{27}\right)^{4/3}$
50. $(-1)^{3/5}$

51. $\left(\dfrac{8}{27}\right)^{-2/3}$

52. $100^{3/2}$

53. $\sqrt{4x^4}$

54. $\sqrt[4]{16y^8}$

55. $\sqrt[3]{ab^2}\sqrt[3]{a^2b}$

56. $\sqrt[3]{8x^6}$

57. $\sqrt{72a^4b^3c^5}$

58. $\sqrt[3]{8x^4b^8}$

59. $\sqrt[5]{32a^{10}b^{15}}$

60. $\sqrt{18a^4b^5}$

61. $\sqrt[3]{\dfrac{8a^9b^6}{c^{12}}}$

62. $\sqrt[5]{\dfrac{4x^6}{x^{-4}y^5}}$

63. $\sqrt{ab}\sqrt{a^3b^3}$

64. $\sqrt[3]{ab}\sqrt[3]{a^2b^2}$

In Exercises 65 through 72 write the expressions using rational exponents rather than radicals.

65. $\dfrac{1}{\sqrt{x}} + \sqrt{x}$

66. $\sqrt{\sqrt{\sqrt{x}}}$

67. $\sqrt[3]{(x+1)^2}$

68. $(\sqrt[3]{(x+1)})^2$

69. $\sqrt[3]{x^2}$

70. $\sqrt{x^3}$

71. $\sqrt{\dfrac{x}{y}}$

72. $\sqrt[3]{\dfrac{x}{y}}$

In Exercises 73 through 84 simplify.

73. $(x^2+1)^{3/4}(x^2+1)^{2/3}$

74. $\dfrac{(x^2+1)^{1/3}}{(x^2+1)^{-3/4}}$

75. $\dfrac{x^{2/3}x^{-1/3}}{x^{2/3}}$

76. $\dfrac{y^{-1/4}y^{-1/2}}{y^{-3/4}}$

77. $\dfrac{a^{2/3}b^{3/4}}{a^{-1/3}b^{-1/4}}$

78. $\dfrac{a^{1/3}b^{-2/5}}{a^{-2/3}b^{1/5}}$

79. $\dfrac{a^{2/3}b^{4/3}}{a^{2/3}b^{1/3}}$

80. $\dfrac{a^{-1/3}b^{-3/4}}{a^{-2/3}b^{1/4}}$

81. $\dfrac{(x+1)^{2/3}(x+2)^{1/3}}{(x+1)^{-1/3}(x+2)^{-2/3}}$

82. $\dfrac{(x+1)^{-1/3}(y+1)^{-2/3}}{(x+1)^{2/3}(y+1)^{-5/3}}$

83. $[(x+1)^3]^{5/3}$

84. $[(y-2)^{1/3}]^6$

In Exercises 85 through 92 rationalize the denominator.

85. $\dfrac{1}{\sqrt{5}}$

86. $\dfrac{1}{\sqrt{7}}$

87. $\dfrac{1}{\sqrt{2}+1}$

88. $\dfrac{1}{3-\sqrt{5}}$

89. $\dfrac{1}{\sqrt[3]{5}}$

90. $\dfrac{3}{\sqrt[4]{3}}$

91. $\dfrac{1+\sqrt{3}}{1-\sqrt{3}}$

92. $\dfrac{2-\sqrt{5}}{2+\sqrt{5}}$

In Exercises 93 through 96 rationalize the numerator.

93. $\dfrac{\sqrt{5}-\sqrt{2}}{3}$

94. $\dfrac{\sqrt{7}-\sqrt{3}}{4}$

95. $\dfrac{\sqrt{10}-2}{6}$

96. $\sqrt{10}-3$

A.2 POLYNOMIALS AND RATIONAL EXPRESSIONS

■ Polynomials ■ Factoring ■ Rational Expressions

Polynomials

We begin with the following definition.

Polynomial

A polynomial is an expression of the form

$$a_nx^n + a_{n-1}x^{n-1} + \cdots + a_1x + a_0,$$

where n is a nonnegative integer and each coefficient a_i is a real number.

If $a_n \neq 0$, the polynomial is said to have **degree n.**

Lower order polynomials have special names. A polynomial such as 3 has degree equal to *zero* and is called a **constant.** A polynomial such as $x - \frac{1}{2}$ has degree equal to *one* and is called linear. A polynomial such as $4x^2 - 2x + 4$ has degree equal to *two* and is called a quadratic. A polynomial such as $2x^3 - 5x^2 - \frac{3}{2}x - 6$ has degree equal to *three* and is called a **cubic.**

Since for any real number x a polynomial is a real number, polynomials must follow the same rules as any real number.

EXAMPLE 1

Simplify the following. (a) $(x^3 + 2x^2 + 4x - 7) + (2x^3 - x^2 + 2x + 10)$,
(b) $(x^4 - 2x^2 + 3) - (x^4 + x^3 - x - 4)$, (c) $(x^2 - 2)(3x^3 - x + 1)$

SOLUTIONS ■

(a) $(x^3 + 2x^2 + 4x - 7) + (2x^3 - x^2 + 2x + 10)$

$$= (1 + 2)x^3 + (2 - 1)x^2 + (4 + 2)x + (-7 + 10)$$

$$= 3x^3 + x^2 + 6x + 3$$

(b) $(x^4 - 2x^2 + 3) - (x^4 + x^3 - x - 4)$

$$= (1 - 1)x^4 - x^3 - 2x^2 + x + (3 + 4)$$

$$= -x^3 - 2x^2 + x + 7$$

(c) $(x^2 - 2)(3x^3 - x + 1) = x^2(3x^3 - x + 1) - 2(3x^3 - x + 1)$

$$= (3x^5 - x^3 + x^2) + (-6x^3 + 2x - 2)$$

$$= 3x^5 + (-1 - 6)x^3 + x^2 + 2x - 2$$

$$= 3x^5 - 7x^3 + x^2 + 2x - 2 \quad ■$$

We now list some products that occur frequently. They can be checked by multiplying the terms on the left side.

Special Products

1. $(a + b)(a - b) = a^2 - b^2$
2. $(a + b)^2 = a^2 + 2ab + b^2$
3. $(a - b)^2 = a^2 + 2ab - b^2$
4. $(ax + b)(cx + d) = acx^2 + (ad + bc)x + bd$
5. $(x + y)^3 = x^3 + 3x^2y + 3xy^2 + y^3$
6. $(x - y)^3 = x^3 - 3x^2y + 3xy^2 - y^3$
7. $(a - b)(a^2 + ab + b^2) = a^3 - b^3$
8. $(a + b)(a^2 - ab + b^2) = a^3 + b^3$

Factoring

We will often find it useful to **factor** a polynomial, that is, write the polynomial as a product of two or more nonconstant polynomials, each of which is as low an order as possible.

For example, using the first special product listed above, we see that

$$x^2 - 1 = (x + 1)(x - 1).$$

Although, in general, factoring a polynomial can be extremely difficult, many polynomials can be factored using the other special products listed above. The following are some examples.

EXAMPLE 2

Factor (a) $4x^2 - 9$, (b) $8x^3 - 27$, (c) $81x^4 - 16$, (d) $4x^{-4} - 9x^{-6}$.

SOLUTIONS ■

(a) Use Formula 1 with $a = 2x$ and $b = 3$. Then

$$4x^2 - 9 = (2x)^2 - 3^2$$

$$= (2x + 3)(2x - 3)$$

(b) Use Formula 7 with $a = 2x$ and $b = 3$.

$$8x^3 - 27 = (2x)^3 - (3)^3$$

$$= (2x - 3)[(2x)^2 + (2x)(3) + (3)^2]$$

$$= (2x - 3)(4x^2 + 6x + 9)$$

(c) Use Formula 1 twice as follows.

$$81x^4 - 16 = (9x^2)^2 - (4)^2$$
$$= (9x^2 + 4)(9x^2 - 4)$$
$$= (9x^2 + 4)(3x + 2)(3x - 2)$$

(d) There are two ways of factoring this expression. In calculus one normally factors out the highest negative power of x first. Doing this gives

$$4x^{-4} - 9x^{-6} = x^{-6}(4x^2 - 9)$$
$$= x^{-6}(2x + 3)(2x - 3) \quad \blacksquare$$

Rational Expressions

A **rational expression** is a sum, difference, product, or quotient of terms, each of which is the quotient of polynomials. Thus the following are rational expressions.

$$\frac{x^2 + 1}{x^3 - x - 3} \qquad 2x^3 + \frac{5x + 1}{x^2 - 2}$$

Rational expressions must satisfy the same rules as real numbers.

Rules for Rational Expressions

Let $a(x)$, $b(x)$, $c(x)$, $d(x)$ be rational expressions with $b(x) \neq 0$ and $d(x) \neq 0$. Also $c(x) \neq 0$ in Rule 5. Then

Rule 1 $\dfrac{a(x)}{b(x)} + \dfrac{c(x)}{b(x)} = \dfrac{a(x) + c(x)}{b(x)}$

Rule 2 $\dfrac{a(x)}{b(x)} + \dfrac{c(x)}{b(x)} = \dfrac{a(x) - c(x)}{b(x)}$

Rule 3 $\dfrac{a(x)d(x)}{b(x)d(x)} = \dfrac{a(x)}{b(x)}$

Rule 4 $\dfrac{a(x)}{b(x)} \cdot \dfrac{c(x)}{d(x)} = \dfrac{a(x)c(x)}{b(x)d(x)}$

Rule 5 $\dfrac{a(x)}{b(x)} \div \dfrac{c(x)}{d(x)} = \dfrac{a(x)}{b(x)} \cdot \dfrac{d(x)}{c(x)}$

EXAMPLE 3   Rule 1

Simplify $\dfrac{x^2 + x + 1}{s^2 + 1} - \dfrac{x + 1}{x^2 + 1}$.

SOLUTION ∎

$$\frac{x^2 + x + 1}{x^2 + 1} - \frac{x + 1}{x^2 + 1} = \frac{(x^2 + x + 1) - (x + 1)}{x^2 + 1} = \frac{x^2}{x^2 + 1} \quad \blacksquare$$

EXAMPLE 4   Rule 3

Simplify $\dfrac{x^2 - 1}{x^2 + 4x + 3}$.

SOLUTION ∎

$$\frac{x^2 - 1}{x^2 + 4x + 3} = \frac{(x + 1)(x - 1)}{(x + 1)(x + 3)} = \frac{x - 1}{x + 3} \quad \blacksquare$$

EXAMPLE 5   Rule 4

Simplify $\dfrac{x^2 + 3x + 2}{2x^2 - x - 1} \cdot \dfrac{x - 1}{x + 1}$.

SOLUTION ∎

$$\frac{x^2 + 3x + 2}{2x^2 - x - 1} \cdot \frac{x - 1}{x + 1} = \frac{(x + 1)(x + 2)(x - 1)}{(x - 1)(2x + 1)(x + 1)} = \frac{x + 2}{2x + 1} \quad \blacksquare$$

EXAMPLE 6   Rule 5

Simplify $\dfrac{x^3 - 2x^2 + x}{x^2 + 2x} \div \dfrac{x - 1}{x^2 - 4}$.

SOLUTION ∎

$$\frac{x^3 - 2x^2 + x}{x^2 + 2x} \div \frac{x - 1}{x^2 - 4} = \frac{x^3 - 2x^2 + x}{x^2 + 2x} \cdot \frac{x^2 - 4}{x - 1}$$

$$= \frac{x(x^2 - 2x + 1)}{x(x + 2)} \cdot \frac{x^2 - 4}{x - 1}$$

$$= \frac{(x - 1)^2 (x - 2)(x + 2)}{(x + 2)(x - 1)}$$

$$= (x - 1)(x - 2) \quad \blacksquare$$

Least Common Multiple. In order to combine the two fractions

$$\frac{1}{x^2 - 2x + 1} - \frac{1}{x^3 - x},$$

we must first factor, obtaining

$$\frac{1}{(x - 1)^2} - \frac{1}{x(x - 1)(x + 1)}.$$

We now take as the least common denominator, the product of each of the factors, taking the power of each factor as the highest power that appears. Thus the least common denominator for the above example is $x(x - 1)^2(x + 1)$. Thus

$$\frac{1}{(x - 1)^2} - \frac{1}{x(x - 1)(x + 1)} = \frac{1}{(x - 1)^2} \frac{x(x + 1)}{x(x + 1)} - \frac{1}{x(x - 1)(x + 1)} \frac{x - 1}{x - 1}$$

$$= \frac{x(x + 1) - (x - 1)}{x(x - 1)^2(x + 1)}$$

$$= \frac{x^2 + x - x + 1}{x(x - 1)^2(x + 1)}$$

$$= \frac{x^2 + 1}{x(x - 1)^2(x + 1)}$$

The following is a problem that comes up in calculus.

EXAMPLE 7

Simplify $\dfrac{\dfrac{1}{x + h} - \dfrac{1}{x}}{h}$.

S O L U T I O N ∎

$$\frac{\dfrac{1}{x+h}-\dfrac{1}{x}}{h} = \frac{1}{h}\left(\frac{1}{x+h}-\frac{1}{x}\right)$$

$$= \frac{1}{h}\left(\frac{1}{x+h}\frac{x}{x}-\frac{1}{x}\frac{x+h}{x+h}\right)$$

$$= \frac{1}{h}\frac{x-(x+h)}{x(x+h)}$$

$$= \frac{1}{h}\frac{-h}{x(x+h)}$$

$$= \frac{-1}{x(x+h)} \quad ∎$$

Rationalizing the Numerator. In calculus one often must rationalize the numerator.

EXAMPLE 8

Rationalize $\dfrac{\sqrt{x}-2}{x-4}$.

S O L U T I O N ∎

$$\frac{\sqrt{x}-2}{x-4} = \frac{\sqrt{x}-2}{x-4}\frac{\sqrt{x}+2}{\sqrt{x}+2}$$

$$= \frac{(\sqrt{x})^2-(2)^2}{(x-4)(\sqrt{x}+2)}$$

$$= \frac{(x-4)}{(x-4)(\sqrt{x}+2)}$$

$$= \frac{1}{\sqrt{x}+2} \quad ∎$$

EXERCISE SET A.2

In Exercises 1 through 14 simplify.

1. $(x^2 + 2x - 1) + (2x^2 + x + 3)$

2. $(x^2 - x + 2) + (-x^2 - x - 1)$

3. $(-2x^2 + x - 2) + (-x^2 - 2x + 1)$

4. $(-3x^2 - x + 1) + (x^2 + x - 1)$

5. $(x^4 + x^2 + x + 1) - (x^3 + 2x - 3)$

6. $(x^4 - 2x^3 - 1) - (2x^4 + x^2 + x - 1)$

7. $(2x^3 - 2x^2 + 1) - (x^3 - x^2 - x - 1)$

8. $(-x^3 - x + 1) - (x^3 - x^2 + 2)$

9. $(x + 1)(x^2 + 2x + 3)$ **10.** $(x + 2)(x^2 - x - 2)$

11. $(2x - 1)(x^2 - x + 3)$ **12.** $(2x - 3)(x^2 - 2)$

13. $(x^2 + 3)(x^2 + 1)$ **14.** $(x^2 + 2)(x^2 + x + 2)$

In Exercises 15 through 58 factor.

15. $x^3 - 27$ **16.** $x^3 - 8$

17. $x^3 - \dfrac{1}{8}$ **18.** $x^3 - \dfrac{1}{64}$

19. $27x^3 - 8$ **20.** $8x^3 - 125$

21. $\dfrac{1}{8}x^3 - 27$ **22.** $\dfrac{1}{64}x^3 - 27$

23. $\dfrac{1}{64}x^3 - \dfrac{1}{27}$ **24.** $\dfrac{1}{125}x^3 - \dfrac{1}{8}$

25. $a^3x^3 - b^3$ **26.** $a^3x^3\dfrac{1}{b^3}$

27. $x^2 - 16$ **28.** $x^2 - 81$

29. $16x^2 - 1$ **30.** $81x^2 - 1$

31. $x^4 - \dfrac{1}{16}$

32. $x^4 - \dfrac{1}{81}$

33. $\dfrac{1}{16} x^4 - 1$

34. $\dfrac{1}{81} x^4 - 1$

35. $x^4 - 4$

36. $x^4 - 9$

37. $\dfrac{1}{4} x^4 - 1$

38. $\dfrac{1}{9} x^4 - 1$

39. $\dfrac{x^2 - x - 2}{x^2 - 3x + 2}$

40. $\dfrac{x^2 + 3x + 2}{x^2 + 5x + 6}$

41. $\dfrac{x^3 + x^2}{x - x^3}$

42. $\dfrac{x^2 - 4}{x^2 + x - 6}$

43. $\dfrac{x^2 - 1}{x^2 - 4} \cdot \dfrac{x - 2}{x + 1}$

44. $\dfrac{x^2 - 2x + 1}{x^2 + 3x + 2} \cdot \dfrac{x + 1}{x - 1}$

45. $\dfrac{x^3 - 1}{x^2 - 16} \cdot \dfrac{x - 4}{x^2 + x + 1}$

46. $\dfrac{x^4 - 1}{x + 2} \cdot \dfrac{x}{x^2 - 1}$

47. $\dfrac{x^2 - 1}{x + 1} \div \dfrac{x - 1}{2x + 1}$

48. $\dfrac{2x^2 + x - 1}{x^2 - 2x + 1} \div \dfrac{2x - 1}{x - 1}$

49. $\dfrac{x^2 + 1}{x + 1} \div \dfrac{x}{x + 1}$

50. $\dfrac{a^3 - b^3}{a^2 - b^2} \div \dfrac{1}{a + b}$

51. $\dfrac{x}{x + 1} - \dfrac{1}{x + 1}$

52. $\dfrac{x + 1}{x^2 + 1} - \dfrac{2x + 3}{x^2 + 1}$

53. $\dfrac{x}{x^2 - 1} - \dfrac{1}{x^2 - 1}$

54. $\dfrac{x^2}{x^4 - 8} + \dfrac{4}{x^4 - 8}$

55. $\dfrac{x^2}{x^2 - 1} - \dfrac{x}{x - 1}$

56. $\dfrac{1}{x^2 + x - 2} - \dfrac{1}{x^2 - 4}$

57. $\dfrac{x^2}{x^4 - 16} - \dfrac{1}{x^2 + 4}$

58. $\dfrac{1}{x - 3} + \dfrac{1}{x^2 - 9} - \dfrac{1}{x^2 - 2x - 3}$

In Exercises 59 through 66 rationalize the denominator.

59. $\dfrac{1}{\sqrt{x} + 1}$

60. $\dfrac{1}{\sqrt{x} - 1}$

61. $\dfrac{1}{\sqrt{x} + \sqrt{y}}$

62. $\dfrac{1}{\sqrt{x} - \sqrt{3}}$

63. $\dfrac{y - 9}{\sqrt{y} - 3}$

64. $\dfrac{a - b}{\sqrt{a} - \sqrt{b}}$

65. $\dfrac{a^2 - 1}{\sqrt{a} - 1}$

66. $\dfrac{x^2 - y^2}{\sqrt{x} - \sqrt{y}}$

In Exercises 67 through 70 rationalize the numerator.

67. $\dfrac{\sqrt{x} - 1}{x - 1}$

68. $\dfrac{\sqrt{x} - \sqrt{y}}{x - y}$

69. $\dfrac{\sqrt{x + h} - \sqrt{x}}{h}$

70. $\dfrac{\sqrt{x} - \sqrt{x - h}}{h}$

A.3 EQUATIONS

- Linear Equations ■ Quadratic Equations ■ Polynomial Equations
- Linear Equations with Absolute Values

Linear Equations

The polynomial $ax + b$ is called a **linear** expression and the equation $ax + b = 0$ is called a **linear equation**.

If $a \ne 0$, the linear equation $ax + b = 0$ can always be solved as follows.

$$ax + b = 0$$

$$ax = -b$$

$$x = -\dfrac{b}{a}$$

EXAMPLE 1

Solve $\dfrac{2x}{3} - \dfrac{x - 1}{3} = \dfrac{1}{4} - \dfrac{1}{2}\left(3x - \dfrac{3x - 2}{3}\right)$.

SOLUTION ■ First remove the parenthesis.

$$\dfrac{2x}{3} - \dfrac{x - 1}{3} = \dfrac{1}{4} - \dfrac{3}{2}x + \dfrac{3x - 2}{6}$$

Remove the fractions by multiplying by 12.

$$4(2x) - 4(x - 1) = 3 - 6(3x) + 2(3x - 2)$$

$$8x - 4x + 4 = 3 - 18x + 6x - 4$$

Move all x-terms to the left and all constant terms to the right.

$$8x - 4x + 18x - 6x = 3 - 4 - 4$$

$$16x = -5$$

$$x = \frac{5}{16} \quad \blacksquare$$

Quadratic Equations

If $a \neq 0$, the polynomial $ax^2 + bx + c$ is called a **quadratic expression**. The equation

$$ax^2 + bx + c = 0, \qquad a \neq 0,$$

is called a **quadratic equation**. We shall always assume that a, b, c, and x are *real* numbers.

Some quadratic equations can be solved easily by factoring.

EXAMPLE 2

Solve (a) $x^2 - 4 = 0$, (b) $9x^2 - 16 = 0$, (c) $3x^2 + 7x - 6 = 0$.

SOLUTIONS ■

(a) To solve $x^2 - 4 = 0$, first recognize the expression $x^2 - 4$ as a difference of squares $a^2 - b^2$ with $a = x$ and $b = 2$. Then factoring gives

$$0 = x^2 - 4 = x^2 - 2^2 = (x + 2)(x - 2).$$

The only way the product of two expressions can be zero is if at least one of them is zero. Thus either $x + 2 = 0$ or $x - 2 = 0$. This in turn implies that $x = -2$ or $x = 2$. Thus the quadratic equation $x^2 - 4 = 0$ has two solutions, $x = -2$ and $x = 2$.

(b) To solve $9x^2 - 16 = 0$ first recognize the expression $9x^2 - 16$ as a difference of squares $a^2 - b^2$ with $a = 3x$ and $b = 4$. Factoring then gives

$$0 = 9x^2 - 16 = (3x)^2 - 4^2 = (3x + 4)(3x - 4)$$

$$\text{implies } 3x + 4 = 0 \quad \text{or} \quad 3x - 4 = 0$$

$$x = -\frac{4}{3} \quad \text{or} \quad x = \frac{4}{3}$$

(c) To solve $3x^2 + 7x - 6 = 0$, first factor $3x^2 + 7x - 6$ into a product of two linear expressions. Since the first term is $3x^2$ and the last term is negative, we try $(3x+ \quad)(x- \quad)$ or $(3x- \quad)(x+ \quad)$. Now 6 can be written as $6 = 3 \cdot 2$ or $6 = 6 \cdot 1$. Using trial and error with the first way gives the following.

$$(3x - 3)(x + 2) = 3x^2 + 3x - 6 \qquad \text{wrong}$$

$$(3x + 3)(x - 2) = 3x^2 - 3x - 6 \qquad \text{wrong}$$

$$(3x + 2)(x - 3) = 3x^2 - 7x - 6 \qquad \text{wrong}$$

$$(3x - 2)(x + 3) = 3x^2 + 7x - 6 \qquad \text{correct}$$

Thus we have $3x^2 + 7x - 6 = (3x - 2)(x + 3)$. Therefore

$$0 = 3x^2 + 7x - 6 = (3x - 2)(x + 3)$$

implies $3x - 2 = 0$ or $x + 3 = 0$

$$x = \frac{2}{3} \quad \text{or} \quad x = -3$$ ∎

The procedures used in the last example do not always work. A procedure that does always work is to use the quadratic formula.

Solutions of the Quadratic Equation

The solutions of the quadratic equation

$$ax^2 + bx + c = 0, \quad a \neq 0$$

are given by the quadratic formula

$$x = \frac{-b \pm \sqrt{b^2 - 4ac}}{2a}$$

The term $b^2 - 4ac$ is called the **discriminant**.

The quadratic formula breaks down into three cases.

Case 1. If $b^2 - 4ac > 0$, there are the two solutions

$$\frac{-b \pm \sqrt{b^2 - 4ac}}{2a}$$

Case 2. If $b^2 - 4ac = 0$, there is one solution given by

$$-\frac{b}{2a}.$$

Case 3. If $b^2 - 4ac < 0$, there are no solutions in the set of real numbers.

EXAMPLE 3

Solve $2x^2 - 3x - 1 = 0$.

SOLUTION ■ Use the quadratic formula with $a = 2$, $b = -3$, $c = -1$ and obtain

$$x = \frac{-(-3) \pm \sqrt{(-3)^2 - 4(2)(-1)}}{2(2)}$$

$$= \frac{3 \pm \sqrt{17}}{4}$$ ∎

Polynomial Equations

We now consider solving polynomial equations of any order. We have just seen how any quadratic equation can be solved using the quadratic formula. It would be convenient if any polynomial equation such as $x^4 + 3x^3 - 2x^2 - 8 = 0$ could be solved using some corresponding ''polynomial formula.'' Unfortunately, no such formula exists for polynomials of degree four or higher except for very special cases. We can solve polynomial equations in the cases when the polynomial can be completely factored. We now consider such cases.

EXAMPLE 4

Solve (a) $4x^4 - 36x^2 = 0$, (b) $2x^5 - 2x = 0$, (c) $10x^3 - 5x^2 - 5x = 0$.

SOLUTIONS ■

(a)
$$\begin{aligned} 0 = 4x^4 - 36x^2 &= 4x^2(x^2 - 9) &&\text{common factor} \\ &= 4x^2(x + 3)(x - 3) &&\text{difference of squares} \end{aligned}$$

From this factorization we can immediately obtain the solutions $x = 0$, $x = -3$, and $x = 3$.

(b)
$$\begin{aligned} 0 = 2x^5 - 2x &= 2x(x^4 - 1) &&\text{common factor} \\ &= 2x(x^2 + 1)(x^2 - 1) &&\text{difference of squares} \\ &= 2x(x^2 + 1)(x + 1)(x - 1) &&\text{difference of squares} \end{aligned}$$

From this factorization we can immediately obtain the solutions $x = 0$, $x = -1$, and $x = 1$.

(c)
$$\begin{aligned} 0 = 10x^3 - 5x^2 - 5x &= 5x(2x^2 - x - 1) &&\text{common factor} \\ &= 5x(2x + 1)(x - 1) &&\text{factoring a quadratic} \end{aligned}$$

From this factorization we can immediately obtain the solutions $x = 0$, $x = 1$, and $x = -1/2$. ■

Linear Equations with Absolute Value

We now give the definition of absolute value.

Absolute Value

$$|x| = \begin{cases} x & \text{if } x \geq 0 \\ -x & \text{if } x < 0 \end{cases}$$

REMARK. You may have learned to find the absolute value by "dropping the sign." Doing this for algebraic expressions will be disastrous. For example $|-x|$ is x if x is positive but $-x$ if x is negative. Thus $|-x|$ is sometimes equal to x and sometimes to $-x$.

It follows immediately that the absolute value has the following property.

A Property of Absolute Value

For any $a \geq 0$,

$$|x| = a \quad \text{if, and only if,} \quad x = a \quad \text{or} \quad x = -a.$$

EXAMPLE 5

Solve $|2x - 3| = 2$.

SOLUTION ■ This will be true if $2x - 3 = 2$ or if $2x - 3 = -2$. Thus

$$2x - 3 = 2 \quad \text{or} \quad 2x - 3 = -2$$
$$2x = 5 \quad \text{or} \quad 2x = 1$$
$$x = \frac{5}{2} \quad \text{or} \quad x = \frac{1}{2} \quad ■$$

EXERCISE SET A.3

In Exercises 1 through 10 solve.

1. $x - 2(x - 3) = 4 - 2(x + 1)$

2. $3(x - 1) - 4(x - 2) - 3 = -(3x - 1)$

3. $-(2x - 1) + 3(x + 4) - 4 = 2(4x - 1)$

4. $(4x - 5) - 3(2x + 1) - 4 = 3 + 2(x - 1) - 3(2x + 1)$

5. $\dfrac{1}{2} x - \dfrac{x - 2}{4} - 1 = 2(x - 2)$

6. $\dfrac{x - 1}{2} - \dfrac{2x - 3}{4} - \dfrac{1}{2} = \dfrac{1}{2}(3x - 1)$

7. $\dfrac{2x - 3}{3} - \dfrac{2x + 1}{2} - \dfrac{2}{3} = \dfrac{1}{3}(4x + 3)$

8. $\dfrac{7}{12} x - \dfrac{2x - 1}{3} + \dfrac{5}{6} = \dfrac{1}{2}(5x - 2)$

9. $\dfrac{1}{2}\left(3x - \dfrac{1}{2}\dfrac{x - 1}{2}\right) = 2x - \dfrac{3}{2}$

10. $\dfrac{1}{2}\left(\dfrac{3x - 2}{3} - \dfrac{1}{3}\dfrac{2x - 1}{2}\right) = \dfrac{1}{3}\left(\dfrac{2x - 3}{2} + \dfrac{3}{2}\right)$

In Exercises 11 through 50 set the quadratic expression equal to zero and solve by factoring.

11. $y^2 - 4$

12. $x^2 - 9$

13. $4y^2 - 9$

14. $36x^2 - 25$

15. $\dfrac{1}{4} x^2 - 1$

16. $\dfrac{1}{9} y^2 - 1$

17. $x^2 - \dfrac{1}{4}$

18. $y^2 - \dfrac{1}{9}$

19. $25x^2 - 9$

20. $16x^2 - 25$

21. $\dfrac{1}{4} y^2 - 9$

22. $\dfrac{1}{16} y^2 - \dfrac{1}{9}$

23. $4y^2 - \dfrac{1}{9}$

24. $9y^2 - \dfrac{1}{4}$

25. $a^2x^2 - b^2$

26. $a^2y^2 - \dfrac{1}{b^2}$

27. $x^2 + 4x + 4$

28. $x^2 + 4x + 3$

29. $x^2 - 5x + 4$

30. $x^2 + 2x - 8$

31. $2x^2 + 3x - 2$

32. $2x^2 - 3x + 1$

33. $3x^2 + 7x + 2$

34. $6x^2 + 5x + 1$

35. $6x^2 + x - 1$

36. $6x^2 + x - 1$

37. $10x^2 - 12x + 2$

38. $10x^2 - 11x - 6$

39. $6x^2 + 13x + 6$

40. $6x^2 - 5x - 6$

41. $12x^2 - 11x + 2$

42. $12x^2 - x - 6$

43. $9x^2 - 6x + 1$

44. $x^2 - 6x - 9$

45. $9x^2 + 12x + 4$

46. $4x^2 + 20x + 25$

47. $x^2 - x + \dfrac{1}{4}$

48. $x^2 - \dfrac{1}{2} x + \dfrac{1}{16}$

49. $4x^2 + 2x + \dfrac{1}{4}$

50. $9x^2 - 3x + \dfrac{1}{4}$

In Exercises 51 through 58 set the polynomial expression equal to zero and solve by factoring.

51. $x^4 - 16$

52. $x^4 - 81$

53. $16x^4 - 81$

54. $81x^4 - 16$

55. $x^5 + x^4 - 2x^3$

56. $x^7 + x^6 - 6x^5$

57. $2x^6 - 162x^2$

58. $3x^7 - 48x^3$

In Exercises 59 through 72 set each of quadratic expressions equal to zero. If there is a solution, solve by using the quadratic formula.

59. $2x^2 + 4x + 2$

60. $2x^2 - 5x + 3$

61. $5x^2 - 7x + \dfrac{1}{5}$

62. $4x^2 - 9x + 2$

63. $-x^2 + x + 1$

64. $-2x^2 - 3x + 2$

65. $-3x^2 - 4x + 5$

66. $-2x^2 - 3x + 4$

67. $-x^2 + 4x - 1$

68. $-2x^2 + 5x - 2$

69. $-3x^2 + 7x - \dfrac{2}{3}$

70. $-x^2 + 3x - 2$

71. $x^2 + x + 1$

72. $x^2 + 2x - 2$

In Exercises 73 through 84 solve.

73. $|2x - 1| = 1$

74. $|2x - 3| = 2$

75. $|3x - 1| = 5$

76. $|5x - 2| = 4$

77. $|2 - 3x| = 2$

78. $|3 - 2x| = 1$

79. $|4 - x| = 5$

80. $|3 - 5x| = 3$

81. $\left|\dfrac{1}{2} - x\right| = \dfrac{2}{3}$

82. $\left|x - \dfrac{1}{3}\right| = \dfrac{1}{4}$

83. $\left|\dfrac{1}{2} x - \dfrac{2}{3}\right| = \dfrac{1}{6}$

84. $\left|\dfrac{1}{3} x - \dfrac{2}{5}\right| = \dfrac{1}{2}$

A.4 INEQUALITIES

■ Linear Inequalities ■ Inequalities with Absolute Values ■ Nonlinear Inequalities

Linear Inequalities

We begin this section with some fundamental properties of inequalities.

Properties of Inequalities

1. $a > b$ implies $a + c > b + c$.
2. $a > b$, $c > 0$ implies $ac > bc$.
3. $a > b$, $c < 0$ implies $ac < bc$.

EXAMPLE 1

Solve the inequality $5x - 4 < 2x + 5$.

SOLUTION ■

$$5x - 4 < 2x + 5$$
$$5x - 2x < 5 + 4$$
$$3x < 9$$
$$x < 3$$

Figure A.1

Thus all numbers *less* than 3 is the solution set. This set can also be represented using the interval notation $(-\infty, 3)$. See Figure A.1. ■

EXAMPLE 2

Solve the inequality $-2 \leq 6 - 2x < 8$.

SOLUTION ■

$$-2 \leq 6 - 2x \qquad \text{and} \qquad 6 - 2x < 8$$
$$2x \leq 6 + 2 \qquad\qquad -2x < 8 - 6 = 2$$
$$2x \leq 8 \qquad\qquad x > -1$$
$$x \leq 4 \qquad\qquad x > -1$$

Figure A.2

Thus x must simultaneously satisfy $x \leq 4$ and $x > -1$. We can describe this solution set by the interval notation $(-1, 4]$. See Figure A.2. ■

Inequalities with Absolute Values

Properties of Inequalities with Absolute Values

For any $a < 0$,

1. $|x| < a$ if and only if $-a < x < a$.
2. $|x| > a$ if and only if either $x > a$ or $x < -a$.

EXAMPLE 3

Solve the inequality $|2x + 4| < 2$.

SOLUTION ■

According to Property 1, this is equivalent to

$$-2 < 2x + 4 < 2.$$
$$-2 < 2x + 4 \qquad \text{and} \qquad 2x + 4 < 2$$
$$-2x < 6 \qquad\qquad 2x < -2$$
$$x > -3 \qquad\qquad x < -1$$

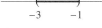

Figure A.3

In interval notation, the solution set is $(-3, -1)$. See Figure A.3. ■

EXAMPLE 4

Solve the inequality $|3 - 4x| > 5$.

SOLUTION ■ According to Property 2, this is equivalent to

$$2 - 4x > 5 \quad \text{or} \quad 3 - 4x < -5$$

$$-4x > 2 \qquad\qquad -4x < -8$$

$$x < -\frac{1}{2} \qquad\qquad x > 2$$

In interval notation, the solution set is $(-\infty, -\frac{1}{2})$ together with $(2, +\infty)$. See Figure A.4. ■

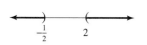

Figure A.4

Nonlinear Inequalities

EXAMPLE 5

Solve the inequality $x^2 - 3x < -2$.

SOLUTION ■ This is equivalent to

$$x^2 - 3x + 2 < 0.$$

Factoring the left side, this is equivalent to

$$(x - 1)(x - 2) < 0.$$

The only way the product of the two terms $(x - 1)$ and $(x - 2)$ can be negative is for one to be positive and the other negative. Thus we need to carefully look at the signs of these two factors. Clearly $x - 1 < 0$ if $x < 1$ and $x - 1 > 0$ if $x > 1$. In the same way $x - 2 < 0$ if $x < 2$ and $x - 2 > 0$ if $x > 2$. We then indicate this with a **sign chart** of these two factors as shown in Figure A.5. The sign of $(x - 1)(x - 2)$ is then obtained by taking the product of the signs of the individual factors. Thus on the interval $(-\infty, 1)$ the sign of both of the factors is negative, so the sign of the product is positive. On the interval $(1, 2)$ one factor is negative while the other factor is positive. Thus the sign of the product is negative. On the interval $(2, +\infty)$ both factors are positive so that the sign of the product is also positive.

From this we see that the product $(x - 1)(x - 2)$ is negative only when $1 < x < 2$. Thus the solution set is $(1, 2)$. ■

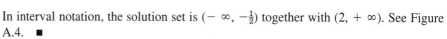

Figure A.5

EXAMPLE 6

Solve the inequality $2x^2 - x \geq 3$.

SOLUTION ■ This is equivalent to

$$2x^2 - x - 3 \geq 0.$$

Factoring the left side, this is equivalent to

$$(x + 1)(2x - 3) \geq 0.$$

$$
\begin{array}{c}
 \quad\quad\quad -1 \quad\quad \tfrac{3}{2} \\
\hline
(x+1) \quad - - - \ + + + \ + + + \\
(2x-3) \quad - - - \ - - - \ + + + \\
(x+1)(2x-3) \quad + + + \ - - - \ + + +
\end{array}
$$

Figure A.6

Figure A.6 gives the sign chart of the two factors $(x + 1)$ and $(2x - 3)$. It is suggested that the first factor be the one with the least zero. This seems to be the least confusing and most systematic, but is not mandatory. In the last line is given the product of the signs in the previous two lines. In this way we find all places where the product $(x + 1)(2x - 3)$ is greater or equal to zero. This is the set $(-\infty, -1]$ together with the set $[\tfrac{3}{2}, +\infty)$. ∎

EXAMPLE 7

Solve the inequality $\dfrac{x^2 - 2}{x} < 1$.

SOLUTION ∎ Notice that we cannot have $x = 0$.
We must reduce the inequality to one with zero on the right. Thus

$$\frac{x^2 - 2}{x} < 1$$

$$\frac{x^2 - 2}{x} - 1 < 0$$

$$\frac{x^3 - 2}{x} - \frac{x}{x} < 0$$

$$\frac{x^2 - x - 2}{x} < 0$$

$$\frac{(x + 1)(x - 2)}{x} < 0$$

The sign of the quotient will be determined by the signs of the three factors $(x + 1)$, x, and $(x - 2)$. We note the signs of the three factors in the sign chart given in Figure A.7. The sign chart of the quotient given in the last line in Figure A.7 is obtained by multiplying the individual signs on the previous three lines. Thus since the sign of all three factors is negative on $(-\infty, -1)$, the product is negative. On the interval $(-1, 0)$, two of the factors have negative signs and one has positive sign, thus the product is positive. We see from Figure A.7 that the sign of the quotient is negative on the interval $(-\infty, -1)$ together with $(0, 2)$. The "N" in the sign chart indicates that the quotient is not defined at $x = 0$. ∎

$$
\begin{array}{c}
\quad\quad -1 \quad\quad 0 \quad\quad\quad 2 \\
\hline
(x+1) \quad - - - \ + + + \ + + + \ + + + \\
x \quad - - - \ - - - \ + + + \ + + + \\
(x-2) \quad - - - \ - - - \ - - - \ + + + \\
\frac{(x+1)(x-2)}{x} \quad - - - \ + + + \ - - - \ + + +
\end{array}
$$

Figure A.7

EXAMPLE 8

Solve the inequality $\dfrac{x^2 - 4}{x^2 + 1} \le 0$.

SOLUTION ■ We factor and obtain

$$\frac{x^2 - 4}{x^2 + 1} \le 0$$

$$\frac{(x + 2)(x - 2)}{x^2 + 1} \le 0$$

We notice that the factor $x^2 + 1$ is *always* positive. Figure A.8 shows the sign chart of the three factors. The sign of the quotient is shown on the last line, where we see that the quotient is less than or equal to zero on $[-2, 2]$. ■

	-2	2
$(x + 2)$	$- - -$ $+ + +$	$+ + +$
$(x - 2)$	$- - -$ $- - -$	$+ + +$
$(x^2 + 1)$	$+ + +$ $+ + +$	$+ + +$
$\dfrac{(x + 2)(x - 2)}{(x^2 + 1)}$	$+ + +$ $- - -$	$+ + +$

Figure A.8

EXERCISE SET A.4

Solve the following.

1. $4x - 1 < x + 3$
2. $9x + 3 > 5x + 7$
3. $3x - 5 > 5x + 7$
4. $3x - 2 < 7x - 10$
5. $5x - 3 \le 2x - 6$
6. $2x - 7 \ge 4x + 3$
7. $3x + 7 \ge 5x - 7$
8. $3 - x \le 2 - 3x$
9. $\dfrac{5x - 1}{3} < \dfrac{2x + 1}{6}$
10. $\dfrac{2x - 1}{2} > -\dfrac{x - 1}{4}$
11. $1 - x \ge \dfrac{x - 1}{2}$
12. $\dfrac{x}{2} - \dfrac{1}{3} \le \dfrac{x + 3}{4}$
13. $|x + 1| < 3$
14. $|x + 2| \le 4$
15. $|2x - 1| \le 1$
16. $|3 - 2x| < 6$
17. $|x - 1| > 2$
18. $|x + 2| \ge 3$
19. $|3x - 2| \ge 6$
20. $|6 - x| > 3$
21. $|2x + 3| < \dfrac{1}{6}$
22. $\left|\dfrac{x}{2} - 3\right| \le 4$
23. $\dfrac{|2x - 5|}{3} \ge 5$
24. $\left|\dfrac{1}{2} - 2x\right| > 4$
25. $(x - 1)(x - 3) < 0$
26. $(x + 1)(x - 2) > 0$
27. $(x + 2)(x - 3) \ge 0$
28. $(x + 2)(x + 1) \le 0$
29. $x^2 - 4 < 0$
30. $x^2 - x - 6 > 0$
31. $x^2 + 2x - 3 \ge 0$
32. $x^2 + 2x \le 0$

33. $2x^2 - 5x < -2$
34. $2x^2 + x > 3$
35. $3x^2 + 7x \ge 6$
36. $6x^2 + 3 \le 11x$
37. $\dfrac{x - 1}{x + 2} < 0$
38. $\dfrac{x + 2}{x + 3} > 0$
39. $\dfrac{x - 1}{x - 2} \ge 0$
40. $\dfrac{x - 2}{x + 3} \le 0$
41. $\dfrac{x}{x + 2} > \dfrac{1}{x + 2}$
42. $\dfrac{2x}{x + 3} < \dfrac{3}{x + 3}$
43. $\dfrac{1}{x - 1} \le \dfrac{x}{1 - x}$
44. $\dfrac{2x}{x - 4} \ge \dfrac{1}{x - 4}$
45. $x \ge \dfrac{1}{x}$
46. $x \le \dfrac{4}{x}$
47. $\dfrac{x^2 + 2}{x} \ge 3x$
48. $\dfrac{x^2 - 3}{x} < 2x$
49. $\dfrac{(x - 2)(x - 3)}{x - 1} < 0$
50. $\dfrac{x^2 - 1}{x - 3} > 0$
51. $\dfrac{x^2 - 1}{x - 2} \ge 0$
52. $\dfrac{x^2 - 2x + 1}{x} \le 0$
53. $\dfrac{x^2 + 2x + 1}{x} \le 0$
54. $\dfrac{x^2 + 1}{x - 2} \ge 0$
55. $\dfrac{x^2 - x - 2}{x^2} < 0$
56. $\dfrac{x^2 - 1}{x^2 + 1} > 0$

A.5 THE CARTESIAN COORDINATE SYSTEM

■ Cartesian Coordinate System   ■ Distance Between Two Points   ■ Creating Equations   ■ Graphs

René Descartes 1596–1650

It is sometimes said that Descartes' work *La Geometrie* marks the turning point between medieval and modern mathematics. He demonstrated the interplay between algebra and geometry, tying together these two branches of mathematics. Due to poor health as a child, he was always permitted to remain in bed as long as he wished. He maintained this habit throughout his life and did his most productive thinking while lying in bed in the morning. Descartes was a great philosopher as well as a mathematician and felt that mathematics should be a model for other branches of study. The following is a quote from his famous *Discours Sur la Methode*: "The long chain of simple and easy reasonings by means of which geometers are accustomed to reach conclusions of their most difficult demonstrations led me to imagine that all things, to the knowledge of which man is competent, are mutually connected in the same way, and that there is nothing so far removed from us as to be beyond our reach, or so hidden that we cannot discover it, provided only we abstain from accepting the false for the true, and always preserve in our thoughts the order necessary for the deduction of one truth from another."

APPLICATION: GETTING TO THE CHURCH ON TIME

A prospective groom is located at point O in Figure A.9 and must drive his car to his wedding taking place in 25 minutes at the church located at point C. The church at C is 13 miles east and 4 miles north of O. The only highway goes through a town located at B which is 10 miles due east of O. From B the highway goes straight to C. From O to B is a nice open stretch of highway, and he can average 50 miles per hour. The stretch from B to C has traffic lights and some congestion, and he can average only 25 miles per hour on this leg of the trip. Can he get to the church on time? See Example 2 for the answer.

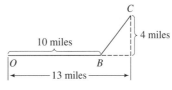

Figure A.9

Cartesian Coordinate System

It is very useful to have a geometric representation of the real numbers called the **number line**. The number line, shown in Figure A.10, is a (straight) line with an arbitrary point selected to represent the number 0. This point is called the **origin**. If the line is horizontal, the number 1 is placed an arbitrary distance to the right of the origin. The distance from the origin to the number 1 then represents the unit length. Each positive real number x then lies x units to the right of the origin, and each negative real number x lies $-x$ units to the left of the origin. In this manner each real number

Figure A.10

corresponds to exactly one point on the number line, and each point on the number line corresponds to exactly one real number.

We use the standard notation shown in the table.

Interval Notation	Definition	Geometric Representation
(a, b)	$\{x \mid a < x < b\}$	
$[a, b]$	$\{x \mid a \leq x \leq b\}$	
$(a, b]$	$\{x \mid a < x \leq b\}$	
$[a, b)$	$\{x \mid a \leq x < b\}$	
$[a, +\infty)$	$\{x \mid a \leq x\}$	
$(a, +\infty)$	$\{x \mid a < x\}$	
$(-\infty, b)$	$\{x \mid x < b\}$	
$(-\infty, b]$	$\{x \mid x \leq b\}$	

The Cartesian coordinate system (named after René Descartes) permits us to express data or relationships between two variables as geometric pictures. Representing numbers and equations as geometric pictures using the Cartesian coordinate system yields deeper insight into the equations and data since we can see them graphically.

To describe a **Cartesian coordinate system** we begin with a horizontal line called the **x-axis** and a vertical line called the **y-axis** both drawn in the plane. The point of intersection of these two lines is called the **origin**. The plane is then called the **xy-plane** or the **coordinate plane** (Fig. A.11).

For each axis select a unit of length. The units of length for each axis need not be the same. Starting from the origin as zero, mark off the scales on each axis. For the x-axis, positive numbers are marked off to the right of the origin and negative numbers to the left. For the y-axis, positive numbers are marked off above the origin and negative numbers below the origin (Fig. A.12).

Each point P in the xy-plane is assigned a pair of numbers (a, b) as shown in Figure A.13. The first number a is the horizontal distance from the point P to the y-axis and is called the **x-coordinate**, and the second number b is the vertical distance to the x-axis and is called the **y-coordinate**.

Conversely, every pair of numbers (a, b) determines a point P in the xy-plane with x-coordinate equal to a and y-coordinate equal to b. Figure A.14 indicates some examples.

We use the standard convention that $P(x, y)$ is the point P in the plane with Cartesian coordinates (x, y).

Warning. The symbol (a, b) is used to denote both a point in the xy-plane and an interval on the real line. The context always indicates in which way this symbol is being used.

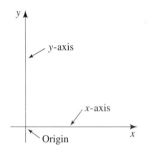

Figure A.11

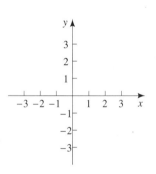

Figure A.12

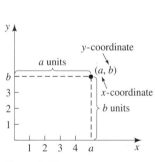

Figure A.13

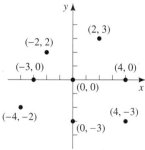

Figure A.14

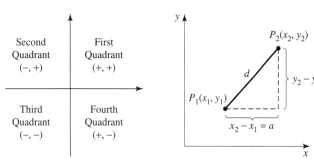

Figure A.15   **Figure A.16**

The x-axis and the y-axis divide the plane into four **quadrants**. The **first quadrant** is all points (x, y) with both $x > 0$ and $y > 0$. The **second quadrant** is all points (x, y) with both $x < 0$ and $y > 0$. The **third quadrant** is all points (x, y) with both $x < 0$ and $y < 0$. The **fourth quadrant** is all points (x,y) with both $x > 0$ and $y < 0$ (Fig. A.15).

Distance Between Two Points

We wish now to find the distance d between two given points $P_1(x_1, y_1)$ and $P_2(x_2, y_2)$ in the xy-plane shown in Figure A.16.

Recall that the Pythagorean theorem says that $d^2 = a^2 + b^2$ in Figure A.16. Then

$$d^2 = |x_2 - x_1|^2 + |y_2 - y_1|^2 = (x_2 - x_1)^2 + (y_2 - y_1)^2$$

Taking square roots of each side gives the following:

Distance Formula

The distance $d = d(P_1, P_2)$ between the two points $P_1(x_1, y_1)$ and $P_2(x_2, y_2)$ in the xy-plane is given by

$$d(P_1, P_2) = \sqrt{(x_2 - x_1)^2 + (y_2 - y_1)^2}$$

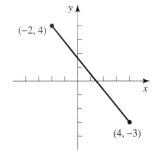

Figure A.17

EXAMPLE 1   Finding the Distance Between Two Points

Find the distance between the two points $(-2, 4)$ and $(4, -3)$ shown in Figure A.17.

SOLUTION ■ We have $(x_1, y_1) = (-2, 4)$ and $(x_2, y_2) = (4, -3)$. Thus,

$$\begin{aligned} d(P_1, P_2) &= \sqrt{(x_2 - x_1)^2 + (y_2 - y_1)^2} \\ &= \sqrt{[(4) - (-2)]^2 + [(-3) - (4)]^2} \\ &= \sqrt{(6)^2 + (-7)^2} = \sqrt{85} \quad ■ \end{aligned}$$

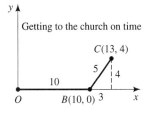

Figure A.18

EXAMPLE 2   Using the Distance Between Two Points

A prospective groom is located at point O in Figure A.18 and must drive his car to his wedding located at the church at point C in 25 minutes. The church at C is located 13 miles east and 4 miles north of O. The only highway goes through a town located at B, which is 10 miles due east of O. From O to B is a nice open stretch of highway, and he can average 50 miles per hour. The stretch from B to C has traffic lights and some congestion, and he can average only 25 miles per hour on this leg of the trip. Can he get to the church on time?

SOLUTION ■ The distance from O to B is 10 miles, and the distance from B to C is $\sqrt{(13-10)^2 + (4-0)^2} = 5$ miles. Distance is the product of velocity with time, $d = v \cdot t$. Thus, $t = d/v$. Then the time to travel the given route is

$$\frac{10}{50} + \frac{5}{25} = \frac{2}{5}$$

of an hour, or 24 minutes. This is less than 25 minutes, so he can make his wedding on time. ■

Creating Equations

In order to solve any applied problem, we must first take the problem and translate it into mathematical terms. Doing this requires creating equations.

EXAMPLE 3   Creating an Equation

A small shop makes two styles of shirts. The first style costs \$7 to make and the second \$9. The shop has \$252 a day to produce these shirts. If x is the number of the first style produced each day and y is the number of the second style, find an equation that x and y must satisfy.

SOLUTION ■ Create a table that includes the needed information. See Table A.1. Since x are produced at a cost of \$7 each, the cost in dollars of producing x of these shirts is $7x$. Since y are produced at a cost of \$9 each, the cost in dollars of producing y of these shirts is $9y$. This is indicated in Table A.1. Since the cost of producing the first style plus the cost of producing the second style must total \$252, the equation we are seeking is

$$7x + 9y = 252 \quad ■$$

TABLE A.1

Style	First	Second	Both
Cost of each shirt	\$7	\$9	
Number of each style produced	x	y	
Total cost	$7x$	$9y$	252

Graphs

Given any equation in x and y (such as $y = 2x - 1$), the **graph of the equation** is the set of all points (x, y) such that x and y satisfy the equation.

Graph of an Equation

The **graph of an equation** is the set

$$\{(x, y) | x \text{ and } y \text{ satisfy the equation}\}$$

For example the point $(0, -1)$ is on the graph of the equation $y = 2x - 1$, because when x is replaced with 0 and y replaced with -1 the equation is satisfied. That is, $(-1) = 2(0) - 1$.

To find the graph of an equation, plot points until a pattern emerges. More complicated equations require more points.

TABLE A.2

x	−3	−2	−1	0	1	2	3
y	−7	−5	−3	−1	1	3	5

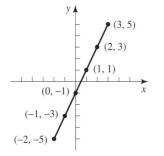

Figure A.19

EXAMPLE 4   Determining a Graph of an Equation

Sketch the graph of $y = 2x - 1$, that is, find the set $\{(x,y) | y = 2x - 1\}$.

SOLUTION ■ We take a number of values for x, put them into the equation and solve for the corresponding values of y. For example if we take $x = -3$, then $y = 2(-3) - 1 = -7$. Table A.2 summarizes this work. The points (x, y) found in Table A.2 are graphed in Figure A.19. We see that the graph is a straight line. ■

EXAMPLE 5   Determining the Graph of an Equation

Sketch the graph of $x = 2$, that is, find the set $\{(x, y) | x = 2\}$.

SOLUTION ■ Points, such as $(2, -1)$, $(2, 1)$, and $(2, 2)$, are all on the graph. The graph is the vertical line shown in Figure A.20. ■

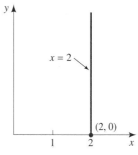

Figure A.20

EXAMPLE 6   Determining the Graph of an Equation

Sketch the graph of $y = x^2$, that is, find the set $\{(x,y) | y = x^2\}$.

SOLUTION ■ We take a number of values for x, put them into the equation, and solve for the corresponding values of y. For example if we take $x = -2$, then $y = (-2)^2 = 4$. Table A.3 summarizes this work. The points (x, y) found in Table A.3 are graphed in Figure A.21. ■

TABLE A.3

x	−3	−2	−1	0	1	2	3
y	9	4	1	0	1	4	9

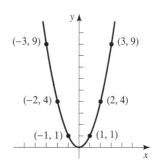

Figure A.21

SELF-HELP EXERCISE SET A.5

1. Plot the points $(-2, 5)$, $(-2, -4)$, and $(3, -2)$.

2. Find the distance between the two points $(-4, -2)$ and $(3, -4)$.

3. Sketch the graph of the equation $7x + 9y = 252$ found in Example 3.

EXERCISE SET A.5

In Exercises 1 through 8 draw a number line, and indentify.

1. $(-1, 3)$ **2.** $[-2, 6]$

3. $(2, 5)$ **4.** $[-3, -1]$

5. $(-2, +\infty)$ **6.** $[2, +\infty)$

7. $(-\infty, 1.5]$ **8.** $(-\infty, -2)$

In Exercises 9 through 20 plot the given point in the xy-plane.

9. $(1, 2)$ **10.** $(-3, -4)$

11. $(-2, 1)$ **12.** $(2, -3)$

13. $(1, 0)$ **14.** $(-2, 0)$

15. $(0, -2)$ **16.** $(0, 4)$

17. $(-2, -1)$ **18.** $(2, 2)$

19. $(3, -4)$ **20.** $(-3, 0)$

In Exercises 21 through 32 find the set of points (x, y) in the xy-plane that satisfies the given equation or inequality.

21. $x \geq 0$ **22.** $y \geq 0$

23. $x < 0$ **24.** $y < 0$

25. $x = 2$ **26.** $x = -1$

27. $y = 3$ **28.** $y = -2$

29. $xy > 0$ **30.** $xy < 0$

31. $xy = 0$ **32.** $x^2 + y^2 = 0$

In Exercises 33 through 40 find the distance between the given points.

33. $(2, 4), (3, 2)$ **34.** $(-2, 5), (2, -3)$

35. $(-2, 5), (-3, -1)$ **36.** $(1, 3), (1, 6)$

37. $(-3, -2), (4, -3)$ **38.** $(-5, -6), (-3, -2)$

39. $(a, b), (b, a)$ **40.** $(x, y), (0, 0)$

In Exercises 41 through 50 sketch the graphs.

41. $y = -2x + 1$ **42.** $y = -2x - 2$

43. $y = 3x - 2$ **44.** $y = 2x + 3$

45. $y = 3$ **46.** $x = 4$

47. $y = 3x^2$ **48.** $y = -x^2$

49. $y = x^3$ **50.** $y = \sqrt{x}$

Applications

51. Transportation. A truck travels a straight highway from O to A and then another straight highway from A to B. How far does the truck travel?

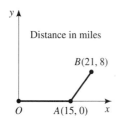

52. Transportation. If the truck in the previous exercise travels 30 miles per hour from O to A and 60 miles per hour from A to B, how long does the trip take?

53. Navigation. A ship leaves port and heads due east for 4 hours with a speed of 3 miles per hour, after which it turns due north and maintains this direction at a speed of 2.5 miles per hour. How far is the ship from port 6 hours after leaving?

54. Navigation. A ship leaves port and heads due east for 1 hour at 5 miles per hour. It then turns north and heads in this direction at 4 miles per hour. How long does it take the ship to get 13 miles from port?

55. Oil Slick. The shore line stretches along the x-axis as shown in the figure, with the ocean consisting of the first and second quadrants. An oil spill occurs at the point $(2, 4)$ and a coastal town is located at $(5, 0)$ where the numbers are in miles. If the oil slick is approaching the town at a rate of $\frac{1}{2}$ mile per day, how many days until the oil slick reaches the coastal town located at $(5, 0)$?

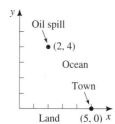

56. Transportation Charges. A furniture store is located 5 miles east and 14 miles north of the center of town. You live 10 miles east and 2 miles north of the center of town. The store advertises free delivery within a 12-mile radius of the store. Do you qualify for free delivery?

57. Profits. A furniture store sells chairs at a profit of $100 each and sofas at a profit of $150 each. Let x be the number of chairs sold each week and y the number of sofas sold each week. If the profit in one week is $3000, write an equation that x and y must satisfy. Graph this equation.

58. Revenue. A restaurant has two specials: steak and chicken. The dinner with steak is $13 and with chicken is $10. Let x be the number of steak specials served and y the number of chicken specials. If the restaurant made $390 in sales on these specials, find the equation that x and y must satisfy. Graph this equation.

59. Nutrition. An individual needs about 800 mg of calcium daily in his diet but is unable to eat any dairy products due to the large amounts of cholesterol in such products. This individual does, however, enjoy eating canned sardines and steamed broccoli. Let x be the number of ounces of canned sardines consumed each day and let y be the number of cups of steamed broccoli. If there is 125 mg of calcium in each ounce of canned sardines and 190 mg in each cup of steamed broccoli, what equation must x and y satisfy if this person is to obtain his daily need of calcium from these two sources?

60. Costs. A contractor builds ranch and split-level style homes. The ranch costs $130,000 to build and the split-level $150,000. Let x be the number of ranch-style homes built and y the number of split-level homes. If the contractor has $1,360,000 to build these homes, find the equation that x and y must satisfy. Graph this equation.

61. Biology. A fisheries biologist decides to set a gill net on his next trip to a remote mountain lake. The net must stretch

between points A and B shown in the figure. Since he must backpack to the lake he wants to carry no more net than necessary. He walks off the distance a from C to B and the distance b from A to C being careful that AC is perpendicular to AB. Find an equation that gives the distance c from A to B in terms of a and b.

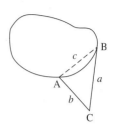

SOLUTIONS TO SELF-HELP EXERCISE SET A.5

1. The points are plotted in the figure.

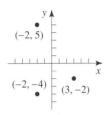

2. The distance between the two points $(-4, -2)$ and $(3, -4)$ is given by

$$\sqrt{[3-(-4)]^2 + [-4-(-2)]^2} = \sqrt{7^2 + (-2)^2} = \sqrt{53}$$

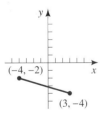

3. First solve for y and obtain $y = -\frac{7}{9}x + 28$. Now take a number of values for x, put them into the equation and solve for the corresponding values of y. For example if we take $x = 9$, then $y = -\frac{7}{9}(9) + 28 = 21$. The table summarizes this work. The points (x, y) found in the table are graphed in the figure. We see that apparently the graph is a straight line.

x	0	9	18	27	36
y	28	21	14	7	0

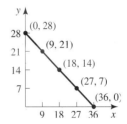

B.1 BASIC GEOMETRIC FORMULAS

- **Pythagorean Theorem**
  $c^2 = a^2 + b^2$

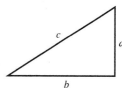

- **Rectangle**
  Area = ab
  Perimeter = $2a + 2b$

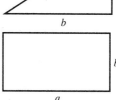

- **Parallelogram**
  Height = h
  Area = ah
  Perimeter = $2a + 2b$

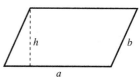

- **Triangle**
  Height = h
  Area = $\frac{1}{2} hc$
  Perimeter = $a + b + c$

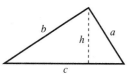

- **Trapezoid**
  Area = $\frac{1}{2}(a + b)h$

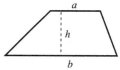

- **Circle**
  Radius = r
  Diameter = $d = 2r$
  Area = πr^2
  Circumference = $2\pi r$
  $\pi = 3.14159$

- **Rectangular Solid**
  Volume = abc
  Total surface area = $2ab + 2ac + 2bc$

- **Right Circular Cylinder**
  Radius of base = r
  Height = h
  Volume = $\pi r^2 h$
  Total surface area = $2\pi r^2 + 2\pi r h$

- **Sphere**
  Radius = r
  Diameter = $d = 2r$
  Volume = $\frac{4}{3}\pi r^3$
  Total surface area = $4\pi r^2$

- **Right Circular Cone**
  Radius of base = r
  Height = h
  Volume = $\frac{1}{3}\pi r^2 h$
  Total surface area = $\pi r^2 + \pi r \sqrt{r^2 + h^2}$

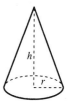

B.2 TABLE OF INTEGRALS

Integrands Containing $(ax + b)$

1. $\displaystyle\int x(ax + b)^n \, dx = \frac{(ax + b)^{n+1}}{a^2}\left[\frac{ax + b}{n + 2} - \frac{b}{n + 1}\right]$
$(n \neq -1, -2)$

2. $\displaystyle\int \frac{x}{ax + b} \, dx = \frac{x}{a} - \frac{b}{a^2} \ln |ax + b|$

3. $\displaystyle\int \frac{x}{(ax + b)^2} \, dx = \frac{1}{a^2}\left[\ln |ax + b| + \frac{b}{ax + b}\right]$

4. $\displaystyle\int \frac{x^2}{ax + b} \, dx = \frac{1}{a^3}\left[\frac{1}{2}(ax + b)^2 - 2b(ax + b) + b^2 \ln |ax + b|\right]$

5. $\displaystyle\int \frac{x^2}{(ax + b)^2} \, dx = \frac{1}{a^3}\left[ax + b - \frac{b^2}{ax + b} - 2b \ln |ax + b|\right]$

6. $\displaystyle\int \frac{1}{x(ax + b)} \, dx = \frac{1}{b} \ln \left|\frac{x}{ax + b}\right| \quad (b \neq 0)$

7. $\displaystyle\int \frac{1}{x^2(ax + b)} \, dx = -\frac{1}{bx} + \frac{a}{b^2} \ln \left|\frac{ax + b}{x}\right| \quad (b \neq 0)$

8. $\displaystyle\int \frac{1}{x(ax + b)^2} \, dx = \frac{1}{b(ax + b)} - \frac{1}{b^2} \ln \left|\frac{ax + b}{x}\right| \quad (b \neq 0)$

Integrands Containing $\sqrt{ax + b}$

9. $\displaystyle\int x\sqrt{ax + b} \, dx = \frac{2}{a^2}\left[\frac{(ax + b)^{5/2}}{5} - \frac{b(ax + b)^{3/2}}{3}\right]$

10. $\displaystyle\int x^2\sqrt{ax + b} \, dx = \frac{2}{a^3}\left[\frac{(ax + b)^{7/2}}{7} - \frac{2b(ax + b)^{5/2}}{5} + \frac{b^2(ax + b)^{3/2}}{3}\right]$

11. $\displaystyle\int \frac{x}{\sqrt{ax + b}} \, dx = \frac{2ax - 4b}{3a^2}\sqrt{ax + b}$

12. $\displaystyle\int \frac{1}{x\sqrt{ax + b}} \, dx = \frac{1}{\sqrt{b}} \ln \left|\frac{\sqrt{ax + b} - \sqrt{b}}{\sqrt{ax + b} + \sqrt{b}}\right| \quad (b > 0)$

Integrands Containing $a^2 \pm x^2$

13. $\displaystyle\int \frac{1}{a^2 - x^2} \, dx = \frac{1}{2a} \ln \left|\frac{x + a}{x - a}\right|$

14. $\displaystyle\int \frac{x}{a^2 \pm x^2} \, dx = \pm\frac{1}{2} \ln |a^2 \pm x^2|$

15. $\displaystyle\int \frac{1}{x(a^2 \pm x^2)} \, dx = \frac{1}{2a^2} \ln \left|\frac{x^2}{a^2 \pm x^2}\right|$

Integrands Containing $\sqrt{a^2 - x^2}$

16. $\displaystyle\int \frac{x}{\sqrt{a^2 - x^2}}\, dx = -\sqrt{a^2 - x^2}$

17. $\displaystyle\int \frac{1}{x\sqrt{a^2 - x^2}}\, dx = -\frac{1}{a} \ln \left| \frac{a + \sqrt{a^2 - x^2}}{x} \right|$

18. $\displaystyle\int \frac{1}{x^2\sqrt{a^2 - x^2}}\, dx = -\frac{\sqrt{a^2 - x^2}}{a^2 x}$

19. $\displaystyle\int \frac{1}{(a^2 - x^2)^{3/2}}\, dx = \frac{1}{a^2} \frac{x}{\sqrt{a^2 - x^2}}$

20. $\displaystyle\int \frac{x}{(a^2 - x^2)^{3/2}}\, dx = \frac{1}{\sqrt{a^2 - x^2}}$

21. $\displaystyle\int \frac{\sqrt{a^2 - x^2}}{x}\, dx = \sqrt{a^2 - x^2} - a \ln \left| \frac{a + \sqrt{a^2 - x^2}}{x} \right|$

Integrands Containing $\sqrt{x^2 \pm a^2}$

22. $\displaystyle\int \frac{1}{\sqrt{x^2 \pm a^2}}\, dx = \ln \left| x + \sqrt{x^2 \pm a^2} \right|$

23. $\displaystyle\int \frac{x}{\sqrt{x^2 \pm a^2}}\, dx = \sqrt{x^2 \pm a^2}$

24. $\displaystyle\int \frac{x^2}{\sqrt{x^2 \pm a^2}}\, dx = \frac{1}{2} x\sqrt{x^2 \pm a^2} \mp \frac{1}{2} a^2 \ln \left| x + \sqrt{x^2 \pm a^2} \right|$

25. $\displaystyle\int \frac{1}{x\sqrt{x^2 + a^2}}\, dx = -\frac{1}{a} \ln \left| \frac{a + \sqrt{x^2 + a^2}}{x} \right|$

26. $\displaystyle\int \frac{1}{x^2\sqrt{x^2 \pm a^2}}\, dx = \mp \frac{\sqrt{x^2 \pm a^2}}{a^2 x}$

27. $\displaystyle\int \frac{1}{(x^2 \pm a^2)^{3/2}}\, dx = \pm \frac{1}{a^2} \frac{x}{\sqrt{x^2 \pm a^2}}$

28. $\displaystyle\int \frac{x}{(x^2 \pm a^2)^{3/2}}\, dx = -\frac{1}{\sqrt{x^2 \pm a^2}}$

29. $\displaystyle\int \frac{x^2}{(x^2 \pm a^2)^{3/2}}\, dx = -\frac{x}{\sqrt{x^2 \pm a^2}} + \ln \left| x + \sqrt{x^2 \pm a^2} \right|$

30. $\displaystyle\int x\sqrt{x^2 \pm a^2}\, dx = \frac{1}{3} (x^2 \pm a^2)^{3/2}$

31. $\displaystyle\int \sqrt{x^2 \pm a^2}\, dx = \frac{1}{2} x\sqrt{x^2 \pm a^2} \pm \frac{1}{2} a^2 \ln \left| x + \sqrt{x^2 \pm a^2} \right|$

32. $\displaystyle\int \frac{\sqrt{x^2 + a^2}}{x}\, dx = \sqrt{x^2 + a^2} - a \ln \left| \frac{a + \sqrt{x^2 + a^2}}{x} \right|$

33. $\displaystyle\int \frac{\sqrt{x^2 \pm a^2}}{x^2}\, dx = -\frac{\sqrt{x^2 \pm a^2}}{x} + \ln \left| x + \sqrt{x^2 \pm a^2} \right|$

Integrands Containing $(ax^2 + bx + c)$

34. $\displaystyle\int \frac{1}{ax^2 + bx + c}\, dx$

$$= \frac{1}{\sqrt{b^2 - 4ac}} \ln\left|\frac{2ax + b - \sqrt{b^2 - 4ac}}{2ax + b + \sqrt{b^2 - 4ac}}\right| \quad (b^2 - 4ac > 0)$$

Integrands Containing $(ax^n + b)$

35. $\displaystyle\int \frac{1}{x(ax^n + b)}\, dx = \frac{1}{nb} \ln\left|\frac{x^n}{ax^n + b}\right| \quad (n \neq 0,\ b \neq 0)$

36. $\displaystyle\int \frac{1}{x\sqrt{ax^n + b}}\, dx = \frac{1}{n\sqrt{b}} \ln\left|\frac{\sqrt{ax^n + b} - \sqrt{b}}{\sqrt{ax^n + b} + \sqrt{b}}\right| \quad (b > 0)$

Integrands Containing Exponentials and Logarithms

37. $\displaystyle\int xe^{ax}\, dx = \frac{1}{a^2}(ax - 1)e^{ax}$

38. $\displaystyle\int x^n e^{ax}\, dx = \frac{1}{a} x^n e^{ax} - \frac{n}{a}\int x^{n-1} e^{ax}\, dx$

39. $\displaystyle\int \frac{1}{b + ce^{ax}}\, dx = \frac{1}{ab}[ax - \ln(b + ce^{ax})] \quad (ab \neq 0)$

40. $\displaystyle\int x^n \ln|x|\, dx = \frac{1}{n + 1} x^{n+1}\left[\ln|x| - \frac{1}{n + 1}\right] \quad (n \neq -1)$

41. $\displaystyle\int (\ln x)^n\, dx = x(\ln x)^n - n\int (\ln x)^{n-1}\, dx$

Miscellaneous Integral

42. $\displaystyle\int \sqrt{\frac{x + a}{x + b}}\, dx = \sqrt{(x + b)(x + a)} + (a - b)\ln\left|\sqrt{x + b} + \sqrt{x + a}\right|$

B.3

Area Under Standard Normal Curve to the Left of $z = \dfrac{x - \mu}{\sigma}$

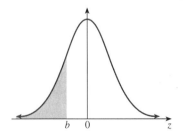

z	.00	.01	.02	.03	.04	.05	.06	.07	.08	.09
−3.4	.0003	.0003	.0003	.0003	.0003	.0003	.0003	.0003	.0003	.0002
−3.3	.0005	.0005	.0005	.0004	.0004	.0004	.0004	.0004	.0004	.0003
−3.2	.0007	.0007	.0006	.0006	.0006	.0006	.0006	.0005	.0005	.0005
−3.1	.0010	.0009	.0009	.0009	.0008	.0008	.0008	.0008	.0007	.0007
−3.0	.0013	.0013	.0013	.0012	.0012	.0011	.0011	.0011	.0010	.0010
−2.9	.0019	.0018	.0017	.0017	.0016	.0016	.0015	.0015	.0014	.0014
−2.8	.0026	.0025	.0024	.0023	.0023	.0022	.0021	.0021	.0020	.0019
−2.7	.0035	.0034	.0033	.0032	.0031	.0030	.0029	.0028	.0027	.0026
−2.6	.0047	.0045	.0044	.0043	.0041	.0040	.0039	.0038	.0037	.0036
−2.5	.0062	.0060	.0059	.0057	.0055	.0054	.0052	.0051	.0049	.0048
−2.4	.0082	.0080	.0078	.0075	.0073	.0071	.0069	.0068	.0066	.0064
−2.3	.0107	.0104	.0102	.0099	.0096	.0094	.0091	.0089	.0087	.0084
−2.2	.0139	.0136	.0132	.0129	.0125	.0122	.0119	.0116	.0113	.0110
−2.1	.0179	.0174	.0170	.0166	.0162	.0158	.0154	.0150	.0146	.0143
−2.0	.0228	.0222	.0217	.0212	.0207	.0202	.0197	.0192	.0188	.0183
−1.9	.0287	.0281	.0274	.0268	.0262	.0256	.0250	.0244	.0239	.0233
−1.8	.0359	.0352	.0344	.0336	.0329	.0322	.0314	.0307	.0301	.0294
−1.7	.0446	.0436	.0427	.0418	.0409	.0401	.0392	.0384	.0375	.0367
−1.6	.0548	.0537	.0526	.0516	.0505	.0495	.0485	.0475	.0465	.0455
−1.5	.0668	.0655	.0643	.0630	.0618	.0606	.0594	.0582	.0571	.0559
−1.4	.0808	.0793	.0778	.0764	.0749	.0735	.0722	.0708	.0694	.0681
−1.3	.0968	.0951	.0934	.0918	.0901	.0885	.0869	.0853	.0838	.0823
−1.2	.1151	.1131	.1112	.1093	.1075	.1056	.1038	.1020	.1003	.0985
−1.1	.1357	.1335	.1314	.1292	.1271	.1251	.1230	.1210	.1190	.1170
−1.0	.1587	.1562	.1539	.1515	.1492	.1469	.1446	.1423	.1401	.1379
−0.9	.1841	.1814	.1788	.1762	.1736	.1711	.1685	.1660	.1635	.1611
−0.8	.2119	.2090	.2061	.2033	.2005	.1977	.1949	.1922	.1894	.1867
−0.7	.2420	.2389	.2358	.2327	.2296	.2266	.2236	.2206	.2177	.2148
−0.6	.2743	.2709	.2676	.2643	.2611	.2578	.2546	.2514	.2483	.2451
−0.5	.3085	.3050	.3015	.2981	.2946	.2912	.2877	.2843	.2810	.2776
−0.4	.3446	.3409	.3372	.3336	.3300	.3264	.3228	.3192	.3156	.3121
−0.3	.3821	.3783	.3745	.3707	.3669	.3632	.3594	.3557	.3520	.3483
−0.2	.4207	.4168	.4129	.4090	.4052	.4013	.3974	.3936	.3897	.3859
−0.1	.4602	.4562	.4522	.4483	.4443	.4404	.4364	.4325	.4286	.4247
−0.0	.5000	.4960	.4920	.4880	.4840	.4801	.4761	.4721	.4681	.4641
0.0	.5000	.5040	.5080	.5120	.5160	.5199	.5239	.5279	.5319	.5359
0.1	.5398	.5438	.5478	.5517	.5557	.5596	.5636	.5675	.5714	.5753
0.2	.5793	.5832	.5871	.5910	.5948	.5987	.6026	.6064	.6103	.6141
0.3	.6179	.6217	.6255	.6293	.6331	.6368	.6406	.6443	.6480	.6517
0.4	.6554	.6591	.6628	.6664	.6700	.6736	.6772	.6808	.6844	.6879
0.5	.6915	.6950	.6985	.7019	.7054	.7088	.7123	.7157	.7190	.7224
0.6	.7257	.7291	.7324	.7357	.7389	.7422	.7454	.7486	.7517	.7549
0.7	.7580	.7611	.7642	.7673	.7704	.7734	.7764	.7794	.7823	.7852
0.8	.7881	.7910	.7939	.7967	.7995	.8023	.8051	.8078	.8106	.8133
0.9	.8159	.8186	.8212	.8238	.8264	.8289	.8315	.8340	.8365	.8389
1.0	.8413	.8438	.8461	.8485	.8508	.8531	.8554	.8577	.8599	.8621
1.1	.8643	.8665	.8686	.8708	.8729	.8749	.8770	.8790	.8810	.8830
1.2	.8849	.8869	.8888	.8907	.8925	.8944	.8962	.8980	.8997	.9015
1.3	.9032	.9049	.9066	.9082	.9099	.9115	.9131	.9147	.9162	.9177
1.4	.9192	.9207	.9222	.9236	.9251	.9265	.9278	.9292	.9306	.9319

z	.00	.01	.02	.03	.04	.05	.06	.07	.08	.09
1.5	.9332	.9345	.9357	.9370	.9382	.9394	.9406	.9418	.9429	.9441
1.6	.9452	.9463	.9474	.9484	.9495	.9505	.9515	.9525	.9535	.9545
1.7	.9554	.9564	.9573	.9582	.9591	.9599	.9608	.9616	.9625	.9633
1.8	.9641	.9649	.9656	.9664	.9671	.9678	.9686	.9693	.9699	.9706
1.9	.9713	.9719	.9726	.9732	.9738	.9744	.9750	.9756	.9761	.9767
2.0	.9772	.9778	.9783	.9788	.9793	.9798	.9803	.9808	.9812	.9817
2.1	.9821	.9826	.9830	.9834	.9838	.9842	.9846	.9850	.9854	.9857
2.2	.9861	.9864	.9868	.9871	.9875	.9878	.9881	.9884	.9887	.9890
2.3	.9893	.9896	.9898	.9901	.9904	.9906	.9909	.9911	.9913	.9916
2.4	.9918	.9920	.9922	.9925	.9927	.9929	.9931	.9932	.9934	.9936
2.5	.9938	.9940	.9941	.9943	.9945	.9946	.9948	.9949	.9951	.9952
2.6	.9953	.9955	.9956	.9957	.9959	.9960	.9961	.9962	.9963	.9964
2.7	.9965	.9966	.9967	.9968	.9969	.9970	.9971	.9972	.9973	.9974
2.8	.9974	.9975	.9976	.9977	.9977	.9978	.9979	.9979	.9980	.9981
2.9	.9981	.9982	.9982	.9983	.9984	.9984	.9985	.9985	.9986	.9986
3.0	.9987	.9987	.9987	.9988	.9988	.9989	.9989	.9989	.9990	.9990
3.1	.9990	.9991	.9991	.9991	.9992	.9992	.9992	.9992	.9993	.9993
3.2	.9993	.9993	.9994	.9994	.9994	.9994	.9994	.9995	.9995	.9995
3.3	.9995	.9995	.9995	.9996	.9996	.9996	.9996	.9996	.9996	.9997
3.4	.9997	.9997	.9997	.9997	.9997	.9997	.9997	.9997	.9997	.9998

Answers to Odd-Numbered Exercises

CHAPTER 1

Section 1.0

1.

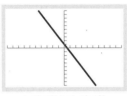

[−10, 10] by [−10, 10]

3.

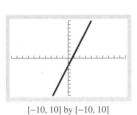

[−10, 10] by [−10, 10]

5.

[−10, 10] by [−10, 10]

7.

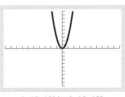

[−10, 10] by [−10, 10]

9.

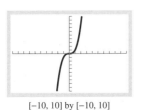

[−10, 10] by [−10, 10]

11. $(8, 9)$, $(−7, −8)$ **13.** $(7, 14)$

15. $[−1, 6]$ by $[−3, 2]$ **17.** $[1, 6]$ by $[−2, 5]$

19. (b)   **21.** (b)

23. (b)   **25.** (b)

27. (a) $100x + 150y = 3000$; (b) $−\frac{2}{3}$; (c) 18. If 3 chairs are sold, then 18 sofas need to be sold for the profit to be $3000; (d) 30

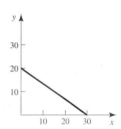

29. (a) $125x + 190y = 820$; (b) $−125/190$; (c) 3. If 2 ounces of sardines are consumed, then 3 cups of broccoli need to be eaten to obtain 820 mg of calcium; (d) 6.56

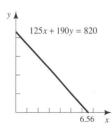

31. 1963.

[1958, 1967] by [60, 75]

13. $\frac{1}{2}$, −4, 2

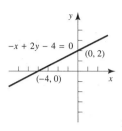

15. undefined, 3, no y-intercept

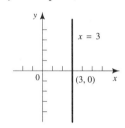

17. 0, no x-intercept, 4

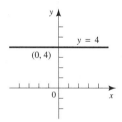

19. $\frac{1}{2}$, 1, $-\frac{1}{2}$

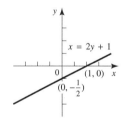

21. 5, 3

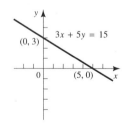

23. 5, −4

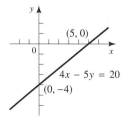

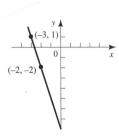

5. 0

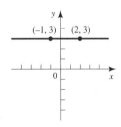

7. undefined

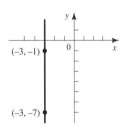

9. 2, $\frac{1}{2}$, −1

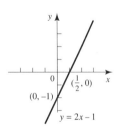

11. −1, 1, 1

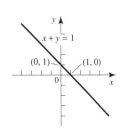

25. $y = -3x + 7$

27. $y = -3$

29. $y = -x + 2$

31. $x = -2$

33. $y = -\dfrac{b}{a}x + b$

35. $y = -\dfrac{2}{5}x + \dfrac{11}{5}$

37. $y = 0$

39. $x = 1$

41. $y = 2x + 5$

43. $y = -\frac{5}{3}x + 4$

45. $y = -\frac{2}{5}x + 4$

47. No

49. $0.08x + 0.05y = 576$, \$160

51. $y = 0.012x - 0.12$, 0.36 kilograms

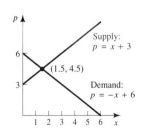

53. Each additional pound of milk per cow per year requires an additional 3 cents in cost per cow per year. \$30

55. $L - 140 = \dfrac{182}{79}(l - 60)$, $\dfrac{182}{79}$, each 79 mm of increase in the tail brings a 182-mm increase in the total length.

57. For each additional fly on the front legs of the bull there are (on average) 2.27 additional total flies on the bull; no flies on the front legs, implies 10.95 flies (on average) on the bull.

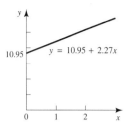

59. If $b \neq 0$, then $ax + by = c$ if and only if $y = -\dfrac{a}{b} \cdot x + \dfrac{c}{b}$. This is the straight line with slope $-\dfrac{a}{b}$ and y-intercept $\dfrac{c}{b}$. If $b = 0$, then by assumption $a \neq 0$, and we have $x = \dfrac{c}{a}$ which is a vertical line. Therefore $ax + by = c$ is a straight line if a and b are not both zero. Conversely, every line is the graph of either the equation $y - y_1 = m(x - x_1)$ if and only if $-mx + y = y_1 - mx_1$ with $a = -m$, $b = 1$ and $c = y_1 - mx_1$, or the equation $x = x_0$ if m fails to exist, in which case $a = 1$, $b = 0$ and $c = x_0$. Every line is therefore the graph of a linear equation.

Section 1.2

1. $C = 3x + 10{,}000$

3. $R = 5x$

5. $P = 2x - 10{,}000$

7. $x = 2$

9. $x = 20$

11. $(1.5, 4.5)$

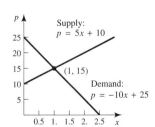

13. $(1, 15)$

15. $V = -5000t + 50{,}000$, \$45,000, \$25,000

17. $p = -0.005x + 22.5$

19. (a) $3x + 4y = 120$; (b) $m = -\frac{3}{4}$, making four more bookcases requires making three fewer desks.

21. (a) $y = 1000 + 0.05x$, $y = 1500 + 0.04x$; (b) \$50,000

23. $t = \dfrac{25}{23}$

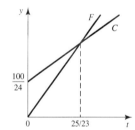

25. $y = 6x + 23$, \$4.5 million

27. outside

29. (a) $1.3333x$; (b) $c = 4000 + 1.3333x$

31. $(200, 86)$

33. $C = 5x + 1000$

35. $C = 3x + 2000$

37. $R = 11x$

39. (a) manual; (b) automatic

	Total Costs	
Volume	**Manual**	**Automatic**
1,000	\$ 2,600	\$8020
10,000	\$17,000	\$8200

41.

	Cost Per Unit	
Volume	Manual	Automatic
1,000	$2.60	$8.02
10,000	$1.70	$0.82
100,000	$1.61	$0.10

$\overline{C}_m = \dfrac{1000}{x} + 1.6$, $\overline{C}_a = \dfrac{8000}{x} + 0.02$. The cost per unit for a manual machine tends toward $1.60 and becomes greater than the cost per unit for an automatic machine, which tends toward $0.02.

43. Fixed cost, zero, negative of the fixed cost

Section 1.3

1. yes   **3.** no

5. yes

7. (1) domain: $\{1, 4\}$, range: $\{2, 7\}$; (2) domain: $\{3, 5\}$, range: $\{7\}$

9. all real numbers x except $x = 1$

11. all $x \geq 0$ **13.** all $x > 1$

15. all real numbers x except $x = 2, 3$

17. all real numbers   **19.** $\{x | x \geq -2, \ x \neq 0\}$

21. all real numbers   **23.** all real numbers

25. domain: $[0, 2]$, range: $[0, 1]$

27. domain: $[0, 2]$, range: $\{1, 2\}$

29. domain: $[0, 6]$, range: $[0, 1]$

31. $5, 1, 3, 6\sqrt{3} + 3, 2x^6 + 3, \dfrac{2}{x^3} + 3$

33. $\frac{1}{2}, \frac{2}{3}, -1, \dfrac{1}{\sqrt{2}}, \dfrac{1}{x + 3}, \dfrac{x + 1}{x + 2}$

35. $3, 0 \ \sqrt{2x^2 - 2}, \sqrt{2/x - 1}, \sqrt{-2x - 1}, -\sqrt{2x - 1}$

37. $1, -1, |x|, x^2$ **39.** $1, 2, 2, 2$

41. $0, 1, 0, 1$ **43.** yes

45. no   **47.** no

49. yes   **51.** $y = -\frac{2}{5}x + 3$

53. no   **55.** $y = \sqrt[3]{x}$

57. no

59.

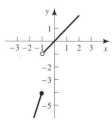

61.

63.

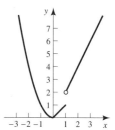

65.

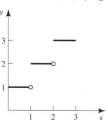

67. all real numbers

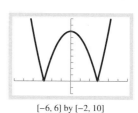

$[-6, 6]$ by $[-2, 10]$

69. $[-3, 3]$

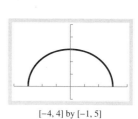

$[-4, 4]$ by $[-1, 5]$

71. $[0, 5) \cup (5, \infty)$

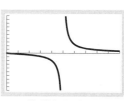

$[0, 10]$ by $[-10, 10]$

73. $A(r) = \pi r^2$ **75.** $V = 3x^3$

77. $d(t) = 60t$ **79.** $R(x) = 60x$

81. $C(x) = \begin{cases} x & \text{if } 0 \le x \le 500 \\ 500 + 2(x - 500) & \text{if } x > 500 \end{cases}$

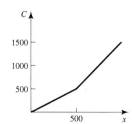

83. $T(x) = \begin{cases} 0 & \text{if } 0 \le x \le 50 \\ 0.01(x - 50) & \text{if } x > 50 \end{cases}$

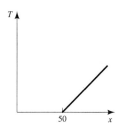

85.

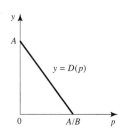

87. 109.6, 70.5

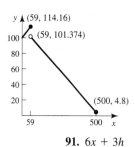

89. 3   **91.** $6x + 3h$

93. $-\dfrac{1}{x(x + h)}$

95. The two functions are the same.

97. $f(a + b) = m(a + b) = ma + mb = f(a) + f(b)$

99. $3f(x) = 3 \cdot 3^x = 3^{x+1} = f(x + 1)$, $f(a + b) = 3^{a+b} = 3^a \cdot 3^b = f(a) \cdot f(b)$

Section 1.4

1.

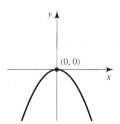

3.

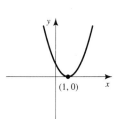

5.

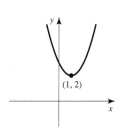

7.

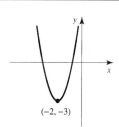

9.

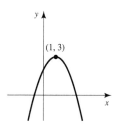

11.

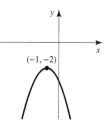

13.

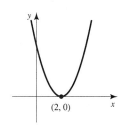

(2, 0)

15.

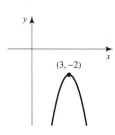

(3, −2)

17.

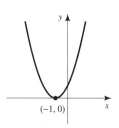

(−1, 0)

19.

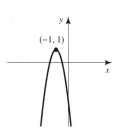

(−1, 1)

21.

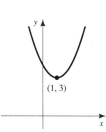

(1, 3)

23.

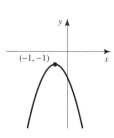

(−1, −1)

25. (1, −1)   **27.** (2, 2)

29. $x = 5$ **31.** $x = 1.5$

33. $x = 500$

35. $P(x) = -2x^2 + 20x - 42$, maximum when $x = 5$, break-even quantities are $x = 3, 7$

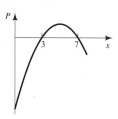

37. $P(x) = -x^2 + 15x - 36$, maximum when $x = 15/2$, break-even quantities are $x = 3, 12$

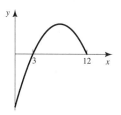

39. Break-even quantities are $x = 0.9$ and 2.2. Maximum at $x = 1.55$

41. Break-even quantities are approximately $x = 1.936436$, 7.663564. Maximum at $x = 4.8$.

43. 20.25   **45.** 50×50

47. 110   **49.** $r = 0.50$

51. 63.8, 2.95116

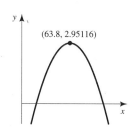

(63.8, 2.95116)

53. 3 pounds, the relative yield of vegetables is maximized

55. 11666.67, 58.46   **57.** 35

59. Both $p(x) = x^2$ and $q(x) = -x^2$ are quadratics, but $p(x) + q(x) = 0$ and is not a quadratic. Thus, the sum of two quadratics need not be a quadratic.

61. Both $f(x)$ and $g(x)$ are increasing for $x < x_1$, thus the peak of $f + g$ must occur after x_1. Both $f(x)$ and $g(x)$ are decreasing for $x > x_2$, thus the peak of $f + g$ must occur before x_2.

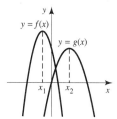

$y = f(x)$

$y = g(x)$

x_1 x_2

63. Since we expect profits to turn negative if too many products are made and sold (the price would have to be very low to unload so many products), a should be negative.

65. The graph of $p = mx + e$ should slope downward, requiring $m < 0$ and $e > 0$. Since $R = xp = mx^2 + ex$ and $R = ax^2 + bx$, $a = m < 0$ and $b = e > 0$.

67. $R(x) = mx^2 + bx = m\left(x + \dfrac{b}{2m}\right)^2 - \dfrac{b^2}{4m}$. Thus, revenue peaks at $x = -\dfrac{b}{2m} > 0$. $P(x) = R(x) - C(x) =$ $m\left(x + \dfrac{b - d}{2m}\right)^2 - \left[e + \dfrac{(b - d)^2}{4m}\right]$. Thus, profit peaks at $x = -\dfrac{b - d}{2m}$. Since $-\dfrac{1}{2m} > 0$ and $b > b - d$, we have $-\dfrac{b}{2m} > -\dfrac{b - d}{2m}$. That is, profit peaks before revenue peaks.

Section 1.5

1. degree: 8, domain: $(-\infty, \infty)$, leading coefficient: 1

3. degree: 9, domain: $(-\infty, \infty)$, leading coefficient: 1000

5. $(-\infty, 2) \cup (2, \infty)$ **7.** $(-\infty, \infty)$

9. all real numbers except zero

11.
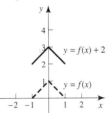

13.
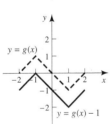

15.
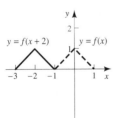

17.
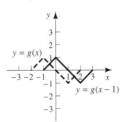

19.
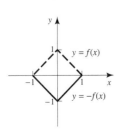

21.
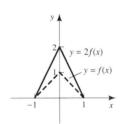

23.
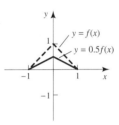

25.
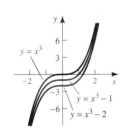

27.
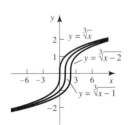

29.
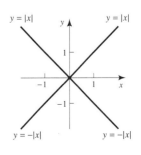

31.

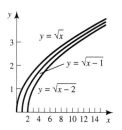

33.

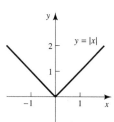

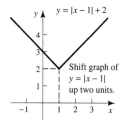

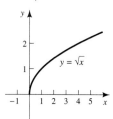

35.

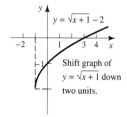

37.

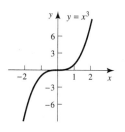

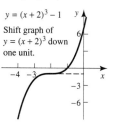

39.

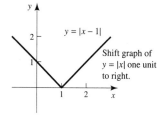

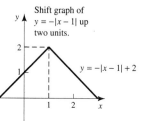

41. $y = 0.5|x| + 3$ **43.** $y = 2|x^3| - 3$

45. $y = -|x - 2| + 3$ **47.** no

49. 0.129934793, 1.186621394, 10.8367459, V increases by a factor of $2^{3.191} \approx 9.13$ when H doubles.

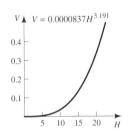

51. S is reduced by a factor of $(\frac{1}{2})^{0.284} \approx 0.82$

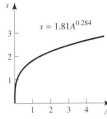

57. (a) is the graph of $y = x^{12}$ since $(-x)^{12} = x^{12}$. (The graph is symmetrical about the y-axis.); (b) is the graph of $y = x^{11}$

59. The graph of $y = f(cx)$ with $0 < c < 1$ is the graph of $y = f(x)$ stretched horizontally by a factor of $1/c$.

61. (a) is the graph of y_1; (b) is the graph of y_2.

Section 1.6

1. (a) 11; (b) 3; (c) 28; (d) 7/4

3. $x^2 + 2x + 4, -x^2 + 2x + 2, (2x + 3)(x^2 + 1), \dfrac{2x + 3}{x^2 + 1}$.
All the domains are $(-\infty, \infty)$.

5. $3x + 4, x + 2, 2x^2 + 5x + 3, \dfrac{2x + 3}{x + 1}$. The domain is $(-\infty, \infty)$ in the first three cases and $(-\infty, -1) \cup (-1, \infty)$ in the last case.

7. $\sqrt{x + 1} + x + 2, \sqrt{x + 1} - x - 2, \sqrt{x + 1}(x + 2), \dfrac{\sqrt{x + 1}}{x + 2}$. Domains are all $[-1, \infty)$.

9. $2x + 1 + \dfrac{1}{x}, 2x + 1 - \dfrac{1}{x}, \dfrac{2x + 1}{x}, x(2x + 1)$. Domains are all $(-\infty, 0) \cup (0, \infty)$.

11. $[-1, \infty)$ **13.** $[\frac{5}{2}, \infty)$

15. (a) 125; (b) -1

17. $6x - 3, 6x + 1$. Domains are all $(-\infty, \infty)$.

19. x, x. Domains: $(-\infty, \infty), (-\infty, \infty)$

21. (a) 5; (b) 54; (c) 7; (d) 16

23. $f(x) = x^5, g(x) = x + 5$

25. $f(x) = \sqrt[3]{x}, g(x) = x + 1$

27. $f(x) = |x|, g(x) = x^2 - 1$

29. $f(x) = \dfrac{1}{x}, g(x) = x^2 + 1$

31. (a) 0; (b) -1; (c) 0; (d) 1

33. $9t + 10$. The firm has a daily start-up cost of \$10,000 and costs of \$9000 per hour.

35. $R(x) - C(x)$, which is the profit $P(x)$

37. $V(t) = \dfrac{4\pi(30 - 2t)^3}{3}$

39. $g(x) = 40x, f(r) = 5.5r, y = f(g(x)) = (f \circ g)(x) = 40(5.5)x = 220x$. There are 220 yards in a furlong.

41. $(-\infty, \infty)$ **43.** $[-5, \infty)$

45. $(-\infty, 5]$

Section 1.7

1.

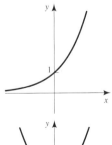

3.

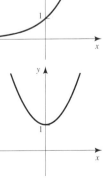

5.

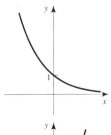

7.

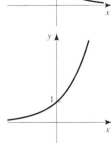

9. 2

11. $\frac{3}{4}$

13. $\frac{3}{4}$

15. $-\dfrac{45}{23}$

17.

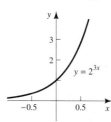

19.

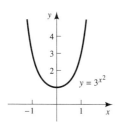

21.

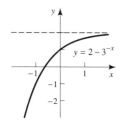

23.

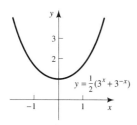

25.

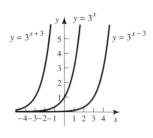

27.

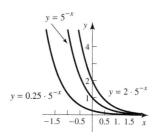

29.

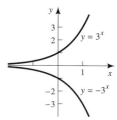

31. -1

33. -3

35. $-\frac{1}{2}$

37. $-\frac{1}{2}$

39. 1

41. -1

43. 0, 1

45. (a) \$1080; (b) \$1082.43; (c) \$1083.00; (d) \$1083.22;
(e) \$1083.28

47. (a) \$3262.04; (b) \$7039.99; (c) \$14,974.46; (d) \$31,409.42;
(e) \$93,050.97; (f) \$267,863.55

49. (a) 8.16%; (b) 8.243%; (c) 8.300%; (d) 8.322%; (e) 8.328%

51. (a) \$1784.31; (b) \$1664.13; (c) \$1655.56

53. \$31,837.58

55. \$26,897.08

57. \$2957 billion

59. \$1104.79, no

61. 6.66%, 6.82%, the second bank

63. \$67,297.13

65. 11.58 years

67. 8.2%

69. 16.82

71. (a) y_2; (b) y_1

73. (a) y_2; (b) y_1

75. $(0, \infty)$, $(-\infty, 0)$

Section 1.8

1. 10^2

3. $10^{-\pi}$

5. $3^{5/3}$

7. 4

9. -1

11. 2π

13. 8

15. 81

17. $2 \log x + \frac{1}{2} \log y + \log z$

19. $\frac{1}{2} \log x + \frac{1}{2} \log y - \log z$

21. $\frac{1}{2}(\log x + \log y - \log z)$

23. $\log x^2 y$

25. $\log \dfrac{\sqrt{x}}{\sqrt[3]{y}}$

27. $\log \dfrac{x^3 y}{\sqrt[3]{z}}$

29. $0.2 \log (0.6)$

31. $-0.5 \log 3$

33. $\dfrac{1 - \log 2}{3}$

35. $\pm\sqrt{\log 4}$

37. $10^{-1.5} - 7$

39. 2.5

41.

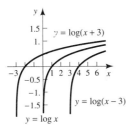

43.

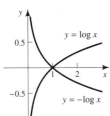

47. 22.5 years

49. 24.8 years

51. $\log_3 x = \dfrac{\log_{10} x}{\log_{10} 3}$

53. No. Since for $y > 0$, $\log_1 x = y$ if and only if $x = 1^y = 1$ has no solution if $x \neq 1$.

55. y_2

Chapter 1 Review

1–4.

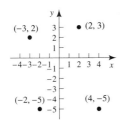

5.

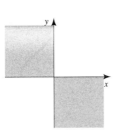

6. $\sqrt{34}$

7.

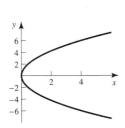

8. $(8, 9)$, $(9, 15)$

9. $[-1, 8]$ by $[-5, 3]$

10. b

11. b

12. $-\dfrac{7}{3}$

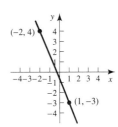

13. $\dfrac{3}{2}$

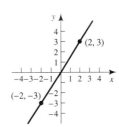

14. Slope undefined

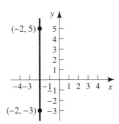

15. 3, $-\dfrac{2}{3}$, 2

16. 0, no x-intercept, -2

17. slope undefined, 3, no y-intercept

18. $y - 5 = -2(x + 2)$

19. $y = -x - 2$

20. $y + 5 = 3(x + 3)$ **21.** $y - 7 = \frac{1}{4}(x + 3)$ **34.**

22. a, b

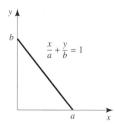

23. yes   **24.** no

25. yes   **26.** no

27. no   **28.** yes

29. $\{x | x \neq 3\}$ **30.** $\{x | x > 5\}$

31. $1, \frac{1}{3}, \dfrac{1}{a^2 - 2a + 2}, \dfrac{1}{x^2 + 2x + 2}$

32.

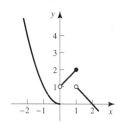

33.

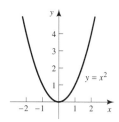

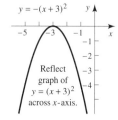

35.

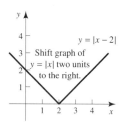

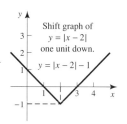

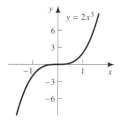

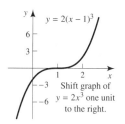

36.

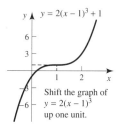

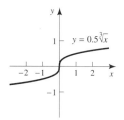

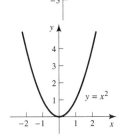

Shift the graph of $y = 0.5\sqrt[3]{x}$ one unit to the left.

$y = 0.5\sqrt[3]{x+1}$

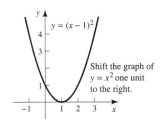

Shift the graph of $y = 0.5\sqrt[3]{x+1}$ down two units.

$y = 0.5\sqrt[3]{x+1} - 2$

37.

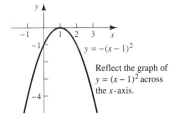

$y = x^2$

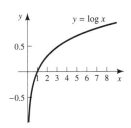

$y = (x-1)^2$

Shift the graph of $y = x^2$ one unit to the right.

$y = -(x-1)^2$

Reflect the graph of $y = (x-1)^2$ across the x-axis.

38.

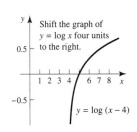

$y = \log x$

Shift the graph of $y = \log x$ four units to the right.

$y = \log (x-4)$

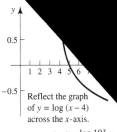

Reflect the graph of $y = \log (x-4)$ across the x-axis.

39.

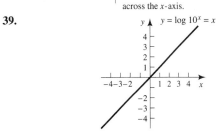

$y = \log 10^x = x$

40. $x^3 + 2x + 3$. Domain: $(-\infty, \infty)$

41. $x^3 - 2x + 5$. Domain: $(-\infty, \infty)$

42. $(x^3 + 4)(2x - 1)$. Domain: $(-\infty, \infty)$

43. $\dfrac{x^3 + 4}{2x - 1}$. Domain: $(-\infty, 0.50) \cup (0.50, \infty)$

44. $9x^2 + 6x + 3$, $3x^2 + 7$. Domains: $(-\infty, \infty)$, $(-\infty, \infty)$

45. $\sqrt{x}$, $\sqrt{x-1} + 1$. Domains: $[0, \infty)$, $[1, \infty)$

46.

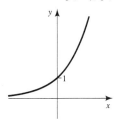

47.

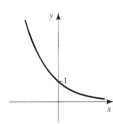

48.

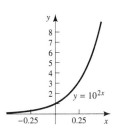

$y = 10^{2x}$

49.

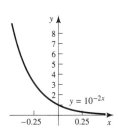

$y = 10^{-2x}$

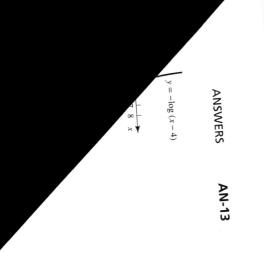

$y = -\log (x - 4)$

71. 10, 20, 30

72. For each unit distance away from the pocket margin, the proportion of dead roots decrease by 0.023 unit.

73. For 10,000 more units of shoot length, there will be 275 more buds.

74. $c = -6x + 12$

75. 6.84 years, 508.29 pounds

76. 103.41 pounds, 4294.39 pounds

77. 149.31, 313.32, 657.48

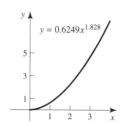

$y = 0.6249x^{1.828}$

78. $10x^2$

53. 3

54. 10   **55.** 0.50

56. 4   **57.** 5

58. 1.17   **59.** $C = 6x + 2000$

60. $R = 10x$ **61.** $P = 4x - 2000$, 500

62. $x = 150$ **63.** $x = 1000$

64. 150 miles   **65.** 42 minutes

66. $p - 1 = -\dfrac{1}{200,000}(x - 100,000)$

67. 12,000   **68.** $4.5x + 5.5y = 15$

69. (a) Boston; (b) Houston

70.

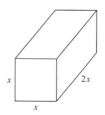

79.

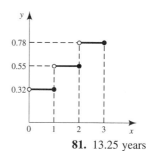

Quadruped	l	h	$l{:}h^{2/3}$
Ermine	12 cm	4 cm	4.762
Dashshund	35 cm	12 cm	6.677
Indian tiger	90 cm	45 cm	7.114
Llama	122 cm	73 cm	6.985
Indian elephant	153 cm	135 cm	5.814

80. $4\pi t^2$ **81.** 13.25 years

82. $k = 0.02$ **83.** in the year 2022

CHAPTER 2

Section 2.1

1. 2, 2, 2   **3.** 2, 2, 2

5. 1, 2, does not exist

7.

x	0.9	0.99	0.999	→	1	←	1.001	1.01	1.1
f(x)	4.7	4.97	4.997	→	5	←	5.003	5.03	5.3

9.

x	1.9	1.99	.1.999	→	2	←	2.001	2.01	2.1
f(x)	2.859	3.881	3.988	→	4	←	4.012	4.121	5.261

11.

x	1.9	1.99	1.999	→	2	←	2.001	2.01	2.1
f(x)	−10	−100	−1000	→	no	←	1000	100	10

13.

x	3.9	3.99	3.999	→	4	←	4.001	4.01	4.1
f(x)	3.9748	3.9975	3.9997	→	4	←	4.0002	4.0025	4.0248

15. 1, 0, does not exist

17. 1, 1, 1

19. 0, does not exist, does not exist

21. -1, 1, does not exist

23. None of the limits exist.

25. 4, does not exist, does not exist

27. 13   **29.** 13

31. -3 **33.** $-\frac{3}{4}$

35. 1   **37.** 4

39. 2   **41.** -2

43. does not exist   **45.** 4

47. (a) 0.32, 0.55, does not exist; (b) 0.55, 0.55, 0.55

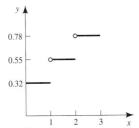

49. 1, 10, 100, 1000. $C(x)$ becomes large without bound

51. 1

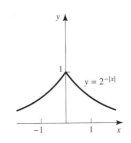

x	-0.1	-0.01	-0.001	$\to$ 0 $\leftarrow$	0.001	0.01	0.1
$2^{-\lvert x\rvert}$	0.933	0.993	0.999	$\to$ 1 $\leftarrow$	0.999	0.993	0.933

53. impossible

55.

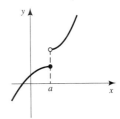

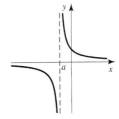

57. no. For example:
$$f(x) = \begin{cases} 1 & x < 0 \\ 0 & x \geq 0 \end{cases} \text{ and } g(x) = \begin{cases} 0 & x < 0 \\ 1 & x \geq 0 \end{cases}$$

59. For very small farms of, say, less than an acre, the owner has the additional time to give additional nurture to the plants and to protect them from disease, insects, and animals.

Section 2.2

1. 0, 2, 3, 4   **3.** nowhere

5. 0   **7.** 1

9. 0   **11.** everywhere

13. everywhere except $x = \pm 1$

15. positive   **17.** negative

19. $\frac{2}{3}$ **21.** 0

23. ∞ **25.** 1

27. 5   **29.** 1

31. 0   **33.** 2

35. 0   **37.** $10^x = e^{(\ln 10)x}$

39. $r \to c$ **41.** 100%

43. $A(t) = \begin{cases} \$100,000 & 0 < t < 0.25 \\ \$102,000 & 0.25 \leq t < 0.50 \\ \$104,040 & 0.50 \leq t < 0.75 \\ \$106,120.80 & 0.75 \leq t < 1 \end{cases}$

the quarter-year marks, if you wished to cash this certificate, you would wait until the beginning of a quarter to do so.

45. not continuous at $x = 100,000$ and $x = 1,000,000$

47. \$1061.84, 6.18%

49. The second bank, the second effective annual interest rate is greater than the first one.

51. $D = e^{-3.3195} K^{1.0629} \approx 0.0362 K^{1.0629}$

53. $S = \dfrac{c}{e^{-a-bt} + 1}, \quad S \to c$

55. 30 days

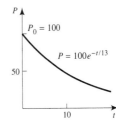

57. $y(1.25) \approx -1.45 < 0$ and $y(2) = 17 > 0$ implies y has a root in $(1.25, 2)$

59. The graph of $y = p(x)$ looks like the graph of $y = x^6$ for large $\lvert x\rvert$.

61.

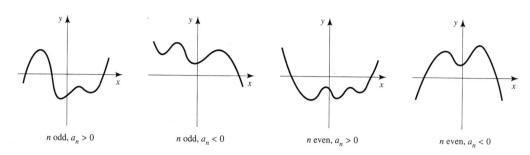

n odd, $a_n > 0$ n odd, $a_n < 0$ n even, $a_n > 0$ n even, $a_n < 0$

63. As x becomes large without bound, any polynomial (other than a constant) must become either positively or negatively large without bound. Thus, a nonconstant polynomial cannot be asymptotic to any horizontal line. Obviously, the indicated polynomial is not a constant.

65. Every polynomial is continuous at every point. The indicated function in the graph is not continuous at the point where the graph ''blows up'' and therefore cannot be the graph of any polynomial.

67. Let the polynomial be $p(x) = a_n x^n + \cdots + a_1 x + a_0$. Suppose $a_n > 0$. Then as x becomes large without bound, $p(x)$ becomes large without bound and therefore must be positive at some point b. Since n is odd and $a_n > 0$, as x becomes negatively large without bound, $p(x)$ becomes negatively large without bound and therefore must be neg-

ative at some point a. By the constant sign theorem, $y = p(x)$ must have at least one zero in (a, b). A similar analysis applies if $a_n < 0$.

69. From Exercise 67, the polynomial $f(x) = x^n - c$ must have at least one zero if n is odd, that is, there exists an $x \in (-\infty, \infty)$ such that $x^n - c = 0$, or $x = \sqrt[n]{c}$.

71. $\log_{10} x = \dfrac{\ln x}{\ln 10}$

73. $0.5P_0 = y = P_0 e^{-kT}$ implies $0.5 = e^{-kT}$ implies $\ln 0.5 = -kT$ implies $\ln 2 = kT$

75. $H \to 0$ as $D \to 0$

77. (a) $x \geq 0$ implies $e^{-kx} \leq e^0 = 1$; $b < 1$ implies $0 < be^{-kx} < 1$ implies $0 < 1 - be^{-kx} < 1$, therefore $y = m(1 - be^{-kx}) < m$. Finally, $\lim\limits_{x \to \infty} 2(1 - 0.5e^{-x}) = 2$.

Section 2.3

1. 96

3. 96

5. 0

7. −64

9. $y - 48 = 32(x - 1)$

11. 2

13. 2

15. −8

17. 0

19. −16

21. 3

23. $B - C, A - B$ and $D - E, C - D$

25. positive: A and D, negative at B, zero at C

27. 0.5

29. $y - 6 = 7(x - 2)$

35. $y \approx 7.39x - 7.39$

37. $y \approx 0.25x + 0.39$

39. $y \approx 0.25x + 1$

41. $0, -8.5, \frac{7}{3}, 12$

43. $15.9, \dfrac{13}{15}, \dfrac{-327}{15}$

45. $4, 0, -4$

47. $8, 0, -2$

49. (a) 800, meaning that by the first week, the sales are increasing at the rate of 800 sales per week; (b) 0, meaning that by the fifth week, the sales stop increasing; (c) −200, meaning that by the sixth week, the sales are decreasing at the rate of 200 sales per week

51. $0, 4, 16$

53. 1.42

57. largest: 3^x; smallest: 2^x

59. 1.099, yes

Section 2.4

1. 5

3. $2x$

5. $6x + 3$

7. $-6x^2 + 1$

9. $\dfrac{-1}{(x + 2)^2}$

11. $\dfrac{-2}{(2x - 1)^2}$

13. $\dfrac{1}{2\sqrt{t + 1}}$

15. $\dfrac{1}{\sqrt{2t + 5}}$

17. (1) $y - 2 = 5(x - 1)$; (3) $y - 5 = 2(x - 1)$; (5) $y - 5 = 9(x - 1)$

19. $y = x - 1$

21. everywhere except $x = 0$

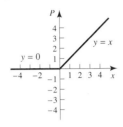

23. everywhere except $x = 1$

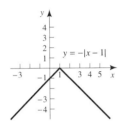

25. everywhere except $x = 1$

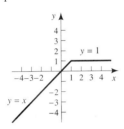

27. everywhere in $(-1, 1)$ except $x = 0$

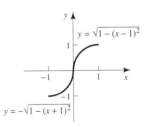

29. 3

31.

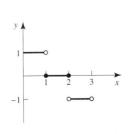

33.

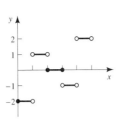

35.

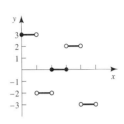

41. $\frac{2}{3}, \frac{2}{5}$

43. $-25, -4$

45. $-0.15, \dfrac{-3}{125}$

49. $\lim\limits_{h \to 0} \dfrac{f(0 + h) - f(0)}{h} = \lim\limits_{h \to 0} \dfrac{\sqrt[3]{h^2} - 0}{h} = \lim\limits_{h \to 0} h^{(2/3)-1} =$
$\lim\limits_{h \to 0} \dfrac{1}{h^{1/3}} = \infty$

51. Graph of $y = f(x)$ appears to have corner at $x = 0$. Closer examination reveals $f'(0) = 0$.

53.

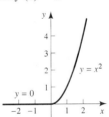

h	-0.1	-0.01	-0.001	$\to 0 \leftarrow$	0.001	0.01	0.1
$\dfrac{f(0 + h) - f(0)}{h}$	0	0	0	$\to 0 \leftarrow$	0.001	0.01	0.1

Since $\lim\limits_{h \to 0^-} \dfrac{f(0 + h) - f(0)}{h} = \lim\limits_{h \to 0^-} \dfrac{0 - 0}{h} = 0,$ and
$\lim\limits_{h \to 0^+} \dfrac{f(0 + h) - f(0)}{h} = \lim\limits_{h \to 0^+} \dfrac{h^2 - 0}{h} = \lim\limits_{h \to 0^+} h = 0,$ then
$f'(0) = \lim\limits_{h \to 0} \dfrac{f(0 + h) - f(0)}{h} = 0$

55. $f'(0) = 1$

57. $f'(1.57) \approx -1$

Section 2.5

1. 3.9375

3. −0.05

5. 0.95

7. $f'(1) \approx \frac{1}{3}$

9. $f'(2) \approx 8.8$

11. $f'(1) \approx 1$

13. 1.0333

15. 4.2401

17. −0.40

19. −1.6

21. −0.015

23. 288,000

25. 1.6π

27. 90.4, 94.7

29. Initially there appears to be a corner at $x = 0$. After zooming several times about $(0, \sqrt{0.0001})$ the graph becomes rounded. After more zooming, the graph becomes a horizontal line. Thus, $f'(0) = 0$.

31. Δ area $\approx 2\pi l \cdot \Delta r$

33. 0.02

Chapter 2 Review

1. $x = -1$: 1, 1, 1; $x = 0$: 2, 1, does not exist; $x = 1$: 2, does not exist, does not exist; $x = 2$: 1, 1, 1

2. −1, 0, 1

3. −1, 0, 1, 2

4. 12.7

5. π^3

6. 0

7. −3

8. −4

9. 1

10. −7

11. 3

12. 7

13. Does not exist

14. Does not exist

15. 2

16. 3

17. 1.5

18. Does not exist

19. 2

20. 0

21. Does not exist (∞)

22. 0

23. 1

24. 0

25. 0

26. $4x + 1$

27. $2x + 5$

28. $\dfrac{-2}{x^3}$

29. $\dfrac{-2}{(2x + 5)^2}$

30. $\dfrac{1}{2\sqrt{x + 7}}$

31. $\dfrac{1}{\sqrt{2x - 1}}$

32. $y - 9 = 12(x - 2)$

33. $y - 10 = 3(x - 4)$

34. $y - 6 = 11(x - 1)$

35. $y - \dfrac{1}{3} = -\dfrac{1}{54}(x - 9)$

36. $f'(1) \approx 0.20$

37. $f'(8) \approx 0.33$

38. $f'(2) \approx -0.19$

39. $f'(4) \approx 0.19$

40. 1.01

41. 3.9

42. 0.11

43. 2.503

44.

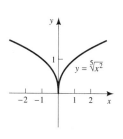

$y = \sqrt[5]{x^2}$

h	−0.1	−0.01	−0.001	→ 0 ←	0.001	0.01	0.1
$\dfrac{f(0 + h) - f(0)}{h}$	−3.98	−15.85	−63.10	→ No ←	63.10	15.85	3.98

$f'(0)$ does not exist.

45. $\lim\limits_{h\to 0} \dfrac{f(h) - f(0)}{h} = \lim\limits_{h\to 0} \dfrac{\sqrt[5]{h^2} - 0}{h} = \lim\limits_{h\to 0} h^{2/5 - 1} = \lim\limits_{h\to 0} h^{-3/5} = \infty$. Therefore $f'(0)$ does not exist.

46. Yes, since $\lim\limits_{x\to 2} f(x) = f(2)$

47. Since $\lim\limits_{h\to 0^-} \dfrac{f(h) - f(0)}{h} = \lim\limits_{h\to 0^-} \dfrac{0 - 0}{h} = 0$, and $\lim\limits_{h\to 0^+} \dfrac{f(h) - f(0)}{h} = \lim\limits_{h\to 0^+} \dfrac{h^3 - 0}{h} = \lim\limits_{h\to 0^+} h^2 = 0$, then $\lim\limits_{h\to 0^-} \dfrac{f(h) - f(0)}{h} = \lim\limits_{h\to 0^+} \dfrac{f(h) - f(0)}{h} = \lim\limits_{h\to 0} \dfrac{f(h) - f(0)}{h} = f'(0) = 0$

48. $0.20

49. $i \to \infty$

50. 0

51. Everywhere except the integer points of minute.

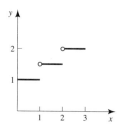

52. 200, 0, −200

53. −2, the distance s is decreasing at the rate of 2 ft/sec when $t = 3$. 0, the distance s stops decreasing when $t = 4$. 2, the distance s is increasing at the rate of 2 ft/sec when $t = 5$.

54. $\Delta V \approx 12.57$

55.

n	1	2	3	4	5
$n!$	1	2	6	24	120
$\sqrt{2\pi n}(n/e)^n$	0.922	1.919	5.836	23.506	118.019

56. $\ln 10/k$

57. (a) If $A > B$, $\lim\limits_{t\to\infty} x(t) = (a - 1)A + B$; (b) If $A < B$, $\lim\limits_{t\to\infty} x(t) = aA$

Section 3.1

1. 0

3. 0

5. $23x^{22}$

7. $1.4x^{0.4}$

9. ex^{e-1}

11. $\frac{4}{3}x^{1/3}$

13. $-5.1x^{-2.7}$

15. $\frac{0.12}{\sqrt{x}}$

17. $2e^x$

19. $\pi + \frac{2}{x}$

21. $-1 - 2x$

23. $\frac{2}{\pi}x + \pi$

25. $-1 - 2x - 3x^2 - \frac{10}{x}$

27. $5x^4 - 5e^x$

29. $0.7x - 0.3e^x$

31. $-\frac{1}{x} - 0.02x - 0.09x^2$

33. $\frac{1}{6}(e^x + 2x - 2)$

35. $9x^2 - \frac{9}{x^4}$

37. $2x + \frac{1}{x^2}$

39. $4x^{-1} + 9x^{-4} + 8x$

41. $-81t^2 - e^{t-3}$

43. $-t^{-2} - 2t^{-3} - 3t^{-4}$

45. $\frac{1}{3}t^{-2/3} + \frac{1}{3}t^{-4/3}$

47. $5u^{2/3} - 2u^{-1/3}$

49. $4u + 3$

51. $\frac{1}{2}u^{-1/2} - \frac{9}{2}u^{-5/2}$

53. $x^{-1/2} + \frac{1}{6}x^{-3/2}$

55. $\pi^2 x^{\pi-1} + \frac{2}{x}$

57. $\sqrt{2}x^{\sqrt{2}-1} + \sqrt{2}x^{-\sqrt{2}-1}$

59. $-\frac{3}{2}x^{-5/2} - \frac{4}{5}x^{-9/5} + \frac{1}{2x}$

61. $\frac{2}{3}x^{-1/3}$

63. $-\frac{2}{5}x^{-7/5}$

65. $y = x + 2$

67. $y - 2 = -2(x - 1)$

69. $y = -3(x + 1)$

71. $x = 0$

73. $x = \pm\sqrt{3}$

75. $-4x + 24$; 12, every unit increase in x at $x = 3$ results in $12 increase in revenue; -16, $16 decrease in revenue is expected if a single unit increase in x is performed at $x = 10$.

77. $-4.4x + 24 - 0.001e^x$; 10.78, every un.. ...
$x = 3$ results in $10.78 increase in profit; -42.03, $42.05 decrease in profit is expected if a single unit increase in x is performed at $x = 10$.

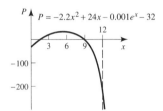

79. $6t^2 - 4$

81. 230.26

83. $5.62L$

85. $0.24(L + 0.0039L^2)$

87. $\frac{292.6T^{0.75}}{(273)^{1.75}P} \approx \frac{0.016}{P}T^{0.75}$

89. (a) -1.36 and 0.92
(b) -1.36 and 0.92
(c) Solutions in part (b) equal to values in part (a)
(d) Exact solution cannot be found.

91. (a) ±0.88
(b) ±0.88
(c) Solutions in part (b) equal to values in part (a)
(d) Exact solution cannot be found.

93. $L_1: y = -4x - 3$, $L_2: y = 4x - 3$. Solve $4x - 3 = -4x - 3$ and obtain $x = 0$.

95. no. For example:
$$f(x) = \begin{cases} 1 & x \le a \\ 0 & x > a \end{cases} \text{ and } g(x) = \begin{cases} 0 & x \le a \\ 1 & x > a \end{cases}$$

97. $1 + 2x - 3\cos x$

99. $\frac{5}{x} - 2\sin x$

Section 3.2

1. $x^3e^x + 3x^2e^x$

3. $\frac{\ln x + 2}{2\sqrt{x}}$

5. $e^x(x^2 + 2x - 3)$

7. $(4e^x - 5x^4)\ln x + \frac{4e^x - x^5}{x}$

9. $e^x(\sqrt{x} + 1) + \frac{e^x + 1}{2\sqrt{x}}$

11. $\left(\frac{2x-3}{x}\right)\left(\frac{x^2+1}{x}\right) + (2x - 3\ln x)\left(\frac{x^2-1}{x^2}\right)$

13. $\frac{1/x - \ln x}{e^x}$

15. $-\frac{1}{x(\ln x)^2}$

17. $\frac{2}{(x+2)^2}$

19. $\frac{8}{(x+5)^2}$

23. $\dfrac{-1}{(2x-1)^2}$

41. $\dfrac{2}{5} \cdot \dfrac{I^2 + 100{,}000I}{(I + 50{,}000)^2}$

43. $\dfrac{7.12(2x - 13.74)}{(13.74x - x^2)^2}$

27. $\dfrac{(2 - 2u - u^2)}{(u^2 + 2)^2}$

45. (a) $2e^{2x}$; (b) $3e^{3x}$; (c) ne^{nx}

31. $\dfrac{-8u^3 + (3u - 1)e^u - 1}{3u^{2/3}(u^3 - e^u - 1)^2}$

47. $f'(x) = (x - a)[2g(x) + (x - a)g'(x)]$ implies $f'(a) = 0$. The equation of the line tangent to $y = f(x)$ at $x = a$ is $y - f(a) = f'(a)(x - a)$ or $y = 0$, which is the x-axis.

$\dfrac{9u^2 + 2u}{(u^2 + 3)^2}$

$\dfrac{1 - x^2}{(1 + x^2)^2}$

35. $\dfrac{0.1}{(10 - 0.1x)^2}$

49. $\sin x + x \cos x$

51. $e^x(\cos x - \sin x)$

37. $\dfrac{1000e^t}{(1 + e^t)^2}$

39. $\dfrac{15 - 6t^2}{(2t^2 + 5)^2}$

53. $\dfrac{x \cos x - \sin x}{x^2}$

55. $1 + \tan^2 x$

Section 3.3

1. $14(2x + 1)^6$

3. $8x(x^2 + 1)^3$

29. $e^x \ln x \left(\ln x + \dfrac{2}{x} \right)$

31. $-12(x + 1)^{-5}$

5. $3e^x\sqrt{2e^x - 3}$

7. $\tfrac{1}{2}\sqrt{e^x}$

9. $\dfrac{1}{\sqrt{2x + 1}}$

11. $4x(x^2 + 1)^{-2/3}$

33. $\dfrac{xe^x - 3}{(x + 3)^4}$

35. $\tfrac{1}{4}(\sqrt{x} + 1)^{-1/2}x^{-1/2}$

13. $-\dfrac{e^x}{(e^x + 1)^2}$

15. $\dfrac{6x^2}{(x^3 + 1)^2}$

37. $-x(1 + x^2)^{-3/2}$

39. $4000\left(1 + \tfrac{1}{12}i\right)^{47}$

17. $\dfrac{x^2(1 - \ln x) + 1}{x(x^2 + 1)^{3/2}}$

41. $\dfrac{4\pi cr^2}{3} \left(1 + \dfrac{c}{r} n^{1/3} \right)^2 n^{-2/3}$

19. $15(2x + 1)(3x - 2)^4 + 2(3x - 2)^5$

21. $\dfrac{4(\ln x)^3}{x}$

43. $\dfrac{an}{2\sqrt{n - b}} + a\sqrt{n - b}$

45. $\dfrac{1}{8\sqrt{\sqrt{\sqrt{x} + 1} + 1}\sqrt{\sqrt{x} + 1}\sqrt{x}}$

23. $e^x(x^2 + 1)^7(x^2 + 16x + 1)$

25. $10(1 - x)^4(2x - 1)^4 - 4(1 - x)^3(2x - 1)^5$

47. Write as the product $(e^x + 1)(x + 3)^{-3}$

27. $12\sqrt{x}(4x + 3)^2 + \tfrac{1}{2}x^{-1/2}(4x + 3)^3$

49. $3(\sin x)^2 \cos x$

51. $-\dfrac{\sin x}{2\sqrt{\cos x}}$

Section 3.4

1. $4e^{4x}$

3. $-3e^{-3x}$

37. $\dfrac{2x + 1}{2(x^2 + x + 1)\sqrt{\ln |x^2 + x + 1|}}$

5. $4xe^{2x^2+1}$

7. $\dfrac{e^{\sqrt{x}}}{2\sqrt{x}}$

39. $\dfrac{1}{x(x + 1)}$

41. $-4xe^{-x^2} \ln |x| + \dfrac{2e^{-x^2}}{x}$

9. $x^2e^x + 2xe^x$

11. $e^x x^4(5 + x)$

43. $2x \ln x + x$

13. $-xe^{-x} + e^{-x}$

15. $2x^3e^{x^2} + 2xe^{x^2}$

45. $-\dfrac{2x \ln x}{(x^2 + 1)^2} + \dfrac{1}{x(x^2 + 1)}$

17. $\dfrac{e^x}{2\sqrt{e^x + 1}}$

19. $\dfrac{1}{2} e^{\sqrt{x}} + \dfrac{e^{\sqrt{x}}}{2\sqrt{x}}$

21. $\dfrac{1 + e^x - xe^x}{(1 + e^x)^2}$

23. $\dfrac{8}{(e^{2x} + e^{-2x})^2}$

47. $\dfrac{11}{x}(\ln x)^{10}$

49. $\dfrac{3}{x}$

25. $\dfrac{3e^{3x}}{2\sqrt{e^{3x} + 2}}$

27. $5^x \ln 5$

51. $\dfrac{e^x}{(e^x + 1) \ln 10}$

53. $e^{-3x}(1 - 3x), \tfrac{1}{3}$

55. $-2xe^{-x^2}, 0$

57. $xe^x(2 + x), 0, -2$

29. $\dfrac{3^{\sqrt{x}} \ln 3}{2\sqrt{x}}$

31. $3^x + x3^x \ln 3$

59. $xe^{-x}(2 - x), 0, 2$

61. $\ln x^2 + 2, \dfrac{1}{e}$

33. $\dfrac{3x^2 + 2x}{x^3 + x^2 + 1}$

35. $4x \ln |x| + 2x$

63. $\dfrac{1 - 2 \ln x}{x^3}, \sqrt{e}$

65. $e^{-2x}(1 - 2x), \tfrac{1}{2}$

67. $\dfrac{2x}{x^2 + 5}$, $(0, \infty)$

69. $R'(t) = \dfrac{A_0}{b - 1}(-kbe^{-kt}) < 0$

71. $e^{e^{e^x}} \cdot e^{e^x} \cdot e^x$

73. **(a)** 500

    (b) $1488e^{-0.024t}(1 + 124e^{-0.024t})^{-2}$

    (c) $p'(t) > 0$

    (d) The graph increases and is asymptotic to $y = 500$

    (e) The two graphs are about the same for the first 50 years.

75. Let $u = f(x)$, then $\dfrac{d}{dx}[\sin f(x)] = \dfrac{d}{dx}(\sin u$

$\dfrac{d}{du}(\sin u) \cdot \dfrac{du}{dx} = (\cos u)\dfrac{du}{dx} = [\cos f(x)]f'(x)$, and

$[\cos f(x)] = \dfrac{d}{dx}(\cos u) = \dfrac{d}{du}(\cos u) \cdot \dfrac{du}{dx} =$

$(-\sin u)\dfrac{du}{dx} = -[\sin f(x)]f'(x).$

77. $-3x^2 \sin x^3$ **79.** $\dfrac{\cos(\ln x)}{x}$

81. $-\sin(\sin x) \cdot \cos x$

Section 3.5

1. 0.60   **3.** 10

5. 30   **7.** 3

9. 83   **11.** 2

13. (a) $\frac{1}{4}$ inelastic; (b) 1 unit; (c) 9 elastic

15. (a) $\frac{1}{7}$ inelastic; (b) 1 unit; (c) 3 elastic

17. (a) $\frac{1}{3}$ inelastic; (b) $\frac{1}{2}$ inelastic; (c) $\dfrac{10}{11}$ inelastic

19. (a) $\dfrac{2}{1.001}$ elastic; (b) $\dfrac{2}{11}$ inelastic

21. (a) 2 elastic; (b) 2 elastic   **23.** (a) 15; (b) 15

25. (a) none; (b) none   **27.** $-x + 31$

29. $-2x + 11, 3, 8$

31. **(a)** $0.03x^2 - 2x + 50$

    (b) 22, the cost of the 21st item is approximately \$22; 18, the cost of the 41st item is approximately \$18; 38, the cost of the 61st item is approximately \$38.

    (c) The marginal cost first decreases, and then at some point, begins to increase.

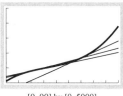

[0, 90] by [0, 5000]

Chapter 3 Review

1. $6x^5 - 6x$ **2.** $\dfrac{24}{5}x^{1/5} + 2x^{-3/5}$

3. $-6x^{-4} - 2x^{-5/3}$ **4.** $2.6x^{0.3} - 7.2x^{-3.4}$

5. $2\pi^2 x + 2$ **6.** $3\pi^3 x^2 + 2e^2 x$

7. $\frac{2}{3}x^{-1/3}$ **8.** $-\frac{5}{3}x^{-8/3}$

9. $(x + 3)(3x^2 + 1) + (x^3 + x + 1)$

10. $(x^2 + 5)(5x^4 - 3) + 2x(x^5 - 3x + 2)$

11. $\sqrt{x}(4x^3 - 4x) + \frac{1}{2}x^{-1/2}(x^4 - 2x^2 + 3)$

12. $\sqrt[3]{x}(3x^2 - 3) + \frac{1}{3}x^{-2/3}(x^3 - 3x + 5)$

13. $x^{3/2}(2x + 4) + \frac{3}{2}x^{1/2}(x^2 + 4x + 7)$

14. $x^{-7/2}(3x^2 + 2) - \frac{7}{2}x^{-9/2}(x^3 + 2x + 1)$

15. $\dfrac{2}{(x + 1)^2}$ **16.** $\dfrac{(2x + 1)\frac{1}{2}x^{-1/2} - 2x^{1/2}}{(2x + 1)^2}$

17. $\dfrac{x^4 + 3x^2}{(x^2 + 1)^2}$ **18.** $\dfrac{3x^2 - 4x - 5}{(3x - 2)^2}$

19. $\dfrac{-2x^5 + 4x^3 + 2x}{(x^4 + 1)^2}$ **20.** $30(3x + 1)^9$

21. $40x(x^2 + 4)^{19}$ **22.** $\frac{5}{4}(\sqrt{x} - 5)^{3/2}x^{-1/2}$

23. $\dfrac{2x + 1}{2\sqrt{x^2 + x + 1}}$

24. $5x^3(x^2 + x + 2)^4(2x + 1) + 3x^2(x^2 + x + 2)^5$

25. $5(x^2 + 3)(x^4 + x + 1)^4(4x^3 + 1) + 2x(x^4 + x + 1)^5$

26. $-4(x^2 + 1)(x^4 + x + 1)^{-5}(4x^3 + 1) + 2x(x^4 + x + 1)^{-4}$

27. $\frac{3}{4}(x^{3/2} + 1)^{-1/2}x^{1/2}$ **28.** $\frac{1}{6}(x^{1/2} + 1)^{-2/3}x^{-1/2}$

29. $-7e^{-7x}$ **30.** $6x^2e^{2x^3+1}$

31. $e^x(x^2 + 3x + 3)$ **32.** $\dfrac{2x}{x^2 + 1}$

33. $\dfrac{e^{x^2}}{x} + 2xe^{x^2}\ln x$ **34.** $\dfrac{2x + e^x}{x^2 + e^x}$

35. $\frac{1}{5}x^{-4/5}$ **36.** $\frac{2}{3}x^{-1/3}$

37. $-3x^{-4}$ **38.** $\frac{1}{2}(x^{-1/2} - x^{-3/2})$

39. 1.01   **40.** 3.9

41. 0.11   **42.** 2.503

43. (a) 606.11, at $x = 2$ items sold, one more item sold results in a \$606.11 profit; (b) 329.14, at $x = 3$ items sold, one more item sold results in a \$329.14 profit; (c) -225.98, at $x = 4$ items sold, one more item sold causes loss of \$225.98

items sold, one more item

ving at 56 ft/sec in positive
st at the instant $t = 2$; -152,
ng at 152 ft/sec in negative

rate of change of the number
ven by $3.2572t^{1.395}$

he age of x, the instantaneous
of the eggs is given by $7.08x^2$
e quadratic formula indicates

— 41.00λ + 0σ...

that $f'(x)$ is never zero and since $f'(0) > 0$, $f'(x) > 0$ for
all x. This means that *Ostrinia furnacalis* lays more eggs
on older plants.

47. $f'(x) = -8.094 \ln(1.053) \cdot (1.053)^{-x} \approx -0.418(1.053)^{-x}$.
$f'(x) < 0$ for all x means that more immature sweet potato
whiteflies cause smaller tomato leaves.

48. 25

49. (a) $\frac{2}{3}$ inelastic; (b) 1 unit; (c) $\frac{3}{2}$ elastic

50. $-\dfrac{900}{(3x + 1)^2}$

CHAPTER 4

Section 4.1

1. relative maximum on about May 30 and June 12; relative
minimum on about June 5

3. relative maximum at about $x = 1300$ and about $x = 1850$;
relative minimum at about $x = 1400$ and $x = 1950$

5. relative maximum at about $x = 2500$ and $x = 3300$; relative
minimum at about $x = 2800$

7. critical value at $x = 3$
increasing on $(3, \infty)$
decreasing on $(-\infty, 3)$
no relative maxima
relative minimum at $x = 3$

9. critical values at $x = 2, 4$
increasing on $(-\infty, 2)$, $(4, \infty)$
decreasing on $(2, 4)$
relative maximum at $x = 2$
relative minimum at $x = 4$

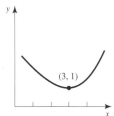

11. critical values at $x = 1, 3, 5$
increasing on $(1, 3)$, $(5, \infty)$
decreasing on $(-\infty, 1)$, $(3, 5)$
relative maximum at $x = 3$
relative minima at $x = 1, 5$

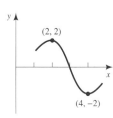

13. critical values at $x = 1, 3, 5$
increasing on $(-\infty, 1)$, $(3, \infty)$
decreasing on $(1, 3)$
relative maximum at $x = 1$
relative minimum at $x = 3$

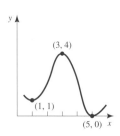

15. critical values at $x = -1, 1, 2$
increasing on $(-\infty, 1)$, $(2, \infty)$
decreasing on $(1, 2)$
relative maximum at $x = 1$
relative minimum at $x = 2$

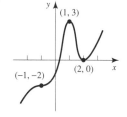

17. critical values at $x = -1, 1, 2$
increasing on $(-1, 1)$, $(2, \infty)$
decreasing on $(-\infty, -1)$, $(1, 2)$
relative maximum at $x = 1$
relative minima at $x = -1, 2$

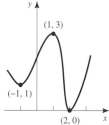

19. critical value at $x = 2$
increasing on $(2, \infty)$
decreasing on $(-\infty, 2)$
no relative maxima
relative minimum at $x = 2$

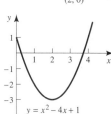

21. no critical values
increasing on $(-\infty, \infty)$
never decreasing
no relative maxima
no relative minima

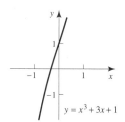

23. critical value at $x = 0$
increasing on $(0, \infty)$
decreasing on $(-\infty, 0)$
no relative maxima
relative minimum at $x = 0$

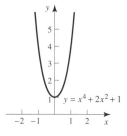

25. critical value at $x = 1$
increasing on $(-\infty, 1)$
decreasing on $(1, \infty)$
relative maximum at $x = 1$
no relative minima

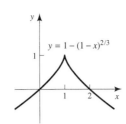
$y = 1 - (1 - x)^{2/3}$

27. critical values at $x = -1, 1$
increasing on $(-\infty, -1), (1, \infty)$
decreasing on $(-1, 1)$
relative maximum at $x = -1$
relative minimum at $x = 1$

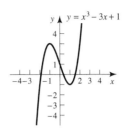
$y = x^3 - 3x + 1$

29. critical values at $x = \frac{1}{2}, \frac{3}{2}$
increasing on $(-\infty, \frac{1}{2}), (\frac{3}{2}, \infty)$
decreasing on $(\frac{1}{2}, \frac{3}{2})$
relative maximum at $x = \frac{1}{2}$
relative minimum at $x = \frac{3}{2}$

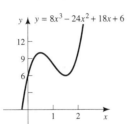
$y = 8x^3 - 24x^2 + 18x + 6$

31. critical values at $x = -2, 0, 2$
increasing on $(-2, 0), (2, \infty)$
decreasing on $(-\infty, -2), (0, 2)$
relative maximum at $x = 0$
relative minimums at $x = -2, 2$

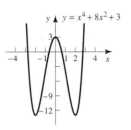
$y = x^4 + 8x^2 + 3$

33. critical values at $x = -1, 1$
increasing on $(-1, 1)$
decreasing on $(-\infty, -1), (1, \infty)$
relative maximum at $x = 1$
relative minimum at $x = -1$

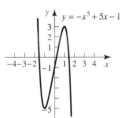
$y = -x^5 + 5x - 1$

35. critical values at $x = -2, 0, 2$
increasing on $(-\infty, -2), (2, \infty)$
decreasing on $(-2, 2)$
relative maximum at $x = -2$
relative minimum at $x = 2$

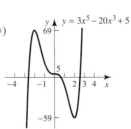
$y = 3x^5 - 20x^3 + 5$

37. critical value at $x = 1$
increasing on $(1, \infty)$
decreasing on $(0, 1)$
no relative maxima
relative minimum at $x = 1$

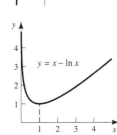
$y = x - \ln x$

39. critical value at $x = 0$
increasing on $(-\infty, 0)$
decreasing on $(0, \infty)$
relative maximum at $x = 0$
no relative minima

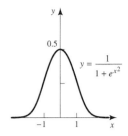
$y = \dfrac{1}{1 + e^{x^2}}$

41. critical values at $x = 0, 2$
increasing on $(0, 2)$
decreasing on $(-\infty, 0), (2, \infty)$
relative maximum at $x = 2$
relative minimum at $x = 0$

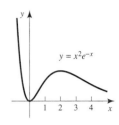
$y = x^2 e^{-x}$

43. $f'(x) = 3ax^2 + b$ is positive if both a and b are positive, and negative if both a and b are negative.

51. $x = 500$ **53.** $x = 20$

55. $x = 3$

57. Solving $\bar{R} = R'$ for x gives $100x - 20x^2 = 200x - 60x^2$ implies $40x^2 - 100x = 0$ implies $x = 2.5$, which is the value where $\bar{R}$ is maximized.

59. $(0, \infty)$

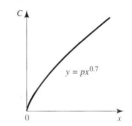
$y = px^{0.7}$

61. $\sqrt{bk}$

63. 1.044   **65.** 30. 24,400

67. $I/2$

69. $f'(x) = -55.232e^{-0.64x} < 0$ for all real x, $y \to 9.1$

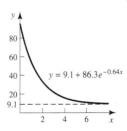
$y = 9.1 + 86.3e^{-0.64x}$

71. $x = 3.59$

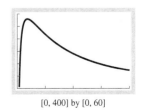
[0, 400] by [0, 60]

73. $f'(x) = \dfrac{-0.0899 \ln 10 \cdot 10^{3.8811 - 0.1798\sqrt{x}}}{\sqrt{x}} < 0$ for all $x >$ $0, y \to 0$

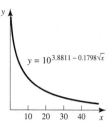

$y = 10^{3.8811 - 0.1798\sqrt{x}}$

75. neither

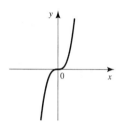

77. relative maximum

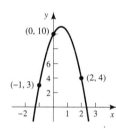

79. $f(0) < 0$ and $f'(x) > 0$ on $(-\infty, 0)$ implies $f(x) < 0$. Hence, f has no root on $(-\infty, 0)$. $f(0) < 0$ and $f'(x) < 0$ on $\left(0, \dfrac{12}{13}\right)$ implies $f(x) < 0$. Hence, f has no root on $\left(0, \dfrac{12}{13}\right)$. $f\left(\dfrac{12}{13}\right) < 0, f(2) > 0$ and $f'(x) > 0$ on $\left(\dfrac{12}{13}, \infty\right)$ implies f has one and only one root on $\left(\dfrac{12}{13}, \infty\right)$.

81. Let $g(x) = e^x - (1 + x)$. Since $g'(x) < 0$ on $(-\infty, 0)$, $g(x) > g(0) = 0$ on $(-\infty, 0)$. Since $g'(x) > 0$ on $(0, \infty)$, $g(x) > g(0) = 0$ on $(0, \infty)$. Thus, $g(x) \geq 0$ for all x, that is, $e^x \geq 1 + x$ for all x.

83. $h'(x) = f'(x)g(x) + f(x)g'(x) < 0$ since $f'(x) > 0, g(x) < 0, f(x) < 0$ and $g'(x) > 0$ ·

85. (b)   **87.** (c)

89. (e)   **91.** $b^2 < 3c$

93. $\cos x$ is positive on $\left(0, \dfrac{\pi}{2}\right)$ and $\left(\dfrac{3\pi}{2}, 2\pi\right)$ and negative on $\left(\dfrac{\pi}{2}, \dfrac{3\pi}{2}\right)$. Therefore, $\sin x$ is increasing on $\left(0, \dfrac{\pi}{2}\right)$ and $\left(\dfrac{3\pi}{2}, 2\pi\right)$ and decreasing on $\left(\dfrac{\pi}{2}, \dfrac{3\pi}{2}\right)$.

95. Let $f(m) = 1 - \left(\dfrac{m}{b}\right)^{x-1}$. Notice that $f(0) = 1$ and $f(b) = 0$. Since $f'(m) = \dfrac{1-x}{b}(m/b)^{x-2}$ and $x > 1, f'(m) < 0$ on $(0, b)$. Thus, $f(m)$ decreases from 1 to 0 on $(0, b)$. Let $g(m) = \left(1 - \dfrac{k}{m}\right)(x - 1)$. Since $x > 1$ and $b > k, g(m) > 0$ on (k, b). Notice $g(k) = 0$. Also, $g'(m) = \dfrac{k(x-1)}{m^2} > 0$ on (k, b). Thus, $g(m)$ is increasing on (k, b). The graph of $y = g(m)$ then must intersect the graph of $y = f(m)$ at exactly one point on (k, b).

97. 0.32

99. 5.982

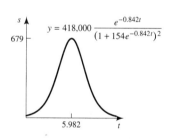

$y = 418,000 \dfrac{e^{-0.842t}}{(1 + 154e^{-0.842t})^2}$

Section 4.2

1. increasing on $(0, 50)$, decreasing on $(50, 100)$

3. 40

5. increasing on $(0, 25)$, decreasing on $(25, 100)$

7. 16,000 pounds, 6,912,000 pounds

9. 41,613, 4116

11. msy = 243, mvp $\approx$ 2.09, cc $\approx$ 12.91

Section 4.3

1. $80x^3 + 2$

3. $2(x + 1)^{-3}$

5. $-(2x + 1)^{-3/2}$

7. $240x^2, 480x$

9. $-6x^{-4}, 24x^{-5}$

11. $\dfrac{8}{27}x^{-7/3}, -\dfrac{56}{81}x^{-10/3}$

13. no inflection value; concave down on $(0, 4)$

15. inflection value at $x = 2$; concave up on $(2, 4)$; concave down on $(0, 2)$

17. inflection value at $x = 2$; concave up on $(2, 4)$; concave down on $(0, 2)$

19. no inflection value; concave up on $(2, 4)$; concave down on $(0, 2)$

21. 2. Concave up on $(2, \infty)$, concave down on $(-\infty, 2)$

23. 1, 3, 5. Concave up on $(1, 3)$ and $(5, \infty)$, concave down on $(-\infty, 1)$ and $(3, 5)$

25. 0, 4. Concave up on $(-\infty, 0)$ and $(4, \infty)$, concave down on $(0, 4)$

27. -2. Concave up on $(-2, \infty)$, concave down on $(-\infty, -2)$

29. $-2, 0, 2$. Concave up on $(-2, 0)$ and $(2, \infty)$, concave down on $(-\infty, -2)$ and $(0, 2)$

31. critical value at $x = 0$
inflection value at $x = 0$
increasing on $(-\infty, \infty)$
decreasing nowhere
concave up on $(0, \infty)$
concave down on $(-\infty, 0)$

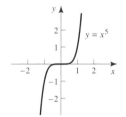

33. critical value at $x = 4$
relative minimum at $x = 4$
no inflection values
increasing on $(4, \infty)$
decreasing on $(-\infty, 4)$
concave up on $(-\infty, \infty)$
concave down nowhere

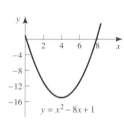

35. critical values at $x = 0, 2$
relative maximum at $x = 0$
relative minimum at $x = 2$
inflection value at $x = 1$
increasing on $(-\infty, 0)$, $(2, \infty)$
decreasing on $(0, 2)$
concave up on $(1, \infty)$
concave down on $(-\infty, 1)$

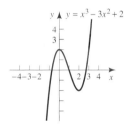

37. critical values at $x = 1, 3$
relative maximum at $x = 3$
relative minimum at $x = 1$
inflection value at $x = 2$
increasing on $(1, 3)$
decreasing on $(-\infty, 1)$, $(3, \infty)$
concave up on $(-\infty, 2)$
concave down on $(2, \infty)$

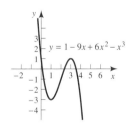

39. critical values at $x = 0, 1$
relative maximum at $x = 0$
relative minimum at $x = 1$
inflection value at $x = \frac{3}{4}$
increasing on $(-\infty, 0)$, $(1, \infty)$
decreasing on $(0, 1)$
concave up on $(\frac{3}{4}, \infty)$
concave down on $(-\infty, \frac{3}{4})$

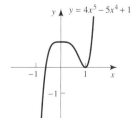

41. no critical values
no inflection values
increasing on $(0, \infty)$
decreasing nowhere
concave up nowhere
concave down on $(0, \infty)$

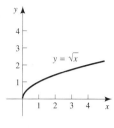

43. no critical values
no inflection values
increasing nowhere
decreasing on $(-\infty, 1)$
concave up nowhere
concave down on $(-\infty, 1)$

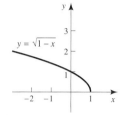

45. critical value at $x = 2$
inflection value at $x = 2$
increasing on $(-\infty, \infty)$
decreasing nowhere
concave up on $(2, \infty)$
concave down on $(-\infty, 2)$

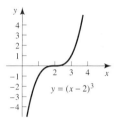

47. critical value at $x = 0$
relative minimum at $x = 0$
no inflection values
increasing on $(0, \infty)$
decreasing on $(-\infty, 0)$
concave up on $(-\infty, \infty)$
concave down nowhere

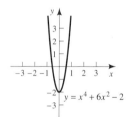

49. critical value at $x = 0$
relative minimum at $x = 0$
no inflection values
increasing on $(0, \infty)$
decreasing on $(-\infty, 0)$
concave up on $(-\infty, \infty)$
concave down nowhere

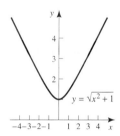

51. critical value at $x = -1$
relative minimum at $x = -1$
inflection value at $x = -2$
increasing on $(-1, \infty)$
decreasing on $(-\infty, -1)$
concave up on $(-2, \infty)$
concave down on $(-\infty, -2)$

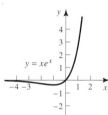

53. critical values at $x = 0, 2$
relative maximum at $x = 2$
inflection values at $x = 0, \frac{4}{3}$
increasing on $(-\infty, 2)$
decreasing on $(2, \infty)$
concave up on $(0, \frac{4}{3})$
concave down on $(-\infty, 0)$, $(\frac{4}{3}, \infty)$

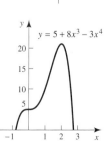

55. critical values at $x = -2, 0$
relative maximum at $x = -2$
relative minimum at $x = 0$
inflection values at $x = -2 \pm \sqrt{2}$
increasing on $(-\infty, -2), (0, \infty)$
decreasing on $(-2, 0)$
concave up on $\left(-\infty, -2 - \sqrt{2}\right)$,
$\left(-2 + \sqrt{2}, \infty\right)$
concave down on $\left(-2 - \sqrt{2}, -2 + \sqrt{2}\right)$

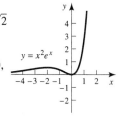

57. Since $f''(x) = 2a$, the sign of $f''(x)$ and a are the same.

59. $f''(x) = 12ax^2 + 2b$

61. Since $f'(x) = 3x^2 - 3b$, the critical values are $x = \pm\sqrt{b}$. Since $f''(x) = 6x$, $f(-\sqrt{b})$ is a relative maximum, and $f(\sqrt{b})$ is a relative minimum.

63. -1, relative minimum

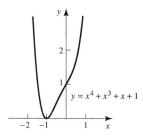

65. -1.221, relative maximum, 0.724, relative minimum

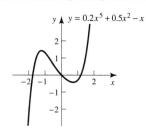

67. -0.384, relative minimum, 0.442, relative maximum, 2.942, relative minimum

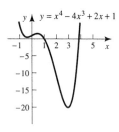

69. -1.108, relative minimum, 0.270, relative maximum, 0.838, relative minimum

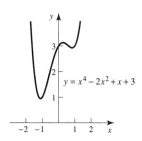

71. 6

73. (a) Increasing on $(0, \infty)$; (b) Concave up on $(0, 10)$, concave down on $(10, \infty)$, inflection point at $(10, 25)$; (c)

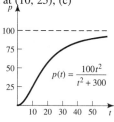

75. increasing on $(0, \infty)$
concave up on $(0, 3)$
concave down on $(3, \infty)$
inflection point at $\left(3, \dfrac{103}{1080}\right)$ point of diminishing returns at $t = 3$

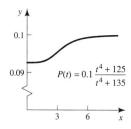

77. 2   **79.** $r = \frac{1}{2}$

81. (a) $y' = -0.175ax^{-1.175}$, $y'' \approx 0.206ax^{-2.175}$. (b) The average amount of needed labor decreases by a factor of 88.6%. Less labor is needed for production since workers become more efficient in performing the repeated manual tasks.

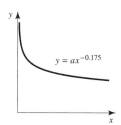

83. (c)   **85.** (d)

87. For any function $g(x)$, if $g'(a)$ exists, then $g(x)$ is continuous at $x = a$. Since $f'' = (f')'$ and $f''(a)$ exists, then $f'(x)$ is continuous at $x = a$. This certainly implies that $f'(a)$ exists. This in turn implies that $f(x)$ is continuous at $x = a$.

89. (a); (b) 4; (c) 2

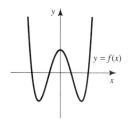

91. (a); (b) yes; (c) odd

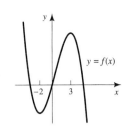

$f'(0) = \lim\limits_{h \to 0} \dfrac{f(h) - f(0)}{h} = 0.$ Since $f'(x) = 0$ for $x < 0$

and $f'(x) = 3x^2$ for $x \ge 0$, we have $\lim\limits_{h \to 0^-} \dfrac{f'(h) - f'(0)}{h} =$

$\lim\limits_{h \to 0^-} \dfrac{0 - 0}{h} = 0,\ \lim\limits_{h \to 0^+} \dfrac{f'(h) - f'(0)}{h} = \lim\limits_{h \to 0^+} \dfrac{3h^2 - 0}{h} =$

$\lim\limits_{h \to 0^+} 3h = 0.$ Thus, $f''(0) = \lim\limits_{h \to 0} \dfrac{f'(h) - f'(0)}{h} = 0.$

93. $g''(x) = \dfrac{-f''(x)[f(x)]^2 + 2f(x)[f'(x)]^2}{[f(x)]^4} < 0$ since $f''(x) > 0$

and $f(x) < 0$ for all x.

97. (a) $-\sin x,\ -\cos x$ **99.** approximately 5.5

95. $\lim\limits_{h \to 0^-} = \dfrac{f(h) - f(0)}{h} = \lim\limits_{h \to 0^-} \dfrac{0 - 0}{h} = 0,$

$\lim\limits_{h \to 0^+} \dfrac{f(h) - f(0)}{h} = \lim\limits_{h \to 0^+} \dfrac{h^3 - 0}{h} = \lim\limits_{h \to 0^+} h^2 = 0.$ Thus,

Section 4.4

1. critical values at $x = 0, -3$
inflection values at $x = 0, -2$

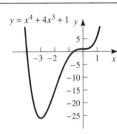

9. no critical values
no inflection values
asymptotes: $x = -2, y = 1$

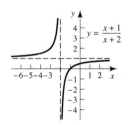

3. critical values at $x = -4, 0, 4$
inflection values at $x = \pm\dfrac{4}{\sqrt{3}}$

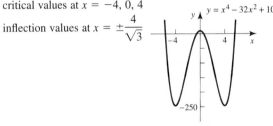

11. critical value at $x = 0$
inflection values at $x = \pm\dfrac{1}{\sqrt{3}}$
asymptote: $y = 0$

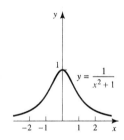

5. critical values at $x = 0, \pm2\sqrt{2}$
inflection values at $x = -2, 0, 2$

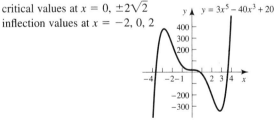

13. critical values at $x = \pm3$
no inflection values
asymptote: $x = 0$

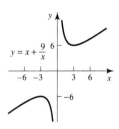

7. no critical values
no inflection values
asymptotes: $x = 1, y = 0$

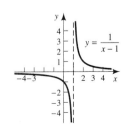

15. critical values at $x = \pm1$
no inflection values
asymptote: $x = 0$

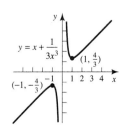

17. critical values at $x = 0, 2$
no inflection values
asymptote: $x = 1$

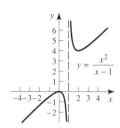

$y = \dfrac{x^2}{x - 1}$

19. critical values at $x = \pm 1$
inflection values at $x = 0, \pm\sqrt{3}$
asymptote: $y = 0$

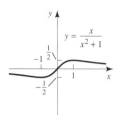

$y = \dfrac{x}{x^2 + 1}$

21. critical value at $x = 2$
inflection value at $x = 3$
asymptotes: $x = 0, y = 0$

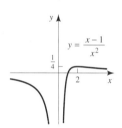

$y = \dfrac{x - 1}{x^2}$

23. critical value at $x = 4$
inflection value at $x = 6$

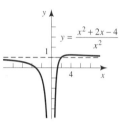

$y = \dfrac{x^2 + 2x - 4}{x^2}$

25. critical values at $x = 0, 1$
inflection value at $x = -\frac{1}{2}$

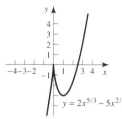

$y = 2x^{5/3} - 5x^{2/3}$

27. critical values at $x = 0, 1$
no inflection values

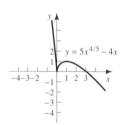

$y = 5x^{4/5} - 4x$

29. critical values at $x = \pm\dfrac{\sqrt{2}}{2}$

inflection values at $x = 0, \pm\sqrt{\dfrac{3}{2}}$

asymptote: $y = 0$

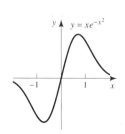

$y = xe^{-x^2}$

31. critical value at $x = 0$
inflection values at $x = \ln(\sqrt{2} \pm 1)$
asymptote: $y = 0$

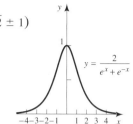

$y = \dfrac{2}{e^x + e^{-x}}$

33. critical value at $x = \sqrt{e}$
inflection value at $x = e^{5/6}$
asymptotes: $x = 0, y = 0$

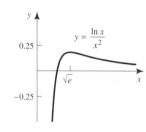

$y = \dfrac{\ln x}{x^2}$

35. no critical value and no inflection
value ($f(0)$ does not exist).
asymptotes: $x = 0, y = 0$

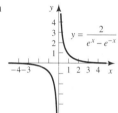

$y = \dfrac{2}{e^x - e^{-x}}$

37. no critical values
no inflection values
asymptote: $x = 100$

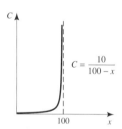

$C = \dfrac{10}{100 - x}$

39. no critical values
no inflection values
asymptote: $n = 9$

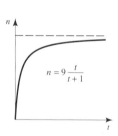

$n = 9\dfrac{t}{t + 1}$

41. critical value at $x = \sqrt{\dfrac{a}{b}}$

asymptote: $x = 0$

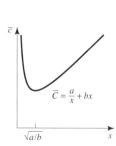

$\bar{C} = \dfrac{a}{x} + bx$

43. $5 \ln 20 \approx 14.98$, 10   **45.** 5.4 years

47. critical value at $x = 0$

inflection values at $x = \pm \dfrac{b}{\sqrt{3}}$

asymptote: $y = 0$

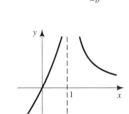

49.

51.

53. (c)   **55.** (a)

57. $P'(t) = \dfrac{akLe^{-kt}}{(1 + ae^{-kt})^2} > 0$ for all $t \geq 0$

59. $P''(t) = ak^2L \dfrac{e^{-kt}(ae^{-kt} - 1)}{(1 + ae^{-kt})^3} = 0$ if and only if $ae^{-kt} - 1$

$= 0$. This implies $t = \dfrac{1}{k} \ln a$. Since $P''(t) > 0$ on

$\left(1, \dfrac{1}{k} \ln a\right)$ and $P''(t) < 0$ on $\left(\dfrac{1}{k} \ln a, \infty\right)$, $c = \dfrac{1}{k} \ln a$ is

an inflection point.

61. Since $e^{-kc} = e^{-\ln a} = \dfrac{1}{a}$, $P(c) = \dfrac{L}{1 + ae^{-kc}} = \dfrac{L}{1 + a/a} = \dfrac{L}{2}$

63. Since $P_0 = \dfrac{100}{1 + a}$, $a = \dfrac{100}{P_0} - 1$. Since $Q = P(T) = $

$\dfrac{100}{1 + ae^{-kT}} = \dfrac{100}{1 + (100/P_0 - 1)e^{-kT}}$, $e^{-kt} = $

$\dfrac{P_0(100 - Q)}{Q(100 - P_0)}$. This implies $k = -\dfrac{1}{T} \ln \dfrac{P_0(100 - Q)}{Q(100 - P_0)} = $

$\dfrac{1}{T} \ln \dfrac{Q(100 - P_0)}{P_0(100 - Q)}$

65. (a) $f'(t) = ake^{-kt} > 0$ for all $t \geq 0$. (b) $\lim_{t \to \infty} f(t) = $

$L - \lim_{t \to \infty} ae^{-kt} = L - 0 = L$. (c) $f''(t) = -ak^2e^{-kt} < 0$

for all $t \geq 0$

67.

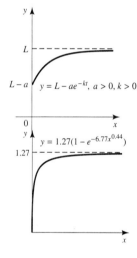

Section 4.5

1. absolute minimum in June
absolute maximum in November

3. absolute minimum in September
absolute maximum in April

5. absolute minimum 1934, absolute maximum in 1910

7. (a) absolute minimum at $x = 2$, absolute maximum at $x = 4$; (b) absolute minimum at $x = 1$, absolute maximum at $x = 2$; (c) absolute minimum at $x = -2, 1$, absolute maximum at $x = -1, 2$

9. (a) absolute minimum at $x = 0$, absolute maximum at $x = -2$; (b) absolute minimum at $x = 2$, absolute maximum at $x = 1$; (c) absolute minimum at $x = 2$, absolute maximum at $x = -2$

11. absolute minimum at $x = 1$, absolute maximum at $x = -1$

13. absolute minimum at $x = 0$, absolute maximum at $x = 2$

15. absolute minimum at $x = 1$, absolute maximum at $x = 3$

17. absolute minimum at $x = -3, 0$, absolute maximum at $x = 2$

19. absolute minimum at $x = -2, 2$, absolute maximum at $x = -3, 3$

21. absolute minimum at $x = 2$, absolute maximum at $x = 8$

23. absolute minimum at $x = \frac{1}{2}$, no absolute maximum

25. absolute minimum at $x = 0$, no absolute maximum

27. absolute minimum at $x = 0$, no absolute maximum

29. (a) absolute minimum at $x = 1$, absolute maximum at $x = \frac{1}{2}$; (b) no absolute minimum, no absolute maximum; (c) absolute minimum at $x = 1$, no absolute maximum; (d) no absolute minimum, absolute maximum at $x = -1$

31. 8, 8   **33.** 10, 10

35. 5, 20

37. 10, −10

39. $2a^2$

41. $\left(\dfrac{c}{2a}, \dfrac{c}{2b}\right)$

43. 200 × 200

45. $x = 1.5$

47. 10 × 20

49. 100 × 200

51. 40

53. $450

55. 4 × 4 × 2, 48

57. 6 × 6 × 9

59. 4 runs of 4000 each

61. C is $10\sqrt{3}$ feet to the right of D.

63. Cut at $\dfrac{48}{3\sqrt{3} + 4} \approx 5.22$, and use the shorter piece for the square.

65. $\dfrac{500}{9 + \sqrt{3}} \approx 46.6$ yards

Section 4.6

1. −2

3. $-\frac{3}{5}$

5. $\frac{9}{2}$

7. $-\frac{2}{7}$

9. −2

11. $-\frac{1}{2}$

13. $\dfrac{16}{31}$

15. $-\frac{1}{2}$

17. −3

19. $-\dfrac{15}{7}$

21. $-\frac{3}{4}$

23. $-\frac{5}{4}$

25. $-\dfrac{\sqrt[4]{al}}{4}$

27. 6

29. 6

31. 16

33. 12

35. −1

37. −1

39. −2

41. −200

43. 400

45. 1.6 miles per hour

47. −3 feet per second

49. $\dfrac{125}{16\pi}$ yards per minute

51. 40 square feet per second

53. Since $f(g(x)) = x$, $\dfrac{df[g(x)]}{dx} = 1$ and $f'(g(x)) \cdot g'(x) = 1$. When $x = c$, $g(c) = y_0$, we have $f'(g(c)) \cdot g'(c) = 1$. Then $f'(y_0) \cdot g'(c) = 1$ and $f'(y_0) = \dfrac{1}{g'(c)}$ since $g'(c) \neq 0$

55. $f(g(x)) = \sqrt[3]{g(x)} = \sqrt[3]{x^3} = x$, $c = 2$, $y_0 = g(c) = c^3 = 8$. Then $f'(y_0) = \dfrac{1}{3} y_0^{-2/3} = \dfrac{1}{12}$ and $g'(c) = 3c^2 = 12$. Therefore $f'(y_0) = \dfrac{1}{g'(c)}$

Section 4.7

1. (1) $f(1) = -1 < 0$, $f(2) = 7 > 0$; (2) $f'(x) = 3x^2 + 1 > 0$ on $[1, 2]$ (3) 1.2478

3. (1) $f(1) = -1 < 0$, $f(2) = 15 > 0$; (2) $f'(x) = 4x^3 + 1 > 0$ on $[1, 2]$ (3) 1.1654

5. (1) $f(1) = 6 > 0$, $f(2) = -2 < 0$; (2) $f'(x) = -(1 + 3x^2) < 0$ on $[1, 2]$ (3) 1.8338

7. (1) $f(1) = -3 < 0$, $f(2) = 14 > 0$; (2) $f'(x) = 12x^2(x - 1) > 0$ on $(1, 2]$ (3) 1.5628

9. 1.7321

11. 2.2240

13. 1.1225

15. −0.5627, 0.5627

17. 0.1206, 2.3473, 3.5321

19. 0.5671

21. 1.3098

23. −0.4

25. −1.1813, 0.7415

27. −1.879385, 0.347296, 1.532089

29. 2.074341

31. Exercise 15: 0.562694 and −0.562694; Exercise 17: 0.120615, 2.347296, 3.532089

33. $x_n \to -\infty$

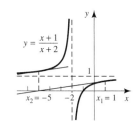

$y = \dfrac{x + 1}{x + 2}$

$x_2 = -5$ −2 $x_1 = 1$ x

35. Notice that $x_2 \approx -8.7$. Then Newton's method converges to the root $x \approx -3.98$.

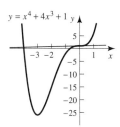

$y = x^4 + 4x^3 + 1$

37. 0.7391

Chapter 4 Review

1. critical values at 3, 5, 7, 11, 13, 15; increasing on (3, 7), (11, 13), (15, 18); decreasing on (1, 3), (7, 11), (13, 15)

2. critical values at −2, 1; increasing on $(-\infty, -2)$, $(1, \infty)$; decreasing on (−2, 1); relative maximum at $x = -2$; relative minimum at $x = 1$

3. critical values at −4, 2, 5; increasing on $(-\infty, -4)$, $(5, \infty)$; decreasing on (−4, 5); relative maximum at $x = -4$; relative minimum at $x = 5$

4. critical value at $x = -\frac{3}{4}$
relative minimum at $x = -\frac{3}{4}$
increasing on $(-0.75, \infty)$
decreasing on $(-\infty, -0.75)$
concave up on $(-\infty, \infty)$
concave down nowhere

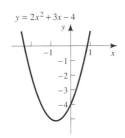

5. critical value at $x = 1$
relative maximum at $x = 1$
increasing on $(-\infty, 1)$
decreasing on $(1, \infty)$
concave up nowhere
concave down $(-\infty, \infty)$

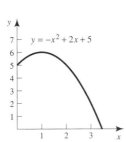

6. critical value at $x = -1$
relative maximum at $x = -1$
increasing on $(-\infty, -1)$
decreasing on $(-1, \infty)$
concave up nowhere
concave down $(-\infty, \infty)$

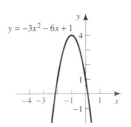

7. critical value at $x = -2$
relative minimum at $x = -2$
increasing on $(-2, \infty)$
decreasing on $(-\infty, -2)$
concave up on $(-\infty, \infty)$
concave down nowhere

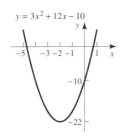

8. critical values at $x = -1, 1$
relative maximum at $x = -1$
relative minimum at $x = 1$
increasing on $(-\infty, -1)$, $(1, \infty)$
decreasing on (−1, 1)
concave up on $(0, \infty)$
concave down on $(-\infty, 0)$
inflection value at $x = 0$

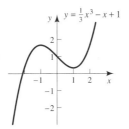

9. critical values at $x = -1, 1$
relative maximum at $x = 1$
relative minimum at $x = -1$
increasing on (−1, 1)
decreasing on $(-\infty, -1)$, $(1, \infty)$
concave up on $(-\infty, 0)$
concave down on $(0, \infty)$
inflection value at $x = 0$

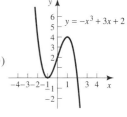

10. critical values at $x = 0, 3$
relative minimum at $x = 3$
increasing on $(3, \infty)$
decreasing on $(-\infty, 3)$
concave up on $(-\infty, 0)$, $(2, \infty)$
concave down on (0, 2)
inflection values at $x = 0, 2$

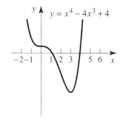

11. critical values at $x = 0, 12$
relative maximum at $x = 0$
relative minimum at $x = 12$
increasing on $(-\infty, 0)$, $(12, \infty)$
decreasing on (0, 12)
concave up on $(9, \infty)$
concave down on $(-\infty, 9)$
inflection value at $x = 9$

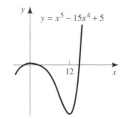

12. critical values at $x = -3, 0, 3$
relative maximum at $x = 0$
relative minimum at $x = -3, 3$
increasing on (−3, 0), $(3, \infty)$
decreasing on $(-\infty, -3)$, (0, 3)
concave up on $\left(-\infty, -\sqrt{3}\right)$, $\left(\sqrt{3}, \infty\right)$
concave down on $\left(-\sqrt{3}, \sqrt{3}\right)$
inflection values at $x = \pm\sqrt{3}$

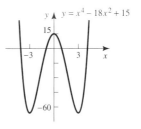

13. critical values at $x = -3, 0, 3$
relative maximum at $x = 3$
relative minimum at $x = -3$
increasing on (−3, 3)
decreasing on $(-\infty, -3)$, $(3, \infty)$
concave up on $\left(-\infty, -\dfrac{3}{\sqrt{2}}\right)$, $\left(0, \dfrac{3}{\sqrt{2}}\right)$
concave down on $\left(-\dfrac{3}{\sqrt{2}}, 0\right)$, $\left(\dfrac{3}{\sqrt{2}}, \infty\right)$
inflection values at $x = 0, \pm\dfrac{3}{\sqrt{2}}$

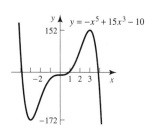

$y = -x^5 + 15x^3 - 10$

20. critical value at $x = 0$
relative minimum at $x = 0$
increasing on $(0, \infty)$
decreasing on $(-\infty, 0)$
concave up on $(-\infty, \infty)$
concave down nowhere
no inflection values

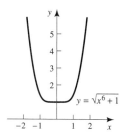

$y = \sqrt{x^6 + 1}$

14. critical value at $x = 2$
relative minimum at $x = 2$
increasing on $(-\infty, 0), (2, \infty)$
decreasing on $(0, 2)$
concave up on $(-\infty, 0), (0, \infty)$
concave down nowhere
no inflection values
asymptote at $x = 0$

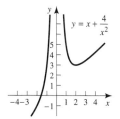

$y = x + \dfrac{4}{x^2}$

21. critical value at $x = 9$
increasing nowhere
decreasing on $(-\infty, \infty)$
concave up on $(9, \infty)$
concave down on $(-\infty, 9)$
inflection value at $x = 9$

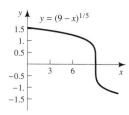

$y = (9 - x)^{1/5}$

15. critical values at $x = -1, 1$
relative minimum at $x = -1, 1$
increasing on $(-1, 0), (1, \infty)$
decreasing on $(-\infty, -1), (0, 1)$
concave up on $(-\infty, 0), (0, \infty)$
concave down nowhere
no inflection values
asymptote at $x = 0$

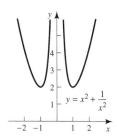

$y = x^2 + \dfrac{1}{x^2}$

22. critical value at $x = 0$
relative maximum at $x = 0$
increasing on $(-\infty, 0)$
decreasing on $(0, \infty)$
concave up on $\left(-\infty, -\dfrac{1}{\sqrt{2}}\right), \left(\dfrac{1}{\sqrt{2}}, \infty\right)$
concave down on $\left(-\dfrac{1}{\sqrt{2}}, \dfrac{1}{\sqrt{2}}\right)$
inflection values at $x = \pm\dfrac{1}{\sqrt{2}}$
asymptote at $y = 0$

16. critical value at $x = -\sqrt[3]{2}$
relative maximum at $x = -\sqrt[3]{2}$
increasing on $\left(-\infty, -\sqrt[3]{2}\right), (0, \infty)$
decreasing on $\left(-\sqrt[3]{2}, 0\right)$
concave up nowhere
concave down on $(-\infty, 0), (0, \infty)$
no inflection values
asymptote at $x = 0$

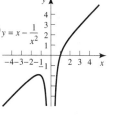

$y = x - \dfrac{1}{x^2}$

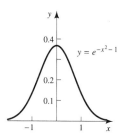

$y = e^{-x^2 - 1}$

17. no critical values
increasing on $(0, \infty)$
decreasing on $(-\infty, 0)$
concave up on $(-\infty, -3), (3, \infty)$
concave down on $(-3, 0), (0, 3)$
inflection values at $x = \pm 3$
asymptote at $x = 0$

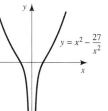

$y = x^2 - \dfrac{27}{x^2}$

23. critical value at $x = 0$
relative minimum at $x = 0$
increasing on $(0, \infty)$
decreasing on $(-\infty, 0)$
concave up on $\left(-\dfrac{1}{\sqrt{2}}, \dfrac{1}{\sqrt{2}}\right)$
concave down on $\left(-\infty, -\dfrac{1}{\sqrt{2}}\right), \left(\dfrac{1}{\sqrt{2}}, \infty\right)$
inflection values at $x = \pm\dfrac{1}{\sqrt{2}}$
asymptote at $y = 1$

18. no critical values
increasing nowhere
decreasing on $(-\infty, 5), (5, \infty)$
concave up on $(5, \infty)$
concave down on $(-\infty, 5)$
no inflection values
asymptotes at $x = 5, y = 1$

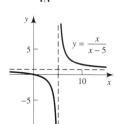

$y = \dfrac{x}{x - 5}$

19. no critical values
increasing nowhere
decreasing on $(-\infty, 5), (5, \infty)$
concave up on $(5, \infty)$
concave down on $(-\infty, 5)$
no inflection values
asymptotes: $x = 5, y = 1$

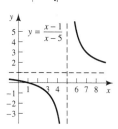

$y = \dfrac{x - 1}{x - 5}$

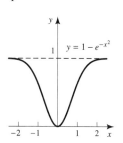

$y = 1 - e^{-x^2}$

24. critical value at $x = 0$
relative minimum at $x = 0$
increasing on $(0, \infty)$
decreasing on $(-\infty, 0)$
concave up on $(-1, 1)$
concave down on $(-\infty, -1)$, $(1, \infty)$
inflection values at $x = \pm 1$

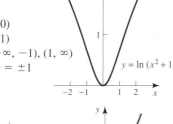

$y = \ln(x^2 + 1)$

25. no critical values
increasing on $(-\infty, \infty)$
decreasing nowhere
concave up on $(-\infty, \infty)$
concave down nowhere
no inflection values
asymptotes at $y = 0$

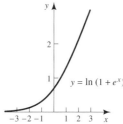

$y = \ln(1 + e^x)$

26. absolute maximum at $x = 1$; absolute minimum at $x = -1$

27. absolute maximum at $x = -2$; absolute minimum at $x = 2$

28. no absolute maximum; absolute minimum at $x = 1$

29. absolute maximum at $x = 1$; no absolute minimum

30. absolute maximum at $x = 2$; absolute minimum at $x = -1$

31. absolute maximum at $x = 3$; absolute minimum at $x = 0$

32. absolute maximum at $x = 0$; absolute minimum at $x = 2$

33. absolute maximum at $x = 1$; absolute minimum at $x = 2$

34. absolute maximum at $x = 1$; absolute minimum at $x = 2$

35. absolute maximum at $x = 4$; absolute minimum at $x = 1$

36. $-\dfrac{11}{30}$

37. $-\dfrac{5}{16}$

38. -1

39. 0

40. $f''(x) = 12ax^2 + 2b$. $f'' > 0$ if $a, b > 0$ and $f'' < 0$ if $a, b < 0$

41. relative minimum at $x = a$; relative maximum at $x = b$

42. $g''(x) = e^{f(x)}[(f'(x))^2 + f''(x)] \geq 0$ since $e^{f(x)} > 0$ and $f''(x) > 0$ for all x

43. $g''(x) = \dfrac{f''(x)f(x) - [f'(x)]^2}{[f(x)]^2} < 0$ since $f > 0$ and $f'' < 0$ for all x

44. increasing on $\left(\sqrt{\dfrac{a}{b}}, \infty\right)$; decreasing on $\left(0, \sqrt{\dfrac{a}{b}}\right)$

45. 10

46. Solving $\overline{C} = C'$ for x we obtain $x = 10$, which is the value where the average cost is minimized.

47. $\overline{C}$ is minimized at the value where $\overline{C}' = 0$, that is $\overline{C}' = \dfrac{C'(x) \cdot x - C(x)}{x^2} = 0$. This implies $C'(x) = \dfrac{C(x)}{x} = \overline{C}(x)$

48. $\dfrac{N}{k} \cdot \dfrac{1-a}{2-a}$

49. $\sqrt[3]{\dfrac{a}{2b}}$

50. Build only a square enclosure.

51. 70

52. 72

53. $r = \dfrac{3}{\sqrt[3]{\pi}}$, $h = \dfrac{12}{\sqrt[3]{\pi}}$

54. 1 run of 1000 cases each

55. When $P'(x) = 0$, $x = \dfrac{1}{|p'(x)|}(p(x) - a)$. Thus, increasing a decreases x. 180

56. 180

57. $\dfrac{\sqrt{6}}{3} \approx 0.82$ feet per second

58. 0, 70.13, everywhere

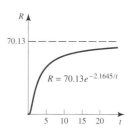

$R = 70.13e^{-2.1645/t}$

59. everywhere

60. $bk \leq a$

61. (b)

62. (f)

63. (a)

64. (e)

65. (d)

66. (c)

67. $f(x)$ has a relative minimum at $x = 1$, since f decreases to the left of 1 and increases to the right of 1.

68. $f(x)$ has a relative maximum at $x = 1$, since f increases to the left of 1 and decreases to the right of 1.

69. $f(x)$ has a relative maximum at $x = 1$, since the graph is like that shown in Exercise 68

70. $f(x)$ has a relative minimum at $x = 1$ since the graph is like that shown in Exercise 67

71. $f(x)$ has a relative minimum since $f''(1) > 0$

72. $f(x)$ has a relative maximum since $f''(1) < 0$

CHAPTER 5

Section 5.1

1. $\dfrac{x^{100}}{100} + C$

3. $\dfrac{-x^{-98}}{98} + C$

9. $\dfrac{\sqrt{2}}{5}y^{5/2} + C$

11. $\frac{3}{5}u^{5/3} + C$

5. $5x + C$

7. $-\frac{5}{2}y^{-2} + C$

13. $2x^3 + 2x^2 + C$

15. $\dfrac{x^3}{3} + \dfrac{x^2}{2} + x + C$

17. $\dfrac{10\sqrt{2}}{11} u^{1.1} - \dfrac{1}{21} u^{2.1} + C$

19. $\dfrac{t^3}{3} - t + C$

21. $\frac{2}{3}t^{3/2} - \frac{3}{8}t^{8/3} + C$

23. $t^6 - t^4 + t + C$

25. $x + 3 \ln|x| + C$

27. $\pi x + \ln|x| + C$

29. $\frac{2}{3}t^{3/2} + 2t^{1/2} + C$

31. $e^x - \frac{3}{2}x^2 + C$

33. $5e^x + C$

35. $5e^x - 4x + C$

37. $x + \ln|x| + C$

39. $R(x) = 30x - 0.25x^2$

41. $p(x) = 2x^{-1/2} + 1$

43. $C(x) = 100x - 0.1e^x + 1000.1$

45. $C(x) = -2x^3 + 20x^{3/2}$ **47.** $-16t^2 + 30t + 15$

49. $-16t^2 + 10t + 6$

51. approximately \$3,200,000

53. $x - 0.25x^2 + 0.2$

55. $0.28x + 0.025x^2 - \dfrac{1}{30000}x^3$

57. $\frac{1}{2}e^{2x} + C$

59. $\frac{1}{3}\sin 3x + C$

Section 5.2

1. $\dfrac{2}{11}(3 + 1)^{11} + C$

3. $-\dfrac{1}{16}(3 - x^2)^8 + C$

5. $\frac{4}{5}(2x^2 + 4x - 1)^{5/2} + C$

7. $\dfrac{(3x^4 + 4x^3 + 6x^2 + 12x + 1)^4}{48} + C$

9. $\frac{4}{3}(x + 1)^{3/2} + C$

11. $\frac{1}{3}(x^2 + 1)^{3/2} + C$

13. $\frac{3}{4}(x^2 + 1)^{2/3} + C$

15. $2(x^{1/3} + 1)^{3/2} + C$

17. $\frac{2}{3}(\ln x)^{3/2} + C$

19. $-e^{1-x} + C$

21. $-\frac{1}{2}e^{1-x^2} + C$

23. $2e^{\sqrt{x}} + C$

25. $\frac{1}{3}\ln|3x + 5| + C$

27. $-\frac{1}{6}(x^2 + 3)^{-3} + C$

29. $-\ln(e^{-x} + 1) + C$

31. $\frac{1}{2}\ln(e^{2x} + e^{-2x}) + C$

33. $\frac{1}{2}\ln|\ln|x|| + C$

35. $p(x) = \dfrac{4}{3x^2 + 1} + 9$

37. $R(x) = 20x^2 - x^4$

39. $C(x) = 2\ln(x^2 + 1) + 1000$

41. 6 tons

43. $20{,}000e^{2.2} + 80{,}000 \approx 260{,}500$

45. $\dfrac{\ln 7}{2} \approx 0.97$ **47.** \$16,703.20

49. Let $u = x^5 + x^4 + 1$, then $du = x(5x^3 + 4x^2)\,dx$ and $\displaystyle\int (x^5 + x^4 + 1)^2(5x^3 + 4x^2)\,dx = \int u^2 \cdot \dfrac{du}{x}$. There is no way to eliminate the factor $\dfrac{1}{x}$, thus the method fails.

51. $-\cos x^2 + C$ **53.** $\frac{1}{3}\sin x^3 + C$

Section 5.3

1. (a); (b) 5.2, 6.8, 1.6

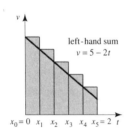

left-hand sum
$v = 5 - 2t$

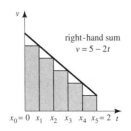

right-hand sum
$v = 5 - 2t$

3. (a); (b) 3.52, 1.92, 1.6

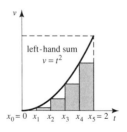

left-hand sum
$v = t^2$

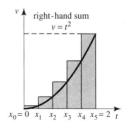

right-hand sum
$v = t^2$

5. (a); (b) 10.32, 7.12, 3.2

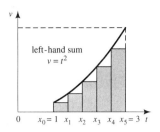

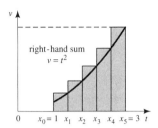

7. (a); (b) 20, 15, 5

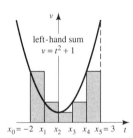

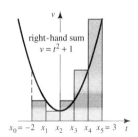

9. (a); (b) 8.38, 6.15, 2.23

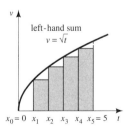

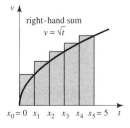

11. 16

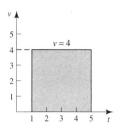

13. 0.5

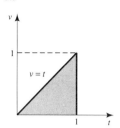

15. 6

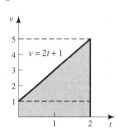

17. 4   **19.** 1

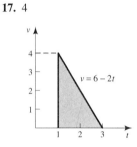
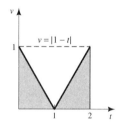

21. t^2

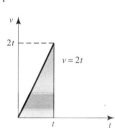

23. Use rectangles with base of length 1 unit and obtain lower bound of 15 and upper bound of 24.

25. 135, 95   **27.** 0.01, 0.001

29. 10, 100

31. The first runner is ahead after 1 minute since the first runner has a larger velocity during every moment of the first minute. The second runner is ahead after 5 minutes. The second runner gets behind the first runner at the 1-minute mark a distance equal to the area A_1 between the two curves on [0, 1]. From the 1-minute mark to the end of the fifth minute the second runner gains a distance equal to the area A_2 between the two curves on [1, 5]. Notice that $A_2 > A_1$. Thus, at the end of the fifth minute the second runner is ahead of the first runner by a distance equal to $A_2 - A_1$.

Section 5.4

1.

n	Left-hand sum	Right-hand sum
10	0.610509	0.710509
100	0.661463	0.671463
1000	0.666160	0.667160

3.

n	Left-hand sum	Right-hand sum
10	0.202500	0.302500
100	0.245025	0.255025
1000	0.249500	0.250500

5.

n	Left-hand sum	Right-hand sum
10	4.920000	5.720000
100	5.293200	5.373200
1000	5.329332	5.337332

7.

n	Right-hand sum	Left-hand sum
10	1.034896	1.168229
100	1.091975	1.105309
1000	1.097946	1.099279

9.

n	Left-hand sum	Right-hand sum
10	1.633799	1.805628
100	1.709705	1.726888
1000	1.717423	1.719141

11.

n	Left-hand sum	Right-hand sum
10	0.351221	0.420535
100	0.382824	0.389756
1000	0.385948	0.386641

13. $n = 10$, left-hand sum gives lower estimate of 0.687405, right-hand sum gives upper estimate of 0.787405

15. $n = 160$, left-hand sum gives lower estimate of 3.950156, right-hand sum gives upper estimate of 4.050156

17. $n = 80$, left-hand sum gives lower estimate of 5.283125, right-hand sum gives upper estimate of 5.383125

19.

n	Left-hand sum	Right-hand sum
10	10.560000	10.560000
100	10.665600	10.665600
1000	10.666656	10.666656

21.

n	Left-hand sum	Right-hand sum
10	2.622043	2.700019
100	2.636816	2.644614
1000	2.639529	2.640309

23. 20   **25.** 6

27. $\frac{9}{2}$ **29.** $\frac{\pi}{2}$

31. $\frac{\pi}{2}$

33. left-hand sun is 0.747140; right-hand sum is 0.746508; left-hand sum must be greater since function is decreasing on $[0, 1]$

35. no difference

37. MSUM is 1.09861214. Correct answer to 8 decimal places is 1.09861229. Left-hand sum is 1.09927925. Right-hand sum is 1.09794592.

39. $N = 22$

Section 5.5

1. 15   **3.** 0

5. 2   **7.** 14

9. $\frac{11}{8}$ **11.** $\frac{(e^{-2} - e^{-4})}{2}$

13. $\frac{1}{2} \ln 3$ **15.** 0

17. $-\frac{1}{16}$ **19.** $\frac{26}{3}$

21. 0   **23.** $\frac{1}{2} \ln \frac{5}{3}$

25. 6   **27.** 5

29. 0   **31.** 0

33. $\frac{1}{3}$ **35.** (a) 5; (b) 7

37. Lower estimate is 13, upper estimate is 21. Take $\Delta t = 1$.

39. 948, 928

41. $2(e^2 - e) \approx 9.34$ million barrels

43. 71   **45.** about 11 years

47. about 5281 gal

49. (a) $V(t) = \int -k(T - t)\, dt = -kT \int dt + k \int t\, dt = -kTt + \frac{k}{2} t^2 + C.$ $I = V(0)$ implies $C = I$. Thus, $V(T) = -kT^2 + \frac{k}{2} T^2 + I = I - \frac{kT^2}{2}$. (b) Solve $V(t) = \frac{I}{2}$ for t: $\frac{k}{2} t^2 - kTt + I = \frac{I}{2}$ implies $t^2 - 2Tt + \frac{I}{k} = 0$ implies $t =$

$$\frac{2T \pm \sqrt{4T^2 - 4l/k}}{2} = T \pm \sqrt{T^2 - \frac{l}{k}}. \text{ Since } T > t > 0,$$

the solution is $t = T - \sqrt{T^2 - \dfrac{l}{k}}$

51. $\dfrac{74}{3}$ thousands   **53.** $3194.53

55. $8 million   **57.** About 10.23

59. $101\frac{1}{3}$

61. (a) First store. Revenue from the first store during the first year is $R(1) - R(0) = \int_0^1 R_1'(t)\, dt$, which is also the area between the graph of the curve $y = R_1'(t)$ and the x-axis on the interval $[0, 1]$. This is greater than the area under the graph of $y = R_2$ and the x-axis on $[0, 1]$.

(b) Second store. Sales for the second store less sales of the first store during the three-week period is $[R_2(3) - R_2(0)] - [R_1(3) - R_1(0)]$ equals

$$\int_0^3 R_2'(t)\, dt - \int_0^3 R_1'(t)\, dt =$$

$$\int_1^3 [R_2'(t) - R_1'(t)]\, dt$$

$$- \int_0^1 [R_1'(t) - R_2'(t)]\, dt$$

The first integral on the right-hand side of the preceding equation is the area under the graph of $y = R_2'(t)$ and above the graph of $y = R_1'(t)$ on $[1, 3]$ and is greater than the second integral, which is the area under the graph of $y = R_1'(t)$ and above the graph of $y = R_2'(t)$ on $[0, 1]$.

63. $\int_b^a f(x)\, dx$ is a real number. $\int f(x)\, dx$ is a function of x. For example: $\int_0^1 x\, dx = 0.5$ and $\int x\, dx = \frac{1}{2}x^2 + C$

65. $\dfrac{a}{t}\left(\dfrac{1}{m} - \dfrac{1}{m + tb}\right)$

67.

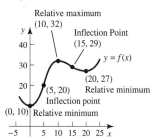

Section 5.6

1. 3.75

3. $10e - 20$

5. $\dfrac{1}{6}$

7. $\frac{1}{2}(e^4 - e^2) - 1$

9. $\frac{9}{2}$

11. $\frac{1}{2}$

13. $5\frac{1}{3}$

15. $\frac{8}{3}$

17. $\dfrac{125}{6}$

19. $\dfrac{125}{6}$

21. $e^2 + e - 2$

23. 16

25. 1.5

27. $\frac{1}{2}$

29. 1

31. 8

33. $\dfrac{23}{2}$

35. 20.75

37. x-coordinates of intercepts are $x = -1.53$, $x = 0$, $x = 1.28$. Upper estimate is 5.5821; lower estimate is 5.5815.

39. x-coordinates of intercepts are $x = -1.80$, $x = 0$, $x = 1.19$. Lower estimate is 5.7052; upper estimate is 5.7069.

41. $\frac{1}{3}$

43. $\dfrac{1}{12}$

45. 0.1

47. $\dfrac{1}{300}$

49. $4.8 billion

51. $\dfrac{26}{3}$

53. 2

55. The coefficient of inequality has become larger.

57. 15

Section 5.7

1. (a) $100; **(b)** $200(1 - e^{-0.50}) = $78.69; **(c)** $200(e^{0.50} - 1) = $129.74

3. (a) $2000(e - 1) = $3436.56; **(b)** $2000; **(c)** $2000e = $5436.56

5. (a) $400(1 - e^{-0.50}) = $157.39; **(b)** $\dfrac{$2000(1 - e^{-1.5})}{15} = $103.58; **(c)** $\dfrac{$2000(e - e^{-0.50})}{15} = $281.57

7. (a) $50; **(b)** $26.42; **(c)** $71.83

9. (a) $333.33; (b) $160.60; (c) $436.56

11. (a) yes, since $P_V(10) = \$948,180.84$; (b) no, since $P_V(10) = \$776,869.84$

13. (a) yes, since $P_V(5) \approx \$316$ thousand; (b) no, since $P_V(5) \approx \$303$ thousand

15. $\dfrac{128}{3}$ **17.** $0.9 - 0.1 \ln 10$

19. $\frac{1}{2}$ **21.** $\frac{2}{9}$

23. 8, 16, where equilibrium point is (4, 8)

25. $\dfrac{128}{3}$, 8, where equilibrium point is (4, 4)

27. 34,000   **29.** 3000

Chapter 5 Review

1. $\dfrac{x^{10}}{10} + C$ **2.** $\frac{2}{3}x^{3/2} + C$

3. $-2y^{-1} + C$ **4.** $\dfrac{\sqrt{3}}{3}\, y^3 + C$

5. $2x^3 + 4x^2 + C$ **6.** $x^2 - 3 \ln|x| + C$

7. $\frac{2}{5}t^{5/2} + 2t^{1/2} + C$ **8.** $\dfrac{e^{3x}}{3} + C$

9. $-e^{-x} - \dfrac{1}{5}e^{-5x} + C$ **10.** $\dfrac{2}{21}(2x+1)^{21} + C$

11. $(2x^3 + 1)^{10} + C$ **12.** $\frac{2}{3}(x^3+1)^{1/2} + C$

13. $0.5 \ln|x^2 + 2x| + C$ **14.** $\dfrac{e^{x^2+5}}{2} + C$

15. $-\frac{5}{7}(x^2+1)^{-7} + C$ **16.** $-2e^{-\sqrt{x}} + C$

17. $\ln(e^x + 1) + C$ **18.** $\dfrac{(\ln x)^4}{4} + C$

19. $n = 10$: 1.77, 1.57; $n = 100$: 1.6767, 1.6567; $n = 1000$: 1.667667, 1.665667; $\int_0^1 (2x^2 + 1)\, dx = \frac{5}{3}$

20.

n	Left-hand sum	Right-hand sum
10	1.2025	1.3025
100	1.2450	1.2550
1000	1.2495	1.2505

$\int_0^1 (x^3 + 1)\, dx = 1.25$

21. $\dfrac{9\pi}{4}$

22. 1.283. The area of the rectangles is the left-hand sum and is greater than the area under the curve.

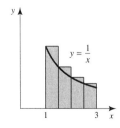

23. 0.95. The area of the rectangles is the right-hand sum and is less than the area under the curve.

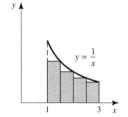

24. 52   **25.** -8

26. 8   **27.** $\dfrac{e^4 - e^2}{2} + \ln 2$

28. $\dfrac{2^{10} - 1}{5}$ **29.** $\dfrac{e^9 - e}{3}$

30. $\dfrac{\ln 5}{2}$ **31.** $3 - \sqrt{3}$

32. $\frac{1}{2}\int_0^2 e^{t^4}\, dt$ **33.** 0

34. $0.50 \ln 3$ **35.** 9

36. $\frac{8}{3}$ **37.** $\frac{8}{3}$

38. 1   **39.** $\dfrac{20}{3}$

40. 72   **41.** $\dfrac{32}{3}$

42. 9   **43.** $\frac{2}{3}$

44. 13   **45.** $4 \ln 2 - 2 \ln 3$

46. $\dfrac{38}{3}$ **47.** $x^2 - 10x$

48. $x^3 + x + 1000$ **49.** $98,419.70

50. 1.96 if $s(0) = 0$ **51.** 32, 26

52. 6906   **53.** 10.82

54. 17,183   **55.** $x_0 = 2$, $p_0 = 6$

56. $\dfrac{16}{3}$ **57.** 6

58. $L = \frac{3}{5}$

59. $\$250(1 - e^{-0.4}) = \$82.42$

60. $\dfrac{\$250}{3}(1 - e^{-1.2}) = \$58.23$

61. $\dfrac{\$250}{3}(e^{0.8} - e^{-0.4}) = \$129.59$

62. $\dfrac{\$5000}{9}(1 - e^{-0.9}) = \$329.68$ thousand

CHAPTER 6

Section 6.1

1. $xe^{2x} - \frac{1}{2}e^{2x} + C$

3. $\frac{3}{4}e^4 + \frac{1}{4}$

5. 2

7. $2x(1 + x)^{1/2} - \frac{4}{3}(1 + x)^{3/2} + C$

9. $\frac{x}{11}(1 + x)^{11} - \frac{1}{132}(1 + x)^{12} + C$

11. $-\frac{x}{x + 2} + \ln|x + 2| + C$

13. $\frac{2}{9}x^3(1 + x^3)^{3/2} - \frac{4}{45}(1 + x^3)^{5/2} + C$

15. $\frac{x^2}{22}(1 + x^2)^{11} - \frac{1}{264}(1 + x^2)^{12} + C$

17. $\frac{1}{2}x^2e^{2x} - \frac{1}{2}xe^{2x} + \frac{1}{4}e^{2x} + C$

19. $\frac{3}{2}e^4 + \frac{1}{2}$ **21.** 111

23. \$26,306.67   **25.** \$36.47

27. $\frac{1000}{e - 1} \approx 582$

29. Let $u = (\ln x)^n$ and $dv = dx$

31. Let $u = f(x)$ and $dv = dx$

Section 6.2

1. $\frac{2}{9}\left[\frac{(3x + 2)^{5/2}}{5} - \frac{2(3x + 2)^{3/2}}{3}\right] + C$

3. $\frac{x}{2}\sqrt{x^2 - 9} - \frac{9}{2}\ln|x + \sqrt{x^2 - 9}| + C$

5. $\ln\left|\frac{\sqrt{1 - x} - 1}{\sqrt{1 - x} + 1}\right| + C$ **7.** $\frac{1}{3}\ln\left|\frac{2x + 1}{2x + 4}\right| + C$

9. $\frac{1}{3}(x^2 - 9)^{3/2} + C$ **11.** $\frac{1}{3x} + \frac{1}{9}\ln\left|\frac{x - 3}{x}\right| + C$

13. $-\frac{x}{\sqrt{x^2 + 16}} + \ln|x + \sqrt{x^2 + 16}| + C$

15. $\frac{1}{3}\ln\left|\frac{x^3}{2x^3 + 1}\right| + C$ **17.** $-\frac{\sqrt{1 - 4x^2}}{x} + C$

19. $\frac{1}{18}x\sqrt{9x^2 - 1} + \frac{1}{54}\ln|x + \frac{1}{3}\sqrt{9x^2 - 1}| + C$

21. $x[(\ln x)^3 - 3(\ln x)^2 + 6\ln x - 6] + C$

23. $\frac{1}{2}(e^x + 1)^2 - 2(e^x + 1) + \ln(e^x + 1) + C$

25. $-e^{-2x}(\frac{1}{2}x^2 + \frac{1}{2}x + \frac{1}{4}) + C$

27. $R(x) = \frac{2(x + 1)^{5/2}}{5} - \frac{2(x + 1)^{3/2}}{3} + \frac{4}{15}$

29. $\frac{10t}{9\sqrt{t^2 + 9}}$ **31.** 4.8

33. 327

35. $y_3(x) = -1$. Thus, $y_1(x) = y_2(x) - 1$

Section 6.3

1. $T_2 = 5$, $S_2 = 4$, exact $= 4$

3. to four decimal places $T_2 = 1.1667$, $S_2 = 1.1111$, exact $= 1.0986$

5. to four decimal places $T_2 = 0.6830$, $S_2 = 0.7440$, exact $= 0.7854$

7. to six decimal places $T_2 = 1.571583$, $S_2 = 1.475731$, exact $= 1.462652$

9. $T_4 = 4.25$, $S_2 = S_4 = 4$, exact $= 4$

11. to four decimal places $T_4 = 1.1167$, $S_4 = 1.1$, exact $= 1.0986$

13. to four decimal places $T_4 = 0.7489$, $S_4 = 0.7709$, exact $= 0.7854$

15. to six decimal places $T_4 = 1.490679$, $S_4 = 1.463711$, exact $= 1.462652$

17. $T(100) = 4.0004$, $S(100) = 4$, exact $= 4$

19. $T(100) = 1.098642$, $S(100) = 1.098612294$, exact to nine decimal places $= 1.098612289$

21. $T(100) = 0.785104$, $S(100) = 0.785283$, exact to six decimal places $= 0.785398$

23. $T(100) = 1.462697$, $S(100) = 1.462652$, exact to six decimal places $= 1.462652$

25. (a) $T_4 = 5.8$; (b) $S_4 \approx 5.87$

27. (a) $T_4 = 50$; (b) $S_4 = 52$

29. (a) $T_6 = 4.325$; (b) $S_6 = 4.35$

31. (a) $\frac{1}{75}$; (b) 0

33. (a) 0.32; (b) 0.00043

35. \$78,818

37. Yes. When the graph of the function is concave down, the line segment connecting the two endpoints of any subinterval lies below the curve. Thus, the trapezoid rule always gives an underestimate. When the graph of the function is concave up, the line segment connecting the two endpoints of any subinterval lies above the curve. Thus, the trapezoid rule always gives an overestimate.

39. Here $\Delta x = \dfrac{b - 0}{2} = \dfrac{b}{2}$, then $S_2 = \dfrac{b}{6}\left[f(0) + 4f\left(\dfrac{b}{2}\right) + f(b)\right] = \dfrac{b}{6}\left(0 + 4 \times \dfrac{b^3}{8} + b^3\right) = \dfrac{b^4}{4}$ and $\displaystyle\int_0^b x^3\, dx = \dfrac{1}{4}x^4 \Big|_0^b = \dfrac{b^4}{4}$. Thus $S_2 = \displaystyle\int_0^b x^3\, dx$

41. left-hand sum $= f(x_0)\,\Delta x + f(x_1)\,\Delta x + \cdots + f(x_{n-1})\,\Delta x$ and right-hand sum $= f(x_1)\,\Delta x + f(x_2)\,\Delta x + \cdots + f(x_n)\,\Delta x$. Thus, $\dfrac{\text{left-hand sum} + \text{right-hand sum}}{2} =$

$$\dfrac{f(x_0) + f(x_1)}{2}\,\Delta x + \dfrac{f(x_1) + f(x_2)}{2}\,\Delta x + \cdots + \dfrac{f(x_{n-1}) + f(x_n)}{2}\,\Delta x = T_n$$

Section 6.4

1. $\frac{1}{2}$

3. diverges

5. 100

7. diverges

9. 1

11. $-\frac{1}{2}$

13. $\frac{1}{3}$

15. $-\frac{1}{2}$

17. 0

19. 0

21. approximately 667 tons

23. $20 million

25. 2000 tons

27. 200 million

29. 1 thousand

31. $1250

33. $555.56

35. (a) no, since $P_V(\infty) =$ $100,000; (b) yes, since $P_V(\infty) =$ $100,000

37. (a) no, since $P_V(\infty) =$ $600,000; (b) yes, since $P_V(\infty) =$ $600,000

39. $\int_{-\infty}^{\infty} x\, dx = \int_{-\infty}^{0} x\, dx + \int_0^{\infty} x\, dx$. Since $\int_{-\infty}^{0} x\, dx$ diverges, $\int_{-\infty}^{\infty} x\, dx$ diverges. Notice that $\lim\limits_{a \to -\infty}\int_a^0 f(x)\, dx + \lim\limits_{b \to \infty}\int_0^b f(x)\, dx \neq \lim\limits_{a \to \infty}\int_{-a}^a f(x)\, dx$

41. Since $\displaystyle\int_1^{\infty} \dfrac{1}{\sqrt{x}}\, dx = \lim_{b \to \infty}\int_1^b x^{-1/2}\, dx = \lim_{b \to \infty}\left(2\sqrt{b} - 2\right)$ diverges, and $f(x) \geq \dfrac{1}{\sqrt{x}}$ on $[1, \infty)$, we have $\displaystyle\int_1^{\infty} f(x)\, dx \geq \int_1^{\infty} \dfrac{1}{\sqrt{x}}\, dx$. Thus, $\displaystyle\int_1^{\infty} f(x)\, dx$ diverges

43. $\displaystyle\int_{-\infty}^c f(x)\, dx + \int_c^{\infty} f(x)\, dx = \lim_{a \to -\infty}\int_a^c f(x)\, dx + \lim_{b \to \infty}\int_c^b f(x)\, dx = \lim_{a \to -\infty}\int_a^0 f(x)\, dx + \int_0^c f(x)\, dx + \int_c^0 f(x)\, dx + \lim_{b \to \infty}\int_0^b f(x)\, dx = \lim_{a \to -\infty}\int_a^0 f(x)\, dx + \lim_{b \to \infty}\int_0^b f(x)\, dx = \int_{-\infty}^{\infty} f(x)\, dx$

Chapter 6 Review

1. $\dfrac{1}{25}(5x - 1)e^{5x} + C$

2. $\dfrac{2}{27}\sqrt{5 + 3x}\,(3x - 10) + C$

3. $\frac{1}{6}x^6(\ln x - \frac{1}{6}) + C$

4. $-\dfrac{x}{1 + x} + \ln|1 + x| + C$

5. $-\frac{1}{5}(9 - x^2)^{3/2}(x^2 + 6) + C$

6. $2\sqrt{x}(\ln x - 2) + C$

7. $\frac{1}{2}x - \frac{3}{4}\ln|2x + 3| + C$

8. $\dfrac{1}{4(x + 4)} - \dfrac{1}{16}\ln\left|\dfrac{x + 4}{x}\right| + C$

9. $-\dfrac{\sqrt{4 - x^2}}{4x} + C$

10. $\sqrt{9 - x^2} - 3\ln\left|\dfrac{3 + \sqrt{9 - x^2}}{x}\right| + C$

11. $\frac{1}{2}x\sqrt{x^2 + 1} + \frac{1}{2}\ln|x + \sqrt{x^2 + 1}| + C$

12. $x - \ln(1 + e^x) + C$

13. 10

14. diverges

15. $\frac{2}{3}$

16. diverges

17. (a) 45; (b) $\dfrac{100}{3}$; (c) correct $= 32$

18. to four decimal places: (a) 1.5375; (b) 1.425; (c) correct $= 1.3863$

19. to four decimal places: (a) 35.3125; (b) 32.0833; (c) correct $= 32$

20. to four decimal places: (a) 1.4281; (b) 1.3916; (c) correct $= 1.3863$

21. $555.6 thousand

CHAPTER 7

Section 7.1

1. (a) 13; (b) 10; (c) 12

3. (a) 2; (b) 13; (c) 2

5. (a) $\frac{1}{3}$; (b) -1; (c) $\frac{1}{3}$

7. (a) $\sqrt{6}$; (b) 3; (c) 2

9. (a) 1; (b) $\frac{1}{2}$; (c) $\frac{1}{2}$

11. All x and y

13. $\{(x, y)|x \neq -y\}$

15. $\{(x, y)|x^2 + y^2 \leq 16\}$

17. 3

19. plane with intercepts $(0, 0, 4)$, $(0, 6, 0)$, $(12, 0, 0)$

21. plane with intercepts $(0, 0, 2)$, $(0, 4, 0)$, $(10, 0, 0)$

23. horizontal plane three units above xy-plane

25. sphere of radius 6 centered at $(0, 0, 0)$

27. (f)   **29.** (c)

31. (d)

33. (a) $z_0 = 0$: $x = 0 = y$; (b) $z_0 = 1$: $x^2 + y^2 = 1$; (c) $z_0 = 4$: $x^2 + y^2 = 4$; (d) $z_0 = 9$: $x^2 + y^2 = 9$; (e) surface is a bowl

35. (a) $z_0 = 1$: $x = 0 = y$; (b) $z_0 = 0$: $x^2 + y^2 = 1$; (c) $z_0 = -3$: $x^2 + y^2 = 4$; (d) $z_0 = -8$: $x^2 + y^2 = 9$; (e) surface is a mountain

37. $S(x, y) = 2x^2 + 4xy$, $S(3, 5) = 78$

39. $R(x, y) = 1400x + 802y - 4xy - 12x^2 - 0.5y^2$, 80,400

41. $P(x, y) = -15,000 + 1350x + 801.5y - 4xy - 12x^2 - 0.5y^2$, 60,300

43. $A(m, t) = 1000(1 + 0.08/m)^{mt}$, \$1489.85

45. (a) $\{(x, y)|y \neq 0\}$; (b) 200; (c) 50

47. (a) $\{(d, P)|P \geq 0\}$; (b) 24π; (c) 8π

49. When $y = mx$, $\lim\limits_{x \to 0} f(x,mx) = \lim\limits_{x \to 0} \dfrac{mx^3}{x^4 + m^2x^2} = \lim\limits_{x \to 0} \dfrac{mx}{x^2 + m^2} = \dfrac{0}{0 + m^2} = 0$ for all m. But that does not imply that f is continuous at $(0, 0)$.

Section 7.2

1. $f_x = 2x$, $f_y = 2y$, 2, 6

3. $f_x = 2xy - 3x^2y^2$, $f_y = x^2 - 2x^3y$, -8, -3

5. $f_x = \dfrac{\sqrt{y}}{2\sqrt{x}}$, $f_y = \dfrac{\sqrt{x}}{2\sqrt{y}}$, $\dfrac{1}{2}$, $\dfrac{1}{2}$

7. $f_x = \dfrac{xy^2}{\sqrt{1 + x^2y^2}}$, $f_y = \dfrac{x^2y}{\sqrt{1 + x^2y^2}}$, 0, 0

9. $f_x = 2e^{2x+3y}$, $f_y = 3e^{2x+3y}$, $2e^5$, $3e^5$

11. $f_x = ye^{xy} + xy^2e^{xy}$, $f_y = xe^{xy} + x^2ye^{xy}$, $2e$, $2e$

13. $f_x = \dfrac{1}{x + 2y}$, $f_y = \dfrac{2}{x + 2y}$, 1, 2

15. $f_x = \dfrac{e^{xy}}{x} + ye^{xy} \ln x$, $f_y = xe^{xy} \ln x$, 1, 0

17. $f_x = \dfrac{-1}{x^2y}$, $f_y = \dfrac{-1}{xy^2}$, $-\dfrac{1}{2}$, $-\dfrac{1}{4}$

19. $f_x = \dfrac{y^2 - x^2 + 2xy}{(x^2 + y^2)^2}$, $f_y = \dfrac{y^2 - x^2 - 2xy}{(x^2 + y^2)^2}$, $\dfrac{1}{25}$, $-\dfrac{7}{25}$

21. $f_{xx} = 2y^4$, $f_{yy} = 12x^2y^2$, $f_{xy} = f_{yx} = 8xy^3$

23. $f_{xx} = 4e^{2x-3y}$, $f_{yy} = 9e^{2x-3y}$, $f_{xy} = f_{yx} = -6e^{2x-3y}$

25. $f_{xx} = -\frac{1}{4}y^{1/2}x^{-3/2}$, $f_{yy} = -\frac{1}{4}y^{-3/2}x^{1/2}$, $f_{xy} = f_{yx} = \frac{1}{4}x^{-1/2}y^{-1/2}$

27. $f_{xx} = 2e^y$, $f_{yy} = x^2e^y$, $f_{xy} = f_{yx} = 2xe^y$

29. $f_{xx} = \dfrac{y^2}{(x^2 + y^2)^{3/2}}$, $f_{yy} = \dfrac{x^2}{(x^2 + y^2)^{3/2}}$, $f_{xy} = f_{yx} = -xy(x^2 + y^2)^{-3/2}$

31. $f_x = yz$, $f_y = xz$, $f_z = xy$

33. $f_x = \dfrac{x}{\sqrt{x^2 + y^2 + z^2}}$, $f_y = \dfrac{y}{\sqrt{x^2 + y^2 + z^2}}$, $f_z = \dfrac{z}{\sqrt{x^2 + y^2 + z^2}}$

35. $f_x = e^{x+2y+3z}$, $f_y = 2e^{x+2y+3z}$, $f_z = 3e^{x+2y+3z}$

37. $f_x = \dfrac{1}{x + 2y + 5z}$, $f_y = \dfrac{2}{x + 2y + 5z}$, $f_z = \dfrac{5}{x + 2y + 5z}$

39. complementary   **41.** competitive

43. competitive

45. $A_r = 1000t(1 + r/12)^{12t-1}$. This gives the rate of change of the amount with respect to changes in the interest rate r.

47. $\dfrac{\partial V}{\partial r} = 4\pi r^2\left(1 + \dfrac{c}{r} N^{1/3}\right)^2$. This gives the rate of change of the detection volume with respect to changes in the detection radius r. $\dfrac{\partial V}{\partial N} = \dfrac{4\pi cr^2}{3}\left(1 + \dfrac{c}{r} N^{1/3}\right)^2 N^{-2/3}$. This gives the rate of change of the detection volume with respect to changes in the number of prey N.

49. $\dfrac{\partial L}{\partial r} = -2p\pi r\left(1 - \dfrac{p\pi r^2}{n}\right)^{n-1}$. This gives the rate of change of the amount of light with respect to changes in the radius r of the leaves. $\dfrac{\partial L}{\partial p} = -\pi r^2\left(1 - \dfrac{p\pi r^2}{n}\right)^{n-1}$. This gives the rate of change of the amount of light with respect to changes in the total density p of the leaves.

51. $\dfrac{\partial A}{\partial S}$ $(S, D) = 0.11297S^{0.43}D^{-0.73}$, $\dfrac{\partial A}{\partial D}$ $(S, D) = -0.05767S^{1.43}D^{-1.73}$

53. $\dfrac{\partial x}{\partial A} > 0$ means that there is more demand when advertising

is at a higher level. $\dfrac{\partial x}{\partial P} < 0$ means that there is less demand when advertising is at a higher price.

55. (a) The instantaneous rate of change of f with respect to x at the point $x = 1$ and $y = 2$ is 3. (b) The instantaneous rate of change of f with respect to y at the point $x = 1$ and $y = 2$ is -5.

Section 7.3

1. relative minimum at $(5, -2)$

3. relative maximum at $(2, 1)$

5. saddle at $(1, 1)$

7. relative maximum at $(-1, -2)$

9. relative minimum at $(0, 0)$

11. saddle at $(0, 2)$

13. saddle at $(0, 0)$, relative minimum at $(1, 1)$

15. saddle at $(0, 0)$, relative maximum at $\left(\dfrac{1}{6}, \dfrac{1}{12}\right)$

17. relative maximum at $(1, 1)$

19. saddle at $(0, 0)$

21. Since $f(x, y) = x^4y^4 \geq 0$ and $f(0, 0) = 0$, f has a relative minimum at $(0, 0)$. Since $f_{xx} = 12x^2y^4$, $f_{yy} = 12x^4y^2$, and $f_{xy} = 16x^3y^3$, $\Delta(0, 0) = 0$.

23. If $y = x$, then $f(x, y) = f(x, x) = x^3x^3 = x^6$ and $f(x, x)$ increases as x increases. But if $y = -x$, $f(x, -x) = -x^6$ and $f(x, -x)$ decreases as x increases. Since $f_{xx} = 6xy^3$, $f_{yy} = 6x^3y$, and $f_{xy} = 9x^2y^2$, $\Delta(0, 0) = 0$.

25. increasing in all cases

27. increasing in the third case and decreasing in all the others

29. $x = 1, y = 2, z = 4$ **31.** base 4×4 and height 6

33. $x = 200, y = 100$ **35.** $x = 18, y = 8$

37. $x = 9, y = 4$

39. $x = 200, y = 100, z = 700$

41. $x = 1600, y = 1000$ **43.** $x = 2100, y = 0$

45. $N \approx 2.463, P \approx 2.399$

47. (1) $x = \left(\dfrac{P}{a}\right)^{1/b} y^{1 - 1/b}$; (2) $C = C(y) = p_1 \left(\dfrac{P}{a}\right)^{1/b} y^{1 - 1/b} + p_2y$; (3) $C'(y) = 0$ implies $y = \left(\dfrac{p_1}{p_2}\right)^b \left(\dfrac{1}{b} - 1\right)^b \dfrac{P}{a}$; (4) $x = \left(\dfrac{p_1}{p_2}\right)^{b-1} \left(\dfrac{1}{b} - 1\right)^{b-1} \dfrac{P}{a}$; (5) $C = C(P) = \left(\dfrac{1}{b} - 1\right)^b \left(\dfrac{p_1}{p_2}\right)^b \left(\dfrac{p_2}{1 - b}\right) \dfrac{P}{a}$

Section 7.4

1. minimum at $x = 1, y = 3$

3. maximum at $x = -3, y = -1$

5. minimum at $x = 1, y = 1$

7. maximum at $x = -1, y = -3$

9. minimum at $x = 2, y = 2$

11. maximum at $x = 4, y = 4$

13. minimum at $x = 4, y = 2, z = 1$

15. maximum at $x = 2, y = 2, z = 1$

17. 10 and 10 **19.** 12, 12, and 12

21. $r = 2, h = 4$ **23.** $x = 200, y = 100$

25. $6 \times 6 \times 9$ **27.** $x = 2000, y = 4000$

29. $x = 1, y = 2, z = 4$

31. $x = 200, y = 100, z = 700$

33. Let $F(x, y, \lambda) = p_1x + p_2y + \lambda f(x, y) - \lambda P_1$, then

$\begin{cases} 0 = F_x = p_1 + \lambda f_x \\ 0 = F_y = p_2 + \lambda f_y \end{cases}$ implies $\begin{cases} \lambda = -\dfrac{p_1}{f_x} \\ \\ \lambda = -\dfrac{p_2}{f_y} \end{cases}$

Thus, $\lambda = -\dfrac{p_1}{f_x} = -\dfrac{p_2}{f_y}$ implies $\dfrac{f_x}{p_1} = \dfrac{f_y}{p_2}$

Section 7.5

1. $y = 0.50x + 0.50$, 0.5

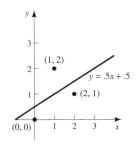

3. $y = 1.1x + 0.1$, 0.95

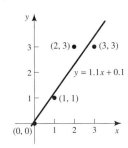

5. $y = -0.9x + 4.5$, -0.92

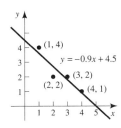

7. $y = -0.7x + 3.4$, -0.90

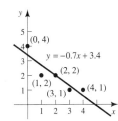

9. $y = 5x + 5.4$, 35.4 (or rounded to 35)

11. $y = x + 2.5$, $5.5 million

13. $y = 0.1006x - 167.12$, 0.98, there is a strong positive correlation between the two variables.

15. (a) $y = 0.0858x + 1.0401$, 0.96; (b) $Y = 0.7426X - 1.2039$, 0.98; (c) model b. (d) For model (b), $y = e^{-1.2039}x^{0.7426}$ and $\dfrac{dy}{dx}$ is a decreasing function. Thus, there is diminishing marginal returns. (e) $y = L - ae^{-kx}$ or even $y = \dfrac{a}{1 + be^{kx}}$, $k > 0$.

Section 7.6

1. 1

3. -6

5. $\dfrac{1}{15}$

7. $\dfrac{-11}{1250}$

9. $\dfrac{1}{20}$

11. $\dfrac{-1}{20}$

13. 0.05

15. $\dfrac{e}{10} \approx 0.2718$

17. 6.05

19. 5.036

21. 3.11

23. 8.9

25. 24.57

27. 1.08

29. $2000

31. 118.75

33. 0.70 ft^3

35. 0.9573 m^2

37. 0.7116

Section 7.7

1. $\dfrac{15}{4}$

3. $\frac{3}{4}$

5. 8

7. 2.5

9. 1

11. 35

13. $\dfrac{4}{15}\left(2\sqrt{2} - 1\right)$

15. $\ln 2 \ln 3$

17. $(e^2 - 1)(e^3 - 1)$

19. $\dfrac{7}{24}$

21. $\dfrac{e^4 - 1}{2}$

23. $\frac{1}{2}$

25. $\frac{1}{3}$

27. $\frac{1}{8}$

29. $\frac{1}{4}$

31. $\dfrac{1}{40}$

33. $\dfrac{1}{24}$

35. $-\frac{2}{3}$

37. $\frac{7}{6}$

39. $\frac{3}{2} - \ln 2$

41. 4

43. 19,000

45. $\dfrac{3627}{532}$

47. $e - 1$

49. 10,000

Chapter 7 Review

1. $-1, 7, 4$

2. $0, \frac{1}{5}, -\frac{8}{5}$

3. (1) all x and y; (2) $\{(x, y)|y \neq 0\}$

4. $2\sqrt{11}$

5. $z_0 = 16$: $x^2 + y^2 = 0$ or the single point $(0, 0)$; $z_0 = 12$; $x^2 + y^2 = 4$, circle of radius 2 centered at $(0, 0)$; $z_0 = 7$: $x^2 + y^2 = 9$, circle of radius 3 centered at $(0, 0)$

6. $f_x = 1 + 2xy^5, f_y = 5x^2y^4$

7. $f_x(2, 1) = 5, f_y(2, 1) = 20$

8. $f_x = -ye^{-xy} + \dfrac{1}{x}, f_y = -xe^{-xy} + \dfrac{1}{y}$

9. $f_{xx} = 2y^5, f_{xy} = f_{yx} = 10xy^4, f_{yy} = 20x^2y^3$

10. $f_{xx} = y^2e^{-xy} - \dfrac{1}{x^2}, f_{xy} = f_{yx} = e^{-xy}(xy - 1), f_{yy} = x^2e^{-xy} - \dfrac{1}{y^2}$

11. $f_x = -yze^{-xyz} + y^2z^3, f_y = -xze^{-xyz} + 2xyz^3,$
$f_z = -xye^{-xyz} + 3xy^2z^2$

12. relative minimum at $(0, 0)$

13. relative maximum at $(0, 0)$

14. relative maximum at $(-2, -1)$

15. relative minimum at $(2, -3)$

16. saddle at $(0, 0)$, relative maximum at $(-1, -1)$

17. saddle at $(-1, -1)$, relative minimum at $(3, 3)$

18. saddle at $(5, 3)$, relative maximum at $(5, -1)$

19. saddle at $(0, 0)$, relative maximum at $(\frac{2}{3}, \frac{1}{9})$

20. since $f_{xx}(0, 0) = 2a$ and $\Delta(0, 0) = 4(ac - b^2)$, the result follows

21. minimum at $(4, -3)$

22. maximum at $(\frac{1}{4}, \frac{1}{2})$

23. 2.8

24. 3.2

25. 4.99

26. 24

27. 80

28. $\dfrac{5}{12}$

29. $\dfrac{7}{24}$

30. $\frac{4}{5}$

31. $\dfrac{3}{20}$

32. $\frac{1}{2}$

33. $\dfrac{13}{20}$

34. $\dfrac{13}{20}$

35. $\frac{1}{6}$

36. 80

37. 480

38. $f_K = 2K^{-0.80}L^{0.80}, f_L = 8K^{0.20}L^{-0.20}$

39. $f_{KK} = -1.6K^{-1.8}L^{0.80}, f_{KL} = 1.6K^{-0.80}L^{-0.20}, f_{LL} = -1.6K^{0.20}L^{-1.20}$

40. 1091

41. $C_x = 2x - y, C_y = 2y - x$

42. $C_{xx} = 2, C_{xy} = C_{yx} = -1, C_{yy} = 2$

43. $A_p = e^{rt}, A_r = Pte^{rt}, A_t = Pre^{rt}$

44. $2 \times 2 \times 4$

45. $x = 10, y = 12$

46. $x = 11, y = 14$

47. 11,000 of regular and 9000 of mint

48. 500 of each

49. 170 cubic inches

50. $\dfrac{\partial h}{\partial e} > 0$ means that the individual firm has a higher harvest rate with a higher effort rate. $\dfrac{\partial h}{\partial N} < 0$ means that more participating firms result in less of a harvest. $\dfrac{\partial h}{\partial X} > 0$ means that larger resource stock size results in more of a harvest.

CHAPTER 8

Section 8.1

11. $y = x^3$

13. $y = e^x - 1$

15. $y = 10e^{x^2}$

17. $y = 100e^{x^3 + x^2}$

19. $y = e^{1 - 1/x}$

21. $y = x^4 + 3$

23. $y = 0.5e^{2x} + 2.5$

25. $y = e^{x^2}$

27. $R(x) = -0.5e^{-x^2} + 0.5$

29. $\dfrac{dR}{dS} = \dfrac{k}{S}$

31. $\dfrac{dy}{dt} = -ky^2$ **33.** $\dfrac{dy}{dt} = k(a - y)(b - y)$

35. $\dfrac{dA}{dt} = 5 - \dfrac{1}{40}A$

37. The general solution of the differential equation $y' = f(x, y)$

is a function $y(x, C)$ for which each solution of the differential equation can be obtained by selecting an appropriate value of C. A particular solution is a solution with a specific value of C. For example, the general solution of $y' = y$ is $y = y(x, C) = Ce^x$. A particular solution is $y = 2e^x$.

Section 8.2

1. $y = 3e^{x^4}$ **3.** $y = \dfrac{2}{2 - x^2}$

5. $y = 1 - \ln(1 - ex)$ **7.** $y = x$

9. $y = \sqrt[3]{1.5x^2 + 3x + 1}$ **11.** $y = x$

13. $\frac{1}{3}y^3 + y = \frac{1}{2}x^2 + x + 12$

15. $x^2 + y^2 = C$

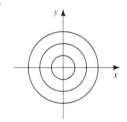

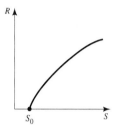

17. (a) $R = k \ln |S| + C$; (b) $R(S) = k \ln \left| \dfrac{S}{S_0} \right|$

19. $y(t) = \dfrac{y_0}{ky_0 t + 1} \cdot t = \dfrac{1}{ky_0}$ **21.** $y = \dfrac{ab(1 - e^{k(b-a)t})}{a - be^{k(b-a)t}}$

23. 417 million   **25.** 15, 42

27. about 23 wolves   **29.** $y = Cx^k$

31. $\dfrac{dA}{dt} = 5 - \dfrac{1}{40}A$, $A(0) = 10$, $A(t) = 200 - 190e^{-t/40}$

Section 8.3

1. (b)   **3.** (c)

5. (d)

7. $y_1 = 1.2000$
$y_2 = 1.4400$
$y_3 = 1.7280$
$y_4 = 2.0736$
$y_5 = 2.48832$
error $= |y_5 = \phi(1)| \approx 0.2300$

9. $y_1 = 0.0000$
$y_2 = 0.0400$
$y_3 = 0.1280$
$y_4 = 0.2736$
$y_5 = 0.4883$
error $= |y_5 - \phi(1)| \approx 0.2300$

11. $y_1 = 0.2000$
$y_2 = 0.4333$
$y_3 = 0.6952$
$y_4 = 0.9821$
$y_5 = 1.2912$
error $= |y_5 - \phi(1)| \approx 0.0951$

13. $y_1 = 1.2000$
$y_2 = 1.4000$
$y_3 = 1.6000$
$y_4 = 1.8000$
$y_5 = 2.0000$
error $= |y_5 - \phi(1)| \approx 0.0000$

15. $y_1 = 0.6000$
$y_2 = 0.4416$
$y_3 = 0.3480$
$y_4 = 0.2850$
$y_5 = 0.2395$
error $= |y_5 - \phi(3)| \approx 0.0462$

17.

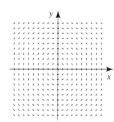

19.

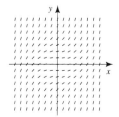

21.

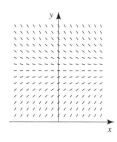

23.

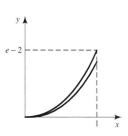

25.

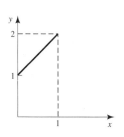

27. For a solution starting at $(1, 7)$ to get into the region $y < 5$ it must cross the line $y = 5$ and have *negative* slope at some point. Any solution in the region $y < 5$, however, must have *positive* slope.

29. Consider the solution curve through $(2, -2)$. The slope appears to be zero. As the solution curve moves to the right, the slope becomes positive. The solution curve then rises. The further it rises the more positive the slope, and thus, the faster it rises.

31. Consider the solution curve through a point in the third quadrant. The solution curve rises, and when near the x-axis, slows somewhat. After crossing the x-axis with slope about 1, the curve then rises ever faster.

33. Consider the solution curve through the point $(-1, -2)$. The slope is positive. As the curve moves up and to the right, however, the slope decreases. The curve must cross the y-axis at some point below the x-axis. After crossing the y-axis, the curve then drops, and approaches $y = -x$ asymptotically.

Section 8.4

1.

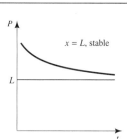

3.

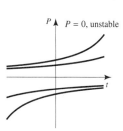

5.

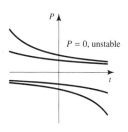

7.

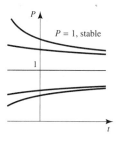

9.

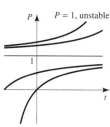

11.

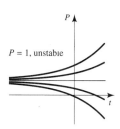

13.

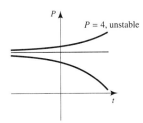

$P = 4$, unstable

15.

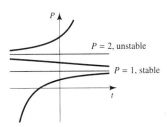

$P = 2$, unstable

$P = 1$, stable

17.

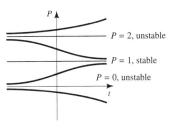

$P = 2$, unstable

$P = 1$, stable

$P = 0$, unstable

19.

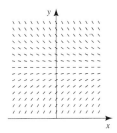

21.

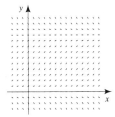

Section 8.5

1. $E \geq k$

3. The population always heads for 0.

5. The figure indicates one constant solution ($P = 1$), which is not stable.

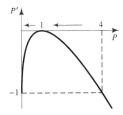

7. The figure indicates two constant solutions, the first not stable, the second stable.

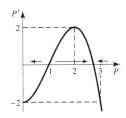

9. The figure indicates no constant solution.

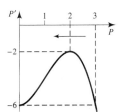

11. The figure indicates two constant solutions, the first not stable, the second stable.

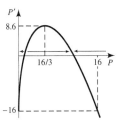

13. $E < 1$, stable

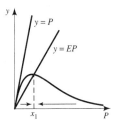

15. Let $f(P) = 2\sqrt{P} - P$. Then $f'(P) = 1/\sqrt{P} - 1$. The graph rises until $P = 1$ and then falls, with the tangent to the curve at $P = 0$ vertical. Thus, any line $y = EP$ with $E > 0$ intersects the curve in exactly one place. Let the P-coordinate of this point of intersection be P^*. If $P < P^*$, $f(P) > P^*$, whereas if $P > P^*$, $f(P) < P^*$. Thus, the constant solution $P = P^*$ is stable.

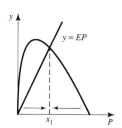

17. Let $f(P) = P^2(3 - P)$. (See the graph.) Note that $f'(0) = 0$. If $E < E^*$, then $y = EP$ intersects the curve $y = f(P)$ in two places. As the figure indicates, P_1 is unstable, whereas P_2 is stable.

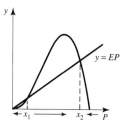

19. $0 < E < 2.25$

Chapter 8 Review

3. $y = 2x^3 e^{-3x} + 5e^{-3x}$

4. $y = 2x^5 + C$

5. $y = 2x^3 + 2x^2 + C$

6. $y = Ce^{2x^5}$

7. $\frac{1}{2}y^2 = x^5 + C$

8. $y^{-3} = x^{-1} + C$

9. $e^y = \frac{1}{2}e^{x^2} + C$

10. Constant solution $y = 3$. When $y > 3$, solutions decrease and head asymptotically to $y = 3$. When $y < 3$, solutions increase and head asymptotically to $y = 3$.

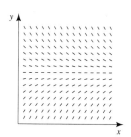

11. Constant solution $y = -0.5$

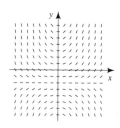

12. $y_0 = 1$, $y_1 = 1.6$, $y_2 = 2.568$, $y_3 = 4.1408$, $y_4 = 6.69728$, $y_5 = 10.843648$

13. $y_0 = 1$, $y_1 = 1.3$, $y_2 = 1.691$, $y_3 = 2.2023$, $y_4 = 2.87199$, $y_5 = 3.749587$, $y_6 = 4.899463$, $y_7 = 6.405302$, $y_8 = 8.375892$, $y_9 = 10.952660$, $y_{10} = 14.319459$

14. $P(t) = \dfrac{h}{k} + \left(P_0 - \dfrac{h}{k}\right)e^{kt}$

16. $P = 0$.

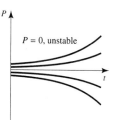

17. $P = 0$.

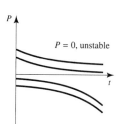

18. $P = 1, 3$.

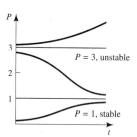

19. $P = 0, 2, 5$.

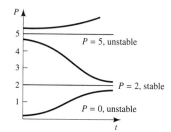

20. $0 < h < 1$. First is unstable, second is stable.

21. $0 < E < 2$. Stable

CHAPTER 9

Section 9.1

1. $\dfrac{\pi}{12}$

3. $\dfrac{4\pi}{3}$

5. $-\dfrac{7\pi}{3}$

7. $75°$

9. $420°$

11. $-720°$

13. (1) I; (3) III; (5) IV

15. $\dfrac{7\pi}{2}$

17. -2π

19.

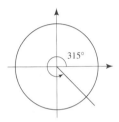

21.

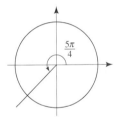

23.

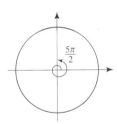

25.

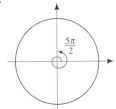

$495°,\ -225°$

27.

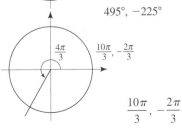

$\dfrac{10\pi}{3},\ -\dfrac{2\pi}{3}$

29.

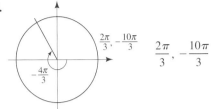 $\dfrac{2\pi}{3},\ -\dfrac{10\pi}{3}$

31. (a) 2.5; (b) $\left(\dfrac{450}{\pi}\right)^{\circ} \approx 143.2°$

33. (a) 2; (b) $\left(\dfrac{360}{\pi}\right)^{\circ} \approx 114.6°$

35. 2

37. $\dfrac{\pi}{2}$

39. (a) $4000 \times \dfrac{\pi}{180} \approx 69.81$ miles; (b) $4000 \times \dfrac{\pi}{4} \approx 3141.6$ miles

41. degrees: $0.25 \times \dfrac{180}{\pi} \approx 14.3$; radian: 0.25

43. (a) π miles; (b) $\dfrac{1761}{1760}\pi$ miles

45. ≈ 75 feet

47. (a) 28.27 feet; (b) 6.67

49.

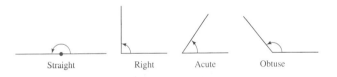

Straight   Right   Acute   Obtuse

51.

Section 9.2

1. $\sin \alpha = \dfrac{\sqrt{15}}{4}$, $\cos \alpha = \dfrac{1}{4}$

3. $\sin \gamma = -\dfrac{2\sqrt{2}}{3}$, $\cos \gamma = -\dfrac{1}{3}$

5. $\sin \theta = \dfrac{1}{2}$, $\cos \theta = \dfrac{\sqrt{3}}{2}$

7. $\sin \theta = \dfrac{\sqrt{3}}{2}$, $\cos \theta = \dfrac{1}{2}$

9. $\sin \theta = -\dfrac{1}{2}$, $\cos \theta = -\dfrac{\sqrt{3}}{2}$

11. $\sin \theta = -\dfrac{\sqrt{3}}{2}$, $\cos \theta = -\dfrac{1}{2}$

13. $\sin \theta = 0$, $\cos \theta = -1$

15. $\sin \alpha = \dfrac{\sqrt{5}}{3}$, $\cos \alpha = -\dfrac{2}{3}$

17. $\sin \gamma = -\dfrac{\sqrt{21}}{5}$, $\cos \gamma = \dfrac{2}{5}$

19. -1

21. 0 **23.** -1

25. $x = 50\sqrt{2}$, $y = 50\sqrt{2}$ **27.** $x = 2\sqrt{3}$, $r = 4$

29. $\sin \alpha = \dfrac{4}{5}$, $\cos \alpha = \dfrac{3}{5}$ **31.** $\sin \gamma = \dfrac{4}{5}$, $\cos \gamma = \dfrac{3}{5}$

33.

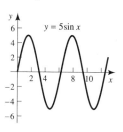

35.

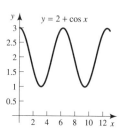

37.

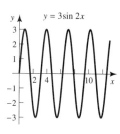

39.

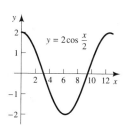

41.

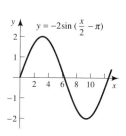

43. $\dfrac{20}{3}\sqrt{3} \approx 11.55$ feet   **45.** $100/\sqrt{3} \approx 57.7$ feet

47. $1000\sqrt{2} \approx 1414.21$ feet

49. Maximum demand occurs in January; minimum demand occurs in July.

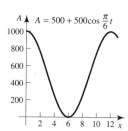

51. Temperatures range from $-10°$ to $90°$F. Day 20 is the coldest. Day 202 is the hottest.

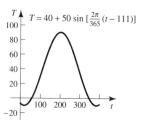

53. When the prey population drops, the predator population decreases because of lack of food. When the prey population increases, the predator population increases because of abundance of food. The lag is a reaction time related to gestation period of the prey.

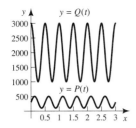

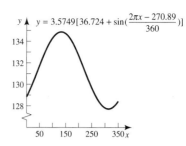

55. Most at 133°. Least at 313°. The maximum warmth occurs from the sun at 130°, the minimum at 310°.

57. 2 radians ≠ 2 degrees, $\sin 2 \approx 0.909$, $\sin 2° \approx 0.0349$

59. $y = -2 \sin x$ **61.** $y = 1 + 2 \cos 2x$

Section 9.3

1. $5 \cos 5x$

3. $-(2x + 1) \sin (x^2 + x + 1)$

5. $3e^{3x} \cos e^{3x}$ **7.** $-2x \dfrac{\sin x^2}{\cos x^2}$

9. $\sin^2 x \cos^3 x (3 \cos^2 x - 4 \sin^2 x)$

11. $\dfrac{3 \cos 3x \sin x - \sin 3x \cos x}{\sin^2 x}$

13. $-\sin x \cos(\cos x)$

15. $y - \dfrac{1}{2} \sqrt{2} = \dfrac{3}{2} \sqrt{2}\left(x - \dfrac{\pi}{12}\right)$

17. Maximum is $\sqrt{2}$ at $x = \dfrac{\pi}{4}$, minimum is $-\sqrt{2}$ at $x = \dfrac{5\pi}{4}$

19. Maximum is 2π at $x = 2\pi$, minimum is 0 at $x = 0$

21. Maximum at $\dfrac{\pi}{4}$, minimum at $\dfrac{5\pi}{4}$. Inflection values at $\dfrac{3\pi}{4}$ and $\dfrac{7\pi}{4}$

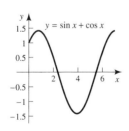

23. Maximum when $x = 2\pi$, minimum when $x = 0$. Inflection value at $x = \pi$

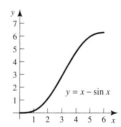

25. 0.48 miles/minute   **27.** $18\sqrt{2}$ feet

29. Predator: minimum population = 100, maximum population = 500. Prey: minimum population = 1000, maximum population = 3000

31. Maximum of 90°F on July 20, minimum of −10° on January 20.

33. 4

35. Follows since $1° = \dfrac{\pi}{180}$ radians

Section 9.4

1. $-\dfrac{1}{5} \cos 5x + C$ **3.** $2 \sin x^5 + C$ **15.** $\dfrac{1}{2}$ **17.** $\sqrt{2} - \cos 1 - \sin 1$

5. $\dfrac{2}{3} \sin^{3/2} x + C$ **7.** $-\ln(2 + \cos x) + C$ **19.** $\sqrt{2} + 1$ **21.** $\dfrac{2a}{\pi}$

9. $-\cos(\ln x) + C$ **11.** $x \sin x + \cos x + C$ **23.** 300   **25.** 40

13. $-x^2 \cos x + 2x \sin x + 2 \cos x + C$

Section 9.5

1. $\tan 0 = 0$, $\sec 0 = 1$, $\cot 0 = $ Undefined, $\csc 0 = $ Undefined

3. $\tan \dfrac{3\pi}{2} = $ Undefined, $\sec \dfrac{3\pi}{2} = $ Undefined, $\cot \dfrac{3\pi}{2} = 0$, $\csc \dfrac{3\pi}{2} = -1$

5. $\tan \dfrac{\pi}{4} = 1$, $\sec \dfrac{\pi}{4} = \sqrt{2}$, $\cot \dfrac{\pi}{4} = 1$, $\csc \dfrac{\pi}{4} = \sqrt{2}$

7. $\tan \dfrac{\pi}{3} = \sqrt{3}$, $\sec \dfrac{\pi}{3} = 2$, $\cot \dfrac{\pi}{3} = \dfrac{\sqrt{3}}{3}$, $\csc \dfrac{\pi}{3} = \dfrac{2\sqrt{3}}{3}$

9. $\tan \theta = 2$, $\sec \theta = \sqrt{5}$, $\cot \theta = \dfrac{1}{2}$, $\csc \theta = \dfrac{\sqrt{5}}{2}$

11. $\tan \theta = \dfrac{3}{2}$, $\sec \theta = -\dfrac{\sqrt{13}}{2}$, $\cot \theta = \dfrac{2}{3}$, $\csc \theta = -\dfrac{\sqrt{13}}{3}$

13. $\tan \theta = \dfrac{1}{2}$, $\sec \theta = \dfrac{\sqrt{5}}{2}$, $\cot \theta = 2$, $\csc \theta = \sqrt{5}$

15. $\tan \theta = \dfrac{2}{3}$, $\sec \theta = \dfrac{\sqrt{13}}{3}$, $\cot \theta = \dfrac{3}{2}$, $\csc \theta = \dfrac{\sqrt{13}}{2}$

17. $15 \tan^4 3x \cdot \sec^2 3x$ **19.** $\sec^2 x(1 + 2x \tan x)$

21. $\dfrac{-2x^2 \csc^2 x^2 - \cot x^2}{x^2}$

23. $-\dfrac{2}{x} \csc(\ln 5x^2) \cdot \cot(\ln 5x^2)$

25. $2xe^{\tan x^2} \sec^2 x^2$ **27.** $\sec x \cdot \tan x \cdot \sec^2(\sec x)$

29. $(-\sin x + \sec^2 x) \cos(\cos x + \tan x)$

31. $3 \tan 3x + C$ **33.** $e^{\tan x} + C$

35. $\dfrac{1}{3} \tan^3 x + C$ **37.** $-\dfrac{1}{5} \ln |\cos 5x| + C$

39. $4 \ln |\sin x^2| + C$ **41.** $\ln |\sec x^2 + \tan x^2| + C$

43. ≈ 2.73 miles   **45.** 24 minutes

47. $\tan(-\theta) = \dfrac{\sin(-\theta)}{\cos(-\theta)} = \dfrac{-\sin \theta}{\cos \theta} = -\tan \theta$

49.
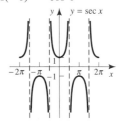

51. $\dfrac{\tan \beta_1}{\tan \beta} = \dfrac{-CD/CE}{-BC/CE} = \dfrac{CD}{BC} = \cos \alpha$

Chapter 9 Review

1. $\dfrac{\pi}{2}$ **2.** $\dfrac{2\pi}{3}$

3. $\dfrac{3\pi}{2}$ **4.** $\dfrac{5\pi}{2}$

5. $-\dfrac{5\pi}{4}$ **6.** $\dfrac{5\pi}{12}$

7. $165°$ **8.** $315°$

9. $-450°$ **10.** $900°$

11. $99°$ **12.** $30°$

13. $\sin \theta = \dfrac{4}{5}$, $\cos \theta = \dfrac{3}{5}$, $\tan \theta = \dfrac{4}{3}$, $\sec \theta = \dfrac{5}{3}$, $\cot \theta = \dfrac{3}{4}$, $\csc \theta = \dfrac{5}{4}$

14. $\sin \theta = -\dfrac{4}{\sqrt{41}}$, $\cos \theta = \dfrac{5}{\sqrt{41}}$, $\tan \theta = -\dfrac{4}{5}$, $\sec \theta = \dfrac{\sqrt{41}}{5}$, $\cot \theta = -\dfrac{5}{4}$, $\csc \theta = -\dfrac{\sqrt{41}}{4}$

15. $\sin \theta = \dfrac{4}{\sqrt{41}}$, $\cos \theta = \dfrac{5}{\sqrt{41}}$, $\tan \theta = \dfrac{4}{5}$, $\sec \theta = \dfrac{\sqrt{41}}{5}$, $\cot \theta = \dfrac{5}{4}$, $\csc \theta = \dfrac{\sqrt{41}}{4}$

16. $\sin \theta = \dfrac{1}{\sqrt{2}}$, $\cos \theta = \dfrac{1}{\sqrt{2}}$, $\tan \theta = 1$, $\sec \theta = \sqrt{2}$, $\cot \theta = 1$, $\csc \theta = \sqrt{2}$

17.
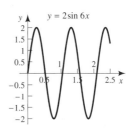

18.
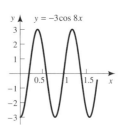

19.
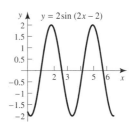

20.

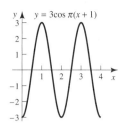

$y = 3\cos \pi(x + 1)$

21. $3\pi \cos \pi x$

22. $-\dfrac{4}{3} \sin \dfrac{x}{3}$

23. $4x \sec^2(x^2 + 1)$

24. $-\csc^2(x + 1)$

25. $2 \sec^2 x \tan x$

26. $\sec x \cdot \csc x \cdot (\tan x - \cot x)$

27. $27x^2 \sin^2 x^3 \cdot \cos x^3$

28. $-\dfrac{\sin x}{2\sqrt{\cos x + 1}}$

29. $2x(\tan x^2 + x^2 \sec^2 x^2)$

30. $\sec e^x \cdot \tan e^x \cdot e^x$

31. $(\cos x - \sin x)e^{\sin x + \cos x}$

32. $-\dfrac{1}{x} \csc^2(\ln |x|)$

33. $\dfrac{x \cos x - \sin x}{x^2}$

34. $\dfrac{2 \cos x + 1}{(2 + \cos x)^2}$

35. $\cos x \cdot \sec^2(\sin x)$

36. $-\dfrac{1}{8} \cos 8x + C$

37. $2 \sin \dfrac{x}{2} + C$

38. $2 \tan \dfrac{x}{2} + C$

39. $\dfrac{1}{3} \sec 3x + C$

40. $-6 \cot x + C$

41. $-2 \ln |\cos x| + C$

42. $\sin e^x + C$

43. $-e^{\cos x} + C$

44. $\sin(\ln x) + C$

45. $\ln(3 + \sin x) + C$

46. $\dfrac{1}{3 + \cos x} + C$

47. $\dfrac{x}{2}[\sin(\ln x) - \cos(\ln x)] + C$

48. Maximum when $x \approx 7.7$, minimum when $x \approx 12.1$

49. $-KLI_0 \dfrac{\sin \theta}{\cos^2 \theta} e^{-KL/\cos \theta}$

CHAPTER 10

Section 10.1

1. $p_4(x) = 1 - x + \dfrac{1}{2}x^2 - \dfrac{1}{6}x^3 + \dfrac{1}{24}x^4$

3. $p_4(x) = \sqrt{2} + \dfrac{\sqrt{2}}{4}x - \dfrac{\sqrt{2}}{32}x^2 + \dfrac{\sqrt{2}}{128}x^3 - \dfrac{5\sqrt{2}}{2048}x^4$

5. $p_3(x) = 1 - x + x^2 - x^3$

7. $p_4(x) = \ln 2 + \dfrac{1}{2}x - \dfrac{1}{8}x^2 + \dfrac{1}{24}x^3 - \dfrac{1}{64}x^4$

9. $p_5(x) = 1 - \dfrac{1}{2}x^2 + \dfrac{1}{24}x^4$

11. $p_2(x) = 1 + 2(x - 1) + (x - 1)^2$

13. $p_4(x) = 1 - (x - 1) + (x - 1)^2 - (x - 1)^3 + (x - 1)^4$

15. $p_3(x) = e^2 + 2e^2(x - 1) + 2e^2(x - 1)^2 + \dfrac{4}{3}e^2(x - 1)^3$

17. $p_4(x) = 1 - \dfrac{1}{2}\left(x - \dfrac{\pi}{2}\right)^2 + \dfrac{1}{24}\left(x - \dfrac{\pi}{2}\right)^4$

19. $p_n(x) = 0$ if $n < 4$, and $p_n(x) = x^4$ if $n \geq 4$

21. $p_0(x) = 3$, $p_1(x) = 3 + 2x$, and $p_n(x) = 3 + 2x + x^2$ if $n \geq 2$

23. $p_n(x) = 1 - x + \dfrac{1}{2}x^2 - \dfrac{1}{6}x^3 + \cdots + \dfrac{(-1)^n}{n!}x^n$

25. $p_n(x) = 1 + x + x^2 + \cdots + x^n$

27. $p_n(x) = \begin{cases} x - \dfrac{1}{3!}x^3 + \cdots + \dfrac{(-1)^{(n-1)/2}}{n!}x^n & \text{if } n \text{ is odd} \\[3mm] x - \dfrac{1}{3!}x^3 + \cdots + \dfrac{(-1)^{n/2-1}}{(n-1)!}x^{n-1} & \text{if } n \text{ is even} \end{cases}$

29. $p_0(x) = a^4$, $p_1(x) = a^4 + 4a^3(x - a)$, $p_2(x) = a^4 + 4a^3(x - a) + 6a^2(x - a)^2$, $p_3(x) = a^4 + 4a^3(x - a) + 6a^2(x - a)^2 + 4a(x - a)^3$, and $p_n(x) = a^4 + 4a^3(x - a) + 6a^2(x - a)^2 + 4a(x - a)^3 + (x - a)^4$ for all $n \geq 4$

31. $p_n(x) = e^a + e^a(x - a) + \dfrac{e^a}{2}(x - a)^2 + \cdots + \dfrac{e^a}{n!}(x - a)^n$

33. 0818733

35. 2.234

37. 3.872986

39. The graphs of $y = \sin x$ and $y = p_3(x)$ are indistinguishable on $[-1, 1]$

41. The graph of $y = e^{-x}$ is indistinguishable from that of $y = p_5(x)$ on $[-1, 1.5]$

43. The graph of $y = e^{-0.5x^2}$ is indistinguishable from the graph of $y = p_2(x)$ on $[-0.5, 0.5]$

45. 2.3785

47. 1.5671

49. 1.733

51. 6.6

53. 3.2

55. 1.9167

57. $a_0 < 0$, $a_1 < 0$, $a_2 > 0$

59. $a_0 > 0$, $a_1 < 0$, $a_2 < 0$

61. $p_2(x) = 3 + (x - 1) + 2(x - 1)^2$; $p_2(x) = f(x)$

63. $p_n(x) = P(x)$

Section 10.2

1. 0.00026

3. 5.2×10^{-8}

5. 0.0016

7. 0.0002

9. 5.7×10^{-6}

11. 0

13. 0.00032

15. 0.032

17. 0.08

19. 5

21. 7

23. 4

25. 5

27. 0.125

29. 0.0078

31. 4

33. 0.47

35. 3.42

37. 0.52

39. 0.13

41. $f(x) = \sin x = p_3(x) + R_3(x)$. Since $\left|f^{(4)}(x)\right| \leq 1$, $\left|R_3(x)\right|$
$\leq \frac{1}{4!}|x|^4$. Since $\lim\limits_{x \to 0} \frac{p_3(x)}{x} = 1$, we have

$$\left|\frac{\sin x}{x} - \frac{p_3(x)}{x}\right| = \left|\frac{R_3(x)}{x}\right| \leq \frac{1}{4!}|x|^3$$

$$\left|\lim\limits_{x \to 0} \frac{\sin x}{x} - 1\right| = 0 \text{ implies } \lim\limits_{x \to 0} \frac{\sin x}{x} = 1$$

43. Let $f(x) = e^x - 1 = p_1(x) + R_1(x)$, $p_1(x) = x$, $\lim\limits_{x \to 0} \frac{p_1(x)}{x} =$

1. On $[-1, 1]$, $\left|R_1(x)\right| \leq \frac{e}{2!}|x|^2$, thus we have

$$\left|\frac{e^x - 1}{x} - \frac{p_1(x)}{x}\right| = \left|\frac{R_1(x)}{x}\right| \leq \frac{e}{2!}|x|$$

$$\left|\lim\limits_{x \to 0} \frac{e^x - 1}{x} - 1\right| = 0 \text{ implies } \lim\limits_{x \to 0} \frac{e^x - 1}{x} = 1$$

Section 10.3

1. 5, 8, 11, 14, 17

3. $0, \frac{1}{3}, \frac{1}{2}, \frac{3}{5}, \frac{2}{3}$

5. $2, 2, \frac{8}{3}, 4, \frac{32}{5}$

7. $-\frac{1}{2}, \frac{1}{4}, -\frac{1}{8}, \frac{1}{16}, -\frac{1}{32}$

9. $-1, 1, -1, 1, -1$

11. $2n + 3$

13. 3^n

15. $(-1)^{n+1}2^n$

17. $\frac{n}{n + 1}$

19.

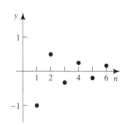

21.

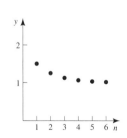

23.

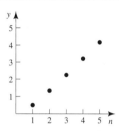

25. Converges to 3

27. Converges to 2

29. Converges to 0

31. Diverges

33. Converges to 0

35. Diverges

37. Converges to 0

39. Converges to 0

41. Converge to 0

43. Converges to 0

45. Converges to 2

47. Converges to 1

49. The concentration becomes zero in the long-term.

51. The profits tend to $10 million in the long-term.

53. e

55. $y_n = 3^{1/n}$, $\ln y_n = \ln 3^{1/n} = \frac{1}{n} \ln 3$, thus $\lim\limits_{n \to \infty} \ln y_n = 0$.
$\lim\limits_{n \to \infty} 3^{1/n} = \lim\limits_{n \to \infty} e^{\ln y_n} = e^0 = 1$. Therefore $\lim\limits_{n \to \infty} 3^{(1/2)^n} = 1$

Section 10.4

1. $S_n = 2n$. Diverges

3. $S_n = \frac{1}{2} - \frac{1}{n + 2}$. Converges to $\frac{1}{2}$

5. Converges to $\frac{5}{4}$

7. Converges to 3

9. Diverges

11. Converges to 10,000

13. Converges to 6

15. Converges to $-\dfrac{5}{6}$

17. Diverges

19. Diverges

21. Converges to -7

23. Converges to $\dfrac{1}{2}$

25. Converges to $\dfrac{1}{20}$

27. $\dfrac{7}{9}$

29. $\dfrac{3}{11}$

31. $\dfrac{377}{110}$

33. \$5

35. \$12,007

37. Let $S_n = \Sigma_{k=1}^n a_k$ and $S'_n = \Sigma_{k=1}^n ca_k = cS_n$. $\Sigma_{k=1}^\infty ca_k =$
$\lim\limits_{n\to\infty} S'_n = c\left(\lim\limits_{n\to\infty} S_n\right) = c\,\Sigma_{k=1}^\infty a_k$

39. $\Sigma_{k=1}^\infty a_k$ is convergent implies $\lim\limits_{k\to\infty} a_k = 0$ implies $\lim\limits_{k\to\infty} \dfrac{1}{a_k} =$
$\infty \ne 0$. Therefore $\sum\limits_{k=1}^\infty \dfrac{1}{a_k}$ diverges.

41. Since $x^k = x \cdot x^{k-1}$, $\Sigma_{k=1}^\infty x^k$ is a geometric series with $a =$
x and $r = x$. If $|r| = |x| < 1$, then $\sum\limits_{k=1}^\infty x^k = \dfrac{a}{1-r} =$
$\dfrac{x}{1-x}$

43. $\lim\limits_{n\to\infty} R_n = \dfrac{D}{e^{bT} - 1}$

Section 10.5

1. Divergent

3. Divergent

5. Convergent

7. Divergent

9. Divergent

11. Divergent

13. Convergent

15. Divergent

17. Convergent

19. Convergent

21. Divergent

23. Convergent

25. Divergent

27. Convergent

29. Convergent

31. Divergent

33. Convergent

35. Convergent

37. Convergent

39. Divergent

41. Convergent

43. \$2.50, \$5.00

45. \$12,007, \$13,792

47. From Figure 10.4 with $f(x) = \dfrac{1}{x}$, $\sum\limits_{k=1}^n \dfrac{1}{k} \le 1 + \displaystyle\int_1^n \dfrac{dx}{x} =$
$1 + \ln n$

49. $\Sigma_{k=1}^\infty a_k$ converges implies $\lim\limits_{k\to\infty} a_k = 0$. Thus, there exists
K such that $k > K$ implies $0 \le a_k < 1$. Then $a_k^2 \le a_k$ for
$k > K$ and therefore $\Sigma_{k=1}^\infty a_k^2$ converges since $\Sigma_{k=1}^\infty a_k$ does.

51. Compare $\Sigma_{k=1}^\infty e^{-k^2}$ with $\Sigma_{k=1}^\infty ke^{-k^2}$, which converges by
integral test.

Section 10.6

1. Convergent

3. Divergent

5. Divergent

7. Convergent

9. Convergent

11. Convergent

13. Convergent

15. Divergent

17. Convergent

19. Convergent

21. Convergent

23. Convergent

25. Convergent

27. Divergent

29. Divergent

31. Converges absolutely

33. Converges absolutely

35. Converges conditionally

37. Converges absolutely

39. Converges conditionally

41. (a) $0 \le b_k \le |a_k|$ for all k. Thus $\Sigma_{k=1}^\infty b_k$ converges if
$\Sigma_{k=1}^\infty |a_k|$ converges. (b) Let $b_k = -c_k$, then $0 \le b_k \le |a_k|$
for all k. Thus $\Sigma_{k=1}^\infty b_k$ converges implies that $\Sigma_{k=1}^\infty c_k =$
$-\Sigma_{k=1}^\infty b_k$ converges.

43. $\Sigma_{k=1}^\infty |a_k|$ converges implies $\Sigma_{k=1}^\infty a_k$ converges. Since
$\left|\Sigma_{k=1}^n a_k\right| \le \Sigma_{k=1}^n |a_k|$,

$$\lim_{n\to\infty} \left|\sum_{k=1}^n a_k\right| \le \lim_{n\to\infty} \sum_{k=1}^n |a_k| \quad \text{or} \quad \left|\sum_{k=1}^\infty a_k\right| \le \sum_{k=1}^\infty |a_k|$$

45. $\Sigma_{k=1}^\infty a_k \cos n$ converges absolutely since $|a_k \cos n| \le a_k$
and $\Sigma_{k=1}^\infty a_k$ converges. But absolute convergence implies
convergence.

Section 10.7

1. $\sum\limits_{k=0}^\infty \dfrac{3^k}{k!} x^k$. Let $b > 0$ and arbitrary. For $x \in (-b, b)$,
$|f^{(n+1)}(x)| \le 3^{n+1}e^{3|b|}$, $|R_n(x)| \le e^{3|b|} \dfrac{|3x|^{n+1}}{(n+1)!} \to 0$ as
$n \to \infty$

3. $\sum\limits_{k=0}^\infty \dfrac{(-1)^k}{(2k+1)!} x^{2k+1}$. For all x, $|f^{(n+1)}(x)| \le 1$, and $|R_n(x)|$
$\le \dfrac{|x|^{n+1}}{(n+1)!} \to 0$ as $n \to \infty$

5. $\sum_{k=0}^{\infty} \frac{(-1)^k 3^{2k}}{(2k)!} x^{2k}$. For all x, $|f^{(n+1)}(x)| \le 3^{n+1}$, and $|R_n(x)|$
$\le \frac{|3x|^{n+1}}{(n+1)!} \to 0$ as $n \to \infty$

7. $\sum_{k=0}^{\infty} \frac{(-1)^k}{(2k)!} \left(x - \frac{\pi}{2}\right)^{2k}$. For all x, $|f^{(n+1)}(x)| \le 1$, and
$|R_n(x)| \le \frac{|x|^{n+1}}{(n+1)!} \to 0$ as $n \to \infty$

9. $\sum_{k=0}^{\infty} \frac{(-1)^{k+1}}{(2k)!} (x - \pi)^{2k}$. For all x, $|f^{(n+1)}(x)| \le 1$, and
$|R_n(x)| \le \frac{|x|^{n+1}}{(n+1)!} \to 0$ as $n \to \infty$

11. $\sum_{k=0}^{\infty} \frac{1}{4^{k+1}} x^k$. $(-4, 4)$

13. $\sum_{k=1}^{\infty} \frac{(-1)^{k+1}}{k} 2^k x^k$. $(-0.5, 0.5)$

15. $\sum_{k=0}^{\infty} (-1)^k (x - 1)^k$. $(0, 2)$

17. $\sum_{k=1}^{\infty} \frac{(-1)^{k+1}}{k} (x - 1)^k$. $(0, 2)$

19. $\ln 2 + \sum_{k=1}^{\infty} \frac{(-1)^{k+1}}{k \cdot 2^k} (x - 1)^k$. $(-1, 3)$

21. $\sum_{k=0}^{\infty} \frac{1}{6^{k+1}} x^k$. For $x \in (-3, 3)$, $|f^{(n+1)}(x)| \le \frac{(n+1)!}{3^{n+2}}$ and
$|R_n(x)| \le \frac{(n+1)!}{3^{n+2}} \frac{|x|^{n+1}}{(n+1)!} = \frac{1}{3} \left|\frac{x}{3}\right|^{n+1} \to 0$ as $n \to \infty$

23. $\sum_{k=1}^{\infty} -\frac{1}{k} x^k$. For $x \in (-0.5, 0.5)$, $|f^{(n+1)}(x)| \le 2^{n+1} n!$ and
$|R_n(x)| \le 2^{n+1} n! \frac{|x|^{n+1}}{(n+1)!} = \frac{1}{n+1} |2x|^{n+1} \to 0$ as $n \to \infty$

25. $f'(0) = 1$, $f''(0) = \frac{2}{9}$, $f^{(5)}(0) = \frac{5!}{3^5}$

27. Yes

29. (a) $e^{3x} = \sum_{k=0}^{\infty} \frac{1}{k!} (3x)^k = \sum_{k=0}^{\infty} \frac{3^k}{k!} x^k$. (b) $\cos 3x = \sum_{k=0}^{\infty} \frac{(-1)^k}{(2k)!} (3x)^{2k} = \sum_{k=0}^{\infty} \frac{(-1)^k \cdot 3^{2k}}{(2k)!} x^{2k}$. (c) If the Taylor series for $f(x)$ at $x = 0$ is $\sum_{k=0}^{\infty} a_k x^k$, then the Taylor series for $f(cx)$ at $x = 0$ is $\sum_{k=0}^{\infty} a_k (cx)^k$.

Chapter 10 Review

1. $p_4(x) = 1 - 2x + 2x^2 - \frac{4}{3}x^3 + \frac{2}{3}x^4$

2. $p_3(x) = 1 + \frac{1}{4}(x - 1) - \frac{3}{32}(x - 1)^2 + \frac{7}{128}(x - 1)^3$

3. $p_4(x) = \ln 5 + (x - 1) - \frac{1}{2}(x - 1)^2 + \frac{1}{3}(x - 1)^3 - \frac{1}{4}(x - 1)^4$

4. $p_3(x) = 1 - 2(x - 1) + 3(x - 1)^2 - 4(x - 1)^3$

5. $p_n(x) = 1 - 3x + \frac{9}{2}x^2 - \cdots + \frac{(-3)^n}{n!} x^n$

6. 1.04663   **7.** 0.18227

8. 2.44×10^{-4} **9.** 0.002048

10. 13   **11.** 2, 4, 8, 16, 32

12. $(-1)^{n+1} 2^n$

13.

14. Converges to 2   **15.** Converges to 0

16. Diverges   **17.** Diverges

18. Converges to 501   **19.** Converges to 12

20. Diverges

22. Converges to $\frac{1}{6}$

24. Diverges

26. Diverges

28. Diverges

30. Converges

32. Diverges

34. Diverges

36. Diverges

38. Converges absolutely

21. Converges to 5

23. Converges

25. Converges

27. Converges

29. Diverges

31. Converges

33. Converges

35. Converges

37. Converges conditionally

39. $\sum_{k=0}^{\infty} \frac{(-5)^k}{k!} x^k$

40. $\ln 2 + \sum_{k=1}^{\infty} \frac{(-1)^{k+1}}{k} (x - 1)^k$

41. $\sum_{k=0}^{\infty} \frac{1}{5^{k+1}} x^k$ **42.** $\sum_{k=0}^{\infty} 2^k (x - 2)^k$

43. Let $b > 0$ be arbitrary. Then for $x \in (-b, b)$, $|f^{(n+1)}(x)| \le 5^{n+1} e^{5b}$ and $|R_n(x)| \le e^{5b} 5^{n+1} \frac{|x|^{n+1}}{(n+1)!} = e^{5b} \frac{|5x|^{n+1}}{(n+1)!} \to 0$ as $n \to \infty$

44. For $x \in (-2.5, 2.5)$, $|f^{(n+1)}(x)| \le \frac{(n+1)!}{(2.5)^{n+2}}$ and $|R_n(x)| \le \frac{(n+1)!}{(2.5)^{n+2}} \frac{|x|^{n+1}}{(n+1)!} = \frac{1}{2.5} \left|\frac{x}{2.5}\right|^{n+1} \to 0$ as $n \to \infty$

45. There will be 100 moose in the long-term.

46. $3.33, $10.00

CHAPTER 11

Section 11.1

1. (a) $\{(1, 1), (1, 2), (1, 3), (1, 4), (2, 1), (2, 2), (2, 3),$
$(2, 4), (3, 1), (3, 2), (3, 3), (3, 4), (4, 1), (4, 2), (4, 3),$
(4, 4)\} (b)

x	2	3	4	5	6	7	8
$P(X = x)$	$\frac{1}{16}$	$\frac{2}{16}$	$\frac{3}{16}$	$\frac{4}{16}$	$\frac{3}{16}$	$\frac{2}{16}$	$\frac{1}{16}$

(c) $\dfrac{3}{16}$ (d) 0.625 (e) 5 (f) 1.58

3. $z = .30$

5. $-\$0.22$

7. $-\$1/19$

9. $-\$1/19$

11. $0.80 or $-\$0.20$ if you account for the cost of the $1.00 ticket

13. $1/6

15. $-.30, 2.01, 1.418$

17. 1.5, 3.850, 1.962

19. 79%

21. (a) .04 (b) .96

23. $1800, $100

25. 2.5

27. $300, $255, the first

29.

Event	Lives for One Year	Dies within the Next Year
Random Variable x	100	-4900
$P(X = x)$	99	.01

$50. If the insurance company were to sell a large number of the same insurance policies under the same conditions, they would expect a $50 return per such policy.

31. A: .83, 1.121, 1.059. B: 1.3, 1.810, 1.345. Salesman B sells the most, but Salesman A is more consistent.

33. Males: 431.8, 13.36, 3.655. Females: 421, 6.0, 2.499. Males.

35. Suppose the experiment has been repeated N times. Then

$$\frac{f(\{e_1, e_2, \cdots, e_T\})}{N} = \frac{f(e_1)}{N} + \frac{f(e_2)}{N} + \cdots + \frac{f(e_T)}{N}$$

For very large N, $\dfrac{f(\{e_1, e_2, \cdots, e_T\})}{N} \approx p(\{e_1, e_2, \cdots, e_T\})$,

$\dfrac{f(e_1)}{N} + \cdots + \dfrac{f(e_T)}{N} \approx p(e_1) + \cdots + p(e_T)$. The result then follows.

37. Since $E \cup E^c = S$ and $E \cap E^c = \phi$, $1 = p(S) = p(E \cup E^c)$
$= p(E) + p(E^c)$. The result then follows.

Section 11.2

1. $\dfrac{2}{3}$

3. 3

5. .25

7. 1

9. 1

11. 3

13. Not always positive

15. Not always positive

17. $\displaystyle\int_1^\infty f(x)\,dx = 0.50 \neq 1$

19. $\displaystyle\int_1^\infty f(x)\,dx$ diverges

21. $\displaystyle\int_{-\infty}^\infty f(x)\,dx$ diverges

23. (a) $\dfrac{1}{16}$ (b) $\dfrac{3}{16}$ (c) 1

25. (a) $\dfrac{1}{2}$ (b) $\dfrac{11}{32}$ (c) $\dfrac{1}{2}$

27. 0.6

29. (a) $\dfrac{1}{6}$ (b) $\dfrac{18}{30}$ (c) $\dfrac{4}{30}$ (d) 0 (e) 1

31. (a) $\dfrac{15}{16}$ (b) $\dfrac{1}{16}$ (c) $\dfrac{1}{2}$

33. (a) $1 - e^{-1.5} \approx 0.777$ (b) $e^{-1.5} \approx 0.223$ (c) $e^{-3} - e^{-6} \approx 0.0473$

35. (a) $\dfrac{2}{3}$ (b) $\dfrac{1}{3}$ (c) $\dfrac{3}{10}$

37. (a) $1 - e^{-.1} \approx .095$ (b) $e^{-.1} \approx .905$ (c) $e^{-.1} - e^{-.2} \approx .086$

39. $f(x) \geq 0$ for all $x \in R$. Since $f(-x) = f(x)$, $\displaystyle\int_{-\infty}^\infty f(x)\,dx = 2\int_0^\infty f(x)\,dx$. Using the Simpson's rule with $n = 50$, we obtain $2\displaystyle\int_0^5 f(x)\,dx \approx 0.99998456$, $2\displaystyle\int_0^6 f(x)\,dx \approx 0.99999993$, $2\displaystyle\int_0^{10} f(x)\,dx \approx 1.00000000$. Thus

$$\lim_{b \to \infty} 2\int_0^b f(x)\,dx = 1$$

41. $f(x) = \dfrac{4}{\pi[(x-1)^2 + 1]} > 0$ on $(1, 2)$. Using the Simpson's rule with $n = 50$, we obtain $\displaystyle\int_1^2 f(x)\,dx \approx 0.99999999999919$

Section 11.3

1. (a) 3 (b) $\dfrac{4}{3}$ (c) $\dfrac{2}{\sqrt{3}}$ (d) 3

3. (a) $\dfrac{8}{3}$ (b) $\dfrac{8}{9}$ (c) $\dfrac{2\sqrt{2}}{3}$ (d) $2\sqrt{2}$

5. (a) $\dfrac{1}{2}$ (b) $\dfrac{1}{20}$ (c) $\dfrac{1}{2\sqrt{5}}$ (d) 0.5

7. (a) 1000 (b) 1,000,000 (c) 1000 (d) 1000 ln 2

9. (a) $\dfrac{4}{3}$ (b) $\dfrac{2}{9}$ (c) $\dfrac{\sqrt{2}}{3}$ (d) $\sqrt[4]{2}$

11. $\dfrac{1}{2}(a+b)$

13. Since $\mu = \sigma = \dfrac{1}{k}$,

$$\int_{\mu-\sigma}^{\mu+\sigma} ke^{-kx}\,dx = \int_0^{2/k} ke^{-kx}\,dx = -e^{-kx}\Big|_0^{2/k} = 1 - e^{-2}$$

15. Since $\mu = \dfrac{1}{k}$,

$$\int_\mu^\infty ke^{-kx}\,dx = \lim_{b\to\infty}\int_{1/k}^b ke^{-kx}\,dx = \lim_{b\to\infty} e^{-kx}\Big|_{1/k}^b$$
$$= \lim_{b\to\infty}(e^{-1} - e^{-kb}) = e^{-1}$$

17. (a) $\dfrac{10}{3}$ minutes (b) $\dfrac{50}{9}$ (c) $\dfrac{5}{3}\sqrt{2}$

19. (a) 5 thousand gallons (b) 5 (c) $\sqrt{5}$

21. (a) $\dfrac{2}{3}$ thousand gallons (b) $\dfrac{2}{9}$ (c) $\dfrac{\sqrt{2}}{3}$

23. (a) 0.5 thousand dollars (b) .75 (c) .87

25. (a) 1.6571 months (b) 0.7151 (c) 0.8456

27. (a) 50 days (b) 2500 (c) 50

29. (a) 0.5 thousand dollars (b) .75 (c) $\dfrac{\sqrt{3}}{2}$

31. (a) $\dfrac{58}{35}$ (b) $\dfrac{848}{245} - \left(\dfrac{58}{35}\right)^2$ (c) .8456

33. (a) .6 (b) No. $\displaystyle\int_0^{x_m} f(x)\,dx = x_m^3(4 - 3x_m) = \dfrac{1}{2}$ can not be solved exactly.

Section 11.4

1. .6915

3. .3821

5. .0668

7. 7881

9. .0919

11. .1498

13. .9332

15. .0228

17. .1151

19. .2417

21. .9544

23. Since $f'(x) = -xe^{-.5x^2}$, $f''(x) = (x^2 - 1)e^{-.5x^2}$. $f' > 0$ and f is increasing on $(-\infty, 0)$, $f' < 0$ and f is decreasing on $(0, \infty)$. Since $f''(0) = -1 < 0$, a maximum is at $x = 0$.

25. Since $f''(x) = (x + 1)(x - 1)e^{-.5x^2}$, $f'' > 0$ and f is concave up on $(-\infty, -1)$ and $(1, \infty)$, $f'' < 0$ and f is concave down on $(-1, 1)$.

27. 1.24%

29. (a) 15.87% (b) 25.14% (c) 25.86%

31. 62.93%   **33.** 6.68%

35. (a) 2.28% (b) 50% (c) 47.72%

37. .9332

39.

Interval	Area
$[-1, 1]$	.19874802
$[-2, 2]$	.73853626
$[-3, 3]$	.97070874
$[-4, 4]$	.99886554
$[-5, 5]$	.99998451
$[-6, 6]$	.99999992

Chapter 11 Review

1. (a) .016, .248, .628, .088, .016, .044 (b) .892 (c) .264

2. (a) $\dfrac{5}{30}$ (b) $\dfrac{15}{30}$ (c) $\dfrac{20}{30}$ **3.** .60, 0, .75

4. (a) .09 (b) .91

5. (a)   (b) .20 (c) .35 (d) .30 (e) .70

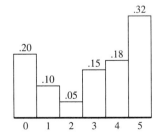

6. 2.97

7. $1.30 or $0.30 if the cost of the $1.00 ticket is subtracted.

8. $400

9. Wins: $\mu = 28$, $\sigma^2 = 21.2$, $\sigma = 4.60$. Loses: $\mu = 12.6$, $\sigma^2 = 13.44$, $\sigma = 3.67$

10. 3.709, 1.926   **11.** .25

12. .50   **13.** 2

14. (a) $\dfrac{1}{8}$ (b) $\dfrac{7}{8}$ (c) 0   **15.** 1.5

16. $\sigma^2 = 0.75$, $\sigma = \dfrac{\sqrt{3}}{2}$ **17.** 1.5

18. 4   **19.** (a) $\sigma^2 = 2$ (b) $\sigma = \sqrt{2}$

20. $3\sqrt{2}$

21. (a) .8944 (b) .1056 (c) .0531

22. (a) .4207 (b) .5793 (c) .0395

23. (a) $1 - e^{-2} \approx .8647$ (b) $e^{-2} \approx .1353$ (c) $e^{-1} - e^{-2} \approx .2325$

24. (a) .0472 (b) .9528 (c) .1556

25. (a) 2.28% (b) 97.72% (c) 95.44%

INDEX